CHEMICAL WARFARE AGENTS
TOXICOLOGY AND TREATMENT
SECOND EDITION

CHEMICAL WARFARE AGENTS
TOXICOLOGY AND TREATMENT

SECOND EDITION

Editors

Timothy C. Marrs, OBE MD DSc FRCP FRCPath
Edentox Associates, Edenbridge, Kent, UK
National Poisons Information Service Birmingham Centre, City Hospital, Birmingham, UK
University of Central Lancashire, Preston, UK

Robert L. Maynard, CBE FRCP FRCPath
Health Protection Agency, Chilton, UK

Frederick R. Sidell MD
Institute of Chemical Defense, Aberdeen Proving Ground, Maryland, USA

BICENTENNIAL 1807 WILEY 2007 BICENTENNIAL

John Wiley & Sons, Ltd

Telephone (+44) 1243 779777

Email (for orders and customer service enquiries): cs-books@wiley.co.uk
Visit our Home Page on www.wileyeurope.com or www.wiley.com

Other Wiley Editorial Offices

John Wiley & Sons Inc., 111 River Street, Hoboken, NJ 07030, USA

Jossey-Bass, 989 Market Street, San Francisco, CA 94103-1741, USA

Wiley-VCH Verlag GmbH, Boschstr. 12, D-69469 Weinheim, Germany

John Wiley & Sons Australia Ltd, 33 Park Road, Milton, Queensland 4064, Australia

John Wiley & Sons (Asia) Pte Ltd, 2 Clementi Loop #02-01, Jin Xing Distripark, Singapore 129809

John Wiley & Sons Canada Ltd, 6045 Freemont Blvd, Mississauga, ONT, Canada L5R 4J3

Wiley also publishes its books in a variety of electronic formats. Some content that appears in print may not be available in electronic books.

Anniversary Logo Design: Richard J. Pacifico

Library of Congress Cataloguing-in-Publication Data

Chemical warfare agents : toxicology and treatment / editors, Timothy
C. Marrs, Robert L. Maynard, Frederick R. Sidell. – 2nd ed.
p. ; cm.
Rev. ed. of: Chemical warfare agents / Timothy C. Marrs, Robert L. Maynard, Frederick R. Sidell. c1996.
Includes bibliographical references and index.
ISBN: 978-0-470-01359-5 (alk. paper)
1. Chemical agents (Munitions) – Toxicology. 2. Antidotes.
I. Marrs, Timothy C. II. Maynard, Robert L. III. Sidell, Frederick R. IV. Marrs, Timothy C. Chemical warfare agents.
[DNLM: 1. Chemical Warfare Agents. 2. Wounds and Injuries – chemically induced.
3. Wounds and Injuries – therapy. QV 663 C5178 2007]
RA648.M37 2007
615.9 – dc22 2006022729

British Library Cataloguing in Publication Data

A catalogue record for this book is available from the British Library

ISBN 978-0-470-01359-5

Typeset in 10/12pt Times by Aptara Inc., New Delhi, India
Printed and bound in Great Britain by Antony Rowe Ltd, Chippenham, Wiltshire
This book is printed on acid-free paper responsibly manufactured from sustainable forestry in which at least two trees are planted for each one used for paper production.

CONTENTS

LIST OF CONTRIBUTORS

David J. Baker
SAMU de Paris, Service du Professeur Pierre Carli, Hôpital Necker Enfants – Malades, 149 Rue de Sevres, 75743 Paris, France

Bryan Ballantyne
871 Chappell Road, Charleston, WV 25304, USA

Chantal Bismuth
Hôpital Fernand Widal, 200 Rue du Faubourg-Saint-Denis, 75010 Paris, France

Robin M. Black
Defence Science and Technology Laboratory, Porton Down, Salisbury, Wiltshire, SP4 0JQ, UK

Sally M. Bradberry
National Poisons Information Service (Birmingham Centre) and West Midlands Poisons Unit, City Hospital, Birmingham, B18 7QH, UK

Robert P. Chilcott
Chemical Hazards and Poisons Division, Centre for Radiation, Chemical and Environmental Hazards, Health Protection Agency, Chilton, Didcot, Oxfordshire, OX11 0RQ, UK

Philippa Edwards
Chemical Hazards and Poisons Division, Centre for Radiation, Chemical and Environmental Hazards, Health Protection Agency, Chilton, Didcot, Oxfordshire, OX11 0RQ, UK

Peter A. Eyer
Walther-Straub-Institute of Pharmacology and Toxicology, Ludwig-Maximilians University, Goethestrasse 33, D-80336 Munich, Germany

Alan H. Hall
Toxicology Consulting and Medical Translating Services, Inc., (TCMS, Inc.), Elk Mountain, WY 82324, USA

Veronique D. Hauschild
US Environmental Protection Agency, Washington, DC 20460, USA

Mathew Hotopf
Institute of Psychiatry, King's Centre for Military Health Research, Weston Education Centre, King's College London, Cutcombe Road, London, SE5 9RJ, UK

Shinichi Ishimatsu
St Luke's International Hospital, Akashicho 9-1 Chio, Tokyo 104-850, Japan

Joseph F. King
US Army Environmental Center, Aberdeen Proving Ground, MD 21010-5400, USA

Roger D. Kingdon
Physical Sciences Department, Defence Science and Technology Laboratory, Porton Down, Salisbury, Wiltshire, SP4 0JQ, UK

J. Michael Lord
Department of Biological Sciences, University of Warwick, Coventry, Warwickshire, CV4 7AL, UK

Hiroshi Maekawa
Advanced Emergency Medical Center, Juntendo Shizuoka Hospital, Nagoka 1129, Izunokuni City, Shizuoka 410-2295, Japan

Timothy C. Marrs
Edentox Associates, Pinehurst, Four Elms Road, Edenbridge, Kent TN8 6AQ, UK; National Poisons Information Service (Birmingham Centre) and West Midlands Poisons Unit, City Hospital, Birmingham B18 7QH, UK; Lancashire School of Health and Postgraduate Medicine, University of Central Lancashire, Preston, PR1 2HE, UK

George Mathers
Defence Science and Technology Laboratory, Porton Down, Salisbury, Wiltshire, SP4 0JQ, UK

Robert L. Maynard
Health Protection Agency, Chilton, Didcot, Oxfordshire OX11 0RQ, UK

John H. McDonough
Pharmacology Branch, Research Division, US Army Medical Research Institute of Chemical Defense, 3100 Ricketts Point Road, Aberdeen Proving Ground, MD 21010-5400, USA

Kunihisa Miura
Department of Anesthesiology and Pain Medicine, Juntendo University, Hongo 2-1-1, Bunkyo-City, Tokyo 113-8421, Japan

Nancy B. Munro
Life Sciences Division, Oak Ridge National Laboratory, PO Box 2008, MS6201, Oak Ridge, TN 37831-6201, USA

Toshio Naito
Department of General Medicine, Juntendo University Hospital, Hongo 2-1-1, Bunkyo-City, Tokyo 113-8421, Japan

Tomohisa Nomura
Department of Emergency and Disaster Medicine, Juntendo University, Hongo 2-1-1, Bunkyo-City, Tokyo 113-8421, Japan

Daan Noort
TNO Defence, Security and Safety, PO Box 45, Rijswijk 2280 AA, The Netherlands

Sumie Okumura
Advanced Emergency Medical Center, Juntendo Shizuoka Hospital, Nagoka 1129, Izunokuni City, Shizuoka 410-2295, Japan

Tetsu Okumura
Department of Emergency and Disaster Medicine, Juntendo University, Hongo 2-1-1, Bunkyo-City, Tokyo 113-8421, Japan

Graham S. Pearson
Department of Peace Studies, Pemberton Building, University of Bradford, Bradford, West Yorkshire, BD7 1DP, UK

Paul Rice
Defence Science and Technology Laboratory, Porton Down, Salisbury, Wiltshire, SP4 0JQ, UK

Gron Roberts
GR Associates, Chappel, Essex, C06 2EE UK

Leah Scott
Biomedical Sciences Department, Defence Science and Technology Laboratory, Porton Down, Salisbury, Wiltshire, SP4 0JQ, UK

Åke Sellström
Swedish Defence Research Institute, S-90182 Umeå, Sweden

Tsung-Ming Shih
Pharmacology Branch, Research Division, US Army Medical Research Institute of Chemical Defense, 3100 Ricketts Point Road, Aberdeen Proving Ground, MD 21010-5400, USA

Frederick R. Sidell
Deceased (2006); formerly, US Army Medical Research Institute of Chemical Defense, Aberdeen Proving Ground, MD 21010-5400, USA

Manabu Sugita
Department of Emergency and Disaster Medicine, Juntendo University, Hongo 2-1-1, Bunkyo-City, Tokyo 113-8421, Japan

Kouichiro Suzuki
Emergency Department, Kawasaki Medical School, Matsushima 577, Kurashiki, Okayama 701-0192, Japan

Toshishige Suzuki
Department of Emergency and Disaster Medicine, Juntendo University, Hongo 2-1-1, Bunkyo-City, Tokyo 113-8421, Japan

Nobukatsu Takasu
St Luke's International Hospital, Akashicho 9-1 Chio, Tokyo 104-850, Japan

Yasuo Takeuchi
Department of Emergency and Disaster Medicine, Juntendo University, Hongo 2-1-1, Bunkyo-City, Tokyo 113-8421, Japan

Sylvia S. Talmage
Life Sciences Division, Oak Ridge National Laboratory, PO Box 2008, MS6201, Oak Ridge, TN 37831-6201, USA

Jennifer G. Treleaven
Royal Marsden Hospital NHS Trust, Downs Road, Sutton, Surrey, SM2 5PT, UK

J. Allister Vale
National Poisons Information Service (Birmingham Centre) and West Midlands Poison Unit, City Hospital, Birmingham, B18 7QH, UK

Stephen Walker
Physical Sciences Department, Defence Science and Technology Laboratory, Porton Down, Salisbury, Wiltshire, SP4 0JQ, UK

Annetta P. Watson
Life Sciences Division, Oak Ridge National Laboratory, PO Box 2008, MS6201, Oak Ridge, TN 37831-6201, USA

Simon Wessely
Institute of Psychiatry, King's Centre for Military Health Research, Weston Education Centre, King's College London, Cutcombe Road, London SE5 9RJ, UK

Anthony Wetherell
Defence Science and Technology Laboratory, Porton Down, Salisbury, Wiltshire, SP4 0JQ, UK

Franz Worek
Bundeswehr Institute of Pharmacology and Toxicology, University of Munich, Neuherbergstrasse 11, D-80937 Munich, Germany

PREFACE

The first edition of 'Chemical Warfare Agents, Toxicology and Treatment' was published in 1996. It appeared to meet a need and we were pleased with the reception given to it by reviewers. Since the first edition was written a number of important and tragic events have occurred. These have led to attention being focused in certain areas: we have tried to respond to this. The first Gulf war increased the attention given to cyclosarin (GF), previously considered a nerve agent of secondary importance, while further development work has been done on newer oxime nerve agent treatments, such as HI-6. In addition, a considerable amount of work has been carried out on the skin effects of sulphur mustard. The terrorist incidents using nerve agents, which took place in 1994 and 1995 in Japan, kindled a considerable amount of interest in other countries and gave rise to a number of symposia, such as the seminar on responding to the consequences of chemical and biological terrorism held at Bethesda, Maryland, in July 1995. Subsequent events, such as the 9/11 attacks in New York and Washington and the Bali, Madrid and London tube bombings, none of which involved chemical agents, have increased the attention given to the possibility of the use of chemicals by Al-Qaida and other groups.

The possibility of the terrorist use of chemical weapons means that the management of civilian casualties has to be considered. Previously, management of chemical casualties has generally been in the context of military personnel, who may be protected physically and in some cases, by pharmacological preparations, against chemical warfare agents, and who will in any case usually be young and physically fit. Civilian casualties, by contrast, may include the infirm, the elderly and children. In addition, armed forces may have in place procedures for dealing with chemical attacks, whereas, until recently, that was not the case for civilians. Most western countries now have in place some procedures to deal with civilian casualties in the event of a terrorist attack using chemicals. However, many problems remain, including, for example, the need for mass decontamination after an incident. This and other topics receive special attention in this second edition.

This new edition of 'Chemical Warfare Agents, Toxicology and Treatment' is a multi-author book and many of the chapters have been extensively rewritten or are completely new. There are new chapters on each group of antidotes for nerve agents, skin lesions produced by mustard gas, on the toxicology of ricin, decontamination and disarmament. It is now about ninety years since chemical weapons were used on a large scale during World War I. That these weapons still pose a threat to both civilians and military personnel says little for mankind's socio-political progress. It is our hope that this book will go some way to allaying this threat.

It is with great regret that we record that our friend and colleague, Dr Fred Sidell, passed away during the preparation of this second edition. Dr Sidell had a unique knowledge of our field, based on long experience of experimental work both in the USA and in the UK. His wise counsel and advice was very valuable to us during the preparation of the first edition of this book. He was unfortunately not well enough to contribute to the new edition but we have reprinted his chapter recording the history of studies involving exposure of volunteers to nerve agents unchanged. We hope this may stand as a small memorial to his work in this area.

The preparation and production of a multi-author book obviously requires a great deal of work by many people, notably the authors of the individual chapters, and we thank them. We would also like to record our thanks to the patient staff of John Wiley and Sons Ltd (notably our Content Editor, Jon Peacock), to Paul Nash for his excellent index, the typesetters (Aptara Inc. of New Delhi, India) and the printers (Antony Rowe Ltd of Chippenham, England).

Tim Marrs, Edenbridge, Kent, UK and
Bob Maynard, Chilton, UK

1 OPINIONS OF CHEMICAL WARFARE

Robert L. Maynard

Health Protection Agency, Chilton, UK

Chemical warfare should be abolished among nations, as abhorrent to civilisation. It is a cruel, unfair and improper use of science. It is fraught with the gravest danger to non-combatants and demoralises the better instincts of humanity (General Pershing, US Army, 1970).

I claim, then, that the use of mustard gas in war on the largest possible scale would render it less expensive of life and property, shorter and more dependent upon brains than upon numbers (Haldane, 1925).

In no future war will the military be able to ignore poison gas. It is a higher form of killing (Haber, 1919).

It is more difficult than might be supposed to say why poison gas seems to be such an immoral weapon to so many people (Fotion and Elfstrom, 1986).

Chemical warfare (CW) has attracted opprobrium since it was first used on a large scale during World War I (WWI). Since 1919, efforts to persuade governments to abandon chemical weapons have been made by members of many sections of society – including the military – and yet, in the early 1990s, the number of countries believed to have, or to be procuring, the means to wage a chemical war is accepted to have never been higher. Third World countries, in particular, have become interested in chemical weaponry, and the use of such weapons by Iraq in the Iran–Iraq conflict has attracted considerable Third World interest, yet disappointingly little and rather low-key, censure from Western nations.

Over the past 90 years, chemical weapons have found only a few defenders. These have included such distinguished scientists as F. Haber and J. B. S. Haldane, military men including Brigadier Fries and Lt Col. Prentiss of the US Chemical Corps, military physicians including Col. E. B. Vedder (US), and military historians, including B. H. Lidell Hart. Several of these writers had first-hand experience of the extensive use of chemicals during WWI, and despite, or perhaps because of this, have advocated strongly the advantages and *humanity* of chemical weapons when compared with high explosives and fragmentation devices.

Such advocacy has had little effect, and the general repugnance for chemical weapons felt by the public and expressed by politicians and pressure groups has grown. During the late 1980s, fresh attempts to produce a ban on chemicals were made. At the same time as these efforts were underway, the threat of terrorists acquiring access to chemical weapons increased.

As stated above, the general public and politicians have long reserved a special dislike and level of criticism for chemical weapons. CW devices are frequently referred to as immoral, cruel, inhumane, a debasement of science, unfair, etc. It is certainly true that many people hold such views. In enquiring into the origins of these views, writers have found difficulty in identifying precisely what it is about chemical weapons that people so dislike and disapprove of – as compared, of course, with other lethal weapon

Chemical Warfare Agents: Toxicology and Treatment (2nd Edition)
Edited by Timothy C. Marrs, Robert L. Maynard and Frederick R. Sidell

systems. The level of dislike encountered should not be underestimated: indeed, many more people in the UK probably disapprove of the use of CW by Germany, during WWI, than disapprove of the use of the much more destructive nuclear weapons by the USA during World War II (WWII). The roots of the disapproval are tangled and involve perceptions of where right lies in a conflict, of the use of particular weapons likely to hasten the end of the conflict, and a feeling that enemies who commit atrocities should be severely punished.

Of all of the means of killing and waging war available to mankind, only biological warfare (BW) attracts more dislike and obloquy than CW. Notions of awful plagues spreading through the people of continents and possibly the world, have been increased by popular authors and journalists, and little credence is given to anyone who argues that limited BW or even CW could be waged. Such opinions are seen as a form of warmongering, or of a descent into madness, and are vigorously opposed.

HISTORY OF CHEMICAL WARFARE

Throughout history, man has sought more efficient means of killing his fellow man. Stones, clubs, spears, arrows, gunpowder, muskets, rifles, high explosives, machine guns, tanks, warships, warplanes, rockets and nuclear weapons form an apparently unending catalogue of increasing military sophistication in destroying one's enemies, while exposing oneself, or the majority of one's forces, to decreasing risk. No means of stemming this tide of weapons has been discovered by accident and, until comparatively recent times, no efforts to find such means had been made. Accompanying this development of hardware, based on the production of physical disruption of men or materials, has been a much less marked development of chemical means of attacking people. Very little effort has been made to devise chemical means of attacking inanimate objects and military smokes, and more recently, infrared screening smokes are the only significant examples of the use of chemicals to frustrate military equipment.

Chemical weapons probably began with smoke and flame. The lighting of bonfires and the hurling of various concoctions of pitch and sulphur (Greek Fire) date from classical times. Irritant smokes were described by Plutarch, hypnotics by the Scottish historian Buchanan, compounds capable of producing incessant diarrhoea by classical Greek authors (Robinson, 1971a), and a whole range of preparations, including arsenical compounds and those containing the saliva of rabid dogs (a remarkably prescient notion) by Leonardo da Vinci (The Reprint Society, 1938). During these remote periods, chemistry and chemical technology were in their infancy, and the use of chemical weapons probably had only a marginal effect on the outcome of wars. Such weapons would, perhaps, have had some terror-inducing effect rather along the lines of all secret weapons, possessed, or alleged to be possessed, by one force but not the other. These early examples of chemical weapons should be distinguished from the early use of poison as a means of removing small numbers of one's enemies. Poisoning probably dates from very earliest times and Indian sources (Robinson, 1971a) from the fourth century BC reveal the use of alkaloids and toxins, including abrin (a compound closely related to ricin, the compound used to kill G. Markow in 1978). Aconite has a long history, and murderous Indian courtesans were reported to coat their lips with an impermeable substance and then apply aconite as a form of lipstick. One kiss, or probably several kisses and a bite, from such ladies was said to mean death. (Some sources allege the lethal dose of pure aconite to be as low as 7 mg.) Poisonous snakes and spiders have also been used as sources of strong poisons. Early 'researchers' in the old 'poison lore' (poison-lehre) recorded the effects of their preparations in detail, and the death of Britannicus (brother of Nero) is particularly well documented. Some practitioners undertook clinical experimental work and Madame de Brinvilliers, who poisoned most of her relatives and developed powders known as 'Les poudres des succession', experimented upon hospital patients in Paris to assay the strengths and determine the effects of her preparations (Blyth and Blyth, 1920). From the fifteenth to seventeenth centuries, poisoning was rife in Italy, and it is probably this period, when the postmortem detection of poisoning was all but impossible, that a deep dislike of poison, as a means of killing and achieving one's goals, stems.[1] This condemnation of poisoning does not seem to have extended to the military use of chemicals before the nineteenth century, but suggestions that fumes of sulphur (sulphur dioxide) should be used as a weapon, made by Admiral Sir Thomas Cochrane (10th Earl of Sunderland) in 1855, were treated with disdain by the British Military Establishment (Robinson, 1971a).

During WWI, chemicals were used on a vast scale: 12 000 tons of mustard gas alone, and in all, 113 000 tons of chemicals were used (Prentiss, 1937). Considerable attention has been paid to the identification of the first uses of CW during WWI and, therefore, the identity of the original transgressor of the Hague declarations of 1899 and 1907 (see below). It is likely that during the latter part of 1914, both Germany and France made use of non-lethal tear-gases, but on 22 April 1915 Germany launched a massive chlorine gas cloud attack, producing 15 000 Allied wounded and 5000 Allied dead. This attack came as a great surprise to the Allied Governments, and not until 25 September 1915 could British forces launch their own chlorine cloud attacks. Phosgene was introduced by Germany late in 1915, and sulphur mustard on 12 July 1917 (Prentiss, 1937). Again, Allied Governments were surprised, although scientists had advised of the likely military values of this compound, and not until September 1918 (two months before the Armistice) could British forces fire any mustard gas shells.

It is difficult to assess the military importance of CW during WWI. Enthusiasts for CW, e.g. Prentiss (1937), stressed the efficiency of the weapon, whereas other writers, including Fritz Haber's son (Haber, 1986), pointed out that once adequate protective equipment became available, the efficacy and efficiency of CW fell sharply away.

[1] It is also true to say that punishments for convicted poisoners at the time were particularly severe: boiling alive and the forced drinking of vast quantities of water being preferred methods (Blyth and Blyth, 1920).

During the period immediately following WWI, public opinion swung firmly against chemical warfare. The reasons for this are considered below. Minor uses of chemicals were alleged during the Russian revolution (White Russian forces used British devices in 1919). Such weapons were also alleged to have been used by the Spanish in Morocco in 1925, and by Chinese forces in the early 1930s, but not until the Italian campaign in Ethiopia (1935–1936) were chemicals used again on a large scale: mustard gas was deployed with great effects against native Ethiopian troops (General Pershing, US Army, 1970). Arguments regarding the ethics of such use are considered at length below.

During WWII, and despite, or perhaps because of, a great deal of civil defence effort (in the UK, Air Raid Precautions (ARPs) were originally intended to defend against aerial gas attacks rather than attacks with high explosives), no proven incidents involving CW were reported in Europe. Japan did use CW against Chinese forces in the late 1930s and early 1940s. Reasons for the non-use of CW by Germany remain obscure but there can be no doubt of German superiority in the CW field: the development of so-called nerve gases tabun, sarin and soman in Germany between 1936 and 1944 came as an almost complete surprise to Allied scientists in 1945 (Holmstedt, 1959). By 1945, some 12 000 tons of tabun had been made in Germany; none had been synthesised by Allied workers. Work on related compounds was certainly underway in the UK and the USA, but the breakthrough to nerve agents had not been made by these countries by 1945.

From 1945 until the 1980s, only two varieties of CW agents were used to any significant extent: lachrymators (CS: tear-gas) and herbicides (e.g. Agent Orange) by US forces in Vietnam. Allegations of CW use had been plentiful during the period and seem particularly convincing regarding Egyptian use of CW in the Yemen (1963–1967). During the 1980s, extensive use was made of mustard gas, and latterly nerve agents (probably tabun) by Iraqi forces during the Iran–Iraq conflict. In one incident at Halabja, some 5000 Iranians and Kurds were reported to have died as a result of a gas attack. Death on this scale, as a result of the use of CW, had not been known since the original German chlorine cloud attack of 1915.

Accusations of the use of chemical weapons in South East Asia have been made and 'Yellow Rain' enjoyed a brief period of public attention. Despite intensive investigations, conclusive evidence of the deliberate use of trichothecene mycotoxins seems to be lacking. Allegations, rather better supported, of the use of CW by Soviet Forces in Afghanistan, have also been made, and a range of effects inexplicable in terms of known agents have been widely reported. Probably most recently, i.e. since 1985, allegations of the use of chemical weapons by Cuban or possibly other forces, in Angola have been made, and again, hard-to-explain effects have been reported. During 1988, reports of Libya constructing a chemical weapons production plant, and the publication in the March 1989 issue of *Scientific American* (Scientific American, 1989) of what appears to be a satellite photograph of the plant, raised fears that terrorists might soon be able to acquire chemical weapons from sympathetic countries and a wave of new chemical terror might be unleashed.

These reported uses of CW and other matters have led to a general impression that while the superpowers may be approaching some measure of agreement regarding the undesirability of CW, Third World countries are not, and terrorists might well take up such weapons. Phrases such as 'the poor man's nuclear weapons' are often used in the press, and certainly a marked new interest in CW substances and how they might be produced has been generated.

THE PUBLIC PERCEPTION OF CW AND THE PERCEPTIONS OF NATIONAL LEADERS

It is difficult to separate the position of chemical weapons as regards international law from the public view of such devices. One might assume public opinion to be one of the forces which form international law and it is probably true that international law, whether based on a treaty or upon alleged customary practice, would have little force unless it *was* supported by public opinion. Equally, public opinion, often based on poor information and very subject to propaganda, may be hardened and possibly formed, by a well-publicized breach of international law. In the field of CW, it is difficult to be definitive regarding the following question:

> *Does the illegality of CW depend upon treaties set up as a result of public opinion that CW is unethical, and therefore, undesirable, or is the general public view that CW is unethical and presumably, therefore, undesirable, based upon the very existence of such treaties declaring this method of war illegal?*

An attempt to answer this question may be made by tracing the development of public opinion and international law regarding CW and attempting to separate examples of political expediency from military desires and ethical thinking. This will be followed by a closer examination of the role of ethical thinking in forming international law regarding CW and particularly of how the concept of the 'Just War' has been applied to the topic.

As said already, chemical warfare has a long, if patchy, history and was waged on a large scale for the first time during WWI. It is, therefore, remarkable that efforts to prevent the use of chemicals in warfare were made *before*, albeit not very long before, WWI. The first Hague Conference held in 1899 formulated a resolution stating:

> *The Contracting Powers agree to abstain from the use of all projectiles, the sole object of which is the diffusion of asphyxiating or deleterious gases.*

The precise wording regarding 'sole object......' resulted from an observation that all high-explosive (HE) shells tended to generate noxious gases in some quantity and that such gases, in confined spaces, could be dangerous. The exact wording was later used by Germany to justify her use of shells containing shrapnel balls embedded in an irritant chemical powder, on 27 October 1914, on the grounds that the diffusion of asphyxiating or deleterious gases was not the *sole* purpose of such munitions.

While the proposition received general assent, the USA stood out against it, and a statement by the Naval Delegate (Captain A. H. Mahan) expressed clearly a view that CW could not be regarded as an unusually appalling method of waging war. He said:

> *It was illogical and not demonstrably humane, to be tender about asphyxiating men with gas, when all are prepared to admit that it was allowable to blow the bottom out of an ironclad at midnight, throwing four or five hundred men into the sea to be choked by water, with scarcely the remotest chance of escape* (Mahan, 1937).

This particularly objective remark has remained the basis of thinking of those who have advanced the case of chemical warfare during the twentieth century. The problem of CW was, however, remote in 1899, and comparatively little attention was paid to the US position.

The Second Hague Conference of 1907 attempted to codify what were regarded as a number of unwritten rules of warfare, although chemical warfare, in the modern sense, was not considered. Article XXIII of the 1907 Hague Convention regarding 'Laws and Customs of War on Land' deals with a closely related issue:

> *In addition to the prohibitions provided by special conventions, it is especially forbidden to employ poison or poisoned weapons.*

The USA clarified its position on poison and poisoned weapons in 'The Rules of Land Warfare', War Department Document No. 468, paragraph 177, wherein it was made clear that "poison" applied to the contamination of water sources and the depositing therein of the carcases of animals. This codification by the USA represented an easing of the position adopted by that country in 1863 (General Orders War Department No. 100):

> *The use of poison in any manner, be it to poison wells or food or arms, is wholly excluded from modern warfare.*

It might be deduced from these several US statements that little conviction existed among the US senior military staff regarding the unethical nature of chemical warfare – they were, in fact, attempting to 'keep their options open'.

Such a deduction is supported by a US State Department opinion given at the time of the First Hague Conference:

> *The expediency of restraining the inventive genius of our people in the direction of devising means of defense is by no means clear, and considering the temptations to which men and nations may be exposed in time of conflict, it is doubtful if an international agreement to this end would prove effective* (US State Department, 1937).

In terms of the formation of international law regarding CW, the USA may be seen as opposing the formation of *conventional* law when no *customary law* existed.

The UK view in the pre-WWI period seems to have been much more clearly against the use of CW. This view was advanced most strongly in traditional military circles and the suggestion, made in 1855 by Sir Thomas Cochrane, that fumes of burning sulphur should be used were met with the comment from the War Ministry that:

> *an operation of this nature would contravene the laws of a civilized warfare* (General Pershing, US Army, 1970).

The USA's refusal to adhere to the 1899 Hague Declaration meant that unanimity at the conference, a requirement of UK agreement, was impossible, and the UK did not so agree. In 1907, the UK did 'adhere to the declaration' and became bound by it. France agreed with the 1899 Hague declaration, as did Germany, Italy, Russia and Japan.

The use of chlorine gas on a large scale by the Germans in 1915 took Allied Forces and governments by surprise and a vigorous propaganda campaign deploring the use of such means of waging war was put in hand in the UK. This campaign, supported by lurid reports in the national press, mobilized public opinion against the German use of chemical weapons. Julian Perry Robinson (Robinson, 1971b) has studied the propaganda disseminated during the period following the German use of chlorine in detail, and has drawn attention to the following description of gas casualties from *The Times* of 30 April 1915, 'The Full Story of Ypres: a New German Weapon' (described as being contributed by 'an authority beyond question'), as revealing the approach taken:

> *Their faces, arms, hands were of a shiny grey–black colour, with mouths open and lead glazed eyes, all swaying slightly backwards and*

forwards trying to get breath. It was a most appalling sight all those poor black faces, struggling, struggling for life what with the groaning and noise of the effort for breath.... The effect the gas has is to fill the lungs with a watery frothy matter, which gradually increases and rises till it fills up the whole lungs and comes up to the mouth; then they die; it is suffocation; slow drowning, taking in some cases one or two days.

Germany responded with propaganda of the opposite kind:

These shells[2] *are not more deadly than the poison of the English explosives,*[3] *but they take effect over a wider area, produce a rapid end, and spare the torn bodies the tortures of pain and death.*

Despite the launching of a propaganda attack on the German use of chlorine, the British Government failed to protest vigorously that Germans had breached at least the spirit, if not the words, of the Hague Declaration of 1899. Protests were made when, later, Germany made use of chemical-containing shells. The British Government was faced in 1915 by a very difficult problem: there was no doubt that general opinion was against the use of chemical weapons, but such weapons were clearly effective (15 000 wounded, 5000 dead and widespread panic induced during the first German gas attack) and seen by some as a means of breaking the deadlock of trench warfare. The British Government took the view that, unfortunately and against its better instincts, chemicals would have to be used by British forces and began the manufacture of such weapons. Senior British Military opinion remained firmly against chemical weapons: Sir John French (12 July 1915) referring to the German use of gas:

As a soldier, I cannot help expressing the deepest regret and some surprise that an Army which, hitherto, has claimed to be the chief exponent of the chivalry of war, should have stooped to employ such devices against brave and gallant foes... (French, 1915).

British opinion then was generally against the German use of chemicals, but accepted that as such weapons had been used and were seen to be effective, Britain would have no option but to reply in kind. The use of gas by British forces was at first represented as a reprisal against the German use of gas. However, gas use soon escalated and British use far passed any reasonable definition of a reprisal. By the close of WWI, chemical warfare seemed to have been accepted as a regrettable fact of military life, and the existence of *customary law* against its use would have been difficult to demonstrate.

The US Government responded more radically than had the British Government when American troops suffered badly from chemical attacks during the early days after their entry into WWI. A Chemical Warfare Service (CWS) was set up and with eventually 4873 men under the command of 210 officers, formed a force more committed to chemical warfare than any in the British Army. At the end of WWI, the CWS was markedly reduced and all but abolished; however, a vigorous and effective propaganda campaign ensured its survival. Articles published in chemical journals extolled the efforts of the CWS and one editorial closed:

....Bestir yourselves, chemists of America! The country glories in the services you have already rendered it in peace and war. Opportunity for further service now presents itself. Whether we will it or not, gas will determine peace or victory in future wars. The Nation must be fully prepared! (Journal of Industrial and Engineering Chemistry, 1919).

DEVELOPMENT OF PUBLIC OPINION AFTER WW1

After WWI, public opinion regarding chemical weapons was influenced by a number of factors which may be worth considering briefly.

The general revulsion towards war which followed WWI – the 'war to end war' – was marked and triggered a pacifist movement which

[2] The date of this quotation is 29 April 1915. 'Through German Eyes, poisonous gases, a quick and painless death' *The Times (London)*. Interestingly, at this time, no 'shells' had been used, and the German Government was arguing that their use of gas clouds produced from cylinders and *not by shells* did not infringe the Hague Declaration of 1899.

[3] Refers to fumes produced on explosion of Lyddite: a picric-acid-based explosive.

remained active during the period 1926–1934. CW had been the subject of much official propaganda during WWI and was seen by activists in the pacifist movements as the very embodiment of all that was evil about modern war. The view was driven deeply into the public mind by the work of several of the 'War Poets', who painted an awful picture of men dying of the effects of chlorine and being blinded by mustard gas. The efforts of the League of Nations and the holding of a series of conferences and meetings were based upon these general feelings that if warfare was evil, then CW was *particularly* evil. The feelings of revulsion described above were fuelled by post-war propaganda designed not only to remind people of the horrors of CW, but also to point out the alleged likely greater horrors if further development of this type of warfare were to be allowed. It should be said that such revelations came from several sources, one being the American CWS. Anxious to preserve its existence, the CWS undertook a series of what would be today regarded as 'leaks' designed to alert the public to the dangers of new chemical 'super-weapons'.

Chlorovinyl dichlorarsine or lewisite, was often the subject of these disclosures, and this compound acquired a particularly bad reputation. Claims that this compound – 'The Dew of Death' – was 'invisible', that if inhaled it 'killed at once', that 'three drops upon the skin would kill', and that the compound 'falling like rain from nozzles attached to an aeroplane, (it could) kill practically everyone in an area over which the aircraft passed' were made (Prentiss, 1937; The Times, 1921a; The Times, 1921b). This theme was adopted by popular authors and 'Sapper'[4] dwelt at length upon CW and lewisite in several of his anti-Bolshevik and anti-German 'Bulldog Drummond' romances. In England, a more cautious line was generally taken by supporters of the need for further work on CW, but, in an attempt to raise interest and support the work, the President of the Society of the Chemical Industry described a new CW agent against which gas masks offered no protection and which would 'stop a man' at a concentration of one in five million. 'Stop a man' is an emotive and imprecise phrase. To many people, the phrase would suggest a lethal effect, although in the case of the compound being considered (adamsite) severe lachrymation would be a much more likely response.

These efforts were supported vigorously by the Allied chemical industries, which, in retrospect, cannot but have been eager for their expansion and the destruction of the world's leading chemical industry: that of Germany. The extent of the German domination of the world chemical industry prior to WWI is often not appreciated. In 1913, Germany produced 85.91% of the world's dyes. Britain produced 2.54% and the USA 1.84% (Prentiss, 1937). The six great German chemical firms had banded together prior to WWI to form Interessen Gemeinschaft Farben (IG Farben: community of interests: dyes) and completely dominated the production of organic chemicals. Fritz Haber, the chemist who organised the German use of chemicals during WWI, was closely associated with IG Farben; Schrader, the developer of nerve agents in the late 1930s, was later one of its leading chemists. British and American writers saw IG Farben, as other saw the Krupps armaments empire, as a dangerous threat and made efforts to have it dismantled and Allied industries expanded in its stead. This line of argument was most effectively pursued by Victor Lefebure in his book *The Riddle of the Rhine* (Lefebure, 1921). Lefebure's arguments have been examined by more recent writers, and despite pointing out that he may have had some vested interest in seeing the decline of IG Farben (he was an employee of Imperial Chemical Industries), most agree his was a good case, well argued. The line taken by commentators such as Lefebure in the UK and Vedder and others in the USA was not a pacifist one: they accepted CW as a likely fact of war and wished to ensure that neither the USA nor the UK should be found again at so great a disadvantage as they had been at the time of the first large scale German CW attack in 1915.

In addition to those pointing out the horrors of CW was a small group of scientists and others who supported further work on CW for quite a different reason: they thought it was a *more* human way to wage war than conventional means. The leading proponent of this view was J. B. S. Haldane who had had first-hand experience of

[4] 'Sapper': pseudonym of Herman Cyril McNeil (1888–1937).

gas warfare as an officer in the Black Watch recalled from France to assist his father Professor J. S. Haldane (brother of Viscount Haldane) in undertaking research in CW. J. B. S. Haldane was frequently exposed to chlorine gas and various lachrymators and irritants. In 1925, he published a series of lectures on CW as a book entitled *Callinicus: A Defence of Chemical Warfare* (Haldane, 1925).[5] Referring to the lack of understanding of CW on the part of army officers:

> *The chemical and physiological ideas which underlie gas warfare require a certain effort to understand, and they do not arise in the study of sport as is the case with those undertaking shooting and motor transport.*

These attacks, combined with Haldane's growing, and carefully cultivated, reputation as an iconoclast, led to his opinions being widely disregarded.

Of medical officers involved in treating CW casualties, most stressed the very unpleasant nature of the effects of these weapons while not comparing them with injuries produced by other weapons systems. Many reports and papers in medical journals stressed the seriousness of mustard gas burns, though hardly any compared these burns with commonplace thermal burns. A few distinguished medical writers, again with extensive first-hand experience of CW, *did* publish comparative material, but this also was largely ignored. Of these authors, Lt Col. E. B. Vedder (Vedder, 1925) is probably the best known. His book remains an indispensable and accurate account of those aspects of chemical warfare of interest to the physician. In Vedder's introduction, he acknowledges the 'considerable prejudice against the use of gas in warfare' and identifies three reasons for this prejudice:

1. That the Germans had used it first and were considered by many to have violated the Hague Declarations.
2. That it was a new weapon and all new weapons seemed to be instinctively disliked.
3. That CW was seen as barbarous and inhumane.

How such comments were received is difficult to establish some 80 years later: they did not, however, seem to convince many that CW was a new and inherently *more* acceptable form of warfare. It is also possible that by drawing attention to chemical warfare, public revulsion was increased.

There was then, after WWI, a preponderance of opinion in favour of the view that CW was a particularly unpleasant form of warfare and that those who supported it, with the exception of industrialists perceived to be concerned with preserving national security, were either cranks or simply deluded. These general feelings were reflected by the attempt by the USA to place some limits on the development of a variety of weapon systems by calling the 'Washington Conference on the Limitation of Armaments' in 1921–1922. CW was referred to a subcommittee which considered the development of this form of warfare during WWI and concluded that it was impracticable to attempt to limit the use of chemicals any more than that of explosives, and stated, regarding chemical weapons:

> *....there can be no limitation on their use against the armed forces of the enemy, ashore or afloat* (Brown, 1968).

As Perry Robinson has pointed out, this view was in accord with the negotiating position of the US delegation (Robinson, 1971b). However, when this was discussed at the conference, the US delegation strongly opposed the subcommittee report. This reversal of position seems to have been the result of intense lobbying and the activities of an advisory subcommittee to the US delegation which had been formed by C. E. Hughes, a US Senator. This group reported that:

> *The frightful consequences of the use of toxic gases, if dropped from airplanes on cities, stagger the imagination...If lethal gases were used in.... bombs (of the size used against cities in the war), it might well be that such permanent and serious damage would be done.... in the depopulation of large sections of the country as to threaten, if not destroy, all that has been granted during the painful centuries of the past.*

[5] Haldane chose the title of the book in honour of the Syrian Callinicus, the alleged inventor of a pitch-tar and sulphur mixture called 'Greek Fire'. In his preface, he pointed out that 'Callinicus' means 'He who conquers in a noble or beautiful manner'!

The ludicrous nature of this assertion, given the level of technology available in 1921, need not be laboured, although it may be that members of the advisory subcommittee were more concerned about possible, although unlikely, future developments than the current position. The view of the advisory subcommittee was supported by a public opinion poll taken in the USA: 366 975 people wanted abolition; 19 wanted retention under the terms of the recommendations of the original subcommittee. The advice of the original subcommittee was rejected, the advice of the advisory subcommittee accepted and the treaty ratified by most nations, including the USA and UK. France did not ratify the treaty, although not for any reasons of disagreement regarding CW, but rather for some regarding submarine warfare.

By 1922, there seemed to exist a widespread consensus that CW was an especially undesirable form of warfare and one which should be prohibited by international convention. This desire to prohibit by treaty had been interpreted by some jurists as evidence of a perceived *lack* of any customary law prohibiting the use of chemical weapons.

The general enthusiasm for the prohibition of CW was followed by the signing of the Geneva Protocol (Stockholm International Peace Research Institute, 1971). This protocol called for an acceptance of earlier conventions regarding chemical warfare and extended them to include bacteriological warfare. The wording regarding chemicals, taken from the 1922 Washington Treaty, prohibiting the use of:

> *asphyxiating, poisonous or other gases, and all analogous liquids, materials or devices,*

was particularly wide-ranging and could be interpreted as embracing tear-gases, herbicides, incapacitants and other non-lethal compounds. As written, the protocol was a very clear reflection of public and national perceptions regarding CW in the 1920s. In the UK, the Geneva Protocol was largely accepted by the military community; in the USA, it was not and intense pressure was brought to bear on the Senate Committee on Foreign Relations. The CWS and the US chemical industry orchestrated this pressure, and by December 1926 had so persuaded the Senate Committee on Military Affairs that the Chairman said:

> *I think it is fair to say that in 1922, there was much of hysteria and much of misinformation concerning chemical warfare. I was not at all surprised at the time that the public generally – not only in this country but in many other countries – believed that something should be done to prohibit the use of gas in warfare. The effects of that weapon had not been studied at the time to such an extent to permit information about it to reach the public. There were many misconceptions as to its effects and as to the character of warfare involved in its use* (Robinson, 1971b).

The results of a survey of 3500 American physicians were quoted, demonstrating that there was an informed consensus that, in comparison with more conventional weapons, gas caused less suffering, both during exposure and during its aftereffects. This view was supported by The Association of Military Surgeons, The American Legion, The Veterans of Foreign Wars of the United States, the Reserve Officers Association of the United States and the Military Order of the World War (Robinson, 1971b). This pressure was irresistible and the USA delayed ratification of the Geneva Protocol until 1975 – and then agreed with reservations. Other countries also delayed ratification: UK, 1930; Germany, 1929; Ethiopia, 1935; USSR, 1928. France ratified the Protocol in 1926. Many of those ratifying entered reservations, usually including the following:

1. The protocol could only be regarded as binding as regards states which had signed, ratified or acceded to it.
2. The protocol would *ipso facto* cease to be binding regarding a state failing to adhere to it.

These reservations seem like examples of simple common sense or prudence, but have been interpreted by some jurists as implying a lack of conviction amongst the contracting parties as regards the 'unthinkability' of chemical warfare: in that, they allow for the possibility of waging CW under certain circumstances.

During the period following the agreement (although not ratification) of the Geneva Protocol, public opinion, in general, remained against chemical weapons as a means of waging war.

The International Committee of the Red Cross (ICRC) conferences held in the late 1920s agreed resolutions deploring chemical warfare. The ICRC also attempted to encourage work in the field of civil defence against CW and pointed out that the level of protection available to the public in many countries was lamentably low. In 1929, *The Times* reported the ICRC as offering a prize for the best mustard gas detector and of planning competitions for design of civilian anti-gas equipment (The Times, 1929). One of the few countries to take such advice seriously was Russia, and in 1928, a simulated CW attack by 30 aeroplanes on Leningrad was undertaken. *The Times* reported a comment from *Izvestia* to the effect that the public regarded the powder bombs as more 'an ordinary street sight' than a 'serious exercise'! (The Times, 1928) The public perception that gas attacks were so horrible to contemplate that no civilized country would indulge in such attack was widespread at this time.

In 1935–1936, this perception changed. Italian forces used mustard gas against Ethiopian troops to considerable effect. The Italians argued for their legitimate use of a weapon prohibited by international law in terms of its use being by way of a reprisal (and therefore seen as exempt from the usual requirements of international law) for the behaviour of Ethiopian troops in torturing and mutilating prisoners of war. This defence, or justification, of the use of CW was seen by the Italian government as acceptable, despite the fact that Italy had signed the Geneva Protocol in 1925 and ratified it in 1928. Interestingly though, Italy had not entered the common reservations regarding the treaty ceasing to be binding under certain conditions and relied on the *customary* international law that reprisals may involve the use of means normally regarded as illegal. The maximum extent of such reprisals has never been specified, but it seems certain that Italy exceeded the spirit of the rules of war concerning such reprisals. It was noted in the UK and the USA that a fascist country, Italy, had used chemical weapons. Evidence was accumulating that Germany, under Hitler, was becoming interested both in chemical weapons and in strategic bombing as the means of waging war. Wickham Steed wrote pointing out the dangers of German developments in these directions and stressed the dangers of 'aerochemical attacks on civilians' (Steed, 1934). Efforts followed, in the UK, to develop air raid precautions against such attacks, and the *Air Raid Precautions Handbook No. 1*, published in 1936, was subtitled 'Personal Protection against Gas' (Ministry of Defence, 1936). Calculations undertaken by the UK and elsewhere, revealed that CW attacks on civilians in cities were, in fact, not likely to be particularly effective: high-explosive (HE) and, particularly, incendiary bombing, was calculated to be likely to be very much more effective. However, ARP measures turned on the dangers of a gas attack, and by 1939, only babies in the UK had not been issued with respirators. The dangers of a CW attack were emphasized, although in a 'we can cope if we are careful' way designed not to induce panic, and measures for dealing with HE bombing were taught at the same time.

Public opinion before WWII was firmly against the use of chemical weapons, although it is clear that some national leaders took a different view, and Winston Churchill returned to his long-held belief in the merits of CW in a series of wartime statements – mainly not intended for public consumption. In 1940, Churchill asked for an assessment of the value of mustard gas as an anti-invasion measure, and by 1944 his view had become characteristically clear:

> *It may be several weeks or even months before I shall ask you to drench Germany with poison gas, and if we do it, let us do it one hundred per cent. In the meanwhile, I want the matter to be studied in cold blood by sensible people and not by that particular set of psalm-singing uninformed defeatists which one runs across now here, now there* (Gilbert, 1991).

It seems clear from this and remarks made earlier in his career that Churchill was prepared to contemplate the use of CW in what he saw as a just cause. The discovery of the very toxic nerve agents by Schrader in Germany just before and during WWII was a major surprise, as already stated, to the Allied Powers, and the acquisition by the Soviet Union of the major German nerve agent production plants at Dühernfurt on the River Oder, now Dyhernfurth, Silesia, Poland, immediately after WWII, was seen as a particularly sinister development. The Soviet Union was understood to have acquired the technology to produce nerve agents in considerable quantities and programmes of aggressive CW research were put in hand in both the USA and the UK. In the UK, an offensive, as compared with a defensive, policy of research and development in the CW field was abandoned in 1956. The reasons for this abandonment are difficult to establish, as many documents relating to the decision remain unavailable.

US policy has changed over the period: chemical weapons were produced until 1969 (Meselson and Robinson, 1980). President Nixon, in 1974, agreed with Secretary Brezhnev of the USSR to consider a joint initiative on the prohibition of chemical weapons. This was confirmed by President Ford in 1976 and bilateral negotiations began in Geneva in that year. As after WWI, the US Chemical Army Corps (which replaced the CWS) was reduced in strength (from *ca.* 4000 to *ca.* 2000) and, as before, began a campaign of lobbying. This, combined with the concept of putting pressure on the Soviet Union by *improving* US weapons at the same time as arguing for their bilateral removal, led in 1981 to President Reagan permitting the production of so-called 'binary weapons'. This programme has been subject to considerable criticism both inside the USA and within other NATO countries. In particular, attempts to store chemical weapons in Europe certainly met strong opposition. During the 1970s and early 1980s, public opinion regarding CW underwent little change. The proponents of CW argued for possession of weapons on the grounds that strong defence was vital and that the Soviet Union's pronouncements regarding its desire to abandon a capacity to conduct a chemical war should not be trusted. The opponents of CW have continued to argue that the use of such weapons would be reprehensible and their possession inevitably increased the likelihood of their use. Very little debate regarding the ethical aspects of chemical warfare took place during this period and articles arguing in favour of chemical warfare, such as that by Lidell Hart, are scarce (Lidell Hart, 1960). The view espoused by Haldane that CW was a good deal *more* desirable than conventional war seemed to have gone into eclipse.

Public opinion regarding CW changed little during the Iran–Iraq conflict in the 1980s and standard condemnations of the use of chemical weapons were made by Western political leaders. These condemnations reveal no new thinking on, or rethinking of, the ethical aspects of chemical warfare but followed the usual assumption of the undesirability of CW and pointed out that it was prohibited under the terms of the Geneva Protocol of 1925. Iraq, the main user of CW during the conflict, had ratified the Geneva Protocol in 1931.

In summary then, public opinion regarding chemical warfare seems to have been formed in the years following WWI. It was markedly influenced by propaganda regarding the peculiar unpleasantness of this type of warfare, spread both by those anxious to build up a national CW capability and those anxious to achieve a prohibition of the future use of chemical weapons. Although others argued in favour of chemical warfare, they seem to have had little influence and public opinion remains against these weapons.

CHEMICAL WARFARE EXAMINED FROM THE STANDPOINT OF THE CUSTOMARY RULES OF WAR

It is widely accepted by civilized nations that war should only be engaged in to achieve certain acceptable ends. Perceptions as to what comprise acceptable ends have changed in the course of history, and, for a long period, the expansion of one's state was seen as a perfectly acceptable purpose with rulers engaging in wars to acquire more land or certain facilities, e.g. access to coasts, to water supplies and to better farm lands. Such reasons for going to war would not be seen as acceptable today. The opinion that wars should not be fought for reasons of expediency is embodied in the first of the often-quoted seven conditions which need to be satisfied before a war can be regarded as a *just war* (McKenna, 1960):

1. The war must be declared by the legitimate public authority and the action must aim at universal good; it must not be a matter of expediency.

2. The seriousness of the injury to be suffered (should war not be declared) must be proportional to the damages war will cause.
3. The seriousness of the injury to be suffered at the hands of the aggressor must be real and immediate.
4. There must be a reasonable chance of winning the war.
5. War may be initiated only as a last resort to redressing grievances.
6. A war may be prosecuted legitimately only insofar as the responsible agents have a right intention.
7. The particular means employed to wage the war must themselves be moral.

Points 1–6 refer to the *ius ad bellum*: the necessity of a just cause for a just war. Point 7 refers to the *ius in bello*: the necessity of just means of fighting for a just war.

The seven conditions have also been defined by J. J. Haldane (Haldane, 1987a)[6] in rather simpler terms:

1. The war must be made by a lawful authority.
2. The war must be waged for a morally just cause, e.g. self-defence.
3. The warring state must have a rightful intention, i.e. to pursue the just cause.
4. The war must be the only means of achieving the just end.
5. There must be a reasonable prospect of victory.
6. The good to be achieved must be greater than the probable evil effects of waging war.
7. The means of war must not themselves be evil: either by being such as to cause gratuitous injuries or deaths, or by involving the intentional killing of innocent civilians.

Before considering the ethical aspects of CW in more detail, it may be as well to define three terms sometimes used in discussing this question.

Ethics There are a number of definitions of 'ethics'. The Shorter Oxford English Dictionary (SOED) defines 'ethics' as 'the, or a, science of morals'. This definition and those like it have led many to confuse ethics with morals or ethical behaviour with moral behaviour. Russell (1927) attempted to separate ethics from morals by defining ethically correct behaviour as transcending the requirements of any particular religious doctrine, eg, ethical behaviour could be defined as behaviour designed to increase the level of happiness of people.

Morals Morals, or more easily moral codes, are considerably easier to understand than ethics or ethical codes. A moral code applies to behaviour and not to objects. In the Christian faith, basic moral standards of behaviour are laid down in the decalogue: 'Thou shalt not kill', etc. Moral teaching lays down rules of behaviour, for use in fairly specific cases.

Casuistry Casuistry is the science of determining, on an ethical or moral basis, the correct action an individual should take under a given set of circumstances. As Russell has pointed out, casuistry has acquired a bad reputation and is sometimes seen as a branch of, or analogous to, sophistry. Casuistry attempts to define ethically and morally correct actions under a given set of circumstances, whereas sophistry is 'the use or practice of specious reasoning as an art of dialectic exercise' (SOED) Casuistry led to the very precise, but often probably pointless arguments of medieval philosophers, and was often regarded as a destructive technique: 'Casuistry destroys, by distinctions and exceptions all morality' (Bolingbroke (SOED)). Unfortunately, condition 7 of the 'just war code' requires a casuistic approach, and it will be demonstrated below that, on some occasions, to wage chemical warfare might be an ethically and morally correct act, but under other circumstances it might not.

An example of the difficulty encountered in applying an ethically and morally correct approach in war is provided by the case of the spy in time of war. Spying is held to be prohibited by generally accepted rules of war as unethical and yet all countries capable of so doing engage in spying in peacetime and to such an extent as is possible in wartime. For a country perceived to be fighting a *just war*, it is difficult to regard the activities of its spies as unethical or immoral; of course, it is likely the other country involved would also

[6] J. J. Haldane is currently Senior Lecturer in Moral Philosophy at the University of St Andrews, Scotland, UK and should not be confused with Professor J. S. Haldane (1860–1936) who undertook research on CW during WWI with his son, Professor J. B. S. Haldane (1892–1964).

perceive its actions as just and have the same difficulty regarding its spies.

The importance of defining morally acceptable behaviour in terms of specific religious philosophies should not be overlooked. Different religions define different morally acceptable codes of conduct, and one cannot argue that use of a chemical weapon by an enemy is morally wrong unless he subscribes to one's own religious beliefs and one has demonstrated that those beliefs lead to such use being held to be immoral. The difference between ethics and morals arises at this point.

One might hold an act by an enemy to be unethical although not immoral. The assumption here is that an ethical code which transcends all religious or moral codes, could, in principle at least, be defined.

Ethical and moral thinking is often seen as leading to the definition of legal principles. Some have attempted to distinguish between the results of ethical and legal thinking, and J. J. Haldane (Haldane, 1987b) has defined the difference between legal principles and objective moral (he uses 'objective moral' and 'ethical' interchangeably) principles as:

> *Law issues from the will of legislators, and binds those who fall within its scope, either by knowingly consenting to be bound by it or else by acting in ways which imply their implicit acceptance of the rules of society. Morality, if it is objective, has a foundation independent of the human will and has a logically different kind of force upon us. What has been willed (i.e. a legal principle) can be opposed and revoked; what is the objective moral case, apart from human attitudes to it, remains such, however much as one or all may wish it were not so.*

The critical difference between legal and ethical principles controls the way in which activities may be construed. By observing patterns of activity, one can arrive at a 'descriptive norm' of behaviour. This is an adequate basis for a legal principle. On the other hand, observing how people act is logically of no value whatsoever in defining how they *should* act: i.e. in defining a norm of behaviour *prescriptively*. Knowing that Germany and the UK used chemical warfare during WWI and that Germany and the UK did not use CW during WWII is of no assistance in defining how they *should* have behaved or how they *should* behave in the future. Defining an ethically acceptable pattern of behaviour is very much more difficult than defining a legally acceptable one; although if it can be done it will likely last longer and likely be more effective. J. J. Haldane has said in considering the establishment of constraints on the use of biological (although his argument applies with equal force to chemical) weapons 'only morality *could* provide an inescapable external constraint'.

Some authorities, including Sims (1987), have been concerned about the origins of attitudes and beliefs regarding chemical, biological and nuclear warfare. It has often been said (see above) that revulsion regarding CW is an instinctive response. Such responses are generally seen as automatic, dependent upon genetic factors and cultural patterns and as being independent of reason. On the other hand, it has been held that a prerequisite for a satisfactory ethical or moral belief is that it should be based upon a rational and demonstrably true judgement. Given that instinctive responses may not necessarily be rationally based, it has been argued that an ethical judgement founded upon an instinctive belief is unsatisfactory and may be unconvincing. J. J. Haldane has considered the following question (my paraphrasing) as regards BW, although the thrust of his answer applies equally to CW:

> *Does the possibility of one's revulsion from induced plague being an instinctive response exclude it from being a moral judgement?*

He commented:

> *This line of thought instantiates the fallacy of supposing that if one has a 'genetic' explanation of why someone is in a certain psychological state, e.g. because of their nature or upbringing, the question of the rationality or truth does not arise. This is fallacious because the origin of a belief is logically independent of its content. Thus, whether an expression of revulsion is instinctive or the product of reflection does not touch upon the issue of whether it is a genuine moral judgement, or whether, if it is such, it is justified and true. These questions can only be answered by looking at the content of the expression.*

This seem to me to be a most important comment and one which strikes in two directions.

The view that CW is unethical *by virtue* of the widespread instinctive revulsion it engenders is dismissed, as is the concern that an ethical judgement based upon such an instinctive revulsion is inevitably flawed. One might conclude that the existence of a marked instinctive revulsion for CW is of no relevance in assessing the ethical or moral propriety of this form of warfare.

Certain standards of acceptable behaviour can be defined, indeed have been defined, and comprise the basis of the generally accepted rules of conduct of war. These standards, or perhaps concepts, are helpful in deciding whether a given use of a weapon would generally be regarded as ethically acceptable. The concepts are sometimes described, in part, as the 'laws of chivalry'. These concepts have certainly existed for very many years and date from a period of limited rather than total warfare: the warfare of armies facing each other on a battlefield. The rules of chivalry include prohibitions of:

1. Treachery.
2. Aggravating the suffering of disabled men.
3. The rendering of death inevitable.
4. The use of poison and poisoned weapons.

These prohibitions all relate to actions and are moral or ethical judgements. However, the list is internally inconsistent in that while (1), (2) and (3) define general aspects of behaviour, (4) specifies a particular weapon system or means of killing. These prohibitions form part of the customary rules of war and as such form a part of 'The Rules of War as Recognized by Civilised Nations'. The broad origins of these rules, some detailed aspects of the rules and the implications of their origins will be considered next.

ORIGINS OF THE RULES OF WAR

The generally accepted rules of war, like much other international law, may arise from several sources. These may be divided into the following:

1. Customary international law which defines how states have behaved over a period of time. Customary law need not make any appeal to ethical thinking; indeed, it may not be rationally based and it is certainly subject to change.
2. Treaty law or conventional law. This is often introduced to reinforce customary law or to create new international law – for example, a boundary change or the agreement between nations to abandon the use of a weapon system.
3. Ethical concepts generally held in common by civilised nations even though they may follow different religions, and, therefore, adhere to different moral codes. Chivalrous behaviour during war is often said to fall into this category.
4. General principles of law as recognized by civilized countries. Some would add the general principles of international law.

Bearing in mind these various sources, five major principles of law as concerns conduct of warfare, may be considered (General Pershing, US Army, 1970):

1. The principle of military necessity. This states that subject to other principles, e.g. of humanity and chivalry, a participant is justified in applying any amount and any kind of force to compel the complete submission of the enemy within the least possible expenditure of life, time and money.
2. The principle of humanity, prohibiting the employment of any kind or degree of violence that is not actually necessary for the purpose of the war.
3. The principle of chivalry, implying concepts of honourable behaviour and mutual respect between forces.
4. The principle of reprisal – or the right to take action which would normally be considered illegal in order to demonstrate that illegal activities on the part of one's opponent will not be tolerated.
5. The principle of self-defence, allowing action to be taken against an actual or impending attack.

Principles (1) and (2) may be considered together: these principles imply much of the accepted rules of how a war should be conducted. They embody the concept that *just wars* are fought to compel one's opponents to amend their actions and not for their annihilation. Under this principle, the disproportionate application of force, destruction of people and property in excess of that needed to achieve the military objective and deliberate extension of the duration of the war, are prohibited. No mention is made of individual weapon systems. How CW conforms to these requirements will depend entirely upon how it is used. The use of chemicals *may* represent a use of force disproportionate to the end in view. However, the use of chemical weapons *does not necessarily* represent such a violation of a principle of war, e.g. the use of one shell loaded with nerve agent during a tank engagement could not be defined as a use of disproportionate force.

A principle often invoked by those who oppose the use of chemical weapons is that of chivalry, and it is here that we approach the centre of many peoples' objections to chemical warfare. The principles of chivalrous behaviour seem very long-lived and their major prohibitions have already been stated.

Along with the prohibitions listed above, the necessity of providing warning of attacks is generally accepted in theory, but ignored in practice, particularly as regards air raids. To the prohibitions, another could be added – embraced in part by the prohibition of the conception of deliberately aggravated suffering – weapons producing suffering but no military advantage should clearly be banned.

Having considered the origins of international law as they affect the conduct of wars, one can consider the commonly made criticisms of chemical warfare and ascribe weight to them.

ANALYSIS OF OFTEN-ASSERTED OBJECTIONS TO CHEMICAL WARFARE IN TERMS OF ETHICS AND THE RULES OF WAR

A great many criticisms have been made of chemical warfare; some of these will be considered below.

Chemical warfare is uncontrollable

If chemical warfare were uncontrollable, then the principle of humanity would clearly be breached. It is important to understand what is meant by uncontrollability of chemical warfare: two possible interpretations exist:

1. The weapons themselves cannot be controlled. This is untrue, as a very great deal is known of the likely spread of chemicals on a battlefield and modern weapons allow more precise targeting than older systems.
2. The second interpretation of uncontrollability is that use would escalate out of control. Given the rules of war, this is no more likely to occur than a loss of control in the use of conventional weapons, e.g. escalation of bombing of civilian populations.

Chemical weapons produce unpredictable effects

This is untrue. Compared with the random chance (real unpredictability) of effects of fragmenting munitions, the effects of chemical weapons may be predicted with considerable accuracy.

Chemical weapons produce inevitable death

This is untrue. During WWI, the mortality in mustard gas casualties was of the order of 2%; over 8% of gunshot wounds resulted in death. High concentrations of lethal agents, such as nerve agents, would likely produce inevitable death, but again this is a criticism of mode of use rather than the weapon.

Chemical weapons produce excessive suffering

Great emphasis was placed during WWI and afterwards on the suffering of gas casualties. The reasons for this propaganda drive have already been discussed. The painting by Sargent showing the shuffling line of blinded gas casualties has made a great impression on public opinion. The picture does not reveal that by three weeks later all, or almost all, of the casualties would have completely recovered their sight. On the other hand, men blinded by shell splinters do not as a general rule recover their sight. Of course, some chemical weapons have been designed to produce pain and so disable the casualty. Phosgene oxime (one of the 'nettle' gases) works in this way. J. B. S. Haldane made the point that injuries, which often became infected, from shell fragments produced a great deal more pain than the majority of chemical injuries. This is supported by the frequent need for strong analgesics – morphine, etc. – in cases of severe physical injury and the observation that such measures are much more rarely required by CW casualties.

Chemical weapons would spread to and devastate civilian populations

During WWI, down-wind drift of gases – particularly chlorine and phosgene (gases released from cylinders) – did occur and civilians were affected. The number of civilians affected was comparatively small, but the fact that women and children had been affected was stressed in the anti-CW propaganda. Drifting contamination is a problem in chemical warfare and is dependent upon wind and local weather conditions. The objection to CW based on its spread to local civilians is, in fact, not an objection to CW, but an objection, again, to a particular mode of use or to its use on a particular occasion. CW can be used in such a way – precisely targeted weapons discharging compounds of low volatility, producing mainly local contamination – that civilians need not be injured.

The question of civilians being injured during war extends beyond the use of chemical weapons. The concept that *no* civilian casualties should be produced dates from the period of very limited wars involving fairly small armies which chose to fight at some distance from civilian populations. In modern warfare – and particularly in so-called 'total war' – civilians will, inevitably, be injured; indeed, some military actions, e.g. bombing of armaments factories and railway terminals, may always be expected to involve the production of civilian casualties. The ethical position of killing civilians has been considered by Catholic and other theologians in terms of the 'double effect' concept (Krikus, 1964). This concept states that if civilians are killed as a result of a justified (just) military engagement, then the deaths of the civilians do not make the act which killed them evil. This argument has been used in the nuclear field (Ramsay, 1961), where the killing of civilians as a side-effect of a justifiable attack on a military target has been *not* regarded as evil. It should be said that this argument has struck many as dangerous – as indeed all arguments based on an assertion of 'necessity' strike many as dangerous:

> *So spake the Fiend, and with necessity,*
> *The tyrant's plea, excus'd his devilish deeds*
> (Milton, *Paradise Lost*).

However, it raises a very interesting ethical question regarding the establishment of an absolute view of the rights and wrongs of a conflict. As said earlier, the concept of the 'just war' is based upon the assertion that one can determine a just cause and then pursue that cause using just means. The just cause is seen as transcending simple gains of territory or influence: one must be, in some way, fighting against evil and it has to be assumed that one's definition of evil is not only correct but would also be admitted to be correct by one's opponents were they able to take an objective view of their actions. There is nothing within the concept of the *just war* which allows the outcome of the conflict to determine whose was the just cause. This represents a change in thinking from times when a military engagement was seen as an acceptable means of settling a claim, i.e. an assertion of the rightness of one's cause, perhaps to an area of territory. This change in concept has much to do with the charges of unchivalrous behaviour often brought against users and proponents of chemical warfare.

Chemical weapons are in some way an unchivalrous means of conducting conflict

The concept of chivalrous behaviour embraces a wide range of attitudes and beliefs regarding how one should behave in conflict. The concept of respect for one's enemy, the belief that one is not *per se* fighting to kill one's enemy but to make his rulers change their ways and the belief that 'fair play' has some role to play, even in war, all make up a part of the concept of chivalry. The concepts are held to be infringed if one side behaves in a treacherous fashion, i.e. having made an agreement, breaks that agreement. This is a sensible concept based as it is upon the notion that normal life, i.e. after a war, as well as the rather abnormal life experienced during a war, will be dependent upon one's capacity to trust others. Clearly, a country which has made an agreement not to use chemical weapons and then uses them, e.g. Iraq in the 1980s, could be considered to have acted treacherously. The validity of the allegation is in no part dependent upon the nature of the treacherous act and the use of chemical weapons cannot *per se* be described as an act of treachery.

The definition of what constitutes a treacherous act is sometimes considerably extended to include acts carried out in a clandestine manner and for which no warning is provided. The idea of a gas seeping along trenches, maiming and killing, conjures up a picture of clandestine activity. There is some dependence here upon the difficulty of detecting an attack by an invisible gas and this is seen again as a hallmark of a clandestine operation. Surprise is a fundamental concept of military tactics and warning of an attack upon troops is not required under generally accepted rules of war. Warning of attacks which could involve injury to civilians has been held to be necessary, to allow evacuation of noncombatants, but in these days of rockets and air attacks, this requirement is seldom regarded as mandatory.

Concepts of chivalry are also held to be infringed if one acts in such a way as does not require some measure of courage, ideally as much courage as shown by one's opponents, and one is therefore not prepared to demonstrate one's willingness to place one's life at risk. Weapons which strike from long range and weapons operated by civilians (in particular, scientists) attract censure on these grounds. In my view, it is here that we are very close indeed to, or perhaps at, the heart of the widespread objections to chemical warfare. When matchlock muskets were introduced, it was observed that 'Soldiers were struck down by abominable bullets which had been discharged by cowardly and base knaves, who would never have dared to meet true soldiers face to face' (Vedder, 1925). J. B. S. Haldane noted that Chevalier Bayard (described by contempories as 'sans peur et sans raproche') was the soul of courtesy to captured knights, and even bowmen, but invariably put to death musketeers or other users of gunpowder who fell into his hands. During WWI, Lord Kitchener described the German use of chemical warfare as an indication of the extent of depravity to which Germany would sink in an attempt to compensate for a lack of courage on the part of her forces (Kitchener, 1915). J. B. S. Haldane excoriated such a view, describing it as 'bayardism' and observed that it represented 'one of the most hideous forms of sentimentalism which has ever supported evil upon earth – the attachment of the professional soldier to cruel and obsolete killing machines' (Haldane, 1925).

Chivalrous behaviour also in some way depends upon a balance of forces and capacities to wage war. Two well-matched armies, composed of equally well-armed and equally brave troops, is the model for a conflict where chivalrous behaviour is likely to occur. The idea that scientists of one side could give that side an advantage which no amount of courage and bravery on the part of their opponents could counter, particularly if the weapon system providing such an advantage was cheap to produce, is often seen as an example of unchivalrous conduct.

A last point usually included in definitions of chivalrous behaviour is one relating to the use of poisons and poisoned weapons. This is almost invariably seen as wrong – particularly among civilized nations. The poisoned weapon is seen as being designed to make death inevitable or to produce unnecessary and aggravated suffering and, along with the expanding (dum-dum) bullet and the barbed lance, has long been prohibited. As regards poison, a rational and deep-rooted fear certainly exists in most people. Presumably, this dates from the early exposure of mankind to naturally occurring poisons, and the subsequent difficult-to-explain deaths of the individuals so exposed instilled this fear in our earliest ancestors. The concepts already mentioned of the lack of need for physical courage on the part of the poisoner and the fact that such an attack may evade detection, also add to the particular dislike of this method of killing. In the first section of this account, the intense level of activity of poisoners in the Middle Ages was stressed. The view that poisoning is unchivalrous may stem from the very pragmatic view that unless poisoning were outlawed, no ruler would be safe, and no war could ever be won, for a late 'assault by poison' could reverse any victory acquired on the field of battle. This prohibition, while rational and sensible, does not bear upon the large-scale use of chemicals during war, but rather upon the small-scale use of chemicals during peace.

Terrorism

In 1994 and 1995 there were two instances of the use of sarin against civilian populations by terrorists, both in Japan. The first was in Matsumoto, a mountain town in central Japan. Seven people died. There were about 200 casualties who suffered symptoms which were typical of organophosphate poisoning. Treatment was with atropine and diazepam. Oxime reactivators were not used for fear that the poisoning was caused by a carbamate. In March 1995 a similar incident happened on the Tokyo underground subway system. There was a large number of casualties and the incident was similar to the one described above, but the cause of the poisoning was recognized earlier as sarin, and PAM methiodide was used in addition to atropine. There was a total of eight deaths. Miosis, headache and dyspnoea were the most constant signs. The miosis was a very severe and long-lasting one and it was associated with marked discomfort. Cholinesterase levels, which were initially very low, seemed to recover very quickly (within hours) in those successfully resuscitated. Psychological and other sequelae seem to have been few and the hospital staff were rarely affected, despite a lack of individual physical protective clothing (see Chapter 13).

CONCLUSIONS

The use of chemical weapons is prohibited by a number of international treaties or conventions. This illegality is said to confirm and codify a customary prohibition based on concepts of ethics and morals. Despite these prohibitions, and unlike other weapon systems also prohibited, the number of countries which have acquired or which are acquiring chemical weapons continues to increase. It might be argued that this is explained, in part, by the lack of force of international agreements resulting from their being based upon false assertions and their being established largely as a result of vigorous propaganda exercises in the aftermath of a world war. Current international moves to abandon chemical weapons may well be brought to a successful conclusion but, with the lack of a sound and persuasive argument against the use of such weapons, it remains very unwise to assume that clandestine production of such weapons will not be attempted and that they will not be used during a war. In addition, the lack of a sound ethical compulsion to desist from using such weapons may very likely lead to their proliferation and use in the so-called Third World conflicts, particularly among nations with no memory of WWI.

REFERENCES

Blyth AW and Blyth MW (1920). *Poisons: Their Effects and Detection.* London: Charles Griffin & Co., Ltd.

Brown FJ (1968). *Chemical Warfare: A Study in Restraints.* Princeton: Greenwood Press.

Fotion N and Elfstrom G (1986). *Military Ethics.* London: Routledge and Kegan Paul.

French J (1915). Second Battle of Ypres. *The Times,* 12 July, p. 9.

General Pershing, US Army (1970). In: *Legal Limits on the Use of Chemical and Biological Weapons* (A Van W Thomas and AJ Thomas, eds). Dallas: Southern Methodist University Press, 1970.

Gilbert M (1991). *Churchill. A Life,* pp. 782–783. London: Heinemann.

Haber F (1919). On the occasion of his being presented with the Nobel Prize for Chemistry. In: *A Higher Form of Killing* (R Harris and J Paxman, eds). London: Chatto and Windus, 1982.

Haber LF (1986). *The Poisonous Cloud – Chemical Warfare in The First World War.* Oxford: Clarendon Press.

Haldane JBS (1925). *Callinicus: A Defence of Chemical Warfare.* London: Kegan Paul, French, Trubner & Co., Ltd.

Haldane JJ (1987a). Defence policy, the just war and the intention to deter. *Defence Anal*, **3**, 51–61.

Haldane JJ (1987b). *Ethics and Biological Warfare Arms Control*. October Issue, p. 24.

Holmstedt B (1959). Pharmacology of organophosphorus cholinesterase inhibitors. *Pharmacol Rev*, **11**, 567–688.

Journal of Industrial and Engineering Chemistry (1919). Beware the Ide[a]s of March! (Editorial). **11**, 814–816.

Kitchener H (1915). Letter from General H Kitchener to Sir J French (copy in possession of author).

Krikus RJ (1964). On the morality of chemical and biological war. *Conflict Resolution*, Vol. IX.

Lefebure V (1921). *The Riddle of the Rhine*. London: W. Collins Sons & Co., Ltd.

Lidell Hart BH (1960). *Deterrent of Defence – A Fresh Look at the West's Military Position*. London: Stevens and Sons, Ltd.

Mahan AH (1937). In: *Chemicals in War* (AM Prentiss, ed.), p. 686. New York: McGraw-Hill Book Company, Inc.

McKenna JC (1960). Ethics and war: a catholic view. *Am Polit Sci Rev*, **54**, 647–658.

Meselson M and Robinson JP (1980). Chemical warfare and chemical disarmament. *Sci Am*, **242**, 34–43.

Ministry of Defence (1936). *Air Raid Precautions Handbook No. 1*. London: HMSO.

Prentiss AM (1937). *Chemicals in War*. New York: McGraw-Hill Book Company, Inc.

Ramsay P (1961). *War and the Christian Conscience*. Durham: Duke University Press.

Robinson JP (1971a). The rise of CB weapons. In: *The Problem of Chemical and Biological Warfare*, Chapter 2. New York: Stockholm International Peace Research Institute.

Robinson JP (1971b). Popular attitudes towards CBW 1919–1939. In: *The Problem of Chemical and Biological Warfare*, Chapter 3. New York: Stockholm International Peace Research Institute.

Russell B (1927). *An Outline of Western Philosophy*. London: George Allen and Unwin, Ltd.

Scientific American (1989). March Issue, p. 8.

Sims NA (1987). *Morality and Biological Warfare Arms Control*. May Issue.

Stockholm International Peace Research Institute (1971). *The Problem of Chemical and Biological Warfare*, Appendix 5. New York: Stockholm International Peace Research Institute.

The Reprint Society (1938). *The Notebooks of Leonardo da Vinci*. London: The Reprint Society.

The Times (1921a). *A Rain of Death*. 14 March, p. 11.

The Times (1921b). *A Deadly War Gas*. 1 September, p. 9.

The Times (1928). *Sham Air Attack on Leningrad: Gas Masks and Bombs*. 11 June, p. 14.

The Times (1929). *Civil Population and Gas Warfare*. 20 May, p. 9.

US State Department (1937). Statement regarding the First Hague Conference, 1899. In: *Chemicals in War* (AM Prentiss, ed.), p. 685. New York: McGraw-Hill Book Company, Inc.

Vedder EB (1925). *The Medical Aspects of Chemical Warfare*. Baltimore: Williams and Wilkins Co.

ANNEX

Terrorism: the possible use of chemicals

In the period since the preceding chapter was originally written, the threat of large-scale chemical warfare seems to have receded. Countries are destroying stockpiles of chemical weapons and even in those countries known at one time to have a capacity and perhaps willingness to wage chemical warfare, for example, Iraq, destruction of these weapons appears to have been completed. It is, of course, possible that a number of countries hold stocks of chemical weapons and that they might, under certain circumstances of warfare, use them. The general trend towards national disarmament as regards chemical weapons is discussed by Pearson in Chapter 28.

Although the likelihood of large-scale use of chemicals against armed forces has receded, the possibility that terrorist groups might use chemicals as weapons to injure civilians has increased. The use of the nerve agent sarin in Tokyo in 1995 has demonstrated that well-organized and well-funded groups can make very deadly compounds and can devise effective methods for their dissemination.

That terrorist groups are willing to injure and kill civilians is, of course, not a new development. The use of explosives and firearms for this purpose is long established and on September 11 2001 Al Qaeda hijacked two airliners and flew them into the World Trade Center (the Twin Towers in New York). The Pentagon, the Headquarters of the US Armed Forces was also attacked. A detailed account of the use of sarin on the Tokyo subway is provided in Chapter 13. Such attacks have given rise to a public outcry: the fact that unarmed and undefended civilians have been attacked has been roundly condemned by all commentators.

The ethical aspects of such attacks are, however, confused and discussion is inhibited by their appalling consequences. It is assumed that there can be no ethical justification for such attacks – the use of chemicals seems, if anything, even more reprehensible than the use of explosives and firearms. As was discussed in the preceding section, this revulsion of the use of chemicals is not well-based and it would be foolish to assume that terrorist groups would be inhibited by general condemnation. On the contrary, it may be that terrorist groups would see chemicals as a low-cost and effective alternative to more conventional means of attack. It is certainly unlikely that terrorist groups will feel bound by international conventions regarding the use of chemicals in warfare. This sombre conclusion leads at once to the need to prepare to deal with the consequences of attacks with chemicals on civilian populations. Prevention for such attacks is clearly a priority and many countries are now actively engaged with this task. As in the case of attacks using conventional weapons, it is sensible to accept that a guarantee of complete prevention is difficult to achieve and thus preparations to deal with the consequences of such an attack are necessary. Many countries, including the UK and USA, are taking steps in this direction and the acquisition of protective equipment for rescue workers, of decontamination facilities for casualties and antidotes to oppose the effects of chemicals have all been publicized. Information has been made available to the public and to the medical profession. In the UK, such information is available on official websites (Health Protection Agency website http://www.hpa.org.uk; Department of Health website http://www/dh/gov.uk). A detailed account of measures that should be taken in responding to such an attack or incident is provided in Chapter 11.

2 THE PHYSICOCHEMICAL PROPERTIES AND GENERAL TOXICOLOGY OF CHEMICAL WARFARE AGENTS

Robert L. Maynard

Health Protection Agency, Chilton, UK

Chemical warfare (CW) agents may be encountered as solids, liquids or gases. Some agents, e.g. sulphur mustard, may appear as solids under North European winter conditions (freezing point 14.4°C), as a liquid at a wide range of temperatures (boiling point 219°C) or as a vapour evaporating from the liquid phase.

CW agents may also be encountered as mixtures or solutions of one agent in another, or of an agent in a solvent. The mixing of lewisite with sulphur mustard has been undertaken to lower the vapour pressure and freezing point of the mustard and hence to increase its persistence, without reducing the effective CW payload of weapon systems. Sulphur mustard has also been mixed with phenyldichlorarsine, the mixture being referred to as Winterlost, i.e. winter mustard.

GASES AND VAPOURS

All gases may be liquefied (converted to the liquid state) by applying sufficient pressure at an appropriate temperature. Each gas is characterized by a critical temperature. At temperatures greater than the critical temperature, the gas cannot be liquefied by the application of pressure. Gases characterized by critical temperatures greater than room temperature are sometimes referred to as *vapours*. Each gas is also characterized by the pressure needed to produce liquification at the critical temperature. This pressure is the critical pressure. Table 1 provides the critical temperatures and pressures for some common gases.

This table contains some surprises: phosgene, which is certainly usually described as a gas, can be liquefied at room temperature as can chlorine. Substances with low critical temperatures, such as oxygen, hydrogen and nitrogen, used to be described as 'permanent gases' to distinguish them from substances such as chlorine and phosgene. Describing phosgene as a vapour rather than as a gas because its critical temperature is greater than room temperature leads to confusion.

The term 'vapour' is sometimes used to describe the gaseous phase of a substance that exists as a liquid at room temperature. Thus, mustard vapour, nerve agent vapour and vapour hazard are terms used in the chemical warfare area. The use of such terms is not, however, standardized or universal and we still speak of mustard gas and nerve gas. There is much to be said for adopting a simple approach: there are three phases of matter: solid, liquid and gas; introducing the term vapour as a variant of gas does not seem to be helpful.

Chemical Warfare Agents: Toxicology and Treatment (2nd Edition)
Edited by Timothy C. Marrs, Robert L. Maynard and Frederick R. Sidell © 2007 John Wiley & Sons, Ltd

Table 1. Critical temperatures and pressures for some common gases

Gas	t_c (°C)	p_c (atm)
Air	−140.7	37.2
Chlorine	144	76.1
Helium	−267.9	2.26
Nitrogen	−147.1	33.5
Oxygen	−118.8	49.7
Carbon dioxide	31.1	73.0
Hydrogen cyanide	183.5	53.2
Hydrogen sulphide	100.4	88.9
Phosgene	181.7	56.0

t_c: critical temperature
p_c: critical pressure

UNITS USED TO DESCRIBE THE CONCENTRATIONS OF GASES

Two systems are in general use: the mass-per-unit-volume system and the volume-fraction system. The former gives concentrations expressed typically as μg or mg per m^3 or per litre(l); the latter gives concentrations expressed as parts per billion or per million, ppb or ppm. When using concentrations to compare the toxicological effects of different toxic gases, the volume-fraction system is more satisfactory, it being unlikely that the relative toxicological effects are dependent on the relative mass of the individual molecules but upon their relative proportions in inhaled air and, of course, on the toxicological properties of the molecules. The systems are easily convertible.

Beginning with the accepted fact that at standard temperature and pressure (STP) 1 gram molecule (GMW) of any gas occupies 22.41 l, we can argue:

$$1\,\text{l} = \frac{\text{GMW}}{22.41}\,\text{g}$$

$$1\,\text{ml} = \frac{\text{GMW}}{22.41}\,\text{mg}$$

$$1\,\text{ml/m}^3 = \frac{\text{GMW}}{22.41}\,\text{mg/m}^3$$

$$1\,\text{ml/m}^3 = 1\,\text{ppm} \quad (1\text{m} = 100\,\text{cm},\ 1\,\text{m}^3 = 10^6\text{cm}^3)$$

$$1\,\text{ppm} = \frac{\text{GMW}}{22.41}\,\text{mg/m}^3$$

$$x\,\text{ppm} = x\,\frac{\text{GMW}}{22.41}\,\text{mg/m}^3$$

This is sometimes expressed as:

$$\text{Concentration in mg/m}^3 \text{ is the concentration in ppm} \times \frac{\text{MW}}{\text{mv}}$$

where mv = molecular volume, i.e. 22.4 l at STP.

The mv can be corrected to ambient conditions:

$$\text{mv (ambient)} = 22.41 \times \frac{T}{273} \times \frac{760}{P}1$$

where T is the ambient temperature expressed in degrees absolute and p is the ambient pressure in mmHg. If p is expressed in millibars, then 1013 should be substituted for 760 in the above equation.

Ozone provides an example, GMW = 48 g.

Accepting that $48 \approx 2 \times 22.41$, the concentration expressed in ppm is, numerically, about half that exposure in mg/m^3:

$$\text{i.e.}100\,\text{ppm} = 100 \times 2 = 200\,\text{mg/m}^3$$

Note that the concentration expressed as for e.g. ppb, is *by volume*. For liquids suspended as an aerosol, expressing the concentration on a mass per unit volume basis is a natural approach. However, the following approach is sometimes used.

Air is a mixture of about 80% nitrogen and 20% oxygen. The molecular weights of nitrogen and oxygen are similar and we might assume, in 'round terms' that the GMW of air was 28 g.

$$\text{Thus 1 l of air weighs } \frac{28}{22.41}\,\text{g}$$

or 1 m^3 of air weighs

$$\frac{28}{22.41} \times 1000\,\text{g} = \text{about 1 kg.}$$

If the concentration of some liquid suspended as an aerosol is 1 μg/m^3, its concentration can, therefore, be expressed as about 1 μg/kg or about 1 ppb. Note that this concentration is expressed as ppb *by mass*.

Calculations of the mass of a toxic material needed to contaminate a large volume of air to a lethal level are often ignored by those suggesting that a specific compound might be used effectively on a large scale. Consider sarin (GB). Let us assume that we wished to contaminate an area of 1 km^2 to a depth of 10 m with GB at a concentration of 100 mg/m^3.

The volume to be contaminated is:

$$1000 \times 1000 \times 10\,\mathrm{m}^3 = 10^7\,\mathrm{m}^3$$

Thus the mass of GB required is:

$$\begin{aligned} 10^7 \times 10^2 &= 10^9\,\mathrm{mg} \\ &= 10^6\,\mathrm{g} \\ &= 10^3\,\mathrm{kg} \\ &= 1000\,\mathrm{kg} \end{aligned}$$

This is a very large mass, about 1 ton.

BEHAVIOUR OF LIQUID AGENTS

Classical CW agents are, in the main, volatile liquids at ordinary temperatures (phosgene is an exception: a gas at ordinary temperatures). The degree of volatility varies, of course, from compound to compound. The relationship between the liquid and the vapour phase is particularly important in explaining the effect of temperature on the damage likely to be produced by exposure to an agent and to calculations of the persistency of agents. The relationship between the liquid phase and gas (or vapour) phase of a volatile substance is defined by the vapour pressure of that substance.

If a sample of a volatile liquid is placed in an enclosed space, evaporation will take place: molecules leave the liquid phase and enter the gas phase at the surface of the liquid. Molecules also leave the gas phase and re-enter the liquid phase. Equilibrium is reached when molecules leave and re-enter the liquid phase at the same rate. In the equilibrium state, the air above the liquid is said to be saturated with vapour, and that vapour to exert the saturated vapour pressure (SVP) characteristic of the liquid at the given temperature.

The SVP of a liquid defines its volatility: the higher the SVP, the greater the volatility of the compound. When the SVP is equal to the atmospheric pressure, the liquid boils. The SVP of water at 100°C = 760 mmHg. It is common knowledge that water boils at a lower temperature at high altitude than at sea level. This is because the SVP becomes equal to the reduced atmospheric pressure at high altitude at a temperature of less than 100°C. Of course, air is not always saturated with water vapour. When saturation is complete, the pressure exerted by the vapour is the SVP. The relative humidity (RH) of air defines the extent of saturation. Air at an RH of 80% contains sufficient water vapour to exert a pressure equal to 80% of the SVP at the specified temperature.

$$\mathrm{RH} = \frac{\text{Actual vapour pressure}}{\text{SVP}} \times 100\%$$

The SVP of a substance is dependent upon temperature: the greater the temperature, the greater the SVP. The SVP of a substance at a particular temperature may be determined by use of Regnault's equation:

$$\log_{10} p = A + B/(273 + t)$$

where A and B are constants which vary from compound to compound and t is the temperature in degrees Celsius. Values of A and B may be calculated from determinations of the boiling points of the substance, t_1 and t_2, at two different pressures, p_1 and p_2. A pair of simultaneous equations is produced:

$$\begin{aligned} \log_{10} p_1 &= A + B/(273 + t_1) \\ \log_{10} p_2 &= A + B/(273 + t_2) \end{aligned}$$

The equations may be solved for A and B. The equation describing the relationship between SVP and temperature for sulphur mustard is:

$$\log_{10} p = 8.3937 - 2734.5/(273 + t)$$

Standard values for the constants A and B for a number of CW agents are given in Table 2.

The SVP of a compound is dependent only upon temperature; it is important to remember that it is independent of barometric pressure. Of course, the gas and liquid phases of a substance must be in contact for these rules to apply.

Table 2. Constants for calculating vapour pressure using the shorter form of Regnault's equation (after Sartori, 1939)

Compound	A	B
Phosgene	7.5595	−1326
Chloropicrin	8.2424	−2045.1
Cyanogen bromide	10.3282	−2457.5
Dichlordiethylene sulphide	8.3937	−2734.5
Methyl dichlorarsine	8.6944	−2281.7
Diphenyl chloroarsine	7.8930	−3288

If a quantity of water is introduced into an evacuated container at 37°C, evaporation will occur and the pressure in the container will rise to 47 mmHg: the SVP of water is 47 mmHg at 37°C.

If a quantity of water is introduced into a closed container of dry air at 37°C and atmospheric pressure (760 mmHg), the water will evaporate until the pressure exerted by the water vapour in the container equals the SVP of water at 37°C, i.e. 47 mmHg. The pressure in the container will be 760 + 47 = 807 mmHg.

If, on the other hand, the contents of the container were maintained at atmospheric pressure, by allowing the container to expand, then the water vapour would still come to exert a pressure of 47 mmHg, the other gases exerting a pressure of 713 mmHg (changes assumed to take place isothermally). The gases other than water vapour would obey Dalton's law of partial pressures and would each exert a partial pressure in accordance with the volume proportion occupied by the gas in question, e.g.

$$\text{partial pressure of oxygen } PO_2 = 0.2093 \times 713 = 149.2\,\text{mmHg}$$

This is an important fact in respiratory physiology: the PO_2 of dry air is 159.1 mmHg but the PO_2 of moist air in the trachea is 149.2 mmHg.

The concentration of a substance in the vapour phase (saturated vapour concentration, SVC) may be calculated from the SVP of the substance. This may be done to a useful level of accuracy by the simple application of the ideal gas law. Table 3 specifies the symbols and units of the parameters used in the calculation.

Consider sulphur mustard at 40°C:

$$\begin{aligned} \text{SVP} &= 0.45\,\text{mmHg} \\ &= 0.45 \times 101\,325/760\,\text{N m}^{-2} \end{aligned}$$

Table 3. Terms, symbols and units needed for the calculation of SVC from the ideal gas law

Symbol	Term	Units
P	Pressure	N m^{-2}
V	Volume	m^3
n	Number of moles of gas present per m^3	—
T	Absolute temperature	K
R	Gas constant: 8.3143	N m K^{-1}mol^{-1}

By the ideal gas law:

$$\begin{aligned} PV &= nRT \\ n &= \frac{PV}{RT} \\ n &= 0.45 \times \frac{101\,325}{760} \times \frac{1}{8.3143} \times \frac{1}{313}\,\text{mol m}^{-3} \end{aligned}$$

The gram molecular weight of sulphur mustard = 159 g

$$\begin{aligned} 1\,\text{gmol} &= 159\,\text{g} \\ \therefore \text{SVC} &= \frac{0.45 \times 101\,325 \times 159}{760 \times 8.3143 \times 313} = 3.67\,\text{g m}^{-3} \end{aligned}$$

i.e. in the units most commonly used in thinking about CW agents, at 40°C the maximum concentration of sulphur mustard (the saturated vapour concentration) = 3670 mg m^{-3}.

This result can also be obtained by arguing as follows:

1 mol of any gas at STP occupies 22.4 l
At 40°C assuming barometric pressure
= 760 mmHg,
1 mol of sulphur mustard would occupy
22.4 × (273 + 40)/273 = 25.68 l

It is important to understand what the last statement means: 159 g of sulphur mustard vapour constrained to occupy 25.68 l would exert a pressure of 760 mmHg. However, we know that the SVP of sulphur mustard at 40°C = 0.45 mmHg, i.e. the 159 g of sulphur mustard exert a pressure of 0.45 mmHg rather than 760 mmHg. The volume occupied by the 159 g of sulphur mustard vapour must then be:

$$25.68 \times 760/0.45 = 43\,370\,\text{l}$$

i.e. 43 370 l of air saturated with sulphur mustard at 40°C contain 159 g of sulphur mustard,

or 1 l of air saturated with sulphur mustard at 40°C contains 159/43 370 = 0.003 67 g of sulphur mustard,

or 1 m^3 of air saturated with sulphur mustard at 40°C contains 3.67 g or 3670 mg of sulphur mustard,

i.e. the SVC of sulphur mustard is 3670 mg m^{-3}.

Use is also made of the SVP in calculating the persistency (or persistence) of CW agents. Classically, the persistence of a CW agent has been compared to that of water at 15°C, and

Table 4. Persistences of some war gases (from Sartori, 1939)

	Temperature (°C)								
	−10	−5	0	+5	+10	+15	+20	+25	+30
Phosgene	0.014	0.012	0.01	0.008	—	—	—	—	—
Chloropicrin	1.36	0.98	0.72	0.54	0.4	0.3	0.23	0.18	0.14
Trichloromethyl chloroformate	2.7	1.9	1.4	1.0	0.7	0.5	0.4	0.3	0.2
Lewisite	96	63.1	42.1	28.5	19.6	13.6	9.6	6.9	5
Dichloroethyl sulphide (liquid)	—	—	—	—	—	103	67	44	29
Dichloroethyl sulphide (solid)	2400	1210	630	333	181	—	—	—	—
Bromobenzyl cyanide	6930	4110	2490	1530	960	610	395	260	173

its value, S, indicates that the compound would take S times as long as water to evaporate at 15°C.

S is given by:

$$S = \frac{p_1}{p}\sqrt{\frac{M_1 T}{M T_1}}$$

where:

S = persistence
p_1 = vapour pressure of water at 15°C (288° K) = 12.7 mmHg
p = vapour pressure of substance in question at T° K
M_1 = molecular weight of water = 18
M = molecular weight of substance in question
T = Absolute temperature
T_1 = Absolute temperature corresponding to 15°C, i.e. 288° K

Table 4 shows the persistence of some CW agents.

Differences between values quoted by different authorities for the SVP of sulphur mustard have led to a range of values being quoted for its persistence. The value given in Table 4 is of more use as a comparison with those of the other agents listed than as an absolute value.

BEHAVIOUR OF SOLUTIONS OF AGENTS

The above discussion has been concerned, in the main, with the vapour pressures produced by pure liquids. However, two other conditions should be considered:

1. Solutions of a volatile liquid in a solvent.
2. Solutions of a gas or gases in a solvent.

Solutions of a volatile liquid in a solvent

If a volatile liquid A is dissolved in a solvent it will continue to exert a vapour pressure, although this will be less than that which would be exerted at the same temperature by a pure sample of A. The change in vapour pressure, for ideal solutions (i.e. a solution in which the cohesive forces are identical to those which would obtain in the pure samples of the separate components of the solution), is defined by Raoult's Law:

> *The partial vapour pressure of A in a solution, at a given temperature, is equal to the vapour pressure of pure A, at the same temperature, multiplied by the mole fraction of A in the solution.*

If both solute A and solvent B are volatile, then the total vapour pressure will be the sum of that exerted by A and by B.

In practice, many solutions of liquids in liquids do not act as ideal solutions and the total vapour pressure above the solution may be greater than or less than that predicted by Raoult's law.

Solution of a gas or gases in a solvent

The amount of a gas which will dissolve in a volume of solvent is dependent upon a number of factors:

- The partial pressure of the gas to which the solvent is exposed. Henry's Law states 'The mass of gas dissolved by a given volume of liquid at a constant temperature is proportional to the pressure of the gas.'
- The solubility of the gas in the liquid: the Bunsen coefficient defines the volume of gas which dissolves, at STP, in a unit volume of solvent at one atmosphere gas pressure.
- The temperature at which the solution is formed. The solubility of most gases decreases as the temperature rises.

Gases dissolved in liquids are often described as exerting a partial pressure 'in the liquid'; for example, the partial pressure of oxygen in arterial blood is 100 mmHg. This means that the amount of oxygen in physical solution in a unit volume of blood is equal to that which would be dissolved in a unit volume of blood equilibrated with oxygen at a pressure of 100 mmHg, or with air containing oxygen exerting a partial pressure of 100 mmHg.

It is important to remember that though the *partial pressure* of a gas in a liquid may be low, the *content* of gas in the solution may be high. Comroe (1974) has pointed out that 'It is possible . . . to have more millilitres of a very soluble gas in 1 litre of liquid than in 1 litre of a gas mixture on top of it at equilibrium (equal partial pressures)'.

BEHAVIOUR OF AEROSOLS

As well as appearing as liquid, e.g. splashes or large droplets, CW agents may be encountered as aerosols. An aerosol has been defined by Muir (1972):

> *The word aerosol is a general name referring to any atmosphere containing particles which remain airborne for a reasonable length of time and is used to describe all particles that can be inhaled whether they are therapeutic, industrial, or of natural origin such as bacteria, fungi and pollens.*

A more formal definition is:

> *An aerosol is a colloid system in which the continuous phase (dispersion medium) is a gas.*

Green and Lane (1964) have pointed out that the term 'aerosol' was coined by Professor F. G. Donnan towards the end of World War I. The term was intended to describe the sort of system exemplified by the irritant arsenical smokes then being developed. It should be recalled that although the term aerosol was intended as a counterpart to the term 'hydrosol', used to describe liquid colloid suspensions, the analogy has always been imperfect in that an aerosol is inherently unstable in comparison with a hydrosol. Green and Lane included dusts of small particle size, smokes and some mists within their use of the term aerosol: their criterion being that the particle size should be sufficiently small to 'confer some degree of stability, at any rate as far as sedimentation is concerned' (Green and Lane, 1964).

A common example of an aerosol is provided by fog. If the temperature of a mass of air containing water vapour falls below the temperature of saturation, condensation of water onto the surface of dust particles will occur, and fog will be formed. Table 5 shows the water content of air saturated with water at different temperatures.

If air saturated with water vapour at 30°C is cooled to 10°C, then some 20 g of water will be condensed from each cubic metre of air. Under suitable conditions the water forms fog. If only the layer of air adjacent to the ground cools, e.g. during the night, dew forms. The dew point is defined as the temperature of saturation of the air with water vapour. For example, air 60% saturated with water at 40°C will contain approximately 33 g of water per m^3. Such a content is equivalent to 100% saturation at 32°C. If the air is cooled, water will begin to condense as the temperature falls below 32°C, i.e. the dew point of the air is 32°C.

Table 5. SVP and water content of air at various temperatures

Temperature (°C)	Vapour pressure (mmHg)	Water content ($g\ m^{-3}$)
0	4.57	4.87
10	9.14	9.36
20	17.36	17.15
30	31.51	30.08
40	54.87	50.67

Table 6. Survival times (s) of water droplets of varying size in air at varying relative humidity (from Florey, 1962)

Droplet diameter (μm)	Relative humidity (%) 0	25	50	75
1	0.29×10^{-4}	0.39×10^{-4}	0.58×10^{-4}	1.18×10^{-4}
2	1.16×10^{-4}	1.55×10^{-4}	2.3×10^{-4}	4.7×10^{-4}
5	7.3×10^{-4}	9.7×10^{-4}	1.5×10^{-3}	2.9×10^{-3}
10	2.9×10^{-3}	3.9×10^{-4}	5.8×10^{-3}	1.2×10^{-2}
20	1.16×10^{-2}	1.6×10^{-2}	2.4×10^{-2}	4.7×10^{-2}
90	0.25	0.3	0.5	1.0

FATE OF LIQUID AEROSOLS

It may be shown that the vapour pressure exerted by a convex surface is greater than that exerted by a plane surface (Starling, 1935). This effect of the curvature of the surface is referred to as the *Kelvin effect*, and is of importance in considering the stability of small liquid droplets. If a very small droplet occurs in an atmosphere in which the vapour pressure is the maximum associated with a plane surface (SVP), evaporation from the surface of the droplet will still occur as a result of the vapour pressure exerted by the droplet surface exceeding the SVP. As the droplet becomes smaller, so the curvature of the surface increases and evaporation continues. This effect in part explains the observation that a dust-free vapour does not form droplets at a temperature below the normal temperature of condensation (the dew point). The Kelvin effect must be taken into account when considering the stability of small liquid aerosol droplets in the airways.

Table 6 shows the time taken (in seconds) for water droplets of different sizes to evaporate completely (Florey, 1962).

If, on the other hand, a droplet of a watery solution which exerts a SVP less than that of pure water is exposed to air saturated with water vapour, it will grow in size. Such growth is described as 'hygroscopic growth'. Hygroscopic growth is a complex topic but may be usefully approached in a series of stages.

All solutions considered in this section are assumed to be watery and dilute.

Consider a small droplet of water exposed to an excess of dry air. Evaporation from the surface of the droplet will occur. As the droplet becomes very small, the rate of evaporation will increase as a result of the Kelvin effect.

Consider a small droplet of water exposed to air saturated with water vapour. Evaporation from the surface of the droplet will occur as a result of the Kelvin effect.

Consider a droplet of solution of a non-volatile solute exerting a vapour pressure of less than the air to which it is exposed. The droplet will grow as a result of water condensing onto its surface. The concentration of the solution comprising the droplet will fall and the vapour pressure exerted by that solution will rise. Once the vapour pressure exerted by the solution equals the vapour pressure of the air, the droplet is described as *stable*. One might wonder, given what has been said regarding the Kelvin effect, how a droplet could ever be described as stable. The Kelvin effect is, however, decreasingly important as the droplet grows, and exerts so small an effect on droplets of > 15 μm diameter that these droplets are, in practice, stable.

Given that the droplet is of a stable size, i.e. in equilibrium at a given relative humidity, the concentration of a given solute in the droplet may be calculated. Cocks and Fernando (1982) calculated that droplets containing a 20% solution of sulphuric acid would be in equilibrium at a relative humidity of 88% and a temperature of 37°C.

The equilibrium size of a droplet may be related to its original size by means of an equation derived by Ferron:

$$\frac{d_e}{d_0} = \left\{ \frac{\rho_0}{\rho_e} [1 + M_w i H / M_0 (K - H)] \right\}^{1/3}$$

where:

d_e = equilibrium diameter
d_0 = initial diameter
ρ_0 = initial density of droplet
ρ_e = equilibrium density of droplet
M_w = molecular weight of solute
M_0 = molecular weight of water
H = relative humidity
i = number of ions produced on dissociation of a molecule of solute
K = constant determined by the Kelvin equation:

$$K = \exp\left(\frac{4\sigma M_w}{RT\rho_e d_e}\right)$$

where, in addition:

σ = surface tension
R = the gas constant
T = absolute temperature

Evaporation is accompanied by local cooling and condensation by local warming. The addition of mass to droplets and the conduction of latent heat away from the surface of the droplet or particle are, as has been pointed out by Pritchard (1987), the determining processes of particle growth. Under conditions of turbulent airflow, conduction of heat away from the surface of droplets or particles is enhanced and hygroscopic growth rates exceed those found under still conditions or under conditions of laminar airflow. The implications of this will be considered later.

The importance of droplet growth under the close-to-saturated conditions obtaining in the respiratory tract is considerable. The pattern of deposition of particles or droplets in the respiratory tract is dependent upon particle or droplet size, and growth of particles or droplets will affect the pattern of deposition. This will also be considered further below.

MATHEMATICAL DESCRIPTION OF AEROSOLS

In this section the term 'particles' should be taken to mean both solid particles and liquid droplets.

Although aerosols containing particles of uniform size (monodisperse aerosols) can be prepared experimentally, naturally occurring aerosols contain particles of varying size. Such aerosols are described as *polydisperse*. If the particles found in such an aerosol are sampled and measured, then the distribution of particle size may be determined. Studies by Green (1927) demonstrated that the distribution of particle size tended to be skewed, there being many more small than large particles. Drinker (1925) showed that the typical skewed distribution curve could be transformed into a normal or Gaussian distribution curve by plotting the logarithm of the particle diameter against the percentage of the total aerosol made up by particles of different sizes. This represented a great step forward in thinking about particle size in aerosol clouds: its importance cannot be overestimated. To understand the terms used to describe particle size, a revision of the theory of normal and log-normal distribution curves may be helpful.

The normal distribution curve is described by the equation:

$$n = \frac{\sum n}{\sigma\sqrt{2\pi}} e^{[-(x-m)^2/2\sigma^2]}$$

where n is the frequency of observations of the value x, $\sum n$ is the total number of observations, m is the arithmetic mean and σ is the standard deviation.

The arithmetic mean is given by:

$$m = \frac{\sum(nx)}{\sum n}$$

and the standard deviation by:

$$\sigma = \sqrt{\frac{\sum[n(x-m)^2]}{\sum n}}$$

Similarly, in the log-normal distribution the mean (now the logarithmic or geometric mean) m_g is given by:

$$\log m_g = \frac{\sum(n \log x)}{\sum n}$$

and the logarithmic or geometric standard deviation by:

$$\log \sigma_g = \sqrt{\frac{\sum[(n(\log x - \log m_g)^2]}{\sum n}}$$

In the case of a normal distribution, the arithmetic mean, the median and the mode are

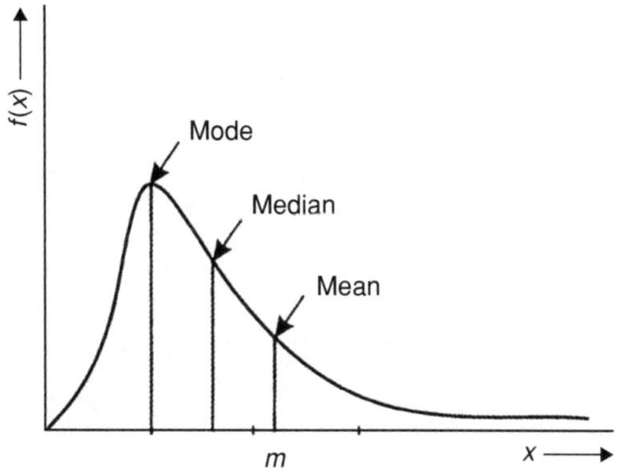

Figure 1. Log-normal distribution showing relationship between mode, mean and median

identical. Similarly, in a log-normal distribution, the geometric mean, the geometric median and the geometric mode are identical. Of course, in a log-normal distribution the arithmetic mean, the arithmetic median and the arithmetic mode all differ one from another. This is shown in Figures 1 and 2.

The geometric mean (m_g) of a log-normal distribution may be shown to be identical with the arithmetic median of that distribution. The arithmetic mean and arithmetic mode of a log-normal distribution are related to the geometric mean. The following equations define the relationships:

$$\log \text{median} = \log m_g$$
$$\log \text{mean} = \log m_g + 1.1513 \log^2 \sigma_g$$
$$\log \text{mode} = \log m_g - 2.3026 \log^2 \sigma_g$$

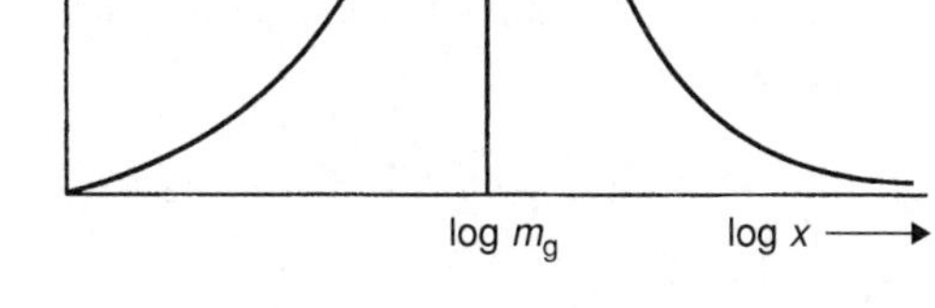

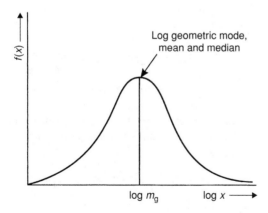

Figure 2. Log-normal distribution plotted with a log scale on the x-axis. Note that the distribution is now normalized, with the geometric mean, median and mode coinciding

The constants preceding the $\log^2 \sigma_g$ terms are the result of conversion from natural logarithms to logarithms of base 10 ($\ln x = 2.3026 \log x$).

It should be understood that log mean is the log of the arithmetic mean and not the logarithmic mean. To avoid confusion we shall use the term geometric mean in place of logarithmic mean.

Let the geometric mean (m_g) of the distribution of particle sizes in an aerosol be 1.0 μm. Let the geometric standard deviation (σ_g) be 2.0 μm. The arithmetic median (usually referred to simply as the median) of the distribution is then 1.0 μm.

The arithmetic mean (usually referred to as the mean) is given by:

$$\log \text{mean} = \log m_g + 1.1513 \log^2 \sigma_g$$
$$\log \text{mean} = 0 + (1.1513 \times 0.3010 \times 0.3010)$$
$$\log \text{mean} = 0.1043$$
$$\text{mean} = a \log 0.1043$$
$$\text{mean} = 1.27\ \mu\text{m}$$

(Note that $\log^2 \sigma_g = (\log \sigma_g)^2$ and not $\log \sigma_g^2$)

Similarly, the arithmetic mode (usually referred to as the mode) is given by:

$$\log \text{mode} = \log m_g - 2.3026 \log^2 \sigma_g$$
$$\log \text{mode} = 0 - (2.3026 \times 0.3010 \times 0.3010)$$
$$\log \text{mode} = -0.2086$$
$$\text{mode} = 0.619\ \mu\text{m}$$

These values are shown in Figure 3 from Raabe (1970).

The scale on the y-axis of Figure 3 has been devised to standardize this sort of plot. The individual values on the scale on the y-axis are dependent upon the range of particle sizes in the aerosol. If all of the particles are between 0 and 1.0 μm in diameter, then the figures on the y-axis may exceed 1.0. If, on the other hand, there is a wide range of particle sizes, the figures on the scale on the y-axis will be very much less than 1.0. Hinds (personal communication) has commented that the values on the fraction per micrometre scale do not convey any immediate intuitive or physical meaning. In Hinds' account, a distribution is shown in which particle size ranges from 0 to 50 μm (Hinds, 1982). The values on the fraction per micrometre scale range from 0 to 0.1. In Figure 3, the distribution of particle size is less broad, 0–6 μm, and the

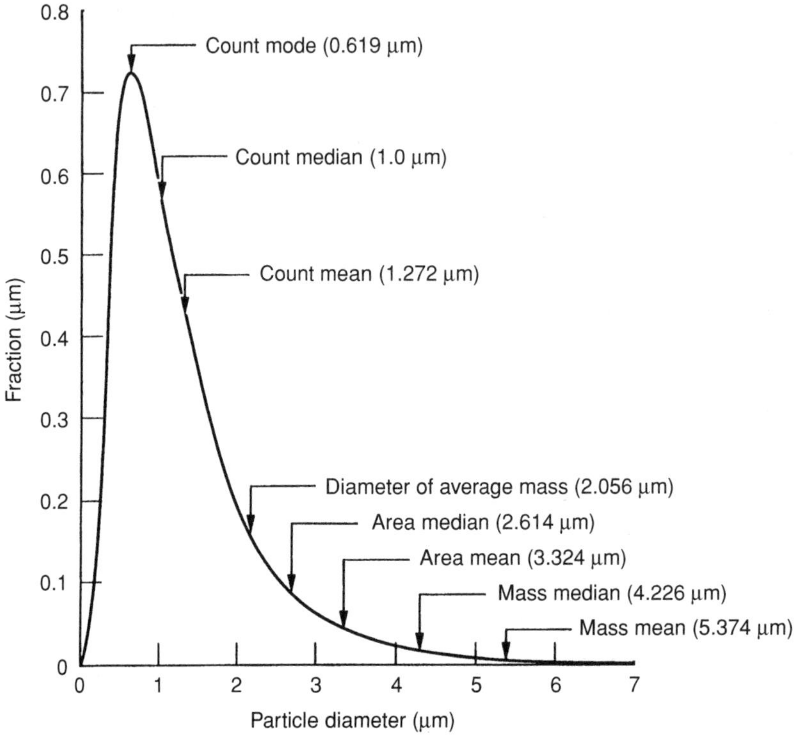

Figure 3. An example of the log-normal distribution function in normalized linear form for CMD = 1.0 and $\sigma_g = 2.0$, showing the mode, median and mean of the size distribution, the surface area distribution median and mean diameters, the mass distribution median and mean diameters, and the diameter of average mass. Reproduced with permission from Raabe OG (1970). Generation and characterization of aerosols. In: *Inhalation Carcinogenesis* (MG Hanna, P Nettersheim and JR Gilbert, eds), pp. 123–172. Proceedings of a Biology Division, Oak Ridge National Laboratory Conference. Oak Ridge, TN, USA: US Atomic Energy Commission

values on the scale on the y-axis range from 0 to 0.8.

The second great advance in the understanding of the distribution of particle sizes in aerosol clouds was reported by Hatch and Choate (1929). These workers showed that if the logarithm of particle diameter was plotted against the fractional percentage of total mass contributed by particles of different sizes, then a log-normal distribution of identical geometric standard deviation to that of the distribution obtained by plotting the logarithm of particle diameter against percentage frequency was obtained. Raabe (1970) summarized this:

Any characteristic of the particles in a population which is proportional to the q*th power of the diameter can also be described by a log-normal distribution of the same geometric standard deviation as the size distribution and with a median diameter given by:*

$$\ln D_q = \ln \mathrm{CMD} + q(\ln \sigma_g)^2$$

Here, CMD stands for count median diameter, i.e. m_g in the equations given above.

Mass is proportional to volume and hence to the third power of diameter of a spherical particle. Hence, we see:

$$\ln \mathrm{MMD} = \ln \mathrm{CMD} + 3 \ln^2 \sigma_g$$

or to the base 10:

$$\log \mathrm{MMD} = \log \mathrm{CMD} + 6.9078 \log^2 \sigma_g$$

where MMD = mass median diameter.

Similarly, considering surface area:

$$\ln \text{AMD} = \ln \text{CMD} + 2\ln^2 \sigma_g$$

$$\log \text{AMD} = \log \text{CMD} + 4.6052 \log^2 \sigma_g$$

MMD and area median diameter (AMD) are the geometric means of the mass and area distributions, respectively. In the same way that the arithmetic mean and mode were calculated from m_g (count median diameter), so equivalent values can be calculated from the MMD and AMD.

In the example given above, (CMD (m_g) = 1.0 μm; $\sigma_g = 2.0$ μm):

$$\log \text{MMD} = \log m_g + 6.9078 \log^2 \sigma_g$$

$$\log \text{MMD} = 0 + 0.6259$$

$$\text{MMD} = 4.226\ \mu\text{m}$$

$$\log \text{AMD} = \log m_g + 4.6052 \log^2 \sigma_g$$

$$\text{AMD} = 2.614\ \mu\text{m}$$

$$\log \text{mass mean diameter} = \log \text{MMD} + 1.1513 \log^2 \sigma_g$$

$$\text{mass mean diameter} = 5.374\ \mu\text{m}$$

$$\log \text{mass mode diameter} = \log \text{MMD} - 2.3026 \log^2 \sigma_g$$

$$\text{mass mode diameter} = 2.6136\ \mu\text{m}$$

$$\log \text{area mean diameter} = \log \text{AMD} + 1.1513 \log^2 \sigma_g$$

$$\text{area mean diameter} = 3.324\ \mu\text{m}$$

$$\log \text{area mode diameter} = \log \text{AMD} = 2.3026 \log^2 \sigma_g$$

$$\text{area mode diameter} = 1.617\ \mu\text{m}$$

The mass mode (or modal) diameter and the area mode (or modal) diameter are not much used in describing aerosols; the other calculated parameters are shown in Figure 3.

In addition to the parameters discussed above, Hatch and Choate (1929) gave equations which allowed the calculation of the parameters of 'average particle size' introduced by Green (1927). These parameters sound confusingly similar to, but are in fact quite different from, those described above. The following discussion is based on Hinds' outstandingly clear account (Hinds, 1982).

Consider 100 spheres, say apples, of different sizes.

Weigh the 100 apples and divide by 100.

The value obtained is the average mass.

Taking the density of an apple as 1, the diameter of a hypothetical apple of average mass could be calculated.

The value obtained is the 'diameter of average mass'.

This approach is described as 'unweighted': no account has been taken of the distribution of masses and a very misleading result would be obtained if the 100 spheres were not in fact apples but 99 grapes and one pumpkin. However, the method entails no more than counting the number of spheres and weighing all of the spheres together.

Consider, alternatively, measuring the diameters of all of the spheres.

Divide the sum of the diameters by 100.

The value obtained is the *arithmetic mean diameter*. An approximation could be reached by allocating the spheres into categories defined by size, and counting the number in each category counted. Using standard statistical terminology, the arithmetic mean diameter would be given by:

$$\bar{d} = \frac{\sum n_i d_i}{N}$$

In this process, the characteristic diameter for each group is weighted by n_i/N, or the fraction of the total number in that size group.

The same approach could be taken for mass and the characteristic diameters of each group weighted by m_i/M, the fraction of the total mass in the size group. Let the average diameter calculated in this way be the mass mean diameter d_{mm}; this will be given by:

$$d_{\text{mm}} = \frac{\sum m_i d_i}{m}$$

The mass mean diameter will not equal the diameter of average mass. Hinds' definition (Hinds, 1982) is useful:

> *In the calculation of the diameter of average mass, a coarse and fine particle are given equal representation in the averaging process but the quantity averaged is the mass. In calculating the mass mean diameter, the quantity averaged is the diameter, but it is weighted according to its mass contribution.*

Similarly, the diameter of an average surface should not be confused with surface mean diameter.

Equations for calculating 'average diameters' are given in Table 7, taken from Hatch (1933).

Table 7. Mathematical definitions of the average diameters of non-uniform particulate substances in terms of the parameters of the distribution curves by count and by weight (from Hatch, 1933)

Average diameter	Symbol	Mathematical definition	Equivalent logarithmic value in terms of statistical parameters of distribution curves: By count (M_g and σ_g)	By weight (screen analysis) (M'_g and σ'_g)
Geometric mean	M_g	$\text{antilog}\left(\frac{\sum n \log d}{\sum n}\right)$	$\log M_g$	$\log M'_g - 6.9078 \log^2 \sigma'_g$
Arithmetic mean	δ	$\left(\frac{\sum nd}{\sum n}\right)$	$\log M_g + 1.1513 \log^2 \sigma_g$	$\log M'_g - 5.7565 \log^2 \sigma'_g$
Specific surface	d_s	$\left(\frac{\sum nd^{-1}}{\sum n}\right)^{-1}$	$\log M_g - 1.1513 \log^2 \sigma_g$	$\log M'_g - 8.0591 \log^2 \sigma'_g$
Surface area	Δ	$\left(\frac{\sum nd^2}{\sum n}\right)^{1/2}$	$\log M_g + 2.3026 \log^2 \sigma_g$	$\log M'_g - 4.6052 \log^2 \sigma'_g$
Volume	D	$\left(\frac{\sum nd^3}{\sum n}\right)^{1/3}$	$\log M_g + 3.4539 \log^2 \sigma_g$	$\log M'_g - 3.4539 \log^2 \sigma'_g$
Surface area per unit volume[a]	D^3/Δ^2	$\left(\frac{\sum nd^3}{\sum nd^2}\right)$	$\log M_g + 5.7565 \log^2 \sigma_g$	$\log M'_g - 1.1513 \log^2 \sigma'_g$

Note:
[a] This diameter gives the specific surface for the sample as a whole; it should not be confused with d_s, the diameter of the hypothetical particle having an *average* specific surface.

The diameter of average mass (D) of the distribution discussed above may be calculated as follows:

$$\log D = \log m_g + 3.4539 \log^2 \sigma_g$$
$$D = 2.056\ \mu\text{m}$$

This parameter is shown in Figure 3. The reader may think it fortunate that diameters of average mass and surface are little used in inhalation toxicology.

If the log-normal distribution of particle size is plotted so that the cumulative frequency appears on the y-axis and the log of the particle diameter on the x-axis, an 'ogive' is produced (Figure 4). This curve may be converted to a straight line by manipulating the scale on the y-axis. If this scale is distributed according to the normal distribution curve, then a straight line will be obtained when the data shown in Figure 4 are plotted (Figure 5). Such a plot is described as a 'log-probability plot' and will be discussed in more detail when the use of 'probit analysis' is considered.

It is worth considering how the geometric standard deviation of the distribution may be derived from this sort of plot. This may be approached by recalling the normal distribution curve obtained when a log-normal distribution is plotted as shown in Figure 6. It will be recalled that in a normal distribution, 68% of the observations lie between +1 and −1 standard deviations of the mean. In Figure 6, it is shown that 68% of the

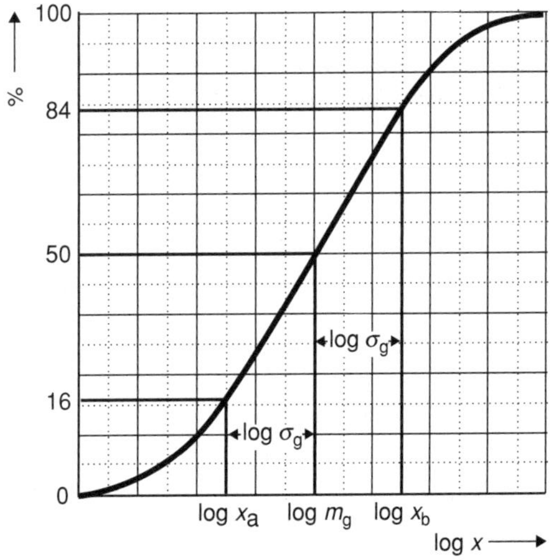

Figure 4. Cumulative plot of a log-normal distribution

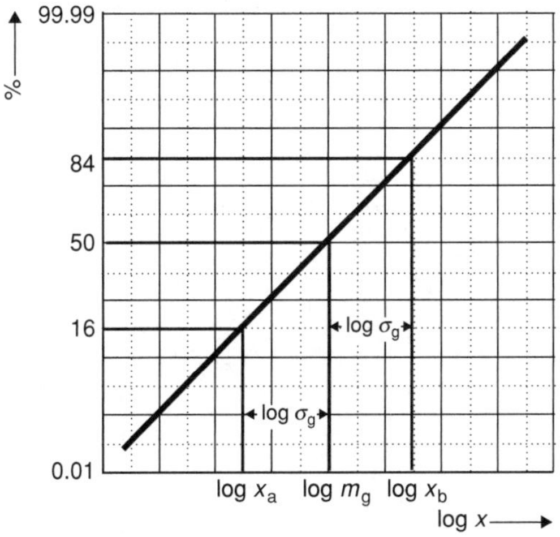

Figure 5. Log-normal distribution

observations fall between log x_a and log x_b.

log x_b − log m_g (or, equally, log m_g − log x_a) is the standard deviation of this distribution of log x

antilog (log x_b − log m_g) is the geometric standard deviation of the distribution of x

antilog (log x_b − log m_g) $= x_b/m_g$

antilog (log m_g − log x_a) $= m_g/x_a$

and thus the geometric standard deviation, σ_g, is given by:

$$\sigma_g = \frac{x_b}{m_g} = \frac{m_g}{x_a}$$

Figures 4 and 5 show how σ_g may be derived from the cumulative frequency plots of the distribution of log x.

The reader will see that in Figures 4, 5 and 6 the logarithms of x_a, m_g and x_b have been plotted on an arithmetic scale. Log-probability paper could have been used, with x_a, m_g and x_b being plotted on a logarithmic axis. In Figure 4, the scale on the y-axis is arithmetic. In Figure 5, the scale on the y-axis has been manipulated to conform to a normal distribution curve. In practice, log-normal graph paper is prepared with the probability scale on the x-axis and the logarithmic scale on the y-axis: examples may be found in Hinds' account (Hinds, 1982).

From Figure 5 it will be seen that:

$$\sigma_g = \text{antilog}(\log x_b - \log m_g) = x_b/m_g$$
$$= 84\%\ \text{size}/50\%\ \text{size}$$

σ_g is often defined in this way.

σ_g may also be calculated as follows:

$$\sigma_g = \frac{xb}{m_g} = \frac{m_g}{xa}$$

$$\sigma_g^2 = \frac{xb}{m_g} \times \frac{m_g}{xa} = \frac{xb}{xa}$$

$$\sigma_g = \sqrt{\frac{xb}{xa}}$$

The geometric standard deviation controls the width of the distribution curve plotted as in Figure 6 and the slope of the distribution curve plotted as in Figure 5.

To simplify plotting of the data as shown in Figure 5, special log-probability graph paper is used. As an example, the data shown in Table 8 have been plotted on this type of graph paper in Figure 7, while σ_g has been calculated from:

$$\sigma_g = \sqrt{\frac{84\%\ \text{size}}{16\%\ \text{size}}} = 2.08$$

$$m_g = 4.45\ \mu\text{m}$$

It has already been stated that σ_g is identical for plots of the distribution of any parameter related to particle diameter, e.g. mass or surface area. Log-probability plots of count distribution,

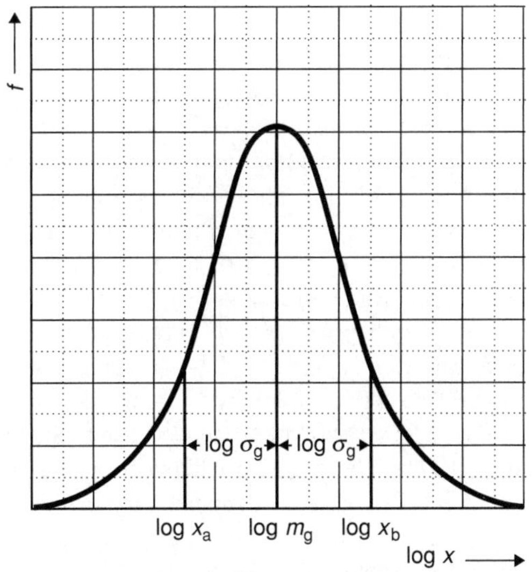

Figure 6. Log-normal distribution

Table 8. Distribution of particle size in a sample of particles (after Drinker and Hatch, 1954)

Group size (filar units)[a]	Upper limit of group (μm)[b]	Group frequency, *f*	Number < upper limit of group size	% < upper limit of group size
1–2	2.3	28	28	18
2–3	3.3	32	60	39
3–4	4.4	20	80	52
4–5	5.5	15	95	62
5–6	6.6	14	109	71
6–7	7.7	8	117	76
7–8	8.8	9	126	82
8–9	9.9	5	131	85
9–10	11.0	4	135	88
10–11	12.1	5	140	91
11–12	13.2	3	143	93
12–13	14.3	2	145	94
13–14	15.4	1	146	95
14–15	16.5	0	146	95
15–16	17.6	2	148	96
16–17	18.7	0	148	96
17–18	19.8	1	149	97
18–19	20.9	0	149	97
19–20	22.0	2	151	98
20–21	23.1	0	151	98
21–22	24.2	0	151	98
22–23	25.3	1	152	99
23–24	26.4	2	154	100

Notes:
[a] By calibration, one filar unit = 1.1 μm.
[b] Note the upper limit of group size is used in plotting cumulative frequency distribution curves. The group midpoints are used in plotting histograms and non-cumulative distribution curves.

surface area distribution and mass distribution against particle diameter would, therefore, be expected to generate parallel lines. This is the case, as shown in Figure 8 (modified from Menzel and Amdur, 1986).

The data upon which these curves are based were obtained by use of a cascade impactor, which sizes particles according to their aerodynamic characteristics. The aerodynamic diameter of a particle is the diameter of a spherical particle of unit density which, when falling, reaches the same terminal velocity as the particle in question. This will be discussed further below.

From Figure 8 it will be seen that only 10% of the particles in the aerosol were greater than 1 μm in diameter but that these accounted for about 80% of the total mass of the particles in the aerosol. More dramatically, 50% of the mass was made up by particles of diameter greater than 1.39 μm (mass median diameter = 1.39 μm) but these comprised only about 4% of the particles in the aerosol.

DEPOSITION OF AIRBORNE PARTICLES IN THE RESPIRATORY TRACT

A proportion of all inhaled particles deposit in the respiratory tract. Study of the pattern of deposition of particles along the airways and the development of mathematical models to predict deposition has been continuous since the work of Findeisen (1935). The latter devised a model of the lung, consisting of a regularly, dichotomously branching series of tubes divided into nine compartments, the last (representing the alveoli) numbering 5×10^7. He considered the pattern of deposition of spheres ranging in size from 0.03 to 30 μm in diameter. A simple model of the respiratory cycle was also used (Stober *et al*, 1993). The model predicted complete deposition of spheres of diameter greater than 3 μm, a minimum of deposition (in fact about 35%) for spheres of diameter 0.1–0.3 μm, and a high total deposition for very small particles.

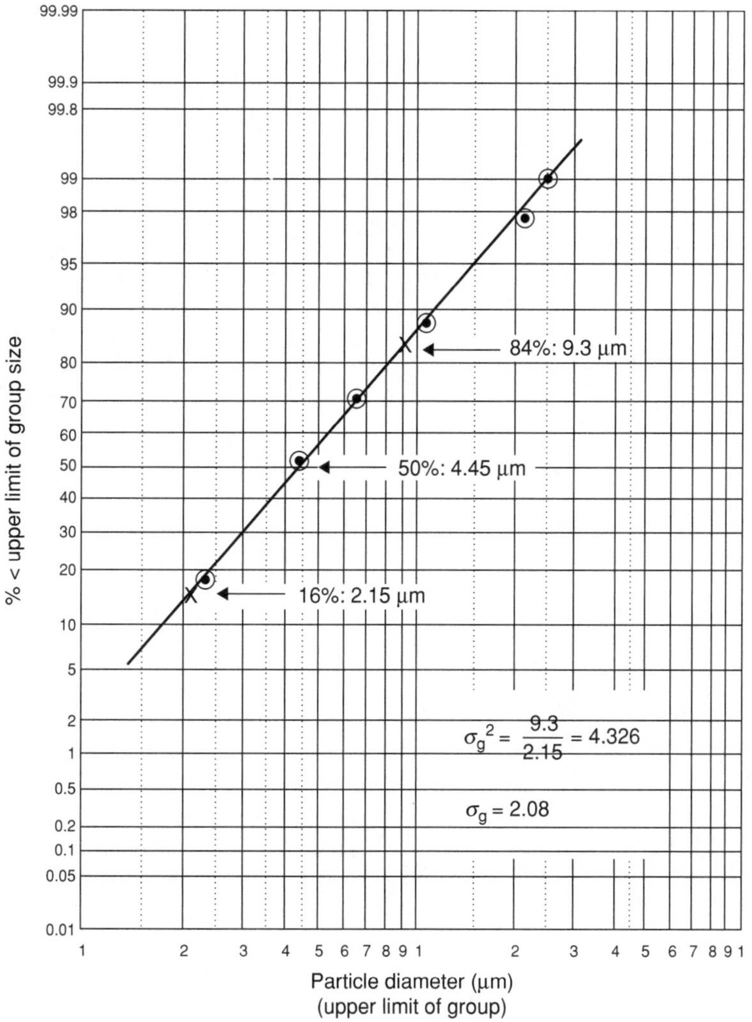

Figure 7. Data from Table 8 plotted on log-probability paper

Findeisen was working in a new field; he quoted only two references in his paper, both by Albert Einstein.

More complex mathematical models have been constructed by Landahl (1950), Hatch and Hemeon (1948) and Altshuler (1959). Although the mathematical sophistication of the modelling has increased, the assumptions made regarding pulmonary anatomy have remained fairly simplistic.

A number of well-known graphs have been drawn to illustrate the predicted deposition of particles of different size in the respiratory tract. Figure 9 is taken from Muir (1972), and shows peak alveolar deposition at between 1 and 5 µm diameter.

Interest in particle deposition after World War II was fuelled by concern about possible deposition of radioactive particles. In 1965, The International Radiological Protection Commission (IRPC) Task Force on Lung Dynamics submitted its report: 'Deposition and retention models for internal dosimetry of the human respiratory tract (Task Group on Lung Dynamics, 1966). The Task Force, comprising Bates, Fish, Hatch, Mercer and Morrow, undertook a detailed examination of the area and developed a deposition model using methods of calculation similar to those used

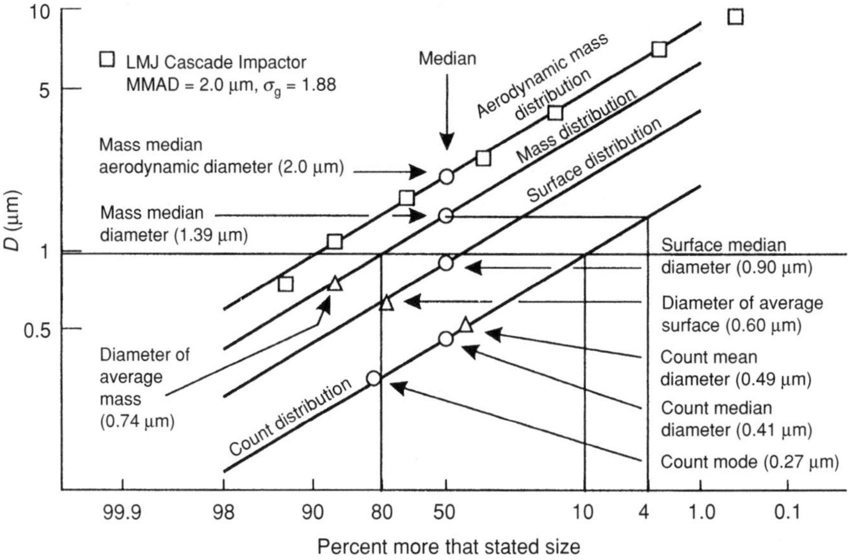

Figure 8. Plot of size distribution of an aerosol on log-probability paper. Curves are shown which characterize aerosol size in regard to various parameters

by Findeisen. The model was, however, more refined in terms of physiological parameters than that used by Findeisen. The respiratory tract was divided into three zones: nasopharyngeal (NP), tracheobronchial (TB) and pulmonary (P). The results of the deposition calculations for a range of particle sizes are shown in Table 9.

The pattern of deposition expected at a respiratory rate of 15 breaths per minute and a tidal volume of 750 ml is shown in Figure 10.

The effect of hygroscopic growth of particles was considered by the Task Force and is considered in detail below.

Having predicted the deposition of particles of different sizes, the Task Force calculated the predicted pattern of deposition of particles from polydisperse aerosols of known count median diameter and σ_g. The results of these most important calculations are shown in Table 10. The importance of this table cannot be overestimated:

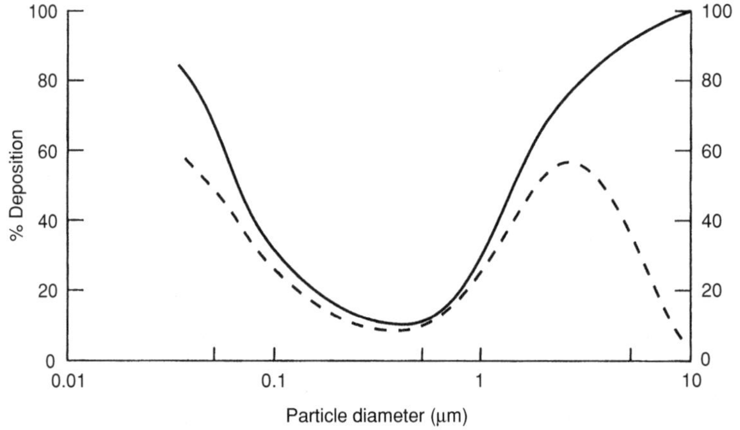

Figure 9. Percentage deposition of inhaled aerosols as a function of particle size. Mouth breathing at rest. Total deposition (—), alveolar deposition (— —). Reproduced with permission from DCF Muir (ed.) (1972). *Clinical Aspects of Inhaled Particles*. William Heinemann books.

Table 9. Deposition of unit density spheres. Reproduced with permission from Task Group on Lung Dynamics (1966). Deposition and retention models for internal dosimetry of the human respiratory tract. *Health Physics*, **12**, 173–207. © Lippincott, Williams & Wilkins

Tidal volume (cm^3)	Location	Diameter of sphere (μm)									
		0.01	0.06	0.20	0.60	1.0	2.0	3.0	4.0	6.0	10.0
750	NP	0	0	0	0	0.036	0.406	0.552	0.654	0.799	0.992
	TB	0.307	0.068	0.027	0.020	0.027	0.051	0.071	0.084	0.091	0.007
	P	0.506	0.585	0.281	0.204	0.250	0.346	0.308	0.238	0.103	0.002
1450	NP	0	0	0	0	0.275	0.522	0.665	0.773	0.923	1.00
	TB	0.256	0.051	0.017	0.019	0.027	0.050	0.064	0.069	0.043	0
	P	0.676	0.711	0.334	0.215	0.242	0.330	0.250	0.150	0.033	0
2150	NP	0	0	0	0.068	0.371	0.607	0.736	0.844	1.0	1.0
	TB	0.208	0.035	0.015	0.021	0.030	0.056	0.067	0.062	0	0
	P	0.746	0.653	0.294	0.209	0.226	0.285	0.195	0.092	0	0

Note:
NP, nasopharyngeal; TB, tracheobronchial; P, pulmonary.

not so much because the predictions should be regarded as immutable, but because of the insight it provides into the factors controlling the deposition of particles in the respiratory tract. The data are plotted in Figure 11.

Figure 12 clearly illustrates minimal overall deposition of particles of 0.1–1.0 μm diameter. (In Figure 12, the activity median aerodynamic diameter (relevant to radioactive particles) has been plotted.)

The models described above and more recent models based upon more realistic anatomical models of the respiratory tract (Weibel, 1963; Horsfield and Cumming, 1967,1968) are based upon the calculations of the relative importance of a number of mechanisms of particle deposition. These are discussed in the following section.

MECHANISMS OF PARTICLE DEPOSITION

Sedimentation

Particles of density (ρ_{part}) greater than that of air (ρ_{air}) sediment under the force of gravity. As particles fall, they accelerate until the resistive forces due to motion through the air equal the force applied by gravity. Once this occurs, the

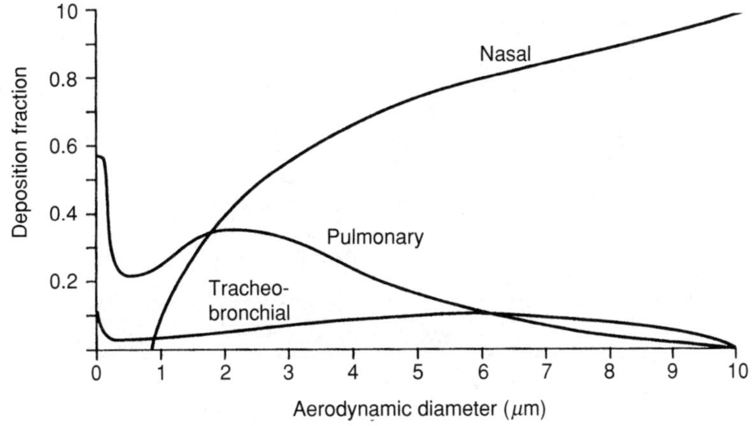

Figure 10. Deposition as a function of particle size for 15 respirations per minute, 750 cm^3 tidal volume. Reproduced with permission from Task Group on Lung Dynamics (1966). Deposition and retention models for internal dosimetry of the human respiratory tract. *Health Physics*, **12**, 173–207. © Lippincott, Williams & Wilkins

Table 10. Computed deposition of log-normal aerosols. Reproduced with permission from Task Group on Lung Dynamics (1966). Deposition and retention models for internal dosimetry of the human respiratory tract. *Health Physics*, **12**, 173–207. © Lippincott, Williams & Wilkins

MMAD (μm)	CMAD (μm)	Sigma (σ_g)	NP	TB (per cent of inspired dust)	P
0.020	0.018	1.2	0.00	21.3	68.3
0.020	0.012	1.5	0.00	19.7	65.4
0.020	0.005	2.0	5.20[a]	20.7[a]	62.7[a]
0.020	0.002	2.5	6.70[a]	22.1[a]	59.0[a]
0.20	0.181	1.2	0.00	2.06	36.4
0.20	0.122	1.5	0.01	2.37	39.1
0.20	0.047	2.0	0.78	2.91	41.2
0.20	0.016	2.5	2.36	3.61	42.3
0.20	0.005	3.0	4.09	4.24	42.8
2.00	1.221	1.5	51.1	4.70	27.2
2.00	0.473	2.0	50.7	4.30	23.6
2.00	0.161	2.5	50.4	3.90	21.8
2.00	0.054	3.0	50.2	3.61	21.0
2.00	0.006	4.0	50.1	3.30	20.6
20.0	18.10	1.2	99.9	0.00	0.00
20.0	1.611	2.5	97.2	0.81	1.70
20.0	0.535	3.0	95.6	1.03	2.60
200.0	181.0	1.2	86.0	0.00	0.00

Note:
[a] Aerosol mass below 0.01 μm is presumed to experience an equal deposition in the three compartments: this is an estimated division reflecting the increased deposition probabilities of very small particles in the airways.

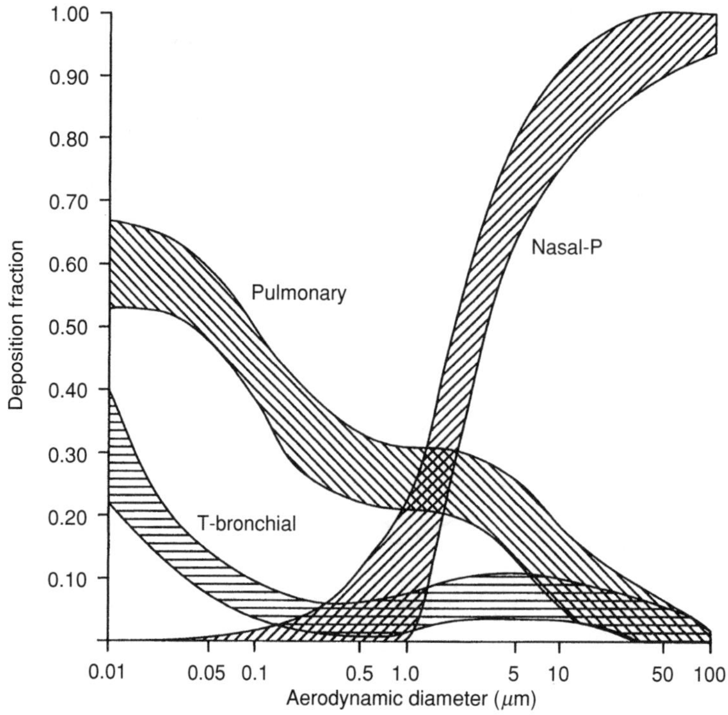

Figure 11. Shaded areas indicate variability of deposition for a given mass median (aerodynamic) diameter in each compartment when the distribution parameter, σ_g, varies from 1.2 to 4.5, and the tidal volume is 1450 ml. Reproduced with permission from DCF Muir (ed.) (1972). *Clinical Aspects of Inhaled Particles*. William Heinemann books

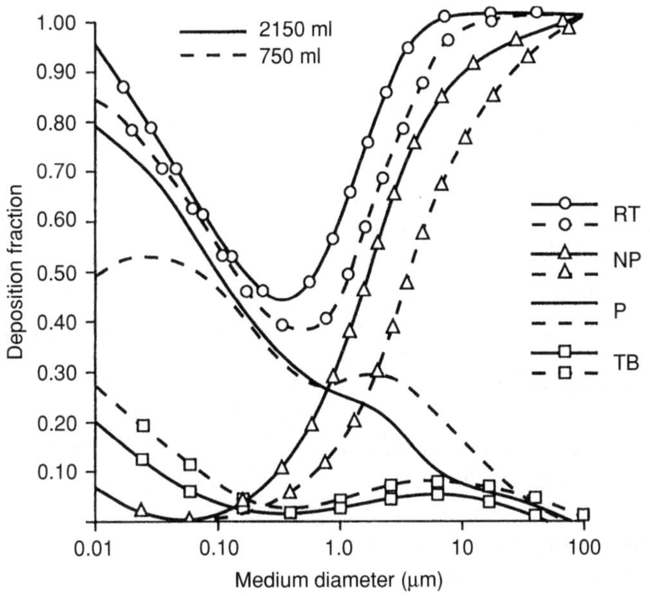

Figure 12. Two ventilatory states, i.e. 750 ml and 2150 ml tidal volume (~ 11 and ~ 32 1 min^{-1} volumes, respectively), are used to indicate the order and direction of change in compartmental deposition which are induced by such physiological factors. Note the crossover in the 'P' curves at approximately 0.8 μm diameter (AMAD). Reproduced with permission from Task Group on Lung Dynamics (1966). Deposition and retention models for internal dosimetry of the human respiratory tract. *Health Physics*, **12**, 173–207. © Lippincott, Williams & Wilkins

particle continues to fall at a constant velocity: the terminal velocity. Terminal velocity (V_t) may be calculated as follows.

Stokes (Starling, 1935) showed that the force exerted by gravity upon a falling body could be given by:

$$F_g = \frac{4}{3}\pi r^3(\rho_{part} - \rho_{air})g$$

where r is the radius of the particle, and g is the acceleration due to gravity.

He also showed that the resistive forces due to the motion of the body through the air could be given by:

$$F_{res} = 6\pi\eta r V$$

where η is the coefficient of viscosity of air and v is the velocity of the body.

For small particles where particle diameter is similar to the mean free path of gas molecules (λ), Cunningham's correction factor C_c must be applied:

$$F_{res} = \frac{6\pi\eta r V}{C_c}.$$

When $F_g = F_{res}$, then the particle will have reached its terminal velocity:

$$\frac{6\pi\eta r V_t}{C_c} = \frac{4}{3}\pi r^3(\rho_{part} - \rho_{air})g$$

$$V_t = \frac{4}{3}\pi r^3 \frac{C_c}{6\pi\eta r}(\rho_{part} - \rho_{air})g$$

$$V_t = \frac{4}{18\eta} r^2 C_c(\rho_{part} - \rho_{air})g$$

Note that V_t or the rate of sedimentation is proportional to the square of the radius of the particle.

Let:

$$d = \text{particle diameter}$$
$$d = 2r$$
$$d^2 = 4r^2$$
$$r^2 = d^2/4$$

Then:

$$V_t = (\rho_{part} - \rho_{air})\frac{C_c g d^2}{18\eta}$$

Cunningham's correction factor is given by:

$$C_c = 1 + 2.52\frac{\lambda}{d}$$

Valberg (1985) stated that the equation for terminal velocity quoted above was valid for particles of unit density of diameter 0.1–40 μm, settling in air. Correction factors allowing predictions to be made for particles of 0.001–200 μm diameter have been provided by Davies (1966).

AERODYNAMIC DIAMETER

Inhaled particles vary both in shape and density and these factors affect their capacity to be deposited by sedimentation. The behaviour of such particles can be determined by converting their actual diameter(s) to their aerodynamic diameter(s). What does this mean? Imagine a low-density particle of irregular shape – this will be characterized by a certain terminal velocity as it settles in air. The aerodynamic diameter of the particle is defined as the actual diameter of a spherical particle of unit density with the same terminal velocity.

Aerodynamic diameter is defined by the following equation:

$$d_a = d_e\left(\frac{P_p}{P_0\chi}\right)^{\frac{1}{2}}$$

where d_a is the aerodynamic diameter, P_p is the particle density and P_0 is the 'standard' particle density (= 1.0 g/cm^3); d_e and χ are new terms.

In the above, d_e is the equivalent volume diameter, i.e. the diameter of a sphere having the same volume as an irregular particle. As Hinds (1982) points out: 'The equivalent volume diameter can be thought of as the diameter of the sphere that would result if an irregular particle were melted to form a droplet!'; d_e is calculated from microscopic measurement of the actual particles being considered, while χ is the dynamic shape factor which is included to allow for the effects of shape on terminal velocity. For example, talc dust is characterized by a dynamic shape factor (χ) of 1.88, sand particles by 1.57, etc. Spheres have a dynamic shape factor of 1.0 while cubes have a dynamic shape factor of 1.08.

The following example is provided by Hinds (1982).

Consider a quartz particle of $d_e = 20$ μm and $P_p = 2700$ kg/m^3:

$$d_a = d_e\left(\frac{P_p}{P_0\chi}\right)^{\frac{1}{2}} \quad \text{where } \chi = 1.36$$

Thus:

$$d_a = 20\left(\frac{2700}{1000 \times 1.36}\right)^{\frac{1}{2}} = 28.2\ \mu\text{m}$$

Note that for all particles with a density greater than that of water and with a dynamic shape factor > 1.0, d_a will be greater than d_e.

If the particle being considered is, in fact, a sphere then:

$$d_a = d_p\left(\frac{P_p}{P_0}\right)^{\frac{1}{2}}$$

Note that for a sphere, $d_p = d_e$.

If very small particles are being considered, the *Cunningham correction factor* (slip correction factor, C_c) must be applied and:

$$d_a = d_p\left[\frac{C_c(d_p)}{C_c(d_a)}\right]^{\frac{1}{2}}\left(\frac{P_p}{P_0}\right)^{\frac{1}{2}}$$

Impaction

If the airstream in which a particle is travelling suddenly changes direction, e.g. at the bifurcation of an airway, force will be applied to the particle, causing it to move across the airstream. Should the particle encounter the wall of the tube in which the air is flowing during its journey across the stream, it will be deposited by impaction. Of course, the particle does not move at right angles across the stream, but follows a curved trajectory. If the particle was initially at the centre of the stream then it would come to rest (with regard to the direction of flow of the airstream), assuming it had not impacted on a wall, at some distance away from the centre of the stream, and continues along with the stream in this new position. The distance travelled (χ) before the particle comes to rest with regard to the airstream is given by Valberg (1985) by:

$$\chi = \left(\frac{u \sin\theta}{g}\right)V_t$$

where u is the velocity before deflection and θ is the angle of deflection of the airstream; V_t is the terminal velocity of the particle, as calculated above. Note that here again, the deposition is dependent upon V_t and hence upon the square of the diameter (or radius) of the particle. Valberg (1985) calculated that a sphere of unit density and 1.0 μm diameter travelling in an airstream at 1 m s^{-1} (typical of the velocity in a major bronchus) would move 1.7 μm away from its previous stream line when the direction of airflow changed by 30°. This is a small movement. Deposition of particles in the airways as a result of impaction relates to those particles close to the walls of the airway, is proportional to the square of the particle radius and will be most efficient at bifurcations of the airways.

Impaction will also be most effective as a means of particle deposition when the air velocity is high. As air moves down the respiratory tract, the airflow velocity falls: impaction is most important in the larger airways. The effect of the falling velocity of airflow is, as pointed out by Muir (1992), offset to some extent by the reduction in airway diameter: 'the fraction of particles entering the tenth generation of airways removed by impaction is similar to that removed in the third generation'.

The phenomenon of impaction has been thoroughly analyzed both theoretically and experimentally and forms the basis of one of the most commonly used methods of particle sizing. Hinds' account of impactors should be consulted for details (Hinds, 1982). In essence, impactors direct a jet of particle-containing air via a nozzle towards a plate. The plate causes a 90° change in the direction of flow, and some particles will impact upon the plate. Control of flow rate and nozzle diameter allows size-control of the particles deposited.

Diffusion

The molecules of a gas are in constant motion and collide with aerosol particles. If these particles are small, they are disturbed by the impact of the gas molecules and move in an irregular fashion described as *Brownian motion*. Particles moving in this way may encounter the walls of the airways and thus be deposited. The root-mean-square displacement after time t is given by:

$$\Delta = \sqrt{6Dt}$$

where D is the diffusion coefficient of the particle, and is given by:

$$D = \frac{KTC_c}{3\pi \eta d}$$

when T is the absolute temperature, K the Boltzmann constant, η the gas viscosity, C_c the Cunningham correction factor and d the diameter of the particle. It should be noted that the smaller the particle, the greater will be the displacement in a given period.

Valberg (1985) presented a useful table comparing displacement due to Brownian motion and that due to sedimentation (Table 11). This table makes clear that for very small particles diffusion will be the *more* important mechanism of deposition.

Interception

Interception occurs when long fibres travelling in the airstream impact upon airway walls as a result of not being able to bend with changes in direction of airflow. It is an important means of deposition of fibres of materials such as asbestos but is not important in a CW context.

Electrostatic precipitation

Particles and surfaces may be described as charged if an excess of negative or positive charges exists at their surface. Charged particles are attracted to surfaces which carry the opposite charge to that upon the particles. The lining of the airways is uncharged. 'Image charge' can be induced at the surface by the approach of a charged particle and deposition is enhanced in comparison with that of uncharged particles. Valberg (1985) pointed out that electrostatic effects were likely to be important if a particle carried more than 10 charges upon its surface. Charged particles lose charge in the atmosphere and only in the case of 'highly charged, freshly generated particles' is deposition in the respiratory tract likely to be significantly dependent upon electrostatic precipitation.

Table 11. Root-mean-square Brownian displacement in 1s compared with distance fallen in air in 1 s of unit density particles of different diameters.[a] Reproduced with permission from Valberg PA (1985). Determination of retained lung dose from toxicology of inhaled materials. In: *Handbook of Experimental Pharmacology* Vol. 75 (HP Witschi and JD Brain, eds), pp. 57–91. Berlin: Springer-Verlag. © 1985 Springer-Verlag

Aspect	Diameter (μm)	Brownian displacement in 1 s (μm)	Distance fallen in 1 s (μm)
Settling greater in 1 s	50	1.7	70 000
	20	2.7	11 500
	10	3.8	2 900
	5	5.5	740
	2	8.8	125
	1	13.0	33
Diffusion greater in 1 s	0.5	20	9.5
	0.2	37	2.1
	0.1	64	0.81
	0.05	120	0.35
	0.02	290	0.013
	0.01	570	0.0063

Note:
[a] Temperature 37°C; gas viscosity 1.9×10^{-5} Pa s. Appropriate correction factors were applied for motion outside the range of validity of Stokes' law.

In summary, there are a number of mechanisms by which particles are deposited in the respiratory tract. Impaction becomes less significant as one moves along the airways, sedimentation plays an increasingly important role and, in the terminal parts of the airway, diffusion of small particles is important.

HYGROSCOPIC GROWTH OF AND NEUTRALIZATION OF ACIDIC DROPLETS IN THE AIRWAYS

It was indicated above that particles or droplets (hereafter referred to as particles) of inorganic salts could grow in the airways as a result of condensation of water upon their surfaces. If particles were acidic or alkaline on inhalation, dilution with water will move the pH towards 7. In addition, ammonia, produced mainly by bacterial action in the mouth, is absorbed by particles and neutralization of acidic particles will occur.

Cocks and Fernando (1982) undertook a detailed theoretical analysis and computer simulation of the growth of droplets on passing from ambient air to the conditions obtaining in the lung. Predicted patterns of growth are shown in Figure 13. The growth of particles of two initial sizes was modelled. Droplets of 0.1 and 1.0 μm radius containing 20%, 40% and 60% solutions of sulphuric acid were considered. These solutions have vapour pressures corresponding to different relative humidities: 88%, 57% and 16%, respectively. The Kelvin effect was taken into account in calculating these equivalent relative humidities: thus, the vapour pressure (measured at a plane surface) of a 20% solution of sulphuric acid would be less than that exerted by water vapour at an RH of 88% but a droplet of 20% sulphuric acid and radius 0.1 μm is stable at an RH of 88% because it exerts a slightly greater vapour pressure than would be exerted by 20% sulphuric acid at a plane surface.

It was assumed that the RH in the respiratory tract was 99.5%. It was discovered that smaller droplets would grow more rapidly than larger droplets and thus approached equilibrium size more rapidly. It was also noted that: 'comparing droplets of the same initial composition, those of smaller initial radii will attain a lower value of r/r_0 (equilibrium radius/initial radius) given sufficient time than larger droplets'. This is due to the greater significance of the Kelvin effect, which retards the growth of small droplets.

Figure 14 illustrates the need for precise knowledge regarding the RH in the respiratory

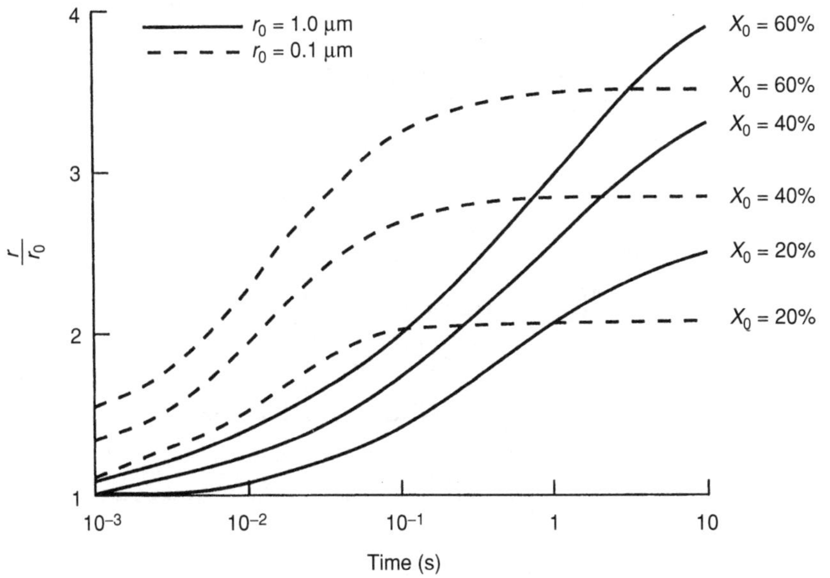

Figure 13. Growth curves for sulphuric acid droplets: r, equilibrium radius; r_0, initial radius; X_0 initial concentration of sulphuric acid. Reprinted from *Journal of Aerosol Science*, **13**, AT Cocks and RP Fernando, The growth of sulphate aerosols in the human airways, pp. 9–19, 1982, with kind permission from Elsevier Science Ltd

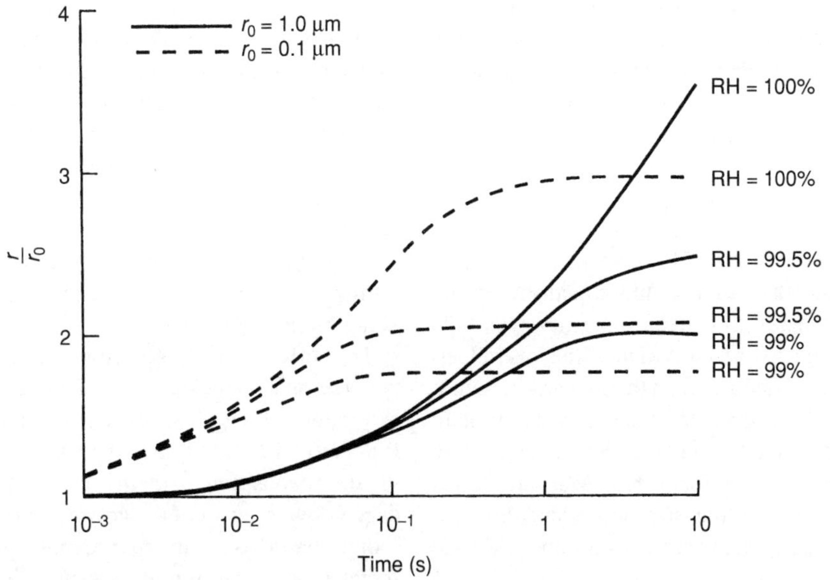

Figure 14. The effect of the assumed relative humidity (RH) in the lung on the growth of sulphuric acid droplets: r, equilibrium radius; r_0, initial radius. Reprinted from *Journal of Aerosol Science*, **13**, AT Cocks and RP Fernando, The growth of sulphate aerosols in the human airways, pp. 9–19, 1982, with kind permission from Elsevier Science Ltd

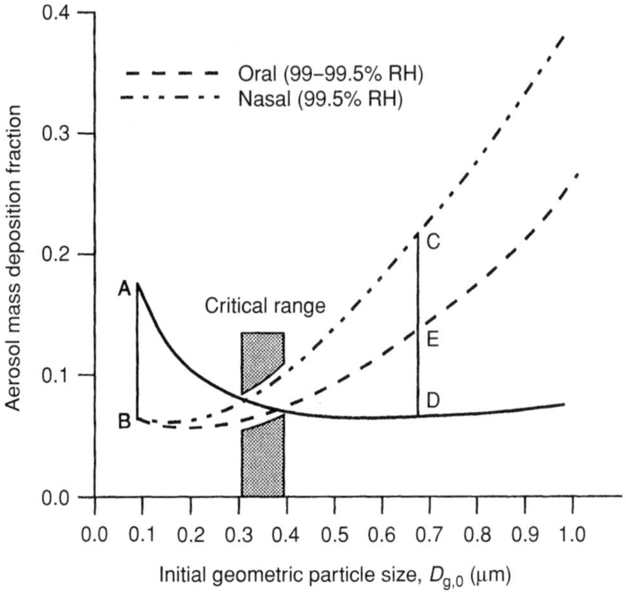

Figure 15. The influence of inhaled particle size and breathing mode on ammonium sulphate aerosol deposition: (—) non-hygroscopic particle behaviour included for comparison; minute volume = 30 l. RH = relative humidity Reproduced with permission from Martonen TB (1985). Ambient sulphate deposition in man: modelling the influence of hygroscopicity. *Environmental Health Perspectives*, **63**, 11–24

tract when predicting particle growth. Droplets containing 20% sulphuric acid of 0.1 and 1.0 μm radius were modelled growing under conditions of differing RH. Note that at humidities of 99.5% and 99% all particles approach an equilibrium radius.

As small particles grow, their diffusion-dependent deposition in the peripheral lung becomes less efficient. Conversely, as large particles grow, their sedimentation and impaction-dependent deposition in the upper airways would be expected to increase. Thus, the overall effect of hygroscopic growth on particle deposition in the respiratory tract may be difficult to predict: an effect on location of deposition might, however, be expected. Martonen *et al* (1985) illustrated the size-dependent differential effect of particle growth on deposition see (Figure 15).

This figure shows the mass deposition fraction plotted against the initial geometric particle size. Martonen *et al* (1985) argued that a critical particle size could be defined: D_c. For hygroscopic particles initially smaller than this size, hygroscopic growth would be expected to reduce deposition compared with that of non-hygroscopic particles. The extent of the reduction in deposition is shown for a particle of initial diameter 0.1 μm by AB. For hygroscopic particles of greater initial diameter than D_c, hygroscopic growth would increase deposition: the effect on particles of 0.7 μm initial diameter is shown by ED for oral breathing and CD for nasal breathing. Note that more growth occurs during nasal breathing, the inspired air being better humidified.

The effects of hygroscopic growth upon the regional deposition of particles in the respiratory tract have been considered by Pritchard (1987). Figure 16 shows the shift in the deposition pattern of particles produced by hygroscopic growth. For hygroscopic particles, minimum deposition is associated with an initial particle diameter of 0.1 μm. For non-hygroscopic particles, the size of minimum deposition was closer to 0.5 μm. In these studies, the hygroscopic particles considered were sodium chloride: such particles would show maximal hygroscopic growth which would not be matched by other particles.

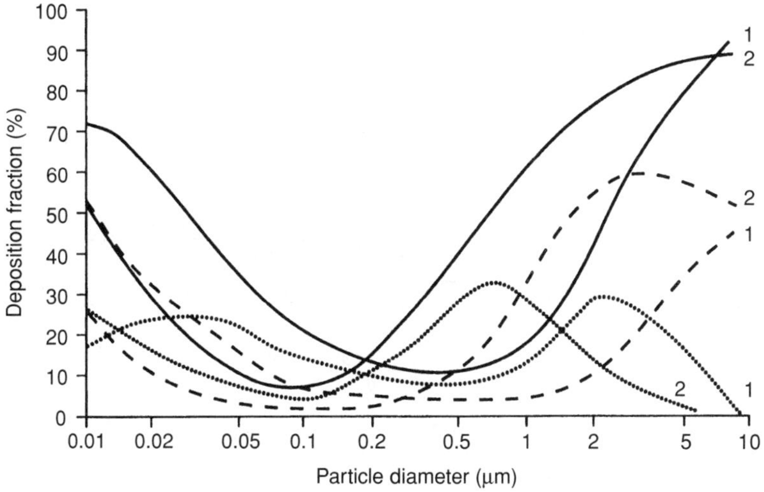

Figure 16. Total and regional distribution of hygroscopic particles in the lung. Oral breathing: tidal volume, 500 ml; breathing rate, 13.7 breaths per minutes. 1, non-hygroscopic; 2, hygroscopic. Solid line, total deposition; dotted line, deposition in pulmonary region; dashed line, deposition in tracheobronchial region. Reproduced from Pritchard JN (1987). Particle growth in the airways and the influence of airflow. In *A New Concept in Inhalation Therapy* (SP Newman, F Moren and GK Crompton, eds), Medicom

The Task Group of the IRPC considered the effects of particle growth upon likely deposition in the respiratory tract and pointed out that for particles of high density, hygroscopic growth would have less effect than might be predicted from a simple consideration of the effects of an increase in particle diameter (Task Group on Lung Dynamics, 1966). Addition of water to a dense particle will reduce particle density and offset the effect of increasing real diameter on the aerodynamic diameter. The ratio of aerodynamic diameter of the particle after hygroscopic growth to that of the original dry particle is given by:

$$\frac{D_{AS}}{D_{AC}} = \left[\frac{\rho_s C_s}{\rho_c C_c}\right]^{\frac{1}{2}} \frac{D_s}{D_c}$$

where ρ_c and ρ_s are the densities of the dry particle and the droplet, respectively, D_{AS} and D_{AC} are the aerodynamic diameters of the droplet and the original particle, respectively, D_s and D_c are the actual diameters, and C_s and C_c are the respective Cunningham's correction factors.

It is clear that droplets may change their composition as they pass along the airways. Little work has been reported on this as regards effects on CW agents, but extensive and informative studies have been undertaken on acid droplets. Droplets of sulphuric acid grow as they pass along the airways. It has been calculated that a particle of less than 0.1 μm initial diameter would reach 99% of its equilibrium diameter on being exposed to the air of the respiratory tract for as little as 0.1 s. This should be compared with the 10 s needed for a droplet of 1.0 μm initial diameter to reach 99% of equilibrium diameter.

Consider a droplet of sulphuric acid growing in the airstream in the respiratory tract: assuming the equilibrium radius of a particle to be 3 times the initial radius, the equilibrium volume will be 27 times the initial volume. Given also that no sulphuric acid is added to the particle during growth, then the sulphuric acid concentration will fall during particle growth by a factor of 27.

Let the initial hydrogen ion molar concentration $= x_0$M. Then:

equilibrium hydrogen ion

$$\text{concentration} = x_0/27\ \text{M}$$

$$\text{Initial pH} = -\log x_0$$

$$\begin{aligned}\text{Equilibrium pH} &= -\log x_0/27\\ &= -(\log x_0 - \log 27)\\ &= -\log x_0 + 1.4314\end{aligned}$$

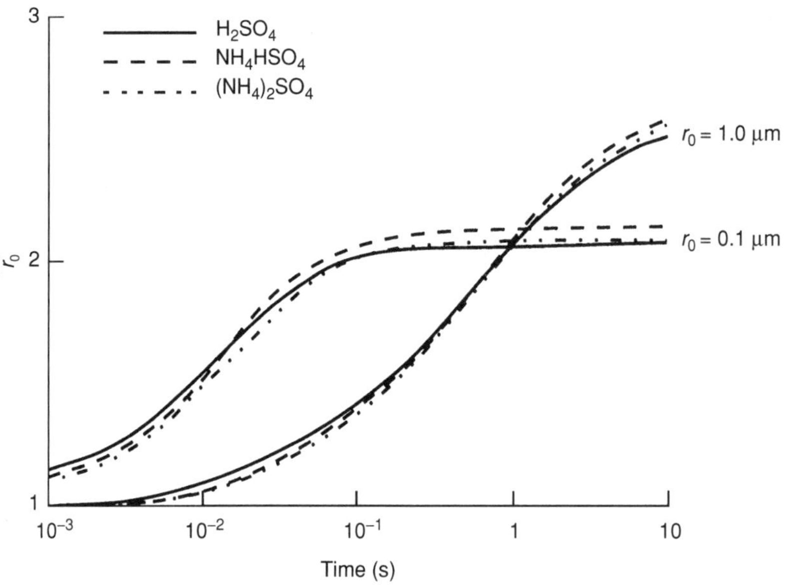

Figure 17. The effect of neutralization by ammonia on the growth of sulphuric acid droplets: r, equilibrium radius; r_0, initial radius. Reprinted from Cocks and Fernando (1982). The growth of sulphate aerosols in the human airways. *J Aerosol Sci*, **13**, 9–19, with kind permission from Elsevier Science Ltd

The dilution factor of 27 will increase the pH of the droplet by approximately 1.4.

NEUTRALIZATION OF ACID DROPLETS BY AMMONIA IN THE RESPIRATORY TRACT

Ammonia is produced in the mouth as a result of bacterial metabolism. Kupprat *et al* (1976) measured ammonia concentrations of between 210 and 700 μg m^{-3} in expired air during quiet mouth breathing. Larson *et al* (1977) recorded values of 10–50 μg m^{-3} during quiet nasal breathing.

Acid droplets absorb ammonia and become partially neutralized as they pass along the respiratory tract. Cocks and McElroy (1984) extended the work already discussed (Cocks and Fernando, 1982) on the hygroscopic growth of particles to model the growth and neutralization of acid aerosols. Before considering the results of the Cocks and McElroy (1984) study, two difficulties identified and dealt with by these authors should be examined:

1. One might ask whether droplets of sulphuric acid which have been neutralized by ammonia and converted into droplets of a solution of ammonium sulphate would grow hygroscopically at the same rate as the original droplets of acid. Cocks and Fernando (1982) modelled the growth of droplets of ammonium bisulphate and ammonium sulphate and showed that the expected growth patterns were very similar to that of sulphuric acid droplets. The results of this modelling exercise are shown in Figure 17.
2. Acid aerosols are often found in association with high concentrations of acidic gases, such as sulphur dioxide. It might, therefore, be asked whether significant depletion of nasal or oral ammonia by sulphur dioxide might occur and thus impair droplet neutralization. This was modelled by Cocks and McElroy (1984) who concluded that at a sulphur dioxide concentration of 100 μg m^{-3} only some 2% of the ammonia likely to be present would be removed in 10 s. The effect may be neglected.

In the main modelling exercise Cocks and McElroy (1984) modelled the following.

Acid loading of the atmosphere: 1000 *and* 100 μ gm^{-3}

It should be noted that given constant relative humidity and all the acid being assumed to be contained in droplets of a given size, e.g. 0.1 μm diameter, there will be 10 times as many droplets present per unit volume of air at the higher loading than at the lower loading.

Ammonia concentrations

These were 500 μg m^{-3} corresponding to oral levels and 50 μg m^{-3} corresponding to nasal levels.

Initial droplet composition

As explained earlier, droplets of sulphuric acid of differing size and acid concentration are in stable equilibrium at different relative humidities. The authors calculated that acid concentration and RH under conditions of stable equilibrium were related as shown in Table 12.

Let us assume that all the acid is present in 5 μm diameter droplets.

Let us also assume that the acid loading of the air is 1000 μg m^{-3}.

Now consider a relative humidity of 99.97%. The droplets will contain acid at a concentration of 7×10^{-3} M.

Now consider a relative humidity of 60%. The droplets will contain acid at a concentration of 5.08 M.

Table 12. Sulphuric acid concentrations in stable droplets at varying ambient relative humidity

RH (%)	[H_2SO_4] M
99.97	7.0×10^{-3}
99.5	0.139
80	3.29
60	5.08

In both cases, total acid loading is constant. This could only be accomplished if there were many more droplets present per unit volume of air at a relative humidity of 99.97% as compared with air at a relative humidity of 60%.

The ratio of the number of droplets present in the two cases is given by:

$$(5.08/7) \times 10^{-3} = 725.7$$

That is, at an atmospheric loading of x μg m^{-3} with the acid contained in droplets of diameter y μm, there will be 725.7 times as many droplets present in a unit volume of air at a relative humidity of 99.97% than at 60%.

Initial droplet diameter

Diameters of 0.1, 0.5, 5.0 and 15 μm were considered. Not all droplet diameters were modelled for each value of relative humidity.

The following were calculated for 0.1, 0.3, 1.0, 3.0 and 10 s of growth.

Reduction in droplet acidity (H)

$$H = [H^+]/[H^+]_0$$

Extent of neutralization (N) *expressed as a percentage*

$$N = [NH_4{}^+]/(2([HSO_4{}^-] + [SO_4{}^{2-}])) \times 100$$

Complete neutralization is reached when [$NH_4{}^+$] is twice the concentration of sulphur(VI)-containing molecules. This is complex: it may be approached as follows.

When ammonia in solution reacts with sulphuric acid, two $NH_4{}^+$ ions are produced for each $SO_4{}^{2-}$ ion, one $NH_4{}^+$ ion is produced for each $HSO_4{}^-$ ion and a further $NH_4{}^+$ ion is produced in converting the acidic $HSO_4{}^-$ ion to a neutral $SO_4{}^{2-}$ ion.

Neutrality is reached when the concentration of $NH_4{}^+$ ions is twice the sum of the concentrations of the $SO_4{}^{2-}$ ions and the $HSO_4{}^-$ ions, i.e.

$$[NH_4{}^+] = 2([HSO_4{}^-] + [SO_4{}^{2-}])$$

Table 13. Data from the Cocks–McElroy model of droplet growth and neutralization

$[H_2SO_4]$ ($\mu g\ m^{-3}$)	$[NH_3]$ ($\mu g\ m^{-3}$)	Droplet diameter (μm)		Time (s) 0.1	0.3	1	3	10
1000	500	5	H	0.822	0.535	0.087	5.2×10^{-4}	
RH = 99.5%			V	1.0	1.0	1.0	1.0	
$[H_2SO_4]_0 = 0.139$ M			N	11.7	32.3	81.9	100	

The contribution of droplet growth to the change in acidity (V)

$$V = V_0/V$$

It will be appreciated that a complex matrix of results was generated by this modelling exercise. From the detailed tables of results presented by the authors, certain particularly interesting results have been selected for more detailed consideration here.

For the results shown in Table 13, the model has included sufficient ammonia to bring about complete neutralization of the acid droplets in the air. As the initial relative humidity was high (99.5%), no hygroscopic growth of the particles occurred. Complete neutralization of 5 μm droplets would be expected in 3 s. The capacity of the ammonia present to neutralize all of the acid present should be contrasted with the extent of neutralization possible had the ammonia concentration been 50 $\mu g\ m^{-3}$.

Calculation of the extent of neutralization possible given an acid loading of 1000 $\mu g\ m^{-3}$ and an ammonia concentration of 50 $\mu g\ m^{-3}$:

$$\underline{H_2SO_4:} \qquad \underline{\text{GMW} = 98\text{g}}$$

$$\begin{aligned} 1\ \mu g &= 1/98 \times 10^{-6}\ \text{M} \\ &= 2/98 \times 10^{-6}\ \text{Eq} \\ 1000\ \mu g &= 2/98 \times 10^{-6} \times 10^{3}\ \text{Eq} \\ &\sim 20 \times 10^{-6}\ \text{Eq} \end{aligned}$$

$$\underline{NH_3:} \qquad \underline{\text{GMW} = 17\text{g}}$$

$$\begin{aligned} 50\ \mu g &= 1/17 \times 10^{-6} \times 50\ \text{Eq} \\ &= 2.94 \times 10^{-6}\ \text{Eq} \end{aligned}$$

$$\begin{aligned} \text{Neutralization ratio} &= 2.94/20 = 0.147 \\ &= 14.7\% \end{aligned}$$

The results shown in Table 13 should be compared with those in Table 14. Here, again, no hygroscopic growth occurred but neutralization was very rapid as a result of the high surface to volume ratio of the small droplets.

The results shown in Table 15 are more difficult to interpret. Neutralization is slow because of the high acid content of each droplet. However, considerable hygroscopic growth occurs and droplet acidity is more affected by droplet growth than by droplet neutralization. It should be noted that acidity has fallen to 40% of its original value by 0.1 s despite the fact that neutralization has only reached 0.42%.

For Table 16, the same conditions as shown in Table 15 are simulated except that the initial droplet diameter is set at 1.0 μm. Here, the droplets are neutralized more rapidly as a result of the more favourable surface-to-volume ratio and also grow more rapidly than the 5 μm droplets. Almost complete neutralization and reduction of droplet acid concentration to low levels is achieved by 1.0 s.

These studies show clearly that substantial changes in size and composition of aerosol droplets may occur during passage along the respiratory tract. It is not surprising that hygroscopic growth may produce significant changes in patterns of deposition of particles. Interestingly, hygroscopic growth is in part dependent upon the pattern of airflow in the airway. This will be considered when the general effects of increased ventilation upon particle deposition have been considered.

Muir (1972) reviewed the effects of variations in the pattern of breathing upon total particle deposition and on the distribution of the deposited particles along the respiratory tract: the effects are complex. Inertial deposition of particles is dependent upon the velocity of the particles and hence upon the velocity of the airstream. Sedimentation and diffusion-dependent deposition

Table 14. Data from the Cocks–McElroy model of droplet growth and neutralization

$[H_2SO_4]$ ($\mu g\ m^{-3}$)	$[NH_3]$ ($\mu g\ m^{-3}$)	Droplet diameter (μm)		Time (s) 0.1	0.3	1	3	10
1000	500	1.0	*H*	5.2×10^{-4}				
RH = 99.5%			*V*	1.0				
$[H_2SO_4]_0 = 0.139$ M			*N*	100				

Table 15. Data from the Cocks–McElroy model of droplet growth and neutralization

$[H_2SO_4]$ ($\mu g\ m^{-3}$)	$[NH_3]$ ($\mu g\ m^{-3}$)	Droplet diameter (μm)		Time (s) 0.1	0.3	1	3	10
1000	500	5	*H*	0.400	0.241	0.126	0.063	0.011
RH = 60%			*V*	0.398	0.241	0.134	0.081	0.048
$[H_2SO_4]_0 = 5.08$–5.17 M			*N*	0.42	1.43	5.43	19.0	61.5

are dependent upon the time available for particle displacement to bring particles into contact with the walls of the airways. Deposition of large particles in the upper airways might be expected to rise under conditions of increased airflow. Increased rates of airflow increase the likelihood of turbulent flow occurring and this again increases the likelihood of particles on the walls of the airways. Turbulent flow occurs in the airways when the Reynold's number exceeds about 1000. Muir (1972) summarized the effects of changing airflow as follows.

> *For a given minute ventilation rapid shallow breathing reduces overall particle deposition and, in particular, reduces the fraction of the aerosol penetrating to the alveoli. On the other hand slow, deep, breathing increases the deposition of the aerosol in the depths of the lung....*

Muir went on to point out that during exercise minute volume increases as a result of increases in both tidal volume and breathing rate. Increases in tidal volume would lead to particles being drawn further into the lung and would offset the reduced fractional deposition which might be expected had only the respiratory rate increased. Muir and Davies (1967) confirmed, using 0.5 μm diameter particles, that fractional deposition remained constant in the resting and exercising subject. Dennis (1971) in a study using larger particles (1.0–3.0 μm diameter) showed an increase in the deposition fraction with minute volume. Given that the deposition fraction does not fall during exercise, it follows that total particle deposition will increase with minute volume.

During exercise, oral breathing becomes increasingly important and the reduction in fractional deposition in the nose will increase the fraction of the inspired aerosol delivered to the lung.

Pritchard (1987) has drawn attention to the increased rate of removal of latent heat from

Table 16. Data from the Cocks–McElroy model of droplet growth and neutralization

$[H_2SO_4]$ ($\mu g\ m^{-3}$)	$[NH_3]$ ($\mu g\ m^{-3}$)	Droplet diameter (μm)		Time (s) 0.1	0.3	1	3	10
1000	500	1.0	*H*	0.09	0.028	2.9×10^{-5}	2.0×10^{-5}	1.8×10^{-5}
RH = 60%			*V*	0.115	0.07	0.043	0.032	0.029
$[H_2SO_4]_0 = 5.08$–5.17 M			*N*	14.3	44.6	99.9	99.9	99.9

Table 17. The influence of breathing pattern on ammonium bisulphate deposition in the lung[a] Reproduced from Pritchard JN (1987). Particle growth in the airways and the influence of airflow. In: *A New Concept in Inhalation Therapy* (SP Newman, F Moren and GK Crompton, eds) Medicom

		Flow = 15 l min^{-1} Particle size			Flow = 60 l min^{-1} Particle size		
		0.1 μm	0.5 μm	1.0 μm	0.1 μm	0.5 μm	1.0 μm
Deposition under oral breathing %							
Region:	Upper	0	1	2	0	2	8
	Middle	2	2	5	1	4	14
	Lower	7	8	21	4	6	12
Total		9	11	28	5	12	34
Deposition under nasal breathing %							
Region:	Upper	0	2	**8**	0	8	**30**
	Middle	2	3	**19**	1	7	**16**
	Lower	7	8	**34**	4	5	**19**
Total		9	13	**7**	5	20	**65**

Note:
[a] Upper region, airway generation 0–5; middle region, airway generation 6–10; lower region, airway generation 11–15.

particles undergoing hygroscopic growth under conditions of turbulent airflow. As growth rate is dependent upon the rate of removal of latent heat, conditions of turbulent flow increase hygroscopic growth rates. Given that inertial impaction is dependent upon the square of particle diameter, increased hygroscopic growth will increase particle deposition. Of course, this effect may be offset in the upper airways by the reduced particle residence time imposed by increased flow rates: deposition in the smaller airways would then be likely to be increased. Pritchard (1987) using data from Martonen *et al* (1985) produced Table 17. The significant effects are illustrated by comparison of the sets of figures printed in bold typeface. The marked increase in deposition of 1.0 μm particles in the upper airway under conditions of nasal breathing and high flow rate is obvious. Interestingly, the fractional deposition ratio between the upper and lower airways is reversed under these conditions as compared with low-flow-rate nasal breathing.

GENERAL CONCEPTS CONCERNING THE TOXICITY OF CW AGENTS

The toxicity of a compound may be defined in a number of ways. The descriptor until recently used in general toxicology and still widely used when thinking about CW agents is the LD_{50}. This specifies the dose of the compound in question which would be expected to kill 50% of a group of animals of the same species, i.e. the median lethal dose. The statement 'the LD_{50} of lewisite is 2 mg kg^{-1}' is, as it stands, meaningless. Only when the route of administration and the species are specified does the statement become informative. Computer-stored toxicology databases often list LD_{50} data with the essential qualifications added, e.g.

$$\text{lewisite, } LD_{50} \text{ (subcut, dog)} = 2 \text{ mg kg}^{-1}$$

LD_{50} values are derived from a study of the toxicity of the given compound. This involves exposing groups of animals to different doses of the test compound and noting the incidence of deaths in the groups by a given time. Hence, a further refinement is added to the LD_{50}: the time at which the count of decedents was made. This should be added to the statement of the LD_{50}.

The LD_{50} may be derived graphically from a plot of the probit transform of the incidence of death against the logarithm of the dose administered. This is the classical quantal assay.

The basis of the probit transform lies in the observation that, for a wide range of toxic compounds, if the incidence of mortality is plotted against the logarithm of the dose of the test

Table 18. Hypothetical data relating dose to mortality[a]

Dose, d_i	Log (dose)	Mortality, m_i Obs	(%)	Probit	Tolerance, $m_i - m_{i-1}$ (%)
0.3	−0.3	0/40	0	—	0
1.0	0.0	1/40	2.5	3.04	2.5
1.5	0.176	4/410	10	3.72	7.5
2.0	0.3	9/40	22.5	4.24	12.5
2.5	0.4	15/40	37.5	4.68	15
3.0	0.48	21/40	52.5	5.06	15
3.5	0.54	25/40	62.5	5.32	10
4.0	0.60	29/40	72.5	5.60	10
4.5	0.65	32/40	80	5.84	7.5
5.0	0.70	34/40	85	6.04	5
5.5	0.74	36/40	90	6.28	5
6.0	0.78	37/40	92.5	6.44	2.5
6.5	0.81	38/40	95	6.64	2.5
7.0	0.85	39/40	97.5	6.96	2.5
7.5	0.88	40/40	100	—	2.5

Note:
[a] d = dose; d_i = ith dose; m = mortality; m_i = mortality at the ith dose.

compound the distribution curve obtained approximates to the cumulative normal distribution curve. As the steps taken in moving from this observation to the 'probit slope' are a little complicated, an example, starting with hypothetical data and working through to the probit slope, is given below.

Consider the hypothetical data given in Table 18. These data may be plotted in a number of ways. If the percentage mortality is plotted against dose, the curve shown in Figure 18 is obtained. If the percentage mortality is plotted against log dose, the curve shown in Figure 19 is obtained. This conforms closely to a cumulative normal distribution curve.

To understand the next manipulation the normal distribution curve should be recalled (Figure 20). The probability of an observation falling between any two values of x may be described by the proportion of the whole area under the curve occupied by the area bounded by the two values. For example, the probability that measurements of x will fall between a and b in Figure 20 is given by the proportion of the whole area under the curve occupied by the shaded area. This may, of course, be thought of in terms of the standard deviation: 68.2% of observations fall within +1 and −1 standard deviations of the mean; in fact, a and b in Figure 20 are standard deviations. Ninety-five per cent of observations lie between +1.96 and −1.96 SDs, and 99% of observations lie between +3 and −3 SDs of the mean. Intermediate values may be obtained from statistical tables of the normal distribution curve.

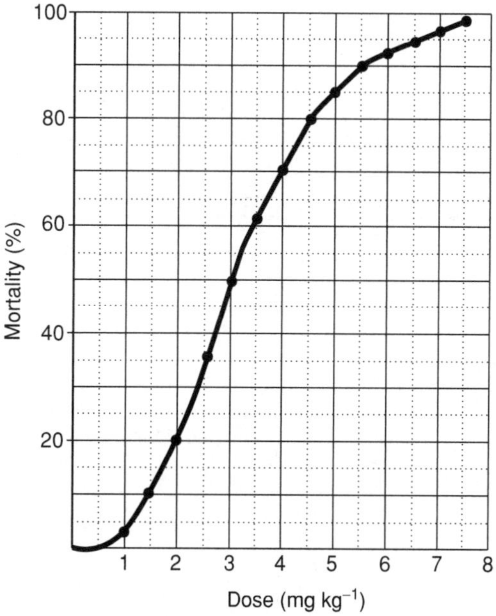

Figure 18. Data from Table 18 plotted arithmetically

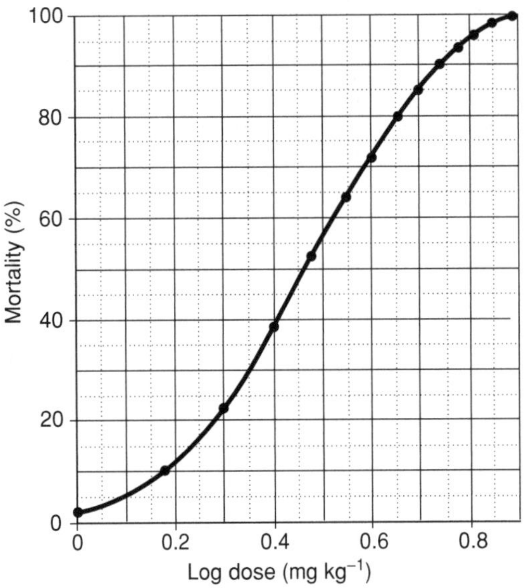

Figure 19. Data from Table 18 plotted using the log of the dose. This could have been shown on semi-log paper by simply plotting the dose on the x-axis

The scale along the x-axis of Figure 20, in units of standard deviation, i.e.

$$\frac{x - \bar{x}}{\text{SD}}$$

is described as the scale of normal equivalent deviation and the values −3, −2, −1, 0, +1,

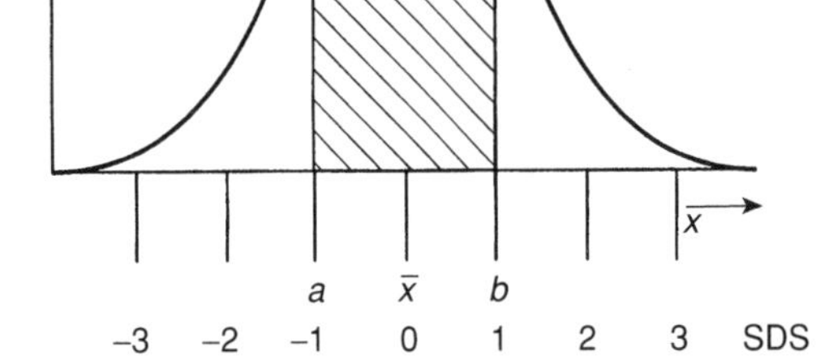

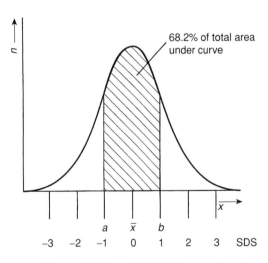

Figure 20. Normal distribution curve

Table 19. Relationships between NEDs and probits

NEDs	−3	−2	−1	0	1	2	3
Probits	2	3	4	5	6	7	8

+2 and +3 are called NEDs or normal equivalent deviates. If 5 is added to each value of the NED, the equivalent 'probit' values are obtained. Finney (1952) commented on the origin of these terms: 'Bliss (1934) first proposed the name "probit" for his modification of Gaddum's normal equivalent deviate [NED], which he increased by 5 so as to simplify the arithmetical procedure by avoiding negative values'. Probabilities for NED > 5 are seldom encountered: NED $= -5$ is equivalent to about 1 in a million. Table 19 shows the equivalence between NEDs and probits.

Recalling how the percentage of observations lay on the normal distribution curve in terms of NEDs allows one to deduce how they will lie in terms of probits. This is shown in Figure 21.

More accurate calculation of probits from the equation of the normal distribution curve has been undertaken and tables are available.

If instead of percentage mortality the corresponding probits are plotted against log dose, a straight line is obtained (Figure 22). The median lethal dose, i.e. the dose at which 50% of the animals would be expected to die, may simply be read off the graph. In practice, the actual dose is plotted on a logarithmic scale on the x-axis and the percentage mortality on the y-axis on a scale modified so as to correspond to the probit scale. Such special graph paper is described as *log-probability paper*, which has been mentioned before in this chapter, and is widely used in toxicology. Figure 23 shows the same data as discussed above plotted on this paper.

A question often asked by students of toxicology is: 'Why does the probit versus log-dose plot so often come out as a straight line?'. This remarkably simple question is difficult to answer. It may be approached in two stages.

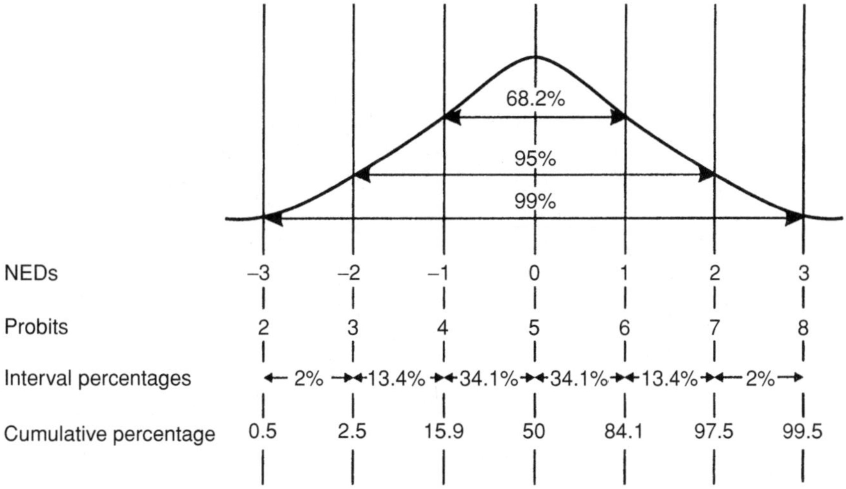

Figure 21. Normal distribution curve showing relationships between NEDs, probits and areas under the curve

Given that a log-normal distribution curve may be used to describe the relationship regarding tolerance of the species in question to compound x, then the probit–log dose plot must yield a straight line. This is because the probit scale is related to the percentage mortality scale (i.e. the scale on the y-axis) in precisely the same way as the percentage mortality is related to the log dose (i.e. the scale on the x-axis).

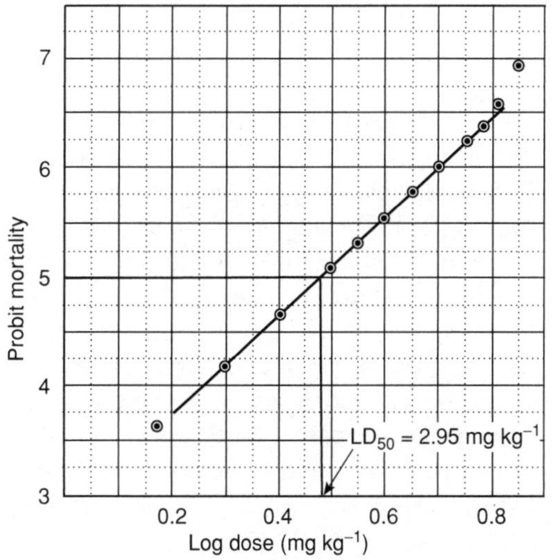

Figure 22. Data from Table 18 plotted showing the log of dose on the x-axis and the mortality probit on the y-axis

This may be summarized as follows:

normal distribution normal distribution

↓ ↓

probits ⟷ % mortality ⟷ log dose

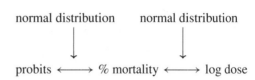

It is no surprise, then, that probits, which are derived from percentage mortality via the normal distribution curve, and log dose, which is related to percentage mortality by the normal distribution curve, should be related by a straight line.

An algebraic analogy may help:

Let $x = k \log y$: a non-linear relationship

Let $z = k' \log y$: a similar non-linear relationship

Then $x/z = k \log y / k' \log y$

$x = k/k'z$: a linear relationship

The second part of the explanation is much less satisfactory: why percentage mortality is related to log dose via the normal distribution curve is unknown.

Many CW agents are encountered as gases. Establishing the dose of a gas is difficult compared with that provided by, say, an intravenous injection. This has led to the definition of exposure rather than dose and to the concentration × time product. Exposure may be thus measured in

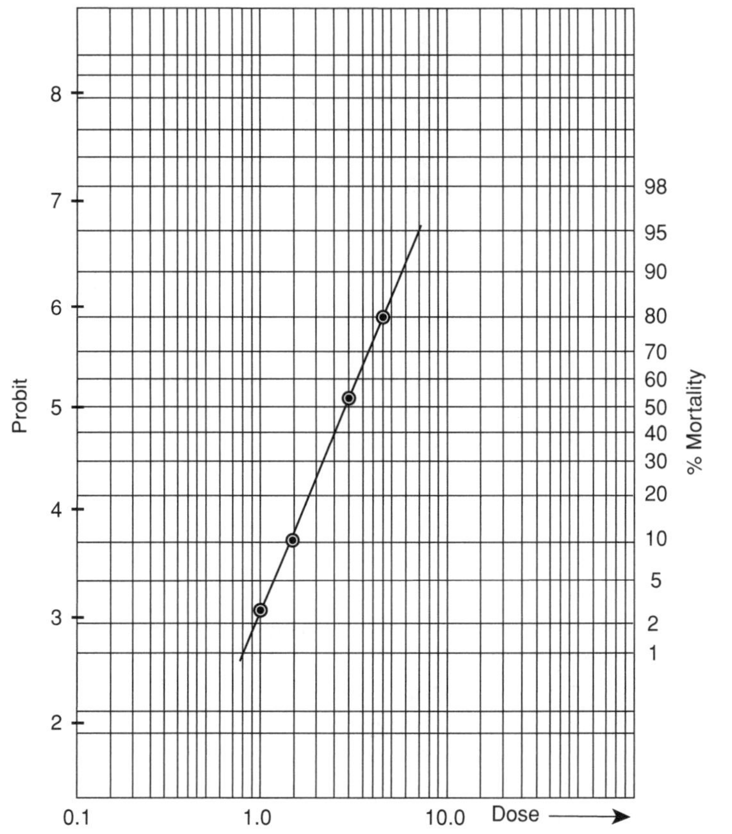

Figure 23. Data from Table 18 plotted on 'log-probability' paper

terms of:

concentration (mg m^{-3}) × duration of exposure (min)

and the median lethal exposure determined experimentally as the LCt_{50}.

The units of *Ct* are mg m^{-3} min or, more euphonically but less obviously, mg min m^{-3}. The toxicity of the nerve agent sarin in humans is often given as 100 mg min m^{-3}.

The fundamental difference between LD_{50} and LCt_{50} is that the former relates to dose and the latter to exposure. It cannot be stressed too strongly: *exposure does not equal dose.*

In addition to the descriptors already identified for use with the LD_{50} value, species and time at which decedents are counted, another must be added regarding LCt_{50} (route need not normally be considered: LCt_{50} usually refers to inhalation, although in experimental studies absorption across the skin might be being studied).

A further additional factor describing the physiological conditions associated with the LCt_{50} must also be added. For example, sarin (GB) is well absorbed across the lung and some 90% of inhaled GB is absorbed. This is effectively independent of respiratory rate and depth. Thus, a person breathing 50 1 min^{-1} will take in five times as much GB as a person breathing 10 1 min^{-1}. Thus, in an identical exposure, the dose absorbed by the former will be five times that absorbed by the latter. In terms of dose, the same dose will be inhaled by an exercising man although he is exposed, for the same period, to only one-fifth of the concentration of GB as a non-exercising man. Thus, the LCt_{50} exercising $= 1/5 LCt_{50}$ at rest.

For GB:

$$\text{Human } LCt_{50} \text{ resting} = 100 \text{ mg min m}^{-3}$$
$$\text{Human } LCt_{50} \text{ exercising} = 20 \text{ mg min m}^{-3}$$

Table 20. Toxicity of the nerve agent sarin (GB)[a]

	C (mg m^{-3})	t (min)	Ct (mg min m^{-3})	Expected % mortality
a	10 000	0.01	100	?
b	1 000	0.1	100	? < 50
c	100	1	100	50
d	10	10	100	50
e	1	100	100	50
f	0.1	1 000	100	? < 50
g	0.01	10 000	100	?
h	0.001	100 000	100	?
i	0.0001	1 000 000	100	?

Note:
[a] The values shown in rows (c), (d) and (e) are approximately correct. Data in other rows have been derived on the assumption that the LCt is constant; this assumption is discussed in the text.

One more complicating factor must now be considered: the effect of duration of exposure. Consider the data shown in Table 20. In each of the rows, the Ct product $= 100$ mg min m^{-3}. In the rows (c), (d) and (e), one might feel reasonably confident that the mortality likely to occur would be about 50%, i.e. an LCt_{50} exposure. At the outer rows, one becomes distinctly less confident and it is unlikely that continuous exposure of a person to 0.0001 mg m^{-3} GB for 1.9 years would have the same effect as exposure to 10 mg m^{-3} for 10 min. Interestingly, the US permissible exposure limit for GB (expressed as an 8-h time-weighted average (TWA), is 0.0001 mg m^{-3}: such an exposure, on a daily, life-long basis, would be regarded as safe.

The concept that a constant Ct product might be expected to produce a constant effect was propounded by F. Haber, who defined:

$$Ct = W$$

where W is the 'product of mortality' (Todlichkeitsprodukt) or 'lethal index'.

The Haber relationship may be represented by a hyperbola. During World War I it was found that the toxicity of many CW agents deviated from the relationship proposed by Haber, and attempts to refine the real relationship were made. The equation:

$$(C - e) \times t = W$$

takes into account elimination (e) of the compound by metabolic processes.

Table 21. Relationship between duration of exposure (time) and concentration and LCt_{50} for hydrogen cyanide

Time (min)	Concentration (mg m^{-3})	LCt_{50} (mg min m^{-3})
0.25	2400	660
1.0	1000	1000
10	200	2000
15	133	4000

One of the most interesting examples of deviations from the Haber relationship is provided by hydrogen cyanide. Consider the data in Table 21. At low concentrations the body is able to metabolize cyanide, and thus as long as the concentration is low a substantial total exposure may be tolerated. The LCt_{50} rises with the duration of exposure. This increase in LCt_{50}, with time, is intuitively understandable and fairly common; the converse, a fall of LCt_{50} with time, is rare but may occur.

Because of the ease of assessing lethality, many toxicity studies have concentrated upon determination of LD_{50} or LCt_{50}. Of more importance in the military context are the incapacitating effects of compounds, and the ID_{50} (median incapacitating dose) or the ICt_{50} are important descriptors of a compound's toxicity. These are, of course, much more difficult to determine in animals and the relevance of such determinations to likely effects in humans is open to doubt. Of even more military importance are the parameters ICt_5 and ICt_{10}, although to determine these with any accuracy is yet more difficult and great caution should be exercised when trying to determine these parameters from studies designed to identify the median exposures. The reliability of predictions made from the probit slope outside the < 20% to > 80% region is low.

Haber's law, or rule, has continued to fascinate toxicologists: as recently as 1999, Hans-Peter Witschi published on this subject (Witschi, 1999) and Miller and coworkers have since published a most interesting paper showing that the relationship described by Haber was but one of a family of such relationships (Miller *et al.*, 2000). The following brief exploration of these relationships is closely based on Miller's paper.

As stated above, the concept $C \times t = k$ is generally associated with Haber. Miller examined data relating to the acute toxicity of chlorine in mice, using the LC_{50} concentration at a range of values of t and found excellent agreement between the data and a simple power law function:

$$C^{\alpha}t = k$$

He fitted the following equation:

$$y = 59709\,x^{-0.983\,78}(R^2 = 0.969\,44)$$

Here, $y = t$ and $x = C$.

Note that the exponent of C is very nearly 1.0.

An equally good fit was found for data relating to a 1% mortality from exposure to chlorine among rats. That the toxicity of chlorine follows Haber's Rule so closely is unsurprising: Haber studied chlorine closely.

Interestingly, Miller also showed that data relating to the delayed acute toxicity in rats of the dioxin compound, HpCDD (1,2,3,4,6,7,8-heptachlorodibenzo-*p*-dioxin), also followed Haber's Rule. If time to death from 'wasting' is plotted against dose ($\mathrm{mg\,kg^{-1}}$) the data are closely fitted by a curve of the equation:

$$y = 97.144x^{-0.978\,08}(R^2 = 0.947\,95)$$

Again, the exponent of x (meaning dose) is very close to unity.

An equation such as:

$$t = kC^{-\alpha}$$

lends itself to a log-log transform:

$$\log t = -\alpha \log C + \log K$$

Miller fitted such an equation to the dioxin data (but using time to 10% mortality) and produced:

$$y = 1.8102 + -0.9351x(R^2 = 0.859\,23)$$

In this equation, $y = \log t$ ($t =$ time to 10% mortality) and $x = \log$ (dose) ($\mathrm{\mu g\,kg^{-1}}$)

However, not all dose–response curves follow Haber's rule. If the dioxin data referred to above are considered again and time is plotted on the x-axis, with dose needed to produce a 10% mortality on the y-axis, then the data are well fitted by an equation:

$$y = 7.3165x^{-0.233\,88}(R^2 = 0.846\,79)$$

The exponent of x is now far from unity.

Note that in the simple Haber relationship, the exponents of both C and t are unity.

It is possible that the exponents of *both* C and t differ from unity. This creates problems for a two-dimensional representation of the data. Why is this? Miller approached the problem along the following lines:

Begin with Haber's rule, $C \times t = k$.

In this equation, the exponents of C and t are unity.

Let us introduce an exponent of the concentration, $C^{\alpha}t = k$.

The introduction of the exponent generates a power function and the terms of the equation can be divided into parameters and variables. This is important, as parameters are often confused with variables: C and t are variables, while α and k are parameters.

We might think that we could write a more general equation, including exponents for both C and t:

$$C^{\alpha}t^{\beta} = k$$

However, now we are in difficulties. Miller points out that this equation is over-parameterized, i.e. we have more parameters (α, β and k) than variables (C and t). Why is this such a problem?

Miller explains:

$$C^{\alpha}t^{\beta} = k$$

If $\alpha = \beta = 1$, this is Haber's rule.

Let us now write:

$$C^{\alpha}t^{\beta} = k$$

as:

$$C^{\alpha} = kt^{-\beta}$$

and thus:

$$C = k^{1/\alpha}t^{-\beta/\alpha}$$

or:

$$C = k^{1}t^{-\gamma}$$

where:

$$\gamma = \beta/\alpha$$

Now comes the critical step: various combinations of β and α can produce the same value of γ. In a two-dimensional plot, these various combinations of β and α will not be distinguishable. This is what is meant by describing the equation as over-parameterized: it cannot be solved to yield unique values of α and β.

Miller suggests an admirable way forward from this impasse – move from a two-dimensional plot to a three-dimensional plot.

Let us assume that for a series of combinations of concentration (C) and duration of exposure (t) to some toxic compound we know the percentage mortality. A probit model can be fitted to the data and an equation in the following form produced:

$$y = m + \alpha \ln C + \beta \ln t \quad \text{(where ln = log to the base } n\text{)}$$

This equation contains three variables (y, C and t) and three parameters (m, α and β) and is thus not over-parameterized and can yield unique values of the parameters that, as Miller points out, 'hold across all levels of responses'. Miller's paper should be consulted for a further exploration of these concepts and illustrative plots in three dimensions.

ABSORPTION OF GASES

The kinetics of gas absorption across the lung is simplified by dividing gases into those which react with tissue components in the lung and those which do not. The latter include volatile organic compounds, such as the anaesthetic gases, while the former include common air pollutants, such as ozone and gases of interest in chemical warfare studies, e.g. chlorine, phosgene and the mustards. Detailed accounts of the kinetics of both types of gases have been provided by workers at the Chemical Industry Institute of Toxicology in the United States (Medinsky *et al*, 1999; Kimbell and Miller, 1999). Only a brief summary is provided here. Before beginning, it is worth noting that water-soluble gases, such as sulphur dioxide, are absorbed efficiently in the nasal passages and at low concentrations little actually reaches the lungs. Comparatively insoluble gases, such as ozone, penetrate to the alveolar level.

Volatile organic compounds

The following properties are important in controlling the uptake of gases by the mucosa of the respiratory tract, this being the first stage of absorption. These factors are: convection, diffusion, dissolution, solubility, partitioning and reaction. These are discussed in detail by Medinsky *et al* (1999). Uptake from the lung into the blood is a comparatively simple process which depends largely on the rate at which air containing the gas is delivered to the alveolar region of the lung, the partition coefficient of the gas (i.e. the ratio of the concentrations of the gas in the two media, or phases, of importance, i.e. the blood and the air at equilibrium) and the rate at which absorbed material is carried away from the lung in the blood. The diffusion of gases, such as nitrous oxide (N_2O), is largely limited by pulmonary blood flow as the partial pressure of N_2O in blood rises rapidly to equal that in the alveoli. Diffusion of gases, such as carbon monoxide, on the other hand are not limited in this way but by the characteristics of the blood–air barrier. This is because the blood has a large 'capacity' to absorb carbon monoxide: the gas binds avidly to haemoglobin and this ensures that the partial pressure of carbon monoxide in the blood remains low. Use is made of these differences in factors controlling the diffusion of N_2O and carbon monoxide by respiratory physiologists: N_2O is used to measure pulmonary blood flow; carbon monoxide to measure the diffusing capacity of the lung. Diffusing capacity is defined as the rate of transfer of carbon monoxide from air to blood (in ml per minute) per unit difference in the partial pressure of carbon monoxide between the alveoli and the blood. It will be apparent that maintaining the P_{CO} in blood at a very low level appreciably simplifies the calculations. Indeed, it was this factor that led to carbon being used as a measure of diffusing capacity instead of the, perhaps, more obvious choice of oxygen.

Determining the increasing partial pressure of oxygen in pulmonary capillary blood presented great difficulties: accounts by Comroe *et al* (1973) and Lumb (2000) should be consulted for details.

It is worth noting in passing that the diffusion of gases is controlled by partial-pressure gradients rather than by concentration gradients. In a system, such as the lung where the solubility of a specified gas is constant (e.g. taken as solubility in water), the gas will diffuse down its partial-pressure gradient which will also be its concentration gradient in terms of direction. However, if two phases are present and the gas solubility differs between these phases the picture becomes more complicated. Gas will diffuse down a partial-pressure gradient even if this movement is against its concentration gradient. One should always consider the partial-pressure gradient in deciding the direction of diffusion. Then, one should consider concentration in deciding how much gas actually diffuses in unit time. A very valuable discussion of these problems has been provided by Dejours (1975). This point is of importance in discussing the relative properties of oxygen and carbon monoxide. Krogh's constant of diffusion describes the rate of diffusion of a gas in terms of nmol s^{-1} cm^{-1} $torr^{-1}$, i.e. how much gas diffuses per second across unit area per unit difference in partial pressure (Krogh, 1915). Expressed in this way, carbon dioxide diffuses about 20 times as rapidly as does oxygen. It will be recalled that Fick's first law of diffusion defines a coefficient of diffusion as:

$$\frac{Mx}{\Delta t} = \dot{M}x = DxA\frac{\Delta Cx}{E}$$

where $Mx/\Delta t$, i.e. $\dot{M}x$ = rate of diffusion of x, Dx is the coefficient of diffusion, A is the area across which diffusion occurs, ΔCx is the concentration gradient and E is the thickness of the barrier.

Thus, we can see that the larger the area and the thinner the diffusion barrier, the more rapidly gas diffuses. The reader will recall that the area of the alveolar surface is about 70 m^2 and the thickness of the air–blood barrier is less than 0.5 μm. The above equation can be rearranged as follows:

$$Dx = \dot{M}x\left(\frac{1}{A}\right)\left(\frac{E}{\Delta Cx}\right)$$

If the units of the terms on the right-hand side of this equation are considered in terms of dimensions we may proceed as follows.

Taking as usual, M for mass, L for length and T for time:

$$MT^{-1}L^{-2}LM^{-1}L^{3}$$

(concentration being, of course, represented by ML^{-3}, and the reciprocal of concentration is $M^{-1}L^{3}$).

Thus, the units of the diffusion coefficient Dx are $L^2.T^{-1}$, usually given as cm^2 s^{-1}. This is an example of the units of a term not being immediately transparent!

Krogh (1915) defined his coefficient of diffusion in terms of partial pressure and we must recall that partial pressure is related to concentration by the solubility constant, βx, of the gas in the medium being considered, i.e.

$$Cx = Px\beta x$$

We may then write:

$$\dot{M}x = DxA\frac{\beta x\Delta Px}{E}$$

and:

$$Dx\beta x = \dot{M}x\left(\frac{1}{A}\right)\left(\frac{E}{\Delta Px}\right)$$

Dimensional analysis is now complicated by the pressure term. Pressure is defined as force per unit area, and force is defined as mass times acceleration. Thus, the dimensional units of pressure are $MLT^{-2}L^{-2}$ or $ML^{-1}T^{-2}$. Thus, the units of $Dx\beta x$ can be defined as:

$$MT^{-1}L^{-2}LM^{-1}LT^{2} \quad \text{or} \quad T$$

The reader will be relieved to know that Krogh's constant of diffusion is not defined in such an unhelpful way but, rather, in the more obvious form of nmol s^{-1} cm^{-1} $torr^{-1}$. This formulation can be easily derived:

$$Dx\beta x = \dot{M}x\left(\frac{1}{A}\right)\left(\frac{E}{\Delta Px}\right) = \left(\frac{\text{nmol}}{\text{s}}\right)\left(\frac{1}{\text{cm}}\right)\left(\frac{1}{\Delta Px}\right)$$

Modelling of the uptake of non-reactive gases has been extensive. The simplest model assumes that at, equilibrium, the rate at which the gas leaves the alveolar space is equal to the rate at which it enters the alveolar space.

Let us define some terms:

$\dot{V}_A$ = alveolar ventilation rate
C_i = concentration of gas in inspired air
C_A = concentration of the gas in alveolar air
$\dot{Q}_c$ = rate of blood flow through pulmonary capillaries
C_v = concentration of gas in venous blood
C_a = concentration of gas in arterial blood.

Let us now look at a simple model. The rate of inhalation of gas is given by $\dot{V}_A C_i$.

The rate at which gas is arriving at the alveolar space via the blood is given by $\dot{Q}_c C_v$.

The rate at which gas is being removed from the alveolar space via expired air is given by $\dot{V}_A C_A$.

The rate at which gas is being removed from the alveolar space by the blood is given by $\dot{Q}_c C_a$.

Thus, the total inflow rate to the alveolar space is given by:

$$\dot{V}_A C_i + \dot{Q}_c C_v$$

and the total outflow rate from the alveolar space is given by:

$$\dot{V}_A C_A + \dot{Q}_c C_a$$

At equilibrium, the total inflow rate to the alveolar space equals the total outflow rate from the alveolar space.

C_A can be replaced by C_a/P_b where P_b is the blood/air partition coefficient.

Thus, we can write:

$$\dot{V}_A C_i + \dot{Q}_c C_v = \dot{V}_A \frac{Ca}{Pb} + \dot{Q}_c C_a$$

and:

$$C_a\left(\frac{\dot{V}}{P_b} + \dot{Q}_c\right) = \dot{V}_A C_i + \dot{Q}_c C_v$$

thus:

$$C_a = \dot{V}_A C_i + \dot{Q}_c C_v \bigg/ \frac{\dot{V}_A}{P_b} + \dot{Q}_c$$

Medinsky *et al* (1999) point out that this equation, derived by Haggard (1924), is incorporated into many models for predicting the arterial concentration of organic gases. The same authors point out that the model ignores metabolism of the compound by the lung and such factors as flow-restricted uptake and diffusion-limited storage in lung tissues. Their account should be consulted for further details and references to the more advanced literature.

Reactive gases

Modelling the uptake of reactive gases by the lung is much more complex and the results less intuitively obvious than those applicable to non-reactive gases. In particular, attention has to be paid to the amount of gas that is taken up by airway-lining fluid, the tissue of the lung and the blood. It will be obvious that a great deal of anatomical information regarding, for example, lining fluid composition and thickness would be needed in the construction of such models. The account by Miller and his colleagues (Miller *et al*, 1993) should be consulted for details. These workers have modelled the uptake of gases in each generation of the airways in experimental animals and in man. Ozone has been studied in detail. One important product of this work has been the capacity to distinguish between 'net-dose' and 'tissue-dose' at each airway generation. Interestingly, although net (i.e. total) ozone uptake falls slowly as the terminal bronchioles are approached, the tissue-dose actually rises rapidly and reaches a peak expressed as $\mu g\, m^{-2}$ per unit inhaled concentration (μg O_3 cm^{-2} min^{-1})/(μg ambient O_3 m^{-3}). However, the tissue-dose then falls away rapidly as the alveolar zone is entered by the model. Several factors control this phenomenon: the thinning of mucus-rich lining fluid as the terminal airways are approached and the rapid increase in surface area once the alveolar zone is reached. The tissue-dose modelling fits nicely with the location of maximal damage seen in experimental animals exposed to ozone. The models have been extended to deal with human adults and children under conditions of rest and exercise.

The need for such work to be applied to gases of interest in the chemical warfare field is obvious.

PARTICLES AS VECTORS FOR CHEMICAL WARFARE AGENTS

Since World War I, the suggestion that inert particles could be used as vectors for CW agents has recurred from time to time. Lefebure (1921) recorded that a prisoner of a German gas battalion had reported the use of small pumice granules impregnated with phosgene. This was apparently an attempt to increase the persistence of phosgene under field conditions. Lefebure also considered that particulate clouds might penetrate respirators designed to absorb gases and, if combined with a lethal gas, present a significant threat. Much of the World War I development of arsenical smokes was based upon this concept. In part, these ideas did not require the particles to act as vectors: a highly irritant particulate capable of penetrating the respirator and inducing coughing or vomiting would lead to removal of the respirator and possible exposure to a lethal compound, such as phosgene. Attempts to develop particulate material impregnated with more-or-less volatile CW agents persisted into the 1930s and 1940s. Mustard-gas-impregnated dust was studied extensively in Germany in this period. Dautrebande (1962), commenting on the role of particles as vectors, noted:

> *For example, when submitted to mustard gas vapours (25 mg/m^3 for 30 min rats do not develop signs of general toxicity or of respiratory distress; simply, in the following days, they exhibit irritation of the accessible mucosae (eyes, nose, ears, rectum) and are all completely cured after three weeks. On the other hand, when submitted for the same length of time (30 min) to a concentration of vesicant vapours of 5 mg/m^3 only (instead of 25 mg/m^3) in the presence of submicronic inert carbon-black particles, they develop an acute pulmonary oedema and most of them die rapidly, usually in less than 6 h.*

This idea that the toxicity of sulphur mustard could be enhanced by the presence of particles was put forward again during the Iran-Iraq war, when chemical weapons were used on a large scale, and again in the Gulf War, when chemical weapons were *not* used. The concept is worth considering in some detail.

The toxicity of CW agents could be enhanced by absorption onto particles by the following mechanisms:

1. Sulphur mustard is absorbed mainly in the upper respiratory tract. The damage produced leads to incapacitation but is seldom fatal. The adsorption of sulphur mustard onto and slow release from small particles could lead to bypassing of the upper respiratory tract and the production of more serious damage in the gas-exchanging part of the lung.
2. Release of toxic materials from particles could lead to increased local concentrations of these substances and thus increased, though very localized, damage.
3. The duration of tissue exposure to toxic chemicals might be increased by slow release from particles.

However, adsorption of toxic gases by particles without subsequent release would be expected to reduce toxicity. For substances which act systemically and do not produce their effect by damage to the lung, little is likely to be gained by adsorption onto particles. In addition, dilution of a toxic material with inert particles inevitably reduces the weapon payload of that toxic material.

Detailed experimental studies of the effects of adsorbing toxic materials onto particles have been undertaken, or at least reported, by only a very few workers. Probably the most detailed analysis reported is that of Goetz (1961), who worked largely from rather unsatisfactory data collected by LaBelle *et al* (1955). Goetz's analysis is considered below.

The surface area of a sphere is given by:

$$4\pi r^2 = \pi d^2$$

and its volume by:

$$\tfrac{4}{3}\pi r^3 = \tfrac{1}{6}\pi d^3 = 0.52d^3$$

Let the mass of particles per unit volume of aerosol be C_p g 1^{-1}, and let particle density be ρ. The mass of one particle is then:

$$\frac{\pi d^3 \rho}{6}$$

The number of particles per litre is given by:

$$\frac{6C_\mathrm{p}}{\pi d^3 \rho}$$

The total surface area of particles, per litre of aerosol, is given by:

$$A_\mathrm{p} = \frac{6C_\mathrm{p}}{\pi d^3 \rho}\pi d^2$$

$$A_\mathrm{p} = \frac{6C_\mathrm{p}}{d\rho}$$

Let t = mean duration of survival of a group of animals exposed to toxic substance T in the absence of particles.

Let t_A = mean duration of survival of a group of animals exposed to toxic substance T in the presence of particles at a concentration of C_p g 1^{-1}.

Let α', the relative synergistic effect of the aerosol, be defined as:

$$\alpha' = 1 - \frac{t_\mathrm{A}}{t}$$

If α' is negative, then the particles have an attenuating effect on the toxicity of T.

If α' is positive, then the particles have an intensifying effect upon the toxicity of T.

Let the specific effectiveness or synergistic potential of the particles be defined in terms of unit surface area by:

$$\alpha = \frac{\alpha'}{A_\mathrm{p}}$$

$$\alpha = \frac{\alpha'}{A_\mathrm{p}} = \frac{\rho d}{6C_\mathrm{p}}\left[1 - \frac{t_\mathrm{A}}{t}\right]$$

This allows comparison of the effects of different particles upon the toxicity of substance T.

Data from LaBelle (1955) with calculated values of α are shown for a range of toxic gases and a variety of particles in Table 22. In common with much of Goetz's paper (Goetz, 1961), the table and the associated calculations are difficult to follow: α seems to be expressed as a percentage. Values for A_p in the rows relating to nitric acid should be taken as the same as those shown for acrolein, and the original table contained a misprint: for acrolein and NaCl, $t_A = 71$ and *not* 0.71.

Goetz (1961) summarized these findings in a shorter table (Table 23). It will be seen that the nature of the particle-toxicant interaction is dependent both upon the identity of the toxic material and that of the particles. Goetz reached the following conclusions:

1. In general, the intensifying action of particles decreases with decreasing volatility of the toxic substance.
2. The porous particles, e.g. clay, act as intensifiers for very volatile materials such as formaldehyde but as attenuators for nitric acid.
3. Glycols and glycerol showed little effect.

Goetz (1961) proceeded to develop a theoretical framework to explain the above observations. His argument is complex and the mathematical analysis, although rewarding, is bedevilled by a number of misprints. In essence, his argument was that deposition of gas molecules from the gas phase on the respiratory surface would lead (unless gas concentrations were very high) to only a low percentage coverage of the surface, i.e. molecules would be widely spaced out on the surface. He argued that the irritant effect of the gas would be dependent upon the percentage areal coverage at any small part of the total respiratory surface. This might be disputed.

If, on the contrary, deposition of gas molecules occurred from the surface of a well-laden particle which had impacted upon the surface, then a much greater local areal coverage of the respiratory surface might be obtained. Goetz concluded from his model that synergism between toxicants and particles could occur but only in a comparatively narrow range of gas concentrations: he suggested that this was why the effect had not been more commonly observed in experimental work (Goetz, 1961).

It is a matter of regret that other authors have not followed up the work of Goetz. Experimental

Table 22. Experimentally derived constants for calculation of synergistic potential.[a] Reprinted from *International Journal of Air and Water Pollution*, **4**, A Goetz, On the nature of the synergistic action of aerosols, pp. 168–184, 1961, with kind permission from Elsevier Science Ltd

Parameter	Triethylene glycol	Ethylene glycol	Glycerol	Mineral oil	Silica gel	Clay	Dicalite	Celite	NaCl	Control
ρ (g cm^{-3})	1.1	1.1	1.3	0.8	1.4	2.5	2.1	2.1	2.1	
d (10^{-4} cm)	1.8	2.0	2.0	2.1	2.7	3.3	3.3	2.9	2.6	—
Formaldehyde										
A_p (cm^2)	67	79	29.6	50	4.9	7.0	3.65	3.55	25.3	
t_A(min)	71	(168)	114	72	(145)	(157)	118	102	114	147
α	+0.775	(−0.177)	+0.78	+1.02	(+ 0.20)	(−0.71)	+ 5.5	+ 8.72	+ 0.91	—
Acrolein										
A_p (cm^2)	11.5	13.6	5.1	8.5	0.79	1.16	0.61	0.59	4.25	
t_A(min)	(73)	(106)	(94)	69	65	(78)	(91)	(99)	71	87
α	(+1.39)	(−1.62)	(−1.37)	+2.47	+31.7	(+ 0.86)	(− 8.2)	−23.8.	+4.22	—
Nitric acid										
t_A (min)	142	151	(103)	72	177	198	254	188	220	118
α	−1.74	−2.06	(+ 3.9)	+4.5	−63	−59	−189	−100	−20.2	—

Note:
[a] Data in parentheses are of no statistical significance.

Table 23. Synergistic potential of various aerosol types[a]. Reprinted from *International Journal of Air and Water Pollution*, **4**, A Goetz, On the nature of the synergistic action of aerosols, pp. 168–184, 1961, with kind permission from Elsevier Science Ltd, The Boulevard, Langford Lane, Kidlington OX5 IGB, UK

Feature	Formaldehyde	Acrolein	Nitric acid
Intensifying action	*Celite* Dicalite *Mineral oil* NaCl *Glycerol* *Triethylene glycol*	Silica gel NaCl Mineral oil	Mineral oil
Attenuating action	None	(Celite) (Dicalite)	*Dicalite* *Celite* *Silica gel* *Clay* *NaCl* *Ethylene glycol* *Triethylene glycol*

Note:
[a] In italics, 'highly significant' data; in parentheses, data of no statistical significance if $\alpha > 5$.

studies by Amdur *et al.* (1988) have shown that very fine metal-containing particles may enhance the effects of low concentrations of sulphur dioxide upon the guinea pig lung. Interestingly, the most sensitive index of effect was a change in the diffusing capacity of the guinea pig lung for carbon monoxide. This measure of lung function reflects the state of the diffusion barrier at the alveolar level and suggests that the sulphur dioxide, which is usually removed by the upper respiratory tract, penetrated to, and had effects upon, the distal lung. It would be interesting to know whether the combination of sulphur mustard with similar particles would have the same effect.

REFERENCES

Altshuler B (1959). Calculation of regional deposition of aerosol in the respiratory tract. *Bull Math Biophys*, **21**, 257–270.

Amdur MO, Chen LC, Guty J *et al.* (1988). Speciation and pulmonary effects of acidic SO_2formed on the surface of ultrafine zinc with aerosols. *Atmos Environ*, **22**, 557–560.

Cocks AT and Fernando AP (1982). The growth of sulphate aerosols in the human airways. *J Aerosol Sci*, **13**, 9–19.

Cocks AT and McElroy WJ (1984). Modelling studies of the concurrent growth and neutralization of sulfuric acid aerosols under conditions in the human airways. *Environ Res*, **35**, 79–96.

Comroe JH (1974). *Physiology of Respiration*, 2nd Edition. Chicago: Year Book Medical Publishers.

Comroe JH, Forster RE, Dubois AB *et al.* (1973). *The Lung: Clinical Physiology and Pulmonary Function Tests*, 2nd Edition. Chicago: Year Book Medical Publishers, Inc.

Dautrebande L (1962). *Microaerosols*. New York: Academic Press.

Davies CN (ed.) (1966). *Aerosol Science*. New York: Academic Press.

Dejours P (1975). *Principles of Comparative Respiratory Physiology*, Oxford: North-Holland Publishing Company.

Dennis WL (1971). The effect of breathing rate on the deposition of particles in the human respiratory system. In: *Inhaled Particles, III* (WH Walton, ed.), pp. 91–102. Old Woking: Unwin.

Drinker P (1925). The size-frequency and identification of certain phagocytosed dusts. *J Ind Hyg*, **7**, 305.

Drinker P and Hatch T(1954). *Industrial Dust*, 2nd Edition. New York: McGraw-Hill Book Company, Inc.

Findeisen W (1935). Über das Absetzen kleiner in der Luft suspendierter Teilchen in der menschlichen Lunge bei der Atmung. *Pfüger's Arch Gesamte Physiol*, **236**, 367–379.

Finney DJ (1952). *Probit Analysis: a Statistical Treatment of the Sigmoid Response Curve*, 2nd Edition. Cambridge: Cambridge University Press.

Florey H (1962). *General Pathology*, 3rd Edition. London: Lloyd Luke (Medical Books).

Goetz A (1961). On the nature of the synergistic action of aerosols. *Int J Air Water Poll*, **4**, 168–184.

Green H (1927). The effect of non-uniformity and particle shape on 'average particle size'. *J Franklin Inst*, **204**, 713–729.

Green HL and Lane WR (1964). *Particulate Clouds: Dusts, Smokes and Mists*, 2nd Edition. London: E and HF Spon.

Haggard HW (1924). The absorption, distribution and elimination of ethyl ether: II. Analysis of the mechanism of the absorption and elimination of such a gas or vapour as ethyl ether. *J Biol Chem*, **49**, 753–770.

Hatch T (1933). Determination of 'average particle size' from the screen-analysis of non-uniform particulate substances. *J Franklin Inst*, **215**, 27–38.

Hatch T and Choate SP (1929). Statistical description of the size properties of non-uniform particulate substances. *J Franklin Inst*, **207**, 369–387.

Hatch T and Hemeon WCL (1948). Influence of particle size in dust exposure. *J Ind Hyg Toxicol*, **30**, 172.

Hinds WC (1982). Particle size statistics. In: *Aerosol Technology. Properties, Behavior and Measurement of Airborne Particles*, pp. 69–103. New York, Chichester, Brisbane, Toronto, Singapore: John Wiley & Sons, Inc./Ltd.

Horsfield K and Cumming G (1967). Angles of branching and diameters of branches in the human bronchial tree. *Bull Math Biophys*, **29**, 245–259.

Horsfield K and Cumming G (1968). Morphology of the bronchial tree in man. *J. Appl Physiol*, **24**, 373–383.

Kimbell JS and Miller FJ (1999). Regional respiratory tract absorption of inhaled reactive gases. In: *Toxicology of the Lung*, 3rd Edition (DE Gardner, JD Crapo and RO McClellan, eds), pp. 557–597. London: Taylor and Francis.

Krogh M (1915). The diffusion of gases through the lungs of man. *J Physiol*, **49**: 271–300.

Kupprat I, Johnson R and Hertig B (1976). Ammonia: a normal constituent of expired air during rest and exercise. *Fed Proc Fed Am Soc Exp Biol*, **35**, 478.

LaBelle CW, Long JE and Christofano EE (1955). Synergistic effects of aerosols. Particulates as carriers of toxic vapours. *AMA Arch Ind Health*, **11**, 297–304.

Landahl HD (1950). On the removal of air-borne droplets by the human respiratory system. The lung. *Bull Math Biophys*, **12**, 43–56.

Larson TV, Covert DS, Frank R *et al.* (1977). Ammonia in the human airways: neutralization of inspired acid sulphate aerosols. *Science*, **197**, 161–163.

Lefebure V (1921). *The Riddle of the Rhine*. London: W. Collins & Company.

Lumb AB (2000). *Nunn's Applied Respiratory Physiology*, 5th Edition. London: Butterworth-Heinemann.

Martonen TB, Barnett AE and Miller FJ (1985). Ambient sulphate aerosol deposition in man: modelling the influence of hygroscopicity. *Environ Health Perspect*, **63**, 11–24.

Medinsky MA, Bond JA, Schlosser PM *et al.* (1999). Mechanisms and models for respiratory tract update of volatile organic chemicals. In: *Toxicology of the Lung*, 3rd Edition (DE Gardner, JD Crapo and RO McClellan, eds), pp. 483–512. London: Taylor and Francis.

Menzel DB and Amdur MO (1986). Toxic responses of the respiratory system. In: *Casarett and Doull's Toxicology: the Basic Science of Poisons*, 3rd Edition (LJ Casarett, J Doull, CD Klaasen and MO Amdur, eds), pp. 330–358. London, New York: Macmillan.

Miller FJ, Schlosser PM, Janszen DB, (2000). Haber's Rule: a special case in a family of curves recording concentration and duration of exposure to a fixed level of response for a given endpoint. *Toxicology*, **149**: 21–34.

Miller FJ, Overton JH, Kimbell JS *et al.* (1993). Regional respiratory-tract absorption of inhaled reactive gases. In: *Toxicology of the Lung*, 2nd Edition (DE Gardner, JD Crapo and RO McClellan, eds), pp. 485–525. New York: Raven Press.

Muir DCF (1972). *Clinical Aspects of Inhaled Particles*. London: William Heinemann Medical Books.

Muir DCF and Davies CN (1967). The deposition of 0.5 μm diameter aerosols in the lungs of man. *Ann Occup Hyg*, **10**, 161–174.

Pritchard JN (1987). Particle growth in the airways and the influence of airflow. In: *A New Concept in Inhalation Therapy* (SP Newman, F Morén and GK Crompton, eds), pp. 3–24. Proceedings of an International Workshop on a New Inhaler, 21–22 May 1987, London, UK. London: Medicom.

Raabe OG (1970). Generation and characterization of aerosols. In: *Inhalation Carcinogenesis* (MG Hanna, P Nettershein and JR Gilbert, eds), pp. 123–172.

Proceedings of a Biology Division, Oak Ridge National Laboratory Conference. Oak Ridge, TN, USA: US Atomic Energy Commission.

Sartori M (1939). *The War Gases*. London: J & A Churchill Ltd.

Starling SG (1935). *Mechanical Properties of Matter*. London: Macmillan and Company.

Stober W, McClellen RO and Morrow PE (1993). Approaches to modelling disposition of inhaled particles and fibres in the lung. In: *Toxicology of the Lung*, 2nd Edition (DE Gardner, JD Crapo and RO McClennan, eds), pp. 527–602. New York: Raven Press.

Task Group on Lung Dynamics (1966). Deposition and retention models for internal dosimetry of the human respiratory tract. *Health Phys*, **12**, 173–207.

Valberg PA (1985). Determination of retained lung dose from toxicology of inhaled materials. In: *Handbook of Experimental Pharmacology*, Vol. 75 (H-P Witschi and JD Brain, eds), pp. 57–91. Berlin: Springer-Verlag.

Weibel ER (1963). *Morphometry of the Human Lung*. Berlin: Springer-Verlag.

Witschi H-P (1999). Some notes on the history of Haber's Law. *Toxicol Sci*, **50**, 164–168.

3 DISPERSION AND MODELLING OF THE SPREAD OF CHEMICAL WARFARE AGENTS

Roger D. Kingdon and Steven Walker

Dstl, Salisbury, UK

The views expressed in this chapter are those of the authors and should not be construed as the position or policy of the Ministry of Defence unless so designated by other authorised documents.

INTRODUCTION

It is one thing to hold up a sealed flask containing a few millilitres of nerve agent and proclaim that it has the *potential* to kill thousands of people; it is quite another thing to *actually* bring that number of people into contact with the nerve agent such that their lives will be in jeopardy. The difference between these two situations, which can be very great, resides in the means of distribution and delivery of the agent. Chemical warfare agents (CWAs) may be distributed and delivered using any or all available media, including the water supply, the food supply and through dispersion in the atmosphere. Since its large-scale exploitation during World War I, atmospheric dispersion has been the primary means of distributing and delivering CWAs and accordingly this is the topic to be discussed in this chapter.

In considering the atmospheric dispersion of CWAs, it is necessary to take into account both the physical mechanisms involved and the means by which those mechanisms are understood. That is, it is not sufficient to describe how CWAs are dispersed in the atmosphere; we must also understand how we arrived at that description. The reason for this is the sheer complexity of the physical mechanisms involved and the necessity to handle the resulting proliferation of models in a controlled fashion. Atmospheric dispersion depends upon a wide variety of processes, including:

- The vertical stratification of temperature and wind speed in the atmospheric boundary layer.
- The relative motion and the mixing of CWAs in the atmosphere, taking into account their density, momentum, temperature and bulk-release properties.
- Different behaviour at different spatio–temporal scales, ranging from small-scale concentration fluctuations in the turbulence inertial range to large-scale air movements in synoptic events affecting the entire troposphere.
- The effect of obstacles, *i.e.* hills, buildings and trees.

In order to describe this wide variety of processes, it is necessary to formulate a correspondingly wide variety of mathematical approximations, each of which sufficiently complex that it is necessary to implement them as computational models. Each model embodies its own unique set of approximations, which can be expected to be mutually consistent; however, there may be inconsistencies between the different sets

Chemical Warfare Agents: Toxicology and Treatment (2nd Edition)
Edited by Timothy C. Marrs, Robert L. Maynard and Frederick R. Sidell

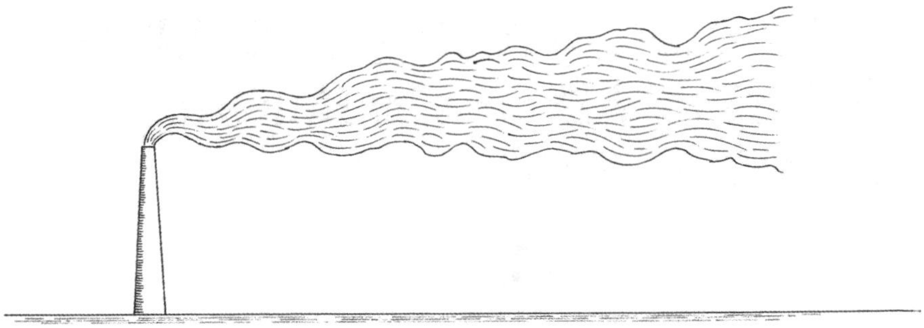

Figure 1. The characteristic *coning* form of a smoke-plume in a neutrally stable atmosphere (after Pasquill, 1962, p. 181)

of approximations underlying different models. That is, it might be expected that situations arise where different models predict different outcomes which appear to contradict one another. In order to handle these situations effectively, it is necessary to employ a clear and robust 'concept of use' defining how the models are to be operated and their results interpreted.

This chapter describes, first, the observed phenomena associated with atmospheric dispersion and, secondly, the range of models used to describe these physical processes, and their concept of use in relation to the atmospheric dispersion of CWAs.

DISPERSION IN THE ATMOSPHERE

Much can be learnt from the direct observation of atmospheric dispersion phenomena. Many observations are of the dispersion of a plume of visible tracer (e.g. smoke or water vapour) from a chimney or 'stack', such that it is quite the convention to describe dispersion in this context. Accordingly, stack-plume observation is used here to help identify the main processes of atmospheric dispersion pertinent to the distribution and delivery of CWAs. In the following, these processes are described under four headings:

- Atmospheric stability
- Relative motion and mixing of CWAs in the atmosphere
- Scale behaviour
- The effect of obstacles.

Atmospheric stability

The classic text describing atmospheric dispersion phenomena is Pasquill's *Atmospheric Diffusion* (Pasquill, 1962). On page 181, Pasquill includes a number of diagrams (attributed to Church (1949) and the United States Weather Bureau (1955)) illustrating the characteristic forms of smoke-plumes from chimneys under different conditions of atmospheric stability. These diagrams have been copied here as Figures 1–5.

Figure 1 shows the typical shape of a smoke-plume in a neutrally stable atmosphere. In this case, there are no resultant vertical gradients of temperature or wind speed, so that the dominant dispersing process is the prevailing horizontal airflow. Under this influence, the smoke is advected downwind and the resulting smoke-plume has a characteristic form of a cone. This shape is known as a 'Gaussian plume' because the cross-section of its time-averaged smoke concentration has a normal or Gaussian distribution (Figure 6). This is an important observation as the associated mathematical model – the 'Gaussian plume approximation' – can be formulated on a much-reduced set of variables, principally, the time-dependent Gaussian variances describing the extent of the plume in three dimensions (i.e. longitudinal, lateral and vertical with respect to the advection vector). Accordingly, it is convenient to describe the effects of all dispersion processes in terms of their differential influence on this classic Gaussian plume.

Figure 2 shows the typical shape of a smoke-plume in an unstable atmosphere. In this case, there is a vertical temperature gradient such that

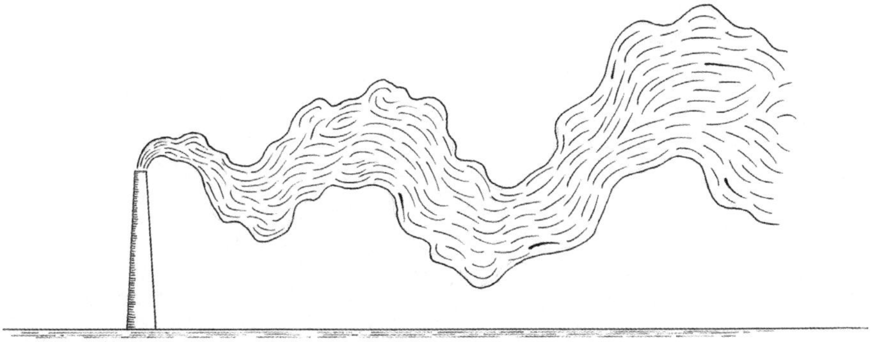

Figure 2. The characteristic *looping* form of a smoke-plume in an unstable atmosphere (after Pasquill, 1962, p. 181)

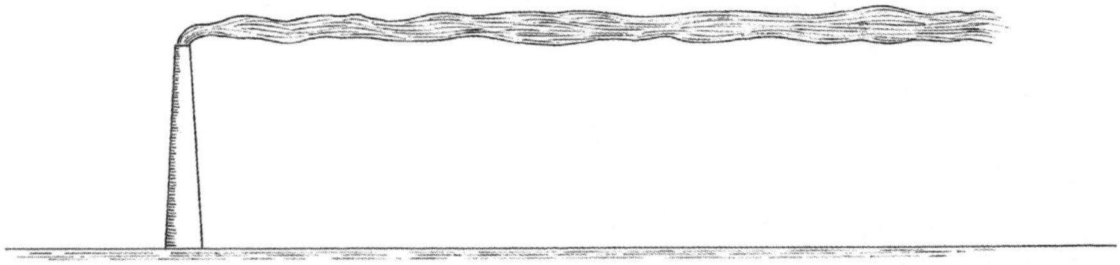

Figure 3. The characteristic *fanning* form of a smoke-plume in a stable atmosphere (after Pasquill, 1962, p. 181)

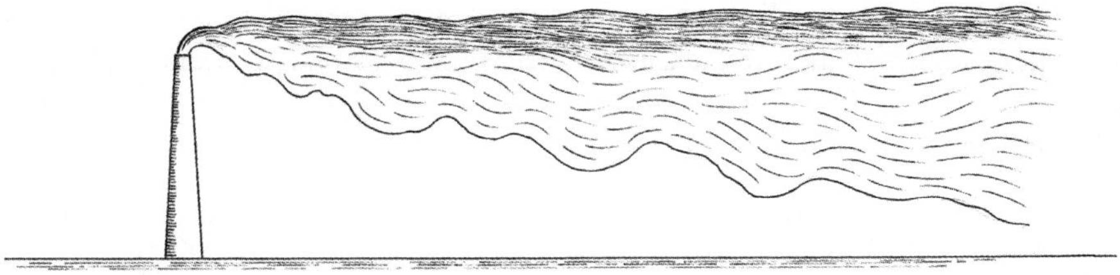

Figure 4. The characteristic *fumigation* form of a smoke-plume in a stable above/unstable below atmosphere (after Pasquill, 1962, p. 181)

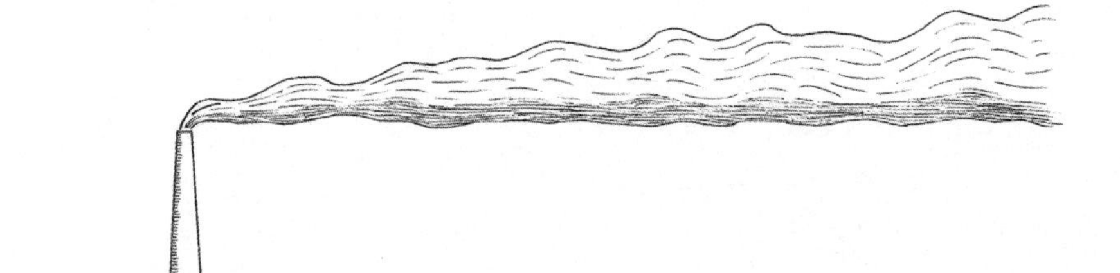

Figure 5. The characteristic *lofting* form of a smoke-plume in an unstable above/stable below atmosphere (after Pasquill, 1962, p. 181)

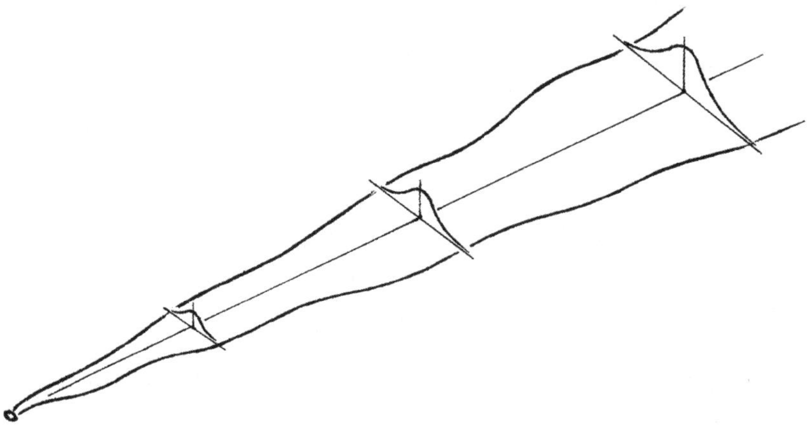

Figure 6. The Gaussian plume. The cross-section of the time-averaged tracer concentration has a normal or Gaussian distribution (after Gifford, 1959)

the air near the ground is hotter and less dense than the air aloft, and the resulting convection currents entrain the smoke in a characteristic looping form. In comparison with its profile for neutral stability, the unstable-atmosphere plume may be supposed to retain a Gaussian cross-section at any point downwind, but the location of the plume will be different, as will the variances describing the extent of the plume. When modelling this plume it is often convenient to include the large-scale fluctuations in the estimation of the plume variances, so that the resulting Gaussian plume is a 'hazard area envelope' within which the actual plume is to be found (Figure 7).

Figure 3 shows the typical shape of a smoke-plume in a stable atmosphere. In this case, there is a vertical temperature gradient such that the air aloft is hotter and less dense then the air near the ground – a so-called 'temperature inversion' – and the resulting *absence* of ambient vertical motion causes the plume to fan out under the influence of the prevailing horizontal airflow. In comparison with its profile for neutral stability, the flattened cone of the stable-atmosphere plume may be supposed to retain a

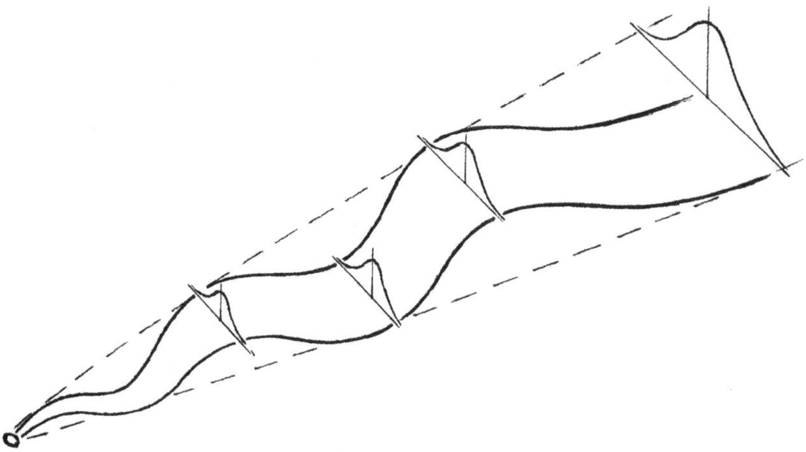

Figure 7. The meandering Gaussian plume. The cross-section of the time-averaged tracer concentration has a normal or Gaussian distribution, as does the 'hazard area envelope' within which the actual plume is to be found. The latter is obtained by including the large-scale fluctuations in the time-average of the tracer concentration (after Gifford, 1959)

Gaussian cross-section at any point downwind, but the variances describing the extent of the plume in each direction will be different.

For temperate climates, neutrally stable atmospheres are the norm, with unstable conditions arising on warm summer afternoons and stable conditions arising on clear, cold nights. Occasionally, the atmospheric stability can enter a period of diurnal variation, in which there is a transition from stable to unstable during the early morning and from unstable to stable in the evening. These transitional states can result in fumigation (Figure 4) and lofting (Figure 5), respectively. In principle, these plumes could be described in terms of the Gaussian plume approximation, but in practice their transience makes validation of the resulting models very difficult. Instead, transitional states are typically modelled as either stable or unstable.

The stability of the atmosphere is clearly a major influence on the dispersion of a smoke-plume. If the plume comprised a CWA rather than a harmless tracer, then the atmospheric stability would have an important effect on the extent of the downwind hazard area and on the concentrations of agent experienced within that hazard area. Atmospheric stability can be determined by direct observation – as in Figures 1–5 – or by estimation of the Monin–Obukhov lengthscale, L (m), which is a height proportional to the height above a surface at which thermal effects first dominate shear (momentum) effects (Pasquill, 1962), as defined in Equation (1):

$$L = -\frac{u_*^3 c_p \rho T}{kgH} \qquad (1)$$

where:

u_* = turbulence friction velocity (m s^{-1})
c_p = specific heat capacity at constant pressure (J kg^{-1} K^{-1})
ρ = density of fluid (kg m^{-3})
T = mean temperature of an adiabatic atmosphere (K)
k = von Kármán constant
g = acceleration due to gravity (m s^{-2})
H = upward turbulent flux of heat (J m^{-2} s^{-1}).

Table 1 gives the equivalence between the Monin–Obukhov lengthscale and the Pasquill stability classes.

Table 1. The equivalence between the Monin–Obukhov lengthscale, L (expressed in terms of the dimensionless ratio z/L, where z (m) is the height above the surface) and the Pasquill stability classes (Jacobson, 1999)

z/L	Pasquill stability class
> 0	Stable
$= 0$	Neutral
< 0	Unstable

Relative motion and mixing of CWAs in the atmosphere

The dispersion of CWAs in the atmosphere depends both on the properties of the atmosphere (as discussed above) and on the properties of the CWAs, in particular, their density, momentum, temperature and bulk quantity at the point of release. These four factors are discussed in turn.

'Classic' CWAs – i.e. those used during First World War I – were chosen to be denser than air, so that they would remain in high concentrations close to the ground and in the trenches. Indeed, the prevalence of trench warfare on the Western Front meant that CWAs were chosen primarily on the basis of density, rather than other factors, such as toxicity. When released to the atmosphere, tracer gases or droplets experience vertical buoyancy forces depending upon the difference between their net density and the density profile of the atmosphere. More-dense tracers descend, while less-dense tracers ascend through the atmosphere. As the CWAs move through the atmosphere, they mix with the ambient air, a process known as *entrainment*. When the mixed volume of tracer and entrained air has the same density as the surrounding atmosphere, then there is no net buoyancy and further vertical motion ceases. Figures 8 and 9 show the effects of buoyancy and entrainment on tracer plumes that are more-dense and less-dense than the atmosphere, respectively. In these cases, the atmosphere is neutrally stable, such that there is no density gradient in the atmosphere itself, and the plumes continue to move downwards or upwards, albeit at a decreasing rate owing to entrainment. Figure 10 shows the effect of buoyancy and entrainment on a lighter-than-air tracer released in a stable atmosphere, for which the density decreases with

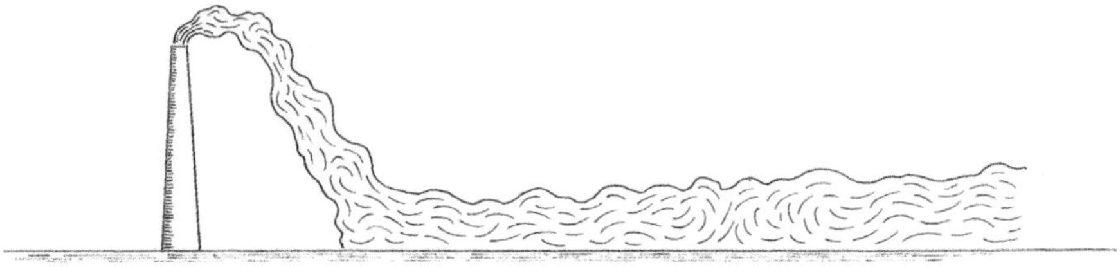

Figure 8. The effect of buoyancy and entrainment on a heavier-than-air tracer released in a neutrally stable atmosphere. The vertical descent of the plume is halted only by the presence of the ground (after Hunt, 1975)

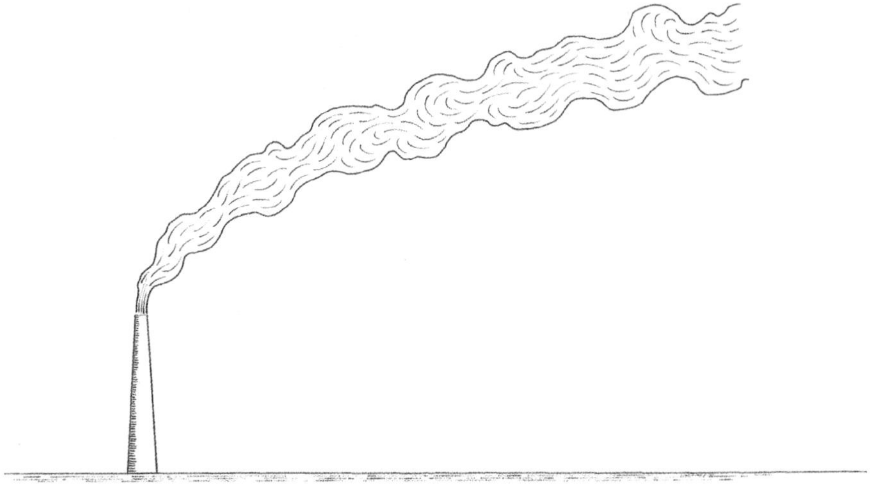

Figure 9. The effect of buoyancy and entrainment on a lighter-than-air tracer released in a neutrally stable atmosphere (after Hunt, 1975)

height. In this case, the vertical motion of the plume ceases when its entrained density is the same as that of the surrounding air. Further dispersion in the vertical direction is as a result of the mixing induced by the prevailing horizontal airflow only.

If a CWA is released to the atmosphere with net momentum – as the result of a pressure difference in a chimney, or an explosion, for example – then the CWA's initial movement and associated mixing is considerably enhanced. After a transient period, however, the initial impulse

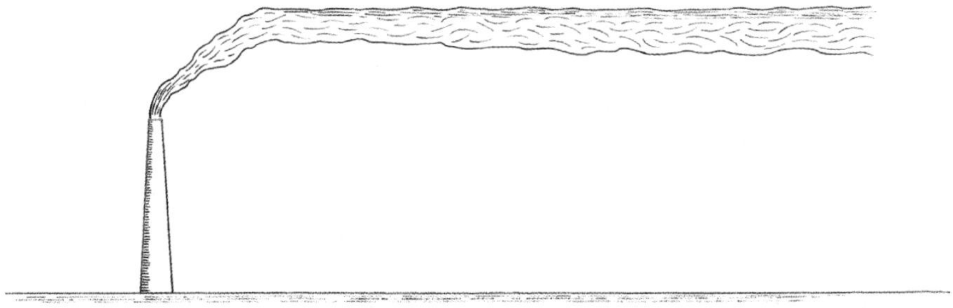

Figure 10. The effect of buoyancy and entrainment on a lighter-than-air tracer released in a stable atmosphere, for which the density decreases with height (after Hunt, 1975)

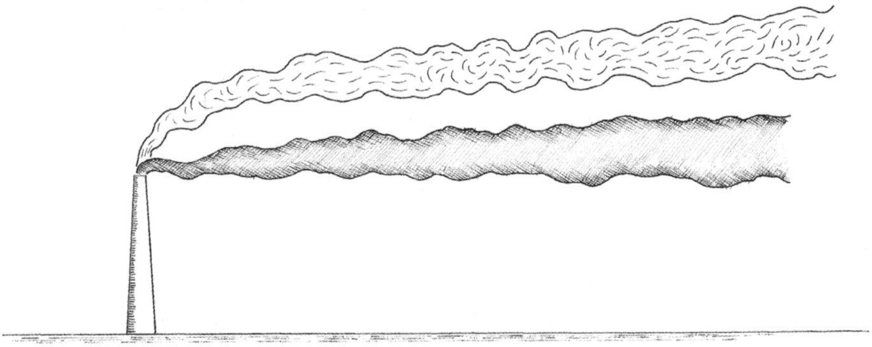

Figure 11. The effect of additional initial vertical momentum on a tracer released in a neutrally stable atmosphere (after Hunt, 1975)

of the CWA is absorbed into the ambient flow and there is no further effect on the subsequent dispersion of the plume. Figure 11 shows the effect of the release of a tracer with additional initial vertical momentum into a neutrally stable atmosphere having the same density as the tracer. Comparing with Figure 1, it can be seen that the additional momentum affects the initial plume only. Accordingly, modellers typically include additional initial momentum through the simple expedient of increasing the spatial extent of the release and adjusting its location, so that the effects of the initial transient period are taken into account in the definition of the source term release.

Most CWAs behave (to a first approximation) as ideal gases, so that an increase in temperature is matched by a corresponding decrease in density. Therefore, heating a CWA (e.g. through combustion of another material) will cause it to rise through the atmosphere, and cooling a CWA (e.g. through expansion) will cause it to fall. As the CWA mixes with the ambient air it exchanges heat, the result being an equilibriation process similar to the entrainment of gases of different densities, albeit with different exchange rates. Indeed, these processes are inextricably linked, and generally referred to as *buoyancy*. For an atmospheric dispersion model to have general applicability it must take buoyancy into account. Buoyancy does not necessarily invalidate the Gaussian plume approximation, but it does involve the introduction of a number of correction factors for the advection and dispersion of the Gaussian plume.

If a fluid that is denser or less buoyant than air is released in bulk quantity, then it can move along the ground as a 'gravity current', encountering minimal retardation and maintaining a forward speed that is truly alarming. The most spectacular demonstration of this effect is in the generation of pyroclastic flows following a volcanic eruption. Fortunately, this particular phenomenon has yet to be exploited for the dispersion of CWAs. However, this example highlights the need to take account of the bulk quantity of the released material, which depends on the delivery system. CWA delivery systems are as varied as the human imagination: cruise missiles, ballistic missiles, submunitions, bombs, shells, crop sprayers, improvised explosive devices and even mail envelopes have been considered or actually used at some time or another. All of these systems must be taken into account when defining the CWA 'source term' for the atmospheric dispersion model.

Scale behaviour

Simple dispersion behaviour – as described above – is scale-independent, so that, for example, Gaussian variances estimated from small-scale experiments in a wind tunnel are applicable in the prediction of the atmospheric dispersion of smoke from a large chimney. However, the specification of a dispersion *environment* (the planetary boundary layer (PBL), terrain, obstacles, closed spaces, etc.) and particular *assets* (i.e. people or property that we might wish to protect) introduces absolute spatial and temporal scales.

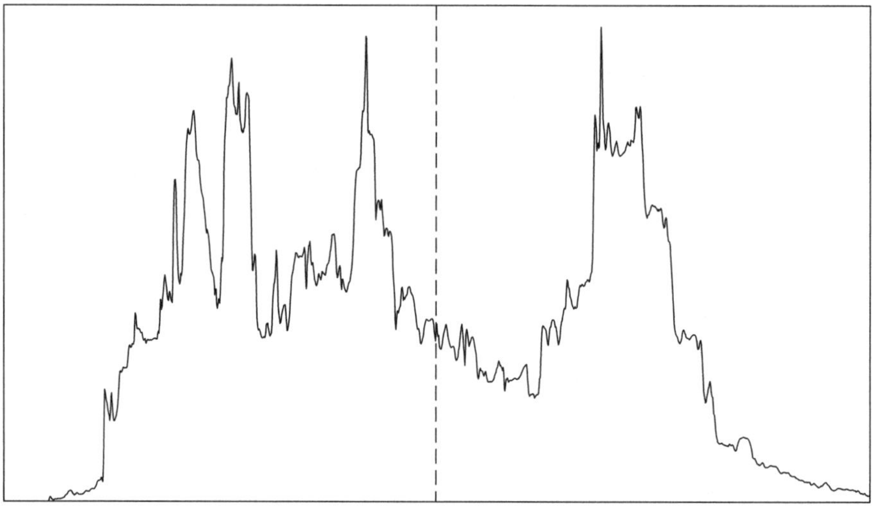

Figure 12. A typical cross-section of the instantaneous tracer concentration, showing small-scale concentration fluctuations (based on field experiments undertaken at Dstl, Porton Down).

Depending upon these scales, different complex dispersion phenomena come into play, and these can have a greater impact on the estimation of the hazard than the scale-independent simple dispersion behaviour. A comprehensive description of scale dispersion behaviour is beyond the scope of this publication, but the following paragraphs identify the main themes:

- small-scale concentration fluctuations in the turbulence inertial range;
- integral-scale plume meander;
- obstacle-scale perturbations (discussed separately, below);
- PBL-scale effects, e.g. sea breezes, urban heat islands and katabatic flows;
- large-scale air movements in synoptic events affecting the entire troposphere.

The Gaussian plume illustrated in Figure 6 represents the cross-section of a *time-averaged* tracer concentration. That is, if time-series concentration measurements taken at a number of points across the plume were separately averaged over their duration, then one would expect to obtain a Gaussian profile. However, at any one time the *instantaneous* concentration profile would look very different. Figure 12, a typical instantaneous concentration cross-section, shows the small-scale concentration fluctuations resulting from the interaction of coherent structures (eddies) of all scales within the turbulence inertial range. The importance of these concentration fluctuations depends upon the nature of the CWA and the nature of the asset. In assessing the hazard associated with a CWA, one is typically concerned with estimating lethality, which is usually expressed in terms of dose, which is concentration integrated through time. That is, in the usual case it is the time-integrated concentration that determines the outcome of the hazard assessment, not the instantaneous concentration, and the small-scale concentration fluctuations have no impact. Increasingly, however, analysts are called upon to assess *incapacitation* as well as *lethality*, and in such cases it is necessary to know the characteristics of the concentration fluctuations, since for several CWAs incapacitation is a function of peak concentration rather than dose.

The turbulence inertial range is a range of spatial scales (and corresponding temporal scales) within which it is possible to predict the broad statistical properties of the fluctuations. Relating these statistical expressions to reliable measures of health effects remains an area of research that is largely uncharted. Less problematic is the assessment of effects relating to the top end of the turbulence inertial range, delineated by the *integral scale*. Fluctuations at this scale result in the meander of the plume, e.g. Figure 7. Often,

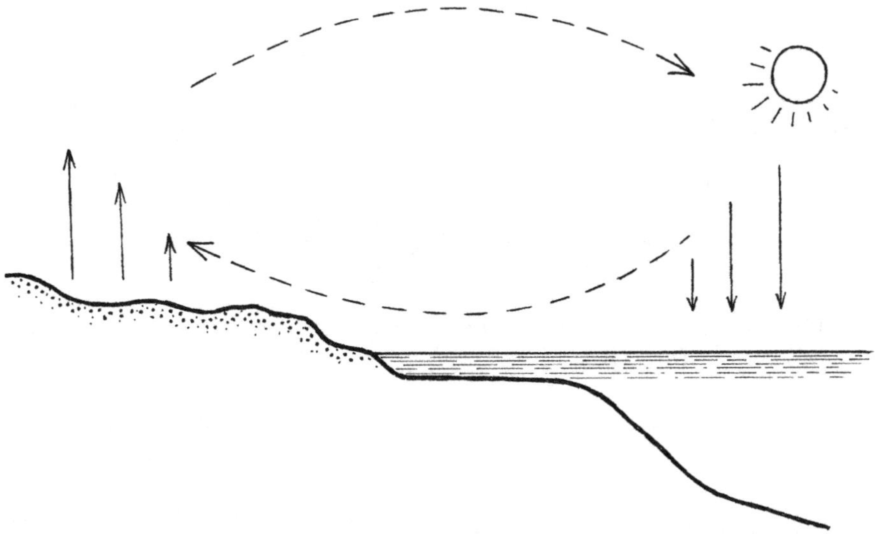

Figure 13. The sea breeze. On a sunny day the land heats up faster than the sea, as the sea has a larger heat capacity and is a better conductor of heat. The air over the land is heated and rises, generating an onshore sea breeze (after ARIC, 2006)

the statistical description of these large-scale fluctuations can be inferred from the stability of the atmosphere or from other meteorological parameters.

Once a dispersing plume is comparable in size to the depth of the PBL – about 1 km in the case of neutral stability – then it is necessary to take into account various meteorological processes operating on that scale. These include sea breezes (Figure 13), urban heat islands (Figure 14) and katabatic flows (Figure 15). These processes introduce significant modifications to the prevailing airflow predicted on the basis of large-scale air movements. Often it is necessary to include their effects explicitly, utilizing knowledge of the local conditions in addition to the results of PBL airflow models and measurements. This is the domain of the meteorologist rather than the dispersion modeller.

Very large releases of material into the atmosphere, such as occurred following an explosion at the Chernobyl nuclear reactor in 1986, can result in plumes that extend above the PBL into the free troposphere. The movement of this material is determined by global airflow patterns and by large-scale synoptic events affecting the entire troposphere. That is, on this scale it is *advection* rather than *dispersion* that largely determines the movement of tracer material. The prediction and mapping of global airflows is of

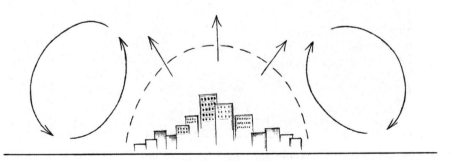

Figure 14. The urban heat island. Space heating and other human activity generate temperature gradients that can give rise to thermal convection cycles (after Oke, 1987)

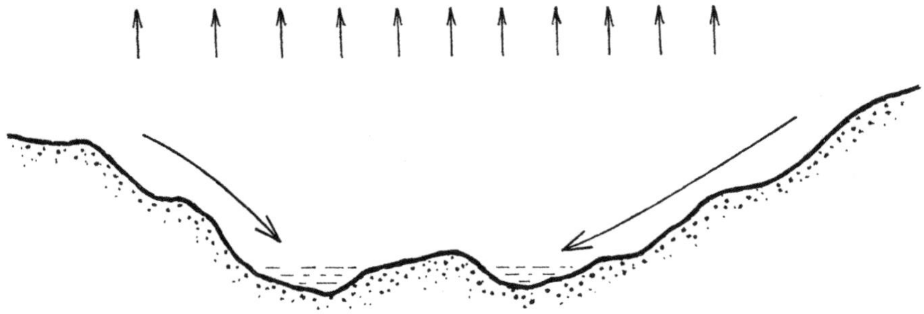

Figure 15. Katabatic airflows. On clear nights surface cooling generates a flow of cold, dense air down hillside slopes (after the ARIC, 2006)

primary importance: again, this is the domain of the meteorologist.

The effect of obstacles

For the release of CWAs into the atmosphere, we are mainly concerned with effects on scales ranging from metres to a few kilometres. In this range, the influence of obstacles such as hills, buildings and trees becomes significant.

Hilly terrain will channel and redirect the prevailing airflow, and with it any tracer that is advected by this wind. This can be seen from Figure 16, which gives examples of tracer dispersion associated with flow over (a) flat terrain and (b) hilly terrain. In (b), the modified flow pattern and the resulting tracer dosage contours have been calculated using a linear airflow model, FACTS (Griffiths, 2000), coupled to the Urban Dispersion Model (UDM) (Hall *et al.*, 2003). Linear airflow models are adequate for small, smooth hills, but larger or steeper obstacles introduce significant distortions to the flat-earth vertical pressure and temperature profiles. These

(a)

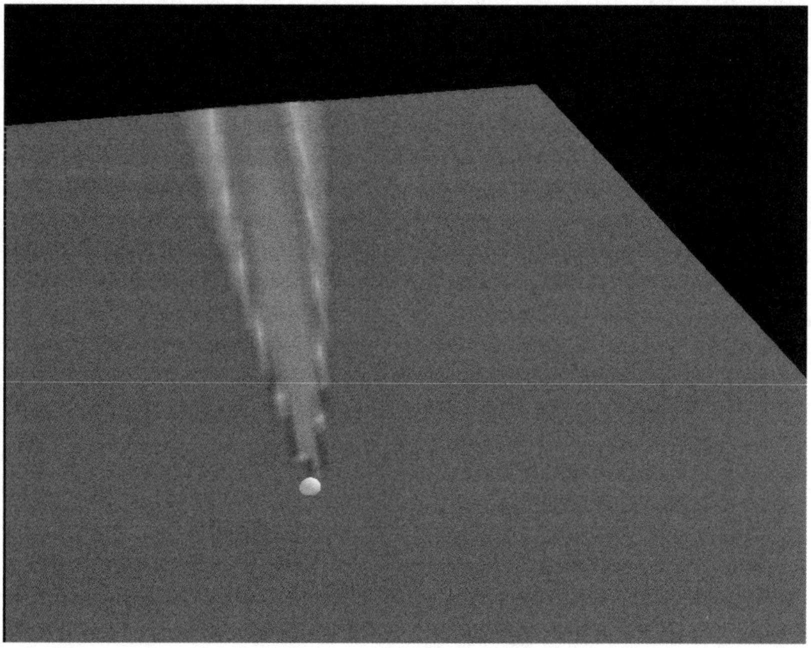

Figure 16. FACTS (Griffiths, 2000)/UDM (Hall *et al.*, 2003) simulations of the release of a tracer over (a) flat terrain, (b) hilly terrain and (c) hilly terrain with buildings, all other conditions being equal. The contours are of dosage, i.e. tracer concentration integrated through time

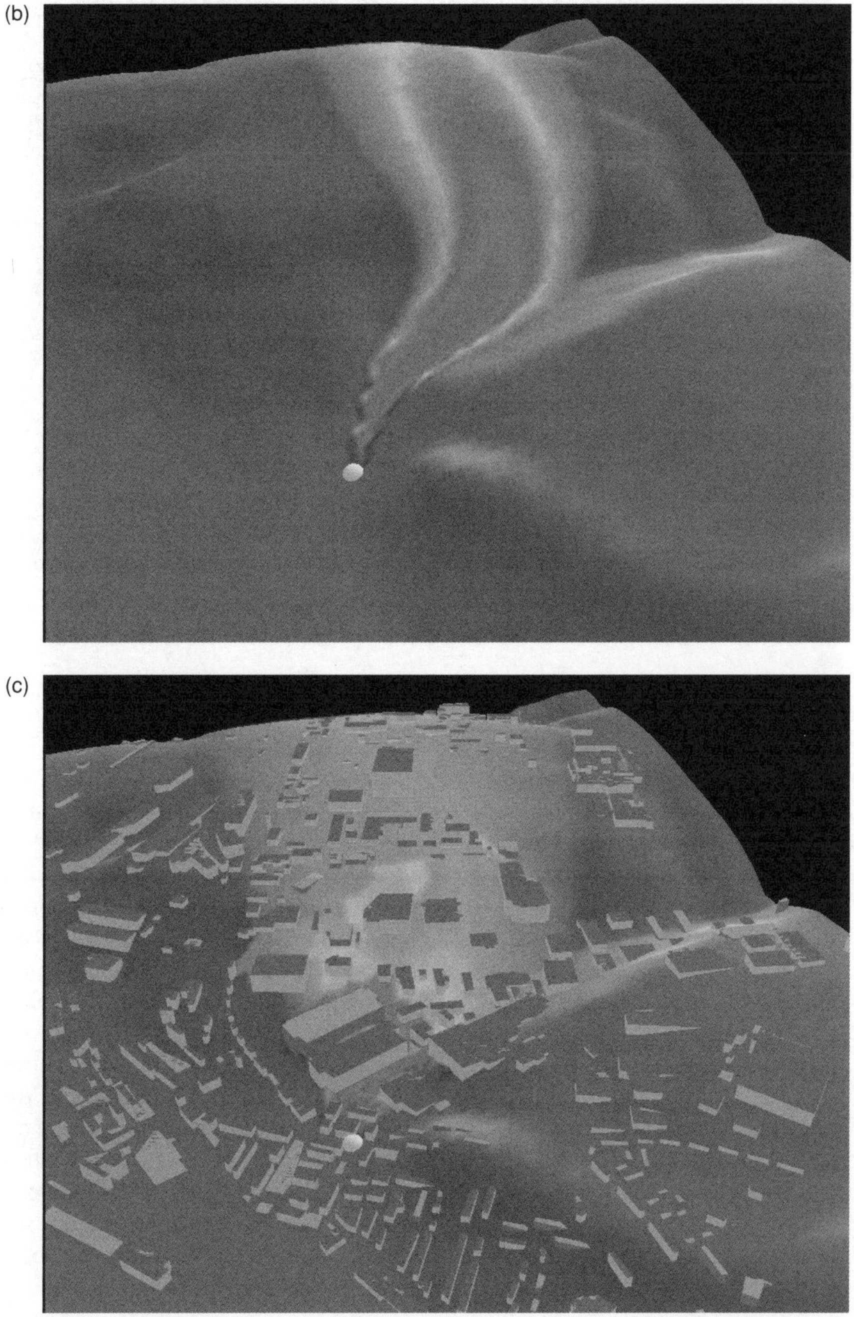

Figure 16. *(cont.)*

modified atmospheric gradients trigger complex mesoscale effects such as katabatic airflows (Figure 15), and they require the application of flow models that are considerably more sophisticated than the simple linear approach.

Figure 16(c) illustrates the effect of adding buildings to the hilly terrain. In addition to the channelling of the airflow by the hills, the buildings entrain the dispersed tracer, resulting in a broader hazard area that persists for a longer

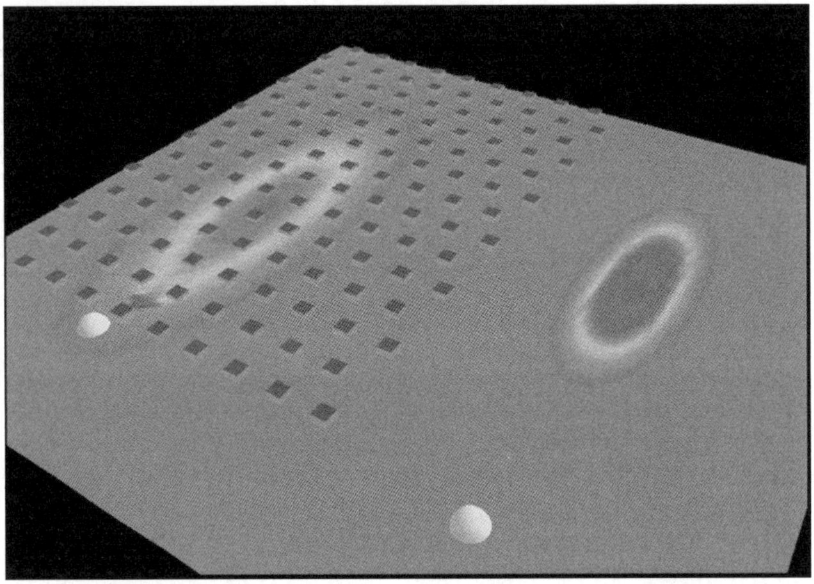

Figure 17. UDM (Hall *et al.*, 2003) simulation of releases of tracer over a regular urban array and over flat terrain. The buildings enhance the dispersion of the tracer and retard its downwind advection

period of time. Figure 17 shows a clearer picture of this effect. In this instance, the obstacles are considerably smaller than the plume width, and their effect is to modify the tracer plume rather than introduce wholly new features. Indeed, in Figure 17 the regular urban array acts as a large-scale 'Galton board', introducing random fluctuations in such a manner that a downwind cross-section of the tracer concentration is guaranteed to have a Gaussian distribution, albeit with variances specific to the particular urban geometry. Accordingly, in such cases it is possible to account for the effect of buildings simply by including correction factors in the expressions for the variances of the Gaussian plume. However, this 'roughness length' approach is not applicable in cases where the tracer plume is of a similar size to the obstacles. The flow patterns around an isolated obstacle can be extremely complex, with the generation of numerous features including horseshoe vortices, entrainment vortices, and a turbulent wake region (Hosker, 1981; Hunt, 1982). While these flow patterns rarely generate dispersion plumes having a comparable complexity, nevertheless they do introduce dispersion features that are unique to the situation, e.g. the obstacle can act as a diffuser (Figure 18(b)) or a channel (Figure 18(c)). As the development of the UDM has shown, it is possible to include these dispersion features within the framework of the Gaussian plume approximation – see below.

Similar to isolated obstacles, trees are efficient diffusers of tracers. Trees have the additional feature that their leaf canopy affords some degree of shelter from CWAs released to the atmosphere. This shelter is effected both through the physical barrier presented by the mass of the leaf canopy – which is only effective if there is significant deposition of the CWA to the canopy itself – and through the aerodynamic barrier presented by the mass of fairly-still air under the canopy. Conversely, any CWA dropped beneath the leaf canopy will be sheltered from the prevailing above-canopy airflow and so it will disperse slowly and will remain in high concentrations for long periods of time. In effect, as far as dispersion is concerned, a leafy avenue or an area of woodland will behave similarly to a large room with leaky windows and skylights. Thus, one is likely to have more success simulating dispersion effects in and around groups of trees using an adapted indoor dispersion model rather than the Gaussian plume approximation.

(a)

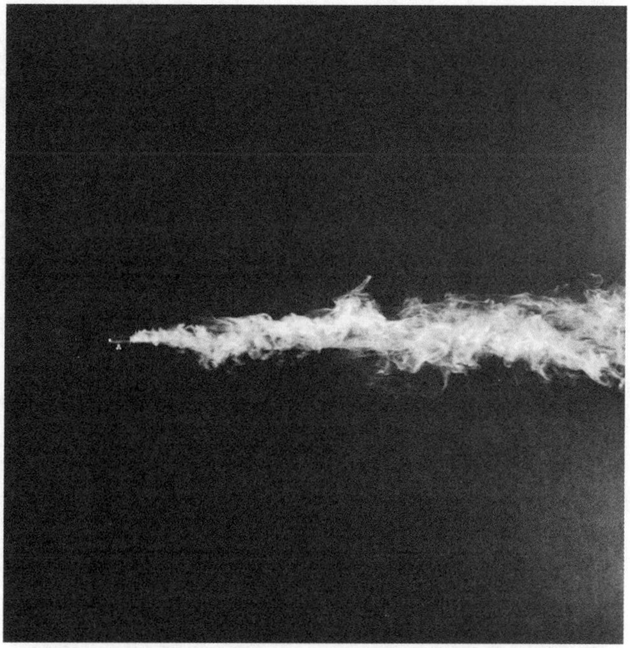

(b)

Figure 18. Plan-view photographs of different smoke plumes: (a) control experiment with no obstacle; (b) oblong obstacle at 0° angle of incidence. The size of the downwind plume is determined more by the size of the obstacle than the size of the initial plume source, and in this sense the obstacle acts as a *diffuser*; (c) oblong obstacle at 45° angle of incidence. The obstacle *channels* the plume at an angle to the prevailing airflow (wind tunnel experiments undertaken by Dstl, Porton Down)

(c)

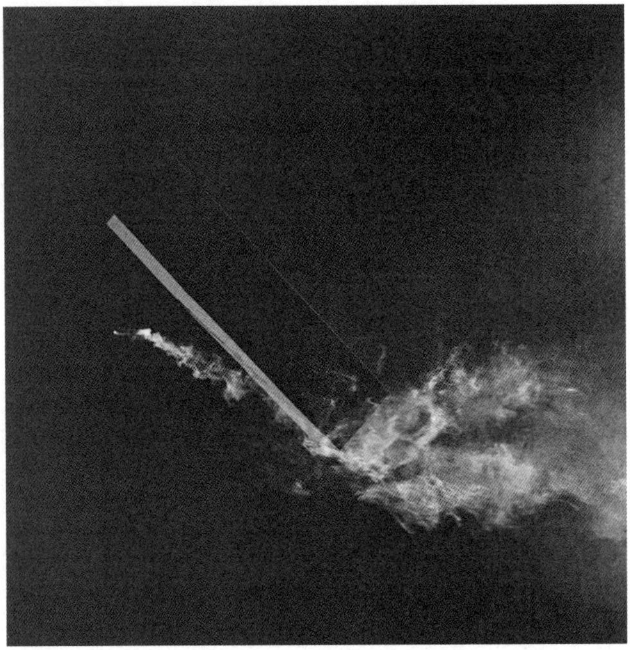

Figure 18. *(cont.)*

Similarly, semi-confined spaces such as courtyards and street canyons may be best treated as 'leaky rooms' and simulated using an adapted indoor dispersion model.

MODELLING ATMOSPHERIC DISPERSION

By necessity, all modelling involves (i) the identification of key physical parameters that together can give an appropriate description of the system under scrutiny, and (ii) the formulation of mathematical equations by which those parameters might be calculated. In the case of fluid flow, the key physical parameters are the mass and momentum of the system, and the corresponding mathematical equations are the continuity equation (expressing the conservation of mass) and the Navier–Stokes equation (expressing the rate of change of momentum). In the following we adopt the usual convention of referring to the Navier–Stokes equation only, presupposing the existence of the ubiquitous continuity equation. Following this approach, one would model the atmospheric dispersion of a CWA by solving the Navier–Stokes equation to give the flow field, and then extending this solution to take account of the dispersion of heat and 'species' (i.e. total mass of CWA). In practice, however, atmospheric dispersion modelling rarely involves the direct solution of the Navier–Stokes equation in this manner. Instead, it is often the case that models such as the Gaussian plume approximation incorporate phenomenological descriptions of dispersion which are difficult to relate to the underlying equations of motion. The following paragraphs discuss the origins of these different models and how they relate to one another:

- The Navier-Stokes equation
- Atmospheric flow models
- Atmospheric dispersion models.

Finally, as discussed in the Introduction, a concept of use for modelling the dispersion of CWAs is established.

The Navier–Stokes equation

Fluid flow is defined by the continuity equation, expressing the conservation of mass, and the Navier–Stokes equation, expressing Newton's

Second Law of Motion applied to a fluid. These are given as Equations (2) and (3), respectively (Batchelor, 1967):

$$\frac{\partial \rho}{\partial t} + \nabla (\rho \mathbf{u}) = 0 \quad (2)$$

where:

ρ = density of fluid (kg m^{-3})
t = time (s)
μ = velocity vector (m s^{-1}).

$$\rho \frac{Du_i}{Dt} = \rho F_i - \frac{\partial p}{\partial x_i} + \frac{\partial}{\partial x_j}\left[2\mu\left(e_{ij} - \tfrac{1}{3}e_{ii}\delta_{ij}\right)\right] \quad (3)$$

where:

ρ = density of fluid (kg m^{-3})
D/Dt = material derivative, i.e. rate of change at a point moving with the fluid locally (s^{-1})
u_i = velocity component (m s^{-1})
t = time (s)
F_i = body force component per unit mass of fluid (m s^{-2})
p = pressure (N m^{-2})
x_i = spatial component (m)
μ = viscosity of fluid (kg m^{-1} s^{-1})
e_{ij} = rate-of-strain tensor, $0.5(\partial u_i/\partial x_j + \partial u_j/\partial x_i)$ (s^{-1})
δ_{ij} = Kronecker delta tensor.

Some of these terms may be eliminated, depending upon the scale of the motion and the complexity of the spatial domain. In a few instances, these simplifications allow the Navier–Stokes equation to be solved analytically, e.g. the parabolic velocity profile associated with laminar flow in a straight duct can be derived from Equations (2) and (3) with appropriate simplifying assumptions and boundary conditions (Batchelor, 1967). However, most practical applications demand a numerical solution of the Navier-Stokes equation. This involves discretizing Equations (2) and (3) on a computational grid that represents the spatial domain in which the flow takes place. Solution involves iterating to convergence an estimated flow solution defined at each point on the computational grid, such that the resulting set of discretized equations is self-consistent, corresponding to a solution where the values of density and momentum at each point on the grid are consistent with neighbouring values, within a specified margin of error. This numerical solution technique (which typically requires several tens or hundreds of iterations in order to reach a converged solution) is known as computational fluid dynamics (CFD).

Given a well-defined flow system, it is fairly straightforward to discretize and solve the associated equations of fluid flow in the manner described above. However, it is not in the nature of CFD that the resulting computer code will be at all suitable for adaptation to other systems, even those that appear to be quite similar. Often, it is quicker to create each solution program from scratch, rather than adapting an existing code; however, this is simply not a practical proposition in the usual circumstance where one wishes to scope the properties of a flow system by solving for various combinations of its defining parameters. Fortunately, there exist commercial CFD software packages (e.g. 'Fluent', available from Fluent Inc., and 'CFX', available from ANSYS Inc.) that implement efficient gridding techniques and numerical solvers, and these are sufficiently flexible that a wide range of flow systems can be set up and solved without the need to create a 'bespoke' computer code.

In providing the means by which fluid flow problems may be solved in a flexible, efficient manner, commercial CFD packages go a long way towards addressing related problems involving the movement of heat and/or species with the flow. As well as the Navier–Stokes equation, such problems require the solution of the equations of conservation of energy and species. A CFD package having the capability to solve for mass, momentum, energy and species is a very powerful tool and a suitable basis for further developments, in particular:

- modelling species reaction, combustion and decay;
- modelling multiple species, multi-phase flow and phase changes;
- modelling thermodynamic processes, including the transfer of heat through conduction and convection, and ideal gas behaviour;
- modelling the advection and dispersion of particles, taking into account their removal through deposition onto surfaces;

- modelling aerosol processes, including condensation, coagulation, scavenging, reaerosolization and evaporation.

Why, then, is CFD not the first tool of choice for the atmospheric dispersion modeller? The 'catch' is that commercial CFD packages have been developed primarily for engineering applications (e.g. to calculate the flow past a streamlined vehicle or through a turbine), where very accurate flow solutions are required at relatively small scales. Accordingly, a great deal of effort has gone into the creation of gridding techniques for complex geometries and the parameterization of accurate boundary layer profiles on the small scale. Neither of these features is likely to be of interest to the atmospheric dispersion modeller, whose spatial regime is much simpler, and whose boundary-layer depth is much greater, than the engineer's. Much more is known about the centimetre-scale boundary-layer profile around an airfoil than the kilometre-scale boundary-layer profile around the earth. In order to be applicable to atmospheric flow and dispersion calculations, CFD packages need to be equipped with a range of PBL profiles. In application, it must be possible to select from this range according to the known meteorological conditions, with the selected profiles being used to define the solution region, such that spatial and temporal variations in the PBL depth are taken into account. Until such a capability is available and demonstrated to be reliable in routine applications, atmospheric dispersion analysts will continue to use models whose relation with the Navier–Stokes equation is often tenuous, but whose relation to observed phenomena is much more secure. These models are discussed in the following paragraphs.

Atmospheric flow models

The wide variety of atmospheric flow models may be conveniently categorised as follows:

- linear models
- Lagrangian models
- discrete dynamical models
- Eulerian or zonal models.

Most of these models can be derived from the Navier–Stokes equation through including appropriate closure approximations. This is not always the case, however, and even where such a derivation exists one often finds that it is a '*post hoc* rationalization', i.e. the derivation is formulated after the selection of modelling approximations, and it exists simply to lend an appearance of respectability. Of more importance are the modelling approximations themselves, that typically reflect observations of persistent trends or structures occurring at particular physical scales. For example, observations of varying atmospheric pressures and associated cloud formations have been explained in terms of structures in the atmosphere known as frontal systems. An idealized frontal system may be thought of as a 'model' encapsulating a set of modelling approximations. It might be possible to formulate these approximations and relate them to the Navier–Stokes equation, but this seems scarcely relevant, given that the frontal system model is known to be a useful and reliable tool for predicting the weather. That is, the main justification for using such a model is not that it can be derived from the Navier–Stokes equation, but that it can be used for predicting physical behaviour at a particular scale. Accordingly, the following brief descriptions of the different types of atmospheric flow models emphasize *application* rather than *derivation*.

Linear models, e.g. FACTS (Griffiths, 2000) and FLOWSTAR, developed by CERC Ltd., solve linearized versions of the Navier–Stokes equation. Such models are fast and stable in operation. Typically, they are applicable for calculating the modification to prevailing horizontal airflow due to the presence of surface features, e.g. small, smooth hills and/or surface roughness.

Lagrangian models, e.g. NAME, developed by the UK Meteorological Office, and LODI, developed by Lawrence Livermore National Laboratory, simulate the advection of infinitesimal tracer 'particles' by the prevailing airflow. Dispersion is simulated through the inclusion of random perturbations to the trajectories of the particles. The quality of the results from a Lagrangian model depends entirely on the quality of the input windfield. Accordingly, early simulations were restricted to large scales, e.g. the prediction of the free troposphere advection of airborne particles from Chernobyl. As the availability of high-resolution windfield data improves, so will the applicability of Lagrangian models to local scales. Lagrangian models are particularly suited to the simulation of atmospheric chemistry, their 'particles' acting as vehicles for chemical species having very slow reaction rates. However, Lagrangian models can suffer from poor sampling of the atmosphere, owing to the 'bunching' of the particle trajectories. This problem is overcome in the 'semi-Lagrangian' approach (Jacobson, 1999), in which particles are advected *backward* in time from a regular (Eulerian) grid. The particles are ascribed the chemical characteristics of their earlier locations, and then returned forward in time along their trajectories to their regular positions, with the timestep determined by the limiting chemical reaction rate. Semi-Lagrangian models are not mass-conservative and therefore their chemical compositions need to be normalized before each backward trajectory calculation. Another variant on the Lagrangian theme is the straight-line trajectory model, in which the chemical composition at a 'receptor' location is calculated as the windrose-weighted average of contributions from each of a number of incoming straight-line trajectories. Since windroses are calculated as long-term averages (e.g. quarterly or annually), this method is unsuitable for the simulation of chemical reaction schemes involving rapid transients or nonlinear effects.

Discrete dynamical models operate on a similar principle to Lagrangian models, except that the 'particles' are ascribed dynamical properties affecting their interaction, enabling the simulation of particular flow features. Dynamical models may be formulated empirically in relation to observed phenomena, or analytically in relation to the Navier–Stokes equation. Examples of empirical models are idealized frontal systems, and the Gaussian plume approximation. As discussed already, the justification for using such models is that they accurately express observed phenomena, and their relationship to the Navier–Stokes equation and to CFD is of less importance. Examples of analytical models are the Random Vortex Method (RVM) (Chorin, 1973) and Lattice-Gas Automata (LGA) (Frisch *et al.*, 1986). In the RVM, the Lagrangian 'particles' are 'vortices' which generate the basic features of turbulent flow through their interaction. LGA 'particles' move with an integer timestep on a regular grid and interact according to local rules. While both models can be shown to reproduce Navier–Stokes fluid flow behaviour (for a limited range of applications), neither model does anything that has not already been bettered by CFD; in this respect, RVM and LGA are representative of their type.

Eulerian or zonal models (e.g. the Urban Airshed Model® (UAM®), developed by Systems Applications International, and the COMIS indoor dispersion model, available from Lawrence Berkeley National Laboratory) divide the flow region into a number of boxes or zones, and simulate advection by transferring material between adjacent boxes. Thus, Eulerian models conserve mass but they do not take into account the rate of change of momentum. Species concentrations are averaged within each box, resulting in considerable 'numerical diffusion' if the boxes are large. Eulerian models are suitable for systems in which the outcome is dominated by non-dynamical factors, e.g. atmospheric chemistry, in the case of UAM, or building layout and ventilation, in the case of COMIS.

Atmospheric dispersion models

Commonly, when people refer to 'atmospheric dispersion models' they mean implementations of the Gaussian plume approximation. Initially, this was simply a matter of convention, resulting from the fact that the Gaussian plume approximation is exclusively based on observations of atmospheric dispersion, independent of the underlying flow structures. However, the appelation has persisted because the Gaussian plume approximation has proved successful *in practice*. While other techniques have been used for specialized dispersion applications (e.g. zonal models for indoor dispersion), the modeller's first tool of choice for the simulation of *atmospheric* dispersion is the Gaussian approach. Even where empirical parameterizations exist that involve little or no computer simulation, e.g. ATP45 dispersion templates (NATO, 1988), it is certain that the Gaussian plume approximation will have been called upon at some stage in their development, either in their initial parameterization or in the subsequent validation process.

There exist a number of different formulations of the Gaussian plume approximation.

Software packages include AERMOD, available from the US Environmental Protection Agency, and ADMS, developed by CERC, Ltd. The most general formulation of the Gaussian plume approximation is the 'Gaussian puff method', a Lagrangian approach in which the 'particles' are puffs of tracer that can disperse as they are advected by the prevailing airflow. In this case, the continuous Gaussian plume is recovered as the superposition of the ensemble of puffs. The HPAC modelling toolkit developed by the US Defense Threat Reduction Agency uses a Gaussian puff model, SCIPUFF (Sykes *et al.*, 2004), for its atmospheric dispersion calculations. Equation (4) defines the Gaussian profile of a puff of tracer:

$$c(\boldsymbol{x}) = \frac{Q}{\sqrt{8\pi^3\,|\sigma|}} \exp\left[-\frac{(x_i - \bar{x}_i)(x_j - \bar{x}_j)}{2\sigma_{ij}}\right] \qquad (4)$$

where:

c = puff tracer concentration (kg m^{-3})
$\boldsymbol{x}$ = location vector (m)
Q = spatially-averaged puff tracer concentration (kg m^{-3})
σ = Gaussian spread parameters, expressed as a tensor of second moments (m).

The Gaussian puff method has a number of advantages:

- Analytic expressions for the Gaussian spread parameters σ may be derived from the Navier–Stokes equation, through the implementation of second-order closure approximations (Sykes *et al.*, 2004).
- The analysis also gives consistent expressions for the Gaussian variances defining the 'hazard area envelope' resulting from a meandering plume, e.g. Figure 7.
- Discrete puffs are a convenient mechanism for the inclusion of particle characteristics (size, mass, etc.) and so simple particle dynamics may be included (e.g. puffs containing larger particles will undergo gravitational settling at a different rate than puffs containing smaller particles).
- Puff splitting and merging algorithms may be included, in order to improve the sampling of the variability of the prevailing airflow.

An unexpected benefit of the inclusion of a puff-splitting mechanism is that it facilitates the simulation of dispersion in urban areas. The UDM (Hall *et al.*, 2003), a semi-empirical Gaussian puff model for the simulation of dispersion in the urban environment, implements a number of mechanisms including puff splitting in order to account for the effects of buildings. Figure 19 illustrates how the UDM simulates the entrainment of a dispersed tracer within the wakes of buildings. The incident puff is split into two puffs, one of which is entrained behind the building. The splitting ratio, which depends on the relative dimensions of the puff and the building, is determined empirically from the results of wind-tunnel experiments. Other urban flow and dispersion effects, such as building diffusion and channelling (Figure 18), may be taken into account using similar mechanisms and parameterization techniques.

Concept of use for modelling the dispersion of CWAs

Through the preceding discussion it is apparent that atmospheric dispersion is a complex phenomenon involving a wide variety of physical processes. It is apparent also that there are a wide variety of approaches to modelling atmospheric dispersion. With this in mind, it is possible to formulate a number of key requirements for the present application, viz. modelling the dispersion of CWAs:

- Atmospheric dispersion modelling requires an accurate and reliable description of the weather, including descriptions of the prevailing airflow, the atmospheric stability and the effects of flow structures at the spatial and temporal scales of interest.
- Spatial scales range from metres to a few kilometres.
- Temporal scales range from seconds (for the estimation of incapacitation) to hours.
- It must be possible to model the effects of dense gases, buoyancy and terrain structures, such as hills, buildings and trees.
- It must be possible to characterize the CWA source term, taking into account the features of the delivery system.

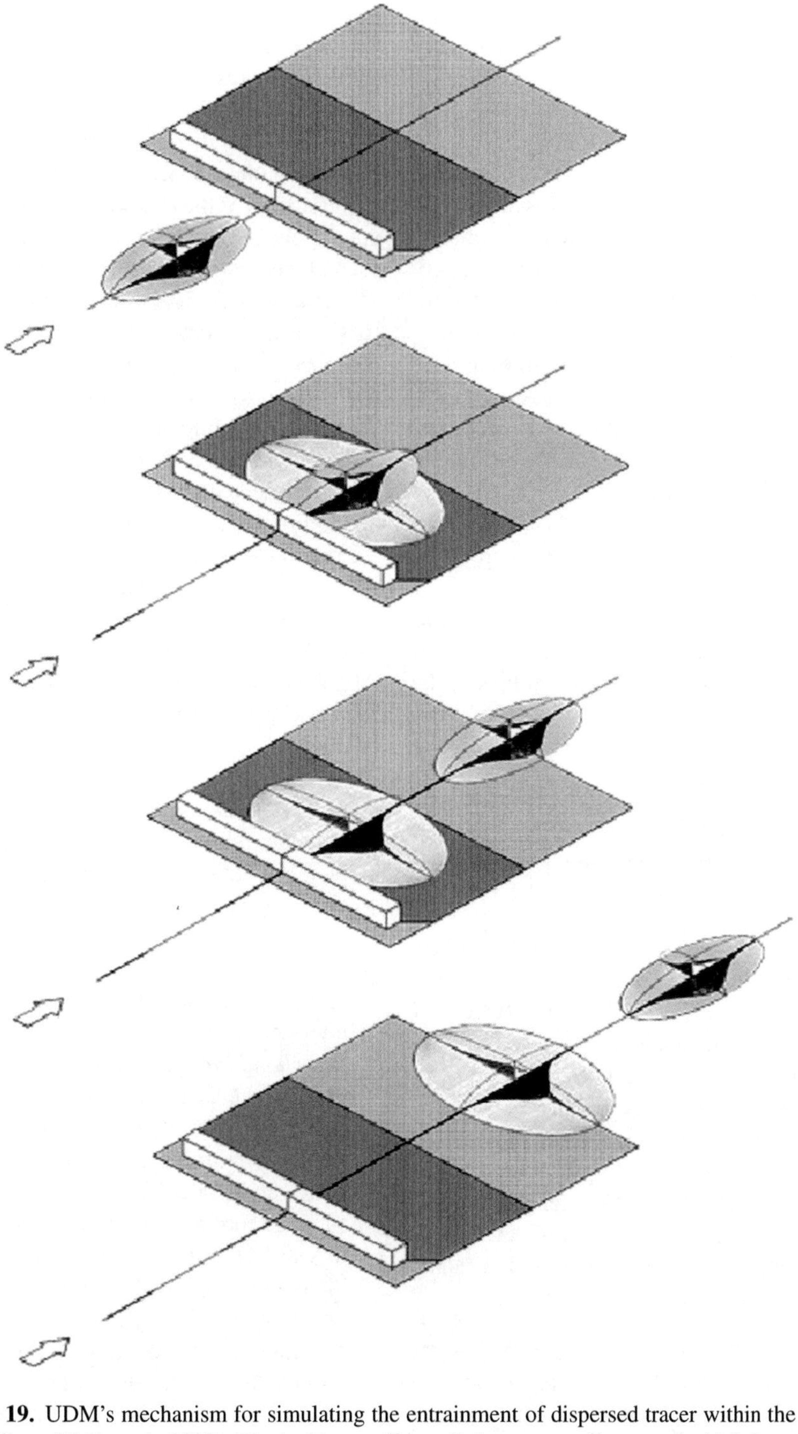

Figure 19. UDM's mechanism for simulating the entrainment of dispersed tracer within the wakes of buildings (Hall *et al.*, 2003). The incident puff is split into two puffs, one of which is entrained behind the building

- To be of any use in an emergency situation, the initial modelling response must be prompt, e.g. within 10 minutes of a reported release of a CWA.
- In a continuing (or anticipated) emergency situation, there is the opportunity to amend or elaborate advice based on modelling, but this further advice must be consistent with the initial response and it must lead to an improved understanding of the situation (and not merely add to the confusion through 'information overload').
- Modelling in support of CWA hazard assessments must be able to demonstrate compliance with statutory limits and procedures, which may involve assessments of health effects and other impacts.
- All modelling must be 'verifiable', i.e. amenable to checking and re-running if required.
- All modelling must be 'robust', i.e. conclusions drawn from model simulations must be qualified by statements (i) setting out the underlying modelling assumptions and their possible pitfalls, and (ii) defining the range of applicability of the conclusions.

In addition, it is possible to identify a number of features that are not 'key requirements' but would be 'nice to have':

- Modelling dispersion source term processes, e.g. combustion, explosions.
- Modelling the incapacitation effects of concentration fluctuations.
- Modelling aerosol and particle processes.
- Modelling chemical reactions.
- CWA atmospheric dispersion modelling should be consistent with similar studies for biological and radiological agents. The easiest way to implement this consistency is to use the same dispersion models and the same datasets for weather, terrain, assets, etc.

How are these diverse and challenging requirements to be met? There is a growing acceptance that no one model can fulfil all of these requirements, and that what is needed is a 'general-purpose' dispersion modelling 'toolkit' comprising a variety of models and datasets that between them cover the required range of applicability. Thus, the toolkit might include pre-calculated dispersion templates for prompt response, a Gaussian plume model for continuing response and hazard assessment, a Gaussian puff model for dispersion in urban environments, an adapted indoor dispersion model for simulating dispersion in semi-confined spaces and so on. The differing requirements of these models for meteorological data might be met by including several datasets, each at the appropriate scale of resolution.

The toolkit approach has major implications for the *modus operandi* or 'concept of use' for atmospheric dispersion modelling. Whereas expertise in the operation of a single model requires a deep understanding of its underlying algorithms, expertise in the operation of a toolkit of models requires a deep understanding of the range of applicability of each model. In practical situations, the role of the single-model expert is little more than that of a glorified data-entry clerk, while the role of the toolkit expert is that of a skilled artisan, selecting and applying the right tool for the job. This change of role affects the concept of use for CWA dispersion modelling in a number of ways:

- *Model development*. While single-model experts rarely admit that their models are complete, toolkit experts must be able to use whatever models are available, regardless of their state of development. Thus, the toolkit approach changes the focus of research from model development to model acquisition, integration and use.
- *Data handling*. The toolkit approach depends upon the existence of a database of common parameter values, defining CWA source terms, meteorology, terrain, operational scenarios and so on. This database must be in a standard format (e.g. a relational database) so that different toolkit models can be applied interchangeably at short notice.
- *Training*. Emergency exercises and the development of a database of operational scenarios provide the best means for honing the toolkit expert's understanding of the applicability of models.
- *Uncertainty handling*. Situations will arise in which equally applicable models predict different outcomes. The toolkit expert can

use this spread in predictions to estimate the margin of error in the advice to decision-makers.

- *Control and reporting*. Toolkit experts need to be able to reconcile the outputs from different models. Typically, this is carried out by a variety of informal methods ranging from automated information fusion to open debate. In an operational situation, such 'methods' would be deeply worrying to the decision-makers, who prefer to receive advice that is coherent, consistent and, above all, uncontested. The legitimate needs of experts and decision-makers can be satisfied by providing each with a separate environment – the 'expert panel' and the 'operations room', respectively – and restricting official communications between these environments to a standard reporting format that forces the experts to come to a consensus. Equally, this mechanism forces decision-makers to properly formulate their requests for advice from the experts, and to ensure the adequate provision of appropriate information. This mechanism has the additional advantage that it is 'self-documenting' and assists post-operation analysis and review.

These measures are recommended to the CWA dispersion modelling community.

ACKNOWLEDGEMENTS

Many thanks to Professor Rex Britter for his useful comments.

REFERENCES

Atmospheric Research and Information Centre (ARIC) (2006). *Weather and Climate Teaching Pack*. Manchester Metropolitan University, Manchester, UK: ARIC [http://www.ace.mmu.ac.uk/].

Batchelor GK (1967). *An Introduction to Fluid Dynamics*. Cambridge, UK: Cambridge University Press.

Chorin AJ (1973). Numerical study of slightly viscous flow. *J Fluid Mech*, **57**, 785–796.

Church PE (1949). Dilution of waste stack gases in the atmosphere. *Ind Eng Chem*, **41**, 2753.

Frisch U, Hasslacher B and Pomeau Y (1986). Lattice-gas automata for the Navier–Stokes equation. *Phys Rev Lett*, **56**, 1505–1508.

Gifford FA (1959). Statistical properties of a plume dispersion model. *Adv Geophys*, **6**, 117–138.

Griffiths IH (2000). *Dispersion in Complex Terrain*, Report number DERA/CBD/CR000268. Porton Down: DERA.

Hall DJ, Spanton AM, Griffiths IH *et al.* (2003). *The Urban Dispersion Model (UDM)*, Version 2.2 Technical Documentation Release 1.1, Report number Dstl/TR04774. Porton Down: Dstl.

Hosker RP (1981). *Methods for Estimating Wake Flow and Effluent Dispersion near Simple Block-Like Buildings*, Technical Memorandum ERL ARL-108. Washington, DC: US Department of Commerce, National Oceanic and Atmospheric Administration.

Hunt JCR (1975). *Air Pollution Dispersion*. Milton Keynes, UK: Open University (film).

Hunt JCR (1982). Mechanisms for dispersion of pollution around buildings and vehicles. In: *Proceedings of BMFT/TÜV Colloquium on Exhaust Gas Air Pollution caused by Motor Vehicle Emissions*, pp. 235–266. Cologne, Germany: Verlag TÜV Rheinland GmbH.

Jacobson MZ (1999). *Fundamentals of Atmospheric Modelling*. Cambridge, UK: Cambridge University Press.

North Atlantic Treaty Organization (1988). *STANAG 2103. Reporting Nuclear Detonation, Biological and Chemical Attacks, and Predicting and Warning of Associated Hazards and Hazards Area (ATP-45)*. NATO Brussels.

Oke TR (1987). *Boundary Layer Climates*, 2nd Edition. London: Methuen and Company.

Pasquill F (1962). *Atmospheric Diffusion*. London: Van Nostrand Company, Ltd.

Sykes RI, Parker SF and Henn DS (2004). *SCIPUFF*, Version 2.1 Technical Documentation, ARAP Report number 728. Princeton, NJ, USA: Titan Corporation.

United States Weather Bureau (1955). *Meteorology and Atomic Energy*. Washington, DC: US Government Printing Office.

4 THE FATE OF CHEMICAL WARFARE AGENTS IN THE ENVIRONMENT

Sylvia S. Talmage[1], Nancy B. Munro[1], Annetta P. Watson[1], Joseph F. King[2] and Veronique Hauschild[3]

[1]*Oak Ridge National Laboratory, Oak Ridge, TN, USA*
[2]*US Army Environmental Center, Aberdeen Proving Ground, MD, USA*
[3]*US Environmental Protection Agency, Washington, DC, USA*

INTRODUCTION

Information on the environmental fate of chemical warfare agents (CWAs) is needed in order to facilitate destruction of aging stockpiles, cleanup of non-stockpile sites and waste areas and formulate response and recovery from possible warfare use by rogue nations or terrorists. Emphasis is placed on chemicals that were part of chemical stockpiles in several countries (Carnes and Watson, 1989; OPCW, 2005), with information on sites of manufacture and disposal as available. In this chapter, CWA degradation products will be evaluated by assessing their likelihood of formation, as well as characterizing their chemistry, toxicity and persistence.

Most persistent compounds are characterized by low vapor pressure, low water solubility and low rates of natural abiotic and microbial degradation. However, some moderately water-soluble compounds resistant to hydrolysis or biodegradation may persist in dry soil or leach into groundwater where they may persist for long periods. The fate of chemical agents in the environment depends largely on their chemical and physical properties (discussed in Chapter 2). Meteorological conditions (temperature, humidity and soil moisture), as well as bulk size, also influence the rate of degradation and dissipation. For example, agents present in bulk amounts will dissipate much more slowly than chemicals dispersed as aerosols or vapors. Some of the agent chemical and physical properties are reiterated here as needed to explain environmental persistence or fugacity. In addition, available information on the chemical and physical properties of the degradation products is presented.

Fauna may be exposed to chemicals at contaminated sites by the oral, dermal or inhalation route. Compounds that are lethal to 50% of tested animals (LD_{50} or LC_{50} values) at < 50 mg/kg, < 50 mg/m^3 and < 200 mg/kg following a single (acute) exposure are considered highly toxic by the oral, inhalation and dermal routes, respectively (O'Bryan and Ross, 1988). Compounds with LD_{50} or LC_{50} values of 50–500 mg/kg, 50–500 mg/m^3 and 200–500 mg/kg for the respective routes are considered moderately toxic, and compounds with values higher than these ranges exhibit a low order of toxicity. Chronic toxicity values by the respective categories and routes of exposure are generally an order of magnitude

Chemical Warfare Agents: Toxicology and Treatment (2nd Edition)
Edited by Timothy C. Marrs, Robert L. Maynard and Frederick R. Sidell

lower. For aquatic organisms, an acute LC_{50} value of < 1 mg/l is considered to be of high acute toxicity, while a chronic LC_{50} of < 0.1 mg/l is considered to be of high chronic toxicity. Where available, toxicity values for terrestrial and aquatic species are included in order to evaluate potential environmental impact.

From an environmental standpoint, the primary warfare agents of concern include the vesicant (blister) agents, sulfur mustard (H, HD and HT) and lewisite (an organic arsenical; agent L) and four nerve agents-VX, GA (tabun), GB (sarin) and GD (soman). Agent GF (*O*-cyclohexyl methylfluorophosphonate; cyclosarin) is currently considered of little to limited strategic interest (Sidell, 1997), is not part of the US domestic chemical unitary munitions stockpile and will not be evaluated further here. While commercially produced for weaponry use by the German Ministry of Defense in the 1940s, there is no available record of nerve agent weapon deployment by German forces during WWII (Sidell, 1997; Robinson, 1997). Sulfur mustard was deployed on the battlefield during WWI, by Japanese military forces in Manchuria in the years leading up to WWII, and by other state parties during certain regional conflicts of the 20th Century (IOM, 1993; Watson *et al.*, 1989; Harris and Paxman, 2002). Lewisite was also produced for use by Japanese forces in Manchuria and in the same facility that manufactured sulfur mustard agent under operation command by the Japanese Army (Watson *et al.*, 1989). Multiple countries have produced and stockpiled chemical warfare agents (OPCW, 2005).

Nitrogen mustards have not been stockpiled in most countries and are not further discussed here other than to say that their chemical properties resemble those of sulfur mustard (IOM, 1993). The toxicity and environmental fate of the nitrogen mustard agents have been reviewed previously (Marrs *et al.*, 1996; Munro *et al.*, 1999). Based on chemical properties, the persistence of nitrogen mustards range from environmentally persistent (tris[2-chloroethyl]amine [HN3]) to moderately persistent (ethyl bis[2-chloroethyl]amine [HN1] and methyl bis[2-chloroethyl]amine [HN2]). The nitrogen mustards break down into water-soluble, less toxic compounds. Cyanogen chloride (agent CK), although highly toxic as a vapor, is both highly volatile and subject to hydrolysis and is therefore considered nonpersistent. Cyanogen chloride is one of several dual-use compounds thought to be located at one or more nonstockpile sites in the continental USA (Opresko *et al.*, 1998). Phosgene (carbonyl chloride; agent CG) was also deployed as a war gas on WWI battlefields, is hydrolyzed slowly (to CO_2 and HCl) in freshwater, and rapidly in seawater, and does not present the problem of environmental persistence. As another dual-use compound, phosgene is presently manufactured and used as a chemical intermediate for end uses such as the manufacture of dyes. It is thought to be located at one or more nonstockpile sites in the continental USA (Opresko *et al.*, 1998).

Emphasis in this review is on those potential degradation products resulting from agent contact with soil, water or the atmosphere after unintended release from historically buried chemical weapons and wastes, or potential spills during munition demilitarization and disposal activities. The principal degradation processes include photolysis, hydrolysis, oxidation and microbial degradation. Volatilization is an important mechanism for the transfer of some CWAs from soil and water to air. Decontamination procedures, considered in Chapter 7, may incorporate some or all of these processes.

Most degradation data for these compounds are collected from laboratory studies or field studies of soil. The compound-specific evaluations presented below summarize information on CWA environmental degradation, degradation product toxicity and any available guidelines for site restoration based on degradation product exposure or environmental screening. Some degradation products are derived from several CWAs (e.g. methyl phosphonic acid from agents VX, GB or GD) or used in commercial industrial processes (e.g. thiodiglycol; Lundin, 1991), while others are unique products of agent-specific degradation e.g. *S*-(2-diisopropylaminioethyl) methylphosphonothioic acid or EA 2192; a product of VX hydrolysis.

Of the many possible degradation products listed in available literature and formed under multiple (sometimes rare) and site-specific conditions of pH, temperature, moisture, sunlight and soil type, there are relatively few products of sufficient persistence or potential toxicity to pose concern. These are summarized in Table 1, along with information characterizing their persistence, chemical–physical properties and chronic toxicity parameters. Detailed information characterizing agent-specific degradation products, impurities and stabilizers are provided in the Appendix to this chapter. It should be noted that, for any given CWA, not every degradation product or impurity listed in the Appendix tables is likely to be present following a CWA release. Impurities are a consequence of the manufacturing processes and conditions (which vary) at the time of munition fill, and degradation reactions and their yields are governed by site-specific conditions at the time of agent release, as well as environmental conditions following release.

It is noteworthy that additional studies are being conducted on the fate of chemical agents deposited on outdoor surfaces such as concrete, asphalt, sand, grass and brackish water. The 'Chemical and Biological Warfare Agent Fate Research Program' (US Army, 2002; US Air Force, 2003; Murdock *et al.*, 2004) is an international effort involving studies conducted in the USA, UK (Porton Down), The Netherlands and the Czech Republic. In addition to the use of multiple surfaces, open-air testing is designed to simulate droplet deposition under multiple-attack scenarios and to incorporate various meteorological conditions.

Finally, although there are no official standards for cleanup and re-entry of contaminated sites, guidelines developed by several groups have been proposed and accepted. Although these guidelines were developed to protect human health, it is believed that such guidelines would be protective of indigenous species at contaminated sites. The US Army Center for Health Promotion and Preventive Medicine (US Army CHPPM, 1999; Bausum, 1998; Enslein, 1984; Reddy *et al.* 2005) has estimated chronic exposure levels or reference doses (RfDs) for agent degradation products using experimental data from structurally related compounds and/or QSAR using commercially available TOPKAT7 software (Bausum, 1998). Inhalation reference concentrations (RfCs) were calculated from the corresponding RfDs. Where exposure guidelines for degradation products are not available, estimated RfDs for the parent compounds (Opresko *et al.*, 1998,2001; Bakshi *et al.*, 2000) summarized in this chapter have been used for screening levels for soil at contaminated sites. Estimated reference doses, calculated with established procedures, are central to any site-specific estimation of potential risks posed by residual levels of these compounds in the environment, as the toxic hazard from a CWA release is primarily a consequence of exposure to the parent agent and not to CWA degradation products (which are of generally lower toxicity, with few exceptions; Munro *et al.*, 1999). Sound logic and objective, site-specific, risk assessment should govern remediation decisions following actual or potential CWA release; elements of a recommended decision logic include the following:

- Acknowledging that the potential risks posed to human health and the environment by release of a parent compound are the primary basis for determining the level or stringency of remediation and regulation.
- Remediation for chemical warfare agent releases should mirror the basic logic and requirements applied to similar toxic materials generated by private industry.

HISTORICAL ASPECTS

Prior to and during WWII, various agents were manufactured, stockpiled and/or placed into munitions in Germany, the USA, the UK, Japan, Italy and the former USSR. Under the conditions of world war, some production sites were heavily contaminated by manufacturing and waste-disposal practices in common use at the time. Improper incineration, destruction and disposal of agents and weaponry after the war also left some sites contaminated. Additional sites have been used for stockpiling. Although most such legacy sites have controlled public access, they remain an area of concern.

German Ministry of Defense

Records indicate that the German Ministry of Defense produced 25 000 tonnes of HD, 2000 tonnes of nitrogen mustard, 12 000 tonnes of tabun, 5900 tonnes of phosgene and several

Table 1. Principal chemical warfare agent degradation products with associated chronic toxicity guidelines[a]

Agent/synonyms (CAS number)	Degradation process	Degradation product or impurity (CAS number)	Persistence[b]/ chemical–physical properties	Chronic toxicity values of degradation products[c]
Sulfur mustard (H, HD) (505-60-2)	Hydrolysis	Thiodiglycol (111-48-8)	Moderate nonvolatile miscible with water resistant to hydrolysis biodegradable	RfD: 400 μg/kg/day RfC: 469 $\mu g/m^3$
Lewisite (dichloro-(2-chlorovinyl)arsine) (541-25-3)	Hydrolysis, dehydration	Lewisite oxide[d] (3088-37-7)	High water-insoluble potential oxidation in soil	RfD: 0.1 μg/kg/day
VX (*O*-ethyl-*S*-(2-(diisopropylamino)ethyl) methylphophonothioate) (50782-69-9)	Hydrolysis	EA 2192[e] (73207-98-4)	Moderate low volatility high water solubility resistant to hydrolysis	RfC: 6×10^{-4} μg/kg/day RfC: 7×10^{-4} $\mu g/m^3$
	Hydrolysis	Ethyl methylphosphonic acid (EMPA) (1832-53-7)	Moderate low volatility water-soluble resistant to hydrolysis biodegradable[f]	RfD: 28 μg/kg/day RfC: 34 $\mu g/m^3$
	Hydrolysis	Methyl phosphonic acid (MPA) (993-13-5)	High low volatility resistant to photolysis resistant to hydrolysis high water solubility mobile in soils resistant to biodegradation	RfD: 20 μg/kg/day[g] RfC: 24 $\mu g/m^3$
GA (tabun; ethyl *N*,*N* -dimethylphosphoramidocyanidate) (77-81-6)	Hydrolysis	None of potential concern	—	—

GB (sarin; isopropyl methylphosphonofluoridate) (107-44-8)	Hydrolysis	Isopropyl methylphosphonic acid (IMPA) (1832-54-8)	High low vapor pressure water-soluble resistant to hydrolysis resistant to biodegradation	RfD: 100 μg/kg/day RfC: 110 μg/m³
	Hydrolysis	Methyl phosphonic acid (MPA) (993-13-5)	High low volatility resistant to photolysis resistant to hydrolysis high water solubility mobile in soils resistant to biodegradation	RfD: 20 μg/kg/day[g] RfC: 24 μg/m³
	Impurity	Diisopropyl methylphosphonate (DIMP) (1445-75-6)	High low volatility water-soluble resistant to hydrolysis slow biodegradation	RfD: 80 μg/kg/day[h]
GD (soman; pinacolyl methylphosphonofluoridate) (96-64-0)	Hydrolysis	Methyl phosphonic acid (MPA) (993-13-5)	High low volatility resistant to photolysis resistant to hydrolysis high water solubility mobile in soils resistant to biodegradation	RfD: 20 μg/kg/day RfC: 24 μg/m³

Notes:
[a] Degradation products were selected on the basis of environmental persistence and toxicity. Production and yield depends on site-specific conditions.
[b] Persistence depends on environmental conditions; in general, moderate persistence indicates weeks to months and high persistence indicates months to years.
[c] Estimated by Bausum *et al.* (1999), unless otherwise noted: RfD, reference dose; RfC, reference concentration. RfD estimate for thiodiglycol as derived in Reddy *et al.* (2005); RfC estimate derived by assuming inhalation rate of 20 m^3/day, body weight of 70 kg and extrapolation uncertainty factor of 3 (method of US Army CHPPM, 1999, App.F).
[d] Under continually moist conditions, the hydrolysis product 2-chlorovinyl arsonous acid (CAS 85090-33-1), a probable vesicant, may be present.
[e] Retains the toxic mechanism of action of the parent CW agent.
[f] Disappearance from soil may be due to a combination of hydrolysis and biodegradation.
[g] US EPA (1996).
[h] US EPA Integrated Risk Information System (IRIS) online database.

tonnes of various tear gases and irritants during WWII (Glasby, 1997). HD was produced only near the town of Halle in former East Germany (Stephan, 2005). Much of the equipment used to manufacture HD was destroyed by the Germans before the end of WWII. The silver-lined production vessels were buried and covered with calcium hypochlorite for detoxification. When the US Army arrived, they destroyed some of the chemical by incineration; later, the Russians blew up the plant. Anecdotal information indicates that the vessels were later recovered for their silver content; this recovery effort resulted in chemical burns to the workers.

Sources differ, but it is generally agreed that some 60 000 tonnes of chemical warfare agents (nearly 300 000 tonnes of weaponry) were found in Germany's arsenal after the war (HELCOM, 1994; Stock, 1996; Glasby, 1997). Captured weapons within the US-occupied sector were transferred to five former ammunition depots – Frankenberg, Wildflecken, Grafenwöhr, Schierling and St. Georgen (Frondorf, 1996). These weapons were either incinerated, buried in flooded mines, dumped at sea, or transported abroad for study or stockpiling (Glasby, 1997). The bulk of the munitions was dumped at sea. Although the plan was to drop the weapons and agents in the Atlantic Ocean at a depth of 4000 ft, ships equipped for this purpose were not available. As a result, weapons were either dumped or old ships containing weaponry were scuttled in shallower depths close to land.

Quantities of chemicals disposed of at sea or incinerated by allied countries following WWII can only be estimated as records are incomplete. After WWII, some 40 000 tonnes of munitions containing approximately 15 000 tonnes of chemical agents in wooden crates were dumped in the Baltic Sea (Kaffka, 1996). Disposal occurred in three locations: Lille Boelt between Sweden and Denmark, east of Bornholm and the Gotland basin (Granbom, 1994, 1996). However, material was either dropped en route or has drifted, as some material has been caught in nets by trawling fishermen, and some material has drifted to the west coast of Poland. Another disposal site was the Skagerrak area of the North Sea, off the Swedish and Norwegian coasts (Granbom, 1996). Disposed materiel near the Skagerrak included about 270 000 tonnes dropped by the British and Americans and 30 000 tonnes submerged near the Gottland and Bornholm islands by the Soviet Army (Konkov, 1996). Ocean disposal also took place in the North Sea (Stock, 1996). Sixty sea dump sites, located around the world, were used by the US alone (Frondorf, 1996).

United Kingdom

During WWII, the United Kingdom produced 40 000 tonnes of mustard gas (Glasby, 1997). The primary production site for all chemical agents was Porton Down on the southern edge of the Salisbury Plain in Wiltshire. This facility was established in 1916 in response to the use of chemical weapons by Germany during WWI (Harris and Paxman, 2002). Following WWII, the plant was decommissioned; silver reaction vessels were filled with caustic soda and allowed to react for several years before being taken apart for silver recovery. Several additional sites were used either to store chemical weapons prior to the end of the war or to dispose of chemicals after the war (Perera and Thomas, 1986). CWA munitions that did not undergo ocean disposal were burned. After WWII, the British Ministry of Defense is reported to have sunk barges loaded with captured German chemical agents and unused British chemical agents off the west coast of Ireland (Fokin and Babievsky, 1996). HD was also sunk in the Bay of Biscay.

United States

Industrial production of sulfur mustard was discontinued in 1968; at that time, the US stockpile contained some 17 000 tonnes (ATSDR, 2003). The US unitary chemical weapons stockpile includes sulfur mustard which was originally present in various munitions and tonne containers at the Aberdeen Proving Ground in Maryland, Deseret Chemical Depot in Utah, Anniston Army Depot in Alabama, the Umatilla Depot Activity in Oregon, Pine Bluff Arsenal in Arkansas and Tooele Army Depot in Utah (DOD, 1996). HD, H and HT are stored in various containers and munitions at these and several other nonstockpile sites. Destruction of the remaining stockpile, either by hydrolysis or incineration, is presently underway. Incineration of nearly 4000 tonnes of chemical weapons was begun in April 2005 at the Pine Bluff Arsenal in Arkansas, USA. These materials include mustard and nerve agents in rockets, land mines and bulk containers (Ault, 2005). In March 2005, the last tonne container of sulfur mustard at the Aberdeen Proving Ground was drained and neutralized (Public Affairs, Chemical Materials Agency, 2005). During the 1940s–1960s, some sulfur mustard was dumped in the coastal waters of the USA (Brankowitz, 1987).

Lewisite was produced in limited quantity at the Rocky Mountain Arsenal (RMA) near Commerce City, Colorado, and was stored only at the Deseret Chemical Depot, near Tooele, Utah (US Army, 1988). Arsenic is present in soil at the RMA as a consequence of lewisite manufacture, as well as from later commercial manufacture of insecticides and herbicides (Corwin *et al.*, 1999).

The US unitary chemical weapons stockpile also includes the nerve agents GA, GB, and VX, stored in either tonne containers or munitions at the Anniston Army Depot in Alabama, Bluegrass Army Depot in Kentucky, Deseret Chemical Depot in Utah, Newport Chemical Depot in Indiana, Pine Bluff Arsenal in Arkansas and the Umatilla Chemical Depot in Oregon (DOD, 1996). Agent VX is present in missiles and projectiles at five army depots or arsenals in the USA and in tonne containers at the Deseret and Newport Chemical Activity (DOD, 1996; US Army, 1996a). GA was produced in limited quantities in the USA. Production of GB, the major nerve agent of the USA, took place only at the RMA from 1953 to 1957 (Robson, 1981). The RMA site was also the location for filling of munitions with GB from bulk stocks. GD nerve agent and the nitrogen mustards were not stockpiled as part of the US chemical weapons inventory. Nonstockpile chemical material, defined as waste, contaminated containers and buried weapons, is found at 88 locations in 33 States, often at multiple sites, the District of Columbia and the Virgin Islands (Opresko *et al.*, 1998).

Former Soviet Union

There is limited information on storage and dumping of agents in the former Soviet Union. Fedorov (1966) reports that the Volga River valley was the primary area of manufacture of V- and G-agents. During the 1940–1945 war years, the Soviet Union produced primarily HD (77 400 tonnes) but also 20 600 tonnes of lewisite, 11 100 tonnes of hydrogen cyanide, 8300 tonnes of phosgene and 6100 tonnes of adamsite (an arsenical irritant). Some of this materiel was put into weapons; some was stockpiled. From 1958–1987, 12 000 tonnes of GB, 5000 tonnes of GD and 15 500 tonnes of V-gas were produced. Materiel destroyed after WWII included 60 000–65 000 tonnes of HD and ~14 000 tonnes of lewisite. Materiel was buried, incinerated or dumped into the surrounding seas; wastes were often discharged directly into rivers. Many production sites throughout the former Soviet Union remain contaminated and subject to ongoing problems. For example, in 1965, a dam at the former Stalingrad production site broke, allowing nerve-agent-containing effluents to flow into the Volga River. An extensive fishkill resulted. Information on official dumping in the seas is limited, but available records indicate dumping in the Baltic (captured German weapons), White, Barents and Black Seas and the seas of Okhotsk and Japan over a period of 50 years (Fedorov, 1996; Surikov, 1996; Yufit *et al.*, 1996).

Japanese Army

Between 1927 and 1945, the Japanese Army operated a chemical warfare agent factory on Okuno-jima, an island of the Inland Sea (Wada *et al.*, 1962; Watson *et al.*, 1989; Yamakido *et al.*, 1996). At peak capacity (1937), this facility produced sulfur mustard (450 tonnes/month), lewisite (50 tonnes/month), diphenylcyanarsine (sneezing gas, 50 tonnes/month), hydrocyanic acid (50 tonnes/month), chloroacetophenone (tear gas, 25 tonnes/month) and phosgene (unreported). After the war, the factory was closed and the remaining agents were disposed of at sea.

Recent incidents

Materiel placed into areas of active fishing or in close proximity to densely populated coastlines may pose an ecological as well as human health problem when it is retrieved in fishing nets or washes up on nearby coasts (Kaffka, 1996; Assennato *et al.*, 1997). In early 2004, nearly 100 explosive devices were recovered from driveways in Sussex County, Delaware, USA. These munitions, dumped at sea years ago, were found in driveways of residents who purchased dredged clam shells for paving material. In one such case, three members of a military explosives disposal team retrieving a leaking WWI-era shell suffered extensive blistering (Chase, 2004). In January of 1993, WWI chemical weapons were recovered from a construction site in Washington, DC (Smartt and Kopp, 2005). The site had been a former chemical testing area operated by the American University during WWI.

In Albania, unsecured stockpiles of sulfur mustard, lewisite and adamsite were recently found (Warrick, 2005). It is thought that these 16 tonnes of chemicals were acquired during the mid-1970s by the then-current regime. These stockpiles have now been secured and plans are underway to incinerate the chemicals onsite.

As late as 1994, *démineurs* (France's bomb-disposal experts) were still collecting unexploded WWI bombs and grenades in the French countryside (Webster, 1994). Although most of the devices pose only an explosive hazard, a small percentage still contain mustard agent.

In the 1980s, Iraq used mustard gas, initially as a defensive agent, against Iranian troops (Harris and Paxman, 2002; Dunn, 1986). By the mid-1980s, Harris and Paxman (2002) report that GB and VX were added to the Iraqi arsenal. These agents, singly or in combination, were deployed by Iraqi forces against Kurdish civilians.

TERRORISTS EVENTS

The Chemical Weapons Convention, whereby nations eschewed programs of development of chemical warfare agents by international treaty, entered into force in 1997 (OPCW, 2005). Unfortunately, such treaties cannot eliminate development and use of CWAs by a rogue nation or a terrorist organization.

The nerve agent sarin (GB) can be manufactured easily and inexpensively. GB was the first CWA to be used by terrorists in modern times when it was deployed in attacks against Japanese citizens (see Chapters 13). In 1994, several hundred people were exposed in the city of Matsumoto (Morita *et al.*, 1995); another incident involved near-simultaneous releases at several points within the Tokyo subway system in March of 1995 (Okumura *et al.*, 1996). In the latter case, casualties were limited due to the purity of the GB (30%) and the passive method of dispersal (Pangi, 2002). The Japanese Self Defense Force, as well as the Tokyo Municipal Fire Department in collaboration with the National Police Agency, was involved in decontamination of subway cars and stations. The subway system was back in public service within 24 h post-agent release.

SULFUR MUSTARD (H, HD)

Environmental degradation

Sulfur mustard, although usually referred to as 'mustard gas', is actually a liquid that boils at 217°C. Most available information characterizes the distilled or purified form of sulfur mustard (HD; bis(2-chloroethyl)sulfide) in contrast to the undistilled form (H) which contains numerous impurities. Agent HT was made by an older manufacturing process and contains about 60% HD and ~40% agent T (bis(2-(2-chloroethylthio)ethyl)ether) to lower the freezing point of the blend for deployment in cold-weather operations. The HT blend also contains a variety of impurities. In some cases, solvents such as benzene or thickeners such as polyacrylates were added to HD to change the melting point.

The fate of HD in the environment is determined by its chemical and physical properties. Observations of persistence and degradation products from both field and laboratory studies support the characterization of HD as a *persistent agent*. As noted, sulfur mustard is a liquid at ambient temperatures; the vapor pressure is low (0.11 mmHg at 25°C), but sufficient for mustard to be in the air immediately surrounding droplets of the liquid. The primary dissipation mechanism for HD from soil is evaporation. Sulfur mustard vapor is 5.5 times heavier than air, and evaporation or volatilization from surfaces or soils is projected to require days at temperatures above its freezing point (Puzderliski, 1980). Below its freezing temperature, HD is a solid, and evaporation as well as hydrolysis would be slower. Photodegradation does not appear to be a significant mode of dissipation (Rewick *et al.*, 1986). Water solubility is low, and based on a density greater than that of freshwater and seawater, HD would sink in water where it would persist for considerable periods of time while retaining blister-forming properties (Sanches *et al.*, 1993). Low water solubility inhibits flow into groundwater where it is not normally found (Rosenblatt *et al.*, 1975; Bartelt-Hunt *et al.*, 2006), i.e. once dissolved in groundwater, HD rapidly hydrolyzes. In a field-conducted study, HD droplets applied at a surface density of approximately 1 g/m^2 to soil plots appears to have hydrolyzed fairly rapidly based on measurements that thiodiglycol (the final degradation product) was detected throughout the 5-day post-application sampling period (McGuire *et al.*, 1993).

The primary environmental fate mechanism followed by stored or buried HD is hydrolysis. Although HD is rapidly hydrolyzed (a half-life of 4 to 8 min at 25°C in distilled water has been reported [Bartlett and Swain, 1949]), the overall process of hydrolytic destruction is limited by the very low water solubility of HD. Intermediate hydrolysis products and/or water-insoluble thickeners that can coat or encapsulate droplets of mustard retard hydrolysis. Because of low water solubility and formation of intermediate products, bulk amounts of HD may persist undispersed under water for some time. However, HD dispersed as droplets or mist, as in the case of an aerial attack, is expected to hydrolyze rapidly in humid air.

Sulfur mustard hydrolysis is 'surface-controlled', with products formed at the HD–water interface diffusing into the bulk water phase (Rosenblatt *et al.*, 1975,1995; Small, 1984; MacNaughton and Brewer, 1994; Yang *et al.*, 1992). The hydrolysis mechanism is complex and, depending on the availability of water, occurs by two routes, both of which lead to formation of thiodiglycol (TDG) and hydrochloric acid (Figure 1). In a dilute aqueous solution, dissolved HD is rapidly converted first to a sulfonium ion and then to the hemi-mustard and TDG. In the presence of insufficient water to initially dissolve all available HD, several sulfonium ion aggregates (TDG–mustard aggregates) are formed at the water–HD interface. These aggregates may shield the bulk of the material and contribute to environmental persistence. Although the reactions shown in Figure 1 are reversible, the conditions required to produce reversible hydrolysis would not normally be encountered in the environment (Small, 1984). In several laboratory studies, the hydrolysis half-life of dissolved HD ranged from 158 min at 0.6°C to ~1.5 min at 40°C and did not vary appreciably in the typical environmental pH range (Small, 1984). The reported half-life in seawater is 15 min at 25°C, 49 min at 15°C and 175 min at 5°C (Small, 1984; Stock, 1996). The final degradation product, TDG, is miscible with water and readily biodegradable by bacteria and fungi present in the environment (Harvey *et al.*, 1996; Pham *et al.*, 1996; Itoh *et al.*, 1997; Lee and Allen, 1998). In fact, most of the degradation products of HD are biodegradable under anaerobic conditions (Sklyar *et al.*, 1999). The overall hydrolysis reaction yielding TDG is:

$$\text{Cl–CH}_2\text{–CH}_2\text{–S–CH}_2\text{–CH}_2\text{–Cl} + 2\text{H}_2\text{O} \longrightarrow \text{HO–CH}_2\text{–CH}_2\text{–S–CH}_2\text{–CH}_2\text{–OH} + 2\text{HCl}$$

In addition to the major degradation products outlined in Figure 1, numerous degradation products, impurities and secondary reaction products have been identified from both controlled laboratory studies and field studies of contaminated sites. The analysis of materiel in stored tonne containers of HD (US Army, 1996b; NRC, 1996; Amr *et al.*, 1996; Harvey *et al.*, 1996) supports the observation of slow hydrolysis during storage and verifies the identification of chemicals from controlled laboratory studies. For example, HD stored for approximately 40 years in tonne containers was 89.2% pure. The major impurity (4.7%) was 1,2-bis-(2-chloroethylthioethane), also known as compound Q or sesquimustard. Other impurities and degradation products present in the greatest amounts included hexachloroethane, 1,4-dithiane (estimated at 40 lb/container), 2-chloroethyl 3-chloropropyl sulfide and 2-chloroethyl 4-chlorobutyl sulfide. A residue in the bottom of the containers was composed primarily of 1-(2-chloroethyl)-

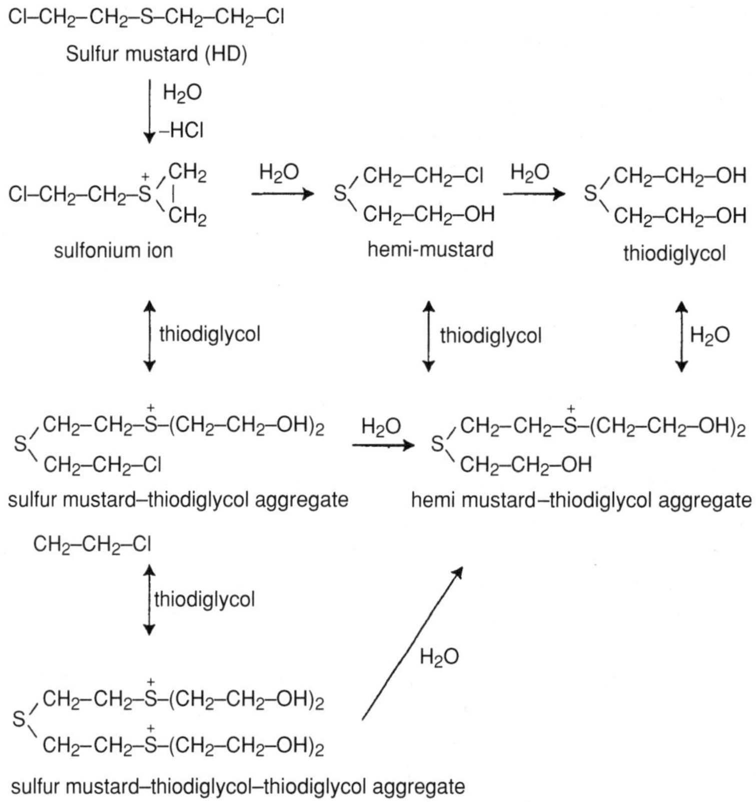

Figure 1. Primary hydrolysis pathways of agent HD in the environment

1-thiona-4-thiane chloride (Q-sulfonium), formed by reaction of HD with the metal on the inside of the tonne containers or from a reaction with metal impurities during the agent manufacturing process (Amr *et al.*, 1996). HD may form metal complexes with storage containers (HD–$FeCl_2$) or with the metal sulfides present in soil (Small, 1984). The degradation products, impurities and stabilizers of HD are listed in the Appendix to this chapter (Table A1), along with synonyms and Chemical Abstract Service (CAS) numbers. Inorganic degradation products and well-characterized organic entities, such as ethanol and isopropyl alcohol, are not listed.

Few data on the environmentally relevant chemical and physical properties of HD degradation products are available. In the absence of such data, Small (1984) calculated physical properties for degradation products of HD that are predicted to be stable in the environment. These properties are the octanol/water partition coefficient (K_{ow}; the affinity of a compound for the organic phase of the environment and thus the affinity to bioaccumulate), the soil adsorption coefficient (K_{oc}) which is a measure of adsorption to the organic fraction of soil or sediment, water solubility and vapor pressure (Table 2). These calculated physical properties may differ by an order of magnitude from actual values. The physical properties for the parent sulfur mustard compound are included for comparison purposes. Data were also available on 1,4-oxathiane (Berkowitz *et al.*, 1978). Based on a similar chemical structure, the behavior of 1,4-oxathiane in the environment should be similar to that of 1,4-dithiane, with 1,4-oxathiane being more volatile and more water-soluble. No dissociation constants were located for the chemicals shown in Table 2.

As can be seen from the values in Table 2, several of the compounds are moderately to highly soluble in water (> 1 g/l). Vapor pressures are generally low and indicate little volatility, with the possible exception of the vinyl sulfides and 1,2-dichloroethane. Log K_{oc} values of

Table 2. Physical properties of sulfur mustard degradation products[a,b]

Compound	Water solubility (g/l)	log K_{ow}	log K_{oc}	Vapor pressure (mmHg)
Thiodiglycol	Miscible	−0.77	0.96	0.000 02
2-Chloroethyl vinyl sulfide	1.4	1.11	1.98	5.8
Divinyl sulfide	2.5	0.85	1.84	6.0
Mustard sulfoxide	93	−0.85	0.91	0.65
Mustard sulfone	11	−0.51	1.11	0.96
2-Chloroethyl vinyl sulfoxide	160	−1.11	0.77	0.064
Vinyl sulfoxide	280	−1.37	0.63	0.92
2-Hydroxyethyl vinyl sulfide	5.0	0.53	1.66	3.8
2-Chloroethyl vinyl sulfone	78	−0.77	0.96	0.023
Divinyl sulfone	140	−1.03	0.82	0.09
1,4-Dithiane	3.0	0.77	1.80	0.80
1,4-Oxathiane[c]	167	0.60	ND[d]	3.9
1,2-Dichloroethane	11	1.48	2.18	8.5

Notes:

[a] Values were calculated by Small (1984), except data on 1,4-oxathiane.

[b] For comparison purposes, the water solubility, log K_{ow}, log K_{oc} and vapor pressure for sulfur mustard are 1.0 g/l, 1.37, 2.12 and 0.1 mmHg, respectively.

[c] Berkowitz *et al.* (1978).

[d] ND, no data.

approximately two or less indicate that little soil adsorption will occur. Small (1984) used the calculated physical properties to derive physical indices, including persistence or removal by several processes. The derivations were based on water solubility and, for illustrative purposes, assumed a fractional soil organic carbon content (f_{oc}) of 0.02. According to Small (1984), the *leaching index*, which he defined as the number of leachings required to reduce a compound to one-tenth of its initial concentration, was high for sulfur mustard and 1,2-dichloroethane, indicating little leaching, whereas values for 2-chloroethyl vinyl sulfide, divinyl sulfide, 2-hydroxyethyl vinyl sulfide and 1,4-dithiane were intermediate, indicating a moderate amount of leaching. The volatility potential estimates (the loss of a compound from soil), ranged from practically none for TDG to 2.3 days for divinyl sulfide and 1.8 years for HD. Calculated Henry's law constants, a measure of volatilization from surface water, indicate that TDG, 2-chloroethyl vinyl sulfoxide, vinyl sulfoxide, 2-chloroethyl vinyl sulfone and divinyl sulfone are essentially nonvolatile, whereas divinyl sulfide and 1,2-dichloroethane would rapidly volatilize. The other compounds were calculated to be of intermediate volatility.

Two common degradation products of HD that persist in the environment are 1,4-oxathiane and 1,4-dithiane. 1,4-Oxathiane is formed by dehydrohalogenation of partially hydrolyzed mustard while 1,4-dithiane is a thermal degradation product of mustard formed by dechlorination. Formation of 1,4-dithiane occurs very slowly at ambient temperatures. 1,4-dithiane photo-oxidizes to sulfoxides and sulfones (Deardorff *et al.*, 1992). 1,4-Oxathiane in soil may also be formed by rearrangement of 2-hydroxyethyl vinyl sulfide (Small, 1984). Both 1,4-oxathiane and 1,4-dithiane are groundwater contaminants in the RMA area of the USA (Sanches *et al.*, 1993) and, along with TDG, have been used as indicators of HD contamination (Tompkins *et al.*, 1998). Historically, concentrations of 1,4-dithiane as high as 9 mg/l have been found at RMA (Deardorff *et al.*, 1992).

It is interesting to note that the environmental stability of several of the degradation products allows forensic monitoring to ascertain potential use or presence of agents in warfare incidents. For example, TDG, 2-hydroxyethyl vinyl sulfide, 1,4-dithiane (which may also be found in the starting mustard), 1,4-oxathiane and divinyl sulfide were among compounds identified in soil, munition fragments and wool samples associated with a chemical warfare incident in Iraq. A total of 23 mustard-related compounds were tentatively identified, along with the explosives 2,4,6-trinitrotoluene and tetryl (Hay and Roberts, 1990;

Black *et al.*, 1993). In addition to sulfur mustard (the primary component), sesquimustard and Agent T, 15 additional components, including dehydrochlorination products, were identified in an Iran/Iraq soil sample suspected to have been contaminated with mustard (D'Agostino and Provost, 1988). It should be noted that while TDG can be used as a starting material for the production of HD, it is also a high-production volume chemical, used in the textile industry in the USA.

Degradation product toxicity

The acute and chronic toxicity of HD degradation products has been reviewed (Munro *et al.*, 1999). Incomplete hydrolysis yields products of varying, but lower toxicity than that of the parent compound. For example, the sulfoxides and sulfones have mammalian oral LD_{50} values of ~100 mg/kg and are moderate skin irritants, whereas TDG, 1,4-oxathiane and 1,4-dithiane have LD_{50} values ranging from 2830 to 6610 mg/kg (Munro *et al.*, 1999) and are considered of low toxicity. The initial degradation products that retain the 2-chloroethylsulfide moiety, such as mustard- and hemimustard-TDG aggregates, mustard sulfone and divinyl sulfone retain the vesicant property. 1,2-Bis(2-chloroethylthio)ethane is moderately to highly acutely toxic by the inhalation route (Robinson, 1967; Vocci *et al.*, 1963). The hemisulfur mustard forms DNA adducts, but both 1,4-dithiane and 1,4-oxathiane are negative in mutagenicity tests. Rodent acute oral LD_{50} values of several thousand mg/kg indicate low toxicity for both 1,4-dithiane and 1,4-oxathiane (Mayhew and Muni, 1986). The generally low K_{ow} values (< 2) for the mustard degradation products listed in Table 2 indicate a low potential to partition to the lipid phase in organisms and thus little potential to bioconcentrate. 1,4-Dithiane and the final degradation products, TDG and hydrochloric acid (in dilute form), are practically nontoxic to vegetation and aquatic organisms (Smyth *et al.*, 1941; Inamori *et al.*, 1990).

Concern regarding the potential hazards posed by sulfur mustard munitions historically buried at sea has been expressed. The literature contains multiple reports of incidents in which fishermen that have snagged damaged or corroded mustard munitions in their nets exhibit blisters and chemical burns as well as ocular effects (Assennato *et al.*, 1997; Wulf *et al.*, 1985). Although the munitions have been in the sea for years, these effects are due to sulfur mustard and not degradation products. The exposure of fishermen in the Adriatic Sea most probably stems from the 'Disaster at Bari' (Infield, 1988). During WWII, Bari Harbor was the site of a German bombing raid that resulted in explosion of sulfur mustard munitions stored in the hold of a transport ship at anchor; over 1000 military personnel and Italian civilians died from contact with mustard agent released to the air and water during the raid (Alexander, 1947; Gage, 1946). Debris from this incident was later sunk at sites in the Adriatic Sea.

In addressing the ecotoxicological implications of HD dumped in the Baltic Sea, Muribi (1997) reported on the acute toxicity of HD to the invertebrate *Daphnia magna* in brackish water at 19.5°C. Exposure to 0.5 mg/l HD (the highest concentration tested) for 48 h did not induce any visible effects. Additional toxicity studies with salt- and freshwater organisms, reviewed by Muribi (1997) and Munro *et al.* (1999), indicate that dissolved sulfur mustard is generally not acutely toxic at concentrations below 1 mg/l (close to its water solubility value). For some algae, a concentration of 1 mg/l may be acutely toxic (Stock, 1996). Thus, the environmental action of HD is limited by its low water solubility.

An ecotoxicological investigation of the Skagerrak area off the Swedish coast was undertaken in 1992 (Granbom, 1996). Analysis of sediment samples taken in the area where ships loaded with chemical agents and weapons were scuttled showed low levels of HD, up to 190 parts per thousand (ppt).

Degradation product guidelines for restoration

Although health-based exposure guidelines, such as reference doses and environmental screening levels, have been derived for the parent agents (NRC, 2003; Opresko *et al.*, 1998; US Army CHPPM, 1999), exposure guidelines have been estimated for only a single HD degradation product, TDG. Using a subchronic oral

Table 3. Current range of estimated health-based screening levels for residential and industrial soil[a,b]

Agent	Residential soil value (mg/kg)	Industrial soil value (mg/kg)
HD	0.01–0.55	0.3–14
Lewisite	0.3–7.8	3.7–7.8
VX	0.042–0.047	1.1–1.2
GA	1.2–3.1	68–82
GB	0.5–1.6	32–41
GD	0.22–0.31	5.2–8.2

Notes:
[a] Values were estimated based on several US EPA chronic risk assessment methodologies and RfD estimates from Opresko *et al.* (1998,2001).
[b] Source: US Army CHPPM, 1999.

study with the rat, an RfD of 0.4 mg/kg/day was estimated (Reddy *et al.*, 2005). This compares to an estimated RfD for HD of 7×10^{-6} mg/kg/day (Opresko *et al.*, 1998). Health-based environmental screening levels for HD for residential soil range from 0.01 to 0.55 mg/kg of soil (Table 3).

LEWISITE – DICHLORO-(2-CHLOROVINYL)ARSINE

Environmental degradation

With a vapor pressure of 0.58 mmHg at 25°C, lewisite is considered non-volatile. However, it is more volatile than HD and may be used as a moderate irritant vapor over greater distances than HD (Watson and Griffin, 1992). Based on its UV absorption band (Rewick *et al.*, 1986), some photodegradation may take place in the atmosphere. Hydrolysis may also occur in the gas phase (MacNaughton and Brewer, 1994).

Lewisite is practically insoluble in water, but the small amount that dissolves is rapidly hydrolyzed (Rosenblatt *et al.*, 1975), resulting in the formation of the water-soluble dihydroxy arsine or 2-chlorovinyl arsonous acid. Lewisite in solution becomes essentially 100% 2-chlorovinyl arsonous acid (Major, 1998):

$$\text{Cl–CH=CH–AsCl}_2 + 2\text{H}_2\text{O} \longleftrightarrow \text{Cl–CH=CH–As(OH)}_2 + 2\ \text{HCl}$$

With the removal of water, 2-chlorovinyl arsonous acid forms lewisite oxide (2-chlorovinyl arsenous oxide). Formation of lewisite oxide and polymerized lewisite oxide is essentially a dehydration reaction:

$$\text{Cl–CH=CH–As(OH)}_2 \longleftrightarrow \text{H}_2\text{O} + \text{Cl–CH=CH–AsO} + (\text{Cl–CH=CH–AsO})_n$$

Once formed, lewisite oxide and polymerized lewisite oxide are relatively insoluble in water. Once dry, the oxide will probably not readily redissolve or form the acid in the environment. The degradation products (and impurities) of lewisite are listed in the Appendix, Table A2.

Lewisite is easily hydrolyzed by soil moisture, and minerals present in the soil would speed the process (Cooper, 1990). Alkaline soils would neutralize lewisite. Slow oxidation in soil results in 2-chlorovinyl arsonic acid (Rosenblatt *et al.*, 1975). There are no specific data on microbial degradation, but suggested pathways of microbial degradation in soil include epoxidation of the C=C bond and reductive dehalogenation and dehydrohalogenation (Morrill *et al.*, 1985). The latter pathways result in toxic metabolites due to the epoxy bond and arsine group. Depending on the environmental conditions, various inorganic arsenic compounds can be formed in the course of complete lewisite mineralization; inorganic arsenic compounds are found in areas of past lewisite releases, although the limited quantity of lewisite present in the USA would suggest limited areas of risk there. Movement of arsenic to groundwater at contaminated sites is retarded by adsorption to soil (Corwin *et al.*, 1999). Even if lewisite is completely degraded, the toxic element arsenic (combined with metals or as salts) would remain.

Degradation product toxicity

Lewisite has vesicant properties as well as systemic toxicity. Because lewisite rapidly hydrolyzes to 2-chlorovinyl arsonous acid upon contact with moist surfaces, the toxic properties of lewisite may actually be those of 2-chlorovinyl arsonous acid (Fowler *et al.*, 1991). Percutaneous LD_{50} values in small mammals range from 5 to 24 mg/kg (Marrs *et al.*, 1996). Lewisite oxide most likely also has vesicant properties (US Army, 1974). The vesicant potency of lewisite degradation products compared to lewisite is unknown. With a lowest lethal oral dose of 50 mg/kg (RTECS, 2005), 2-chlorovinyl arsonic acid is moderately toxic to small mammals. The impurity bis(2-chlorovinyl)arsine is reported to have a toxicity comparable to that of lewisite (Rosenblatt *et al.*, 1975).

Lewisite vapor is extremely phytotoxic and has been implicated in the death of vegetation in lewisite shell target areas (Armstrong *et al.*, 1928). In aquatic situations, available data indicate that lewisite degradation products are less toxic than lewisite. At 200 mg/l, 2-chlorovinyl arsonic acid was only slightly toxic to aquatic organisms (Price and von Limbach, 1945), whereas the threshold for lethality of lewisite to several species of fish is <2.0 mg/l (Buswell *et al.*, 1944). When lewisite solutions were allowed to age for several days, there was a decrease in the toxicity to aquatic organisms (Price and von Limbach, 1945). Lewisite and its major degradation products, although extremely toxic, do not bioaccumulate through food chains, whereas arsenic, the result of complete mineralization, does.

Degradation product guidelines for restoration

Bausum *et al.* (1999) estimated a RfD for lewisite oxide of 0.1 µg/kg/day (Table 1) for use in developing environmental restoration guidelines. The estimated RfD for lewisite (1×10^{-4} mg/kg/day) is appropriate when the presence of lewisite, 2-chlorovinyl arsonous acid or lewisite oxide is known to be present in the environmental medium of concern (Opresko *et al.*, 2001). Otherwise, the RfD for inorganic arsenic (0.3 µg/kg/day) may be applied (US EPA, 2005a) because lewisite in environmental media is eventually degraded to inorganic arsenic. Estimated health-based soil values for lewisite in residential soil range from 0.3 to 7.8 mg/kg of soil (Table 3). No guidelines are available for cleanup or protection of the environment for lewisite degradation products. Because lewisite in environmental media is degraded to inorganic arsenic, the RfD for inorganic arsenic (0.3 µg/kg/day) may be applied (US EPA, 2005a). Estimated health-based soil values for residential soil range from 0.3 to 7.8 mg/kg of soil. In developing these soil screening levels, the persistent degradates 2-chlorovinyl arsonous acid, lewisite oxide and arsenic were considered.

NERVE AGENTS

The nerve agents are alkylphosphonic acid esters. They are generally divided into V agents, the primary one being VX (*O*-ethyl *S*-(2-diisopropylaminoethyl) methylphosphonothioate), and G agents, the principal ones being GA (tabun, ethyl-*N*,*N*-dimethylphosphoramidocyanidate), GB (sarin; isopropyl methylphosphonofluoridate) and GD (soman; pinacolyl methylphosphonofluoridate). V agents, such as VX, contain a sulfur atom and are alkylphosphonothiolates. GA contains a cyanide group, while GB and GD contain a fluorine substituent group and are methylphosphonofluoridate esters. These nerve agents contain a C–P bond which is almost unique, is not found in organophosphate pesticides, is very resistant to hydrolysis. The cyanide–phosphorus bond is relatively easily hydrolyzed.

The nerve agents are viscous, clear liquids when relatively pure. Agent VX is the least volatile (vapor pressure of 0.0007 mmHg at 20°C), while GB (vapor pressure of 2.10 mmHg at 20°C) is the most volatile. The nerve agents are all anticholinesterases; they inactivate cholinesterases, the enzymes responsible for the destruction of acetylcholine, which, in turn, terminates a nerve impulse. When cholinesterases are inactivated, the resulting prolonged neurotransmission is manifest as tremors, paralysis or other symptoms depending on the affected portion of the nervous system.

VX (*O*-ethyl *S*-(2-(diisopropylamino)ethyl) methylphosphonothioate)

ENVIRONMENTAL DEGRADATION

Agent VX is a persistent, odorless and colorless to amber-colored liquid. VX is far less volatile (10.5 mg/m^3 at 25°C) than the G-agents and does

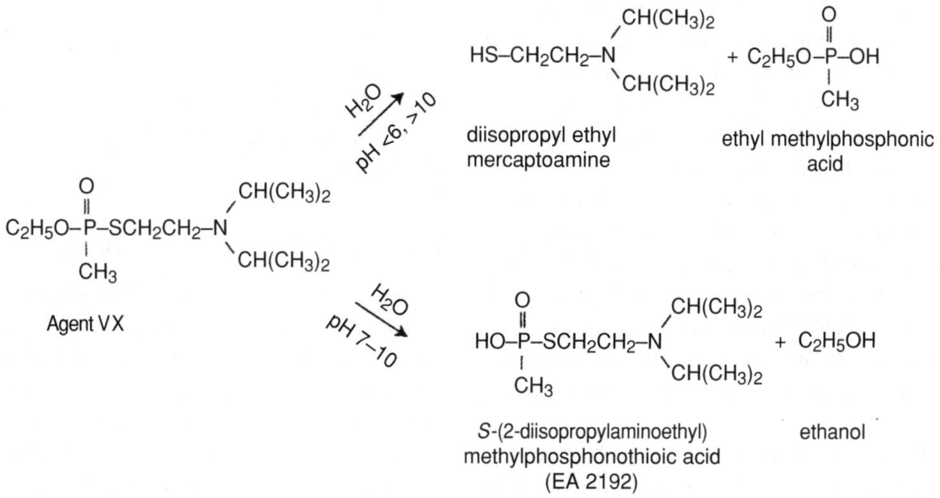

Figure 2. Primary hydrolysis pathways of agent VX in the environment

not evaporate readily. A Henry's law constant of 3.5×10^{-9} atm m^3/mol (Small, 1984) indicates that VX is essentially non-volatile from water. VX is moderately persistent on bare ground and may remain in significant concentrations for 2–6 days, depending on temperature, organic carbon content of the soil and moisture (Sage and Howard, 1989). Dissipation is the result of a combination of processes including evaporation, hydrolysis and microbial degradation. Although VX adsorbs to soil (Verweij and Boter, 1976), Small (1984), in reviewing field and closed-container studies of VX persistence, estimated that 90% of initially applied VX in soil would be lost in ~15 days. In the laboratory, unstabilized VX of 95% purity decomposed at a rate of 5% per month at 22°C (US Army, 1992). All four major degradation products (Figure 2), as well as methylphosphonic acid (MPA), were detected within three days following deposition of VX on soil at low concentrations (McGuire *et al.*, 1993).

VX is soluble in water, and, although relatively resistant to hydrolysis (Franke, 1982), the rate of hydrolysis is still reasonably fast relative to groundwater flow. Figure 2 is a description of the primary hydrolysis pathways of VX in the environment. The hydrolysis pathway is somewhat dependent on environmental pH. In acid and alkaline conditions, cleavage of the P–S bond predominates, resulting in formation of ethyl methylphosphonic acid (EMPA) and diisopropylethyl mercaptoamine (DESH). The latter compound can be oxidized to bis(2-diisopropylaminoethyl) disulfide ($(DES)_2$; EA 4196) or react with the diisopropyl ethyleneimmonium ion $(CH_2)_2N^+(C_3H_7)_2$ to form bis(2-diisopropylaminoethyl) sulfide. In a solution of 0.01 M VX and aqueous 0.1 M NaOH, VX was hydrolyzed to EMPA and diisopropylaminoethyl methyl thiolophosphonate (EA 2192) ions in a ratio of 87% to 13%, respectively; under these conditions, the half-life of VX was 31 min (Yang *et al.*, 1993). In another experimental study, trace amounts of VX on concrete surfaces (a neutral to alkaline surface) degraded with a half-life of 2–3 h at room temperature. Degradation was by cleavage of the P–S and S–C bonds, with the major degradation product being DESH (Williams *et al.*, 2005). This result is consistent with alkaline hydrolysis within water films associated with the concrete surface.

At neutral to alkaline pH values (7 to 10), the above pathway competes with dealkylation of the ethoxy group (cleavage of the O–P bond, followed by addition of a hydroxyl group), the latter pathway yielding the environmentally stable EA 2192 and ethanol. Although MPA can theoretically be formed slowly by hydrolysis of EMPA, this has not been demonstrated to occur in aqueous solutions (Kingery and Allen, 1995).

Numerous additional hydrolysis and reaction products and impurities have been identified from sampling and analysis of stored tonne containers of VX (Appendix, Table A3) (US Army, 1996b; NRC, 1996; Amr *et al.*, 1996). During 30–40 years of storage at the Newport Chemical Activity, Indiana, most of the nerve agent remained intact (90.5 to 94.8% pure). VX was formulated with a small percent of the stabilizers, diisopropyl carbodiimide or dicyclohexylcarbodiimide, to protect it against decomposition from trace amounts of water. During the ensuing storage period, some of the stabilizers hydrolyzed.

Few data are available from which to further characterize the fate of the major VX degradation products in the environment. The measured and estimated chemical and physical properties of VX degradation products are listed in Table 4. The sulfur-containing EA 2192, although soluble in water, is relatively stable to hydrolysis (Michel *et al.*, 1962; Szafraniec *et al.*, 1990). The highly water-soluble metabolite EMPA is resistant to hydrolysis, but disappears from soil fairly rapidly. Its metabolite, MPA, is only slowly biodegraded. Analysis for EMPA is not difficult, and the presence of EMPA would confirm VX contamination, as well as contamination by potentially toxic metabolites, such as EA 2192. Other than dimerization and reaction with the ethyleneimmonium ion, information on the fate of DESH was not found. However, both laboratory and field studies show that VX and its products, including DESH (but not MPA), disappear from soil fairly rapidly (Small, 1984; Verweij and Boter, 1976; Kaaijk and Frijlink, 1977). Small's (1984) review of field studies conducted at Carroll Island, MD, USA, showed that VX sprayed on soil decreased by about three orders of magnitude within 17 to 52 days. In an area of field tests at Dugway Proving Ground, where soil levels before 1969 were as high as 6 mg/g, no VX was detected (detection limit 0.4 μg/g) 10 years later. The degradation product, MPA, was detected at concentrations ranging from 14.9 to 23 μg/g and was distributed uniformly through a 120-cm depth.

Additional information characterizing aquatic degradation of agent VX, as well as hydrolysis products and stabilizers, may result from the ongoing CDC and Region II EPA evaluation of treatment and disposal for the caustic VX hydrolysate resulting from the approved process for demilitarizing bulk VX stockpiled at the Newport Chemical Agent Disposal facility in Indiana (DHHS, 2005).

DEGRADATION PRODUCT TOXICITY

Although information is limited about other VX hydrolysis products, none display the high acute toxicity of EA 2192, and most for which information is available can be characterized as having low to moderate acute lethality (Michel *et al.*, 1962; Munro *et al.*, 1999). DESH, for example, is much less toxic than VX. No toxicity data were found for EMPA, but it is structurally similar to isopropyl methylphosphonic acid (IMPA) and is expected to have the same low to moderate toxicity as IMPA and MPA. MPA, with an oral LD_{50} in the rat of 5000 mg/kg (Williams *et al.*, 1987), is practically nontoxic. EA 2192, with an oral LD_{50} of 630 μg/kg in the rat, is considered highly toxic; it also retains anticholinesterase properties. Because of its environmental stability, it is the degradation product of most concern. Since it is a solid at environmental temperatures, EA 2192 vapor inhalation is not a likely route of uptake (Rosenblatt *et al.*, 1995). Under normal exposure conditions, EA 2192 does not readily penetrate the skin or have blister-forming effects (Michel *et al.*, 1962). However, it retains some of the anticholinesterase activity of VX.

All of the parent nerve agents are highly toxic to aquatic organisms (Munro *et al.*, 1999). Acute ecotoxicity information was found only for the degradation product, MPA. The 48- to 96-h LC_{50} values of several thousand mg/l for daphnids and fish indicates that MPA has low environmental toxicity.

In his ecotoxicological investigation of the Skagerrak area off the Swedish coast, Granbom (1996) reported on the placement of cages of crabs and mussels near scuttled ships. The organisms were checked for decreases in the enzyme activity of acetylcholinesterase. After a 2–3-week exposure, the retrieved animals appeared healthy and no statistically significant effect on cholinesterase activity was observed.

Table 4. Physical properties of agent VX degradation products[a]

Compound	Water solubility (mg/l)	log K_{ow}	log K_{oc}	pK_a(25°C)[b]	Vapor pressure (mmHg)[b]
Ethyl methylphosphonic acid (EMPA)	1.8×10^5	−1.15	0.75	2.00, 2.76[c]	3.6×10^{-4}
S-(2-Diisopropylaminoethyl) methylphosphonothioic acid (EA 2192)	Infinitely soluble	0.96	1.90	11.05	ND
bis(2-Diisopropylaminoethyl) sulfide	1.2	4.47	3.81	ND	2.7×10^{-7}
bis(2-Diisopropylaminoethyl) disulfide	9.5	3.48	3.28	ND	5.9×10^{-9}
Ethyl methylphosphonothioic acid	1.1×10^3	1.26	2.06	1.85	4.3×10^{-2}
Diisopropylaminoethanol	1.5×10^3	1.08	1.96	10.08[d]	1.8
Methylphosphonic acid (MPA)	$> 1.0 \times 10^6$	−2.28	0.15	2.38	2×10^{-6}[e]
Diethyl dimethylpyrophosphonate	$> 1.0 \times 10^6$	−2.12	0.23	ND	ND

Notes:
[a] Modified from Small (1984), except where otherwise indicated.
[b] ND, no data.
[c] Bossle *et al*. (1983).
[d] Estimated (HSDB, 2005).
[e] Howard and Meylan (1997).

DEGRADATION PRODUCT GUIDELINES FOR RESTORATION

Provisional RfDs and RfCs for 25 degradation products, impurities and stabilizers (Bausum, 1998) are listed in Table 5. These values can be compared with the estimated oral RfD for VX of 0.0006 μg/kg/day (Opresko *et al.*, 1998). Where data were unavailable for a degradation product, but the mechanism of action for the chemical was the same or predicted to be the same as that of VX, the value for VX was assigned. Thus, although both EA 2192 and bis(*S*, *S*-(diisopropylaminoethyl) methylphosphonodithiolate are less toxic than VX, the conservative value of 0.0006 μg/kg/day value was proposed. Health-based soil screening levels for the parent compound range from 0.042 to 0.047 mg/kg of soil (US Army CHPPM, 1999).

GA (Tabun; ethyl *N,N*-dimethylphosphoroamidocyanidate)

ENVIRONMENTAL DEGRADATION

Agent GA, with a vapor pressure of 0.037 mmHg, is more volatile than VX and will evaporate, but no data on concentration or fate in the atmosphere were located. GA dissolves rapidly and is subject to hydrolysis, with hydrolysis to *O*-ethyl *N*,*N*-dimethylamido phosphoric acid and hydrogen cyanide as the principal pathway under neutral environmental conditions (Figure 3). The initial reaction is fairly rapid; hydrolysis of *O*-ethyl *N*,*N*-dimethylamido phosphoric acid to dimethyl phosphoramidate and then phosphoric acid is much slower. Although this latter pathway predominates under neutral and basic conditions, phosphorocyanidate may also be formed from dimethyl phosphoramidate. Under acidic conditions, hydrolysis to ethylphosphoryl cyanidate and dimethylamine occurs. The final phosphorus product by all pathways is phosphoric acid. Although theoretically possible, there is little likelihood of formation of a detectable amount of MPA from GA (Sanches *et al.*, 1993).

Potential GA degradation products and impurities are listed in the Appendix, Table A4. Impurities may account for up to 28% of the volatile organic content of munitions-grade GA. The principal impurity is diethyl dimethylphosphoramidate, up to 12% of the munition fill (D'Agostino *et al.*, 1985,1989,2003; D'Agostino and Provost 1992a,b). Many of the phosphorus-containing products are likely to be degraded to phosphoric acids. Unique contaminants and impurities are generally present in trace amounts and would be of minimal environmental concern. As noted in Table A4 (see the Appendix), many of the degradation products and impurities were identified in soil below a leaking container of GA. Several species of bacteria, such as *Alteromonas* sp. and *Flavobacterium* sp., are capable of degrading G-type agents (Cheng *et al.*, 1999).

DEGRADATION PRODUCT TOXICITY

With the exception of the major impurity, diethyl dimethylphosphoramidate, no data characterizing toxicity of major degradation products and impurities were located. An intramuscular LD_{50} of 440 mg/kg in the mouse (Grechkin *et al.*, 1977) indicates a moderate to low order of toxicity for this impurity. Other degradation products, such as triethyl phosphate, have been well characterized in the past. No data were located on the toxicity of unique degradation products or impurities to aquatic organisms.

DEGRADATION PRODUCT GUIDELINES FOR RESTORATION

No exposure guidelines for Agent GA degradation products were found. The estimated oral reference dose for GA for humans is 0.04 μg/kg/day (Opresko *et al.*, 1998). Estimated health-based soil screening values for residential soil range from 1.2 to 3.1 mg/kg of soil (US Army CHPPM, 1999).

GB (sarin; isopropyl methylphosphonofluoridate)

ENVIRONMENTAL DEGRADATION

Agent GB is considered nonpersistent in the environment, as it is volatile, soluble in water and subject to acidic and basic hydrolysis. GB is the most volatile of the standard-threat G-agents, with an

Table 5. Estimated reference doses (RfDs)[a] and reference concentrations (RfCs)[a] for agent VX degradation and related products[b]

Degradation product (formula; CAS number)[c]	RfD (μg/kg/day)	RfC (μg/m³)
Ethyl methylphosphonic acid (EMPA) ($C_3H_9PO_3$; 1832-53-7)	28	34
Diisopropyl ethyl mercaptoamine ($C_8H_{19}NS$; 5842-07-9)	3.8	4.6
S-(Diisopropylaminoethyl) methylphosphonothioic acid (EA2192) ($C_9H_{22}NPO_2S$; 73207-98-4)	0.0006	0.0007
bis(2-Diisopropylaminoethyl) sulfide ($C_{16}H_{36}N_2S$; 110501-56-9)	8.6	10.3
bis(2-Diisopropylaminoethyl) disulfide ($C_{16}H_{36}N_2S_2$; 65332-44-7)	6.6	7.9
O-Ethyl methylphosphonothioic acid ($C_3H_9O_2PS$; 18005-40-8)	7	8.5
2-Diisopropylaminoethanol ($C_8H_{19}NO$; 96-80-0)	8.4	10
Methylphosphonic acid (MPA) (CH_5O_3P; 993-13-5)	20	24
bis(*S*, *S*-(2-Diisopropylaminoethyl) methylphosphonodithiolate ($C_{17}H_{39}N_2PS_2$; 169493-13-4)	0.0006	0.0007
2-(Diisopropylamino) ethyl ethyl sulfide ($C_{10}H_{23}NS$; NA)	8.6	10.3
Diethyl methylphosphonate ($C_5H_{13}O_3P$; 683-08-9)	29	35
1,2-bis(Ethyl methylphosphonothiolo) ethane (NA; NA)	5.0×10^{-2}	6.0×10^{-2}
O-(2-Diisopropylaminoethyl) *O*′-ethyl methylphosphonate ($C_{11}H_{26}NO_2P$; 71840-26-1)	0.025	0.03
O,*O*-Diethyl *P*, *P*′-dimethyldiphosphonothionate (NA; NA)	0.05	0.06
O,*O*-Diethyl methylphosphonothioate ($C_5H_{13}O_2PS$; 6996-81-2)	12	13
O-Ethyl methylethylphosphinate (NA; NA)	14	17
Diethyl dimethylpyrophosphonate ($C_6H_{16}O_5P$; 32288-17-8)	0.05	0.06
O,*S*-Diethyl methylphosphonothioate ($C_5H_{13}O_2PS$; 2511-10-6)	0.017	0.02
Diisopropylamine ($C_6H_{15}N$; 108-18-9)	4.3	5.1
N,*N*-Diisopropylmethylamine ($C_7H_{17}N$; 10342-97-9)	0.56	2
N,*N*-Diisopropylethylamine ($C_8H_{19}N$; 7087-68-5)	0.56	2

(*cont.*)

Table 5. *(cont.)*

Degradation product (formula; CAS number)[c]	RfD (μg/kg/day)	RfC (μg/m^3)
Diisopropyl carbodiimide ($C_7H_{14}N_2$; 693-13-0)	0.25	0.3
Dicyclohexyl carbodiimide ($C_{13}H_{22}N_2$; 538-75-0)	0.25	0.3
1,3-Diisopropylurea ($C_7H_{16}N_2O$; 4128-37-4)	0.87	1.04
1,3-Dicyclohexylurea ($C_{13}H_{24}N_2O$; 2387-23-7)	6.7	8

Notes:
[a] RfD estimates are based on data from structurally related chemicals or from QSAR estimates using the TOPKAT computer program. RfC values were calculated from RfD values.
[b] Sources: Bausum (1998), Bausum *et al.* (1999).
[c] NA, not available.

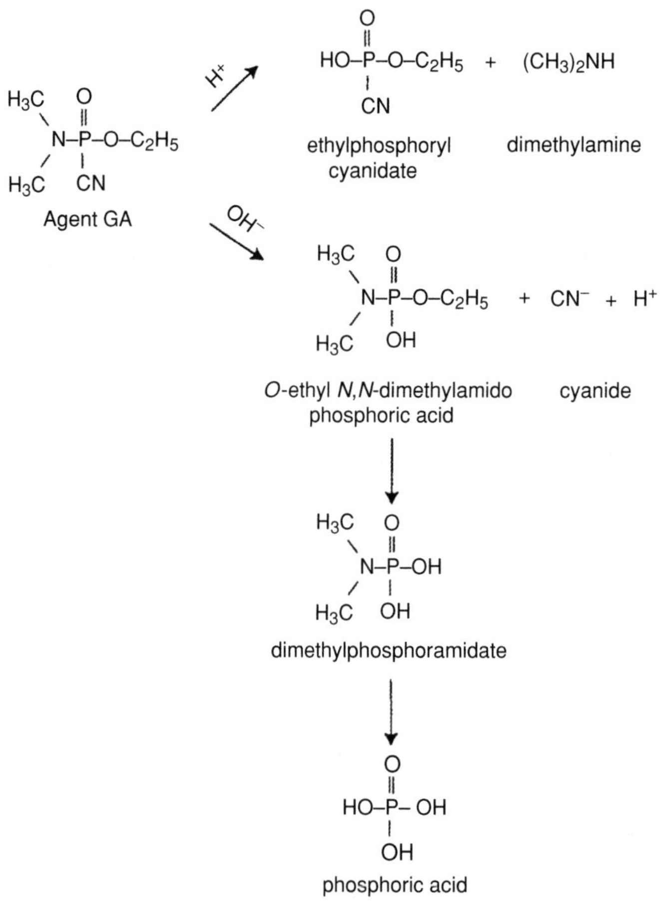

Figure 3. Primary hydrolysis pathways of agent GA in the environment

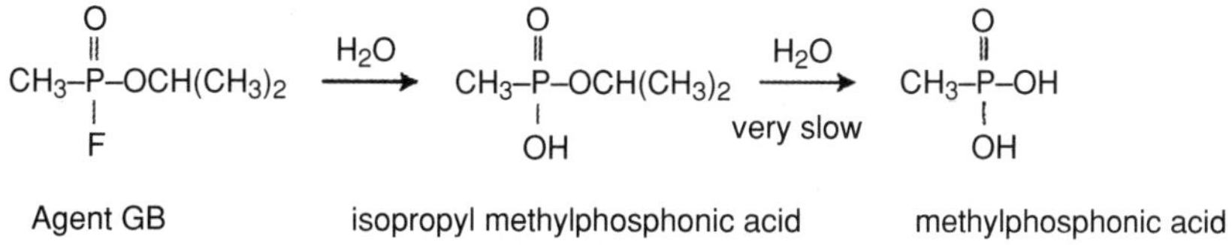

Figure 4. Primary hydrolysis pathway of agent GB in the environment

evaporation rate similar to that of water (Rosenblatt *et al.*, 1995). Small (1984) used a surface-deposition model to calculate a volatilization half-life of 7.7 h for GB. No information on fate in the atmosphere was located, although Kingery and Allen (1995) state that nerve agents can be degraded by photolysis and/or radical oxidation. The low calculated Henry's law constant of 5.4×10^{-7} atm m^3/mol, based on high water solubility, indicates slow to essentially no volatilization from water.

The fate mechanisms of GB in soil includes hydrolysis, evaporation and leaching; the phosphonic acid hydrolysis products are subject to biodegradation. Depending on temperature, $> 90\%$ of GB added to soil may be lost in 5 days (Small, 1984). As shown by field studies under snow in Norway, low temperatures would increase persistence. In this setting, approximately 55% was removed by evaporation within 5 h and 15% was removed by hydrolysis. Hydrolysis products and several impurities were present up to four weeks later (NMFA, 1982–1983; Johnsen and Blanch, 1984). Hydrolytic half-lives are highly dependent upon pH and temperature. Hydrolytic half-lives are shorter in acidic and basic solutions than at a neutral pH. At 20°C and the pH of natural waters where the half-life is a maximum, estimates of the half-life range from 461 h (pH 6.5) to 46 h (pH 7.5) (Clark, 1989). At 25°C, the half-life ranges from 237 h (pH 6.5) to 24 h (pH 7.5). A half-life of 8300 h at 0°C and a pH of 6.5 was estimated. Durst *et al.*, (1988) have documented a half-life of 3 s at a pH of 12.

GB hydrolyzes first through the loss of fluoride, producing isopropyl methylphosphonic acid (IMPA) and hydrofluoric acid, and second, more slowly through the loss of the isopropanol to produce MPA (Rosenblatt *et al.*, 1995; MacNaughton and Brewer, 1994; Kingery and Allen, 1995) (Figure 4). The same products are produced under acidic conditions. Alkaline hydrolysis is predicted to result in isopropanol, methylfluorophosphonic acid and, with the loss of fluoride, MPA (Clark, 1989). This latter pathway has not been confirmed in other studies. IMPA is subject to microbial degradation by acclimated bacteria (Zhang *et al.*, 1999).

Breakdown of GB results in only a few degradation products (see the Appendix, Table A5) and these are predicted to be relatively nontoxic (Rosenblatt *et al.*, 1995). The hydrolysis products are acids, and their presence increases the rate of hydrolysis. The rate of hydrolysis under natural conditions is accelerated by the presence of certain hydroxy cations in solution, such as $Cu(OH)^+$, $Ca(OH)^+$ and $Mn(OH)^+$ (Epstein and Rosenblatt, 1958). Metal cations, such as copper and manganese, in seawater also increase the rate of hydrolysis (Epstein, 1974).

GB manufactured in the USA was at least 93% pure. Impurities may include diisopropyl methylphosphonate (DIMP), a stable byproduct of manufacture, and methylphosphonic difluoride. DIMP is usually present at 2–3% in isopropyl methylphosphonate waste and is present in sampling wells, both on and off the RMA (Rosenblatt *et al.*, 1995; Aulerich *et al.*, 1979). In 1974, DIMP concentrations of 0.5 μg/l (the limit of detection) to 44 000 μg/l were found in the groundwater near the Arsenal (Robson, 1981).

Stabilizers and byproducts of GB manufacture may also be present at contaminated sites. The stabilizers N, N'-diisopropyl carbodiimide (1.5%) and/or tributylamine were added to weapons-grade GB (Epstein *et al.*, 1977; Rosenblatt *et al.*, 1995). N, N'-diisopropyl carbodiimide breaks down into N, N'-diisopropylurea as noted in Table A5 in the Appendix.

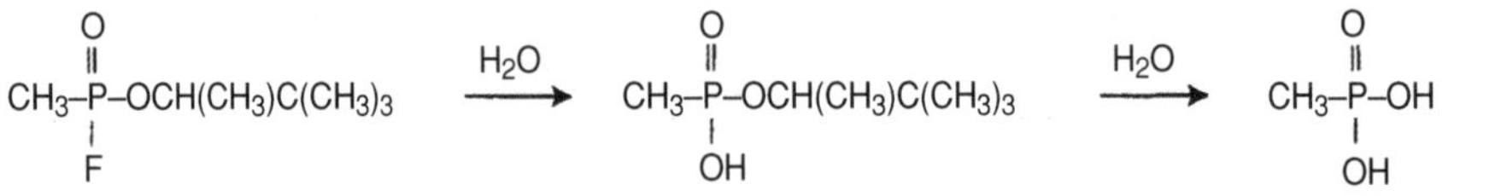

Agent GD — pinacolyl methylphosphonic acid — methylphosphonic acid

Figure 5. Primary hydrolysis pathway of agent GD in the environment

DEGRADATION PRODUCT TOXICITY

The degradation product IMPA is of low acute and subchronic toxicity (Munro *et al.*, 1999). Mammalian oral LD_{50} values are ~5600 mg/kg. In a rodent study, 300 ppm in drinking water for 90 days was a 'no-effect level' (Mecler, 1981). The RfD for IMPA is 100 μg/kg/day (US EPA, 2005b). DIMP, which is 90% metabolized to IMPA within 24 h in mammalian species, has demonstrated low acute and negligible chronic and reproductive toxicity (ATSDR, 2005; Hart, 1976,1980; Bucci *et al.*, 1997). DIMP has a slight phytotoxic effect, but is of low toxicity to birds (Aulerich *et al.*, 1979) and aquatic organisms including fish and daphnids (Bentley *et al.*, 1976).

DEGRADATION PRODUCT GUIDELINES FOR RESTORATION

Of the several GB degradation products, the US EPA Integrated Risk Information system (IRIS) has developed a RfD for only one, DIMP (RfD of 80 μg/kg/day). Calculated RfD estimates are also available for MPA (estimated RfD of 20 μg/kg/day) and IMPA (estimated RfD of 100 μg/kg/day). Calculated RfC estimates have also been generated for IMPA (estimated RfC of 110 $\mu g/m^3$) and MPA (estimated RfC of 24 $\mu g/m^3$) (see Table 1). No other environmental restoration guidelines for GB degradation products have been located. The estimated oral RfD for GB is 0.02 μg/kg/day (Opresko *et al.*, 1998,2001). Health-based environmental screening levels for GB for residential soil range from 0.5 mg/kg (vapor inhalation only from soil) to 1.6 mg/kg (ingestion) of soil (US Army CHPPM, 1999).

GD (soman; pinacolyl methylphosphonofluoridate)

ENVIRONMENTAL DEGRADATION

Agent GD is less volatile than GB, evaporating at one-fourth the rate of water (Rosenblatt *et al.*, 1995). Added thickeners, such as methyl methacrylate, retard evaporation. The volatility of GD, intermediate between that of GA and GB, is high enough to make it a vapor hazard. Some volatilization from water may occur.

Like the other G-agents, GD is subject to hydrolysis (Yang *et al.*, 1992). GD hydrolyzes first through the loss of fluoride, and second, more slowly through the loss of the alkoxy group (Figure 5). Thus, the primary hydrolysis product is pinacolyl methylphosphonic acid which slowly hydrolyzes, with the release of pinacolyl alcohol, to MPA (Clark, 1989; MacNaughton and Brewer, 1994; Kingery and Allen, 1995). GD stored at pH 6 for 8 weeks had a pinacolyl methylphosphonic acid/MPA ratio of 250 (Hambrook *et al.*, 1971). Qualitatively, the hydrolysis of GD is similar to that of GA; however, the reaction rate is 5 times slower than that of GA, and GD has an estimated half-life of about 60 h at pH 6 and 25°C (Hambrook *et al.*, 1971). The reaction is both acid- and base-catalyzed, resulting in a hydrolysis curve similar to that of GA (Clark, 1989). At a pH >10, hydrolysis to pinacolyl methylphosphonic acid occurs within a few minutes (Yang *et al.*, 1992). Because an acid is produced, the pH will decrease, lessening the rate of hydrolysis. However, this effect would be small in the environment. The hydrolysis products and impurities are listed in the Appendix, Table A6.

Under actual field conditions, the measured decay of GD is rapid (McGuire *et al.*, 1993). The Chemical and Biological Defence Establishment at Porton Down (Salisbury, UK) conducted outdoor experiments in which GD was deposited on 1 m^2 plots of soil at an areal concentration of 10 g/m^2. Samples were collected immediately (day '0') and during the following 3 days. The initial decomposition was from hydrolysis of the P–F bond as evidenced in the day '0' samples. The phosphonate partial ester and MPA peaked in the day '1' samples and declined subsequently. Pinacolyl alcohol was also detected.

Short-term field trials have been performed with GD droplets applied to silica sand at a contamination density of 5.21 g/m^2 and laboratory experiments with GD liquid applied to soil (US Air Force, 2003; Murdock *et al.*, 2004). These studies have demonstrated potential for GD 'off-gassing' due to volume displacement following actual and simulated rain events within hours after initial agent application. With repeated and simulated rain events, GD off-gassing from soil declined to nondetectable concentrations at 73.3 h post-application (following five simulated rain events).

DEGRADATION PRODUCT TOXICITY

Other than MPA (discussed under VX), no data on the toxicity of major degradation products were found. The thickener, methyl methacrylate, is an irritant, but has a low acute oral and inhalation toxicity (Thiokol/Ventron, 1980; Deichman, 1941; Spealman *et al.*, 1945; Klimkina *et al.*, 1976). Although not teratogenic, exposure of pregnant rats to high concentrations, i.e. 110 mg/l for approximately an h/day during days 6 through 15 of gestation, caused embryo- and feto-toxicity (Nicholas *et al.*, 1979; Luo *et al.*, 1986). There are no data available on the toxicity of degradation products to aquatic organisms.

DEGRADATION PRODUCT GUIDELINES FOR RESTORATION

Other than for MPA (see discussion under VX), no guidelines for environmental cleanup of degradation products are available. The estimated oral RfD for GD is 0.004 μg/kg/day (Opresko *et al.*, 1998,2001). Health-based environmental screening levels for GD for residential soil range from 0.22 to 0.31 mg/kg of soil (US Army CHPPM, 1999).

SUMMARY

Under environmental conditions, CWAs (vesicant agents, sulfur mustard (H, HD and HT) and lewisite (L); nerve agents, GA, GB, GD and VX) can undergo multiple-degradation processes such as hydrolysis, oxidation, dehydration and photolysis. These baseline degradation reactions can vary in rate and completeness, depending upon reaction temperature and pH, as well as the presence of free radicals and catalysts. Knowledge of these baseline reaction parameters has formed the basis for many modern decontamination procedures.

Examination of degradation product data evaluated in this analysis indicates that, in most cases and given sufficient time, agent dispersed (or leaked or spilled) in humid air, moist soil or in marine or freshwater bodies would result in eventual degradation and reaction yields of less toxic compounds (when compared to parent agent properties). Notable exceptions are lewisite oxide and EA 2192, a toxic product of VX hydrolysis. Degradation reactions exhibit reaction rates and half-times governed by site-specific physical and chemical conditions, including temperature and pH.

Many CWAs are sparingly soluble; thus, concentrations of soluble CWAs are not likely to accumulate or be distributed in water. The more soluble compounds, such as GA, degrade to non-toxic compounds. Sulfur mustard in bulk form would persist and be toxic (as the agent), but its low solubility would retard formation and dispersal of toxic HD concentrations in fresh or seawater. At cold seabed temperatures less than or equal to the HD freezing point of 13–14°C, ocean-buried sulfur mustard would be frozen, slow to solubilize, and thus exhibit a theoretically long, but finite half-life. Sulfur mustard agent is also known to be persistent for extended periods when the agent becomes encapsulated by the naturally occurring polymeric products of surface hydrolysis. Such encapsulated HD has been found associated with continuous sources of sulfur mustard, such as buried (leaking) bulk containers.

Low concentrations of the G-agents, VX, or lewisite in the environment would generally degrade within a period of days or weeks to months. Some degradation products may persist in the environment for extended periods. Persistent degradation products include TDG from HD, lewisite oxide and elemental arsenic from lewisite and EA 2192, EMPA and MPA from VX. MPA is also a degradate of both GB and GD. Of these degradation products, only EA 2192 and lewisite oxide possess high mammalian toxicity. Little is known of encapsulated HD, which retains vesicant properties and persists in the environment. Thiodiglycol is practically nontoxic. Methylphosphonic acid and its derivative, EMPA (and by extension IMPA), are also of low oral toxicity. These key degradation products, along with associated chronic reference values, are summarized in Table 1.

All products known to result from degradation of the evaluated chemical warfare agents listed above are summarized in the Appendix to this chapter (Tables A1–A6). These extensive tables are provided for completeness and should not be interpreted or applied as checklists of all products for which monitoring is required prior to site clearance. It bears repeating that the degradation reactions and yields are determined by site-specific conditions, and not every degradation product identified in the appendix will be present at any given CWA release site.

The current evaluation concludes that toxic hazards associated with a potential CWA release, regardless of origin (e.g. controlled storage, transport, demilitarization activity, deliberate terrorist event) are primarily a consequence of exposure to the toxic parent compound, and *not* to a degradation product. Lewisite oxide and EA 2192 are exceptions, but the degradation reaction yields for these products is not 1:1: thus, concentrations of these specific degradation products present at a release site would be less than the parent agent concentration contained in the release. It is acknowledged that munitions containing quantities of highly toxic CWA can be redistributed by winds and tides to sites distant from the original disposal area; any agent existing as munition fill and thus protected from the degradation reactions described above should be managed and treated as military-grade parent agent.

ACKNOWLEDGEMENTS

This work was prepared under two Interagency Agreements (IAGs): IAG No. 2134-K006-A1 with the US Army Environmental Center of Aberdeen Proving Ground, MD, and IAG No. 2207-M135-A1 with the US Army Center for Health Promotion and Preventive Medicine of Aberdeen Proving Ground, MD (the latter as support to the Chemical Stockpile Emergency Preparedness Program). The Oak Ridge National Laboratory (ORNL) is managed and operated by UT-Battelle, LLC., for the US Department of Energy under contract DE-AC05-00OR22725.

The views expressed in this paper do not necessarily represent official Federal agency position or policy. Mention of trade names or commercial products does not constitute endorsement or recommendation of use.

APPENDIX

Agent degradation products

Details of the degradation products, impurities, stabilizers, etc. for the various evaluated chemical warfare agents discussed in this chapter are given in Tables A1–A6.

Table A1. Degradation products and impurities of sulfur mustard agent (HD)[a,b]

Name/Synonym	Formula	CAS number[c]	Source
Hemi-sulfur mustard (CH) Mustard chlorohydrin 2-Hydroxyethyl 2-chloroethyl sulfide 2-((2-Chloroethyl)thio)ethanol	C_4H_9ClOS	693-30-1	Hydrolysis of sulfur mustard
Thiodiglycol (TDG) 2,2′-Thiobisethanol 2,2′-Thiodiethanol Thiodiethylene glycol	$C_4H_{10}O_2S$	111-48-8	Hydrolysis of sulfur mustard
bis(2-Hydroxyethyl)-2-(2-chloroethylthio) ethyl-sulfonium chloride Sulfur mustard–thiodiglycol aggregate HD–TDG or H1TG	$C_8H_{18}ClO_2S_2{\cdot}Cl$	64036-91-5	Hydrolysis of sulfur mustard
bis(2-Hydroxyethyl)-2-(2-hydroxyethylthio)-ethyl sulfide Hemi-mustard–thiodiglycol aggregate CH–TDG	$C_8H_{19}S_2O_3$	64036-92-6	Hydrolysis of sulfur mustard
bis-2(bis(2-Hydroxyethyl)-sulfonium ethyl)-sulfide dichloride Sulfur mustard–thiodiglycol–thiodiglycol aggregate HD–TDG–TDG or H2TG	$C_{12}H_{28}O_4S_3{\cdot}2Cl$	64036-79-9	Hydrolysis of sulfur mustard
Mustard sulfoxide 1,1′-Sulfinylbis(2-chloroethane) bis(2-Chloroethyl) sulfoxide	$C_4H_8Cl_2OS$	5819-08-9	Oxidation of sulfur mustard
Mustard sulfone 1,1′-Sulfonylbis(2-chloroethane) bis(2-Chloroethyl) sulfone	$C_4H_8Cl_2O_2S$	471-03-4	Oxidation of sulfur mustard
2-Chloroethyl vinyl sulfide 2-Chloroethylthio ethene	C_4H_7ClS	81142-02-1	Dechlorination of sulfur mustard
Divinyl sulfide Ethylthioethene	C_4H_6S	627-51-0	Dechlorination of sulfur mustard
2-Chloroethyl vinyl sulfoxide	C_4H_7ClOS	40709-82-8	Dechlorination of sulfur mustard
Vinyl sulfoxide sulfone Divinyl sulfoxide 1,1-Sulfinylbis ethene	C_4H_6OS	1115-15-7	Dehydrochlorination of mustard
2-Chloroethyl vinyl sulfone	$C_4H_7ClO_2S$	7327-58-4	Dechlorination of mustard sulfone
Divinyl sulfone	$C_4H_6O_2S$	77-77-0	Dechlorination of mustard sulfone
1,4-Dithiane Diethylene disulfide	$C_4H_8S_2$	505-29-3	Impurity, thermal decomposition, dechlorination of sulfur mustard, present in tonne containers
1,4-Oxathiane 1,4-Thioxane	C_4H_8OS	15980-15-1	Dechlorination of hemi-mustard, present in tonne containers

(cont.)

Table A1. *(cont.)*

Name/Synonym	Formula	CAS number[c]	Source
bis(2-(2-Chloroethylthio)ethyl) ether Agent T	$C_8H_{16}Cl_2OS_2$	63918-89-8	Impurity of sulfur mustard
1,2-bis(2-Chloroethylthio) ethane Compound Q Sesqui-mustard	$C_6H_{12}Cl_2S_2$	3563-36-8	Impurity, present in tonne containers
1,8-Dichloro-3-oxa-6-thiaoctane	$C_6H_{12}Cl_2OS$	NA	Impurity
Tetrachloroethylene	C_2Cl_4	127-18-4	Present in tonne containers
Hexachloroethane	C_2Cl_6	67-72-1	Present in tonne containers
2-Chloroethyl 3-chloropropyl sulfide	$C_5H_{10}Cl_2S$	71784-01-5	Present in tonne containers
2-Chloroethyl 4-chlorobutyl sulfide	$C_6H_{12}Cl_2S$	114811-35-7	Present in tonne containers
2-Chloroethyl (2-chloroethoxy)ethyl sulfide	$C_6H_{12}Cl_2OS$	114811-38-0	Present in tonne containers
bis(2-Chloroethyl) disulfide HD disulfide	$C_4H_8Cl_2S_2$	1002-41-1	Impurity, present in tonne containers
bis(2-Chloropropyl) sulfide	$C_6H_{12}Cl_2S$	22535-54-2	Impurity, present in tonne containers
1,2-Dichloroethane	$C_2H_4Cl_2$	107-06-2	Impurity, present in tonne containers
1,2,5-Trithiepane	$C_4H_8S_3$	6576-93-8	Present in tonne containers
1,2,3,4-Tetrathiane	$C_2H_4S_4$	NA	Present in H
1,1,2,2-Tetrachloroethane	$C_2H_2Cl_4$	79-34-5	Present in tonne containers
bis(2-Chloroethyl) trisulfide HD trisulfide	$C_4H_8Cl_2S_3$	19149-77-0	Impurity, present in tonne containers
1,2,3-Trithiolane	$C_2H_4S_3$	NA	Impurity, present in tonne containers
HD tetrasulfide	$C_4H_8Cl_2S_4$	NA	Present in tonne containers
2-Methyl 1-propene Methylpropene Isobutylene	C_4H_8	115-11-7	Present in tonne containers
Thiirane Ethylene sulfide	C_2H_4S	420-12-2	Present in tonne containers
2-Chlorobutane *sec*-Butyl chloride	C_4H_9Cl	78-86-4	Present in tonne containers
Trichloroethylene	C_2HCl_3	79-01-6	Present in tonne containers
Q Sulfonium 1-(2-Chloroethyl) 1,4-dithanium chloride 1-(2-Chloroethyl)-1-thiona-4-thiane chloride	$C_6H_{12}ClS_2 \cdot Cl$	30843-67-5	Residue in tonne containers

Notes:

[a] For any given release of sulfur mustard agent, degradation products and yield are determined by site-specific environmental conditions. Not all products or impurities listed in this table are associated with every HD release.

[b] Sources: Rosenblatt *et al.* (1995); NRC (1996); Amr *et al.* (1996); D'Agostino and Provost (1988); MacNaughton and Brewer (1994); Kingery and Allen (1995); Yang *et al.* (1992).

[c] NA, not available.

Table A2. Degradation products and impurities of lewisite (L)[a]

Name/synonym	Formula	CAS number[b]	Source
2-Chlorovinyl arsonous acid (CVA) 2-Chlorovinyl arsenous acid 2-Chloroethenyl arsonous acid Dihydroxy(2-Chlorovinyl)arsine 2-Chloroethenyl dihydroxyarsine	$C_2H_4AsClO_2$	85090-33-1	Hydrolysis of Lewisite
2-Chlorovinyl arsenous oxide 2-Chlorovinyl arsenic oxide 2-Chlorovinyl arsine oxide 2-Chloroethenyl arsinic oxide Lewisite oxide	C_2H_2AsClO	3088-37-7	Hydrolysis or dehydration of 2-chlorovinyl arsonous acid
Lewisite oxide polymer	$(C_2H_2AsClO)_n$	NA	Polymer of lewisite oxide
2-Chlorovinyl arsonic acid 2-Chloroethenyl arsonic acid	$C_2H_4AsClO_3$	64038-44-4	Oxidation of lewisite oxide
bis(2-Chlorovinyl)chloroarsine Lewisite 2	$C_4H_4AsCl_3$	40334-69-8	Impurity
tris(2-Chlorovinyl)arsine Lewisite 3	$C_6H_6AsCl_3$	40334-70-1	Impurity
Arsenic trichloride	$AsCl_3$	7784-34-1	Impurity

Notes:
[a] Sources: Rosenblatt *et al.* (1975); Franke (1982); Clark (1989); Goldman and Dacre (1989); Sanches *et al.* (1993).
[b] NA, not available.

Table A3. Degradation products, impurities, and stabilizers of agent VX[a]

Name/synonym	Formula[b]	CAS number[b]	Source
Ethyl methylphosphonic acid (EMPA) Ethyl hydrogen methylphosphonate	$C_3H_9O_3P$	1832-53-7	Hydrolysis of VX
Diisopropyl ethyl mercaptoamine Diisopropylamino ethyl mercaptan (DESH, DIAEM) Diisopropylamino ethylthiolate 2-(Diisopropylamino) ethane thiol Thiolamine	$C_8H_{19}NS$	5842-07-9	Hydrolysis of VX
Diisopropylaminoethyl methyl thiolophosphonate *S*-(2-Diisopropylaminoethyl) methylphosphonothioic acid *S*-(2-Diisopropylaminoethyl) methylphosphonothioate EA 2192	$C_9H_{22}NO_2PS$	73207-98-4	Hydrolysis of VX
bis(2-Diisopropylaminoethyl) sulfide ($(DE_2)S$) *N*,*N*′-Thiodi-2,1-ethanediyl)bis(*N*-(1-methylethyl)-2-propanamine VX sulfide	$C_{16}H_{36}N_2S$	110501-56-9	Reaction of DESH with ethyleneimmonium ion
bis(2-Diisopropylaminoethyl) disulfide ($(DES)_2$) *N*,*N*′(Dithio-2,1-ethanediyl)bis(*N*-(1-methylethyl)-2-propanamine VX disulfide EA 4196	$C_{16}H_{36}N_2S_2$	65332-44-7	Dimerization of DESH, air oxidation of DESH
Ethyl methylphosphonothioic acid *O*-Ethyl methylphosphonothioate *O*-Ethyl methylthiophosphonate	$C_3H_9O_2PS$	18005-40-8	Hydrolysis of VX, VX precursor for some processes
2-Diisopropylaminoethanol *N*,*N*-Diisopropylamino ethanol	$C_8H_{19}NO$	96-80-0	Hydrolysis of VX
Methylphosphonic acid (MPA)	CH_5O_3P	993-13-5	Hydrolysis of EMPA
bis(*S*,*S*-(2-Diisopropylaminoethyl)) methylphosphonodithiolate *S*,*S*-bis(2-(bis(1-Methylethyl)amino)ethyl) methylphosphono-dithioic acid 'bis'	$C_{17}H_{39}N_2OPS_2$	169493-13-4	Impurity formed during manufacture
Diisopropylaminoethyl sulfide	$C_8H_{19}NS$	NA	Present in tonne containers
2-(Diisopropylamino) ethyl ethyl sulfide	$C_{10}H_{23}NS$	NA	Present in tonne containers
Diethyl methylphosphonate *O*,*O*-Diethyl methylphosphonate	$C_5H_{13}O_3P$	683-08-9	Present in tonne containers, degradation product and impurity
1-(2-Chloroethyl-1,4 dithianium chloride[c] 1-(2-Chloroethyl)-1-thiona-4-thiane chloride Q-sulfonium	$C_6H_{12}ClS_2\$Cl$	30843-67-5	Present in tonne containers

(*cont.*)

Table A3. (*cont.*)

Name/synonym	Formula[b]	CAS number[b]	Source
1,2-bis(Ethyl methylphosphonothiolo) ethane	$C_8H_{18}O_4P_2S_2$	NA	Present in tonne containers
O-(2-Diisopropylaminoethyl) *O′*-ethyl methylphosphonate 2-(bis(1-Methylethyl)amino)ethyl methylphosphonic acid	$C_{11}H_{26}NO_3P$	71840-26-1	Present in tonne containers
O-Ethyl *S*-2-(diisopropylamino)ethyl methylphosphonothioate	$C_{11}H_{26}NO_2PS$	50782-69-9	Present in tonne containers and stored glass containers
O,*O*-Diethyl dimethylpyrophosphonothioate	NA	NA	Present in tonne containers
O,*O*-Diethyl methylphosphonothioate	$C_5H_{13}O_2PS$	6996-81-2	Present in tonne containers
O-Ethyl methylethylphosphinate	NA	NA	Present in tonne containers
Diethyl dimethyl pyrophosphonate (pyro) Diethyl dimethyldiphosphonate	$C_6H_{16}O_5P_2$	32288-17-8	Impurity, anhydride of EMPA, present in tonne containers
O,*S*-Diethyl methylphosphonothioate	$C_5H_{13}O_2PS$	2511-10-6	Impurity, present in tonne containers
N,*N*-Diisopropylmethylamine	$C_7H_{17}N$	10342-97-9	Present in tonne containers
N,*N*-Diisopropylethylamine	$C_8H_{19}N$	7087-68-5	Present in tonne containers
2-Diisopropylamino ethyl vinyl sulfide	NA	NA	Impurity, present in tonne containers
O,*O′*-Diethyl *P*,*P′*dimethyldiphosphonothioate	$C_6H_{14}O_4P_2S$	NA	Impurity, present in tonne containers
Diisopropylamine	$C_6H_{15}N$	108-18-9	Present in tonne containers
Diisopropyl carbodiimide	$C_7H_{14}N_2$	693-13-0	Stabilizer
Dicyclohexyl carbodiimide *N*,*N′*-Methanetetraylbiscyclohexaneamine	$C_{13}H_{22}N_2$	538-75-0	Stabilizer
1,3-Diisopropylurea *N*,*N′*-Diisopropylurea *N*,*N′bis*-(1-Methylethyl) urea	$C_7H_{16}N_2O$	4128-37-4	Hydrolysis of diisopropyl carbodiimide
N,*N′*-Dicyclohexylurea	$C_{13}H_{24}N_2O$	2387-23-7	Hydrolysis of dicyclohexyl carbodiimide
1,9-bis(Diisopropyl amino)-3,4,7-trithianonane	$C_{18}H_{40}N_2S_3$	110501-59-2	Present in stored glass containers

Notes:
[a] Sources: Clark (1989); Rosenblatt *et al.* (1995); NRC (1996); Amr *et al.* (1996); MacNaughton and Brewer (1994); Kingery and Allen (1995); Epstein *et al.* (1974); Rohrbaugh (1998); D'Agostino *et al.* (1987); Creasy *et al.* 1999.
[b] NA, not available.
[c] Isolated as chloride salt; CAS number of parent compound is 199982-97-3.

Table A4. Degradation products and impurities of agent GA[a]

Name/Synonym	Formula[b]	CAS number[b]	Source
O-Ethyl-*N*,*N*-dimethylamido phosphoric acid Ethyl-*N*,*N*-dimethyl phosphoramidate (EDPA) Ethyl hydrogen dimethylphosphoramidate	$C_4H_{12}NO_3P$	2632-86-2	Hydrolysis of GA, identified in soil[c]
Dimethylphosphoramidate *N*,*N*-Dimethylphosphoramidate Dimethyl phosphoramidic acid	$C_2H_8NO_3P$	33876-51-6	Hydrolysis of GA
Dimethylphosphoramide cyanidate Phosphoramidocyanidic acid	$C_3H_7N_2O_2P$	63917-41-9	Hydrolysis of GA
Phosphorocyanidate Phosphorisocyanatidous acid	CH_2NO_3P	23852-43-9	Hydrolysis of GA
Ethylphosphoryl cyanidate	$C_3H_6NO_3P$	117529-17-6	Hydrolysis of GA
Dimethylamine	C_2H_7N	124-40-3	Hydrolysis of GA
Ethyl phosphoric acid	$C_2H_7O_4P$	NA	Hydrolysis of GA
Phosphoric acid	H_3O_4P	7664-38-2	Hydrolysis of GA
O,*O*-Diethyl *N*,*N*-dimethylphosphoramidate Diethyl dimethylphosphoramidate	$C_6H_{16}NO_3P$	2404-03-7	Major impurity, identified in soil[c]
O-Ethyl bis(*N*,*N*-dimethyl)phosphordiamidate Ethyl tetramethylphosphorodiamidate	$C_6H_{17}N_2O_2P$	2404-65-1	Identified in soil[c]
bis(*N*,*N*-Dimethyl)phosphoramidocyanidate tetramethylphosphorodiamidic cyanide	$C_5H_{12}N_3OP$	14445-60-4	Impurity, identified in soil[c]
bis(Ethyl dimethylphosphoramidic) anhydride	NA	NA	Identified in soil[c]
Dimethyl phosphoric ethyl dimethyl-phosphoramidic anhydride	NA	NA	Identified in soil[c]
Ethyl dimethylphosphoramidic tetramethyl-phosphorodiamidic anhydride	NA	NA	Identified in soil[c]
bis(Ethyl dimethylamidophosphonyl) Dimethylamidophosphonate	NA	NA	Identified in soil[c]
O-Ethyl *N*,*N*-dimethyl phosphoramidic chloride	NA	2510-93-2	Present in sample
Triethyl phosphate	$C_6H_{15}O_4P$	78-40-0	Identified in soil[c]
Diethyl hydrogen phosphate	$C_4H_{11}O_4P$	NA	Identified in soil[c]
Ethyl dihydrogen phosphate	$C_2H_7O_4P$	NA	Identified in soil[c]
N,*N*-Dimethylphosphoramidic dichloride Dimethylaminophosphoryl dichloride	NA	683-85-2	Impurity, identified in soil[c]
Ethyl isopropyl dimethyl phosphoramidate	$C_7H_{18}NO_3P$	NA	Impurity, munitions-grade GA

Notes:
[a] Sources: MacNaughton and Brewer (1994); Sanches *et al.* (1993); D'Agostino *et al.* (1985,1989); D'Agostino and Provost (1992a,b); Creasy *et al.* (1997).
[b] NA, not available.
[c] Compounds identified in soil contaminated by a leaking container of GA (D'Agostino and Provost 1992a).

Table A5. Degradation products, impurities and stabilizers of agent GB[a]

Name/synonym	Formula	CAS number	Source
Isopropyl methylphosphonic acid (IMPA)	$C_4H_{11}PO_3$	1832-54-8	Hydrolysis of GB
Methylphosphonic acid (MPA)	$C_4H_{11}PO_3$	993-13-5	Hydrolysis of GB
Diisopropyl methylphosphonate (DIMP)	$C_7H_{17}PO_3$	1445-775-6	Impurity
Methylphosphonic difluoride	CH_3F_2OP	676-99-3	Potential impurity
Diisopropyl carbodiimide	$C_7H_{14}N_2$	676-99-3	Stabilizer
N,N′-Diisopropylurea	$C_7H_{16}N_2O$	4128-37-4	Hydrolysis of diisopropyl carbodiimide
Tributylamine	$C_{12}H_{27}N$	102-82-9	Stabilizer
Dibutylchloramine	$C_8H_{18}ClN$	999-33-7	Stabilizer

Note:
[a] Sources: Small (1984); Rosenblatt *et al.* (1995); MacNaughton and Brewer (1994); Sanches *et al.* (1993).

Table A6. Degradation products, impurities, and thickener of agent GD[a]

Name/synonym	Formula	CAS number	Source
Pinacolyl methylphosphonic acid	$C_7H_{17}O_3P$	616-52-4	Hydrolysis of GD
Methylphosphonic acid (MPA)	CH_5O_3P	993-13-5	Hydrolysis of GD
Pinacolyl alcohol	$C_6H_{14}O$	464-07-3	Hydrolysis of pinacolyl methylphosphonic acid
Dipinacolyl methylphosphonate	$C_{13}H_{29}O_3P$	7040-58-6	Impurity
Methyl pinacolyl methylphosphonate	$C_8H_{19}O_3P$	7040-59-7	Impurity
Methyl methylphosphonofluoridate	$C_2H_6FO_2P$	353-88-8	Impurity
Methyl methacrylate	$C_5H_8O_2$	80-62-6	Thickener

Note:
[a] Sources: Rosenblatt *et al.* (1995); MacNaughton and Brewer (1994); Kingery and Allen (1995); Sanches *et al.* (1993); Johnsen and Blanch (1984).

REFERENCES

Alexander SF (1947). Medical report of the Bari Harbor mustard casualties. *Milit Surg*, **101**, 1–17.

Amr AT, Cain TC, Cleaves DJ *et al.* (1996). Preliminary Risk Assessment of Alternative Technologies for Chemical Demilitarization, MTR 96W0000023. McLean, VA: Mitretek Systems.

Armstrong GC, Wells HB, Wilkes AE *et al.* (1928). *Comparative Test with Mustard Gas (HS) Lewisite (M1), Methyldicloroarsine (MD) and Methyldifluorarsine (MD2) in 75 mm Shell Fired Statically in Collaboration with Chemical Division.* EAMRD 95. Edgewood Arsenal, MD, USA: Department of the Army, Medical Research Division.

Assennato G, Ambrose F and Sivo D (1997). Possibili effetti a lungo termine sull'apparato esposizone ad iprite tra pescatori (Possible long-term effects on the respiratory tract of the sulfur mustard exposure among fisherman). *Medic Lavoro (Milan)*, **88**, 148–154.

ATSDR (Agency for Toxic Substances and Disease Registry) (2003). *Toxicological Profile for Sulfur Mustard.* Atlanta, GA: Agency for Toxic Substances and Disease Registry, US Department of Health and Human Services.

ATSDR (Agency for Toxic Substances and Disease Registry) (2005). *Toxicological Profile for Diisopropyl Methylphosphonate.* Atlanta, GA: Agency for Toxic Substances and Disease Registry, US Department of Health and Human Services.

Aulerich RJ, Coleman TH, Polin D *et al.* (1979). *Toxicology Study of Diisopropyl Methylphosphonate and Dicyclopentadiene in Mallard Ducks, Bobwhite Quail and Mink.* AD-A087-257/2. East Lansing, MI: Michigan State University.

Ault L (2005). Incineration pace beginning to pick up at arsenal. *Pine Bluff Commercial.* Online edition (http://www.pbcommerical.com/articles), retrieved April 6, 2005.

Bakshi KS, Pang SNJ and Snyder R (eds) (2000). Review of the US Army's health risk assessments for oral exposure to six chemical-warfare agents. *J Toxicol Environ Health*, **A59**, 281–526.

Bartlett PD and Swain CG (1949). Kinetics of hydrolysis and displacement reactions of β, β'-dichlorodethylsulfide (mustard gas) and of β-chloro-β'-hydroxydiethyl sulfide (mustard chlorohydrin). *J Am Chem Soc*, **71**, 1406–1415.

Bartelt-Hunt, Barlaz MA, Knappe DRU *et al.* (2006). Fate of chemical warfare agents and toxic industrial chemicals in landfills. *Environ Sci Technol*, **40**, 4219–4225.

Bausum HT (1998). *Toxicological and related data for VX, suggested breakdown products and additives; suggested RfD and RfC values.* Prepared by the US Army Center for Health Promotion and Preventive Medicine, Aberdeen Proving Ground, MD for the Newport (IN) Chemical Agent Disposal Facility.

Bausum HT, Reddy G and Leach GJ (1999). *Suggested interim estimates of the reference dose (RfD) and reference concentration (RfC) for certain key breakdown products of chemical agents.* Report to the USACHPPM Chemical Standards Working Group, December 10, 1998. Aberdeen Proving Ground, MD: US Army Center for Health Promotion and Preventive Medicine.

Bentley RE, LeBlanc GA, Hollister TA *et al.* (1976). *Acute Toxicity of Diisopropylmethyl Phosphonate and Dicyclopentadiene to Aquatic Organisms.* Wareham, MA: EG&G Bionomics.

Berkowitz JB, Goyer MM, Harris JC *et al.* (1978). *Literature Review – Problem Definition Studies on Selected Chemicals. Final Report. Vol. II. Chemistry, Toxicology and Potential Environmental Effects of Selected Organic Pollutants.* AD B052964. Fort Detrick, MD: US Army Medical Bioengineering Research and Development Laboratory.

Black RM, Clarke RJ, Cooper DB *et al.* (1993). Application of headspace analysis, solvent extraction, thermal desorption and gas chromatography–mass spectrometry to the analysis of chemical warfare samples containing sulphur mustard and related compounds. *J Chromatogr*, **637**, 71–80.

Bossle PC, Martin JJ, Sarver EW *et al.* (1983). High-performance liquid chromatography analysis of alkyl methylphosphonic acids by derivatization. *J Chromatogr*, **267**, 209–212.

Brankowitz WR (1987). *Chemical weapons movement: history compilation.* ADA193348. Aberdeen Proving Ground, MD: Office of the Program Manager for chemical munitions (demilitarization and binary).

Bucci TJ, Mercieca MD, Perman V *et al.* (1997). *Study No. TP-001, Two-Generation Reproductive Study in Mink Fed Diisopropyl Methylphosphonate (DIMP), Final Report.* Frederick, MD: Pathology Associates International.

Buswell AM, Price CC, Prosser CL *et al.* (1944). *The Effect of Certain Chemical Warfare Agents in Water on Aquatic Organisms.* Report No. OSRD 3589. Washington, DC: National Defense Research Command Office, Office of Scientific Research and Development.

Carnes SA and Watson AP (1989). Disposing of the U.S. chemical weapons stockpile: An approaching reality. *J Am Med Assoc*, **262**, 653–659.

Chase R (2004). Air Force personnel injured by vintage ordnance. Associated Press website (http://www.ap.org/), Friday, July 23.

Cheng TC, DeFrank JJ and Rastogi VK (1999). *Alteromonas* prolidase for organophosphorus G-agent decontamination. *Chem–Biol Interact*, **119/120**, 455–462.

Clark DN (1989). Review of Reactions of Chemical Agents in Water, AD-A213 287. Fort Belvoir, VA: Defense Technical Information Center.

Cooper WA (1990). *A Data Base on Persistent CW Agents on Terrain.* AD-B149 283. Aberdeen Proving Ground, MD: Chemical Research, Development and Engineering Center.

Corwin DL, David A and Goldberg S (1999). Mobility of arsenic in soil from the Rocky Mountain Arsenal area. *J Contam Hydrol*, **39**, 35–58.

Creasy WR, Stuff JR, Williams B *et al.* (1997). Identification of chemical-weapons-related compounds in decontamination solutions and other matrices by multiple chromatographic techniques. *J Chromatogr A*, **774**, 253–263.

Creasy WR, Brickhouse MD, Morrissey KM *et al.* (1999). Analysis of chemical weapons decontamination waste from old ton containers from Johnston Atoll using multiple analytical methods. *Environ Sci Technol*, **33**, 2157–2162.

D'Agostino PA and Provost LR (1988). Capillary column isobutane chemical ionization mass spectrometry of mustard and related compounds. *Biomed Environ Mass Spectrom*, **15**, 553–564.

D'Agostino PA and Provost LR (1992a). Determination of chemical warfare agents, their hydrolysis products and related compounds in soil. *J Chromatogr*, **589**, 287–294.

D'Agostino PA and Provost LR (1992b). Mass spectrometric identification of products formed during degradation of ethyl dimethylphosphoramidocyanidate (tabun). *J Chromatogr*, **598**, 89–95.

D'Agostino PA, Hansen AS, Lockwood PA *et al.* (1985). Capillary column gas chromatography–

mass spectrometry of tabun. *J Chromatogr*, **347**, 257–266.

D'Agostino PA, Provost LR, Visentini J (1987). Analysis of O-ethyl S-[diisopropylamino)ethyl] methylphosphonothiolate (VX) by capillary column gas chromatography-mass spectrometry. *J. Chromatogr*, **402**, 2221–232.

D'Agostino PA, Provost LR and Looye KM (1989). Identification of tabun impurities by combined capillary column gas chromatography–mass spectrometry. *J Chromatogr*, **465**, 271–283.

D'Agostino PA, Hancock JR and Chenier CL (2003). Mass spectrometric analysis of chemical warfare agents and their degradation products in soil and synthetic samples. *Eur J Mass Spectrom*, **9**, 609–618.

Deardorff MB, Das BR and Roberts WC (1992). Health Advisory for 1, 4-dithiane, PB93-117026. Washington, DC: US Environmental Protection Agency.

Deichman WB (1941). Toxicity of methyl, ethyl, and *n*-butyl methacrylate. *J Ind Hyg Toxicol*, **23**, 343–351.

DHHS (Department of Health and Human Services) (2005). *Review of the US Army Proposal for Off-Site Treatment and Disposal of Caustic VX Hydrolysate from the Newport Chemical Agent Disposal Facility*. Atlanta, GA: Centers for Disease Control and Prevention, DHHS.

DOD (Department of Defense) (1996). US Chemical weapons stockpile information declassified, News Release, January 22, 1996. Washington, DC: Office of Assistant Secretary of Defense (Public Affairs), 9 pp.

Dunn P (1986). The chemical war: Iran revisited (1986). *NBC Defense Tech Int*, **1**, 32–39.

Durst HD, Sarver EW, Yurow HW *et al.* (1988). Hydrolysis Reaction Equations for Chemical Agents, Rep. CRDEC-TR-009. MD: Chemical Research, Development and Engineering Center, Aberdeen Proving Ground.

Durst, HD, Sarver EW, Yurow HW *et al.* (2002). Support for the delisting of decontaminated liquid chemical surety materials as listed hazardous waste from the specific sources (State), MD02 in COMAR 10.51.02.16-1. Aberdeen Proving Ground, MD: U.S. Army Chemical Research Development and Engineering Center.

Enslein K (1984). Estimation of toxicological endpoints by structure-activity relationships. *Pharmacol Rev*, **36**, 131S–135S.

Epstein J (1974). Properties of GB in water. *J Am Water Works Assoc*, **66**, 31–37.

Epstein J and Rosenblatt DH (1958). Kinetics of some metal ion-catalyzed hydrolyses of isopropyl methyl phosphonofluoridate (GB) at 25°C. *J Am Chem Soc*, **80**, 3596–3598.

Epstein J, Callahan JJ, Bauer VE (1974). The kinetics and mechanisms of hydrolysis of phosphonothiolates in dilute aqueous solution. *Phosphorus*, **4**, 157–163.

Epstein J, Davis GT, Eng L *et al.* (1977). Potential hazards associated with spray drying operations. *Environ Sci Technol*, **11**, 70–75.

Fedorov LA (1996). Pre-convention liquidation of Soviet chemical weapons. In: *Sea-Dumped Chemical Weapons. Aspects, Problems and Solutions* (AV Kaffka, ed.), pp. 17–27. Dordrecht, The Netherlands: Kluwer Academic Publishers.

Fokin AV and Babievsky KK (1996). Chemical 'echo' of the wars. In: *Sea-Dumped Chemical Weapons. Aspects, Problems and Solutions* (AV Kaffka, ed.), pp. 29–33. Dordrecht, The Netherlands: Kluwer Academic Publishers.

Fowler WK, Stewart DC and Weinberg DS (1991). Gas chromatographic determination of the lewisite hydrolysate, 2-chlorovinylarsonous acid, after derivatization with 1,2-ethanedithiol. *J Chromatogr*, **558**, 235–246.

Franke S (1982). *Textbook of Military Chemistry*, Vol 1, USAMIIA-HT-039-82, AD B062913. Alexandria, VA: Defense Technical Information Center.

Frondorf MJ (1996). Special study on the sea disposal of chemical munitions by the United States. In: *Sea-Dumped Chemical Weapons: Aspects, Problems and Solutions* (AV Kaffka, ed.), pp. 35–47. Dordrecht, The Netherlands: Kluwer Academic Publishers.

Gage, EL (1946). Mustard gas (dichloroethyl sulfide) burns: in clinical experiences. *West Virginia Med J*, **42**, 180–185.

Glasby GP (1997). Disposal of chemical weapons in the Baltic Sea. *Sci Total Environ*, **206**, 267–273.

Goldman M, Dacre JC (1989). Lewisite: its chemistry, toxicology, and biological effects. *Rev Environ Contam Toxicol*, **110**, 75–115.

Granbom PO (1994). Ett faktahäfte om dumpad C-ammunition i Skagerrak och Östersjön, 1994-05-01. Umeå, Sweden: Foersvarets Forskningsanstalt, OA ABC-skydd.

Granbom PO (1996). Investigation of a dumping area in the Skagerrak. In: *Sea-Dumped Chemical Weapons: Aspects, Problems and Solutions* (AV Kaffka ed.), pp. 41–49. Dordrecht, The Netherlands: Kluwer Academic Publishers.

Grechkin NP, Grishina LN, Neklesova ID *et al.* (1977). Synthesis and physiological activity of some amidophosphates. *Pharm Chem J*, **11**, 38–41.

Hambrook JL, Howells DJ and Utley D (1971). Degradation of phosphonates. Breakdown of soman (*O*-pinacolyl-methylphosphonofluoridate) in wheat plants. *Pest Sci*, **2**, 172–175.

Harris R and Paxman J (2002). *A Higher Form of Killing: The Secret History of Chemical and Biological Warfare*. New York: Random House Trade Paperbacks.

Hart ER (1976). Mammalian Toxicological Evaluation of DIMP and DCPD, Final Report, AD-AO58 323. Kensington, MD: Litton Bionetics, Inc.

Hart ER (1980) Mammalian Toxicological Evaluation of DIMP and DCPD (Phase II), AD-AO82 685. Kensington, MD: Litton Bionetics, Inc.

Harvey SP, Szafraniec, LL, Beaudry WT *et al.* (1996). HD Hydrolysis/Biodegradation Toxicology and Kinetics. ERDEC-TR-382. Edgewood Research, Development and Engineering Center, Aberdeen Proving Ground, MD, ADA319798. Springfield, VA: National Technical Information Center.

Hay A and Roberts G (1990). The use of poison gas against the Iraqi Kurds: analysis of bomb fragments, soil, and wool samples. *J Am Med Assoc*, **263**, 1065.

HELCOM (1994). Report on chemical munitions dumped in the Baltic Sea, Report to the 16th Meeting of the Helsinki Commission, March 8–11, 1994, 1994:43. Helsinki, Finland: HELCOM CHEMU.

Howard PH and Meylan WM (eds) (1997). *Handbook of Physical Properties of Organic Chemicals*. New York: Lewis Publishers.

HSDB (Hazardous Substances Data Base) (2005). Diisopropylaminoethanol, online file. Washington, DC: National Library of Medicine.

Inamori Y, Ohno Y, Nishihata S *et al.* (1990). Phytogrowth-inhibitory and anti-bacterial activities of 2,5-dihydroxy-1,4-dithiane and its derivatives. *Chem Pharm Bull*, **38**, 243–245.

Infield G (1988). *Disaster at Bari*. New York: Bantam Books.

IOM (Institute of Medicine) (1993). *Veterans at Risk: The Health Effects of Mustard Gas and Lewisite* (CM Pechura and D Rall, eds), Committee to Survey the Health Effects of Mustard Gas and Lewisite. Institute of Medicine, Washington, DC: National Academy Press.

Itoh N, Yoshida M, Miyamoto T *et al.* (1997). Fungal cleavage of thioether bond found in Yperite. *FEBS Lett*, **412**, 281–284.

Johnsen BA and Blanch JH (1984). Analysis of snow samples contaminated with chemical warfare agents, *Arch Belg Med Soc*, Supplement. In: *Proceedings World Congress on New Compounds for Biological and Chemical Warfare: Toxicological Evaluation*, pp. 22–30.

Kaaijk J and Frijlink C (1977). Degradation of *S*-2-diisopropylaminoethyl *O*-ethyl methylphonothioate in soil: sulfur-containing products. *Pest Sci*, **8**, 510–514.

Kaffka AV (ed.) (1996). *Sea-Dumped Chemical Weapons: Aspects, Problems and Solutions*. Dordrecht, The Netherlands: Kluwer Academic Publishers.

Kingery AF and Allen HE (1995) The environmental fate of organophosphorus nerve agents: a review. *Toxicol Environ Chem*, **47**, 155–184.

Klimkina NV, Ekhina RS and Sergeev AN (1976). Experimental substantiation of the maximum permissible concentration of methyl and butyl ethers of methacrylic acid in water bodies. *Gigie Sanitar (Moscow)*, **41**, 6–11.

Konkov VN (1996). The technological problems with sea-dumped chemical weapons from the standpoint of defense conversion industries. In: *Sea-Dumped Chemical Weapons: Aspects, Problems and Solutions* (AV Kaffka, ed.), pp. 87–91. Dordrecht, The Netherlands: Kluwer Academic Publishers.

Lee KP and Allen HE (1998). Environmental transformation mechanisms of thiodiglycol. *Environ Toxicol Chem*, **17**, 1720–1726.

Lundin SJ (1991). *Verification of dual-use chemicals under the Chemical Weapons Convention: The case of thiodiglycol*. Stockholm, Sweden: Stockholm International Peace Research Institute.

Luo S-Q, Gang B-Q and Sun S-Z (1986). Study on embryotoxicity and fetotoxicity in rats by maternal inhalation of low level methyl methacrylate. *Toxicol Lett*, **31**(Supplement), 80.

MacNaughton MG and Brewer JH (1994). *Environmental Chemistry and Fate of Chemical Warfare Agents*, SWRI Project 01-5864. San Antonio, Texas: Southwest Research Institute.

Major M (1998). Memorandum for MCHB-TS-EHRARCP (August 5, 1998). Aberdeen Proving Ground, Maryland: US Army Center for Health Promotion and Preventive Medicine.

Marrs TC, Maynard RL and Sidell, FR (1996). *Chemical Warfare Agents: Toxicology and Treatment*. Chichester, UK: John Wiley & Sons, Ltd.

Mayhew DA and Muni IA (1986). *Dermal, Eye and Oral Toxicological Evaluations. Phase II. Acute Oral LD Determinations of Benzothiazole, Dithiane and Oxathiane*, AD-A172 647/0/XAB. Woburn, MA: Bioassay Systems Corporation.

McGuire RR, Haas JS and Eagle RJ (1993). *The Decay of Chemical Weapons Agents under Environmental Conditions*, UCRL-ID-114107, Lawrence

Livermore National Laboratory Report. Lawrence, CA: Lawrence Livermore National Laboratory.

Mecler FJ (1981). *Mammalian Toxicological Evaluation of DIMP and DCPD (Phase 3-IMPA)*, AD-A107574. Fort Detrick, MD: US Army Medical Research and Development Command.

Michel HO, Epstein J, Plapinger RR *et al.* (1962). *EA 2192: A Novel Anticholinesterase*, CRDLR 3135. Army Chemical Center, MD: US Army Chemical Research and Development Laboratories.

Morita H, Yanagisawa N, and Nakajima T (1995). Sarin poisoning in Matsumoto, Japan. *Lancet*, **346**, 290–293.

Morrill LG, Reed LW and Chinn KSK (1985). *Toxic Chemicals in the Soil Environment*, Vol 2, *Interaction of Some Toxic Chemicals/Chemical Warfare Agents and Soils*, TECOM Project 2-CO-210-049 (DTIC: AD-A158 215). Stillwater, OK: Oklahoma State University.

Munro NB, Talmage SS, Griffin GD *et al.* (1999). The sources, fate, and toxicity of chemical warfare agent degradation products. *Environ Health Perspect*, **107**, 933–974.

Murdock P, Kilpatrick W and Love S (2004). *Technical Note: Response to questions regarding agent fate testing: Effects of water on previously contaminated surfaces*. AFRL/HEPC CBD Chemical and Biological Defense Team. Dayton, OH: Wright-Patterson Air Force Base.

Muribi M (1997). *Toxicity of Mustard Gas and Two Arsenic Based Chemical Warfare Agents on Daphnia magna (for the evaluation of the ecotoxicological risk of the dumped chemical warfare agents in the Baltic Sea.)*, Report FOA-R-97-00430-222-SE; PB97-206825. Springfield, VA: National Technical Information Service.

Nicholas CA, Lawrence WH and Autian J (1979). Embryotoxicity and fetotoxicity from maternal inhalation of methyl methacrylate monomer in rats. *Toxicol Appl Pharmacol*, **50**, 451–458.

NMFA (1982–1983). *Verification of a Chemical Weapons Convention: Sampling and Analysis of Chemical Warfare Agents under Winter Conditions*, Parts 1 and 2. Oslo, Norway: Royal Norwegian Ministry of Foreign Affairs.

NRC (1996). *Review and Evaluation of Alternative Chemical Disposal Technologies. Panel on Review and Evaluation of Alternative Disposal Technologies*. Washington, DC: National Research Council, National Academies Press.

NRC (2003). *Acute Exposure Guideline Levels for Selected Airborne Chemicals, Vol. 3*. Washington, DC: National Research Council, National Academies Press.

O'Bryan TR and Ross RH (1988). Chemical scoring system for hazard and exposure identification. *J Toxicol Environ Health*, **1**, 119–134.

Okumura T, Takasu N, Ishimatu S *et al.* (1996). Report on 640 victims of the Tokyo subway sarin attack. *Ann Emerg Med*, **28**, 129–135.

OPCW (Organisation for the Prohibition of Chemical Weapons) (2005). Online database: http://www.opcw.org/index.html, retrieved March, 2005.

Opresko DM, Young RA, Faust RA *et al.* (1998). Chemical Warfare agents; estimating oral reference doses. *Rev Environ Contam Toxicol*, **156**, 1–183.

Opresko DM, Young RA, Watson AP *et al.* (2001). Chemical warfare agents: Current status of oral reference doses. *Rev Environ Contam Toxicol*, **172**, 65–85.

Pangi R (2002). Consequence management in the 1995 sarin attacks on the Japanese subway system. In: *Studies in Conflict and Terrorism*. Routledge, UK: Taylor & Francis Group.

Perera J and Thomas A (1986). Mustard gas at the bottom of the garden. *New Sci*, **13**, 18–19.

Pham MQ, Harvey SP, Weigand WA *et al.* (1996). Reactor comparisons for the biodegradation of thiodiglycol, a product of mustard gas hydrolysis. *Appl Biochem and Biotechnol* **57/58**, 779–789.

Price CC and von Limbach B (1945). Further data on the toxicity of various CW agents to fish, OSRD No. 5528. Washington DC: National Defense Research Committee, Office of Scientific Research and Development.

Public Affairs, Chemical Materials Agency (2005). ABCDF Processes Final Batch of Mustard Agent Drained from last Aberdeen Container. *CBIAC Newslett*, **6**(2), 5.

Puzderliski A (1980). The persistence of sarin and Yperite drops in soil. *Pregl Naucno-Tehnick Radova Inform*, **30**, 47–54.

Reddy G, Major AA and Leach GJ (2005). Toxicity assessment of thiodiglycol. *Internat J Toxicol*, **24**, 435–442.

Rewick RT, Shumacher ML and Haynes DL (1986). The UV absorption spectra of chemical agents and simulants. *Appl Spectros*, **40**, 152–156.

Robinson JP (1967). Chemical warfare. *Sci J*, **3**, 33–40.

Robson SG (1981). Computer simulation of movement of DIMP-contaminated groundwater near the Rocky Mountain Arsenal, Colorado. In: *Permeability and Groundwater Contaminant Transport: A Symposium, ASTM Spec Tech Publ*, **746**, 209–220.

Rohrbaugh DK (1998). Characterization of equimolar VX-water reaction product by gas chromatography-mass spectrometry. *J Chromatogr*, **809**, 131–139.

Rosenblatt DH, Miller TA, Dacre JC *et al.* (1975). *Problem Definition Studies on Potential Environmental Pollutants. II. Physical, Chemical, Toxicological, and Biological Properties of 16 Substances*. Technical report 7509; AD AO30428. Fort Detrick, MD: US Army Medical Bioengineering Research and Development Laboratory.

Rosenblatt DH, Small MJ, Kimmell TA *et al.* (1995). *Agent Decontamination Chemistry: Technical Report*. US Army Test and Evaluation Command (TECOM) Technical Support, Phase 1. Prepared for Environmental Quality Office, US Army Test and Evaluation Command. Argonne, IL: Argonne National Laboratory.

RTECS (2005). 2-Chlorovinyl arsonic acid. Registry of Toxic Effects of Chemical Substances, MEDLARS Online Information Retrieval System, National Library of Medicine, retrieved February 21, 2005.

Sage GW and Howard PH (1989). *Environmental Fate Assessments of Chemical Agents HD and VX*. CRDEC-CR-034. MD: Aberdeen Proving Ground.

Sanches ML, Russell CR and Randolf CL (1993). Chemical Weapons Convention (CWC) Signature Analysis, DNA-TR-92-73; AD B171788. Alexandria, VA: Defense Technical Information Center.

Sidell FR (1997). Nerve agents. In: *Medical Aspects of Chemical and Biological Warfare* (FR Sidell, ET Takafuji and DE Franz, eds), pp. 129–179. Washington, DC: Office of the Surgeon General, TBMM Publications, Borden Institute, Walter Reed Army Medical Center.

Sklyar VI, Mosolova TP, Kucherenko IA *et al.* (1999). Anaerobic toxicity and biodegradability of hydrolysis products of chemical warfare agents. *Appl Biochem Biotechnol*, **81**, 107–117.

Small MJ (1984). *Compounds formed from the chemical decontamination of HD, GB and VX and their environmental fate*. Technical report 8304; AD A149515. Fort Detrick, MD: U.S Army Medical Bioengineering Research and Development Laboratory.

Smartt J and Kopp C (2005). Waves of change: Army transformation at Aberdeen Proving Ground reaches Technical Escort Unit. *CBIAC Newslett*, **6**(1), 1, 12–15.

Smyth HF Jr, Seaton J and Fischer L (1941). The single dose toxicity of some glycols and derivatives. *J Ind Hyg Toxicol*, **23**, 259–268.

Spealman CR, Main RJ, Haag HB *et al.* (1945). Monomeric methyl methacrylate. Studies on toxicity. *Ind Med*, **14**, 292–298.

Stephan U (2005). Personal communication from Professor Ursula Stephan, Gefahrstoff-Büro, Halle, Germany.

Stock T (1996). Sea-dumped chemical weapons and the chemical weapons convention. In: *Sea-Dumped Chemical Weapons: Aspects, Problems and Solutions* (AV Kaffka, ed.), pp. 49–66. Dordrecht, The Netherlands: Kluwer Academic Publishers.

Surikov BT (1996). How to save the Baltics from ecological disaster. In: *Sea-Dumped Chemical Weapons: Aspects, Problems and Solutions* (AV Kaffka, ed.), pp. 67–70. Dordrecht, The Netherlands: Kluwer Academic Publishers.

Szafraniec LJ, Szafraniec LL, Beaudry WT *et al.* (1990). *On the Stoichiometry of Phosphonothiolate Ester Hydrolysis*. ADA-250773. Aberdeen Proving Ground, MD: US Army Chemical Research, Development and Engineering Center.

Thiokol/Ventron Division (1980). Material Safety Data Sheet, Methylphosphonic Acid. Danvers, MA: Thiokol/Ventron Division.

Tompkins BA, Sega GA and MacNaughton SJ (1998). The quantitation of sulfur mustard by-products, sulfur-containing herbicides and organophosphonates in soil and concrete. *Anal Lett*, **3**, 1603–1622.

US Air Force (2003). Fate of Agent: Czech Republic Field Trials: Results from 2000 and 2001, Trial Data, Vol. 2, AFRL-HE-WP-TR-2003-0055. Dayton, OH: Wright Patterson Air Force Base.

US Army (1974). *Chemical Agent Data Sheets, Vol. 1*. Technical Report, EO-SR-74001 (Edgewood Arsenal Special Report), AD B028222, pp. 167–179. Aberdeen Proving Ground, MD: Department of the Army, Edgewood Arsenal.

US Army (1988). *Final Programmatic Environmental Impact Statement for the Chemical Stockpile Disposal Program*. Aberdeen Proving Ground, MD: US Department of the Army.

US Army (1992). *Material Safety Data Sheets: Lethal Nerve Agents GA, GB, VX*. Department of the Army, Aberdeen Proving Ground, MD: Edgewood Research, Development and Engineering Center.

US Army (1996a). Newport Chemical Activity Concept Design Package for VX Neutralization Followed by Off-Site Biodegradation. April 4, 1996. Product Manager for Alternative Technologies and Approaches. Aberdeen Proving Grounds, MD: US Army Program Manager for Chemical Demilitarization.

US Army (1996b). Cumulative Sample Analysis Data Report, HD Ton Container Survey Results. March

14, 1996. Aberdeen Proving Ground, MD; US Army Program Manager for Chemical Demilitarization.

US Army (2002). *Chemical and Biological Warfare Agent Fate Research Program.* Department of Def The Pentagon, Arlington, VA: Joint Science and Technology Panel for Chemical and Biological Defense.

US Army CHPPM (1999). *Derivation of Health-Based Environmental Screening Levels for Chemical Agents: A Technical Evaluation.* MD: Aberdeen Proving Ground.

US EPA (1996). Risk Assessment Issue Paper for: Derivation of an Oral RfD for Methylphosphonic Acid (CASRN 993-13-5), Paper No. 96-024. Cincinnati, OH: NCEA, U.S. Environmental Protection Agency.

US EPA (2005a). Arsenic. IRIS online database: http://www.epa.gov/iriswebp/iris/index.html. Washington, DC: US Environmental Protection Agency, retrieved February 21, 2005.

US EPA (2005b). Isopropyl methyl phosphonic acid. IRIS online database: http://www.epa.gov/iriswebp/iris/index.html. Washington, DC: US Environmental Protection Agency, retrieved April 18, 2005.

Verweij A and Boter HL (1976). Degradation of *S*-2-diisopropylaminoethyl *O*-ethyl methylphosphonothioate in soil: phosphorus-containing products. *Pest Sci*, **7**, 355–362.

Vocci FJ, Ballard TA, Yevich P *et al.* (1963). Inhalation toxicity studies with aerosols of sesqui-mustard. *Toxicol Appl Pharmacol*, **5**, 677–684.

Wada S, Nishimoto Y, Katsuta S *et al.* (1962). Review of Okuno-jima poison gas factory regarding occupational environment. *Hiroshima J Med*, **11**, 75–78.

Warrick J. (2005). Albania's chemical cache raises fears about others. Washington Post Online file: washingtonpost.com, January 9, 2005.

Watson AP and Griffin GD (1992). Toxicity of vesicant agents scheduled for destruction by the Chemical Stockpile Disposal Program. *Environ Health Perspect*, **98**, 259–280.

Watson AP, Jones TD and Griffin GD (1989). Sulfur mustard as a carcinogen: application of relative potency analysis to the chemical warfare agents H, HD and HT. *Regulat Toxicol Pharmacol*, **10**, 1–25.

Webster D (1994). The soldiers moved on. The war moved on. The bombs stayed. *Smithsonian Mag*, February, 26–37.

Williams RT, Miller WR III and MacGillivray AR (1987). *Environmental Fate and Effects of Tributyl Phosphate and Methyl Phosphonic Acid.* CRDEC-CR-87103: NTIS AD-A184 959/5. Aberdeen Proving Ground, MD: US Army Armament Munitions Chemical Command, Chemical Research, Development and Engineering Center; West Chester, PA: Roy F Weston.

Williams JM, Rowland B, Jeffery MT *et al.* (2005). Degradation kinetics of VX on concrete by secondary ion mass spectrometry. *Langmuir*, **21**, 2386–2390.

Wulf HC, Aasted A, Darre E, Niebuhr E (1985). Sister chromatid exchanges in fishermen exposed to leaking mustard gas shells. *Lancet*, Mar 23; **1(8430)**, 690–691.

Yamakido M, Ishioka S, Hiyama K *et al.* (1996). Former poison gas workers and cancer: incidence and inhibition of tumor formation by treatment with biological response modifier N-CWS. *Environ Health Perspect*, **104** (Supplement 3), 485–488.

Yang YC, Baker JA and Ward JR (1992). Decontamination of chemical warfare agents. *Chem Rev*, **92**, 1729–1743.

Yang YC, Szafraniec LL and Beaudry WT (1993). Perhydrolysis of nerve agent VX. *J Org Chem*, **58**, 6964–6965.

Yufit SS, Miskevich IV and Shtemberg ON (1996). Chemical weapons dumping and White Sea contamination. In: *Sea-Dumped Chemical Weapons: Aspects, Problems and Solutions* (AV Kaffka, ed.), pp. 157–166. Dordrecht, The Netherlands: Kluwer Academic Publishers.

Zhang Y, Autenrieth RL, Bonner JS *et al.* (1999). Biodegradation of neutralized sarin. *Biotechnol Bioeng*, **64**, 221–231.

5 BIOLOGICAL MARKERS OF EXPOSURE TO CHEMICAL WARFARE AGENTS

Robin M Black[1] and Daan Noort[2]

[1] Dstl, Salisbury, UK
[2] TNO Defence, Security and Safety, Rijswijk, The Netherlands

INTRODUCTION

The analysis of biomedical samples, such as urine and blood, can provide qualitative and quantitative information on exposure to CW agents. Detection of metabolites and covalent adducts provides forensic evidence in cases of allegations of military or terrorist use of CW agents. The methodology may also be used for diagnostic purposes to ensure administration of appropriate medical countermeasures, and for monitoring exposure in workers engaged in demilitarization and other defensive activities. Some of the simpler analytical methods have been applied in toxicokinetic studies. This chapter reviews the biological fate of CW agents, the metabolites and adducts that may be used as biological markers of exposure, and analytical methods for their detection. The emphasis is placed on the biological fate of CW agents and their biological markers of exposure. Analytical methods are summarized; they have been reviewed in greater detail elsewhere (Noort *et al.*, 2002a; Black and Noort, 2005; Noort and Black, 2005), together with methods for unchanged agents in biological fluids. In cases of allegations of CW use, battlefield use of riot control agents (RCAs), which contravenes the Chemical Weapons Convention, may be an important issue. The biological fate of RCAs is therefore also included.

BIOLOGICAL REACTIONS OF CW AGENTS

The first requisite in analyzing biomedical samples is knowledge of the biological fate of CW agents. Most are reactive electrophiles, i.e. they have atoms which are electron-poor that react with electron-rich, nucleophilic sites on other molecules. This chemical reactivity is a key component in the biochemical mechanisms of action of vesicants, nerve agents and phosgene. Because they are chemically reactive, CW agents generally have short lifetimes in the body, although the more lipophilic agents may be partially protected by sequestration into fatty tissues. The major fraction of an absorbed dose is rapidly metabolized and eliminated as metabolites, predominantly in the urine but with a small fraction in the faeces. The remainder of the absorbed dose is accounted for by covalent reactions with nucleophilic sites on macromolecules, such as proteins and DNA.

The most abundant free nucleophiles in the body are water and the tripeptide glutathione (γ-Glu–Cys–Gly), which has a reactive thiol group associated with the cysteine residue. Reactions with these nucleophiles, either chemically or mediated by enzymes, are the starting point for most metabolic pathways of electrophiles and account for most of the metabolites excreted in urine. A host of nucleophilic sites also exist on proteins,

Chemical Warfare Agents: Toxicology and Treatment (2nd Edition)
Edited by Timothy C. Marrs, Robert L. Maynard and Frederick R. Sidell

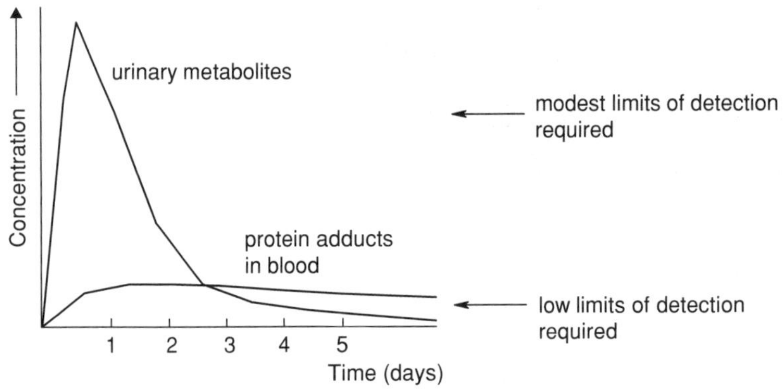

Figure 1. Typical profiles for urinary metabolites and protein adducts in blood in relation to required limits of detection

DNA and other macromolecules (Van Welie *et al.*, 1992). Examples are cysteine (SH), serine and tyrosine (OH), lysine and *N*-terminal valine (NH_2), and aspartic and glutamic acid (CO_2H) amino acid residues on proteins, and NH and PO_3^- residues on DNA. Haemoglobin and albumin are the most abundant proteins in blood, and covalently bound residues on these proteins provide biological markers of exposure to many electrophiles, including CW agents. Nerve agents are somewhat more selective in that they specifically target a serine OH group in the enzymes acetylcholinesterase (AChE) and butyrylcholinesterase (BuChE). Sulphur and nitrogen mustards react with guanine NH residues in DNA, which are present in every tissue in the body.

As described below, urinary metabolites have been identified for vesicants, nerve agents, 3-quinuclidinyl benzilate (BZ), hydrogen cyanide and the RCAs, CS, CR and capsaicin. Protein adducts have been identified for vesicants, nerve agents and phosgene, and DNA adducts for sulphur and nitrogen mustards. With the rapid advances being made in proteomics and metabonomics, new biological markers of exposure will undoubtedly be identified in the near future.

FREE METABOLITES AS BIOLOGICAL MARKERS

Free metabolites occur primarily in urine. Urinary excretion may account for up to ~ 90% of an absorbed dose of agent. Unlike blood, urine does not require invasive collection and is often the easiest biomedical fluid to obtain. It is a simpler, though more variable, matrix than blood, and does not have the complications of coagulation or red cell lysis. It should be frozen as soon as possible after collection to minimize chemical or microbial degradation of metabolites. Most urinary metabolites are relatively simple molecules, the synthesis of analytical standards is not too demanding, and in many cases analytical methods can be performed using benchtop instrumentation, e.g. gas chromatography–mass spectrometry (GC–MS) or gas chromatography–tandem mass spectrometry (GC–MS–MS). The major disadvantage of urinary metabolites as biological markers of exposure is demonstrated by the typical excretion profile shown in Figure 1. Up to ~ 90% of the total amount excreted is usually eliminated within the first 48–72 h following exposure (with percutaneous exposure, the curve may be somewhat flatter and broader). Depending on the exposure level, the detection of urinary metabolites within this time period is usually not too demanding, as illustrated by the detection of hydrolysis products of sarin following terrorist attacks in Tokyo and Matsumoto (Minami *et al.*, 1997; Nakajima *et al.*, 1998). After about 72 h, there usually follows a prolonged elimination of low concentrations of metabolites, and more sensitive methods are required to detect them. Experience has shown that in cases of allegations of CW use, particularly in remote conflicts, samples may be collected days or even weeks after the alleged exposure. Urine collected 13 days after exposure is the longest period after which metabolites have been reported in human CW casualties, and concentrations were < 1 ng/ml (Black and Read, 1995a). In some cases, e.g. the hydrolysis products of sulphur mustard, nitrogen mustard (HN-3) and tabun, metabolites may not be unequivocal indicators of exposure because of their occurrence from other sources. Although urine is the preferred sample, free metabolites may be detected in blood if collected within a few days of exposure. This was illustrated by samples collected after an assassination with the nerve agent VX (Tsuchihashi *et al.*, 1998) and the terrorist release of sarin in the Tokyo subway (Noort *et al.*, 1998). Saliva is another possible sample for detecting metabolites (e.g. Timchalk *et al.*, 2004) but so far has received little attention in the context of CW agents.

PROTEIN AND DNA ADDUCTS AS BIOLOGICAL MARKERS

Blood is the most important source of protein adducts, but requires medically trained personnel for its collection and stringent safety precautions in handling. The major advantage of protein adducts is that they are potentially much longer-lived biological markers than urinary metabolites. Provided the adduct is stable to chemical or metabolic degradation, it may persist for the lifetime of the protein. Approximate half-lives of albumin, haemoglobin and BuChE are 21 days, 42 days and 5–16 days, respectively. As shown schematically in Figure 1, although the concentrations of adducts in blood are initially much lower than those of urinary metabolites, after several days most adducts remain close to their initial concentration while those of urinary metabolites have declined substantially. The major disadvantages of protein adducts are that acquisition of analytical standards can be costly and demanding, and many of the methods require liquid chromatography–tandem mass spectrometry (LC–MS–MS), which is generally more costly and experimentally more challenging than GC–MS and GC–MS–MS.

Haemoglobin and albumin are the two most abundant proteins in blood, and for reactions that are purely chemical (i.e. non-enzymatic) usually provide the most abundant biomarkers. In the case of nerve agents, enzymatic reactions with the enzymes AChE and BuChE provide sensitive biomarkers. Haemoglobin and albumin are water-soluble proteins that have a large number of hydrophilic and nucleophilic amino acid residues on the periphery of the molecule. An important difference is that an agent must penetrate the red cell membrane to react with haemoglobin, and this will depend partially on its physical properties. Albumin has a cysteine residue (34) which is much more accessible than those in haemoglobin. Ideally, blood should be separated into red cell and plasma fractions and stored frozen shortly following collection, but this is not always possible.

Many electrophilic toxicants, including ethylene oxide, methyl bromide and acrylamide, react with haemoglobin and albumin (Törnqvist *et al.*, 2002). It is therefore not surprising that one or both of these proteins form adducts with vesicants, nerve agents and phosgene, at least *in vitro*. Adducts with haemoglobin have been identified for sulphur mustard (Noort *et al.*, 1996; Black *et al.*, 1997b), lewisite (Fidder *et al.*, 2000) and phosgene (Noort *et al.*, 2000). Adducts with albumin have been identified for sulphur mustard (Noort *et al.*, 1999) and nitrogen mustard HN-2 (Noort *et al.*, 2002b), some nerve agents (Black *et al.*, 1999; Harrison *et al.*, 2006) and phosgene (Noort *et al.*, 2000). Adducts with AChE and BuChE have been identified for the range of nerve agents (Fidder *et al.*, 2002; Elhanany *et al.*, 2001). Another possible source of protein adducts is skin. Sulphur mustard forms adducts with keratin aspartic and glutamic acid residues (Van der Schans, G.P. *et al.*, 2003). Hair is also a possible source of protein adducts but, unlike albumin and haemoglobin, most of the nucleophilic sites are on the interior of the protein with hydrophobic residues on the exterior.

DNA adducts provide biomarkers of exposure for sulphur and nitrogen mustards. However, turnover of these adducts appears to be much shorter than the lifetime of natural DNA due to the intervention of repair mechanisms.

ANALYTICAL METHODS

Analytical methods for biomedical samples require careful sample preparation, i.e. selective concentration of the analyte from the matrix, and a sensitive and selective method for detection. Analytes may be extracted directly from blood, plasma or urine using liquid–liquid extraction, but more commonly by solid-phase extraction (SPE) using a variety of solid phases. Early methods tended to use C_{18}- or C_8-bonded silica cartridges, or anion exchange for acidic analytes such as phosphonic acids. In recent years, there has been an increasing tendency to use hydrophobic–hydrophilic polymeric cartridges, which are often more efficient for extracting partially polar analytes such as thiodiglycol (TDG). In a few examples, solid-phase microextraction has been used, which in favourable cases may provide very low limits of detection (Wooten *et al.*, 2002). Affinity SPE using molecularly imprinted polymers has been applied to phosphonic acids in urine (Meng and Qin, 2001).

Mass spectrometry is generally the only spectrometric technique that universally provides the requisite combination of sensitivity and specificity for analysis at low parts per billion (ppb) in urine and blood. Nuclear magnetic resonance (NMR) spectrometry, although a very specific technique, still does not have the sensitivity required for trace analysis in complex matrices. Mass spectrometry is usually used in combination with gas or liquid chromatography, or less commonly with capillary electrophoresis. For analysis down to mid–low ppb levels, single stage GC–MS or LC–MS may provide adequate limits of detection, but for detection at low-sub ppb levels with a high degree of confidence tandem mass spectrometry is required. Immunoassays are useful for the rapid screening of multiple samples but, unless antibodies for the analyte are available, they require a considerable effort to develop.

Most metabolites, particularly those derived from hydrolysis, are polar and require derivatization prior to analysis by GC–MS. A disadvantage of this approach is that derivatization can be a major source of error in trace analysis. The metabolites normally require isolation from aqueous media before derivatization, and this can result in loss of analyte. In addition, the presence of large amounts of extraneous material in an extract, plus residual traces of water, may suppress derivatization. An advantage of derivatization is that it can be used to enhance detection. Conversion to perfluorinated derivatives and detection by negative-ion chemical ionization mass spectrometry (NICI-MS) provides the lowest limits of detection for TDG and phosphonic acids. LC–MS (usually LC–MS–MS) provides an alternative that can be applied directly to concentrated aqueous solutions of metabolites, in most cases without the need to derivatize. LC–MS–MS can provide very low limits of detection, but for many of the simpler polar metabolites, such as TDG and phosphonic acids, lowest limits of detection have been obtained with derivatization and GC–MS–MS. However, LC–MS instrumentation is still improving, particularly with the wider use of capillary columns.

A number of strategies can be applied to the detection of protein adducts. Some proteins can be selectively isolated from blood by affinity SPE; alternatively, blood may be fractionated using precipitation techniques. Detection of the entire protein adduct is possible with modern MS techniques, e.g. using electrospray or desorption ionization methods, but limits of detection are usually modest with considerable chemical background. A common approach is to selectively digest the protein with enzymes, such as trypsin or pepsin, to produce short-chain peptides, and detect the alkylated or phosphylated peptide using LC–MS–MS. Alternatively, the protein may be digested to its constituent amino acids using the protease from *Streptomyces griseus* (Pronase) or 6 M hydrochloric acid, although these methods tend to produce a large chemical background. In some cases, particularly where ester linkages are formed, the bound moiety may be displaced from the protein by hydrolysis, nucleophiles such as fluoride ion, or chemical derivatization, and detected using a simpler methodology, usually GC–MS.

An important aspect of trace analysis is that rigorous quality control must be included in analytical protocols if the results are to withstand international scrutiny. In cases of allegations of CW use, a chain of custody must be maintained and any positive analysis must be preceded by

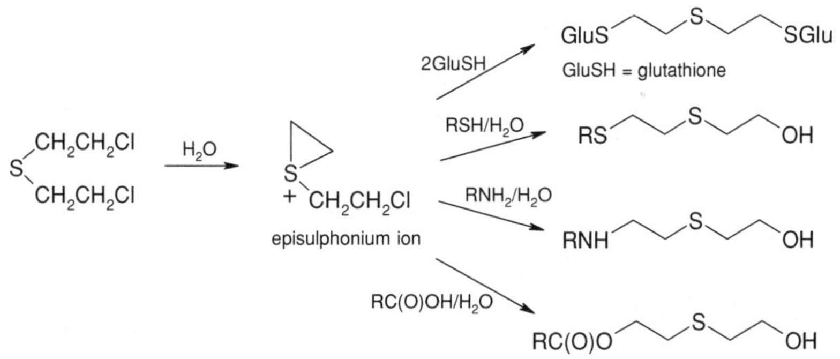

Figure 2. Sulphur mustard adduct formation with various nucleophiles

a negative control sample taken through the entire analytical procedure to demonstrate that no cross-contamination of equipment has occurred. Acceptable criteria for the trace-level detection of biological markers of CW agent poisoning are currently being discussed with the Organization for the Prohibition of Chemical Weapons.

SULPHUR MUSTARD

Sulphur mustard remains one of the CW agents of most concern because of its ease of synthesis, advantageous physical properties and the dual hazard that it presents from skin contact and lung and eye damage from exposure to vapour. More biomarkers have been identified for sulphur mustard than for any other CW agent, in part due to its relatively indiscriminate reactions with nucleophiles. Research into retrospective identification of poisoning was stimulated by the extensive use of sulphur mustard in the Iraq–Iran conflict and against the Kurdish population in Iraq.

Biological fate

DISTRIBUTION AND METABOLISM

Studies with ^{35}S-sulphur mustard showed that radioactivity was distributed rapidly throughout the tissues following intravenous (IV) administration in the rat (Maisonneuve *et al.* 1993). Typically, 50–80% of an absorbed dose is eliminated in urine following IV, intraperitoneal (IP) or percutaneous (PC) administration in the rat (Hambrook *et al.*, 1992). Much lower amounts of radioactivity are excreted in faeces, typically 5–15% of the dose. Radioactivity persisted in the blood for > 6 weeks following IV administration, associated mainly with the haemoglobin, indicative of covalent adduct formation (Hambrook *et al.*, 1993). A small percentage of radioactivity was retained in the skin following percutaneous exposure, presumably also bound to macromolecules.

The chemical and metabolic reactions of sulphur mustard are dominated by reactions with nucleophiles at its two electrophilic carbon atoms, plus oxidation of the electron-rich sulphur atom. Nucleophilic reactions proceed by an internal SN_1 type mechanism, via the episulphonium ion shown in Figure 2.

The formation of the episulphonium ion is rate-limiting and occurs very rapidly in polar solvents (Bartlett and Swain, 1949). In a competitive environment, the episulphonium ion reacts preferentially with 'soft' nucleophiles such as thiols, although it will react with a broad range of soft and hard nucleophiles. Under physiological conditions, reaction with the cysteinyl thiol function in glutathione (probably chemical rather than enzyme-mediated) competes with hydrolysis and reactions with nucleophilic sites on macromolecues. Metabolism studies in the rat with ^{35}S- and ^{13}C-labelled sulphur mustard showed that the initial reaction products with glutathione are metabolized by two divergent pathways to produce mercapturic acids (*N*-acetylcysteine conjugates) and β-lyase metabolites (methylthio/methylsulphinyl conjugates) (Black *et al.*, 1992a). The mustard sulphur atom in most metabolites is oxidized to sulphoxide or sulphone. The various permutations of

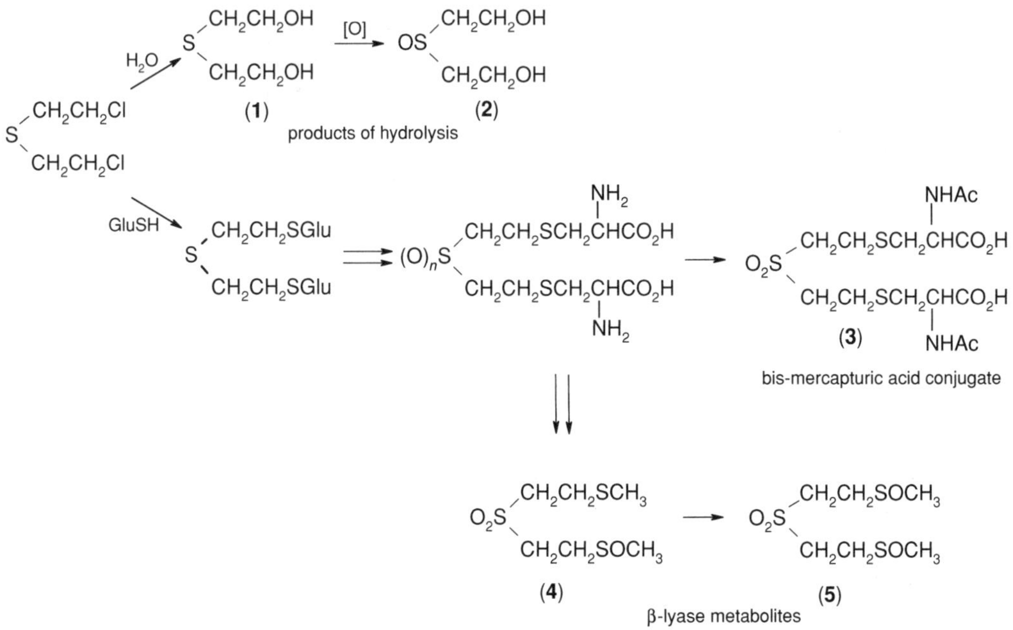

Figure 3. Urinary metabolites of sulphur mustard that have been investigated as biomarkers

reactions with nucleophiles, on one or both electrophilic carbon atoms, combined with three possible oxidation states of the sulphur atom, produce a large number of metabolites (at least 20 by HPLC using radioactivity detection). Figure 3 shows 5 of the 9 metabolites identified (**1**–**5**), which are the ones that have so far been investigated as biological markers. Not shown are three mono-mercapturic acid conjugates and mustard sulphoxide, which is much less reactive than mustard and was a very minor metabolite. These metabolites were identified by offline mass spectrometry following isolation by preparative HPLC, before the introduction of modern LC–MS systems that use atmospheric pressure ionization. Many more metabolites (e.g. derived from the partial hydrolysis product, hemi-mustard) could be identified using current instrumentation. Metabolism by the mercapturic acid pathway to the bis-*N*-acetylcysteine conjugate of mustard sulphone (**3**) is consistent with early studies by Roberts and Warwick (1963), in which TLC and dilution assays were used for tentative identification. The β-lyase metabolites **4** and **5** are formed through cleavage of the S–C bond in an intermediate bis-cysteinyl conjugate by the enzyme β-lyase, followed by *S*-methylation (Bakke and Gustafsson, 1984). This metabolic pathway is believed to be mediated predominantly by gut flora. Of the metabolites derived from simple hydrolysis, thiodiglycol sulphoxide (TDGO) (**2**) was the major one, with much lower amounts of TDG (**1**). A separate metabolism study of TDG in the rat showed that > 90% is excreted as TDGO (Black *et al.*, 1993). An early metabolism study using IV administration in the rat reported excretion of the bis-glutathione conjugate of sulphur mustard (Davison *et al.*, 1961).

In a quantitative study of urinary excretion in rats following PC exposure to sulphur mustard, 3.7–13.6% of the dose was excreted as products of hydrolysis, mainly as TDGO, and 2.5–5.3% as β-lyase metabolites, but with considerable variation between animals (Black *et al.*, 1992b). The excretion of β-lyase metabolites showed a sharper decline than that of hydrolysis products, suggesting that the prolonged excretion of the latter results from TDG being slowly liberated from adducts with macromolecules, e.g. from esters formed with aspartic and glutamic acid residues. A satisfactory analytical method for the bis-mercapturic acid conjugate (**3**) was not then available.

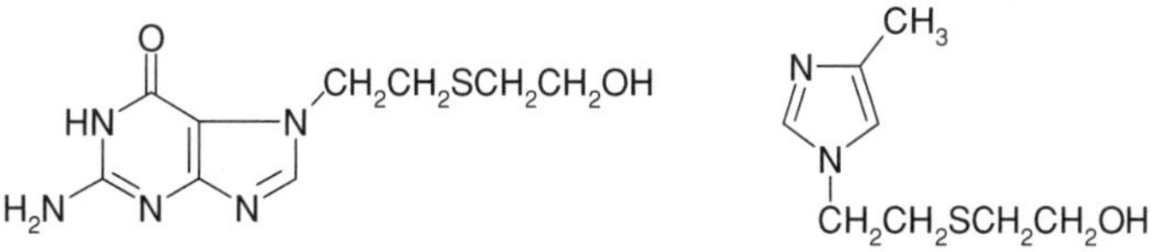

(6) (7)

Figure 4. Additional metabolites of sulphur mustard

Two other urinary metabolites have been reported in animal studies (Figure 4). N7-(2-Hydroxyethylthioethyl)guanine (**6**), derived from the breakdown of alkylated DNA, was detected in rats (Fidder *et al.*, 1996). The unusual metabolite (**7**), assumed to result from reaction with a histidine residue, was identified in the pig following percutaneous administration (Sandelowsky *et al.*, 1992).

REACTIONS WITH PROTEINS

Haemoglobin

Incubation of human blood with ^{35}S-sulphur mustard, and analysis of tryptic digests by LC–MS–MS, identified alkylation on 6 histidine residues, 3 glutamic acid residues, and both of the *N*-terminal valines (Noort *et al.*, 1996; Black *et al.*, 1997b). N1 and N3 histidine adducts were the most abundant. Alkylated cysteine, aspartic acid, lysine and tryptophan were also detected in Pronase digests, although the cysteine, which is somewhat hindered in haemoglobin, may have originated from glutathione. In all cases, and with albumin and keratin, the adducts possess a $CH_2CH_2SCH_2CH_2OH$ residue resulting from a single alkylation with hydrolysis at the second electrophilic carbon atom.

Albumin

Sulphur mustard alkylates the cysteine-34 residue in human serum albumin (Noort *et al.*, 1999, 2004b), which is known to react with a number of electrophiles. This residue is much more accessible than the cysteines in haemoglobin, and its reactivity is promoted by its relatively low pK_a, resulting from intramolecular stabilization of the thiolate anion. Aspartic and glutamic acid residues on albumin also react with sulphur mustard to give 2-hydroxyethylthioethyl esters (Capacio *et al.*, 2004). Adducts with albumin histidine residues have not been reported.

Keratin

The skin is a primary target of sulphur mustard. Exposure of human callus to ^{14}C-sulphur mustard showed that a significant part (15–20%) of the radioactivity was covalently bound to keratin (Van der Schans, G.P. *et al.*, 2003). Approximately 80% of the bound radioactivity could be released by alkaline hydrolysis, indicating it to be bound as esters of glutamic and aspartic acid residues.

REACTION WITH DNA

It has long been known that the predominant interaction of sulphur mustard with DNA is alkylation of N7 of deoxyguanosine residues (Brookes and Lawley, 1960; Fidder *et al.*, 1994). This and crosslinking actions are assumed to be responsible for the carcinogenic activity of sulphur mustard. Depurination of the resulting N7-(2-hydroxyethylthioethyl)-2′-deoxyguanosine releases N7-(2-hydroxyethylthioethyl)guanine (**6**), which can be detected in tissue, blood and urine samples (Fidder *et al.*, 1996). Sulphur mustard may also react with the phosphate groups in DNA, although no evidence for this has been presented.

Analytical methods

URINARY METABOLITES

Sensitive analytical methods have been developed for TDG, TDGO, the two β-lyase metabolites (**4, 5**) and the bis-*N*-acetylcysteine conjugate (**3**). TDG is best analyzed by GC–MS

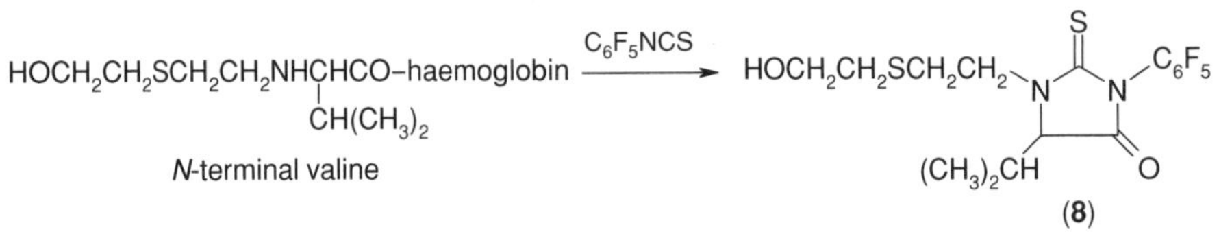

Figure 5. Selective cleavage of alkylated *N*-terminal valine with pentafluorophenyl isothiocyanate

or GC–MS–MS after derivatization. The bis-pentafluorobenzoyl derivative in combination with NICI-MS (Black and Read, 1988,1995a), and the bis-heptafluorobutyryl derivative in combination with electron ionization or positive-ion chemical ionization (Jakubowski *et al.*, 1990; Boyer *et al.*, 2004; Riches *et al.*, 2007), provide the most sensitive methods, with limits of detection (LODs) down to ~ 0.1 ng/ml. Conversion to the bis-trimethylsilyl or *tert*-butyldimethylsilyl derivative is widely used in environmental analysis but these give higher LODs in urine (Ohsawa *et al.*, 2004). TDGO is most easily analyzed after reduction to TDG with titanium trichloride (Black and Read, 1991). This is because of the difficulty of isolation of this very polar metabolite from aqueous media, and the different reactions involving the sulphoxide function that may occur on derivatization (Black and Muir, 2003). TDG and TDGO can be analyzed by LC–MS but detection limits have been too high for biomedical sample analysis. The β-lyase metabolites are readily analyzed by GC–MS and GC–MS–MS, provided that the sulphoxide groups, as with TDGO, are reduced with titanium trichloride (Black *et al.*, 1991; Black and Read 1995a; Young *et al.*, 2004). This produces the single analyte $O_2S(CH_2CH_2SCH_3)_2$, which is easily extracted and analyzed (LOD ~ 0.1 ng/ml). Thus, TDG, TDGO and the two β-lyase metabolites can be analyzed as two analytes, TDG and $O_2S(CH_2CH_2SCH_3)_2$, from the same aliquot of urine treated with titanium trichloride (Black and Read, 1995a; Boyer *et al.*, 2004). Some TDG and TDGO is excreted as glucuronides, as indicated by increased levels after treatment of urine with glucuronidase. LC–MS–MS, using positive electrospray ionization (ESI), also provides a sensitive method for the β-lyase metabolites, detecting them individually as the original metabolites (Read and Black, 2004a). No satisfactory GC–MS method has been developed for the bis *N*-acetylcysteine conjugate (**3**), probably because of thermal instability, but LC–MS–MS using negative electrospray ionization provides acceptable detection limits (Read and Black, 2004b).

PROTEIN ADDUCTS

Haemoglobin

Sensitive methods have been reported for sulphur mustard adducts with *N*-terminal valine and histidine, and for TDG released from aspartic acid and glutamic acid residues. In the case of *N*-terminal valine, the alkylated amino acid can be selectively cleaved using a method developed for other alkylating agents (Törnqvist *et al.*, 1996). Reaction of haemoglobin with pentafluorophenyl isothiocyanate (a fluorinated Edman reagent) releases the alkylated amino acid as the hydantoin (**8**) (Figure 5), which is further derivatized to its heptafluorobutyryl derivative and analysed by GC–MS/GC–MS–MS (Noort *et al.*, 1996, 2004a; Black *et al.*, 1997a). This provides a relatively simple method. Alkylated *N*-terminal valine could be detected up to 90 days following administration of sulphur mustard (4.1 mg/kg IV) to a marmoset (Noort *et al.*, 2002a; Benschop *et al.*, 2000). TDG can be cleaved from aspartic and glutamic acids residues on haemoglobin and analyzed by GC–MS (Capacio *et al.*, 2004). TDG released from aspartic and glutamic residues in whole blood proteins could be detected up to 45 days following administration of sulphur mustard (1 mg/kg IV) to an African Green monkey. Alkylated histidine is more problematic; haemoglobin is digested to its constituent amino acids with 6 M HCl and the alkylated histidine converted to its fluorenylmethoxycarbonyl (Fmoc) derivative for analysis by LC–MS–MS (Noort *et al.*, 1996; Black *et al.*, 1997a).

Albumin

A sensitive method was developed for detecting the alkylated cysteine residue based on affinity isolation of the protein, digestion with Pronase, and detection of the alkylated tripeptide Cys*–Pro–Phe (Noort *et al.*, 1999, 2004b). This method is more sensitive than those for alkylated haemoglobin, based on incubations of plasma with sulphur mustard.

DNA ADDUCTS

N7-(2-Hydroxyethylthioethyl)guanine (**6**) is analyzed in urine using LC–ESI–MS–MS. The adduct was detected in the urine of guinea pigs exposed to sulphur mustard, although concentrations declined rapidly after 36–48 h (Fidder *et al.*, 1996). HPLC (Ludlum *et al.*, 1994), ^{32}P-postlabelling (Niu *et al.*, 1996) and ELISA (Van der Schans, G.P. *et al.*, 1994, 2004) methods have also been developed.

Human exposures

A limited number of biomedical samples became available from CW casualties of sulphur mustard poisoning during the Iraq–Iran conflict, and from attacks on Kurdish communities in Iraq. Additional samples have since been obtained from accidental exposures in the laboratory and from old munitions.

URINARY METABOLITES

A summary of positive analyses is shown in Table 1. Results for β-lyase metabolites from Iranian and Kurdish casualties (Black and Read, 1995a) clearly show the advantages of having sub-ng/ml detection limits when samples are collected several days after the exposure. In the case of two Kurdish casualties, where urine was collected 13 days after exposure, levels of β-lyase metabolites (0.3 and 0.1 ng/ml) were close to the LOD. The excretion of TDG was monitored for 14 days following an accidental laboratory exposure to sulphur mustard (Jakubowski *et al.*, 2000). The casualty developed blisters on hands and arms (< 1% of body area) and erythema on his face and neck (< 5% of body area). A maximum excretion of 20 μg per day was observed between days 3 and 4 (maximum concentration 65 ng/ml), with < 10 ng/ml after 7 days. A problem with TDG and particularly TDGO is the occurrence of very low background levels in human (and animal) urine (Wils *et al.*, 1985, 1988; Black and Read, 1988, 1991, 1995a). Recent screening of 105 human samples from subjects with no known exposure to sulphur mustard indicated detectable levels of TDG normally < 2.5 ng/ml. When combined with TDGO analysis, the geometric mean was 3.4 ng/ml, and as high as 20 ng/ml (Young *et al.*, 2004). The source of these background levels is unknown, but TDG is used in inks and dyes for fabrics; there may also be a dietary source. For this reason, β-lyase metabolites are regarded as the more definitive biological markers of sulphur mustard poisoning, because these have not been detected in non-exposed individuals (Black and Read, 1995a; Young *et al.*, 2004). The relative amounts of β-lyase and hydrolysis products appear to be quite variable. In two casualties accidentally exposed to a WWI munition, hydrolysis products were more abundant in the urine of one casualty and β-lyase metabolites were more abundant in the other (Black and Read, 1995b). The bis-*N*-acetylcysteine conjugate (**3**) was detected in the same samples, but only at levels close to the LOD, and 15 years following collection, when a sensitive method had been developed (Read and Black, 2004b). Further analyses for this metabolite in human urine are required to determine its importance. In the rat it is a major metabolite.

PROTEIN AND DNA ADDUCTS

Adducts with *N*-terminal valine, histidine and aspartic/glutamic acid residues have been detected in samples from human casualties. *N*-Terminal valine adducts were detected in blood from Iranian casualties collected up to 26 days after exposure (Benschop *et al.*, 1997; Black *et al.*, 1997a), and in blood collected 2–3 days after exposure from the two subjects (see above) accidentally exposed from a WWI munition (Black *et al.*, 1997a). In the case of a sample collected 26 days after exposure, adduct concentration corresponded with that found in human blood after incubation with 0.9 μM sulphur mustard. The

Table 1. Analyses of samples from human casualties of deliberate or accidental exposure to sulphur mustard.

Analyte[a]	Casualties (number)	Time of sample collection	Approximate concentration	Remarks
TDG	Iranian CW (25)	Up to ~ 10 days	1–100 ng/ml (24/25) 330 ng/ml (1/25)	Mean in control samples was 5 ng/ml; highest 20 ng/ml. TDG was converted to sulphur mustard with concentrated HCl (Wils *et al.*, 1985, 1988)
Albumin-cysteine adduct			≡ 0.4–1.8 μM	Expressed as concentration of sulphur mustard that induced the same level of adduct on incubation with human blood (Noort *et al.*, 1999)
TDG + TDGO	Iranian CW (5)	5–10 days	27–69 ng/ml (3/3)	Believed to include samples from some of the casualties above
β-Lyase metabolites			0.5–5 ng/ml (4/5) 220 ng/ml (1/5)	Patient with 220 ng/ml died (Black and Read, 1995a)
Hb-valine adduct			0.3–0.8 ng/ml (4/4)	Calculated on the basis of released amino acid adduct. Control < 0.15 ng/ml
Hb-histidine adduct			0.7–2.5 ng/ml	Control < 0.3 ng/ml (Black *et al.*, 1997a)
β-Lyase metabolites	Kurdish CW (2)	13 days	0.3, 0.1 ng/ml	TDG + TDGO levels within those of control urine (Black and Read, 1995a)
Hb-valine adduct	Iranian CW (2)	22, 26 days	≡ 0.9, 0.9 μM	
DNA adduct			≡ 0.22, 0.16 μM ≡ 0.43, 0.25 μM	First figure refers to lymphocytes, second figure to granulocytes; determined by immunoassay (Benschop *et al.* 1997)
TDG	Accidental (2)	2–3 days	2, 2 ng/ml	Exposed from a WWI munition. Control urine < 1 ng/ml.
TDG + TDGO			69, 44 ng/ml	Control urine ~ 11 ng/ml
β-Lyase metabolites (**4**,**5**)			42, 56 ng/ml	Control urine < 0.1 ng/ml (Black and Read, 1995b)
β-Lyase metabolite (**4**)			15, 17 ng/ml	Analyzed by LC–MS–MS after 15 years storage (Read and Black, 2004b)
β-Lyase metabolite (**5**)			30, 34 ng/ml	
Bis-conjugate (**3**)			1, 1 ng/ml	Analysed after 15 years storage; some degradation may have occurred (Read and Black, 2004b)
Hb-valine adduct			0.3 ng/ml (1/1)	Calculated on basis of released amino acid adduct; control blood < 0.15 ng/ml
Hb-histidine adduct			2.5 ng/ml (1/1)	Control blood < 0.3 ng/ml (Black *et al.*, 1997a)
TDG	Accidental (1)	Up to 14 days	maximum 20 μg/day, ~ 65 ng/ml	Laboratory exposure. Concentrations < 10 ng/ml after day 7; 0.243 mg excreted over 14 days (Jakubowski *et al.*, 2000)

Note:
[a] metabolites in urine, protein and DNA adducts in blood

results were corroborated by immunochemical analysis for the DNA adduct in lymphocytes from the same blood samples (Benschop *et al.*, 1997). The histidine adduct was detected at slightly higher concentrations than the valine adduct in those samples analyzed for both (Black *et al.*, 1997a). In samples collected up to 10 days after exposure, signal-to noise ratios for β-lyase metabolites were generally greater than those obtained for adducts, but after a longer period adducts are the superior biomarkers.

NON-METABOLIZED SULPHUR MUSTARD

In most circumstances, absorbed sulphur mustard should be fully metabolized. It is, however, highly lipophilic and can partition into fatty tissues. High concentrations of sulphur mustard were reported in various organs and tissues, particularly abdominal fat, removed *post mortem* from a deceased casualty of severe sulphur mustard poisoning (Drasch *et al.*, 1987). Hair may also be a source of unchanged sulphur mustard. A United Nations (1986) investigation reported the detection of sulphur mustard in hair removed from a CW casualty.

NITROGEN MUSTARDS

Biological fate

METABOLISM

Three nitrogen mustards, HN-1, HN-2 and HN-3 (Figure 6), are included in Schedule 1 of the CWC. Most of the limited information on metabolism relates to HN-2, which has been used as an anti-cancer drug (mechlorethamine).

In vitro studies of HN-2 incubated with rat and rabbit liver homogenates indicated *N*-demethylation to be a significant pathway (up to 7% in 120 min), as judged by the generation of formaldehyde (Trams and Nadkarni, 1956). Loss of one of the CH_2CH_2X substituents also occurred, as indicated by the generation of acetaldehyde. In aqueous media, nitrogen mustards are hydrolysed to the corresponding ethanolamines (**9**) (Figure 6) and, like TDG, these are excretion products in rodents following exposure. No metabolites derived from conjugation of nitrogen mustards with glutathione have yet been reported.

Ethanolamines were determined quantitatively following PC administration of nitrogen mustards in the rat. Excretion in urine up to 48 h accounted for < 0.1% of the applied doses of HN-1 and HN-2, and up to ~ 0.3 % of HN-3 (absorbed doses were not determined) (Lemire *et al.*, 2004). The ethanolamines appeared to be excreted unconjugated, as treatment of the urine with β-glucuronidase had no effect.

Metabolism studies with ethanolamines have been undertaken in the context of their use as industrial chemicals. Following PC and IV administration of *N*-methyldiethanolamine in the rat, a major fraction of the absorbed dose was excreted as unidentified urinary metabolites, with some unchanged *N*-methyldiethanolamine (Leung *et al.*, 1996). Triethanolamine was excreted predominantly unmetabolized in mice following both IV and percutaneous administration (Stott *et al.*, 2000).

PROTEIN ADDUCTS

Similar to sulphur mustard, HN-2 binds covalently with the cysteine-34 residue of human serum albumin *in vitro*. Although no detection following *in vivo* exposures has been reported, analogous albumin adducts have been demonstrated in cancer patients being treated

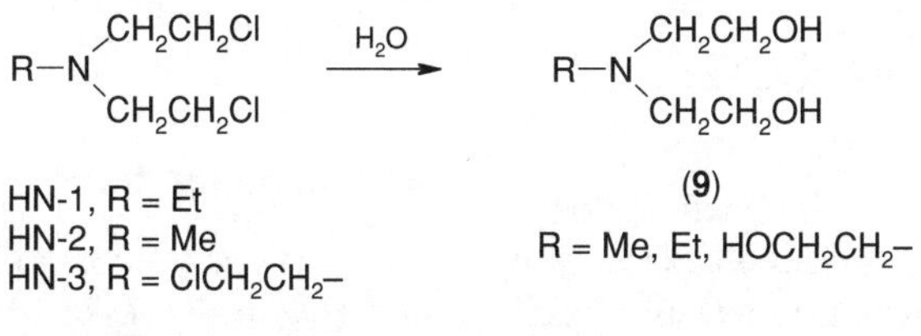

Figure 6. Nitrogen mustards and their hydrolysis products

with the anti-cancer nitrogen mustard drugs melphalan and cyclophosphamide (Noort *et al.*, 2002b). HN-2 alkylation of histidine residues in haemoglobin has been indicated (Fung *et al.*, 1975) but no definitive studies have been reported. Nornitrogen mustard, $HN(CH_2CH_2Cl)_2$, reacts with *N*-terminal valines in haemoglobin *in vitro* (Thulin *et al.*, 1996).

DNA ADDUCTS

The major mono-alkylated adduct of HN-2 with DNA is *N*-[2-(hydroxyethyl)-*N*-(2-(7-guanyl)ethyl]methylamine, analogous to the adduct formed with sulphur mustard. Alkylation of DNA in water with HN-2 gave four principal products, derived from mono-alkylation of guanine at N-7 and adenine at N-3, and from crosslinking of guanine to guanine or guanine to adenine (Osborne *et al.*, 1995). The ratio of alkylation at N-7 of guanine to N-3 of adenine was 86:14.

Analytical methods

METABOLITES

The di- and triethanolamine hydrolysis products of nitrogen mustards are best analyzed in urine by LC–ESI-MS–MS (Lemire *et al.*, 2003, 2004), with preconcentration using strong cation-exchange SPE (LODs between 0.4 and 3 ng/ml). Application of this method to 120 human urine samples found no background levels above the LODs for *N*-methyl and *N*-ethyldiethanolamines, but a high occurrence (47%) of triethanolamine, ranging from < 3 ng/ml to ~ 6500 ng/ml. Triethanolamine is therefore not an appropriate biological marker of exposure for HN-3. Triethanolamine is widely used in commercial products, including household detergents and cosmetics.

PROTEIN AND DNA ADDUCTS

The albumin adduct with HN-2 can be detected by LC–MS–MS using a similar methodology to that used for the sulphur mustard adduct (Noort *et al.*, 2002b). HPLC with UV detection (LOD, 10 ng/ml) has been reported for the HN-2 N7–guanine adduct (Sperry *et al.*, 1998), but does not appear to have been applied to human samples.

LEWISITE

Biological fate

Weapons-grade lewisite consists of lewisite I, $(CHCl–CH)AsCl_2$ (90%) and lewisite II, $(CHCl–CH)_2AsCl$ ($\sim$10%), with very small amounts of the non-vesicant lewisite III, $(CHCl–CH)_3As$ (which imparts the characteristic geranium-like odour). Biomedical sample analysis has been directed only at products derived from lewisite I.

METABOLISM

No detailed metabolism studies have been reported for lewisite I. It is rapidly hydrolysed to 2-chlorovinylarsonous acid (CVAA), which is excreted in the urine of experimental animals (Jakubowski *et al.*, 1993; Logan *et al.*, 1999). It appears to be excreted over a relatively short period.

PROTEIN ADDUCTS

Trivalent arsenic has a high affinity for thiol groups, a characteristic that is exploited in derivatization for GC–MS analysis. Consistent with this reactivity, lewisite I reacts *in vitro* with cysteine residues in haemoglobin, forming a crosslink. On incubation of human blood with ^{14}C-lewisite I, 93% of the total radioactivity was found in the erythrocytes, with 25–50% associated with globin (Fidder *et al.*, 2000). The residual radioactivity in the erythrocytes was probably bound to glutathione. LC–MS–MS of tryptic digests indicated the presence of several binding sites, and specifically identified a crosslink between the cysteine-93 and cysteine-112 residues of β-globin. Binding to albumin was not observed.

Analytical methods

CVAA is analyzed by GC–MS after derivatization (Figure 7); Like lewisite I, CVAA reacts readily with mono- and dithiols at ambient

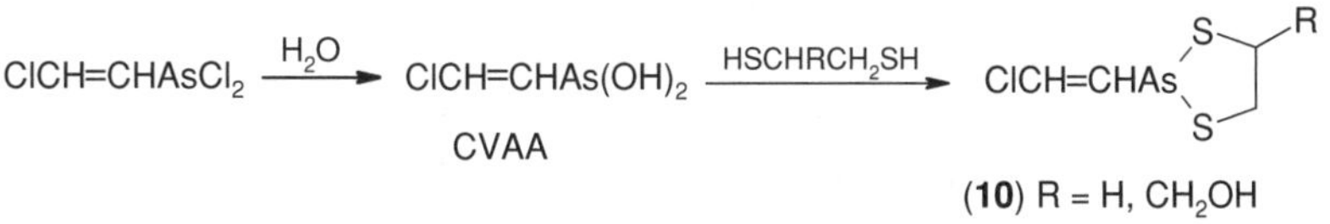

Figure 7. Hydrolysis of lewisite I to chlorovinylarsonous acid and its derivatization to 1,3-dithioarsenolines (**10**) with 1,2-dithiols

temperature. 1,2-Ethanedithiol (Jakubowski *et al.*, 1993; Logan *et al.*, 1999), 1,3-propanedithiol (Wooten *et al.*, 2002) and 2,3-dimercaptopropan-1-ol (British Anti-Lewisite, BAL) (Fidder *et al.*, 2000) have been used for biomedical sample analysis. Unlike most derivatizing reagents, which are reactive electrophiles, thiols can derivatize *in situ* in the biomedical fluid or aqueous extract. In the case of BAL as a derivatizing agent, the free hydroxyl is subsequently converted to its heptafluorobutyryl derivative. CVAA can be concentrated from urine by C_{18} SPE, either before or after derivatization, or by solid phase microextraction after derivatization. The latter provides a very sensitive method. LC–MS of CVAA gives ill-defined peaks unless it is oxidized to the pentavalent arsonic acid (Black and Muir, 2003).

Lewisite bound to haemoglobin cysteine residues is displaced by conversion to the BAL derivative shown in Figure 7 (Fidder *et al.*, 2000). Exposure of guinea pigs to lewisite (0.25 mg/kg, subcutaneous) could be demonstrated by whole blood analysis (i.e. adducts + CVAA) up to at least 240 h.

There have been no reported human exposures to lewisite for which biomedical samples have been analyzed.

NERVE AGENTS

The three types of nerve agent known to have been weaponized are typified by sarin (GB), VX and tabun (GA). Soman (GD) and cyclosarin (GF) are less volatile phosphonofluoridate analogues of sarin, and RVX or R-33 is a Russian analogue of VX with broadly similar properties. Tabun differs from the other nerve agents in that it does not possess a *P*-methyl substituent, which has important implications for retrospective identification. A large number of human biomedical samples were collected following terrorist attacks with sarin in Matsumoto City in 1994 and the Tokyo subway in 1995, and these have contributed substantially to the development of analytical methods.

Biological fate

DISTRIBUTION AND METABOLISM

Following systemic administration, nerve agents are rapidly distributed throughout the tissues, partly as unchanged agent, partly as metabolites and partly bound irreversibly to serine esterase-type enzymes and other proteins. Approximately 90% of subcutaneous doses (0.075 mg/kg) of sarin and cyclosarin in the rat were excreted in the urine within 24 h, as phosphonic acids resulting primarily from enzymatic hydrolysis (Shih *et al.*, 1994). Soman was eliminated more slowly with a biphasic elimination curve; approximately 50% was excreted within the first 24 h, rising to 62% after 7 days. The first phase of elimination is due to rapid enzymatic hydrolysis of the inactive P(+) isomers by phosphorylphosphatases; the second phase is from slower hydrolysis of the active P(-) isomers (Benschop and De Jong, 2001). VX, which is a poor substrate for phosphorylphosphatases, disappears from the blood at a much slower rate than sarin and soman. Toxicologically relevant levels of VX were present in guinea pig blood 10–20 h following IV administration ($2 \times LD_{50}$) compared to < 2 h for GB or GD (Van der Schans, M.J. *et al.*, 2003). Following PC exposure, which is the main battlefield hazard of VX, blood levels of agent may be present for significantly longer.

The metabolism of nerve agents is dominated by hydrolysis. This occurs to a small extent by chemical reaction with water, but is predominantly mediated by

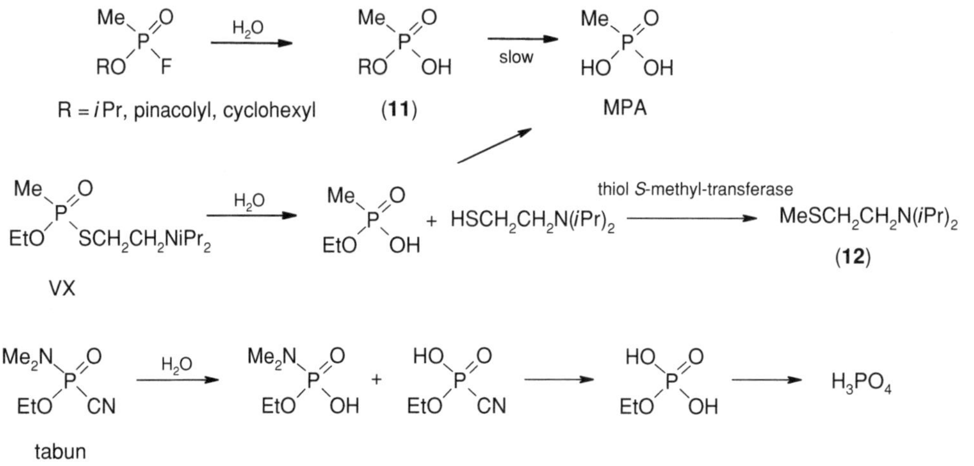

Figure 8. Metabolic pathways of nerve agents

phosphorylphosphatases. The main hydrolytic pathways that occur in the body are shown in Figure 8. Alternative hydrolytic pathways are possible for VX (through P–O and O–C cleavage) but these have not been reported in animal studies. For phosphonofluoridates (sarin, soman and GF) and V agents, the major metabolites are alkyl methylphosphonic acids (**11**). Further hydrolysis to methylphosphonic acid (MPA) may proceed slowly, as indicated by analyses of victims of the Tokyo subway attack (Nakajima *et al.*, 1998). In rats, only traces of MPA were observed in urine (Shih *et al.*, 1994).

Tabun hydrolyses by two pathways, through P–N and P–CN cleavage. It has not yet been established which of these predominates in animals or humans. Further hydrolysis produces ethyl phosphoric acid and eventually phosphate, both of which are ubiquitous from other sources.

Hydrolysis of VX produces 2-diisopropylaminoethanethiol, $HSCH_2CH_2N(i\text{Pr})_2$. A metabolite (**12**), derived from enzymatic *S*-methylation of this hydrolysis product, was identified in human plasma following an assassination with VX (Tsuchihashi *et al.* 1998). Experiments in rats confirmed the rapid metabolic formation of (**12**) from $HSCH_2CH_2N(i\text{Pr})_2$ (Tsuchihashi *et al.*, 2000). This metabolite has not been reported in urine. By analogy with β-lyase metabolites of sulphur mustard, it might be excreted as a sulphoxide.

PROTEIN ADDUCTS

Acetyl and butyrylcholinesterase

The biochemical target for nerve agents is the enzyme AChE. Nerve agents inhibit the enzyme by catalysed phosphylation of a serine hydroxyl group within the active site. The phosphylated enzyme regenerates extremely slowly, unless reactivation is accelerated by particular nucleophiles such as oximes or fluoride. Nerve agents react similarly with the related enzyme BuChE (Figure 9), which acts as a scavenger of nerve agents; its main physiological function has yet to be elucidated. In rodent species and the rabbit, carboxylesterases (which are also serine proteases) act as additional scavengers. Adducts with the range of nerve agents have been identified by LC–MS–MS (Fidder *et al.*, 2002; Elhanany *et al.*, 2001).

AChE and BuChE, which have half-lives of 5–16 days, provide excellent biomarkers, but with one disadvantage. With certain nerve agents, particularly soman, a rapid secondary reaction occurs within the active site, in which the phosphyl moiety is dealkylated. This process, known as 'ageing', leads to loss of structural information on the inhibitor, for example with soman the pinacolyl group is lost (Figure 9). It also results in a negatively charged phosphyl moiety, which is resistant to reactivation by oximes and fluoride ion. Half-times quoted for ageing of human red blood cell AChE *in vitro* are: soman, 2–6 min;

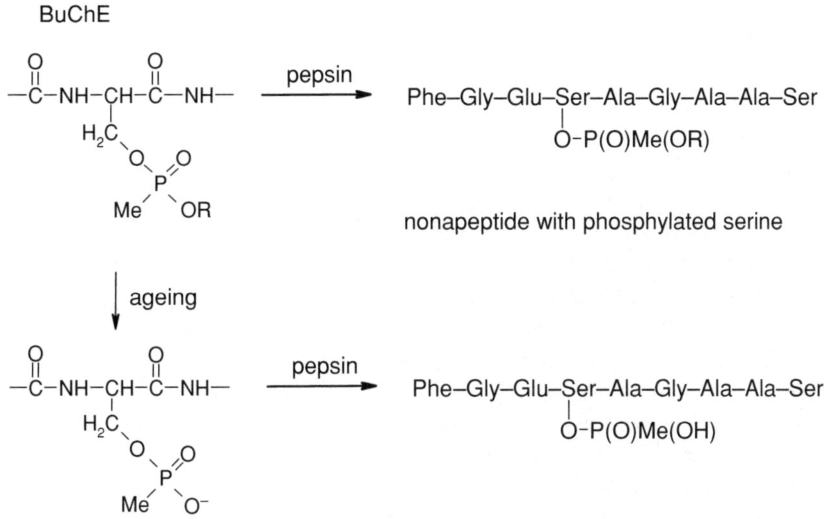

Figure 9. Nerve agent adducts with BuChE and their digestion with pepsin

sarin, 3 h, 5 h; tabun, 13 h, > 14 h; GF 40 h, 7.5 h; VX, 48 h (Dunn *et al.*, 1997).

Adducts with albumin

Nerve agents also react with a tyrosine residue associated with the albumin fraction in blood (Black *et al.*, 1999) (Figure 10). Analysis of tryptic digests from plasma incubated with sarin identified a phosphylated tripeptide, MeP(O)(O^{i}Pr)–Tyr–Thr–Lys, consistent with the protein being albumin (tyrosine residue 411), although this sequence is common and occurs in other proteins. Before the advent of modern mass spectrometry, diisopropyl fluorophosphate was reported to bind to a tyrosine residue in bovine serum albumin (Murachi, 1963). The reaction with tyrosine is assumed to be purely chemical, and therefore at low exposure levels a catalytic reaction with ChE should predominate. At higher exposure levels, tyrosine adducts are formed at significant concentrations, both in human blood *in vitro* and in guinea pigs *in vivo*. Detectable levels of tyrosine adducts occur at BuChE inhibition levels of 10–20% for soman, GF and tabun, and at 70% for sarin in incubates with human plasma *in vitro* (Harrison *et al.*, 2007). For example, when tabun was incubated with human plasma, levels of adduct remained below the limit of detection until ~15% of the BuChE was inhibited; there

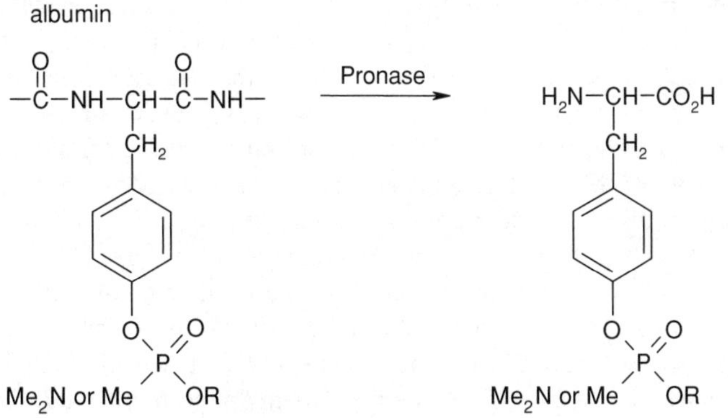

Figure 10. Adducts of G agents with a tyrosine residue on albumin

then followed a steep rise in the concentration of the adduct. Adducts with all four agents have been detected in guinea pigs exposed to 0.5 or 2 × LD_{50} doses. The tyrosine adducts do not appear to age rapidly and can still be detected after therapeutic treatment with oximes (which should substantially reduce the amounts of non-aged phosphylated AChE and BuChE). Non-aged adducts with soman and tabun were detected 7 days after exposure to 5 × LD_{50} doses after treatment with oxime, atropine and anticonvulsant. Consistent with the reaction being entirely chemical, a tyrosine adduct with the less reactive VX was observed in blood *in vitro* only at high concentrations. The probable formation of tyrosine adducts in a non-human primate is supported by the fluoride regeneration of soman (2.4 ng/ml) from plasma 3 days after an exposure to 2 × LD_{50} (intramuscular), when all of the inhibited BuChE should have aged (Adams *et al.*, 2004). These studies also showed that agent could be regenerated from incubates of agent with human serum albumin. Albumin was recently shown to bind a biotinylated organophosphorus agent in mice, with diisopropyl phosphate and some organophosphorus pesticides competing for the same site (Peeples *et al.*, 2005).

Analytical methods

METABOLITES

There are numerous methods for the analysis of alkyl methylphosphonic acids in urine and blood, mostly using GC–MS and GC–MS–MS. These have been recently reviewed (Black and Muir, 2003; Black and Noort, 2005) and only a representative selection is summarized here. Isolation from urine is usually achieved by hydrophobic SPE (C_{18}, C_8 or polymeric) at low pH, or by anion-exchange SPE. Phosphonic acids require derivatization for GC–MS analysis, and at least four different derivatives have been applied to biomedical samples. Silylation (trimethylsilyl or *tert*-butyldimethylsilyl) are commonly used for environmental analysis and were used in the analyses of samples from casualties of the Matsumoto and Tokyo terrorist attacks (e.g. Minami *et al.*, 1997; Nakajima *et al.*, 1998). In these cases, the first samples were collected within hours of the exposure and the detection limits obtainable with silylation were sufficient. A GC–MS–MS method using methyl esters has been developed for high-throughput analysis (Baar *et al.*, 2004). The lowest limits of detection, down to at least 0.1 ng/ml, have been obtained by conversion to pentafluorobenzyl esters and detection by negative-ion chemical ionization (Shih *et al.*, 1991; Fredriksson *et al.*, 1995; Miki *et al.* 1999). Although these methods were initially developed using research-grade mass spectrometers, simple ion-trap instruments can achieve comparable detection limits (Riches *et al.*, 2005). LC–MS–MS is less sensitive but LODs of 1–4 ng/ml were achieved for isopropyl methylphosphonic acid in serum and applied successfully to samples from Japanese casualties (Noort *et al.*, 1998). The metabolite (**12**) derived from VX can be simply extracted from serum and analysed by GC–MS (Tsuchihashi *et al.*, 1998).

PROTEIN ADDUCTS

Historically, the inhibition of BuChE in plasma has been used to monitor exposure to nerve agents using the classical Ellman colorimetric method (Ellman *et al.*, 1961), or modifications thereof (Worek *et al.*, 1999). This method is used routinely in occupational health monitoring, and could be used to rapidly screen casualties. Its disadvantages are that detection of low-level exposure requires previous baseline measurements, and the assay is non-specific with regard to the inhibitor.

Mass spectrometric detection of phosphylated BuChE provides a much more specific biomarker of exposure. BuChE is usually preferred to AChE because it is much more abundant in blood plasma than AChE is in erythrocytes. A versatile method of detection involves isolation of BuChE from plasma using affinity SPE, digestion with pepsin, and LC–MS–MS detection of a phosphylated nonapeptide encompassing the active site serine (Fidder *et al.*, 2002). In the case of aged residues, this method will only identify part of the structure of the nerve agent although, provided it is a phosphonofluoridate or V agent, a methylphosphonyl MeP(O)-residue will be a clear indication that a nerve agent has been used (there are no pesticides in use that include a Me–P(O) moiety; fonofos has Et-P(O)).

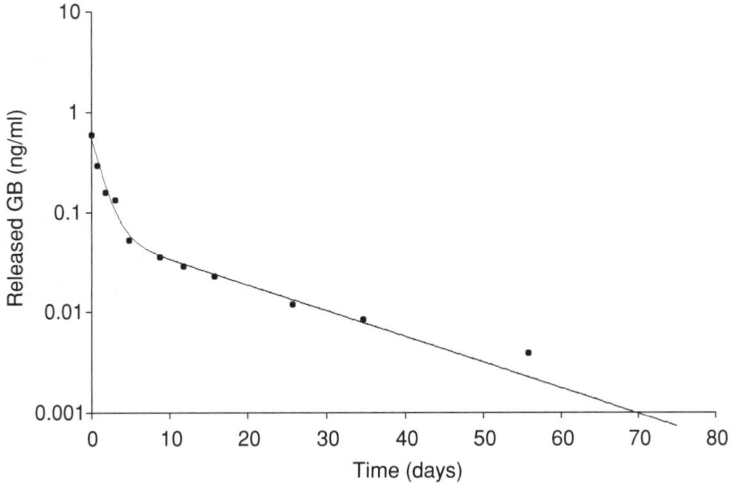

Figure 11. Fluoride-induced regeneration of sarin from sarin-inhibited BuChE in plasma of a rhesus monkey after IV administration. The rhesus monkey received a final dose of 0.7 μg/kg (taken from Van der Schans, M.J. *et al.*, 2004). Reprinted from *Archives of Toxicology*, **78**: 2004, 508–524, 'Retrospective detection of exposure to nerve agents', by M. J. Van der Schans *et al.*, original copyright notice with kind permission of Springer Science and Business Media

An alternative, and experimentally less demanding method, displaces the organophosphorus moiety as a fluoridate (e.g. with sarin-inhibited BuChE the original nerve agent is regenerated), either from plasma BuChE (Polhuijs *et al.*, 1997) or red cell AChE (Jakubowski *et al.*, 2004). The alkyl methylphosphonyl fluoridate is readily extracted and detected using GC–MS (or GC–FPD). Fluoride reactivation is currently the most sensitive method for detecting ChE inhibition by nerve agents, enabling detection in blood samples with < 1 % BuChE inhibition (Degenhardt *et al.*, 2004). The method has been demonstrated to be effective for GA, GB, GF and VX. Figure 11 shows detection up to 56 days following exposure of atropinized rhesus monkeys to a dose producing an initial 40% inhibition of BuChE (Van der Schans *et al.*, 2004).

In the case of sarin-inhibited ChE, the phosphyl moiety has also been displaced as isopropyl methylphosphonic acid using trypsin digestion and alkaline phosphatase (Nagao *et al.*, 1997; Matsuda *et al.*, 1998).

Albumin adducts can be detected by LC-MS-MS as phosphylated tyrosine, after digestion of the albumin fraction with Pronase and SPE clean-up (Harrison *et al.*, 2000b).

Human exposures

The only incidents of nerve agent poisoning where biomedical samples have been reported are those resulting from terrorist disseminations of sarin in Matsumoto City (1994) and the Tokyo subway (1995), plus an assassination using VX. Positive analyses are summarized in Table 2. In contrast to the CW incidents involving sulphur mustard, most of the biomedical samples were collected within hours of the incidents.

METABOLITES

Very high levels of isopropyl methylphosphonic acid (*i*PrMPA) and methylphosphonic acid (MPA) were detected in the urine of a Matsumoto casualty with low blood AChE activity and rendered unconscious (Nakajima *et al.*, 1998). Urine was collected over a 7-day period. Estimated concentrations of *i*PrMPA declined from 760 ng/ml to 10 ng/ml and MPA from 140 ng/ml to below the LOD. A crude estimate of the total exposure was 2.8 mg. *i*PrMPA was also detected in urine collected over 7 days from casualties of the Tokyo attack (Minami *et al.*, 1997). Concentrations were not reported but the estimated exposures were 0.13–0.25 mg

Table 2. Analysis of samples from casualties of Japanese terrorist incidents, exposed to sarin or VX

Analyte/matrix	Casualties (number)	Time of sample collection	Approximate concentration	Remarks
*i*PrMPA/urine	Matsumoto (1)	< 1–7 days	760 ng/ml (day 1) 80 ng/ml (day 2) 10 ng/ml (day 7)	Severe casualty. Quantitation estimated on basis of response to MPA standard. Total excretion estimated as 2.1 mg
MPA/urine			140 ng/ml (day 1) 20 ng/ml (day 3)	Total excretion estimated as 0.45 mg (Nakajima *et al.*, 1998)
*i*PrMPA/urine	Tokyo (not given)	~ 1 h–7 days	See remarks	In one severely intoxicated patient, maximum excretion was at 12 h, expressed as 3.4 μl/g creatinine (Minami *et al.*, 1997)
*i*PrMPA/MPA from blood AChE	Tokyo (4)	Few days, *post-mortem*	Not given	Full-scan MS spectra obtained of TMS derivatives (Nagao *et al.*, 1997)
MPA from brain AChE			Not given	*Post-mortem* tissue samples stored for ~2 years. No *i*PrMPA was detected (Matsuda *et al.*, 1998)
*i*PrMPA/serum	Matsumoto, Tokyo (18)	1.5–2.5 h	< 1–136 ng/ml	Analysis by LC–MS–MS (Noort *et al.*, 1998)
Sarin from BuChE/serum			0.2–4.1 ng/ml (12/18)	Fluoride reactivation used (Polhuijs *et al.*, 1997)
BuChE/serum			10–20 pM	Analysed as peptic nonapeptide (Fidder *et al.*, 2002)
Ethyl MPA/serum	Assassination (1)	1 h	1250 ng/ml	PC exposure on neck, died after 10 days, analyzed after 6 months storage (Tsuchihashi *et al.*, 1998)
VX metabolite (**12**)			143 ng/ml	

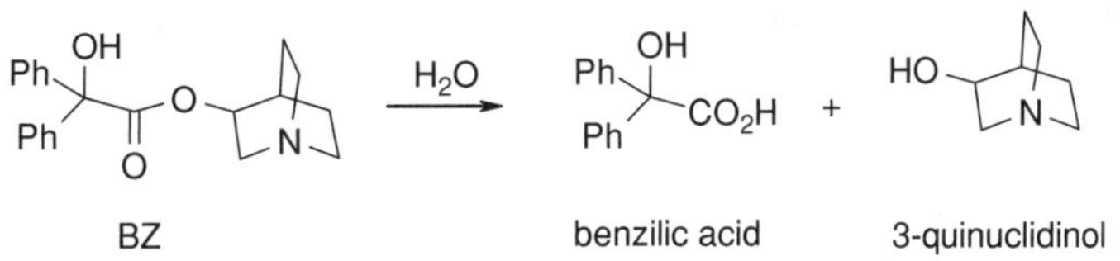

Figure 12. BZ and its hydrolysis products

of sarin in a comatose patient and 0.016–0.032 mg in less severely intoxicated patients. *i*PrMPA was detected by LC–MS–MS in serum collected within 2 h of hospitalization at concentrations of 3–136 ng/ml in 4 casualties of the Matsumoto incident and 2–100 ng/ml in 13 casualties of the Tokyo attack (Noort *et al.*, 1998). High levels of *i*PrMPA correlated with low levels of BuChE activity. Ethyl methylphosphonic acid and 2-diisopropylaminoethyl methyl sulphide (**12**) were detected by GC–MS and GC–MS–MS in the serum of a subject assassinated by application of VX from a disposable syringe to the neck (Tsuchihashi *et al.*, 1998, 2000).

PROTEIN ADDUCTS

Application of fluoride reactivation to serum samples of casualties of the Matsumoto and Tokyo incidents yielded sarin concentrations in the range 0.2–4.1 ng/ml serum (Polhuijs *et al.*, 1997). Hydrolytic displacement of the phosphonyl residue identified *i*PrMPA at levels sufficient for full-scan mass spectra to be obtained from samples collected from casualties who died (Nagao *et al.*, 1997); MPA was also identified. MPA was detected in formalin-fixed brain tissues some two years later using a similar procedure (Matsuda *et al.*, 1998). The phosphonylated peptic nonapeptide from BuChE was identified in serum samples from several casualties of the subway attack (Fidder *et al.*, 2002).

QUINUCLIDINYL BENZILATE, BZ

BZ, 3-quinuclinidinyl benzilate, is an antagonist at central and peripheral muscarinic cholinergic receptors. It produces both mental and physical incapacitation in humans.

Biological fate

No metabolism studies have been reported but it has been assumed that its hydrolysis products, benzilic acid and 3-quinuclidinol (Figure 12), would be excreted in urine.

Unlike the vesicants and nerve agents, BZ is not a reactive electrophile (other than with regard to hydrolysis) and no adducts are known to be formed either directly or indirectly through reactive metabolites.

Analytical methods

Little attention has been paid to the analysis of biomarkers of BZ. A method for detecting BZ, benzilic acid and 3-quinuclidinol in urine, isolated the metabolites (at different pH values) by SPE (C_{18} for BZ and benzilic acid, Florisil for 3-quinuclidinol) (Byrd *et al.*, 1987, 1988). They were converted to their trimethylsilyl derivatives and analyzed by GC–MS. Urine was pretreated with glucuronidase, although there was no evidence for the formation of glucuronide conjugates. LODs were in the range 0.5–5 ng/ml. LC–MS–MS methods are currently being developed.

PHOSGENE

Biological fate

METABOLISM

No detailed metabolism studies of phosgene have been reported, although some relevant information is available from its occurrence as an active metabolite (through P450-dependent oxidation) of chloroform and carbon tetrachloride. Phosgene is a highly reactive electrophile with a half-life < 1 s in water. It reacts chemically with two molecules of glutathione to form diglutathionyl dithiocarbonate, GluS–C(O)–SGlu, (Fabrizi *et al.*, 2001) and with cysteine to form 2-oxothiazolidine-4-carboxylic acid (Kubic and Anders, 1980). Diglutathionyl dithiocarbonate has been detected in bile following IP administration of chloroform and carbon tetrachloride to phenobarbital-treated rats (Pohl *et al.*, 1981). However, it has not been established if significant amounts of these or related compounds are excreted in urine following exposure to phosgene. As an active metabolite of chloroform, phosgene reacts with the polar heads of phospholipids (Di Consiglio *et al.*, 2001).

PROTEIN ADDUCTS

Consistent with its broad reactivity with nucleophiles, ^{14}C-phosgene binds to haemoglobin and albumin upon *in vitro* exposure of human blood (Noort *et al.*, 2000). Following Pronase digestion of globin, LC–MS–MS identified the pentapeptide O–C–(Val–Leu)–Ser–Phe–Ala, apparently derived from hydantoin formation between the N-terminal valine and leucine residues of the 1–5 *N*-terminal amino acids of α-globin. This adduct was rejected as a suitable biomarker of exposure because a peptide with similar properties was detected at trace levels in control samples, possibly formed by reaction of α-globin with carbon dioxide. LC–MS–MS of tryptic and V8 protease digests of human serum albumin identified crosslinking (via a urea linkage) of the lysine residues 195 and 199. ^{14}C-Chloroform has been shown to bind *in vivo* to histones, which are lysine-rich nuclear proteins (Diaz Gomez and Castro, 1980). In a model chemical system, representing amino acid residues 1–23 of human histone H2B, phosgene was shown to bind predominantly to lysine residues leading to cross linkages (Fabrizi *et al.*, 2003).

Analytical methods

Micro-LC-MS–MS enabled the adducted tryptic fragment from albumin to be detected in human blood exposed to ≥ 1 μM phosgene *in vitro*. The biomarker has not been demonstrated in animals exposed *in vivo*.

HYDROGEN CYANIDE

Biological fate

DISTRIBUTION AND METABOLISM

Hydrogen cyanide (pK_a, 9.22) is distributed in the body as hydrogen cyanide rather than cyanide ion. Inhaled hydrogen cyanide passes immediately into the systemic circulation and is distributed rapidly throughout the tissues (World Health Organization, 2004). The major proportion of cyanide in blood concentrates in the red cells bound to methaemoglobin, which acts as a cyanide sink. Cyanide is metabolized predominantly in the liver by conversion to thiocyanate (^{-}SCN), through the action of the mitochondrial enzyme rhodanese, which catalyses the transfer of a sulphur atom from thiosulphate. Cyanide is excreted primarily as thiocyanate in urine. Approximately 15% of absorbed cyanide reacts with cystine to form 2-iminothiazolidine-4-carboxylic acid, which is in equilibrium with its tautomer 2-aminothiazoline-4-carboxylic acid (**13**) (Figure 13).

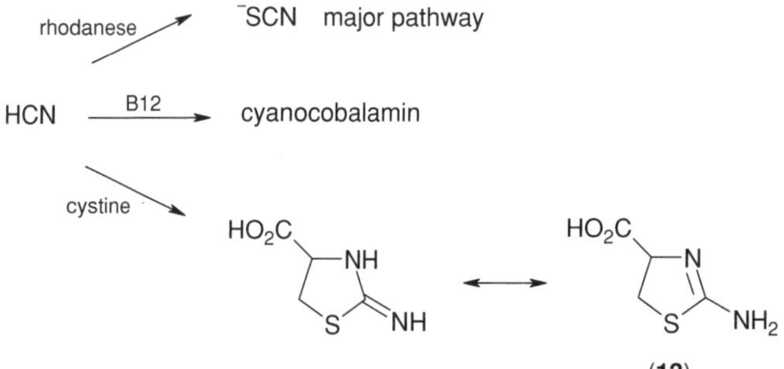

(**13**)

Figure 13. Metabolism of hydrogen cyanide

A small percentage reacts with vitamin B12, appearing as cyanocobalamin in urine.

Analytical methods

Cyanide can be regenerated from methaemoglobin by treatment with acid. There are many methods for detecting free and bound cyanide, and thiocyanate in blood; these have recently been reviewed (Black and Muir, 2003; Black and Noort, 2005). They have been applied mostly to the determination of cyanide levels in smokers and fire victims rather than cases of deliberate poisoning. 2-Aminothiazoline-4-carboxylic acid has been analysed in urine using LC with fluorescence detection after conversion to *N*-carbamylcysteine (Lundquist *et al.*, 1995), and by GC–MS after conversion to its tris-trimethylsilyl derivative (Logue *et al.*, 2005). The analyte was concentrated from urine on a cation exchange resin.

Human exposures

Retrospective identification of cyanide poisoning in a CW context would be complicated by exposure from other sources, which include cigarette smoke, smoke from fires and some foods, e.g. cyanogenic glycosides in bitter almonds, fruit seeds and a number of plants. Quoted blood concentrations in non-smokers vary from a few ng/ml to >100 ng/ml. In nine fire victims, the concentrations determined were 687 ± 597 ng/ml (Ishii *et al.*, 1998). In smokers, cyanide levels in blood may rise to ~ 500 ng/ml. 2-Aminothiazoline-4-carboxylic acid was detected in the urine of moderate cigarette smokers at concentrations between < 44–162 ng/ml (Lundquist *et al.*, 1995).

RIOT CONTROL AGENTS

The major riot control agents, or 'aids to arrest', in current use are 2-chlorobenzylidene malononitrile (CS), 1-chloroacetophenone (CN) and capsaicin, *N*-(4-hydroxy-3-methoxybenzyl)-8-methyl-6-nonenamide (or pepper spray); the potent and persistent irritant dibenz[b,f]1:4-oxazepine (CR) has rarely been used (Olajos and Salem, 2001). Methods for the retrospective identification of exposure to RCAs have not yet been developed; the known metabolic pathways are summarized below.

CS

Biological fate

METABOLISM

Chemically, CS is a moderately reactive electrophile due to the presence of two electron-withdrawing nitrile groups attached to the olefinic bond. The site of reaction with nucleophiles is the olefinic carbon adjacent to the aromatic ring. Reactions are SN_2-like, i.e. CS reacts with nucleophiles directly in a bimolecular fashion. CS reacts quite rapidly with water when in solution (half-life ~ 14 min, 25°C, pH 7.4) to give 2-chlorobenzaldehyde and malonitrile. Reactions are much faster with thiols and amines. CS reacts rapidly with glutathione and plasma protein, although the reaction products have not been characterized (Cucinell *et al.*, 1971).

The metabolism of CS is dominated by two pathways, the major one resulting from initial hydrolysis to 2-chlorobenzaldehyde and malononitrile (Figure 14), and the minor one from reduction to 2-chlorobenzylmalononitrile (dihydro-CS) (Figure 15) (Brewster *et al.*, 1987). 2-Chlorobenzaldehyde (**14**) is further metabolized by hepatic oxidation to 2-chlorobenzoic acid (**15**), which is conjugated with glycine to form the major urinary metabolite 2-chlorohippuric acid (**16**). A minor metabolic pathway of 2-chlorobenzaldehyde is through reduction to 2-chlorobenzyl alcohol (**17**), which is excreted as glucuronide (**18**) and the *N*-acetylcysteine conjugate (**19**). The formation of the latter is postulated to occur via a sulphate intermediate (Rietveld *et al.*, 1983). In these metabolic pathways, three of the carbon atoms of CS are lost as malononitrile, which is further metabolized to cyanide and thiocyanate. In the second metabolic pathway (Figure 14), CS is reduced to the hydrolytically more stable dihydro-CS (**20**), which is metabolized by hydrolysis with decarboxylation and excreted as the glycine conjugate (**21**), or by simple hydrolysis to the corresponding carboxamide

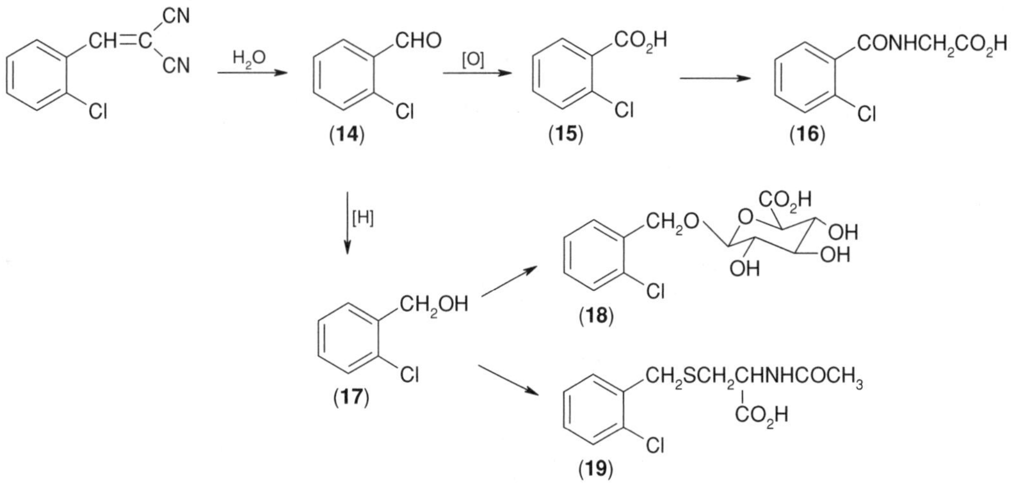

Figure 14. Metabolic pathway of CS initiated by hydrolysis

(**22**) and carboxylic acid (**23**). In this pathway, the carbon skeleton is either fully retained or loses two carbon atoms. No protein adducts of CS have been reported.

Human exposures

There have been no reported cases where exposure to CS has been confirmed by biomedical sample analysis. In human volunteer trials, with low concentrations of aerosolized CS, neither CS or 2-chlorobenzaldehyde were detected in the blood of six volunteers shortly after termination of the exposure (Leadbeater, 1973). A disadvantage of the major metabolic pathway is that three carbons are lost in the initial hydrolysis. It has yet to be shown if background levels of metabolites from this pathway exist in non-exposed individuals, resulting from environmental, dietary or drug exposure, e.g. chlorobenzoic acid is used as a preservative for glues and paints, and as an intermediate in the manufacture of fungicides and dyes. The alternative pathway, originating from reduction to dihydro-CS, should produce more definitive biomarkers but they account for a very low percentage of the dose in the rat.

CN

CN is a less reactive electrophile than CS. Its reaction rate constant with phosphate buffer at pH 7.2 is approximately two orders of magnitude less than that for CS (Cucinell *et al.*, 1971). It

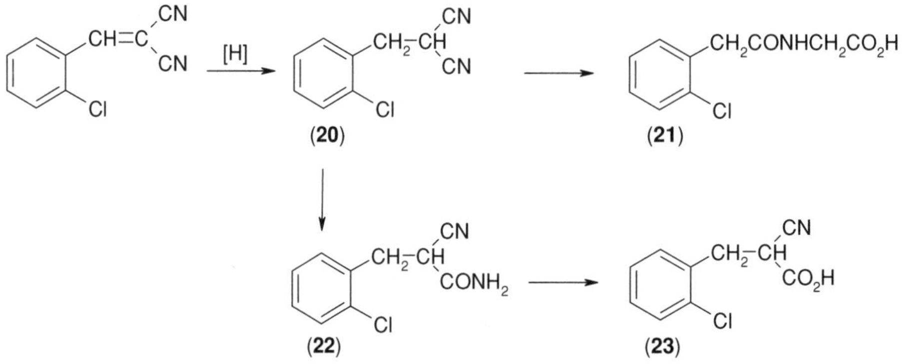

Figure 15. Metabolic pathway of CS initiated by reduction

reacts faster with sulphur-based nucleophiles than with water, but still at a slower rate than CS. The reaction products of CN with nucleophiles have been poorly characterized. Although it has been in use much longer than CS, no metabolism studies appear to have been reported. It will inhibit a number of sulphydryl-based enzymes such as lactic dehydrogenase (Mackworth, 1948), and so it may form covalent adducts with blood proteins.

CR

METABOLISM

Chemically CR is much less reactive than CS or CN, and is hydrolyzed relatively slowly. In the rat, following IV and intragastric administration of ^{14}C-CR, most of the radioactivity (59–93%) was excreted in the urine, of which > 90% was excreted in the first 24 h (French *et al.*, 1983a). Similar patterns were observed in the guinea pig and rhesus monkey after intragastric administration. The major metabolic pathway (Figure 16) is initiated by oxidation of the azomethine moiety to lactam (**24**). This is followed by hydroxylation at aromatic positions 4, 7 or 9 to give hydroxy-lactams (**25–27**), with excretion predominantly as sulphate conjugates in the rat and guinea pig, and as the non-conjugated hydroxy-lactams in the rhesus monkey. C-7 is the major site of hydroxylation in all three species, the 7-hydroxy-lactam and its sulphate conjugate accounting for up to ~ 60% of the administered dose. Deuterium-labelling studies indicated that hydroxylation occurs via an arene epoxide intermediate (cf. benzene) (Harrison *et al.*, 1978), but no mercapturic acid conjugates have been identified derived from reaction of a putative epoxide intermediate with glutathione.

Minor pathways arise from hydrolytic or oxidative cleavage of the azomethine moiety and reduction to dihydro-CR (French *et al.*, 1983b). No samples from human exposures have been reported.

Capsaicin

Capsaicin is the major pungent component of Oleoresin Capsicum (OC), commonly known as pepper spray. OC is extracted from dried ripe chilli peppers and is a variable mixture of many compounds. Related irritants (capsaicinoids) present in the mixture include dihydrocapsaicin, nordihydrocapsaicin, homocapsaicin, homodihydrocapsaicin and nonivamide. The latter is used as a synthetic substitute for pepper spray.

METABOLISM

The metabolism of capsaicin and dihydrocapsaicin has been studied in the context of food and their use in skin creams for the treatment of arthritic pain and inflammation. Capsaicin (Figure 17) is a much more complex molecule than other RCAs, and offers more functional groups and other sites for metabolism. ω-Hydroxycapsaicin (**28**) was detected in the urine

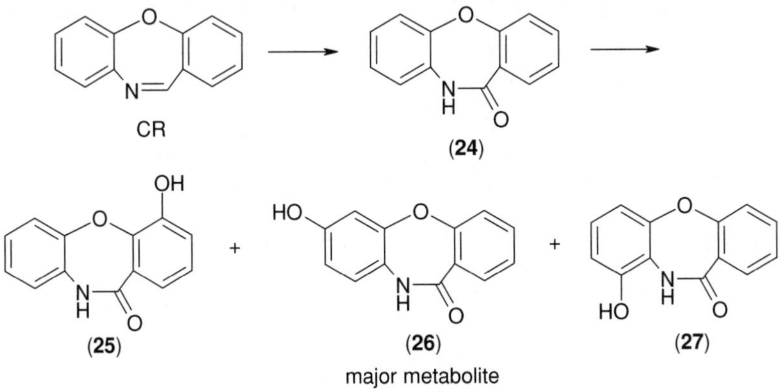

Figure 16. Major metabolic pathway of CR, with excretion occurring mainly as sulphates in the rat and guinea pig and mainly as non-conjugated species in the rhesus monkey

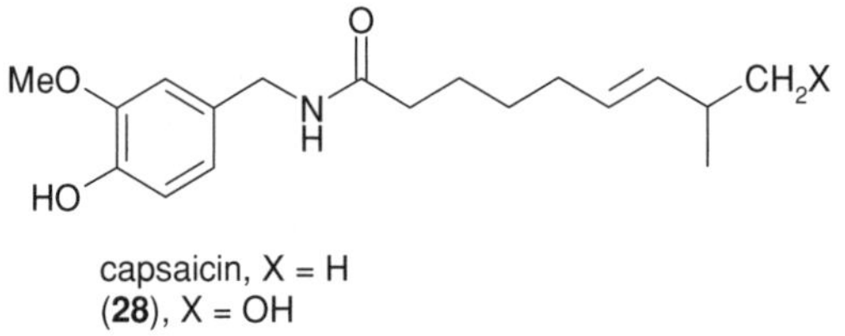

Figure 17. Capsaicin and its urinary metabolite ω-hydroxycapsaicin

of rabbits following IV administration of capsaicin (Surh *et al.*, 1995). Following oral administration of dihydrocapsaicin (20 mg/kg) in the rat, ∼ 75% of the dose was eliminated in the urine as unchanged dihydrocapsaicin plus eight metabolites (Kawada and Iwai, 1985). The metabolites were all derived from an initial hydrolytic cleavage of the carboxamide function and consisted of free metabolites (14.5% of the total dose) and glucuronide conjugates (60.5%). The following were identified: dihydrocapsaicin (8.7%), vanillamine (**29**) (4.7%), vanillin (**30**) (4.6%), vanillyl alcohol (**31**) (37.6%) and vanillic acid (**32**) (19.2%) (Figure 18). Dihydrocapsaicin-hydrolyzing enzyme activity was found in various organs of the rat, but particularly in the liver and gut.

The other product of hydrolysis is initially 8-methylnonanoic acid, but it is not clear to what extent this is metabolized.

One of the concerns for capsaicins as food components is the potential for the aromatic phenolic moiety to be oxidised by P450 enzymes to potentially carcinogenic epoxide, phenoxy radical or quinone-type electrophilic intermediates. Detailed studies of the metabolism of capsaicin *in vitro*, by recombinant cytochrome P450 enzyme and hepatic and lung microsomes, were reported by Reilly *et al.* (2003) and Reilly and Yost (2005). Metabolites were identified that were derived from aromatic and alkyl hydroxylation, aryl *O*-demethylation, alkyl dehydrogenation and an additional ring oxygenation. Addition of GSH to microsomal incubations with capsaicin trapped several reactive intermediates as GSH adducts, although these have not been reported as metabolites in animal studies.

Human exposures

Confirming an exposure to capsaicin is likely to be complicated by the presence of background levels of metabolites. Capsaicin and dehydrocapsaicin are commonly ingested as hot

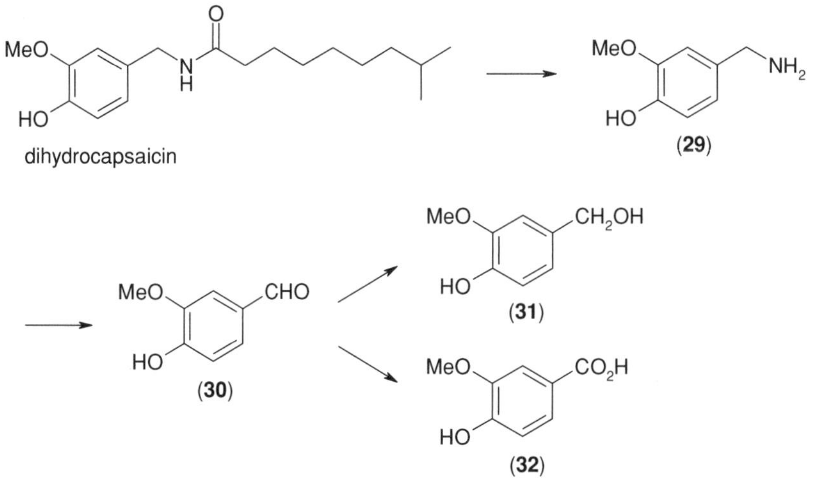

Figure 18. Metabolites of dihydrocapsaicin identified in the rat (IG administration)

chilli spices. Vanillin, an intermediate in the hydrolytic pathway, is used extensively as a food additive.

SUMMARY

Free metabolites and protein adducts for sulphur mustard and nerve agents, and DNA adducts for sulphur mustard, have been identified in experimental animals. Their validity as biomarkers of exposure has been demonstrated in samples collected from human casualties of accidental or deliberate exposure. Metabolites derived from hydrolysis have been detected in animal studies for nitrogen mustards and lewisite, plus adducts with either albumin or haemoglobin. These have not been demonstrated in human samples. In the case of HN-3, high and variable levels of the hydrolysis product, triethanolamine, in normal human urine precludes its use for confirming an exposure. No suitable biomarker for phosgene has yet been identified in animal studies although a protein adduct has been identified *in vitro*. Background levels of cyanide from sources such as cigarette smoke, fire smoke and certain food constituents make confirmation of cyanide exposure difficult.

Detailed metabolism studies have been reported for the RCAs, CS and CR, and to a lesser extent, capsaicin, but sensitive analytical methods for the metabolites have yet to be developed. The formation of covalent adducts with proteins has been little studied, although observations have suggested that CS and CN react with proteins. In the case of CS and capsaicin, major metabolites are derived from an initial hydrolysis with loss of some of the carbon skeleton, and it needs to be established if background levels of these metabolites occur in non-exposed individuals.

Analytical methods continue to be improved, and in some cases adapted for less costly instrumentation. At present, expertise in biomedical sample analysis for CW agents is restricted to a small number of laboratories. Recent interest shown by the OPCW, and the concern for terrorist use of CW, may encourage a larger number of laboratories to acquire expertise.

REFERENCES

Adams TK, Capacio BR, Smith JR *et al.* (2004). The application of the fluoride reactivation process to the detection of sarin and soman nerve agent exposures in biological samples. *Drug Chem Toxicol*, **27**, 77–91.

Bakke J and Gustafsson J-Å (1984). Mercapturic acid pathway metabolites of xenobiotics: generation of potentially toxic metabolites during entereohepatic circulation. *TIPS*, **5**, 517–521.

Barr JR, Driskell WJ, Aston LS (2004). Quantitation of metabolites of the nerve agents sarin, soman, cyclosarin, VX and Russian VX in human urine using isotope-dilution gas chromatography-tandem mass spectrometry. *J Anal Toxicol*, **28**, 372–378.

Bartlett PD and Swain CG (1949). Kinetics of hydrolysis and displacement reactions of β,β′-dichlorodiethyl sulfide (mustard gas) and of β-chloro-β′-hydroxydiethyl sulfide (mustard chlorohydrin). *J Am Chem Soc*, **71**, 1406–1415.

Benschop HP, Van der Schans GP, Noort D *et al.* (1997). Verification of exposure to sulfur mustard in two casualties of the Iran–Iraq conflict. *J Anal Toxicol*, **21**, 249–251.

Benschop HP, Noort D, Van der Schans GP *et al.* (2000). *Diagnosis and dosimetry of exposure to sulfur mustard: development of standard operating procedures; further exploratory research on protein adducts.* Final Report Cooperative Agreement DAMD17-97-2-7002, NTIS number: ADA381035/XAB. Springfield, VA, USA: NTIS.

Benschop HP and De Jong LPA (2001). Toxicokinetics of nerve agents. In: *Chemical Warfare Agents: Toxicity at Low Levels* (SM Somani and JA Romano, eds), pp. 25–81. Boca Raton, FL, USA: CRC Press.

Black RM and Read RW (1988). Detection of trace levels of thiodiglycol in blood, plasma and urine using gas chromatography–electron capture negative ion chemical ionisation mass spectrometry. *J Chromatogr*, **449**, 261–270.

Black, RM and Read, RW (1991) Methods for the analysis of thiodiglycol sulphoxide, a metabolite of sulphur mustard, in urine using gas chromatography–mass spectrometry. *J Chromatogr*, **558**, 393–404.

Black RM, Clarke RJ and Read RW (1991). Analysis of 1,1′-sulphonylbis[2-(methylsulphinyl)ethane] and 1-methylsulphinyl-2-[2-(methylthio)ethylsulphonyl]ethane, metabolites of sulphur mustard, in urine using gas chromatography–mass spectrometry. *J Chromatogr*, **558**, 405–414.

Black RM, Brewster K, Clarke RJ *et al.* (1992a).

Biological fate of sulphur mustard, 1,1-thiobis(2-chloroethane): isolation and identification of urinary metabolites following intraperitoneal administration to rat. *Xenobiotica*, **22**, 405–418.
Black RM, Hambrook JL, Howells DJ *et al.* (1992b). Biological fate of sulfur mustard, 1,1′-thiobis(2-chloroethane). Urinary excretion profiles of hydrolysis products and β-lyase metabolites of sulfur mustard after cutaneous application in rats. *J Anal Toxicol*, **16**, 79–84.
Black RM, Brewster K, Clarke RJ *et al.* (1993). Metabolism of thiodiglycol (2,2′-thiobis-ethanol): isolation and identification of urinary metabolites following intraperitoneal administration to rat. *Xenobiotica*, **23**, 473–481.
Black RM and Read RW (1995a). Improved methodology for the detection and quantitation of urinary metabolites of sulphur mustard using gas chromatography–tandem mass spectrometry. *J Chromatogr B*, **665**, 97–105.
Black RM and Read RW (1995b). Biological fate of sulphur mustard, 1,1-thiobis(2-chloroethane): identification of β-lyase metabolites and hydrolysis products in human urine. *Xenobiotica*, **25**, 167–173.
Black RM, Clarke RJ, Harrison JM *et al.* (1997a). Biological fate of sulphur mustard: identification of valine and histidine adducts in haemoglobin from casualties of sulphur mustard poisoning. *Xenobiotica*, **27**, 499–512.
Black RM, Harrison JM and Read RW (1997b). Biological fate of sulphur mustard: *in vitro* alkylation of human haemoglobin by sulphur mustard. *Xenobiotica*, **27**, 11–32.
Black RM, Harrison JM and Read RW (1999). The interaction of sarin and soman with plasma proteins: the identification of a novel phosphylation site. *Arch Toxicol*, **73**, 123–126.
Black RM and Muir R (2003). Derivatization reactions in the chromatographic analysis of chemical warfare agents and their degradation products. *J Chromatogr A*, **1000**, 253–281.
Black RM and Noort D (2005). Methods for the retrospective detection of exposure to toxic scheduled chemicals. Part A: analysis of free metabolites. In: *Chemical Weapons Convention Related Analysis* (M Mesilaakso, ed.), pp. 403–431. Chichester, UK: John Wiley & Sons Ltd.
Boyer AE, Ash D, Barr D *et al.* (2004). Quantitation of the sulfur mustard metabolites 1,1′-sulfonylbis[2-(methylthio)ethane] and thiodiglycol in urine using isotope-dilution gas chromatography–tandem mass spectrometry. *J Anal Toxicol*, **28**, 327–332.
Brewster K, Harrison JM, Leadbeater J *et al.* (1987). The fate of 2-chlorobenzylidene malononitrile (CS) in rats. *Xenobiotica*, **17**, 911–924.
Brookes P and Lawley PD (1960). The reaction of mustard gas with nucleic acids *in vitro* and *in vivo*. *Biochem J*, **77**, 478–484.
Byrd GD, Sniegoski LT and White VE (1987). *Development of a confirmatory chemical test for exposure to 3-quinuclidinyl benzilate (BZ)*. Report 1987 Order No. AD-A191746. US Army Medical Research & Development Command, Fort Dietrick, Maryland (*Chem Abstr*, **110**, 70602).
Byrd GD, Sniegoski LT and White VE (1988). Determination of 3-quinuclidinyl benzilate in urine. *J Res Nat Bur Stand (US)*, **93**, 293–295.
Capacio BR, Smith JR, DeLion MT *et al.* (2004). Monitoring of sulfur mustard exposure by gas chromatography–mass spectrometry: analysis of thiodiglycol cleaved from blood proteins. *J Anal Toxicol*, **28**, 306–311.
Cucinell SA, Swentzel KC, Biskup R *et al.* (1971). Biochemical interactions and metabolic fate of riot control agents. *Fed Proc*, **30**, 86–91.
Davison C, Rozman RS and Smith PK (1961). Metabolism of bis-β-chloroethyl sulfide (sulfur mustard gas). *Biochem Pharmacol*, **7**, 65–74.
Degenhardt CEAM, Pleijsier K, Van der Schans MJ *et al.* (2004). Improvements of the fluoride reactivation method for the verification of nerve agent exposure. *J Anal Toxicol*, **28**, 364–371.
Diaz Gomez MI and Castro JA (1980). Covalent binding of chloroform metabolites to nuclear proteins – no evidence for binding to nucleic acids. *Cancer Lett*, **9**, 213–218.
Di Consiglio E, De Angelis G, Testai E *et al.* (2001). Correlation of a specific mitochondrial phospholipid–phosgene adduct with chloroform acute toxicity. *Toxicology*, **159**, 43–53.
Drasch G, Kretschmer E, Kauert G *et al.* (1987). Concentrations of mustard gas [bis(2-chloroethyl)sulfide] in the tissues of a victim of vesicant exposure. *J Forensic Sci*, **32**, 1788–1793.
Dunn MA, Brennie EH and Sidell FR (1997). Pretreatment for nerve agent exposure. In: *Medical Aspects of Chemical and Biological Warfare* (FR Sidell, ET Takafuji and DR Franz, eds), pp. 181–196. Washington, DC, USA: Office of the Surgeon General at TMM Publications, Borden Institute, Walter Read Army Medical Center.
Elhanany E, Ordentlich A, Dgany O *et al.* (2001). Resolving pathways of interaction of covalent inhibitors with the active site of acetylcholinesterases: MALDI-TOF/MS analysis of various nerve agent phosphyl adducts. *Chem Res Toxicol*, **14**, 912–918.

Ellman GL, Courtney KD and Anders V (1961). A new and rapid colorimetric determination of acetylcholinesterase activity. *Biochem Pharmacol*, **7**, 88–95.

Fabrizi L, Taylor GW, Edwards RJ *et al.* (2001). Adducts of the chloroform metabolite phosgene. In: *Biological Reactive Intermediates VI* (Dansette PM, Snyder R, Delaforge M, eds), pp. 129–132. New York, NY, USA and Amsterdam, The Netherlands: Kluwer Academic/Plenum Publishers.

Fabrizi L, Taylor GW, Cañas B *et al.* (2003). Adduction of the chloroform metabolite phosgene to lysine residues of human histone H2B. *Chem Res Toxicol*, **16**, 266–275.

Fidder A, Moes GWH, Scheffer AG *et al.* (1994). Synthesis, characterization and quantitation of the major adducts formed between sulfur mustard and DNA of calf thymus and human blood. *Chem Res Toxicol*, **7**, 199–204.

Fidder A, Noort D, de Jong LPA *et al.* (1996). N7-(2-Hydroxyethylthioethyl)-guanine: a novel urinary metabolite following exposure to sulphur mustard. *Arch Toxicol*, **70**, 854–855.

Fidder A, Noort D, Hulst AG *et al.* (2000). Biomonitoring of exposure to lewisite based on adducts to haemoglobin. *Arch Toxicol*, **74**, 207–214.

Fidder A, Noort D, Hulst AG *et al.* (2002). Retrospective detection of exposure to organophosphorus anti-cholinesterases: mass spectrometric analysis of phosphylated human butyrylcholinesterase. *Chem Res Toxicol*, **15**, 582–590.

Fredriksson S-Å Hammarström L.-G, Henriksson L *et al.* (1995). Trace determination of alkyl methylphosphonic acids in environmental and biological samples using gas chromatography/negative-ion chemical ionization mass spectrometry and tandem mass spectrometry. *J Mass Spectrom*, **30**, 1133–1143.

French MC, Harrison JM, Inch TD *et al.* (1983a). The fate of dibenz[*b*, *f*]-1,4-oxazepine (CR) in the rat, rhesus monkey and guinea-pig. Part I. Metabolism *in vivo*. *Xenobiotica*, **13**, 345–359.

French MC, Harrison JM, Newman J. *et al.* (1983b). The fate of dibenz[*b*, *f*]-1,4-oxazepine (CR) in the rat. Part III. The intermediary metabolites. *Xenobiotica*, **13**, 373–381.

Fung LW-M, Ho C, Roth EF *et al.* (1975). The alkylation of hemoglobin S by nitrogen mustard. *J Biol Chem*, **250**, 4786–4789.

Hambrook JL, Harrison JM, Howells DJ *et al.* (1992). Biological fate of sulphur mustard (1,1′-thiobis(2-chloroethane)): urinary and faecal excretion of ^{35}S by rat after injection or cutaneous application of ^{35}S-labelled sulphur mustard. *Xenobiotica*, **22**, 65–75.

Hambrook JL, Howells DJ and Schock C (1993). Biological fate of sulphur mustard (1,1′-thiobis(2-chloroethane)): uptake, distribution and retention of ^{35}S in skin and in blood after cutaneous application of ^{35}S-sulphur mustard in rat and comparison with human blood *in vitro*. *Xenobiotica*, **23**, 537–561.

Harrison JM, Clarke RJ, Inch TD *et al.* (1978). The metabolism of dibenz[b,f]-1,4-oxazepine (CR): *in vivo* hydroxylation of 10,11-dihydrodibenz[b,f]-1,4-oxazepin-11-(1OH)-one and the NIH shift. *Experientia*, **34**, 698–699.

Harrison JM, Read RW, Williams NH *et al.* (2007). Phosphylated tyrosine as a biomarker of exposure to organophosphorus nerve agents. *Arch Toxicol*, in press.

Ishii A, Seno H, Watanabe-Suzuki K, *et al.* (1998). Determination of cyanide in whole blood by capillary gas chromatography with cryogenic oven trapping. *Anal Chem*, **70**, 4873–4876.

Jakubowski EM, Woodard CL, Mershon MM *et al.* (1990). Quantification of thiodiglycol in urine by electron ionization gas chromatography–mass spectrometry. *J Chromatogr*, **528**, 184–190.

Jakubowski EM, Smith JR, Logan TP *et al.* (1993). Verification of lewisite exposure: quantification of chlorovinyl arsonous acid in biological samples. In: *Proceedings of the 1993 Medical Defense Bioscience Review*, 10–13 May, pp. 361–368. Aberdeen Proving Ground, MD, USA: US Army Medical Research Institute of Chemical Defense.

Jakubowski EM, Sidell FR, Evans RA *et al.* (2000). Quantification of thiodiglycol in human urine after an accidental sulfur mustard exposure. *Toxicol Methods*, **10**, 143–150.

Jakubowski EM, McGuire JM, Evans RA *et al.* (2004). Quantitation of fluoride ion released sarin in red blood cell samples by gas chromatography–chemical ionization mass spectrometry using isotope dilution and large-volume injection. *J Anal Toxicol*, **28**, 357–363.

Kawada T and Iwai K (1985). *In vivo* and *in vitro* metabolism of dihydrocapsaicin, a pungent principle of hot pepper, in rats. *Agric Biol Chem*, **49**, 441–448.

Kubic VL and Anders MW (1980). Metabolism of carbon tetrachloride to phosgene. *Life Sci*, **26**, 2151–2156.

Leadbeater L (1973). The absorption of *o*-chlorobenzylidene malononitrile (CS) by the respiratory tract. *Toxicol Appl Pharmacol*, **25**, 101–110.

Lemire SW, Ashley DL and Calafat AM (2003). Quantitative determination of the hydrolysis products of nitrogen mustards in human urine by liquid chromatography–electrospray ionization tandem mass spectrometry. *J Anal Toxicol*, **27**, 1–6.

Lemire SW, Barr JR, Ashley DL *et al.* (2004). Quantitation of biomarkers of exposure to nitrogen mustards in urine from rats dosed with nitrogen mustards and from an unexposed human population. *J Anal Toxicol*, **28**, 320–326.

Leung H-W, Ballantyne B and Frantgz SW (1996). Pharmacokinetics of *N*-methyldiethanolamine following acute cutaneous and intravenous dosing in the rat. *J Toxicol-Cutaneous Ocular Toxicol*, **15**, 343–353.

Logan TP, Smith JR, Jakubowski EM *et al.* (1999). Verification of lewisite exposure by the analysis of 2-chlorovinyl arsonous acid in urine. *Toxicol Methods*, **9**, 275–284.

Logue BA, Kirschten NP, Petrikovics I *et al.*(2005). Determination of the cyanide metabolite 2-aminothiazoline-4-carboxylic acid in urine and plasma by gas chromatography–mass spectrometry. *J Chromatogr B*, **819**, 237–244.

Ludlum DB, Austin-Ritchie P, Hagopian M *et al.* (1994). Detection of sulfur mustard-induced DNA modifications. *Chem-Biol Interact*, **91**, 39–49.

Lundquist P, Kagedal B, Nilsson L *et al.* (1995). Analysis of the cyanide metabolite 2-aminothiazoline-4-carboxylic acid in urine by high-performance liquid chromatography. *Anal Biochem*, **228**, 27–34.

Mackworth JR (1948). The inhibition of thiol enzymes by lachrymators. *Biochem J*, **42**, 82–90.

Maisonneuve A, Callebat I, Debordes L *et al.* (1993). Biological fate of sulphur mustard in rat: toxicokinetics and disposition. *Xenobiotica*, **23**, 771–180.

Matsuda Y, Nagao M, Takatori T *et al.* (1998). Detection of sarin hydrolysis product in formalin-fixed brain tissues of victims of the Tokyo subway terrorist attack. *Toxicol Appl Pharmacol*, **150**, 310–320.

Meng ZH and Qin L (2001). Determination of degradation products of nerve agents in human serum by solid phase extraction using molecularly imprinted polymer. *Anal Chim Acta*, **435**, 121–127.

Miki A, Katagi M, Tsuchihashi H *et al.* (1999). Determination of alkyl methylphosphonic acids, the main metabolites of organophosphorus nerve agents, in biofluids by gas chromatography–mass spectrometry by liquid–liquid–solid-phase-transfer-catalyzed pentafluorobenzylation. *J Anal Toxicol*, **23**, 86–93.

Minami M, Hui D-M, Katsumata M *et al.* (1997). Method for the analysis of the methylphosphonic acid metabolites of sarin and its ethanol-substituted analogue in urine as applied to the victims of the Tokyo sarin disaster. *J Chromatogr B*, **695**, 237–244.

Murachi T (1963). A general reaction of diisopropyl phosphorofluoridate with proteins without direct effect on enzymic activities. *Biochim Biophys Acta*, **71**, 239–241.

Nagao M, Takatori T, Matsuda Y *et al.* (1997). Definitive evidence for the acute sarin poisoning diagnosis in the Tokyo subway. *Toxicol Appl Pharmacol*, **144**, 198–203.

Nakajima T, Sasaki K, Ozawa H *et al.* (1998). Urinary metabolites of sarin in a patient of the Matsumoto sarin incident. *Arch Toxicol*, **72**, 601–603.

Niu T, Matijasevic Z, Austin-Ritchie P *et al.* (1996). A ^{32}P-postlabeling method for the detection of adducts in the DNA of human fibroblasts exposed to sulfur mustard. *Chem Biol Interact*, **100**, 77–84.

Noort D, Verheij ER, Hulst AG *et al.* (1996). Characterization of sulfur mustard induced structural modifications in human hemoglobin by liquid chromatography–tandem mass spectrometry. *Chem Res Toxicol*, **9**, 781–787.

Noort D, Hulst AG, Platenburg DHJM *et al.* (1998). Quantitative analysis of *O*-isopropyl methylphosphonic acid in serum samples of Japanese citizens allegedly exposed to sarin: estimation of internal dosage. *Arch Toxicol*, **72**, 671–675.

Noort D, Hulst AG, De Jong LPA *et al.* (1999). Alkylation of human serum albumin by sulfur mustard *in vitro* and *in vivo*: mass spectrometric analysis of a cysteine adduct as a sensitive biomarker of exposure. *Chem Res Toxicol*, **12**, 715–721.

Noort D, Hulst AG, Fidder A *et al.* (2000). *In vitro* adduct formation of phosgene with albumin and hemoglobin in human blood. *Chem Res Toxicol*, **13**, 719–726.

Noort D, Benschop HP and Black RM (2002a). Biomonitoring of exposure to chemical warfare agents: a review. *Toxicol Appl Pharmacol*, **184**, 116–126.

Noort D, Hulst AG and Jansen R (2002b). Covalent binding of nitrogen mustards to the cysteine-34 residue in human serum albumin. *Arch Toxicol*, **76**, 83–88.

Noort D, Fidder A, Benschop HP *et al.* (2004a). Procedure for monitoring exposure to sulfur mustard based on modified Edman degradation of globin. *J Anal Toxicol*, **28**, 311–315.

Noort D, Fidder A, Hulst AG *et al.* (2004b). Retrospective detection of exposure to sulfur mustard: improvements on an assay for liquid chromatography–tandem mass spectrometry analysis of albumin/sulfur mustard adducts. *J Anal Toxicol*, **28**, 333–338.

Noort D and Black RM (2005). Methods for the retrospective detection of exposure to toxic scheduled chemicals. Part B: analysis of covalent adducts to proteins and DNA. In: *Chemical Weapons Convention Related Analysis* (M Mesilaakso, ed.), pp. 433–451. Chichester, UK: John Wiley & Sons Ltd.

Ohsawa I, Kanamori-Kataoka M, Tsuge K *et al.* (2004). Determination of thiodiglycol, a mustard gas hydrolysis product, by gas chromatography–mass spectrometry after *tert*-butyldimethylsilylation. *J Chromatogr B*, **1061**, 235–241.

Olajos EJ and Salem H (2001). Riot control agents: pharmacology, toxicology, biochemistry and chemistry. *J Appl Toxicol*, **21**, 355–391.

Osborne MR, Wilman DE and Lawley PD (1995). Alkylation of DNA by the nitrogen mustard bis(2-chloroethyl)methylamine. *Chem Res Toxicol*, **8**, 316–320.

Peeples ES, Schopfer LM, Duysen EG *et al.* (2005). Albumin, a new biomarker of organophosphorus toxicant exposure, identified by mass spectrometry. *Toxicol Sci*, **83**, 303–312.

Pohl LR, Branchflower RV, Highet RJ *et al.* (1981). The formation of diglutathionyl dithiocarbonate as a metabolite of chloroform, bromotrichloromethane and carbon tetrachloride. *Drug Metab Disp*, **9**, 334–339.

Polhuijs M, Langenberg JP and Benschop HP (1997). New method for retrospective detection of exposure to organophosphorus anticholinesterases: application to alleged sarin victims of Japanese terrorists. *Toxicol Appl Pharmacol*, **146**, 156–161.

Read RW and Black RM (2004a). Analysis of the sulfur mustard metabolite 1,1′-sulfonylbis[2-*S*-(*N*-acetylcysteinyl)ethane] in urine by negative ion electrospray liquid chromatography–tandem mass spectrometry. *J Anal Toxicol*, **28**, 352–356.

Read RW and Black RM (2004b). Analysis of β-lyase metabolites of sulfur mustard in urine by electrospray liquid chromatography–tandem mass spectrometry. *J Anal Toxicol*, **28**, 346–351.

Reilly CA, Ehlhardt WJ, Jackson DA *et al.* (2003). Metabolism of capsaicin by cytochrome P450 produces novel dehydrogenated metabolites and decreases cytotoxicity to lung and liver cells. *Chem Res Toxicol*, **16**, 336–349.

Reilly CA and Yost GS (2005). Structural and enzymatic parameters that determine alkyl dehydrogenation/hydroxylation of capsaicinoids by cytochrome p450 enzymes. *Drug Metab Dispos*, **33**, 530–536.

Riches J, Morton I, Read RW *et al.* (2005). The trace analysis of alkyl alkylphosphonic acids in urine using gas chromatography-ion trap negative ion tandem mass spectrometry. *J Chromatogr B*, **816**, 251–258.

Riches J, Read RW and Black RM (2007). Analysis of the sulphur mustard metabolite thiodiglycol in urine using isotope-dilution gas chromatography-ion trap tandem mass spectrometry. *J Chromatogr B*, **845**, 114–120.

Rietveld EC, Delbressine LPC, Waegemaekers THJM *et al.* (1983). 2-Chlorobenzylmercapturic acid, a metabolite of the riot control agent chlorobenzylidine malononitrile (CS) in the rat. *Arch Toxicol*, **54**, 139–144.

Roberts JJ and Warwick GP (1963). Studies of the mode of action of alkylating agents-VI. The metabolism of bis-2-chloroethylsulphide (mustard gas) and related compounds. *Biochem Pharmacol*, **12**, 1329–1334.

Sandelowsky I, Simon GA, Bel P *et al.* (1992). N^1-(2-Hydroxyethylthioethyl)-4-methylimidazole (4-met-1-imid-thiodiglycol) in plasma and urine: a novel metabolite following dermal exposure to sulphur mustard. *Arch Toxicol*, **66**, 296–297.

Shih ML, Smith JR, McMonagle JD *et al.* (1991). Detection of metabolites of toxic alkyl methylphosphonates in biological samples. *Biol Mass Spectrom*, **20**, 717–723.

Shih ML, McMonagle JD, Dolzine TW *et al.* (1994). Metabolite pharmacokinetics of soman, sarin and GF in rats and biological monitoring of exposure to toxic organophosphorus agents. *J Appl Toxicol*, **14**, 195–199.

Sperry ML, Skanchy D and Marino MT (1998). High-performance liquid chromatographic determination of *N*-[2-(hydroxyethyl)-*N*-(2-(7-guaninyl)ethyl]methylamine, a reaction product between nitrogen mustard and DNA and its application to biological samples. *J Chromatogr B*, **716**, 187–193.

Stott WT, Waechter JM, Rick DL *et al.* (2000). Absorption, distribution, metabolism and excretion of intravenously and dermally administered triethanolamine in mice. *Food Chem Toxicol*, **38**, 1043–1051.

Surh YJ, Ahn SH, Kim KC *et al.* (1995). Metabolism of capsaicinoids: evidence for aliphatic hydroxylation and its pharmacological implications. *Life Sci*, **56**, PL305–PL311.

Thulin H, Zorcec V, Segerback D *et al.* (1996). Oxazolidonylethyl adducts to hemoglobin and DNA following nornitrogen mustard exposure. *Chem-Bio Interact*, **99**, 263–275.

Timchalk C, Poet TS, Kousba AA *et al.* (2004). Noninvasive biomonitoring approaches to determine dosimetry and risk following acute chemical exposure: analysis of lead or organophosphate insecticide in saliva. *J Toxicol Environ Health Curr Issues*, **67**, 635–650.

Törnqvist M, Mowrer J, Jensen S *et al.* (1996). Monitoring of environmental cancer initiators through hemoglobin adducts by a modified Edman degradation method. *Anal Biochem*, **154**, 255–266.

Törnqvist M, Fred C, Haglund J *et al.* (2002). Protein adducts: quantitative and qualitative aspects of their formation, analysis and applications. *J Chromatogr B*, **778**, 279–308.

Trams EG and Nadkarni MV (1956). Studies on the *N*-dealkylation of nitrogen mustard and triethylenediamine by liver homogenates. *Cancer Res*, **16**, 1069–1075.

Tsuchihashi H, Katagi M, Nishikawa M *et al.* (1998). Identification of metabolites of nerve agent VX in serum collected from a victim. *J Anal Toxicol*, **22**, 383–388.

Tsuchihashi H, Katagi M, Tatsuno M *et al.* (2000). Determination of metabolites of nerve agent *O*-ethyl-*S*-2-diisopropylaminoethyl methylphosphonothioate (VX). In: *Natural and Selected Synthetic Toxins* (Tu AT and Gaffield W, eds), ACS Symposium Series, 745, pp. 369–386. Washington, DC, USA: American Chemical Society.

United Nations (1986). Report of the mission dispatched by the Secretary-General to investigate allegations of the use of chemical weapons in the conflict between the Islamic Republic of Iran and Iraq, Report No. S/17911, 12 March. New York, NY, USA: United Nations.

Van der Schans GP, Scheffer AG, Mars-Groenendijk RH *et al.* (1994). Immunochemical detection of adducts of sulfur mustard to DNA of calf thymus and human white blood cells. *Chem Res Toxicol*, **7**, 408–413.

Van der Schans GP, Noort D, Mars-Groenendijk RH *et al.* (2003). Immunochemical detection of sulfur mustard adducts with keratins in the stratum corneum of human skin. *Chem Res Toxicol*, **15**, 21–25.

Van der Schans GP, Mars-Groenendijk R, De Jong LPA *et al.* (2004). Standard Operating Procedure for immunoslotblot assay for analysis of DNA/sulfur mustard adducts in human blood and skin. *J Anal Toxicol*, **28**, 316–319.

Van der Schans MJ, Lander BJ, Van der Wiel H *et al.* (2003). Toxicokinetics of the nerve agent (+/−)-VX in anesthetized and atropinized hairless guinea pigs and marmosets after intravenous and percutaneous administration. *Toxicol Appl Pharmacol*, **191**, 48–62.

Van der Schans MJ, Polhuijs M, Van Dijk C *et al.* (2004). Retrospective detection of exposure to nerve agents: analysis of phosphofluoridates originating from fluoride-induced reactivation of phosphylated BuChE. *Arch Toxicol*, **78**, 508–524.

Van Welie RTH, van Dijck RGJM and Vermeulen NPE (1992). Mercapturic acids, protein adducts and DNA adducts as biomarkers of electrophilic chemicals. *Crit Rev Toxicol*, **22**, 271–306.

Wils ERJ, Hulst AG, de Jong AL *et al.* (1985). Analysis of thiodiglycol in urine of victims of an alleged attack with mustard gas. *J Anal Toxicol*, **9**, 254–257.

Wils ERJ, Hulst AG and van Laar J (1988). Analysis of thiodiglycol in urine of victims of an alleged attack with mustard gas, Part II. *J Anal Toxicol*, **12**, 15–19.

Wooten JV, Ashley DL and Calafat AM (2002). Quantitation of 2-chlorovinylarsonous acid in human urine by automated solid-phase microextraction-gas chromatography–mass spectrometry. *J Chromatog B*, **772**, 147–153.

Worek F, Mast U, Kiderlen D *et al.* (1999). Improved determination of acetylcholinesterase activity in human blood. *Clin Chim Acta*, **288**, 73–90.

World Health Organization (2004). *Hydrogen cyanide and cyanides: human health aspects*. Concise International Chemical Assessment Document 61. New York, NY, USA: World Health Organization, Geneva.

Young CL, Ash D, Driskell WJ *et al.* (2004). A rapid, sensitive method for the quantitation of specific metabolites of sulfur mustard using isotope-dilution gas chromatography–tandem mass spectrometry. *J Anal Toxicol*, **28**, 339–345.

6 RESPIRATORY PROTECTION

Anthony Wetherell and George Mathers

Dstl, Salisbury, UK

INTRODUCTION

Respirators are arguably the first, and most important, line of personal defence against chemical (and biological) agents as, operationally, these agents all have their greatest and most rapid effects via the respiratory system. Respirators are generally of two types: those that filter ambient air (air-filtering/air-purifying respirators) and those that provide breathable air from containers (closed-circuit breathing apparatus). This chapter deals with military air-filtering respirators, as that is the type mostly used by military personnel to protect against chemical (and biological) warfare agents. The chapter covers the history of respirators, respirator structure, function and design issues, effects on the wearer and respirator testing methods.

HISTORY OF RESPIRATORS

First World War

The first documented use of respiratory protective devices dates from the German chlorine gas attack against the Ypres Salient in 1915, when Allied soldiers covered their mouths, noses, and sometimes their eyes, with handkerchiefs, woollen scarves and any other fabric at hand. Some soldiers soaked the fabric in water, and went on, as the attacks continued, to use moist earth, soda solution and sodium thiosulphate (photographers' 'hypo').

The British then developed mouth muffs: multi-layered pads of cotton soaked in sodium carbonate and sodium thiosulphate,[1] followed by the 'Black Veil Respirator', which consisted of cotton waste soaked in sodium thiosulphate, washing soda, water and glycerine (to keep it all moist), held together by black veiling material. Following this came various types of 'Hypo Helmet': hoods made of muslin/flannel material, with skirts to tuck under the collar, and incorporating mica, celluloid and, finally, glass eyepieces and rubber 'flapper' exhale valves. The filter material was soaked in the same chemicals as before at first, and then supplemented by alkaline sodium phenolate solution (carbolic acid) and later by hexamethylene tetramine (urotropine) to provide protection against phosgene.

Examination of a German respirator led to development of the 'Large Box Respirator', consisting of an oronasal mask of multi-layered muslin soaked in a solution of sodium zincate and urotropine, a rubber mouthpiece to hold between the teeth, a noseclip and a set of rubber goggles. This respirator also had a filter, containing layers of charcoal, soda lime and potassium permanganate, connected to the mask by a hose. The Large Box Respirator was followed by the 'Small Box Respirator', which had a facepiece that covered the mouth, nose and eyes, was made of fabric coated with gum rubber, and incorporated celluloid lenses and a rubber 'flutter' exhale valve. This respirator continued in use throughout the rest of the First World War.

[1] The official History of World War I records that the women of England responded to an appeal published in the national press and produced these simple pads of cotton wool or lint in large numbers.

Chemical Warfare Agents: Toxicology and Treatment (2nd Edition)
Edited by Timothy C. Marrs, Robert L. Maynard and Frederick R. Sidell

The French developed the 'M-2 mask', consisting of 20 layers of muslin impregnated with 'Greasene' and another 20 impregnated with 'Complexene'. The mask had twin circular celluloid lenses and a weatherproof flap for rain protection, but no outlet valve or filter canister. The M-2 was followed by the 'Tissot mask', with a thin, moulded rubber facepiece incorporating eyepieces, a flapper-type rubber exhale valve and a flexible, fabric-covered hose running over the shoulder to a filter canister worn in a harness on the back. The Tissot mask also ducted air over the eyepieces to help demist them: a feature still used in most present-day respirators.

The Germans, like the British, also first used pads soaked in sodium carbonate and sodium thiosulphate, but then went on to develop a full face mask (which inspired the British 'Box' designs). This was followed by a respirator, known simply as 'The German Mask', with a facepiece of natural rubber (later of oiled leather when rubber became unavailable), twin circular eyepieces with outer lenses of celluloid and inner lenses of coated gelatin, and a filter canister that screwed directly into the facepiece: another design feature still used in most present-day respirators.

The USA benefited from the earlier experience of the British and French, and developed a series of respirators based mostly on the British Small Box Respirator and called the 'American Small Box Respirator', the 'Corrected English Mask' and the 'RFK mask' (the initials of three personnel in the Chemical Service), together with the French M-2 and Tissot respirators.

Inter-war and Second World War

After WWI, respirators followed the general style of elastomeric facepieces, with eyepieces/visors and exhale valves, and filters connected by tubes or mounted directly on the facepiece. The first all-rubber, full-face respirator was issued to the Royal Navy in 1922 and the Army in 1924. The facepiece was moulded in natural rubber and its outer surface was covered in stockinette to aid the 'wicking out' and surface evaporation of chemical agents. The respirator had two-inch diameter glass eyepieces and an expiratory valve fitted in the region of the nosepiece. The filter canister was mounted on the wearer's body and connected to the facepiece by a corrugated hose; the Navy version had a longer hose than the Army version.

The realization that it was more economical to have a single respirator design for all three services led to the introduction of the 'General Service' (GS) respirator, familiar to all those who served in the Second World War. Over 25 million of these respirators were made in the UK alone, and issued to all Commonwealth forces and Civil Defence Corps. The GS respirator consisted of a rubber facepiece with stockinette covering, flat circular eyepieces, a diver-type outlet valve, an elastic webbing harness and a filter canister fitted with an inlet valve. The canister, worn on the body and attached to the facepiece by a hose, was filled with activated charcoal granules, with a wool–asbestos mixture added later, and impregnated charcoal added even later.

In 1942, to meet the need for a lighter respirator more suited to a war of movement, the 'Light-Type respirator' was introduced. This had a canister, filled with activated, impregnated charcoal and a wool–resin mixture, directly attached to the left cheek of the facepiece, and, for the first time, a speech transmitter was incorporated.

Post-Second World War to present day

Improvements in materials, methods of manufacture and charcoal performance continued to be made to the Light-Type respirator into the 1950s, when work began on a replacement with higher protection levels to meet the emerging threat from biological agents. The new British respirator, the 'S6', began to be issued in 1962, and was considered to be the best of its time. The facepiece was of natural rubber, with an airbag faceseal, also of natural rubber, but a softer compound to give improved comfort and a better seal. The eyepieces were of toughened glass, curved to improve all-round vision, and there was provision for vision-corrective spectacle inserts. Airflow management included the much-copied inner airguide to direct inhaled air over the eyepieces to reduce misting. Early filter canisters were filled as for the Light-Type, but later the resin–wool mixture was replaced with a higher-efficiency pleated paper element.

Figure 1. UK respirators from WW1 to present day

The S6 was replaced by the 'S10' in 1986, to meet the changing requirements of new weapon and communications systems, and improved serviceability and reduced manufacturing costs. The S10 profited from the availability of new polymer materials, and had a butyl rubber facepiece with a reflex faceseal and eyepieces with improved compatibility with optical systems.

The S10 is due to be replaced by the 'General Service Respirator (GSR)' in 2006/2007, to meet the continued changing requirements. The GSR design is yet to be finalized, but will incorporate further improvements to protection and equipment compatibility, with improved vision and speech, and reduced breathing resistance.

While developments described above were underway in the UK, the USA developed a variety of respirators for various purposes, including the 'M9' and 'M17' for general purpose and infantry use, followed by the 'M40', and the 'M24' and 'M25', followed by the 'M42' and 'M43', for tank crews. All these respirators are similar in the way they operate, but their designs differ according to the requirements of particular user groups. These respirators are to be replaced by the new 'M50 Joint Services General Purpose Mask' by 2006. A selection of UK respirators from World War I to the present day is shown in Figure 1.

RESPIRATOR STRUCTURE, FUNCTION AND DESIGN ISSUES

Modern-day, air-filtering, military respirators consist of a facepiece (mask), on which are mounted various sub-assembly components such as filters, faceseals, visors/eyepieces, valves, airguides, speech-enhancing modules and drinking systems.

Respirator design is concerned firstly to give protection, and secondly to reduce the physiological and psychological burdens on the user and to improve compatibility with other equipment. There are two major design issues. The first is that there is not enough space: the facepiece must have as low a profile as possible to maximize compatibility with other equipment, but still has to accommodate all the sub-assembly components, even on the smallest size respirator. The second issue is that anything done to improve one feature will almost certainly impair one or more of the other features. Thus, all respirator design is a compromise to maximize protection while providing the users with the optimum design to meet their operational requirements.

Materials

Facepiece materials must be chemically resistant and impermeable to agents ('chemically hard') while being flexible (especially at the seals), comfortable, non-toxic to the skin, easily workable, inexpensive and have a long shelf-life. Eyepieces and visors must also be transparent. It is difficult to combine all of these features in one material; in particular, flexible and chemically hard materials are typically difficult to work with and uncomfortable next to the skin. On the other hand, easily workable and comfortable materials tend to absorb agent (nerve agents are good solvents), which will then 'off-gas', e.g. in a collective protection system. In extreme cases, the agent can penetrate the material itself. Chemical hardness is of primary importance and most facepieces are made of chlorobutyl rubber.

Rigid sub-assembly components, e.g. filter canister bodies, and mountings for valves, eyepieces and speech modules, were originally made of metal but tend more now to be made of polymers. In the S10, polyacetal is used as it is not only chemically hard but also has mechanical properties that allow snap-fitting, which reduces production costs and allows easy replacement of parts.

Facepiece

The facepiece, in addition to protecting the face, is the carrier for the functional, sub-assembly components. The size and location of these requires very careful consideration to ensure not only that they all work properly together, but also that they do not interfere with the use of other equipment, e.g. optical, communications and weapons systems. For example, reducing the 'snout' profile will minimize its intrusion into the visual field and allow better access to confined spaces, but will reduce the space available for other components. Keeping the cheek profiles smooth will allow better use of shoulder-held and fired weapons, but means that components, such as speech modules and filters, cannot be mounted on the cheeks. Reducing the brow profile will allow better fitting of combat helmets, but could affect the faceseal efficiency in that area.

Sizing is dealt with below, but it should be mentioned here that fitting the whole range of face sizes is especially challenging, as the smallest size facepiece can be 25% smaller than the largest, yet the sizes of the functional components remain the same, and all the components must still be accommodated.

Faceseal

The faceseal is normally moulded as an integral part of the facepiece; it seals the edge of the facepiece to the face and has to achieve total surface contact with the skin. The faceseal is the most vulnerable leakage path, and the most difficult to deal with; it has to accommodate a wide variety of face sizes, shapes and movements (e.g. speaking) and must also cope with face secretions, soiling, user-applied creams such as camouflage cream, and beard growth. Various face seals designs are possible, each with advantages and disadvantages. The main types are:

- Single-skin seals, where the edge of the facepiece forms the seal. These provide the least protection, but are simple and cheap to manufacture.
- Airbag seals: a closed tube-like structure attached around the edge of the facepiece. These are more expensive as they need separate moulding from the facepiece and also need to incorporate a valve or air bleed so that the pressure inside the airbag can be matched with ambient pressure, e.g. at high and low altitudes. However, there are three main advantages. First, they can be made of a softer rubber compound, which improves comfort. Secondly, they inflate slightly when the wearer breathes in and the pressure inside the respirator falls; this gives a tighter seal when it is most needed and improves protection. Thirdly, they are more conformable to different face shapes and movements, which improves protection and comfort.
- Reflex seals, where the edge of the facepiece is turned back in. These are cheaper than airbag seals to make and require only a single-piece moulding, but they involve complex tooling

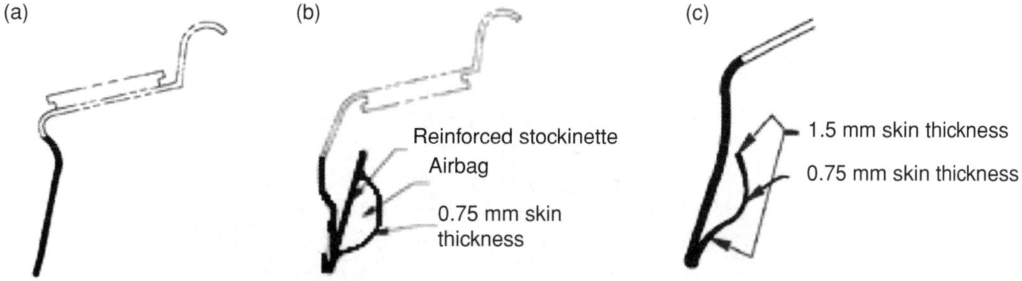

Figure 2. Some typical faceseals: (a) single-skin seal, Light Type respirator; (b) airbag seal, S6 respirator; (c) reflex seal, S10 respirator

and are less comfortable and conformable to face variations.

- Double-bladed seals, usually formed as two parallel reflex seals. These can be very compliant and tolerant of face variations within a given size, but they are difficult to manufacture and their sizing is critical.

Whatever the type, faceseals must not extend beyond the hairline, otherwise trapped hair will cause leaks, and all faceseals need a cross-sectional skin contact of about 10 mm to be effective (not easy to achieve at the temples). If the respirator incorporates an oronasal mask designed to seal to the skin, then this, too, needs to seal properly without compromising the main faceseal. In addition, the topography of the oronasal sealing area is complex, particularly around the nose, and good fits across the whole user population are difficult to achieve. Some typical faceseals are shown in Figure 2.

Filter canisters

Filter canisters are usually the largest single component of a respirator, apart from the facepiece itself. The main design challenge is to make the canister(s) as small as possible to reduce weight, bulk, intrusion into the visual field, and fouling or knocking in confined spaces that could breach the faceseal, all while still providing the required protection. The level of protection required is determined by the anticipated threat, which in turn determines the types and amounts of filter material and, hence, the size of the canister.

Canister bodies are usually made from an aluminium alloy or chemically hard plastic such as glass-filled nylon. Most canisters are cylindrical in shape, with a standard NATO STANAG 4155 screw thread in the centre of one end, so that different nations' canisters are interchangeable. This design is perhaps the simplest and easiest to manufacture, and the standard screw thread provides a positive, leak-free mounting. The canisters on many non-NATO respirators also follow this pattern. However, the canisters can be bulky in themselves, and the standard mounting causes the canister to 'stand off' from the facepiece. This further increases the overall bulk of the respirator, restricting access to confined spaces, and the canister intrudes into the visual field.

Other shapes are possible, such as ovoids or ellipsoids. These can remove some of the bulk of the canister from the visual field, but they can be more difficult to fill properly during manufacture, the airflow patterns through the filter material can be more complex, and the canister orientation can be wrong when fully mounted using a screw thread. Other types of attachment, e.g. bayonet fitting, could ensure correct canister orientation, reduce canister 'stand-off', and could also make changing canisters easier. However, they can involve complex designs to ensure that the canister is secure and the mounting is gas-tight, they tend to be more rigid such that the faceseal is more likely to breach if the canister is knocked, and they would not conform to the NATO standard.

Most canisters are mounted on the facepiece, but they can be mounted elsewhere, e.g. on the chest (as in some WWI designs) or on the back of the helmet or neck. However, while these locations might help some wearers in some jobs,

Figure 3. Sectioned view of the S10 filter canister

they would hinder others. Some personnel, e.g. infantry, are so encumbered with equipment that there is no spare space on their bodies, and a tube connecting the canister to the facepiece would be vulnerable to damage.

Most military respirators have a single filter canister, mounted on the right or left cheek or on the front of the facepiece, to allow use of shoulder-fired weapons. Some respirators have two canisters – one on each cheek. The advantages of two canisters are improved protection with decreased breathing resistance, the respirator is more balanced side-to-side, and breathing can continue while changing canisters during wear (provided a shut-off valve is incorporated and functions when a canister is removed). The disadvantages are that two mountings are required which take up even more valuable space on the facepiece, overall bulk may not be reduced and can even be increased, canister changing can take twice as long, left- and right-handed canisters may be needed which increases logistic problems, downward view can be impeded on both sides of the respirator, and access to shoulder-held and fired weapons can be obstructed whichever shoulder is used. A sectional view of the S10 filter canister is shown in Figure 3.

Filters

The filter can consist of up to three elements. The outer element is a particulate filter, often made of glass fibre paper, pleated to increase surface area. The inner element is a vapour adsorbent, usually activated granular charcoal. The third element comprises various chemicals impregnated on to the charcoal, such as copper, chromium, silver and triethylenediamine (TEDA), to react with volatile chemical agents such as hydrogen cyanide and cyanogen chloride that are poorly adsorbed.

Valves and deadspaces

All respirators have an outlet (exhalation, expiratory) valve and most have an inlet (inhalation, inspiratory) valve, both usually of the mushroom (or umbrella) type. Outlet valves allow exhaled air to exit the respirator, but prevent ambient air from entering; they are usually mounted on the respirator 'snout' mid-line, and made of chemically hard material as they are exposed to ambient air. Inlet valves are mounted just inboard of the filter to prevent humid, exhaled air going back into the filter (filters can lose efficiency when

Figure 4. Some typical inlet/outlet mushroom (umbrella) valves

wet), and can be made from more compliant and less chemically hard materials as they are unlikely to become seriously contaminated.

The outlet valve housing often incorporates a deadspace (not to be confused with respiratory deadspace – see below) just outboard of the valve. This is a small, partially enclosed volume that traps some of the exhaled air next to the valve surface, to ensure that any back leakage into the respirator is of clean air. There are various designs of deadspace including simple, open-ended structures, double valves and complex spirals or volutes designed to ease the exit of exhaled air while restricting the entry of contaminated air. The deadspace in the UK S10 respirator is in the form of a folded acoustic horn that also functions as a speech module. A selection of some typical inlet/outlet mushroom (umbrella) valves is given in Figure 4.

Airflow management

Most respirators incorporate airguides to ensure that the airflow inside the respirator is directed where it is needed. Air enters through the filter, is first directed over the eyepiece(s) to help prevent misting, and then into the oronasal cavity, from which it is inhaled. Exhaled air is usually directed straight out of the outlet valve (some being retained in the outlet valve deadspace). Some respirators have an oronasal mask that seals to the oronasal region of the face, inside the main face mask, with valves to separate the oronasal space from the eye-space. Airflow in the UK S6 respirator is illustrated in Figure 5.

Eyepiece/visor

Eyepieces and visors are usually designed to provide the best compromise between a wide visual field, eye relief (stand-off from the eyes), and compatibility with optical equipment, such as weapon sights and binoculars. Eyepieces are usually made from polycarbonate, for ballistic protection, and treated to resist scratching.

A great deal of design effort is put into making eyepieces that are distortion-free and give as wide a field of view as possible. Clarity of vision is not normally a problem as design guidelines and materials with good transparency are available. However, scratching of eyepieces can distort vision, and so anti-scratch coatings are needed, and clarity may also be an issue with flexible, transparent materials.

Visual acuity can also be impaired by eyepiece misting, caused mostly by exhaled air escaping from the oronasal space into the eyespace, but exacerbated by sweat. Misting tends to occur when the air management system is not well designed and/or the wearer is breathing hard. Some people tend to suffer from misting more than others, perhaps because of mismatch between their facial features and the airguides. Anti-misting coatings can help, but they are not very durable and must be re-applied at intervals.

The visual field allowed by the respirator can be improved by enlarging the eyepieces, using a single visor instead of separate eyepieces, and/or by reducing the eye–eyepiece stand-off. Bigger eyepieces and single visors will also make more of the wearer's face visible, helping with recognizability and speech intelligibility (non-verbal communication), and also help reduce the 'claustrophobic' effect. Users tend to like single visors for these reasons, but there are some factors to be considered. Firstly, one-piece visors, or bigger eyepieces, will reduce the space available for other components, but this may be an acceptable trade-off. Secondly, if not designed correctly, one-piece visors or bigger eyepieces could reflect more light (possibly betraying tactical positions) and could cause a local 'greenhouse effect' which could increase the heat load. Thirdly,

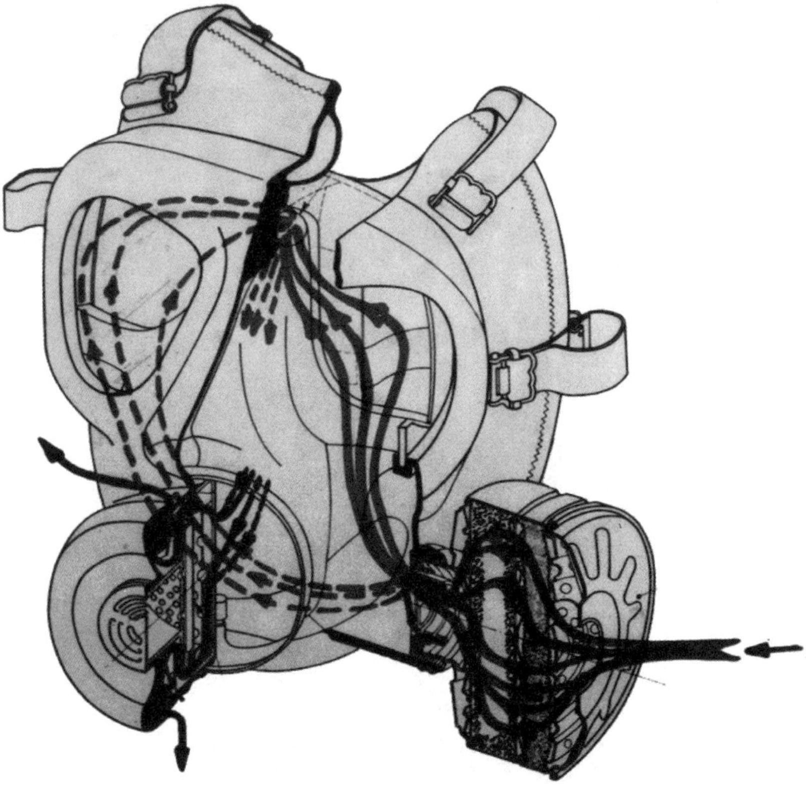

Figure 5. Airflow in the UK S6 respirator (similar for the S10 respirator)

one-piece visors need to be flexible so that the respirator can conform to different face shapes and movements, and so far there have been no materials with the required transparency, flexibility and impact resistance. However, materials technology is improving rapidly, and visors with the required characteristics are now becoming available. Fourthly, the improvement in visual field may be more imagined than real. It is true that visual field will be improved, but it will always be restricted to some extent by the faceseal boundary and the respirator snout and canister(s). Also, binocular vision immediately in front of the nose will be improved, but people normally do not view objects so closely. Nevertheless, an imagined improvement will make the respirator more acceptable to the users, and this raises an interesting question: respirator designers try to provide real benefits, and can include imagined benefits where there is no cost in real terms, but what should the designer do if there is a conflict between real and imagined benefits?

Speech module

Most respirators have some sort of speech module to help overcome the muffling and attenuating effect and improve speech intelligibility. Speech modules are normally mounted on the respirator mid-line for direct speech, or one cheek of the facepiece for indirect speech, e.g. using telephones or radios. Direct speech modules are usually circular or annular in shape and incorporate a thin sheet of polyester such as 'Mylar', about 5 cm in diameter (the minimum for acceptable speech intelligibility) tensioned on a frame and protected by grilles inside and outside the respirator.

The primary speech module of the S10 respirator does not have a diaphragm; instead, the exhale valve housing is in the shape of a folded acoustic horn, to serve both as a speech module and a deadspace. The S10 also has a secondary speech module, that does incorporate a small diaphragm, to which radio microphones can be attached.

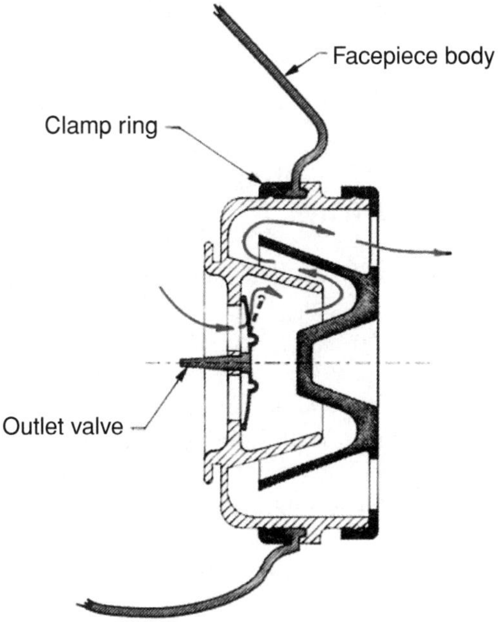

Figure 6. The UK S10 primary speech module/outlet valve/deadspace assembly

The secondary speech module is normally on the right cheek to allow the wearer to use right shoulder-fired weapons, but can be exchanged with the filter canister to allow left shoulder firing.

The design issues in this area are concerned with improving speech intelligibility. Bigger speech modules would transmit speech better, but would take up more space. Most of the sound leaves the respirator via the exhale valve, which needs only to be open slightly for sound transmission to increase markedly, but the valve tends to vibrate and distort the sound. This can be remedied by asymmetrically biassing or weighting the valve to dampen the vibration, but this can affect valve operation. Electronic components in the form of microphones, amplifiers and signal processors may also help, but microphones need to be positioned carefully to capture the speech and withstand the large pressure changes involved. In addition, electronic components require power, which increases the logistic burden. A schematic of the UK S10 primary speech module/outlet valve/deadspace assembly is shown in Figure 6.

Drinking facility

Removing the respirator in a contaminated environment is obviously undesirable but maintaining fluid intake is important. This has led to the incorporation of drinking facilities into respirator design. Most consist of a tube leading from outside to inside the respirator, and a tap, usually where the tube passes through the facepiece. The tube outside the respirator terminates in a valve designed to fit to a water canteen, and is stored in a holder or clip when not in use. The tube inside the respirator is normally stored to one side; turning the tap also moves the inside tube to where it can reach the wearer's mouth.

There are three main design issues. The first is to ensure that the inner and outer tubes are long enough to reach the mouth and the water container without kinking, but not so long that they become cumbersome. The second is to ensure that the outer tube can be removed from its storage fitting and the tap can be used easily by people wearing protective gloves that impair tactility and dexterity. The third is to ensure that the fluid flow rate is sufficient to cope with requirements in hot environments and/or heavy physical exercise. In addition, if respirators need to be worn for long periods, then there would be a need to provide not only water but also foodstuff. This could be achieved by using nutrient fluids, but the drinking facility would need to be designed so that food particles would not be trapped and decay.

Sizing

Sizing is a very important issue as respirators must provide the required protection across the whole of the user population. A large number of sizes would provide better fits for more people, but would incur unacceptable manufacturing and logistic costs. At the other extreme, a 'one-size-fits-all' respirator might reduce manufacturing and logistic problems, but would simply not be feasible: the variation in face size from the smallest female to the largest male is about 5 cm for both face length and width. Thus, a respirator that fits the largest face would be far too big and cumbersome on small faces.

Traditionally, respirators are sized and fitted using two facial dimensions: the distance

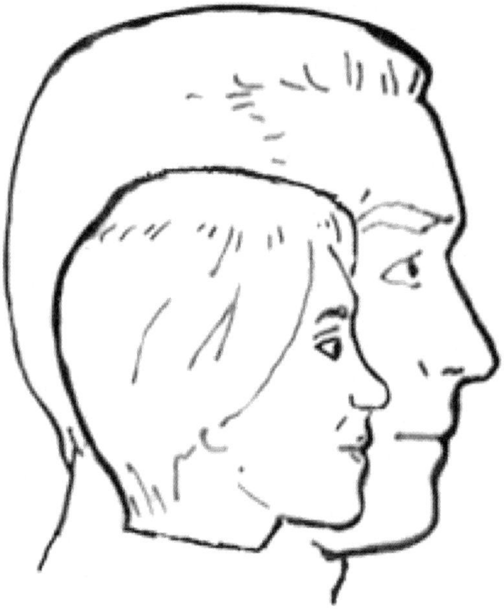

Figure 7. Comparative sizes of the smallest female and the largest male head

between the widest points of the cheekbones (bizygomatic breadth) and the distance from the tip of the chin to the deepest point of the nasal root depression (nasion-menton length). The latter is the more important as it determines the position of the eyes relative to the eyepieces: an important design parameter to maximize the visual field. The bizygomatic breadth and the nasion-menton length tend to be correlated in terms of overall face size, but it is recognized that there are significant numbers of people with long, thin faces or short, wide faces.

Other facial features that need to be considered include prominent cheekbones, often with sunken cheeks and/or temples, low cheekbones, narrow, receding, lumpy, broad or square chins, heavy brows, deep eye sockets and scars or deep lines in the skin.

Requirements for more comfort and better fit (individually and across the user population) mean that bizygomatic breadth and nasion-menton length alone are probably no longer sufficient, and that more facial measurements are needed. These measurements are not always available, and even when they are, there can be problems in determining the variation and range of each facial dimension in terms of all of the other dimensions. Comparative sizes of the smallest female and the largest male head are shown in Figure 7.

EFFECTS OF WEARING RESPIRATORS

The effects of wearing respirators can be divided into two: the burden placed directly on the user and compatibility with other equipment. User burdens are divided into three: physiological, psychological and ergonomic. Equipment compatibility is really an ergonomic issue, but is usually considered in its own right because of its importance to the military user.

There are two main physiological burdens, thermoregulatory and respiratory, and two main types of psychological burden, 'hard' effects on attention, cognition and perceptual-motor function, and 'soft' effects on conation, feelings and subjective state. There is a large literature on the effects of wearing respirators, with and without protective clothing, and for brevity, this section covers only the effects of respirators, either alone or in terms of their contribution to full protection. In addition, for brevity, this section contains only a selection of the large number of references.

Thermoregulatory effects

Respirators feel hot to wear because the impermeable material of the facepiece and faceseal prevents heat loss, causing heat build-up and sweating, which can cause problems with heavy work and/or hot environments. Sweat can also cause irritation and soreness where it collects around the chin (the lowest point).

These problems could be reduced by using materials that are impermeable to agent and air but permeable to water, and preferably with high heat conductivity, but such materials are not yet available. Cooling systems could be used, but some, e.g. condensation refrigerators, thermionic coolers and Peltier devices, require power, which creates logistic problems. Furthermore, with current technology, having to carry the weight of these systems would generate more metabolic heat than the systems could remove. Other systems, e.g. heat pipes, do not require power, but the technology is not yet sufficiently developed to make these feasible. In addition, all such systems would generate and/or dissipate extra heat, which would increase the wearers' heat signatures and risk of being detected under battlefield conditions. Phase-change materials (such as heptadecane and octadecane) that change from solid to liquid and absorb heat at 'near-skin' temperatures, might help; they are already commercially available, but are of limited use in that once melted, they absorb no more heat and their cooling effect ends.

Siting exhale valve assemblies at the lowest point on the facepiece would allow sweat to drain away, and would also help take the exhale valve assembly out of the visual field. This is done in some respirators, but there are three disadvantages. First, valves do not perform well when half-submerged. Secondly, drying sweat leaves a residue that can affect the valve seating and increase leakage. Thirdly, like chin filters, low-mounted valves can get in the way, and get dirty, when the user is lying prone or crawling. In addition, it is important to remember that sweat must evaporate in order to cool the skin; simply removing sweat might help stop its build-up, but it would not improve cooling.

Respiratory effects

There are three respiratory effects. First, inhalation resistance is imposed by the filter canister, inhalation valve and, to some extent, the airflow pattern. Secondly, exhalation resistance is imposed by the exhale valve and to some extent the exhale valve deadspace design. Thirdly, the internal volume of the respirator increases the respiratory deadspace, which can cause build-up of carbon dioxide.

Moderate respiratory resistance causes slowing and deepening of respiration, and high resistances cause rapid, shallow breathing (Sharp and Ernsting, 1988). These effects are exacerbated by exercise, and generally result in a fall in 'minute volume' and a failure to eliminate carbon dioxide adequately (Ernsting, 1965; Deno *et al.*, 1981; Dressendorfer *et al.*, 1977).

Wearing respirators affects respiratory parameters, such as ventilatory volumes and tidal volumes, both at rest and during exercise (Stannard and Russ, 1948; Bartlett *et al.*, 1972; Askanazi *et al.*, 1980; Jetté *et al.*, 1990). Some reports say that wearing respirators decreases exercise endurance (Van Huss *et al.*, 1967; Craig *et al.*, 1971; Flook and Kelman 1973; Stemler and Craig 1977; Lerman *et al.*, 1983). Other reports say that wearing respirators has little or no effect on exercise endurance (Demedts and Anthonisen, 1973; Dressendorfer *et al.*, 1977; Gamberale *et al.*, 1978; Deno *et al.*, 1981; Chamberlain *et al.*, 1997). There is some evidence that exhalation resistance may be more important than inhalation resistance in determining exercise tolerance (Flook and Kelman, 1973).

With current filter and adsorption technology, inhalation resistance may only be reduced by reducing the amount of filter material, which will reduce protection, or by increasing the surface area, which will make the filter bigger, reducing access to confined spaces and increasing the chance of fouling and knocking. Exhalation resistance can be reduced to some extent by better valve and deadspace design and more flexible materials.

There is an argument that breathing resistance should not be reduced to zero, even if it could be. Wearers expect that a respirator will impose some breathing resistance and may be anxious if there is none: they would not know whether the respirator is working properly. In addition, some exhale resistance is important, not only for psychological reasons, but also to improve protection: the build-up of pressure on exhalation, before the valve opens, is necessary to resist the ingress of agent via the face-seal. Other psychological factors also need to be taken into account: there is some evidence that wearing a facepiece alone, without a canister, can degrade work performance (Craig *et al.*, 1970).

The internal volume of the respirator adds to the respiratory deadspace, which can cause build-up of carbon dioxide, which in turn can reduce exercise endurance. The deadspace can be reduced by reducing the overall volume of the respirator, which would also improve equipment compatibility and access to confined spaces. However, it would further reduce the space available for other components, could exacerbate feelings of 'claustrophobia' and, if it is reduced to the point at which the internal features of the respirator begin to touch the face and lips, it can cause sufficient distraction as to impair exercise tolerance (Kelm *et al.*, 2000).

Wearing respirators has little effect on physiological functions, such as heart rate, blood pressure and body temperature, whether the wearer is resting (Shephard, 1962; Ryman *et al.*, 1987) or exercising (Spioch *et al.*, 1962; Gee *et al.*, 1968; Hermansen *et al.*, 1972; Flook and Kelman, 1973; Stemler and Craig, 1977; Harber *et al.*, 1982; Dukes-Dobos and Smith, 1984; Harber *et al.*, 1984; Kelm *et al.*, 2000; Wetherell, 2000).

Perceptual-motor and cognitive effects

The most important perceptual-motor effects of respirators are on vision and speech intelligibility. It is common experience that respirators restrict the visual field, depending not only on the eyepiece shape, size and 'eye-relief' (distance from eye to eyepiece), but also on intrusion of filter canisters and other components into the visual field. Respirators can also impair visual acuity if the eyepieces are not well-designed or become misted.

Respirators muffle and distort speech because the facepiece material absorbs sound, the facepiece moulding forms resonant cavities and the faceseal restricts movement of the lower jaw (Mozo and Peters, 1984; Johnson and Sleeper, 1985; Nelson and Mozo, 1985; Fine and Kobrick, 1987; Taylor and Orlansky, 1987; Wetherell and Gwyther, 1993). Voice modules help, but do not resolve the problem completely.

Respirators can impair attention (Spioch *et al.*, 1962; Kobrick and Sleeper, 1986; Barba *et al.*, 1987; Van de Linde, 1988; Caretti, 1995), but can also, at least for a few hours, improve performance on tasks requiring close attention (Wetherell, 1989; Zimmerman *et al.*, 1991), probably by removing potentially distracting stimuli, focusing attention and/or increasing psychological arousal.

Anecdotal reports suggest that respirators impair thinking and reasoning, but close inspection reveals that most of the effects are due to sensory and/or motor impairment, to other factors such as environmental temperature and humidity, or to distraction caused by factors such as discomfort, breathing resistance, feelings of isolation and anxiety.

Conative effects, feelings and subjective state

Nobody likes wearing respirators; it is common experience that wearers feel isolated and disinclined to communicate, unhappy, anxious, impatient, frustrated, irritable and uncomfortable (Warren *et al.*, 1988; Caretti, 1995).

Some wearers can suffer more extreme reactions, including respiratory distress, heightened anxiety and hyperventilation, leading to vomiting, claustrophobia and convulsions (Morgan, 1983; Muza, 1988). Wearing full chemical and biological protection can also produce dramatic reactions, mostly attributable to respirators, including panic, confusion, shortness of breath, claustrophobia and even hallucinations (Jones *et al.*, 1985; Brooks *et al.*, 1983; Carter and Cammermeyer, 1985a,b,1989). These feelings can contribute significantly to impaired military effectiveness (Brooks *et al.*, 1983; House, 1985; Rauch *et al.*, 1986).

Ergonomics

Ergonomic issues are often forgotten, but they can impose serious burdens. Respirators have to be carried, and take up space and weight needed for other items such as food, water, weapons and ammunition. Respirators require attention and maintenance, including changing canisters, careful cleaning, decontamination when necessary, and correct disassembly and re-assembly, all of which require training. If an attack occurs, the respirator must be found (usually at the bottom of a haversack) and donned quickly and properly, which also requires training. During wear, the respirator may have to be lifted from the face for eating, taking medical countermeasures, e.g. tablets, or special decontamination, and re-seated properly. The drinking tube needs to be operated properly, often while wearing protective NBC gloves, and to deliver liquid at the required rate.

The burdens are not just on the wearers, but also on those supporting the wearers. Faces must be measured accurately and the correct size must be issued, fitted and tested at intervals to ensure the required protection is achieved and maintained. This requires training of instructors. If damaged, respirators must be returned for repair or disposal, and a new one issued – this requires extra manpower and logistics organization.

Equipment compatibility

The ability to use equipment while wearing a respirator has already been mentioned. The problem is that the respirator's bulk can obstruct the wearer's access to other equipment, particularly equipment that comes into direct contact with parts of the respirator. The main respirator components involved are the eyepieces/visor and the filter canister(s), and the main pieces of equipment are shoulder-held and fired weapons, optical equipment, such as weapon sights, binoculars, night vision aids, and communications equipment.

Equipment compatibility works both ways: as a respirator may interfere with the operation of equipment, so the use of equipment may interfere with the respirator. For example, a respirator may prevent the wearer from obtaining a good 'sight picture' through a rifle sight, but pushing against the rifle to obtain a good sight picture, and firing the weapon, may distort the respirator and cause breaching of the face-seal.

RESPIRATOR PROTECTION TESTING

The protection afforded by a respirator is usually expressed as the protection factor (PF), i.e. the ratio of the concentration of agent outside to that inside the respirator. Respirators are tested on 'real' people using simulated agents; most testing is carried out in the laboratory, but workplace/field testing is increasingly being used.

Test agents

Simulated agents most used include aerosols of sodium chloride (common salt), corn oil, sunflower oil, dioctyl phthalate (DOP) and poly alpha olefins (usually poly 1-decene, also known as 'Emery Oil'). Bacteria and viruses are sometimes used: mostly spores of Bacillus subtilis var Niger (formerly known as B globigii [BG]) and the MS2 bacteriophage. Methyl salicylate is used both as a vapour and an aerosol, often containing very small amounts of ultraviolet tracers such as 'Tinopal' (used in fabrics and soap powders to enhance 'whiteness' of clothing) to aid detection. Some smokes and mists used to train firefighters and to create special effects in discotheques are also used (the smokes generally contain 1,2-propane diol and 1,2,3-propane triol).

Measurement methods

Typical methods used to measure concentrations of test compounds include flame photometry, light scattering and ambient particle counting, although other methods including time-of-flight measurement and measurement of inward air leakage during reduced internal pressure are sometimes used.

Flame photometry involves burning the sample in a hydrogen flame and measuring the characteristic emission of the agent used, e.g. sodium chloride. The method was developed at the Dstl, Porton Down, UK in the 1960s, and is now used extensively throughout the world; it is specified in British Standards (BS) (1969) and is also specified as a routine laboratory test in the BS (1992).

A salt aerosol is generated in an enclosure, usually up to a concentration of 13 mg m^{-3}, with a particle mass median diameter of 0.3 to 1.3 μm, depending on the aerosol generator used. The subject dons the respirator, enters the enclosure and performs a set of simple exercises while the salt concentration inside and outside the respirator are measured by flame photometry. The Dstl, Porton Down uses a specially built high-sensitivity flame photometer to measure very low salt concentrations (and hence to confirm very high PFs).

Light scattering is similar to the effect by which a flashlight beam can be seen in a dusty atmosphere. A light is shone at the aerosol, usually an oil, and the scattered light is focused onto a photomultiplier, which generates voltages proportional to the amount of light present. Light scattering can be used to measure particle size (amount of light scattered), number (number of scattering events) and even shape (proportion of light scattered to each detector).

Particle-counting methods count particles present in ambient air and do not require the subject to be exposed to any test aerosol. The most commonly used method is condensation nuclei counting (CNC) in which a vapour, usually isopropyl alcohol, is condensed onto particles to grow them to a size that can be 'seen' and counted, by a laser and/or a light-scattering system.

Air leakage measurements use a system of controlled negative pressure in which air is pumped out of the respirator to create a fixed negative pressure (usually 1 in water gauge) and the airflow need to maintain this pressure is measured. The airflow is taken as the same as the leakage rate into the respirator. The method requires wearers to hold their breath and the filter canister to be 'blanked off' (otherwise, air will 'leak' through the filter).

Respirator testing procedures and exercises

Subjects should not be tested immediately after donning the respirator, as it takes 5 to 10 min, depending on the ambient temperature, for the respirator face-seal to warm up and settle fully on to the face.

Subjects undergoing respirator protection tests are often required to perform simple exercises. Different testing and regulatory authorities require different exercises, but they typically include the following:

- breathing normally;
- breathing deeply;
- nodding the head slowly up and down;
- shaking the head slowly from side to side;
- bending forward and nodding and/or shaking the head slowly;
- turning the head as far as possible to look over the shoulder;
- making various facial movements, e.g. smiling, frowning, miming, speaking;
- reading aloud from a prepared text, e.g. the 'Rainbow Passage' (Fairbanks, 1960) which includes a variety of labial, stop, fricative, velar and dental sounds;
- Walking on a treadmill (to increase breathing rate and depth, jog the respirator and possibly induce sensible sweating).

THE FUTURE

Respirators are continually being improved as design methods, materials technology and test methodology improve.

New materials are under development which combine physical and chemical hardness with easy workability. Some materials have selective permeability, e.g. to water vapour, which could help reduce the heat load on the face. Some materials are transparent and flexible, but still provide good ballistic protection, allowing the use of panoramic visors that flex and allow the facepiece to conform better to different face shapes and movements. New, more permanent hydrophilic and hydrophobic coatings are also under development to help reduce eyepiece/visor misting.

New filter materials are under development to improve filtration while reducing breathing resistance, and also to allow more conformable and ergonomic canister designs. New particle filter materials include 'electret' (captures particles by electrostatic attraction) and polytetrafluoroethylene (PTFE). New adsorbents include new forms of carbon, including resin-immobilized and monolithic carbons, carbon fibres (to combine particle and adsorbent filtration), together with other types of adsorbent, such as silicas and zeolytes. New adsorbents and impregnants are also under development to protect against toxic industrial chemicals that present an increasing risk to military forces.

New developments in face scanning using three-dimensional cameras and lasers will help design better-fitting respirators and determine optimum sizing regimens. Face-scanning techniques, together with recent developments in computer-aided design and rapid prototyping, allow the possibility of individually fitted, 'bespoke' respirators. However, allowance must be made to accommodate face movements, e.g. speaking, smiling and frowning, and changes in face size and shape, e.g. through losing or gaining weight.

It has long been recognized that respirator tests performed in the laboratory can indicate what level of protection a respirator can achieve, but do not reflect the PF obtainable in the workplace or the field. Systems are now being developed that can be fitted onto personal load carriage equipment, and used to measure the PF while personnel are performing their normal duties. Protection factors are measured by CNC particle counters, and recorded onto data loggers or computers, either worn on the body or, via radio telemetry, sited remotely. Other features can also be incorporated, e.g. helmet-mounted video cameras to relate PFs to activities, internal mask pressure monitors, accelerometers to monitor movement in the x-, y- and z-planes, global positioning systems to monitor subject location and RF modems to relay data to base stations.

Finally, the chemical (and biological) threat has changed since the end of the Cold War, mainly in its uncertainty, and in many respects the threat is now greater. While the threat continues, respirators will continue to be one of the first and most important lines of defence, and work will continue to improve protection while reducing user-burden and interference with the use of other equipment.

REFERENCES

Askanazi J, Silverberg PA, Foster RJ *et al.* (1980). Effects of respiratory apparatus on breathing pattern. *J Appl Physiol*, **48**, 577–580.

Barba CA Stamper DA, Penetar DM *et al.* (1987). *The effects of the M17A2 mask on human pursuit tracking performance.* Report No. 250. San Francisco, CA, USA: Letterman Army Institute of Research.

Bartlett HL, Hodgson JL, Kollias J *et al.* (1972). Effect of respiratory value dead space on pulmonary function at rest and during exercise. *Med Sci Sports*, **4**, 132–137.

Brooks FR, Ebner DG, Xenakis SN *et al.* (1983). Psychological reactions during chemical warfare training. *Mil Med*, **148**, 232–235.

Caretti DM (1995). *Signal detection over the visual field during wear of various respirators and respirator configurations.* ERDEC Report TR-386. Edgewood, MA, USA: Edgewood Research, Development and Engineering Center.

Carter BJ and Cammermeyer M (1985a). Biopsychological responses of medical unit personnel wearing chemical defence ensemble in a simulated chemical warfare environment. *Mil Med*, **150**, 239–249.

Carter BJ and Cammermeyer M (1985b). Emergence of real casualties during simulated chemical warfare training. *Mil Med*, **150**, 657–663.

Carter BJ & Cammermeyer M (1989). Human responses to simulated chemical warfare training in US Army Reserve personnel. *Mil Med*, **154**, 281–288.

Chamberlain S, Cook JM, Spayer J *et al.* (1997). *Physiological and psychological effects of breathing resistance on the respirator wearer.* Report CBD/MC/TR97/158. Porton Down, UK: Defence Evaluation and Research Agency, Chemical and Biological Defence Sector.

Craig FN, Blevins WV and Cummings EG (1970). Exhausting work limited by external resistance and inhalation of carbon dioxide. *J Appl Physiol*, **29**, 847–851.

Craig FN, Blevins WV and Froehlich HL (1971). *Training to improve endurance in exhausting work of men wearing protective masks: a review and some preliminary experiments.* Report 4535. Edgewood, MA, USA: Edgewood Arsenal.

Demedts M and Anthonisen NR (1973). Effects of increased external airway resistance during steady-state exercise. *J Appl Physiol*, **35**, 361–366.

Deno NS, Kamon E and Kiser DM (1981). Physiological response to resistance breathing during short and prolonged exercise. *Am Ind Hyg Assoc J*, **42**, 616–623.

Dressendorfer RH, Wade CE and Bernauer EM (1977). Combined effects of breathing resistance and hyperoxia on aerobic work tolerance. *J Appl Physiol*, **42**, 444–448.

Dukes-Dobos RJ and Smith R (1984). Effects of respirators under heat/work conditions. *Am Ind Hyg Ass J*, **45**, 399–404.

Ernsting J (1965). The physiological requirements of aircraft oxygen systems. In: *A Textbook of Aviation Physiology* (JA Gillies, ed.), pp. 209–289. London, UK: Pergamon Press.

Fairbanks G (1960). *Voice and Articulation Drill Book.* New York, NY, USA: Harper and Row.

Fine BJ and Kobrick JL (1987). Effect of heat and chemical protective clothing on cognitive performance. *Aviat Space Environ Med*, **58**, 149–154.

Flook V and Kelman GR (1973). Submaximal exercise with increased inspiratory resistance to breathing. *J Appl Physiol*, **35**, 379–384.

Gamberale F, Holmer I, Kindblom AS *et al.* (1978). Magnitude perception of added inspiratory resistance during steady-state exercise. *Ergonomics*, **21**, 531–538.

Gee JBL, Burton G, Vassalo C *et al.* (1968). Effects of external airway obstruction on work capacity and pulmonary gas exchange. *Am Rev Resp Disease*, **98**, 1003–1012.

Harber P, Tamimie J, Bhattacharya A *et al.* (1982). Physiological effects of respirator dead space and resistance loading. *J Occ Med*, **24**, 681–684.

Harbar P, Tamimie J, Emory J *et al.* (1984). Effects of exercise using industrial respirators. *Am Ind Hyg Ass J*, **45**, 603–609.

Hermansen L, Vokac Z and Lereim P (1972). Respiratory and circulatory response to added air flow resistance during exercise. *Ergonomics*, **15**, 15–24.

House GL (1985). Leadership challenges on the nuclear battlefield. *Mil Rev*, **3**, 60–69.

Jette M, Thoden J and Livingstone S (1990). Physiological effects of inspiratory resistance on progressive work capacity. *Eur J Appl Physiol*, **60**, 65–70.

Johnson RF and Sleeper LA (1985). Effects of thermal stress and chemical protective clothing on speech intelligibility. In: *Proceedings of the 29th Meeting of the Human Factors Society*, pp. 541–545, Santa Monica, CA, USA.

Kelm DM, Puxley KPM and Withey WR (2000). *The contribution of head and respiratory protection to the human factors burden of individual protective equipment.* Report CHS/PPD/TR000259. Centre for Human Sciences, Farnborough, UK: Defence Evaluation and Research Agency.

Kobrick JL and Sleeper LA (1986). Effects of wearing chemical protective clothing in the heat on signal detection over the visual field. *Aviat Space Environ Med*, **57**, 144–148.

Lerman Y, Shefer A, Epstein Y *et al.* (1983). External inspiratory resistance of protective respiratory devices: effects on physical performance and respiratory function. *Am J Ind Med*, **4**, 733–740.

Morgan WP (1983). Psychological problems associated with the wearing of industrial respirators: a review. *Am Ind Hygiene Ass J*, **44**, 671–676.

Mozo BT and Peters LJ (1984). *Effects of chemical protective and oxygen masks on attenuation and intelligibility when worn with the SPH-4 helmet.* Report 84-5. Fort Rucker, AL, USA: US Army Aeromedical Research Laboratory.

Muza SR (1988). Biomedical aspects of NBC masks and their relation to military performance. In: *Handbook on Clothing: Biomedical Effects of Military Clothing and Equipment Systems*, Chapter 11. Nato, Brussels, Belgium: Research Study Group 7 on Biomedical Research Aspects of Military Protective Clothing.

Nelson WR and Mozo BT (1985). Effects of XM-40 chemical protective mask on peal-ear attenuation and speech intelligibility characteristics of the SPH-4 aviator helmet, Report 85-2. Fort Rucker, AL, USA: US Army Aeromedical Research Laboratory.

Jones PD, Stokes JW, Newhouse PA *et al.* (1985). Neuropsychiatric casualties of chemical, biological and nuclear warfare. In: *Psychiatry: The State of the Art* (P Pichot, P Berner, R Wolf, K Thau, eds.), pp. 539–543, New York, NY, USA: Plenum Publishing.

Nielsen R, Gwosdow AR, Berglund LG *et al.* (1987). The effect of temperature and humidity levels in a protective mask on user acceptability during exercise. *Am Ind Hyg Assn J*, **48**, 639–645.

Rauch TM, Munro I, Tharion W *et al.* (1986). Subjective symptoms, human endurance, and cognitive interventions, Report TR-86-1. Colorado Springs, CO, USA: Proceedings of Psychology in the Department of Defense, US Air Force Academy.

Ryman DH, Englund CE, Kelly TL *et al.* (1987). Mood, symptom, fatigue, sleepiness and vital sign changes in mask and protective suit under non-exercising conditions. In: *Proceedings of the 6th Medical and Chemical Defense Bioscience Review*, pp. 669–672, US Army Medical Research Institute of Chemical Defense, MA, USA.

Sharp J and Ernsting J (1988). Hypoxia and hyperventilation. In: *Aviation Medicine*, 2nd Edition (J Ernsting and P King, eds), pp. 45–59. London, UK: Butterworths.

Shephard RJ (1962). Ergonomics of the respirator. In: *Design and Use of Respirators* (CN Davies, ed.), pp. 51–56. New York, NY, USA: MacMillan.

Spioch FM, Kobza R and Rump S (1962). The effects of respirators on the physiological reactions to physical effort. *Acta Physiol Polon*, **13**, 637–649.

Stannard JN and Russ EM (1948). Estimation of critical dead space in respiratory protective devices. *J Appl Physiol*, **1**, 326–332.

Stemiler FW and Craig FN (1977). Effects of respiratory equipment on endurance in hard work. *J Appl Physiol: Resp Environ Exercise Physiol*, **42**, 28–32.

Taylor HL and Orlansky J (1987). *The effects on human performance due to wearing protective clothing for chemical warfare: implications for training.* Report P-2026. Alexandria, VA, USA: Institute for Defence Analysis.

Van de Linde FJG (1988). Loss of performance while wearing a respirator does not increase during a

22.5 hour wearing period. *Aviat Space Environ Med*, **59**, 273–277.

Van Huss WD, Hartman F, Craig FN *et al.* (1967). Respiratory burden of the field protective mask under exercise load (abstract). *Fed Proc*, **26**, 721.

Warren RH, Poole PM and Busamralc A (1988). *The effects of microencapsulation on sensory-motor and cognitive performance: relationship to personality characteristics and anxiety*. Report TR 89/015. Natick, MA, USA: US Army Natick Research, Development and Engineering Center.

Wetherell A (1989). *Effects of wearing the S6 respirator for 6 hours on the cognitive and psychomotor performance of male and female subjects.* Technical Note 989. Porton Down, UK: Chemical and Biological Defence Establishment.

Wetherell A (2000). *The contribution of the respirator and its components to the human factors load of NBC individual protective equipment*. Report CBD/CR000485. Porton Down, UK: Defence Evaluation and Research Agency, Chemical and Biological Defence Sector.

Wetherell A and Gwyther R (1993). *Effects of IPE on speech transmission and reception*. Unpublished report. Porton Down, UK: Chemical and Biological Defence Establishment.

Zimmerman WJ, Eberts C, Salvendy G *et al.* (1991). Effects of respirators on performance of physical, psychomotor and cognitive tasks. *Ergonomics*, **34**, 321–324.

7 RESPONDING TO CHEMICAL TERRORISM: OPERATIONAL PLANNING AND DECONTAMINATION

Gron Roberts[1] and Robert L. Maynard[2]

[1] *GR Associates, Chappel, UK*
[2] *Health Protection Agency, Chilton, UK*

OPERATIONAL PLANNING

Introduction

A great deal has been written about the properties and health effects of many kinds of chemicals – and the treatment of their victims – but there is much less published advice available for those operational planners or responders facing the practical challenges of dealing effectively with a deliberate release in a non-military setting. This chapter provides an overview of the basic factors that need to be considered when developing contingency arrangements for the initial health response and aims to provide general advice and information for those who might find themselves planning for or directing those operations. A summary of current thinking on casualty decontamination is also provided.

Chemical terrorism is a real threat

With some 11 million chemical substances known to man and over 70 000 in regular use, providing a rapid, safe and effective health response to an incident involving any accidental or deliberate release is a real and significant operational challenge for any emergency medical system. Unintentional discharges have occurred in many parts of the world and the release of large quantities of methyl isocyanate in Bhopal in December 1984 was a stark reminder of the potential impact of industrial accidents.

The release of the nerve gas sarin by the Aum Shinrikyo cult in Matsumoto and Tokyo – in June 1994 and March 1995, respectively (see Chapter 13) – alerted the world to the very real chemical threat from terrorist groups. The determination and sophistication of subsequent terrorist attacks, the access such groups seem to have to the necessary funds, expertise and materials and their avowed intent to cause mass casualties and fatalities all emphasise that the possibility of a deliberate chemical release in a civilian setting can no longer be ignored.

The need for planning: failing to plan is planning to fail

Victims of chemical contamination need prompt and effective treatment, both at the scene of the incident and in hospital. Ambulance or emergency medical services and hospital emergency rooms provide the front line health response and most will already have tried and tested contingency arrangements for accidental chemical releases – often referred to as HAZMAT or HAZCHEM (hazardous materials and hazardous chemicals) protocols. Whether for an accidental or deliberate chemical release contingency arrangements need to address the following **key objectives**:

- the safety, protection and decontamination of responding staff;

Chemical Warfare Agents: Toxicology and Treatment (2nd Edition)
Edited by Timothy C. Marrs, Robert L. Maynard and Frederick R. Sidell

Table 1. Differences between accidental and deliberate releases

Accidental releases	Deliberate releases
Usually single scene	May be multi-scene
Usually one substance	Possibly multi-substance
Known or identified substance	Unknown/unidentified agent(s)
Not intended to cause harm	Designed to cause harm
Identifiable source and quantity	Unknown source or quantity
Information available	Information not readily available
Site-specific safety plans in place	No specific safety plans in place
Limited affected area[a]	Wider affected area
Limited consequential disruption	Widespread consequential disruption
Generally few casualties[a]	Potentially many casualties
Chemical effects likely to be main cause of injury	Chemical effects could be accompanied by physical injuries

Note:
[a] The release of toxic methyl isocyanate in Bhopal, India in December 1984 is a noteworthy exception.

- rapid hazard identification and containment on site as far as possible;
- quick and effective initial patient triage, rescue and decontamination;
- basic life support, assessment and advanced treatment – including the administration of any available countermeasures at scene and in hospital;
- transfer to hospital for secondary decontamination and treatment;
- implementing rapid measures to protect health facilities;
- providing specialist care in an appropriate setting.

As a general rule, building on normal response arrangements usually provides the best foundation for developing any contingency plans, but arrangements for dealing with a deliberate chemical releases will need to recognize and reflect some important features not seen in accidental settings. A number of such differences are summarized in Table 1.

The scale and nature of the problems that a major deliberate release presents to emergency responders makes managing such incidents disproportionately complex and demanding. Planning assumptions need to be varied significantly and specific contingency arrangements for a deliberate release are essential as any approach based simply on dealing with them as *super-hazmat* incidents is unlikely to represent an operationally viable or safe approach.

Success requires a team effort

Teamwork is always the key to a successful response to any major incident. Additional threats to the safety of the responding personnel and the potential need for patient decontamination are just two of the compounding factors that add to the complexity of dealing with a large-scale chemical release. The emergency services responding to any such incident need to have a high degree of confidence in each other's capabilities and awareness of each other's operating systems. Therefore, the importance of developing integrated multi-agency contingency plans, clearly defining organizational roles and responsibilities and practising regularly together cannot be over emphasized.

The value of military knowledge and experience

Because the use of poisonous substances in warfare has a long history (see Chapter 1), there has been a considerable amount of military research into the toxicological hazard posed by various agents, their optimal offensive use, the most effective protection against them and how best to decontaminate and deal with victims. Specific research and experience in a civilian setting is very much more limited and serious interest much more recent, although it has increased significantly following sarin releases in Matsumoto and Tokyo.

Civil emergency planners have naturally looked to military research and information as the basis of their planning, but access has often been restricted and relevant knowledge and experience is only gradually becoming more available to civilian responding agencies. Such knowledge – particularly the military's experience in developing personal protection, casualty treatment and decontamination – provides a very useful basis for civilian planning, but the aims and operational challenges facing civilian responders differ in some significant and important respects from those facing military planners. Civilian planners particularly need to bear in mind that:

- the overriding military objective is to protect troops while maintaining their ability to fight in the type of protective equipment that allows them to perform the basic tasks necessary to continue the battle;
- there is likely to be more accurate intelligence available on probable agents and appropriate countermeasures in a military setting;
- the military's ability to detect, identify, measure and prepare for a chemical threat is likely to be much greater in the initial phases;
- military personnel are generally young, fit, disciplined and well trained to use protective equipment to operate in demanding environments.

The efficacy of the agent that might be used, or the delivery system that may be adopted by a terrorist group is likely to be somewhat less sophisticated than military options. However, an attack on an unprotected population is also likely to produce many casualties, lead to much more severe exposure across a wider range of age groups, create more panic and result in operating conditions that are likely to be more chaotic than those encountered in a military setting. For all those reasons, the applicability of military assumptions, doctrines and procedures needs critical examination and careful adaptation before adoption in civilian operations.

Risk assessment

Unless some pre-warning is given, an emergency call to a deliberate chemical release is unlikely to specify the true extent or nature of the risk. Recognizing a potential hazard is the first major safety challenge. Rapid risk assessments and prudent measures to protect the safety of initial responding crews are required.

Time to initial intervention is likely to be a key determinant of mortality and morbidity. More than in most other emergency scenarios, the initial responders arriving at the scene are going to be under acute pressure to *do something*. Emergency services face a very real tension between their duty to protect their own staff and their basic *raison d'être* – to protect and assist the public. If responding agencies adopt an entirely risk-free approach, this is likely to be to the detriment of some potential survivors, and striking the right balance will be critical to operational effectiveness and will almost certainly require justification at a subsequent inquiry.

Vigilant and well-trained despatch-centre staff can be the first line of defence. All those receiving calls from the public in emergency despatch-centres should receive specific chemical awareness training and be routinely alert to the possibility of chemical involvement. Warning questions and triggers should be built-in to the call-taking algorithms that are now widely used and despatch protocols should require that all relevant information – particularly any doubts or suspicions – be passed immediately to all first responding units.

In responding to an unclassified emergency call, operational staff should be routinely alert to the possibility of chemical contamination. Their initial training and continuing education programmes should remind them of the general warning signs of chemical deployment and place particular emphasis on the need for a dynamic risk assessment. Whenever the cause of any incident is not immediately obvious, **if in doubt – stay out** is a useful general rule. That can be supported by simple protocols such as the *Safety Triggers for Emergency Personnel* (STEP) rules adopted by many UK emergency services as the basis of an initial risk assessment (Table 2).

When approaching the scene of any suspicious incident, staff should also be aware of and alert to general indicators of a possible toxic release such as:

Table 2. Safety triggers for emergency personnel (STEP) 1, 2, 3

STEP 1	One casualty down	• Nothing unusual – adopt normal approach
STEP 2	Two casualties down	• Approach with **caution** • Consider all possible causes • Report arrival – update despatch
STEP 3	Three plus casualties down	• **Do not approach** the scene • Withdraw and contain • Provide situation report • Protect self and others • Await specialist help

- **Environmental** – Discoloration, withering or dead foliage or grass, localized haze or hazy conditions, unusual or unexplained odours, deposits, liquids or droplets.
- **People** – Many displaying similar signs or symptoms, such as loss of consciousness, difficulty in breathing, convulsions, nausea, miosis, headache, disorientation or skin blistering.
- **Animals** – Dead or dying animals, birds or fish.

Protecting responding personnel

The type of personal protective equipment suitable for health staff is outside the scope of this chapter, but a decision on the level of protection necessary at an incident is one of the most critical initial command decisions and one that has to strike the right balance between protecting staff and helping the public. Although designs are improving, the more protection such equipment offers the wearer, the more time it generally tends to take to put on. Wearing such equipment also limits the rescue time available, makes communication particularly difficult and hinders the mobility and effectiveness of the wearer – particularly when attempting to carry out patient assessment or treatment.

Scene management

Law enforcement agencies will usually attempt to control access to and egress from the site by establishing and enforcing cordons to limit and control movement. An outer cordon will normally surround the entire incident and an inner cordon protects the site itself with fire agencies usually assuming overall responsibility for co-ordination and safety within this inner zone. To maintain safety and contain/limit the spread of contamination the site of a chemical incident will usually be divided into a contaminated (hot), decontamination (warm) and clean (cold) zones (Ambulance Service Association) (Figure 1).

In practice, the initial boundaries between these 'zones' are unlikely to be clearly defined or fixed and will need rapid adjustment should conditions change or contamination spread. Limiting that spread by maintaining the integrity of these 'zone' boundaries is an important objective and immediate controls should be instigated to ensure that access/egress points are properly designated and managed, that rescue personnel wear protective equipment appropriate to the risk in each zone and that no contaminated person (or equipment) is allowed to cross the clean/dirty line, i.e. to enter the cold zone.

The basic concept of operations envisages that contaminated casualties who are rescued or self-evacuate from the heavily contaminated 'hot zone' should be rapidly undressed, triaged (Fisher *et al.*, 1999), given basic life support treatment and decontaminated in the 'warm zone' before being passed to the 'cold zone' for fuller assessment, treatment and, if necessary, subsequent transfer to hospital.

Health operations in the 'hot zone'

Any chemical release will have its maximum harmful effect against unprotected individuals, so those who self-evacuate and the rapidly rescued will generally have the best survival and

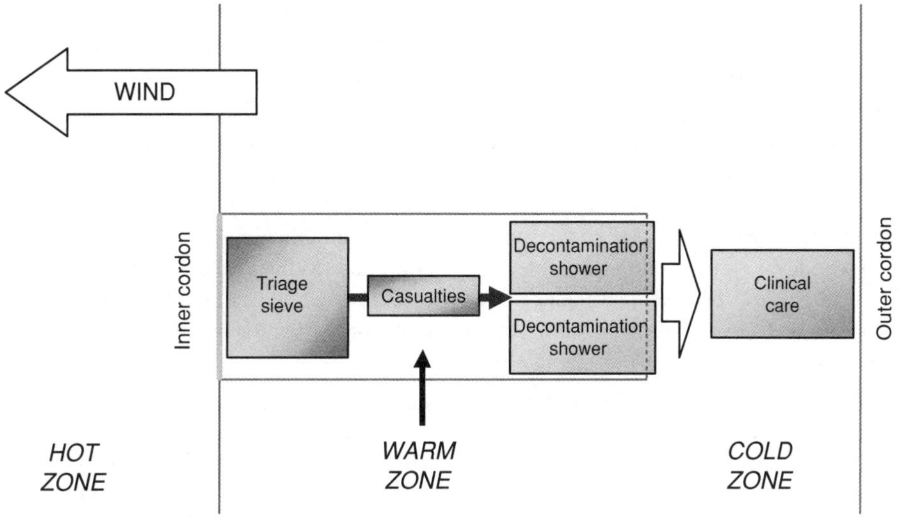

Figure 1. Cordons and zonings (Ambulance Service Association, 2004). Reproduced with kind permission of the UK Ambulance Service Association, 2004

recovery prospects. Planners and incident 'commanders' should bear in mind that by the time rescue efforts are mounted survivors are generally unlikely to be found in environments or circumstances that demand the wearing of self-contained respiratory protection. Health plans should also recognize that the level of meaningful assessment or treatment that can be carried out in the most heavily contaminated area (hot zone) will be very limited – not least by the fact that the patients themselves will remain exposed and unprotected and therefore at significant risk.

Rapid rescue from the hot zone represents the most practical and effective operational approach and in most major releases the most effective health response is likely to be focused in the more lightly contaminated ('warm') and uncontaminated ('cold') zones. There – and at receiving hospitals – the main emphasis should be on providing the type of respiratory and physical protection for staff so that secondary contamination is prevented. Unprotected staff should not come into contact with undecontaminated patients.

Most health personnel are likely to find wearing and working in fully self-contained protective equipment physically demanding and the limitations frustrating. Employing organizations are also likely to find the investment required in training and providing practice exercises for larger numbers of staff disproportionately high but a limited presence of trained and protected staff in specialist teams may be justified. The benefits often claimed from sending such personnel into the 'hot zone' – reassurance and clinical support for other rescuers, early assessment of casualty numbers/signs/symptoms and providing earlier information to the health system – need to be weighed carefully against the cost, additional risks and physical demands involved. Further operational research designed to quantify such benefits is urgently needed.

Identifying/classifying the agent(s) used by terrorists

Identifying the chemical(s) involved and measuring its concentration would obviously be helpful in informing early tactical decision-making, but specialist advice is unlikely to be readily available at that stage and early sampling difficult or impossible. Despite the increasing availability of a widening and improving range of detection devices, these are still not fully 'soldier-proof', results usually require a degree of interpretation under field conditions and they are also prone to recording false positives. Identifying the actual agent(s) involved may not prove possible until more specialist support arrives and so responders are likely to be reliant on environmental indicators and/or the signs and symptoms displayed by victims. In practice, information gathered during clinical examination may be most useful. This is sometimes referred to by the paramilitary term as MEDINT (standing for Medical Intelligence).

This potential lack of ability to precisely identify agents and measure their concentrations may not be as much of an obstacle as might be supposed as most first responders will have limited choices in the level of protection available, few specific antidotes and limited treatment options during the early stages of dealing with an incident. Symptom-based protocols are likely to be the most effective basis for making initial diagnostic decisions and of enabling emergency treatment to begin and to be continued until the suspected agent can be identified.

Patient and mass decontamination

The past few years have seen an increase in research into procedures and protocols for patient decontamination and the development of a range of equipment to facilitate that process at the scene and in hospitals. A detailed account of recent developments in this field is provided below. Depending on the type of agent released, speedy decontamination can be a critical part of preventing or limiting harm and controlling the spread of contamination.

Decontaminating 'non-ambulatory' patients is very staff-intensive and time-consuming but should generally be the primary focus of health effort at the scene. Handling the equipment and decontaminating patients lying on rescue stretchers or back-boards effectively, while not exacerbating any injuries, involves considerable physical effort and requires practice. Specifically trained teams are likely to provide the most effective operational approach. Dealing with those who are 'ambulatory' is a quicker process as they should normally be able to carry out their own decontamination under supervision and mass decontamination should generally be a joint emergency service responsibility.

A victim receiving instructions to undress, to wait in a decontamination line while being told that all is well by rescue personnel who are themselves wearing protective equipment is hardly likely to be reassured and may well become uncooperative. Significant delays in decontamination are likely to present major problems and there is a danger that the very existence of specialist equipment, and containment protocols might lead to inflexibility and the 'best' may become the enemy of the 'good'. Realistic contingency plans will recognize that providing sufficient capacity for rapid mass decontamination will be critical to success and incident commanders should be encouraged to consider pragmatic solutions – such as the use of sports facility showers, swimming pools or car park sprinkler systems.

Hospital planning

Major incidents cause fear and disorientation. The natural response to a sudden emergency is to move away and past experience indicates many people will leave the scene by any means available, especially in an urban environment. That was clearly demonstrated following the 1995 sarin attack in Tokyo where over 4000 people made their own way by taxi or car to hospitals or doctors' offices throughout the city while under 10% were transported by ambulance (Kulling *et al.*, 2000).

It has to be anticipated that in the chaos that will inevitably follow a terrorist attack, many victims are likely to bypass the management arrangements at the scene before controls can be established. With an increased threat of chemical terrorism, the potential for such a release is no longer confined to areas surrounding manufacturing plants, storage facilities or during transport of chemicals. Any hospital – with or without an emergency department – has to consider the prospect that it could find itself faced with potentially contaminated casualties who have 'self-evacuated' and their major incident plans need to recognize this.

All hospital plans should include provisions for a proportionate response to the arrival of contaminated casualties, with procedures for rapid 'lock-down' and establishing entry and exit restrictions. Most ambulance service plans already provide for sending an immediate alert to all hospitals designated to receive casualties whenever a major incident is declared. Although patients may have already self-evacuated before the emergency services arrive, ambulance plans should also provide for warning **all** hospitals in the general area if chemicals are involved.

Prudent hospital major incident plans should also include a framework for protecting staff, holding 'self-presenters' outside clinical areas and providing facilities for basic decontamination. Additionally, those hospitals with emergency departments should develop robust and resilient arrangements for holding and dealing with contaminated casualties apart from other patients, training and protecting their staff, facilities for primary or secondary decontamination, obtaining expert advice and providing definitive care for such patients.

Training, protocols and exercises

Responding to any major incident is stressful and the pressures of dealing with a terrorist chemical attack will be compounded by staff concerns for their own safety and the wider chaos such an event will prompt. A disciplined response is essential for safety and responders are only likely to perform well under such demanding circumstances if they are very familiar with operational procedures, confident in their own skills and abilities – as well as those of their co-responders – have trained regularly and been given previous opportunities to practice under realistic conditions.

General chemical awareness and safety should be an integral part of the basic training of all emergency service staff and more specific courses provided for those who are expected to carry out specific tasks. Specific courses should include a realistic understanding of risk, risk-assessment and personal protection and consist mainly of realistic practice to ensure familiarity with equipment and build confidence. Effective response plans should also be translated into operational protocols and role-specific action cards.

Summary

No emergency service or health care establishment can afford to ignore the potential and very real threat now posed by chemical terrorism. Such incidents provide different and exceptional challenges for which traditional major incident-response plans are unlikely to prove adequate. An effective whole health system response requires multi-agency pre-planning, a particular focus on teamwork, adequate attention to risk assessment, proportionate provision for staff protection, disciplined working at the scene, adequate arrangements for rapid patient decontamination and protocols for clinical care, both at the scene and in hospitals. No single operational 'blueprint' is likely to provide for all potential circumstances but particular attention to staff training and realistic practice are likely to be the most critical success factors.

DECONTAMINATION

Chemical weapons may be encountered as gases, liquids or solids. Contamination of individuals with liquids, or solids, may lead to absorption of the chemical agents across the body surface: removal of agents from the body is an obvious priority. Contamination of equipment and buildings mean that these cannot be used without taking protective measures until they, too, have been decontaminated. In the military setting, rapid decontamination to enable military activity to continue is essential and much effort has been put into the development of decontaminants. Some agents are fairly easily removed, while others, including thickened agents (thickened soman [GD] or thickened sulphur mustard) present much more difficult problems and scraping away the viscous material may be necessary. Some equipment is resistant to chemicals and aggressive decontaminants; for example, strong bleach solutions or caustic alkalis can be used: such substances are clearly unsuitable for decontamination of the skin and may be very damaging if they contaminate the eyes. In the civilian setting, mass decontamination of people may be necessary and this poses some logistic problems. Showering a few hundred potentially contaminated and probably very frightened civilians, outdoors on a cold day is a formidable task and yet this is likely to be necessary should an effective attack on a civilian population by, perhaps, terrorists occur. Decontamination following industrial accidents is also sometimes necessary and plans for this are well developed as part of the major chemical incident procedures developed in many countries. The following sections focus on the decontamination of patients: some may be casualties, while others may have been exposed to lower amounts of chemicals and may be decontaminated on a precautionary basis. Decontaminants for use on skin and on the eyes are considered separately.

Objectives of decontamination

Decontamination is undertaken to remove toxic materials from the body. In general, the term 'decontamination' is used to imply removal of materials from the body surface. Internal decontamination is sometimes possible and the use of activated charcoal to remove toxic substances from the gut and haemodialysis to remove substances from the blood fall into this category. Internal decontamination will not be considered here: it is touched upon, as appropriate, in the chapters dealing with the management of casualties exposed to specific agents. Removing material from the body surface and eyes prevents absorption which may lead to damage both to the skin and eyes and the systemic effects of the absorbed agent. Nerve agents, for example, do not damage the skin *per se* but they cause severe systemic effects and rapid removal from the skin can be life-saving. Sulphur mustard or lewisite, on the other hand, cause severe damage to the skin and eyes, as well as significant systemic effects: again, rapid removal from the skin and eyes is essential. That removal should be rapid is critically important: a patient with significant skin contamination by a nerve agent is in grave danger of death; a patient with eye contamination by lewisite is in grave danger of losing his sight unless the agent is rapidly removed. Rapidity of decontamination is thus a key objective.

Contaminated patients pose a hazard to their attendants. Secondary contamination by contact with liquids or solids, e.g. powders, is an obvious risk, while 'off-gassing' of volatile agents from clothing is, perhaps, less obvious. However, it was noted in the Tokyo subway incident (see Chapter 13) that hospital staff developed miosis as a result of exposure to sarin (GB) given off by patients' clothing. The patients had themselves not been exposed to liquid sarin but only to sarin vapour and yet enough had been adsorbed onto to their clothing to present a significant secondary hazard. Staff dealing with undecontaminated casualties need to wear protective equipment: gloves, appropriate clothing and masks. This reduces their capacity to provide clinical treatment – indeed, their capacity to undertake an adequate clinical examination of patients may be considerably impaired. It is essential that before chemical casualties come into contact with unprotected clinical staff they, the casualties, are fully decontaminated. This essential requirement imposes a delay in the chain of evacuation – the delay must be minimized by appropriate decontamination facilities and training.

First steps in decontamination

Civilians exposed to chemical warfare agents present very different problems from those presented by contaminated military personnel. In the latter case, the soldier may be well protected by his special clothing and respirator and decontamination is focused on the need to allow him to continue with his military duties – to continue to fight. Thus, he needs to be able to decontaminate himself so that he can, later, remove his protective equipment and possibly use it again. He must also be able to decontaminate his equipment, including his personal weapon, vehicles etc. In the civilian setting, this is not the case. Civilians exposed to chemical agents as a result of military or terrorist activity are likely to be regarded as casualties and to be in need of assistance from trained staff, for example, from the Emergency Services. Civilians are unlikely to be provided with personal decontamination kits. This places them at a disadvantage as compared with military personnel: a delay before decontamination can begin is likely and if exposure to chemical weapons, such as nerve agents or sulphur mustard, has occurred this will increase the likelihood of casualties dying or being severely injured. Delay can occur during evacuation of contaminated civilians to decontamination facilities and this is especially likely should contamination occur in an enclosed environment to which access by rescuer staff is likely to be both delayed and dangerous. That unprotected and untrained civilians are likely to be at greater risk than military personnel is hardly surprising. Rapid removal of civilian casualties from a contaminated environment and then rapid removal of their clothing are thus the first steps in decontamination.

Decontaminants

Decontaminants can act by:

- removing contaminants, unchanged, from the body surface;
- reacting with contaminants to produce non-toxic products.

Removal may be achieved by dilution, for example, by water, or by adsorption of the chemical by a material such as fullers' earth. Washing the chemical from the body surface, for example, by the physical effects of rubbing with a wet sponge or cloth and by dilution, is perhaps the most familiar form of decontamination. Many chemicals are hydrolyzed by water though the reaction may be slow at normal pH levels. Changing the pH of the medium can greatly increase the rate of hydrolysis.

Adsorption is an effective means of removal of chemicals but delayed release of the chemical from the adsorptive medium may be a problem. Providing substances that react with the toxic contaminant to produce non-toxic products has been the objective of much research in the field of decontamination. This has been pursued in laboratories tasked with providing means of defence against chemical weapons. Details of the research undertaken are less generally available than one might wish but Yang and her colleagues, in 1992, published an outstanding review of the area (Yang *et al.*, 1992). We have drawn extensively on this review in the following paragraphs. Recent work by Raber and her colleagues has also provided much valuable information (Raber *et al.*, 2001; Raber and McGuire, 2002).

Before considering the chemistry of decontamination, it may be useful to define the characteristics of the ideal decontaminant. It should be noted that the ideal decontaminant has not yet been discovered: it is, in fact, a theoretical construct or concept rather like the 'alkahest' (the universal solvent) sought by the alchemists in mediaeval times. However, the ideal decontaminant might have the following characteristics:

(i) efficacy – it would destroy the toxic material;
(ii) speed of action – destruction of the toxic material would occur rapidly;
(iii) portability – it would be easily portable, i.e. light in weight;
(iv) low toxicity – it would not damage those materials (skin and eyes in our case) to which it was applied;
(v) cost – it would be cheap to make;
(vi) shelf-life – it would have a long shelf-life;
(vii) independence of environmental conditions – it would function well at all environmental temperatures;
(viii) low capacity to spread toxic chemicals on the surface to which it is applied – this is a less obvious point but one of the drawbacks of using water alone to remove sulphur mustard is that the chemical agent is poorly miscible with water and spreading can occur;
(ix) low environmental toxicity – run-off of decontaminants or products of decontamination should not cause damage to flora or fauna.

Each of these factors may achieve different levels of importance depending on circumstances of use. In a military setting, portability is critically important: soldiers cannot carry gallons of water for use as a personal decontaminant. However, in a civilian setting, the ready access to clean water may make water an attractive choice. Similarly, in the military setting a personal decontamination kit that deteriorates rapidly may be less useful than, for example, a pack filled with fullers' earth that retains its efficacy indefinitely. In all cases, speed of action and efficacy will be important. It will be understood that corrosive materials are unsuitable for personal decontamination and may also damage delicate equipment: thus, a decontaminant suitable for use on the exterior of an armoured fighting vehicle may not be suitable for use either on the skin or on electronic equipment.

Chemical aspects of decontamination

As indicated above, decontaminants may be reactive or non-reactive (adsorbents and diluents).

Table 3. Decontaminants composed of hypochlorite (modified from Yang et al., 1992)

Decontaminant	Composition
Bleach	2–6 wt% NaOCl in water
HTH (high-test hypochlorite)	$Ca(OCl)_2$ Cl + $Ca(OCl)_2$ as a solid powder or a 7% aqueous slurry
STB (super-tropical bleach)	$Ca(OCl)_2$ + CaO as a solid powder or as 7, 13, 40 and 70 wt% aqueous slurries
Dutch powder	$Ca(OCl)_2$ + MgO
ASH (activated solution of hypochlorite)	0.5% $Ca(OCl)_2$ + 0.5% sodium dihydrogen phosphate buffer + 0.05% detergent in water
SLASH (self-limiting activated solution of hypochlorite)	0.5% $Ca(OCl)_2$ + 1.0% sodium citrate + 0.2% citrate acid + 0.05% detergent in water

Reactive decontaminants are likely to be preferable, under some circumstances, because of their rapidity of action and the complete removal of the toxic material that they offer. Yang *et al.* (1992) identified two key reactions that are made use of in decontamination, i.e. hydrolysis, and oxidation. Two sub-categories – catalytically-enhanced and enzyme-enhanced hydrolysis were also identified. These authors discussed a variety of media – aqueous, organic and solid – in which these reactions could occur.

Hydrolysis

Hydrolysis, in this context, implies the reaction of a compound with hydroxyl ions, OH^-, provided by water. Such reactions may be slow at neutral pH but much faster in the presence of higher concentrations of hydroxide ions. Chemicals that are soluble in water undergo more rapid hydrolysis than insoluble compounds. Thus, the rate of hydrolysis of sulphur mustard is slow. Nerve agents such as sarin (GB) and soman (GD) are rapidly hydrolyzed at high pH values. Yang *et al.* (1992) quote Epstein *et al.* (1956) as suggesting that hydrolysis is catalyzed by the hypochlorite ion found in solutions of bleach. The hydrolysis of sarin or soman is shown in the following, generalized, equation:

$$RO\text{-}\underset{CH_3}{\overset{O}{\overset{\|}{\underset{|}{P}}}}\text{-}F + OH^- \longrightarrow RO\text{-}\underset{CH_3}{\overset{O}{\overset{\|}{\underset{|}{P}}}}\text{-}OH + F^- \quad (1)$$

The nerve agent has been converted into the corresponding phosphonic acid.

Yang *et al.* (1992) point out that this base-catalyzed hydrolysis can lead to the formation of toxic products, for example, in the case of the nerve agent VX. In this case a very toxic compound with the following formula can be produced:

$$HO\text{-}\underset{CH_3}{\overset{O}{\overset{\|}{\underset{|}{P}}}}\text{-}SCH_2CH_2N[CH(CH_3)_2]_2 \quad (2)$$

Oxidation

Oxidation of chemical warfare agents by solutions or solid preparations of bleach was the first form of chemical decontamination to be studied. Preparations of bleach were introduced during World War I and a list of preparations (modified from Yang *et al.* (1992)) is given in Table 3.

All of these mixtures may be used on skin or on equipment although the concentrated preparations should be avoided in clinical settings. In 2002, the Swedish Defence Research Agency (2002) advised the use of a mixture of chlorinated lime and magnesium oxide (cf. Dutch powder in Table 3) as a skin decontaminant. It was noted that this was effective against thickened agents due to its capacity to bind the sticky preparations and to make them easier to remove. The authors noted that the powder had an irritating effect on skin and recommended early bathing following its use.

Reactions of some chemical warfare agents with bleach (oxidation reactions) are illustrated in Equations (3) and (4).

Sulphur mustard (H)

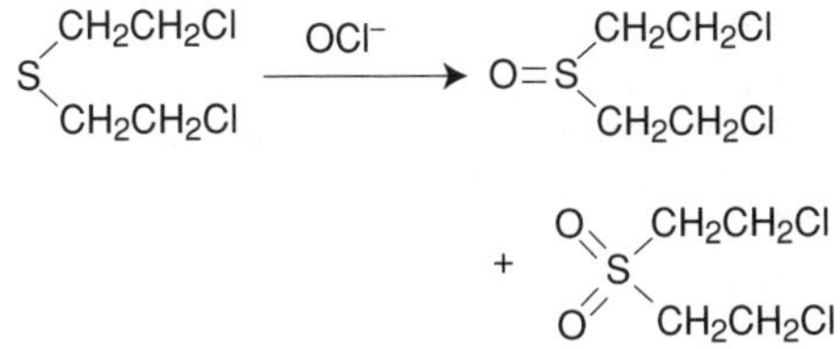

and other products

(3)

VX

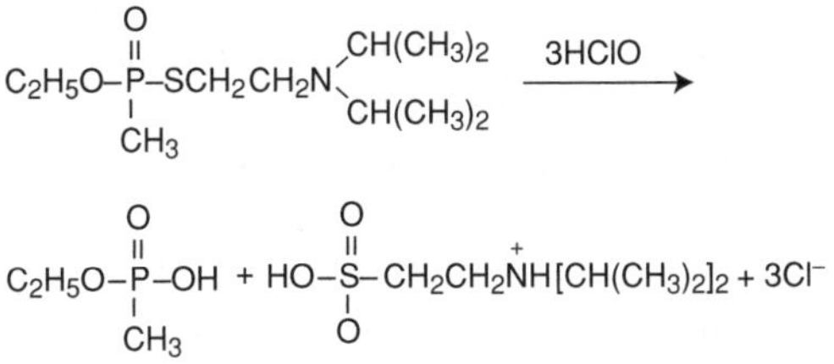

(4)

Other oxidants have been developed. The commercial product 'Oxone' is described as a component of a preparation 'L-Gel', described by Raber and McGuire (2002). This very effective system is active against chemical warfare and biological warfare agents but the focus of Raber and McGuire's (2002) paper is on decontamination of buildings and materials, rather than people and no details of studies of effects on skin are reported. Many other oxidizing agents, including potassium permanganate and hydrogen peroxide, have been studied as decontaminants. Cabal *et al.* (2003) describe foam-making decontaminants containing cationic and non-ionic tensides (the latter being effective detergents) and hydrogen peroxide.

Catalysis

Both hydrolysis and oxidation can be speeded up by catalysts. Wagner-Jauregg *et al.* (1955) and Gustafson and co-workers (1959,1962,1963) have investigated the use of copper as a catalyst for the hydrolysis of nerve agents and sulphur mustard. Iron has been studied as a catalyst of oxidative decontamination: the reader is referred to Yang *et al.* (1992) for detailed comments. However, catalysts have not found a clear application in clinical decontamination.

Enzymes

Nerve agents can be decontaminated by enzymatically accelerated hydrolysis. Hoskin and others have investigated the use of organophosphorus acid anhydrases produced by squid and also by bacteria (Hoskin *et al.*, 1966; Hoskin and Long, 1972; Hoskin and Roush, 1982; Amitai *et al.*, 2003; Ghanem and Raushel, 2005). Cloning of bacterial genes for these anhydrase enzymes has been undertaken (Phillips *et al.*, 1990; Rowland *et al.*, 1991; Dumas *et al.*, 1989; Cheng *et al.*, 1999). Here again, these preparations have not been established for clinical use although further work is underway in this area. Bacterial enzymes have been incorporated in foams for surface and soil decontamination (LeJeune and Russell, 1999).

Reactive decontaminants in use today

(a) Bleach and other compounds generating hypochlorous acid

As stated above, preparations of bleach were the first chemical decontaminants developed during World War I. In addition to the preparations listed in Table 3, a number of non-aqueous preparations have been produced. Of these, diethylenetrimine ($H_2NCH_2CH_2NHCH_2CH_2NH_2$) is the best known: it is a constituent of the decontaminant DS2 adopted in the USA in 1960 (Jackson, 1960; Richardson, 1972). DS2 also contains sodium hydroxide and ethylene glycol monoethyl ether. This mixture was found to be more effective than the other bleach preparations at low temperatures. DS2 is corrosive and is not suitable for skin contamination – protective clothing is necessary when it is used. Yang *et al.* (1992) report the discovery of a Soviet personal decontamination kit in 1973. Analyses of its contents led to the development of the USA M258 kit in the 1970s and the M258A1, M280 and M291 systems in the 1980s. Details of these systems as developed for skin decontamination are presented in Table 4 – based on 'Table 2' of Yang *et al.* (1992).

Table 4. Field decontamination equipment and systems (modified from Yang *et al.*, 1992)

Item name	Description	Decontaminants	Applications
M258A1, decontamination kit, personal	Consists of foil-packaged pairs of towelettes in a plastic carrying case	I. Water, phenol, NaOH, ethanol and ammonia II. Water, ethanol, chloramine-B and $ZnCl_2$	Skin and individual equipment
M291, skin decontamination kit	Consists of six foil-packaged non-woven fibre pads filled with XE-555 resins	2.8 g of XE-555 resins of a total water content of 25 wt%	Skin

It will be noted that the M291 kit involves a solid (resin) decontamination medium. This is attractive for military use in that the kit provides adsorptive and reactive decontamination, is light in weight and thus easily portable. The resin is a styrene/divinyl copolymer and provides a complex combination of ion-exchange and adsorptive properties. The resin system was found to be less irritating to the skin than the earlier M258A1 water-based system. Preparations based on chloramine and on an *N*-chloro-oxidant ('Fichlor', sodium *N*, *N*-dichloroisocyanate) have also been developed. Yang *et al.* (1992) listed three problems associated with the use of aqueous preparations of bleach as a decontaminant. These are:

(i) decline in the active chlorine content with time – solutions should be used fresh;
(ii) large amounts of the decontaminant are needed;
(iii) bleach is corrosive.

The last point is important in dealing with patients: contamination of the eyes with bleach is a potentially serious problem. This is considered below. The *N*-chloro-compounds (chloroamines and 'Fi-clor') are more stable, more effective and are less irritating than the hypochlorites. Chloramine has been used as a wound disinfectant and as a surgical antiseptic. *Martindale* (Reynolds, 1996) notes that the maximum allowable concentration in cosmetics, in the UK, is 0.2%. This text also notes that ingestion of chloramine can cause vomiting, cyanosis and respiratory failure: fatalities have occurred. Such effects are very unlikely following the sensible use of dilute preparations. Sodium dichloroisocyanurate ('Fi-clor') is used as a sterilizing agent for babies' feeding bottles and as an antibacterial treatment for water in swimming pools. A good account of another chloramide decontaminant (S-320) developed for use against sulphur mustard has been provided by Shih *et al.* (1999).

(b) The US military decontaminants have been discussed above. Such complex systems for personal decontamination are not in use in all armed forces.

(c) Canadian workers have developed reactive decontaminants which contain oximates. One form, described by Sawyer *et al.* (1991) comprised potassium 2,3-butanedione monoximate in polyethylene glycol methylether 50. Reaction of this decontaminant (reactive skin decontaminant, RSO) with nerve agent was rapid. The possibility that potentially toxic reaction products (phosphorylated oximes) might be absorbed and cause undesirable effects was considered by Sawyer *et al.* (1991) in the paper referred to above. These authors studied the effects of the reaction products of the interaction between the decontaminant and a number of nerve agents (GA, GB, GD and VX) by intraperitoneal injection in rats and in chick embryo neuronal cell cultures. The authors reported that, depending on the decontaminant–OP molar mixing ratio, very significant reductions in OP anticholinesterase activity were produced. Lethality fell dramatically in the rats injected with a 1:10 mol ratio of OP/decontaminant. The original paper should be consulted for details, especially as regards the interesting differences of the OP/decontaminant mixtures depending on the OP component actually used.

Nanotechnology is a fast growing field and the reaction between nanocrystals of magnesium oxide and OP compounds have been reported by Rajagopalan *et al.* (2002). The large and reactive surface area (expressed on a surface area per unit mass basis) of such particles allows for rapid reaction with OP compounds. This is a promising advance that may find an application in skin decontaminants: nanoparticles are already used as a component of sun-screening lotions.

(d) Fullers' earth

Fullers' earth is a form of aluminium silicate (a clay) with a high adsorption capacity when prepared as a dry powder. It was used for removing grease from wool – a process described as 'fulling': hence its name. Note the lack of an upper-case 'F', except as the first word in a sentence and the position of the apostrophe – it was the earth used by 'fullers'. Several minerals are commonly described as fullers' earth: bentonite (sodium montmorillonite clays) and attapulgite clays which are not related chemically to bentonite but which have similar absorptive properties (Parkes, 1994). Long-term exposure to fullers' earth can cause pneumoconiosis but this is irrelevant to its use as a decontamintant. Fullers' earth is used as a component of emulsion paints, cosmetics and baby powders. It is an effective decontaminant for chemical warfare agents and was long the personal decontaminant issued to members of the UK Armed Forces. It is light in weight and easy to use. Areas of liquid contamination should be patted with cloth pads impregnated with fullers' earth or covered with fullers' earth dispersed from a plastic puffer bottle. The contaminated fullers' earth presents little risk in open-air conditions although off-gassing of chemical from the powder could be a problem indoors. Fullers' earth should not be allowed to enter wounds or the eyes. Fullers' earth that has been exposed to chemicals should not be allowed to enter the collective protection environment.

Water with or without detergents

Water is probably the decontaminant most likely to be used in a civilian setting. Thorough washing of contaminated areas of skin and of the eyes contaminated by liquid chemicals is needed. As discussed below, this can be done, simply, by means of showers. For an emergency, any source of water should be used: the artificial sprinklers in multi-storey car parks might represent an option. Adding detergent to the water increases the rate of dissolution of some chemicals and this is generally recommended (http://www.hpa.org.uk). It should be noted that detergents damage the barrier function of the stratum corneum of the epidermis (Scheuplein and Blank, 1971) and rinsing away with clean water is recommended.

Organic solvents: paraffin, solvents and petrol (gasoline)

A detailed survey of means of removing sulphur mustard from the skin was undertaken during World War I: the results of wide practical experience are reported by Vedder (1925). In addition to bathing in the hot soapy water after exposure to mustard gas vapour, washing contaminated areas of skin with solvents such as kerosene, acetone or absolute alcohol was recommended. It was argued that this could wash mustard out of the skin (not just from the surface). Rubbing with solvents, removing contaminated solvent and replacing with fresh solvents, for up to thirty minutes, was recommended. This method is not recommended today and it is believed that solvents could hasten the penetration of chemicals through the skin, although it is to deny the great experience of doctors caring for thousands of mustard gas casualties during World War I.

Current recommendations for decontamination of civilians exposed to chemicals as a result of release of such substances by terrorists

DECONTAMINATION OF EYES

All experts in the field of chemical warfare agents toxicology agree that very rapid removal of liquid mustards (sulphur or nitrogen) and lewisite from the eye is critically important. Water, dilute (0.9%) sodium chloride solution or sodium bicarbonate solution (1.26%) should be used. In an emergency, only water is likely to be available.

The patient's eye should be opened and irrigated thoroughly. Any delay in decontamination may lead to severe and possibly permanent eye damage. Contact lenses should be removed prior to decontamination.

DECONTAMINATION OF THE SKIN

It is unlikely that precise information about the areas of the body contaminated with chemicals will be available in an emergency setting – thus, decontamination of the entire body – not just, for example, the hands and face, should be undertaken. All clothing should be removed and placed in sealed polythene bags. Jewellery, watches, money and other personal possessions should be collected and placed in labelled polythene bags. Two approaches to decontamination are feasible: the 'bucket and sponge' method taking one patient at a time, or the mass decontamination approach with attendants supervising the passage of a queue of people through some communal shower unit. Patients who have been injured or are unable to help themselves should be decontaminated by trained attendants using a system in which a stretcher is placed on a table equipped with rollers that allows transfer of the decontaminated patient to a clean stretcher, an ambulance and thus to hospital. Some authorities have argued for the addition of liquid soap to the water used in automated decontamination units. This will speed the process of material but the interaction of water and soap leading to foaming when passed through a jet should be considered. Whatever means of decontamination are used it should be thorough. This means that each patient will need some minutes of attention: up to fifteen or so minutes may be needed to deal with an incapacitated patient. Whatever system is used, plans for dealing with waste water and any decontaminants used should be devised and practiced in advance of an incident occurring. Waste water may be collected inside an area protected by inflatable 'bunds' (bund is a word of Hindu origin – a man-made embankment or damn). In some countries, the Fire Brigade is equipped to provide such water-collection facilities. If waste water is to be allowed to run into drains, the local water company should be informed.

The logistical problems of stripping and showering a large number of people, including babies, children and the elderly, are clearly formidable. Suitable towels and clothing must be provided, modesty should be preserved and consideration given to how people are to make their way home. Some resistance among those not believing themselves to be contaminated should be expected. In addition, the security forces may wish to interview members of the crowd queuing for decontamination and facilities for fast decontamination of such individuals should be provided.

Not all authorities support the use of water-based decontamination but for use on a mass decontamination scale it is difficult to imagine any other method that is likely to work as well as washing with water.

Some thought has been given to the question of whether cold or warm water should be used. The view held currently by many authorities is that *warm* water should be used: showering people is likely to be difficult enough without the added shock of cold water. The argument against the use of warm water is that is causes dilatation of superficial blood vessels and thus could increase the rate of absorption of contaminants. In practice – as long as a plentiful supply of water is used – this is unlikely to be a problem: contaminants will be rapidly removed.

DECIDING ON ADEQUACY OF DECONTAMINATION PROCEDURES

In the military environment, chemical detectors are often deployed to confirm that decontamination has been completed before casualties are passed via an air-lock into a 'Collective Protection Environment'. The reason for this is obvious: staff in the Collective Protection Environment do not wear personal protective equipment (PPE) and must be protected by rigorous maintenance of an uncontaminated environment. The 'Chemical Agent Monitor' (CAM) is particularly suited for detecting vapour given off by unremoved areas of liquid contamination. However, proper monitoring of patients takes several minutes and in a mass-casualty or mass-decontamination setting this may not be feasible. Reliance must, therefore, be placed on the adequacy of the decontamination process itself.

This clearly imposes some risk, as well as a large responsibility on those in charge of the decontamination process. It is essential that staff are properly trained and that they adhere to the methods that they have been taught, even as pressure from those awaiting decontamination builds up. The need for staff from the security services, for example, the Police, to supervise queues of people and to maintain order is obvious.

REFERENCES

Ambulance Service Association (2004). *The National Decontamination Provider Manual* – Issue 1. http://www.asa.uk.net

Amitai G, Adani R, Hershkovit P *et al.* (2003). Degradation of VX and sulfur mustard by enzymatic haloperoxidation. *J Appl Toxicol*, **23**, 225–233.

Cabal J, Kassa J and Severa J (2003). A comparison of the decontamination efficacy of foam-making blends based on cationic and non-ionic tensides against organophosphorus compounds determined *in vitro* and *in vivo*. *Human Exp Toxicol*, **22**, 507–514.

Cheng T-C, DeFrank JJ and Rastogi VK (1999). *Alteromonas* prolidase for organophosphorus G-agent decontamination. *Chem-Biol Interact*, **119–120**, 455–462.

Dumas DP, Caldwell SR, Wild JR and Raushel FM (1989). Purification and properties of the phosphotriesterase from *Pseudomonas diminuta*. *J Biol Chem*, **264**, 19659–19665.

Epstein J, Bauer VE, Melvin S *et al.* (1956). The chlorine-catalyzed hydrolysis of isopropyl methylphosphonofluoridate (Sarin) in aqueous solution. *J Am Chem Soc*, **78**, 4068–4071.

Fisher J, Morgan-Jones D, Murray V *et al.* (1999). *Chemical Incident Management*, Section 2, Figure 4. London, UK: The Stationery Office.

Ghanem E and Raushel FM (2005). Detoxification of organophosphate nerve agents by bacterial phosphotriesterase. *Toxicol Appl Pharmacol*, **207** (Supplement 2), S459–S470.

Gustafson RL and Martell AE (1959). Hydrolytic tendencies of metal chelate compounds. V. Hydrolysis and dimerization (II) chelates of 1,2-diamines. *J Am Chem Soc*, **81**, 525–529.

Gustafson RL and Martell AE (1962). A kinetic study of the copper(II) chelate-catalyzed hydrolysis of isopropyl methylphosphonofluoridate (Sarin). *J Am Chem Soc*, **84**, 2309–2316.

Gustafson RL, Chaberek S and Martell AE (1963). A kinetic study of the cipper (II) chelate catalyzed hydrolysis of diisopropyl phosphofluoridate. *J Am Chem Soc*, **85**, 598–601.

Hoskin FC and Long RJ (1972). Purification of a DFP-hydrolyzing enzyme from squid head ganglion. *Arch Biochem Biophys*, **150**, 548–555.

Hoskin FC and Roush AH (1982). Hydrolysis of nerve gas by squid-type diisopropyl phosphorofluidate hydrolyzing enzyme on agarose resin. *Science*, **215**, 1255–1257.

Hoskin FCG, Rosenberg P and Brain M (1966). Re-examination of the effect of DFP on electrical and cholinesterase activity of squid giant nerve axon. *Proc Natl Acad Sci*, **55**, 1231–1235.

Jackson JB (1960). *Development of Decontamination Solution DS2*, CWLR, 2368.

Kulling P (2000). *The Terrorist Attack with Sarin in Tokyo*, Article No. 2000-00-040. Stockholm, Sweden: KAMEDO, Swedish National Board of Health and Welfare.

LeJeune KE and Russell AJ (1999). Biocatalytic nerve agent detoxification in fire fighting foams. *Biotechnol Bioeng*, **62**, 659–665.

Parkes WR (1994). *Occupational Lung Disorders*. London, UK: Butterworth-Heinemann.

Phillips JP, Xin JH, Kirkby K *et al.* (1990). *Proc Natl Acad Sci*, **87**, 8155–8159.

Raber E and McGuire R (2002). Oxidative decontamination of chemical and biological warfare agents using L-Gel. *J Hazard Mater*, **B93**, 339–352.

Raber E, Jin A, Noonan K *et al.* (2001). Decontamination issues for chemical and biological warfare agents: how clean is clean enough? *Int J Environ Health Res*, **11**, 128–148.

Rajagopalan S, Kopez O, Decker S *et al.* (2002). Nanocrystalline metal oxides as destructive adsorbents for organophosphorus compounds at ambient temperatures. *Chem Eur J*, **8**, 2602–2606.

Reynolds JEF (ed.) (1996). *Martindale: The Extra Pharmacopoeia*. London, UK: Royal Pharmaceutical Society.

Richardson GA (1972). *Development of a Package Decontamination System*. EACR-1310-17, Final Report, Contract DAA15-71-C. Dayton, OH, USA: Monsanto Research Corporation.

Rowland SS, Speedie MK and Pogell BM (1991). Purification and characterization of a secreted recombinant phosphotriesterase (parathion hydrolase) from *Streptomyces lividans*. *Appl Environ Microbiol*, **57**, 440–444.

Sawyer TW, Parker D, Thomas N *et al.* (1991). Efficacy of an oximate-based skin decontaminant against organophosphate nerve agents determined *in vivo* and *in vitro*. *Toxicology*, **67**, 267–277.

Scheuplein RJ and Blank IH (1971). Permeability of the skin. *Physiol Rev*, **51**, 702–747.

Shih ML, Korte WD, Smith JR *et al.* (1999). Analysis and stability of the candidate sulfur mustard decontaminant S-330. *J Appl Toxicol*, **19**, S89–S95.

Swedish Defence Research Agency (2002). *Chemical Weapons – Threat, Effects and Protection*. Umeå, Sweden: FOI.

Vedder EB (1925). *The Medical Aspects of Chemical Warfare*. Baltimore, MD, USA: Williams and Wilkins Company.

Wagner-Jauregg T, Hackley BE, Lies TA *et al.* (1955). Model reactions of phosphorus-containing enzyme inactivators. IV. The catalysis of certain metal salts and chelates in the hydrolysis of diisopropyl fluorosphate. *J Am Chem Soc*, **77**, 922–929.

Yang Y-C, Baker JA and Ward JR (1992). Decontamination of chemical warfare agents. *Chem Rev*, **92**, 1729–1743.

8 TOXICOLOGY OF ORGANOPHOSPHATE NERVE AGENTS

Timothy C. Marrs

Edentox Associates, Edenbridge, UK

INTRODUCTION

The organophosphate (OP) anticholinesterases include the chemical warfare nerve agents, a variety of OP pesticides (Ballantyne and Marrs, 1992) and, less well-known, a natural compound, anatoxin-As (Dittmann and Wiegand, 2005). The active ingredients of OP pesticides are used as drugs in human medicine, e.g. malathion to treat head-lice and metrifonate/trichlorfon in tropical medicines. OPs are also used in veterinary medicine notably as ectoparasiticides (Beesley, 1994). The chemical warfare nerve agents have a much higher mammalian acute toxicity, particularly via the percutaneous and inhalation routes, than the OP pesticides. Additionally many chemical warfare agents are phosphonofluoridates and phosphonothioates, while many pesticides are = S type-phosphorothioates. Qualitatively, the anticholinesterase toxicology of the OP nerve agents and pesticides is similar and in general treatment strategies are alike so that much information on OP pesticides is relevant to nerve agent toxicology.

History

OP compounds were intensively investigated in Germany in the 1930s. The German conglomerate IG Farbenindustrie looked at a number of these compounds for use as insecticides and a programme of synthesis of a large number of compounds was undertaken. Tabun and sarin, OPs of little use as insecticides but of very high mammalian toxicity, were synthesized by Schrader in 1937 and a small pilot production plant was set up at Münster-Lager. Later, at Dühernfurt near Breslau in Prussian Silesia (now Bzerg Dolny and Wrocław in Poland), a production plant for these agents was established and both tabun and sarin were manufactured in quantity: in the case of tabun, 12 000 tonnes were produced. Soman, another nerve agent, was also synthesized in Germany during the war, but only manufactured in small quantities. Strangely, perhaps in view of the large stocks held by Germany, the nerve agents were not used in World War II (UK Ministry of Defence, 1972) and appear to have been dumped in the Baltic sea. The V agents similarly arose out of studies of putative insecticides and VX has been variously reported as being first synthesized at the Chemical Defence Experimental Establishment (CDEE) at Porton Down, UK (now Dstl, Porton Down) or by Ghosh at Imperial Chemical Industries in 1952 (SIPRI, 1971). Other nerve agents were developed subsequently and stocks have been held by a number of countries, including USA, the former USSR and successor states, the UK, France and Iraq. However, nerve agents have rarely been used in warfare; the only notable instance of use being by Iraq against that country's own Kurdish population (le Chêne, 1989). There have also been allegations of use of OP nerve agents during the Iran/Iraq war. The nerve agent sarin, in an impure form, was used in two terrorist attacks in Japan, respectively, in Matsumoto 1994 (Okudera, 2002) and Tokyo in 1995 (Nagao *et al.*,

Chemical Warfare Agents: Toxicology and Treatment (2nd Edition)

Table 1. Possible targets for nerve agents in warfare.[a] Adapted and reproduced from Chapter 34, 'Organophosphorus compounds as chemical warfare agents', by Maynard RL and Beswick F, in *Clinical and Experimental Toxicology of Organophosphates and Carbamates* (B Ballantyne and TC Marrs, eds), 1992, with the kind permission of the publishers, Elsevier, and the authors

	Type of agent			
Target	Sarin	Soman	VX	Delivery system
Rear areas				
Airports/airfields	—	L	L	Aircraft (bombs, cluster spray bombs, spray tanks, missiles)
Seaports	—	L	L	Aircraft (bombs, cluster spray bombs, spray tanks, missiles)
Railways, especially junctions	—	L	L	Aircraft (bombs, cluster)
Headquarters and communication centres	—	L	L	Aircraft (bombs, missiles)
Storage sites	—	L	L	Aircraft (bombs, cluster missiles)
Troop concentrations	—	L	L	Aircraft (bombs, spray tanks)
Forward areas				
Nuclear delivery weapons, other key weapons and systems	L	—	L	Multiple rocket launchers, aircraft (bombs, rockets)
Defence positions	L	—	L	Multiple rocket launchers, artillery, mortars, aircraft (bombs), rockets
Own flanks	L	—	L	Mines
Own defence front generally	L	—	L	Artillery, mortars, mines
To produce casualties, to harass and reduce combat efficiency	L	—	L	Multiple rocket launchers, artillery, mortars, aircraft (bombs, rockets)
To deny ground	—	L	L	Aircraft (spray, mines)
Harass civilian populations	L	—	—	Aircraft (bombs, rockets, sprays)

Note:
[a] L, likely use.

1997), as was, almost certainly, VX, but this agent was used for assassination of individuals (Nozaki *et al.*, 1995).

Use

Possible roles of OP nerve agents in warfare are outlined in Table 1. Until the use of sarin in Matsumoto and Tokyo, the use of chemical weapons as terrorists' tools had been considered unlikely. Since then, and the emergence of al-Qaida, this has been reconsidered and Table 2 gives some of the roles whereby terrorists might make use of chemical weapons. Note that the synthesis of tabun is easier than the other G agents, so that tabun is more likely to be used in terrorist scenarios (see below). The reason is that tabun has no fluorine in its structure. Incorporation of the fluorine leaving group requires the use of hydrofluoric acid during the synthesis and this is, of course, corrosive to glass. Early bulk synthesis of nerve agents with fluorine leaving groups was carried out using special apparatus made or lined with pure silver. Such a process is inevitably costly, although the difficulties did not deter a Japanese terrorist group (the Aum Shinrikyo). It is of interest that the nerve agent likely to have been used by Iraq against the Kurds was tabun,

Table 2. Possible terrorist targets for nerve agents

Major target	Specific target
Air transport	Airport terminals/interiors of planes
Railways/subway systems	Stations, interiors of carriages
Road transport	Freeways/motorways, especially intersections and service stations
Public meeting places	Concert halls, major sporting events, political meetings, churches and synagogues
Financial centres	Headquarters of financial institutions, exchanges, e.g. trading floors
Energy supply	Power stations, gas terminals, oil terminals
Communications	Television and radio broadcasting stations, telephone exchanges

the first nerve agent to be synthesized on a large scale.

Structure and physical properties

The nerve agents comprise a group of OPs of high acute mammalian toxicity. They are derivatives of phosphoric or phosphonic acids (more often the latter) and contain two alkyl groups (R and R′) and a leaving group. The general formulae of the OP nerve agents is similar to the OP pesticides:

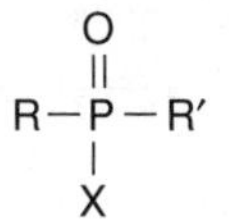

The nerve agents are traditionally divided into the G agents and V agents (Table 3). According to Watson *et al.* (2005), G stands for German and V for venom. In the case of the G agents the leaving group is often a fluorine atom and, exceptionally in GF (cyclosarin), one of the alkyl groups is replaced by a cyclohexyl group. Soman is distinguished by the fact that one of its alkyl groups is a bulky pinacolyl group, while tabun does not contain a fluorine atom and is a cyanidate. The G agents include GB (sarin, isopropyl methylphosphonofluoridate), GD (soman, pinacolyl methylphosphonofluoridate), GA (tabun, ethyl *N*,*N*-dimethylphosphoramidocyanidate) and GF (cyclosarin, cyclohexyl methylphosphonofluoridate). The V agents are phosphonothioates of the P–O type in which the leaving group is linked to phosphorus through a sulphur atom, except for VG which is a phosphorothioate. The V agents are exemplified by VX (*O*-ethyl *S*- (2-(diisopropylamino)ethyl) methylphosphonothioate).

All of the nerve agents are colourless liquids, although impure agents may be yellow to brown in colour. However, these compounds differ amongst themselves in physical properties (Table 4), for example, the V agents are much less volatile than the G agents, the latter being volatile liquids (tabun less volatile than sarin or soman). Soman may be thickened to increase persistence (Marrs and Maynard, 2001). VX is a non-volatile liquid with the result that the VX (unless aerosolized) is not an inhalation hazard, a fact that may significantly affect its role in warfare. Tabun is said to have a fruity odor, while the other agents are said to be odorless. Sarin may be mixed with tributylamine (sarin type I) or diisopropylcarbodiimide (sarin type II), to prevent spontaneous hydrolysis.

As is discussed below, more information on cyclosarin (GF, cyclohexyl methylphosphonofluoridate) has accumulated recently, because, during operations ‘Desert Shield’ and ‘Desert Storm’, it was discovered that cyclosarin was among the nerve agents that Iraq possessed. Thus, some limited information is now available on the physical properties of this nerve agent (gulflink, 2005). The material is a liquid at room temperature with a boiling point of 239°C and a freezing point of −30°C. The vapour pressure is 0.044 mmHg at 20°C and the volatility is 438 mgm^{-3} at 20°C.

TOXICOLOGY

Until recently, attention has almost entirely been given to the acute toxicity of nerve agents and, in the military context, this is still the most

(O-ethyl-S (2-(diisopropylamino) ethyl) me

Table 3. Formulae of nerve agents. Adapted and reproduced from Chapter 34, 'Organophosphorus compounds as chemical warfare agents', by Maynard RL and Beswick F, in *Clinical and Experimental Toxicology of Organophosphates* (B Ballantyne and TC Marrs, eds), 1992, with the kind permission of the publishers, Elsevier, and the authors

Abbreviation	Common name	Proper name
GA	Tabun	Ethyl *N*, *N*-dimethylphosphoramidocyanidate 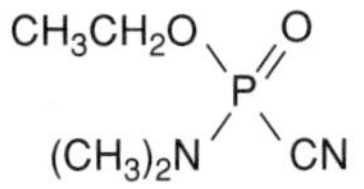
GB	Sarin	Isopropyl methylphosphonofluoridate 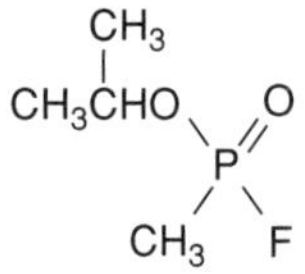
GD	Soman	Pinacolyl methylphosphonofluoridate 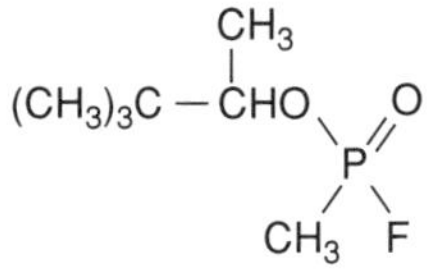
GE	—	Isopropyl ethylphosphonofluoridate 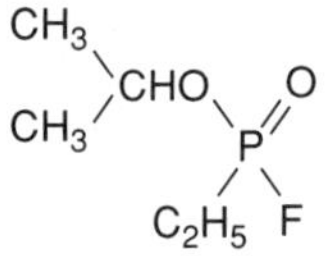
GF	Cyclosarin	Cyclohexyl methylphosphonofluoridate 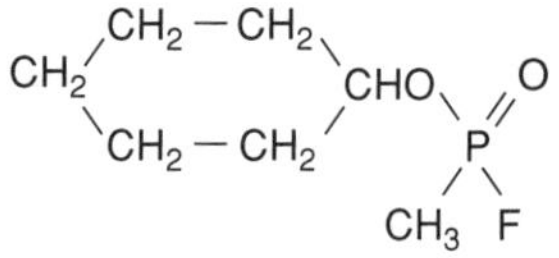
VX	—	*O*-Ethyl-*S*-[2(diisopropylamino)ethyl]methylphosphonothioate 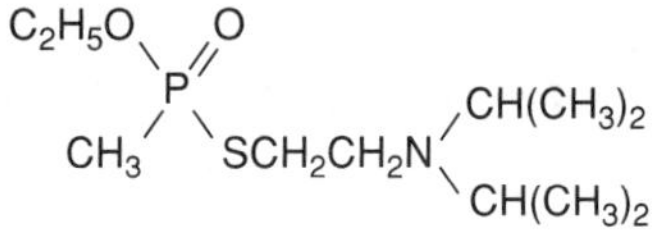
VE	—	*O*-Ethyl-*S*-[2-(diethylamino)ethyl]ethylphosphonothioate 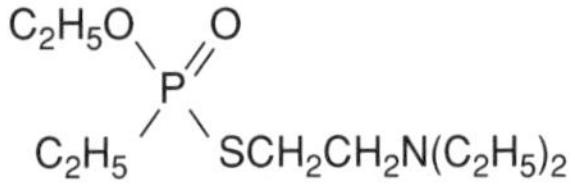
VG	—	*O*,*O*-Diethyl-*S*-[2-(diethylamino)ethyl] phosphorothioate 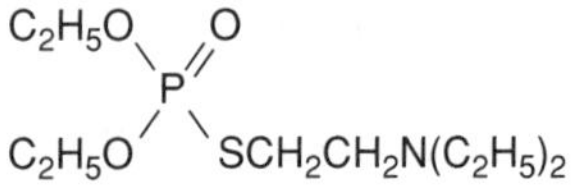
VM	—	*O*-Ethyl-*S*-[2-(diethylamino)ethyl] methylphosphonothioate 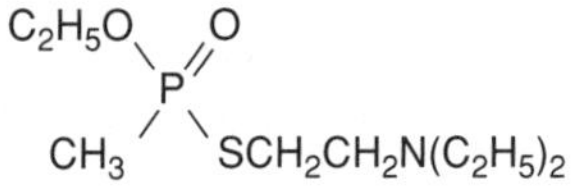

Table 4. Physico-chemical properties of nerve agents. Adapted and reproduced from Chapter 34, 'Organophosphorus compounds as chemical warfare agents', by Maynard RL and Beswick F, in *Clinical and Experimental Toxicology of Organophosphates* (B Ballantyne and TC Marrs, eds), 1992, with the kind permission of the publishers, Elsevier, and the authors

Property	Tabun, GA	Sarin, GB	Soman, GD	VX	GF
Molecular weight, MW (Da)	162.3	140.1	182.18	267.36	180.14
Specific gravity at 25°C	1.073	1.0087	1.022	1.0083	1.133[a]
Boiling point °C	246	147	167	300	—
Melting point °C	−49	−56	−80	−20	−12

Vapour pressure (VP) and Volatility (Vol)[b]	°C	Tabun, GA VP (mmHg)	Tabun, GA Vol (mg m^{-3})	Sarin, GB VP (mm Hg)	Sarin, GB Vol (mg m^{-3})	Soman, GD VP (mm Hg)	Soman, GD Vol (mg m^{-3})	VX VP (mmHg)	VX Vol (mm m^{-3})	GF VP (mm Hg)	GF Vol (mg m^{-3})
	0	0.004	38	0.52	4 279	0.044	470.9	—	—	0.006	63
	10	0.013	119.5	1.07	8 494	0.11	1 135.5	—	—	0.017	173
	20	0.036	319.8	2.10	16 101	0.27	2 692.1	0.00044	5.85[c]	0.044	438[d]
	25	0.07	611.3	2.9	21 862	0.40	3 921.4	0.0007	10.07	0.068	659
	30	0.094	807.4	3.93	29 138	0.61	5 881.4	—	—	0.104	991
	40	0.23	1912.4	7.1	60 959	—	—	—	—	0.234	2159
	50	0.56	4512.0	12.3	83 548	2.60	23 516.0	—	—	0.501	4480

Notes:

[a] Temperature, 20°C.

[b] Volatility = concentration of saturated vapour at specified temperature. Volatility calculated from $PV = nRT$.

$$\text{Vol} = \frac{VP \times 101\,325 \times \text{MW}}{760 \times 8.3143 \times A} = \frac{\text{VP} \times \text{MW} \times 16.035}{A}$$

where A is the Absolute temperature.

[c] Some authorities quote values as low as 0.1–1.0 mg m^{-3}.

[d] Where a range is given, this may be because materials of different purity have been studied.

important consideration. It is also the case when considering casualties from terrorist use. Two other exposure patterns have also to be considered, (1) the long-term effects of low dose exposure, particularly in relation to manufacturing workers and individuals involved in clean-up of contaminated sites, and (2) the delayed and long-term effects after recovery from acute exposure. It has been pointed out (Reutter, 1999) that the toxicity of chemical weapons is unchanged but our perception of toxicity has since these compounds were developed in the 1940s and 1950s.

Other scenarios that need to be considered include contamination of food and public places and, in such situations, levels to which it is necessary to decontaminate to ensure public safety. A further consideration of importance is the toxicity of nerve agent degradation products (see review by Munro *et al.*, 1999) and, particularly in the case of terrorist use, manufacturing contaminants.

Acute toxicity

The toxic actions of all of the nerve agents are very similar, although some differences have been observed. There are differences between the OPs with respect to their relative central and peripheral effects (Ligtenstein, 1984; Misulis *et al.*, 1987) and, within the central nervous system (CNS), some differences between the OPs may exist. For example, in the case of soman, there is some indication that there are differences in the sensitivity of acetylcholinesterase within the nervous system (Sellström *et al.*, 1985). Acute toxicity figures are available for the nerve agents tabun, sarin, soman and VX in many species (Table 5). The acute toxic dose in humans in not known with any exactitude as poisoning with nerve agents has only rarely been observed in man (Sidell, 1974; Maynard and Beswick, 1992). Where poisoning has been observed in man, detailed information on dose has usually not been available and this was the case with the sarin casualties in Japan, where the material was, in any case, impure. Therefore, lethal doses and likely clinical effects in man have to be inferred from experimental poisoning in animals and from OP pesticide poisoning. The cases that have been studied, together with low-dose volunteer studies with nerve agents, do not suggest any major differences between man and other animals in clinical response to nerve agents.

Acute toxicity figures for nerve agents other than tabun, sarin, soman and VX are scanty, with the exception of cyclosarin. Since the 1st Gulf War, a considerable amount of work has been carried out on the treatment of poisoning with cyclosarin and data have also become available on the acute toxicity of this compound. In studies by Anthony *et al.* (2004), the LCt50s in the rat for inhaled cyclosarin were of a similar order of magnitude to the LCt50s for sarin (Table 6). Although cyclosarin is a powerful acetylcholinesterase inhibitor, rapid ageing of the inhibited enzyme as seen with soman, does not occur, while spontaneous reactivation does (Worek *et al.*, 1998) (see below). Qualitatively, the toxic effects of cyclosarin seem similar to other nerve agents (Young and Koplovitz, 1995).

It should be noted that the nerve agents have one or more chiral centres and the toxicity of the enantiomers may differ dramatically from one to another and the racemic mixtures (Spruit *et al.*, 2000).

Mechanism of toxicological actions

The action of OP nerve agents on the nervous system results from their effects on enzymes, particularly esterases. The most notable of these esterases is acetylcholinesterase. The active site of acetylcholinesterase comprises a catalytic triad of serine, histidine and glutamic acid residues and other important features of the enzyme are a 'gorge' connecting the active site to the surface of the protein and a peripheral anionic site (Bourne *et al.*, 1995,1999; Sussman *et al.*, 1991; Thompson and Richardson, 2004), The OPs phosphylate[1] the serine hydroxyl group in the active site of the enzyme.

The binding to acetylcholinesterase of acetylcholine (the natural substrate of the enzyme) results in acetylation of the serine at the active site of the enzyme, with loss of the choline moiety. The reaction for acetyl choline can be envisaged as shown below, with E being the

[1] Phosphylation includes phophorylation and phosphonylation, the latter being more common with nerve agents.

Table 5. Comparative acute toxicity of nerve agents. Adapted and reproduced from Chapter 34, 'Organophosphorus compounds as chemical warfare agents', by Maynard RL and Beswick F, in *Clinical and Experimental Toxicology of Organophosphates* (B Ballantyne and TC Marrs, eds), 1992, with the kind permission of the publishers, Elsevier, and the authors

Species	Route[a]	Term	Unit (duration)	Tabun	Sarin	Soman	VX
Man	pc	LD_{50}	mg kg^{-1}	—	28[b]	—	—
	pc	LCLO	µg kg^{-1}	—	—	—	86[c]
	pc	LDLO	mg kg^{-1}	23[b]	—	18[b]	—
	inhal	LDLO	mg m^{-3}	150[b]	—	70[b]	—
	inhal	LD_{50}	mg m^{-3}	—	70[b]	—	—
	inhal	ECLO	µg m^{-3}	—	90[d]	—	—
	iv	TDLO	µg kg^{-1}	14[b]	—	—	—
	iv	TDLO	µg kg^{-1}	—	—	—	1.5[e]
	oral	TDLO	µg kg^{-1}	—	2[f]	—	4[e]
	sc	LDLO	µg kg^{-1}	—	—	—	30[g]
	im	TDLO	µg kg^{-1}	—	—	—	3.2[g]
Rat	pc	LD_{50}	mg kg^{-1}	18[h]	—	—	—
	inhal	LC_{50}	mg m^{-3} (10 min)	304[h]	150[d]	—	—
	iv	LD_{50}	µg kg^{-1}	66[h]	39[i]	44.5[j]	—
	oral	LD_{50}	µg kg^{-1}	3700[h]	550[h]	—	12[m]
	sc	LD_{50}	µg kg^{-1}	193[j]	103[k]	75[l]	—
	im	LD_{50}	µg kg^{-1}	800[f]	108[n]	62[n]	—
	im	LD_{50}	µg kg^{-1}	130[G]	—	—	—
	ip	LD_{50}	µg kg^{-1}	—	218[i]	98[o]	—
Mouse	pc	LD_{50}	mg kg^{-1}	1[h]	1.08[h]	—	—
	inhal	LC_{50}	mg m^{-3} (30 min)	15[h]	5[p]	1[p]	—
	iv	LD_{50}	µg kg^{-1}	150[h]	113[q]	35[r]	—
	sc	LD_{50}	µg kg^{-1}	250[s]	60[p]	40[p]	22[s]
	sc	LD_{50}	µg kg^{-1}	—	319[q]	—	—
	sc	LD_{50}	µg kg^{-1}	—	172[E]	—	—
	im	LD_{50}	µg kg^{-1}	440[q]	222[q]	—	—
	ip	LD_{50}	µg kg^{-1}	—	420[u]	393[u]	50[c]
Dog	pc	LD_{50}	mg kg^{-1}	30[h]	—	—	—
	inhal	LC_{50}	mg m^{-3} (10 min)	400[h]	100[h]	—	—
	iv	LD_{50}	µg kg^{-1}	84[v]	19[v]	—	—
	oral	LD_{50}	µg kg^{-1}	200[p]	—	—	—
	sc	LD_{50}	µg kg^{-1}	284[l]	—	12[w]	—
	sc	LD_{50}	µg kg^{-1}	—	120[H]	14[H]	10[H]
Monkey	pc	LD_{50}	µg kg^{-1}	9300[h]	—	—	—
	inhal	LC_{50}	mg m^{-3} (10 min)	250[h]	100[h]	—	—
	sc	LD_{50}	µg kg^{-1}	—	—	13[x]	—
	im	LD_{50}	µg kg^{-1}	—	22.3[y]	9.5[z]	—
Cat	inhal	LC_{50}	mg m^{-3} (10 min)	250[h]	100[v]	—	—
	iv	LD_{50}	µg kg^{-1}	—	22[h]	—	—
Rabbit	pc	LD_{5}	µg kg^{-1}	2500[h]	925[h]	—	—
	inhal	LC_{50}	mg m^{-3} (10 min)	840[h]	120[h]	—	—
	iv	LD_{50}	µg kg^{-1}	63[h]	15[A]	—	—
	oral	LD_{50}	µg kg^{-1}	16300[h]	—	—	—
	sc	LD_{50}	µg kg^{-1}	375[B]	30[C]	20[w]	14[D]
	ip	LD_{50}	µg kg^{-1}	—	—	—	66[D]

(cont.)

Table 5. *(cont.)*

Species	Route[a]	Term	Unit	Tabun	Sarin	Soman	VX
Guinea Pig	pc	LD_{50}	$mg\,kg^{-1}$	35[h]	—	—	—
	inhal	LC_{50}	$mg\,m^{-3}$ (2 min)	393[h]	—	—	—
	sc	LD_{50}	$\mu g\,kg^{-1}$	120[C]	—	—	—
	sc	LD_{50}	$\mu g\,kg^{-1}$	—	30[B]	24[C]	8.4[C]
Hamster	sc	LD_{50}	$\mu g\,kg^{-1}$	245[F]	95[B]	—	—
Farm	pc	LD_{50}	$\mu g\,kg^{-1}$	1100[h]	—	—	—
Animal	inhal	LC_{50}	$mg\,m^{-3}$ (14 min)	400[h]	—	—	—
Chickens	sc	LD_{50}	$\mu g\,kg^{-1}$	—	—	50[w]	—
	ip	LD_{50}	$\mu g\,kg^{-1}$	—	—	71[o]	—
Frog	ip	LD_{50}	$\mu g\,kg^{-1}$	—	—	251[o]	—

Notes:
[a] pc, percutaneous; inhal, inhalation; iv, intravenous; sc, subcutaneous; im, intramuscular; ip, intraperitoneal. [b] Robinson (1967). [c] WHO (1970). [d] Rengstorff (1985). [e] Sidell (1974). [f] Grob and Harvey (1958). [g] National Academy of Sciences (1982). [h] Gates and Renshaw (1946). [i] Fleisher *et al.* (1963). [j] Pazdernik *et al.* (1983). [k] Brimblecombe *et al.* (1970). [l] Bosković *et al.* (1984). [m] Jovanovic (1982). [n] Schoene *et al.* (1985). [o] Chattopadhyay *et al.* (1986). [p] Lotts (1960). [q] Schoene and Oldiges (1973). [r] Brezenoff *et al.* (1984). [s] Maksimović *et al.* (1980). [t] Fredriksson (1957). [u] Clement (1984). [v] O'Leary *et al.* (1961). [w] Berry and Davies (1970). [x] Clement *et al.* (1981). [y] D'Mello and Duffy (1985). [z] Lipp (1972). [A] Wills (1961). [B] Coleman *et al.* (1968). [C] Gordon and Leadbeater (1977). [D] Leblic *et al.* (1984). [E] Inns *et al.* (1992). [F] Coleman *et al.* (1966). [G] Cabal *et al.* (2004). [H] Weger and Scinicz (1981).

enzyme, AX acetylcholine, EAX is a reversible Michaelis–Menten complex and A is acetate:

$$E + AX \xrightarrow{k_{+1}} EAX \xrightarrow{k_2} \begin{matrix} EA \\ +X \end{matrix} \xrightarrow{k_3} E + A$$

where k_{+1}, k_{-1}, k_2 and k_3 are rate constants. Inhibition with OPs takes place by a process analogous to the reaction of acetylcholine with the enzyme, by a reaction in which the leaving group of the OP is lost so that the esterase is phosphylated at the hydroxyl group of the serine residue instead of acetylated. In the above equation, AX would be the OP nerve agent, EAX a reversible Michaelis–Menten complex of the enzyme and nerve agent, EA the phosphylated enzyme, X the leaving group and A, with nerve agents, (typically) an alkoxy alkylphosphonate. Reactivation of nerve agent-inhibited acetylcholinesterase occurs by hydrolysis of the alkoxy alkylphosphonyl or phosphoryl enzyme, resulting in dephosphylation and the rates of phosphylation are very variable, which partly accounts for differences in acute toxicity between the nerve agents.

The affinity with which a substrate such as acetylcholine or an inhibitor such as an OP binds to acetylcholinesterase is described by the dissociation constant for the complex EAX, K_D. This is equal to k_{-1}/k_{+1}. For inhibitors whose

Table 6. Acute toxicity of cyclosarin (GF)

Species	Sex	Route[c]	LD_{50} ($\mu g\,kg^{-1}$) or LCt_{50} ($mg\,min\,m^{-3}$)
Rat	Male	inhal	371 (10 min)[b] 396 (1 h)[b] 585 (4 h)[b]
Rat	Female	inhal	253 (10 min)[b] 334 (1 h)[b] 533 (4 h)[b]
Rat	Male	im	80[c]
Mouse	Male	sc	243[d]
Guinea pig	Male	sc	44[e]
Monkey, rhesus	Male	im	46.6[f]

Notes:
[a] Inhal, inhalation; sc, subcutaneous; im, intramuscular.
[b] Anthony *et al.* (2004).
[c] Kassa and Cabal (1999).
[d] Clement (1992).
[e] Lundy *et al.* (1992).
[f] Koplovitz *et al.* (1992).

complexes with acetylcholinesterase reactivate slowly, such as OPs, k_3 can be ignored and the reaction with acetylcholinesterase can be described by a bimolecular rate constant, k_i as follows:

$$E + AX \xrightarrow[k_i]{} X + EA$$

Some useful relationships can then be derived, e.g. $k_i = k_2/K_D$ (Main and Iverson, 1966). In addition, $k_i = \ln 2/I_{50}$ (Aldridge, 1950) which allows easy estimation of k_i (The I_{50} is the concentration of inhibitor, which inhibits the enzyme by 50%). These constants have been measured for many OP chemical warfare agents and also pesticides (e.g. Gray and Dawson, 1987). The hydrolysis reaction for acetylated acetylcholinesterase is fast (Koelle, 1992), in the region of 100 μs (Lawler, 1961; O'Brien, 1976). The key to the powerful anticholinesterase effects of OPs is what happens after inhibition by these compounds. In the case of OPs, hydrolysis of the phosphylated serine residue is much slower[2] than the acetylated analogue.

Rates of phophorylation or phosphonylation (inhibition) are a function of the whole OP molecule [see Maxwell and Lenz (1992) for a discussion of the structure-activity effects that underlie cholinesterase inhibition]. Rates of reactivation (hydrolysis) are determined by the structure left attached to the active-site serine residue after loss of the leaving group. Reactivation by hydrolysis of the phosphylated enzyme, which is a nucleophilic displacement reaction, occurs at a clinically significant rate, with nearly all important OPs but always much slower than when the enzyme is acetylated by its natural substrate: thus the enzyme deacetylates in μs, but dephosphorylates in a matter of hours to days, In general, spontaneous reactivation is faster, the smaller the alkyl groups; thus dimethyl phosphoryl acetylcholinesterase reactivates faster than the diethyl analogue and diisopropyl phosphorfluoridate (DFP)-inhibited enzyme hydrolyses extremely slowly, in reality so slowly that the binding of the organophosphate to the enzyme has been described as irreversible (WHO, 1986) (for a tabulation of the $t_{1/2}$s for spontanous reactivation of various phosphylated acetylcholinesterases, see Wilson *et al.*, 1992). While the active site of the enzyme is phosphylated it is of course unavailable for hydrolysis of acetylcholine. With some organophosphorylated and organophosphonylated acetylcholinesterases (but notably not most OP insecticides), hydrolysis is very slow. With soman, the situation is further complicated by an additional reaction known as ageing. This consists of monodealkylation of the dialkylphosphyl enzyme, creating a much more stable monoalkylphosphyl enzyme, the reactivation rate of which is negligible (see review by Curtil and Masson, 1993). On the basis of studies using acetylcholinesterase from *Torpedo californica*, Millard *et al.* (1999) suggested a number of non-covalent forces that might stabilize the aged enzyme, thereby preventing reactivation. Rates of ageing depend on the structure of the inhibited acetylcholinesterase produced with each nerve agent. Soman produces an inhibited acetylcholinesterase, where the active site serine is phosphylated with a pinacoloxy methylphosphonyl structure. This ages very rapidly by loss of the large pinacolyl group, with the result that reactivation of inhibited acetylcholinesterase does not occur to any clinically significant extent. Talbot *et al.* (1988) studied *in vitro* and *in vivo* ageing rates of soman-inhibited erythrocyte cholinesterase in several animal species. *In vitro*, $t_{1/2}$s were (mean ± sem) 8.0 ± 0.82 min for guinea pigs, 1.1 ± 0.08 min for marmosets, 1.4 ± 0.11 min for cynomolgus monkeys and 0.88 ± 0.03 min for squirrel monkeys. *In vivo* $t_{1/2}$s were 8.6 ± 0.94 min for the rat, 7.5 ± 1.7 min for the guinea pig and 0.99 ± 0.10 min for the marmoset. The ageing $t_{1/2}$ in human erythrocytes is known to be rapid (1.3 min) for soman-inhibited enzyme (Harris *et al.*, 1978). Incidentally, this suggests that the guinea pig would not be a good model for humans in antidotal experiments in soman poisoning and that such experiments could only be undertaken realistically in primates. Because of the rapid rate of ageing of the soman-inhibited acetylcholinesterase, recovery of function depends on resynthesis of acetylcholinesterase (Gray,

[2] The term irreversible is often used for this reaction. This should not be taken to mean that reactivation of acetylcholinesterase by hydrolysis does not occur. The term is used because the OP is not recovered intact upon reactivation of the enzyme (Chambers, 1992).

1984). Substances which produce complexes with acetylcholinesterase which age virtually instantaneously have been synthesized, for example crotylsarin, *O*-(2-butenyl) methylphosphonofluoridate, which has been used as an experimental tool in the investigation of the action of oximes (van Helden *et al.*, 1994). With most other OPs, including most other nerve agents, ageing occurs, by the same mechanism as soman (monodealkylation or, in the case of tabun, possibly P–N scission (Barak *et al.*, 2000)), but more slowly and some spontaneous reactivation may take place. Where treatment is instituted late, or where there has been repeated exposure, significant amounts of enzyme may be in the aged state (for tabulations of ageing $t_{1/2}$s of various phosphylated acetylcholinesterases, see Wilson *et al.*, 1992 and Maynard, 1999). The slow ageing process which occurs in regard to tabun and sarin has been taken by some to indicate that treatment with oxime acetylcholinesterase reactivators can be safely delayed. Nothing could be further from the truth. Ageing has little to do with the toxic effects of organophosphates: these are, of course, dependent on the accumulation of acetylcholine at essential sites and the patient will always benefit from treatment being instituted as soon as possible.

Tabun-inhibited enzyme (an *O*-ethyl *N*,*N*-dimethylamidophosphoro derivative) is slow to reactivate and Heilbronn (1963) found no detectable spontaneous reactivation with human acetylcholinesterase. The reasons for this are becoming more clear. Tabun binding of mouse acetylcholinesterase causes conformational changes in the enzyme that may stabilize the enzyme–inhibitor complex even without ageing of the complex (Ekstrom *et al.*, 2006).

ACCUMULATION OF ACETYLCHOLINE

In normal circumstances, acetylcholine is hydrolyzed almost immediately by acetylcholinesterase close to receptors in the synaptic cleft at sites of action (see Bowman, 1993). The normal function of acetylcholinesterase is to hydrolyze acetylcholine in the synaptic cleft, parasympathetic effector organ or neuromuscular junction, in order to terminate transmission of a nerve impulse: failure of acetylcholinesterase activity results in accumulation of acetylcholine (Burgen and Hobbiger, 1951), which in turn causes enhancement and prolongation of cholinergic effects and also depolarization blockade. At the neuromuscular junction, where accumulation of acetylcholine initially causes fasciculation, continued accumulation produces flaccid type paralysis due to depolarization blockade.

CHOLINERGIC NEUROTRANSMISSION

Fully to undertand the effects of esterase inhibitors, such as OPs, it is necessary to discuss the cholinergic neurotransmission system. The essential features in this system are a synthetic enzyme, choline acetyltransferase (ChAT), the neurotransmitter itself (acetylcholine), a hydrolytic enzyme, acetylcholinesterase and specialized receptors, with which the acetylcholine interacts. Acetylcholine is one of a number of neurotransmitters in the nervous systems of mammals. It is synthesized from acetyl coenzyme A and choline by the enzyme ChAT in the perikaryon of cholinergic neurones and transported to nerve terminals, with ChAT existing in both soluble and membrane-bound forms. The ChAT gene also encodes another protein besides ChAT, the vesicular acetylcholine transporter. This transporter is responsible for transporting acetylcholine from the neuronal cytoplasm to the synaptic vesicles (Oda, 1999). At the nerve terminals, acetylcholine is released in response to an action potential, thereby triggering opening of voltage-gated calcium channels in the presynaptic terminal. The acetylcholine thus released crosses the synaptic cleft. Contact between the acetylcholine and specialized receptors (see below) at the proximal end of the post-ganglionic nerve fibre results in localized depolarization of the postsynaptic membrane and generation of a nerve impulse in the postganglionic nerve fibre. Transmission is similar at parasympathetic nerve endings except that the acetylcholine stimulates receptors in parasympathetic effector organs. At the neuromuscular junction, muscle fibre depolarization is initiated at the motor end plate following a similar sequence of events. In the cases of both autonomic ganglia and muscle, it is necessary for the depolarization to exceed a threshold in order to produce postjunctional activity, e.g. initiate an action potential in a postsynaptic nerve cell or muscle fibre.

CHOLINERGIC RECEPTORS

Cholinergic receptors are divided into muscarinic and nicotinic on the basis of their sensitivity to pharmacological stimulation by muscarine and nicotine, respectively; muscarinic receptors are found in parasympathetic effector organs, and, prejunctionally at the neuromuscular junction, whereas nicotinic receptors are found at autonomic ganglia and the neuromuscular junction. More or less specific antagonists are available; thus, atropine antagonizes muscarinic agonists at muscarinic sites but has little effect at the neuromuscular junction, whereas the reverse is true of tubocurarine. The two types of receptor are fundamentally different – muscarinic receptors using a second messenger system, whereas nicotinic receptors are ligand-gated ion channels.

MUSCARINIC RECEPTORS

Five sub-types of muscarinic receptor, designated M_1 to M_5, have been cloned. Each receptor has a serpentine structure that spans the cell membrane seven times. The receptor is coupled via guanosine triphosphate-binding proteins (G-proteins) to the enzymes adenylate cyclase or phospholipase C. M_I receptors, activation of which leads to an increase in Ca^{2+} conductance of local ligand-gated Ca^{2+} channels, are found in neuronal tissue. These calcium channels are ligand-gated, unlike those in the presynaptic or prejunctional terminals which are voltage-gated. Influx of calcium ions via open channels is both electrically driven by the negative intracellular potential and concentration driven as the intracellular free Ca^{2+} concentration is about 100 $nmol\,l^{-1}$ while the extracellular concentration is about 1.2×10^6 $nmol\,l^{-1}$ (Ganong, 1991). Activation of M_2 receptors leads to a decrease in Ca^{2+} conductance. M_2 receptors may play a prejunctional role in modulating the activity at the neuromuscular junction. M_2 receptors are also found in the heart and mediate the depressor effects of parasympathetic activity via the vagus nerve. The distribution and some functions of the other sub-types have also been characterized (Eglen and Whiting, 1990; Jones, 1993). M_3 receptors are found in glandular tissue and M_4 receptors are found in the striatum while the M_5 subtype is present in the hippocampus and brainstem.

NICOTINIC RECEPTORS

Nicotinic receptors, found at all autonomic system ganglia and at the skeletal neuromuscular junction have been well characterized at a molecular level largely because the same type of receptor is found in the fish (*Torpedo californica)* electric organ. The structure of the nicotinic acetylcholine receptor has been reviewed (Kaminski and Ruff, 1999). These receptors are part of a 'superfamily' of ligand-gated ion channels with γ-aminobutyric acid A ($GABA_A$) receptors and glycine receptors (Ortells and Lunt, 1995). Each receptor comprises five sub-units arranged around a channel that passes through the cell membrane, but there are some differences between the receptors found at the neuromuscular junction and those in neuronal tissue, in structure and in inhibition characteristics (the former are inhibited by α-bungarotoxin). In neuronal tissue, the receptors are composed of α and β sub-units and a number of each type of sub-unit have been cloned (Boyd, 1997). The muscle receptor in adult innervated muscle contains 5 sub-units, 2 α, 1 β 1 δ and 1 ε; in denervated or embryonic muscle the ε is replaced by a γ sub-unit. Binding of acetylcholine leads to a widening of the channel and an increase in Na^+ conductance. Influx of Na^+ ions leads to depolarization of the cell membrane (see Lefkowitz *et al.*, 1996).

Clinical Effects

The clinical effects of nerve agents are, to a large extent, those of acetylcholine accumulation and the effects of all of the nerve agents are similar. Those differences that have been observed are presumably due to a combination of different rates of inactivation and reactivation of the enzymes, together with different rates of ageing of the inhibited enzyme and differences in absorption, distribution, metabolism and

Table 7. Main effects of nerve agents at various sites in the body

Receptor	Target organ	Symptoms and signs
Central	Central nervous system	Giddiness, anxiety, restlessness, headache, tremor, confusion, failure to concentrate, convulsions, respiratory depression, respiratory arrest
Muscarinic	*Glands*	
	Nasal mucosa	Rhinorrhea
	Bronchial mucosa	Bronchorrhea
	Sweat	Sweating
	Lachrymal	Lachrymation
	Salivary	Salivation
	Smooth muscle	
	Iris	Miosis
	Ciliary muscle	Failure of accommodation
	Gut	Abdominal cramp, diarrhoea, involuntary defecation
	Bladder	Frequency, involuntary micturition
	Heart	Bradycardia
Nicotinic	Autonomic ganglia	Sympathetic effects, including pallor, tachycardia, hypertension
	Skeletal muscle	Weakness, fasciculation

excretion of the nerve agent. It is noteworthy that in the case of soman, kinetic differences have been recorded between different stereoisomers (Benschop *et al.*, 1987).

The symptoms and signs of nerve agent poisoning may be divided into three groups, muscarinic, mediated by muscarinic receptors in parasympathetic effector organs, nicotinic, mediated by nicotinic receptors in autonomic ganglia and the neuromuscular junction, and effects in the central nervous system, which are mediated by receptors of both types. Acute effects of OP nerve agents are given in Table 7. The muscarinic symptoms and signs result from increased activity of the parasympathetic system and include bronchorrhoea, salivation, constriction of the pupil of the eye (miosis),[3] abdominal colic and bradycardia (Grob and Harvey, 1953). Nicotinic effects at autonomic ganglia can produce pallor, tachycardia and hypertension. The clinical effects in the cardiovascular system depend on whether muscarinic or nicotinic effects predominate; bradycardia or tachycardia may occur and, in some cases arrhythmias (see below). At the neuromuscular junction, nicotinic signs include muscle fasciculation and later paralysis. If the patient survives the acute cholinergic syndrome, the effects of nerve agents are largely reversible, although, as discussed above, with soman recovery may be very slow and in certain circumstances there may be long-term changes in the CNS (see below). Where death occurs, it is caused by respiratory paralysis, which may be central or due to the anticholinesterase action at the neuromuscular junction (Chang *et al.*, 1990).

[3] Miosis is the term used to describe the constriction of the pupil. The term is often assumed to be derived from the same Greek root as meiosis, i.e. a diminution. This is in fact not the case. Miosis, and the earlier variant myosis, are derived from the Greek root myein (or muein): to close, blink (Shorter Oxford English Dictionary, Webster's Third New International Dictionary).

NON-ANTICHOLINESTERASE EFFECTS

Clinically, the most important effects of OP nerve agents are anticholinesterase actions. However, it should not be forgotten that OPs bind to a variety of enzymes, including esterases other than acetylcholinesterase, e.g. carboxylesterase, (long-chain fatty acid hydrolase), serine peptidases, amidases and proteases and others (see Lockridge and Schopfer, 2005). Moreover, there is some evidence that some OP anticholinesterases can act directly on muscarinic and nicotinic receptors (Bakry *et al.*, 1988; Silveira *et al.*, 1990; Mobley, 1990; Eldafrawi *et al.*, 1992). More complex direct effects at cholinergic receptors may be important: Rocha *et al.* (1999) showed that VX, at concentrations that had little effect on cholinesterase, interacted directly with presynaptic muscarinic receptors, causing a selective inhibition of the evoked release of GABA. This would increase excitatory neurotransmission and these workers hypothesized that this might account for the ability of VX to induce convulsions.

It has also been demonstrated that OPs, including nerve agents, can affect neurotransmission pathways other than cholinergic ones, for example γ-aminobutyric acid (GABAergic) receptors as well as glutamatergic, dopaminergic, somatostatinergic and noradrenergic systems (Lau *et al.*, 1988; Fletcher *et al.*, 1989; Fosbraey *et al.*, 1990; Naseem, 1990; Smallbridge *et al.*, 1991; Chechabo *et al.*, 1999; Tonduli *et al.*, 1999; Rocha *et al.*, 1999). It is likely that most if not all such perturbations are secondary to effects on cholinergic systems, but there is evidence that recruitment of non-cholinergic systems is responsible, at least in part, for seizure activity (Solberg and Belkin, 1997). Furthermore, the efficacy of some experimental treatments not directed at the cholinergic neurotransmission system suggests a role for other neurotransmission systems. Thus, Braitman and Sparenborg (1989) found that a potent antagonist at the *N*-methyl-D-aspartate (NMDA)-type glutamate receptor (MK-801), together with other antidotes, protected against seizures induced in guinea pigs by soman. Moreover, Filliat *et al.* (1999) found that learning deficits produced in rats by soman were reduced by antagonists at α-amino-3-hydroxy-5-methyl-4-isoxazolepropionic acid (AMPA)-type glutamate receptors and NMDA-type glutamate receptors, together with atropine. It has also been suggested that the protective action of caramiphen aginst OP poisoning may be, in part, due to the ability of this drug, which is also an anticholinergic agent, to modulate the NMDA receptor (Raveh *et al.*, 1999)

Kovacic (2003) suggested that oxidative stress might play a part in the toxic manifestations of OPs.

It is possible that some of the CNS effects of nerve agents are secondary to changes in the blood brain barrier. There is little evidence that nerve agents will cause OP-induced delayed polyneuropathy at doses likely to be encountered and survived by man (see below).

EFFECTS ON SPECIFIC ORGANS

The eye

Unlike OP insecticides, the G-type nerve agents are most likely to be encountered as vapours and eye effects occur early. Nerve agents produce miosis (constriction of the pupil); this produces a feeling that the surroundings are dim, or that illumination has been reduced. The onset is rapid and may last for several days. Miosis is a very sensitive indicator of exposure to nerve agents and van Helden *et al.* (2004a) concluded, on the basis of studies in guinea pigs and marmosets with sarin, that miosis would occur during low-level exposure at levels that would not be detectable by the currently fielded alarm systems, assuming that humans are of similar sensitivity to these experimental species.

Spasm of the ciliary muscle may impair accommodation and is associated with severe headache. Long-lasting miosis, associated with eye pain, was a notable clinical sign in the Tokyo Subway (underground railway) terrorist attack with sarin and the same was true of the sarin attack at Matsumoto (Nohara and Segawa, 1996).

Dilatation of subconjunctival blood vessels occurs and the eye becomes bloodshot. After exposure to high concentrations of nerve agent, the eyes take on a glassy appearance: the appearance is sometimes compared to that of a glass marble. The lachrymal glands do not seem to be much affected by exposure to nerve agent vapour and tearing is not a reliable early sign of exposure.

The Respiratory Tract

The upper respiratory tract contains two components that are under cholinergic control. These are the mucous glands and smooth muscle. The response to nerve agents by the former is an increase of secretions resulting in bronchorrhea and rhinorrhea (runny nose) and, in severe cases, foaming around the nose may occur. The effect on smooth muscle is to produce bronchospasm. Respiratory function may also be affected by the action of nerve agents on the neuromuscular junction of the (striated) muscles of respiration and by central effects on the respiratory centre.

Heart

The acute effect on the heart depends on the relative predominance of muscarinic or nicotinic effects. Bradycardia or tachycardia may occur, as well as arrhythmias, including atrioventricular block and various ventricular arrhythmias. An arrhythmia characteristic of poisoning with OP pesticides is Torsade de Pointes (Ludomirski *et al.*, 1982), while soman and other OPs have been reported to produce histopathological changes in the myocardium in both experimental animals and humans (McDonough *et al.*, 1989; Singer *et al.*, 1987; Pimentel and Carrington da Costa, 1992; Koplovitz *et al.*, 1992; Britt *et al.*, 2000). A few patients in the Matsumoto nerve agent attack had arrhythmias (Okudera, 2002).

Nervous system and skeletal muscle

Systemically, nerve agents produce fasciculation and then blockade at the neuromuscular junction with weakness and paralysis. Paralysis of the muscles of respiration (see above) may interfere with respiration and is potentially life-threatening. Some of these effects may be mediated by direct actions at the receptor-ion channel complex. Soman and sarin were reported by Goldstein *et al.* (1987) to alter muscle spindle function.

Separate from the acute effects of anticholinesterases upon the neuromuscular junction (discussed previously), two further syndromes involving the neuromuscular system have been associated with OP poisoning. These are (1) the intermediate syndrome (IMS), and (2) organophosphate-induced delayed polyneuropathy (OPIDP).

INTERMEDIATE SYNDROME (IMS)

Senanayake and Karalliedde (1987) described a new form of neurotoxicity following intoxication by organophosphorus insecticides. As it occurred after the acute syndrome and before the onset of classical OPIDN, they called it the 'Intermediate Syndrome'. This phenomenon consists of marked weakness of the proximal skeletal musculature (including the muscles of respiration and neck muscles) and cranial nerve palsies. IMS comes on 1–4 days after acute poisoning; respiratory support is often necessary and, if it is provided, recovery occurs within 4–18 days. Although the IMS has not been described in cases of accidental nerve agent poisoning, it is likely that it would occur, at least in some cases. The pathogenesis of IMS is not known with certainty, although it appears to be a post-junctional non-depolarizing blockade. IMS is probably a consequence of cholinergic overactivity at the neuromuscular junction (NMJ), perhaps through down-regulation of post-junctional nicotinic receptors (i.e. a reduced density of functioning nicotinic receptors), with a possible contribution from prejunctional muscarinic receptors (Marrs *et al.*, 2005). It has also been suggested that failure to hydrolyze acetylcholine may reduce the supply of choline, which is a substrate for ChAT, the enzyme that synthesizes acetylcholine. Furthermore, connection has been suggested between the IMS and OP-induced myopathy (Senanayake and Karalliedde, 1992). Myopathy associated with OPs was first observed many years ago and has been observed histologically in experimental animals with the nerve agents tabun, soman, and sarin (Preusser, 1967; Ariens *et al.*, 1969; Gupta *et al.*, 1987a,b; Bright *et al.*, 1991; Hughes *et al.*, 1991; Koplovitz *et al.*, 1992; Britt *et al.*, 2000). The changes characteristic of OP-induced myopathy seem to be initiated by calcium influx (Leonard and Salpeter, 1979) as a consequence of acetylcholine accumulation at the neuromuscular junction (Marrs *et al.*, 1990; Inns *et al.*, 1990). In general the distribution of the myopathy parallels the distribution of muscles affected by IMS, although that was not true of the study in rhesus macaques by Britt *et al.* (2000). If it is these histological changes that underlie IMS and, as they have been observed with chemical warfare nerve agents, the development of the syndrome can be anticipated in the recovery phase of nerve agent poisoning in at least some cases. However, it should be noted that oximes are reported to protect against the myopathy, whereas there have been reports that, in insecticide poisoning, oximes have not always protected against IMS. Furthermore, the time-course of the myopathy observed in experimental animals is not similar to the time-course for the development of IMS. Karalliedde *et al.* (2006) concluded that IMS arose from down-regulation of the nicotinic acetylcholine receptor, caused by acetylcholine accumulation, and that myopathy and IMS are not causally related, but have a common origin in acetylcholine accumulation.

NERVE AGENT-INDUCED DELAYED POLYNEUROPATHY

Organophosphate-induced delayed polyneuropathy (OPIDP) is a symmetrical sensory-motor neuropathy, with both central and peripheral components, tending to be most severe in long axons and occurring 7–14 days after exposure to OPs. In severe cases, it is an extremely disabling condition, the central component being irreversible. Inhibition of neuropathy target esterase (NTE), an esterase of unknown function present at several sites, including neurones, where it is an integral membrane protein (Glynn, 2000), appears to be necessary for OPIDP to develop (see reviews by Somani and Husein, 2000 and Senanayake, 2001). This is followed by an ageing reaction similar to that described for soman with acetylcholinesterase above (Johnson, 1975). By contrast with IMS, it is unlikely that nerve agents possess the capability to cause OPIDP. In experimental studies, nerve agents do not usually bring about OPIDP (Anderson and Dunham, 1985; Crowell *et al.*, 1989; Johnson *et al.*, 1985; Goldstein *et al.*, 1987; Parker *et al.*, 1988; Henderson *et al.*, 1992). Gordon *et al.* (1983) reported studies in hens (Ross white or Sussex) *in vivo,* protected with physostigmine, atropine sulphate and pralidoxime mesilate (P2S): OPIDP (as assessed clinically) with NTE inhibition was only seen at 30–60 × LD_{50} for sarin, but not at 38 × LD_{50} for soman or 82 × LD_{50} for tabun. Furthermore, in the case of soman, Johnson *et al.* (1985) showed that only a tiny proportion of inhibited NTE from hen brain and spinal cord underwent ageing. Crowell *et al.* (1989) studied the potential for sarin and soman to cause OPIDP. Sarin type I[4] was given by gavage at single doses of 61, 200 and 400 μg kg^{-1}, sarin type II at doses of 70, 140 and 280 μg kg^{-1} and soman at doses of 3.5, 7.1 and 14.2 μg kg^{-1} to groups of five Leghorn hens protected from acute cholinergic toxicity with atropine. There were appropriate vehicle controls and tri-*o*-cresyl phosphate was used as a positive control. NTE was measured in brain, spinal cord and lymphocytes. With sarin type I, the highest dose was lethal within 24 h and the treatment did not significantly depress NTE in any tissue. The highest dose of sarin II decreased lymphocyte NTE to 33% of the control value. In the brain and spinal cord, significant depression of NTE was not seen at any dose. Soman did not produce significant depression of NTE. In all cases tri-*o*-cresyl phosphate produced significant depression of NTE. Despite this, there have been reports that sarin may produce histopathological changes in rodents similar to those found in OPIDP (see Somani and Husain, 2001).

Studies *in vitro* also suggest that the development of OPIDP is unlikely to be a clinical problem in humans. Thus, Gordon *et al.* (1983) compared the inhibitory potency of the nerve agents, sarin, soman, tabun and VX, and some other OP compounds, including diisopropyl phosphorofluoridate (DFP), mipafox and related compounds, against NTE and AChE. The ratio I_{50} for inhibition of acetylcholinesterase/I_{50} for inhibition of NTE was 0.0056 for sarin, 0.0012 for soman, 0.0005 for tabun and 10^{-6} for VX. In the case of the neuropathic OPs, the ratio was 1.13 for DFP and 1.8 for mipafox. Moreover, structure–activity considerations lend no support to suggestions that nerve agents would be neuropathic (Aldridge *et al.*, 1969).

Thus, the reason for the expectation that nerve agents would be non-neuropathic in man is that concentrations of nerve agent required to produce AChE inhibition are low, by comparison with concentrations required for inhibition of

[4] Types I and II refer to the stabilizers used.

NTE. Nevertheless, OPIDP was one hypothesis for Gulf War Syndrome (Institute of Medicine, 1996).

CENTRAL NERVOUS SYSTEM

OPIDP, which has a central component, was discussed above. The nerve agents produce a wide range of effects upon the central nervous system, ranging from anxiety and emotional lability at low doses, to convulsions and respiratory paralysis at higher ones. It is important to note that doses considerably below lethal doses can markedly degrade performance of tasks in behavioral studies (Brimblecomb, 1974; Wolthuis and Vanwersch, 1984; D'Mello and Duffy, 1985; D'Mello, 1993; DiGiovanni, 1999) and there is evidence that the decremental effects of exposure to single doses of nerve agents may be prolonged (McDonough *et al.*, 1986). Clearly, this is of importance as it is likely that military performance of personnel would be impaired: affected servicemen might not only lose the motivation to fight but also lose the ability to defend themselves and be unable to carry out the complex tasks frequently required in the modern armed forces. Effects on skilled personnel, such as pilots and navigators, would be particularly disabling. In a terrorist attack, rescuers may be affected and their skills impaired.

Long-term and delayed effects

Previously, experimental work with nerve agents has been concentrated on the acute effects. However with the possibility of mass casualties from terrorist outrages, the possibility of persistent neurobehavioral and psychiatric effects, such have sometimes been observed with excessive exposure to OP pesticides, needs to be considered. Such effects have included headache, anxiety, agitation, insomnia, irritability, impaired memory and difficulty in concentrating (Annau, 1992; Jamal, 1995; Lader, 2001; Feldman, 1999).

The long-term effects of low dose exposure and the delayed effects of acute exposure are often conflated. The latter are easier to understand. A number of studies have been performed to investigate subtle changes in humans exposed to various sorts of OPs, and with differing patterns of exposure, and these have been reviewed (Ray, 1998a; Committee on Toxicity of Products in Food, Consumer Products and the Environment, 1999; Karalliedde *et al.*, 2000; Romano *et al.*, 2001). Many of the studies reviewed refer to OP insecticides. It is difficult to summarize this work as the patterns of exposure have been very varied. Of course, long-term low dose exposure and acute high dose exposure are two ends of a spectrum of different exposure scenarios and many intermediate patterns of exposure are possible. Studies have also been undertaken in experimental animals (including primates) and some of these are discussed below.

Delayed effects of high dose exposure

The anticholinesterase effects induced by nerve agents in the CNS are reversible and histopathological changes are usually exiguous in fatalities and in animals dying during acute toxicity studies. Thus Anzueto *et al.* (1986) found that, in baboons, given intravenous infusions of soman, only minimal CNS histopathological changes were observed. The absence of specific changes hinders diagnosis at autopsy; moreover, it implies that complete recovery after successful treatment is probable. However, the situation may be different after survival of near-lethal doses and where humans survive, it is probable that functional deficits would be observed. It is also likely that were humans to survive for an appreciable period before dying, histopathological changes would be seen in the CNS. In animals, after substantial doses of soman, the initial histopathological changes seen are edema, particularly astrocytic and perivascular hemorrhages. Neuronal degeneration and necrosis, sometimes diffuse, may be observed, together with more discrete infarcts, with necrosis of all cell types. Such changes may be detected particularly in the hippocampus and piriform cortex (McLeod, 1985). This picture, which does not seem to correlate with areas in which the blood-brain barrier is compromised, may proceed to an encephalopathy. Most commonly, this affects the cortex, hippocampus and thalamic nuclei, a distribution that suggests anoxia is a likely cause (McLeod, 1985; McDonough *et al.*, 1986,1989). The hypothesis that hypoxia secondary to convulsions is the cause, is supported by the fact that the nerve agent-induced effects have been correlated with seizure activity (Anzueto *et al.*, 1986; Britt *et al.*, 2000) and that anticonvulsant γ-aminobutyric acid ($GABA_A$) agonists, such as diazepam or midazolam, alleviate the effects (Martin *et al.*, 1985; Anderson *et al.*, 1997). Nevertheless, the attribution of nerve-agent-induced pathological changes to hypoxia and/or convulsions is not the view of all (Petras, 1981) and in an *in vitro* ultrastructural study using rat rat hippocampal slices, Lebeda *et al.* (1988) found morphological differences between the effects produced by hypoxia and soman. OP poisoning is indeed sometimes associated with long-term CNS changes, both in experimental animals and in humans (Holmes and Gaon, 1956; Korsak and Sato, 1977; Rosenstock *et al.*, 1991) and this is likely to be the case both with pesticides and nerve agents (see reviews by Marrs and Maynard, 1994 and Ray, 1998a).

Human data on nerve agent exposure is scanty compared to that on OP pesticides. A notable study (on workers) involving nerve agents was carried out by Duffy *et al.* (1979). Behavioral signs and subtle EEG changes were noted after exposure to sarin, severe enough to cause symptoms and clinical signs. Burchfiel *et al.* (1976) described changes, which persisted for a year in the EEGs of rhesus monkeys after a single large dose or repeated small doses of sarin. The lower dose used was 1 μg, given by ten weekly injections (total dose 10 μg = 7% LD50). This result, which is somewhat surprising, must be interpreted cautiously as the group size was small (three per group) and similar but not identical electroencephalographic changes were seen after the organochlorine cyclodiene pesticide, dieldrin, namely a relative increase in ß activity. Subjects from volunteer programmes have been studied in the USA (National Academy of Sciences, 1982) and here there was no clear evidence of CNS effects or any effect on mortality. In a 1- and 2-year follow-up of survivors of the Matsumoto sarin exposure, epileptiform EEG abnormalities were found in four out of six severely poisoned subjects (Sekijima *et al.*, 1997). In a study of 18 victims of the Tokyo subway sarin incident, 6–8 months after exposure, the event-related and visual evoked potentials were significantly prolonged compared to matched controls, although no subject had obvious clinical abnormality (Murata *et al.*, 1997). A case-report from Japan noted specific interference with memory in a victim of the sarin exposure in Tokyo: this persisted to 6 months after exposure and extended to the period in a retrograde fashion to 70 days before exposure (Hatta *et al.*, 1996). In a suspected case of VX poisoning, antegrade and retrograde amnesia was observed (Nozaki *et al.*, 1995). In a study of members of a rescue team, who attended the Matsumoto sarin exposure, all symptoms of the affected personnel had resolved a year after exposure (Nakajima *et al.*, 1997). In a cross-sectional study of rescue staff and policemen approximately 3 years after they were exposed to sarin on the Tokyo subway, effects were observed on memory (Nishiwaki *et al.*, 2001). Yokoyama *et al.* (1998) reported vestibulocerebellar effects 6–8 months after exposure to sarin in the Tokyo incident. This comprised low-frequency sway, which was more severe in females. Of course, in the aftermath of a terrorist use of a nerve agent, psychological effects are to be expected in addition to any effect from damage to the CNS (DiGiovanni, 1999).

The implications for treatment are, at this time, uncertain; the most logical course would be to avoid convulsions and/or anoxia as much as possible during treatment of acute nerve agent poisoning.

Low dose exposure

A variety of effects in human and animals have been attributed to OP exposure but it is less clear whether low doses, particularly subconvulsive ones, can bring about long-term changes in CNS function. If that were the case, it is clearly desirable to know if there is a threshold dose below which effects are not observed. Effects observed after long-term low dose exposure to OPs are less easy to explain than delayed effects of high dose exposure, but it should be noted that OPs, including nerve agents, can react with targets other than cholinesterases (see above). Richards *et al.* (1999) found two novel OP-reactive sites in rat brain homogenates by inhibition of ^{3}H-DFP labeling using several OP insecticides or their oxons. One of the sites had a molecular mass of 85 kDa and Richards *et al.* (2000) purified the 85 kDa site from porcine brain and characterized it as an acylpeptide hydrolase (see reviews by Ray, 1998a,b).

After reviewing the available data, both in humans and in experimental animals, Moore (1998) concluded that the available data indicated that exposure to nerve agents at doses producing no clinical signs or symptoms of acute toxicity did not produce chronic illness (see also Brown and Brix, 1998). Pearce *et al.* (1999) were unable to find significant persisting effects of single im doses of sarin producing 36.4 to 67.1% erythrocyte acetylcholinesterase inhibition in marmosets on EEG or the CANTAB automated test battery. There was a marginal increase of beta 2 EEG activity attributable to an affect in one subject: Haley (2000) speculated that this might be due to a polymorphism for the enzyme paraoxonase 1 (PON1) in one marmoset. van Helden *et al.* (2004b) have shown effects on the EEG from sarin vapor at concentration × time values (Cts) of 0.2 and 0.1 mg min m^{-3} (time of exposure was 5 h) in vehicle and pyridostigmine-treated marmosets; these effects were persistent. Furthermore, Bajgar *et al.* (2004) found persistent (4 weeks) effects on a few behavioural parameters in guinea pigs given single inhalation exposures to soman. The concentrations used were 1.2, 1.5 and 2.7 mg m^{-3} and exposure was for 1 h. Only butyrylcholinesterase was inhibited at the lowest concentration at 1 day, while red cell cholinesterase was 66% control at the mid-concentration and 28% control at the highest concentration, both at 1 day. Brain acetylcholinesterase measured at 4 weeks was only inhibited > 20% at the highest concentration and then only in the basal ganglia, not the other areas in which enzyme activity was estimated. Significant increases in neuroexcitability score were seen at all test concentrations, compared to controls at 4 weeks, although the differences were small in magnitude. Unsurprisingly, larger differences between test groups and controls were seen at 1 day. The same group (Kassa *et al.*, 2004) reported a study in rats, in which exposure to sarin by inhalation was studied. Exposure was for 1 h at concentrations of 0.8 $\mu g\,l^{-1}$, which produced red cell acetylcholinesterase depression of < 20% compared to controls, 1.25 $\mu g\,l^{-1}$, which produced red cell acetylcholinesterase depressions of 20–30%, and 2.5 $\mu g\,l^{-1}$, which produced red cell acetylcholinesterase depression of 40–50%. Group sizes were 10. Exposures were on single occasions, except that exposure to the middle concentration was carried out repeatedly and additionally in a separate group of animals. A functional observational test battery (FOB) of behaviour was carried out at 3 months. Significant abnormality was seen after single exposure at the highest concentration in respect of a number of observations (gait disorder and score, mobility score, activity and stereotypy and at the mid-concentration only in respect of the last – more changes were seen after repeated exposure). Changes in spatial discrimination in the Y-maze were seen at all doses shortly after exposure at all test concentrations, but at the highest concentration this persisted for 3 weeks. This study could be interpreted as suggesting a single exposure threshold for long-term effects at 20–30% erythrocyte acetylcholinesterase depression in the rat.

Sleep disruption was seen in mice during a study on soman by Crouzier *et al.* (2004). The dose used was 50 $\mu g\,kg^{-1}$ bw sc.

Visual field defects persisting for 1 year, which had disappeared 17 months after exposure, were found in a single subject exposed to sarin in Matsumoto (Sekijima *et al.*, 1997).

Reproductive toxicity, developmental toxicity and developmental neurotoxicity

On the basis of a study with VX (Goldman *et al.*, 1988), there was little evidence that OP nerve agents presented a reproductive hazard at doses below those toxic to the parents. On the basis of studies with sarin (Denk, 1975; La Borde and Bates, 1986), there is little evidence that OP nerve agents have a potential for teratogenesis or fetoxicity at doses below those toxic to the dams. There is no direct evidence that exposure to nerve agents has produced developmental toxicity in humans; data on exposure of pregnant women to nerve agents are unsurprisingly extremely scanty. Four pregnant women were exposed to sarin in the Tokyo exposure at 9–36 weeks' gestation (Ohbu *et al.*, 1997). All had normal offspring. Nevertheless, because of the profound effects of

OPs on neurotransmission, there is a strong theoretical basis for concern that effects which, in adults would be reversible, would in the embryo, fetus and neonate be irreversible. With pesticidal OPs, there are data in animals to suggest a potential for developmental neurotoxicity. Thus, studies have suggested the possibility that chlorpyrofos may have effects on developing organisms (Tang *et al.*, 1999; Qiao *et al.*, 2001) and a body of work is accumulating on OP pesticides in response to the development of a standardized developmental neurotoxicity test in the rat (US EPA, 1998; OECD, 2003; Fenner-Crisp *et al.*, 2005). However, formal developmental neurotoxicity tests on OP nerve agents have not been done and the effects have to be inferred from knowledge on OP pesticides [see reviews by Kitos and Suntornwat (1992), Slotkin (2005) and Makris (2005)]. There is little doubt that, with the appropriate pattern of exposure to pregnant animals, including humans, nerve agents would be developmental neurotoxicants. Further infants and children also have developing nervous systems and these individuals may be at greater risk of long-term effects on the nervous system than adults (Guzelian *et al.*, 1992). Nevertheless, with pesticides, it has been suggested that most exposures of pregnant women have not shown adverse effect on the offspring, presumably because exposure is of short duration and occurs only once (McElhatton, 1987; Minton and Murray, 1988) and it seems likely that that would be the case with nerve agents.

Delayed effects outside the CNS

The nerve agents are generally considered to be acute toxicants and their delayed effects have been relatively little investigated. Li *et al.* (1998) reported an increase in sister chromosome exchange (SCE) in the lymphocytes of victims of the Tokyo sarin disaster. Because of the probability that the victims were exposed to by-products of sarin synthesis, diisopropyl methylphosphonate, diethyl methylphosphonate and isopropyl ethyl methylphosphonate, these were also studied. The frequency of SCE was determined in human lymphocytes exposed to these by-products: all three compounds increased the frequency of SCE compared with controls. Sarin and soman did not produce unscheduled DNA synthesis in isolated rat hepatocytes but decreased repair synthesis was seen with sarin (Klein *et al.*, 1987). Decreased excision repair capacity has also been seen with the insecticide diazinon, and so this may be a more general effect of OPs.

EFFECT OF ROUTE OF EXPOSURE

Vapour

Onset is rapid and the eyes and respiratory system are most affected. Low-level exposure causes tightness of the chest, rhinorrhea and salivation. Dimming of vision due to miosis, eye pain and headache then follow. On examination, the pupils are constricted and the conjunctivae hyperaemic. These effects may last several hours after cessation of exposure and the headache and visual problems several days. In severe cases, salivation and rhinorrhea are more marked, and wheezing and dyspnea are prominent. Other effects, such as abdominal pain, vomiting, involuntary defecation and micturition, weakness, fasciculation and convulsion, follow depending on the degree of systemic absorption. Death may occur from respiratory failure.

The attack with sarin on the Tokyo Subway (underground railway) has added considerably to our knowledge of the clinical effects of sarin vapor. The symptoms observed were largely as expected, namely cough, difficulty in breathing and tightness of the chest, bradycardia and eye pain (Masuda *et al.*, 1995; Nozaki *et al.*, 1995; Ohbu *et al.*, 1997) (see Chapter 13).

Skin contamination

Local effects at the site of contamination include sweating and local fasciculation. Fasciculation may spread to involve whole muscle groups, while the onset of systemic symptoms and signs is generally slower than after vapour exposure.

Ingestion

Ingestion of nerve agent may occur from contaminated food or water. Colicky pain occurs, together with nausea, vomiting, diarrhoea and involuntary defecation, would be expected.

SUB-GROUPS PARTICULARLY LIABLE TO NERVE AGENTS

Most work on nerve agents has been directed at management of poisoning in military personnel, particularly soldiers; soldiers are generally physically fit and young. With the possibility of terrorist use of chemical weapons in public places, other groups need to be considered. Developmental neurotoxicity and nerve agents has already been discussed (see above) and in this respect (and possibly others) the fetus, infant and young child might represent a susceptible subgroup. It has already been noted that four pregnant women were exposed to sarin in the Tokyo disaster and that all had normal offspring. Nevertheless, it must be recognized that OP pesticides are embryo- and fetocidal and OP pesticide exposure of mothers has been associated with effects including cardiac defects in offspring (see review by Pelkonin *et al.*, 2005). In a mass-casualty situation with a single high-dose exposure, it is likely that the outcome for the embryo/fetus would be largely determined by the outcome for the mother.

The elderly also represent a potentially susceptible sub-group to toxicants (Stevenson, 1990) including nerve agents; the reasons for this are numerous but may include impaired organ function, most notably, in acute intoxication, poor cardiovascular and respiratory status. Furthermore, decreased hepatic and renal function may impair metabolism and excretion of nerve agents (and also antidotes). Other groups, those with chronic conditions such as diabetes, may represent challenges particularly in management.

In healthy people, variation in activity of metabolizing enzymes, including that due to polymorphisms, may cause differences in susceptibility. One enzyme that has received a lot of attention is paraoxonase 1 (PON1). The Q-type allozyme (Gln_{192}) of PON1 hydrolyzes sarin and soman more effectively than type R (Arg_{192}) (Costa *et al.*, 1999,2005). The clinical significance of the PON1 192 polymorphism remains to be established in respect of nerve agents (see reviews by Costa *et al.*, 2002,2003,2005). A connection between this polymorphism and Gulf war syndrome has been suggested based on epidemiological studies (Haley *et al.*, 1999) as it has with ill-health in UK sheep farmers and shepherds, using OP sheep dips (Cherry *et al.*, 2005). It should be noted that the PON1 192 polymorphism (and the 55 polymorphism) have been associated with ill-health outcomes independent of OP exposure, including coronary heart disease (Shih *et al.*, 2002); this may confound some of the studies mentioned above. There are several variants of butyrylcholinesterase, and individuals are known who completely lack plasma butyrylcholinesterase activity (Østergaard, 1992). It has been suggested that in nerve agent poisoning butyrylcholinesterase acts as a 'sink' for nerve agents. If that is the case it would be expected that those with low or absent butyrylcholinesterase activity would be more susceptible to nerve agents.

LABORATORY INVESTIGATIONS

Measurement of cholinesterase activity

The cornerstone of laboratory diagnosis of nerve agent poisoning is measurement of enzyme activity. Numerous methods are available for determination of both acetylcholinesterase and butyrylcholinesterase, most of which measure catalytic activity (see St Omer and Rottinghaus (1992) and Swaminathan and Widdop (2001). Of these, the Ellman method (Ellman *et al.*, 1961) and its subsequent modifications are most widely used. Plasma butyrylcholinesterase activity correlates badly with brain acetylcholinesterase activity and is best thought of as a marker of poisoning rather than a prognostic indicator, although in certain circumstances it may have a role in cholinergic neurotransmission (See review by Casida and Quistad, 2004). Even activity of the preferred red cell acetylcholinesterase correlates poorly with central nervous acetylcholinesterase activity, seriously limiting the use of the former to assess the severity of poisoning (Jimmerson *et al.*, 1989; Karalliedde, 2002). The reasons for this are complex. Butyrylcholinesterase is a different enzyme than acetylcholinesterase, with different substrate specificities and with different rates of inhibition, reactivation and ageing from acetylcholinesterase. Butyrylcholinesterase is predominantly synthesized in the liver and activity is affected by conditions such as liver disease and a large number of other factors, including the existence of inherited variants with different or even no enzymatic activity (Østergaard *et al.*, 1992; Swaminathan and Widdop, 2001). These factors present difficulties with using butyrylcholinesterase in the diagnosis and management of nerve agent poisoning (see discussion by Brock and Brock, 1993). The activity of butyrylcholinesterase in an individual poisoned with an OP is, at any time, a function of inhibition, spontaneous reactivation and resynthesis, with the added possibility of ageing, all occurring concurrently. It should be noted that the rate constants for these processes are different for butyrylcholinesterase and acetylcholinesterase in any organism, as the two enzymes are not the same gene product. Acetylcholinesterase activity is the sum of the same processes, but there are two important additional considerations, i.e. (1) it is the same gene product as acetylcholinesterase in the nervous system (the enzyme of interest), and (2) resynthesis cannot occur in the erythrocyte and the $t_{1/2}$ for recovery here is consequently the same as the half-life of the red blood cell, while resynthesis can occur at other sites, and may not be insubstantial (Wehner *et al.*, 1985). Additionally, the enzymes in the plasma and red blood cell may be more accessible to inhibitor than that in the nervous system, especially the central nervous system (for discussion, see Marrs, 2001 and Karalliedde, (2002). It is unlikely that a peak erythrocyte acetylcholinesterase depression of 30% or less would produce clinical signs or symptoms, and below 40% any signs would be expected to be very mild.

George *et al.* (2003) have studied the use of antisera to distinguish between native and inhibited acetylcholinesterase, providing a basis for measuring biomarkers of OP, including nerve agent, exposure.

Alkylphosphates

OPs, both pesticides and nerve agents, lose their leaving groups when they react with acetylcholinesterase. Hydrolysis of the reaction product produces an alkylphosphate or alkylphosphonate. Methods have been developed for measuring the VX–acetylcholinesterase hydrolysis product, *O*-ethyl methylphosphonic acid, and the sarin–acetylcholinesterase hydrolysis product, *O*-isopropoxy methylphosphonic acid (IMPA) (Noort *et al.*, 1998) (for review, see Noort *et al.*, 2002). IMPA and methylphosphonic acid were detected in patients from the Matsumoto sarin exposure (Nakajima *et al.*, 1998). An alternative is to release isopropyl methylphosphonylserine or methylphosphonoserine from the active site of inhibited erythrocytic acetylcholinesterase and measure the sarin hydrolysis product and this was done on some victims of the Tokyo sarin exposure (Nagao *et al.*, 1997). Another approach is to release the inhibitor bound to butyrylcholinesterase within the enzyme–inhibitor complex, using fluoride and measure the resulting phosphofluoridate using gas chromatography (van der Schans *et al.*, 2004).

Black *et al.* (1999) found that sarin and soman bind to a tyrosine residue in human plasma and suggested that this could form the basis of a biomarker of exposure to these agents.

DIAGNOSIS POST-MORTEM

There may be signs of the acute cholinergic crisis preceding death, e.g. excess secretions in the form of foam around the nose. Furthermore, devices may have detected nerve agents in the atmosphere or on surfaces. Diagnosis of acute nerve agent poisoning at autopsy would be made difficult because of the paucity of histopathological changes that would be present in the nervous system. Cerebral edema may be present. One of the few histopathological changes to be expected *post-mortem* would be changes in skeletal muscle, sometimes called 'segmental myopathy'. If biological fluids could be obtained, marked inhibition of acetyl- and butyrylcholinesterase activity would be present and, depending on the nerve agent involved, urinary alkylphosphates might be present if the patient has survived any length of time (see above). Histochemistry has been used to diagnose OP pesticide poisoning (Petty, 1958) and could doubtless be used to detect nerve agent poisoning, although it is, at most, semiquantitative (Marrs and Bright, 1992). Methods may be available to measure the nerve agent itself in body fluids. Measurement of cholinesterase activity and demonstration of inhibition is not, of course, specific for particular types of OP, whereas alkylphosphates may be (see review by Ballantyne, 1992).

REFERENCE DOSES/HEALTH-BASED GUIDANCE VALUES FOR NERVE AGENTS

In the past, the notion of reference doses for nerve agents would have been thought a little preposterous. However, chronic reference doses/acceptable daily intakes have been established for chemical warfare agents, *inter alia* the nerve agents, soman, sarin, tabun and VX for exposure by the oral route. This was at the behest of the US Army, which asked the National Research Council to review the US Army's interim reference doses (Bakshi *et al.*, 2000; Subcommittee on Chronic Reference Doses for Selected Chemical-Warfare Agents, 2000a–g). This was chiefly a result of the consideration of environmental contamination at various sites in the USA, but such reference doses could also be useful in deciding detection limits that would be needed where food was contaminated with nerve agents, either accidentally or deliberately. As many effects of nerve agents are not route- or pathway-specific, these reference doses could probably be applied to other routes of exposure. The data used to calculate these reference doses, in comparison with dossiers available for regulated substances, such as pesticides, are deficient. There was a lack of long-term studies, and with some nerve agents, multigeneration and developmental toxicity studies. Furthermore, there were problems with study design, in terms of numbers of animals and good laboratory practice. Allowance was made for these problems by the use of extra uncertainty factors.

CONCLUSIONS

The organophosphate nerve agents continue to pose major problems in chemical defence. Important measures in prevention of military casualties include adequate detection measures, physical protection, chemical prophylaxis and proper treatment. In the terrorist context, proper emergency planning is key. Treatment of poisoning by organophosphorus nerve agents is dealt with in other chapters in this volume.

REFERENCES

Aldridge WN (1950). Some properties of specific cholinesterase with particular reference to the mechanism of inhibition of diethyl *p*-nitrophenyl thiophosphate (E605) and analogues. *Biochem J*, **46**, 451–460.

Aldridge WN, Barnes JM and Johnson MK (1969). Studies on delayed neuropathy produced by some organophosphorus compounds. *Ann NY Acad Sci*, **160**, 314–322.

Anderson RJ and Dunham CB (1985). Electrophysiologic changes in peripheral nerve following repeated exposure to organophosphorus agents. *Arch Toxicol*, **58**, 97–101.

Anderson DR, Harris LW, Chang F-C *et al.* (1997). Antagonism of soman-induced convulsions by midazolam, diazepam and scopolamine. *Drug Chem Toxicol*, **20**, 115–131.

Annau Z (1992). Neurobehavioral effects of organophosphorus compounds. *Organophosphates, Chemistry, Fate and Effects* (JE Chambers and PE Levi, eds), pp. 419–432. San Diego, CA, USA: Academic Press.

Anthony JS, Haley M, Manthei J *et al.* (2004). Inhalation toxicity of cyclosarin (GF) vapour in rats as a function of exposure concentration and duration: potency comparison to sarin (GB). *Inhalation Toxicol*, **16**, 103–111.

Anzueto A, Berdine GG, Moore GT *et al.* (1986). Pathophysiology of soman intoxication in primates. *Toxicol Appl Pharmacol*, **86**, 56–68.

Ariens AT, Wolthuis OL and van Bentham RMJ (1969). Reversible necrosis at the end plate region in striated muscles of the rat poisoned with cholinesterase inhibitors. *Experientia*, **1**, 57–59.

Bajgar J, Ševelová L, Krejčová G *et al.* (2004). Biochemical and behavioural effects of soman vapors in low concentrations. *Inhal Toxicol*, **16**, 497–507.

Bakri NMS, El-Rashidi AH, Eldefrawi AT *et al.* (1988). Direct actions of organophosphate anticholinesterases on nicotinic and muscarinic actylcholine receptors. *J Biochem Toxicol*, **3**, 235–239.

Bakshi KS, Pang SNJ and Snyder R (2000). Introduction to the special issue. *J Toxicol Environ Health A*, **59**, 283–288.

Ballantyne B (1992). Forensic aspects of acute anticholinesterase poisoning. In: *Clinical and Experimental Toxicology of Organophosphates and Carbamates* (B Ballantyne and TC Marrs, eds), pp. 618–622. Oxford: Butterworth-Heinemann.

Ballantyne B and Marrs TC (1992). Overview of the biological and clinical aspects of organophosphates and carbamates. In: *Clinical and Experimental Toxicology of Organophosphates and Carbamates* (B Ballantyne and TC Marrs, eds), pp. 1–14. Oxford: Butterworth-heinemann.

Barak D, Ordentlich A, Kaplan D *et al.* (2000). Evidence for P–N bond scission in phosphoramidate nerve agent adducts of human acetylcholinesterase. *Biochemistry*, **8**, 1156–1161.

Beesley WN (1994). Sheep dipping, with special reference to the UK. *Pesticides Outlook*, February, 16–29.

Benschop HP, Bijleveld EC, de Jong LPA *et al.* (1987). Toxicokinetics of the four stereoisomers of the nerve agent soman in atropinized rats – influence of a soman simulator. *Toxicol Appl Pharmacol*, **90**, 490–500.

Berry WK and Davies DR (1970). Use of carbamates and atropine in the protection of animals against poisoning by 1,2,2-trimethylpropyl phosphonofluoridate. *Biochem Pharmacol*, **19**, 927–934.

Black RM, Harrison JM and Read RW (1999). The interaction of sarin and soman with plasma proteins: the identification of a novel phosphonylation site. *Arch Toxicol*, **73**, 123–126.

Bosković B, Kovacević V and Jovanović D (1984). 2-PAM chloride, H16 and HGG 12 in soman and tabun poisoning. *Fund Appl Toxicol*, **4**, 106–115.

Bourne Y, Taylor P and Marchot P (1995). Acetylcholinesterase inhibition by fasciculin, crystal structure of the complex. *Cell*, **83**, 503–512.

Bourne Y, Taylor P, Bougis PE *et al.* (1999). Crystal structure of mouse acetylcholinesterase. A peripheral site-occluding loop in a tetrameric assembly. *J Biol Chem*, **274**, 2963–2970.

Bowman WC (1993). Physiology and pharmacology of neuromuscular transmission, with special reference to the possible consequences of prolonged blockade. *Intensive Care Med*, **19**, S45–S53.

Boyd RT (1997). The molecular biology of neuronal nicotinic acetylcholine receptors. *Crit Rev Toxicol*, **27**, 299–318.

Braitman DJ and Sparenborg S (1989). MK-801 protects against seizures induced by the cholinesterase inhibitor soman. *Brain Res Bull*, **23**, 145–148.

Brezenoff HE, McGee J and Knight V (1984). The hypertensive response to soman and its relation to brain acetylcholinesterase inhibition. *Acta Pharmacol Toxicol*, **55**, 270–277.

Bright JE, Inns RH, Tuckwell NJ *et al.* (1991). A histochemical study of changes observed in the mouse diaphragm after organophosphate poisoning. *Human Exp Toxicol*, **10**, 9–14.

Brimblecombe RW (1974). *Drug Actions in Cholinergic System*, pp. 64–132. New York: MacMillan.

Brimblecombe RW, Green DM, Stratton JA *et al.* (1970). The protective actions of some anticholinergic drugs in sarin poisoning. *Brit J Pharmacol*, **39**, 822–830.

Britt JO, Martin JL, Okerberg CV *et al.* (2000). Histopathologic changes in the brain, heart and skeletal muscle of rhesus macaques ten days after exposure to soman (an organophsophorus nerve agent). *Comp Med*, **50**, 133–139.

Brock A and Brock V (1993). Factors affecting interindividual variation in human plasma cholinesterase activity: body weight, height, sex, genetic polymorphisms and age. *Arch Environ Contam Toxicol*, **24**, 93–99.

Brown MA and Brix KA (1998). Review of health consequences from high-, intermediate- and low-level exposure to organophosphorus nerve agents. *J Appl Toxicol*, **18**, 393–408.

Burchfiel JL, Duffy FH and Sim van M (1976). Persistent effects of sarin and dieldrin upon the primate electroencephalogram. *Toxicol Appl Pharmacol*, **35**, 365–379.

Burgen ASV and Hobbiger F (1951). The inhibition of cholinesterases by alkylphosphates and alkylphenophosphates. *Br J Pharmacol Chemother*, **6**, 593–605.

Cabal J, Kuca K and Kassa J (2004). Specification of the structure of oximes able to reactivate tabun-inhibited acetylcholinesterase. *Basic Clin Pharmacol Toxicol*, **95**, 81–86.

Casida JE and Quistad GB (2004). Organophosphate toxicology: safety aspects of nonacetylcholinesterase secondary targets. *Chem Res Toxicol*, **17**, 983–997.

Chambers HW (1992). Organophosphorus compounds: an overview. In: *Organophosphates, Chemistry, Fate and Effects*. (JE Chambers and PE Levi, eds), pp. 3–18. San Diego, CA, USA: Academic Press.

Chang F-CT, Foster RE, Beers ET *et al.* (1990). Neurophysiological concomitants of soman-induced respiratory depression in awake, behaving guinea pigs. *Toxicol Appl Pharmacol*, **102**, 233–250.

Chatthopadhay DP, Dighe SK, Nashikkar AB *et al.* (1986). Species differences in the *in vitro* inhibition of grain acetylcholinesterase and carboxyl esterase by mipafox, paraoxon and soman. *Pest Biochem Physiol*, **26**, 202–208.

Chechabo SR, Santos MD and Albuquerque EX (1999). The organophosphate sarin, at low concentrations, inhibits the evoked release of GABA in rat hippocampal slices. *Neurotoxicology*, **20**, 871–882.

Cherry NM, Durrington PN, Mackness B *et al.* (2005). *Genetic variation in susceptibility to chonic effects of organophosphate exposure*. Health and Safety Executive Research Report 408. London, Edinburgh, Belfast, UK: Her Majesty's Stationary Office.

Clement JG (1984). Role of aliesterase in organophosphate poisoning. *Fund Appl Toxicol*, **4**, S96–S105.

Clement JG (1992). Efficacy of various oximes against GF (cyclohexylmethylphosphonofluoridate) poisoning in mice. *Arch Toxicol*, **66**, 143–144.

Clement JG, Hand BT and Shiloff JD (1981). Differences in the toxicity of soman in various strains of mice. *Fund Appl Toxicol*, **1**, 419–420.

Coleman IW, Little PE, Patton GE *et al.* (1966). Cholinolytics in the treatment of anticholinesterase poisoning IV. The effectiveness of five binary combinations of cholinolytics with oximes in the treatment of organophosphorus poisoning. *Can J Physiol Pharmacol*, **44**, 743–764.

Coleman IW, Patton GE and Bannard RA (1968). Cholinolytics in the treatment of anticholinesterase poisoning V. The effectiveness of parpanit with oximes in the treatment of organophosphorus poisoning. *Can J Physiol Pharmacol*, **46**, 109–117.

Committee on Toxicity of Products in Food, Consumer Products and the Environment (1999). *Organophosphates*, Report of the Committee on Toxicity of Products in Food, Consumer Products and the Environment. London, UK: Department of Health.

Costa LG, Li WF, Richter RJ *et al.* (1999). The role of paraoxonase (PON1) in the detoxication of organophosphates and its human polymorphism. *Chem-Biol Interact*, **119/120**, 429–438.

Costa LG, Li W-F, Richter RJ *et al.* (2002). PON1 and organophosphate toxicity. In: *Paraoxonase (PON1) in Health and Disease: Basic and Clinical Aspects* (LG Costa and CE Furlong, eds), pp. 165–184. Dordrecht, The Netherlands: Kluwer Academic Publishers.

Costa LG, Cole TB and Furlong CE (2003). Polymorphisms of paraoxonase (PON1) and their significance in clinical toxicology of organophosphates. *J Toxicol Clin Toxicol*, **41**, 37–45.

Costa LG, Cole TB, Vitalone A *et al.* (2005). Paraoxonase polymorphisms and toxicity of organophosphates. In: *Toxicology of Organophosphate and Carbamate Compounds*. (R Gupta, ed.) pp. 247–255. San Diego, CA, USA: Academic Press.

Crouzier D, Baille le Crom V, Four E *et al.* (2004). Disruption of mice sleep stages induced by low doses of organophosphorus compound soman. *Toxicology*, **199**, 59–71.

Crowell JA, Parker RM, Bucci TJ *et al.* (1989). Neuropathy terget esterase in hens after sarin and soman, *J Biochem Toxicol*, **4**, 15–20.

Curtil C and Masson P (1993). Le vieillissement des cholinesterase après inhibition par les organophosphorés. *Ann Pharmacoceut Franç*, **51**, 63–77.

Denk JR (1975). *Effect of GB on mammalian germ cells and reproductive parameters*, Final report EB-TR-74087, AD-A006503. Aberdeen Proving Ground, Fort Detrick, MD, USA: US Army Medical Research and Development Command.

DiGiovanni C (1999). Domestic terrorism with chemical or biological agents: psychiatric aspects. *Am J Psychiatr*, **156**, 1500–1505.

Dittmann E and Wiegand C (2005). Cyanobacterial toxins – occurrence, biosynthesis and impact on human affairs. *Mol Nutr Food Res*, **50**, 7–17.

D'Mello GD (1993). Behavioural toxicity of anticholinesterases in humans and animals – a review. *Human Exp Toxicol*, **12**, 3–7.

D'Mello GD and Duffy EAM (1985). The acute toxicity of sarin in marmosets (*Callithrix jacchus*): a behavioral analysis. *Fund Appl Toxicol*, **5**, S169–S174.

Duffy FH, Burchfiel JL, Bartels PH *et al.* (1979). Long-term effects of an organophosphate upon the human electroencephalogram. *Toxicol and Appl Pharmacol*, **47**, 161–176.

Eglen RM and Whiting RL (1990). Heterogeneity of vascular muscarinic receptors. *J Autonom Pharmacol*, **10**, 233–245.

Ekstrom F, Akfur C, Tunemalm AK *et al.* (2006). Structural changes of phenylalanine 338 and histidine 447 revealed by the crystal structures of tabun-inhibited murine acetylcholinesterase. *Biochemistry*, **10**, 74–81.

Eldefrawi AT, Jett D and Eldefrawi ME (1992). Direct actions of organophosphorus anticholinesterases on muscarinic receptors. In: *Organophosphates, Chemistry, Fate and Effects* (JE Chambers and PE Levi, eds), pp. 268–323. San Diego, CA, USA: Academic Press.

Ellman GL, Courtney D and Andres V (1961). A new and rapid colorimetric determination of acetylcholinesterase activity. *Biochem Pharamcol*, **7**, 88–95.

Feldman RG (1999). *Occupational and Environmental Neurotoxicology*, p. 432. Philadelphia, PA, USA: Lippincott-Raven.

Fenner-Crisp P, Adams J, Balbus J *et al.* (2005). Application of developmental neurotoxicity testing to public health protection. *Neurotoxicol Teratol*, **27**, 371.

Filliat P, Baubichon D, Burckhart M-F *et al.* (1999). Memory impairment after soman intoxication in rat: correlation with central neuropathology. Improvement with anticholinergic and antiglutamatergic therapeutics. *Neurotoxicol*, **20**, 535–550.

Fleisher JH, Harris LW, Prudhomme C *et al.* (1963). Effects of *p*-nitrophenyl phosphonate (EPN) on the toxicity of isopropyl methyl phosphonofluoriate (GB). *J. Pharmacol Exp Ther*, **139**, 390–396.

Fletcher HP, Noble S and Spratto GR (1989). Effect of the acetylcholinesterase inhibitor, soman, on plasma levels of endorphin and adrenocorticotrophic hormone (ACTH). *Biochem Pharmacol*, **38**, 2045–2046.

Fosbraey P, Wetherell JR and French MC (1990). Neurotransmitter changes in guinea-pig brain regions following soman intoxication. *J Neurochem*, **54**, 72–79.

Fredriksson T (1957). Pharmacological properties of methyl fluorophosphomylcholines – two synthetic cholinergic drugs. *Arch Int Pharmacol Ther*, **113**, 101–104.

Ganong WF (1991). *Review of Medical Physiology*, 15th Edition. Paramus, NJ, USA: Prentice Hall International Ltd.

Gates M and Renshaw BC (1946). Fluorophosphates and other phosphorus-containing compounds. In: *Summary Technical Report of Division 9*, Vol. 1, Parts I, II, pp. 131, 155. Washington, DC, USA: Office of Scientific Research and Development.

George KM, Schule T, Sandoval LE *et al.* (2003). Differentiation between acetylcholinesterase and the organophosphate-inhibited form using antibodies and the correlation of antibody recognition with reactivation mechanism and rate. *J Biol Chem*, **278**, 45512–45518.

Glynn P (2000). Neural development and neurodegeneration: two faces of neuropathy target esterase. *Prog Neurobiol*, **61**, 61–74.

Goldman M, Wilson BW, Kawakami TG *et al.* (1988). *Toxicity studies on agent VX*, Final Report, DTIC AD-A201397. Fort Detrick, MD, USA: US Army Medical Research and Development Command.

Goldstein BD, Fincher DR and Searle JR (1987). Electrophysiological changes in the primary sensory neuron following subchronic soman and sarin: aletations in sensory receptor function. *Toxicol Appl Pharmacol*, **91**, 55–64.

Gordon JJ, Inns RH, Johnson MK *et al.* (1983). The delayed neuropathic effects of nerve agents and some other organophosphate compounds. *Arch Toxicol*, **51**, 71–82.

Gordon JJ and Leadbeater LL (1977). The prophylatic use of (1-methyl 2-hydroxy-iminomethylpyridinium methanesulfonate) (P2S) in the treatment of organophosphate poisoning. *Toxicol Appl Pharmacol*, **40**, 109–114.

Gray AP (1984). Design and structure–activity relationships of antidotes to organophosphorus anticholinesterase agents. *Drug Metab Rev*, **15**, 557–589.

Gray PJ and Dawson RM (1987). Kinetic constants for the inhibition of eel and rabbit brain acetylcholinesterase by some organophosphates and carbamates of military significance. *Toxicol Appl Pharmacol*, **91**, 140–144.

Grob D and Harvey AM (1953). The effects and treatment of nerve gas poisoning. *Am J Med*, **14**, 52–63.

Grob D and Harvey AM (1958). Effects in man of the anticholinesterase compound sarin (isopropyl methyl phosphonofluoridate). *J Clin Invest*, **37**, 350–368.

gulflink (2005). http:/www.gulflink.osd.mil/library/randrep/mr1018.5.appb.pdf, accessed 14th December 2005.

Gupta RC, Patterson GT and Dettbarn W-D (1987a). Acute tabun toxicity; biochemical and histochemical consequences in brain and skeletal muscles of rats. *Toxicology*, **46**, 329–341.

Gupta RC, Patterson GT and Dettbarn W-D (1987b). Biochemical and histochemical alterations following acute soman intoxication in the rat. *Toxicol Appl Pharmacol*, **87**, 393–402.

Guzelian PS, Henry CJ and Olin SS (1992). *Similarities and Differences between Children and Adults: Implications for Risk Assessment*. Washington, DC, USA: ILSI Press.

Haley RW (2000). PON1 and low dose sarin in marmosets. *J Psychopharm*, **14**, 87–88.

Haley RW, Billecke S and La Du BN (1999). Association of low PON1 type Q (type A) arylesterase activity with neurological symptom complexes in Gulf war veterans. *Toxicol Appl Pharmacol*, **157**, 227–233.

Harris LW, Heyl WC, Stitcher DL *et al.* (1978). Effects of 1,1′-oxydimethylene bis(4-tert-butylpyridinium chloride (SAD-128) and decamethonium on reactivation of soman and sarin-inhibited cholinesterase by oximes. *Biochem Pharmacol*, **27**, 757–761.

Hatta K, Miura Y, Asukai N *et al.* (1996). Amnesia from sarin poisoning. *Lancet*, **347**, 1343.

Heilbronn E (1963). *In vitro* reactivation and aging of tabun-inhibited blood cholinesterases: studies with *N*-methylpyridinium-2-aldoxime methane sulphonate and *N*, *N'*-trimethylene bis(pyridinium-4-aldoxime) dibromide. *Biochem Pharmacol*, **12**, 25–36.

Henderson JD, Higgins RJ, Dacre JC *et al.* (1992). Neurotoxicity of acute and repeated treatments of tabun, paraoxon, diisopropyl fluorophosphate and isofenphos to the hen. *Toxicology*, **72**, 117–129.

Holmes JH and Gaon MD (1956). Observations on acute and multiple exposure to anticholinesterase agents. *Trans Am Clin Chem Assoc*, **68**, 86–103.

Hughes JN, Knight R, Brown RFR *et al.* (1991). Effects of experimental sarin intoxication on the morphology of the mouse diaphragm: a light and electron microscopical study. *Int J Exp Path*, **72**, 195–209.

Inns RH, Tuckwell NJ, Bright JE *et al.* (1990). Histochemical demonstration of calcium accumulation in muscle fibres after experimental organophosphate poisoning. *Human Exp Toxicol*, **9**, 245–250.

Institute of Medicine (1996). *Health Consequences of Service during the Persian Gulf War: Recommendations for Research and Information Systems.* Washington, DC, USA: National Academy Press.

Jamal GA (1995). Long-term neurotoxic effects of organophosphate compounds. *Adv Drug React Toxicol Rev*, **14**, 85–99.

Jimmerson VR, Shih T-M and Mailman RB (1989). Variability in soman toxicity in the rat: correlation with biochemical and behavioral measures. *Toxicology*, **57**, 241–254.

Johnson MK (1975). Organophosphorus esters causing delayed neurotoxic effects. *Arch Toxicol*, **34**, 259–288.

Johnson MK, Willems JL, de Bisschop HC *et al.* (1985). Can soman cause delayed neuropathy? *Fund Appl Toxicol*, **5**, S180–S181.

Jones SV (1993). Muscarinic receptor subtypes: modulation of ion channels. *Life Sci*, **52**, 457–464.

Jovanović D (1982). The effect of bis-pyridirium oximes on neuromuscular blockade induced by highly toxic organophosphates in the rat. *Arch Int Pharmacol Ther*, **262**, 231–241.

Kaminski HJ and Ruff RL (1999). Structure and kinetic properties of the acetylcholine receptor. In: *Myasthenia and Myasthenic Disorders* (AG Engel, ed.), pp. 40–64. New York, Oxford University Press.

Karalliedde L (2002). Cholinesterase estimations revisited: the clinical relevance. *Eur J Anaesthesiology*, **19**, 313–316.

Karalliedde L, Wheeler H, Maclehose R *et al.* (2000). Possible intermediate and long-term health effects following exposure to chemical warfare agents. *Public Health*, **114**, 238–248.

Karalliedde L, Baker D and Marrs TC (2006). Organophosphate-induced intermediate syndrome: its aetiology and the relationship with myopathy. *Toxicol Rev*, **25**, 1–14.

Kassa J and Cabal J (1999). A comparison of the efficacy of actylcholinesterase reactivators against cyclohexylmethylphosphonofluoridate (GF agent) by *in vitro* and *in vivo* methods. *Pharmacol Toxicol*, **84**, 41–45.

Kassa J, Krejčová G, Skopek F *et al.* (2004). The influence of sarin on various physiological functions in rats following single or repeated low-level exposure. *Inhal Toxicol*, **16**, 517–530.

Kitos PA and Suntornwat O (1992). Teratogenic effect of organophosphorus compounds. In: *Organophosphates, Chemistry, Fate and Effects* (JE Chambers and PE Levi, eds), pp. 387–417. San Diego, CA, USA: Academic Press.

Klein AK, Nasr ML and Goldman M (1987). The effects of *in vitro* exposure to the neurotoxins sarin (GB) and soman (GD) on unscheduled DNA synthesis by rat hepatocytes. *Toxicol Lett*, **38**, 239–249.

Koelle GB (1992). Pharmacology and toxicology of organophosphates. In: *Clinical and Experimental Toxicology of Organophosphates and Carbamates* (B Ballantyne and TC Marrs eds), pp. 33–37. Oxford, UK: Butterworth-Heinemann.

Koplovitz I, Gresham VC, Dochterman LW *et al.* (1992). Evaluation of the toxicity, pathology and treatment of cyclohexylmethylphosphonofluoridate (CMPF) poisoning in rhesus monkeys. *Arch Toxicol*, **66**, 622–628.

Korsak RJ and Sato MM (1977). Effects of chronic organophosphate pesticide exposure on the central nervous system. *Clin Toxicol*, **11**, 83–95.

Kovacic P (2003). Mechanism of organophosphates (nerve gases and pesticides) and antidotes: electron transfer and oxidative stress. *Curr Med Chem*, **16**, 2705–2709.

La Borde JB and Bates HK (1986). *Developmental toxicity study on agent GB-DCSM types I and II in CD rats and NZW rabbits*, Final Report. Fort Detrick, MD, USA: US Army Medical Research and Development Command.

Lader M (2001). The effects of organophosphates on neuropsychiatric and psychological function. In:

Organophosphates and Health (L Karalliedde, S Feldman, J Henry and T Marrs, eds), pp. 175–198. London, UK: Imperial College Press.

Lau W-M, Freeman SE and Szilagyí M (1988). Binding of some organophosphorus compounds at adenosine receptors in guinea pig brain membranes. *Neurosci Lett*, **94**, 125–130.

Lawler HC (1961). Turnover time of actylcholinesterase. *J Biol Chem*, **236**, 2296–2301.

Lebeda FJ, Wierwille RC, VanMeter WG *et al.* (1988). Acute ultrastructural alterations in duced by soman and hypoxia in rat hippocampal CA3 pyramidal neurones. *NeuroToxicol*, **9**, 9–22.

Leblic C, Cox HM and le Moan L (1984). Etude de la toxicité, de l'eserine, VX et le paraoxon, pour établir un modéle mathematique de l'extrapolation à être humain. *Arch Blg Med Soc Hyg Trav Med* (Supplement), 226–242.

le Chêne E (1989). *Chemical and Biological warfare – Threat of the Future*, Mackenzie Paper. Toronto, Canada: The Mackenzie Institute.

Lefkowitz RJ, Hoffnman BB and Taylor P (1996). Drugs acting at synaptic and neuroeffector junctional sites. In: *Goodman and Gilman's Pharmacological Basis of Therapeutics*, 9th Edition (JG Hardman, LE Limbird, PB Molinoff, RW Rudden and AG Gilman, eds), pp. 105–140. New York, NY, USA: McGraw-Hill.

Leonard JP and Salpeter MM (1979). Agonist induced myopathy at the neuromuscular junction is mediated by calcium. *J Cell Biol*, **82**, 811–819.

Li Q, Minami M, Clement JG *et al.* (1998). Elevated frequency of sister chromatid exchanges in lymphocytes of victims of the Tokyo sarin disaster and in experiments exposing lymphocytes to by-products of sarin synthesis. *Toxicol Lett*, **98**, 95–103.

Ligtenstein DA (1984). On the synergism of the cholinesterase reactivating bispyridinium-aldoxime HI-6 and atropine in the treatment of organophosphate intoxications in the rat, *MD Thesis*. Leyden, the Netherlands: University of Leyden.

Lipp SA (1972). Effect of diazepam upon soman induced seizure activities and convulsions. *Electroenceph Clin Neurophysiol*, **32**, 557–560.

Lockridge O and Schopfer LM (2005). Biomarkers of organophosphate exposure. In: *Toxicology of Organophosphate and Carbamate Compounds* (R Gupta, ed.), pp. 703–711. San Diego, CA, USA: Academic Press.

Lotts von K (1960). Zur Toxikologie and pharmakologie organischer Phosphosäurester. *Dtsch Gesundheitsweren*, **15**, 2179–2133.

Ludomirsky A, Klein HO and Sarelli P (1982). Q-T Prolongations and polymorphous ('torsade de pointes') ventricular arrhythmias associated with organophosphorus insecticide poisoning. *Am J Cardiol*, **49**, 1654–1658.

Lundy PM, Hansen AS, Hand BT *et al.* (1992). Comparison of several oximes against poisoning by soman, tabun and GF. *Toxicology*, **72**, 99–105.

Main AR and Iverson F (1966). Measurement of the affinity and phosphorylation onstants governing irreversible inhibition of cholinesterases by diisopropyl phosphoofluoridate. *Biochem J*, **100**, 525–531.

Makris S (2005). Regulatory considerations in developmental neurotoxicity of organophosphorus and carbamate pesticides In: *Toxicology of Organophosphate and Carbamate Compounds* (R Gupta, ed.), pp. 633–642. San Diego, CA, USA: Academic Press.

Maksimović M, Bosković B, Rodović L *et al.* (1980). Antidotal effects of bis-pyridinium 2-mono oxime carbonyl derivatives in intoxication with highly toxic organophosphorus compounds. *Acute Pharm Jagoslav*, **30**, 151–160.

Marrs TC (2001). Organophosphates: history, chemistry, pharmacology. In: *Organophosphates and Health* (L Karalliedde, S Feldman, J Henry and T Marrs, eds), pp. 1–37. London, UK: Imperial College Press.

Marrs TC and Bright JE (1992). Histochemical localization of cholinesterase in anticholinesterase poisoning. In: *Clinical and Experimental Toxicology of Organophosphates and Carbamates* (B Ballantyne and TC Marrs, eds), pp. 28–43. Oxford, UK: Butterworth-Heinemann.

Marrs TC and Maynard RL (1994). Neurotoxicity of chemical warfare agents. In: *Handbook of Clinical Neurology*, Part 1, Volume 64. (PJ Vinken, GW Bruyn and FA de Wolff, eds), pp. 223–247. Amsterdam, The Netherlands: Elsevier.

Marrs TC and Maynard RL (2001). Organophosphorus chemical warfare agents. In: *Organophosphates and Health*. (L Karalliedde, S Feldman, J Henry and T Marrs, eds), pp. 83–108. London, UK: Imperial College Press.

Marrs TC, Bright JE, Inns RH *et al.* (1990). Histochemical demonstration of calcium influx into mouse diaphragms induced by sarin. *Toxicologist*, **10**, 132.

Marrs TC, Karalliedde L and Baker D (2005). Intermediate Syndrome, does it exist? Can it be prevented?. *Toxicology*, **213**, 208.

Martin LJ, Doebbler JA, Shih T-M *et al.* (1985). Protective effect of diazepam pretreatment on soman-induced brain lesion formation. *Brain Res*, **325**, 287–289.

Masuda N, Takatsu M, Morinari H *et al.* (1995). Sarin poisoning on the Tokyo subway. *Lancet*, **345**, 1446.

Maxwell DM and Lenz DE (1992). Structure–activity relationships and anticholinesterase activity. In: *Clinical and Experimental Toxicology of Organophosphates and Carbamates*. (B Ballantyne and TC Marrs, eds), pp. 47–58. Oxford, UK: Butterworth-Heinemann.

Maynard RL (1999). Toxicology of chemical warfare agents. In: *General and Applied Toxicology*. (B Ballantyne, TC Marrs and T Syversen, eds), pp. 2079–2109. Basingstoke, UK: MacMillan.

Maynard RL and Beswick FW (1992). Organophosphorus compounds as chemical warfare agents. In: *Clinical and Experimental Toxicology of Organophosphates and Carbamates*. (B Ballantyne and TC Marrs, eds), pp. 373–385. Oxford, UK: Butterworth-Heinemann.

McDonough JH, Jaax NK, Crowley RA *et al.* (1989). Atropine and/or diazepam therapy protects against soman-induced neural and cardiac pathology. *Fund Appl Toxicol*, **13**, 256–276.

McDonough JH, Smith RF and Smith CD (1986). Behavioral correlates of soman-induced neuropathology: deficits in DRL acquisition. *Behav Toxicol Teratol*, **8**, 179–187.

McElhatton P (1987). Personal communication cited in Minton NA and Murray VSG (1988). A review of organophosphate poisoning. *Med Toxicol*, **3**, 350–375.

McLeod CG (1985). Pathology of nerve agents: perspectives on medical management. *Fund Appl Toxicol*, **5**, S10–S16.

Millard CB, Kryger G, Ordentlich A *et al.* (1999). Crystal structures of aged phosphonylated acetylcholinesterase: nerve agent reaction products at the atomic level. *Biochemistry*, **38**, 7032–7039.

Minton NA and Murray VSG (1988). A review of organophosphate poisoning. *Med Toxicol*, **3**, 350–375.

Misulis KE, Clinton ME, Dettbarn W-D *et al.* (1987). Differences in central and peripheral neural actions between soman and diisopropyl fluorophosphate, organophosphorus inhibitors of acetylcholinesterases. *Toxicol Appl Pharmacol*, **89**, 391–398.

Mobley PL (1990). The cholinesterase inhibitor soman increases inositol triphosphate in rat brain. *Neuropharmacology*, **29**, 189–191.

Moore DH (1998). Health effects of exposure to low doses of nerve agent – a review of present knowledge. *Drug Chem Toxicol*, **21**(Supplement 1), 123–130.

Munro NB, Talmage SS, Griffin GD *et al.* (1999). The sources, fate, and toxicity of chemical warfare agent degradation products. *Environ Health Perspect*, **107**, 933–974.

Murata K, Araki S, Yokoyama K *et al.* (1997). Asymptomatic sequelae to acute sarin poisoning in the central and autonomic nervous system 6 months after the Tokyo subway attack. *J Neurol*, **244**, 601–666.

Nagao M, Takatori T, Matsuda Y *et al.* (1997). Definitive evidence for the acute sarin poisoning diagnosis in the Tokyo subway. *Toxicol Appl Pharmacol*, **144**, 198–203.

Nakajima T, Sato S, Morita H *et al.* (1997). Sarin poisoning of a rescue team in the Matsumoto sarin incident in Japan. *Occ Environ Med*, **54**, 697–701.

Nakajima T, Sasaki K, Ozawa H *et al.* (1998). Urinary metabolites of sarin in a patient of the Matsumoto sarin incident. *Arch Toxicol*, **72**, 601–603.

Naseem SM (1990). Effect of organophosphates on dopamine and muscarinic receptor binding in rat brain *Biochem Int*, **20**, 799–806.

National Academy of Sciences (1982). *Possible Long-Term Health Effects of Short-Term Exposure to Chemical Agents*, Volume 1, *Anticholinesterases and Anticholinergics*. Washington, DC, USA: National Academy Press.

Nishiwaki Y, Maekawa K, Ogawa Y *et al.* (2001). Effects of sarin on the nervous system in resuce team staff members and police officers 3 years after the Tokyo sarin attack. *Environ Health Perspect*, **109**, 1169–1173.

Nohara M and Segawa K (1996). Ocular symptoms due to organophosphorus gas (Sarin) poisoning in Matsumoto. *Brit J Ophthalm*, **80**, 1023.

Noort D, Hulst AG, Platenburg DHJM *et al.* (1998). Quantitative analysis of *O*-isopropyl methylphosphonic acid in serum samples of Japanese citizens exposed to sarin: estimation of internal dosage. *Arch Toxicol*, **72**, 671–675.

Noort D, Benschop HP and Black RM (2002). Biomonitoring of exposure to chemical warfare agents: a review. *Toxicol Appl Pharmacol*, **184**, 116–126.

Nozaki H, Aikawa N, Fujishima S *et al.* (1995). A case of VX poisoning and the difference from sarin. *Lancet*, **346**, 698–699.

O'Brien RD (1976). Acetylcholinesterase and its inhibition. In: *Insecticides Biochemistry and Physiology*. (CF Wilkinson, ed.), pp. 271–296. New York, NY, USA: Plenum Press.

Oda Y (1999). Choline acetyltransferase: the structure, distribution and pathologic changes in the central nervous system. *Pathol Int*, **49**, 921–937.

OECD (2003). *Developmental Neurotoxicity Study*, Draft New Guideline 426, September. Paris, France: Organization for Economic Co-operation and Development.

Ohbu S, Yamashina A, Takasu N (1997). Sarin poisoning on Tokyo subway. *Southern Med J*, **90**, 587–593.

Okudera H (2002). Clinical features on nerve gas terrorism in Matsumoto. *J Clin Neurosci*, **9**, 17–21.

O'Leary TF, Kunkel AM and Jones AH (1961). Efficacy and limitations of oxime – atropine treatment of organophosphorus anticholinesterase poisoning. *J Pharmacol Exp Ther*, **132**, 50–52.

Ortells MO and Lunt GG (1995). Evolutionary history of the ligand-gated ion-channel superfamily of receptors. *Trends Neurosci*, **18**, 121–127.

Østergaard D, Jensen FS and Viby-Morgensen J (1992). Pseudocholinesterase deficiency and anticholinesterase toxicity. In: *Clinical and Experimental Toxicology of Organophosphates and Carbamates* (B Ballantyne and TC Marrs, eds), pp. 520–527. Oxford, UK: Butterworth-Heinemann.

Padzernik TL, Cross R, Nelson S *et al.* (1983). Soman-induced depression of grain activity in TAB-pretreated rats : 2-dooxyglucose study. *Neurotoxicity*, **4**, 27–34.

Parker RM, Crowell JA, Bucci TJ *et al.* (1988). Negative delayed neuropathy study in chickens after treatment with isopropyl methylphosphonofluoridate (sarin, type 1). *Toxicologist*, **8**, 248.

Pearce PC, Crofts HS, Muggleton NG *et al.* (1999). The effects of acutely administered low dose sarin on cognitive behaviour and the electroencephalogram in the common marmoset. *J Psychopharmacol*, **13**, 128–135.

Pelkonin O, Vähäkangas K and Gupta RC (2005). Placental toxicity of organophosphate and carbamate pesticides. In: *Toxicology of Organophosphate and Carbamate Compounds* (R Gupta, ed.), pp. 463–479. San Diego, CA, USA: Academic Press.

Petras JM (1981). Soman neurotoxicity. *Fund Appl Toxicol*, **1**, 242.

Petty CS (1958). Histochemical proof of organic phosphate poisoning. *Arch Pathol*, **66**, 458–463.

Pimentel JH and Carrington da Costa RB (1992). Effects of organophosphates on the heart. In: *Clinical and Experimental Toxicology of Organophosphates and Carbamates* (B Ballantyne and TC Marrs, eds), pp. 145–148. Oxford, UK: Butterworth-Heinemann.

Preusser H-J (1967). Die Ultrastructur der motorischen Endplatte im Zwerchfell der Ratte und Veranderungen nach Inhibierung der Acetylcholinesterase. *Zeit Zellforsch*, **80**, 436–457.

Qiao D, Seidler FJ and Slotkin TA (2001). Developmental neurotoxicity of chlorpyrifos modelled *in vitro*: comparative effects of metabolites and other cholinesterase inhibitors on DNA synthesis in PC12 and C6 cells. *Environ Health Perspect*, **109**, 909–913.

Raveh L, Chapman S, Cohen G *et al.* (1999). The involvement of the NDMA receptor complex in the protective effect of anticholinergic drugs against soman poisoning. *Neurotoxicology*, **20**, 551–510.

Ray D (1998a). *Organophosphate Esters*, Report prepared for the Department of Health. Leicester, UK: Institute for Environment and Health.

Ray D (1998b). Chronic effects of low level exposure to anticholinesterases – a mechanistic review. *Toxicol Lett*, **102/103**, 527–533.

Rengstorff HH (1985). Accidental exposure to sarin : vision effects. *Arch Toxicol*, **56**, 201–203.

Reutter S (1999). Hazards of chemical weapons release during war: new perspectives. *Environ Health Perspect*, **107**, 985–990.

Richards P, Johnson M, Ray D *et al.* (1999). Novel targets for organophosphorus compounds. *Chem-Biol Interact*, **119–120**, 503–511.

Richards P, Johnson M and Ray D (2000). Identification of acylpeptide hydrolase as a sensitive site for reaction with organophosphorus compounds and a potential target for cognitive enhancing drugs. *Mol Pharmacol*, **58**, 577–583.

Robinson JP (1967). Chemical warfare. *Sci J*, **3**, 33–40.

Rocha ES, Santos MD, Chechabo SR *et al.* (1999). Low concentrations of the organophosphate VX affect spontaneous and evoked transmitter release from hippocampal neurons: toxicological relevance of cholinesterase-independent actions. *Toxicol Appl Pharmacol*, **159**, 31–40.

Romano JA, McDonough JH, Sheridan R *et al.* (2001). Health effects of low-level exposure to nerve agents. In: *Chemical Warfare Agents: Toxicity at Low Levels* (SM Somani and JA Romani, eds), pp. 1–24. Boca Raton, FL, USA: CRC Press.

Rosenstock L, Keifer M, Daniell WE *et al.* (1991). Chronic central nervous system effects of acute organophosphate pesticide poisoning. *Lancet*, **338**, 223–227.

Schoene K and Oldiges H (1973). Efficacy of pyridinium salts against tabun and sarin poisoning *in vivo* and *in vitro*. *Arch Int Pharmacodyn Ther*, **204**, 110–123.

Schoene K, Hochrainer D, Oldiges H *et al.* (1985). The protective effect of oxime pretreatment upon the inhalative toxicity of sarin and soman in rats. *Fund Appl Toxicol*, **5**, 584–588.

Sekijima Y, Morita H and Yanagisawa N (1997). Follow-up of sarin poisoning in Matsumoto. *Ann Int Med*, **127**, 1042.

Sellström Å, Algers G and Karlsson B (1985). Soman intoxication and the blood-brain barrier. *Fund Appl Toxicol*, **5**, S122–S126.

Senanayake N (2001). Organophosphorus-induced delayed polyneuropathy. In: *Organophosphates and Health* (L Karalliedde, S Feldman, J Henry and T Marrs, eds), pp. 159–173. London, UK: Imperial College Press.

Senanayake N and Karalliedde L (1987). Neurotoxic effects of organophosphorus insecticides. An intermediate syndrome. *New Eng J Med*, **316**, 761–763.

Senanayake N and Karalliedde L 1992). The intermediate syndrome in anticholinesterase neurotoxicity. In: *Clinical and Experimental Toxicology of Organophosphates and Carbamates* (B Ballantyne and TC Marrs, eds), pp. 126–134. Oxford, UK: Butterworth-Heinemann.

Shih DM, Reddy S and Lusis AJ (2002). CHD and atherosclerosis: human epidemiological studies and transgenic mouse models. In: *Paraoxonase (PON1) in Health and Disease: Basic and Clinical Aspects* (LG Costa and CE Furlong, eds), pp. 93–123. Dordrecht, The Netherlands: Kluwer Academic Press.

Sidell FR (1974). Soman and sarin: clinical manifestations and treatment of accidental poisoning. *Clin Toxicol*, **7**, 1–17.

Silveira CLP, Eldefrawi AT and Eldefrawi ME (1990). Putative M2 muscarinic receptors of rat heart have high affinity for organophosphorus anticholinesterases. *Toxicol Appl Pharmacol*, **103**, 474–481.

Singer AW, Jaax NK, Graham JS *et al.* (1987). Cardiomyopathy in soman and sarin intoxicated rats. *Toxicol Lett*, **36**, 243–249.

SIPRI (1971). Stockholm International Peace Research Institute, *The Problem of Chemical and Biological Warfare*, Volume 1, *The Rise of CB Weapons*. Stockholm, Sweden: Almqvist and Wiksell.

Slotkin TA (2005). Developmental neurotoxicity of organophosphorus: a case study of chlorpyrifos. In: *Toxicology of Organophosphate and Carbamate Compounds* (R Gupta, ed.), pp. 293–314. San Diego, CA, USA: Academic Press.

Smallbridge RC, Carr FE and Fein HG (1991). Diisopropylfluorophosphate (DFP) reduces serum prolactin, thyrotropin, luteinizing hormone, and growth hormone and increases adrenocorticotropic and corticosterone in rats: involvement of dopaminergic and somatostatinergic as well as cholinergic pathways. *Toxicol Appl Pharmacol*, **108**, 284–295.

Solberg Y and Belkin M (1997). The role of excitotoxicity in organophosphorus nerve agents central poisoning. *Trends Pharm Sci*, **18**, 183–185.

Somani SM and Husein K (2001). Low-level nerve agent toxicity under normal and stressful conditions. In: *Chemical Warfare Agents: Toxicity at Low Levels* (SM Somani and JA Romano, eds), pp. 83–120. Boca Raton, FL, USA: CRC Press.

Spruit HET, Langenberg JP, Trap HC *et al.* (2000). Intravenous and inhalation toxicokinetics of sarin stereoisomers in atropinized guinea pigs. *Toxicol Appl Pharmacol*, **169**, 249–254.

Stevenson IH (1990). Susceptibility of different age groups to toxic damage. In: *Basic Science in Toxicology* (GN Volans, J Sims, FM Sullivan and P Turner, eds), pp. 404–411. London, UK: Taylor & Francis.

St Omer VEV and Rottinghaus GE (1992). Biochemical determination of cholinesterase activity in biological fluids. In: *Clinical and Experimental Toxicology of Organophosphates and Carbamates* (B Ballantyne and TC Marrs, eds), pp. 15–27. Oxford, UK: Butterworth-Heinemann.

Subcommittee on Chronic Reference Doses for Selected Chemical-Warfare Agents (2000a). Summary. *J Toxicol Clin Toxicol A*, **59**, 283–288.

Subcommittee on Chronic Reference Doses for Selected Chemical-Warfare Agents (2000b). Introduction. *J Toxicol Clin Toxicol A*, **59**, 289–295.

Subcommittee on Chronic Reference Doses for Selected Chemical-Warfare Agents (2000c). Derivation of reference doses. *J Toxicol Clin Toxicol A*, **59**, 297–301.

Subcommittee on Chronic Reference Doses for Selected Chemical-Warfare Agents (2000d). Evaluation of the army's interim reference dose for GA. *J Toxicol Clin Toxicol A*, **59**, 303–311.

Subcommittee on Chronic Reference Doses for Selected Chemical-Warfare Agents (2000e). Evaluation of the army's interim reference dose for GB. *J Toxicol Clin Toxicol A*, **59**, 313–321.

Subcommittee on Chronic Reference Doses for Selected Chemical-Warfare Agents (2000f). Evaluation of the army's interim reference dose for GD. *J Toxicol Clin Toxicol A*, **59**, 323–330.

Subcommittee on Chronic Reference Doses for Selected Chemical-Warfare Agents (2000g). Evaluation of the army's interim reference dose for VX. *J Toxicol Clin Toxicol A*, **59**, 331–338.

Sussman JL, Harel M, Frolow F *et al.* (1991). Atomic structure of acetylcholinesterase from Torpedo calfornica: a prototypic acetylcholine-binding protein. *Science*, **253**, 872–879.

Swaminathan R and Widdop B (2001). Biochemical and toxicological investigation related to OP compounds. In: *Organophosphates and Health* (L Karalliedde, S Feldman, J Henry and T Marrs, eds), pp. 357–406. London, UK: Imperial College Press.

Talbot BG, Anderson DR, Harris LW *et al.* (1988). A comparison of *in vivo* and *in vitro* rates of aging of soman-inhibited erythrocyte acetylcholinesterase in different animal species. *Drug Chem Toxicol*, **11**, 289–305.

Tang J, Carr RL and Chambers JE (1999). Changes in rat brain cholinesterase activity and muscarinic receptor density during and after repeated oral exposure to chlorpyrifos in early postnatal development. *Toxicol Sci*, **51**, 265–272.

Thompson CM and Richardson RJ (2004). In: *Pesticides, Toxicology and International Regulation* (TC Marrs and B Ballantyne, eds), pp. 89–127. Chichester, UK: John Wiley & Sons Ltd.

Tonduli LS, Testylier G, Pernot Marino I *et al.* (1999). Triggering of soman-induced seizures in rats: multiparametric analysis with special correlation between enzymatic, neurochemical and electrophysiological data. *J Neurosci Res*, **58**, 464–473.

UK Ministry of Defence (1972). *Medical Manual of Defence against Chemical Agents*, JSP 312 A/24/Gen/4392, pp. 7–12. London, Edinburgh and Belfast, UK: Her Majesty's Stationary Office.

US EPA (1998). United States Environmental Protection Agency, Guidelines for developmental toxicity risk assessment. *Fed Reg*, **56**, 63798–63826 (5 Dec).

Van der Schans M, Polhuijs M, van Dijk C *et al.* (2004). Retrospective detection of exposure to nerve agents: analysis of phosphofluoridates originating from fluoride-induced reactivation of phosphonylated BuChE. *Arch Toxicol*, **78**, 508–524.

van Helden HPM, van der Wiel HJ, Zijlstra JJ *et al.* (1994). Comparison of the therapeutic effects and pharmacokinetics of HI-6, HLö-7, HGG-12, HGG-42 and obidoxime following non-reactivatable acetylcholinesterase inhibition in rats. *Arch Toxicol*, **68**, 224–230.

van Helden HPM, Trap HC, Kuijpers WC *et al.* (2004a). Low-level exposure of guinea pigs and marmosets to sarin vapour in air: lowest-observable-adverse-effect level (LOAEL) for miosis. *J Appl Toxicol*, **24**, 59–68.

van Helden HPM, Vanwersh RAP, Kuijpers WC *et al.* (2004b). Low levels of sarin affect the EEG in marmoset monkeys: a pilot study. *J Appl Toxicol*, **24**, 475–483.

Watson A, Bakshi K, Opresko D *et al.* (2005). Cholinesterase inhibitors as chemical warfare agents: community preparedness guidelines. In: *Toxicology of Organophosphate and Carbamate Compounds*. (R Gupta, ed.), pp. 47–68. San Diego, CA, USA: Academic Press.

Weger N and Szinicz L (1981). Therapeutic effects of new oximes, benactyzine and atropine in soman poisoning: part I. Effects of various oximes in soman, sarin and VX poisoning in dogs. *Fund Appl Toxicol*, **1**, 161–163.

Wehner JM, Smolen A and Smolen TN (1985). Recovery of acetylcholinesterase activity after acute organophosphate treatment of CNS re-aggregate cultures. *Fund Appl Toxicol*, **5**, 1104–1109.

WHO (1970). Technical Report, *Health Aspects of Chemicals and Biological Weapons*, Report of a World Health Organization Group of Consultants. Geneva, Switzerland: World Health Organization.

WHO (1986). *Organophosphorus Insecticides: A General Introduction*, Environmental Health Criteria 63. Geneva, Switzerland: World Health Organization.

Wills JH (1961). Anticholinergic compounds as adjuncts to atropine in preventing lethality by sarin in the rabbit. *J Med Pharm Chem*, **3**, 353–359.

Wilson BW, Hooper MJ, Hansen ME *et al.* (1992). Reactivation of organophosphorus inhibited AChE with oximes. In: *Organophosphates, Chemistry, Fate and Effects* (JE Chambers and PE Levi, eds), pp. 107–137. San Diego, CA, USA: Academic Press.

Wolthuis OL and Vanwersch RAP (1984). Behavioral changes in the rat after low doses of cholinesterase inhibitors. *Fund Appl Toxicol*, **4**, S195–S208.

Worek F, Eyer P and Szinicz L (1998). Inhibition, reactivation and aging kinetics of cyclohexylmethylphosphonofluoridate-inhibited human cholinesterase. *Arch Toxicol*, **72**, 580–587.

Yokoyama K, Araki S, Murata K *et al.* (1998). A preliminary study on delayed vestibulo–cerebellar effects of Tokyo subway sarin poisoning in relation to gender difference: frequency analysis of postural sway. *J Occ Environ Med*, **40**, 17–21.

Young GD and Koplovitz I (1995). Acute toxicity of cyclohexylmethylphosphonofluoridate (CPMF) in rhesus monkeys: serum biochemical and pharmacological changes. *Arch Toxicol*, **69**, 379–383.

9 A HISTORY OF HUMAN STUDIES WITH NERVE AGENTS BY THE UK AND USA

Frederick R. Sidell

Bel Air, MD, USA

The views expressed in this chapter are those of the author and should not be construed as the position or policy of the Department of the Army (USA) unless so designated by other authorized documents.

INTRODUCTION

The toxic organophosphorus cholinesterase-inhibiting substances that are known as nerve agents were first synthesized by a German chemist, Dr Gerhard Schrader, in 1936. Germany manufactured and stockpiled two of these compounds, tabun (GA) and sarin (GB), and developed a third, soman (GD), during World War II. Since the Allies had no knowledge of these agents and no defense against them, it was fortunate that Germany did not use them. Organophosphorus cholinesterase inhibitors were first synthesized in the mid-1800s, but were not used as warfare agents. In fact, English officials had studied and rejected one, diisopropyl phosphorofluoridate (DFP), as a potential chemical agent in the World War II period.

In the closing days of the war in Europe, the USSR captured a manufacturing facility in eastern Germany and moved the facility and personnel to Russia to continue production (Koelle, 1981). In western Germany, the British and the Americans captured a stockpile of munitions containing an unknown chemical agent. Their initial lack of awareness of the properties of this agent, tabun, was indicated in an early British report:

> *Before the substance was identified, 1 mm and 2 mm drops were placed on the skin of the arms of human observers to ascertain whether or not it was a vesicant. The results were negative. A 1 mm drop was also placed in the eye of a rabbit to show whether the substance caused eye damage. The rabbit went into convulsions and died in a few minutes* (MOD1).

This report also described the first of many studies in animals which were conducted before humans were next exposed to these agents.

The potency of the vapor of these compounds was probably noted by Dr Schrader and his staff, since some reports suggest that they had miosis, eye and head discomfort and dyspnea while working with the substances. Before they realized the volatility and potency of tabun, which was the first agent studied, British scientists also suffered miosis and eye and head discomfort (K. W. Wilson, personal communication).

Beginning at the conclusion of World War II in Europe, the US and the UK military establishments began large research and development programs to investigate the effects of nerve agents and to develop them for military use. Human studies formed an integral part of these programs. This chapter is intended to provide a history of these human studies. However, it is incomplete.

Chemical Warfare Agents: Toxicology and Treatment (2nd Edition)

Some reports are not easily found and have undoubtedly been overlooked. A few are still classified and cannot be quoted or referenced directly. However, the classified documents contain no significant data that are not also found in the unclassified reports. Finally, this is intended to be a history of a program, not a scientific review of this literature or a description of the clinical effects of nerve agents in humans, reports of which can be found elsewhere (Sidell, 1992).

In the USA, this work was performed under the direct supervision of the Office of The Surgeon General of the Army. For each study a detailed protocol was written and approved by a medical review board which was chaired by a direct appointee of the Surgeon General. A physician prepared to administer emergency care was present during all studies (although he often was not the investigator). Each agent was thoroughly studied by several routes of administration in at least seven species of animals before it was given to humans. Representatives of outside agencies, e.g. the Food and Drug Administration, reviewed the program on several occasions and reported that procedures and standards exceeded those of many other laboratories performing human studies. In the great majority of the studies mentioned, the subjects were enlisted military personnel. In all instances, these personnel volunteered for the studies without coercion and gave informed consent. The conditions under which studies were performed in the UK were similar if not identical to those in the USA.

Several initial studies will be mentioned in chronological order. Later studies generally addressed a specific aspect of human pharmacology and these will be discussed by topic. The agent VX was synthesized long after studies on the G agents were underway, and its effects were investigated much later. The first section of this chapter discusses studies with the G agents, the second describes studies with VX, and the third provides an overview of accidental exposures.

G AGENTS

General studies

In the first study (MOD2), reported in the mid-1940s, 49 subjects were exposed to tabun vapor at *Cts* of 0.7–21 mg min m^{-3} (one subject with respiratory protection and one eye protected was exposed to a *Ct* of 30 mg min m^{-3}). (*Ct* is the product of concentration of vapor or aerosol and time of exposure [mg m^{-3} × min = mg min m^{-3}].) The effects included miosis, frontal headache, retrobulbar pain, engorgement of vessels in the eye, rhinorrhea, nausea and vomiting and complaints of tightness in the chest and blurring of vision. The signs and symptoms were at their maximal at 24–48 h after exposure. No changes in near or far visual acuity were noted at *Cts* under 30 mg min m^{-3}. Topical atropine in the eye relieved most of the symptoms, including the nausea and vomiting.

This was followed by a study in which 22 infantry soldiers, 2 officers and 5 civilian scientists were exposed to a *Ct* of 28 mg min m^{-3} of tabun vapor (MOD3). The soldiers then performed a number of military field tasks at 2, 5, 24 and 30 h after exposure and also at twilight (about 10 h after exposure). These tasks included assembling and disassembling weapons, shooting a rifle, map reading, use of the compass and slide rule and writing messages. The scientists performed their usual work. All 29 had miosis, disturbances of vision and headache; these effects were most severe at 24 h. Of the 29, 22 vomited, 24 had poor or no sleep, 22 had tiredness or lassitude and 26 felt depressed. Their average far vision changed from 6/6 (m) before exposure to 6/24 (m) and recovered gradually over a week. Rhinorrhea was common, but there is no mention of lower airway complaints. The conclusions were that the agent produced general harassing effects which would cause some military disability, especially if sustained effort were involved, and that the agent is detectable by eye signs, respiratory effects and smell. The miosis was felt to be of relatively little importance 'from a practical point of view'. Topical atropine helped vision, gave a large measure of relief, and improved the general well-being of the subjects.

A similar study with GB followed (MOD4). Subjects received *Cts* of 3.3 or 6.6 mg min m^{-3} and had miosis and head and eye complaints. Respiratory symptoms were not consistent. When the subjects received a *Ct* of 3.3 mg min m^{-3} daily for 3 or 4 consecutive days, the eye effects intensified.

A total of 120 subjects were exposed to sarin vapor at *Cts* of 1, 2, 4 or 6 mg min m^{-3} (Harvey, 1952). At each *Ct*, the time and concentration were changed by a factor of 10; for example, at a *Ct* of 6 mg min m^{-3} the concentrations were 3.0 and 0.3 mg m^{-3} and the times 2 and 20 min respectively. Rhinorrhea was the most commonly noted effect and occurred in all who received a *Ct* of 2 mg min m^{-3} or higher. Miosis was noted in all at a *Ct* of 6 mg min m^{-3}, but was an inconsistent finding at lower *Cts*. The complaint of a tight chest was more common at *Cts* of 2 and 4 than at 6 mg min m^{-3}. At the highest *Ct*, the mean erythrocyte cholinesterase activity after exposure was 73% of the subjects' baseline values for those exposed for 20 min and 87% for those exposed for 2 min.

Eight subjects who inhaled sarin vapor while at rest and nine who inhaled it while exercising had normal neuromuscular function, respiratory function, electrocardiographs and blood cholinesterase after exposure (Freeman *et al.*, 1952). The retained amount of sarin ranged from 0.16 to 0.8 μg kg^{-1} in the resting group and from 0.29 to 0.99 μg kg^{-1} in the exercising group.

In an attempt to relate erythrocyte cholinesterase activity to signs and symptoms, five subjects received a *Ct* of 6.6 mg min m^{-3} in a single exposure, and five received two exposures to a *Ct* of 3.3 mg min m^{-3} 2 or 3 days apart (MOD5). The first group had miosis and eye symptoms, tight chests and rhinorrhea. The two-exposure group had fewer effects after the second exposure, and no cumulative effects were noted. In no subject was the erythrocyte cholinesterase activity beyond his normal range (of five samples before the exposure). This was the first of many observations that there is no relationship between the effects caused by direct contact of the nerve agent vapor (on the eye, nose and airways) and inhibition of erythrocyte cholinesterase activity.

In the early 1950s, Dr David Grob (then associated with The Johns Hopkins Hospital) and associates investigated the clinical pharmacology of sarin in an elaborate series of studies extending over several years. They administered the agent orally, intra-arterially, percutaneously and topically in the eye, and described the clinical and laboratory effects, including those on the electroencephalogram. They noted that the erythrocyte cholinesterase activity was inhibited to a greater degree than the plasma enzyme, but that the plasma enzyme recovered faster. They also noted that atropine was effective at reversing the effects in organs with muscarinic receptor sites. They published their studies in numerous government contract reports. Since these findings are also published in journals that are readily available (Grob and Harvey, 1953,1958; Grob, 1956), they will not be detailed here. Over the several-year's duration of these studies, the secrecy surrounding the compounds decreased. In the first report (Grob and Harvey, 1953) the investigators described the effects of 'nerve gas', but the agent is not mentioned nor are doses given. Later, both the specific agent and the doses were included. These reports contain the familiar table showing the effects of nerve agents that has been reproduced in many places, including military manuals.

Skin exposure

The toxicity of liquid nerve agents on the skin was studied extensively. In the first study reported (Freeman *et al.*, 1953), 5-mg droplets of sarin (6–129 droplets; total doses of 25–550 mg) were placed on the volar aspects of the forearms of 11 subjects. The agent evaporated in 1.5–6 min from the skin when the area was relatively free of hair, and in 6.5–15 min from the skin when placed in areas of heavy hair. The agent was mixed with a dye for visualization, and after evaporation was complete the site was tested for the agent and then decontaminated. Maximal inhibition of erythrocyte cholinesterase activity was 18% (82% of normal). Two subjects (cholinesterase of 82%) had transient diarrhea. Sweating at the site of application continued for as long as 34 days.

A small amount (3.21 mg) of liquid sarin was applied to the skin of six subjects, and about 98% of it was recovered in the air (Marzulli and Williams, 1953). The tendency of sarin to evaporate rather than to penetrate skin was noted in a later, more extensive study (MOD6). In a similar study by two of the same investigators (Freeman *et al.*, 1954), tabun, 50–400 mg in 0.5-mg droplets was placed on the volar aspects of forearm skin of six subjects. The droplets evaporated in 18–167 min, and the maximal inhibition of erythrocyte cholinesterase activity was 34% (in a subject who received 400 μg kg^{-1}). Sweating, a feeling of coolness and blanching occurred at the site after exposure, and localized sweating persisted for 8–95 days. Ambient temperatures were about 24°C, relative humidities about 40%, and wind speed about 1 mile per hour (1.6 km h^{-1}) in these two studies.

In an extensive study (MOD6), sarin in amounts of 30–300 mg as 0.5-mg droplets was placed on the bare forearm skin of 114 subjects at a temperature of 70°F (21°C) and a relative humidity of 70%. In contrast to the studies noted above, in which decontamination was not done until the droplet had been absorbed or had evaporated, the site of application was decontaminated 30 min after exposure. The average inhibition of erythrocyte cholinesterase at the highest dose was 40%, but the variation was large. Four subjects had inhibition of greater than 80%; three of these had no effects and one had generalized sweating with cold clammy skin and vomiting, and was given atropine. Amounts of 100–300 mg (0.5-mg droplets) of sarin were placed on one layer of serge over the forearm skin. The mean inhibition of cholinesterase activity at the highest dose was 48%. Four subjects, each with more than 85% inhibition of erythrocyte cholinesterase, vomited starting 30 min to 26 h after exposure. A fifth had profuse sweating and difficulty in breathing soon after decontamination and were treated. Sarin, 200 mg (0.5-mg droplets), was placed on a layer of serge over a layer of flannel on the forearm skin of 18 subjects. Mean erythrocyte cholinesterase inhibition was 43% (this amount had caused 25% inhibition when placed on serge and 19% on bare skin). One subject with 96% inhibition of erythrocyte cholinesterase had generalized muscular fasciculations and difficulty in respiration 23 min after the start of exposure and was treated. No subject with less than 80% inhibition in erythrocyte cholinesterase activity had any effects from the agent.

In another part of this study, soman, 10–40 mg in 0.5-mg droplets, was placed on the forearms of 32 subjects. The mean erythrocyte cholinesterase inhibition increased from 13% to 27% as the dose increased. GF (another organophosphorus cholinesterase inhibitor and nerve agent), in doses of 5–30 mg in 0.5-mg droplets, produced a mean inhibition of 57% after the highest amount. Several doses of soman and GF, given as single drops, caused 2–10 times more inhibition of cholinesterase than when given as multiple small droplets. The time course of cholinesterase inhibition was not detailed. The amounts of agent needed to inhibit erythrocyte cholinesterase activity by 50% when placed on bare skin were estimated to be 400 mg per person for sarin, 65 mg per person for soman and 30 mg per person for GF.

In a later study (Naitlich, 1965), 0.2–6.1 mg of soman was placed on the forearms of subjects in an open ward. Several subjects and the investigators had eye effects from agent vapor, but there were no systemic effects or inhibition of erythrocyte cholinesterase. The next part of this study took place in a static chamber, with the investigators watching through a window, where 2.0–7.8 mg was applied to the forearm skin of 25 subjects. Most had localized sweating at the site of application. On several days of the study when the ambient temperature was over 80°F (27°C) and relative humidity over 85%, the cholinesterase activity inhibition was slightly greater than on the other days when temperature and humidity were lower. Otherwise, maximal inhibition of erythrocyte cholinesterase activity occurred between 2 and 24 h after exposure, and the maximal inhibition was to 69% of baseline.

Exposure of skin to vapor

One arm of each of four subjects was put into a chamber and exposed to *Ct*s of 1000–8000 mg min m^{-3} of tabun (Krackow and Fuhr, 1949). There were no effects or changes in plasma cholinesterase activity. Later, 16 subjects dressed in shoes, shorts and protective masks were exposed to *Ct*s of 520–2000 mg min m^{-3} of tabun (roughly 5–10 times the estimated LCt_{50} of tabun by inhalation). The cholinesterase activity was inhibited more at 1 h after exposure than at 15 min after exposure, but at neither time was it markedly inhibited. There was a slight but statistically significant difference between the cholinesterase activity of the group exposed to 520–970 mg min

m^{-3} (between 90% and 100% of control) and that of the group exposed to 1320–2000 mg min m^{-3} (80–90%). There were no signs or symptoms, and electrocardiograms and blood pressures were normal. In a similar study, effects were minimal after skin exposure to about 10–15 times the inhalational LCt_{50} of sarin.

Performance studies

The potential effects of nerve agents on intellectual, cognitive or psychomotor functioning were examined in several studies. Sarin was the agent in each study. Twenty subjects who received a *Ct* of 10 mg min m^{-3} had no deterioration in central intellectual capacity, but were impaired slightly on a visual search task, and their rate of learning on a hand–eye task was slower 5 and 24 h after exposure than before exposure. Extra stress caused their performance to deteriorate further, and they were noted to be lethargic and disinclined to be bothered (MOD7).

In a similar study by the same investigator (MOD8). 12 subjects received a *Ct* of 14.7 mg min m^{-3}, and 12 non-exposed subjects were controls. Again, intellectual capacity was not impaired. However, the perceptual span was slightly less and the flicker threshold frequency (FTF) was decreased in the exposed subjects. Their three-dimensional space coordination was markedly impaired. The decrease in FTF was similar to that seen with fatigue.

In a third study (MOD9), eight subjects received a *Ct* of 14.6 mg min m^{-3} (which inhibited their erythrocyte cholinesterase activity by an average of 36%). At 5 and 24 h after exposure, there was no serious intellectual decline, but cognitive function was slightly decreased and subjects had a decrease in manual dexterity/discrimination, possibly because of finger incoordination. Self-rating scales indicated that the subjects had dysphoria and reduced mental alertness.

Twenty-eight subjects who received a rather low *Ct* of 4 mg min m^{-3} with 17% inhibition of erythrocyte cholinesterase activity took a battery of tests, including the Minnesota rate of manipulation, Purdue pegboard, dual pursuit meter, reaction time, a steadiness aiming device and an addition task (Davy and Gossar, 1959). Although their pupil sizes had decreased by at least 2 mm, their performance was not severely impaired on any task and was improved on many.

No changes in performance were found on a heavy pursuit meter task, a light pursuit meter task and a ball bearing picking task in 25 subjects who received 5 μg kg^{-1} of vapor by oral inhalation (eyes not exposed) (MOD10).

Intellectual function was examined in 24 individuals who had been accidentally exposed to sarin (Combs and Freeman, 1954). They were divided into high- and low-dose groups based on their clinical signs and the judgement of the investigators. Subscales of the Wechsler–Bellevue test were used. The only differences between immediate post-exposure tests and baseline tests (tests done several months after exposure when the individuals were considered to have completely recovered) were on the similarities and comprehension scales in the high-dose group. These differences appeared to indicate impairment of judgement and of higher-level verbal concept formation.

A study of behavioral changes in humans after VX is noted in the section on VX.

Physical performance

Several studies examined physical performance after sarin. In the first (MOD11), two groups of 20 subjects each received 5 μg kg^{-1} of sarin using the single-breath technique of oral inhalation. (In the single-breath technique the subject breathed a measured amount of air containing a known amount of agent through a mouthpiece with nose clamped and with no eye exposure. The amount is given as μg kg^{-1}. When the amount is given as a *Ct*, the subject generally was exposed in a chamber with no protection to eyes, nose or mouth. The exact inhaled dose is not known with the latter method.) Because of vapor condensation on the apparatus, the first group did not receive the full amount of sarin, and their mean erythrocyte cholinesterase activity was depressed by only 18%. They rode a bicycle ergometer as fast as possible for 5 min per day for 5 days and inhaled the sarin just before the ride on day 3. The incidence of chest symptoms was lower than usual during the ride, and some felt the ride was easier than usual. Some complained of a tight chest 10–15 min after the ride.

There were no changes in minute volume, oxygen intake or carbon dioxide output. Another group rode the bicycle at a constant rate for 15 min three times a day for 10 days and inhaled sarin on day 8 (mean erythrocyte cholinesterase activity was 65% of baseline). There was a small but insignificant increase in oxygen consumption but no other changes.

In a later study (MOD12), 14 subjects exposed to a *Ct* of 14.7 mg min m^{-3} of sarin (mean erythrocyte inhibition of 50%) carried one-third of their body weight up and down a 40.5-cm step 30 times a minute for 5 min (the Harvard pack test). This was done 45 min and 24 h after exposure. Their fitness index, which includes heart rate and other measures, did not differ from that of a group exposed to agent but not exercised.

Physiology

Needle electrodes were placed in intercostal muscles in 10 subjects who received 70–150 μg of sarin by the single-breath technique of vapor administration (MOD13). Although eight noted chest tightness, none had changes in whole blood cholinesterase, and there were no abnormalities in the electrical pattern from the muscles. It was concluded that muscular abnormalities did not contribute to the sensation of dyspnea.

Venous tone in the legs was examined by a *g*-suit technique in 23 subjects who received 5 μg kg^{-1} of sarin by the single-breath technique of vapor exposure (33% decrease in erythrocyte cholinesterase activity) (MOD14). Although there was a small decrease in the average tidal volume and a small increase in the respiratory rate in these subjects, there was no change in the amount of end-tidal shift of air in the lungs (which would have suggested a change in venous return). This suggested that pressor amines probably would not be useful in therapy of sarin intoxication.

Using a low-frequency critically damped ballistocardiograph, cardiac output was estimated in 25 subjects given 5 μg kg^{-1} of sarin by oral inhalation (which caused a mean 37% decrease in erythrocyte cholinesterase activity). There was a small increase in heart rate, which the investigator felt was a cortical response to the sensation of a tight chest, but no changes in cardiac output were noted (MOD15).

Fifteen subjects were simultaneously acclimatized to heat and cold by spending part of each day in a hot chamber and part in a cold chamber (MOD16). After acclimatization had been achieved (as determined by appropriate techniques), eight of the subjects received sarin by inhalation (a *Ct* of 15 mg min m^{-3}). Physiological studies indicated they did not differ from the unexposed acclimatized group, and it was concluded that the agent did not interfere with acclimatization.

Odor

At a concentration of sarin under 1.5 mg m^{-3}, 8 of 15 people noted an odor, but gave no good description of it and did not think that they would recognize it again (MOD17). In contrast, 14 of 15 described a stronger musty, spicy or fruity odor after sniffing a similar concentration of soman and thought that they would recognize it again.

About 50% of 34 subjects (22 male and 12 female laboratory workers) detected the odor of soman in concentrations of 3.3–7.0 mg m^{-3} and described it as sweet, fruity or nutty (Dutreau *et al.*, 1950). About 65% of the subjects had mild nasal and airway symptoms from the agent.

Twenty people were exposed to GF vapor. At a concentration of 10.4 mg m^{-3}, 35% could smell it, and 65% could detect an odor at a concentration of 14.8 mg m^{-3} (McGrath *et al.*, 1953). There was no agreement on the odor.

Miscellaneous

Dressed subjects wearing the standard US M17 mask, hood and combat uniform (not protective clothing) were exposed for 20 min to concentrations of sarin of 0.25–4.8 mg m^{-3} or for 5 min at concentrations of 8.9–27.1 mg m^{-3}. Absorption through the skin is negligible under these circumstances. While in the vapor atmosphere, they carefully removed their mask and hood to eat or drink. They could take a bite or a sip of liquid in an average of 4 s, and the total unprotected time while they ate rations or drank some water ranged from 0.21 to 1.5 min. Many had miosis and rhinorrhea and complained of chest tightness. The inhibition of whole blood cholinesterase activity ranged from 2% to 32%. Those who only drank expired deeply before removing their mask, effectively clearing it, had no symptoms or physical signs, and averaged 26% inhibition of cholinesterase activity. The investigator concluded that it was safe to eat and drink in a contaminated atmosphere if these activities are done carefully (Cresthull and Oberst, 1964).

Pulmonary

In a study using 150 laboratory employees and enlisted personnel, pulmonary function (vital capacity and maximum breathing capacity) was examined after various *Ct*s up to 6.0 mg min m^{-3} (Cooper and Moloney, 1951). There was a good correlation between degree of bronchoconstriction and *Ct* over the dose range used (0.7). Maximal bronchoconstriction occurred between 15 min and 2 h after exposure, with a tendency to recover by 3 h. Although subjects were directed to breathe through their noses only, one also breathed through his mouth and had more severe symptoms and pulmonary function decrement. This suggested that nasal breathing offers some protection.

Using the same 150 subjects at the same time, another investigator measured airway resistance with an interrupted-flow technique. He found minimal airway obstruction and that only at the highest *Ct* (Clements *et al.*, 1952).

In another study, 105 subjects inhaled *Ct*s ranging from 2.4 to 19.6 mg min m^{-3} (retained doses of 0.1–3.1 μg kg^{-1}). There was a mean erythrocyte cholinesterase inhibition of 30% at a retained dose of 2.2 μg kg^{-1} and a 'rough' correlation of enzyme inhibition with retained dose. However, although measures of pulmonary function varied widely both above and below baseline, none changed consistently with dose. Measures used included resting tidal volume, exercise tidal volume, maximum breathing capacity and vital capacity (MOD18).

Using a single-breath technique of sarin administration, 63 subjects inhaled doses of 0.6–3.3 μg kg^{-1}. Increase in airway resistance was dose-dependent, but it correlated better with the concentration of agent (which varied from 5 to 40 mg m^{-3}) than with *Ct*. At the highest dose, resistance was 2.9 times baseline at 1 min after exposure, 2 times baseline at 15 min, and 1.6 times baseline at 1 h. At lower amounts, resistance was less and improvement was faster. The incidence and severity of airway symptoms paralleled the time course of the increased resistance. Symptoms were maximal immediately after exposure and improved over the next 15 min (MOD19).

Also with a single-breath technique, 25 subjects inhaled 0.5–5.0 μg kg^{-1} of sarin and had fewer symptoms than usually seen after these amounts (MOD20). There was a correlation between the amount retained and erythrocyte cholinesterase inhibition, with a maximal inhibition of 27% at 5.0 μg kg^{-1}. The half-time for inhibition of the enzyme was about 90 s, and inhibition was complete in about 10 min.

The same investigator (MOD21) administered a dose of 0.5 μg kg^{-1} on two occasions to four subjects and a dose of 2.5 μg kg^{-1} on two occasions to 29 others. In each group, the rate of inhalation varied with each exposure. Symptoms depended on the rate of inhalation. A flow rate of 20 l min^{-1} caused a minimal change in pulmonary resistance, and symptoms were related to the tongue and throat as well as chest. At a rate over 40 l min^{-1}, there was a marked increase in pulmonary resistance, and symptoms were confined to the chest. The investigator suggested that at lower rates much of the vapor is absorbed in the upper airways and that at higher rates absorption is primarily in the lower airways. Erythrocyte cholinesterase activity inhibition was 17–20% in both groups.

Others studied the absorption of sarin vapor in rabbits, monkeys and humans (MOD22; Ainswarth and Shephard, 1961). Retention in the nasal tract and trachea was 98% in rabbits and 93% in monkeys, leaving little to be absorbed in the terminal airways. During nasal breathing humans retained 96% of inhaled sarin.

In the first of two related studies by the same investigators (Oberst *et al.*, 1952,1959), subjects retained 82% of orally inhaled sarin while at rest and 81% while exercising on a bicycle ergometer. In the second study, 52 subjects breathing orally retained 87% of inhaled sarin and 39 subjects breathing nasally retained 89% of the agent. When exposed to sarin vapor while exercising and breathing orally, 41 subjects retained 80%. The estimated retained amounts to reduce the erythrocyte cholinesterase activity by 50% were 3.73 μg kg^{-1} for oral inhalation at rest, 4.06 μg kg^{-1} for nasal inhalation at rest, and 4.43 μg kg^{-1} for oral inhalation while exercising. The investigators remarked that because the minute volume was about eight times greater in the exercising subjects, the concentration was decreased by a factor of about eight, and that minute volume must be considered when using the term *Ct* (Oberst *et al.*, 1986; Oberst, 1961).

Using a mathematical model and data from other sources (not stated, but probably from the above studies), an investigator suggested that after oral inhalation of sarin at low flow rates about 50–70% is absorbed in the upper airway and 20–40% in the lungs (lower airways), while about 20% is exhaled. At high flow rates, these are essentially reversed, with about 20% absorption in the upper airway, 50–70% in the lungs (lower airways), and 10–30% exhaled (Brown, 1957).

Military effectiveness

One of the first studies done with a nerve agent in humans concerned soldiers' abilities to conduct military operations after exposure to the agent (MOD3). A similar study was done with 134 enlisted soldiers and five officers exposed to sarin at *Ct*s of 11.8–13.9 mg min^{-3} (MOD23). After exposure, all subjects had miosis, about 75% had rhinorrhea (which decreased by 5 h), about half complained of a headache, and fewer than half noted the sensation of a tight chest. They conducted an approach march, an attack with support weapons, controlled firing, and a twilight advance into position. In addition, night patrols were sent on a compass route. The conclusions included the comments that the men showed more signs of fatigue than normal, that commanders became irritable and unduly flustered, and that accuracy and speed of lower ranks were not up to normal. The investigators felt that most normal military operations could be conducted satisfactorily except for those at night.

In a similar exercise, 169 subjects received *Ct*s of about 12 mg min m^{-3} and then were divided into four groups (MOD24). Signs and symptoms were the same as had been noted in previous studies: eye effects, rhinorrhea and dyspnea. One group then received atropine, one group wore respirators after exposure, one group received atropine and wore respirators, and one group received neither atropine nor respirators after exposure. The conclusions suggested that after exposure to agent and administration of atropine, operations may be hazardous by day and would definitely be dangerous by night because of impaired eyesight and mental alertness. Masking after exposure increased the basic mental harassment and produced distress during intense physical effort. With the combination of masking and atropine, the deleterious effects were additive.

Eye

The first studies on these agents focused on eye effects (MOD2; MOD3), and many later ones concentrated on this organ. In subjects exposed to *Ct*s of 1 to 6 mg min m^{-3}, a linear relationship was noted between *Ct* and decrease in pupil size. At a *Ct* of 1 mg min m^{-3} the decrease was 1.1 mm, and at a *Ct* of 6 mg min m^{-3} the decrease was 3.0 mm. Near and far points did not change (Johns, 1952).

Of a total of 240 subjects, one group had their eyes 'bandaged' (not complete occlusion), other groups had no protection, and another group received 2 mg of atropine intramuscularly 0.5 h before all were exposed to *Ct*s of 5–15 mg min m^{-3} of sarin (MOD25). The *Ct* of 5 mg min m^{-3} caused slower and less complete constriction than the higher *Ct*s. Bandaged eyes constricted more slowly and less completely. Although atropine decreased conjunctival suffusion and complaints of eye pain, it appeared not to affect the miosis.

The sensitivity of human and rabbit eyes to agents was compared in a study in which the agent was administered in air flowing through tightly fitting goggles (in both the humans and the rabbits) (MOD26). By this technique, the ECt_{50} (*Ct* to produce the effect in 50% of the population) for miosis after sarin was 3.13 mg min m^{-3} for humans and 2.33 mg min m^{-3} for the animals. At a concentration of 3.7 mg m^{-3} there was a delay of several minutes for miosis after sarin and soman, but a delay of 30–40 min after VX.

The ECts to cause miosis in the rabbit in a static chamber were 1.32 mg min m^{-3} for sarin, 0.59 mg min m^{-3} for soman, and 0.04 mg min m^{-3} for VX. The investigators commented that VX might cause miosis in concentrations under the limits of field detectors. These data also illustrate the reason that rabbits were felt to be very effective detectors of agent.

Various measures of eye function were examined in 27 volunteers who received *Ct*s of 1.87–15 mg min m^{-3} of sarin (MOD27). Miosis was maximal within 30 min in all except the lowest dose group. The contrast threshold changed by 23% at the highest *Ct* and by 15% in the next highest dose group and correlated well with the square of the pupillary diameter. There was no change in visual acuity and little (under ID) change in refractive state. The near point decreased. The symptoms included pain on convergence with an increase in pain with increased illumination. The conclusion was that an aircrew would function well in daytime even at the highest *Ct* (15 mg min m^{-3}), but that at night the maximal allowable *Ct* should be 2 mg min m^{-3}.

Six subjects were exposed to a *Ct* of 15 mg min m^{-3} of sarin vapor (Moylan-Jones and Price-Thomas, 1973). Visual acuity was unchanged after the exposure (it was actually improved in one subject), despite the subjects' complaints to the contrary. The topical instillation of cyclopentolate markedly decreased near visual ability. The investigators suggested that topical mydriatics be reserved for those needing night vision or those with marked discomfort from the eye changes.

In a series of studies (Rubin and Goldberg, 1957a,b,1958, Rubin *et al.*, 1957), Rubin and associates investigated the effects of sarin on night vision. Inhalation of sarin (in a chamber) with full eye protection caused an elevation of the scotopic threshold (decreased ability to perceive light in darkness) even though miosis was not present. Local instillation of sarin into an eye caused miosis, but not an increase in the scotopic threshold. When one eye was completely protected (miosis did not occur) and the other exposed to sarin vapor which the subject inhaled, the protected eye did not become miotic but there was an increased threshold in both eyes. Finally, after sarin exposure intramuscular administration of atropine sulfate (which enters the central nervous system) reversed the increased threshold, whereas injection of atropine methylnitrate (which does not enter the central nervous system) did not. The investigators concluded that the decreased ability to perceive light is related to central mechanisms, possibly in the retina or visual pathways, rather than to miosis.

In another study (MOD28; Gazzard and Price-Thomas, 1975), miosis was induced by inhalation of sarin or by physostigmine eye drops. Although the degree of miosis was equal, the subjects with sarin exposure were less able to perceive a stimulus of standard luminance. This study concluded, as had the previously described study, that effects on central pathways rather than miosis *per se* were the primary determinants of the agent-induced visual deficit.

However, the amount of elevation of the scotopic threshold did correlate with decrease in pupillary area in 10 subjects who received sarin eye drops (Stewart *et al.*, 1968). The investigators in this study concluded that the decrease in night vision was directly related to miosis alone.

In three workers accidentally exposed to sarin, the return of the ability of the iris to dilate correlated better with return of plasma cholinesterase activity than with return of erythrocyte enzyme activity (Sidell, 1974). The dark-adapted pupil size, measured in complete darkness, did not return to normal for about 6 weeks.

Two other workers accidentally exposed also exhibited slow recovery of pupillary function (Rangstorff, 1985). Visual acuity was temporarily improved by sarin in these presbyopic subjects.

Studies with protection

The oxime 2-pralidoxime chloride (2-PAMCl) was administered by slow intravenous infusion to subjects who had received sarin. The half-time for *in vivo* aging of the sarin–erythrocyte complex seemed to be about 5 h (Sidell and Groff, 1974).

Sarin was given intravenously to counteract the effects of intoxication by a cholinergic blocking compound in two subjects (Ketchum *et al.* 1973). It caused total inhibition of erythrocyte cholinesterase activity in one subject and reversed both the central nervous system and more peripherally mediated effects of the anticholinergic compound in both subjects. There were no effects from sarin.

VX

Intravenous

Human studies with VX began in 1959 when an investigator, Dr Van M. Sim, volunteered to be the first subject. VX, 0.04 μg kg^{-1}, was given intravenously over 30 s with no untoward effects and no change in erythrocyte cholinesterase activity. Three and a half hours later, 0.08 μg kg^{-1} was administered by the same method. He developed a headache, felt sweaty and lightheaded, and complained of abdominal cramps, but blood cholinesterase activity was normal. For regulatory and approval reasons, a year intervened before the study continued.

Again, Dr Sim was the initial subject and received a 30-s injection of 0.225 μg kg^{-1} of VX which caused a decrease in erythrocyte cholinesterase activity to 63% of baseline. After 2 h, an intravenous infusion of 1 μg min^{-1} was begun. A variety of minor effects occurred during the infusion, but at 3.5 h he became pale, stopped talking, appeared 'out of contact', and had profuse salivation and vomiting. The confusional state lasted about 15 min. It was learned later that his erythrocyte cholinesterase activity was 15% of his baseline, (Kimura *et al.*, 1960).

In a continuation of this study, six other subjects received 1 μg kg^{-1} of VX by slow intravenous infusion. Their average erythrocyte cholinesterase activity ranged from 34% to 45% of their control activities.

In a later study, 27 subjects received 1.3–1.5 μg kg^{-1} of VX intravenously over 30 s (Sidell, 1967). By 15 min, 12 subjects (with erythrocyte cholinesterase activities 17–45% of controls) were dizzy, shaky or lightheaded. Of 17 subjects who later had erythrocyte cholinesterase activity of less than 30% of their baseline, five vomited and two were treated for severe nausea. Of eight with enzyme activity less than 25%, four vomited, two were treated for severe nausea, and one complained of nausea. These severe effects occurred more than an hour after agent administration. The time of maximal inhibition of erythrocyte cholinesterase activity was also about an hour after agent administration. Recovery of the enzyme was about 1% per hour for the first day or two.

VX was also administered intravenously as an antidote for intoxication with cholinergic blocking compounds (Sidell *et al.*, 1973). It was extremely effective at reversing both the central and peripheral effects of these compounds. There were no other effects from the agent.

Vapor exposure

The forearm or the entire arm of 29 subjects was exposed to *Ct*s of 7–681 mg min m^{-3} of VX vapor (Cresthull *et al.*, 1963). Blood was taken for whole blood cholinesterase activity once before exposure and once at 20 h after exposure. (Because the VX-inhibited cholinesterase reactivates rapidly during the first hours after inhibition, this may not be a good measure of the maximal amount of inhibition.) The enzyme activity was inhibited in a dose-related fashion from 3% to 43%. There were neither signs nor symptoms.

The heads and necks of medical officers were exposed to VX vapor on 54 occasions (MOD29). (Because the VX-inhibited cholinesterase recovers rapidly, the same few subjects were used repeatedly.) On 19 occasions there was no respiratory protection (*Ct*s of 0.6 mg min m^{-3} to 6.4 mg min m^{-3}), and on 35 occasions the respiratory tract was protected by having the subjects breathe clean air through a tube with the nose clamped (*Ct*s of 0.7 mg min m^{-3} to 25.6 mg min m^{-3}). In the unprotected subjects, some had miosis even though their eyes had been closed; the onset of miosis was about 1–3 h after exposure (and possibly could have been from vapor from the skin exposure). Chest tightness and rhinorrhea occurred from 7 min to 30 min after exposure. There was no significant inhibition in erythrocyte cholinesterase activity immediately after exposure, but there was some inhibition at 15 min after exposure, and maximal inhibition occurred at 4 h. Most of the subjects who were exposed only by skin exposure had miosis within 0.5–4 h after exposure, and two also had nausea. The erythrocyte cholinesterase activity did not change within the first 20 min, and maximal inhibition occurred at 8–12 h. The investigators noted the slower onset of effects from VX vapor compared to sarin vapor (onset of effects is slow even after intravenous administration of VX (Sidell, 1967).

The recovery of VX-inhibited erythrocyte cholinesterase activity was studied after the enzyme had been inhibited by percutaneous or inhalational exposure (in other studies) (MOD30). The enzyme recovered at a rate of 6% per day for the first 3–4 days, and then at 1% per day. A rapid early recovery of the enzyme was noted in several other studies.

Application on the skin

VX, in amounts of 5–35 μg kg^{-1} in small droplets, was applied to the volar aspect of the forearms of 103 subjects. The ambient temperature was 70–80°F (*Ca* 21–27°C) and the relative humidity ranged from 40% to 70% (Sim and Stubbs, 1960). Under these conditions, the amount estimated to produce a 50% inhibition of erythrocyte cholinesterase activity was 34 μg kg^{-1}. Symptoms, which usually occurred later than 6 h after exposure, included local sweating at the site (local muscular fasciculations were seen in only two subjects), a feeling of tiredness or weakness, nausea, vomiting and headaches. The incidence of vomiting was related to inhibition of cholinesterase; after more than 70–80% inhibition of erythrocyte cholinesterase activity, more than half of the subjects vomited. There were few symptoms when cholinesterase activity was about 50% of normal, but symptoms increased in severity and duration as the enzyme activity fell below 30% of baseline activity. When VX was mixed with an amine (1:1), absorption of the agent was enhanced.

Amounts of VX of 5–30 μg kg^{-1} were applied to 19 different skin sites in a large group of subjects (Sim, 1962). The estimated amount to cause a 70% reduction in erythrocyte cholinesterase activity varied by a factor of 26, from 5.1 μg kg^{-1} on the cheek to 132 μg kg^{-1} on the palm. The ear (6.6 μg kg^{-1}), top of the head (10.8 μg kg^{-1}), and forehead (11.2 μg kg^{-1}) were among the more sensitive areas of skin, while the dorsal forearm (93.8 μg kg^{-1}), dorsal foot (94.3 μg kg^{-1}) and knee (102 μg kg^{-1}) were among the least sensitive. Thirty-two subjects had a maximal drop in erythrocyte cholinesterase activity of 70% or more, and 25 of these became sick. The median time of onset of illness was 5 h after application of the agent to the head, 7 h for the extremities, and 10 h for the trunk. When the agent was applied to bearded skin, both the amount of inhibition of cholinesterase and the symptoms were more severe. The oximes 2-PAMCI and pralidoxime mesilate (P2S) reactivated the inhibited cholinesterase and reduced the severity of the symptoms.

Radiolabeled (^{32}P) VX was applied to the palm (26 μg), the back (25 μg) or the forearm (15.7 μg) of several subjects, and the amount in the skin and the amount recovered by washing the site were counted (MOD31). Based on a decrease of whole blood cholinesterase activity of 3%, the investigators concluded that less than 1% had penetrated through the palm skin, 8% the skin of the back, and 15% the skin of the forearm.

Neat VX, 20 μg kg^{-1}, was applied to the skin of four subjects, while four other subjects received 20 μg kg^{-1} mixed 1:1 with octylamine, and four others received neat VX, 35 μg kg^{-1} (Lubash and Clark, 1960). Seven were symptomatic with insomnia, nightmares, lightheadedness, nausea, epigastric discomfort, vomiting and diarrhea. (The whole blood activity was 14–38% of control in these subjects.) The whole blood erythrocyte activity was above 42% of control activity in the five asymptomatic subjects. Plasma and urinary electrolytes, BSP excretion, SGOT, SGPT and serum amylase were all normal following exposure.

One subject received neat VX on one occasion and VX mixed with *n*-octylamine on another [3:1, amine/VX (Vocci *et al*., 1959)]. Absorption of agent was increased by the amine (maximal erythrocyte cholinesterase activity inhibition of 9% with agent alone, and 41% with the mixture).

In two large and elaborate studies, Craig, Cummings and associates studied the penetration of VX through the skin of the cheek and forearm at ambient temperatures of 0°F, 35°F, 65°F and 115°F (*Ca* –18°, 1.7, 18, 46°C) (Cummings and Craig, 1965; Craig *et al*., 1967). Because these studies are reported elsewhere (Craig *et al*., 1977), they will not be described in detail. Among their conclusions were that absorption of agent was faster and more complete at higher temperatures (27% more per 10°F [*Ca* 5.6°C] increase in temperature) and that absorption continued for hours after decontamination of the application site (usually with 5.25% hypochlorite). The onset of effects and maximal inhibition of erythrocyte cholinesterase activity typically occurred several hours after the conclusion of the 3-h exposure (at which time the site was decontaminated).

The arms of three medical officers were covered with an outer layer of battledress serge over flannel and 200 mg of VX (as droplets of 0.8–1.61 μl) was dropped on the serge and allowed to remain for 8 h (MOD32). (The current estimate for the LD_{50} for VX liquid on the skin is 10 mg.) About 85% of the agent was recovered from the serge and about 5% from the flannel. The investigators suggested that the remainder evaporated. The subjects had no symptoms and no decrease in erythrocyte cholinesterase activity.

Later, under the same experimental conditions VX was allowed to remain on the serge in two subjects for 24 h. Erythrocyte cholinesterase activity began to fall within 14 h, and at 24 h, when the material was removed and the arms decontaminated (5.25% hypochlorite), it was 73% in one subject and 81% in the other. Maximal inhibition of the enzyme (34% and 64% of baseline) occurred at 48 h and 24 h after decontamination. The activity of the lower cholinesterase returned at a rate of about 10% per day for the next 3 days, but then slowed to about 1% per day. Both subjects were symptom- and sign-free. (This study should be compared to that in which sarin was placed on serge and flannel and left in place for 30 min.)

Ingestion

Fifty-four subjects drank water contaminated with VX to find the time to incipient toxicity, defined as inhibition of erythrocyte cholinesterase activity of 70% or more (Sim *et al.*, 1964). Doses of VX and types of water were varied. At an amount of 400 μg per 70 kg in distilled water, cholinesterase activity fell to 22% of the control activity in 1 day. With the same amount in distilled water to which tetraglycine hydroperiodide (a water-purifying agent) was added, the mean cholinesterase activity was 17% of control activity in 16 h. At the same dose in water treated with the standard field kit for chemical agent decontamination of water, toxicity occurred on day 4 in four of eight subjects. The same amount of VX added to tap water produced toxicity in six of nine subjects on day 4 also. Doses of 100–125 mg per 70 kg caused incipient toxicity in 6–7 days.

Maximal cholinesterase inhibition occurred in 2–3 h in subjects who drank VX in saline or dextrose in water (Sidell and Groff, 1966). At the same doses, eating before ingesting VX enhanced absorption. When the agent was mixed with tap water instead of with saline or dextrose in water, absorption was less. About 2.5 $\mu g\ kg^{-1}$ of VX caused a 50% inhibition of erythrocyte cholinesterase activity, and 3.8 $\mu g\ kg^{-1}$ produced a 70% inhibition.

When given orally, VX was ineffective in reversing the intoxication caused by a cholinergic blocking component, even though an equipotent amount (the amount causing the same inhibition of erythrocyte cholinesterase activity) given intravenously produced very effective antagonism (Sidell *et al.*, 1973).

Psychological effects

Psychological and behavioral effects were studied in 93 subjects who had been exposed to VX on the skin (Bowers *et al.*, 1964). (Results from skin exposure were reported in Sims and Stubbs (1960) and Sim (1962); in this report the agent is referred to as EA 1701.) About 30% of subjects had symptoms of anxiety, 57% had psychomotor depression and 57% had intellectual impairment after inhibition of whole blood cholinesterase activity by more than 60% (or below 40% of their baseline activity). (As the erythrocyte cholinesterase activity is usually about half that of the whole blood after VX exposure, this suggests that these subjects had about 20% of their control erythrocyte cholinesterase activity.) Fewer than 10% of the subjects with less enzyme inhibition had these effects. Psychological effects usually appeared earlier than physical effects (nausea, vomiting) or appeared in the absence of physical effects. The psychological effects were characterized by difficulty in sustaining attention and a slowing of intellectual and motor processes. There were no illogical or inappropriate trends in language and thinking, no conceptual looseness, and no perceptual distortions.

Odor

Half of a group of subjects smelled VX at a concentration of 3.9 mg m^{-3}. When two different stabilizers were added, the odor was detectable at 1.9 and 1.0 mg m^{-3}. However, the odor was not clearly described (Koon *et al.*, 1959).

Therapeutic studies

Subjects given VX in other studies received the oxime 2-PAMCl (by slow intravenous infusion) at times as long as 48 h after VX administration. Significant reactivation of the inhibited erythrocyte cholinesterase activity occurred at all times, even though the doses of oxime were small (5–25 mg kg^{-1}). Aging of the VX–enzyme complex occurs later than 48 h or possibly not at all (Sidell and Groff, 1974).

An experimental VX-like compound was given intravenously to 30 subjects (Sidell *et al.*, 1965). The dose to cause 70% inhibition of erythrocyte cholinesterase activity was estimated to be 0.99 μg kg^{-1}. Four subjects had symptoms of lightheadedness, anorexia, fatigue and difficulty in sleeping. More severe effects did not occur, although three subjects had erythrocyte cholinesterase activity less than 10% of their baselines. 2-PAMCl, given 3 min after agent administration, did not affect the enzyme activity.

ACCIDENTAL EXPOSURES

During the research, developmental and manufacturing programs there were many accidental exposures to these agents. Almost all were vapor exposures to sarin. Most were considered mild, and few casualties were given therapy. At the height of the program, the sudden onset of 'dim vision', a runny nose and tightness in the chest was not considered the serious matter that it might be today. Individuals have told the author that they continued work under these circumstances rather than seek medical attention. (This occurred despite emphasis by management and medical staff on prompt reporting of even minor symptoms.)

The large majority of reported instances of accidental exposure had one or more of these effects. Very often there was associated pain in the head, a vague sensation of 'not feeling well', nausea, and perhaps some very mild intellectual or cognitive slowing. Sometimes there were no objective signs of exposure, but only complaints of dyspnea, nausea, or 'not feeling well'. In the earlier years this was compounded by lack of cholinesterase monitoring, so this evidence was absent also. Several years ago the author reviewed the records of the medical facility at Edgewood Arsenal dating from the late 1940s. About 200 records had objective evidence of exposure noted on the medical chart (miosis, depressed cholinesterase activity, rhinorrhea, obvious signs of respiratory difficulty), although many times that number of records did not report any objective evidence even though the patient had reported to the facility because of nerve agent exposure. A report published in a medical journal in 1959 described 22 typical cases (Craig and Woodson, 1959).

Four severe exposures have been reported. They will be discussed later.

The largest number of exposures reported at a single facility were those discussed in two separate reports from Rocky Mountain Arsenal (Gaon and Werne; Holmes *et al.*, 1958). Each report tabulates over 300 exposures, but some cases may have been listed in both. Fewer than 4% of the patients were treated, and of those who were treated, some were in a study of whether orally administered atropine or a placebo provided the better therapy. (Atropine appeared to provide more relief in the more severe cases, but in the mild exposures there was no difference between the two.)

Two investigators reported effects in themselves from accidental exposure (Lekou *et al.*, 1966; Jager, 1957). The first report described an investigator and some of his staff who were exposed to soman vapor with ensuing miosis, headache, general malaise and some impairment of intellectual abilities. The second report documents more severe effects. The investigator had serious breathing difficulties and severe muscular spasms, but did not lose consciousness, convulse, or stop breathing. He received large amounts of atropine (45 mg within the first hour) and experienced mental difficulties, including inability to concentrate for weeks afterwards. Although the time of their return varied, all of these scientists resumed their normal work.

The earliest report of accidental exposures described 10 cases (Brown, 1948). Over the next decade other reports described about 200 cases of exposure to G agents (Gammill *et al.*, 1954; Seed, 1952; Craig and Freeman, 1953; Craig and Cornblath, 1953; Grob and Johns, 1956; Brody, 1954) and about 20 cases of exposure to VX-like agents (although none to VX) (Freeman *et al.*, 1956; Bertino *et al.*, 1957).

The first severe exposure was described in 1952 (Ward *et al.*, 1952; Ward, 1962). A physician was suddenly exposed to a large concentration of sarin. Within seconds he lost consciousness and convulsed. A minute or two later he became apneic. The second case was a subject in an experimental program who had a more severe toxic effect from a small amount of sarin on his abraded skin than others had had (Grob, 1956). A third individual had small leak in his protective mask while working in an atmosphere of sarin (Sidell, 1974). He exited the area, suddenly lost consciousness, convulsed, and became apneic. A chemist was pipetting a solution containing soman and got some in and around his mouth (Sidell, 1974). He went to the medical facility, where he lost consciousness in about 15 min after the exposure. Medical assistance was available almost immediately for all four of these individuals, and all survived. Cases one (a physician), three (a technician) and four (a chemist) resumed their normal work after 6–8 weeks with no known decrements. The further effects, if any, in case two were not reported.

CONCLUSION

Over a two-decade period the effects of nerve agents were studied in hundreds of volunteer subjects. This chapter provides a history and overview of these studies and, although incomplete, furnishes a general description of the nature of those investigations. It is a tribute to those who conducted the investigations and to the stringent and well-controlled conditions under which they were conducted that so many subjects were exposed to these toxic materials with so few serious effects.

ACKNOWLEDGEMENTS

I thank Van M. Sim, MD, a leader and major investigator in the US effort in this program, for reviewing this chapter and for his helpful advice and comments.

I also thank Mr Patsy D'Eramo for library assistance in patiently locating and retrieving the old and obscure reports referenced.

REFERENCES

Ainsworth M and Shephard RJ (1961). The intrabronchial distribution of soluble vapours at selected rates of gas flow. In: *Inhaled Particles and Vapors* (CN Davies, ed.), pp. 233–247. New York, NY, USA: Pergamon Press.

Bertino JR, Geiger LE and Sim VM (1957). *Accidental V-agent exposures*. CWLR 2156.

Bowers MB Jr, Goodman E and Sim VR (1964). Some behavioural changes in man following anticholinesterase administration. *J Nerv Ment Dis*, **138**, 383–389.

Broby BB (1954). *Seventy five cases of accidental nerve gas poisoning at Dugway Proving Ground.* DPG-MIB SR-5.

Brown EC Jr (1948). *Effects of G agents on man: clinical observations*. MDR 158.

Brown ES (1957). *Distribution of GB absorbed by oral inhalation*. CWL TM 22-1.

Clements JA, Moore JC, Johnson RP *et al.* (1952). *Observations on airway resistance in men given low doses of GB by chamber exposure*. MLRR 122.

Coombs AY and Freeman G (1954). *Obsevations of the effects of GB on intellectual function in man.* MLRR 310.

Cooper DY and Maloney JV Jr (1951). *The pulmonary effects of inhalation of low concentrations of GB in man*. MLRR 82.

Craig AB and Cornblath M (1953). *Further clinical observations on workers accidentally exposed to 'G' agents*. MLRR 234.

Craig AB and Freeman G (1953). *Clinical observations on workers accidentally exposed to 'G' agents*. MLRR 154.

Craig AB and Woodson GS (1959). Observations on the effects of exposure to nerve gas. I. Clinical observations and cholinesterase depression. *Am J Med Sci*, **238**, 13–17.

Craig FN, Cummings EG, Mounter LA *et al.* (1967). *Penetration of VX applied to the forearm at environmental temperatures of 65°and 115°F*. EATR 4064.

Craig FN, Cummings EG and Sim VM (1977). Environmental temperature and the percutaneous absorption of a cholinesterase inhibitor, VX. *J Invest Dermatol*, **68**, 357–361.

Cresthull P and Oberst FW (1964). Hazard to masked men eating and drinking in a sarin-contaminated atmosphere. *Armed Forces Chem J*, 21–23.

Cresthull P, Koon WS, Musselman NP *et al.* (1963). *Percutaneous exposure of the arm or the forearm of man to VX vapor*. CRDLR 3176.

Cummings EG and Craig FN (1965). *Effect of environmental temperature on the penetration of VX applied to the cheek*. CRDLR 3256.

Davy E and Grosser G (1959). *Some effects of certain chemical agents on motor and mental performance*. CWL SP 2-20.

Dutreau CW, McGrath FP and Bray EH (1950). *Toxicity studies on GD. 1. Median lethal concentration by inhalation in pigeons, rabbits, rats and mice. 2. Median detectable concentration by odor for man*. MDRR 8.

Freeman G, Moore JC, Clanton BR *et al.* (1952). *Observations on the effects of low concentrations of GB on man in rest and exercise*. MLRR 148.

Freeman G, Marzulli FN, Craig AB *et al.* (1953). *The toxicity of liquid GB applied to the skin of man*. MLRR 217.

Freeman G, Marzulli F, Craig AB *et al.* (1954). *The toxicity of liquid GA applied to the skin of man*. MLRR 250.

Freeman G, Hilton KC and Brown ES (1956). *V poisoning in man*. CWLR 2025.

Gammill JF, Gibson JDS and Cutuly E (1954). *Report of mild exposure to Gb in 21 persons*. DPG-MIB SR-1.

Gaon MD and Werne J. *Report of a study of mild exposure to GB at Rocky Mountain Arsenal*. Medical Department US Army. Undated report.

Gazzard MF and Price-Thomas D (1975). A comparative study of central visual field changes induced by sarin vapour and physostigmine eye drops. *Exp Eye Res*, **20**, 15–21.

Grob D (1956). The manifestations and treatment of poisoning due to nerve gas and other organic phosphate anticholinesterase compounds. *Arch Int Med*, **98**, 221–239.

Grob D and Harvey AM (1953). The effects and treatment of nerve gas poisoning. *Am J Med*, **14**, 52–63.

Grob D and Harvey JC (1958). Effects in man of the anticholinesterase compound sarin (isopropyl methyl phosphonofluoridate). *J Clin Invest*, **37**, 350–368.

Grob D and Johns RJ (1956). *Effects of V-agent organic phosphate anticholinesterase compound EA 1508 in man following accidental exposure*. CWLR 2004.

Harvey JC (1952). *Clinical observations on volunteers exposed to concentrations of GB*. MLRR 114.

Holmes JH, Vincent T, Gingrich F *et al.* (1958). *Final progress report*. Contract Progress Report DA-18-108-M-5586.

Jager BV (1957). *Case of GB poisoning*. Contract Progress Report DA 18-108-CML- 5421.

Johns RJ (1952). *The effects of low concentrations of GB on the human eye*. MLRR 100.

Ketchum JS, Sidell FR, Crowell EB Jr *et al.* (1973). Atropine, scopolamine and ditran: comparative pharmacology and antagonists in man. *Psychopharmacology*, **28**, 121–145.

Kimura KK, McNamara BP and Sim VM (1960). *Intravenous administration of VX in man*. CRDLR 3017.

Koelle GB (1981). Organophosphate poisoning – an overview. *Fund Appl Toxicol*, **1**, 129–134.

Koon WS, Cresthull P, Crook JW *et al.* (1959). *Odor detection of VX vapor with and without a stabilizer*. CWLR 2292.

Krackow EH and Fuhr I (1949). *Toxicity of GA vapor by cutaneous absorption for monkey and man*. MDR 179.

Lekov D, Dimitrov V and Mizkow Z (1966). Clinical observations of individuals contaminated by a pinacolic ester of methylfluorophosphine acid (soman). *Voenno Meditsinsko Delo*, **4**, 47–52.

Lubash GD and Clark BJ (1960). *Some metabolic studies in humans following percutaneous exposure to VX*. CRDLR 3033.

Marzulli FN and Williams MR (1953). *Studies on the evaporation, retention and penetration of GB applied to intact human and intact and abraded rabbit skin*. MLRR 199.

McGrath FP, vonBerg VJ and Oberst FW (1953). *Toxicity and perception of GF vapor*. MLRR 185.

MOD1. *The toxicology, symptoms, pathology and treatment of T2104 poisoning in animals*. Unpublished MOD report.

MOD2. *Eye effect of T2104*. Unpublished MOD report.

MOD3. *Tests of visual acuity affecting efficiency following exposure to a harassing dosage of tabun*. Unpublished MOD report.

MOD4. *Report on exposures of unprotected men and rabbits to low concentrations of nerve gas vapour*. Unpublished MOD report.

MOD5. *Cholinesterase as an aid to the early diagnosis of nerve gas poisoning. Part II: The variation of blood cholinesterases in man before and after the administration of very small quantities of G vapour by inhalation*. Unpublished MOD report.

MOD6. *The percutaneous toxicity of the G-compounds*. Unpublished MOD report.

MOD7. *Psychological effects of a G-agent on men*. Unpublished MOD report.

MOD8. *Psychological effects of a G-agent on men; second report*. Unpublished MOD report.

MOD9. *Cognitive and emotional changes after exposure to GB*. Unpublished MOD report.

MOD10. *Studies of psychomotor performance. The effect of GB*. Unpublished MOD report.

MOD11. *Physical performance following inhalation of GB*. Unpublished MOD report.

MOD12. *The effects of a single exposure to GB (sarin) on human physical performance*. Unpublished MOD report.

MOD13. *The intercostal muscles and GB inhalation in man*. Unpublished MOD report.

MOD14. *The action of anticholinesterases and pressor amines on the capacity veins of the leg*. Unpublished MOD report.

MOD15. *Changes in cardiac output following the administration of sarin and other pharmacological agents. Part 3. Human experiments using the low frequency critically damped ballistocardiograph*. Unpublished MOD report.

MOD16. *GB exposure during simultaneous experimental acclimizatization to heat and cold in man*. Unpublished MOD report.

MOD17. *Detection of the 'G' gases by smell*. Unpublished MOD report.

MOD18. *An evaluation of the functional changes produced by the inhalation of GB vapour*. Unpublished MOD report.

MOD19. *Air-way resistance changes in men exposed to GB vapour*. Unpublished MOD report.

MOD20. *The single breath administration of sarin*. Unpublished MOD report.

MOD21. *The intrabronchial distribution of soluble vapours at selected rates of gas flow*. Unpublished MOD report.

MOD22. *The retention of inhaled GB vapour*. Unpublished MOD report.

MOD23. *The effects of a minor exposure to GB on military efficiency*. Unpublished MOD report.

MOD24. *The effects of atropine and wearing respirators on the military efficiency of troops exposed to GB*. Unpublished MOD report.

MOD25. *Effect of pupil size of exposure to GB vapor*. Unpublished MOD report.

MOD26. *Estimation of the concentrations of nerve agent vapour required to produce measured degrees of miosis in rabbit and human eyes*. Unpublished MOD report.

MOD27. *The effects of a chemical agent on the eyes of aircrew*. Unpublished MOD report.

MOD28. *A comparative study of central visual field changes induced by GB (isopropylmethyl phosphonofluoridate) vapours and physostigmine salicylate eyedrops*. Unpublished MOD report.

MOD29. *Human exposure to VX vapour*. Unpublished MOD report.

MOD30. *Recovery of blood cholinesterase in man after exposure to VX*. Unpublished MOD report.

MOD31. *Passage of VX through human skin*. Unpublished MOD report.

MOD32. *Penetration of VX through clothing*. Unpublished MOD report.

Moylan-Jones R and Price-Thomas DA (1973). Cyclopentolate in treatment of sarin miosis. *Br J Pharmacol*, **48**, 309–313.

Neitlich H (1965). Effect of percutaneous GD on human studies. *CRDL TM-2-21*.

Oberst FW (1961). Factors affecting inhalation and retention of toxic vapors. In: *Inhaled Particles and Vapours* (CN Davies, ed.), pp. 249–265. New York, NY, USA: Pergamon Press.

Oberst FW, Koon WS and Crook JW (1952). *Methods for quantitative determination of GB absorbed from inspired air of man during rest and exercise*. MLRR 143.

Oberst FW, Crook JW, Christensen MK *et al.* (1959). *Inhaled GB retention studies in man at rest and during activity*. CWLR 2296.

Oberst FW, Koon WS, Christensen MK *et al.* (1968). Retention of inhaled sarin vapor and its effect on red blood cell cholinesterase activity in man. *Clin Pharmacol Ther*, **9**, 421–427.

Rengstorff RH (1985). Accidental exposure to sarin; vision effects. *Arch Toxicol*, **56**, 201–203.

Rubin LS and Goldberg MN (1957a). *The effect of GB on dark adaption in man. III. The effect of tertiary and quaternary atropine salts on absolute scotopic threshold changes engendered by GB*. CWL TR 2155.

Rubin LS and Goldberg MN (1957b). Effect of sarin on dark adaptation in man: threshold changes. *J Appl Physiol*, **11**, 439–444.

Rubin LS and Goldberg MN (1958). Effect of tertiary and quaternary atropine salts on absolute scotopic threshold changes produced by an anticholinesterase (sarin). *J Appl Physiol*, **12**, 305–310.

Rubin LS, Krop S and Goldberg MN (1957). Effect of sarin on dark adaptation in man: mechanism of action. *J Appl Physiol*, **11**, 445–449.

Seed JC (1952). *An accident involving vapor exposure to a nerve gas. MLRR 146.*

Sidell FR (1967). *Human responses to intravenous VX.* EATR 4082.

Sidell FR (1974). Soman and sarin: clinical manifestations and treatment of accidental poisoning by organophosphates. *Clin Toxicol*, **7**, 1–17.

Sidell FR (1992). Clinical considerations in nerve agent intoxication. In: *Chemical Warfare Agents* (SM Somani, ed.), pp. 156–194. San Diego, CA, USA: Academic Press, Inc.

Sidell FR and Groff WA (1966). *Oral toxicity of VX to humans.* EATR 4009.

Sidell FR and Groff WA (1974). The reactivatibility of cholinesterase inhibited by VX and sarin in man. *Toxicol Appl Pharmacol*, **27**, 241–252.

Sidell FR, Groff WA and Vocci F (1965). *Effects of EA 3148 administered intravenously to humans.* CRDL TM 2-31.

Sidell FR, Aghajanian GK and Groff WA (1973). The reversal of anticholinergic intoxication in man with the cholinesterase inhibitor VX. *Proc Soc Exp Biol Med*, **144**, 725–730.

Sim VM (1962). *Variability of different intact human-skin sites to the penetration of VX.* CRDLR 3122.

Sim VM and Stubbs JL (1960). *VX percutaneous studies in man.* CRDLR 3015.

Sim VM, McClure C Jr, Vocci FJ *et al.* (1964). *Tolerance of man to VX-contaminated water.* CRDLR 3231.

Stewart WC, Madill HAD and Dyer AM (1968). Night vision in the miotic eye. *Can Med Assoc J*, **99**, 1145–1148.

Vocci FJ, Hickman WE, Mehlman MA *et al.* (1959). *The effect of n-octylamine on the penetration of VX by the percutaneous route in man.* CWL TM, 24–22.

Ward JR (1962). Case report: exposure to a nerve gas. In: *Artificial Respiration. Theory and Applications* (JL Whittenberger, ed.), pp. 258–265. New York, NY, USA: Harper and Row.

Ward JR, Gosselin R, Comstock J *et al.* (1952). *Case report of a severe human poisoning by GB.* MLRR 151.

10 NERVE AGENTS: LOW-DOSE EFFECTS

Leah Scott

Dstl, Porton Down, Salisbury, UK

INTRODUCTION

What constitutes 'low' dose and why is it important?

The traditional emphasis of defence research in the area of nerve agents has, understandably, focused on the prevention or mitigation of the effects of exposure to doses that induce lethality or severe physical incapacitation. In consequence, relatively little attention has been paid to the effects of lower doses.

The sequelae of poisoning by nerve agents are highly dependent upon both the level and the characteristics of exposure in terms of route, concentration, duration and frequency. The spectrum of observable signs of poisoning ranges from miosis and transitory disruption of behaviour, with or without minor clinical signs, to the well-known major manifestations of cholinergic over stimulation, involving marked motor and CNS perturbations, which may lead to death or severe and prolonged physical incapacitation which in the absence of medical intervention would result in death.

'Low dose' is sometimes defined in terms of exposure to levels of nerve agent that are described as asymptomatic in that they do not induce overt acute signs or symptoms. While this chapter will focus primarily on the potential effects of such asymptomatic exposures, levels of nerve agent that induce miosis and those which result in transitory disruption of behaviour with minor clinical signs will also be considered.

Nerve agent–induced miosis is generally believed to be a local event that is a consequence of direct contact between nerve agent vapour and the eye. It can occur following exposure to extremely low levels of nerve agent that do not lead to inhibition of erythrocyte cholinesterase and which do not produce other signs or symptoms. Following higher levels of nerve agent exposure, the more marked clinical signs observed are associated with high levels of cholinesterase inhibition.

Figure 1 illustrates the relationships between threshold levels for various indices of nerve agent exposure.

It is important to be aware of both the short-term and long-term effects of exposure to levels of nerve agent which are not acutely life threatening. Widespread debate continues about whether exposure to such levels of nerve agents and other OP anticholinesterases leads to long-term adverse effects on human health. One of the key recommendations of a report by the Committee on Toxicity of Chemicals in Food, Consumer Products and the Environment (1999) (UK) was that further work was needed to clarify the potential long-term effects of OP exposure, especially when acute exposure did not give rise to signs or symptoms of toxicity.

There are a number of military scenarios in which potential exposure to extremely low levels of nerve agents could theoretically occur and these range from a single exposure lasting for some minutes (e.g. in the event of a delay in donning respiratory protection within an area

Chemical Warfare Agents: Toxicology and Treatment (2nd Edition)
Edited by Timothy C. Marrs, Robert L. Maynard and Frederick R. Sidell

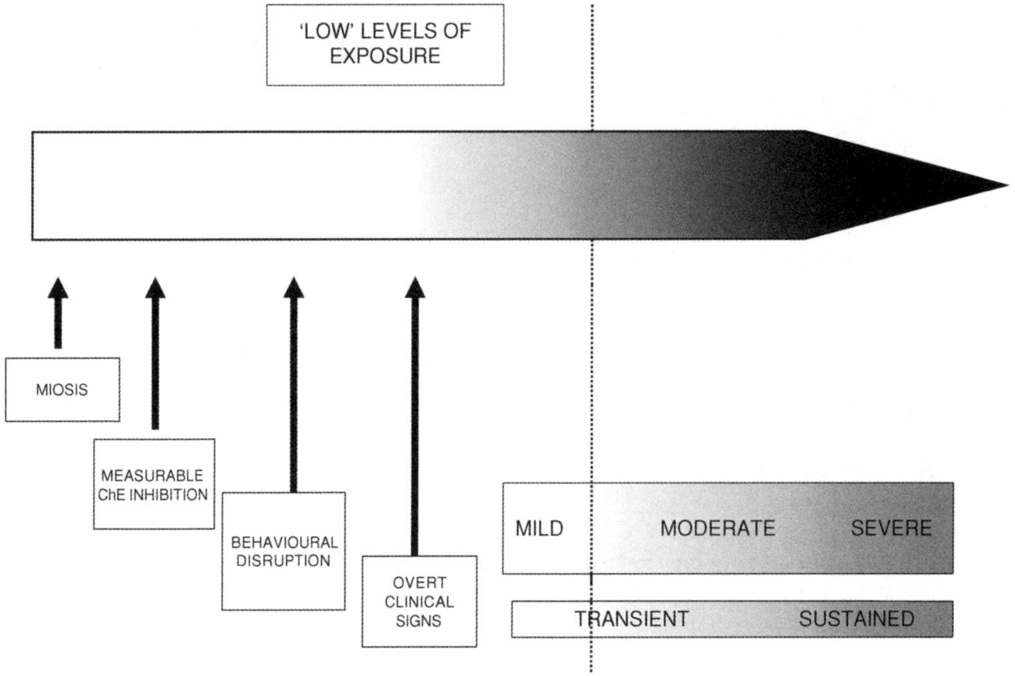

Figure 1. Illustration of the relationships between threshold levels for various indices of nerve agent exposure

of downwind hazard following a nerve agent release), to repeated exposures, either continuous or intermittent and lasting for several days (e.g. while personnel are engaged in activities within some areas of collective protection). In such circumstances, it is necessary (a) to assess whether there are operational implications arising from low level exposures and (b) to consider whether such events have implications in the context of the duty of care owed to service personnel.

In both civilian and military situations, data on the effects of a range of doses of agent are required to inform the need for medical management. Expedient, life-saving therapeutic interventions and subsequent medical management for those exhibiting marked acute signs and symptoms of nerve agent poisoning are already relatively well established (see Chapters 11, 14, 15 and 16), although the longer term sequelae of higher-level exposure supported by medical intervention are not well understood. Much less information is available on the management of those exposed to lower levels of nerve agents. If clinical signs and symptoms are absent, blood cholinesterase profiles would provide limited but not unequivocal evidence of nerve agent exposure.

Nerve agent exposure in man

The potential sources of information on the acute and long-term consequences of human exposures to OPs in general are extremely limited and collecting and interpreting information on the effects of low-level exposures is even more problematical. The main opportunities for gathering such information are as follows: (a) data derived from accidental/occupational exposure to nerve agents and OP pesticides, (b) data from laboratory-based experiments involving volunteers, (c) data collected following the intentional release of sarin by terrorists in Japan in 1994/5, and (d) data from recent and contemporary exposure to OP pesticides, especially in the developing world where deliberate self-poisoning represents a major health problem (see Chapters 9 and 13 and review by Karalliedde *et al.* [2001]).

The acute toxicity of OP compounds is relatively well defined and based upon historical data (e.g. Grob and Harvey, 1953) derived from accidental/occupational exposure to nerve agents and pesticides and experiments in volunteers. Long-term effects, especially those that reflect central nervous system function, are less well characterized. Following on from an investigation, which began in the early 1950s, of industrial and agricultural workers exposed to a diversity of OPs, Metcalf and Holmes (1969) reported EEG, psychological and neurological indices in this population. A subset of this population was also included in EEG studies reported by Duffy *et al.* (1979) and Burchfiel and Duffy (1982), which were undertaken to investigate a number of established accidental exposures to sarin in the late 1960s and the early 1970s in civilian employees of the US government. These individuals had complained of a range of symptoms, including sleep disturbances, memory loss and difficulties in concentration. Following pesticide exposure, a similar range of symptoms has been reported; Sherman *et al.* (1995), for example, describes tension, anxiety, difficulty in concentrating, slowness of recall, mental confusion and sleep disturbances in such circumstances.

The Burchfiel and Duffy (1982) study, which also included an investigation of the effects of sarin in rhesus monkeys (see below), showed statistically significant differences between sarin exposed and control groups. There were increases in the amount of beta-activity in the EEG of the sarin-exposed subjects who also showed significantly increased rapid eye movement (REM) sleep. Similar changes in beta-activity were also observed in rhesus monkeys. Unfortunately, the functional significance of these differences remains unclear.

In more recent times, the Aum Shinrikyo religious cult used sarin in two terrorist attacks in Japan. At Matsumoto, in 1994, overall 600 individuals reported acute signs and symptoms, 58 were admitted to hospital and seven people died (Morita *et al.*, 1995). Following the release of sarin on the Tokyo underground in 1995, approximately 1000 individuals exhibited signs and symptoms and twelve people died (Nagao *et al.*, 1997; IoM, 2000) (see Chapter 13). Not only were substantial numbers of members of the public directly affected, but in both situations, first responders and medical staff were also exposed to sarin as they attended those exhibiting marked signs and symptoms (Nakajima *et al.*, 1997). One to five years following exposure, a proportion of patients reported continuing long-term sequelae reflecting physical and/or psychological disturbances (Yokoyama *et al.*, 1998; Nishiwaki *et al.*, 2001).

It is a common feature of the historical human evidence obtained from accidental/occupational exposures and instances of intentional self-poisoning as well as the more recent studies which have monitored the survivors of the 1995 sarin attack on the Tokyo underground, that, inevitably, there are no definitive data on exposure levels, biochemical sequelae or information on predisposing factors. Moreover, the serious psychological consequences of terrorist events involving large numbers of people represent an additional complication. In such cases, when outcomes such as Post Traumatic Stress Disorder occur, it is extremely difficult, if not impossible, to define the involvement of the toxicological elements *per se*, let alone to ascribe causality. These problems pose enormous challenges for experimental design, statistical comparison and interpretation of the results.

Determining whether exposure to nerve agents or other OPs has occurred is particularly difficult if acute signs and symptoms are absent. In situations such as the terrorist use of sarin in Japan, the priorities following the event will, wholly appropriately, be primarily concerned with saving life. Subsequent medical monitoring will undoubtedly be focused on monitoring the condition of survivors who have exhibited moderate to severe acute effects.

No acute signs or symptoms were exhibited following the destruction of chemical munitions at Khamisiyah, Iraq in 1991, although it was theoretically possible that almost 100 000 US troops could have been exposed to extremely low concentrations of sarin or cyclosarin (e.g. Winkenwerder, 2002; IoM, 2000). The large number of retrospective studies concerning this incident has been extensively reviewed (IoM, 2004).

At present, the scope for laboratory-based studies involving exposure of volunteers to OP compounds is extremely limited. Earlier studies have been summarized previously (NRC, 1982,1983) and Page (2003) reported a retrospective study of US military volunteers from 1955 to 1975, summarized by Sidell in

Chapter 9. The study by Page involved a survey of neuropsychiatric impairment, illness attitudes, peripheral nerve disease, sleep disorders, vestibular dysfunction and reproductive history undertaken at least 25 years following exposure to a range of anticholinesterases ($n = 1339$). No excess mortality from particular conditions was reported but the interpretation of the study outcomes is severely limited by a lack of information on exposure dose history and no clear conclusions could be drawn.

Baker and Sedgwick (1996) reported small changes in subtle electrophysiological measures of muscle electrical activity (jitter) in volunteers 3 h following exposure of volunteers – to low concentrations of sarin (Ct: 5–15 mg min m^{-3}) during the 1980s. The effects were still present one year later but were absent two years after sarin administration. The functional significance of such changes are not well characterized. More recently, Garfitt *et al.* (2002) investigated the effects of both oral and dermal exposure to low doses of the OP sheep dip, diazinon, although these studies were directed towards identification of biochemical markers of exposure rather than the investigation of long-term sequelae.

Instances of deliberate self-poisoning (e.g. Eddleston, 2000) inevitably involve the self-administration of high doses of OPs and are, in consequence, unlikely to represent opportunities for investigation of the effects of exposure which does not give rise to signs and symptoms. These situations, do, however, represent an important opportunity to collect information on the long-term consequences of medical intervention, notwithstanding difficulties in interpretation due to predisposing medical conditions and the fact that exposure to pesticides, rather than nerve agents, are involved.

While inhibition of cholinesterase is an important common feature of poisoning by OP pesticides and nerve agents, there are marked differences in potencies, physicochemical properties and toxicokinetic profiles between these materials. For these reasons, care must always be taken when relating the significant body of literature pertaining to OPs other than nerve agents (both from epidemiological and non-epidemiological sources) to nerve agents.

ANIMAL STUDIES: GENERAL CONSIDERATIONS

For the reasons outlined above, this area of work is heavily reliant upon extrapolation of animal-derived data to man. The sources of animal data have been extensively reviewed by the IoM (2000, 2004). Well-designed animal studies can provide important information on dose–response relationships, although it is recognized that the relative strengths and weaknesses of particular animal models must always be considered. Thus, data from studies in rats where the presence of scavenger enzymes in this species to some extent limits the extrapolation of data to man, have nonetheless provided important pointers for more definitive studies in porcine and non-human primate models. Moreover, as regards studies of effects on the central nervous system, rodent species are unable to perform tasks of the required level of complexity and their polyphasic sleeping patterns are markedly different from human monophasic sleeping patterns. For these reasons, non-human primate models are recommended for closely controlled studies that set out to investigate the more subtle sequelae of exposure to nerve agents and to relate the findings to the pattern of adverse signs and symptoms most frequently reported following nerve agent exposure in man.

In the majority of animal studies which have investigated the effects of low-level exposures, it has been expedient to administer nerve agents by injection. While this is a useful starting point, more definitive studies involving inhalation or percutaneous administration may also be required to address specific questions or model particular situations.

From a methodological perspective, it is particularly challenging to investigate the effects of low-level exposure of long duration in animal models and compromises are inevitable. Nose-only exposures have been used by Benschop *et al.* (1998) to determine the toxicokinetics of soman stereoisomers in atropinized guinea-pigs and subsequently, related studies have been undertaken in marmosets (van Helden *et al.*, 2005). Hulet *et al.* (2006) have investigated the effects of exposure concentration and duration on pupil size in conscious, restrained minipigs. Undoubtedly, the requirement for anaesthesia and/or restraint in such studies represents a significant potential confounder, which needs to be considered in the interpretation and extrapolation of the results. Whole-body exposure systems enable the effects of nerve agents to be studied in conscious unrestrained animals (e.g. Bartosova-Sevelova and Bajgar (2005) who reported on the effects of a 4-h exposure to sarin in rats), although, in such circumstances, there are potential confounding factors associated with agent retained on the fur and which could be subsequently ingested.

ANIMAL STUDIES: SHORT-TERM EFFECTS OF LOW DOSES

Miosis studies

Nerve agent vapour rapidly crosses the conjunctiva and cornea and inhibits local cholinesterases. This leads to over-stimulation of muscarinic receptors of the sphincter muscle of the iris and the ciliary muscle of the lens. Constriction of the pupil and difficulties with accommodation result and the capability of the pupil to dark-adapt, i.e. to dilate, is also severely compromised. The impact on visual ability is operationally significant and as indicated previously there are a number of situations in which miosis might be encountered.

There have long been important questions in inhalation and ocular studies about the relationship between dosage as expressed by the product of exposure concentration and time (see Chapter 2). With a growing appreciation of the importance of exposure duration in defining toxic dosages, the concept of toxic load is becoming more extensively used (ten Berge *et al.*, 1986). The effects of exposure concentration and duration on pupil size have been investigated in rats following sarin administration (Mioduszewski *et al.*, 2002) and these studies have recently been repeated in minipigs (Hulet *et al.*, 2006). This facilitates further investigation of the effects of repeated exposures and may lead to a better understanding of the miotic tolerance, previously reported in rats (Dabisch *et al.*, 2005). This, in turn, may have implications for understanding tolerance observed in rats to multiple exposures of nerve agents at higher dose levels. For example, the effects of repeated exposure at what are described as subacute levels (0.3–0.6 LD_{50}) have been reported by Hulet *et al.* (2002). Doses of up to 0.3–0.4 LD_{50} administered by the subcutaneous route on up to 10 occasions over a 2-week period were remarkably well tolerated although there were statistically significant changes from control in terms of behavioural responsiveness following the higher dose.

Clinical signs and visually guided reaching in marmosets

Figure 2 shows the relationship between levels of inhibition of erythrocyte cholinesterase and performance of a task of visually guided reaching in marmosets. No clinical signs were observed

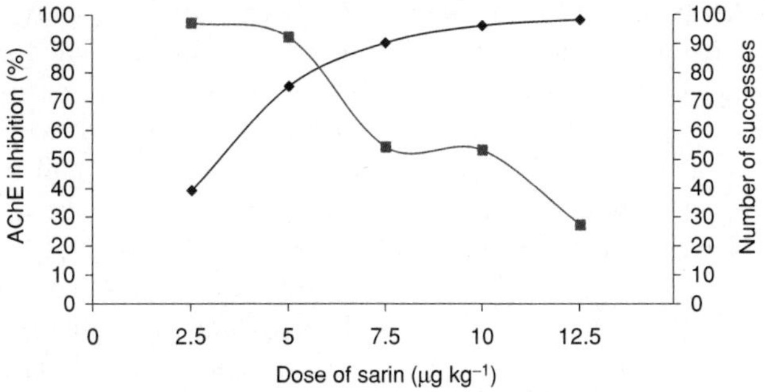

Figure 2. AChE inhibition and visually guided reaching in marmosets 1 h following sarin – dose/effect profiles: ♦, % RBC AChE inhibition; ■, successful retrievals. From D'Mello and Duffy (1985) and French *et al.* (unpublished observations).

following intramuscular administration of 7.5 $\mu g\,kg^{-1}$ sarin, which approximates to 0.3 LD_{50} of sarin and was the lowest dose which induced a marked decrement in a visually-guided reaching task in marmosets (D'Mello and Duffy, 1985). Minor, transient clinical signs were seen at the higher doses (see below).

Do nerve agents improve cognitive function? While investigating the effects of a combination of physostigmine and hyoscine as a pretreatment for nerve agent poisoning, Muggleton *et al.* (2003) made an interesting observation concerning the effects of 0.5 LD_{50} doses of either sarin or soman, on a serial reversal performance. Sequences of cognitive tests were presented which required marmosets, trained to associate one of a pair of stimuli with reward, to switch their responding to the previously unrewarded stimulus. In the absence of pretreatment, the doses of agents used induced minor transient signs of poisoning such as salivation, ataxia, piloerection and tail arching which were no longer evident 30–60 min following nerve agent administration. When a serial reversal task was presented 60 min following nerve agent administration, the majority of animals made no attempt to perform. However, when tested 24 h later, performance was significantly better than control in that animals made fewer errors per subject per reversal. Although the test conditions had not been optimized to facilitate observation of performance improvements, the pattern of responding clearly suggested that these improvements were due to administration of sarin or soman rather than an effect on learning *per se*. This observation was felt to warrant further investigation. If the finding is confirmed, administration of substances, such as nerve agents, which induce cholinesterase inhibition of long duration, may have utility in the management of conditions such as Alzheimer's Disease where the use of reversible cholinesterase inhibitors has been established for some time.

ANIMAL STUDIES: LONG-TERM EFFECTS OF LOW DOSES

Experimental studies in rhesus monkeys suggest that administration of sarin at doses which did not induce overt clinical signs gave rise to small but statistically significant changes in brain electrical activity as gauged by electroencephalography (EEG). Burchfiel *et al.* (1976) demonstrated a significant increase in the beta-component (13–50 Hz) of the EEG spectrum of rhesus monkeys in response to both single and repeated doses of sarin. Unfortunately, in this study, EEG findings were only reported 24 h and 12 months following sarin administration and no indication of level of cholinesterase inhibition was given. The functional significance of these EEG changes is unknown.

More recent studies undertaken at Dstl (Pearce *et al.*, 1999) were designed to determine whether similar sarin-induced changes in EEG would be observed in marmosets up to 15 months following administration. Performance of a challenging cognitive task and levels of erythrocyte cholinesterase were monitored throughout. The behavioural task was included to aid in the interpretation of the functional significance of any EEG changes that might be observed.

The dose of sarin used in the study, which inhibited erythocyte cholinesterase by approximately 50%, and approximated to 1/10 of the LD_{50} for this agent in marmosets, did not give rise to overt signs of poisoning and produced no significant changes in either the EEG or performance of the behavioural task. The inconsistency with the findings of the earlier study in rhesus monkeys in that no consistent changes were observed in the EEG may have resulted from methodological differences between the studies with respect to technical refinements in the later marmoset study. The fact that there was no treatment-related disruption of cognitive performance on sensitive indices of attentional set shifting is strong evidence that at this dose level acute administration of sarin does not give rise to adverse long-term behavioural effects.

Although a dose–response investigation using the long-term low dose marmoset model has not yet been undertaken for sarin, a subsequent study investigated the effects of a range of doses of the OP sheep dip, diazinon (Muggleton *et al.*, 2005). In this study, EEG, performance of a similar cognitive task and sleep were monitored for up to 12 months following doses of diazinon which inhibited erythrocyte cholinesterase by ca. 40–80% and plasma cholinesterase by 85–97% Although there were dose-related short-term changes in sleep architecture, there were no long-term changes in the EEG and no indication of long-term disruption of behaviour.

SUMMARY AND WAY FORWARD

In their 2004 report, which focused on sarin, the IoM review committee considered animal toxicological data relating to neurotoxicity, immunotoxicity, genotoxicity, carcinogenicity and genetic susceptibility. In terms of human data, the committee considered epidemiological studies which primarily investigated associations between neurological and cardiovascular symptoms and signs and *possible* exposure to sarin or cyclosarin.

The committee concluded that there was 'inadequate or insufficient evidence' to determine whether an association exists between exposure to sarin at doses which do not give rise to acute signs and symptoms and subsequent long-term neurological or cardiovascular effects. 'On present evidence, medically significant late sequelae of sarin exposure seem hard to prove' (Sharp, 2006).

There remains a paucity of data on other nerve agents and more complete dose–response relationships from animal studies which optimize extrapolation of animal-derived data to man are needed.

REFERENCES

Baker DJ and Sedgwick EM (1996). Single fibre electromyographic changes in man after organophosphate exposure. *Human Exp Toxicol*, **15**, 369–375.

Bartosova-Sevelova L and Bajgar J (2005). Changes of acetylcholinesterase activity after long-term exposure to sarin vapors in rats. *Human Exp Toxicol*, **24**, 363–367.

Benschop HP, Trap HC, Spruit HET, van der Wiel HJ *et al.* (1998). Low level nose-only exposure to the nerve agent soman: toxicokinetics of soman stereoisomers and cholinesterase inhibition in atropinized guinea pigs. *Toxicol Appl Pharmacol*, **153**, 179–185.

Brown MA and Brix KA (1998). Review of health consequences from high-, intermediate- and low-level exposure to organophosphorus nerve agents. *J Appl Toxicol*, **18**, 393–408.

Burchfiel JL (1976). Persistent effect of sarin and dieldrin on the electroencephalogram of monkey and man. *Toxicol Appl Pharmacol*, **35**, 365–379.

Burchfiel JL and Duffy FH (1982). Organophosphate neurotoxicity: chronic effects of sarin on the electroencephalogram of monkey and man. *Neurobehav Toxicol Teratol*, **4**, 767–778.

Committee on Toxicity of Chemicals in Food, Consumer Products and the Environment (1999). *Organophosphates.* London, UK: Department of Health.

D'Mello GD and Duffy EAM (1985). 'The acute toxicity of sarin in marmosets *(Callithrix jacchus)*: a behavioural analysis. *Fundam Appl Toxicol*, **5**, S169–S174.

D'Mello GD, Lewandowska EAM and Miles SS (1983). Effects of the acetylcholinesterase inhibitor sarin upon a visually guided reaching response in the marmoset. In: *Application of Behavioral Pharmacology in Toxicology* (G Zbinden, V Cuomo, G Racagni and B Weiss eds), pp. 153–156. New York, NY, USA: Raven Press.

Dabisch PA, Burnett DC, Miller DB *et al.* (2005). Tolerence to the miotic effect of sarin vapor in rats after multiple low-level exposures. *J Ocu Pharmacol Ther*, **21**, 182–195.

Dabisch PA, Miller DB, Reutter SA *et al.* (2005). Miotic tolerence to sarin vapor exposure: role of sympathetic and parasympathetic nervous systems. *Toxicol Sci*, **85**, 1041–1047.

Duffy FH, Burchfiel JL, Bartels PH *et al.* (1979). Long-term effects of an organophosphate upon the human electroencephalogram. *Toxicol Appl Pharmacol*, **47**, 161–176.

Eddleston M (2000). Patterns and problems of deliberate self-poisoning in the developing world. *QJ Med*, **93**, 715–731.

Garfitt SJ, Jones K, Mason HJ *et al.* (2002). Exposure to the organophosphate diazinon: data from a human volunteer study with oral and dermal doses. *Toxicol Lett*, **134**, 105–113.

Grob D and Harvey AM (1953). The effects and treatment of nerve gas poisoning. *Am J Med*, **14**, 52–63.

Hulet SW, McDonough JH and Shih TM (2002). The dose–response effects of repeated subacute sarin exposure on guinea-pigs. *Pharmacol Biochem Behav*, **72**, 835–845.

Hulet SW, Sommerville DR, Crosier RB *et al.* (2006). Comparison of low-level sarin and cyclosarin vapor exposure on pupil size of the Göttingen minipig: effects of exposure concentration and duration. *Inhal Toxicol*, **18** 143–153.

IoM (Institute of Medicine) (2000). *Gulf War and Health*, Volume 1, *Depleted Uranium, Pyridostigmine Bromide, Sarin and Vaccines*. Washington, DC, USA: National Academy Press.

IoM (Institute of Medicine) (2004). *Gulf War and Health: Updated Literature Review of Sarin*. Washington, DC, USA: National Academy Press (www.nap.edu).

Karalliede L, Eddlestone M, Murray V (2001). The global picture of organophosphate insecticide poisoning. In: *Organophosphates and Health* (Karalliede L, Feldman S, Henry J, Marrs T eds), pp 431–471. London, UK: Imperial College Press.

Metcalf DR and Holmes JH (1969). EEG, psychological and neurological alterations in humans with organophosphorus exposure. *Ann NY Acad Sci*, **160**, 357–365.

Mioduszewski RJ, Manthei JA, Way RA *et al.* (2002). Interaction of exposure concentration and duration in determining acute toxic effects of sarin vapor in rats. *Toxicol Sci*, **66**, 176–184.

Morita H, Yanagisawa N, Nakajima T *et al.* (1995). Sarin poisoning in Matsumoto, Japan. *Lancet*, **346**(8970), 290–293.

Muggleton NG, Bowditch AP, Crofts HS *et al.* (2003). Assessment of a combination of physostigmine and scopolamine as a pretreatment against the behavioural effects of organophosphates in the common marmoset (*Callithrix jacchus*). *Psychopharmacology*, **166**, 212–220.

Muggleton NG, Smith AJ, Scott EAM *et al.* (2005). A long-term study of the effects of diazinon on sleep, the electrocortigram and cognitive behaviour in common marmosets. *J Psychopharmacol*, **19**, 455–466.

Nagao M, Takatori T, Matsuda Y *et al.* (1997). Definitive evidence for the acute sarin poisoning diagnosis in the Tokyo subway. *Toxicol Appl Pharmacol*, **144**, 198–203.

Nakajima T, Satos S, Morita H *et al.* (1997). Sarin poisoning of a rescue team in the Matsumoto sarin incident in Japan. *Occupational and Environmental Medicine*, **54**(10), 697–701.

Nishiwaki Y, Maekawa K, Ogawa Y *et al.* (2001). Sarin Health Effects Study Group. Effects of sarin on the nervous system in rescue team staff members and police officers three years after the Toyko subway sarin attack. *Environ Health Perspect*, **109**, 1169–1173.

NRC (National Research Council) (1982). *Possible Long-Term Health Effects of Short-Term Exposure to Chemical Agents*, Volume 1, *Anticholinesterases and Anticholinergics*. Washington, DC, USA: National Academy Press.

NRC (National Research Council) (1985). *Possible Long-Term Health Effects of Short-Term Exposure to Chemical Agents*, Volume 3, *Final Report on Current Health Status of Test Subjects*. Washington, DC, USA: National Academy Press.

Page WF (2003). Long-term health effects of exposure to sarin and other anticholinesterases chemical warfare agents. *Milit Med*, **168**, 239–245.

Pearce PC, Crofts HS, Muggleton NG *et al.* (1999). The effects of acutely administered low dose sarin on cognitive behaviour and the electroencephalogram in the common marmoset. *J Psychopharmacol*, **13**, 128–135.

Sharp D (2006). Long-term effects of sarin. *Lancet* Jan 14; **367**(9505), 95–97.

Sherman JD (1995). Organophosphate pesticides – neurological and respiratory toxicity. *Toxicol Ind Health*, **11**, 33–39.

ten Berge WF, Zwart A and Appelman LM (1986). Concentration–time mortality response relationship of irritant and systemically acting vapours and gases. *J Hazard Mater*, **13**, 301–309.

van Helden HPM, Vanwersch RAP, Kuijpers WC *et al.* (2004). Low levels of sarin affect the EEG in marmoset monkeys: a pilot study. *J Appl Toxicol*, **24**, 475–483.

Winkenwerder W (2002). *US demolition operations at Khamisiyah*. Final Report. Washington, DC, USA: Department of Defense (http://www.gulflink.osd.mil/khamisiyah_iii/).

Yokoyama K, Ariaki S, Murata K, Nishikitani M, Okumura T, Ishimatsu S, Takasu N (1998). Chronic neurobehavioural effects of Tokyo subway sarin poisoning in relation to posttraumatic stress disorder. *Arch Environ Health*, **53**, 249–56.

11 MANAGING CIVILIAN CASUALTIES AFFECTED BY NERVE AGENTS

J. Allister Vale[1], Paul Rice[2] and Timothy C. Marrs[1,3]

[1] *National Poisons Information Service, Birmingham, UK*
[2] *Dstl Porton Down, Salisbury, UK*
[3] *Edentox Associates, Edenbridge, UK*

INTRODUCTION

The successful management of patients affected by nerve agent poisoning depends on the clinician:

(i) recognizing the factors which influence the impact of nerve agents;
(ii) appreciating the differences between on-target military attacks against relatively well-protected armed forces and nerve agent attacks initiated by terrorists against a civilian population;
(iii) understanding the mechanisms of nerve agent toxicity and applying this knowledge to the treatment options;
(iv) making an accurate diagnosis and assessment of the severity of intoxication;
(v) maintaining vital body functions and undertaking adequate clinical monitoring;
(vi) minimizing further absorption of the nerve agent to which the casualty has been exposed;
(vii) using atropine, oxime and diazepam optimally.

This chapter provides an overview of the management of civilian casualties from nerve agent poisoning and should be read in conjunction with the detailed reviews in Chapter 14 (McDonough and Shih), Chapter 15 (Eyer and Worek), Chapter 16 (Marrs and Sellström) and Chapter 12 (Baker).

FACTORS INFLUENCING THE IMPACT OF NERVE AGENTS

An understanding of the impact of the physicochemical properties and toxicity of each nerve agent, as well as the prevailing meteorological factors and the route of delivery, is important if the clinical and public health responses to a deliberate release are to be optimized.

Physicochemical properties

PHYSICAL STATE

Is the nerve agent a volatile or non-volatile liquid? Sarin (volatility = 21 862 mg m^{-3} at 25°C) is much more volatile than tabun (611 mg m^{-3} at 25°C), although both agents tend to evaporate from the skin (or other surface) more quickly than VX, which is a non-volatile oily liquid (10.07 mg m^{-3} at 25°C) (Maynard and Beswick, 1992).

VAPOUR PRESSURE

This is a measure of how quickly a nerve agent will evaporate and is increased by a rise in ambient temperature. For example, the vapour pressure for sarin is 0.52 mmHg at 0°C and 2.9 mmHg at 25°C (Maynard and Beswick, 1992), whereas that of tabun is 0.004 mmHg at 0°C and 0.07 mmHg at 25°C (Maynard and Beswick, 1992).

Chemical Warfare Agents: Toxicology and Treatment (2nd Edition)
Edited by Timothy C. Marrs, Robert L. Maynard and Frederick R. Sidell © 2007 John Wiley & Sons, Ltd

VAPOUR DENSITY

Nerve agents with a high vapour density compared to air, such as VX (9.2), stay at ground level and tend to accumulate in low-lying areas.

SOLUBILITY AND STABILITY IN WATER

The solubility of tabun is 9.8 g $(100\ g)^{-1}$, whereas soman has a solubility of 2.1 g $(100\ g)^{-1}$. The stability of nerve agents in water depends on the temperature and pH of the solution. For example, the half-life of sarin is 5.4 h at 25°C and pH 7.0, whereas the half-lives of soman and VX are, respectively, 82 h and 350 days (Somani *et al.*, 1992).

STABILITY

This refers to the ability of a nerve agent to survive dissemination and transport to the site of deployment.

PERSISTENCE

Non-persistent agents, such as sarin, soman and tabun, disperse rapidly after release and present an immediate short duration inhalational hazard but may be made persistent by a 'thickening agent', such as ethyl methacrylate. In contrast, persistent agents, such as VX, continue to be a contact hazard and may be absorbed through the skin or can vaporize over a prolonged period to produce an inhalation hazard.

Toxicity of nerve agents

The LC_{50} by inhalation (30 min exposure) in the mouse is 15 mg m^{-3} (Gates and Renshaw, 1946), 5 mg m^{-3} (Lotts, 1960) and 1 mg m^{-3} (Lotts, 1960), for tabun, sarin and soman, respectively. By the subcutaneous route, the LD_{50} in the rabbit is 375 $\mu g\ kg^{-1}$ (Coleman *et al.*, 1968), 30 $\mu g\ kg^{-1}$ (Gordon and Leadbeater, 1977), 20 $\mu g\ kg^{-1}$ (Berry and Davies, 1970) and 14 $\mu g\ kg^{-1}$ (Leblic *et al.*, 1984) for tabun, sarin, soman and VX, respectively. These data demonstrate the relatively greater toxicity of soman (and VX), the reasons for which will be discussed below (see also Chapter 8).

Route of delivery

The major potential routes of delivery of nerve agents are air (both outdoor and indoor), water (hence the solubility of the nerve agent is important) and food.

Meteorological factors

Meteorological factors are important in the case of air delivery as the wind may disperse volatile agents and a higher ambient temperature increases volatility and decreases persistence. Moreover, some agents may freeze on clothing and then vaporize if carried indoors. Rain tends to dilute the agent, reducing its toxicity and promoting its hydrolysis, thereby leading to its inactivation.

DIFFERENCES BETWEEN ATTACKS AGAINST MILITARY PERSONNEL AND THOSE INITIATED BY TERRORISTS AGAINST A CIVILIAN POPULATION

There are important differences between on-target military attacks against relatively well-protected armed forces and nerve agent attacks initiated by terrorists against a civilian population. In contrast to military personnel, civilians are unlikely to be pre-treated with pyridostigmine and protected by personal protective equipment (PPE). Furthermore, the time after exposure when specific therapy can first be administered to civilians is likely to be delayed. Even conservative estimates suggest a delay between exposure and the first administration of atropine/oxime of at least 30 min. The clinical importance of this delay is discussed below.

MECHANISMS OF NERVE AGENT POISONING AND IMPLICATIONS FOR MANAGEMENT

The fundamental processes involved in acetylcholinesterase (AChE) inhibition, reactivation and 'ageing', are similar for both nerve agents and organophosphorus insecticides, although there are important differences, including the fact that ageing is not a clinically significant problem in organophosphorus pesticide poisoning. Acetylcholine (ACh) binds to AChE, which is found predominantly at the neuromuscular junction and at cholinergic neuroeffector junctions of tissues. The substrate binding domain on AChE contains a glutamate and an aromatic tryptophan that interact with the charged ammonium group of acetylcholine, together with a serine, histidine, and glutamate triad which interacts with the ester group. When acetylcholine binds to AChE, the ester moiety of ACh undergoes a nucleophilic attack by serine, resulting in the hydrolysis of acetylcholine (and formation of choline) and acetylation of serine. The acetylated serine is then hydrolyzed rapidly, thereby regenerating free AChE. Choline is taken up by the presynaptic or pre-junctional nerve terminals and recycled by combination with acetyl CoA, catalyzed by the enzyme choline acetyltransferase, to form more ACh.

The destruction of acetylcholine by AChE is a very rapid reaction. Approximately 10^4 molecules of acetylcholine are hydrolyzed per second by a single AChE molecule. In contrast, hydrolysis of the nerve agent–AChE complex is much slower than hydrolysis of the acetylated enzyme that is produced by complexation with the normal substrate, ACh. The result is that the activity of AChE is inhibited for a prolonged period.

The combination of the nerve agent soman (GD) with AChE to produce soman-inhibited AChE is shown diagrammatically in Figure 1. The final steps (Figure 1(c,d) take place so slowly that no clinically relevant regeneration of the enzyme takes place before a process known as ageing occurs (Figure 1(e)).

Ageing

The ageing process consists of the monodealkylation of the AChE–nerve agent complex. The loss of an alkyl group produces a conformational change that results in the formation of a very stable agent–enzyme complex which is then resistant to spontaneous hydrolysis and reactivation by oximes. The rate of ageing is dependent on the nature of the alkyl group (Wilson *et al.*, 1992). Monodealkylation occurs to some extent with all dialkylphosphorylated AChE complexes but, in general, is only of clinical importance in relation to the treatment of soman poisoning. The human *in vitro* ageing half-life for soman-inhibited AChE is 1.3 min (Harris *et al.*, 1978), for sarin-inhibited AChE 3 h (Davies and Green, 1956; Heilbronn, 1963) and tabun-inhibited AChE 13 h.

In the case of soman, therefore, once ageing has occurred, recovery of enzyme function depends on 're-synthesis' of AChE. As a result, it is important that an oxime is administered as soon after soman exposure as possible so that some reactivation of AChE occurs before all of the enzyme becomes aged. The phenomenon of soman-induced ageing led to the development of carbamate (pyridostigmine) prophylaxis for nerve agent poisoning.

MAKING THE DIAGNOSIS

The diagnosis of nerve agent poisoning is based on the patient's history, clinical presentation and laboratory tests. In a patient with a positive history, characteristic symptoms and depressed erythrocyte cholinesterase activity, diagnosis is not difficult to make. Unfortunately, the history may be unobtainable and the clinical features may not be recognized as such by those clinicians who have no personal experience of diagnosing patients with nerve agent poisoning. Only the number of casualties may prompt consideration of the diagnosis.

CLINICAL FEATURES

Sidell (Sidell, 1974,1992,1997) has reviewed the features and management of nerve agent poisoning. Systemic nerve agent poisoning may follow inhalation, ingestion or dermal exposure, although the onset of systemic toxicity is slower by the latter route. Miosis, which may be painful and last for several days, occurs rapidly following ocular exposure to a nerve agent and appears to be a very sensitive index of exposure (Nozaki *et al.*, 1997). Ciliary muscle spasm may impair

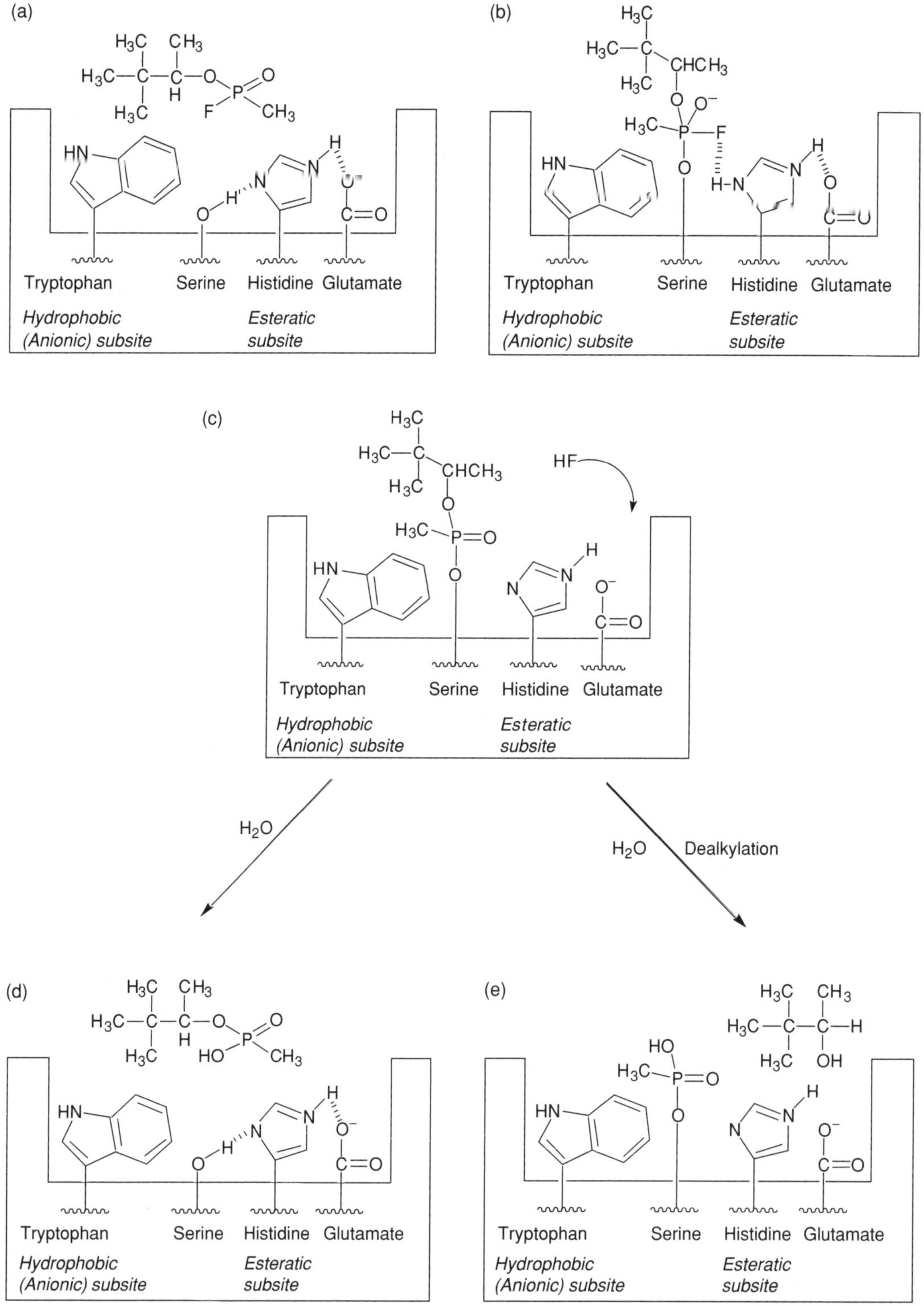

Figure 1. Reaction of soman (GD) with acetylcholinesterase (AChE). (a) Soman and the active site of AChE shown together but not having undergone any reaction. (b) Soman combined with AChE to form an inhibitor–enzyme intermediate. (c) The leaving group (fluoride) has been lost, leaving a complex of soman with AChE. (d) The ester link in the phosphonylated AChE has been hydrolyzed, the enzyme has reactivated and an alkylphosphate has been formed. (e) The link between the large pinacolyl group and phosphorus has been cleaved with the formation of a stable monoalkylphosphonylated complex with AChE and pinacolyl alcohol. The process is known as 'ageing'.

accommodation and conjunctival injection and eye pain may occur. Contact with liquid nerve agent may produce localized sweating and fasciculation, which may spread to involve whole muscle groups. Chest tightness, rhinorrhoea and increased salivation occur within seconds/ minutes of inhalation of a nerve agent. In contrast, ingestion of food or water contaminated with nerve agent may cause abdominal pain, nausea, vomiting, diarrhoea and involuntary defecation, although the onset of symptoms may be delayed.

Miosis may also occur as a systemic feature, although more usually it follows direct exposure. This explains why, for example, modest dermal exposure may produce systemic features but not miosis. Abdominal pain, nausea and vomiting, involuntary micturition and defecation, muscle weakness and fasciculation, tremor, restlessness, ataxia and convulsions may follow dermal exposure, inhalation or ingestion of a nerve agent. Bradycardia, tachycardia and hypertension may occur, dependent on whether muscarinic or nicotinic effects predominate. If exposure is substantial, death may occur from respiratory failure within minutes, whereas mild or moderately exposed individuals usually recover completely, although EEG abnormalities have been reported in those severely exposed to sarin in Japan (Murata *et al.*, 1997; Sekijima *et al.*, 1997).

The presence of miosis is a most important sign and is more helpful in determining the diagnosis of poisoning than measurement of erythrocyte AChE activity soon after presentation. The absence of miosis either raises serious doubt about the diagnosis, if exposure was secondary to the release of vapour, or suggests that exposure has produced only sub-clinical poisoning.

Features and outcome in those exposed to nerve agents in Japan

It is relevant to review the impact of nerve agents (sarin and VX) released on an unprotected civilian population in Japan on six occasions in 1994–1995 (see also Chapter 13). On two successive evenings in Tokyo in the Spring of 1994, Aum Shinrikyo, a politico-religious group, employed sarin in an attempt to assassinate the head of a rival religious sect (Kaplan and Marshall, 1996). However, the sprayer failed to work on the first night and on the second night sarin was blown back into the faces of Aum Shinrykyo members, including their 'Home Affairs Minister', Mr Niimi, who became seriously ill and required the administration of pralidoxime. A third more major release took place on 27 June 1994 in Matsumoto (see below). VX was used to poison a former supporter of Aum Shinrikyo on 2 December 1994 in Tokyo and the victim was in hospital for six weeks after he was found comatose following the attack (Tu, 1999,2000). Ten days later, the Aum Shinrikyo 'Home Affairs Minister' plunged a syringe containing VX into the neck of a 28-year-old man in Osaka who was believed to be a police informer. He suffered a cardiac arrest and died 10 days later; two metabolites of VX were detected in blood taken from the casualty (Tsuchihashi *et al.*, 1998,2000). The final attack undertaken by Aum Shinrikyo was in Tokyo on 20 March 1995; this is described in detail below.

Matsumoto

Some 600 people were exposed to sarin released from a truck using a heater and fan in a residential area of Matsumoto on 27 June 1994 (Okudera *et al.*, 1997). Sarin spread over an elliptical area measuring 800 m by 570 m, affecting most intensely those in a smaller 400 m by 300 m region. The evening was warm, and some residents made the fatal mistake of going to bed with their windows open. Some 600 people were exposed to sarin: 277 individuals with mild symptoms did not seek medical referral, a further 253 sought medical attention at outpatient clinics, 58 residents were admitted to six hospitals and all recovered; seven casualties living close to the sarin release died outside hospital (Morita *et al.*, 1995; Nakajima *et al.*, 1997,1998; Okudera *et al.*, 1997; Okudera, 2002; Smithson, 2004; Suzuki *et al.*, 1997). The death toll probably would probably have been higher had more residents opened their windows and the sarin release been more efficient. The features experienced by 264 casualties are summarized in Table 1 (Morita *et al.*, 1995).

Three weeks after exposure, 129 of 471 casualties still had symptoms such as dysaethesiae of the extremities (Nakajima *et al.*, 1998). Many

Table 1. Features in those exposed to sarin in Japan in 1994 and 1995[a]

Feature	Matsumoto %[b]	Tokyo %[c]
Miosis (pupil diameter < 1.5 mm)	44	99
Decreased visual acuity and miosis	57	N/A
Eye pain	N/A	45
Blurred vision	N/A	40
Nausea	N/A	60
Rhinorrhoea	37	N/A
Breathlessness	25	63
Headache	23	75
Malaise	12	N/A
Low-grade fever	6	N/A
Dysesthesia of the extremities	6	N/A

Notes:
[a] N/A, not available.
[b] $n = 264$; Morita *et al.*, 1995.
[c] $n = 111$; Okumura *et al.*, 1996.

casualties also complained of asthenopia (eyestrain), which was even more frequent at four months (Nakajima *et al.*, 1998). This symptom was still present in some casualties at one year and, in those symptomatic at one year, more than 50% complained of fatigue, asthenia and blurred vision and 40% had shoulder stiffness (Nakajima *et al.*, 1998). Follow up of casualties up to three years after exposure has been carried out (Nakajima *et al.*, 1999; Nohara *et al.*, 1999; Okudera, 2002; Sekijima *et al.*, 1997). Of those casualties with the most severe initial features, four developed epileptiform EEG abnormalities and one developed a sensory neuropathy seven months after exposure (Sekijima *et al.*, 1997).

Eighteen of 52 rescuers were reported to have mild symptoms of poisoning (Nakajima *et al.*, 1997), whereas in another study, eight of 95 rescuers developed mild features of nerve agent poisoning (Okudera *et al.*, 1997).

Tokyo

On 20 March 1995, a terrorist attack occurred in the Tokyo subway system during rush hour. Sarin was placed in five subway cars on three separate lines in plastic bags opened so that the agent, which is liquid under temperate conditions, could evaporate. Over 5000 'casualties' sought medical attention of whom 984 were moderately poisoned and 54 were severely poisoned; 12 died. However, a substantial number of those presenting (some 4000) had no signs of toxicity and 4973 individuals were seen on 'day one' and sent home. In the following 24 h, many more individuals presented, although none had features of nerve agent poisoning.

The majority of casualties were assessed at St. Luke's International Hospital, which is located within 3 km of five of the affected subway stations, although a further 568 patients were assessed at five other hospitals (Kato and Hamanaka, 1996; Masuda *et al.*, 1995; Nozaki *et al.*, 1995; Suzuki *et al.*, 1995; Yokoyama *et al.*, 1995,1996). Doctors at St Luke's were notified at 08:16 hours of an explosion and fire at a nearby subway system (Matsui *et al.*, 1996; Okumura *et al.*, 1996) and, twelve minutes later, the first victims arrived on foot at the emergency department. A casualty in cardiopulmonary arrest was brought in by private car at 08:43 hours. In all, 640 casualties were assessed at St Luke's Hospital on 20 March 1995, with the Chapel being used as the main treatment area (Okumura *et al.*, 1996). Initially (at 09:12 hours), the Fire Department identified acetonitrile as the suspected agent. However, medical staff discounted this diagnosis as all casualties had marked miosis and atropine was therefore administered; pralidoxime iodide was first given at 10:00 hours. Miosis appeared to be a more sensitive early indicator of exposure than erythrocyte acetylcholinesterase activity (Nozaki *et al.*, 1997).

There were inadequate facilities in the Emergency Department at St Luke's to permit a large number of casualties to remove contaminated clothing and to shower; formal decontamination was, therefore, impossible. In addition, the ventilation in the patient reception area was poor. Consequently, some of the medical staff complained of eye or throat pain, nausea, or miosis (Okumura *et al.*, 1996). This was relieved by improving ventilation and by rotation of affected staff to other locations within the hospital. Secondary exposure of medical staff from patients affected by sarin vapour was limited. No medical staff required pharmacological treatment for their signs and symptoms.

The features in 111 patients admitted to St Luke's Hospital are shown in Table 1 (Okumura *et al.*, 1996). Of the 640 patients who presented during the first day to St Luke's Hospital, only two patients died (Ohbu *et al.*, 1997). Within 2 to 4 days, 105 patients, 95% of those who were admitted to the hospital, had been discharged with satisfactory relief of their complaints (Okumura *et al.*, 1996).

ANALYTICAL CONFIRMATION OF THE DIAGNOSIS

The activity of two enzymes may be estimated to diagnose and/or monitor the progress of nerve agent poisoning. These are red cell acetylcholinesterase (AChE) and plasma butyrylcholinesterase (plasma cholinesterase; plasma pseudocholinesterase). Both are surrogates for activity of AChE in the central and peripheral nervous system. A number of factors need to be borne in mind when interpreting cholinesterase activity measurements. AChE and butyrylcholinesterase are different gene products, although there are similarities in their structures (Darvesh *et al.*, 2003). There are differences in the kinetics of the inhibition of the two enzymes by nerve agents, and in the kinetics of reactivation and ageing of the resulting inhibited enzymes (Wilson *et al.*, 1992). Thus, on enzyme kinetic grounds measurement of the activity of plasma butyrylcholinesterase would be expected to be a poorer surrogate than measurement of the activity of red cell AChE as a surrogate for the activity of AChE in nervous tissue.

However, even erythrocyte AChE measurements cannot be expected to be a perfect surrogate for the nervous tissue enzyme: this is because pharmacokinetic factors may result in differential access of the inhibitor to the red cell and to neural structures. A further consideration is that, where nerve agents react with the enzyme to produce a phosphonylated structure that does not spontaneously reactivate, red cells of mammals lack the protein synthetic capability to synthesize new AChE. By contrast, in nervous tissue, after inhibition by OPs whose enzyme–inhibitor complex with AChE does not readily reactivate, activity may reappear relatively quickly. Thus, Wehner *et al.* (1985) observed approximately 30% recovery after 24 h in diisopropylfluorophosphate (DFP)-treated mouse CNS reaggregates, which was clearly due to synthesis *de novo* of AChE. Another consideration in the interpretation of butyrylcholinesterase activity measurements is that the normal range is relatively wide, rendering interpretation in individual patients difficult unless the results of previous estimations in the patient are available (Swaminathan and Widdop, 2001).

Few laboratories can determine the nerve agent responsible for the intoxication and measure either the parent compound or metabolites in body fluids. Furthermore, such measurement has no place in the immediate diagnosis or early management of nerve agent poisoning.

PRINCIPLES OF MANAGEMENT

Maintenance of vital body functions and adequate clinical monitoring

Patients who are moderately or severely poisoned, as shown, for example, by drowsiness, coma, hypotension, severe bronchorrhoea and marked muscle fasciculation, require treatment in a critical care unit as soon as possible as further deterioration may occur and mechanical ventilation may be required.

Bronchorrhoea requires prompt relief with intravenous atropine and supplemental oxygen should be given to maintain $P_aO_2 > 10$ kPa

(75 mmHg). If these measures fail, the patient should be incubated and assisted ventilation (with positive end expiratory pressure) should be instituted.

In severely poisoned patients who are hypotensive, it may be necessary not only to expand plasma volume but also to use a vasopressor (e.g. dopamine titrated to a systolic pressure greater than 90 mmHg) or an inotrope (e.g. dobutamine 2.5–10 $\mu g\,kg^{-1}\,min^{-1}$), to maintain cardiac output). Careful attention must be given to fluid and electrolyte balance and adjustments to infusion fluids made as necessary. Heart rate, blood pressure, ECG and arterial blood gases should be monitored routinely. Cardiac arrhythmias should be treated conventionally and hypoxia must be considered as a possible aetiology.

The management of convulsions and muscle fasciculation with diazepam is discussed below.

Minimizing further absorption of the nerve agent, if exposure is dermal

In principle, after resuscitation and stabilization of the casualty, if exposure is dermal, thorough skin decontamination should be carried out by removing all contaminated clothing and washing affected skin thoroughly with soap and cold water, including exposed areas (e.g. hands, arms, face, neck and hair). This should be done without 'care-givers' themselves being contaminated and casualties becoming hypothermic. However, given the circumstances of likely exposure and the number of casualties, decontamination may be difficult to achieve in practice. The removal and appropriate storage of contaminated clothing may be all that can be done. It is essential that decontamination does not lead to delays in the administration of antidotes to those who are severely poisoned. If exposure is by inhalation, skin decontamination is unnecessary (Sidell, 1997).

Appropriate use of atropine, oximes and diazepam

Impact of a delay in administration of atropine and oxime

In experimental studies, a delay of even 12 min in the administration of atropine and oxime reduced the protection ratio (LD_{50} with treatment/LD_{50} without treatment) substantially, even in the case of nerve agents other than soman (Green *et al.*, 1983; Table 2). While it is important that an oxime is administered as soon after soman exposure as possible, so that some reactivation of AChE occurs before all the enzyme becomes 'aged', early atropine and oxime administration is still clinically important in patients poisoned with other nerve agents, even though 'ageing' occurs more slowly and reactivation occurs relatively rapidly.

Atropine

Atropine competes with ACh and other muscarinic agonists for a common binding site on the muscarinic receptor, thus effectively antagonizing the actions of ACh at muscarinic receptor sites, which would otherwise lead to increased tracheobronchial and salivary secretions, bronchoconstriction and bradycardia. The

Table 2. Effect of delaying therapy (IV atropine, pralidoxime, diazepam) on protection ratios in guinea pigs given pyridostigmine 30 min before nerve agent (Green *et al.*, 1983)

Time of therapy	Protection ratio (95% confidence limits)		
	Sarin	Soman	VX
10 s	93.0 (56–157)	15.0 (9.5–23)	35.0 (21–57)
1 min	34.0 (30–46)[a]	13.0 (7.5–21)	28.0 (16–49)
4 min	12.0 (7.1–19)[a]	7.8 (5.5–11)[a]	12.0 (6.5–22)[a]
8 min	5.4 (3.8–7.7)[a]	4.7 (3.8–5.8)[a]	5.6 (3.8–8.3)
12 min	—	4.3 (3.0–6.2)[a]	—

Note:
[a] Significantly different ($p < 0.05$) from 10 s value.

usefulness of atropine is virtually undisputed, although there is some controversy regarding the dosage schedule. This is discussed in detail in Chapter 14 (McDonough and Shih).

Oximes

Clinically, the main benefit of oximes is to reverse cholinergic effects at peripheral nicotinic sites so that, for example, muscle strength may improve. Oximes are much less effective than atropine at peripheral muscarinic sites and their effects on central nervous system-mediated symptoms and signs may not be clinically significant.

The choice of an oxime should be governed by the following criteria:

(i) The oxime should give adequate protection against all nerve agents. It could be argued that in a military context, where challenge levels of agent on the battlefield against specific target are likely to be high, it is important to select an oxime with the highest possible protection ratio (the factor by which a treatment raises the lethal dose), such as a value of ≥ 5. In contrast, the terrorist use of nerve agents in a civilian population is likely to result in challenge levels where a protection ratio of 2 would have a significant impact on survivability.
(ii) The oxime should be effective against lethality when given after exposure in the absence of pre-treatment with pyridostigmine.
(iii) The oxime should be relatively free from significant acute adverse effects and should have a high therapeutic index (ratio of the LD_{50} to its effective $dose_{50}$ (ED_{50}).
(iv) The oxime should be pharmacologically compatible with other drugs that may be given simultaneously and be chemically compatible if administered in the same injection device.
(v) The oxime should have a shelf life ($< 10\%$ deterioration (t_{90})) of at least five years, preferably in varying environmental conditions. For example, it is known that obidoxime is much more stable than pralidoxime mesilate in solution at high environmental temperatures. The stability of HI-6 as a powder is excellent (estimated shelf life at 25°C of 73 years), but HI-6 has extremely poor stability in solution. For this reason, a number of wet/dry injection devices have been developed, which have the advantage that by separating the solid drug from the liquid vehicle the stability is markedly increased. However, the disadvantage is that it is necessary to dissolve the drug in the vehicle before it can be injected and this may involve a delay in administering the drug. HI-6 methanesulphonate is much more rapidly soluble than HI-6 dichloride (Krummer *et al.*, 2002).

Although the therapeutic combination of oxime and atropine is well established in the treatment of nerve agent poisoning, there is still no international consensus on the choice of oxime, on aspects of dosing and, indeed, some have doubted the worth of oxime therapy altogether. The present authors take the view that with the possible exception of the treatment of GF and soman poisoning, when HI-6 might be preferred, a review of available experimental evidence suggests that there are no clinically important differences between pralidoxime, obidoxime and HI-6 in the treatment of nerve agent poisoning, if pre-treatment with pyridostigmine has not been undertaken. This is discussed further in Chapter 15 (Eyer and Worek).

Benzodiazepines

Benzodiazepines (most often diazepam, sometimes midazolam) may also be of benefit by reducing anxiety and restlessness, reducing muscle fasciculation, arresting seizures, and possibly reducing morbidity and mortality when used in conjunction with atropine and pralidoxime. The use of benzodiazepines in nerve agent poisoning has been reviewed recently (Marrs, 2004) and is discussed in detail in Chapter 16 (Marrs and Sellström).

MANAGEMENT OF NERVE AGENT POISONING *OUTSIDE* HOSPITAL

The release of a nerve agent among a civilian population requires the deployment of special measures and personnel to ensure the rescue of casualties and the rapid administration of antidotes. The arrangements in France are described in Chapter 12 (Baker). Rescue and drug administration should be undertaken by trained staff that are protected by personal protective equipment (PPE) and equipped with pressure demand, self-contained breathing apparatus, to prevent nerve agent exposure in contaminated areas and secondary contamination from casualties, which has been reported (Nozaki *et al.*, 1995; Nozaki *et al.*, 1995). In addition, the use of small portable gas-powered ventilators can assure a high quality of essential ventilatory care at the scene.

The priority is to remove the casualty from further nerve agent exposure and to establish and maintain a clear airway; supplemental oxygen should be given as required. If possible, the victim's contaminated clothing should be removed to reduce further nerve agent absorption. For the reasons stated above, civilian casualties who have been exposed substantially to a nerve agent should receive antidotal treatment as soon as possible after exposure; the rapid parenteral administration of atropine to patients presenting with rhinorrhoea and bronchorrhoea may be life-saving.

It is also recommended that casualties requiring atropine should also receive immediately whichever oxime is available, as it is very unlikely that the identity of the nerve agent will be known before the admission of casualties to hospital. This can be done most conveniently in adults by the administration of the contents of an autoinjector, such as the 'ComboPen' (the UK version contains atropine 2 mg, pralidoxime mesilate 500 mg and the diazepam precursor, avizafone 10 mg), intramuscularly. Severely intoxicated adult casualties may require the administration of the contents of three ComboPens at 5–10 min intervals prior to admission to hospital. In children, the contents of one ComboPen should be administered although alternative administration arrangements will need to be made for very small children.

Casualties receiving antidotes should be moved to hospital as soon as possible. Casualties who do not develop the features of systemic toxicity, notably rhinorrhoea and bronchorrhoea, should be triaged but not given atropine or oxime.

MANAGEMENT OF NERVE AGENT POISONING IN HOSPITAL

In symptomatic patients, intravenous access should be established and blood should be taken for measurement of erythrocyte cholinesterase activity to confirm the diagnosis. If the characteristic features of nerve agent poisoning are present, however, antidotal treatment should not be delayed until the result is available.

If rhinorrhea or bronchorrhea develops, atropine 2 mg in an adult (20 $\mu g\,kg^{-1}$ in a child) should be administered intravenously every 5–10 min until secretions are minimal and the patient is atropinized (dry skin and sinus tachycardia). In all patients receiving atropine, an oxime, such as pralidoxime chloride or mesilate, should be administered in a dose of 30 $mg\,kg^{-1}$ body weight intravenously, followed by an infusion of pralidoxime 8–10 $mg\,kg^{-1}\,h^{-1}$, the infusion rate depending on severity. Alternatively, pralidoxime chloride or mesilate 30 $mg\,kg^{-1}$ body weight intravenously can be given every 4–6 h to patients with systemic features. In the case of GF and soman poisoning, consideration should be given to the use of HI-6, if supplies are available.

The duration of oxime treatment will depend on the presence of features, the clinical response and the erythrocyte AChE activity. It is recommended that the oxime should be administered for as long as atropine is indicated. For the majority of individuals this will be for less than 48 h; the exception would be individuals exposed dermally to VX where a depot of VX might result in prolonged intoxication.

Intravenous diazepam (adult 10–20 mg; child 1–5 mg) is useful in controlling apprehension, agitation, fasciculation and convulsions; the dose may be repeated as required. In some experimental studies, the addition of diazepam to an atropine and oxime regimen has increased survival further (Marrs, 2004).

If ocular exposure has occurred the victim should remove contact lenses, if present, and if they are easily removable. The eyes should be irrigated immediately with lukewarm water or sodium chloride 0.9% solution. Local anesthetic should be applied if ocular pain is present.

PREPARATION AND TRAINING FOR A NERVE AGENT RELEASE

Following the Japanese nerve agent releases, substantial numbers of casualties presented to hospital over a short time period, which stretched the resources available. Hence, each country and every hospital should now have a major accident plan that covers deliberate chemical releases, including nerve agents. This plan should be tested at least annually. It should include arrangements to triage substantial numbers of non-poisoned casualties, as well as those who are severely poisoned and require urgent treatment and admission.

REFERENCES

Berry WK and Davies DR (1970). The use of carbamates and atropine in the protection of animals against poisoning by 1,2,2-trimethylpropyl methylphosphonofluoridate. *Biochem Pharmacol*, **19**, 927–934.

Coleman IW, Patton GE and Bannard RA (1968). Cholinolytics in the treatment of anticholinesterase poisoning. V. The effectiveness of Parpanit with oximes in the treatment of organophosphorus poisoning. *Can J Physiol Pharmacol*, **46**, 109–117.

Darvesh S, Hopkins DA and Geula C (2003). Neurobiology of butyrylcholinesterase. *Nat Rev Neurosci*, **4**, 131–138.

Davies DR and Green AL (1956). The kinetics of reactivation, by oximes, of cholinesterase inhibited by organophosphorus compounds. *Biochem J*, **63**, 529–535.

Gates M and Renshaw BC (1946). *Fluorophosphates and other phosphorus-containing compounds*. In: Summary Technical Report of Division 9, Volume 1, pp. 131, 155. Washington, DC, USA: Office of Scientific Research and Development.

Gordon JJ and Leadbeater L (1977). The prophylactic use of 1-methyl,2-hydroxyiminomethylpyridinium methanesulfonate (P2S) in the treatment of organophosphate poisoning. *Toxicol Appl Pharmacol*, **40**, 109–114.

Green DM, Inns RH and Leadbeater L (1983). Unpublished observations, Porton Down, UK.

Harris LW, Heyl WC, Stitcher DL *et al.* (1978). Effects of 1,1′-oxydimethylene bis-(4-tert-butylpyridinium chloride) (SAD-128) and decamethonium on reactivation of soman- and sarin-inhibited cholinesterase by oximes. *Biochem Pharmacol*, **27**, 757–761.

Heilbronn E (1963). In vitro reactivation and 'ageing' of tabun-inhibited blood cholinesterases: studies with *N*-methylpyridinium-2-aldoxime methane sulphonate and *N*, *N*′-trimethylene bis (pyridinium-4-aldoxime) dibromide. *Biochem Pharmacol*, **12**, 25–36.

Kaplan DE and Marshall A (1996). *The Cult at the End of the World: The Incredible Story of Aum*. London, UK: Hutchison.

Kato T and Hamanaka T (1996). Ocular signs and symptoms caused by exposure to sarin gas. *Am J Ophthalmol*, **121**, 209–210.

Krummer S, Thiermann H, Worek F *et al.* (2002). Equipotent cholinesterase reactivation *in vitro* by the nerve agent antidotes HI 6 dichloride and HI 6 dimethanesulfonate. *Arch Toxicol*, **76**, 589–595.

Leblic C, Cox HM and Le Moan L (1984). Etude de la toxicité de l'eserine, VX et le paraoxon, pour établir un modèle mathematique de l'extrapolation à être humain. *Arch Belg Med Soc Hyg Trav Med* (Supplement), 226–242.

Lotts vK (1960). Zur Toxikologie und pharmakologie organischer Phosphosäurester. *Dtsch Gesundheitsw*, **15**, 2133–2179.

Marrs TC (2004). The role of diazepam in the treatment of nerve agent poisoning in a civilian population. *Toxicol Rev*, **23**, 145–157.

Masuda N, Takatsu M, Morinari H *et al.* (1995). Sarin poisoning in Tokyo subway. *Lancet*, **345**, 1446.

Matsui Y, Ohbu S and Yamashina A (1996). Hospital deployment in mass sarin poisoning incident of the Tokyo subway system – an experience at St. Luke's International Hospital, *Tokyo Jpn Hosp*, **15**, 67–71.

Maynard RL and Beswick FW (1992). Organophosphorus compounds as chemical warfare agents. In *Clinical and Experimental Toxicology of Organophosphates and Carbamates* (B Ballantyne, TC Marrs, eds), pp. 373–385. Oxford, UK: Butterworth-Heinemann.

Morita H, Yanagisawa N, Nakajima T *et al.* (1995). Sarin poisoning in Matsumoto, Japan. *Lancet*, **346**, 290–293.

Murata K, Araki S, Yokoyama K *et al.* (1997). Asymptomatic sequelae to acute sarin poisoning in the central and autonomic nervous system 6 months after the Tokyo subway attack. *J Neurol*, **244**, 601–606.

Nakajima T, Sato S, Morita H *et al.* (1997). Sarin poisoning of a rescue team in the Matsumoto sarin incident in Japan. *Occup Environ Med*, **54**, 697–701.

Nakajima T, Ohta S, Morita H *et al.* (1998). Epidemiological study of sarin poisoning in Matsumoto City, Japan. *J Epidemiol*, **8**, 33–41.

Nakajima T, Ohta S, Fukushima Y *et al.* (1999). Sequelae of sarin toxicity at one and three years after exposure in Matsumoto, Japan. *J Epidemiol*, **9**, 337–343.

Nohara M, Sekijima Y, Nakajima T *et al.* (1999). Ocular manifestations in the follow-up of victims after sarin poisoning in Matsumoto area. *Jpn J Clin Ophthalmol*, **53**, 659–663.

Nozaki H and Aikawa N (1995). Sarin poisoning in Tokyo subway. *Lancet*, **345**, 1446–1447.

Nozaki H, Aikawa N, Shinozawa Y *et al.* (1995). Sarin poisoning in Tokyo subway. *Lancet*, **345**, 980–981.

Nozaki H, Hori S, Shinozawa Y *et al.* (1997). Relationship between pupil size and acetylcholinesterase activity in patients exposed to sarin vapor. *Intensive Care Med*, **23**, 1005–1007.

Ohbu S, Yamashina A, Takasu N *et al.* (1997). Sarin poisoning on Tokyo subway. *South Med J*, **90**, 587–593.

Okudera H (2002). Clinical features on nerve gas terrorism in Matsumoto. *J Clin Neurosci*, **9**, 17–21.

Okudera H, Morita H, Iwashita T *et al.* (1997). Unexpected nerve gas exposure in the city of Matsumoto: report of rescue activity in the first sarin gas terrorism. *Am J Emerg Med*, **15**, 527–528.

Okumura T, Takasu N, Ishimatsu S *et al.* (1996). Report on 640 victims of the Tokyo subway sarin attack. *Ann Emerg Med*, **28**, 129–135.

Sekijima Y, Morita H and Yanagisawa N (1997). Follow-up of sarin poisoning in Matsumoto. *Ann Intern Med*, **127**, 1042.

Sidell FR (1974). Soman and sarin: clinical manifestations and treatment of accidental poisoning by organophosphates. *J Toxicol Clin Toxicol*, **7**, 1–17.

Sidell FR (1992). Clinical considerations in nerve agent intoxication. In: *Chemical Warfare Agents* (SM Somani, ed.), pp. 155–194. San Diego, CA, USA: Academic Press.

Sidell FR (1997). Nerve agents. In: *Textbook of Military Medicine. Part I. Warfare, Weaponry and the Casualty. Medical Aspects of Chemical and Biological Warfare.* First Edition (R Zajtchuk and RF Bellamy, eds), pp. 129–179. Washington, DC, USA: Borden Institute, Walter Reed Medical Center.

Smithson AE (2004). *Ataxia: The Chemical and Biological Terrorism Threat and the US Response*, Henry L. Stimson Center Report. Volume 35. Washington, DC, USA: Henry L Stimson Center.

Somani SM, Solana RP and Dube SN (1992). Toxicodynamics of nerve agents. In: *Chemical Warfare Agents* (SM Somani, ed.), pp. 67–123. San Diego, CA, USA: Academic Press.

Suzuki T, Morita H, Ono K *et al.* (1995). Sarin poisoning in Tokyo subway. *Lancet*, **345**, 980.

Suzuki J, Kohno T, Tsukagosi M *et al.* (1997). Eighteen cases exposed to sarin in Matsumoto, Japan. *Intern Med*, **36**, 466–470.

Swaminathan R and Widdop B (2001). Biochemical and toxicological investigation related to OP compounds. In: *Organophosphates and Health* (L Karalleidde, S Feldman, J Henry, and T Marrs, eds), pp. 357–406. London: Imperial College Press.

Tsuchihashi H, Katagi M, Nishikawa M *et al.* (1998). Identification of metabolites of nerve agent VX in serum collected from a victim. *J Anal Toxicol*, **22**, 383–388.

Tsuchihashi H, Katagi M, Tatsuno M *et al.* (2000). Determination of metabolites of nerve agent *O*-ethyl-*S*-2-diisopropylaminoethyl methylphosphonothioate (VX). In: *Natural and Selected Synthetic Toxins: Biological Implications* (AT Tut, and W Gaffield, eds), ACS Symposium Series, Volume 745, pp. 369–386. Washington, DC, USA: American Chemical Society.

Tu AT (1999). Anatomy of Aum Shrinrikyo's organization and terrorist attacks with chemical and biological weapons. *Arch Toxicol Kinet Xenobiot Metab*, **7**, 45–84.

Tu AT (2000). Overview of sarin terrorist attacks in Japan. In: *Natural and Selected Synthetic Toxins: Biological Implications* (AT Tut and W Gaffield, eds), ACS Symposium Series, Volume 745, pp. 304–317. Washington, DC, USA: American Chemical Society.

Wehner JM, Smolen A, Smolen TN *et al.* (1985). Recovery of acetylcholinesterase activity after acute organophosphate treatment of CNS reaggregate cultures. *Fundam Appl Toxicol*, **5**, 1104–1109.

Wilson BH, Hooper MJ, Hansen ME *et al.* (1992). Reactivation of organophosphorus inhibited AChE with oximes. In: *Organophosphates – Chemistry, Fate and Effects* (JE Chambers and PE Levi, eds), pp. 107–137. San Diego, CA, USA: Academic Press.

Yokoyama K, Ogura Y, Kishimoto M *et al.* (1995). Blood purification for severe sarin poisoning after the Tokyo subway attack. *JAMA*, **274**, 379.

Yokoyama K, Yamada A and Mimura N (1996). Clinical profiles of patients with sarin poisoning after the Tokyo subway attack. *Am J Med*, **100**, 586.

12 THE MANAGEMENT OF CASUALTIES FOLLOWING TOXIC AGENT RELEASE: THE APPROACH ADOPTED IN FRANCE

David J. Baker

SAMU de Paris, Hôpital Necker Enfants – Malades, Paris, France

INTRODUCTION

The past decade has confirmed that terrorist attacks against mass civilian targets are a constant feature of modern life. The attack on the World Trade Center in 2001, the bus bombings in Israel, the tube attacks in London in 2005 and the suicide bombings in Iraq at the time of writing all demonstrate the mass injuries sustained following the use of improvised devices causing explosion or fires in crowded urban areas. Mass casualties have been produced with a range of injuries caused by the effects of blast or burns. However, behind this picture of conventional injuries which stretch emergency resources to the limit remains the constant fear of an urban attack using chemical agents, following the sarin attacks in Matsumoto and Tokyo in 1995 where the emergency services were not only understandably stretched to the limit but themselves became casualties from secondary contamination (Mehta *et al.*, 1990; Morita *et al.*, 1995; Okumura *et al.*, 1996, 1998). The concern has been fuelled by the knowledge that incidents of this nature have already occurred not only in Japan but in the USA in 2001 with targeted delivery of anthrax spores causing some fatalities (Borio *et al.*, 2001).

Despite public anxiety, deliberate attacks using chemical and biological agents have been rare since the end of World War One (Harris and Paxman, 1982). However, accidental release of chemicals and epidemics are not rare and there is much that is relevant from such events to the management of deliberate counterparts. In addition, there is a growing belief that terrorists will strike again using chemical, biological and radiological agents in an urban attack. In response to this fear and to a sometimes irrational appraisal of the numbers of casualties that such attacks will cause, a number of countries have put together response plans which involve an integrated response of all emergency services.

This chapter considers the strategic aspects of management of incidents involving the release of chemical agents in an urban environment, with particular reference to planning and protocols used by the emergency services in France.

MASS DESTRUCTION? CASUALTIES FOLLOWING CHEMICAL AND BIOLOGICAL AGENT RELEASE

Discussion of the management of casualties from chemical attack must be based upon a realistic appraisal of the likely numbers and type of casualties that are produced. Over the past five years public concern has been increased by constant

Chemical Warfare Agents: Toxicology and Treatment (2nd Edition)

Table 1. Classes of HAZMAT compounds

Class	Hazard
1	Explosives
2	Pressurized gas
3	Flammable liquid
4	Flammable solids
5	Oxidizing substances
6	Toxic substances
7	Radioactive substances
8	Corrosive substances
9	Miscellaneous dangerous substance and articles

references in the media to 'weapons of mass destruction' – a term used since the Cold War to describe chemical, biological and nuclear agents. The fear has been reinforced by knowledge of extensive chemical and biological agent production programmes (Alibek and Handelman, 2000; Harris and Paxman, 1982). However, the term has been used without any real critical analysis of its accuracy and validity from the medical standpoint and there is a need to define the possible emergency medical responses to chemical and biological agent injury. While nuclear weapons have clearly been demonstrated to cause massive destruction of material and loss of life, similar destruction (in terms of casualty numbers) has never been shown to be associated with the release of chemical substances or of biological organisms.

Although chemical and biological agents are usually linked together as potential hazards, there are substantial differences between the effects of chemical substances and bacteria or viruses and the casualties caused. Chemical warfare is the deliberate equivalent of the accidental release of chemical substances. In the civil context this is termed hazardous materials release (HAZMAT) (Anon, 1995; Borak *et al.*, 1991; Bronstein and Currance, 1994; Moles, 1999; Organization of Economic Co-operation and Development, 1994) (Table 1 and Figure 1).

Accidental releases may cause substantial injury and loss of life, as in the case of the mass exposure to methyl isocyanate in Bhopal in 1984 where over 5000 people lost their lives (Anon, 1984; Dhara and Dhara, 2002; Mehta *et al.*, 1990). Deliberate release of chemical agents as seen during the World War I (Harris and Paxman, 1982) and also in the Iran–Iraq war of the 1980s (United Nations, 1987) also produced substantial casualties but it is worth noting that the ratio of dead to wounded was lower than that from the use of conventional weapons. In the case of the Iran–Iraq War where modern medical techniques were available to treat casualties from mustard gas and nerve agent attack, the proportion of fatalities among some 27 000 wounded was less than 1% (United Nations, 1987), a figure which is attributed to the battlefield emergency medical facilities available. The Japanese sarin attack in 1995, which used a standard military nerve gas, caused only 12 deaths but a large number of injured persons who were treatable using antidotes and standard medical life-support techniques (Okumura *et al.*, 1996, 1998). However, the chemical attack on Hallubjah in Kurdistan in 1988 where a remote civilian population with little or no medical resources was subjected to a co-ordinated military chemical attack caused a very high number of fatalities (Hammick, 1991). Against this picture, conventional explosive weapons used by terrorists have been shown to cause substantial numbers of wounded and dead, particularly when used in confined spaces. Thus, the Bali bombing in 2002 left over 200 dead, while the terrorist campaigns in England during the past 30 years have caused equally substantial numbers of casualties.

Biological agent attacks are quite different from those using chemical agents (Baker, 2005a); the release of biological agents is essentially a

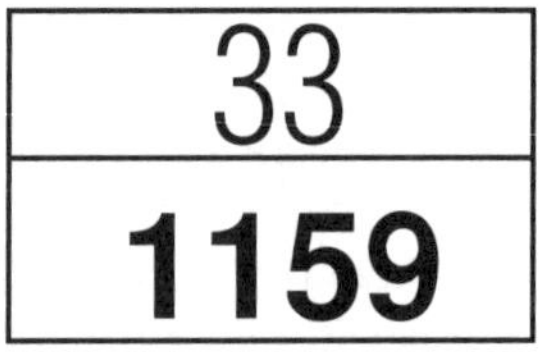

Figure 1. An example of a simple 'Kemmler' (HAZMAT) identification plate. The upper number (repeated to emphasize its nature) is the class number of the compound (flammable liquid), while the lower number is the international identification code for the compound (in this case, isopropyl ether). The exact form of the plate varies according to national practice

deliberately created epidemic. The time taken for the effects of the infection to show (latency) is days to weeks rather than the minutes to hours seen after chemical release. The mode of action of bacterial and viral infection in attacking whole body systems is also different from those chemical agents where the essential target is the respiratory system which may cause death from asphyxia within minutes. The casualty patterns from chemical and biological releases are also different. Chemical agents usually cause a major cluster of casualties within a short space of time. Biological agents cause patients to present with the same type of signs and symptoms at different times following the agent release, particularly when the infectious agent is transmissible from one person to another.

HAZARDS AND THREATS: THE RISKS TO MEDICAL RESPONDERS

It is important to distinguish between *hazard* and *threat*. These terms are often used interchangeably but have distinct meanings. In the context of chemical and biological warfare, a hazard is something that is harmful to human life but a threat implies that a hazard can be delivered (weaponized) and that the assailant has the intention to use the weapon. While many chemical hazards become real threats because they can be weaponized and delivered as bombs, shells or a vapour cloud, many of the feared biological agents are very sensitive to environmental extremes and cannot be easily delivered to infect a large population.

There are a number of characteristics of a chemical agent release which are relevant to the management:

(1) The presentation of the first victims with a common pattern of signs and symptoms (toxidromes) may be the first indication that an attack or release has taken place.
(2) Victims may have different patterns and rates of developing signs and symptoms and may overwhelm the emergency medical services as well as potentially making them secondary casualties.
(3) Specific management in terms of antidotes is different from mass trauma but there is a common pathway in the need for the provision of early life support.
(4) Chemical and radiological attacks (where a radioisotope is released) may produce both physical and toxic casualties as a result of explosive release of the agent.
(5) Chemical and radiological agents pose inherent dangers for emergency responders – the equivalent of the second bomb which became an integral part of terrorist attacks during the 1970s and 1980s and which was designed to harm emergency services specifically.

PROPERTIES OF CHEMICAL AND BIOLOGICAL HAZARDS AND THE CHEMICAL-BIOLOGICAL HAZARD SPECTRUM

Although chemical and biological agents have traditionally been separated, it is appropriate medically to regard them as part of a continuous spectrum of hazards (Baker, 1993). This is shown diagrammatically in Figure 2.

Agents are arranged in ascending order of molecular weight from chemical toxic agents through to self-replicating agents, such as bacteria and viruses on the right. The spectral approach to hazards is useful in emphasizing that agents from different parts of the line act in a similar way

CW agents ⟶ toxins ⟶ BW agents

Figure 2. The chemical–biological spectrum of hazards. Toxic trauma results from exposure to agents which have been classified as chemical or biological. The traditional distinctions can be unified as a spectrum of hazards ranging from low- to high-molecular-weight compounds through to self-replicating organisms. This approach highlights the fact that agents from different parts of the spectrum may produce common final pathophysiological pathways (for example, neuromuscular paralysis, toxic pulmonary oedema and multiple organ dysfunction syndrome) on the body. The failure of the neuromuscular junction by nerve agent anticholinesterases and botulinum toxin is a good example. Bacteria exert their toxic effects through toxins which may affect a variety of somatic systems. The advantage of the spectral approach is that it serves as a reminder that medical management of chemical–biological warfare (CBW) injury should respond primarily to system dysfunction rather than to specific etiologic factors.

The management of chemical releases is determined by the essential properties of the agents concerned (Baker, 2005b). There are four key properties of all chemical and biological agents in the CBW spectrum that determine the way the incident is managed and the care that is given to the patient. Each hazard has (a) toxicity, which may be defined as the damage done to the patient's whole organism and for a respiratory hazard is expressed by Ct_{50} (being the concentration × time which leads to fatality in 50% of an exposed population), (b) latency, which is the time interval between being exposed and for signs and symptoms to appear – given the biological variability of any population at risk this should be expressed as L_{50}, (c) persistency, which is the time that the release agent is physically present at the point of release and is related to the physicochemical properties, and (d) transmissibility, which is the capacity of the agent to be passed from one contaminated or infected person to another. Persistency is a function of the physical characteristics of the agent and local degradation factors such as humidity and temperature. Transmissibility is a function of persistency of contamination on affected patients. Toxicity is usually expressed in terms of LD_{50} or LCt_{50} where C is the concentration of agent inhaled for time (t) required to produce lethality in 50% of the exposed population, while LD_{50} usually relates to toxicity via other routes. In the case of most chemical agents, the inhaled route is normal and so expressions of concentration and time are used. Toxicity and latency determine the way the exposed patient presents and is managed medically, while persistency and transmissibility determine risks to emergency responders and measures that should be taken to decontaminate the patient and his environment and to prevent further secondary casualties.

Haber defined a 'lethality coefficient' as follows:

$$W = C \times t$$

where C is the inhaled concentration of toxic agent and t is the time of exposure. In practice, the absorbed amount of agent depends upon the respiratory minute volume of the exposed person. This is just one factor modifying the expression of toxicity. Others include life-support responses in the case of respiratory failure and the effects of antidotes. Toxicity in humans is usually extrapolated from animal toxicological data. The fact that human toxicity can be radically reduced by emergency medical procedures emphasizes the need for early emergency support in mass chemical incidents.

In the management of chemical releases, toxicity and latency essentially govern the management of the casualty, while persistency and transmissibility govern the management of the incident and the risk to emergency responders.

PROBLEMS IN CHEMICAL INCIDENT MANAGEMENT

There are a number of recognizable problems in the medical management of CBW incidents. These are:

(1) Failure to recognize CBW release.
(2) Failure to implement appropriate HAZMAT procedures to limit the number of secondary casualties.
(3) Release of the agent overwhelming the available emergency and medical resources.
(4) Insufficient planning and training.
(5) Lack of appropriate protective equipment.
(6) Lack of resuscitation skills for early life support.
(7) Lack of specific knowledge for the management of toxic or infectious hazards.
(8) Panic and hysterical reactions from a frightened general public fuelled by irresponsible reporting.

These problems may be minimized by careful planning, equipping and training with the objective of making a CBW release manageable in the same sense as other mass – disasters.

DETECTION, IDENTIFICATION AND MONITORING OF A CHEMICAL ATTACK

In military operations, where CW attack is anticipated as a result of intelligence, detection systems are used which 'alarm' when the presence of one or more identified hazards is detected. A number of physical characteristics of the agents may be used for this, such as ion drift, infrared spectroscopy and mass spectrometry (Box 1).

Box 1: Detection identification and monitoring

Detection *of a CBW agent means detecting its presence in a contaminated environment* and effecting an alarm.

Identification of the released agent in the civil context is usually a subsequent stage, using a number of physicochemical techniques.

Monitoring *is a term used to detect contamination on an exposed patient. This is important to determine the risk of transmission and the effectiveness of decontamination.*

Specific detection techniques include:

- agent-specific chemistry
- generic chemical detection techniques
- mass spectrometry
- ion-current devices
- bioluminescence
- microbial techniques
- chemical-pathological studies on an affected patient after the attack
- internal versus external detection

In the military context, detection and identification are usually combined stages. In the civil setting, however, where there is a wide range of toxic industrial chemicals that may be released accidentally or deliberately, fixed detection systems are less practicable although may be an option for certain high-risk locations such as metro systems.

In a civil setting, the sudden arrival of a large number of victims at an emergency medical facility or calling the ambulance service may be the first indication of an attack. Early detection of an attack may be based only on presenting toxidromes with specific detection only possible after casualties have presented. Detection systems using infrared spectroscopy are in use with a number of emergency services.

Monitoring, which detects contamination on a casualty either before or after decontamination, uses close-range detection such as the chemical agent monitor (CAM) which uses ion-drift technology to monitor contamination at close range on victims and their surroundings.

EMERGENCY MEDICAL ORGANIZATION IN RESPONSE TO A CHEMICAL AGENT ATTACK

Emergency medical organization in response to a chemical attack is based on (1) interaction with the fire and police services who manage the general response to the incident to make sure that attending medical teams are properly protected and are operating within standard procedures, and (2) the provision of emergency medical care to potentially contaminated patients who pose a secondary transmission risk. The basic management of any toxic release is to create three notionally concentric zones, called *hot*, *warm* and *cold*. This approach is a standard internationally and is fundamental to the fire-brigade management of a toxic chemical release (Laurent *et al.*, 1999). The hot zone corresponds to the point of release where the highest toxic concentrations of agent may be found (and where the agent, if persistent, will remain) and is usually entered only by special rescue workers from the fire service. If the identity of the hazard is not known it must be assumed to be toxic, persistent and transmissible and therefore the rescuers are equipped with self-contained suits with their own air supply. These responders bring casualties away from the point of release of agent into a surrounding area, called the warm zone, which has a

Table 2. Classification of personal protection levels

Level of PPE	Description
Level A	• Breathing apparatus with positive pressure • Isolating suit • Gloves, double layer (resistant to chemical substances) • Boots (resistant to chemical substances) • An over-covering system for areas between gloves and arm suite, as well as boots and leg portion of suit
Level B	• Breathing apparatus with positive pressure • Isolating suit • Gloves, double layer (resistant to chemical substances) • Boots (resistant to chemical substances)
Level C	• Mask equipped with a NRBC filter • Suit resistant to chemical substances • Gloves, double layer (resistant to chemical substances) • Boots (resistant to chemical substances)
Level D	• Normal everyday clothing without any protection

lower level of contamination which is due either to transmission by contaminated personnel or as a wind-borne contamination if the released agent has a significant vapour pressure. Here, specially trained medical personnel can operate to sort the casualties into those who have been exposed and those requiring immediate emergency medical treatment. Following decontamination, casualties are moved into the contamination-free cold zone for transfer to hospital and continuing medical care. In practice, the three-zone model for chemical release response may be variable according to the nature of the release. A point release due to an explosion or breach of a containment facility comes closest to the model. However, if the agent is being released continuously (for example, from a breached moving tanker) then the situation becomes a line-release and the concentric model is not applicable. In potential terrorist releases of a chemical agent, there are in practice only two zones, contaminated and non-contaminated. In the three-zone model the decontamination module on the boundary between the warm and cold zone may be effectively only the decontamination facility.

PROTECTION IN CHEMICALLY CONTAMINATED AREAS

The classical HAZMAT procedure where the nature of the released chemical may be unknown holds that fire personnel entering the hot zone should be protected with self-contained level-A suits (see Table 2) having their own air supply.

This is important for chemical releases involving fires and corrosive materials. Filtration respirators are not appropriate in this area because (1) the atmosphere may be oxygen deficient as a result of the fire, and (2) filtration canisters do not remove carbon monoxide, one of the main risks in products of combustion. In a warm zone, however, where oxygen levels are normal and there is no carbon monoxide risk, level-A suits are not required and level-C protection is appropriate This protection uses a lightweight suit with a filtration respirator giving mobility, dexterity and communication potential. It is the type of protection that should be used by emergency medical personnel involved in chemical release incidents (see Box 2).

Box 2: Individual personal protective equipment

The essential difference between the fire service and military approach to working in personal protective equipment (PPE) has been highlighted by the current perceived threat of deliberate toxic and biological release by terrorists in an urban setting. The traditional use of heavy suits and breathing apparatus by fire services for toxic releases had been applied uncritically for all civil chemical and biological releases, particularly in the United States. While this approach leans heavily on the side of safety, the use of such equipment in the warm zone, where ambient concentrations of toxic agents are less than in the hot zone is counterproductive, (a) because of the limited time a wearer can remain safely inside the suit, and (b) because of the severe limitations placed on manoeuverability. There has been a growing realization that early medical care in the form of advanced life support is required for some casualties who are delayed in the warm zone while awaiting decontamination. Since many chemical warfare (CW) and other toxic agents have their life-threatening effects on the respiratory system, early airway and ventilation management is a vital adjunct to antidote therapy. This has lead to consideration of what PPE medical responders should wear.

Military PPE is essentially level-C equipment (see Figure 3). However, it should be noted that protective suits, respirators and filtration cartridges have been developed under conditions of considerable secrecy over the years, working in response to intelligence information about what CW agents are likely to be faced. Data about penetration of agents through the suits and canisters are not openly published. Military PPE developed in the UK has adopted a 'breathing' approach to try to reduce the heat stress load by using layers of material impregnated with activated charcoal. These are designed to allow soldiers to remain operational in persistently contaminated battle zones for considerable periods. In the civilian area, manufacturers have adopted the use of lightweight impermeable suits and data for their chemical resistance are openly available. Similarly data are available for the type and duration of protection offered by the respirator canisters.

Level-C suits are regarded by many emergency medical services (EMSs) in Europe as being suitable for operations by emergency personnel working in the warm zone to provide essential care to patients awaiting and during decontamination (see Figure 4). The weight of the suits and thickness of the gloves

Figure 3. NATO military level-3 protection. Figure supplied by David Baker

Figure 4. SAMU de Paris level-C protective suit. Figure supplied by David Baker

allows manoeuverabilty and dexterity which permit essential medical operations, such as inserting a pharyngeal airway or intubation and ventilation. While level-A and -B suits allow only about 20–30 min of breathing from a self-contained air supply, filtration respirators can be used for considerably longer and the cartridges can be changed inside a contaminated zone using established procedures. Heat stress, which is a serious disadvantage of level-A and -B PPE, is reduced in level-C PPE and this can be improved further by the use of ventilated suits. In the UK, there is still debate about the deployment of EMS personnel in the warm zone. In France, however, plans and equipment and training have been provided to allow the emergency medical service (SAMU) to work in the warm zone. The approach has been to use specially procured level-C suits and a filtration respirator which allows good visual contact and speech transmission (Figure 4). SAMU emergency teams are medically led and the doctor in charge is allowed clinical freedom in his choice of treatment options, rather than following set protocols as in paramedical systems. This improves flexibility of clinical response in the difficult operating circumstances of the hot/warm zone. SAMU personnel have received extensive training in wearing level-C PPE and stocks of equipment are held and maintained in ambulance dispatching centers ready for immediate use by normal-on call emergency medical response teams. Personnel are able to perform all the usual advanced life support actions while wearing PPE, such as endotracheal intubation, artificial ventilation and peripheral intravenous vascular access.

Level-C protection allows medical personnel to provide essential life support for victims of toxic release during decontamination. However, the protection provided will depend on the toxic compound released, the concentration and the duration of time spent in a contaminated zone. More work is needed, in conjunction with designers of military PPE, to assess the suitability of the variety of PPE available on the commercial market for use in urban CW release.

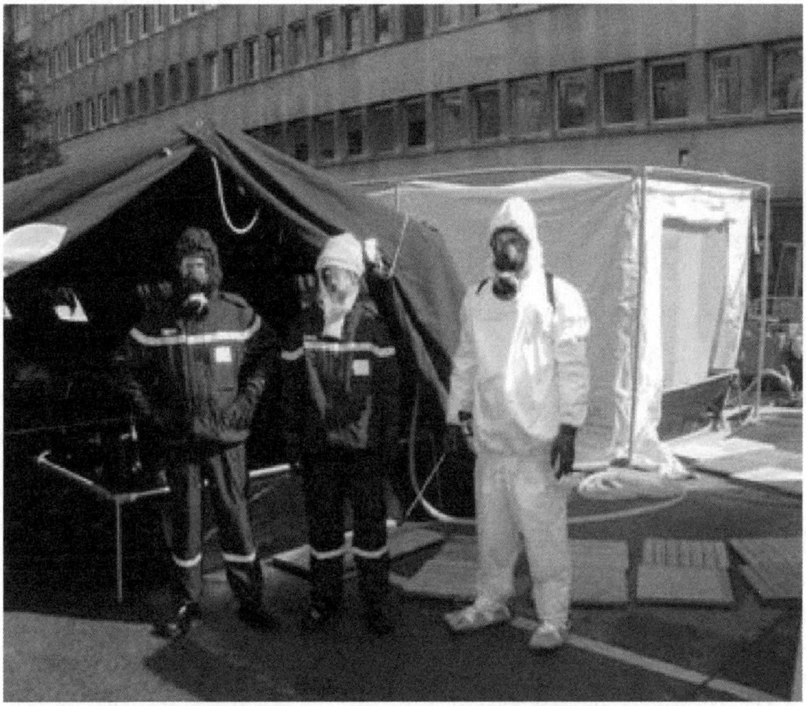

Figure 5. Reception and decontamination facilities at a Paris Plan Biotox reference hospital; the emergency physicians are wearing level-C protection, stored at the hospital, with filtration respirators. Figure supplied by David Baker

DECONTAMINATION OF CASUALTIES

Before any contaminated casualty can be moved on to the cold zone and to hospital medical care, decontamination of the released chemical substance is required. This is only the case if the agent released is persistent. Knowledge of the characteristics may only be available from detection and identification information. As noted above, this may be lacking or delayed in the civil context and therefore full decontamination is the default state in cases of doubt. Decontamination in civil releases is usually done using water or other active decontamination solutions. Multi-stage showering of patients whose clothing has been removed is the basis of this technique. Two points are relevant medically. The first is that in cold climates the decontamination process may lead to significant hypothermia because of the circumstances. Secondly, the decontamination procedure leads to potentially serious delays in providing medical care for the seriously injured. The second factor has lead to plans for the provision of emergency medical care inside the contaminated zone by specially protected, trained and equipped personnel (see Figure 5).

PLANNING THE EMS RESPONSE AND THE MANAGEMENT OF VICTIMS: THE FRENCH APPROACH

Pre-hospital disaster care in France is controlled by two national response plans, called *red* (Plan Rouge: http://en.wikipedia.org/wiki/Plan_rouge) and *white* (Plan Blanc: http://www.sante.gouv. fr/htm/actu/31_030814b.htm; http://en/wikipedia.org/wiki/Plan_blanc). The red plan concerns the rescue and evacuation of victims from a disaster site by the fire and rescue service. The plan provides for an overall on-site commander (COS) who controls a fire and rescue and a

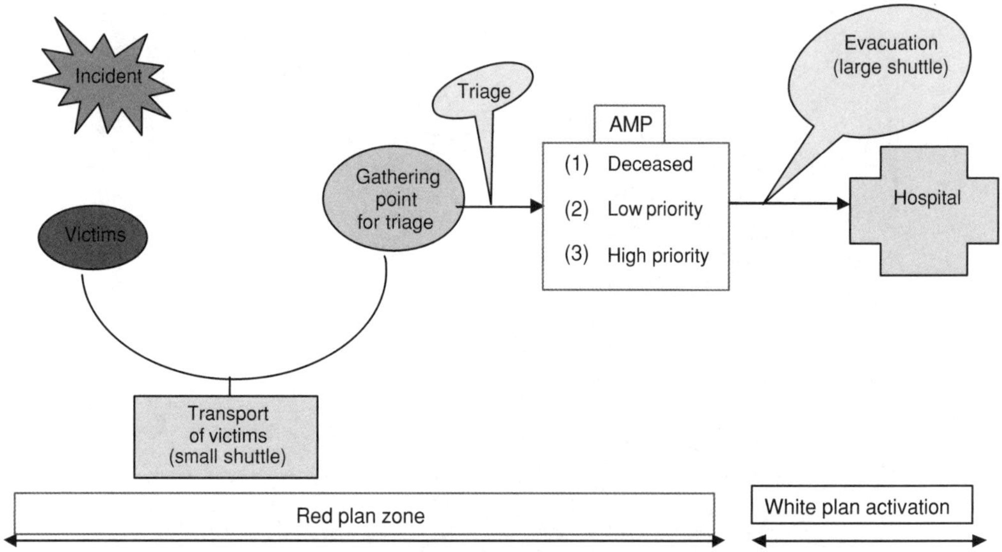

Figure 6. Interaction of the red and white plans for the evacuation of casualties following a disaster. The fire service are responsible for rescue and primary evacuation of casualties as far as the advanced medical post. Here, casualties are triaged and receive primary treatment. The hospital white plan for the management of mass casualties is extended to this point using medically manned mobile intensive care units which can carry out extensive early care before transporting the patient to the most appropriate hospital facility. In the case of a toxic release, this will be to a special 'reference' hospital, manned and equipped to deal with casualties from a chemical–biological release

medical chain. He reports to the Prefect of the Departement (in Paris to the Prefect of Police) and then directly to the Prime Minister. The fire and rescue chain under the control of director of fire and rescue (DSIS) is concerned with managing the cause of the disaster, rescuing victims and providing essential primary medical care using their own internal medical resources (all French fire services have medical units and firefighters are trained to emergency medical technician levels). Firefighters rescue victims in a shuttle operation (*le petit noria*, literally 'small waterwheel') and deliver them to the medical chain at the advanced medical post distant from the site of the disaster (see Figure 6).

The AMP is under the overall control of the director of medical rescue (DSM) who is usually a fire service medical officer (these are a feature of the French system). Running of the AMP is the responsibility of a physician, chosen by the DSM, whose responsibilities include triage, immediate casualty care and evacuation of the patients to designated hospitals (*le grand noria*).

At this point the, white plan begins. This plan is essentially concerned with hospital response to mass casualties caused by disaster. Its provisions include setting-up crisis management cells, recalling personnel, freeing hospital resources and organizing controlled reception of mass casualties. The organization of French emergency medical services allows a particular approach to incidents involving mass casualties. The ambulance services (SAMU) are medically controlled and manned which allows a number of on-site emergency measures which are not easily provided by services led by paramedical staff. Thus, patients receive a large degree of care before arriving at the hospital. In addition, the SAMU and hospital services are operated by the same organization (in Paris this is the AP–HP) which allows integration of planning and a continuum of care from the pre-hospital to the hospital under the control of hospital doctors who also control the SAMU.

In terms of the white plan, this means that the hospital can effectively be extended beyond its

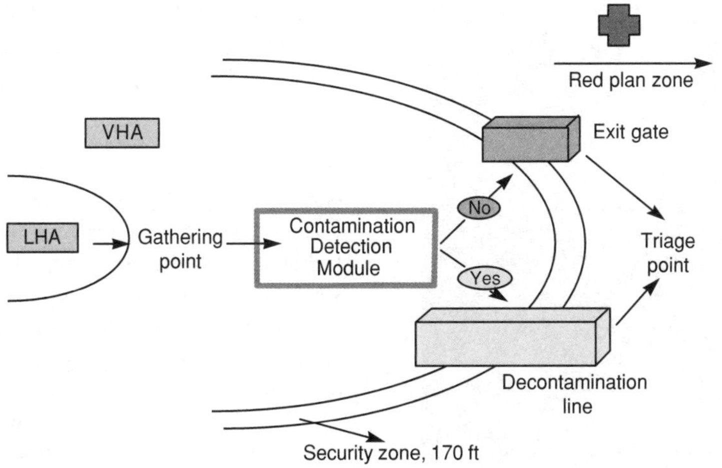

Figure 7. Plan Piratox (red plan) arrangement of HAZMAT zones [LHA, liquid hazard (hot) zone; VHA, vapour hazard (warm) zone]. Triage is conducted (1) to detect contamination, and (2) for medical status in the triage point or the AMP. Later modifications of the plan allow for the provision of early life support (TOXALS) when required inside the warm zone

boundaries by sending the SAMU mobile medical units to the scene of the disaster where they link up with the fire service operating the red plan. In effect, therefore, the AMP is the interface between the on-site and hospital management of casualties.

SPECIAL PLANNING FOR CHEMICAL RELEASES

Following the toxic attack in Japan in 1995, many countries around the world considered planning options in case further terrorist attacks should take place. In France, detailed planning (Plan Piratox) had been put in place in 1987 as a special modification of the red plan to deal with chemical releases (see Figure 7).

The plan was revised in a government circular, dated 26 April 2002 (Circulaire 700 SGDN/PSE/PPS, 2002) with special reference to the management of possible terrorist chemical, biological, radiological and nuclear (CBRN) incidents. At about the same time, a further circular (Circulaire DHOS/HFD No 2002/284, 2002) was issued to provide specific instructions to hospitals for the management of mass casualties from CBRN incidents. This was, in effect, an updating of the white plan. These circulars contain not only organizational flow charts, but also technical annexes which provide definitive medical and scientific data about management of patients exposed to the main toxic hazards.

In addition to these plans, in October 2001 the French Minister of Health announced Plan Biotox as an immediate response to the terrorist anthrax attacks then taking place in the United States (Plan Biotox, 2001). This plan focused on the provision of specific therapy and early life support to be provided in a contaminated zone with the provision of specially trained and equipped SAMU crews capable of operating in a toxic environment and also provision for the reception of contaminated patients at designated hospitals. A key feature of the new plan was the mass purchase of small portable gas-powered ventilators that can assure a high quality of essential ventilatory care at the scene of the emergency and in hospitals that may receive mass casualties (see Figure 8).

In the government circulars of 2002, the role of special receiving hospitals is clearly defined. France is divided administratively into seven ‘defence zones’ (Circulaire DHOS/HFD No 2002/284, 2002). For each of these, an organization for response to CBRN incidents

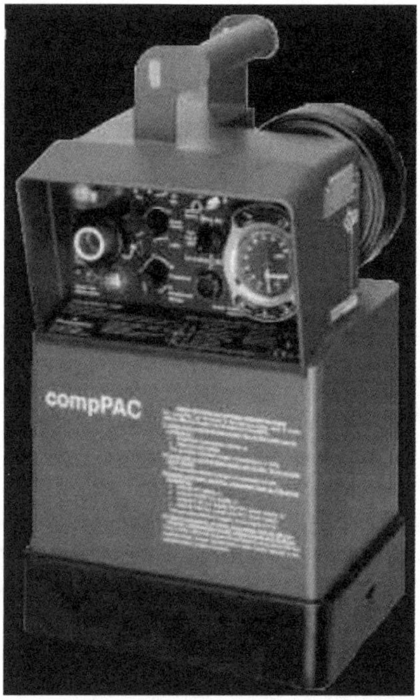

Figure 8. The CompPac ventilator (photograph courtesy of Smiths Medical International Ltd.). This portable pneumatic device can be driven from a variety of power options, including an internal battery, and uses compressed ambient as the ventilating gas. This is filtered through a conventional CBRN respirator canister. The driving gas can be enriched with bottled oxygen given an endurance of over four hours ventilation in a contaminated zone

is under the direct control of the defence zone administrator (Prefect). In each zone, key hospitals (termed 'reference hospitals') are designated to take CB casualties. They are chosen because they contain, as part of their normal daily work, expertise and facilities that can be used quickly in the event of CBRN incidents. These establishments are essentially the head of an extensive referral system which includes the SAMU ambulance services and its constituent stations (SMUR); the SAMU of each region is co-ordinated by the reference hospital. In addition, the reference hospital is linked to specialist poisons centres, infectious disease facilities, nuclear medical facilities and occupational health organizations in order to provide a broad-spectrum approach to CBRN releases.

MANPOWER AND CASUALTY MANAGEMENT LOGISTICS IN THE DECONTAMINATION ZONE

Basic assumptions are important for the provision of protected personnel and the number of casualties that can be treated within a given time following a chemical agent release. Thus, responses may be planned around special emergency medical teams made up of a limited number of protected and equipped personnel. As an example, the SAMU de Paris operates three-person special response teams which include an anaesthetist, an anaesthetic nurse (these work as anaesthetists in their own right in France) and a driver/technician. Six such teams are available for the inner-Paris region. The provision of special ventilators capable of operating in the toxic zone, together with airway equipment and lightweight oxygen cylinders, means that such teams can operate in a contaminated zone for up to six hours. Casualties requiring advanced life support have priority in decontamination and should remain in the hands of the decontamination medical team for as little time as possible before being handed over to other SAMU personnel in the cold zone who have access to large numbers of portable ventilators, as well as to bulk oxygen provided from liquid oxygen stores on site.

The number of casualties receiving life-support will depend upon the skills of the intervention teams. Teams that contain a high proportion of anaesthetists, or nurse anaesthetists, who work on a regular basis within SAMU, will be able to provide rapid and definitive airway management since this speciality is experienced on a daily basis in the management of difficult airways and ventilation. Once the airway has been secured, the patient can be left on a ventilator under the clinical observation of a less-skilled operator while the anaesthetist moves on to manage another case. In this way, each intervention team with automatic ventilators can, in theory, manage several patients with severe respiratory distress at the same time. In practice, there is likely to be a spread of severity of requirement and full life-support will be required only for the most seriously affected. Taking the example of the Paris SAMU teams, the 12 skilled anaesthetic personnel in the 6 special teams could be expected to be able to handle at least 50 cases of severe respiratory failure within the decontamination zone at the same time, with an endurance of up to 6 h of ventilation using filtration ventilation with 'Kevlar' cylinders. In practice, many of the injured will required only oxygen support before decontamination and this can be provided by less-skilled emergency medical technicians from the fire services. There will be a rolling triage procedure which should mean that the most severely affected are managed and decontaminated first before being sent to hospital.

LIFE SUPPORT FOLLOWING CHEMICAL AGENT RELEASE

Life support in medical emergency is usually classified as basic or advanced. Both systems are based upon the ABC approach (the management of airway, breathing and circulation). Whether basic or advanced life-support is provided depends on the skills of the responders and the equipment they use. For the general public, the most familiar use of life-support is resuscitation following cardiac arrest. Thus, holding the head in a position that opens the air passage, mouth-to-mouth ventilation and chest compression are familiar to all those who have completed a first-aid course. In the context of toxic injury, the approach is identical but the means of holding the airway and delivering artificial respiration are more sophisticated. In addition, in toxic injury, cardiac arrest is a possible consequence of respiratory arrest. Thus, the provision of respiratory support is vital if the function of the heart is to continue.

Because of the time-scale of effects, life-support is more likely to be required for chemical rather than biological agent attack. In the past, the approach to the management of patients following release of a chemical agent has been traditionally based upon the use of antidotes. There are many specific antidotes for the recognized hazards but those for nerve agents and cyanide are the best known. In recent years, however, the importance of providing life-support in addition to antidotes following toxic release has become recognized (Baker, 1996,2005a).

Many, perhaps most, chemical agents exert their lethal effects because of effects on the respiratory system and thus the provision of breathing support is essential. This is because toxic agents produce effects that cause blockage of the air passages, depression and failure of the respiratory control centres in the brain or paralysis of the muscles of respiration. To overcome these combined effects, the emergency medical response must include the ability to be able to clear and support the airway and also to be able to ventilate the lungs artificially when there is respiratory deficiency or arrest. This support is now part of a standard response for advanced life support in both conventional and toxic trauma and is termed TOXALS (Baker, 1996; Department of Health, 2003) (see Box 3).

If the chemical agent is not persistent, normal life-support measures can be provided by the emergency services at the casualty collection point. However, problems arise if there is a need for decontamination since this will inevitably produce delays in crossing over to the clean zone. Casualties within this zone who have received a toxic dose of the released hazard will therefore require immediate life-support while they are being decontaminated. Such care can be provided by SAMU teams.

Triage

In the French system, triage in Plans Piratox and Biotox is based on (1) whether the patient is contaminated or not, and (2) the patient's physical status (see Figure 9).

Triage following a chemical agent release is complicated by the latency of onset of signs and symptoms that is a feature of all toxic agents. The principle danger is the time taken to reveal developing toxic pulmonary oedema (Baker, 2005b). In the case of a hazard such as phosgene, there is a dual latency of action with an immediate ('choking') effect on the upper airways which serves as an alert for patients at risk of developing pulmonary oedema later. In the French, system patients are triaged according to whether the emergency is 'absolute' or 'relative.' The absolute category implies that the patient will not

Box 3: TOXALS – General management of casualties contaminated by chemical compounds

Airway – the airway of the casualty must be maintained at all times. In the unconscious casualty, this may involve simple basic airway manœuvres plus suction of the copious secretions associated with chemical poisoning. Occasionally, there may be a requirement for advanced airway management, such as tracheal intubation, to protect the airway from the excessive secretions and to prevent aspiration of regurgitated stomach contents.

Breathing must be carefully observed until full decontamination and recovery have occurred. Supplemental oxygen will speed the recovery from volatile chemical poisoning. If breathing becomes compromised, it must be supported by artificial ventilation with supplemental oxygen using a self-inflating resuscitation bag-valve-mask or automatic ventilator. Entrained air must be filtered when ventilating casualties in a contaminated environment.

Circulation must be carefully observed and monitored. Non-invasive blood pressure, pulse oximetry and ECG monitoring are all useful indicators of circulatory function. The early establishment of intravenous access will aid the administration of fluids and drugs.

Disability should be assessed using the simple AVPU scale (**A**lert, responds to **V**oice, responds to **P**ain, **U**nresponsive). This assessment should be repeated at frequent intervals to assess the progress of the casualty.

Drugs, especially specific antidotes, should be administered.

Exposure of the casualty is essential not only to assess physical damage but to remove all clothes that have been contaminated by the chemical.

Environment – it is important to remember that the primary management described above may be severely limited by the need of the rescuer to wear protective clothing. Therefore, only those skilled in these techniques and trained in protective clothing should enter and treat casualties in a contaminated area. All others should await the casualties' arrival in cold/clean zone, following decontamination.

It should be remembered that the identification of the chemical and therefore its specific antidote might take some time. However this must not delay the basic medical management of the casualty.

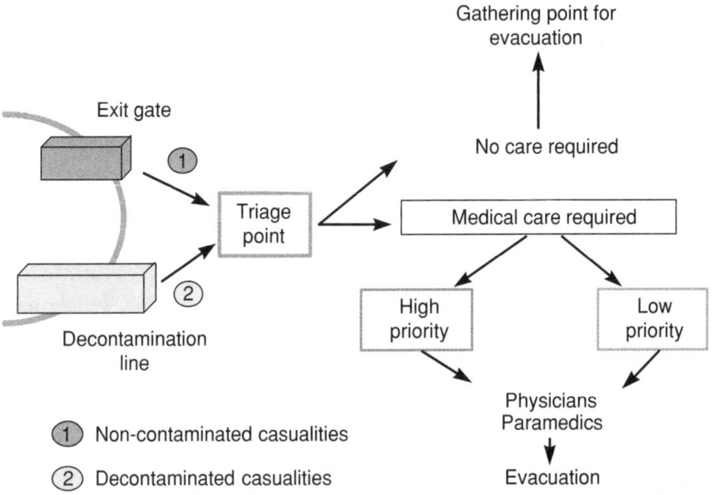

Figure 9. Plan Piratox (modified red plan) management of casualties. Under the French system, emergencies are classified as either 'absolute' or 'relative'. The decision is based upon medical experience

survive unless immediate life-support measures are applied. This is more fluid than the T1-T4 system used in English-speaking countries and is based upon the fact that assessment is made by attending emergency physicians from SAMU who make their decisions on the basis of medical experience.

Casualty management

SAMU operates mobile intensive-care units which are extensively equipped to provide a definitive level of airway, ventilation and circulatory support. These are operated as part of the normal daily emergency response and are manned by anesthesiologists, nurse anaesthetists, physicians and general physicians who have had special postgraduate emergency training. For many emergency medical personnel, casualties from chemical agent release are a relative rarity. Although chemical injury may be unfamiliar, there are many emergency situations such as smoke inhalation, carbon monoxide poisoning and bronchial asthma which provide analogies in clinical management which are useful training for the management of the pathophysiology of chemical releases. In this situation, detailed information is available from reference hospitals and defence-zone regional poisons centres. This is the equivalent of information services such as the Chemical Hazards and Poisons Division of the Health Protection agency in the United Kingdom. Specific information about the use of antidotes and TOXALS is available from the internet (Department of Health, 2003; Plan Blanc: http://en/wikiperid.org/wiki/Plan_blanc). Detailed information about antidote therapy is set out in other chapters of this book.

CONCLUSIONS

Chemical agent release is increasingly likely from both accidental and deliberate causes. Release presents dangers not only for those immediately exposed but also for emergency medical responders. Toxicity of the agent together with latency of onset of effects complicate early medical management. Agents that are physically persistent may be transmitted to responders and dictate the need for both physical protection and decontamination. Life-threatening delays may occur during decontamination, which means that emergency medical support within a contaminated zone by suitably trained and protected responders is essential to break the link between mass injury and potential loss of life.

Planning initiatives such as those in place currently in France ensure that early pre-hospital medical support is available for chemical casualties continuing through to specially designated hospital facilities. Equipment and logistic options exist to ensure a co-ordinated life-support response to multiple casualties. Early and continuing life-support may be required for many casualties, in addition to antidote treatment. Proper planning, training and equipping of emergency medical personnel, together with hospital experience in medical conditions that are analogous to the pathophysiology of chemical attack (Moles and Baker, 1999), mean that the management of casualties from toxic release is as feasible as with casualties from conventional explosive attacks.

REFERENCES

Alibek K and Handelman S (2000). *Biohazard.* London, UK: Arrow Books.

Anon (1984). Calamity at Bhopal. *Lancet*, **137**, 1378–1379.

Anon (1995). *Managing Hazardous Materials Incidents*, Volumes 1–3. Washington, DC, USA: Department of Health and Human Services /Public Health Services Agency for Toxic Substances.

Baker DJ (1993). Chemical and biological warfare agents: a fresh approach. *Jane's Intelligence Rev*, **5**(1).

Baker DJ (1996). Advanced life support for toxic injuries. *Eur J Emerg Med*, **3**, 256–262.

Baker DJ (ed.) (1999a). Special Edition: Toxic Substances. *Resuscitation*, **42**, 101–159.

Baker DJ (1999b). Management of respiratory failure in toxic disasters. *Resuscitation*, **42**, 125–131.

Baker DJ (2003). Management of casualties from terrorist chemical and biological attack: a key role for the anaesthetist. *Br J Anaesth*, **89**, 211–214.

Baker DJ (2005a). Chemical and biological warfare agents: the role of the anesthesiologist. In: *Anesthesia* (ED Miller, ed.), Chapter 62, pp. 2497–2525. New York, NY, USA: Churchill Livingstone.

Baker DJ (2005b). Aspects of critical care following toxic agent release. *Crit Care Med*, **133**, S66–S74.

Borak J, Callan M and Abbott W (1991). *Hazardous Materials Exposure: Emergency Response and Patient Care.*, NJ, USA: Prentice Hall Inc.

Borio L, Frank D and Mani V (2001). Death due to bioterrorism-treated anthrax. Report of two patients. *JAMA*, **286**, 2554–2559.

Bronstein AC and Currance PL (1994). *Emergency Care for Hazardous Materials Exposure*, 2nd Edition. St Louis, MO, USA: Mosby Lifeline.

Circulaire 700 SGDN/PSE/PPS (2002). Relative à la doctrine nationale d'emploi des moyens de secours et de soins face à une action terroriste mettant en oeuvre des matières chimiques, 26 April.

Circulaire DHOS/HFD No 2002/284 (2002). Relative à l'organisation du systeme hospitalier en cas d'afflux de victimes, 3 May (http://www.sante.gouv.fr/htm/pointsur/attentat/circ_020503.pdf).

Department of Health (2003). *Treatment of Poisoning by Selected Chemical Compounds – Blain Report* (First Report by Expert Group on the Management of Chemical Casualties Caused by Terrorist Activity), October 2003 (http://www.dh.gov.uk/PublicationsAndStatistics/Publications/PublicationsPolicyAndGuidance/PublicationsPAmpGBrowsable Document/fs/en?CONTENT_ID=4094986&chk= V5T%2BZB). London, UK: Department of Health.

Dhara VR and Dhara K (2002). The Union Carbide disaster in Bhopal: a review of health effects. *Arch Environ Health*, **57**, 391–404.

Hammick M (1991). All stick and no carrot. *Int Defense Rev*, (December), 1323–1327.

Harris R and Paxman J (1982). *A Higher Form of Killing*. London, UK: Chatto and Windus.

Laurent JF, Richter F and Michel A (1999). Management of victims of urban chemical attack: the French approach. *Resuscitation*, **42**, 141–149.

Matsuda N, Takatsu M, Morinari H *et al.* (1995). Sarin poisoning in the Tokyo subway. *Lancet*, **345**, 1446–1447.

Mehta PS, Mehta AS, Mehta SJ *et al.* (1990). Bhopal tragedy's health effects: a review of methyl isocyanate toxicity. *JAMA*, **264**, 2781–2787.

Moles TM (1999). Emergency medical services systems and HAZMAT major incidents. *Resuscitation*, **42**, 103–107.

Moles TM and Baker DJ (1999). Clinical analogies for chemical injury. *Resuscitation*, **42**, 117–124.

Morita H, Yariagisawa N, Nakaji T *et al.* (1995). Sarin poisoning in Matsumoto, Japan. *Lancet*, **346**, 290–293.

Okumura T, Takasu N, Ishimatsu S *et al.* (1996). Report on 640 victims of the Tokyo subway sarin attack. *Ann Emerg Med*, **28**, 129–135.

Okumura T, Suzuki K, Fukuda A *et al.* (1998). The Tokyo subway Sarin attack. Disaster management, part 2: hospital response. *Acad Emerg*, **5**, 618–624.

Organization for Economic Co-operation and Development (1994). *Health Aspects of Chemical Accidents: Guidance on Chemical Accident Awareness, Preparedness and Response for Health Professionals and Emergency Responders*, OECD Environment Monograph No. 81 (OCDE/GD(94)1). Paris, France: Organization for Economic Co-operation and Development.

Plan Biotox (2001). An introduction, 5 October, (http://www.sante.gouv.fr/htm/dossiers/biotox.intro.htm) (in French).

Plan Blanc. (http://www.sante.gouv.fr/htm/actu/31_030814b.htm) (in French).

Plan Blanc. An introduction (http://en.wikipedia.org/wiki/Plan_blanc) (in English).

Plan Rouge. An introduction (http://en.wikipedia.org/wiki/Plan_rouge) (in English).

United Nations (1987). Report of the Mission Dispatched by the Secretary General to Investigate Allegations of the use of Chemical Weapons in the Conflict between the Islamic Republics of Iran and Iraq, UN document /18852. New York, NY, USA: United Nations.

13 THE DARK MORNING: THE EXPERIENCES AND LESSONS LEARNED FROM THE TOKYO SUBWAY SARIN ATTACK

Tetsu Okumura[1], Tomohisa Nomura[1], Toshishige Suzuki[1], Manabu Sugita[1], Yasuo Takeuchi[1], Toshio Naito[2], Sumie Okumura[3], Hiroshi Maekawa[3], Shinichi Ishimatsu[4], Nobukatsu Takasu[4], Kunihisa Miura[5], and Kouichiro Suzuki[6]

[1]*Department of Emergency and Disaster Medicine, Juntendo University, Tokyo, Japan*
[2]*Department of General Medicine, Juntendo University Hospital, Tokyo, Japan*
[3]*Advanced Emergency Medical Center, Juntendo Shizuoka Hospital, Izunokuni, Japan*
[4]*St Luke's International Hospital, Tokyo, Japan*
[5]*Department of Anesthesiology and Pain Medicine, Juntendo University, Tokyo, Japan*
[6]*Emergency Department, Kawasaki Medical School, Kurashiki, Japan*

INTRODUCTION

The Matsumoto and Tokyo subway sarin attacks were wake-up calls to NBC terrorism. These incidents proved that terrorists could actually deploy chemical weapons and weapons of mass destruction. We have previously analyzed and reported on the Tokyo subway sarin attack from the viewpoint of clinical medicine (Okumura *et al.*, 1996,1998a,1999). Here, by including the findings of the court trials and information related to the attacks that has become available, we review the experiences and lessons learned from the Tokyo subway sarin attack in the hope that doing so will improve measures against chemical terrorism.

SUMMARY OF THE SUBWAY SARIN ATTACK

The subway sarin attack occurred during the Monday morning rush-hour of March 20, 1995, which was the day before the Spring Equinox holiday. The group responsible for the attacks was a cult called *Aum Shinrikyo*, and the purpose of the attack was to disrupt criminal investigations into the cult that were being conducted by the police. According to the police, five terrorists released sarin at a concentration of 35% (the mixture comprised 35% sarin) in five subway cars. The sarin was mixed with hexane and diethylaniline to a final weight of 600 g in nylon/polyethylene bags. The bags were wrapped in newspaper so that they would be inconspicuous, using the newspaper published by the Japanese Communist Party. The terrorists then punctured the bags using the tip of an umbrella before disembarking quickly from the subway cars. The terrorists were exposed to the sarin and experienced the symptoms of miosis, dyspnea and involuntary movements of the hands and feet. Criminal C, a physician and the person responsible for the administration of *Aum Shinrikyo's* own clinic, treated the five terrorists using atropine sulfate and 2-PAM (2-pyridine aldoxime

Chemical Warfare Agents: Toxicology and Treatment (2nd Edition)
Edited by Timothy C. Marrs, Robert L. Maynard and Frederick R. Sidell

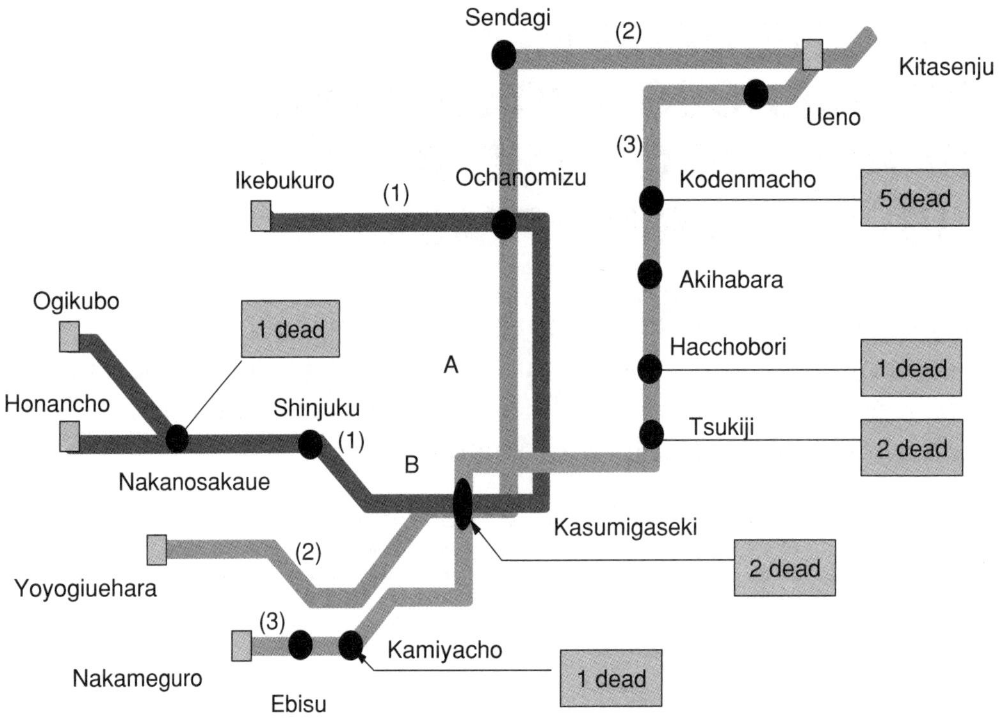

Figure 1. The three affected subways lines and the number of fatalities: (1), Marunouchi line; (2), Chiyoda line; (3), Hibaya line: A, Imperial Palace; B, National Diet (Parliament)

methiodide). Figure 1 shows the three affected subway lines and the number of fatalities, while Figure 2 shows the movements of the terrorist group and the number of fatalities and injuries that resulted from the attack. The number of fatalities was the highest at Kodenmacho Station: 5 people died there. Criminal A brought three bags (the other criminals brought two bags each), and because passengers had kicked the bags onto the platform, a total of three open bags were left to stand on the station platform. In the other subway cars, the subway operator sensed that something was wrong, stopped the train and closed the subway cars, which meant that the bags containing sarin were left inside the cars. Two bags remained unopened, and these bags were recovered and subsequently analyzed. The two fatalities at Kasumigaseki Station on the Chiyoda line were station workers who removed the bags using only cloth gloves and cleaned the floors before slipping into coma and dying. Their actions prevented deaths among the passengers. Another contributing factor was that Criminal C, the physician, only opened one of the two bags of sarin he brought to the Chiyoda line. The number of fatalities and injuries initially reported varied among the agencies involved in dealing with the incident. While all agencies eventually reported twelve fatalities, the number of people with minor and serious injuries has generally been reported at more than 5500. The Fire Department reported 5642 injuries, the Tokyo Metropolitan Police cited 3796 injuries and the subway authority reported a total of 5654 victims, including 12 deaths (10 passengers and 2 station workers), 999 hospitalized individuals (960 passengers and 39 station workers) and 4643 treated individuals (4446 passengers and 197 station workers).

Given that the Tokyo subway sarin attack was the first major chemical terrorist attack in an urban setting, the first responders were unaware of the need to remove the victims from the subway stations as quickly as possible. Some of the victims who had trouble walking were barely able to get out of the subway cars and collapsed on the station platforms, resulting in continued

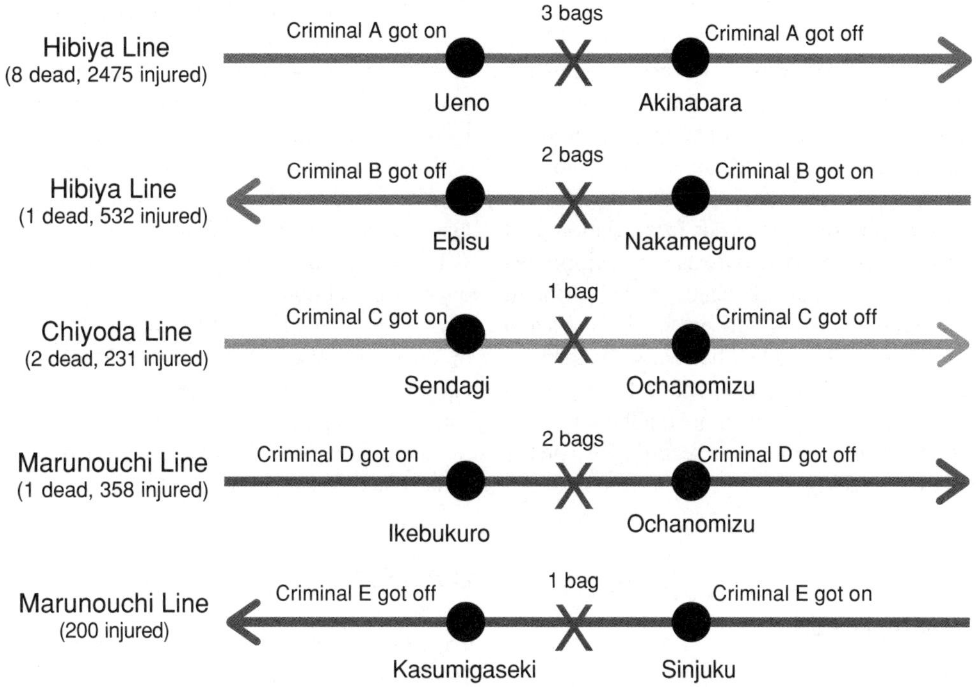

Figure 2. Movements of the terrorist group and the number of fatalities and injuries that resulted from the attack

exposure to sarin. Furthermore, some victims who were eventually brought up to street level were allowed to rest in an area near a vent from the subway system and, again, may have continued to be exposed.

The five criminals stated that they all punctured their bags at 8:00 am. The first call for an ambulance was placed at 8:09 am and reported patients with convulsions at Kayabacho Station. By 8:15 am there were many victims and a false report at around this time of severe causalities caused by an explosion at Tsukiji Station complicated matters further. A total of 19 stations made calls for ambulance services and after 8:30 am, many people went to the surrounding hospitals on their own or with the help of passing drivers. According to a report compiled by the Tokyo Fire Department, a total of 5493 people were treated at 267 medical institutions in Tokyo, including 14 emergency medical centers and 67 hospitals. An additional 17 patients were treated at 11 hospitals outside Tokyo (Ieki, 1997). Another report found that a total of 6185 victims were treated at 294 hospitals (Chigusa, 1995). There was a great deal of confusion, and the number of victims varies depending on the agency involved. The St. Luke's International Hospital treated the greatest number of victims (640 victims on the day of the attacks) because the hospital is located on the Hibiya Line, which had the highest number of victims, and also because some media outlets reported that the hospital had the antidote. Based on the aging process of sarin, 2-PAM should be administered within 4–5 h of exposure, but many victims continued to visit the hospital to receive 2-PAM for approximately one week after the attack.

SYMPTOMS

According to the data for 627 victims treated at the St. Luke's International Hospital, the most common symptom was miosis ($n = 568$, 90.5%), followed by headache ($n = 316$, 50.4%), visual darkness ($n = 236$, 37.6%), eye pain ($n = 235$, 37.5%), dyspnea ($n = 183$, 29.2%), nausea ($n = 168$, 26.8%), cough ($n = 118$, 18.8%), throat pain

($n = 115$, 18.3%) and blurred vision ($n = 112$, 17.9%), respectively (Okumura *et al.*, 1999). As far as severity was concerned, among the doctors who treated the victims, the consensus was that severely affected patients required artificial respiration due to convulsions and respiratory arrest, moderately affected patients exhibited respiratory distress or fasciculation, and mildly affected patients only had eye symptoms. On the day of the attack, chaos made it difficult to differentiate clearly between convulsions and fasciculations. There has been a view that the use of EEG monitoring could have differentiated convulsions from fasciculation, but there was not enough time or manpower to record EEGs on the day of the attack.

MEDICAL TREATMENT AT THE ATTACK SITES

At the time of the attack, there were no established measures to deal with chemical weapons, decontamination was not undertaken, and no antidote was administered at the sites of the attack. During the initial critical hours when doctors could have played an important role, none were present at the sites of the attack. At the request of the Tokyo Fire Department, a total of 73 medical staff – 47 doctors, 23 nurses and three clerks – were dispatched to 4 subway stations for triage and emergency treatment (Chigusa, 1995). However, because they reached the stations after 9:00 am, they did not provide emergency medical treatment such as tracheal intubation.

CLINICAL LABORATORY FINDINGS FROM THE VICTIMS

Sarin poisoning is associated with a marked decrease in plasma cholinesterase. In the Matsumoto sarin attack, Okudera (2002) documented that mean standard values (midpoint of the upper and lower levels) for plasma cholinesterase levels and the mean pupil diameters were $\leq 50\%$ and 0.9 mm in severe cases, 50–100% and 1.3 mm in moderate cases, and $\geq 100\%$ and 2.2 mm in mild cases, respectively. They concluded that patients with a pupil diameter of ≤ 1 mm are severely poisoned and require emergency care such as tracheal intubation, patients with a pupil diameter of 1–2 mm require medical care at medical institutions, and patients with a pupil diameter of ≥ 2 mm do not require emergency care (Research Committee on Nerve Gas Poisoning in Matsumoto, 1995).

However, our findings show that the level of plasma cholinesterase and the degree of miosis did not always correlate with severity of exposure. We believe that level of plasma cholinesterase and pupil diameter should only be used as approximate guides to the level of poisoning and that severity and the need for treatment should be based on clinical findings.

DIAGNOSIS

The Matsumoto sarin attack occurred on June 27, 1994, one year before the Tokyo subway sarin attack, and although the group responsible for the Matsumoto attack was not known at the time of the Tokyo subway attack, emergency doctors in Japan knew that the diagnosis and treatment of sarin poisoning should follow that of accidental organophosphorus poisoning.

One of the diagnostic signs of organophosphorus poisoning, including sarin poisoning, in clinical settings was that many cardiopulmonary arrest victims had miosis. This is unusual given that in cases of cardiac arrest the pupils are generally dilated. The doctors in Matsumoto noted miosis in patients with cardiopulmonary arrest. Consequently, a group of patients presenting with cardiac arrest and miosis is an important clue for suspecting chemical terrorist activity, especially a chemical attack involving use of a nerve agent.

Dr Yanagisawa heads the Shinshu University Hospital where many of the victims of the Matsumoto sarin attack were treated. When he heard that several people had died in Tokyo and that many of the patients had miosis, he telephoned the St. Luke's International Hospital after 10:00 am to exchange information with emergency doctors at the hospital. The doctors reached the conclusion that the victims were probably suffering from sarin poisoning. In general, except for patient referral, communication and sharing information of this nature among medical institutions in Japan is very rare. Dr Yanagisawa broke with tradition because one of his students died in the Matsumoto sarin attack. On the day of the Tokyo subway sarin attack, the deceased student would have graduated from the university and this prompted Dr Yanagisawa to make the call. Shinshu University Hospital faxed the documents compiled at the time of the Matsumoto sarin attack to St. Luke's International Hospital and the other major hospitals in Tokyo.

At the same time, doctors and nurses from the Self Defense Force Central Hospital also suspected sarin poisoning and brought various documents and supplies to St. Luke's International Hospital. By the time most of the patients with cardiopulmonary or respiratory arrest had been resuscitated, results of laboratory tests revealed decreased plasma cholinesterase. Based on the various medical data, clinical and test findings, sarin poisoning was diagnosed with relative ease.

Initially, the Tokyo Fire Department's HazMat Team conducted a rapid field test and reported to medical institutions that the causative agent was acetonitrile. However, this was incorrect. Fortunately, given the number of victims with mild to severe symptoms treated at St. Luke's International Hospital, acetonitrile poisoning was ruled out. However, some medical institutions that handled only severe cases, suspected cyanide poisoning and treated patients with antidotes to cyanide.

In Japan, the Japan Poison Information Center is responsible for managing poisoning-related information. The center has an office in Tsukuba and Osaka to handle enquiries from citizens and medical institutions 24 hours a day, 365 days a year. On the day of the subway sarin attack, the center received a total of 143 calls from medical institutions. At the time when the causative agent was not yet identified, the center received calls describing the symptoms and decreased cholinesterase. Once the report of acetonitrile had been announced, the center dispensed information on acetonitrile. When sarin was identified as the causative agent, the center received calls regarding treatment methods, handling of contaminated clothes and pregnancy (Chigusa, 1995).

TREATMENT

At the St. Luke's International Hospital, after 9:00 am but before sarin poisoning was clearly suspected, atropine sulfate was administered palliatively to treat muscarinic symptoms such as miosis and increased secretion. A total of 515 mg was administered to 640 victims, and while this dose was sufficient, it was considerably lower than the recommended dose in war zones (Okumura *et al.*, 1996,1998b,2005) and may have been related to the fact that the concentration of sarin used in the attack was 35%. The administration of 2-PAM by the St. Luke's hospital at approximately 11:00, appeared to control fasciculation, but the effects associated with sarin poisoning could have been alleviated by time and it is difficult to say whether 2-PAM was therapeutically effective. None of the patients were saved by the use of 2-PAM. In addition to atropine sulfate and 2-PAM, diazepam was used in patients with convulsions.

It was not difficult to perform tracheal intubation in emergency settings in Tokyo. However, according to Okudera (2002), tracheal intubation was very difficult in the Matsumoto sarin attack due to excessive airway secretion and tracheal constriction (Research Committee on Nerve Gas Poisoning in Matsumoto, 1995; Okudera, 2002). The difference between the two attacks might have been the purity and large quantities of the sarin used; in the Matsumoto sarin attack, the purity of sarin was thought to be relatively high (in fact, nobody knows the exact concentration of sarin which was used in Matsumoto) but in the Tokyo subway sarin attack, the purity of sarin was 35%. Dr Sidell was a world authority on chemical weapons. He recommended that if a nerve agent is used, the following measures should be taken in the following order: decontamination, drug, airway, breathing and circulation (DDABC). If A, B and C are carried out first, efforts to ensure sufficient ventilation may be wasted. Atropine sulfate should be administered to suppress airway secretions, and airway spasm should be relieved before ensuring adequate ventilation (Sidell, 2003). Using the Matsumoto attack as a benchmark, clinicians should look for symptoms and signs, such as miosis, increased secretion and excessive nasal discharge, and if many victims exhibit increased parasympathetic activity, a chemical terrorist attack using a nerve agent should be immediately suspected and the decision to administer atropine sulfate must be taken immediately.

At the St. Luke's International Hospital, one of three cardiopulmonary arrest victims and two respiratory arrest patients recovered fully and have been socially rehabilitated. This is important and shows that triage in cases of chemical terrorism must be different from that used in large-scale natural disasters. In the latter case, if respiration does not resume after establishing an airway, then trauma victims are generally labeled with a black tag and not treated further (Hodgetts and Porter, 2004). However, in a chemical terrorism attack caused by a nerve agent, medical care should be actively provided to patients with cardiopulmonary or respiratory arrest: recovery by such patients is clearly possible.

SECONDARY EFFECTS EXPERIENCED BY FIRST RESPONDERS

In the Tokyo subway sarin attack, primary decontamination was not undertaken at the attack sites, and the first responders and healthcare workers did not wear appropriate personal protective equipment (PPE). Because of this, 135 of 1364 firefighters (9.9%) suffered from symptoms associated with secondary exposure. While the number of policemen who experienced secondary exposure has not been released, the proportion who suffered from secondary exposure is expected to be similar. Fortunately, no deaths were officially attributed to secondary exposure. If higher-purity sarin had been used in the Tokyo subway sarin attack, it is possible that some people might have died from secondary exposure. However, a pregnant nurse who was exposed to sarin at the hospital became concerned whether or not her child would be born healthy, and after consulting with her doctor, she had an abortion. Therefore, while this was not officially recorded, one life may be said to have been lost due to secondary exposure.

In the Tokyo subway sarin attack, secondary exposure also affected staff of medical institutions. According to a survey conducted by the St. Luke's International Hospital, 23% of the hospital staff that worked on the day of the attack suffered the effects of secondary exposure. Among the various medical professionals, the incidence of secondary exposure was the highest for nursing assistants (39.3%), followed by nurses (26.5%), volunteers (25.5%), doctors (21.8%) and clerks (18.2%). The incidence of secondary exposure was high when the chance of contact with the victims was high. Among the various areas of the hospital, the incidence of secondary exposure was the highest in the chapel (45.8%), followed by the ICU (38.7%), the outpatient department (32.4%), in the wards (17.7%) and in emergency department (16.7%), respectively. Given that so many victims were brought into the Emergency Department at St Luke's one after another, the automatic door was left open. This meant that the area was well ventilated with fresh air from outside and had the effect of minimizing secondary exposure. Conversely, ventilation in the chapel was not good, and because many victims were placed there, the incidence of secondary exposure was high. Given that the attack occurred in winter, the victims were placed in the chapel and continued to wear the clothes they had on at the time of the attack. This meant that every time someone took their jacket off or moved, the sarin that was trapped between or under their clothes was released and caused secondary exposure. In this manner, secondary exposure occurred because decontamination was not performed. Secondary exposure could have been prevented by, at least, performing dry decontamination.

LONG-TERM DAMAGE

While no conclusive large-scale studies have been undertaken on the long-term effects of sarin poisoning, several preliminary studies have been conducted. Murata and colleagues (1997) reported that six to eight months after the Tokyo subway attack in March 1995, the neurophysiological effects of acute sarin poisoning were investigated in 18 passengers exposed to sarin in the subways to determine whether the focal or functional brain deficits were detectable. The event-related and visual evoked potentials (P300 and VEPs), brainstem auditory evoked potential, and electrocardiographic R-R interval variability (CVRR), together with the score on the post-traumatic stress disorder (PTSD) checklist, were recorded in the sarin cases and in the same number of control subjects matched for sex and age. None of the sarin cases had any obvious clinical abnormalities at the time of testing. The P300 and VEP (P100) latencies in the sarin cases were significantly prolonged compared with the matched controls. In the sarin cases, the CVRR was significantly related to plasma cholinesterase (ChE) levels determined immediately after exposure; the PTSD score was not significantly associated with any neurophysiological data despite the high PTSD score in the sarin cases. These findings suggest that asymptomatic sequelae to sarin exposure, rather than PTSD, persist in the central nervous system beyond the turnover period of ChE. Sarin may have neurotoxic actions as a result of mechanisms in addition to the inhibitory action on brain ChE. Yokoyama and colleagues (1998c) argued: 'It is suggested that a delayed effect on the vestibulo-cerebellar system was induced by acute sarin poisoning; females might be more sensitive than males'. Yokoyama and colleagues (1998a,b) also state that: 'A chronic effect on psychomotor performance was caused directly by acute sarin poisoning; on the other hand, the effects on psychiatric symptoms (General Health Questionnaire) and fatigue (Profile of Mood States) appeared to result from post-traumatic stress disorder induced by exposure to sarin' and that: 'The results suggested delayed effects on psychomotor performance, the higher and visual nervous system and the vestibulo-cerebellar system with psychiatric symptoms resulting from PTSD'. In addition, Nishiwaki and colleagues (2001) documented that: 'Our findings suggest the chronic decline of memory function two years and 10 months to three years and nine months after exposure to sarin in the Tokyo subway attack, and further study is needed'. In both the Matsumoto and Tokyo attacks, a single case of organophosphorus-induced delayed neuropathy (OPIDN) was reported (Sekijima *et al.*, 1997; Himuro *et al.*, 1998). In addition, some of the victims of the Matsumoto attack experienced arrhythmia and reduced cardiac contractile force (Okudera, 2002). These central nervous system, peripheral nervous system and cardiovascular symptoms have been suggested as the after-effects of sarin poisoning. However, most cases are anecdotal reports from small-scale studies. It will therefore be necessary to conduct a large-scale, well-designed scientific study.

Victims have been followed up periodically by St. Luke's International Hospital. A survey was conducted on 660 victims one year after the attack, and valid responses were obtained from 303 victims (Ishimatsu *et al.*, 1996). The results revealed that 45% still had some symptoms. Regarding physical symptoms, 18.5% of the victims still complained of eye problems, 11.9% of easy fatigability, and 8.6% of headache. As for psychological symptoms, 12.9% complained of fear of subways while, 11.6% indicated fears concerning the escape from the attack. Another survey was conducted three years after the attack, and 88% of the respondents still complained of some aftereffects (Okumura *et al.*, 1999). However, these studies lack objectivity; victims with after-effects tended to respond more often and there were limits in eliminating biases.

As part of scientific research funded by the Ministry of Health, Labor and Welfare of Japan, Matsui and colleagues conducted two important studies (Matsui *et al.*, 2002).

The first was a case-control study comparing the victims who were treated at St. Luke's International Hospital and regular patients at the hospital. Using a specially designed survey sheet, different items were assessed and quantified. A statistical significance of $p < 0.05$ was found for the following items: chest pain, eye fatigability, blurred near vision, eye mucus, nightmares, fear, restlessness, lack of concentration and forgetfulness. Furthermore, a statistical significance of $p < 0.01$ was observed for hazy vision, blurred far vision, difficulty in ocular focusing, eye discomfort, flashbacks, fear of the attack sites and avoiding the topics and news related to the attack. Incidence of PTSD was analyzed using different diagnostic criteria and was higher among the victims. Of the various diseases that required treatment, the incidence of eye diseases was significantly higher for the victims, but there was no marked intergroup difference in the incidence of other diseases.

The second study was a cohort study comparing victims who received intervention by medical personnel after the attacks and those who did not receive any intervention. There was no significant intergroup difference in items such as listlessness, diarrhea, blurred far vision, blurred near vision, difficulty in ocular focusing, eye mucus and apathy. However, scores for other items was significantly higher among victims who had not received medical intervention. The incidence of masked PTSD for the victims without intervention was significantly higher ($p = 0.007$). Therefore, whilst the incidence of eye symptoms among victims was significantly higher than it was for the non-victims (unexposed patients), there was no significant intergroup difference in the incidence of eye symptoms with respect to intervention, hence suggesting that eye symptoms are long-term and physical after-effects are associated with sarin poisoning.

In the future, it will be necessary to investigate asymptomatic lesions of the nervous system using high-sensitivity tests such as, single photon emission computed tomography (SPECT) and positron emission tomography (PET).

LESSONS LEARNED AND RECOMMENDATIONS

The findings and data gathered from the Tokyo subway sarin attack are summarized above. The section below will focus on the lessons learned and recommendations.

(1) Because the Tokyo subway sarin attack was the first major chemical terrorist attack in an urban setting, the first responders were not aware of the countermeasures required when dealing with chemical terrorism. No concerted effort was made to remove quickly the victims from the subway system. Some victims were lying on the station platform, and when they were finally brought up to the street level, they were placed near a vent for the subway system. This means that zoning (defining safe areas for casualty holding) was not properly done.

(2) In the Tokyo subway sarin attack, a field test yielded a false-positive result. It is important to remember that field tests can yield false-negative and false-positive results. Results from field tests must be compared and clinical tests need to be undertaken in the event of any discrepancy.

(3) Given the earlier Matsumoto sarin attack experience, sarin poisoning was diagnosed relatively smoothly based on the symptoms and laboratory findings (particularly assays of plasma cholinesterase).

(4) The first clinical sign was contraction of the pupils in patients experiencing cardiopulmonary arrest. Consequently, miosis in a group of cardiopulmonary arrest patients is an important indicator for chemical terrorism.

(5) For triage of the victims exposed to a nerve agent, victims with cardiopulmonary or respiratory arrest should be actively resuscitated while maximizing the available medical resources.

(6) In the Tokyo subway sarin attack, the amount of atropine sulfate administered was markedly smaller than what has been proposed in the past. The reason for this may be that the concentration of sarin used in the attack was low, at 35%. Large quantities of sarin of a higher concentration were thought to have been used in the Matsumoto attack, and intubation was difficult due to airway spasm and excessive airway secretion. Therefore, if the use of a nerve agent is suspected clinically, at a minimum, atropine sulfate should be administered as quickly as possible. It is therefore important to establish a system whereby antidotes can be administered early during pre-hospital care.

(7) While 2-PAM appeared to be effective in clinical settings, there was no clear evidence to support its clinical efficacy. In fact, it is thought that 2-PAM administration did not save any lives.

(8) In instances where a chemical gas is used in a terrorist attack, it is necessary to perform dry decontamination (i.e. change clothes) in order to minimize secondary exposure. Because the Tokyo subway sarin attack occurred in winter, many victims were wearing many layers of clothes. The extent of secondary exposure could have been

minimized if victims had changed their clothes.

(9) Long-term effects of sarin poisoning need to be monitored more closely in the future and a well-designed large-scale study needs to be carried out.

REFERENCES

Chigusa H (1995). The Tokyo Subway sarin attack. In: *Disaster Medicine Learned from the Cases* (T Ukai, Y Takahasshi and M Aono, eds), pp. 98–102. Tokyo, Japan: Nanko-do.

Himuro K, Murayama S, Nishiyama K *et al.* (1998). Distal sensory axonopathy after sarin intoxication. *Neurology*, **51**, 1195–1197.

Hodgetts TJ and Porter C (2004). Triage. In: *Major Incident Management System (MIMS)*, p. 30. London, UK: BMJ Books.

Ieki R (1997). Overview of the Tokyo Subway sarin attack. In: *Organophosphorus Poisoning (Sarin Poisoning)* (R Ieki, ed.), pp. 1–3. Tokyo, Japan: Sindan to Chiryo sha.

Ishimatsu S, Tanaka K, Okumura T *et al.* (1996). Result of the follow up study of the Tokyo subway sarin attack (1 year after the attack). *Kyuukyu-Igakkai-shi*, **7**, 567.

Matsui Y, Ishimatsu S, Kawana N *et al.* (2002). *Official Report of Ministry of Welfare and Labor Science Project: Sequelae in the Tokyo Subway Sarin Attack Victims*. Tokyo, Japan: Ministry of Welfare and Labor.

Murata K, Araki S, Yokoyama *et al.* (1997). Asymptomatic sequelae to acute sarin poisoning in the central and autonomic nervous system 6 months after the Tokyo subway attack. *J Neurol*, **244**, 601–606.

Nishiwaki Y, Maekawa K, Ogawa Y *et al.* (2001). Sarin Health Effects Study Group. Effects of sarin on the nervous system in rescue team staff members and police officers 3 years after the Tokyo subway sarin attack. *Environ Health Perspect*, **109**, 1169–1173.

Okudera H (2002). Clinical features on nerve gas terrorism in Matsumoto. *J Clin Neurosci*, **9**, 17–21.

Okumura T, Takasu N, Ishimatsu S *et al.* (1996). Report on 640 victims of the Tokyo subway sarin attack. *Ann Emerg Med*, **28**, 129–135.

Okumura T, Suzuki K, Fukuda *et al.* (1998a). The Tokyo subway sarin attack: disaster management. Part I. Community emergency response. *Acad Emerg Med*, **5**, 613–617.

Okumura T, Suzuki K, Fukuda *et al.* (1998b). The Tokyo subway sarin attack: disaster management. Part II. Hospital response. *Acad Emerg Med*, **5**, 618–624.

Okumura T, Suzuki K, Fukuda A *et al.* (1998c). The Tokyo subway sarin attack: disaster management. Part III. National and international response. *Acad Emerg Med*, **5**, 625–628.

Okumura T, Suzuki K, Fukuda A *et al.* (1999). Preparedness against nerve agent terrorism. In: *Natural and Synthetic Toxins Biological Implications* (AT Tu and W Gaffield, eds), pp. 356–368. Washington, DC, USA: American Chemical Society.

Okumura T, Hisaoka T, Naito T *et al.* (2005). Acute and chronic effects of sarin exposure from the Tokyo subway incident. *Environ Toxicol Pharmacol*, **19**, 447–450.

Research Committee on Nerve Gas Poisoning in Matsumoto (1995). *Report on the Poisonous Gas Incident in Matsumoto*. Matsumoto, Japan: Matsumoto Regional Medical Council.

Sekijima Y, Morita H and Yanagisawa N (1997). Follow-up of sarin poisoning in Matsumoto. *Ann Intern Med*, **127**, 1042.

Sidell FR (2003). Nerve agents. In: *Advanced Disaster Medical Response Manual for Providers* (SM Briggs, ed.), pp. 62–63. Boston, MA, USA: Harvard Medical International.

Yokoyama K, Araki S, Murata K *et al.* (1998a). A preliminary study on delayed vestibulo-cerebellar effects of Tokyo subway sarin poisoning in relation to gender difference: frequency analysis of postural sway. *J Occup Environ Med*, **40**, 17–21.

Yokoyama K, Araki S, Murata K *et al.* (1998b). Chronic neurobehavioral effects of Tokyo subway sarin poisoning in relation to post-traumatic stress disorder. *Arch Environ Health*, **53**, 249–256.

Yokoyama K, Araki S, Murata K *et al.* (1998c). Chronic neurobehavioral and central and autonomic nervous system effects of Tokyo subway sarin poisoning. *J Physiol*, **92**, 317–323.

14 ATROPINE AND OTHER ANTICHOLINERGIC DRUGS

John H. McDonough and Tsung-Ming Shih

US Army Medical Research Institute of Chemical Defense, Aberdeen Proving Ground, MD, USA

The opinions expressed herein are solely of the authors and not necessarily those of the Department of Defense, the Department of the Army or the Army Medical Research and Materiel Command.

The nerve agents are highly toxic organophosphorous (OP) compounds. The agents of greatest concern, along with their chemical names and two-letter military designations, are tabun (*o*-ethyl *N*,*N*-dimethyl phosphoramidocyanidate; GA), sarin (isopropyl methylphosphonofluoridate; GB), soman (pinacolyl methylphosphonofluoridate; GD), cyclosarin (cyclohexyl methylphosphonofluoridate, GF), VX (*o*-ethyl *S*-2-*N*,*N*-diisopropylaminoethyl methyl phosphonofluoridate) and a Russian V-type agent designated VR (*o*-isobutyl *S*-(2-diethylamino)ethyl methylphosphonothioate). The nerve agents inhibit the cholinesterase (ChE) family of enzymes that includes acetylcholinesterase (AChE) and butyrylcholinesterase (BChE). It is the inhibition of AChE, the enzyme that hydrolyzes the cholinergic neurotransmitter acetylcholine (ACh), that produces the toxic action of nerve agents. Inhibition of BChE activity by itself is not known to produce any toxic effect. Nerve agents bind to the active site of the AChE enzyme, thus preventing it from hydrolyzing ACh. The enzyme is inhibited irreversibly, and the return of esterase activity depends on the synthesis of new enzyme molecules ($\sim 1\%$ per day in humans). All nerve agents penetrate the central nervous system (CNS), with the G-type agents acting more rapidly centrally than the V-type.

ACh is the neurotransmitter at the neuromuscular junction of skeletal and smooth muscle, the preganglionic nerves of the autonomic nervous system, the postganglionic parasympathetic nerves and muscarinic and nicotinic cholinergic synapses within the CNS. Following nerve agent exposure and subsequent inhibition of the AChE enzyme, levels of ACh rapidly increase at the various effector sites, resulting in continuous stimulation. It is this hyperstimulation of the cholinergic system at central and peripheral sites that leads to the toxic signs of poisoning with these compounds, resulting in a syndrome referred to as a *cholinergic crisis*. These signs include miosis (constriction of the pupils), increased tracheobronchial secretions, bronchial constriction, laryngospasm, increased sweating, urinary and fecal incontinence, muscle fasciculations, tremor, convulsions, electrical seizures and loss of respiratory drive from the CNS. The relative prominence and severity of a given toxic sign depend highly on the route and degree of exposure. Ocular and respiratory effects occur rapidly and are most prominent following vapor exposure, while localized sweating, muscle fasciculations and gastrointestinal disturbances are the initial signs following percutaneous exposures and usually develop more gradually. The acute lethal effects of the nerve agents are generally attributed to respiratory failure caused by a combination of effects at both central (loss of respiratory drive) and peripheral (weakness at diaphragm and intercostal muscles) levels. These effects are

Chemical Warfare Agents: Toxicology and Treatment (2nd Edition)
Edited by Timothy C. Marrs, Robert L. Maynard and Frederick R. Sidell

exacerbated by copious secretions, bronchoconstriction, bronchospasm, muscle fasciculations and convulsions, which also contribute to the compromise of respiratory status. Several excellent reference sources provide more detailed discussions of the history, chemistry, physiochemical properties, pharmacology and toxicology of the nerve agents (Koelle, 1963; Sidell, 1997; Taylor, 2001).

PRINCIPLES OF TREATMENT OF NERVE AGENT EXPOSURE

Physical protective measures (e.g. gas masks, gloves and overgarments) and strict decontamination procedures are the most effective means of protection against the toxic action of nerve agents. The USA and North Atlantic Treaty Organization (NATO) also advocate the use of carbamate prophylaxis with pyridostigmine bromide as a way to enhance the therapeutic efficacy of antidote treatments in the case of poisoning with rapidly aging nerve agents, such as soman. Discussion of the use of carbamate pretreatment is covered in more detail in another chapter in this volume.

If intoxication does occur, treatment of nerve agent poisoning is focused along several lines. Prevention or reduction of the toxic signs is accomplished primarily via (a) administration of anticholinergic drugs, atropine sulfate being almost universally used for this purpose, (b) reactivation of agent-inhibited enzyme with oxime reactivators such as pralidoxime chloride (2-PAM Cl), and, when indicated, in cases of more severe poisoning, (c) treatment of convulsions and seizures with the benzodiazepine class of drugs (Army FM 8-285: Treatment of Chemical Agent Casualties, 1995; *Medical Management of Chemical Casualties Handbook*, 2000; Sidell, 1997). It should be noted that different countries have different complements of drugs for treating nerve agent casualties, but the differences are more in the specific drug used rather than in the general treatment approach (anticholinergic, oxime reactivator, anticonvulsant) itself (Moore *et al*., 1995). Virtually all countries use atropine as an anticholinergic treatment compound and some use other synthetic anticholinergic drugs to supplement the effects of atropine. Diazepam, or a water-soluble prodrug form of diazepam (avizafone), is typically used as the benzodiazepine for field treatment. The greatest difference among the countries involves the choice of oxime treatment. The USA uses the chloride salt of the monopyridinium oxime, pralidoxime (2-PAM Cl); the UK uses the methanesulfonate salt, referred to as P2S or pralidoxime mesilate; France uses pralidoxime methylsulfate, known as Contrathion; Japan uses the iodide salt. Other countries favor more potent bispyridinium oximes, such as obidoxime (Toxogonin), trimedoxime (TMB-4) or HI-6.

An anticholinergic drug such as atropine blocks the effects of ACh overstimulation at central and peripheral muscarinic sites. Since it is the muscarinically mediated effects of nerve agent poisoning that are the most life-threatening, atropine or another anticholinergic is the most important life-saving treatment. It provides symptomatic relief of the excessive secretory responses (nose – rhinorrhea, salivary, pulmonary and gastrointestinal), laryngospasm, and bronchoconstriction. Atropine also increases the heart rate and, to a lesser extent, antagonizes the loss of central respiratory drive (Brown and Taylor, 2001). Atropine at high doses and other centrally active anticholinergic drugs are also effective treatments of nerve agent-induced seizures/convulsions (Capacio and Shih, 1991; McDonough and Shih, 1993; McDonough *et al*., 2000; Shih *et al*, 2003). Atropine and other muscarinic anticholinergic drugs are unable to counteract the nicotinic signs of intoxication (e.g. muscle fasciculations, muscle fatigue, weakness). Reversal of the nicotinic signs of intoxication is therapeutically accomplished via oxime reactivation of inhibited AChE (Dawson, 1994; Kassa, 2002). This topic is dealt with in detail in Chapter 15 in this volume. Treatment of nerve agent-induced seizures/convulsions is essential for overall casualty management and prevention of neurological damage (Lemercier *et al*., 1983; Lallement *et al*., 1998; McDonough and Shih, 1997; Shih *et al*., 2003). While there is a major cholinergic component to the initiation and early maintenance of these seizures, benzodiazepine drugs such as diazepam are most commonly used to antagonize this feature of nerve agent poisoning. The use of benzodiazepines for the treatment of nerve agent-induced seizures is discussed in greater detail in Chapter 16.

Since nerve agents can produce rapid lethal effects, military personnel are issued several different automatic injector devices to deliver drugs intramuscularly (IM) for immediate emergency treatment of exposure. In the US military, individuals are issued three MARK I treatment drug kits; each kit contains two autoinjectors, one with 2 mg of atropine and the other with 600 mg of the oxime 2-PAM Cl. The MARK I kits are currently in the process of being replaced by a multichambered autoinjector (drugs in separate chambers) known as Antidote Treatment – Nerve Agent Autoinjector (ATNAA), that delivers atropine (2.1 mg) and 2-PAM Cl (600 mg) as a single injection. Individuals are also issued a single autoinjector (known as CANA – Convulsive Antidote Nerve Agent) containing 10 mg of diazepam. Thus, each soldier carries 6 mg of atropine, 1800 mg of 2-PAM Cl, and 10 mg of diazepam. The armed forces of many other countries provide their service members with autoinjector devices similar to the types described above. The guidelines and training for the use of these treatment drugs are based upon the medical treatment recommendations and doctrines established by the medical services of each country (e.g. Army FM 8-285: Chemical Casualty Care, 1995; Medical *Management of Chemical Casualties Handbook*, 2000). In the USA, the Food and Drug Administration has just recently approved autoinjectors with 0.5 and 1.0 mg atropine pediatric dosage forms for homeland defense use. Guidelines for atropine dosages to use in children poisoned with nerve agent have been recently published (Rotenberg and Newmark, 2003).

The treatment of nerve agent exposure in a military setting poses a unique medical problem. Individuals who have limited emergency medical training must accurately recognize and diagnose the signs and symptoms of a potentially lethal toxic exposure and then promptly administer to themselves or a fellow soldier the necessary treatment drugs in the proper order and the proper dosage. In addition to self-protection, protection of the casualty from further exposure, as well as decontamination and evacuation needs to be considered. This poses a set of tasks that requires constant and realistic training if they are to be accomplished smoothly in the case of a real nerve agent attack.

ANTICHOLINERGIC DRUGS

Atropine and similar anticholinergic drugs are pharmacologically classified as muscarinic receptor antagonists (Brown and Taylor, 2001). Muscarinic receptor antagonists compete with ACh for a common binding site on the muscarinic receptor. This binding prevents ACh from binding to muscarinic cholinergic receptors at neuroeffector sites on smooth muscle, heart muscle, glands, peripheral ganglia and in the CNS. In the context of nerve-agent intoxication, this means that muscarinic receptor antagonists can block the hyperstimulation that is produced by the repeated binding of high levels of ACh at peripheral and central muscarinic cholinergic synapses. Five subtypes of the muscarinic receptor have been identified, designated M_1, M_2, M_3, M_4 and M_5 (Caulfield and Birdsall, 1998). Each receptor subtype couples to a second messenger system through an intervening G-protein. M_1, M_3 and M_5 receptors stimulate phosphoinositide metabolism, while M_2, and M_4 receptors inhibit adenylate cyclase. The relative tissue distribution also differs for each subtype. M_1 receptors are enriched in the forebrain, especially in the hippocampus and cerebral cortex. M_2 receptors are found in the heart and brainstem, while M_3 receptors are found in smooth muscle, exocrine glands and the cerebral cortex. M_4 receptors are most abundant in the striatum; M_5 receptors are most concentrated in the substantia nigra. Studies with muscarinic receptor subtype knockout mice have revealed that specific receptor subtypes are involved in agonist-induced hypothermia (M_2, M_3), tremor (M_2), salivation (M_3, and to a lesser extent, M_1 and M_4), pupil diameter (M_3), dilation of cerebral blood vessels (M_5) and seizures (M_1) (Hamilton *et al.*, 1997; Gomeza *et al.*, 1999; Yamada *et al.*, 2001; Bymaster *et al.*, 2003). Because all of these responses are involved to varying extents in the toxic effects of nerve-agent intoxication, nonspecific anticholinergic drugs, such as atropine or scopolamine, that act at all receptor subtypes, produce the most effective antidotal effects.

The antagonism of ACh by atropine at muscarinic receptor sites is competitive; this means the antagonism produced by a certain level of atropine can be overcome if the concentration of ACh is sufficiently high. ACh levels have been reported to increase from 150 to 300% in various brain regions following nerve-agent exposure in experimental animals (Shih, 1982; Fosbraey *et al.*, 1990; Lallement *et al.*, 1992). Since the increases in ACh are directly at the cholinergic synaptic nerve terminals, this leads to very high ACh concentrations close to the receptors. It is for this reason that atropine, or any other antimuscarinic drug, has to be given at relatively high doses to antagonize the effects of this increased ACh in cases of severe poisoning. This is also why a rapid administration of sufficient atropine or an anticholinergic drug like atropine is essential to quickly blockade the receptors and reverse the toxic muscarinic effects of poisoning.

A second feature that determines the speed of atropine action is the route of administration. The action of 2 mg of atropine sulfate on the human heart rate begins 1, 8 or 20 min after intravenous (IV), intramuscular (IM) or oral administration, and maximal rate increases occur in 6, 35 or 50 min by these routes, respectively (Grob, 1956; Ketchum *et al.*, 1973). The use of autoinjectors to administer atropine IM speeds up uptake of the drug over conventional IM injections when using a needle and syringe. This is due to a spraying effect as the needle plunges into the muscle, resulting in the distribution of drug within a larger muscle area, allowing for more rapid drug uptake (Sidell *et al.*, 1974). A conventional IM injection, using a needle and syringe, results in a depot disposition of drug, leading to slower uptake. Nevertheless, even with autoinjectors, it should be remembered that IM drug administration has an inherent lag-time between drug delivery, onset of effects and peak drug effect. Even though these times may only be minutes, such a delay may be clinically significant in a severely poisoned casualty with compromised respiration and cardiovascular status.

Atropine and other natural (e.g. scopolamine) or synthetic (e.g. benactyzine, biperiden, caramiphen, trihexyphenidyl) antimuscarinic compounds all cause the same constellation of effects, with the greatest difference between the drugs being their abilities to penetrate the CNS and their durations of action (Brown and Taylor, 2001). Atropine has more prominent peripheral effects at low doses than does scopolamine or other synthetic anticholinergics (Ketchum *et al.*, 1973). In contrast, scopolamine and other synthetic anticholinergics produce marked CNS effects at low doses with minimal concominent peripheral effects. Low doses of atropine (2 mg) depress salivation, bronchial secretions and sweating, increase heart rate, produce pupilary dilation and inhibition of lens accommodation for near vision (Headley, 1982; Penetar, 1990; McDonough, 2002). The one side-effect of atropine with the greatest operational impact on military performance is inhibition of sweating and the potential for inducing heat casualties due to the inability to regulate core temperature. This can occur even after a 2 mg dose of atropine, especially with heavy work, a hot environment or the use of chemical protective suits. Larger doses (5 mg) of atropine have more pronounced effects on salivation, bronchial secretions, sweating, heart rate and pupilary dilation, as well as inhibiting parasympathetic control of the urinary bladder and gastrointestinal tract. Still higher doses ($\geq$ 8 mg) inhibit gastric secretion and motility and produce a constellation of CNS effects (restlessness, disorientation, amnesia, hallucinations) best characterized as delirium (Ketchum *et al.*, 1973). At these doses, the EEG is shifted to slower activity, there is a reduction in the voltage and frequency of the alpha-rhythm and rapid eye movement (REM) sleep is depressed (Longo, 1966; Pickworth *et al.*, 1990). One feature of practical significance in using atropine as an immediate antidote for nerve-agent poisoning is that increasing the dose of atropine, or any of the other anticholinergics, beyond a certain maximum dosage will not produce any greater response. The onset of effects will just be more rapid after IM administration and the duration of effects will be longer.

Atropine has been the antidote of choice for treatment of nerve agent intoxication since nerve agents were first discovered and produced during World War II. It was included in the German nerve agent first-aid kits (Comstock and

Krop, 1948) and was determined to be an effective antidote by British scientists at Porton Down who first analyzed the pharmacology and toxicology of tabun obtained from captured German artillery shells (Wilson, personal communication; Sidell, 1997). Since the 1940s, atropine has been adopted as the first-line antidote to counteract nerve-agent poisoning by the armed forces of most countries. It is also almost universally used as the antidote to treat anticholinesterase poisoning by organophosphate or carbamate pesticides (Eddleston *et al.*, 2004a,b). The use of 2 mg as the 'unit dose' of atropine in the MARK I or ATNAA autoinjectors was established because this amount of atropine can reverse the effects of low or moderate exposures to nerve agent. The associated side-effects of this dose are well-tolerated, change in mental status is very unlikely and reasonable military performance can be maintained as long as care is taken to prevent heat injuries (McDonough, 2002). Thus, if exposure were suspected, this dose could be self-administered without significant performance compromise even if it was given inadvertently in the absence of agent exposure (Sidell, 1997).

Several countries use, or have proposed to use, other anticholinergic drugs as adjuncts to atropine for the treatment of nerve-agent poisoning. The common feature of all of these products is that these anticholinergics have much more potent and rapid effects on the CNS than does atropine. Israel uses a mixture of drugs, known as TAB, as their immediate nerve-agent treatment; this mixture contains the oxime TMB-4, atropine and the synthetic anticholinergic benactyzine. From 1975 to 1980, the US military also used TAB. The atropine and benactyzine combination in the TAB mixture is similar in composition to atropine, benactyzine and 2-PAM-combination antidote mixtures investigated by Yugoslov researchers in the early 1970s (Vojvodic and Maksimovic, 1972; Vojvodic *et al.*, 1972). Animal studies have shown that benactyzine is much more potent and rapidly acting in reversing the CNS effects of nerve-agent intoxication than atropine (Jovic and Milosevic, 1970; McDonough *et al.*, 2000). In addition, benactyzine is significantly less potent in inhibiting sweating or producing mydriasis than atropine, and therefore less likely to induce heat casualties in a warm environment or compromise nearvision in the case of accidental use. Military researchers in the Czech Republic have advocated the use of the synthetic anticholinergics benactyzine and trihexyphenidyl, along with the carbamate pyridostigmine, in a prophylactic mixture they have designated as PANPAL (Bajgar *et al.*, 1994). In addition, the Czechs utilize benactyzine and biperiden, as well as atropine, as post-exposure antidotal treatments (Bajgar *et al.*, 1994; Kassa and Bajgar, 1996). For several years, researchers in Israel, the UK and the Netherlands have demonstrated the effectiveness of scopolamine or hyoscine as part of a pretreatment combination with the centrally active carbamate physostigmine against soman poisoning (Meshulam *et al.*, 2001; Philippens *et al.*, 2000; Wetherell, 1994; Wetherell *et al.*, 2002). Most recently, Russian scientists have discussed the use of a synthetic anticholinergic, pentifin, as a potential post-exposure treatment (Petrov *et al.*, 2004). This compound is reported to possess central muscarinic and nicotinic antagonistic activity and is a strong M_1 muscarinic cholinoreceptor blocker. Exact details of how this compound may possibly be used (dose, frequency of administration) were not discussed. While many countries have other anticholinergic drugs to use as adjuncts to atropine for the treatment of nerve-agent poisoning, none of these compounds have been tested or used in human clinical cases of poisoning either with nerve agents or other organophosphate or carbamate pesticides.

Animal studies

Since the 1940s, animal studies have been critical for understanding the biochemical and physiological mechanisms by which nerve agents produce their toxic effects and for evaluation of various drugs to provide effective medical countermeasures. Over those years, there have been numerous studies to determine whether atropine is the most effective anticholinergic drug to treat nerve-agent poisoning. Initial studies by US researchers evaluated the effects of different anticholinergics alone to protect against increasing challenge doses of nerve agents (Wills, 1963). In the 1960s, Canadian researchers (Coleman *et al.*, 1962,1968) performed an extensive series of studies of both tertiary and quaternary anticholinergic drugs, in conjunction with the oxime P2S, to antagonize the lethal effects of sarin in mice and rats. They were trying to identify the specific cholinolytic mechanisms of action that were associated with enhanced protective activity against sarin toxicity. None of the pharmacological tests of peripheral anticholinergic action (mydriatic action, inhibition of ACh-induced spasm in ileum and lung, inhibition of gut motility) predicted protection against sarin-induced toxicity, a finding also confirmed by Brimblecombe *et al.* (1970). Further tests of using the compounds as adjuncts with a standard dose of atropine showed that caramiphen (called Parpanit in the papers) and the glycolate compound G-3063 (4′-*N*-methylpiperidyl 1-phenylcyclopentanecarboxylate HCl) demonstrated significantly enhanced protective activity against sarin lethality. A study by Jovic and Milosevic (1970) examined the protective effects of twelve anticholinergic compounds alone or in conjunction with 2-PAM against poisoning by soman, sarin and tabun, as well as a number of other highly toxic OP compounds in mice. They concluded that anticholinergics with pronounced central effects, specifically caramiphen or benactyzine in their study, could enhance the protective action of atropine, especially against the nerve agents, which they believed have more pronounced central toxic effects than the other OP compounds tested.

Nerve agents are potent convulsant compounds (McDonough and Shih, 1993,1997). The contribution of these seizures to the overall toxicity of these agents and the need to control them as part of total poisoning treatment were just being fully appreciated in the 1970s (Lipp, 1972,1973; Rump *et al.*, 1972,1973). It was Green *et al.* (1977) who first recognized that some anticholinergics have anticonvulsant activity and that this property was related to their enhanced antidotal activity against nerve agents. Subsequent studies by many other groups expanded on this observation, showing that potent centrally acting anticholinergics can (a) antagonize nerve agent-induced seizures (Pazdernik *et al.*, 1983; Samson *et al.*, 1985; Capacio and Shih, 1991; McDonough and Shih, 1993), (b) protect against the development of seizure-related brain damage (Samson *et al.*, 1985; McDonough *et al.*, 1989,1995), and (c) reverse the physical incapacitation associated with nerve-agent intoxication (Leadbeater *et al.*, 1985; Anderson *et al.*, 1994). All of these anticholinergic actions enhance the ability to protect against the lethal effects of nerve agents (Shih *et al.*, 2003). Atropine also displays anticonvulsant action against nerve agent-induced seizures in animal studies, but much higher doses are required compared to the anticholinergics with strong central activity (e.g., scopolamine, benactyzine, trihexyphenidyl) (McDonough *et al.*, 2000; Shih and McDonough, 2000; Shih *et al.*, 2003). It should be noted that, except in this context, anticholinergics are not routinely thought of as anticonvulsants and none is clinically licensed for such an indication.

In studies with rodents (rats, guinea pigs), there is a pronounced time-dependency to the anticonvulsant effects of these anticholinergic compounds. Relatively low doses of these compounds can rapidly terminate seizures when they are administered shortly (5 min) after seizure onset. As the delay between seizure onset and anticholinergic treatment is increased (40 min), some animals become totally refractory to the anticonvulsant effects, whereas others require significantly greater amounts of drug (1-2 log units), and the delay between drug administration and seizure termination is also greatly delayed (McDonough and Shih, 1993; McDonough *et al.*, 2000). It is not known if this same time-dependency is seen in higher species, including non-human primates and man. The reason for such a pronounced shift in anticonvulsant effectiveness of these anticholinergics is thought to be due to the early trigger of seizure activity by the high ACh levels and the later recruitment of non-cholinergic excitatory neurotransmitter systems (i.e. glutamatergic) by the excessive neural activity of the seizure itself (McDonough and Shih, 1997). In that regard, many of these synthetic anticholinergic drugs (e.g. benactyzine, biperiden, procyclidine, trihexyphenidyl) have been shown to have *N*-methyl-D-aspartate (NMDA) antagonistic properties (Olney *et al.*, 1987; McDonough and Shih, 1995), an additional pharmacological feature that has been demonstrated to be beneficial in the treatment of nerve agent-induced seizures (Sparenborg *et al.*, 1992; Carpentier *et al.*, 1994; Lallement *et al.*, 1994,1998).

USE OF ATROPINE IN THE TREATMENT OF NERVE AGENT POISONING IN HUMANS

There are three major sets of clinical reports on the use of atropine in the treatment of humans suffering from severe nerve-agent poisoning. Sidell (1974,1997) described the treatment of a small number of workers at Edgewood Arsenal accidentally exposed to either soman or sarin. There are several clinical reports on the treatment of the victims of the terrorist attacks with sarin in Matsumoto and Tokyo (Masuda *et al.*, 1995; Morita *et al.*, 1995; Nozaki and Aikawa, 1995; Ohbu *et al.*, 1997; Okumura *et al.*, 1996; Sekijima *et al.*, 1995) as well as one individual poisoned with VX (Nozaki *et al.*, 1995). Finally, Newmark (2004) has summarized a series of articles originally published in the Kowsar Medical Journal by Dr Syed Abbas Foroutan that describes his experiences and protocols for treating nerve-agent casualties in a chemical aid station during the 1981–1987 Iran–Iraq war. The Sidell and Japanese reports describe the treatment of either single or limited numbers of severely poisoned casualties in fully equipped clinical emergency room settings. This is in contrast to the chemical-aid-station environment of Dr Syed Abbas Foroutan, where multiple nerve agent and other chemical casualties were treated following attacks, stabilized, and then evacuated further behind the front lines for recovery.

Sidell (1974,1997) described treating two severely poisoned individuals. The first received a vapor exposure to sarin and was seen in the treatment facility 5–10 min after the first symptom. He was cyanotic and convulsing, with labored breathing, fasciculations and copious secretions. He was immediately given 2 mg of atropine IV and 2 mg IM and an additional 2 mg IV several minutes later; 2-PAM and oxygen were also given, and another 2 mg IV of atropine was administered about 20 min after admission to control the return of secretions. Over the next 30 min, the patient deteriorated despite another 2 mg IM of atropine. By 60 min after admission, the patient became apneic, required assisted ventilation and received an additional 3 mg of atropine IV. Respiratory assistance was required for ~ 2 h during which another 1 mg of atropine was given IV to counteract bronchoconstriction. About 2.5 h after admission, the patient began to regain consciousness, began to breathe spontaneously and slowly recovered. In all, the casualty received 14 mg of atropine over ~ 2.25 h after admission, 10 mg of which was given IV. The second patient seen by Sidell was accidentally exposed to a small amount of soman solution orally. He immediately flushed his mouth with water and arrived at the treatment facility within 10 min of the accident. There, he immediately developed signs of intoxication and collapsed. Within a minute of developing symptoms, he was administered 2 mg of atropine IV and received a total of 4 mg IV and 8 mg IM of additional atropine over the next 15 min, along with 2-PAM, as well as oxygen and frequent nasopharyngeal suction. Bronchoconstriction and a decreased respiratory rate and amplitude were noted and the patient developed cyanosis. These signs began to diminish following the atropine therapy; heart rate and blood pressure remained stable, and about 30 min after admission the patient began to regain consciousness. He received additional atropine, 4 mg IV and 4 mg IM, at 14 h and 22 h post-exposure, respectively, to control nausea, vomiting and abdominal pain. This casualty received a total of 14 mg of atropine (6 mg IV) within ~ 20 min after admission, and a total of 22 mg of atropine in the first 24 h.

Treatment of the victims exposed to sarin and VX in the Japanese terrorist incidents is similar in many ways to the descriptions of the 'Sidell casualties'. All victims were seen in hospital emergency rooms where treatment was initiated. In cases of severe poisoning, atropine therapy was given IV, which insures maximum therapeutic effect within minutes. Nozaki and Aikawa (1995) described the treatment of one severely poisoned patient of the Tokyo subway exposures at Keio University Hospital. The patient arrived at the hospital in a coma ~ 1 h after exposure (Glasgow coma scale: E1M1V1), displaying profuse sweating and oral secretions, convulsions, cyanosis and respiratory arrest. Over the next 10 min, the patient was intubated, mechanically ventilated and administered 0.5 mg IV of atropine, 5 mg IV of diazepam and 1000 mg IV of pralidoxime iodide; over the next 25 min, he received an additional 4 mg IV of atropine and 20 mg IV of diazepam. The patient gradually recovered consciousness and was extubated ~ 4.5 h after admission. He was treated with a total dose of 15 mg IV of atropine for two days in addition to the atropine administered as initial treatment. Treatment of five severely exposed patients at St. Luke's International Hospital (Okumura *et al.*, 1996; Ohbu *et al.*, 1997) followed a similar protocol. These patients either presented in a coma and respiratory arrest or were 'drowsy' and then developed generalized convulsions and lapsed into respiratory arrest. Two of these patients were in total cardiopulmonary arrest: one did not respond to cardiopulmonary resuscitation and died, while the other was successfully resuscitated but suffered severe brain damage and died 28 days later. All of the severely poisoned patients required intubation and ventilatory support; IV atropine was given up to total doses of 2–5 mg, along with IV pralidoxime iodide (1000 mg initially, followed by 500 mg h^{-1} for total doses up to 8500 mg); IV diazepam (5–20 mg total dose) was used to control convulsions. St. Luke's International Hospital also treated 105 moderately exposed patients. Miosis and other ocular signs of exposure were present in all patients; dyspnea, nausea, vomiting, muscle weakness, fasciculations and agitation were the other most common signs. These patients received 2 mg IV of atropine, 2000 mg IV of pralidoxime iodide and IV diazepam (dose not noted) for those with fasciculations. The authors note that miosis was severe and persistent (continuing > 1 week) in these patients and was unresponsive to systemic atropine treatment, but did respond to locally applied mydriatic agents.

Nozaki *et al.* (1995) also reported on the treatment of a patient exposed to VX in an attempted murder by the same *Aum Shinrikyo* cult that released sarin in Matsumoto and the Tokyo subway terrorist attacks. Reportedly, VX was sprayed on the victim's back; the man noted impaired vision and then experienced seizures and loss of consciousness. He arrived at the emergency room about 2 h after exposure semi-comatose (Glasgow scale E2M4V2) with profuse oral secretions, sweating, cyanosis, muscle fasciculations and convulsions, and was intubated. Organophosphate poisoning was not suspected at first, and he was treated with dopamine and isoprenaline to increase blood pressure and heart rate and with phenytoin to control the convulsions, but all of these treatments were without effect. At 3.5 h after exposure, he was treated with 2 mg IV of atropine, which immediately increased both heart rate and blood pressure, and 10 mg IV of diazepam, which controlled the convulsions. He was maintained on continuous IV atropine (3 mg day^{-1}) and mechanical ventilation, and 9 days after exposure he became alert and was extubated. The authors stressed the importance of systemic atropine for treating the bradycardia produced by the VX.

Battlefield experience contrasts strongly with the treatment environments of 'Sidell's patients' and with those of the Japanese terrorist attacks. Dr Syed Abbas Foroutan is an Iranian physician who set up and ran a chemical casualty-aid station during the 1981–1987 Iran–Iraq War. As such, he saw sometimes hundreds of nerve-agent casualties following a single Iraqi attack. Several years after the war, he published a series of articles in Farsi in the *Kowsar Medical Journal* describing his experiences and treatment protocols. These articles have been reviewed and summarized by Newmark (2004); he compares Dr Foroutran's experiences and treatment protocols for nerve-agent casualties with both US and NATO nerve-agent care doctrine. Dr Foroutan categorized the conditions of his nerve agent casualties as either 'mild', 'moderate' or 'poor'. Patients in the 'mild' classification consisted mostly of those with ocular symptoms, rhinorrhea and mild dyspnea. Patients in the 'moderate' classification presented with miosis, rhinorrhea, dyspnea, nausea, muscle weakness and fasciculations, but were conscious. The 'poor' classification would correspond in clinical presentation to the severely poisoned patients described by Sidell or those in the Japanese terrorist incidents.

The hallmark of Iranian nerve-agent-casualty treatment doctrine was 'to administer the highest required dosage of atropine in the shortest possible period of time'. This almost total reliance on atropine may have been due to limited availability of oximes (see discussion in Newmark, 2004). Each Iranian soldier carried three 2 mg autoinjectors of atropine for immediate treatment of exposure. However, Dr Foroutan stated that this amount of atropine 'is effective on mild to moderate poisoning and has no effect on severely poisoned patients'. For those severely poisoned or for those who did not receive atropine, he states 'it is imperative that the patient be administered atropine under any and all circumstances' before evacuation back to the emergency unit. In the emergency unit, his atropine treatment protocol for severe exposures was a tradeoff between the need for rapid atropinization and over-atropinization. The patient would be given a test dose of 4 mg of atropine IV, and if in 1 to 2 min there was no sign of atropinization, then the patient would be given 5 mg of atropine IV over the next 5 min while checking pulse rate. An increase in pulse rate of 20–30 beats per min (BPM) was taken as evidence of initial atropinization. The rate at which atropine was given was tied to the pulse rate; if pulse rate decreased below 60–70 BPM, the rate at which atropine was given was increased, and the rate of atropine administration was slowed if the pulse rate was > 110 BPM. The total amount and rate at which atropine was given, according to Dr Foroutan's recommendations, was considerably higher than that recommended by the US or NATO treatment doctrine. He stated that if atropinization was not achieved following the first 4 and 5 mg IV atropine doses, that doses of 25 to 50 mg atropine be given IV every 5 min using the pulse rate guidelines outlined above, up to a total dose of 150 mg. In several very severe cases, he reported administering up to 200 mg IV of atropine in a 10–15 min period in an effort to achieve atropinization. Ease of breathing and drying of respiratory secretions were the end-points he recommended to consider atropine therapy adequate.

The rate and total amounts of atropine administered by Dr Foroutan to severely poisoned nerve-agent casualties differ considerably from the doses and treatment protocols used by Sidell and the physicians treating the Japanese terrorist victims. Dr Foroutan states that using more conservative treatment approaches wastes time and risks lives. In mass casualty situations such as those he faced, with limited oxime available and probably limited available means of mechanical respiration, this is probably a reasonable approach. He also stated that the faster atropinization is achieved, then the lower the chance of cardiopulmonary arrest. While offering no proof of this last statement, animal studies show that the more rapidly normal respiration and cardiac status are restored, then the less likely is the development of cyanosis, hypoxia, lowered cardiac rate and blood pressure, all factors that increase the risk of cardiopulmonary collapse (Lipp, 1976; Lipp and Dola, 1978; personal observation). Several of these factors (lowered cardiac rate, low blood pressure) also work against rapid uptake and distribution of IM administered treatment drugs.

Once atropinization had been accomplished in Dr Foroutan's Chemical Emergency Unit, casualties were evacuated to a recovery unit 100 km behind the front lines. There, he recommended that the atropine dose be reduced to a few milligrams per day, and that atropine 1 to 2 mg every 4 to 6 h be given for up to two days after the resolution of all clinical symptoms, with the dose only to be increased if bradycardia developed.

FIELD TREATMENT OF NERVE AGENT EXPOSURE WITH ANTICHOLINERGICS

It must be remembered that for immediate field treatment most soldiers are equipped with only three 2 mg autoinjectors of atropine (as well as oxime, and possibly anticonvulsant injectors) for administration by themselves or a 'buddy'. Most medical treatment doctrines call for oxime administration only with the first three autoinjectors of atropine. Additional oxime beyond this initial treatment will be administered under direction of a physician at a medical treatment facility. Additional atropine and anticonvulsant treatment is carried by the medic/corpsman in most Western/NATO forces and will be absolutely required in cases of severe poisoning. US medical treatment guidelines call for the administration of the first CANA anticonvulsant (10 mg of diazepam) whenever all three atropine autoinjectors are administered, regardless of the apparent presence or absence of convulsions/seizures. Other countries have slightly different treatment regimens for anticonvulsant use. General guidelines for the treatment of nerve-agent exposure are based on the speed and intensity with which signs and symptoms of poisoning develop, as well as consideration of the probable route of exposure. Therapy should be titrated to relieve distress, minimize the casualty's discomfort and stop or reverse the toxic process. For anticholinergic drugs, this will mean primarily the control of excessive secretions, relief of bronchoconstriction, reversal of dyspnea and the maintenance of adequate oxygenation. Atropine and/or other anticholinergic drugs administered IM as immediate antidotes under most conditions will not counteract miosis or eye pain produced by vapor exposure, nor do they affect muscle fasciculations (Sidell, 1997). As a general rule, consciousness, spontaneous and clear respiration and lack of seizure activity are three indicators of successful therapy and a good clinical outcome for treatment in severe cases.

Knowledge of the suspected route of nerve-agent exposure is crucial in determining treatment guidelines. As a general rule, then the greater the exposure the more rapid the onset of signs, while the longer the time between exposure and onset of effects, then the less severe the effects will eventually be. Signs and symptoms develop relatively rapidly following vapor exposure, and both ocular and respiratory symptoms dominate the clinical picture in the case of adults. In contrast, signs and symptoms develop in a more gradual fashion following dermal exposure to liquid nerve agent, with localized sweating and muscle fasciculations being the first signs, followed by gastrointestinal distress and nausea. With dermal exposures, the development of signs and symptoms may be quite delayed, up to 18 h. Toxic effects can begin hours later, even after thorough decontamination since absorbed dose is determined by how much agent was on the skin and the duration of contact before decontamination. As can be seen from the clinical descriptions of poisoning cited above, rapid onset of miosis, excessive respiratory secretions, difficulty in breathing and, especially, the development of muscle fasciculations are good indications of a moderate to severe vapor exposure that needs to be aggressively treated. Likewise, gastrointestinal signs, accompanied by localized sweating, fasciculations, and dyspnea, are indications of a moderate to severe dermal exposure. Exposure through an open wound is expected to follow a time-course intermediate between the vapor and dermal routes. With either vapor or dermal exposure, diminished cognitive status or loss of consciousness should be taken as a sign of moderate to severe exposure. The general guidelines for atropine treatment following different categories of nerve agent exposure are shown in Table 1. It has been adapted from recommendations found in the Army Field Manual 8-285: Treatment of Chemical Agent Casualties (1995), the *Medical Management of Chemical Agent Casualties Handbook* (2000), and Sidell (1997). A more detailed discussion of these treatment guidelines for vapor and dermal exposures is provided below.

Vapor exposures

The effects of nerve agents occur very quickly following vapor exposure and can reach maximum intensity within minutes, even after the casualty is protected or removed from the vapor. Because toxic effects occur so rapidly, antidotal therapy should be more aggressive for a casualty seen during or immediately after an exposure than for one seen 15 to 30 min after exposure has ended. The effects from an absorbed vapor exposure may continue to progress to a maximum over several minutes, even after the exposure is terminated. Thus, the more aggressive therapy given immediately after onset of effects is due to anticipation of more severe effects in the ensuing minutes. In contrast, a casualty seen 15 to 30 min after vapor exposure is terminated will most probably be displaying maximum signs and not progress further. If the patient is displaying only mild signs at this time (e.g. miosis, rhinorrhea), these signs may resolve without any or with minimal atropine therapy.

For a casualty seen immediately after a vapor exposure, 2 mg of atropine should be given if the only toxic sign is miosis. If any dyspnea is also present, a second 2 mg of atropine

Table 1. General guidelines for immediate atropine treatment of a nerve agent casualty based upon the suspected route and severity of clinical signs of exposure

Route of exposure	Severity of exposure	Signs and symptoms	atropine dose[a,b,c]
Inhalation – vapor	Mild	Miosis; rhinorrhea; mild dyspnea; nausea/vomiting	> 5–10 min since exposure: observation or 1 × 2 mg atropine autoinjector, depending upon severity of dyspnea < 5 min of exposure: 2 × 2 mg atropine autoinjectors
	Moderate	Miosis; rhinorrhea; moderate to severe dyspnea; nausea/vomiting	> 5–10 min since exposure: 1 or 2 × 2 mg atropine autoinjectors < 5 min of exposure: 3 × 2 mg atropine autoinjectors
	Moderately severe	Miosis; rhinorrhea; severe dyspnea; nausea/vomiting; fasciculations	3 × 2 mg atropine autoinjectors; additional atropine (2 mg every 5 min) and ventilatory support may be required
	Severe	Loss of consciousness; convulsions; severe dyspnea/apnea	5 × 2 mg atropine autoinjectors; additional atropine (2 mg every 5 min) and ventilatory support probably required
Dermal	Mild	Localized sweating; fasciculations	1 × 2 mg atropine autoinjector
	Moderate	Localized sweating; fasciculations; nausea/vomiting	1 or 2 × 2 mg atropine autoinjectors
	Moderately severe	Localized sweating; nausea/vomiting; generalized fasciculations; dyspnea	3 × 2 mg atropine autoinjectors; additional atropine (2 mg every 5 min) and ventilatory support may be required
	Severe	Loss of consciousness; convulsions; severe dyspnea/apnea	5 × 2 mg atropine autoinjectors; additional atropine (2 mg every 5 min) and ventilatory support probably required

Notes:
[a] For each of the first three 2 mg autoinjectors of atropine administered, there should also be corresponding autoinjectors of oxime that is delivered in conjunction with the atropine dosing; no further oxime is to be given in the field after the delivery of these first three doses.
[b] If the severity of toxic signs requires that all three 2 mg autoinjectors of atropine need to be administered, then for US forces, anticonvulsant treatment is also given.
[c] Injections administered IM.

(total dose = 4 mg of atropine) should be given. If dyspnea is severe or if any other sign of severe intoxication is present (e.g. fasciculations, collapse/loss of consciousness) then all three 2 mg atropine injectors (total dose = 6 mg of atropine) should be administered and the medic/corpsman alerted to the possibility of the need for additional atropine. For casualties seen 15–30 min after vapor exposures have been terminated, no atropine is required if miosis is the only sign. If mild or moderate dyspnea is present, 2 mg of atropine should be given; if dyspnea is more severe (obvious gasping for breath), then two 2 mg of atropine (total dose = 4 mg of atropine) should be given. Improvement should be noted within 5 to 10 min following such treatment. If dyspnea is severe and more serious signs of intoxication are also present (e.g. fasciculations, collapse/loss of consciousness, convulsions), then all three 2 mg atropine injectors (total dose = 6 mg of atropine) should be administered and the medic/corpsman alerted to the need for additional atropine.

Dermal exposures

Because of the time-lag between exposure and the onset of toxic effects, the treatment of dermal exposure to nerve agent is more problematic. As stated above, toxic effects may develop hours after a dermal exposure, even though thorough decontamination may have been performed, since a toxic dose may have already been absorbed through the skin. Nerve agents penetrate skin at different rates based on moisture, temperature, location on the body and even age and gender of the patient. An asymptomatic person who has had dermal contact with a nerve agent should be kept under medical observation for up to 24 h and be regularly re-evaluated for changes in their condition. This is especially so for contact with agents such as VX or VR, which have greater skin penetrating capabilities than the G-agents. Localized sweating and muscle fasciculations are the first signs of dermal exposure to nerve agents and indicate that the nerve agent has already penetrated the skin. Observation of these signs calls for the administration of 2 mg of atropine. Nausea and vomiting are the other early signs of exposure to liquid nerve agent. If these signs occur soon after a suspected dermal exposure, it is an indication of a severe exposure and the need for more aggressive treatment. If these signs occur within 1 h of a suspected dermal exposure, then two 2 mg of atropine (total dose = 4 mg of atropine) should be administered and the need for further therapy anticipated. If nausea and vomiting occur several hours after suspected exposure, they may be successfully treated with a single 2-mg atropine dose if symptoms do not grow worse, but the casualty needs to be closely monitored. When dyspnea and/or more generalized muscle fasciculations are present in addition to the nausea and vomiting, this requires administration of all three 2 mg atropine autoinjectors (total dose = 6 mg of atropine) and the medic/corpsman alerted to the possibility of the need for additional atropine. The presence of more severe signs, e.g. loss of consciousness convulsions/seizures, requires administration of all three 2 mg atropine autoinjectors (total dose = 6 mg of atropine) and immediate notification of the medic/corpsman for additional atropine.

The severe casualty

A casualty that has been severely exposed by either the vapor or dermal route will probably have altered mental status or be unconscious, have severe dyspnea or apnea, cyanosis, copious secretions, generalized fasciculations and/or periodic convulsions/seizures. Not all signs need be present, but two of the above signs, along with altered mental status, indicate a potentially life-threatening exposure. In addition to therapeutic drugs, such a casualty requires ventilatory support, since the success of therapy under such circumstances depends upon the status of the cardiovascular system. Studies have also shown that rapid IV administration of atropine to animals rendered hypoxic by OP agents can trigger ventricular fibrillation, a potentially fatal cardiac arrhythmia (Kunkel *et al.*, 1973; Wills *et al.*, 1950). Because of this, it has always been recommended that ventilatory support be provided to counteract the hypoxia before or in conjunction with administration of large amounts of atropine (Sidell, 1997). In field treatment, a cricothyroidotomy may be the most practical way to provide an airway for assisted ventilation, while in a medical treatment facility endotracheal intubation should be attempted. Supplemental oxygen is helpful if available, as well as suction to clear secretions from the airway. Resistance to ventilation will decrease and secretions dry up as atropinization is achieved. Ventilatory assistance may be required for only a short time (20–30 min), but in cases of severe poisoning it may be required for hours before the return of spontaneous respiration (Sidell, 1974). Ohbu *et al.* (1997) describe several patients from the Tokyo subway incident in cardiopulmonary or respiratory arrest at time of admission that were successfully resuscitated, provided cardiovascular and respiratory support, and then successfully treated with atropine, oxime and diazepam.

Additional atropine beyond the initial 6 mg carried by the service member will most assuredly be needed promptly in a severely poisoned casualty. Sidell (1997) recommends that an additional 4 mg of atropine be given immediately, for a total initial dose of 10 mg. If the patient is in a medical treatment facility, the atropine should be given IV if that is possible and if the casualty is not hypoxic. If the casualty is in the field, the medic/corpsman should give two additional 2 mg atropine autoinjectors. Under either circumstance, the caregiver should then wait for several minutes to assess the response to this initial 10 mg dose. Additional 2 mg autoinjector doses of atropine should then be given at 3 to 5 min intervals until atropinization is achieved. Signs of successful atropinization include decreased bronchospasm, reduced airway resistance, the drying up of respiratory and salivary secretions and a heart rate $\geq$ 90 beats per minute. All of these effects result in ease of respiration and adequate oxygenation, the keys to successful therapy.

Guidelines for the treatment of pediatric casualties in a potential homeland defense type of situation have been recently published (Rotenberg and Newmark, 2003). They recommend atropine doses scaled to the weight of the child, with 0.05 mg kg^{-1} per dose being the rough unit dose: 2 mg for 40 kg+; 1 mg for 20 kg+; 0.5 for 10 kg+. They recommend that these doses be given every 5–10 min to a moderate or severe pediatric casualty until atropinization is accomplished.

There has recently been increased discussion about the end-points of atropinization and the most efficient means to achieve it. Eddleston *et al.* (2004a) assessed variations in textbook recommendations for early atropinization using model data of atropine dose requirements in patients severely poisoned with OP pesticides. They concluded that a 'dose-doubling strategy', continued 'doubling' of successive doses, would be the most rapid and efficient way to achieve atropinization. Likewise, the treatment regimen used by Dr Foroutan also would result in a rapid atropinization. However, it must be remembered that the guidelines discussed above are for field treatment of a casualty, where a medic/corpsman is the caregiver and respiratory support is limited, while those of Eddleston *et al.* (2004a) and Dr Foroutan refer to treatment at a medical facility by a physician. The end-points of atropinization recommended by Army Field Manual 8-285: Treatment of Chemical Agent Casualties (1995), the *Medical Management of Chemical Agent Casualties Handbook* (2000), Sidell (1997), Dr Foroutan and Eddleston *et al.* (2004a) are very similar: lack of bronchoconstriction, ease of respiration, drying of respiratory secretions and a heart rate > 90 beats per minute. Eddleston *et al.* (2004a) recommended a target heart rate of > 80 beats per minute). Pupil size is not recommended as a therapeutic end-point to judge adequacy of atropinization in nerve-agent casualties since miosis produced by exposure to nerve agent vapor is resistant to systemic atropine treatment (Sidell, 1997).

Once atropinization has been achieved, the casualty needs to be closely monitored and additional atropine provided to maintain atropinization if toxic signs begin to reappear. Although very large amounts of atropine may be required to achieve initial atropinization (15 mg or more in Sidell's cases; up to 200 mg in Foroutan's experience), the need for such large doses has not extended beyond 2–3 h with nerve agents. This statement applies primarily to clinical experiences with vapor exposures to the G-type agents. There is limited clinical experience treating severe percutaneous exposures with V-type agents, where poisoning may very well be more protracted.

Additional atropine may be required beyond the initial atropinization to control reappearance of secretions and continuing nausea and vomiting for the next 6 to 36 h after severe poisoning. This has been typically administered as 1 or 2 mg doses IM (Sidell, 1974) or given by slow IV infusion of 1 to 2 mg h^{-1}. The point at which to discontinue atropine administration is a clinical judgment. With severe intoxication by nerve agents, continuing atropine administration beyond 36 h has not been needed. This is not the case with severe OP or carbamate pesticide poisoning where pesticides may be sequestered in fat tissue and cause continued acute cholinergic crisis for days or even weeks (LeBlanc *et al.*, 1986; Vale *et al.*, 1990).

Atropine, or other anticholinergic drugs, will not reverse all toxic signs of exposure. Muscle fasciculations may persist for hours after the other signs of intoxication have been controlled with atropine. The casualty may feel weak or easily fatigued for days. Some of the sarin-exposed victims of the Matsumoto terrorist attack still complained of fatigue, asthenopia (weakness or easy fatigue of the visual organs), dimness of vision and a general loss in strength up to one year following that incident (Nakajima *et al.*, 1997). Sidell (1974) was the first to call attention to the protracted CNS effects that can linger for several weeks following exposure. A patient may become depressed, withdrawn, have poor sleep accompanied by vivid dreams and have moderate to mild cognitive impairment. These effects are virtually identical to the behavioral changes that are seen in volunteers exposed to moderate doses of DFP, sarin or VX (Grob *et al.*, 1947; Grob and Harvey, 1953,1958; Bowers *et al.*, 1964). These CNS changes noted by Sidell (1974) could be counteracted with moderate doses of scopolamine, and this treatment seemed to assist with recovery, especially achieving restful sleep. This is the only report in the literature that describes the use of anticholinergics to treat the behavioral sequela of a severe nerve-agent exposure.

SUMMARY

Atropine is an anticholinergic drug that binds to muscarinic cholinergic receptors and blocks the stimulating effects of the neurotransmitter ACh. As such, atropine or other antimuscarinic anticholinergic drugs provide life-saving immediate therapy to individuals exposed to nerve agents, OP pesticides or overdoses of other ChE inhibitors. Atropine decreases secretions and reverses the spasm or contraction of smooth muscle. It relieves bronchoconstriction and allows for better air exchange and maintenance of cardiovascular function. Rapid administration of what may seem as very large doses of atropine is required to control the toxic signs of nerve-agent poisoning in severely exposed individuals. Caregivers must be trained to be aggressive in the use of atropine for successful management of nerve agent casualties. Other centrally active anticholinergic drugs show significant promise as either replacements for, or as adjuncts to, atropine therapy based on extensive laboratory tests in animals, but have not, as of yet, been used clinically for this purpose.

REFERENCES

Anderson DR, Gennings C, Carter WH *et al.* (1994). Efficacy comparison of scopolamine and diazepam against soman-induced debilitation in guinea pigs. *Fund Appl Toxicol*, **22**, 588–593.

Bajgar J, Fusek J and Vachek J (1994). Treatment and prophylaxis against nerve agent poisoning. *ASA Newslett*, **99**(4), 10–11.

Bowers MB, Goodman E and Sim VN (1964). Some behavioral changes in man following anticholinesterase administration. *J Nerv Ment Dis*, **138**, 383–389.

Brimblecombe RW, Green DM, Stratton JA *et al.* (1970). The protective actions of some anticholinergic drugs in sarin poisoning. *Br J Pharmacol*, **39**, 822–830.

Brown JH and Taylor P (2001). Muscarinic receptor agonists and antagonists. In: *Goodman and Gilman's The Pharmacological Basis of Therapeutics*, 10th Edition (JG Hardman, LE Limbird and AG Gilman, Eds), pp. 155–173. New York, NY, USA: McGrow-Hill.

Bymaster FP, Carter PA, Yamada M *et al.* (2003). Role of specific muscarinic receptor subtypes in cholinergic parasympathomimetic responses, *in vivo* phosphoinositide hydrolysis and pilocarpine-induced seizure activity. *Eur J Neurosci*, **17**, 1403–1410.

Capacio BR and Shih T-M (1991). Anticonvulsant actions of anticholinergic drugs in soman poisoning. *Epilepsia*, **32**, 604–615.

Carpentier P, Foquin-Tarricone A, Bodjarian N *et al.* (1994). Anticonvulsant and antilethal effects of the phencyclidine derivative TCP in soman poisoning. *Neurotoxicology*, **15**, 837–852.

Caulfield MP and Birdsall NJ (1998). International Union of Pharmacology XVII. Classification of muscarinic acetylcholine receptors. *Pharmacol Rev*, **50**, 279–290.

Coleman IW, Little PE and Bannard RAB (1962). Cholinolytics in the treatment of anticholinesterase poisoning. I. The effectiveness of certain cholinolytics in combination with an oxime for treatment of sarin poisoning. *Can J Biochem Pharmacol*, **40**, 815–826.

Coleman IW, Patton GE and Bannard RAB (1968). Cholinolytics in the treatment of anticholinesterase poisoning. V. The effectiveness of parpanit with oximes in the treatment of organophosphorus poisoning. *Can J Physiol Pharmacol*, **46**, 109–117.

Comstock CC and Krop S (1948). *German first aid kits for treatment of tabun (GA) casualties*. Medical Division Report No. 151. Edgewood Arsenal, MD, USA: Army Chemical Center.

Dawson RM (1994). Review of oximes available for treatment of nerve agent poisoning. *J Appl Toxicol*, **14**, 317–331.

Departments of the Army, Navy, Air Force and Commandant Marine Corps (1995). *Field Manual: Treatment of Chemical Agent Casualties and Conventional Military Chemical Injuries*. Army FM-8-285; Navy NAVMED P-5041; Air Force AFJMAN 44-149; Marine Corps FMFM 11-11. Washington, DC, USA (December).

Eddleston M Buckley NA, Checketts H *et al.* (2004a). Speed of initial atropinisation in significant organophosphorus pesticide poisoning – a systematic comparison of recommended regimens. *J Toxicol Clin Toxicol*, **42**, 852–862.

Eddleston M, Dawson A, Karalliedde L *et al.* (2004b). Early management after self-poisoning with an organophosphorus or carbamate pesticide – a treatment protocol for junior doctors. *Crit Care*, **8**, R391–R397.

Fosbraey P, Wetherell JR and French MC (1990). Neurotransmitter changes in guinea-pig brain regions following soman intoxication. *J Neurochem*, **54**, 72–79.

Gomeza J, Shannon HE, Kostenis E *et al.* (1999). Pronounced pharmacological deficits in M2 muscarinic acetylcholine receptor knockout mice. *Proc Nat Acad Sci USA*, **96**, 1692–1697.

Green DM, Muir AW, Stratton JA *et al.* (1977). Dual mechanisms of antidotal action of atropine-like drugs in poisoning by organophosphorus anticholinesterases. *J Pharm Pharmacol*, **29**, 62–64.

Grob D (1956). The manifestations and treatment of poisoning due to nerve gas and other organic phosphate anticholinesterase compounds. *AMA Arch Intern Med*, **98**, 221–239.

Grob D and Harvey AM (1953). The effects and treatment of nerve gas poisoning. *Am J Med*, **14**, 52–63.

Grob D and Harvey AM (1958). The effects in man of the anticholinesterase compound sarin (isopropyl methyl phosphnonofluoridate). *J Clin Invest*, **37**, 350–368.

Grob D, Harvey AM, Langworthy OR *et al.* (1947). The administration of di-isopropyl fluororphosphate (DFP) to man, III: Effect on the central nervous system with special reference to the electrical activity of the brain. *Bull Johns Hopkins Hosp*, **81**, 257–266.

Hamilton SE, Loose MD, Qi M *et al.* (1997). Disruption of the M1 receptor gene ablates muscarinic receptor-dependent M current regulation and seizure activity in mice. *Proc Nat Acad Sci USA*, **94**, 13311–13316.

Headley DB (1982). Effects of atropine sulfate and pralidoxime chloride on visual, physiological, performance, subjective and cognitive variables in man: A review. *Milit Med*, **147**, 122–132.

Jovic R and Milosevic M (1970). Effective doses of some cholinolytics in the treatment of anticholinesterase poisoning. *Eur J Pharmacol*, **12**, 85–93.

Kassa J (2002). Review of oximes in the antidotal treatment of poisoning by organophosphorus nerve agents. *J Toxicol Clin Toxicol*, **40**, 803–816.

Kassa J and Bajgar J (1996). The influence of pharmacological pretreatment on efficacy of HI-6 oxime in combination with benactyzine in soman poisoning in rats. *Human Exp Toxicol*, **15**, 383–388.

Ketchum JS, Sidell FR, Crowell EB *et al.* (1973). Atropine, scopolamine, and ditran: comparative pharmacology and antagonists in man. *Psychopharmacologia*, **28**, 121–145.

Koelle GB (1963). Cholinesterases and anticholinesterase agents. In: *Handbuch der Experimentellen Pharmakologie*, Volume 15 (O Eichler and A Farah, eds) 989–1027. Berlin, Germany: Springer-Verlag.

Kunkel AM, O'Leary JF and Jones AH (1973). *Atropine-induced ventricular fibrillation during cyanosis caused by organophosphorus poisoning*. Edgewood Arsenal Technical Report 4711. Edgewood Arsenal, MD, USA: Medical Research Laboratory.

Lallement G, Denoyer M, Collet A *et al.* (1992). Changes in hippocampal acetylcholine and glutamate extracellular levels during soman-induced seizures: influence of septal cholinoceptive cells. *Neurosci Lett*, **139**, 104–107.

Lallement G, Pernot-Marino I, Foquin-Tarricone A *et al.* (1994). Modulation of soman-induced neuropathology with an anticonvulsant regimen. *NeuroReport*, **5**, 2265–2268.

Lallement G, Clarencon D, Masqueliez C *et al.* (1998). Nerve agent poisoning in primates: antilethal, antiepileptic and neuroprotective effects of GK-11. *Arch Toxicol*, **72**, 84–92.

Leadbeater L, Inns RH and Rylands JM (1985). Treatment of poisoning by soman. *Fund Appl Toxicol*, **5**, S225–S231.

LeBlanc FN, Benson BE and Gilg AD (1986). A severe organophosphate poisoning requiring the use of an atropine drip. *Clin Toxicol*, **24**, 69–76.

Lemercier G, Carpentier P, Sentenac-Roumanou H *et al.* (1983). Histological and histochemical changes in the central nervous system of the rat poisoned by an irreversible anticholinesterase organophosphorus compound. *Acta Neuropathol (Berlin)*, **61**, 123–129.

Lipp JA (1972). Effect of diazepam upon soman-induced seizure activity and convulsions. *EEG Clin Neurophysiol*, **32**, 557–560.

Lipp JA (1973). Effects of benzodiazepine derivatives on soman-induced seizure activity and convulsions in the monkey. *Arch Int Pharmacodynam Ther*, **202**, 244–251.

Lipp JA (1976). Effect of atropine upon the cardiovascular system during soman-induced respiratory depression. *Arch Int Pharmacodynam Ther*, **220**, 19–27.

Lipp JA and Dola TJ (1978). Effect of atropine upon the cerebrovascular system during soman-induced respiratory depression. *Arch Int Pharmacodynam Ther*, **235**, 211–218.

Longo VG (1966). Behavioral and electroencephalographic effects of atropine and related compounds. *Pharmacol Rev*, **18**, 965–996.

Masuda N, Takatsu M, Morinari H *et al.* (1995). Sarin poisoning in Tokyo subway. *Lancet*, **345**, 1446.

McDonough JH (2002). Performance impacts of nerve agents and their pharmacological countermeasures. *Milit Psychol*, **14**, 93–119.

McDonough JH and Shih T-M (1993). Pharmacological modulation of soman-induced seizures. *Neurosci Biobehav Rev*, **17**, 203–215.

McDonough JH and Shih T-M (1995). A study of the *N*-methyl-D-aspartate antagonist properties of anticholinergic drugs. *Pharmacol Biochem Behav*, **51**, 249–253.

McDonough JH, Jaax NK, Crowley RA *et al.* (1989). Atropine and/or diazepam therapy protects against soman-induced neural and cardiac pathology. *Fund Appl Toxicol*, **13**, 256–276.

McDonough JH, Dochterman LW, Smith CD *et al.* (1995). Protection against nerve agent-induced neuropathology, but not cardiac pathology, is associated with the anticonvulsant action of drug treatment. *Neurotoxicology*, **15**, 123–132.

McDonough JH and Shih T-M (1997). Neuropharmacological mechanisms of nerve agent-induced seizure and neuropathology. *Neurosci Biobehav Rev*, **21**, 559–579.

McDonough JH, Zoeffel LD, McMonagle J *et al.* (2000). Anticonvulsant treatment for nerve agent seizures: anticholinergics versus diazepam in soman-intoxicated guinea pigs. *Epilep Res*, **38**, 1–14.

Medical Management of Chemical Casualties Handbook (2000). 3rd Edition, pp. 102–135. Aberdeen Proving Ground, MD, USA: Chemical Casualty Care Division, USAMRICD.

Meshulam Y, Cohen G, Chapman S *et al.* (2001). Prophylaxis against organophosphate poisoning by sustained release of scopolamine and physostigmine. *J Appl Toxicol*, **21**(Supplement 1), S75–S78.

Moore DH, Clifford CB, Crawford IT *et al.* (1995). Review of nerve agent inhibitors and reactivators of acetylcholinesterase. In *Enzymes of Cholinesterase Family* (DM Quinn, AS Balasubramanian, BP Doctor and P Taylor, eds), pp. 297–304. New York, NY, USA: Plenum Press.

Morita H, Yanaagisawa N, Nakajima T *et al.* (1995). Sarin poisoning in Matsumoto, Japan. *Lancet*, **346**, 290–293.

Nakajima T, Ohta S, Morita H *et al.* (1997). Epidemiological study of sarin poisoning in Matsumoto City, Japan. *J Epidemiol*, **8**, 33–41.

Newmark J (2004). The birth of nerve agent warfare: lessons from Syed Abbas Foroutan. *Neurology*, **62**, 1590–1596.

Nozaki H and Aikawa N (1995). Sarin poisoning in Tokyo subway. *Lancet*, **345**, 1446–1447.

Nozaki H, Aikawa N, Fujishima S *et al.* (1995). A case of VX poisoning and the difference from sarin. *Lancet*, **346**, 698–699.

Ohbu S, Yamashina A, Takasu N *et al.* (1997). Sarin poisoning on Tokyo subway. *Southern Med J*, **90**, 587–593.

Okumura T, Takasu N, Ishimatsu S *et al.* (1996). Report on 640 victims of the Tokyo subway sarin attack. *Ann Emerg Med*, **28**, 129–135.

Olney JA, Price MT, Labruyere J *et al.* (1987). Anti-Parkinsonian agents are phencyclidine agonists and *N*-methyl-aspartate antagonists. *Eur J Pharmacol*, **142**, 319–320.

Pazdernik TL, Cross R, Nelson SR *et al.* (1983). Soman-induced depression of brain activity in TAB-pretreated rats: a 2-deoxyglucose study. *Neurotoxicology*, **4**, 27–34.

Penetar DM (1990). A brief review of atropine effects on physiology and performance. *Drug Develop Res*, **20**, 117–121.

Petrov AN, Sofronov GA, Nechiporenko SP *et al.* (2004). Organophosphorus poison antidotes (in Russian). *Ross Khim Zhur (Russ J Chem)*, **48**, 110–116.

Philippens IH, Melchers BP, Oliver B *et al.* (2000). Scopolamine augments the efficacy of physostigmine against soman poisoning in guinea pigs. *Pharmacol Biochem Behav*, **65**, 175–182.

Pickworth WB, Herning RI, Koeppl B *et al.* (1990). Dose-dependent atropine-induced changes in spontaneous electroencephalogram in human volunteers. *Milit Med*, **155**, 166–170.

Rotenberg JS and Newmark J (2003). Nerve agent attacks on children: diagnosis and management. *Pediatrics*, **112**, 648–658.

Rump S, Grudzinska E and Edelwein Z (1972). Effects of diazepam on abnormalities of bioelectrical activity of the rabbit's brain due to fluostigmine. *Acta Nerv Super (Prague)*, **14**, 176–177.

Rump S, Grudzinska E and Edelwein Z (1973). Effects of diazepam on epileptiform patterns of bioelectric activity of the rabbit's brain induced by fluostigmine. *Neuropharmacology*, **12**, 813–817.

Samson FE, Pazdernik TL, Cross RS *et al.* (1985). Brain regional activity and damage associated with organophosphate induced seizures: effects of atropine and benactyzine. *Proc West Pharmacol Soc*, **28**, 183–185.

Sekijima Y, Morita H, Shindo M *et al.* (1995). A case of severe sarin poisoning in the sarin attack at Matsumoto – one-year follow-up on the clinical findings and laboratory data (in Japanese). *Rin Shinkeij (Clin Neurol)*, **35**, 1241–1245.

Shih T-M (1982). Time course effects of soman on acetylcholine and choline levels in six discrete areas of the rat brain. *Psychopharmacologia (Berlin)*, **78**, 170–175.

Shih T-M and McDonough JH (2000). Efficacy of biperiden and atropine as anticonvulsant treatment for organophosphorus nerve agent intoxication. *Arch Toxicol*, **74**, 165–172.

Shih T-M, Duniho SM and McDonough JH (2003). Control of nerve agent-induced seizures is critical for neuroprotection and survival. *Toxicol Appl Pharmacol*, **188**, 69–80.

Sidell FR (1974). Soman and sarin: Clinical manifestations and treatment of accidental poisoning by organophosphates. *Clin Toxicol*, **7**, 1–17.

Sidell FR (1997). Nerve agents. In: *Textbook of Military Medicine: Medical Aspects of Chemical and Biological Warfare* (R Zajtchuk, ed.), pp. 129–179. Washington, DC, USA: Office of the Surgeon General.

Sidell FR, Markis JE, Groff WA *et al.* (1974). Enhancement of drug absorption after administration by an automatic injector. *J Pharmacokin Biopharm*, **2**, 197–210.

Sparenborg S, Brennecke LH, Jaax NK *et al.* (1992). Dizocilpine (MK-801) arrests status epilepticus and prevents brain damage induced by soman. *Neuropharmacology*, **31**, 357–368.

Taylor P (2001). Anticholinesterase agents. In: *Goodman and Gilman's The Pharmacological Basis of Therapeutics*, 10th Edition (JG Hardman, LE Limbard and AG Gilman, eds), pp. 175–191. New York, NY, USA: McGraw-Hill.

Vale JA, Meredith TJ and Heath A (1990). High dose atropine in organophosphorus poisoning. *Postgrad Med J*, **66**, 881.

Vojvodic VB and Maksimovic M (1972). Absorption and excretion of pralidoxime in man after intramuscular injection of PAM-2Cl and various cholinolytics. *Eur J Clin Pharmacol*, **5**, 58–61.

Vojvodic VB, Jovic R, Rosic N *et al.* (1972). Effect of a mixture of atropine, benactyzine, and pralidoxime on the body and on certain of the elements of the fighting qualities of people – volunteers. *Vojnosan Pregl*, **29**, 103–107.

Wetherell JR (1994). Continuous administration of low dose rates of physostigmine and hyoscine to guinea-pigs prevents the toxicity and reduces the incapacitation produced by soman poisoning. *J Pharm Pharmacol*, **46**, 1023–1028.

Wetherell J, Hall T and Passingham S (2002). Physostigmine and hyoscine improves protection against the lethal and incapacitating effects of nerve agent poisoning in the guinea pig. *Neurotoxicology*, **23**, 341–349.

Wills JH (1963). Pharmacological antagonists of the anticholinesterase agents. In *Handbuch der Experimentellen Pharmakologie*, Volume 15 (O Eichler and A Farah, eds), pp. 883–920. Berlin, Germany: Springer-Verlag.

Wills JH, McNamara BP and Fine EA (1950). Ventricular fibrillation in delayed treatment of TEPP poisoning. *Fed Proc*, **9**, 136.

Wilson KW (late 1970s). Personal Communication. Aberdeen Proving Ground, MD, USA: Biomedical Laboratory, Chemical Systems Laboratories.

Yamada M, Lamping KG, Duttaroy A *et al.* (2001). Cholinergic dilation of cerebral blood vessels is abolished in M5 muscarinic acetylcholine receptor knockout mice. *Proc Nat Acad Sci USA*, **98**, 14096–14101.

15 OXIMES

Peter A. Eyer[1] and Franz Worek[2]

[1]*Ludwig-Maximilians University, Munich, Germany*
[2]*Bundeswehr Institute of Pharmacology and Toxicology, Munich, Germany*

INTRODUCTION

While atropine is highly effective in antagonizing acetylcholine at most peripheral *m*-receptors and partly at central *m*-receptors, atropine is ineffective at *n*-receptors (*m*-muscarinic, *n*-nicotinic). Here, reactivating oximes can be expected to act as specific antidotes. Search for suitable reactivators dates back to the early 1950s, starting with hydroxylamine and hydroxamic acids (Hobbiger, 1963). Later on, ketoximes and aldoximes proved to be more efficacious. Meanwhile, more than 1500 compounds have been tested, of which only a few have been studied for human use. This chapter deals with the two widely licensed drugs pralidoxime and obidoxime, together with trimedoxime (TMB-4) and methoxime (MMB-4), and also two 'Hagedorn –oximes', HI 6 and HLö 7 (for structural formulae, see Figure 1). Trimedoxime, methoxime and HI 6 have already been tested in humans and are available for human use in some countries. HLö 7 has been included because this bispyridinium oxime has probably the broadest spectrum of antidotal action of any pyridinium oxime against nerve agents.

HISTORY OF OXIME DEVELOPMENT

In 1955, Wilson and Ginsburg in the United States and Childs *et al.* in the United Kingdom researched and published independently on the efficacy of the compound 2-pyridine aldoxime methiodide (2-PAM) as a reactivator of phosphorylated cholinesterases. The next compound of interest was trimedoxime, which was synthesized and tested by Poziomek *et al.* in 1958 (Poziomek *et al.*, 1958) and independently by Hobbiger *et al.* (Hobbiger *et al.*, 1958). This bis(pyridinium-4-aldoxime) derivative was clearly superior to 2-PAM, particularly in tabun poisoning, but worsened the outcome in soman-poisoned laboratory animals. A year later, Hobbiger and Sadler (Hobbiger and Sadler, 1959) described methoxime, which was reported to be more effective than trimedoxime in sarin poisoning (Bajgar *et al.*, 1975) and in reactivating soman-inhibited AChE (Harris *et al.*, 1990): methoxime also increased the survival of soman-challenged rabbits (Koplovitz and Stewart, 1992). Obidoxime, a structural analogue of trimedoxime was synthesized by Lüttringhaus and Hagedorn in the early 1960s (Lüttringhaus and Hagedorn, 1964) and studied clinically by Erdmann and von Clarmann (1963). This compound, marketed as Toxogonin®, is one of the most active reactivators against organophosphorus insecticides. However, like trimedoxime, obidoxime was of limited efficacy in soman poisoning. This therapeutic gap was attributed to formation of phosphyloximes (this term is used to cover both phosphoryl and phosphonyl oximes), which are probably more reactive than the parent organophosphorus compounds. This was suggested by Wilson and Ginsburg as early as 1955 (Wilson and Ginsburg, 1955) and later by others (Hackley Jr *et al.*, 1959; Lamb *et al.*, 1965; Schoene, 1973; Nenner, 1974; De Jong and Ceulen, 1978). The stability of phosphyloximes of the 4-pyridinium compounds was clearly greater than that of the 2-pyridinium analogs. This suggestion stimulated the synthesis

Chemical Warfare Agents: Toxicology and Treatment (2nd Edition)
Edited by Timothy C. Marrs, Robert L. Maynard and Frederick R. Sidell

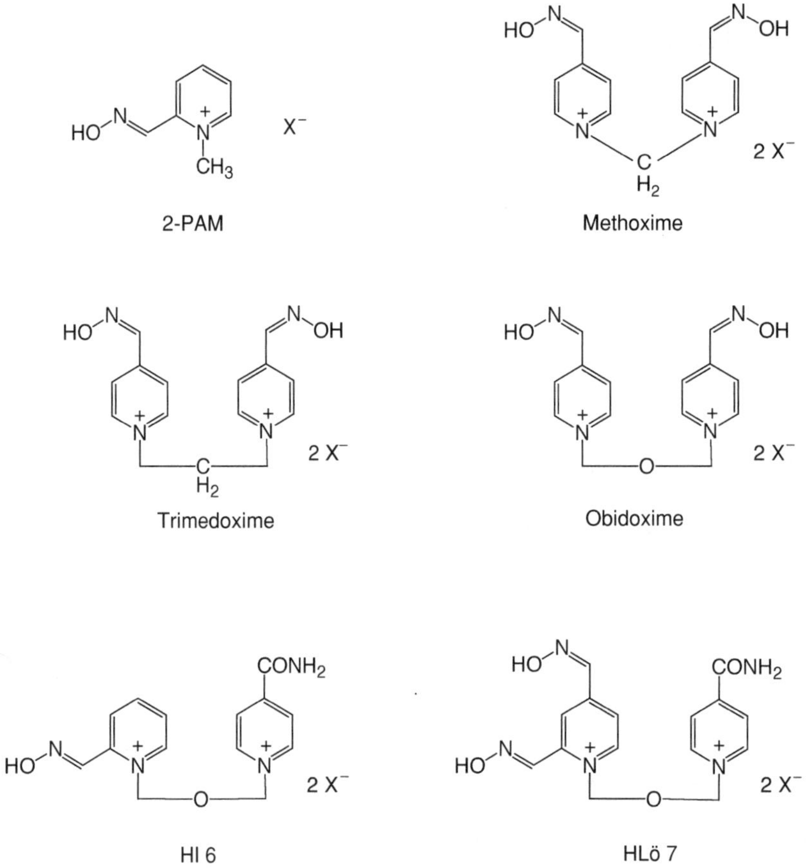

Figure 1. Structural formulae of pyridinium oximes

of hundreds of oxime derivatives in the laboratory of Hagedorn, leading to asymmetric bispyridinium oximes of the H-series and culminating in the development of HI 6 (Stark, 1968; Oldiges and Schoene, 1970) and HLö 7 (Löffler, 1986; De Jong *et al.*, 1989; Eyer *et al.*, 1992). As well as pyridinium oximes, imidazolium and quinuclidinium compounds have been synthesized and tested in the laboratory of Binenfeld in Croatia and have recently been reviewed (Primozic *et al.*, 2004). This list is by no means complete but comprises probably the compounds of broader interest as published in the open literature.

MECHANISM OF OXIME ACTION

There is considerable evidence that the oxime reactivators form a complex of similar orientation to that expected for the phosphylation reaction. Figure 2 depicts the essential features. While reaction (1) is thought to be fully reversible with $K_D = k_{-1}/k_{+1}$, nucleophilic attack of the oximate on phosphorus leads to a pentacoordinate transition state of the intermediate phosphyloxime, a reaction that is facilitated by polarization of the phosphorus oxygen bond in the oxyanion 'hole' of the enzyme. Ideally, the nucleophile and the leaving group, i.e. the reactivated enzyme, would occupy apical positions, assuming an in-line S_N2 displacement reaction (Ashani *et al.*, 1995). However, owing to spatial constraints of the 'gorge' and steric limitations of the bulky oximes, optimal geometry for the interaction is not achieved. Accordingly, reactions with oximes are usually much slower than phosphylation reactions (Taylor *et al.*, 1999).

The efficacy of the reactivator is attributed to the nucleophilicity of the oximate and the decay rate of the phosphyloxime, depending on the

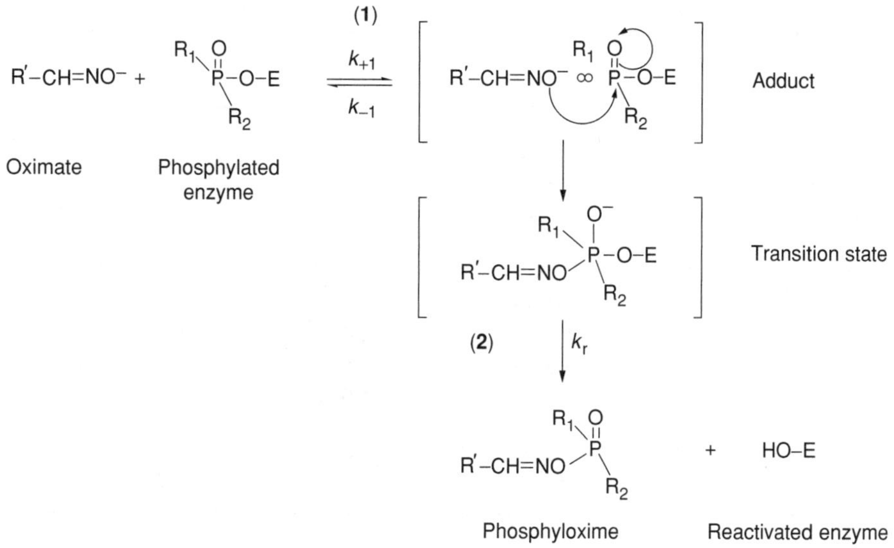

Figure 2. Proposed scheme of chemical reactions involved in the reactivation of phosphylated cholinesterase

structure of the reactivator molecule and on the bound organophosphorus moiety, and finally on the architecture and hence the source of the enzyme. Much has been learnt in recent years from the development of mutant cholinesterases and their structural elucidation by X-ray structural analysis.

It should be recalled that the active site of the cholinesterases lies at the base of a deep 'gorge'. In AChE, this 'gorge' is quite narrow with the walls lined by plentiful aromatic side-chains forming a well-defined choline and acyl 'pocket' (Sussman *et al.*, 1991). The narrow confines of the 'gorge' in AChE determine the orientation of the organophosphorus conjugate and the oxime and hence the reaction rate of phosphylation and reactivation (Kovarik *et al.*, 2004). The spatial constraints of high stereoselectivity explain the particular chiral selectivity of the organophosphonates in their reaction with AChE where the phosphonyl oxygen occupies the oxyanion hole. Thus, the S_p-phosphonates, particularly those with bulky alkoxy groups, as in soman, are usually much more reactive than the R_p-enantiomers (Taylor *et al.*, 1999). Interestingly, the more reactive S_p compounds bound to AChE are usually also more susceptible to oxime reactivation. This finding suggested that the oxyanion hole-induced polarization of the phosphonyl oxygen may be required for efficient reactivation (Taylor *et al.*, 1999). It has been assumed that the route of nucleophilic attack on the phosphorus atom is via the acyl 'pocket' and that enhancing the clearance to this 'pocket' may optimize the attack angle of the oximate (Kovarik *et al.*, 2004). From this, it can be expected that various oximes will have different efficacy in reactivating inhibited AChE, depending on the species and the organophosphate. Hence, optimizing general oxime effectiveness is hampered by the tremendously large permutations of species and organophosphates available for study.

Formation of phosphyloximes

The lower part of Figure 2 deserves additional comment with regard to reaction (2). This reaction is in fact reversible and the free phosphyloxime (POX) is in itself a powerful phosphylating agent. Meanwhile, several phosphyloximes have been prepared by synthetic means (Green and Saville, 1956; Hackley Jr *et al.*, 1959; Portmann *et al.*, 1991; Becker *et al.*, 1997; Leader *et al.*, 1999). The inhibition rate constants of some phosphonyloximes can be compared with those of the parent OPs (Table 1).

Table 1 indicates that phosphyloximes of 2-substituted pyridinium compounds are very

Table 1. Inhibitory activity and stability of phosphyloximes

Parent OP	Inhibition rate constants (M^{-1} min^{-1} $\times 10^6$)	POX from		Decomposition $t_{1/2}$ (min)
Tabun	7.4[a]	2-PAM	11[b]	1.1[c]
		4-PAM	2.5[b]	63[c]
Sarin	27[a]	2-PAM	1.420[b]	1.1[c]
		4-PAM	110[b]	75[c]
Soman	92[a]	2-PAM	1.570[b]	1.5[c]
		4-PAM	237[c]	112[c]
Cyclosarin	490[a]	2-PAM	4.020[b]	1.3[c]
		4-PAM	1.800[b]	84[c]
VX	120[a]	2-PAM	1.940[b]	0.7[c]
		4-PAM	382[b]	62[c]
		Obidoxime	13.000[d]	169[e]
		Trimedoxime	9.120[d]	315[e]

Notes:
[a] Human erythrocyte AChE in 100 mM phosphate, pH 7.4, 37°C (Worek *et al*., 2004).
[b] Recombinant mouse AChE in 10 mM HEPES, pH 7.8, 29°C.
[c] 10 mM HEPES, pH 7.8, 29°C (Ashani *et al*., 2003).
[d] Recombinant mouse AChE in 20 mM HEPES, pH 7.2, 25°C (Luo *et al*., 1999).
[e] In 20 mM HEPES, pH 7.2, 25°C (Luo *et al*., 1999).

unstable with half-lives under physiological conditions of approximately 1 min. The hydrolysis of phosphyloximes is hydroxide ion–catalyzed and hence faster at more alkaline pH (Ashani *et al*., 2003). Moreover, the decomposition has a high energy of activation, which for the diethylphosphoryl oxime of obidoxime amounts to 85 kJ mol^{-1} (Kiderlen *et al*., 2005). The primary reaction product is the phosphylic acid, along with the pyridinium nitrile, as would be expected with a ß-elimination reaction (Hagedorn *et al*., 1972; Ashani *et al*., 2003) that produces cyanide and the pyridone or rearranges into the carboxamide (Kiderlen *et al*., 2005). In contrast to the 2-substituted pyridinium POXs, those of 4-pyridinium aldoximes are remarkably stable. It has been suggested that the acidity of the methin proton is decisive for the stability of the POX, being more influenced by the pyridinium moiety than by the nature of the *O*-alkyl substituent (Hagedorn *et al*., 1972; Ashani *et al*., 2003). The high reactivity of the POX towards AChE, as illustrated in Table 1, is an obstacle to reactivation since a topochemical equilibrium develops where reactivation may soon be stopped by the POX, which is inevitably formed. At AChE concentrations of approximately 3 nM, as found in human blood (Lockridge and Masson, 2000), 20% reactivation yields 0.6 nM POX. POX derivatives with inhibition rate constants of 10^9 M^{-1} min^{-1} exhibit an inhibition velocity (0.36 nM min^{-1}) that approaches the decomposition velocity ($t_{1/2} = 1$ min; 0.42 nM min^{-1}). Hence a biphasic reactivation is to be expected with sarin, soman, cyclosarin and VX-inhibited AChE when reacting with 2-PAM. Clearly, the situation is much worse when obidoxime and trimedoxime are the reactivators. This phenomenon has been observed with sarin–inhibited AChE when reactivated with obidoxime and to a lesser extent with pralidoxime, but virtually not at all with HI 6 (Worek *et al*., 2000). It can be anticipated and has indeed been shown that POX formation greatly interferes in the reactivation of cyclosarin-inhibited AChE by pralidoxime (Worek *et al*., 1998) and with the reactivation of VX-inhibited AChE by obidoxime and trimedoxime (Ashani *et al*., 1995; Luo *et al*., 1998a; Luo *et al*., 1999). Interestingly, HLö 7 with an oxime group both in the 2- and 4-positions (see Figure 1) forms no stable POX that interferes with reactivation (Luo *et al*., 1998b), implying that the oximate vicinal to the pyridinium nitrogen atom is the preferred nucleophile. This

Table 2. Reactivation constants of organophosphate-inhibited human erythrocyte AChE

OP[a]	Reactivation constant	Obidoxime	2-PAM	Methoxime[b]	HI 6	HLö 7
Sarin	k_r	0.937	0.25	0.18	0.677	0.849
	K_D	31.3	27.6	78.4	50.1	24.2
Cyclosarin	k_r	0.395	0.182	0.85	1.3	1.663
	K_D	945.6	3159	370.6	47.2	17.9
n-Butylsarin	k_r	0.24	0.08	ND	0.72	0.7
	K_D	43.3	138	ND	25.6	15.3
DFP	k_r	0.06	0.05	ND	0.02	0.02
	K_D	63.8	847	ND	1935	83.7
VX	k_r	0.893	0.215	0.328	0.242	0.49
	K_D	27.4	28.1	241.9	11.5	7.8
VR	k_r	0.63	0.06	0.853	0.71	0.84
	K_D	106	30.7	144.3	9.2	5.3
Tabun	k_r	0.04	0.01	0.011	0	0.02
	K_D	97.3	706	1252	ND	106.5
Fenamiphos	k_r	0.09	0.02	ND	0.02	0.03
	K_D	615	695	ND	889	240
Methamidophos	k_r	0.84	0.31	ND	0.39	0.49
	K_D	7.5	2.1	ND	9.1	3.3
MFPCh	k_r	0.02	0.004	ND	0.076	0.051
	K_D	1133	2949	ND	1233	606
MFPhCh	k_r	0.009	0.003	ND	0.028	0.011
	K_D	3547	3837	ND	1037	599
MFPßCh	k_r	0.015	0.002	ND	0.09	0.078
	K_D	1658	3131	ND	859	458
Paraoxon	k_r	0.81	0.17	0.36	0.2	0.34
	K_D	32.2	187.3	264.3	548.4	47.8

Notes:
[a] MFPCh, methylfluorophosphonylcholine iodide; MFPhCh, methylfluorophosphonylhomocholine iodide; MFPßCh, methylfluorophosphonyl-ß-methylcholine iodide.
[b] ND, no data; k_r in min^{-1}; K_D in μM.

suggestion fits with the much higher pK_a value of the 4-oximate (8.52) compared to the 2-oximate group (7.04) (Eyer *et al.*, 1992). In conclusion, formation of the more stable phosphyloximes impedes reactivation by obidoxime, trimedoxime and presumably by methoxime, in part by 2-PAM, but apparently not by HI 6 and HLö 7.

KINETICS OF OXIME-INDUCED REACTIVATION OF HUMAN ACHE *IN VITRO*

Determination of reactivation rate constants of OP-inhibited AChE *in vitro* is a versatile tool for quantifying the ability of oximes to remove the phosphyl residue from the enzyme. Numerous studies aimed at estimating the reactivating potency of oximes using different enzymes from different sources and experimental protocols have been performed (Bismuth *et al.*, 1992). Recently, the kinetic constants of AChE from human erythrocyte membranes inhibited by various OPs have been determined under physiological conditions, i.e. pH 7.4 and 37°C (Worek *et al.*, 2004).

Table 2 is a compilation of kinetic data for the reactivation of OP-inhibited AChE by the licensed oximes obidoxime and pralidoxime and the experimental compounds HI 6, HLö 7 and methoxime. These data demonstrate a large variability of the affinity ($1/K_D$) and reactivity (k_r) of the oximes depending on the

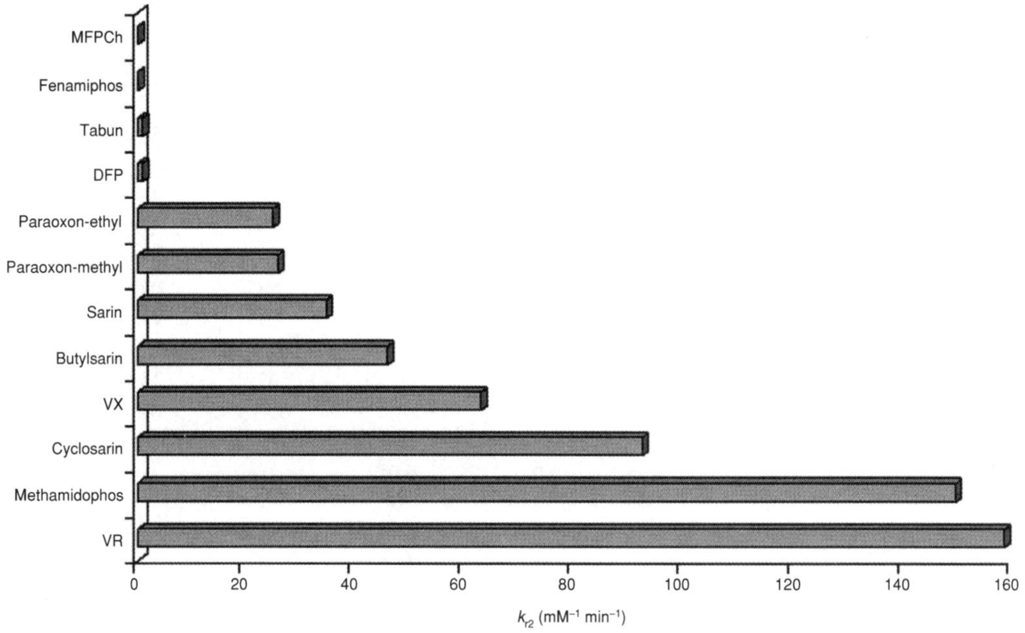

Figure 3. Efficiencies, k_{r2} (mM^{-1} min^{-1}) of the oximes for the reactivation of OP-inhibited human AChE. Data of the most effective oximes are given

structure of the phosphyl moiety. Phosphonylated as well as dimethyl- and diethylphosphoryl AChE were highly susceptible to reactivation by oximes, while human AChE inhibited by diisopropylphosphorofluoridate (DFP), phosphoramidates (tabun, fenamiphos) and phosphonylcholines was rather resistant (Figure 3). The efficiency, i.e. the second-order rate constant ($k_{r2} = k_r/K_D$) for HLö7 is superior for phosphonylated enzymes when compared to the other oximes while obidoxime was most effective with AChE inhibited by organophosphates and phosphoramidates (Table 2).

Experiments using site-directed mutagenesis and computational modelling have shown that oximes appear to have preferred entry routes for attack of the phosphyl moiety and that side-chain substitutions of the active site 'gorge' of AChE may significantly alter the reactivity (Radic and Taylor, 1999; Taylor *et al.*, 1999; Wong *et al.*, 2000; Kovarik *et al.*, 2004). Hence, marked species differences in reactivatability of phosphylated AChE towards oximes are to be expected. In fact, experimental data support the assumption of substantial species differences in reactivation of OP-inhibited AChE (Berry, 1971; De Jong and Wolring, 1984; Su *et al.*, 1986; De Jong *et al.*, 1989; Hanke and Overton, 1991; Clement *et al.*, 1994). In addition, the determination of reactivation rate constants with sarin-, cyclosarin- and VX-inhibited human, guinea-pig, rat and rabbit erythrocyte AChE revealed considerable differences in the efficacy of oximes between human and animal enzymes (Worek *et al.*, 2002). These species differences were particularly evident with the experimental oximes HI 6 and HLö 7 when comparing human and guinea-pig AChE, while minor differences were observed with obidoxime and pralidoxime (Figure 4).

Similar results were obtained with soman-inhibited human, non-human primate and guinea-pig AChE (Figure 5).

EFFICACY OF OXIMES AGAINST NERVE AGENTS IN ANIMALS *IN VIVO*

Numerous studies have been carried out to evaluate the efficacy of oximes in nerve agent poisoning in different species, mainly mice, rats,

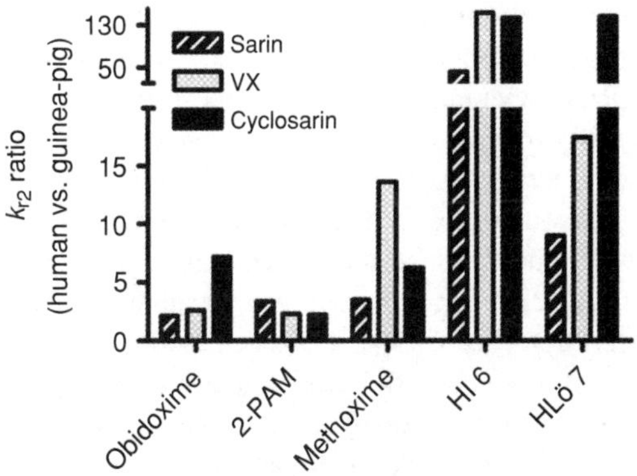

Figure 4. Comparison of the reactivation efficiencies of various oximes on sarin-, VX- and cyclosarin-inhibited human and guinea-pig AChE. The ratios of the second-order rate constants, k_{r2}, between human and guinea-pig AChE are given

guinea-pigs and rabbits. Mostly, oxime-induced increase in survival rate, expressed as the protective ratio of treated versus non-treated animals, was defined as the main endpoint. Unfortunately, a comparison of the data is hampered by interlaboratory differences in species, strains, route of agent and antidote administration, combination with other drugs and dosing of antidotes. Dawson's comprehensive review provides an excellent overview of the different experimental protocols used in the past decades and the large variations in protective ratios, even in a single species (Dawson, 1994). Nevertheless, some general conclusions can be derived from the available animal data. Basic atropine treatment is a prerequisite for successful oxime treatment. Obidoxime and pralidoxime were shown to be effective in sarin- and VX-poisoning in different species but seem to be rather ineffective against cyclosarin, an agent where HI 6 was shown to be a potent antidote. Tabun poisoning is generally considered to be a therapeutic challenge. Obidoxime seems to be superior to pralidoxime and HI 6, but in a single non-human primate study, HI 6 showed a beneficial effect without reactivation of inhibited AChE (Hamilton and Lundy, 1989). Data on HLö 7 are sparse but indicate a promising effect against different nerve agents, including soman and tabun (Eyer *et al.*, 1992).

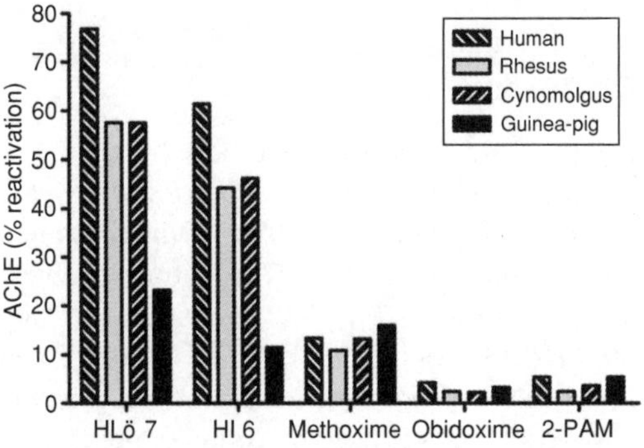

Figure 5. Reactivation of soman-inhibited AChE from different species: reactivation of AChE (%) after incubation of AChE with 300 μM oxime at pH 7.4, 37°C for 10 min

Pyridostigmine pre-treatment increased the antidotal efficacy of oxime–atropine combinations in various studies, especially in soman-poisoned animals (Gordon *et al.*, 1978; Dirnhuber *et al.*, 1979; Inns and Leadbeater, 1983; Leadbeater *et al.*, 1985), but some reports indicate that pyridostigmine reduced the antidotal effect in sarin, VX and VR poisoning (Anderson *et al.*, 1992; Koplovitz and Stewart, 1992; Maxwell *et al.*, 1997).

Rapid aging of inhibited AChE is thought to be the major reason for the low efficacy of oximes in soman poisoning (Fleisher and Harris, 1965). Surprisingly, protective ratios between 6 and 8 were recorded in soman-poisoned mice, rats and guinea-pigs treated with HI 6 and atropine (Kepner and Wolthuis, 1978; De Jong and Wolring, 1984; Maxwell and Brecht, 1991). It should be emphasized that the aging half-time of soman-inhibited AChE varies substantially between species, i.e. 1–2 min in humans and non-human primates (Shafferman *et al.*, 1996), 8–10 min in rats and guinea-pigs (Talbot *et al.*, 1988) and 16 min in rabbits (Boskovic *et al.*, 1968). These marked species differences in aging kinetics of soman-inhibited AChE, which seem to be of minor importance with AChE inhibited by other nerve agents (Harris *et al.*, 1966; Smith and Usdin, 1966; Maxwell *et al.*, 1997; Worek *et al.*, 2004), have to be taken into account when extrapolating animal data to humans.

The notable therapeutic efficacy of HI 6 in soman- and tabun-poisoned non-human primates in the absence of significant AChE reactivation (Hamilton and Lundy, 1989) was attributed in part to a non-reactivating, 'direct' oxime effect (van Helden *et al.*, 1996). This assumption was supported by mechanistic studies *in vivo* and *in vitro* in rats after exposure with rapidly aging OPs (Busker *et al.*, 1991; Melchers *et al.*, 1991; van Helden *et al.*, 1991), but could not be verified in experiments with soman-exposed human intercostal muscle (Tattersall *et al.*, 1998). At any rate, the suspected 'direct' effects were only observed at unrealistically high HI 6 concentrations. Hence, there is at present no convincing evidence that 'direct' oxime effects may be of therapeutic significance in humans.

EFFICACY OF OXIMES AGAINST NERVE AGENTS IN HUMANS *IN VIVO*

There are only a few reports in the open literature on the effect of oximes in nerve agent-exposed humans. Pralidoxime chloride was very effective in reactivating erythrocyte AChE in individuals exposed to sublethal intravenous or oral VX while this oxime was substantially less effective in humans exposed to IV sarin (Sidell and Groff, 1974). Accidental sarin exposure by inhalation resulted in an initial progressive deterioration (coma, apnea) of the patient despite atropine and 2-PAM treatment and substantial recovery of erythrocyte AChE activity (Sidell, 1974). It took several hours until the patient's condition improved. Sidell also reported an accidental oral soman exposure. A lethal dose of diluted soman splashed into and around the mouth of an individual, resulting in coma, bronchoconstriction and respiratory depression, which was successfully treated with repeated atropine injections. 2-PAM (2 g IV) had no effect on inhibited erythrocyte AChE.

STABILITY OF OXIMES AND PHARMACEUTICAL ASPECTS

The stability of the reactivating pyridinium aldoximes in aqueous solution varies widely. In general, decomposition takes place via two pH-dependent mechanisms. At pH values below 3, the oximes are hydrolyzed to the corresponding aldehydes and hydroxylamine. Various states of equilibrium between the oxime and the hydrolysis products were established, depending on pH, temperature and oxime concentration. Generally, concentrated solutions, as used for autoinjectors, are less stable and degradation products sometimes accelerate decomposition. Buffer salts and the material the container is made of (metal, rubber stoppers) may accelerate decomposition (Ellin, 1982; Hartwich, 2004).

Above pH 4, the attack at the oxime group is catalyzed either by water or hydroxide ions, resulting in an intermediate pyridinium nitrile. The latter is highly unstable and decomposes into a pyridone derivative and free cyanide or forms the carboxamide, which further degrades into the carboxy derivative and ammonia. Besides these general routes some additional pathways have been detected, partly depending on the oxime studied.

Pralidoxime

Analysis of aqueous 2-PAM chloride solutions (300 mg ml^{-1}) that were stored at room temperature for 10 years yielded 91% 2-PAM, 6.5% 2-carboxy-1-methylpyridinium chloride, 2.1% 2-aminocarbonyl-1-methylpyridinium chloride and less than 7 μg ml^{-1} cyanide. The pH of the solution had dropped from 4.2 to about 1.0, which was attributed to the 2-carboxy-1-methylpyridinium chloride (Schroeder *et al.*, 1989). For more details of 2-PAM decomposition the reader is referred to the specialist literature (Ellin, 1982; Fyhr *et al.*, 1986; Holcombe, 1986; Utley, 1987).

Obidoxime

Analysis of aqueous obidoxime (250 mg ml^{-1}) stored at approximately 20°C for 19 years yielded 92% obidoxime, 2.5% of the presumed monocarboxy derivative, 0.16% formaldehyde and only traces (0.7 μg mL^{-1}) of cyanide, while the pH had dropped to about 3.5 (Spöhrer and Eyer, 1995). Formation of free formaldehyde may accelerate the decomposition of obidoxime and it has been argued that formaldehyde reacting with the liberated hydroxylamine may shift the equilibrium in favour of the aldehyde and eventually to the carboxylic acid (Rubnov *et al.*,

1999b). The decomposition of obidoxime has been comprehensively studied by Christenson (Christenson, 1968a,b, 1972).

Trimedoxime and methoxime

A similar autocatalytic phenomenon of oxime degradation has not been observed with trimedoxime, yet trimedoxime degraded faster when formaldehyde was intentionally added (Rubnov *et al.*, 1999a). Interestingly, methoxime, which is the monomethyl analogue of trimedoxime, was very much less stable and showed 25% degradation within 3 weeks when stored in 50 mM sodium phosphate, pH 7, at room temperature. During that time, less than 1% of trimedoxime had decomposed. The pronounced lability of methoxime has been attributed to the acidic methylene protons of the bridge, exhibiting rapid deuterium exchange and unusual downfield shifts in the NMR spectrum. Such a behaviour suggests easy hydrolysis involving an S_N2 displacement of the pyridinium ring by a water molecule (Lin and Klayman, 1986).

HI 6 and HLö 7

Dilute aqueous solutions of HI 6 and HLö 7 (pH 2.5) mainly decompose by nucleophilic attack of another oximate molecule on the methylene group of the aminal–acetal bridge, yielding a pyridine(bis)aldoxime, formaldehyde and isonicotinamide; in addition, hydrolysis of the isonicotinamide group was observed (Eyer *et al.*, 1986,1988,1989; Korte and Shih, 1993; Hartwich, 2004). In concentrated aqueous HI 6 solutions (dichloride or dimethanesulfonate, 400 mM, pH 3.2), isonicotinamide deamination was the predominant pathway. Interestingly, HI 6 degradation, and formation of the deamination product, showed an autocatalytic period, pointing to formation of decomposition products acting as catalysts. Formation of free hydroxylamine was considered a candidate. However, free hydroxylamine was always < 2% during the degradation process of HI 6 and did not contribute at this low concentration (Korte and Shih, 1993). On the other hand, the pH of concentrated, unbuffered solutions of HI 6 dropped from 3.4 to 2.2 at the beginning of the decomposition, presumably due to formation of the highly acidic carboxylic acid derivative. The structural analogue, 1-methyl-isonicotinic acid, reportedly has a pK_a value of 1.72 (Black, 1955) and a pK_a value of 2.4 was determined for deaminated HI 6 (Korte and Shih, 1993). Thus, it is plausible that lowering the pH value during decomposition accelerates the degradation rate of HI 6. In fact, aqueous solutions of less pure HI 6 dichloride preparations had a lower initial pH value and decomposed without showing a lag phase (Korte and Shih, 1993).

Besides specific proton catalysis, general acid catalysis by oximes may contribute to the advanced degradation of concentrated pyridinium aldoxime solutions. In fact, addition of 2-PAM to HI 6 accelerated not only the cleavage of the aminal–acetal bridge of the latter (Eyer *et al.*, 1988), but also isonicotinamide deamination (Korte and Shih, 1993). Due to this 'molecular canibalism', concentrated solutions of HI 6 and HLö 7 seem to be condemned to increased degradation and cannot be stockpiled in dissolved form.

For more in-depth information the reader is referred to the specialist literature (Eyer *et al.*, 1986,1988; Mdachi *et al.*, 1990; Korte and Shih, 1993; Hartwich, 2004); HLö 7 (Eyer *et al.*, 1989,1992).

Table 3 shows the approximate shelf-lives of oxime preparations as anticipated for 'ready-to-use' formulations. It is obvious that aqueous solutions of trimedoxime dichloride and obidoxime dichloride and, with certain reservations, 2-PAM chloride, are sufficiently stable and therefore suitable to be used in autoinjectors. From practical and economical considerations it is of minor importance if 10% of active material is lost during storage, provided that no toxic decomposition products arise. 2-PAM solutions subjected to higher temperatures and showing considerable decomposition were less toxic to mice than freshly prepared solutions (Kondritzer *et al.*, 1961; Barkman *et al.*, 1963). In fact, the most toxic product expected, cyanide, was present at less than 0.05%.

This situation changes when pyridinium oximes decompose under physiological conditions (dilute solutions, pH 7.4, 37°C). In fact, pralidoxime iodide (1 $mg\,ml^{-1}$) decomposes

Table 3. Shelf-lives of pyridinium aldoximes at 25°C (10% decomposition)

Compound	Concentration (%)	Initial pH	$t_{90\%}$ (years)	Activation energy (kJ mol^{-1})	Reference
Pralidoxime chloride	50	3.6	3.4	113	Ellin, 1982
(in autoinjectors)	30	4.2	10	—	Schroeder *et al.*, 1989
Trimedoxime dibromide	11.4	3.4	14	90	Rubnov *et al.*, 1999[a]
Obidoxime dichloride	25	2.45	38	110	Rubnov *et al.*, 1999[b]
HI 6 dichloride	37	2.5	0.055	93	Eyer *et al.*, 1988
(in autoinjectors)	13	3.6	0.148	—	Hartwich, 2004
HLö 7 dimesilate	52	2.5	0.02	85	Eyer *et al.*, 1992

much faster, the half-life being 8 days (37°C, pH 7.4) (Ellin *et al.*, 1962). Because the biological half-life of pralidoxime in man is about 75 min (Sidell *et al.*, 1972; Josselson and Sidell, 1978), abiotic transformation is expected to contribute to pralidoxime elimination by only 1%. HI 6 and HLö 7, the least stable bispyridinium compounds, yielded 0.6 equivalents of cyanide per mole of decomposed oxime, provided cyanide was trapped in alkali. Since the decomposition under physiological conditions is rather fast (half-life 12 h, pH 7.4, 37°C) it was calculated that about 4% cyanide (on a molar basis) would be produced from a therapeutic HI 6 dose by abiotic transformation at a biological half-life of 0.9 h, as is found in dogs. Such an amount was found in dogs (50 μmol HI 6/kg IV) when cyanide was trapped in methaemoglobin-containing erythrocytes (Eyer *et al.*, 1987). As the biological half-life of HI 6 in man is about 80 min (Kusic *et al.*, 1985; Clement *et al.*, 1992a,1994), abiotic transformation of HI 6 is expected to contribute by up to 8% of the dose.

To circumvent the problem of limited stability of HI 6 and HLö 7, dry/wet autoinjectors have been developed in which the unstable oxime is dissolved by an atropine-containing diluent in an adjacent chamber upon activation of the device (Schlager *et al.*, 1991; Thiermann *et al.*, 1994,1995). Such devices, filled with 500 mg of HI 6 dichloride and 2 mg of atropine sulfate, are now on the market for military use and were fielded by the Canadian troops in the Second Gulf War (Dawson, 1994). While the pharmaceutical stability may be sufficient and a shelf-life of 5 years has been envisaged (Brodin and Wellenstam, 1996), HI 6 dichloride appears less suitable for dry/wet autoinjectors since it dissolves only slowly in the cold, i.e. below 10°C. In contrast, HI 6 dimethanesulfonate is 5 times more water-soluble and has been proposed as a substitute for HI 6 dichloride (Thiermann *et al.*, 1995a,1996). HI 6 dimethanesulfonate and the dichloride have been shown to be equieffective in reactivating human AChE inhibited by paraoxon, sarin and VX (Krummer *et al.*, 2002). Since methanesulfonate is used as the anion in the preparation of pralidoxime (P2S), used in the United Kingdom, HI 6 dimethanesulfonate may be the better choice compared to the dichloride.

PHARMACOKINETICS OF THE PYRIDINIUM OXIMES

Most pharmacokinetic data stem from volunteer studies and do not necessarily reflect oxime kinetics in severely poisoned patients.

Distribution

With the exception of pralidoxime, which also penetrates cell membranes and is found within red blood cells, the pyridinium oximes, being ionized, usually distribute in the extracellular fluid. The transfer of pralidoxime, however, is slow with a half-life of about 4.5 h (Ellin *et al.*, 1974). By contrast, bispyridinium oximes such as obidoxime, trimedoxime and HLö 7 did not penetrate the red cell membrane to any appreciable extent (Ellin *et al.*, 1974; Spöhrer, 1994). The apparent volumes of distribution are in agreement with these findings: the Vd_{ss} for pralidoxime was $0.7 \pm 0.1\, l\, kg^{-1}$ in human volunteers (Sidell *et al.*, 1972; Josselson and Sidell, 1978), but

$0.17 \pm 0.02\ l\,kg^{-1}$ for obidoxime (Sidell *et al.*, 1972) and $0.25 \pm 0.04\ l\,kg^{-1}$ for HI 6 in healthy young men (Clement *et al.*, 1995). A similar V_d of 0.25 and 0.26 $l\,kg^{-1}$ was found in dogs after IM injection of HI 6 and HLö 7, respectively, administered along with 2 mg of atropine sulfate by dry/wet autoinjectors (Spöhrer *et al.*, 1994). The Vd of methoxime injected IM into rabbits and pigs was 0.27 and 0.36 $l\,kg^{-1}$, respectively (Stemler *et al.*, 1991; Woodard and Lukey, 1991).

In OP-poisoned patients, Willems and coworkers found a volume of distribution for pralidoxime of $2.77 \pm 1.45\ l\,kg^{-1}$ in the ß-phase after 4.4 mg kg^{-1} IV followed by continuous infusion at 2.1 mg $kg^{-1}h^{-1}$ (Willems *et al.*, 1992). In OP-poisoned patients treated with an obidoxime bolus 3.5 mg kg^{-1} IV followed by continuous infusion at 0.45 mg $kg^{-1}h^{-1}$ the volume of distribution at steady state was $0.66 \pm 0.23\ l\,kg^{-1}$ ($n = 27$; unpublished results). An even larger Vd of obidoxime, 0.85 $l\,kg^{-1}$, was reported in a methamidophos-poisoned patient with renal failure (Bentur *et al.*, 1993). These data point to a deep 'compartment' that is filled only slowly.

Autoradiographic studies in rats revealed that radioactivity from radiolabelled pralidoxime, obidoxime, trimedoxime and HI 6 was concentrated in the kidneys and mucopolysaccharide-containing tissue such as intervertebral discs and other cartilaginous structures. The enrichment of cationic substances in proteoglycan-rich tissue is not unusual and most probably based on ionic interactions (Garrigue *et al.*, 1991; Maurizis *et al.*, 1992). In a patient who had received obidoxime for 12 days and died 15 days after parathion poisoning, obidoxime concentrations in cartilage were 100-fold higher than in plasma (unpublished). These data indicate that the chondroitin sulfate-rich tissues may represent the deep 'compartment' from which pyridinium aldoximes are slowly released.

A more important issue is the permeability of the blood–brain barrier to pyridinium oximes. Pralidoxime was not detected in the spinal fluid of a 36-year-old epileptic man after a 1h infusion of pralidoxime iodide (44 mg kg^{-1}) (Jager and Stagg, 1958). On the other hand, pralidoxime chloride (500 mg IV over 15 min) was presumed to have acted centrally in a 3.5-year-old child with parathion poisoning, because of a dramatic improvement of the EEG pattern along with a simultaneous regain of consciousness (Lotti and Becker, 1982). Obidoxime was detected at a concentration of 4 $\mu mol\,l^{-1}$ in the CSF of a parathion-poisoned patient on day 6 after continuous infusion of obidoxime, at a time when the plasma concentration was 17 $\mu mol\,l^{-1}$. Obidoxime was also detected in the frontal cortex of a deceased patient following parathion poisoning with 0.85 nmol g^{-1} wet weight at a plasma concentration of 0.5 nmol ml^{-1}, four days after discontinuation of obidoxime infusion (unpublished). Animal experiments with dogs and rats have shown that HI 6 could be detected in brain and CSF (Ligtenstein and Kossen, 1983; Klimmek and Eyer, 1986), and Clement showed that HI 6 was able to normalize the decreased body temperature observed after sarin poisoning, which he attributed to a central effect (Clement, 1992). All these data indicate that the blood–brain barrier is not entirely impermeable to quaternary pyridinium oximes and may become even more permeable in severe intoxication (Grange-Messenet *et al.*, 1999).

METABOLISM AND EXCRETION

When 1-[^{14}C]methyl-pyridinium aldoxime iodide or radioactive pralidoxime [^{14}C]-labelled in the oxime group was parenterally administrated to rats, 90% of the radioactivity was recovered in urine and 6% in the faeces, irrespective of the position of the label. About 90% of the urinary radioactivity was associated with intact pralidoxime. In addition, some 5% of the dose was excreted as 1-methyl-2-pyridone, indicating some cyanogenesis (Enander *et al.*, 1962). In humans, the 1-methyl-2-cyanopyridinium ion was detected in urine of male volunteers without significantly increased urinary thiocyanate. Since 90% of pralidoxime chloride, 5 mg kg^{-1} IV, was recovered from urine, cyanide formation is probably of no toxicological concern (Garrigue *et al.*, 1990).

Obidoxime is chemically more stable than pralidoxime and no metabolites have been found hitherto. In dogs ($n = 4$) following obidoxime

10 mg kg^{-1} IV, 77% of the dose was recovered unchanged in the urine over 24 h (Spöhrer, 1994). In human volunteers ($n = 5$) who had received 5 mg kg^{-1} obidoxime IV, 70% of the dose was excreted unchanged in the urine in 24 h (Sidell *et al.*, 1972). In OP-insecticide-poisoned patients treated with obidoxime (250 mg IV bolus followed by continuous infusion at 750 mg $(24\ h)^{-1}$, 50–70% of the infused obidoxime could be recovered in urine in steady state conditions (unpublished results). Although the urine chromatograms were screened, it was not possible to identify material indicative of a pyridinium compound (Eyer, unpublished data). Since obidoxime was still eliminated in a patient suffering from complete anuria, some elimination by the biliary/faecal route cannot be ruled out (Spöhrer, 1994; Eyer 1996). Urinary recovery of trimedoxime IM was 80% in healthy volunteers (Vojvodic and Boskovic, 1976).

As mentioned above, the Hagedorn oximes HI 6 and HLö 7, when dissolved in water, are chemically less stable than obidoxime. When 50 mg kg^{-1} IV was administered to rats only 57% of the dose of HI 6 was recovered unchanged in the urine over 24 h. Two metabolites, very probably both with a pyridone structure, were identified comprising some 20% of the dose administered (Ligtenstein *et al.*, 1987). In dogs, following HI 6 38 mg kg^{-1} adminstered IV, 80% of the dose was recovered unchanged in the 6 h urine. On a molar basis, about 4% of the HI 6 was found as cyanide (trapped in methaemoglobin-containing red blood cells) (Eyer *et al.*, 1987). This amount was expected from abiotic transformation of HI 6 via the cyanogenic pathway (see above). In a subsequent study with [^{14}C]carbonyl-labelled HI 6 administered to dogs 38 mg kg^{-1} IV, 94% of the radioactivity was found in the urine excreted within 5 days, 83% being associated with unchanged HI 6. The remainder was found as the pyridone derivative (5%) and as the deaminated carboxylic acid of HI 6 (2%). The presence of free isonicotinamide (3%) and isonicotinic acid (1%) pointed to cleavage of the aminal–acetal bridge (Ladstetter, 1990). Most of this 'metabolism' apparently stems from abiotic transformation and can be expected to occur similarly in humans. When HI 6 dichloride 31 mg kg^{-1} IM was injected into dogs together with atropine sulfate 0.125 mg kg^{-1} using autoinjectors, 70% of the dose was excreted unchanged with the 24 h urine (Spöhrer *et al.*, 1994). Data on HLö 7 are scarce. In one study with dogs ($n = 15$), receiving HLö 7 dimethanesulfonate 13 mg kg^{-1} IM together with atropine sulfate 0.125 mg kg^{-1}, 78% of the dose was excreted unchanged in the urine, within 24 h (Spöhrer *et al.*, 1994). Hence, a metabolic pattern similar to HI 6 can be expected.

Elimination

The elimination rate of the various pyridinium oximes in human volunteers is very similar, with a half-life of around 80 min (Table 4). In OP-intoxicated patients, however, the elimination rate is considerably reduced with a mean elimination half-life of about 160 min. It can be expected that impairment of renal function in severe intoxications is a major contributor of the retarded elimination. Human data on methoxime and HLö 7 elimination are not available. From experiments with pigs it appears that methoxime is eliminated at a slower rate than pralidoxime (157.6 ± 107.9 vs. 86.8 ± 14.7 min) (Stemler *et al.*, 1991). Similarly, trimedoxime elimination in human volunteers was somewhat delayed. After IM administration, plasma concentration fell with a mean half-life of 128 min ($n = 14$) (Vojvodic and Boskovic, 1976).

SAFETY AND TOLERABILITY OF THERAPEUTIC OXIMES

Reactivating oximes have been designed to fit optimally into the active site of the enzyme acetylcholinesterase to maximize dephosphylation. This property inevitably implies competition with the substrate of AChE. The K_i values describing the dissociation constant of the oxime from the substrate-free enzyme is around 300 μM, while the K_{ii} value describing the dissociation constant of the oxime from the enzyme–substrate complex is about one order of magnitude higher (Mast, 1997; Eyer, 2003). That means 10 μM oxime is virtually without effect while 100 μM is expected to inhibit AChE to an appreciable extent. Such a peak concentration may

Table 4. Elimination rate constants of oximes in humans

Oxime	Dose ($mg\,kg^{-1}$)	Route	$t_{1/2el}$ (min)	Reference
Volunteers				
PAM Chloride	2.5–10	IV	73.7 ± 15.2	Creasey and Green, 1959
PAM Chloride	5.0	IV	78.6 ± 7.8	Sidell *et al.*, 1972
PAM Mesilate	5.0	IV	84.6 ± 14.4	Sidell *et al.*, 1972
PAM Chloride	10.0	IM	79	Sidell and Groff, 1971
PAM Chloride	12.5	IM	150 (96[a])	Jovanovic, 1989
Patients				
PAM Chloride	13.2	IM	174 ± 71	Jovanovic, 1989
PAM Methylsulfate	$2.14\ h^{-1}$	IV	206 ± 54	Willems *et al.*, 1992
Volunteers				
Obidoxime dichloride	0.5–1.0	IV	72 ± 9.6	Sidell *et al.*, 1972
Obidoxime dichloride	2.5–10	IM	83	Sidell and Groff, 1970
Obidoxime dichloride	3.5	IM	120	(Erdmann *et al.*, 1965)
Patients				
Obidoxime dichloride	$3.5 + 0.45\ h^{-1}$	IV	129 ± 77	Eyer, 2003
Volunteers				
HI 6 dichloride	3.5	IM	84.9	Kusic *et al.*, 1985
HI 6 dichloride	7	IM	79.9	Kusic *et al.*, 1985
HI 6 dichloride	7	IM	64.4 ± 6.4	Clement *et al.*, 1994
Patients				
HI 6 dichloride	7	IM	120	Kusic *et al.*, 1991

Note:
[a] Recalculated (Eyer, 2003).

be reached by IM injection of two 600 mg P2S autoinjectors (Stemler *et al.*, 1991).

Animal studies

The acute toxicity of the various oximes dealt with in this chapter has been tested in a variety of animal strains of both sexes and using several routes of administration. For more in-depth information, the reader is referred to the review by Marrs (1991). An overview of the LD_{50} data of the various salts of pralidoxime is depicted in Figure 6. The slope of unity indicates that there is no marked sex difference between the LD_{50} values found in male and female animals. The LD_{50} following IM injection is roughly double that with IV administration.

There are no obvious differences in sensitivity among small laboratory animals. Figure 7 compares the LD_{50} data of various oximes after IM administration to various animals. It is clear that, on a molar basis, methoxime has the lowest LD 50 while HI 6 is probably the least toxic oxime.

The oximes apparently do not exhibit direct cytotoxic actions. There had been some concern about the cyanogenic potential of the aldoximes. This is probably not an issue with pralidoxime, obidoxime, trimedoxime and methoxime but

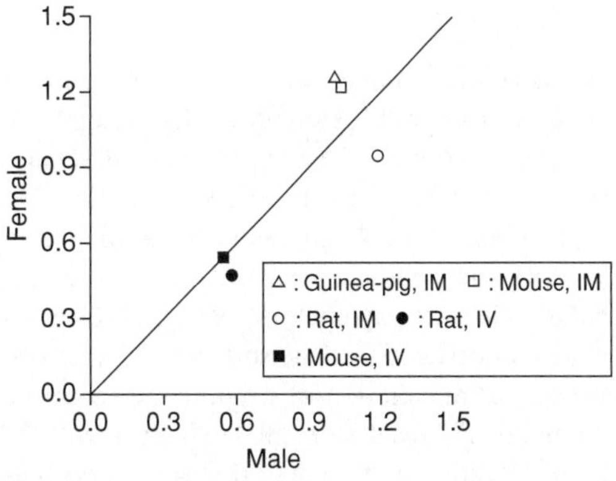

Figure 6. Toxicity of 2-PAM after IM administration into various rodents of both sexes. LD_{50} values are given in $mmol\,kg^{-1}$; a slope of unity has been drawn through the origin (Marrs, 1991; Clement *et al.*, 1992b; Eyer *et al.*, 1992; Kassa and Cabal, 1999)

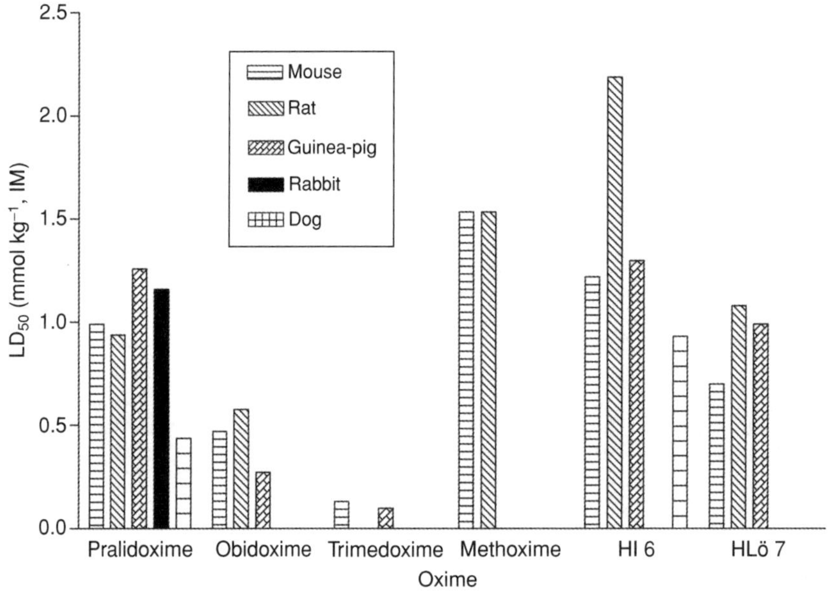

Figure 7. Toxicity of oximes in various species after IM administration. LD_{50} values are means of both sexes where available (for references, see Figure 6)

could pose some problems with the more unstable Hagedorn oximes that degrade rapidly under physiological conditions. Thus, in shock, when HI 6 elimination may be impeded, up to 0.4 mmol cyanide may be liberated from 0.5 g HI 6 within 12 h. This amount corresponds to 10 mg HCN, an amount that would normally be detoxified within 2.5 h (Schulz *et al.*, 1982).

Human studies

When young volunteers ($n = 18$) were given a single dose of pralidoxime iodide 15–30 mg kg^{-1} IV over 2–4 min, they experienced dizziness ($n = 8$), blurred vision ($n = 13$), diplopia ($n = 9$), headache ($n = 4$), impaired accommodation ($n = 4$) and nausea ($n = 3$) (Jager and Stagg, 1958). The same symptoms also were observed with pralidoxime methanesulfonate (Sundwall, 1960). Administration of pralidoxime chloride 30 mg/kg IM resulted in ECG changes with T-wave elevation and increased systolic and diastolic blood pressure. Headache, altered accommodation and vomiting also occurred (Calesnick *et al.*, 1967). In another study, IM injection of pralidoxime chloride up to 600 mg caused no changes in heart rate or blood pressure and only mild pain at the site of injection (Sidell and Groff, 1971). Burning and stinging at the injection site following 0.5–1.0 g pralidoxime chloride IM also was reported by others (Xue *et al.*, 1985). Volunteers experienced dizziness and blurred vision when plasma pralidoxime levels approached 80 μM (Medicis *et al.*, 1996). Rapid injection was also associated with tachycardia, laryngeal spasm, muscle rigidity and transient neuromuscular blockade (Ellenhorn and Barceloux, 1988). Hence it appears prudent to avoid excessive peak plasma concentrations that may elicit toxic symptoms. A particularly serious effect was related to pralidoxime iodide infusion in a coumaphos-poisoned patient who experienced a cardiac arrest in asystole following infusion of 0.4 g over 2 min. He responded quickly to intravenous sodium bicarbonate and adrenaline. However, when the pralidoxime infusion was recommenced, the patients had a second cardiac arrest within 2 min (0.4 g). The reason of this serious adverse event remains obscure (Scott, 1986).

Obidoxime chloride 250mg IM elicited a different profile of adverse effects in humans, including a hot and tight feeling in the orofacial region, along with numbness. In addition, a cold sensation in the rhino-pharyngeal region was noted, similar to the effect of menthol. The symptoms usually developed within 15 min of injection (Erdmann *et al.*, 1965; Sidell and Groff, 1970; Simon *et al.*, 1976). At a higher dose, 10 mg kg^{-1} IM, a dry mouth and hot feelings in the throat were observed in all subjects who also noticed paraesthesia of variable severity. All subjects had a mild-to-moderate transient increase in systolic and diastolic blood pressure along with some tachycardia (increase in heart rate 30 bpm after 10 mg kg^{-1} IM) (Sidell and Groff, 1970). In another study, 11 volunteers of both sexes received obidoxime 250 mg IM to screen for possible hepatotoxicity. No abnormal liver function data were found in the 1 month follow-up period. The volunteers experienced the usual sensations as already described (Boelcke *et al.*, 1970). In severely OP-poisoned patients treated with obidoxime, several grams per day, hepatotoxic effects were occasionally observed, including cholestasis, jaundice and elevated serum aminotransferases (Prinz, 1967; von Gaisberg and Dieterle, 1967; Boelcke *et al.*, 1970). In a recent multicenter study of severely OP-poisoned patients lead from Munich, all patients received a 250 mg obidoxime short infusion followed by continuous infusion of 750 mg obidoxime over 24 h. No patient with dimethoate poisoning ($n = 6$) had pathological liver findings, five patients with oxydemeton-methyl poisoning ($n = 12$) showed transient liver pathology, while eight patients with parathion poisoning ($n = 13$) had abnormal liver function tests. Hence, it cannot be ruled out that the type of OP and the other ingredients of the formulation may contribute to the liver pathology. Cardiac arrhythmias have occasionally been associated with obidoxime therapy: 1/17 had a second-degree atrioventricular block, while 2/17 had atrial fibrillation (Xue *et al.*, 1985). We also observed some transient arrhythmias in our Munich study but did not consider these effects to be related to obidoxime (unpublished data).

Intramuscular and intravenous trimedoxime dichloride 15 or 30 mg kg^{-1} caused a long-lasting fall in systolic and diastolic blood pressure which was also observed after oral trimedoxime (4 g); ECG changes were not reported. Plasma concentration peaked between 6 and 9 μM after oral administration and 100 μM trimedoxime after intravenous administration after which the blood pressure decreased. Chronic oral administration of trimedoxime caused hepatotoxic effects with icterus, elevated serum liver enzymes, petechial bleeding and a prothrombin time of 40 and 50% of normal (Calesnick *et al.*, 1967). It appeared to the investigators that trimedoxime dichloride was six to ten times as toxic as pralidoxime chloride. Palpitations and dilation of superficial capillaries has been reported after only 0.15 g trimedoxime IV (Xue *et al.*, 1985). Short-term IV infusion of trimedoxime dibromide 10 mg kg^{-1} in male and female healthy volunteers over 5 to 8 min ($n = 33$) resulted in sensations of warmth and numbness in the orofacial region, slight dizziness and mild headache in some 80% of the volunteers. Mean blood pressure increased by 10 mmHg and heart rate by 45% during infusion. These changes were not observed with pralidoxime iodide (18 mg kg^{-1}) (Wiezorek *et al.*, 1968). A comparative study of pralidoxime chloride 1 g, obidoxime dichloride 0.25 g and trimedoxime dichloride 0.25 g IM in healthy male volunteers ($n = 14$) showed very similar adverse effects for all three compounds. The oximes invariably produced pain ('burning') at the site of injection and sensations in the facial region (see above). The effects on heart rate and blood pressure were small throughout (Vojvodic and Boskovic, 1976).

In a recent evaluation of pediatric poisoning from trimedoxime and atropine-containing autoinjectors in Israel, no serious side effects were associated with exposure to the oxime. Twenty two children who inadvertently injected more than the recommended, age-adjusted dose (1 mg atropine and 40 mg trimedoxime for children 3–8 years or double that dose for persons > 8 years) did not develop severe adverse reactions, nor did they require medical intervention (Kozer *et al.*, 2005). HI 6 dichloride was tested in a double-blind, placebo controlled, ascending dose-tolerance study (HI 6 + 2 mg atropine sulfate) in 24 healthy male volunteers (Clement and Erhardt, 1994). Doses from 62.5 up to 500 mg were well tolerated by the subjects without serious complaints. There were no clinically significant changes in heart rate or ECG, respiration or blood pressure or visual and mental acuity following 500 mg HI 6 plus atropine IM. The changes in aspartate aminotransferase (AST), creatine phosphokinase (CPK), creatinine and gamma-glutamyl transferase (gamma-GT) following the highest HI 6 dose were considered clinically insignificant. The increase in CPK that has already been observed in an earlier study with 3 volunteers at 500 mg HI 6 (Clement *et al.*, 1992a) was attributed to the IM injection of a hypertonic solution of HI 6 that had produced mild pain at the injection site. Good tolerability of HI 6 500 mg IM administered to 22 healthy servicemen has already been reported earlier (Kusic *et al.*, 1985). In a dimethoate-poisoned patient, HI 6 dichloride was administered by continuous infusion 4 g day^{-1} (20 g total) apparently without adverse effects (Jovanovic *et al.*, 1990). No human data on tolerability of methoxime and HLö 7 were found in the open literature.

In conclusion, a rough estimate of the lowest adverse effect level (LOAEL), on a molar basis, may be provided for adults – pralidoxime chloride: HI 6 dichloride: obidoxime dichloride: trimedoxime dichloride = 2:1.3:0.4:0.25 mmol.

ASSESSMENT OF THERAPEUTIC OXIME EFFECTS BY THEORETICAL MODELS

The relative lack of human data on the treatment of nerve-agent poisoning and the difficulties in extrapolating data from pharmacodynamic animal studies to humans has prompted the development of theoretical models for the estimation of the therapeutic efficacy of antidotes in humans. Until recently, mainly physiologically based pharmacokinetic models of nerve agents (de Jong *et al.*, 1996; Langenberg *et al.*, 1997) and models for the estimation of the protective effect of scavengers were available (Sweeney and Maxwell, 1999,2003; Ashani and Pistinner, 2004). In order to overcome the shortcomings of these methods, a dynamic model was developed by combining toxicokinetics of nerve agents, pharmacokinetics of oximes and enzyme kinetics (inhibition, reactivation and aging) which enables the estimation of AChE changes in different scenarios (Worek *et al.*, 2005).

In doing so, the estimation of the effect of nerve-agent exposure and oxime treatment on AChE activity is hampered by the lack of human toxicokinetic data. Nevertheless, such model calculations may provide valuable information on the ability of different oximes to reactivate human AChE.

Poisoning by cyclosarin is thought to be a therapeutic challenge due to the high inhibitory activity of this agent and the low reactivating efficacy of the oximes that are generally available, obidoxime and 2-PAM (Sidell, 1992; Worek *et al.*, 2004). In fact, the calculation of AChE status after intravenous exposure to cyclosarin and simultaneous intramuscular oxime administration demonstrates a rather small effect of both obidoxime and 2-PAM (Figure 8). Due to the low affinity of these oximes towards cyclosarin-inhibited AChE (see Table 2), an increase in oxime dose, i.e. two or three autoinjector equivalents, has only a small effect. In contrast, HI 6 should be an effective reactivator of cyclosarin-inhibited AChE.

Sidell and coworkers (Sidell and Groff, 1974) showed reasonable efficacy of 2-PAM in VX-exposed humans and, according to *in vitro* reactivation data (see Table 2), adequate reactivation of VX-inhibited AChE by obidoxime, 2-PAM and HI 6 can be expected. However, the high percutaneous toxicity of VX (Craig *et al.*, 1977; Sidell, 1997), its long-lasting penetration through skin and its long persistence in the body (Duncan and Griffith, 1992; Chilcott *et al.*, 2003; van der Schans *et al.*, 2003) may affect the therapeutic efficacy of oximes. Actually, the calculation of AChE status after percutaneous VX exposure and intramuscular oxime administration shows a progressive decrease in AChE activity due to the long persistence of VX and rapid elimination of the oximes, leading to re-inhibition of reactivated enzyme (Figure 9). Obviously, additional oxime injections would be necessary to preserve an adequate proportion of the enzyme in an active state.

CONCLUSIONS

The rapid onset of intoxication by nerve agents requires fast administration of an antidote

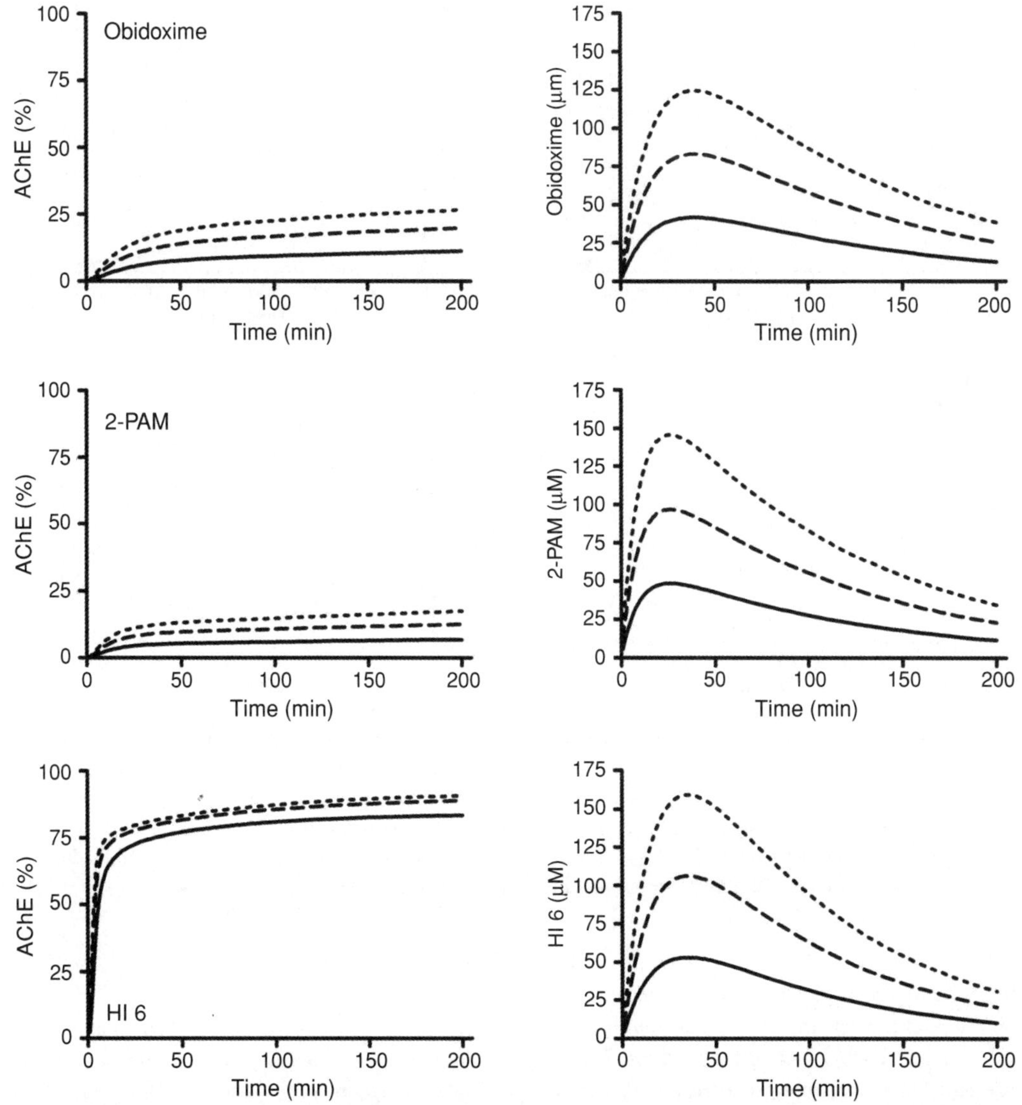

Figure 8. Calculated human AChE activities (left-hand side) after intravenous cyclosarin and simultaneous intramuscular oxime injection. The changes in AChE activity were calculated by using the toxicokinetic data of (–)-sarin (Spruit *et al.*, 2000) and human pharmacokinetic data of the oximes (Worek *et al.*, 2005) for 1, 2 and 3 autoinjector equivalents (AIs) of obidoxime (1 AI = 250 mg), 2-PAM (1 AI = 600 mg) and HI 6 (1 AI = 500 mg) and exposure to 0.8 × LD_{50} cyclosarin ((—), 1 AI; (– – –), 2 AI; (- - - - - -), 3 AI). The calculated plasma oxime concentrations are shown on the right-hand side

'cocktail'. In the military context or with mass casualties, this can be done most conveniently by means of autoinjectors containing a cholinolytic such as atropine and an oxime. In addition a benzodiazepine may be administered separately or with the atropine/oxime combination. Sarin, cyclosarin and VX poisoning respond favourably to HI 6; tabun is notorious for producing poisoning which is quite resistant to oxime reactivation at the usual doses. Here, HI 6 is practically ineffective while trimedoxime and obidoxime may be of some value. The treatment of soman

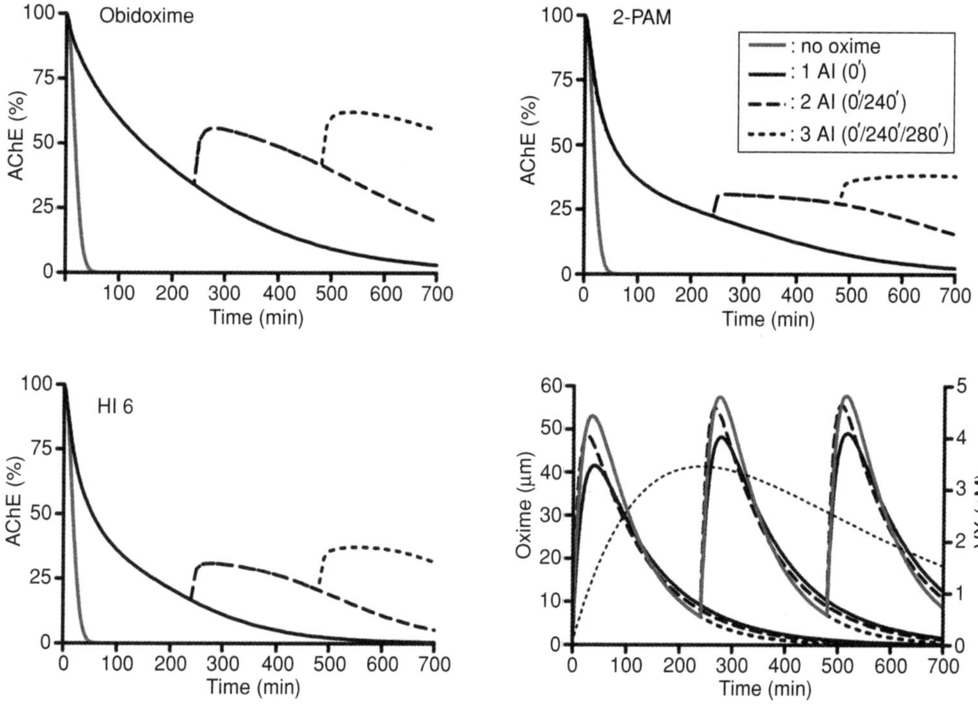

Figure 9. Calculated human AChE activities after percutaneous VX and simultaneous intramuscular oxime injection. The changes in AChE activity were calculated by using plasma VX concentrations, measured in guinea-pigs (van der Schans *et al.*, 2003) and human pharmacokinetic data of the oximes for 250 mg of obidoxime (continuous line), 600 mg of 2-PAM (broken line) and 500 mg of HI 6 ('grey' line) and exposure to 5 × LD_{50} VX (VX concentration given as dotted line, right-hand side, bottom graph). Oxime administration, 1 autoinjector (AI), was assumed to occur at 0, 240 and 480 min

poisoning is particularly fraught with problems because of the rapid aging of human enzyme, and all oximes are therefore presumed to be of little value. The 'direct' effects of HI 6 probably cannot be exploited at therapeutic concentrations of that drug. HLö 7 has apparently the broadest antidotal spectrum, but at present, there is no experience of human safety and tolerance. Methoxime was quite disappointing in terms of reactivation potency and apparently there is a lack of experience of methoxime in human use. Pralidoxime, although well tolerated, is a weak reactivator throughout and requires much higher doses than are delivered by a single autoinjector. Hence, HI 6 may be the favorite, if tabun exposure is less likely. It is to be remembered that the half-life of most oximes is quite short and further administrations at 4-h intervals should be considered. This is of particular concern in case of percutaneous poisoning.

REFERENCES

Anderson DR, Harris LW, Woodard CL *et al.* (1992). The effect of pyridostigmine pretreatment on oxime efficacy against intoxication by soman or VX in rats. *Drug Chem Toxicol*, **15**, 285–294.

Ashani Y and Pistinner S (2004). Estimation of the upper limit of human butyrylcholinesterase dose required for protection against organophosphates toxicity: a mathematically based toxicokinetic model. *Toxicol Sci*, **77**, 358–367.

Ashani Y, Bhattacharjee AK, Leader H *et al.* (2003). Inhibition of cholinesterases with cationic phosphonyl oximes highlights distinctive properties of the charged pyridine groups of quaternary oxime reactivators. *Biochem Pharmacol*, **66**, 191–202.

Ashani Y, Radic Z, Tsigelny I *et al.* (1995). Amino acid residues controlling reactivation of organophosphonyl conjugates of acetylcholinesterase by mono- and bisquaternary oximes. *J Biol Chem*, **270**, 6370–6380.

Bajgar J, Patocka J, Jakl A *et al.* (1975). Antidotal therapy and changes of acetylcholinesterase activity following isopropyl methylphosphonofluoridate intoxication in mice. *Acta Biol Med Germ*, **34**, 1049–1055.

Barkman R, Edgren B and Sundwall A (1963). Self-administration of pralidoxime in nerve gas poisoning with a note on the stability of the drug. *J Pharm Pharmacol*, **15**, 671–677.

Becker G, Kawan A and Szinicz L (1997). Direct reaction of oximes with sarin, soman, or tabun in vitro. *Arch Toxicol*, **71**, 714–718.

Bentur Y, Nutenko I, Tsipiniuk A *et al.* (1993). Pharmacokinetics of obidoxime in organophosphate poisoning associated with renal failure. *Clin Toxicol*, **31**, 315–322.

Berry WK (1971). Some species differences in the rates of reaction of diaphragm particulate acetylcholinesterase with tetraethyl pyrophosphate and pralidoxime. *Biochem Pharmacol*, **20**, 1333–1334.

Bismuth C, Inns RH and Marrs TC (1992). Efficacy, toxicity and clinical use of oximes in anticholinesterase poisoning. In: *Clinical and Experimental Toxicology of Organophosphates and Carbamates* (B Ballantyne and TC Marrs, eds), pp. 555–577. Oxford, UK: Butterworth-Heinemann.

Black ML (1955). The ionization constants of the pyridinium monocarboxylic acids: a reinterpretataion. *J Phys Chem*, **59**, 670–671.

Boelcke G, Creutzfeldt W, Erdmann WD *et al.* (1970). Untersuchungen zur Frage der Lebertoxizität von Obidoxim (Toxogonin®) am Menschen. *Dtsch med Wschr*, **95**, 1175–1178.

Boskovic B, Maksimovic M and Minic D (1968). Ageing and reactivation of acetylcholinesterase inhibited with soman and its thiocholine-like analogue. *Biochem Pharmacol*, **17**, 1738–1741.

Brodin A and Wellenstam K (1996). *Pharmaceutical and technical requirements, design, solutions and documentation of performance for binary autoinjectors*. CBMTS 2. Spiez, Switzerland: NC-Laboratory Spiez.

Busker RW, Zijlstra JJ, van der Wiel HJ *et al.* (1991). Organophosphate poisoning: a method to test therapeutic effects of oximes other than acetylcholinesterase reactivation in the rat. *Toxicology*, **69**, 331–344.

Calesnick B, Christensen JA and Richter M (1967). Human toxicity of various oximes. 2-Pyridine aldoxime methyl chloride, its methane sulfonate salt and 1,1′-trimethylenebis-(4-formylpyridinium chloride). *Arch Environ Health*, **15**, 599–608.

Chilcott RP, Dalton CH, Hill I *et al.* (2003). Clinical manifestations of VX poisoning following percutaneous exposure in the domestic white pig. *Human Exp Toxicol*, **22**, 255–261.

Childs AF, Davies DR, Green AL *et al.* (1955). The reactivation by oximes and hydroxamic acids of cholinesterase inhibited by organo-phosphorus compounds. *Brit J Pharmacol*, **10**, 462–465.

Christenson I (1968a). Hydrolysis of bis(4-hydroxyiminomethyl-1-pyridiniomethyl)ether dichloride (Toxogonin®). I. Decomposition products. *Acta Pharm Suec*, **5**, 23–36.

Christenson I (1968b). Hydrolysis of bis(4-hydroxyiminomethyl-1-pyridiniomethyl)ether dichloride (Toxogonin®). II. Kinetics and equilibrium in acidic solution. *Acta Pharm Suec*, **5**, 249–262.

Christenson I (1972). Hydrolysis of obidoxime chloride (Toxogonin®). III. Kinetics in neutral and alkaline solution. *Acta Pharm Suec*, **9**, 309–322.

Clement JG (1992). Central activity of acetylcholinesterase oxime reactivators. *Toxicol Appl Pharmacol*, **112**, 104–109.

Clement JG and Erhardt N (1994). *In vitro* oxime-induced reactivation of various molecular forms of soman-inhibited acetylcholinesterase in striated muscle from rat, monkey and human. *Arch Toxicol*, **68**, 648–655.

Clement JG, Pierce CH and Houle J-M (1992a). Ascending dose tolerance study of the oxime, HI-6, in man. In: *Proceedings of the 4th International Symposium on Protection against Chemical Warfare Agents*, 287–296, FAO Report A40067-4.6, 4.7. Stockholm, Sweden.

Clement JG, Hansen AS and Boulet CA (1992b). Efficacy of HLö-7 and pyrimidoxime as antidotes of nerve agent poisoning in mice. *Arch Toxicol*, **66**, 216–219.

Clement JG, Madill HD, Bailey D *et al.* (1994). *Clinical study of a new therapy for nerve agent poisoning: Ascending dose tolerance study of HI 6 and atropine*. Suffield Report, pp. 1–32. Suffield. AB, Canada: Defence Research Establishment.

Clement JG, Bailey DG, Madill HD *et al.* (1995). The acetylcholinesterase oxime reactivator HI-6 in man: pharmacokinetics and tolerability in combination with atropine. *Biopharm Drug Disposit*, **16**, 415–425.

Craig FN, Cummings EG and Sim VM (1977). Environmental temperature and the percutaneous absorption of a cholinesterase inhibitor, VX. *J Invest Dermatol*, **68**, 357–361.

Creasey HN and Green AC (1959). 2-Hydroxyiminomethyl-*N*-methyl-pyridiniummethansulfonate (P2S), an antidote to organophosphorus poisoning. Its preparation, estimation, and stability. *J Pharm Pharmacol*, **11**, 485–490.

Dawson RM (1994). Review of oximes available for treatment of nerve agent poisoning. *J Appl Toxicol*, **14**, 317–331.

De Jong LPA and Ceulen DI (1978). Anticholinesterase activity and rate of decomposition of some phosphylated oximes. *Biochem Pharmacol*, **27**, 857–863.

De Jong LPA and Wolring GZ (1984). Stereospecific reactivation by some Hagedorn-oximes of acetylcholinesterases from various species including man, inhibited by soman. *Biochem Pharmacol*, **33**, 1119–1125.

De Jong LPA, Verhagen MAA, Langenberg JP *et al.* (1989). The bispyridinium-dioxime HLö 7. A potent reactivator for acetylcholinesterase inhibited by the stereoisomers of tabun and soman. *Biochem Pharmacol*, **38**, 633–640.

De Jong LPA, Langenberg JP, van Dijk C *et al.* (1996). Development of a physiologically based model for the toxicokinetics of C(+−)P(+−)-soman in the atropinized guinea pig. *Proc Med Def Biosci Rev*, pp. 71–80.

Dirnhuber P, French MC, Green DM *et al.* (1979). The protection of primates against soman poisoning by pretreatment with pyridostigmine. *J Pharm Pharmacol*, **31**, 295–299.

Duncan RC and Griffith J (1992). Screening of agricultural workers for exposure to anticholinesterases. In: *Clinical and Experimental Toxicology of Organophosphates and Carbamates* (B Ballantyne and TC Marrs, eds), pp. 421–429. Oxford, UK: Butterworth-Heinemann.

Ellenhorn MJ and Barceloux DG (1988). *Medical Toxicology: Diagnosis and Treatment of Human Poisoning*. New York, NY, USA: Elsevier.

Ellin RI (1982). Stability of concentrated aqueous solutions of pralidoxime chloride. *J Pharmaceut Sci*, **71**, 1057–1059.

Ellin RI, Carlese JS and Kondritzer AA (1962). Stability of pyridine-2-aldoxime methiodide II. Kinetics of deterioration in dilute aqueous solutions. *J Pharmaceut Sci*, **51**, 141–146.

Ellin RI, Groff WA and Sidell FR (1974). Passage of pyridinium oximes into human red cells. *Biochem Pharmacol*, **23**, 2663–2670.

Enander I, Sundwall A and Sörbo B (1962). Metabolic studies on *N*-methylpyridinium-2-aldoxime – III. Experiments with the ^{14}C-labelled compound. *Biochem Pharmacol*, **11**, 377–382.

Erdmann WD and von Clarmann M (1963). Ein neuer Esterase-Reaktivator für die Behandlung von Vergiftungen mit Alkylphosphaten. *Dtsch med Wschr*, **88**, 2201–2206.

Erdmann WD, Bosse I and Franke P (1965). Zur Resorption und Ausscheidung von Toxogonin nach intramuskulärer Injektion am Menschen. *Dtsch med Wschr*, **90**, 1436–1438.

Eyer P (1996). Optimal oxime dosage regimen, a pharmacokinetic approach. In: *Role of Oximes in the Treatment of Anticholinesterase Agent Poisoning* (L Szinicz, P Eyer and R Klimmek, eds), pp. 33–51. Heidelberg, Germany: Spektrum, Akademischer Verlag.

Eyer P (2003). The role of oximes in the management of organophosphorus pesticide poisoning. *Toxicol Rev*, **22**, 165–190.

Eyer P, Hell W, Kawan A *et al.* (1986). Studies on the decomposition of the oxime HI 6 in aqueous solution. *Arch Toxicol*, **59**, 266–271.

Eyer P, Kawan A and Ladstetter B (1987). Formation of cyanide after i.v. administration of the oxime HI 6 to dogs. *Arch Toxicol*, **61**, 63–69.

Eyer P, Hagedorn I and Ladstetter B (1988). Study on the stability of the oxime HI 6 in aqueous solution. *Arch Toxicol*, **62**, 224–226.

Eyer P, Ladstetter B, Schäfer W *et al.* (1989). Studies on the stability and decomposition of the Hagedorn-oxime HLö 7 in aqueous solution. *Arch Toxicol*, **63**, 59–67.

Eyer P, Hagedorn I, Klimmek R *et al.* (1992). HLö 7 dimethanesulfonate, a potent bispyridinium-dioxime against anticholinesterases. *Arch Toxicol*, **66**, 603–621.

Fleisher JH and Harris LW (1965). Dealkylation as a mechanism for aging of cholinesterase after poisoning with pinacolyl methylphosphonofluoridate. *Biochem Pharmacol*, **14**, 641–650.

Fyhr P, Brodin A, Ernerot L *et al.* (1986). Degradation pathway of pralidoxime chloride in concentrated acidic solution. *J Pharm Sci*, **75**, 608–611.

Garrigue H, Maurizis JC, Nicolas C *et al.* (1990). Disposition and metabolism of two acetylcholinesterase reactivators, pyrimidoxime and HI6, in rats submitted to organophosphate poisoning. *Xenobiotica*, **20**, 699–709.

Garrigue H, Maurizis JC, Madelmont JC *et al.* (1991). Disposition and metabolism of acetylcholinesterase reactivators 2PAM-I, TMB4 and R665 in rats submitted to organophosphate poisoning. *Xenobiotica*, **21**, 583–595.

Gordon JJ, Leadbeater L and Maidment MP (1978). The protection of animals against organophosphate poisoning by pretreatment with a carbamate. *Toxicol Appl Pharmacol*, **43**, 207–216.

Grange-Messenet, Bouchaud C, Jamme M *et al.* (1999). Seizure-related opening of the blood-brain barrier produced by the anticholinesterase compound, soman: new ultrastructural observations. *Cell Mol Biol*, **45**, 1–14.

Green AL and Saville B (1956). The reaction of oximes with isopropyl methylphosphonofluoridate (sarin). *J Chem Soc*, **3**, 3887–3892.

Hackley Jr BE, Steinberg GM and Lamb JC (1959). Formation of potent inhibitors of AChE by reaction of pyridinaldoximes with isopropyl methylphosphonofluoridate (GB). *Arch Biochem Biophys*, **80**, 211–214.

Hagedorn I, Stark I and Lorenz HP (1972). Reaktivierung phosphorylierter Acetylcholin-Esterase – Abhängigkeit von der Aktivator-Acidität. *Angew Chem*, **84**, 354–356.

Hamilton MG and Lundy PM (1989). HI-6 Therapy of soman and tabun poisoning in primates and rodents. *Arch Toxicol*, **63**, 144–149.

Hanke DW and Overton MA (1991). Phosphylation kinetic constants and oxime-induced reactivation in acetylcholinesterase from fetal bovine serum, bovine caudate nucleus, and electric eel. *J Toxicol Environ Health*, **34**, 141–156.

Harris LW, Fleisher JH, Clark J *et al.* (1966). Dealkylation and loss of capacity for reactivation of cholinesterase inhibited by sarin. *Science*, **154**, 404–407.

Harris LW, Anderson DR, Lennox WJ *et al.* (1990). Evaluation of several oximes as reactivators of unaged soman-inhibited whole blood acetylcholinesterase in rabbits. *Biochem Pharmacol*, **40**, 2677–2682.

Hartwich WJ (2004). *Untersuchungen zur schnellen Freigabe von HI6 Dichlorid und HI 6 Dimethansulfonat aus verschiedenen Autoinjektorsystemen*, pp. 1–127. MD Thesis. Munich, Germany: Medical Faculty, Ludwig Maximilians University.

Hobbiger F (1963). Reactivation of phosphorylated acetylcholinesterase. In: *Cholinesterases and Anticholinesterase Agents*, Vol. XV (GB Koelle, ed.), pp. 921–988. Berlin, Germany: Springer-Verlag.

Hobbiger F and Sadler PW (1959). Protection against lethal organophosphate poisoning by quaternary pyridine aldoximes. *Brit J Pharmacol*, **14**, 192–201.

Hobbiger F, O'Sullivan DG and Sadler PW (1958). New potent reactivators of acetycholinesterase inhibited by tetraethyl pyrophosphate. *Nature*, **182**, 1498–1499.

Holcombe DG (1986). Stability of pralidoxime mesylate injections. *Anal Proc*, **23**, 320–321.

Inns RH and Leadbeater L (1983). The efficacy of bispyridinium derivatives in the treatment of organophosphate poisoning in the guinea pig. *J Pharm Pharmacol*, **35**, 427–433.

Jager BV and Stagg GN (1958). Toxicity of diacetylmonoxime and of pyridine-2-aldoxime methiodide in man 15–30 mg/kg 2-PAM: no effects on EEG in a healthy man. *Bull Johns Hopkins Hosp*, **102**, 203–211.

Josselson J and Sidell FR (1978). Effect of intravenous thiamine on pralidoxime kinetics. *Clin Pharmacol Ther*, **24**, 95–100.

Jovanovic D (1989). Pharmacokinetics of pralidoxime chloride. A comparative study in healthy volunteers and in organophosphorus poisoning. *Arch Toxicol*, **63**, 416–418.

Jovanovic D, Randjelovic S and Joksovic D (1990). A case of unusual suicidal poisoning by the organophosphorus insecticide dimethoate. *Human Exp Toxicol*, **9**, 49–51.

Kassa J and Cabal J (1999). A comparison of the efficacy of a new asymmetric bispyridinium oxime BI-6 with presently used oximes and H oximes against sarin by *in vitro* and *in vivo* methods. *Human Exp Toxicol*, **18**, 560–565.

Kepner LA and Wolthuis OL (1978). A comparison of the oximes HS-6 and HI-6 in the therapy of soman intoxication in rodents. *Eur J Pharmacol*, **48**, 377–382.

Kiderlen D, Eyer P and Worek F (2005). Formation and disposition of diethylphosphoryl-obidoxime, a potent anticholinesterase that is hydrolyzed by human paraoxonase PON1. *Biochem Pharmacol*, **69**, 1853–1867.

Klimmek R and Eyer P (1986). Pharmacokinetics and pharmacodynamics of the oxime HI 6 in dogs. *Arch Toxicol*, **59**, 272–278.

Kondritzer A, Ellin RT and Edberg LJ (1961). Investigation of methyl pyridinium-2-aldoxime salts. *J Pharm Sci*, **50**, 109–112.

Koplovitz I and Stewart JR (1992). Efficacy of oxime plus atropine treatment against soman poisoning in the atropinesterase-free rabbit. *Drug Chem Toxicol*, **15**, 117–26.

Korte WD and Shih ML (1993). Degradation of three related bis(pyridinium)aldoximes in aqueous solutions at high concentrations: Examples of unexpectedly rapid amide group hydrolysis. *J Pharm Sci*, **82**, 782–786.

Kovarik Z, Radic Z, Berman HA *et al.* (2004). Mutant cholinesterases possessing enhanced capacity for reactivation of their phosphonylated conjugates. *Biochemistry*, **43**, 3222–3229.

Kozer E, Mordel A, Haim SB *et al.* (2005). Pediatric poisoning from trimedoxime (TMB4) and atropine automatic injectors. *J Pediatr*, **146**, 41–44.

Krummer S, Thiermann H, Worek F *et al.* (2002). Equipotent cholinesterase reactivation *in vitro* by the nerve agent antidotes HI 6 dichloride and HI 6 dimethanesulfonate. *Arch Toxicol*, **76**, 589–595.

Kusic R, Boskovic B, Vojvodic V *et al.* (1985). HI-6 in man: blood levels, urinary excretion, and tolerance after intramuscular administration of the oxime to healthy volunteers. *Fund Appl Toxicol*, **5**, S89–S97.

Kusic R, Jovanovic D, Randjelovic S *et al.* (1991). HI-6 in man: Efficacy of the oxime in poisoning by organophosphorus insecticides. *Human Exp Toxicol*, **10**, 113–118.

Ladstetter B (1990). *Stabilität und metabolisches Schicksal neuer Antidote gegen Organophosphate.* PhD Thesis. Munchen, Germany: Universität München.

Lamb JC, Steinberg GM, Solomon S *et al.* (1965). Reaction of 4-formyl-1-methylpyridinium iodide oxime with isopropyl methylphosphonofluoridate. *Biochemistry*, **4**, 2475–2484.

Langenberg JP, van Dijk C, Sweeney RE *et al.* (1997). Development of a physiologically based model for the toxicokinetics of C(+−)P(+−)-soman in the atropinized guinea pig. *ArchToxicol*, **71**, 320–331.

Leadbeater L, Inns RH and Rylands JM (1985). Treatment of poisoning by soman. *Fund Appl Toxicol*, **5**, S225–S231.

Leader H, Vincze A, Manisterski B *et al.* (1999). Characterization of *O,O*-diethylphosphoryl oximes as inhibitors of cholinesterases and substrates of phosphotriesterases. *Biochem Pharmacol*, **58**, 503–515.

Ligtenstein DA and Kossen SP (1983). Kinetic profile in blood and brain of the cholinesterase reactivating oxime HI-6 after intravenous administration to the rat. *Toxicol Appl Pharmacol*, **71**, 177–183.

Ligtenstein DA, Wils ERJ, Kossen SP *et al.* (1987). Identification of two metabolites of the cholinesterase reactivator HI-6 isolated from rat urine. *J Pharm Pharmacol*, **39**, 17–23.

Lin AJ and Klayman DL (1986). Stability studies of bis(pyridiniumaldoxime) reactivators of organophosphate inhibited acetylcholinesterase. *J Pharm Sci*, **75**, 797–799.

Lockridge O and Masson P (2000). Pesticides and susceptible populations: people with butyrylcholinesterase genetic variants may be at risk. *NeuroToxicology*, **21**, 113–126.

Löffler M (1986). *Quartäre Salze von Pyridin-2,4-Dialdoxim als Gegenmittel für Organophosphat-Vergiftungen* PhD Thesis. Freiburg in Bresgau, Germany: Albert Ludwigs Universität.

Lotti M and Becker CE (1982). Treatment of acute organophosphate poisoning: Evidence of a direct effect on central nervous system by 2-PAM (pyridine-2-aldoxime methyl chloride). *Clin Toxicol*, **19**, 121–127.

Luo C, Ashani Y and Doctor BP (1998a). Acceleration of oxime-induced reactivation of organophosphate-inhibited fetal bovine serum acetylcholinesterase by monoquaternary and bisquaternary ligands. *Mol Pharmacol*, **53**, 718–726.

Luo C, Ashani Y, Saxena A *et al.* (1998b). Acceleration of oxime-induced reactivation of organophosphate-inhibited acetylcholinesterase by quaternary ligands. In: *Structure and Function of Cholinesterases and Related Proteins* (BP Doctor, P Taylor, DM Quinn *et al.*), pp. 215–221. New York, NY, USA: Plenum Press.

Luo C, Saxena A, Smith M *et al.* (1999). Phosphoryl oxime inhibition of acetylcholinesterase during oxime reactivation is prevented by edrophonium. *Biochemistry*, **38**, 9937–9947.

Lüttringhaus A and Hagedorn I (1964). Quartäre Hydroxyiminomethyl-pyridiniumsalze. Das Dichlorid des Bis-[4-hydroxyiminomethyl-pyridinium-(1)-methyl]-äthers ("LüH6"), ein neuer Reaktivator der durch organische Phosphorsäureester gehemmten Acetylcholin-Esterase. *Arzneim Forsch*, **14**, 1–5.

Marrs TC (1991). Toxicology of oximes used in treatment of organophosphate poisoning. *Adverse Drug React Toxicol Rev*, **10**, 61–73.

Mast U (1997). *Reaktivierung der Erythrozyten-Acetylcholinesterase durch Oxime. Ermittlung enzymkinetischer Konstanten und ihre Bedeutung für die Therapie einer Organophosphat-Vergiftung.* MD Thesis. Ludwig-Maximilians–Universität München, Germany.

Maurizis JC, Ollier M, Nicolas C *et al.* (1992). *In vitro* binding of oxime acetylcholinesterase reactivators to proteoglycans synthesized by cultured chondrocytes and fibroblasts. *Biochem Pharmacol*, **44**, 1927–1933.

Maxwell DM and Brecht KM (1991). The role of carboxylesterase in species variation of oxime protection against soman. *Neurosci Biobehav Rev*, **15**, 135–139.

Maxwell DM, Brecht KM and Koplovitz I (1997). Characterization and treatment of the toxicity of *O*-isobutyl *S*-[2-(diethylamino) ethyl]methylphosphonothioate, a structural isomer of VX, in guinea pigs. *J Am Coll Toxicol*, **15** (Supplement 2), S78–S88.

Mdachi RE, Marshall WD, Ecobichon DJ *et al.* (1990). Abiotic transformations and decomposition kinetics of 4-carbamoyl-2′-[hydroxyimino)methyl]-1,1′-(oxydimethylene)-bis(pyridinium chloride) in aqueous phosphate buffers. *Chem Res Toxicol*, **3**, 413–422.

Medicis JJ, Stork CM, Howland MA *et al.* (1996). Pharmacokinetics following a loading plus a continuous infusion of pralidoxime compared with the traditional short infusion regimen in human volunteers. *Clin Toxicol*, **34**, 289–295.

Melchers BPC, Van der Laaken AL and Van Helden HPM (1991). On the mechanism whereby HI-6 improves neuromuscular function after oxime-resistant acetylcholinesterase inhibition and subsequent impairment of neuromuscular transmission. *Eur J Pharmacol*, **200**, 331–337.

Nenner M (1974). Phosphonylierte Aldoxime. Hemmwirkung auf Acetylcholinesterase und hydrolytischer Abbau. *Biochem Pharmacol*, **23**, 1255–1262.

Oldiges H and Schoene K (1970). Pyridinium- und Imidazoliumsalze als Antidote gegenüber Soman- und Paraoxonvergiftungen bei Mäusen. *Arch Toxicol*, **26**, 293–305.

Portmann R, Niederhauser A, Hofmann W *et al.* (1991). 32. Synthesis of 4-(([(isopropyloxy) methylphosphoryloxy]imino)methyl)-1-methyl-pyridinium iodide and its characterisation. *Helv Chim Acta*, **74**, 331–335.

Poziomek EJ, Hackley Jr BE and Steinberg GM (1958). Pyridinium aldoximes. *J Org Chem*, **23**, 714–717.

Primozic I, Odzak R, Tomic S *et al.* (2004). Pyridinium, imidazolium and quinuclidinium oximes: synthesis, interaction with native and phosphylated cholinesterases, and antidotes against organophosphorus compounds. *J Med Chem Def*, **2**, 1–30.

Prinz HJ (1967). Therapie akuter Alkylphosphat-Vergiftungen. *Dtsch Ärztebl*, **36**, 1845–1849.

Radic Z and Taylor P (1999). The influence of peripheral site ligands on the reaction of symmetric and chiral organophosphates with wildtype and mutant acetylcholinesterases. *Chem Biol Interact*, **119**, 111–117.

Rubnov S, Amisar S, Levy D *et al.* (1999a). Stability of trimedoxime in concentrated acidic injectable solutions. *Mil Med*, **164**, 55–58.

Rubnov S, Shats I, Levy D *et al.* (1999b). Autocatalytic degradation and stability of obidoxime. *J Pharm Pharmacol*, **51**, 9–14.

Schlager JW, Dolzine TW, Stewart JR *et al.* (1991). Operational evaluation of three commercial configurations of atropine/HI-6 wet/dry autoinjectors. *Pharm Res*, **8**, 1191–1194.

Schoene K (1973). Phosphonyloxime aus Soman; Bildung und Reaktion mit Acetylcholinesterase *in vitro*. *Biochem Pharmacol*, **22**, 2997–3003.

Schroeder AC, DiGiovanni JH, von Bredow J *et al.* (1989). Pralidoxime chloride stability-indicating assay and analysis of solution samples stored at room temperature for ten years. *J Pharm Sci*, **78**, 132–136.

Schulz V, Gross R, Pasch T *et al.* (1982). Cyanide toxicity of sodium nitroprusside in therapeutic use with and without sodium thiosulfate. *Klin WochenSchr*, **60**, 1393–1400.

Scott RJ (1986). Repeated asystole following PAM in organophosphate self-poisoning. *Anaesth Intens Care*, **14**, 458–60.

Shafferman A, Ordentlich A, Barak D *et al.* (1996). Aging of phosphylated human acetylcholinesterase: catalytic processes mediated by aromatic and polar residues of the active centre. *Biochem J*, **318**, 833–840.

Sidell FR (1974). Soman and sarin: clinical manifestations and treatment of accidental poisoning by organophosphates. *ClinToxicol*, **7**, 1–17.

Sidell FR (1992). Clinical considerations in nerve agent intoxication. In: *Chemical Warfare Agents* (SM Somani, ed.), pp. 155–194. San Diego, CA, USA: Academic Press.

Sidell F (1997). Nerve agents. In: *Medical Aspects of Chemical and Biological Warfare* (F Sidell, ET Takafuji and DR Franz, eds), pp. 130–179. Washington, DC, USA: Walter Reed Army Medical Center.

Sidell FR and Groff WA (1970). Toxogonin: Blood levels and side effects after intramuscular administration in man. *J Pharm Sci*, **59**, 793–797.

Sidell FR and Groff WA (1971). Intramuscular and intravenous administration of small doses of 2-pyridinium aldoxime methochloride to man. *J Pharm Sci*, **60**, 1224–1228.

Sidell FR and Groff WA (1974). The reactivatibility of cholinesterase inhibited by VX and sarin in man. *Toxicol Appl Pharmacol*, **27**, 241–252.

Sidell FR, Groff WA and Kaminskis A (1972). Toxogonin and pralidoxime: Kinetic comparison after intravenous administration to man. *J Pharm Sci*, **61**, 1765–1769.

Simon GA, Tirosh MS and Edery H (1976). Administration of obidoxime tablets to man. Plasma levels and side reactions. *Arch Toxicol*, **36**, 83–88.

Smith TE and Usdin E (1966). Formation of nonreactivatible isopropylmethylphosphonofluoridate-inhibited acetylcholinesterase. *Biochemistry*, **5**, 2914–2918.

Spöhrer U (1994). *HPLC-analytische Untersuchungen zur Pharmakokinetik von Pyridiniumaldoximen*. PhD Thesis. Ludwig-Maximilians-Universität München, Germany.

Spöhrer U and Eyer P (1995). Separation of geometrical syn/anti isomers of obidoxime by ion-pair high-performance liquid chromatography. *J Chromatgr A*, **693**, 55–61.

Spöhrer U, Thiermann H, Klimmek R *et al.* (1994). Pharmacokinetics of the oximes HI 6 and HLö 7 in dogs after i.m. injection with newly developed dry/wet autoinjectors. *Arch Toxicol*, **68**, 480–489.

Spruit HE, Langenberg JP, Trap HC *et al.* (2000). Intravenous and inhalation toxicokinetics of sarin stereoisomers in atropinized guinea pigs. *Toxicol Appl Pharmacol*, **169**, 249–54.

Stark I (1968). *Versuche zur Darstellung eines LüH6 (Toxogonin) überlegenen Acetylcholinesterase-Reaktivators*. Freiburg in Bresgau, Germany: Albert Ludwigs Universität.

Stemler FW, Tezak-Reid TM, McCluskey MP *et al.* (1991). Pharmacokinetics and pharmacodynamics of oxime in unanesthetized pigs. *Fund Appl Toxicol*, **16**, 548–558.

Su C-T, Wang P-H, Liu R-F *et al.* (1986). Kinetic studies and structure–activity relationships of bispyridinium oximes as reactivators of acetylcholinesterase inhibited by organophosphorus compounds. *Fund Appl Toxicol*, **6**, 506–524.

Sundwall A (1960). Plasma concentration curves of *N*-methylpyridinium-2-aldoxime methane sulphonate (P2S) after intravenous, intramuscular and oral administration in man. *Biochem Pharmacol*, **5**, 225–230.

Sussman JL, Harel M, Frolow F *et al.* (1991). Atomic structure of acetylcholinesterase from *Torpedo californica*: A prototypic acetylcholine-binding protein. *Science*, **253**, 872–879.

Sweeney RE and Maxwell DM (1999). A theoretical model of the competition between hydrolase and carboxylesterase in protection against organophosphorus poisoning. *Math Biosci*, **160**, 175–190.

Sweeney RE and Maxwell DM (2003). A theoretical expression for the protection associated with stoichiometric and catalytic scavengers in a single compartment model of organophosphorus poisoning. *Math Biosci*, **181**, 133–143.

Talbot BG, Anderson DR, Harris LW *et al.* (1988). A comparison of *in vivo* and *in vitro* rates of ageing of soman-inhibited erythrocyte anticholinesterase in different animal species. *Drug Chem Toxicol*, **11**, 289–305.

Tattersall JE, Smith AP, Waters K *et al.* (1998). Therapeutic action of HI-6 against soman poisoning in vitro: an interspecies comparison. *Br J Pharmacol*, **125**, 7P.

Taylor P, Wong L, Radic Z *et al.* (1999). Analysis of cholinesterase inactivation and reactivation by systematic structural modification and enantiomeric selectivity. *Chem-Biol Interact*, **119–120**, 3–15.

Thiermann H, Spöhrer U, Klimmek R *et al.* (1994). Operational evaluation of wet/dry autoinjectors containing atropine in solution and powdered HI 6 or HLö 7. *Int J Pharm*, **109**, 35–43.

Thiermann H, Seidl S and Eyer P (1995a). Stand der Entwicklung neuer Autoinjektoren zur Behandlung der Nervenkampfstoff-Vergiftung. *Wehrmed Mschr*, **39**, 189–192.

Thiermann H, Spöhrer U, Klimmek R *et al.* (1995b). Pharmacokinetics of atropine and its combination with HI 6 or HLö 7 in dogs after i.m. injection with newly developed dry/wet autoinjectors. *Przeg Lek*, **52**, 208.

Thiermann H, Seidl S and Eyer P (1996). HI 6 dimethanesulfonate has better dissolution properties than HI 6 dichloride for application in dry/wet autoinjectors. *Int J Pharm*, **137**, 167–176.

Utley D (1987). Analysis of formulations containing pralidoxime mesylate by liquid chromatography. *J Chromatogr*, **396**, 237–250.

van der Schans MJ, Lander BJ, van der Wiel H *et al.* (2003). Toxicokinetics of the nerve agent (+/−)-VX in anesthetized and atropinized hairless guinea pigs and marmosets after intravenous and percutaneous administration. *Toxicol Appl Pharmacol*, **191**, 48–62.

van Helden HPM, de Lange J, Busker RW *et al.* (1991). Therapy of organophosphate poisoning in the rat by direct effects of oximes unrelated to ChE reactivation. *Arch Toxicol*, **65**, 586–593.

van Helden HPM, Busker RW, Melchers BPC *et al.* (1996). Pharmacological effects of oximes: how relevant are they? *Arch Toxicol*, **70**, 779–786.

Vojvodic V and Boskovic B (1976). A comparative study of pralidoxime, obidoxime and trimedoxime in healthy men volunteers and in rats. In: *Medical Protection against Chemical-Warfare Agents* (J Stares, ed.), pp. 65–73. SIPRI, Stockholm, Sweden: Almqvist & Wiksell.

von Gaisberg U and Dieterle K (1967). Organ-Parenchymschäden nach E-605-Vergiftung bzw. hochdosierter Toxogoninbehandlung. *Dtsch Ärztebl*, **64**, 1791–1796.

Wiezorek WD, Kreisel W, Schnitzlein W *et al.* (1968). Eigenwirkungen von Trimedoxim und Pralidoxim am Menschen. *Zeitschr Militärmed*, **4**, 223–226.

Willems JL, Langenberg JP, Verstraete AG *et al.* (1992). Plasma concentrations of pralidoxime methylsulphate in organophosphorus poisoned patients. *Arch Toxicol*, **66**, 260–266.

Wilson IB and Ginsburg S (1955). A powerful reactivator of alkylphosphate-inhibited acetylcholinesterase. *Biochim Biophys Acta*, **18**, 168–170.

Wong L, Radic Z, Brüggemann RJM *et al.* (2000). Mechanism of oxime reactivation of acetylcholinesterase analyzed by chirality and mutagenesis. *Biochemistry*, **39**, 5750–5757.

Woodard CL and Lukey BJ (1991). MMB-4 pharmacokinetics in rabbits after intravenous and intramuscular administration. *Drug Metab Dispos*, **19**, 283–284.

Worek F, Eyer P and Szinicz L (1998). Inhibition, reactivation, and aging kinetics of cyclohexylmethylphosphonofluoridate-inhibited human cholinesterases. *Arch Toxicol*, **72**, 580–587.

Worek F, Eyer P, Kiderlen D *et al.* (2000). Effect of human plasma on the reactivation of sarin-inhibited human erythrocyte acetylcholinesterase. *Arch Toxicol*, **74**, 21–26.

Worek F, Reiter G, Eyer P *et al.* (2002). Reactivation kinetics of acetylcholinesterase from different species inhibited by highly toxic organophosphates. *Arch Toxicol*, **76**, 523–529.

Worek F, Thiermann H, Szinicz L *et al.* (2004). Kinetic analysis of interactions between human acetylcholinesterase, structurally different organophosphorus compounds and oximes. *Biochem Pharmacol*, **68**, 2237–48.

Worek F, Szinicz L, Eyer P *et al.* (2005). Evaluation of oxime efficacy in nerve agent poisoning: development of a kinetic-based dynamic model. *Toxicol Appl Pharmacol*, **209**, 193–202.

Xue SZ, Ding XJ and Ding Y (1985). Clinical observation and comparison of the effectiveness of several oxime cholinesterase reactivators. *Scand J Work Environ Health*, **11**, 46–48.

16 THE USE OF BENZODIAZEPINES IN ORGANOPHOSPHORUS NERVE AGENT INTOXICATION

Timothy C. Marrs[1] and Åke Sellström[2]

[1] *Edentox Associates, Edenbridge, UK*
[2] *Swedish Defence Research Institute, Umeå, Sweden*

INTRODUCTION

Anticonvulsants, including diazepam, other benzodiazepines, barbiturates and phenitoin were studied initially in the management of organophosphorus (OP) ester-induced convulsions for symptomatic reasons. Diazepam is by far the most intensively studied benzodiazepine for nerve-agent poisoning, although two other benzodiazepines need consideration, namely avizafone and midazolam. Nerve-agent poisoning has mainly been a military concern. The scenario involves advanced first aid or immediate medical assistance in the field to a homogenous population of soldiers. Following extended experimental research on the protective efficacy of diazepam (see below), this was introduced for military use against nerve-agent intoxication as tablets of 5 mg. The UK armed forces' autoinjection devices carried a 5 mg diazepam tablet in a detachable cap and the Swedish armed forces issued 5 mg diazepam tablets for prophylactic use for up to three days. Diazepam was introduced in autoinjector devices, in November 1990. Each autoinjector device contained 10 mg of diazepam, and was intended for co-administration with atropine and an oxime reactivator at the onset of severe nerve agent intoxication. Further development of the autoinjectors resulted in the introduction of a water soluble 'pro-drug' version of diazepam, avizafone. Avizafone is a peptide-aminobenzophenone, sometimes called 'pro-diazepam'. Avizafone is compatible with atropine and pralidoxime mesilate when used by intramuscular injection and is included in some autoinjection devices, such as the L4A1 ComboPen©.

Unfortunately, preparedness for treatment of nerve-agent intoxication is no longer only a military concern. Today, the civilian health care community also has to be prepared and such preparation includes the therapeutic use of anticonvulsants. Although the mechanisms of intoxication are the same in both the military and the civilian cases, it is important that the scenarios are not conflated. The military scenario, the only one so far widely studied, involves a relatively homogenous population of trained servicemen, usually physically fit, frequently pre-treated with pyridostigmine. Until recently, nerve-agent poisoning of civilians was given little attention. This dramatically changed after the Halabja, Matsumoto and Tokyo incidents. For example, the search engine 'Google' gave approximately 170 000 hits for the search 'nerve+agent+civilian+treatment' – of these 3670 also mentioned diazepam. When considering the use of OP nerve agents in civilians, it must be remembered that the population will not have been pre-treated and may include the sick, children and old people. Some of these aspects have been addressed; see, for example, the guidelines issued by the Department of Health, UK (2005) and a publication on the treatment

Chemical Warfare Agents: Toxicology and Treatment (2nd Edition)
Edited by Timothy C. Marrs, Robert L. Maynard and Frederick R. Sidell

of children after nerve-agent attack (Rotenberg and Newmark, 2003). In particular, the use of diazepam in civilian populations has been reviewed (Marrs, 2004).

BENZODIAZEPINES

The main site of action of the benzodiazepines is on γ-aminobutyric acid A ($GABA_A$) receptors, although when benzodiazepines are used for the specific indication of nerve-agent poisoning, some other beneficial actions cannot be ruled out. The $GABA_A$ receptor is one of a 'superfamily' of receptors, which also includes the glycine receptor and the nicotinic acetylcholine receptor, and is a ligand-gated ion channel (Ortells and Lunt, 1995). The GABAergic system is the major inhibitory neurotransmission system in the mammalian central nervous system. Benzodiazepines, including diazepam, alter GABA binding at the $GABA_A$ receptor. This action is allosteric and these drugs are not direct GABA agonists (Charney *et al.*, 2001), which may account in part for their safety. Benzodiazepines are often thought of as increasing the effects of GABA (Costa and Guidotti, 1979) and these drugs thereby increase endogenous control of the central nervous system against hyper-excitation (Sellström, 1992). It has been suggested that the various actions of the benzodiazepines may be mediated by different $GABA_A$ receptor subtypes [see reviews by Lüddens and Korpi (1995), Johnston (1996) and Sieghart and Sperk (2002)].

Benzodiazepines, unlike other anticonvulsants, are effective against all seizure types and are the preferred agent of use in status epilepticus (for review, see Rosenow *et al.* (2002)). Untreated seizures or convulsion are dangerous and may have long-term sequelae in the central nervous system (CNS), while nerve-agent intoxication, in particular, may damage the CNS (see below). The molecular mechanisms behind this damage, anoxia and/or excitotoxicity, is an area of research interest (for review, see Solberg and Belkin, 1997). Benzodiazepines are used to alleviate or prevent convulsions in OP nerve-agent poisoning and the use of benzodiazepines may have long-term as well as short-term benefit in severe nerve-agent poisoning. The aim is to prevent convulsions or reduce their duration. The major argument against the use of benzodiazepines as anticonvulsants is the tendency for patients to develop tolerance (Haigh and Feely, 1988). The period of treatment with anticonvulsants in nerve-agent intoxication would, however, be expected to be relatively short, 1-6 days, in most cases (Sellström, 1992). This means that the development of tolerance to the anticonvulsant effect of benzodiazepines normally may be ignored and that benzodiazepines, particularly, diazepam, may thus be advocated as the first-line choice against anticholinesterase-induced convulsions. Diazepam is the most-studied benzodiazepine in nerve-agent poisoning, while midazolam has found some favor in poisons units as an alternative to diazepam in the treatment of OP-insecticide poisoning and should perhaps be considered in nerve-agent poisoning. Midazolam has the advantage of rapid and efficient uptake after intramuscular administration (Mattila *et al.*, 1983).The diazepam pro-drug avizafone has also been studied in nerve-agent poisoning.

Diazepam

The main use of diazepam is as an anxiolytic in anxiety states while the drug also has muscle-relaxant properties (Diamantis and Kretzkin, 1966) and relatively unimportant effects on the cardiovascular system (Baldessarini, 1980). Diazepam is effective in treating convulsions from a variety of causes (Pieri *et al.*, 1981). Diazepam, together with midazolam and lately lorazepam, are preferred drugs in treatment of status epilepticus (for a review, see Rosenow *et al.*, 2002). Long-term use of benzodiazepines, including diazepam, can result in habituation, which may be related to changes in glutamatergic receptors (Allison and Pratt, 2003). Tolerance also develops to the anticonvulsant effects of diazepam (as with other benzodiazepines – see above), for which reason diazepam is not often used for the long-term treatment of epilepsy (for review, see Haigh and Feely, 1988).

PHARMACOKINETICS

In the context of nerve-agent treatment the most likely mode of administration would be by intravenous injection. In healthy human subjects, blood concentrations of 0.4 and 1.2 mg l^{-1} were found 15 min after intravenous bolus doses of 10 and 20 mg, respectively (Hillestad *et al.*, 1974). Other modes of administration have been considered for self-administration, or administration by those not trained in intravenous injection. Such routes include intranasal, intramuscular and the use of suppositories. In humans, absorption is poor after intramuscular injection; plasma levels acheived were equal to only 60% of those reached after the same oral dose (Hillestad *et al.*, 1974). Studies in experimental animals have shown that following oral administration, the bioavailability of diazepam is almost 100% (Mandelli *et al.*, 1978).

Diazepam is largely protein-bound both in humans and in experimental animals. Thus Klotz *et al.* (1976a) found that more than 85% of the plasma diazepam was protein-bound in humans, dogs, rabbits, guinea pigs and rats, and the same is true of desmethyldiazepam, the diazepam metabolite, except in the guinea pigs. The unbound fraction of diazepam in the plasma was observed to be lower in young human subjects than older ones (Viani *et al.*, 1992). On the basis of studies in dogs, rabbits, guinea pigs and rats, as well as man, by Klotz *et al.* (1976a,b), a two-compartment open model was proposed for the elimination in all species investigated, humans included. However, there were major quantitative pharmacokinetic differences between man and the experimental animals, which may complicate interpretation of pharmacodynamic studies. In cats, diazepam is rapidly distributed in the grey matter and more slowly accumulated by the white matter, the effect of repeated dosing suggesting that the white matter acts as a deep compartment (Morselli *et al.*, 1974). Capacio *et al.* (2001) studied the pharmacokinetics of diazepam after IM administration to soman-exposed guinea pigs, pretreated with intramuscular pyridostigmine. The V_d observed in a study was much higher than that observed by Klotz *et al.* (1976a) in guinea pigs solely exposed to diazepam.

The elimination kinetics of diazepam in humans has been described by a two-compartment open model, with a plasma clearance of 26–35 ml min^{-1} after a single intravenous dose (Klotz *et al.*, 1975; Andreasen *et al.*, 1976; Klotz *et al.*, 1976a). The $t_{1/2}$ for diazepam in the β-phase was 1–2 days and somewhat longer for desmethyldiazepam (Klotz *et al*, 1976a; Mandelli *et al.*, 1978). The half-life of diazepam was increased where liver damage was present (Andreasen *et al.*, 1976), and with age (Klotz *et al.*, 1975).

Studies by Klotz *et al.* (1975,1976a,b) suggest that biliary excretion of diazepam is unimportant in man, but there is some evidence (see above) for species differences (Klotz *et al.*, 1975,1976a; van der Kleijn *et al.*, 1971). Urinary excretion of diazepam is mainly in the form of sulphate and glucuronide conjugates (Mandelli *et al.*, 1978). The main metabolic pathway is demethylation and hydroxylation to metabolites with CNS depressant activity in animals and man. These metabolites are desmethyldiazepam and oxazepam.

PHARMACODYNAMICS

Diazepam was one of the earliest benzodiazepines and has been very widely studied and used. In addition, diazepam has been investigated in many experimental animal pharmacodynamic studies as therapy for both OP nerve agent and pesticide poisoning. Most experimental work on the efficacy of diazepam has been carried out using treatment combinations, usually atropine, pyridinium oximes and diazepam. Many studies, in addition, also included the administration of a carbamate, usually pyridostigmine. The action of pyridostigmine is to protect a proportion of the acetylcholinesterase from dialkylphosphonylation, the subsequent monodealkylation (aging) of which produces, with soman, alkylphosphonylated acetylcholinesterase which is refractory to reactivation.

Animal studies that have employed prophylactic protocols, for antidotes that would not be used as prophylaxes should be extrapolated to post-poisoning treatment cautiously. Likewise, those studies involving antidotes used on pyridostigmine-pretreated animals should be extrapolated to non-pretreated human populations with caution.

STUDIES ON DIAZEPAM AND NERVE AGENTS

Lipp (1972) showed in monkeys exposed to soman (15–20 μg kg^{-1} IM) that a combination of diazepam (2–5 mg kg^{-1} IV) and atropine (0.1 mg kg^{-1} IV) was more effective than atropine alone in preventing death and avoiding seizure activity. Atropine, without diazepam, did not abolish seizures. There was some indication that diazepam by itself could provide limited protection. A year later Lipp (1973) and Rump *et al.* (1973) showed diazepam to be effective against seizures in monkeys and rabbits induced by soman and DFP, including when diazepam was used as a pre-treatment. In a study by Johnson and Lowndes (1974), treatment with diazepam (1.0 mg kg^{-1} IV) and atropine (0.5 mg kg^{-1} IV) 20 min before poisoning increased survival to a greater extent than either drug given singly, in rabbits poisoned with soman (15 or 20 μg kg^{-1}).

In an attempt to understand the mechanism whereby diazepam was efficacious, Johnson and co-workers (Johnson and Lowndes, 1974; Johnson and Wilcox, 1975) showed diazepam to counteract the over-activity normally associated with skeletal and heart muscle following soman intoxication. It was also shown, however, that diazepam enhances the respiratory depression produced by soman in the pentobarbitone-anaesthetized rabbit. Bosković (1981) found that atropine and diazepam increased approximately threefold the survival time of rats poisoned with soman, when given 1 min after poisoning. In addition, there are several pharmacodynamic studies of oximes by the same author, in which diazepam (and frequently atropine) were included, but it is often difficult to separate out the effects of the diazepam from the other drugs used (e.g. Bošković *et al.*, 1984).

Lundy *et al.* (1978) showed in rats that diazepam (alone, 2.5 mg kg^{-1} IP) possessed powerful anticonvulsant activity against soman (1 LD_{50} SC, 135 μg kg^{-1}), the diazepam being administered 30 min before the nerve agent. In another study, using a prophylactic protocol, diazepam (3.2 mg kg^{-1} IM) effectively countered the convulsant actions of soman (104 μg kg^{-1} SC) administered 10 min later to rats (Churchill *et al.*, 1987). In spite of the findings by Lipp (1972) and by Johnson and Lowndes (1974), the beneficial effects of diazepam alone on the lethality of OPs are more arguable than those on convulsions. Accordingly, Doebler *et al.* (1985) showed that soman-induced RNA depletion in rats was almost completely prevented by diazepam (2.2 mg kg^{-1} IM) pretreatment 10 min before soman challenge, but lethality was not prevented.

Evidence from animal studies suggests that diazepam might prevent some of the sequelae associated with convulsions. Martin *et al.* (1985) showed that histological changes produced in the central nervous systems of rats by injection of just sub-lethal doses (0.9% LD_{50}) of soman could be prevented with diazepam, injected 10 min previously. Following the study by Martin *et al.* (1985), more attention has been given to the protection afforded by diazepam against OP-induced effects on the CNS. Accordingly, Pazdernik *et al.* (1986) showed that diazepam normalized local cerebral glucose utilization in the soman-intoxicated rat.

In a study in rats, Shih *et al.* (2003) found the intramuscular dose of diazepam for the abolition of seizures to be much smaller when given 30 min before soman than when given 5 min after the onset of seizures (these animals had all been treated with HI–6).

A number of pharmacodynamic studies have been carried out with diazepam in which pre-treatment with carbamates such as pyridostigmine has also been employed. Inns and Leadbeater (1983) carried out studies on the efficacy of a number of pyridinium compounds, including both clinically available drugs and two bispyridinium non-oximes. In a study in which pyridostigmine iodide was injected intramuscularly into guinea pigs 30 min before the guinea pigs were challenged with soman, it was found that a combination of atropine and diazepam was generally more effective than atropine alone. Capacio *et al.* (2001) studied the efficacy of diazepam in treating seizures induced by soman in guinea pigs, implanted with EEG electrodes, the animals having been pre-treated with intramuscular pyridostigmine. One minute after the soman, the guinea pigs were treated with atropine sulphate and pralidoxime chloride). Diazepam terminated seizure activity within 30 min in 52% (33/63) of the animals.

Shih *et al.* (2003) evaluated the potency and rapidity of action of some anticholinergics, including atropine and two benzodiazepines (diazepam and midazolam), for seizures produced by six nerve agents, namely tabun, sarin, soman, cyclosarin, VR, and VX in guinea pigs. The guinea pigs were pre-treated with pyridostigmine bromide 30 min prior to challenge with a supralethal dose (SC) of the various nerve agents. The animals were thereafter treated with atropine sulphate and pralidoxime chloride. Five minutes after the start of seizures on the recorded EEG, animals were treated IM with different doses of the anticholinergics or benzodiazepines. All drugs were capable of terminating seizure convulsions. An important finding of this study was that of the benzodiazepines, midazolam was more than 10-fold more potent than diazepam against all six nerve agents. This study provided further evidence for seizure control being linked with protection against neuronal necrosis.

Studies on OP insecticides provide considerable support for the value of diazepam in experimental poisoning in a variety of animal models. However, such studies must be interpreted in the light of the differences in likely routes of exposure to pesticides and nerve agents and the differences in the pharmacokinetics of the two groups of OPs. With OP pesticides, it seems clear that atropine and diazepam together have a greater effect than each individually, while pyridinium oximes further increase the effectiveness. With pesticides, as with nerve agents, the effect on convulsions of diazepam is clearer than that on lethality (Bokonjić *et al.*, 1987; Rump and Grudzinska, 1974; Rump *et al.*, 1976; Kleinrok and Jagiełło-Wojtowicz, 1977; Gupta, 1984; Krutak-Krol and Domino, 1985; Kassa and Bajgar, 1994) (also see review by Marrs, 2003).

TOXICOLOGY

Very extensive toxicological studies have been carried out on diazepam. Much of the data obtained are only relevant to its use as anxiolytic over long time-periods. The main effects observed in OP poisoning would be unwanted sedation.

Diazepam is a developmental neurotoxin and the significance of this is discussed below in relation to the possible use of diazepam in pregnant women in a terrorist attack. There is a considerable body of evidence from animal studies that exposure *in utero* to diazepam may have adverse effects on the developing foetus (Ryan and Pappas, 1986; Silva and Palermo-Neto, 1999; Livezey *et al.*, 1986; Martire *et al.*, 2002). The effect of diazepam administered to pregnant women has been reviewed (McElhatton, 1994). In nerve-agent poisoning the use of diazepam would have to be balanced against the risk of untreated nerve-agent poisoning (see below).

Avizafone

Avizafone is a lysylglycin derivative of diazepam which is water-soluble and when injected into the body is converted to diazepam and the amino acid lysine. Avizafone, therefore, is more suitable for intramuscular administration than the lipophilic diazepam. The pro-drug, accordingly, has attracted attention as a substitute for diazepam in autoinjector development and is now present together with atropine and pralidoxime as one of the active components of the Nerve Agent Antidote L4A1 (Combopen®). Combopen® contains 10 mg of avizafone.

Avizafone is rapidly converted into diazepam following its injection into the body (Maidment and Upshall, 1990; see also, Lallement *et al.*, 2000 below). Its usefulness in nerve-agent intoxication was predicted from the well-documented effect of diazepam and has been confirmed in a number of studies. Karlsson *et al.* (1990) studied diazepam and avizafone, together with pyridostigmine and atropine, as antidotes to soman poisoning in guinea pigs. Diazepam and avizafone were used at equimolar doses. Avizafone was converted to diazepam and protection provided by avizafone and diazepam was comparable. Clement and Broxup (1993) compared the protection against nerve-agent-induced neuropathology provided by diazepam and avizafone. Their study illustrated the importance of early intervention with either anticonvulsant. Lallement *et al.* (1997) compared two different regimes of treatment against soman intoxication. It was concluded that atropine/HI-6/avizafone protected better than atropine/pralidoxime/diazepam. The result may, however, reflect that HI-6 is a better reactivator of soman-inhibited acetylcholine esterase than pralidoxime (Kassa, 2002). McDonough *et al.* (1999) evaluated the protection provided by avizafone, clonazepam, diazepam, lorazolam, lorazepam and midazolam given following exposure of guinea pigs to soman and pre-treated with pyridostigmine and treated with atropine and pralidoxime chloride. All of the benzodiazepines were effective in protecting against soman-induced seizure. Midazolam was the most potent and rapidly acting compound. Lallement *et al.* (2000) compared the efficacy of diazepam and avizafone to protect the primate brain against soman-induced EEG and pathological changes. Given at equimolar doses, diazepam protected better than avizafone; indeed, at equimolar dosing diazepam in plasma following diazepam injection exceeded diazepam in plasma following avizafone injection by 1:0.7. This is supposedly due to incomplete conversion of the pro-drug to diazepam *in vivo.* It was observed that maximum plasma concentrations of diazepam were achieved more rapidly after avizafone injection than after injection of diazepam itself. In 2004, Lallement *el al.* continued their study, comparing the efficacy of avizafone versus diazepam in protecting mice against soman intoxication (Lallement *et al.*, 2004). The protective effect was studied over 24 h. The combination, atropine/pralidoxime/avizafone, protected 30 % better, from 15 min following soman injection up 6 h following the injection, than did atropine/pralidoxime/diazepam. Following 6 h, there was no difference in the protection provided by the two regimes. Prior lyophilization of atropine/pralidoxime/avizafone, in order to increase its storability, did not reduce this protection. Taysse *et al.* (2003) compared the protection provided by diazepam and avizafone in guinea pigs intoxicated by sarin, pretreated with pyridostigmine and treated with atropine and pralidoxime. Diazepam and avizafone both were effective in preventing sarin-induced neuropathology.

Midazolam

Midazolam, an imidazobenzodiazepine, have several properties that make it a benzodiazepine of interest with regard to seizure control and neuroprotection against nerve agents. It is water-soluble at acid pH, but highly lipid-soluble *in vivo.* Midazolam has a relatively rapid onset of action and high metabolic clearance when compared with other benzodiazepines (For review, see Reves *et al.*, 1985 and Blumer, 1998). Midazolam is better absorbed than diazepam following intramuscular injection (Mattila *et al.*, 1983). Chamberlain *et al.* (1997) compared intramuscular midazolam and intravenous diazepam for the treatment of seizures in children. The study concluded that in spite of the different times of onset after intravenous and intramuscular injections, more total time elapsed for the patients in the diazepam group before they received benefit than the patients in the midazolam group. Accordingly, the midazolam group had a more rapid cessation of seizures. The authors concluded that, given the relative ease of the intramuscular injection of midazolam, this should be a particularly useful route 'in physicians' offices, in a pre-hospital setting and for children with difficult intravenous access'. Together, this makes diazepam the best intramuscular treatment of *status epilepticus* when this is the only available route (Drislane, 1997). Not surprisingly, midazolam is used routinely by many poisons units for the treatment of OP insecticide poisoning.

The protective effect of midazolam in nerve-agent intoxication has been documented, although not to the same extent as diazepam. Krutak-Krol and Domino (1985) were the first to study the protective effect of midazolam against OP acetylcholine esterase intoxication. They found that both diazepam and midazolam protected against paraoxon-induced lethal effects when given together with atropine. Midazolam was more potent than diazepam, which was explained by its better absorption after intramuscular injection (Mattila *et al.*, 1983). McDonough *et al.* (1999) compared the protective efficacy by six benzodiazepines (avizafone, clonazepam, diazepam, lorazolam, lorazepam and midazolam) in the soman-intoxicated guinea pig (see above). The animals were pre-treated with pyridostigmine and treated with atropine and pralidoxime chloride. Although, all benzodiazepines tried were effective in protecting against soman induced seizure, midazolam was the most potent and most rapidly acting compound. Shih *et al.* (2003) compared the efficacy of different anticholinergics, diazepam and/or midazolam to control seizures produced by six different nerve agents (see above). Guinea pigs were pre-treated with pyridostigmine bromide 30 min prior to challenge with a supralethal dose (SC) of the nerve agent, after which the animals were treated with atropine sulphate and pralidoxime chloride. Five minutes after the start of seizures, animals were treated IM with different doses of the anticholinergics or benzodiazepines. As in the previous study, all drugs were capable of terminating seizure convulsions and midazolam was more than 10-fold more potent than diazepam to protect against all six nerve agents.

CLINICAL STUDIES

For obvious reasons, clinical trials of diazepam, midazolam or avizafone, used either alone or in combination with other therapy, have not been carried out in OP nerve-agent poisoning. It has been suggested that an anticonvulsive effect of diazepam can be obtained at plasma levels of 400–500 $\mu g\ l^{-1}$ (1390–1740 $nmol\ l^{-1}$) or above. This corresponds to a single intravenous dose of 10–20 mg in an adult (Hvidberg and Dam, 1976). A single intravenous dose of 20 mg would produce levels above 400 $\mu g\ l^{-1}$ for just over 2 h, based on the data of Hillestad *et al.* (1974).

A few instances of the use of diazepam have been reported in nerve-agent poisoning. In the Tokyo Subway sarin attack, two cases were reported by Okumura *et al.* (1996), in which diazepam was used. In both cases, atropine and pralidoxime methiodide were also administered. In the first case, the patient had convulsions and generalized fasciculations and received 35 mg of diazepam, while the second had fasciculations and received 30 mg of diazepam. Both patients recovered without sequelae. Apart from these instances, the use of nerve agents have been few, the number of case reports of treatment of nerve-agent poisoning has been small and, in warfare and terrorist use, details of exposure dose and treatment are frequently sketchy. As a consequence, it has been largely necessary to extrapolate the efficacy of diazepam in OP nerve-agent poisoning from experiments in animals and the known efficacy of diazepam in human OP insecticide poisoning.

Thus, there are many case-reports and series of the apparently successful treatment of OP insecticide poisoning with benzodiazepines, most often diazepam. As in nerve-agent intoxications, diazepam is used as adjunctive but widely accepted therapy (Karalliedde and Szinicz, 2001), the primary therapy being an anticholinergic such as atropine sulphate and a pyridinium oxime (pralidoxime mesilate or obidoxime). Examples include Barckow *et al.* (1969), Vale and Scott (1974) Yacoub *et al.* (1981), Merrill and Mihm (1982), Marti *et al.* (1985), LeBlanc *et al.* (1986), de Kort *et al.* (1988), Jovanović *et al.* (1990) and Kusić *et al.* (1991). The indication in most cases for the use of diazepam was convulsions, but diazepam has also been used to control muscle fasciculation and agitation.

Intoxicated patients on medication normally have one or several recurrent cholinergic crises (LeBlanc *et al.*, 1986; Menzel and Wessel, 1966; Merrill and Mihm, 1982). In these crises, more severe symptoms, such as fasciculations and convulsions, are often preceded by increased anxiety or restlessness (Barr, 1964,1966; Hopmann and Wanke, 1974; Menzel and Wessel, 1966; Stoeckel and Meinecke, 1966). It has been suggested that such symptoms should be used as indications for the use of an anticonvulsant and that such prophylactic treatment of the cholinergic crisis would ensure less damage to the brain and the muscles. Accordingly, diazepam administration at doses of 5–10 mg intravenously has been recommended in cases of OP poisoning accompanied by anxiety and restlessness, in the absence of convulsions, (Johnson and Vale, 1992; Marrs, 2004).

From experimental studies, diazepam has been reported to provide limited anticonvulsant efficacy in rats when administered late after the onset of seizures (Shih *et al.*, 2003; McDonough *et al.*, 1995). The loss of diazepam's anticonvulsant effects has been attributed to rapid modulation of $GABA_A$ receptors during status epilepticus (Jones *et al.*, 2002; Mazarati *et al.*, 1998). If large doses of diazepam are required to suppress seizure activity, phenytoin should be considered as an alternative (Johnson and Vale, 1992).

INDICATIONS AND DOSAGE FOR USE OF DIAZEPAM

Diazepam is indicated in poisoning by OP nerve agents where convulsions or pronounced muscle fasciculations are present. Diazepam can also be used to treat anxiety. There appears to be no absolute contraindication to its use in OP poisoning, but in the elderly and in patients with respiratory problems, equipment for ventilatory support should be available, if at all possible. Diazepam should not be given to patients with impaired ventilation unless hypoxia is due to severe convulsions. Diazepam may also aggravate hypotension in critically ill patients. Although diazepam is a developmental neurotoxin, acute nerve-agent poisoning and associated uncontrolled convulsions would represent the greater risk to the fetus and should be treated by all recommended means, including diazepam. Thus, the use of diazepam in pregnancy would not be contraindicated in nerve-agent poisoning of pregnant women. Breast feeding should not be done because of the possibility of nerve agent or nerve agent metabolites and lipid-soluble drugs, such as diazepam, being present in the milk.

Dosage and route

Recommendation of doses of relevant benzodiazepines in different patients can be found in various publications, for example, the guidelines issued by the Department of Health, UK (2005) on pre-hospital emergency treatment of deliberate release of OP nerve agents and Rotenberg and Newmark (2003) and Marrs (2004).

Where nerve agent-induced convulsions are present, the standard intravenous dosage règime for diazepam for the treatment of convulsions is used:

Adults: 10–20 mg IV

Children: 0.2–0.3 mg kg^{-1} IV

Elderly: half the adult dose

These doses can be repeated as necessary, or followed by slow intravenous infusion of diazepam. No maximum dose can be recommended as the severity of convulsions may vary. However, if more than 3–5 mg kg^{-1} is needed over 24 h, it should be considered whether insufficient atropine/oxime therapy or inappropriate supportive care (oxygen) may explain convulsions, and whether ventilatory support is necessary. Diazepam potentiates the sedative effects of other sedatives/hypnotics, including alcohol. Diazepam should not be mixed with other drugs in the same infusion solution or in the same syringe as there are numerous incompatibilities.

CONCLUSIONS

Following exposure to nerve agents, rapid administration of an anticonvulsant at the scene may be critical to success. The armed forces rely on various standardized procedures for administration of diazepam. With the aim of achieving as rapid an onset as possible and a safe and standardized dose, an autoinjector for intramuscular use containing an appropriate anticonvulsant, i.e. avizafone, has become the preferred choice. All likely civilian scenarios do not permit a similar ‘one-fits-all’ approach. Medical personnel may be equipped with autoinjectors to use at their discretion where the perceived risk of civilian exposure is high, e.g. when there is a threat of terrorist activity. However, autoinjectors are designed for the young and fit personnel of the armed forces, whereas exposure of children and the old and sick may occur if nerve agents were used against a civilian population. Diazepam is poorly absorbed via the intramuscular route and intramuscular midazolam, therefore, may be the best alternative or complement to the autoinjector in the acute phase for symptomatic treatment of severely intoxicated civilians. In a mass-casualty situation, with poisoning of civilians, it would be extremely difficult to give optimal treatment until the casualties reached hospital.

The later phase of the treatment could be expected to take place in a hospital setting. The patient is, accordingly, under monitoring and the intravenous route and/or rectal route for administration would be available. This means that whenever symptoms such as restlessness, anxiety, seizures and/or convulsions appear, prompt treatment is easier to achieve. Diazepam and midazolam are still valid as good choices, but it may also mean that another benzodiazepine, such as lorazepam, more and more discussed for treatment of status epilepticus, should be studied.

Given the conditions and the requirements for the initial symptomatic treatment of mass casualties, the diazepam derivative avizafone in autoinjectors and/or midazolam for intramuscular injection should be sufficient. In view of lorazepam's reported efficacy in treatment of status epilepticus, it is recommended that more experimental studies are performed on the efficacy of lorazepam, and also in controlling nerve-agent-induced seizures and convulsions.

REFERENCES

Allison C and Pratt JA (2003). Neuroadaptive processes in GABAergic and glutamatergic systems in benzodiazepine dependence. *Pharmacol Ther*, **98**, 171–195.

Andreasen PB, Hendel J, Greisen G *et al.* (1976). Pharmacokinetics of diazepam in disordered liver function. *Eur J Clin Pharmacol*, **10**, 115–120.

Baldessarini RJ (1980). Drugs and the treatment of psychiatric disorders. In: *Goodman and Gilman's, the Pharmacological Basis of Therapeutics* (AG Gilman, LS Goodman and A Gilman, eds), pp. 391–447. New York, NY, USA: Macmillan.

Barckow D, Neuhaus G and Erdmann WD (1969). Zur Behandlung der schweren Parathion – (E 605) – Vergiftung mit dem Cholinesterase-Reaktivator Obidoxim (Toxogonin). *Arch Toxicol*, **24**, 133–146.

Barr MA (1964). Poisoning by anticholinesterase organic phosphates: it significance in anaesthesia. *Med J Aust*, **00**, 792–796.

Barr MA (1966). Further experience in the treatment of severe organic phosphate poisoning. *Med J Aust*, **00**, 490–492.

Blumer JL (1998). Clinical pharmacology of midazolam in infants and children. *Clin Pharmacokinet*, **35**, 37–47.

Bokonjić D, Jovanović D, Jokanović M *et al.* (1987). Protective effects of oximes HI-6 and PAM-2 applied by osmotic minipumps in quinalphos-poisoned rats. *Arch Int Pharmacodyn Ther*, **288**, 309–318.

Bošković B (1981). The treatment of soman poisoning and its perspectives. *Fund Appl Toxicol*, **1**, 203–213.

Bošković B, Kovačević V and Joanović D (1984). PAM-Cl, HI-6, and HGG-12 in soman and tabun poisoning. *Fund Appl Toxicol*, **4**, S106–S115.

Capacio BR, Whalley CE, Byers CE *et al.* (2001). Intramuscular diazepam pharmacokinetics in soman-exposed guinea pigs. *J Appl Toxicol*, **21**, S67–S74.

Chamberlain JM, Altieri MA, Futterman C *et al.* (1997). A prospective, randomized study comparing intramuscular midazolam with intravenous diazepam for the treatment of seizures in children. *Pediatr Emerg Care*, **13**, 92–94.

Charney DS, Mihic SJ and Harris RA (2001). Hypnotics and sedatives. In: *Goodman and Gilman's, the Pharmacological Basis of Therapeutics*, 10th Edition, pp. 399–427. New York, NY, USA: McGraw-Hill.

Churchill L, Padzernik TL, Cross RS *et al.* (1987). Cholinergic systems influence local cerebral glucose use in specific anatomical areas diisopropyl phosphorofluoridate versus soman. *Neuroscience*, **20**, 329–340.

Clement JG and Broxup B (1993). Efficacy of diazepam and avizafone against soman-induced neuropathology in brain of rats. *Neurotoxicology*, **14**, 485–504.

Costa E and Guidotti A (1979). Molecular mechanisms in the receptor actions of benzodiazepines. *Ann Rev Pharmacol Toxicol*, **19**, 531–545.

de Kort WM, Kiestra SH and Sangster B (1988). The use of atropine and oximes in OP intoxications: a modified approach. *Clin Toxicol*, **26**, 199–208.

Department of Health, UK (2005). *Pre-hospital guidelines for the emergency treatment of deliberate release of organophosphorus (OP) nerve agents*. London, UK: Department of Health (www.dh.uk/policyAndGuidance/EmergencyPlanning/DeliberateReleaseChemical).

Diamantis W and Kletzkin M (1966). Evaluation of muscle relaxant drugs by head-drop and by decerebrate rigidity. *Int J Neuropharmacol*, **5**, 305–310.

Doebler JA, Wall TJ, Martin LJ *et al.* (1985). Effects of diazepam on soman-induced brain neuronal RNA depletion and lethality in rats. *Life Sci*, **36**, 1107–1115.

Drislane FW (1997). Status epilepticus. In: *The Comprehensive Evaluation and Treatment of Epilepsy* (SC Schachter and DL Schomer, eds), pp. 149–172. San Diego, CA, USA: Academic Press.

Gupta RC (1984). Acute malathion toxicosis and related enzymatic alterations in Bubulus bubalis: antidotal treatment with atropine, 2-PAM, and diazepam. *J Toxicol Environ Health*, **14**, 291–303.

Haigh JR and Feely M (1988). Tolerance to the anticonvulsant effect of benzodiazepines. *Trends Pharmacol Sci*, **9**, 361–366.

Hillestad L, Hansen T, Melsom H *et al.* (1974). Diazepam metabolism in normal man 1. Serum concentrations and clinical effects after intravenous, intramuscular, and oral administration. *Clin Pharm Therap*, **16**, 479–484.

Hopmann G and Wanke H (1974). (Maximum dose atropine treatment in severe organophosphate poisoning). (Author's translation)). *Dtsch Med Wochenschr*, **99**, 2106–2108.

Hvidberg EF and Dam M (1976). Clinical pharmacokinetics of anticonvulsants. *Clin Pharmacokin*, **1**, 161–188.

Inns RH and Leadbeater L (1983). The efficacy of bispyridinium derivatives in the treatment of organophosphate poisoning in the guinea-pig. *J Pharm Pharmacol*, **35**, 427–433.

Johnson DD and Lowndes HE (1974). Reduction by diazepam of repetitive electrical activity and toxicity resulting from soman. Eur J Pharmacol, **28**, 245–250.

Johnson DD and Wilcox WC (1975). Studies on the mechanism of the protective and antidotal actions of diazepam in organophosphate poisoning. *Eur J Pharmacol*, **34**, 127–132.

Johnson MK and Vale JA (1992). Clinical management of acute organophosphate poisoning: an overview. In: *Clinical and Experimental Toxicology of Organophosphates and Carbamates*, (B Ballantyne and TC Marrs, eds), pp. 528–535. Oxford, UK: Butterworth-Heinemann.

Johnston GAR (1996). $GABA_A$ receptor pharmacology. *Pharmacol Ther*, **69**, 173–198.

Jones DM, Esmaeil N, Maren S *et al.* (2002). Characterization of pharmacoresistance to benzodiazepines in the rat Li-pilocarpine model of status epilepticus. *Epilepsy Res*, **50**, 301–312.

Jovanović D, Randjelović S and Joksović DC (1990). A case of unusual suicidal poisoning by the organophosphorus insecticide, dimethoate. *Human Exp Toxicol*, **9**, 49–51.

Karalliedde L and Szinicz L (2001). Management of organophosphorus compound poisoning. In: *Organophosphates and Health* (L Karalliedde, S Feldman, J Henry *et al.*, eds), pp. 257–293. London, UK: Imperial College Press.

Karlsson B, Lindgren B, Millquist E *et al.* (1990). On the use of diazepam and pro-diazepam (2-benzoyl-4-chloro-*N*-methyl-*N*-lysylglycin anilide), as adjunct antidotes in the treatment of organophosphorus intoxication in the guinea-pig. *J Pharm Pharmacol*, **42**, 247–251.

Kassa J (2002). Review of oximes in the antidotal treatment of poisoning by organophosphorus nerve agent. *J Toxicol Clin Toxicol*, **40**, 803–816.

Kassa J and Bajgar J (1994). Treatment of Stressogenic Effect of Dichlorvos. Treatment of Stressogenic effect of dichlorvous. Sb Ved Pr Lek Fak Karlovy, Univerzity Hradci Králove, Czech Republic.

Kleinrok Z and Jagiełło-Wojtowicz E (1977). Wptyw wybranych lekow przeciwpadaczkowych na toksycznosc ostra estra dwuizopropylowego kwasu fluorofosforowego u myszy. *Ann Univers Mariae Curie-Skłodowska*, **34**, 127–132.

Klotz U, Avant GR, Hoyumpa A *et al.* (1975). The effects of age and liver disease on the disposition and elimination of diazepam in adult man. *J Clin Invest*, **55**, 347–359.

Klotz U, Antonin KH and Bieck PR (1976a). Pharmacokinetics and plasma binding of diazepam in man, dog, rabbit, guinea pig and rat. *J Pharmacol Exper Therap*, **199**, 67–73.

Klotz U, Antonin KH and Bieck PR (1976b). Comparison of the pharmacokinetics of diazepam after single and subchronic doses. *Eur J Clin Pharmacol*, **10**, 121–126.

Krutak-Krol H and Domino EF (1985). Comparative effects of diazepam and midazolam on paraoxon toxicity in rats. *Toxicol Appl Pharmacol*, **81**, 545–550.

Kulkarni YD, Sharma VL, Dua PR *et al.* (1982). Substituted benzophenones as possible anticonvulsants and tranquillizers. *Ind J Pharm Sci*, **44**, 1–4.

Kusić R, Jovanović D, Randjelović S *et al.* (1991). H-6 in man: efficacy of the oxime in poisoning by organophosphorous insecticides. *Human Exp Toxicol*, **10**, 113–118.

Lallement G, Clarencon D, Brochier G *et al.* (1997). Efficacy of atropine/pralidoxime/diazepam or atropine/HI-6/prodiazepam in primate intoxicated by soman. *Pharmacol Biochem Behav*, **56**, 325–332.

Lallement G, Renault F, Baubichon D *et al.* (2000). Compared efficacy of diazepam or avizafone to prevent soman-induced electroencephalographic disturbances and neuropathology in primates: relationship to plasmic benzodiazepine pharmacokinetics. *Arch Toxicol*, **74**, 480–486.

Lallement G, Masqueliez C, Baubichon D *et al.* (2004). Protection against soman-induced lethality of the antidote combination atropine–pralidoxime–pro-diazepam packed as a freeze-dried form. *J Med Chem Def*, **2**, 1–11.

LeBlanc FN, Benson BE and Gilg AD (1986). A severe organophosphate poisoning requiring the use of an atropine drip. *Clin Toxicol*, **24**, 69–76.

Lipp JA (1972). Effect of diazepam upon soman-induced seizure activity and convulsions. *Electroenceph Clin Neurophys*, **32**, 557–560.

Lipp JA (1973). Effect of benzodiazepine derivatives on soman-induced seizure activity and convulsions in the monkey. *Arch Int Pharmacodyn Ther*, **202**, 244–251.

Lipp JA (1974). Effect of small doses of clonazepam upon soman-induced seizure activity and convulsions. *Arch Int Pharmacody Ther*, **210**, 49–54.

Livezey GT, Marczynski TJ and Isaac L (1986). Enduring effects of prenatal diazepam on the behavior, EEG and brain receptors of the adult cat progeny. *Neurotoxicology*, **7**, 319–334.

Lüddens H and Korpi ER (1995). Biological functions of GABAA/benzodiazepine receptor heterogeneity. *J Psychiatr Res*, **29**, 77–94.

Lundy PM, Magor G and Shaw RK (1978). Gamma aminobutyric acid metabolism in different areas of rat brain at the onset of soman-induced convulsions. *Arch Int Pharmacodyn*, **234**, 64–73.

Maidment MP and Upshall DG (1990). Pharmacokinetics of the conversion of peptide-aminobenzophenone pro-drug of diazepam in guinea-pig and rhesus monkeys. *J Biopharm Sci*, **1**, 19–32.

Mandelli M, Tognoni G and Garattini S (1978). Clinical pharmacokinetics of diazepam. *Clin Pharmacokin*, **3**, 72–91.

Marrs TC (2003). Diazepam in the treatment of organophosphorus ester pesticide poisoning. *Toxicol Rev*, **22**, 75–81.

Marrs TC (2004). The role of diazepam in the treatment of nerve agent poisoning in a civilian population. *Toxicol Rev*, **23**, 145–157.

Marti J, Reverte C, Roig M *et al.* (1985). Uso del diacepam en la intoxicacion por insecticidas organofosforados. *An Esp Pediatr*, **23**, 299–301.

Martin LJ, Doebler JA, Shih T-M *et al.* (1985). Protective effect of diazepam pretreatment on soman-induced brain lesion formation. *Brain Res*, **325**, 287–289.

Martire M, Altobello D, Cannizzaro C *et al.* (2002). Prenatal diazepam exposure functionally alters the $GABA_A$ receptor that modulates [^{3}H] noradrenaline release from rat hippocampal synaptosomes. *Dev Neurosci*, **24**, 71–78.

Mattila MA, Suurinkeroinen S, Saila K *et al.* (1983). Midazolam and fat-emulsion diazepam as intramuscular premedication. A double-blind clinical trial. *Acta Anaesthesiol Scand*, **27**, 345–348.

Mazarati AM, Baldwin RA, Sankar R *et al.* (1998). Time-dependent decrease in the effectiveness of antiepileptic drugs during the course of self-sustaining status epilepticus. *Brain Res*, **814**, 179–185.

McDonough JH, Dochterman LW, Smith CD *et al.* (1995). Protection against nerve agent induced neuropathology, but not cardiac pathology, is associated with anticonvulsant action of drug treatment. *Neurotoxicology*, **16**, 123–132.

McDonough JH, McMonagle J, Copeland T *et al.* (1999). Comparative evaluation of benzodiazepines for control of soman-induced seizures. *Arch Toxicol*, **73**, 473–478.

McEllhatton PR (1994). The effects of benzodiazepine use during pregnancy and lactation. *Reprod Toxicol*, **8**, 461–475.

Menzel G, Wessel. G (1962b). Beitrag zur Toxikologie der Alkylphosphate. *Dtsch Gesundheitsw*, **17**, 1297–I304.

Merrill DG and Mihm FG (1982). Prolonged toxicity of organophosphate poisoning. *Crit Care Med*, **10**, 550–551.

Morselli PL, Cassano GB, Placidi GF *et al.* (1974). Kinetics of the distribution of 14C-diazepam and its metabolites in various areas of cat brain. In: *The Benzodiazepines* (S Garattini, E Mussini and LO Randall, eds), pp. 259–270. New York, NY, USA: Raven Press.

Okumura T, Takasu N, Ishimatsu S *et al.* (1996). Report on 640 victims of the Tokyo subway sarin attack. *Ann Emerg Med*, **28**, 129–135.

Ortells MO and Lunt GG (1995). Evolutionary history of the ligand-gated ion-channel superfamily of receptors. *Trends Neurosci*, **18**, 121–127.

Pazdernik TL, Nelson SR, Cross R *et al.* (1986). Effects of antidotes on soman-induced brain changes. *Arch Toxicol* (Supplement), **9**, 333–336.

Pieri L, Schaffner R, Scherschlicht R *et al.* (1981). Pharmacology of midazolam. *Arzneim-Forsch*, **31**, 2180–2201.

Reves JG, Fragen RJ, Vinik HR *et al.* (1985). Diazepam: pharmacology and uses. *Anesthesiology*, **62**, 310–324.

Rosenow F, Arzimanoglou A and Baulac M (2002). Recent developments in treatment of status epilepticus: a review. *Epileptic Disord* (Supplement 2), S41–S51.

Rotenberg JS and Newmark J (2003). Nerve agent attacks on children: Diagnosis and management. *Pediatrics*, **112**, 648–658.

Rump S and Grudzinska E (1974). Investigations on the effects of diazepam in acute experimental intoxications with fluostigmine. *Arch Toxikol*, **31**, 223–232.

Rump S, Grudzinska E and Edelwejn Z (1973). Effects of diazepam on epileptiform patterns of bioelectrical activity or the rabbit's brain induced by fluostigmine mine. *Neuropharmacology*, **12**, 813–817.

Rump S, Faff J, Szymanska T *et al.* (1976). Efficacy of repeated pharmacotherapy in experimental acute poisonings with fluostigmine. *Arch Toxicol*, **35**, 275–280.

Ryan CL and Pappas BA (1986). Intrauterine diazepam exposures: effects on physical and neurobehavioral development in the rat. *Neurobehav Toxicol Teratol*, **8**, 279–286.

Sellström A (1992). Anticonvulsants in anticholinesterase poisoning. In: *Clinical and Experimental Toxicology of Organophosphates and Carbamates* (B Ballantyne and TC Marrs, eds), pp. 578–586. Oxford, UK: Butterworth-Heinemann.

Shih TM, Duniho SM and McDonough JH (2003). Control of nerve agent-induced seizures is critical for neuroprotection and survival. *Toxicol Appl Pharmacol*, **188**, 69–80.

Sieghart W and Sperk G (2002). Subunit composition, distribution and function of $GABA_A$ receptor subtypes. *Curr Top Medicinal Chem*, **2**, 795–816.

Silva FR and Palermo-Neto J (1999). Developmental, neuro and immunotoxic effect of perinatal diazepam treatment in rats. *Immunopharmacol Immunotoxicol*, **21**, 247–265.

Solberg Y and Belkin M (1997). The role of excitotoxicity in organophosphorous nerve agents central poisoning. *Trends Pharmacol Sci*, **18**, 183–185.

Stoeckel H and Meinecke KH (1966). (On a case of occupational poisoning by mevinphos). (Author's translation). *Arch Toxikol*, **21**, 284–288.

Taysse L, Calvet JH, Buee J *et al.* (2003). Comparative efficacy of diazepam and avizafone against sarin-induced neuropathology and respiratory failure in guinea pigs; influence of atropine dose. *Toxicology*, **188**, 197–209.

Vale JA and Scott GW (1974). Organophosphorus poisoning. *Guy's Hosp Rep*, **123**, 13–25.

van der Kleijn E, van Rossum JM, Muskens ETJM *et al.* (1971). Pharmacokinetics of diazepam in dogs, mice and humans. *Acta Pharmacol Toxicol*, **29**, 109–127.

Viani A, Rizzo G, Carrai M *et al.* (1992). The effects of ageing on plasma albumin and plasma protein binding of diazepam, salicylic acid and digitoxin in healthy subjects and patients with renal impairment. *Br J Clin Pharmacol*, **33**, 299–304.

Yacoub M, Skouri H, Amamou M *et al.* (1981). Intoxications aigües par les insecticides organophosphoré. *J Toxicol Med*, **1**, 165–188.

17 PRETREATMENT FOR NERVE AGENT POISONING

Leah Scott

Dstl, Porton Down, Salisbury, UK

INTRODUCTION

Medical countermeasures are aimed at, ideally, preventing but in most cases, and more realistically, mitigating the effects of exposure to chemical and biological agents, which might be encountered in a diversity of military and civilian situations. As such, they need to be, effective, acceptable, practicable and affordable and while, their importance is recognized, it is acknowledged that they are an adjunct, albeit an important one, to detection, physical protection and consequence management measures.

This chapter describes the rationale for pretreatment with the carbamate, pyridostigmine, which is currently licensed and available in a number of countries as an adjunct to post-poisoning therapy regimens. Prospects for improving upon pretreatment with pyridostigmine, such that it is less reliant upon the provision of immediate self-aid or 'buddy aid', will also be discussed. The deterrent value of good medical countermeasures should not be underestimated although the effectiveness of medical countermeasures for poisoning by nerve agents will be dependent upon exposure level, route and the characteristics of individual agents.

Demonstrating the effectiveness of medical countermeasures for poisoning by chemical agents always poses a particular challenge as there is a heavy reliance upon extrapolating from well-characterized animal models to man. An appreciation of the importance of establishing pharmacological equivalence between the selected animal model and man is a key consideration in extrapolating animal-derived data in this context. It is particularly important that the levels of pretreatment and therapy drugs used in animal studies are such that they have a realistic prospect of being administered to humans at comparable doses and the basis for this comparison needs to be understood.

In the case of nerve agent pretreatments, there are known to be marked species differences in the protection afforded by carbamates. Rats, for example, have high circulating levels of carboxylesterase which can act as a scavenger for soman and in consequence are much less sensitive than guinea-pigs which have lower levels of carboxylesterase (Sterri and Fonnum, 1989; Maxwell *et al.*, 1987) Guinea-pigs are now generally agreed to be the best non-primate model for carbamate pretreatment and they have been widely used as a prelude to, and predictor of the outcome of, more definitive studies in non-human primates and man. Neglible levels of plasma carboxylesterase are found in non-human primates and man (Echobichon and Comeau, 1973).

Compared with other agents, there is a very limited window of opportunity for therapeutic intervention following nerve agent poisoning, especially after exposure to nerve agent vapour. Traditional patterns of military deployment enable a pretreatment option to be considered and, although this might have less widespread utility for civilian populations, pretreatments could be considered as a means of providing first responders with additional protection. As is always the case, a well-characterized concept of use is

Chemical Warfare Agents: Toxicology and Treatment (2nd Edition)
Edited by Timothy C. Marrs, Robert L. Maynard and Frederick R. Sidell

required; this needs to identify realistic triggers to start and to stop administration.

For the foreseeable future, it would seem sensible to consider the provision of both the best possible pretreatment options, especially when these are less reliant upon therapeutic intervention, and the best possible therapy options which are less reliant upon pretreatment. Such pretreatment and therapy regimens should be consistent with each other, as well as with subsequent medical management which is required to optimize the benefit of these life-saving interventions.

PRETREATMENT WITH PYRIDOSTIGMINE

Although there are established (and licensed) therapeutic interventions to counteract aspects of nerve agent poisoning, there are significant limitations in relying upon muscarinic antagonism (provided by atropine), anticonvulsive and muscle relaxant properties (provided by avizafone) and reactivation of phosphonylated cholinesterase (provided by an oxime, e.g. pralidoxime mesilate [P2S]). In particular, nerve agents such as soman, which give rise to rapid and, to all intents and purposes, irreversible inhibition of cholinesterase, are intransigent to reactivation by pralidoxime mesilate. An alternative strategy was therefore required to prevent nerve agent-induced lethality.

The concept of using carbamates as pretreatment which protects against OP intoxication was first mooted in 1956 by Koster who demonstrated that cats pretreated with a small dose of eserine (physostigmine) survived poisoning by supralethal doses of diisopropylphosphorofluoridate (DFP) (Koster, 1956). Subsequently, Berry and Davies (1970) demonstrated the effectiveness of a combination of atropine and carbamate pretreatment against soman poisoning. At that time, it was not considered that pretreatments with substances such as atropine, with marked actions on the central nervous system, would be acceptable for human use. For this reason, effort was concentrated upon pretreatment with pyridostigmine, a quaternary carbamate, which would not be expected to cross the blood–brain barrier and affect the central nervous system. Although some early studies in the UK had investigated the benefits of a combined oxime and pyridostigmine pretreatment regimen (Leadbeater *et al.*, 1985), a decision was made to pursue licensure for pyridostigmine bromide as a nerve agent pretreatment because of its established use in the management of myasthenic patients, for which purpose the drug had been licensed for more than 50 years.

Carbamates bind reversibly (Watts and Wilkinson, 1977) to the esteratic site of acetylcholinesterase and thus prevent both the binding of acetylcholine and its subsequent hydrolysis. The concept of their use as a pretreatment is that carbamate occupancy of a proportion of the available acetylcholinesterase, of which there is an superfluity in mammalian systems, renders it inaccessible to circulating nerve agents which are only able to bind to unprotected enzyme; thus, a pool of acetylcholinesterase is protected from nerve agent challenge. Unbound nerve agent is relatively rapidly hydrolyzed while the acetylcholinesterase which has been reversibly inhibited by pyridostigmine spontaneously decarbamoylates, the carbamate inhibitor is subsequently hydrolyzed and the enzyme becomes available once more to hydrolyze acetylcholine. The time-scale of these dynamic interactions is complex and dependent upon the individual characteristics of the carbamate and nerve agent under consideration.

EFFECTIVENESS OF PYRIDOSTIGMINE PRETREATMENT

Pyridostigmine is highly effective in protecting enzymes in the peripheral nervous system and there is clear evidence of its ability to reverse soman-induced neuromuscular block in respiratory muscle in guinea-pigs *in vitro* (French *et al.*, 1979). Although soman induced an irreversible reduction of tetanic tension, pretreatment with pyridostigmine prior to soman exposure, followed by removal of the anticholinesterases from the tissue bath, produced a return of tetanic tension and a small increase in AChE activity. Additional evidence for the protective effect of pyridostigmine at the neuromuscular junction, playing a major role in survival following soman intoxication, was provided by a cross-species study involving respiratory muscle from non-human primates and man (Green and Smith, 1983).

There is compelling evidence from a number of *in vivo* studies in guinea-pigs and non-human primates (Gordon *et al.*, 1978; Inns and Leadbeater, 1983; Dirnhuber *et al.*, 1979; Kluwe *et al.*, 1987) that pyridostigmine pretreatment is an extremely valuable adjunct to post-poisoning therapy in terms of preventing soman-induced lethality as gauged by significant increases in the protective ratios[1] attained in the presence of pretreatment. Protection against tabun is demonstrably better in guinea-pigs pretreated with pyridostigmine than in animals not receiving pretreatment (Koplovitz *et al.*, 1992a) and the protective ratio following pyridostigmine pretreatment supported with therapy is reported to be >5 for cyclosarin (Koplovitz *et al.*, 1992b). In the absence of pretreatment, traditional approaches to therapy (atropine + oxime +/− anticonvulsant) are sufficient to preserve life following exposure to supralethal doses of sarin and VX in guinea-pigs (Koplovitz *et al.*, 1992a). The presence of pyridostigmine pretreatment in these circumstances does not detract from the efficacy of therapy. It has long been acknowledged, however, that, in the absence of supporting therapy, prereatment with pyridostigmine is not effective in preventing or mitigating the effects of nerve agent poisoning.

Reliance upon increases in the protective ratio, while a valuable indicator and comparator, must be considered with some caution as it neglects to take into account the condition of survivors and in many cases, there are marked clinical signs of long duration which are not prevented by a combination of pyridostigmine and supporting therapy. While, in the context of medical countermeasures, preservation of life is, of course, the prime objective, in practical terms it is vitally important to minimize the logistical burden upon the medical support services by preventing or alleviating incapacitation. Thus, there is scope for improving upon the current approach although the tremendous achievement of protecting against lethality following exposure to a range of nerve agents should not be underestimated.

Although a somewhat imperfect measure, in that the precise role of acetylcholinesterase in the blood has yet to be established, the dose of pyridostigmine in animal and human studies is generally gauged and interspecies comparisons made, on the basis of the level of inhibition of cholinesterase in the blood. This is regarded as a functionally meaningful and pragmatic approach. While earlier studies had investigated the effects of higher doses of pyridostigmine, a target of approximately 30% inhibition of erythrocyte cholinesterase is now widely accepted as a level for optimum protection with minimal side effects. In animal studies, higher doses of pyridostigmine give rise to mild gastrointestinal disturbances and loss of muscle tone at levels which induce > 90% inhibition of erythrocyte cholinesterase (unpublished observations).

PRACTICABILITY OF PYRIDOSTIGMINE AS A PRETREATMENT

Pyridostigmine is formulated as 30 mg tablets of pyridostigmine bromide and issued to UK service personnel as the *Nerve Agent Pretreatment Set* (NAPS). NAPS is packaged in blisterpacks of 21 tablets which are designed to be taken at 8 hourly intervals under the direction of the commanding officer. The regimen is designed to provide an average erythrocyte cholinesterase inhibition of ca. 30%, although potentially useful levels of inhibition are achieved after 1–2 h following administration of the first tablet.

In animal studies, pyridostigmine has generally been administered either by subcutaneous or intramuscular injection for studies of acute effects or, in more recent studies, via subcutaneously implanted mini-osmotic pumps which generate sustained release of pyridostigmine for up to 28 days in an attempt to mimic repeated oral administration in field situations. It is acknowledged that such continuous administration is an imperfect, if pragmatic, surrogate for repeated oral administration.

[1] Protective ratio: LD_{50} with treatment/LD_{50} without treatment.

(a)

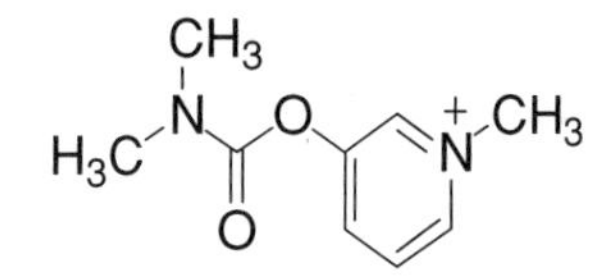

(b)

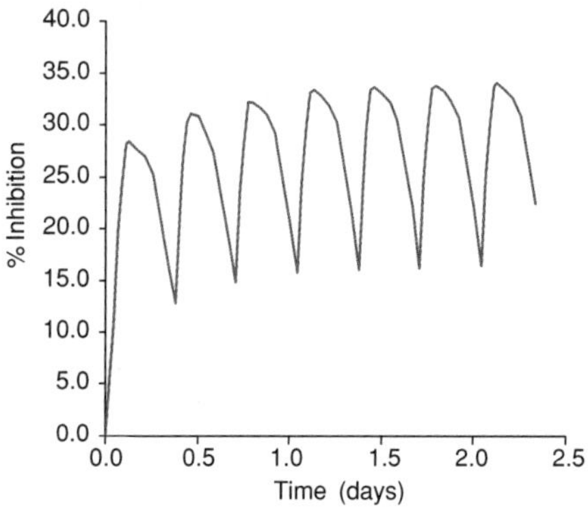

(c)

Information for the Patient on Nerve Agent Pre-Treatment Tablet Set L1A1 (NAPS L1A1)

Why should you read this leaflet?

Please read this leaflet carefully. It tells you what you should know about this pre-treatment when at risk of poisoning by nerve agents.

The leaflet does not contain complete information about this product. If you have any questions about what is written here or wish for any further information please ask your doctor or medical services.

What is Nerve Agent Pre-Treatment Tablet Set L1A1?

Brand Name: Nerve Agent Pre-Treatment Tablet Set L1A1 (NAPS L1A1)

Active ingredient: Each NAPS L1A1 tablet contains 31.5 mg pyridostigmine bromide

Other ingredients: Pregelatinised starch, maize starch, lactose, colloidal silicon dioxide, talc, magnesium stearate

This medicine reduces the action of proteins in the body called cholinesterases.

This helps to counteract the effects of some nerve agents.

What should the product look like?

The round, bioconvex, white to off-white NAPS tablets are marked L1A1 and are in a blister pack in a plastic sleeve, each sleeve contains 21 tablets and is sealed in an aluminium foil sachet. Ten foil sachets are packaged in one cardboard box.

Product Licence and Marketing Authorisation Holder

Secretary of State for Defence
Ministry of Defence
Main Building
Whitehall
London SW1 2HB

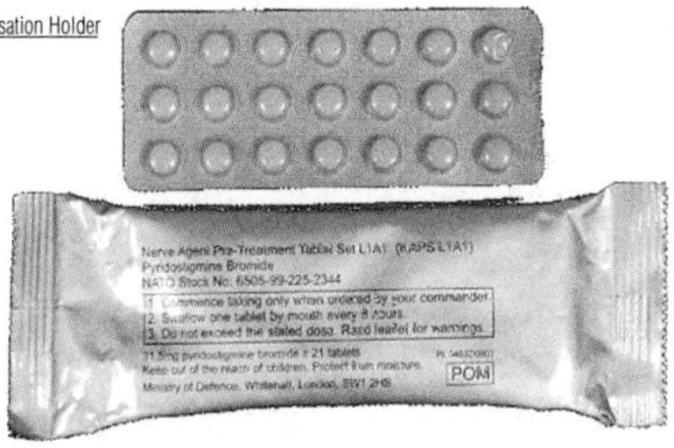

Manufacturer

Pean Pharmaceutical Services Ltd
Units 23 & 24
Tafarnaubach Industrial Estate
Tredegar
Gwent
NP2 3AA

What type of product is Nerve Agent Pre-Treatment Tablet Set?

NAPS is for use as a pre-treatment by members of the armed services and MOD sponsored civilians at risk of poisoning by nerve agents (organophosphorus cholinesterase inhibitors).

Is Nerve Agent Pre-Treatment Tablet Set safe for you?

NAPS is for pre-treatment of active adult service personnel,
Dosage is not applicable to children or the elderly.

NAPS should not be given to subjects with obstructions affecting the digestive system or passing of urine or an allergy to bromides. If you are allergic to any of the ingredients, tell your doctor.

Treatment of subjects with bronchial asthma should be undertaken with extreme caution. Service personnel with bronchial asthma should have been medically downgraded to prevent their deployment outside these areas, and should therefore be close to medical assistance.

Special Warnings and Precautions

Information on recognising symptoms of nerve agent poisoning and the use of NAPS tablets are available in the Survive to Fight booklet issued by the Director of Army Training.

Do not exceed the stated dose.

Effects on ability to drive and to use machines

The recommended treatment with NAPS tablets does not affect ability to drive or use machines.

When to use Nerve Agent Pre-Treatment Tablet Set?

Adult personnel should begin taking it only when ordered to do so in anticipation of nerve agent exposure. One NAPS tablet should be taken by mouth every 8 hours until the order is given to cease taking your medicine. Do not take more than two tablets in an 8 hour period. In the event of a dosage being missed, you should take the next dose as soon as possible. The length of time for which NAPS is taken will be determined by commanders on medical and intelligence advice.

Overdosage

Effects of overdose may include abdominal cramps, diarrhoea, nausea and vomiting, increased peristalsis, salivation, increased bronchial secretions, sweating and constriction of eye pupil (due to muscarinic effects). Further effects may include muscular cramps, localised muscle contractions, general weakness, slow heart action and low blood pressure (due to nicotinic effects).
If respiration is severely depressed, artificial ventilation should be attempted. Atropine sulphate may be given to counteract muscarinic effects.

What side effects may NAPS cause?

The side effects are mainly gastrointestinal and may include nausea, vomiting, diarrhoea and abdominal cramps.

Contact your doctor or medical services if you experience side effects from your medicine, including any that are not mentioned.

Storage Precautions

Packs of NAPS tablets should be stored at room temperature with protection from moisture.
Do not use after expiry indicated on label.

Other important points

Product Licence Number, PL04537/0003
Legal Category: POM
Date of preparation of this leaflet: January 2000

NATO STOCK NO. 6505-99-225-2344

PLT132 2930111

Figure 1. (a) Molecular structure of pyridostigmine. (b) Erythrocyte chlinesterase profile following repeatedoral administration in man. (c) Photograph of blister pack of NAPS, with issued information sheet

ACCEPTABILITY OF PYRIDOSTIGMINE PRETREATMENT

Clearly, when pretreatments are to be administered in anticipation of a nerve agent attack, it is essential to ensure that they are not in themselves incapacitating. Preventing individuals from fulfilling their required duties would be clearly unacceptable. Furthermore, it is important to ensure that pretreatments do not, in themselves, exacerbate the effects of exposure to nerve agents as is theoretically possible.

In the absence of other interventions, administration of this quaternary carbamate, at a dose designed to inhibit erythrocyte cholinesterase by approximately 30%, would not be expected to lead to marked acute effects.

NAPS was issued and deployed used in the 1991 Gulf Conflict and there have been a number of suggestions that it has been responsible for aspects of Gulf Conflict related illnesses (see Chapter 18). In this context, the literature was extensively surveyed by Golomb in 1999. There has been much discussion about the ability of pyridostigmine to cross the blood–brain barrier. As indicated previously, because of its quaternary structure, pyridostigmine would not be expected to cross the blood–brain barrier under normal circumstances, although there has been much discussion about what happens when additional stressors are involved. There are conflicting claims and counterclaims about the effects of pyridostigmine on blood–brain barrier permeability; some groups assert that electric shocks and exercise-induced stress lead to cholinesterase inhibition in the CNS (Friedman *et al*., 1996) although Grauer *et al*. (2000) were unable to replicate these findings. Other groups are adamant that stressors make the blood–brain barrier more resistant to penetration by pyridostigmine (Lallement *et al*., 1998). The, as yet, unresolved controversy undoubtedly reflects the particular challenges associated with the establishment and validation of stress models in small animals and their extrapolation to man.

In the particular context of protecting against the effects of nerve agents, pyridostigmine's limited ability to cross the blood–brain barrier, while, in some respects, an attractive attribute in terms of minimizing potential behavioural disruption when administered as a pretreatment, has a serious flaw in that it is unable to confer protection against the widely acknowledged, if relatively poorly understood, implications of cholinesterase inhibition in the brain. For this reason, the indisputedly centrally active carbamate, physostigmine, is currently under consideration by the UK as an alternative to pyridostigmine pretreatment.

Following the Gulf Conflict, a programme of work was undertaken by the UK Ministry of Defence to investigate whether the multiple vaccinations and /or NAPS used during deployment in the conflict, gave rise to adverse effects. This long-term study did not attempt to establish a model for Gulf Conflict-related illness *per se* but rather set out to refine and extend an existing multifaceted non-human primate model to establish whether the most frequently reported signs and symptoms reported by 'Gulf Veterans', which included inability to concentrate, muscle weakness, sleep disturbances and changes in immunological responsiveness, could be related to administration of the vaccines and/or NAPS. The NAPS elements of the study (which is currently being prepared for publication) will enable definitive statement to be made about the effects of 28 days administration of pyridostigmine on a range of behavioural, electrophysiological and immunological parameters in marmosets up to 18 months following administration (Hornby *et al*., 2006).

FUTURE PROSPECTS: PRETREATMENT WITH PHYSOSTIGMINE AND HYOSCINE

More recently, as mentioned above, effort has been focused on the development of an alternative pretreatment based upon the centrally acting carbamate physostigmine, which has continued to be of interest despite having previously been considered and discounted because of concerns about its potential side-effects and relatively narrow therapeutic window. These concerns have largely been addressed by the concept of co-administration of the potent muscarinic antagonist, hyoscine, which has been shown to reduce the incidence of side-effects of physostigmine administration and has additional anticonvulsive properties in its own right.

EFFECTIVENESS OF PHYSOSTIGMINE/HYOSCINE PRETREATMENT

Continuous administration over 1–13 days of physostigmine, at a dose rate which inhibited erythrocyte acetylcholinesterase by 20–30%, combined with hyoscine, has been shown to be extremely effective in preventing lethality and minimizing soman-induced incapacitation in guinea-pigs (Wetherell, 1994) and soman and sarin-induced incapacitation in non-human primates (Scott *et al.*, 1994). Administration of higher doses of physostigmine alone or in combination with trihexyphenidyl (Lim *et al.*, 1987,1991) or physostigmine and hyoscine had previously been shown to protect against soman and VX poisoning in the guinea-pig (Phillipens *et al.*, 1988,2000; Anderson *et al.*, 1992).

PRACTICABILITY OF PHYSOSTIGMINE/HYOSCINE PRETREATMENT

In addition to the previous misconceptions about the desirability of centrally acting pretreatments, oral dosing with physostigmine has been problematical because of its short half-life. Alternative opportunities for physostigmine administration, such as slow release oral formulations and transdermal delivery systems, have been considered and the latter approach has been deemed to show more promise. Moreover, a useful precedent for transdermal delivery of hyoscine had been established by the existence of a commercially available transdermal patch marketed for the prevention of travel sickness. Meshulam *et al* (1995) demonstrated that physostigmine or a combination of physostigmine and hyoscine, the latter administered via a Scopoderm™ transdermal patch, protected against soman poisoning in the guinea-pig. Transdermal delivery of physostigmine has also been investigated in the treatment of Alzheimer's Disease (Levy *et al.*, 1994).

ACCEPTABILITY OF PHYSOSTIGMINE/HYOSCINE PRETREATMENT

Both physostigmine and hyoscine readily cross the blood–brain barrier and lead to an expected range of effects in the CNS. When administered alone, both drugs, albeit at a much higher dose than are currently being proposed for use as pretreatment, have previously been shown to disrupt performance of a range of cognitive tasks in animals and man. Hyoscine is known to impair aspects of memory (Ridley *et al.*, 1984) and working memory performance (Rusted and Warburton, 1988). Physostigmine has been reported to improve both discrimination learning (Warburton and Brown, 1972) and perceptual function (Wetherell, 1992). When co-administered, pharmacological antagonism between the two drugs mitigates these potentially adverse effects and thus, selection and optimization of dosage regimens continues to represent both an opportunity and a key technical challenge. D'Mello and Sidell (1991) showed that physostigmine-induced emesis occurred at comparable levels of cholinesterase inhibition in marmosets and man, although these represented markedly higher levels of cholinesterase inhibition than are proposed for physostigmine in the context of nerve agent pretreatment. Moreover, physostigmine-induced emesis could be antagonized in a dose-related manner by the co-administration of hyoscine (unpublished observations from this laboratory). Nonetheless, it has been important to demonstrate that the pretreatment regimen does not in itself induce decrements in performance. At doses which are as far as possible, pharmacologically equivalent in marmosets and man, Muggleton *et al.* (2003) showed no effect on a cognitive task in marmosets. Furthermore, the combination of physostigmine and hyoscine used did not exacerbate the effects of 0.5 LD_{50} sarin or soman.

COMPARISON OF PRETREATMENT APPROACHES

The relative strengths and weaknesses of pretreatment with pyridostigmine and physostigmine/hyoscine are exemplified by the results of the following previously unpublished marmoset study which was undertaken to compare the effectiveness of the two pretreatment regimens, supported by triple therapy, against the lethal and incapacitating effects of sarin or soman.

Methods

Sixteen male and sixteen female common marmosets (*Callithrix jacchus*) weighing 280–380 g had previously been trained to perform a task of visually guided reaching which is sensitive to motivational, attentional and motor disturbances (d'Mello *et al.*, 1985). In this task, animals learn to retrieve small items of preferred food from a moving conveyor belt positioned in front of their home cage and performance is gauged by the number of attempts that are made and their effectiveness in retrieving rewards.

Under general anaesthesia, on Day 0, all animals were implanted with a mini-osmotic pump (as described by Muggleton *et al.*, 2003) which delivered either pyridostigmine or physostigmine/hyoscine over 14 days. The dose rates used were selected on the basis of previous studies. The dose of pyridostigmine used would have been expected to inhibit erythrocyte acetylcholinesterase by ca. 30 % and the doses of physostigmine and hyoscine would have been expected to inhibit erythrocyte acetylcholinesterase by ca. 20% and give rise to plasma hyoscine levels of ca. 800 pmol ml^{-1} plasma.

From Day 1, visually guided reaching was assessed throughout administration of both pretreatments. On Day 13, all marmosets were challenged with a dose of either sarin or soman that, in unprotected animals, would have been lethal is less than 30 min. Up to three doses of therapy with atropine, P2S and avizafone, at doses which had been previously shown to be pharmacologically equivalent in marmosets and man (Scott, 1996) were administered by the intramuscular route according to the recommended schedule for field deployment: one ComboPen, in this case, one ComboPen equivalent on the appearance of marked signs of poisoning and afterwards at 15 min intervals if required. The task of visually guided reaching was presented at 1.5, 4, 24 and 48 h following nerve agent administration, unless precluded by signs and symptoms of nerve agent poisoning.

Results

Neither pretreatment regimen disrupted performance of the visually guided reaching task. Following nerve agent challenge, the incidence and duration of clinical signs were such that all animals pretreated with pyridostigmine needed all three ComboPen equivalents. Only one ComboPen equivalent was necessary for other than two of the physostigmine/hyoscine pretreated animals; these required two ComboPen equivalents. All animals survived.

Figure 2 shows the duration of loss of posture observed following nerve agent challenge. The animals which had been pretreated with physostigmine/hyoscine exhibited very few clinical signs following exposure to sarin or soman. Overall, less than half of them did not lose posture; of those which did, all animals had regained posture by 30 min following sarin poisoning and by 45 min following soman poisoning. By contrast, the majority of animals pretreated with pyridostigmine lost posture for prolonged periods. All of the sarin exposed animals regained posture overnight but, in the soman exposed group, 6/8 animals had not regained posture 24 h after nerve agent exposure, and these were humanely killed for ethical reasons at that time.

Figure 2 also shows the performance of the task of visually guided reaching. None of the pyridostigmine-pretreated animals were able to or chose to perform the task 1.5 h following poisoning by either sarin or soman. When presented with the task 4–48 h following soman exposure, the two animals which were in a suitable condition to undertake the task, performed at control levels of performance. At 4 h following poisoning by sarin, three of the pyridostigmine-pretreated animals attempted the task, although performance was poor and seven animals subsequently performed the task at 70% control. By contrast, three of the physostigmine/hyoscine-pretreated animals were sufficiently capable and motivated to attempt the task 1.5 h following sarin exposure at which time they achieved control levels of performance; six animals performed the task at near control levels of performance 4 h following sarin and subsequently, all animals performed the task although the performance of one animal was lower than that of the others. The two physostigmine/hyoscine-pretreated animals

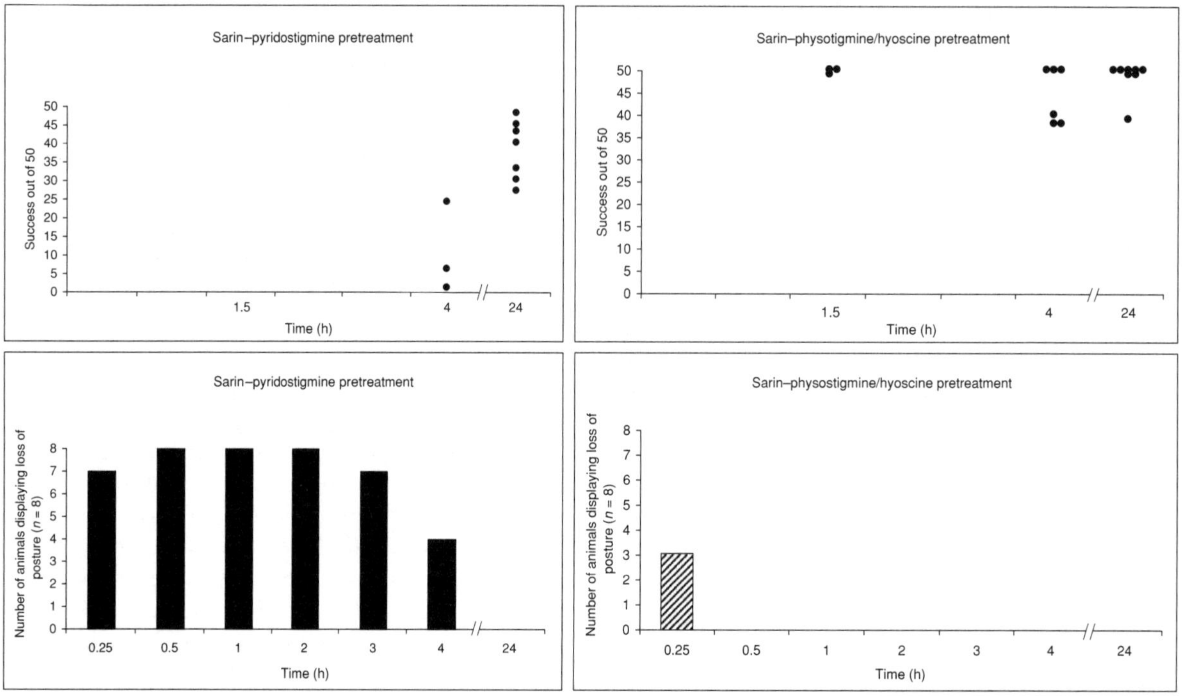

(a) Sarin

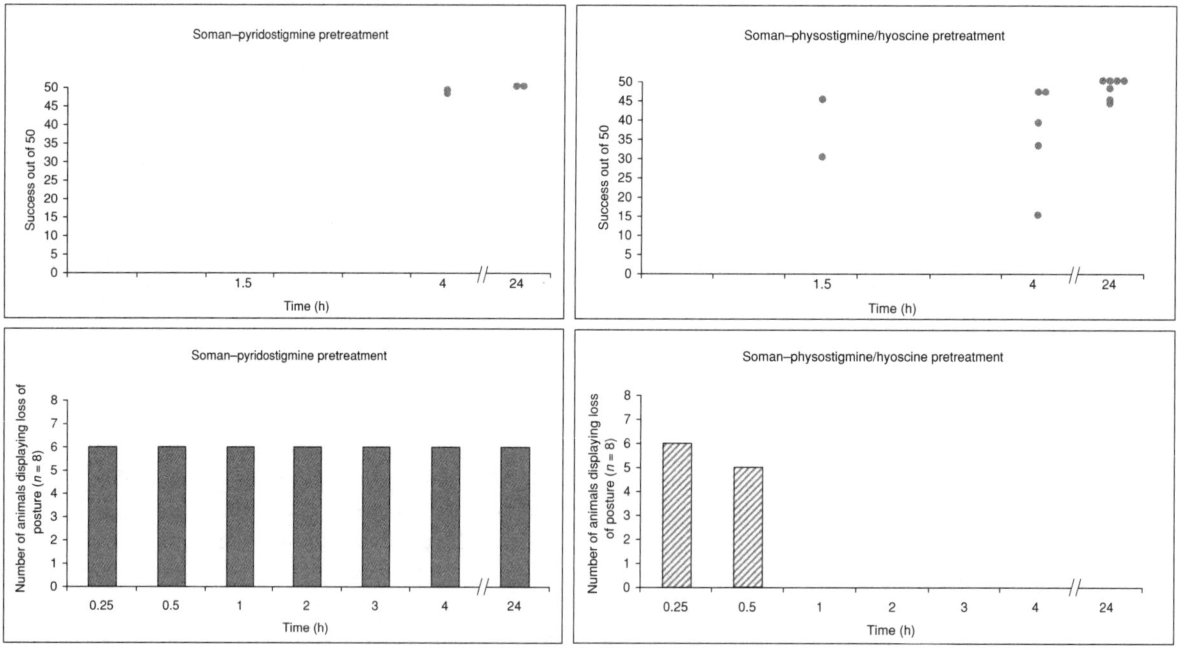

(b) Soman

Figure 2. Duration of loss of posture and performance of a visually guided reaching task in marmosets pretreated with pyridostigmine or physostigmine/hyoscine, and challenged with (a) sarin or (b) soman.

which worked at 1.5 h following soman exposure, performed at 60 and 90% of control. At 4 h following soman, five physostigmine/hyoscine-pretreated animals performed at >50% control and subsequently, 7 animals performed at near control levels.

DISCUSSION AND CONCLUSIONS

On the basis that there were no effects on performance of the task of visually guided reaching, neither pretreatment would be considered to lead to acute decrements in performance in its own right.

If the data on condition post-exposure, as gauged by loss of posture, were to be extrapolated directly to man, the condition of the pyridostigmine-pretreated animals would have placed a significant burden upon the medical support services. That said, it is undoubtedly the case that medical intervention, such has would be expected to be available in military or civilian situations, would be expected to have retrieved the situation for these animals. Pretreatment and therapy are intended to be supported by post-immediate therapy and clinical management although, it is clearly preferable from a logistical viewpoint, that pretreatment alone or in combination with immediate self/'buddy' aid is sufficiently effective to obviate the need for additional intervention. This would seem to be the case if data from the animals in the physostigmine/hyoscine-pretreated groups were extrapolated directly to man. The results of this comparison reinforce the findings of a similar comparison of pretreatment alone against a wider range of nerve agents in guinea-pigs (Wetherall *et al.*, 2002). The results of the behavioural element of the marmoset study summarized here, provides added confidence in the protection afforded against incapacitation.

SUMMARY AND LONGER TERM PROSPECTS

There is now a realistic prospect of introducing a pretreatment which offers major improvement on the currently available in-service provision because of the protection that it affords against lethality and incapacitation and the fact that, depending upon challenge level of nerve agent, pretreatment with physostigmine and hyoscine may obviate immediate post-poisoning therapy and medical management over hours or days.

Nonetheless, broader questions remain about the desirability and practicability of deploying pretreatments for nerve agent poisoning in all of the military and civilian, situations in which exposure to these threats could be encountered. For example, although identifying triggers for instigating administration of pretreatment may be relatively straightforward in certain military scenarios, other situations may be more problematical. It would, therefore, seem wise to continue to aspire to provide the best possible pretreatments, which are less reliant upon supporting therapy as well as the best possible therapies which are less reliant upon pretreatment. In some settings, the distinction between pretreatment and therapy options will inevitably become blurred.

For example, in many of the guinea-pig and non-human primate studies cited above, the experimental design was such that the mini-osmotic pumps used to deliver pretreatments over days, continued to make physostigmine and hyoscine available to the animals for up to 24 h after agent administration. The presence of these two drugs following poisoning, even in the absence of supporting therapy, did not seem to adversely affect the outcome of the studies compared with investigations in which pretreatment was administered acutely. Whether such characteristics can be exploited for bridging traditional approaches to pretreatment and therapy is currently under investigation in studies that are investigating the utility of administering physostigmine and hyoscine, along with various adjuncts, following nerve agent poisoning. Results from preliminary studies in guinea-pigs seem to show considerable promise (Wetherell *et al.*, 2006).

REFERENCES

Anderson DR, Harris LW, Woodard CL *et al.* (1992). The effect of pyridostigmine pretreatment on oxime efficacy against intoxication by soman or VX in rats. *Drug Chem Toxicol*, **15**(4), 285–294.

Berry WK and Davies DR (1970). The use of carbamates and atropine in the protection of animals against poisoning by 1,2,2,-trimethypropyl methylphosphonofluoridate. *Biochem Pharmacol*, **19**, 927–934.

D'Mello GD, Duffy EAM and Miles SS (1985). A conveyor belt task for assessing visuo-motor co-ordination in the marmoset (*Callithrix jacchus*): Effects of diazepam, chlorpromazine, pentobarbital and damphetamine. *Psychopharmacology*, **86**, 125–131.

D'Mello GD and Sidell FR (1991). A model for carbamate and organophosphate-induced induced emesis in humans. *Neurosci Biobehav Rev*, **15**, 179–184.

Dirnhuber P, French MC, Green DM *et al.* (1979). The protection of primates against soman poisoning by pretreatment with pyridostigmine. *J Pharm Pharmacol*, **31**, 295–299.

Echobichon DJ and Comeau AM (1973). Pseudocholinesterases of mammalian plasma: physicochemical properties and organophosphate inhibition in eleven species. *Toxicol Appl Pharmacol*, **24**, 92–100.

French M, Wetherell JR and White PD (1979). The reversal by pyridostigmine of neuromuscular block produced by soman. *J Pharm Pharmacol*, **31**, 290–294.

Friedman A, Kaufer D, Shemer J. *et al.* (1996). Pyridostigmine brain penetration under stress enhances neuronal excitability and induces early immediate transcriptional response. *Nature Med*, **2**, 1382–1384.

Golomb BA (1999). *A Review of the Scientific Literature as it Pertains to Gulf War Illnesses*, Volume 2, *Pyridostigmine Bromide*. Rand Corporation: (http://www.gulflink.osd.mil/library/randrep/pb paper/ or www.rand.or/pubs/monographreports/ MR1018.2/).

Gordon JJ, Leadbeater L and Maidment MP (1978). The protection of animals against organophosphate poisoning by pretreatment with a carbamate. *Toxicol Appl Pharmacol*, **43**, 207–216.

Grauer E, Alkali D, Kapon J *et al.* (2000). Stress does not enable pyridostigmine to inhibit brain cholinesterase after parenteral administration. *Toxicol Appl Pharmacol*, **164**, 301–304.

Green DM and Smith AP (1983). The reversal by pyridostigmine of soman induced neuromuscular blockade in primate respiratory muscle. *Br J Pharmacol*, **79**, 252P.

Harris L, Lennox W, Talbot BG *et al.* (1984). Toxicity of anticholinesterases; interaction of pyridostigmine and physostigmine with soman. *Drug Chem Toxicol*, **7**, 507–526.

Harris LW, Talbot BG, Lennox WJ *et al.* (1991). Physostigmine (alone and together with adjunct) pretreatment against soman, sarin, tabun and VX intoxication. *Drug Chem Toxicol*, **14**, 265–281.

Hornby RJ, Pearce PC, Bowditch AP *et al.* (2006). Multiple Vaccine and Pyridostigmine bromide interactions in the common marmoset *Callithrix Jacchus*: Immunological and endochrinological effects. *Int Immunopharmacol*, **6**, 1765–1779.

Inns RH and Leadbeater L (1983). The efficacy of bispyridinum derivatives in the treatment of organophosphonate poisoning in the guinea-pig. *J Pharm Pharmacol*, **35**, 427–433.

Kluwe WM, Chinn JC, Feder P *et al.* (1987). Efficacy of pyridostigmine pretreatment against acute soman intoxication in a primate model. In: *Sixth Medical Chemical Defense Bioscience Review*, pp. 227–234. Aberdeen Proving Ground, MD, USA: US Army Medical Research Institute of Chemical Defense.

Koplovitz I, Gresham V, Dochterman LW *et al.* (1992a). Evaluation of the toxicity, pathology and treatmenyt of cyclohexylmethylphosphonofluoridate poisoning in rhesus monkeys. *Arch Toxicol*, **66**, 622–628.

Koplovitz I, Harris L, Anderson DR *et al.* (1992b). Reduction by pyridostigmine pretreatment of the efficacy of atropine and 2-PAM treatment of sarin and VX poisoning. *Fund Appl Toxicol*, **18**, 102–106.

Koster R (1956). Synergisms and antagonisms between physostigmine and diisopropyl fluorophosphates in cats. *J Pharmacol Exp Ther*, **88**, 39–46.

Lallement G, Foquin A, Baubuchon D *et al.* (1998). Heat stress, even extreme, does not induce penetration of pyridostigmine into the brain of guinea-pigs. *NeuroToxicology*, **19**, 759–766.

Leadbeater L, Inns RH and Rylands JM (1985). Treatment of poisoning by soman. *Fund Appl Toxicol*, **5**, S225–S231.

Levy A, Brandeis R, Treves TA *et al.* (1994). Transdermal physostigmine in the treatment of Alzheimer's Disease. *Alzheimer Dis Assoc Disord*, **8**, 15–21.

Lim DK, Ito Y, Yu ZJ *et al.* (1987). Prevention of soman toxicity after the continuous administration of physostigmine. *Pharmacol Biochem Behav*, **31**, 633–639.

Lim DK, Hoskins B and Ho IK (1991). Triexyphenidyl enhances physostigmine prophylaxis against soman poisoning in guinea-pigs. *Fund Appl Toxicol*, **16**, 482–489.

Maxwell DM, Brecht KM and O'Neill BL (1987). The effect of carboxylesterase inhibition on interspecies differences in soman toxicity. *Toxicol Lett*, **39**, 35–42.

Meshulam Y, Davidovici R, Wengier A *et al.* (1995). Prophylactic transdermal treatment with physostigmine and scopolamine against soman intoxication in guinea-pigs. *J Appl Toxicol*, **15**(4), 263–266.

Miller SA, Blick DW, Kerenyi SZ *et al.* (1993). Efficacy of physostigmine as a pretreatment for organophosphate poisoning. *Pharmacol Biochem Behav*, **44**, 343–347.

Muggleton NG, Bowditch AP, Crofts HS *et al.* (2003). Assessment of a combination of physostigmine and scopolamine as pretreatment against the behavioural effects of organophosphates in the common marmoset (*Callithrix jacchus*). *Psychopharmacology*, **166**, 212–220.

Phillipens IHCM, Busker RW, Wolthius OL *et al.* (1988). Subchronic physostigmine pretreatment in guinea-pigs: effective against soman and without side effects. *Pharmacol Biochem Behav*, **59**, 1061–1067.

Phillipens IHCM, Melchers BPC, Olivier B *et al.* (2000). Scopolamine augments the efficacy of physostigmine against soman poisoning in guinea-pigs. *Pharmacol Biochem Behav*, **65**, 175–182.

Ridley RM, Bowes PM, Baker HF *et al.* (1984). An involvement of acetylcholine in object discrimination learning and memory in the marmoset. *Neuropsychologia*, **22**, 253–263.

Rusted JM and Warburton DM (1988). The effects of scopolamine on working memory in healthy young volunteers. *Psychopharmacology*, **96**, 145–152.

Scott L (1996). Extrapolation of animal data to man: issues for consideration. In: *Proceedings of the CB Medical Treatment Symposium*, Vol. 2, p. 86. Spiez, Switzerland.

Scott EAM, Hollands RN and Wetherell JR (1994). The current UK medical countermeasures for nerve agent poisoning; Physostigmine and hyoscine as an alternative to pyridostigmine. In: *Proceedings of the CB Medical Treatment Symposium*, Vol. 1, pp. 1.14–1.17.

Sterri S and Fonnum F (1989). Carboxylesterase – the soman scavenger in rodents: heterogeneity and hormonal influence. In: *Enzymes Hydrolysing Organophosphorous Compounds* (E Reiner, WN Aldridge and FCG Hoskinn, eds), pp. 155–164. Chichester, UK: Ellis Harwood.

Warburton DM and Brown K (1972). The facilitation of discrimination performance by physostigmine sulphate. *Psychopharmacology*, **27**, 275–284.

Watts P and Wilkinson RG (1977). The interaction of carbamates with acetylcholinesterase. *Biochem Pharmacol*, **26**, 757–761.

Wetherell A (1992). Effects of physostigmine on stimulus encoding in a memory-scanning task. *Psychopharmacology*, **109**, 198–202.

Wetherell JR (1994). Continuous administration of low dose rates of physostigmine and hyoscine to guinea-pigs prevents the toxicity and reduces the incapacitation produced by soman poisoning. *J Pharm Pharmacol*, **46**, 1023–38.

Wetherell J, Hall T and Passingham S (2002). Physostigmine and hyoscine improves protection against the lethal and incapacitating effects of nerve agent poisoning in the guinea-pig. *NeuroToxicology*, **23**, 341–349.

Wetherell JR, Price M and Mumford H (2006). A novel approach for medical countermeasures to nerve agent poisoning in the guinea-pig. *NeuroToxicology*, **27**(4), 485–491.

18 GULF WAR SYNDROME

Simon Wessely and Matthew Hotopf

Institute of Psychiatry, King's College London, London, UK

THE 1991 GULF WAR

Iraq invaded Kuwait on August 2nd 1990. Shortly after Coalition Forces, led by the United States, began a military deployment known as 'Operation Desert Shield'. On January 17th 1991, an active air campaign began against Iraq, 'Operation Desert Storm', and on February 24th a ground war began, lasting only four days. It was a resounding military success. Iraqi forces were beaten in the field and expelled from Kuwait.

Not only was the campaign a military success, it was also a medical success. It is only during the modern era that deaths from battle have exceeded deaths from disease for most armies. Traditionally, fighting in hostile environments such as the desert has been associated with morbidity and mortality, often substantial, from causes not related to enemy action such as heat stroke, dehydration and infectious disease. There is no evidence for any deaths from those sources among American or British personnel during the Gulf campaign (Hyams *et al.*, 1995). Hence, the military medical authorities must have ended the campaign relieved not to have had to deal with large-scale casualties, and delighted with the success of their preventive measures.

Yet not long after the end of hostilities, this optimistic assessment began to be questioned as reports started to emerge from the United States of unusual clusters of illness in Gulf War veterans. These were either hard to pin down, or, where specific claims were made, as for example of an increase in birth defects, proved to be illusory. However, by 1993 concerns were growing, and the phrase 'Gulf War Syndrome' was starting to be heard, first in the United States, then in the United Kingdom and finally Australia.

GETTING OUR TERMINOLOGIES RIGHT

Strictly speaking, the term Gulf War Syndrome is a misnomer, since no new illness or symptom cluster unique to Gulf War veterans has been identified (see below). Instead it is more accurate to talk about 'Gulf War Illness', or 'Gulf War Illnesses', or even the 'Gulf War health effect'. Having said that, it is likely that it is Gulf War Syndrome that will continue to be the label used to describe the medical legacy of the war, a *fait accompli* belatedly accepted by the UK Ministry of Defence after a 2005 War Pension Tribunal ruling. In addition, it is a definite and by no means minor legacy; 21% of US Gulf veterans now receive some form of disability support from the Veterans' Administration (VA). In 1998, 17% of British Gulf veterans believed they were suffering from Gulf War Syndrome (Chalder *et al.*, 2001). At the time of writing, 10% of British and 21% of US Gulf War veterans are in receipt of some form of war pensions, gratuity or disability payments.

As well as distinguishing between illness and syndrome, we should also distinguish between illness and disease. Kleinman and others (Kleinman, 1988) have emphasized the importance of making a distinction between disease – in which there is pathological evidence of dysfunction – and illness, in which a person reports symptomatic distress and suffering but no pathological

Chemical Warfare Agents: Toxicology and Treatment (2nd Edition)
Edited by Timothy C. Marrs, Robert L. Maynard and Frederick R. Sidell

condition can be found to explain this as yet. At the moment, Gulf veterans are suffering from illness, but not disease (see below). Note that once again the need to be careful with terminology – this time the phrase 'as yet'. The history of medicine gives many examples of illnesses that are subsequently associated with distinct pathologies that are then recognised as diseases, and so it may prove with Gulf War illness. However, it is not all unidirectional – the history of medicine also gives examples of illnesses that are never found to have a basis in pathological changes, and gradually fade away. Few now talk about 'Soldier's Heart' or 'Effort Syndrome', but these were the Gulf War Syndromes of their day (see Jones and Wessely, 2005).

THE EARLY STUDIES

The first co-ordinated response to the problem was to invite any veteran with health problems to come forward for detailed medical evaluation. This began in the United States, with health registries established by both the VA and the Department of Defense (DOD) to provide systematic clinical evaluations, and was then repeated in the United Kingdom with the establishment of the Medical Assessment Programme (MAP).

There have now been several analyses of those attending these programmes, which now amount to over 3000 in the United Kingdom and over 100 000 in the United States. The results have not suggested any unusual pattern of illness – instead the largest diagnostic category has been medically unexplained symptoms and syndromes (Joseph, 1997; Roy *et al.*, 1998; Coker *et al.*, 1999; Lee *et al.*, 2001; Gray *et al.*, 2004). These programmes, however, have served a very important function, since attendance is not random, and is associated with a wide variety of war-related exposures (Smith *et al.*, 2002). The therapeutic effect of a sound assessment, information and provision of services and/or reassurance should not be underestimated.

Scientifically, other than providing evidence of significant concerns in the veteran community, such programmes can give only limited scientific information because of their non-random selection. However, one would expect that if service in the Gulf was associated with either disease new to medical science (as with the first appearance of AIDS at the beginning of the 1980s), or a dramatic elevation of a recognized but hitherto rare condition, then this would have been detected. Neither has happened (Gray *et al.*, 2004).

THE EPIDEMIOLOGICAL STUDIES

The first large epidemiological study of Gulf-related illness was a questionnaire-based study of a random sample of Gulf veterans and appropriate military controls from the state of Iowa (The Iowa Persian Gulf Study Group, 1997). This showed increased rates of symptom reporting in the Gulf cohort. Symptom-defined conditions ranging from chronic fatigue syndrome, depression, post-traumatic stress disorder and others were all elevated.

In the King's study, we compared a random sample of 4246 UK Gulf war veterans, drawn from all three Armed Services and including both serving and non-serving, with similar numbers of non-deployed personnel, and with an active duty control group – namely members of the UK Armed Forces who had served in the difficult and dangerous environment of Bosnia from the start of the UN Peacekeeping Mission (1992–1997) (Unwin *et al.*, 1999).

UK Gulf veterans were between two and three times more likely to report each and every one of the 50 symptoms that were inquired about. Whatever the symptom, the rate was at least twice as high in the Gulf cohort than either the non-deployed cohort, or the Bosnia cohort. Health perception was decreased in the Gulf cohort, but physical functioning was only very slightly different, and still above expected non-military norms. Hence, the Gulf veterans experienced more symptoms, endorsed more conditions, felt worse, but were still physically functioning well as a group, than either the non-deployed cohort or those deployed to an unpleasant and stressful Bosnia (Unwin *et al.*, 1999).

Since then, these findings have been reported from a variety of other population-based studies in the USA, Canada, Australia, Denmark and the United Kingdom, and all show essentially the same findings, the exception being Saudi

Arabia (Gackstetter *et al.*, 2005). In a recent review, Barrett and colleagues were able to conclude that overall Gulf War veterans report two to three times the rates of common symptoms as their non-deployed colleagues (Barrett *et al.*, 2003).

Significantly, all these studies, including our own, are limited by the use of self-report measures. Self-reported symptoms are not, of course, a good guide to findings on clinical examination (McCauley *et al.*, 1999). High rates of reported symptoms do not necessarily reflect high rates of physical disorder. Indeed, if that were the case it would contradict a considerable body of literature on the nature of somatic symptoms in the community.

What exactly is the size of observed increase in morbidity? Symptoms in the community are distributed dimensionally, but if one imposes certain arbitrary but not implausible cut-offs or case definitions, then the excess burden of 'caseness' or ill health, seems to be between 20 to 30% (Cherry *et al.*, 2001a; Steele, 2000; Fukuda *et al.*, 1998; Gray *et al.*, 2002).

However, the evidence also shows that there is no increase in any well-defined physical disorders (Gray *et al.*, 1996; Eisen *et al.*, 2005), with the sole exception of the possibility of an increase in a rare but usually fatal neurological disorder, motor neuron disease (MND), known as amyotrophic lateral sclerosis (ALS) in the USA. A large-scale US study using multiple methods of data collection, including newspaper articles and website appeals, has reported 40 cases in US Gulf veterans, a significant increase, sufficient for the US government to declare the disease service-related (Horner *et al.*, 2003). However, there has been no increase in mortality due to neurological disease, which would be expected given that ALS is a fatal disease. Instead, it remains possible that there has been an 'over-ascertainment' of cases in the Gulf War veterans compared to the controls (Rose, 2004). Irrespective of this, while ALS is a devastating disease for those affected, it is still very rare in veteran populations, and cannot account for anything more than a fraction of the observed increase in morbidity in Gulf veterans, which is not accompanied by evidence of peripheral nervous system involvement.

IS THERE A GULF WAR SYNDROME?

The term Gulf War Syndrome has acquired remarkable media and popular salience, but is there any such thing? A syndrome implies a unique constellation of signs and/or symptoms. For there to be a Gulf War Syndrome, then not only must there be evidence of such a unique constellation, but it must also be found in the context of the Gulf conflict, and not elsewhere (Wegman *et al.*, 1998).

Robert Haley, a Dallas-based epidemiologist was the first to argue in favour of a unique Gulf War Syndrome (Haley *et al.*, 1997), using factor analysis. However, his data came from a much-studied single naval reserve construction battalion, already known to have high rates of illness, had a 41% response rate and a sample size of 249. Haley also did not have a control group, military or non-military. As numerous commentators have pointed out, this makes it difficult to establish whether or not the proposed new syndrome is indeed linked to Gulf service or not (Landrigan, 1997). Undeterred, Robert Haley has continued to pursue the trail of a unique Gulf War Syndrome, and has latterly claimed to have found specific evidence of damage to the basal ganglia and pons (Haley, 2003). He believes the cause of this is exposure to chemical weapons and/or pesticides and has stated that 'there is substantial evidence and general acknowledgement that large numbers of military personnel were repetitively exposed to low environmental levels of the organophosphate chemical nerve agent sarin and pesticides. . . .' (Haley *et al.*, 1999). He believes that exposures to chemical weapons and pesticides has led to specific central and peripheral nerve damage since the end of the conflict, and that this risk will continue to increase over time, leaving a 'Sword of Damocles' hanging over the heads of the veterans . However, a series of expert review committees and panels have failed to be convinced, by the suggestion of widespread use of chemical weapons, while the suggestions of specific localized brain damage needs to be replicated.

Since then, studies that use epidemiological-defined subjects and appropriate controls have generally not found evidence of a unique Gulf War Syndrome. True, clusters of symptoms can be found. For example, there is general agreement that Gulf veterans are affected more by a cluster of symptoms that can be collectively grouped as cognitive/psychological, and most find a factor, labelled by Haley as 'arthromyoneuropathy', but which others prefer to call 'musculo-skeletal'. However, the most important question is not whether such clusters exist, but are they unique to Gulf veterans?

In the King's study we did indeed find evidence to support a particular factor structure to symptoms in the Gulf cohort, but this was no different from the factor structure in the Bosnia or Era controls. The Gulf group had more symptoms experienced at greater intensity, but there was no difference in the way these symptoms could be organized (Ismail *et al.*, 1999; Everitt *et al.*, 2002). A series of controlled US studies draw similar conclusions (Fukuda *et al.*, 1998; Knoke *et al.*, 2000; Doebbeling *et al.*, 2000; Bourdette *et al.*, 2001). Only the VA group find something different in a very large study of deployed and non-deployed veterans. Five of six factors were very similar between the groups, but there was one 'Gulf' factor containing symptoms such as blurred vision, loss of balance, tremors/shaking and speech difficulty (Kang *et al.*, 2002). Those emphasizing this factor reported substantially more Gulf exposures such as DU (depleted uranium), botulism vaccine (not used by UK forces), CARC paint, eating contaminated food and exposure to nerve gas. The possibility that this is influenced by either recall bias or reverse causality cannot be excluded.

The balance of evidence is currently *against* there being a distinct Gulf War Syndrome. In many ways, it is a side-issue that has attracted more interest and polemic than it deserves. The key question is whether or not there is a Gulf health effect, and this is established beyond reasonable doubt. Whether or not this amounts to a unique illness, identifiable only by complex statistical techniques, seems to be a secondary issue. However, having said all that, there seems little doubt that whatever the scientific verdict, the term Gulf War Syndrome is here to stay as a portmanteau term covering all of the medical and social legacy of the conflict.

THE POSITION ELSEWHERE

So far, we have considered the position solely from a US/UK perspective. However, many countries participated in the Coalition Forces, and we are now starting to hear from them as well.

The first non-US country to publish a detailed examination of its Gulf veterans was Canada (Anon, 1998). The results were remarkably consistent with what had already been reported from the USA, and would be reported from the UK, Denmark and most recently finally Australia (Ikin *et al.*, 2004; Kelsall *et al.*, 2004).

Each country's experience has also given examples of natural variations and experiments, which will in time, prove informative. For example, Canada sent three vessels to the Gulf – two used pyridostigmine prophylaxis, and one did not. Yet rates of illness were identical between the three ships (Anon, 1998). Likewise, Danish Gulf veterans also have elevated rates of symptomatic ill health (Ishoy *et al.*, 1999), yet nearly all were only involved in peace-keeping duties after the end of hostilities and neither used pyridostigmine prophylaxis nor received vaccinations against biological agents. Australian health concerns seemed to have surfaced considerably later than those in the USA or UK (McKenzie *et al.*, 2004).

The French experience remains enigmatic. French authorities have consistently denied that any health problems have emerged in their Gulf forces. Likewise, although one or two articles have appeared in the French press about Gulf War Syndrome in other countries, there have been very few media reports of similar stories in France, which largely appeared in 2000 and then seem to have ceased, prompted by the death of a single French Gulf veteran from neoplastic disease (Anon, 2000). If indeed the French were not affected, this would be important epidemiological evidence, since the pattern of Forces protection used by the French differed from that of both the Americans and the British, and in particular anthrax vaccination was not used. One must, however, be cautious about the lack of evidence. No systematic study has yet been reported, although one is underway – and one should recall official denials of any problem in this country prior to the publication of systematic studies.

The cultural pattern of illnesses in the Francophone world also differs from that in the English-speaking world. Illness entities such as chronic fatigue syndrome or multiple chemical sensitivity are hardly acknowledged (Mouterde, 2001; Girault, 2002).

The conclusion so far is therefore that something about Gulf service has affected the symptomatic health of large numbers of those who took part in the campaign from most of the Coalition countries. At the same time, no compelling evidence has emerged to date of neither distinct biomedical abnormalities nor premature mortality.

WHAT MIGHT BE GOING ON?

We now address the question of what does this mean? We argue for the importance of three, related, factors. The first factor we propose is the events of the Gulf War itself. We will suggest that the initial trigger for Gulf-related illnesses were the peculiar hazards of modern warfare and the methods used to protect troops from such hazards. In particular, we will consider the threat of chemical and biological warfare and the methods used to reduce that threat.

The second factor relates to events after the conflict, and the interaction between media, government and military, which served to foster a climate of suspicion and rumour.

The third factor links the particular hazards of modern warfare to Western contemporary societal attitudes on environmental risks and concerns. Many of the hazards encountered during the Gulf campaign had resonances with a common, and often passionate, societal agenda, which gave the narratives of the Gulf veterans a particular resonance and link to wider civilian concerns.

IS GULF WAR ILLNESS THE RESULT OF BIOLOGICAL HAZARDS IN THE GULF?

The most popular explanation for the Gulf Health effect among the media and the public is that the cause of Gulf War Syndrome lies in the particular hazards of that conflict. In particular, most attention has been given to the measures taken to protect the combatants from the threat of chemical and biological warfare (CBW). These included immunizations against biological weapons such as plague and anthrax, and pyridostigmine tablets to protect against exposure to anticholinesterase nerve agents such as sarin. Other hazards included exposure to DU or to the smoke from the oil fires ignited by the retreating Iraqi forces.

Evidence is conflicting. A small group of US Gulf veterans were definitely exposed to DU in the form of shrapnel fragments, and are being intensively monitored. Some subtle changes were soon reported in neuropsychological and neuroendocrine function (McDiarmid *et al.*, 2000). Ten years after exposure urine uranium remains persistently elevated (McDiarmid *et al.*, 2004), but renal function remains normal, which is a crucial observation since renal toxicity is the main long-term hazard of uranium exposure – contrary to the popular perception it is only weakly radioactive. There is some evidence of a mutagenic effect in those with DU shrapnel fragments still in their bodies (McDiarmid *et al.*, 2004) but immune competence was normal. More surprising was the finding that personnel who were involved in the same friendly fire incidents but who did not receive DU shrapnel wounds, also had significant elevations of uranium levels in their urine, presumably via inhalation or contamination (Gwiazda *et al.*, 2004).

Smoke from the burning oil wells received much publicity at the end of the land war, and perhaps for that reason was closely monitored on the spot. However, no sound evidence of any long-term health effects from the smoke has been presented, and that theory has largely disappeared from the literature.

Very relevant to this volume is the fact that pyridostigmine bromide (PB), a reversible inhibitor of acetylcholinesterase, was used as a pre-treatment against exposure to nerve gas. Although side-effects were very frequently reported during its use in the Gulf campaign, these were short lived, and no acute toxicity was observed (Keeler *et al.*, 1991). This is important. Long-term organophosphorus toxicity, which is certainly a hazard, has only been clearly documented in the aftermath of acute toxicity (Fulco *et al.*, 2000). The Canadian experience (*vide supra*) also argues against a prominent role for PB. The latter has been used in civilian practice for the treatment of myasthenia gravis for many years, and in higher doses than used by the Armed Forces, without apparent adverse effect. It has even been used as a treatment for the fatigue associated with post-polio syndrome (Trojan and Cashman, 1995). The extensive cumulative experience with PB, first licensed for use in myasthenia gravis in 1955, in civilian neurological practice argues against an important role for PB *per se* in Gulf-related illness (Anon, 1997b; Ablers and Berent, 2000) (see Chapter 17).

It has also been argued that even though the half-life of PB is measured in hours, a potentially toxic interaction might have resulted from the combination of PB and other agents, most particularly pesticides, and some animal data have been advanced in support (Abou-Donia *et al.*, 1996,2004). In an elegant randomized controlled trial exposing healthy military volunteers to permethrin-impregnated uniforms, DEET-containing skin cream, and oral PB, in a manner consistent with US military doctrine under both restful and stressful conditions, PB levels were higher immediately after stress. However, physical and neurocognitive outcome measures, and self-reported side-effects, did not significantly differ by exposure group, at least not in the short term (Roy *et al.*, 2006).

An alternative approach has been to suggest that host variation may explain differences in individual susceptibility, which is a perfectly plausible hypothesis. Haley has claimed that polymorphisms in the enzyme detoxification pathways for organophosphate compounds are related to symptoms (Haley *et al.*, 1999), which is theoretically plausible (Furlong, 2000) but has not been confirmed in two studies from the UK (Mackness *et al.*, 2000; Hotopf *et al.*, 2003). Finally, animal experiments have failed to confirm any significant adverse delayed neurobehavioural effects from low-dose pyridostigmine alone or in combination with low-dose sarin, as well as confirming that pyridiostigmine did convey some protection against sarin, the purpose for which it is administered (Scremin *et al.*, 2003).

The role of pesticides, and in particular organophosphate pesticides, has also been much discussed. Large quantities of pesticides in various forms were used by all the combatants to reduce the risk of infectious disease. In general, providing these were used appropriately and by trained personnel, little hazard should have resulted (Anon, 1997a). However, whether or not this actually happened in practice is unclear. There have now been numerous expert committee reports on both sides of the Atlantic, but most particularly in the UK on this topic. The conclusions are that there is no disputing the acute toxic effects of organophosphates on the human nervous system, but there remains considerable uncertainty and controversy about the effects of low level chronic exposure (Committee on Toxicity of Chemicals in Food, Consumer Products and the Environment, 1999; Anon, 1997b; United States General Accounting Office, 1997; RCP/RCPsych, 1998). Even when chronic low-dose exposure has been documented, this is not synonymous with evidence of damage to the nervous system (Albers *et al.*, 2004). Information mismanagement has played a part in making this a controversial issue.

Detailed studies of the peripheral nervous system in both US and our own epidemiologically derived samples have failed to find evidence of neuropathy (Bourdette *et al.*, 2001; Davis *et al.*, 2004), including normal findings on the sensitive single-fibre EMG paradigm, which is strong evidence against any chronic peripheral nerve damage (Amato *et al.*, 1997; Sharief *et al.*, 2002). Again the main dissenting opinion has come from Robert Haley in Dallas, even though his own study of his highly selected small population did not reveal any clinically detectable neurological involvement (Haley *et al.*, 1997).

In the USA, but not in the UK, much attention was given to the possibility that troops had been exposed to low levels of the nerve agents sarin and cyclosarin following the probable accidental destruction of an Iraqi arms dump at Khamisiyah. There was no contemporary evidence of chemical weapon detection or clinical evidence of exposure, and since then little evidence has been found that those possibly exposed to the plume thought to have resulted from the incident had any difference in post-war illness (Gray *et al.*,

1999; McCauley *et al.*, 2002). Later studies using better exposure estimate models failed to find any suspicious pattern of hospitalization in those possibly exposed with the exception of a small increase in hospital episodes for cardiac dysrhythmias (Smith *et al.*, 2003), which provides a tantalizing link with the rich historical literature on cardiac dysfunction in service personnel, going back via effort syndrome as far as 'Soldier's Heart' in the US Civil War.

A recent animal study failed to show any adverse effects from low-dose sarin (Pearce *et al.*, 1999).

Irrespective of the health effects of low-dose sarin or cyclosarin, how could such exposure have occurred? Some say that the Iraqi forces did use sarin in the theatre of war, and have been more specific than that, claiming that the results of studies suggest that this happened on day four of the ground war (see http://www.gulflink.org/stories/disasternews/Studies.htm). At a Federal Investigator's meeting in December 2001 one prominent advocate of this theory suggested that the Khamisayah episode was a 'CIA smokescreen' to cover up the real facts about deliberate use of sarin by the Iraqis. Overall, the claim that there was deliberate, but undetected, use of chemical agents by the Iraqis lacks any military or intelligence credibility. We should remember that not only were considerable resources devoted to detection, military health care personnel were on alert for chemical casualties during the war and were particularly attentive to this threat because exposure of health personnel while caring for contaminated patients could cause illness and death.

There is also little evidence of central nervous system damage as indicated by objective evidence of neuropsychological deficits (David *et al.*, 2002). Subjective symptoms of cognitive difficulties are, of course, very common, but just as in the literature of chronic fatigue syndrome, these do not relate very well to objective indices of neuropsychological difficulties, but do relate to symptoms of post-traumatic stress disorder or depression (David *et al.*, 2002; Lindem *et al.*, 2003). In conclusion, 'it is unlikely that the tens of thousands of Gulf War veterans with unexplained health problems are suffering from the results of exposure to neurotoxic chemicals' (Spencer *et al.*, 2001).

Some of the most suggestive evidence comes from studies of the possible effects of the vaccination programme used to protect the Armed Forces against the threat from biological weapons, although this may partly reflect the fact that quantifying exposure to vaccines, although difficult, is not impossible, unlike some of the other postulated hazards.

The United States programme involved immunization against anthrax and botulism, while the United Kingdom chose to protect its Armed Forces against plague and anthrax, with the additional use of pertussis vaccine as adjuvant to speed up the response to anthrax vaccine (Ministry of Defence, 2000). As well as receiving vaccines against potential biological warfare attack, numerous additional routine vaccines were provided. For US personnel, these included boosters of tetanus, diphtheria and oral polio vaccines, plus other measures recommended for the region including meningococcal, typhoid and yellow fever vaccines, and immune globulin against hepatitis A (Ministry of Defence, 2000). Similarly, the UK forces were routinely provided with boosters for tetanus, diphtheria (if required) and oral polio, plus typhoid, yellow fever, cholera, hepatitis A and B (if indicated), meningococcal meningitis A and B. This, with a total of three anthrax, two pertussis and two plague vaccines, would bring the total number of potential vaccines received over a six month period to twenty (Ministry of Defence, 2000).

We found a relationship between receiving both multiple vaccinations in general, and those against CBW agents in particular, and the persistence of symptoms, despite controlling for obvious confounders. The finding that multiple vaccinations in other contexts, including deployment to Bosnia, was not associated with any increased experience of symptoms, suggests some interaction between multiple vaccination and active service deployment to the Gulf (Unwin *et al.*, 1999; Hotopf *et al.*, 2000). The Manchester group likewise found a relationship between reported receipt of multiple vaccinations and subsequent symptoms (Cherry *et al.*, 2001b).

However, what does all this mean? Rook and Zumla argued from a theoretical perspective that the particular medical countermeasures used by the UK Armed Forces, namely the combination of anthrax, pertussis and multiple immunizations, given sometimes under stressful conditions, could bias the immune system towards a Th2-cytokine pattern (Rook and Zumla, 1997). An American study of help-seeking veterans failed to find compelling evidence of any immunological dysfunction (Everson *et al.*, 2002), and nor did a study of Dutch veterans who had served with the UN in Cambodia and who have experienced similar symptoms to Gulf veterans (Soetekouw *et al.*, 1999). We used a nested case-control design to study sick Gulf veterans, well Gulf veterans and sick veterans from other deployments. We found evidence of ongoing Th1 activation associated with ill health in the Gulf veterans detectable some nine years after the conflict, and a biased generation of memory cells secreting the suppressor cytokine, IL-10 (Skowera *et al.*, 2004).

The second countermeasure invoked as Th2-biasing was the shear excess of immunological stimuli represented by the multiplicity of vaccines, administered within a relatively short time-frame. The putative mechanisms through which excess antigen load might lead to Th2 biasing are not immediately obvious, and in the Rook/Zumla original description of the hypothesis, the supporting citations related to excessive stimulus with a single vaccine rather than multiple vaccines. The immunological arbiter of Th polarization is the dendritic cell (DC), which acquires stimuli such as vaccines through pinocytosis, digests them proteolytically and then presents them as short fragments to naïve T cells, along with requisite Th polarizing signals (cytokines, surface molecules), which are generated as a result of the integration of numerous signals relating to the nature of the antigen. Chemical and other manipulations of DCs are known to be able to shape the polarizing signals and lead to skewing, but there is no evidence to date that DCs are altered in the presence of multiple vaccine agents. We have attempted to test this hypothesis *in vitro*, using human DCs exposed to multiple vaccines including anthrax, plague and whole-cell pertussis. Our finding is that DCs summate the signals available into a final, integrated effector response (Skowera *et al.*, 2005). For example, in the presence of three vaccines, of which one is a strong Th1-biasing stimulus, DCs become potent producers of Th1-skewing cytokines and promote Th1 polarization. We could find no evidence that additional stimuli inhibited or subverted the DC response. In summary, the epidemiological evidence for a multiple-vaccine effect on Gulf War-related illness remains a potentially important aetiological lead, but mechanistic studies available at this stage do not identify any immunological basis for it.

RECALL BIAS

There are also some general objections to the view that the cause of the ill-health experienced by Gulf veterans is only the physical and/or toxic hazards of the campaign (Anon, 1997b). Some studies have found that certain symptom patterns are related to certain self-reported exposures (Haley and Kurt, 1997; Wolfe *et al.*, 1998), but others have not (Kroenke *et al.*, 1998; Unwin *et al.*, 1999; The Iowa Persian Gulf Study Group, 1997; Gray *et al.*, 1999,2002). In general, it seems that most exposures are linked to most outcomes, if judged by retrospective self-recall. If one accepts recall as accurate, then the links would be via the aggregate stressors of war, or alternatively if one does not, as a product of recall bias and search after meaning influenced by current health. These are not mutually exclusive – we found when asking veterans identical questions about Gulf War exposures over time, that recall varied according to current health perception. If this has improved, fewer exposures were recalled, while if subjective health had declined, the opposite was true (Wessely *et al.*, 2003). If self-recall of hazard exposure is not stable (Spencer *et al.*, 2001), and is influenced by state markers (Wessely *et al.*, 2003), then this changes the interpretation of studies that find specific links between recall of hazards and specific symptoms, raising the possibility that the symptom came first.

IS GULF WAR ILLNESS A MODERN MANIFESTATION OF POST-CONFLICT ILL-HEALTH?

The argument that the cause of the Gulf health effect lies in the unique nature of modern warfare would be substantially weakened if it could be shown that similar clinical syndromes have arisen after other conflicts which did not involve the particular hazards of the Gulf War.

Interpretable medical records and accounts only commence from the middle of the 19th Century, but from then onwards the literature does contain clinical descriptions of ex-servicemen (and it is always men) with conditions that do show considerable similarities to the Gulf narratives (Hyams *et al.*, 1996). These condition have received many different labels – 'Soldier's Heart', later termed 'Effort Syndrome', owes its provenance to the Crimean and American Civil Wars. Shell shock and neurasthenia dominate the writings of World War I, while 'Agent Orange Syndrome' and 'Post-Traumatic Stress Disorder' emerged after Vietnam.

Hyams' argument rests entirely on a reading of secondary sources, but we have recently concluded a study based on primary sources as well. We began by locating clinical case histories from the Crimean War and Indian Mutiny (Jones and Wessely, 1999) which begin the theme of chronic, unexplained symptoms. This was then continued in a systematic study of UK war pension files from the Boer War, via the World Wars I and II, and ending with clinical files from the Gulf War Medical Assessment Programme (Jones *et al.*, 2002).

The results showed that post-conflict syndromes which show considerable similarities to Gulf War illness have been reported after all of the major conflicts involving the British Armed Forces. The names have changed but there has also been some shift in the symptom patterns recorded, from the debility/weakness picture of Victorian neurasthenia to the more neuropsychiatric (including cognitive and depressive symptoms) of modern times.

Thus, sending young men (and increasingly women) to war does result in some casualties that cannot be explained on a solely physical injury basis, and that the symptoms experienced are similar to those experienced by Gulf War veterans.

Yet the Gulf War was not a particularly stressful conflict in the traditional sense. The active ground war only lasted a few days. Casualties amongst the Coalition Forces were exceptionally light. It would be historically wrong to extrapolate from the prolonged privation, fear and danger of, for example, the trenches or the Pacific War to the Gulf War.

However, it would be equally wrong to claim that Gulf veterans were not exposed to stress or fear of any sort. Once the war was successfully concluded, there were no shortage of commentators telling us how it was inevitable that the Iraqi forces would crumble under the assault of the Coalition Forces, and that there never was a realistic chance that Saddam Hussein would authorize the use of chemical or biological agents. However, that was not how it seemed to those personnel preparing in the desert during the weeks and months before the ground war was launched, nor to the military planners themselves. Plus, some of those stressors have taken their toll.

There is a firm consensus that physical symptomatology is related to stressful exposures in all sorts of circumstances, and going to war is no exception (Ford *et al.*, 2001; Storzbach *et al.*, 2000). We know that rates of formal psychiatric disorder, especially depression and anxiety disorders, have increased about two-fold in Gulf veterans (Stimpson *et al.*, 2003). Likewise, post-traumatic stress disorder (PTSD), is elevated in UK and US Gulf veterans, but not to the extent that could explain the overall increase in ill-health (Ismail *et al.*, 2002) – in other words most symptomatic Gulf War veterans did not have PTSD.

However, PTSD may be a rather limited concept to understand Gulf health. PTSD depends upon a person being exposed to a discrete traumatic event, which in this context means *combat*, and which leads to persistence of the traumatic memory. However, that fails to capture the psychological threat posed by chemical and biological weapons (Stokes and Banderet, 1997; Betts, 1998). Such weapons 'engender fear out of all proportion to their threat' (O' Brien and Payne, 1993) – they are as much, if not more, weapons of psychological as physical warfare (Holloway *et al.*, 1997). Even in training, up to 20% of those who took part in exercises using simulated exposure to irritant gases showed moderate to severe psychological anxiety (Fullerton and Ursano, 1990).

Ignoring the current controversy over weapons of mass destruction, there was no doubting back in 1991 that Iraq possessed such weapons, and had used them extensively during the Iran–Iraq war and against Kurdish civilians. It was anticipated that they would be used in the forthcoming campaign. Countermeasures were untested, and probably insufficient. Effective measures, such as wearing the full Nuclear-Biological-Chemical (NBC) suits were uncomfortable and induced a state of partial sensory deprivation. Surveys during Operation Desert Shield of US Forces confirmed that the threat of CBW was the commonest expressed fear of the coming conflict. The ground war may have only taken a few days, but the deployment itself lasted over many months. During Operation Desert Storm, there were several thousand documented chemical alarm alerts. Subsequently, the consensus of opinion is that none were true positives, and that Iraq did not use its CBW arsenal. However, at the time each alert had to be assumed to be genuine. Thus, even if traditional military stressors were not a prominent feature of the active campaign, a well-founded and realistic anxiety about the threat of 'dread weapons' could still be important. It does not take much imagination to accept the very potent psychological effects of operating in an environment where one could be subject to chemical attack, nor the malign effects of believing, even erroneously, that one has been the victim of such an attack. (Fullerton and Ursano, 1990; Riddle *et al.*, 2003).

Indeed, believing oneself to be exposed to such weapons has been frequently found to be associated with the development of symptoms (Nisenbaum *et al.*, 2000; Unwin *et al.*, 1999), sometimes very strongly (Haley *et al.*, 1997; Proctor *et al.*, 1998; Stuart *et al.*, 2003). The psychological impact generated by the knowledge that such weapons existed, even if they were not used, is substantial.

GULF WAR SYNDROME CAN BE FOUND IN PEOPLE WHO HAVE NEVER BEEN TO THE GULF OR SERVED IN THE ARMED FORCES

Patients with multiple unexplained symptoms, all of them reported in the narratives of Gulf veterans, are also encountered in civilian medical practice and literature. In the popular literature, first-person accounts and patient-orientated literature (in the media and on the Internet) exist with considerable similarities to those of some Gulf veterans. One finds such material under diverse headings such as 'ME', total-allergy syndrome, electrical hypersensitivity, dental amalgam disease, silicon-breast implant disease, hypoglycaemia, chronic Lyme disease, sick building syndrome and many more.

Turning to the professional literature, studies are now reporting that the rates of various symptom-defined conditions originally described in the civilian population are also elevated in the Gulf cohorts. Chief among these are chronic fatigue syndrome (CFS) (Kipen *et al.*, 1999; Kang *et al.*, 2003) and/or multiple chemical sensitivity (MCS) (Black *et al.*, 2000; Reid *et al.*, 2001). These syndromes, which also include fibromyalgia, irritable bowel syndrome (IBS) and others, overlap not only with each other (Wessely *et al.*, 1999), but also with Gulf War illness.

That symptom-based conditions overlap with Gulf War illness is not surprising, given that all of the epidemiological studies confirm that Gulf veterans experience an increased reporting of each and everyone of the symptoms that make up the case-definitions of all of these syndromes found in civilian practice. This does not mean, however, that either CFS/IBS/MCS and the others are all the same, or that Gulf War illness is the same either. It does mean that they all overlap, that discrete boundaries cannot be drawn between them (or if they can, we currently have no idea where these boundaries are). It also means that any explanation of Gulf War illness must explain how similar conditions can be found either in non-deployed military personnel or in civilians as well, who have not been exposed to the threat of chemical weapons.

ATTRIBUTIONS AND EXPLANATIONS

People who feel ill need an explanation for their malaise. Sometimes, doctors can provide such an explanation, while often they cannot (Kroenke and Mangelsdorff, 1989). Plus, in those circumstances, when medicine fails to provide clear answers, people most often turn to their environment to provide those explanations. The choice they make is culturally determined, since it must depend on contemporary and accepted views of health and disease.

It is evident that the range and scope of symptoms, illnesses and conditions blamed on the environment has increased over the course of the last decade or so, reflecting increasing global concern about the effects of chemicals, radiation and infectious diseases, and the collective memories of recent health disasters. The generation that fought the Gulf War was born in a world already sensitised by 'Silent Spring' and the thalidomide tragedy, and came to maturity to a background of the AIDS epidemic, Mad Cow Disease, and numerous well-publicized environmental tragedies, such as Chernobyl, Seveso and Bhopal.

It is the role of environmental attribution that provides a link between the otherwise hard to define new illnesses such as Gulf War Syndrome and health hazards that figure so prominently in the media, such as dental amalgam disease, electromagnetic radiation, ME, organophosphate toxicity, candida, sick-building syndrome, multiple chemical sensitivity and so on. Although the postulated pathophysiological mechanisms are many and varied, all are associated with the presence of multiple unexplained symptoms, and all are in one way or another blamed on some unwelcome external environmental hazard, such as chemicals, pollution, viruses, radiation and so on. Alternatively, it is an internal toxic substance introduced from outside, such as silicon breast implants or dental amalgam. Thus, in civilian life what unites these disparate conditions is not only the clinical evidence of multiple unexplained symptomatology, but also the cognitive schema linking them with ideas of environmental hazard and toxicity. Petrie and Wessely have suggested that all of these new syndromes can be collectively considered as 'illnesses of modernity' (Petrie and Wessely, 2002).

One result of this heightened environmental awareness has been a gradual transformation of popular models of illness and disease. In place of the 'demons and spirits' comes the belief that we as a society are oppressed by mystery gases, viruses and toxins, all of which are invisible, and some of which are as elusive as the 'demons' of old. One can see this in the changing pattern of attributions given by patients with unexplained symptoms (Stewart, 1990). Guy's Poisons Unit, for example, reported that it is only in the last two decades that they have started to see patients with multiple symptoms attributed to environmental poisoning (Hutchesson and Volans, 1989).

Few can deny the heightened anxiety by the public and mass media over the safety of the environment, and the suspicions about the food we eat, the water we drink and the air we breathe. As Barsky points out, 'the world seems generally filled with peril, jammed with other health hazards in addition to disease.... nothing in our environment can be trusted, no matter how comfortable or familiar' (Barsky, 1988).

Likewise, few can doubt the growing strength of the environmental movement. We are far more aware of the risks of our environment than ever before. Activism to combat environmental pollution and toxic waste has been described as a new social movement (Matterson-Allen and Brown, 1990), one which has gradually shifted its focus from its origins in the birth of the ecology movement post-'Silent Spring', which was very much about the challenge to biodiversity posed by pollution, to a new focus on risks to human health. The word 'risk' is itself a quintessentially modern word, and there is even an epidemic of the word 'risk' in the scientific, and most particularly epidemiological, journals (Skolbekken, 1995).

There is a complex relationship between environmental concerns and symptoms. There is no doubt that being exposed to an environmental hazard, such as chemicals, leads to increased fears and concerns (Bowler *et al.*, 1994). This increase occurs whether or not the exposure is real or perceived. These fears, in turn, lead to increased symptom reporting, perhaps via activation of the stress response. The strength of a subject's opinions on environmental matters, pollution, food additives, pesticides, GM food and so on was prospectively linked to increased symptom reporting after pesticide spraying (Petrie *et al.*, 2005). Likewise, those who described themselves as 'very worried' about local environmental conditions were ten times more likely to complain of headaches than those not so concerned (Shusterman *et al.*, 1991). All of these are further amplified by media reporting (Winters *et al.*, 2003).

ATTRIBUTIONS AND EXPLANATIONS: THE GULF WAR AND MODERNITY

Therefore, the particular hazards of the Gulf War have resonance to general societal issues and concerns. There is, for example, a powerful anti-vaccination lobby which receives frequent media coverage, as exemplified by the controversy over whooping cough vaccination, and more latterly MMR vaccination (Jefferson, 2000). Likewise, the intense concerns around the use of DU munitions seem less related to its toxic properties (those of a heavy metal), and more to the powerful emotional impact of its assumed link with radioactivity (actually weak), engendered by the term 'uranium'.

All commentators agree that the Gulf War was the most 'high-tech' military conflict up to that time, the first time the so-called 'Revolution in Military Affairs' (RMA) had been seen in action, and the first demonstration of the overwhelming superiority of American military technology. All of this might be expected to reduce traditional military stressors from more direct, low-tech combat. However, the introduction of new technologies is not without risks of its own – the introduction of new technologies into civilian industries is often accompanied by a rise in non-specific complaints, and is also blamed by many for particular syndromes such as 'repetitive strain injury' (RSI) (Smith and Carayon, 1996). Scandinavian researchers have coined the term 'Techno Stress' to describe this phenomenon (Berg *et al.*, 1992; Arnetz and Wiholm, 1997).

One reason why technological/chemical incidents and disasters have a stronger association with long-term subjective health effects than natural disasters may relate to the differing time-courses of the threats (Havenaar *et al.*, 2002). Technological threats or disasters give rise not only too more health-related fears, plus there is also genuine uncertainty about the long-term risks from such exposures (Havenaar *et al.*, 2002). Hence, it is difficult for experts to confirm or deny such fears, particularly related to possible outcomes which occur endemically in affected communities anyway, such as cancer, miscarriage or reproductive abnormalities. The lack of certainty of long-term health risks, such as seen in the 'Three Mile Island episode' (Prince-Embury and Rooney, 1988), can be applied to many of the non-traditional military hazards of the Gulf campaign, providing a further link between civilian and military health (Bowler and Schwarzer, 1991).

DISTRUST, CONSPIRACY AND CONFIDENCE

The importance of public confidence and political (mis)judgement in shaping health concerns may be illustrated by one US/UK comparison. In the United States, there has been considerable concern and outcry over the role of the accidental discharge of sarin gas at the Khamisiyah arms dump, but this has not been a major issue in the UK. What has been a major issue is the role of exposure to organophosphate pesticides. One reason may be that both issues were accompanied by misinformation. In the UK, it was originally denied that any organophosphate pesticides had been used – a clear misjudgement. This was corrected, but the result was to focus attention on this particular risk, and fuel the cries of 'cover-up'. Something rather similar transpired with regard to Khamisiyah in the USA.

The initial actions of the UK authorities did little to enhance the confidence and trust of the armed forces and the populace. Records that now would give crucial information, such as vaccination records, were destroyed. We do not generally subscribe to conspiracy theories, and instead see this as a low-level decision to get rid of unnecessary paperwork that was no longer of interest. Armies fight wars, not plan epidemiological surveys. However, it handed a weapon to every internet conspiracy buff, who have flocked to the Gulf issues in droves. This was compounded by the delay in commissioning research that might allay fears.

The results of all of these events was a serious lack of trust of governmental and military authorities. This was partly a response to the specific errors related to Gulf War illness, and partly a result of other known misjudgements or denials usually from the Cold War era, and including such events as involuntary experiments carried out on some service personnel during that period. Given that risk communication and management is critically dependent upon a trust between the community that feels exposed, and those responsible for managing that risk (Slovic, 1999), these misjudgements may have been integral to the further development and shaping of Gulf War Syndrome after the conflict.

FRIENDLY FIRE

Most members of professional, volunteer Armed Forces accept that the job entails a certain exposure to physical danger. With that goes an equally inescapable exposure to the risk of psychological injury – the two are closely correlated (Jones and Wessely, 2001). Certain factors also increase this risk. Studies have shown that the psychological consequences of being wounded by your own side are far greater than when it occurs as a result of enemy action. The latter is part of the military contract – the former is not. This is the problem of friendly fire.

Friendly fire traditionally refers to those injured by inadvertently targeted munitions from their own side. However, the Gulf conflict, or more accurately its aftermath, has extended this further. Some of the alleged toxic hazards that have been blamed for ill-health are extends of the friendly-fire concept, since they also originated from our side (Kilshaw, 2004). Of the key exposures most frequently implicated in theories of Gulf War Syndrome, only one, smoke from the burning oil wells, was explicitly the result of enemy action, and it is interesting to note that this is the exposure that has attracted the least coverage and controversy. Perhaps, it is because even while the fires were burning, numerous environmental measurements were taken, and those generally reassuring results were soon disseminated. Alternatively, it was because the pollution from the oil fires was clearly the result of Iraqi hostile action, and not a sin of commission or omission by our own side.

In contrast, any health risks to Coalition forces from DU munitions, for example, can only come from either the UK or US Armed Forces, since these are the only militaries that use DU. Vaccinations to protect against biological weapons are given by one's own side, and thus any side-effects are self-inflicted, even if for a good cause. Anticholinesterase tablets to protect against exposure to nerve agents, such as sarin and cyclosarin, come into the same category, as do the pesticides used to protect against, the threat of insect-borne disease. Sarin nerve agent was possessed by the Iraqis, not the Coalition forces, and the continuing controversy over its role in Gulf ill health seems to be an exception, unless one considers that the cause of exposure, albeit in amounts unlikely to pose any threat to human health, came from the destruction of sarin containing munitions at the Khamisayah site by US forces after the end of hostilities (Gray *et al.*, 1999; McCauley *et al.*, 2002). Similarly, we draw attention to the attempts to sue US contractors who are alleged to have supplied the Iraqi regime with the precursors needed to create sarin and cyclosarin on which basis they are held to be responsible for ill-health in US veterans.

QUESTIONS REMAINING

Despite the best endeavours of a large number of dedicated scientists, we do not possess a simple answer to the question, 'what was the medical legacy of the 1991 Gulf War?', or even simpler 'what is the cause of Gulf War Syndrome?'. It is not for want of trying – serious, albeit belated, efforts have been made in many of the countries that took part in the campaign to try and unravel the problem and address the concerns of veterans. Even if we clearly do not have answers to all of the questions, at least some of the possible causes for long-term ill-health have been investigated, and found not to account for anything more than a small proportion of observed ill-health.

We must face reality. It is now over a decade since the end of the first Gulf War. The possibilities of further direct aetiological research diminish with each year. The chances of finding new evidence on exposures during the conflict is now remote. For what we believe to be largely political reasons, aetiological research continues to be funded – as we write, the US Congress has directly allocated $75 million to a single institution to continue the search, based on the hypothesis that Gulf War Syndrome is the result of low-level exposure to chemical warfare agents (Couzins, 2006). We believe that it is unlikely that such research will generate a eureka moment, or provide evidence that will convince the scientific community. We may be wrong, but it is unlikely.

GULF WAR SYNDROME: THE POST-MODERN ILLNESS

In a characteristically provocative article Muir Gray (Gray, 1999) describes the features of what he calls post-modern medicine – a distrust of science, a readiness to resort to litigation, a greater attention to risk and better access to information (of whatever quality). He also points out, as indeed have many commentators, how consumer and patient values have already replaced paternalistic and professional values, and where doctors used to lead, they now follow. The monolithic role of the doctor has been challenged by lay experts, whose ability to influence public debate and policy increases just as that of the doctor or scientist diminishes – the lay expert may be the survivor of a disaster or the sufferer from a disease (Bury, 1998). The Gulf War veteran may fulfil both roles.

Gulf War Syndrome was perhaps the first major health condition that was constructed almost entirely without the assistance of medicine in any shape or form. Previous syndromes in the military have arisen for many reasons - shell shock was not invented by Myers in his seminal 1915 paper, but without his contribution it is improbable that the term would have gained widespread acceptance (and when professional endorsement was withdrawn, as it was in 1917, the term disappeared from medicine). In modern civilian practice, the success of chronic-fatigue syndrome depended, in part, on the rise in patient consumerism and the reaction against medical paternalism, but key triggers were the paper describing the original Royal Free epidemic that gave rise to the term myalgic encephalomyelitis in the UK or the NIH papers on chronic Epstein Barr virus infection in the USA (Wessely *et al.*, 1998). Without them, CFS would have taken a different course, if it had emerged at all. Sick building syndrome, like Gulf War Syndrome, arose out of confusion, but a key part in its success was the persistent and vocal activities of a handful of doctors and scientists who, in the words of one commentator, unequivocally diagnosed as a pathologic state, something that had not been scientifically demonstrated (Bardana, 1997).

Yet for Gulf War Syndrome, the shape of the syndrome seems to have been determined in the popular and political imaginations long before scientists or doctors had anything to say on the matter. Gulf War Syndrome, we argue, developed without the assistance of science or medicine. Certainly, populist and occasionally maverick scientists have emerged into the limelight of Gulf War Syndrome, and have played roles in subsequent events, but Gulf War Syndrome may be the first truly post-modern illness, in that it developed from the congruence of veterans' narratives, veterans' disquiet and distrust, and a powerful media agenda (Zavestoski *et al.*, 2004). Medical professionals and scientists generally have reacted to events, and not shaped them.

The story of Gulf War Syndrome may well reflect the shape of things to come. Classic conventional warfare between states is becoming increasingly hard to imagine – no state can match the firepower of an advanced military technology, such as the USA. Modern professional militaries could no longer sustain the type of large-scale conventional casualties that were a feature of the two World Wars, and we should be thankful for that. However, as some perceptive commentators have pointed out, reducing the numbers of direct casualties does not appear to have reduced the cost of war, direct or indirect. Harvey Sapolsky uses the example of Gulf War Syndrome to make the case that war is becoming too expensive (Sapolsky, 2003), and he may well have a point. Likewise, the example of Gulf War Syndrome also serves as a reminder the pattern of risks, and threats faced by modern militaries has changed, and that these new hard to understand, hard to measure and hard to manage risks, such as chemicals and chemical warfare agents, can present a more potent challenge to the efficiency of a modern military that the conventional risks of direct trauma (Wessely, 2005).

CONCLUSIONS

So what did happen after the first Gulf War? It is, we argue, a case of 'something old, something new, something borrowed, something blue'. The story began with the experiences of veterans' reporting symptoms. These may have been triggered as an unexpected reaction to measures taken to protect the Armed Forces against modern warfare – 'something new'. These must be added to the unchanging psychological toll of warfare and the legacy of previous post-conflict syndromes – 'something old'. All have been taken up by a powerful media, and shaped into a particular syndrome, under the influence of popular non-military views of health, disease and illness – 'something borrowed', with further impetus coming from the actions, or inactions, of government. Finally, although direct traumatic psychological injury is not prominent, mood disorder also plays a role – 'something blue'.

REFERENCES

Ablers J and Berent S (2000). Controversies in neurotoxicology. *Neurolog Clin*, **18**, 741–763.

Abou-Donia M, Wilmarth K, Abdel-Rahman, A *et al.* (1996). Increased neurotoxicity following concurrent exposure to pyridostigmine bromide, DEET, and chlorpyrifos. *Fund Appl Toxicol*, **34**, 201–222.

Abou-Donia M, Dechkovskaia AM, Goldstein LB *et al.* (2004). Co-exposure to pyridostigmine bromide, DEET, and/or permethrin causes sensorimotor deficit and alterations in brain acetylcholinesterase activity. *Pharmacol Biochem Behav*, **77**, 253–262.

Albers J, Berent S, Garabrant D *et al.* (2004). The effects of occupational exposure to chlorpyrifos on the neurologic examination of the central nervous system: a prospective cohort study. *J Occ Environ Med*, **46**, 367–378.

Amato A, McVey A, Cha, C *et al.* (1997). Evaluation of neuromuscular symptoms in veterans of the Persian Gulf War. *Neurology*, **48**, 4–12.

Anon (1997a). *Gulf War Illnesses: Dealing with the Uncertainities.* London, UK: Parliamentary Office of Science and Technology.

Anon (1997b). *Presidential Advisory Committee on Gulf War Veterans' Illnesses: Final Report.* Washington, DC, USA: US Government Printing Office.

Anon (1998). *Health Study of Canadian Forces Personnel Involved in the 1991 Conflict in the Persian Gulf.* Ottawa, Canada: Goss Gilroy Inc.

Anon (2000). French army wary of first Gulf War syndrome charge. In Paris, France: Reuters.

Arnetz B and Wiholm C (1997). Technological Stress: Psychophysiological Symptoms in Modern Offices. *J Psychosom Res*, **43**, 35–42.

Bardana E (1997). Sick building syndrome – a wolf in sheeps' clothing. *Ann Allergy Asthma Immunol*, **79**, 283–293.

Barrett DG, Doebbeling BN, Clauw DJ *et al.* (2003). Prevalence of symptoms and symptom-based conditions among gulf war veterans: current status of research findings. *Epidemiol Rev*, **24**, 218–227.

Barsky A (1988). *Worried Sick: Our Troubled Quest for Wellness.* Toronto, Canada: Little, Brown & Company.

Berg M, Arnetz B, Liden, S *et al.* (1992). Techo-stress: a psychophysiology study of employees with VDU-associated skin complaints. *J Occup Med*, **34**, 698–700.

Betts R (1998). The new threat of mass destruction. *For Affairs*, **77**, 26–41.

Black D, Doebbeling B, Voelker M *et al.* (2000). Multiple chemical sensitivity syndrome: symptom prevalence and risk factors in a military population. *Arch Inter Med*, **160**, 1169–1176.

Bourdette D, McCauley L, Barkhuizen A *et al.* (2001). Symptom factor analysis, clinical findings and functional status in a population-based case control study of Gulf War unexplained illness. *J Occup Environ Med*, **43**, 1026–1040.

Bowler R and Schwarzer R (1991). Environmental anxiety: Assessing emotional distress and concerns after toxin exposure. *Anx Res*, **4**, 167–180.

Bowler R, Mergler D, Huel, G *et al.* (1994). Psychological psychosocial and psychophysiological sequelae to a community affected by a railroad disaster. *J Traum Stress*, **7**, 601–624.

Bury M (1998). Postmodernity and health. In: *Modernity and Health* (G Scambler and P Higgs, eds), pp. 1–28. London, UK: Routledge.

Chalder T, Hotopf M, Hull L *et al.* (2001). Prevalence of Gulf war veterans who believe they have Gulf war syndrone: questionnaire study. *Br Med J*, **323**, 473–476.

Cherry N, Creed F, Silman A *et al.* (2001a). Health and exposures of United Kingdom Gulf war veterans.

Part 1: The pattern and extent of ill health. *Occup Environ Med*, **58**, 291–298.

Cherry N, Creed F, Silman A *et al.* (2001b). Health and exposures of United Kingdom Gulf war veterans. Part II: The relationship of health to exposure. *Occup Environ Med*, **58**, 299–306.

Coker W, Bhatt B, Blatchley N *et al.* (1999). Clinical findings for the first 1000 Gulf war veterans in the Ministry of Defence's medical assessment programme. *Br Med J*, **318**, 290–294.

Committee on Toxicity of Chemicals in Food (1999). *Organophosphates*. London, UK: Department of Health.

Couzins J (2006). Texas Earmark Allots Millions to Disputed Theory of Gulf War Illness. *Science*, **312**, 668.

David A, Farrin L, Hull L *et al.* (2002). Cognitive functioning and disturbances of mood in UK veterans of the Persian Gulf War: A comparative study. *Psychol Med*, **32**, 1357–1360.

Davis L, Murphy F, Alpern, R *et al.* (2004). Clinical and laboratory assessment of distal peripheral nerves in Gulf War veterans and spouses. *Neurology*, **63**, 1070–1077.

Doebbeling B, Clarke W, Watson, D *et al.* (2000). Is there a Persian Gulf War syndrome? Evidence from a large population-based survey of veterans and nondeployed controls. *Am J Med*, **108**, 695–704.

Eisen SA, Kang HK, Murphy FM *et al.* (2005). Gulf War veterans' health: medical evaluation of a U.S. cohort. *Ann Inter Med*, **142**, 881–890.

Everitt B, Ismail K, David, A *et al.* (2002). Searching for a Gulf War Syndrome Using Cluster Analysis. *Psycholog Med*, **32**, 1371–1378.

Everson M, Shi K, Alreidge P *et al.* (2002). Immunoloical responses are not abnormal in symptomatic Gulf War veterans. *Ann NY Acad Sci*, **966**, 327–343.

Ford J, Campbell K, Storzbach D *et al.* (2001.) Postraumatic stress symptomatology is associated with unexplained illness attributed to Persian Gulf War military service. *Psychosom Med*, **63**, 842–849.

Fukuda K, Nisenbaum R, Stewart G *et al.* (1998). Chronic multisymptom illness affecting air force veterans of the gulf war. *J Am Med Assoc*, **280**, 981–988.

Fulco C, Liverman C and Sox H (eds) (2000). *Gulf War and Health*: Volume 1. *Depleted Uranium, Sarin, Pyridostigmine Bromide, Vaccines*. Washington, DC, USA: Institute of Medicine.

Fullerton C and Ursano R (1990). Behavioral and psychological responses to chemical and biological warfare. *Milit Med*, **155**, 54–59.

Furlong C (2000). PON1 status and neurologic symptom complexes in gulf war veterans. *Genome Res*, **10**, 153–155.

Gackstetter G, Hooper T, Al Qahtani M *et al.* (2005). Assessing the potential health impact of the 1991 Gulf War on Saudi Arabian National Guard soldiers. *Int J Epidem*, **34**, 801–808.

Girault V (2002). Chronic fatigue, an underrated syndrome. *Presse Med*, **31**, 531.

Gray J (1999). Postmodern medicine. *Lancet*, **354**, 1550–1553.

Gray G, Coate B, Anderson C *et al.* (1996). The postwar hospitization experience of US veterans of the Persian Gulf War. *New England J Med*, **335**, 1505–1513.

Gray G, Smith T, Knoke J *et al.* (1999a). The postwar hospitalization experience of Gulf War veterans possibly exposed to chemical munitions destruction at Khamisiyah, Iraq. *Am J Epidem*, **150**, 532–540.

Gray GC, Kaiser KS, Hawksworth AW *et al.* (1999b). Increased postwar symptoms and psychological morbidity among US navy Gulf War veterans. *Am J Trop Med Hyg*, **60**, 758–766.

Gray G, Reed R, Kaiser K *et al.* (2002). Self reported symptoms and medical conditions among 11,868 Gulf War Era veterans. *Am J Epidem*, **155**, 1033–1044.

Gray GG, Kang H, Graham J *et al.* (2004). After more than 10 years of Gulf War veteran medical examinations, what have we learned? *Am J Prevent Med*, **26**, 443–452.

Gwiazda RH, McDiarmid M and Smith D (2004). Detection of depleted uranium in urine of veterans from the 1991 Gulf War. *Health Phys*, **86**, 12–18.

Haley R (2003). Gulf war syndrome: narrowing the possibilities. *Lancet Neurol*, **2**, 272–273.

Haley R and Kurt T (1997). Self-reported exposure to neurotoxic chemical combinations in the Gulf War: a cross-sectional epidemiologic survey. *J Am Med Assoc*, **277**, 231–237.

Haley R, Hom J, Roland, P *et al.* (1997a). Evaluation of neurologic function in Gulf War veterans: a blinded case-control study. *J Am Med Assoc*, **277**, 223–230.

Haley R, Kurt T and Hom J (1997b). Is there a Gulf War syndrome? Searching for syndromes by factor analysis of symptoms. *J Am Med Assoc*, **277**, 215–222.

Haley R, Billecke S and la Du B (1999). Association of low PON1 type Q (type A) arylesterase activity with neurologic symptom complexes in Gulf War veterans. *Toxicol Appl Pharmacol*, **157**, 227–233.

Havenaar J, Cwikel J and Bromet J (eds) (2002). *Toxic Turmoil: Psychological and Societal*

Consequences of Ecological Disasters. New York, NY, USA: Plenum.

Holloway H, Norwood A, Fullerton C *et al.* (1997). The threat of biological weapons: prophylaxis and mitigation of psychological and social consequences. *J Am Med Assoc*, **278**, 425–427.

Horner R, Kamins K, Feussner J *et al.* (2003). Occurence of amyotrophic lateral sclerosis among Gulf War veterans. *Neurology*, **61**, 742–749.

Hotopf M, David A, Hull L *et al.* (2000). The role of vaccinations as risk factors for ill-health in veterans of the Persian Gulf War. *Br Med J*, **320**, 1363–1367.

Hotopf M, Mackness I, Nikolaou V *et al.* (2003). Paraoxonase in Persian Gulf War veterans. *J Occup Environ Med*, **45**, 668–675.

Hutchesson E and Volans G (1989). Unsubstantiated complaints of being poisoned: psychopathology of patients referred to the National Poisons unit. *Br J Psych*, **154**, 34–40.

Hyams K, Hanson K, Wignall F *et al.* (1995). The impact of infectious diseases on the health of US troops deployed to the Persian Gulf during Operations Desert Shield and Desert Storm. *Clin Infect Dis*, **20**, 1497–1504.

Hyams K, Wignall F and Roswell R (1996). War syndromes and their evaluation: from the US Civil War to the Persian Gulf War. *Ann Intern Med*, **125**, 398–405.

Ikin JS, Creamer MC, Forbes AB *et al.* (2004). War-related psychological stressors and risk of psychological disorders in Australian veterans of the 1991 Gulf War. *Br J Psych*, **185**, 116–126.

Ishoy T, Suadicani P, Guldager B *et al.* (1999). State of health after deployment in the Persian Gulf: The Danish Gulf War Study. *Danish Med Bull*, **46**, 416–419.

Ismail K, Everitt B, Blatchley N *et al.* (1999). Is there a Gulf war syndrome? *Lancet*, **353**, 179–182.

Ismail K, Kent K, Brugha T *et al.* (2002). The mental health of UK Gulf war veterans: phase 2 of a two-phase cohort study. *Br Med J*, **325**, 576–579.

Jefferson T (2000). Real or perceived adverse effects of vaccines and the media – a tale of our times. *J Epidem Comm Health*, **54**, 402–403.

Jones E and Wessely S (1999). Chronic fatigue syndrome after the Crimean War and the Indian Mutiny. *Br Med J*, **319**, 1645–1647.

Jones E and Wessely S (2001). Psychiatric casualties of war: an inter and intra war comparison. *Br J Psych*, **178**, 242–247.

Jones E and Wessely S (2005). *From Shell Shock to PTSD: A History of Military Psychiatry*. London, UK: Psychology Press.

Jones E, McCartney H, Everitt B *et al.* (2002). Post-combat syndromes from the Boer War to the Gulf: a cluster analysis of their nature and attribution. *Br Med J*, **324**, 324–327.

Joseph S (1997). A Comprehensive clinical evaluation of 20,000 Persian Gulf War veterans. *Milit Med*, **162**, 149–156.

Kang HM, Lee K, Murphy F *et al.* (2002). Evidence for a deployment-related Gulf War Syndrome by factor analysis. *Arch Environ Health*, **57**, 61–68.

Kang HK, Natelson B, Mahan C *et al.* (2003). Post-traumatic stress disorder and chronic fatigue syndrome-like illness among Gulf War veterans: a population-based survey of 30 000 veterans. *Am J Epidem*, **157**, 141–148.

Keeler J, Hurst C and Dunn M (1991). Pyridostigmine used as a nerve agent pretreatment under wartime conditions. *J Am Med Assoc*, **266**, 693–695.

Kelsall H, Sim M, Forbes, A *et al.* (2004). Symptoms and medical conditions in Australian veterans of the 1991 Gulf War: relation to immunisations and other Gulf War exposures. *Occup Environ Med*, **61**, 1006–1013.

Kilshaw S (2004). Friendly Fire. *Anthropol Med*, **11**, 149–160.

Kipen HM, Hallman W and Natelson BH (1999). Prevalence of chronic fatigue and chemical sensitivities in Gulf registry veterans. *Arch Environ Health*, **54**, 313–318.

Kleinman A (1988). *Patients and Healers in the Context of Culture: An Exploration of the Borderland Between Anthropology, Medicine and Psychiatry*. Berkeley, CA, USA: University of California Press.

Knoke J, Smith TC, Gray G *et al* (2000). Factor analysis of self reported symptoms: Does it identify a Gulf War Syndrome? *Am J Epidem*, **152**, 379–388.

Kroenke K and Mangelsdorff A (1989). Common symptoms in ambulatory care: incidence, evaluation, therapy and outcome. *Am J Med*, **86**, 262–266.

Kroenke K, Koslowe P and Roy M (1998). Symptoms in 18,495 Persian Gulf War Veterans. *J Occup Environ Med*, **40**, 520–528.

Landrigan P (1997). Illness in Gulf War veterans: causes and consequences. *JAMA*, **277**, 259–261.

Lee H, Gabriel R, Bale A *et al.* (2001). Clinical findings of the second 1000 UK Gulf War veterans who attended the Ministry of Defence's Medical Assessment Programme. *J R Army Med Corps*, **147**, 153–160.

Lindem KH, White RF, Proctor SP *et al.* (2003). Neuropsychological Performance in Gulf War Era Veterans Traumatic Stress Symptomatology and Exposure to Chemical-Biological Warfare Agents. *J Psychopathol Behav Assess*, **25**, 105–119.

Mackness B, Durrington P and Mackness M (2000). Low paraoxonase in Persian Gulf War veterans self reporting Gulf War Syndrome. *Biochem Biophys Res Commun*, **276**, 729–733.

Matterson-Allen S and Brown P (1990). Public reaction to toxic waste contamination: analysis of a social movement. *Int J Health Services*, **20**, 484–500.

McCauley L, Joos S, lasarev M *et al.* (1999). Gulf war unexplained illnesses: persistence and unexplained nature of self-reported symptoms. *Environ Res*, **81**, 215–223.

McCauley L, Lasarev M, Sticker D *et al.* (2002). Illness experience of Gulf War veterans possibly exposed to chemical warfare agents. *Am J Prevent Med*, **23**, 200–206.

McDiarmid M, Keogh J, Hooper, F *et al.* (2000). Health effects of depleted uranium on exposed gulf war veterans. *Environ Res*, **82**, 168–180.

McDiarmid M, Engelhrdt S, Oliver M *et al.* (2004). Health effects of depleted uranium on exposed Gulf War veterans: A 10-year follow-up. *J Toxicol Environ Health A*, **67**, 277–296.

McKenzie D, Ikin J, McFarlane A *et al.* (2004). Psychological health of Australian veterans of the 1991 Gulf War: as assessment using the SF-12, GHQ-12 and PCL. *Psycholog Med*, **34**, 1–12.

Ministry of Defence (2000). *British Chemical Warfare Defence During the Gulf Conflict (1990–1991)*. London, UK: Ministry of Defence (www.mod.uk/policy/gulfwar/index.htm).

Mouterde O (2001). Myalgic encephalomyelitis in children. *Lancet*, **357**, 562.

Nisenbaum R, Barrett DH, Reyes M *et al.* (2000). Deployment Stressors and a Chronic Multisymptom Illness Among Gulf War Veterans. *J Nervous Mental Dis*, **188**, 259–266.

O'Brien L and Payne RG (1993). Prevention and management of panic in personnel facing a chemical threat – lessons from the Gulf. *J R Army Med Corps*, **139**, 41–45.

Pearce P, Crofts H, Muggleton N *et al.* (1999). The effects of acutely administered low dose sarin on cognitive behaviour and the electrocephalogram in the common marmoset. *J Psychopharmacol*, **13**, 128–135.

Petrie K and Wessely S (2002). Modern worries and medicine. *Br Med J*, **324**, 690–691.

Petrie KJ, Broadbent E, Kley N *et al.* (2005). Worries about modernity predict symptom complaints following environmental spraying. *Psychosom Med*, **67**, 778–782.

Prince-Embury S and Rooney J (1988). Psychological symptoms of residents in the aftermath of the Three-Mile island nuclear accident in the aftermath of technological disaster. *J Soc Psychol*, **128**, 779–790.

Proctor S, Heeren T, White R *et al.* (1998). Health status of Persian Gulf War veterans: self-reported symptoms, environmental exposures, and the effect of stress. *Int J Epidemiol*, **27**, 1000–1010.

RCP/RCPsych (1998). Organophosphate sheep dip. Clinical management of long-term low-dose exposure. A joint report of the Royal College of Physicians and the Royal College of Psychiatrists. Salisbury, UK: Royal College of Physicians.

Reid S, Hotopf M, Hull L *et al.* (2001). Chronic fatigue syndrome and multiple chemical sensitivity in UK Gulf war veterans. *Am J Epidemiol*, **153**, 604–609.

Riddle JB, Smith T, Ritchie EC *et al.* (2003). Chemical warfare and the Gulf War: a review of the impact on Gulf veterans' health. *Milit Med*, **168**, 600–605.

Rook G and Zumla A (1997). Gulf war syndrome: is it due to a systemic shift in cytokine balance towards a Th2 profile? *Lancet*, **349**, 1831–1833.

Rose M (2004). Gulf war service an uncertain trigger for ALS. *Neurology*, **61**, 730–731.

Roy M, Koslowe P, Kroenke K *et al.* (1998). Signs. symptoms and ill-defined conditions in Persian Gulf War veterans: findings from the comprehensive clinical evaluation program. *Psychosom Med*, **60**, 663–668.

Roy M, Kraus P, Seegers C *et al.* (2006). Pyridostigmine, DEET, permethrin and stress: a double-blind, randomized, placebo-controlled trial to assess harm. Submitted for publication.

Sapolsky H (2003). War needs a warning label. *Breakthoroughs*, **12**, 3–8.

Scremin O, Shih T, Huynh, L *et al.* (2003). Delayed neurologic and behavioral effects of subtoxic does of cholinesterase inhibitors. *J Pharmacol Exp Therapeut*, **304**, 1111–1119.

Sharief M, Pridden J, Delamont R *et al.* (2002). Neurophysiologic evaluation of neuromuscular symptoms in UK Gulf War veterans. A controlled study. *Neurology*, **59**, 1518–1525.

Shusterman D, Lipscomb J, Neutra R *et al.* (1991). Symptom prevalence and odor-worry interaction near hazardous waste sites. *Environ Health Perspect*, **94**, 25–30.

Skolbekken J (1995). The risk epidemic in medical journals. *Social Sci Med*, **40**, 291–305.

Skowera A, Hotopf M, Sawicka E *et al.* (2004). Cellular Immune activation in Gulf War veterans. *J Clin Immunol*, **24**, 60–73.

Skowera A, de Jong EC, Schuitemaker JH *et al.* (2005). Analysis of anthrax and plague bio-warfare vaccine interactions with human monocyte-derived dendritic cells. *J Immunol*, **175**, 7235–7243.

Slovic P (1999). Trust, emotion, sex, politics and science: surveying the risk assessment battlefield. *Risk Anal*, **19**, 689–702.

Smith M and Carayon P (1996). Work organization, stress and cumulative trauma disorders. In: *Beyond Biomechanics: Psychosocial Aspects of Musculoskeletal Disorders in Office Work* (S Moon and S Sauter eds), pp. 23–44. London, UK: Taylor & Francis.

Smith TS, Ryan MAK, Gray GC *et al.* (2002). Ten years and 100 000 participants later: occupational and other factors influencing participation in US Gulf War health registries. *J Occup Environ Med*, **44**, 758–768.

Smith T, Gray G, Weir C *et al.* (2003). Gulf War veterans and Iraqi nerve agents at Khamisayah: post war hospitilisation data revisted. *Am J Epidemiol*, **158**, 457–467.

Soetekouw P, De Vries M, Preijers F *et al.* (1999). Persistent symptoms in former UNTAC soliders are not associated with shifted cytokine balance. *Eur J Clin Investig*, **29**, 960–963.

Spencer P, McCauley L, Lapidus J *et al.* (2001). Self-reported exposures and their association with unexplained illness in a population based case-control study of Gulf War veterans. *J Occup Environ Med*, **43**, 1041–1056.

Steele L (2000). Prevalence and patterns of Gulf War illness in Kansas veterans: Association of symptoms with characteristics of person, place and time of military service. *Am J Epidemiol*, **152**, 992–1002.

Stewart D (1990). The changing faces of somatization. *Psychosomatics*, **31**, 153–158.

Stimpson NT, Weightman AL, Dunstan F *et al.* (2003). Psychiatric disorders in veterans of the Persian Gulf War of 1991. Systematic review. *Br J Psychiat*, **182**, 391–403.

Stokes J and Banderet L (1997). Psychological aspects of chemical defense and warfare. *Milit Psychol*, **9**, 395–415.

Storzbach D, Binder LM, McCauley L *et al.* (2000). Psychological differences between veterans with and without Gulf War unexplained symptoms. *Psychosom Med*, **62**, 726–735.

Stuart J, Ursano R, Fullerton C *et al.* (2003). Belief in exposure to terrorist agents: reported exposure to nerve or mustard gas by Gulf War veterans. *J Nerv Men Dis*, **191**, 431–436.

The Iowa Persian Gulf Study Group (1997). Self-reported illness and health status among Persian Gulf War veterans: a population-based study. *J Am Med Assoc*, **277**, 238–245.

Trojan D and Cashman N (1995). An open trial of pyridostigmine in post poliomyelitis syndrome. *Can J Neurolog Sci*, **22**, 223–227.

United States General Accounting Office (1997). *Gulf War Illnesses: Improved Monitoring of Clinical Progress and Reexamination of Research Emphasis Are Needed.* Washington, DC, USA: United States General Accounting Office.

Unwin C, Blatchley N, Coker W *et al.* (1999). The health of United Kingdom Servicemen who served in the Persian Gulf War. *Lancet*, **353**, 169–178.

Wegman D, Woods N and Bailar J (1998). Invited commentary: how would we know a Gulf War syndrome if we saw one? *Am J Epidemiol*, **146**, 704–711.

Wessely S (2005). Risk, psychiatry and the military. *Br J Psychiat*, **186**, 459–466.

Wessely S, Hotopf M and Sharpe M (1998). *Chronic Fatigue and its Syndromes*. Oxford, UK: Oxford University Press.

Wessely S, Nimnuan C and Sharpe M (1999). Functional somatic syndromes: one or many? *Lancet*, **354**, 936–939.

Wessely S, Unwin C, Hotopf M *et al.* (2003). Is recall of military hazards stable over time? Evidence from the Gulf War. *Br J Psychiat*, **183**, 314–322.

Winters WDS, VanDiest I, Nemery B *et al.* (2003). Media warnings about environmental pollution facilitate the acquisition of symptoms In response to chemical substances. *Psychosom Med*, **65**, 332–338.

Wolfe J, Proctor S, Duncan Davis J *et al.* (1998). Health symptoms reported by Persian Gulf War veterans two years after return. *Am J Ind Med*, **33**, 104–113.

Zavestoski SB, McCormick S, Mayer B *et al.* (2004). Patient activism and the struggle for diagnosis: Gulf War illnesses and other medically unexplained physical symptoms in the US. *Soc Sci Med*, **58**, 161–175.

19 MUSTARD GAS

Robert L. Maynard

Health Protection Agency, Chilton, UK

'Mustard gas' is a misnomer. The compound usually referred to as mustard gas is sulphur mustard: a liquid which boils at 217°C. Both liquid sulphur mustard and the vapour given off are vesicant, i.e. produce blistering. Many mustard compounds other than sulphur mustard have been examined for their potential as chemical warfare agents. During World War II, several hundred mustard-related compounds were synthesized. Of these, the nitrogen mustards and sesqui mustard have attracted some attention. Only sulphur mustard has been used on a large scale in war. Several of the nitrogen mustards have found peaceful uses as antimitotic agents used in the treatment of various cancers. In the following account, mustard gas should be taken as indicating sulphur mustard.

HISTORICAL ASPECTS

The exact date of the first synthesis of sulphur mustard seems to be unknown: 1820, 1822 and 1854 have all been quoted by various writers. West (1920) reported that Despretz recorded the formation of a disagreeably smelling compound produced by the action of ethylene on sulphur chloride. Niemann and Guthrie separately synthesized and described the properties of sulphur mustard in 1860. Niemann's description is accurate and is given below:

> *The characteristic property of this oil is also a very dangerous one. It consists of the fact that the minutest trace which may accidentally come into contact with any portion of the skin, though at first causing no pain, produces in the course of a few hours a reddening and on the following day a severe blister which suppurates for a long time and is very difficult to heal* (West, 1920).

This description written 134 years ago would be difficult to improve upon today. Synthesis of pure sulphur mustard was reported by Victor Meyer (1886).

The development of sulphur mustard as a chemical warfare agent was undertaken by Fritz Haber in Germany during World War I and mustard was used for the first time on the night of 12 July 1917 at Ypres (Prentiss, 1937). British forces were the first to be exposed to mustard gas and during the first three months of use more than 14 000 British casualties were produced. By the end of the war, more than 120 000 British mustard gas casualties had occurred. Some writers have put the total number of mustard gas casualties produced during World War I as high as 400 000. The mortality rate amongst mustard casualties was low: 2–3% (Haldane, 1925).

It is not widely known that British interest in mustard gas preceded its use by German forces. Sir Charles Lovatt Evans worked on chemical warfare agents during World War I and it has been recorded in a biographical memoir (Lovett Evans, 1970) that:

> *in January 1916 Lovatt Evans joined the RAMC and was seconded to work at The Royal Army Medical College of Millbank where Starling, with the rank of Major was in charge of the Anti-Gas Department. There they studied arsine, phosgene, hydrocyanic acid and mustard gas, the last at the suggestion of Harold W Dudley, and Starling advised its use but it was rejected. Lovatt Evans recalled that 'when, some fifteen months later mustard gas was used by the Germans, Starling was infuriated and made a vigorous*

Chemical Warfare Agents: Toxicology and Treatment (2nd Edition)
Edited by Timothy C. Marrs, Robert L. Maynard and Frederick R. Sidell

protest at the highest level; the result of this was that he was promoted to Lt Col and sent to Salonika as Army Chemical Adviser, with nothing particular to do. He did it very well and was awarded a CMG[1]*....'*

The effectiveness of mustard gas was greater than that of all the other chemical warfare agents used during World War I and it came to be known as the 'King of Gases' or the 'King of the Battle Gases'.

Italian forces used sulphur mustard against Ethiopian forces in 1936 and it was used by Japanese forces against Chinese troops during World War II (Robinson, 1971). In the Iran–Iraq war, allegations, which proved well founded, of the use of sulphur mustard by Iraq were made (United Nations Security Council, 1984,1986,1987). In 1986, more than 30 Iranian casualties whose injuries were compatible with exposure to mustard gas were evacuated to London for treatment. The author was thus able to gather first-hand experience of treating mustard gas casualties.

Sulphur mustard has been little used in clinical medicine: the first report of its use being that of Adair and Bagg (1931). Until recently, sulphur mustard was available for use in the treatment of psoriasis: Psoriasin marketed by Malco (Pharmaceutical Society of Great Britain, 1977). For an account of the effects of Psoriazin (sic) on the liver, see Ciszewka-Popiolek *et al.* (1989).

The name 'mustard' was given to the compound by soldiers during World War I because of its smell. Some have described the smell as being similar to that of garlic, mustard, horse radish or leeks. Sulphur mustard should not be confused with mustard oil: allyl isothiocyanate which, interestingly, is also a vesicant (Pharmaceutical Society of Great Britain, 1977). Sulphur mustard is often referred to in military manuals as H – this may be an abbreviated or generalized form of HS said to stand for Hunstoffe. HN is sometimes used to denote nitrogen mustard (see below) and HD for 'Distilled Mustard'. During World War I, German forces referred to sulphur mustard as 'LOST': derived from the first letters of the names of the German chemists involved in its synthesis: Lommel and Steinkopf. The first use of sulphur mustard at Ypres led to French forces referring to it as 'Yperite'. German shells containing sulphur mustard were marked with a yellow cross.

It may be thought that the details of the toxicity of a chemical warfare agent such as sulphur mustard represent a rather esoteric branch of toxicology: they do. However, civilian cases of exposure to sulphur mustard do occur. In 1984, some 23 Baltic fishermen showed effects of exposure to sulphur mustard as a result of mustard gas containing shells being caught in nets (Aasted *et al.*, 1987). More than 50 000 tons of German chemical warfare munitions were dropped into the Baltic at the end of World War II by Allied Forces. It is likely that these shells will continue to be recovered and will remain dangerous for many years yet.

[1] Companion of the Order of St Michael and St George; a British order of knighthood.

PHYSICAL AND CHEMICAL PROPERTIES OF SULPHUR MUSTARD

The chemical structure of sulphur mustard is shown in Figure 1 and the physical characteristics of sulphur mustard are shown in Table 1.

The vapour pressure and volatility of mustard have been discussed in Chapter 2. For reference, Prentiss's table of vapour pressure and volatility is reproduced here in Table 2.

Sulphur mustard is hydrolyzed by water according to the reaction:

$$(CH_2CH_2Cl)_2S + 2H_2O \longrightarrow (CH_2CH_2OH)_2S + 2HCl$$

It should be noted that the vapour of sulphur mustard is 5.5 times as heavy as air and thus the gas will accumulate in shell craters and trenches.

The half-life for the hydrolysis of sulphur mustard by water has been reported as about 3–5 min. Such a figure is in practice misleading: sulphur mustard is rather immiscible with water and mixing is necessary for hydrolysis to take place.

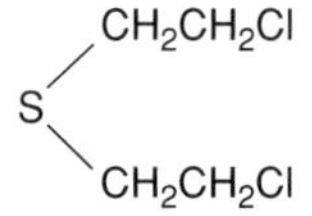

Figure 1. Molecular structure of sulphur mustard

Table 1. Physicochemical characteristics of sulphur mustard

Appearance	Yellow–brown oily liquid
Melting point	14.4°C
Boiling point	228°C (sometimes given as 217°C)
Specific gravity	1.27
Vapour density	5.4
Vapour pressure	
10°C	0.032 mmHg
25°C	0.112 mmHg
40°C	0.346 mmHg
Smell	Garlic, mustard, leeks

Table 2. Vapour pressure and volatility data of sulphur mustard

Temperature		Vapour pressure	Volatility
°C	°F	(mmHg)	(mg l^{-1})
0	32	0.0260	0.250
5	41	0.0300	0.278
10	50	0.0350	0.315
15	59	0.0417	0.401
20	68	0.0650	0.625
25	77	0.0996	0.958
30	86	0.1500	1.443
35	95	0.2220	2.135
40	104	0.4500	3.660

Sulphur mustard placed upon the ground will persist for a long period if protected from wind and rain. Persistence of quantities of sulphur mustard likely to be encountered in chemical warfare for 2–3 days is not unlikely. Increased temperature and pH increase the rate of hydrolysis. Oxidizing agents react with sulphur mustard to produce the corresponding sulphoxide and the sulphone (Figure 2). The sulphone is produced by stronger oxidizing agents, e.g. hypochlorite, and is itself a vesicant and may produce lacrimation and sneezing (described therefore as a 'sternutator' from the Latin sternuo: 'to sneeze'). The sulphoxide is

Dichloroethylsulphoxide Dichloroethylsulphone

Figure 2. Molecular structures of the sulphoxide and sulphone of sulphur mustard

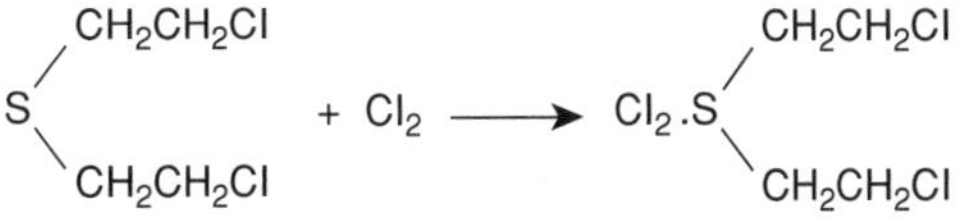

Tetrachloroethylsulphide

Figure 3. Reaction of sulphur mustard with free chlorine

not a vesicant. These effects are not commonly observed by people working with the sulphone because of its low vapour pressure. Sulphur mustard also reacts with free chlorine provided by compounds such as dichloramine, as shown in Figure 3.

Sulphur mustard is soluble in water to the extent of less than 0.1%, but is freely soluble in many common organic solvents including ethanol, ether and chloroform. The vapour given off by a quantity of liquid sulphur mustard has considerable penetrating powers, rapidly passing through clothing and damaging the skin beneath. Substances such as metal glass and glazed tiles are generally impervious to mustard, although painted surfaces may take up vapour. This may be released later from the surface as the local ambient concentration of sulphur mustard vapour falls. Decontamination of painted surfaces is therefore important. Sulphur mustard decomposes at high temperatures to produce toxic compounds which include active lacrimators: disposal of material contaminated with mustard by burning should, therefore, be undertaken with care.

PHYSICAL AND CHEMICAL PROPERTIES OF THE NITROGEN MUSTARDS

There are several nitrogen mustards. Those of relevance here are:

HN_1: *N*-ethyl-2,2′-di(chloroethyl)amine
HN_2: *N*-methyl-2,2′-di(chloroethyl)amine
HN_3: 2,2′,2″-tri(chloroethyl)amine

The chemical structure of the nitrogen mustards are shown in Figure 4.

The physicochemical characteristics of the nitrogen mustards are shown in Table 3.

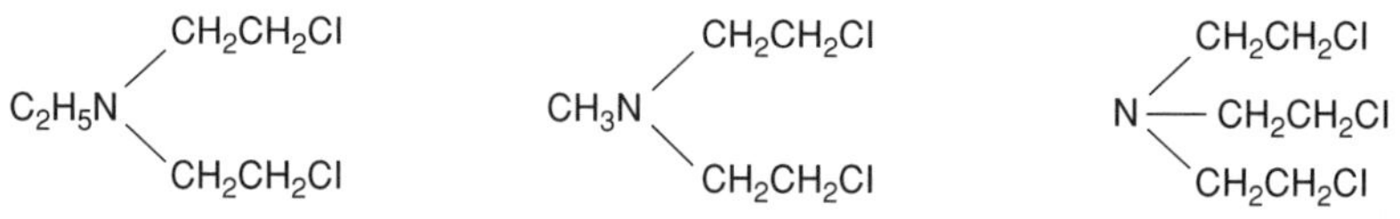

HN_1 HN_2 HN_3

Figure 4. Molecular structures of the nitrogen mustards

Of the nitrogen mustards, HN_2 is familiar as the antimitotic agent mustine hydrochloride or mechlorethamine. From a military standpoint, HN_3 is the major representative of the nitrogen mustards.

LIKELY MODE OF EXPOSURE TO MUSTARD GAS

Sulphur mustard may be used as a chemical warfare agent in a number of ways. It may be delivered by artillery shell, rocket, bomb or aircraft spray. The agent is persistent and under cold conditions long-term contamination of ground may occur. Adequately protected troops would be expected to withstand well an attack with mustard gas. Precautions should be taken to ensure that protected troops do not carry mustard, for example on boots, into designated 'clean areas'.

Contamination of snow by mustard gas presents serious problems. Sulphur mustard freezes at 14°C and detection systems relying on detection of vapour may fail. Carriage of contaminated snow into buildings and the later release of mustard vapour is a very real hazard for troops operating in an arctic climate. It is of interest that the freezing of mustard at the comparatively high temperature of 14°C led to difficulties in the design of artillery shells. Solidification of the contents of a shell were found to alter its ballistic properties and solvents were added to prevent freezing. The use of chlorpicrin for this purpose was developed by American experts. The mixture of phenyldichlorarsine with sulphur mustard to produce 'WINTERLOST' has been mentioned in Chapter 2.

Mustard gas vapour will be carried long distances by wind. Prentiss (1937) reported field tests which showed that:

> *winds not exceeding 12 miles per hour, blowing over a normally saturated terrain, may transfer concentrations of mustard vapour sufficiently strong (0.070 mg per litre) to cause death within 30 min, for 500 to 1000 yards downwind*

Naive troops or those equipped with inadequate protective equipment would not be expected to fare well if attacked with mustard. Because of this, sulphur mustard might prove to be an attractive choice for terrorists planning an attack on a civilian population.

Local contamination with sulphur mustard may extend to exposed water. Liquid mustard tends to sink as a heavy oily layer to the bottom of pools of water, leaving a dangerous oily

Table 3. Physicochemical characteristics of the nitrogen mustards

Characteristic	HN_1	HN_2	HN_3
Appearance	Colourless or yellow oily liquid		
Melting point (°C)	−34	−60	−4
Boiling point (°C)	85	75	138
Specific gravity	1.09	1.15	1.24
Vapour density	5.9	5.4	6.9
Vapour pressure			
10°C	0.0773	0.130	0.002 72
25°C	0.2500	0.427	0.010 90
40°C	0.7220	1.250	0.038 20
Smell	Almost odourless; may smell of fish or soap		

film on the surface. Drinking from contaminated water sources may lead to damage to the gastro-intestinal tract.

ABSORPTION OF SULPHUR MUSTARD, METABOLISM AND EXCRETION

Numerous studies of the penetration of skin by sulphur mustard have been undertaken. Accounts by Cullumbine (1947) and Nagy *et al.* (1946) should be consulted for summaries of early studies. Renshaw (1946) demonstrated that 80% of a sample of liquid sulphur mustard placed upon the skin evaporated. Of that penetrating the surface, about 10% was fixed to the skin with the remainder being absorbed systemically. This finding has been confirmed by recent studies (Papirmeister *et al.,* 1984a,b) of the penetration by sulphur mustard through human foreskin grafted onto athymic nude mice (see below).

The rate of absorption of sulphur mustard across the human respiratory tract is unknown. Cameron *et al.* (1946) demonstrated that some 80% of inhaled mustard was removed by the rabbit nasal mucosa. The corresponding figure for nitrogen mustard was 90%. Absorbed sulphur mustard binds rapidly to protein and disappears from the circulation.

A number of studies of the metabolism and excretion of sulphur mustard were prompted by the use of sulphur mustard in the Iran–Iraq war. Among these, those of Black *et al.* (1993) have been prominent. It was shown that approximately 60% of an intra-peritoneal dose of radiolabelled sulphur mustard was excreted by a rat within 24 h of administration. Some of the products excreted arose by hydrolysis of sulphur mustard but the majority were formed by conjugation with glutathione. The original paper should be consulted for details of the HPLC techniques used.

The relevance of studies of the metabolism of sulphur mustard conducted in rodents to clinical work may be questioned. In 1961, Davison and coworkers reported studies of the metabolism of radiolabelled sulphur mustard in two patients suffering from terminal cancer (Davison *et al.*, 1961). Disappearance of sulphur mustard from the blood was rapid but excretion in urine was delayed as compared with that seen in rodents. This may have been due to kidney damage in these very ill patients. The importance of combination with glutathione was demonstrated in these studies.

TOXICITY OF SULPHUR MUSTARD

It was pointed out in Chapter 2 that the toxicity of chemical warfare agents should not be judged simply in terms of the doses or exposures necessary to cause death. Sulphur mustard is an effective incapacitating agent; indeed its value in World War I rested on this property.

The rat, percutaneous LD_{50} is 9 mg kg^{-1}; corresponding figures for the dog and the rabbit are 20 and 100 mg kg^{-1}, respectively.

The percutaneous LD_{50} is not known in man with any accuracy although death is said to have occurred on exposure to 64 mg kg^{-1}.

Prentiss (1937) gave the following figures for the toxicity of sulphur mustard:

0.15 mg l^{-1} is fatal on 10 min exposure
0.07 mg l^{-1} is fatal on 30 min exposure

Both Vedder (1925) and Prentiss (1937) placed great emphasis on the toxicity of sulphur mustard: Vedder pointed out that:

> *In general 0.07 mgm per litre may be considered as lethal for thirty minutes exposures, and it is therefore approximately five times more toxic than phosgene, which as we have seen is ten times as toxic as chlorine.*

Effects upon the eyes

In humans, the toxicity of mustard as regards effects of the vapour upon the eyes increases as the exposure is increased. In terms of *Ct* (mg min m^{-3}) the following may be used for guidance:

50 mg min m^{-3}	Maximum safe exposure
70 mg min m^{-3}	Mild reddening of the eyes
100 mg min m^{-3}	Partial incapacitation due to eye effects
200 mg min m^{-3}	Total incapacitation due to eye effects

It is important to note that 100 mg min m^{-3} is equivalent to an exposure to 0.0017 mg l^{-1} for

1 h. The odour of sulphur mustard is detectable at about 0.0013 mg l^{-1} (Prentiss, 1937). The risk of incapacitation by conjunctivitis on exposure to levels of sulphur mustard vapour which would be difficult to detect by smell will be appreciated. Indeed, the risk is greater than these figures suggest: after initial exposure to sulphur mustard, the sense of smell seems to be dulled and larger concentrations may be undetected. During World War I, lacrimatory agents were often deployed with sulphur mustard to mask its smell.

Effects upon the skin

The effects of sulphur mustard upon the skin should be considered both in terms of exposure to the agent in liquid and in vapour form.

LIQUID EXPOSURE

Exposure to 50 μg cm^{-2} for 5 min causes slight erythema, while exposure to 250–500 μg cm^{-2} for 5 min leads to blistering. As stated above, much of the applied sulphur mustard evaporates; regarding that which is absorbed, it has been said that as little as 6 μg cm^{-2} can cause blistering (NATO, 1985).

VAPOUR EXPOSURE

The following may be used for guidance:

100–400 mg min m^{-3}	Erythema of skin produced
200–1000 mg min m^{-3}	Leads to blistering
750–1000 mg min m^{-3}	Severe, incapacitating skin burns

The large range of values quoted for the effects of mustard vapour upon the skin takes into account the effects of ambient temperature. It was demonstrated during World War I that at higher ambient temperatures the effects of exposure to mustard vapour were very much more severe than those occurring at lower temperatures. At an ambient temperature of 50°F, exposure to 1000 mg min m^{-3} of vapour may produce the same effects as exposure to 200 mg min m^{-3} of vapour at temperatures in excess of 80°F.

The effects of exposure to sulphur mustard vapour are shown in terms of concentration and time in Figure 5.

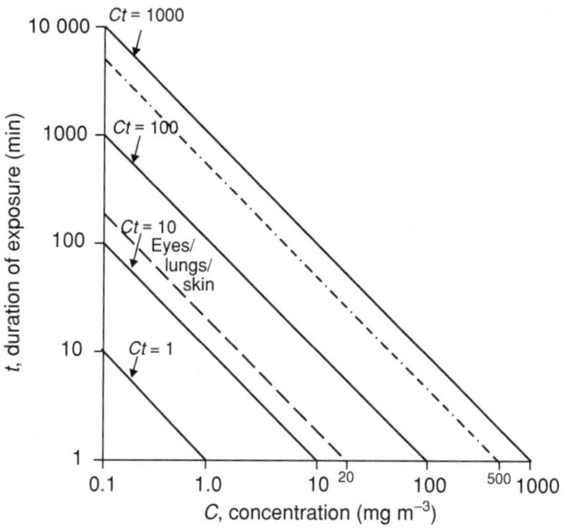

Figure 5. Effects of H. log–log plots of C and t: (– – –) $Ct = 20$ mg min m^{-3}, tolerable level?; (·— —·) $Ct = 500$ mg min m^{-3}, serious skin burns

It will be realized that the figure is intended for guidance only: the likely inaccuracy of the predictions at the extremes of time and concentration will be obvious (see Chapter 2).

TOXICITY OF HN_2 (MUSTINE HYDROCHLORIDE)

Because of its use in clinical practice, more is known about the toxicity of HN_2 than of the other nitrogen mustards. The effects of the accidental administration of 58 mg (instead of 5.8 mg) of HN_2 have been reported (Zaniboni *et al.*, 1988). The patient was a 62 year old woman suffering from Hodgkin's disease. Nine days after the dose, the white cell count had fallen to 140 per mm^3. Haemoglobin fell to 8.2 g per 100 ml. Treatment with blood transfusions restored the red cell count and the patient was discharged on the 25th day after the accident. Some six months later, she was re-admitted with signs of a neurotoxic episode: confusion, ataxia, amnesia, headache, urinary incontinence, hypo-reflexia and papilloedema. A CAT scan revealed hydrocephalus of all cerebral ventricles. Recovery occurred over a period of 1 month. Late neurotoxic effects of nitrogen mustards are rare; early effects on the nervous system are more common (Bethlenfalvay and Bergin, 1972). Early clinical work with sulphur mustard unfortunately does not yield comparable data: the drug tended to be applied directly to superficial tumours or by direct injection into tumours.

GENERAL TOXICOLOGY AND PHARMACOLOGY OF SULPHUR MUSTARD: MECHANISMS OF ACTION

Despite intensive investigation, the exact mechanism by which sulphur mustard produces its characteristic effects upon human skin remain unknown. This is remarkable as so much is now known about this compound. Research on the mechanisms of action of sulphur mustard has accelerated in the 1990s and only a summary of the main lines of research can be provided here. This section focuses on mechanisms; descriptions of effects on particular cell systems and organs follow and further information regarding effects on the skin is provided in Chapter 21. Some details of effects on the bone marrow are provided in Chapter 22, which deals with approaches to the management of mustard-induced bone marrow depression.

Most of the research on the mechanisms of action of sulphur mustard has focused on explaining the effects of this compound on the skin. It has been assumed, perhaps incorrectly, that if this is understood the mechanisms of effect on the cornea and conjunctiva and on the epithelium of the airways will be similar. In outline, the problem seems an easy one: sulphur mustard is an alkylating agent and binds to many intracellular components, including nucleic acids. It seems self-evident that if DNA is damaged, protein synthesis will be impaired, cells will fail to divide and, eventually, will die. This is undoubtedly true. However, this is a superficial analysis and does not explain the delayed appearance of blisters: other compounds that damage the skin, for example, strong acids and alkalis, damage and kill epidermal cells but do not produce blistering. It may be asked whether the processes by which sulphur mustard produces blistering are identical with those which cause blisters to be produced by Lewisite and other arsenical vesicants or even by mustard oil (allyl isothiocyanate). This seems unlikely although the similarity of effect (blistering) suggests that the mechanisms must overlap to some extent: perhaps they differ in the early stages of the process but come together in some final common pathway that leads to epidermal–dermal separation and a marked leak of fluid from dermal papillary capillaries.

Any satisfactory explanation will need to explain the appearance of pyknotic nuclei in the stratum germinativum of the epidermis, separation of hemi-desmosomes from their links with laminin and other compounds of the basal lamina, the accumulation of fluids that leads to the formation of blisters that later collapse as fluid is re-absorbed (and the mechanisms and routes by which it is absorbed) and the very slow healing of the lesions. The biophysics of fluid leakage from dermal capillaries in mustard-damaged skin has been little studied: it is assumed that this is a passive process allowed by separation of capillary endothelial cell junctions rather as in the wheal reaction produced by physical damage (a part of the 'Tripple Response') and by the release of histamine from mast cells in allergic reactions. However, allergic reactions do not produce the large blisters seen in mustard gas casualties and even facial oedema, e.g. angioneurotic oedema is characterized by a marked fluid leak into the tissues but not by blisters. Blistering, itself, is little studied: why a hot iron produces such characteristic blisters but concentrated acid does not, is an unanswered question. It might be asked whether dermal oedema produced by any cause from heart failure to a lack of protein in the blood and thus a low plasma oncotic pressure, could lead to blistering if the epidermal–dermal linkage were to be damaged. This is unknown. However, blistering can occur in other types of poisoning, including barbiturate overdose and exposure to high concentrations of carbon dioxide: again, the mechanisms are unknown. Biophysical research including such obvious steps as measuring blister fluid hydrostatic pressure has not progressed as rapidly as that involving molecular biological techniques. Consider, for example, a hemispherical blister. The surface area is given by the formula $2\pi r^2$ (the surface area of a sphere is given by $4\pi r^2$ and its volume by $4/3\pi r^3$). The area of the base is given by πr^2 – thus the blister surface is exactly twice the area of its base. Yet all of the blister surface is derived from the epidermis that previously covered its base. If we assume that the volume of the epidermal tissue that has been stretched has remained constant, then it must have been reduced in thickness by a factor of 2. The average thickness of the epidermis is 0.06 mm (excluding the palms and soles which do not blister on exposure to mustard). The blister roof is thus likely to be only about 0.03 mm in thickness and the layers of keratinocytes that comprise this layer must have either been reduced or the individual cells stretched. We know little about the forces needed to stretch these cells. If we knew more of these forces we would be able to relate them to the only source of pressure present in a blister: the pressure of the fluid leaking from the capillaries. This pressure will be increased if lymphatic drainage of the damaged area is impaired and the usual process of protein-washout from the interstitium is rendered inoperative: this is discussed further in Chapter 21. No data on lymphatic drainage from mustard damaged skin seem to be available. Some information on capillary leakage has been provided by Vogt *et al.* (1984): this is discussed in a following section but it is noted that the mechanisms underlying the early and reversible increase in capillary permeability and the later more sustained increase, are unknown at a molecular level.

The biochemical reactions of sulphur mustard have been studied in detail. Alkylation: the formation of covalent linkages between an alkylating agent, e.g. sulphur mustard, and nucleophilic molecules so that an alkyl group comes to be attached, is well understood (Bowman *et al.*, 1986). The mode of action of nitrogen mustard have been investigated in more detail than that of sulphur mustard and excellent accounts are available (Goodman and Gilman, 1980; Fox and Scott, 1980). Sulphur mustard and the nitrogen mustards are bifunctional alkylating agents in that their molecules possess two chains each capable of undergoing the cyclization reactions necessary before alkylation can occur. Bifunctionality allows cross-linking of strands of nucleic acids: this is believed by some to be a key process in the toxic effects of the sulphur mustard molecule. The reactions occurring in cross-linking of DNA are shown in Figure 6.

The binding of ethylenesulphonium ions, in the case of sulphur mustard, and ethyleneimmonium ions, in the case of nitrogen mustards, to DNA produces a range of effects including:

- alkylated guanine residues tend to form base-pairs with thymine, rather than cytosine, leading to coding errors and hence to inaccuracies in protein synthesis;
- alkylated guanine residues may be excized from the DNA molecule (see below);
- cross-linking of two alkylated guanine residues leads to a major disruption in DNA functioning.

Papirmeister and his colleagues have studied the effects of sulphur mustard on DNA in great detail (an excellent account is provided in Papirmeister *et al.*, 1985). The authors pointed out that the major sites of DNA alkylation by sulphur mustard (HD: distilled mustard) are:

- monofunctional adducts at the N_7 position of guanine (about 60%);
- monofunctional adducts at the N_3 position of adenine (about 16%);
- bifunctional adducts involving the N_7 positions of guanine residues in the two chains of DNA or of adjacent residues in a single chain: 16% (i.e. interstrand and intrastrand cross-links).

Papirmeister and his colleagues showed that these alkylated purines (guanine and adenine are purines, cytosine and thymine are pyrimidines) are rapidly removed from DNA: a depurination process that may occur spontaneously or as a result of activation of endonucleases. Interestingly, it was shown that alkylation of the N_3 site of adenine had the greatest 'sensitizing effects' on endonucleases derived from extracts of bacterial and mammalian cells (Papirmeister *et al.*, 1970; Lindahl, 1979). For further details of endonuclease activation, the reader should refer to the original papers. Depurination leads to apurinic sites appearing in DNA and these are attacked by apurinic endonucleases. These specific endonucleases lead to breaks in the DNA chain: a critical step in the pathway proposed. Papirmeister summed up the process as:

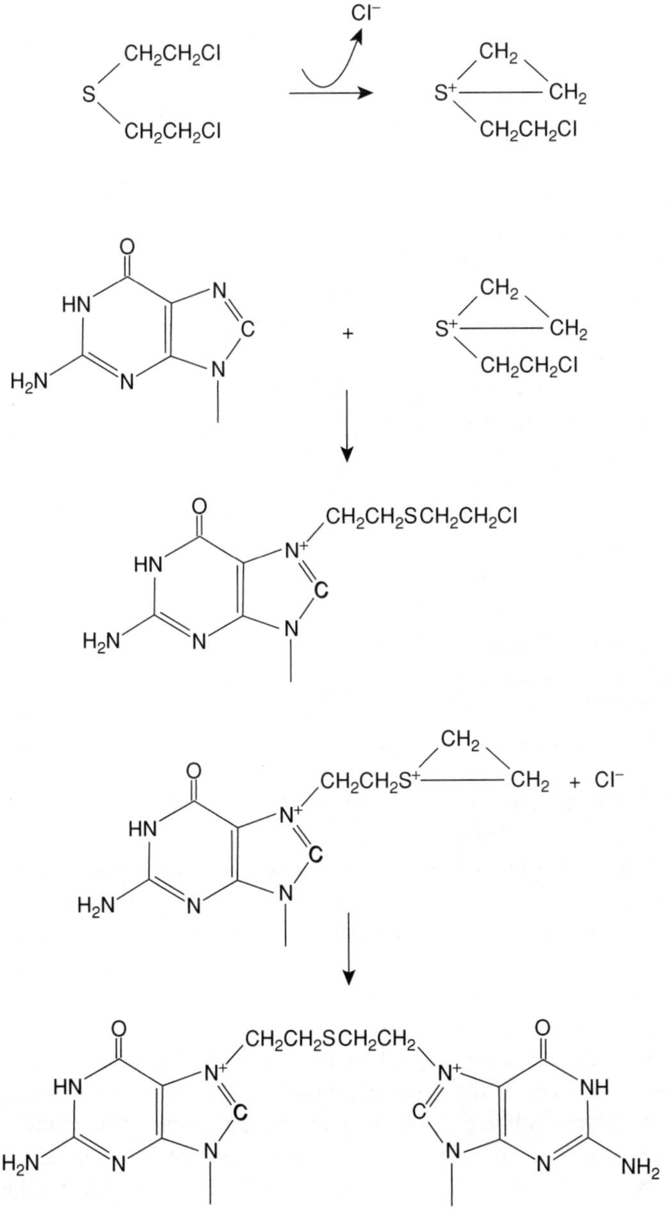

Figure 6. Reactions occuring in the cross-linking of DNA

- the vast majority of DNA breaks are not produced directly by the alkylating agent but by sensitization to enzymatic breakage by apurinic endonucleases;
- apurinic sites are produced by enzyme induced and/or spontaneous; depurination of major monoadducts: N_3 alkyladenine and N_7 alkylguanine;
- apurinic endonucleases convert apurinic sites to DNA breaks, creating good substrates for further attack by exonucleases.

So far, so good but how does this link with a blister formation? It might well be thought that cells with damaged DNA are likely to be impeded as regards protein synthesis and division, and might simply die, but why should such damage lead to the characteristic blistering produced by sulphur mustard?

It had been discovered by Dixon and Needham (1946) that sulphur mustard interfered with carbohydrate metabolism. Studies in rats showed that the Respiratory Quotient of skin fell from 0.9 to 0.56 following exposure to sulphur mustard, suggesting that the metabolism of carbohydrate was significantly impaired. In particular, the anaerobic formation of lactic acid from glucose was impaired: glycolysis had been inhibited. Dixon and Needham (1946) showed that in more than twenty vesicants glycolysis was inhibited: interestingly, many of the vesicants were not closely related chemically. They also showed that hydrogen cyanide, not generally regarded as a vesicant, produced marked vesication when frogs were perfused with prussic acid at a M/100 concentration: a concentration known to inhibit glycolysis. In passing, it may be noted that the 'perfused frog' was one of very few animal models shown by Dixon and Needham (1946) to produce blisters in response to sulphur mustard. These authors noted that mustard did not inhibit glycolysis of hexosediphosphate (fructose-1,6-diphosphate: the fourth molecule in the glycolysis pathway: glucose, glucose-6-phosphate, fructose-6-phosphate, fructose-1,6-diphosphate) and thus argued that sulphur mustard inhibited the initial phosphorylation of glucose, i.e. the hexokinase enzyme was being inhibited. Further work using tissue extracts showed that this was true. They supported their findings with a table which compared the vesicant properties of a range of compounds with their capacity to inhibit the enzyme hexokinase *in vitro* (see Table 4).

Table 4. Vesicant properties of a range of compounds compared with their capacity to inhibit the enzyme hexokinase *in vitro* (adapted from Dixon and Needham, 1946)

Substance	Vesicancy	% Inhibition of hexokinase *in vitro*
$S(CH_2CH_2Cl)_2$	++	80
$OS(CH_2CH_2Cl)_2$	—	0
$S(CHClCH_3)_2$	—	0
$S(CH_2Cl)_2$	—	0
$S(CH_2CH_2CH_2Cl)_2$	—	0
$O_2S(CH{=}CH_2)_2$	+	60
$OS(CH{=}CH_2)_2$	—	0
$S(CH_2CH_2OH)_2$	—	0
$C_2H_5SCH_2CH_2Cl$	+	45
$C_6H_5SCH_2CH_2Cl$	+[a]	40
$CH_3N(CH_2CH_2Cl)_2$	+	65
$N(CH_2CH_2Cl)_3$	+	70
$ClCH{=}CHAsCl_2$	++	100
$ClCH_2CH_2AsCl_2$	—	0
$CH_3CH_2AsCl_2$	+	45
C_2H_5I	—	30
CH_3Br	++[a]	90
BAL	—	0
FCH_2COOCH_3	—	0
$(C_2H_7)_2FPO_3$	—	0

Note:
[a] When evaporation is prevented.

The close correspondence between vesicancy and inhibition of hexokinase is clear; also that the strong vesicants included compounds as different as $S(CH_2CH_2Cl)_2$, $ClCH–CHAsCl_2$ and CH_3Br.

Papirmeister and colleagues (1985) have followed another line of argument. They have shown that activation of the DNA repair enzyme, poly(ADP-ribose) polymerase (PARP) leads to the consumption of the cofactor NAD^+. They argued that depletion of this cofactor explained the inhibition of glycolysis reported by Dixon and Needham (1946) and by Renshaw (1946). Other workers have studied the enzyme PARP in a variety of cell systems (Bhat *et al.*, 2000; Hinshaw *et al.*, 1999; Kraker and Moore, 1998; Meier and Millard, 1998; Meier *et al.*, 2000) and have confirmed its activation in cells exposed to sulphur mustard.

Hinshaw *et al.* (1999) have shown that activation of PARP may be linked to ATP-dependent changes in microfilament architecture in cells. The PARP activation hypothesis is thus well supported although Lin *et al.* (1994) failed to find a link between depletion of NAD^+ and cell damage in rat keratinocyte cultures. These authors used nicotinamide to prevent a fall in NAD^+ levels but found that this did not prevent the decline in DNA content used as an index of cytotoxicity. This is a particularly important finding as Papirmeister *et al.* (1985) have also used nicotinamide and niacin to increase levels of NAD^+ in cells. They noted, however, that nicotinamide was an inhibitor of PARP but that niacin was not. They argued that niacin would allow DNA repair without concomitant depletion of NAD^+. They also noted that niacin deficiency (pellagra) was associated with changes in skin histology similar to that caused by exposure to sulphur mustard.

The hypothesis that led from alkylation of purines in DNA to inhibition of glycolysis thus seems soundly based. It was suggested that a lack of NAD^+ would prevent the glyceraldehyde-3-phosphate to 1,3-diphosphoglyceric acid step in glycolysis. This is an early step and its inhibition prevents formation of pyruvic acid (the entry point for the citric acid cycle) and the formation from pyruvic acid of lactic acid. However, recent authors have not dwelt on Dixon and Needham's finding that hexokinase, the first-step enzyme in the glycolytic pathway, is inhibited. On the contrary, it has been argued that a deficiency of NAD^+ would lead to the hexose monophosphate shunt (the pentose phosphate pathway) being activated (Papirmeister *et al.*, 1985; Hayashi and Ueda, 1982). This pathway does not require the cofactor NAD^+; on the contrary, the cofactor $NADP^+$ is the key cofactor here. If glucose-6-phosphate accumulates (as it would if glycolysis were inhibited at a later, NAD^+-dependent, stage – see above) the hexose monophosphate shunt is activated. However, this, of course, would not occur if Dixon and Needham's (1946) finding of hexokinase inhibition is correct: glucose-6-phosphate would not be accumulating in the presence of sulphur mustard. Further work to resolve this point is needed. The activation of the hexose monophosphate shunt has been shown to be associated with release of a range of proteases including plasminogen activator (Schnyder and Baggiolini, 1980). Thus, a link between the indirect effects of sulphur mustard on glucose metabolism and substances capable of triggering inflammatory cascades and damaging structural proteins may have been found. This is the essence of the hypothesis put forward by Papirmeister *et al.* (1985) and it represents a major step forward in understanding the mechanisms of action of sulphur mustard.

Other lines of study have become prominent in this area in the past ten to fifteen years. We might list these as:

- the apoptosis theory;
- the free radicals/lipid peroxidation theory;
- the cytokine-inflammatory mediators theory;
- the sites of action of sulphur mustard in the epidermis–dermis adhesion complex theory.

This is not suprising – the explosion of interest in apoptosis and in cytokines was bound to spill over into this area: it does, however, pose a problem as regards summarizing the findings of work reported so far.

Apoptosis

Rosenthal *et al.* (1998) pointed out that proteolytic cleavage of PARP was known to be associated with apoptosis (Kaufmann *et al.*, 1993; Neamati *et al.*, 1995; Nicholson *et al.*, 1995; Tewari *et al.*, 1995). Rosenthal *et al.* (1997) argued that the reversible stage of apoptosis was associated with activation, followed by breakdown, of PARP. Calcium ions have been identified as a key factor in controlling apoptosis (Mol and Smith, 1996; Ray *et al.*, 1995). It seems that an increase in intracellular calcium concentration triggers apoptosis. Intracellular concentrations of calcium ions are affected and controlled by a range of mechanisms: oxidative stress is known to be important (Orrenius *et al.*, 1989) as is, in Rosenthal's words, 'activation of protein tyrosine kinase, leading to the activation of phospholipase C, the formation of IP3 (inositol triphosphate) and Ca^{2+} mobilization'. Calmodulin (one of several intracellular calcium binding proteins) has also been shown to be important: this area is complicated and the reader is referred to Ray *et al.* (1995) and Shi *et al.* (1989) for details. These signalling pathways lead to activation of caspases: a group of cysteine proteases. Caspase-3 attacks proteins that are important in maintaining cell structure, including PARP. Rosenthal *et al.* (1997) showed that sulphur mustard induced terminal differentiation and apoptosis in the basal cells of the epidermis. Expression of fibronectin (known to be involved in basal cell–basal lamina adhesion) was reduced. The authors referred to PARP as the 'death substrate' – a singular description of an enzyme playing an important part in controlling apoptosis. Kan *et al.* (2003) have also shown that sulphur mustard induces apoptosis in the hairless guinea pig skin model. The importance of DNA damage as an early step in the pathway leading to caspase activation was stressed by Lodhi *et al.* (2001) and several authors (Kadar *et al.*, 2000; Rosenthal *et al.*, 1998) have stressed the importance of calmodulin in the process. In their latest papers, Rosenthal and colleagues have delved more deeply into the complexities of the apoptosis process and have shown that sulphur mustard up-regulates expression of the Fas receptor (tumour necrosis factor receptor superfamily, member 6 gene encodes the Fas receptor) or its ligand resulting in recruitment of the, again splendidly named, Fas-associated death domain (FADD). A variety of caspases (3,7 and 8) appear to be involved (Rosenthal *et al.*, 2000). Very interestingly, the authors reported that expressing a dominant negative form of FADD, i.e. FADD-DN, in human skin grafted to athymic nude mice, led to a reduction in vesication on exposure to sulphur mustard. The reader is referred to the original paper for details. The apoptosis theory is thus developing well – that sulphur mustard can induce apoptosis seems clear although the exact mechanisms by which it does are undoubtedly complex. Understanding these complexities may lead to ideas for treatment of skin injuries caused by sulphur mustard.

Production of free radicals and lipid peroxidation: links with cytokine production

Molecules which are free radicals are characterized, generally, by an unpaired electron in their outer electron shell. Thus the superoxide ion $O_2^{\bullet -}$ and the hydroxyl radical $OH^{\bullet}$ are free radicals. The unpaired electron makes such molecules highly reactive – indeed chain reaction can be initiated by such molecules reacting with lipid molecules to produce lipo-peroxides. The chemistry of free radicals contains some surprises. Oxygen itself has two unpaired electrons in its outer electron shell and is thus a 'double' free radical. Fortunately, the two unpaired electrons have parallel orbits and this, apparently, confers stability on the molecule. The two unpaired electrons are responsible for the paramagnetic properties of oxygen – properties essentially unique to the naturally occurring oxygen molecule. We should also note that the hydroxyl radical $OH^{\bullet}$ is a very different molecule, in terms of reactivity with biological molecules, than its cousin, the hydroxyl ion OH^-. An interesting series of articles on free radicals and their role in a range of pathological processes can be found in the *Annual Review of Physiology*, Volume 48, 1986. This includes an excellent introduction by W. A. Pryor and an outstanding contribution by Lewis Smith on the role played by free radicals in the toxicology of the herbicide paraquat. Pryor notes in passing that undiluted gas-phase cigarette smoke contains more than 10^{17} reactive free radicals 'per puff': testimony both to the toxicity of cigarette smoke and to the defensive mechanisms in the airways! Free radicals are generated constantly in the body as a side-product of the use of oxygen. Superoxide dismutase (SOD) (an enzyme that is found in forms that contain copper or zinc in the cytoplasm and in a form that contains manganese in mitochondria) catalyses the conversion of the superoxide radical to hydrogen peroxide:

$$O_2^{\bullet -} + O_2^{\bullet -} + 2H^+ \longrightarrow H_2O_2 + O_2$$

Hydrogen peroxide is broken down to water and more oxygen by the enzyme catalase:

$$H_2O_2 + H_2O_2 \longrightarrow 2H_2O + O_2$$

In addition, a large number of free radical-scavenging molecules, including glutathione, are found in cells.

$$2GSH + H_2O_2 \longrightarrow GSSG + 2H_2O$$

The glutathione disulphide (GSSG) is reduced to glutathione by glutathione reductase, with NADPH acting as a cofactor. For more details of these reactions, including the role of myeloperoxidase in forming hypochlorous acid, the reader is referred to the articles previously mentioned in the *Annual Review of Physiology*.

One particular relevant source of free radicals is provided by the xanthine oxidase-catalysed conversion of oxygen to the superoxide radical. This process may be initiated by tumour necrosis factor (TNFα). Arroyo *et al.* (1995) have shown that sulphur mustard can induce human monocytes to release TNFα: thus, a link between sulphur mustard and oxygen free radical formation has been identified. Polymorphonuclear leukocytes respond to the presence of bacteria by an 'oxidative burst' in oxygen free radical production. Levitt *et al.* (2003) have shown that this process is primed by sulphur mustard. Work by Elsayed *et al.* (1992) has also linked sulphur mustard with free radical production. Free radicals are known to be linked via transcription factors to up-regulation of production and release of inflammatory cytokines. The family of known cytokines is now large, over 20 separate interleukin species have been described, and the number of interacting pathways that have been postulated and, in some cases, demonstrated is very large indeed. Sulphur mustard has been linked with the production of a number of such cytokines: IL8, IL6 and TNFα (Sabourin *et al.*, 2002); IL-1ß and IL6 (Sabourin *et al.*, 2000); IL6 (Arroyo *et al.*, 2001); TGFß-1, a fibrogenic cytokine (Aghanouri *et al.*, 2004); IL8 (Lardot *et al.*, 1999) IL-1α (Pu *et al.*, 1995); IL-8, IL-6, TNFα and IL-1ß (Dillman *et al.*, 2004). The last paper is particularly interesting as it demonstrates that down-regulation of the signalling molecule p38 MAP kinase inhibits the effect of sulphur mustard on these cytokines. Further references to the effects of sulphur mustard on cytokine production may be found in Dillmann's paper published in 2004.

Such interest in free radical production and the knock-on effects on production of inflammatory mediators has led to a search for compounds capable of blocking these pathways. Eldad *et al.* (1998) have shown that SOD given prophylactically reduced the effects of sulphur mustard in a guinea pig skin model: no effect was produced when administration of SOD was delayed until after exposure to sulphur mustard. *N*-acetylcysteine (NAC) increases production of glutathione and has been studied by Atkins *et al.* (2000) as a possible protective agent against the effects of sulphur mustard. This work has shown that binding of the NK-κB transcription factor, known to be linked with cytokine production, to its consensus sequence was increased by sulphur mustard. The molecular biology of these processes is complicated: it has been shown, for example, that NAC itself can activate NK-κB although it seems to inhibit this effect of sulphur mustard. Free radicals are now accepted as a possible mediator of the effects of sulphur mustard.

Disturbance of basal cell–basal lamina adhesion complex

The reader might be forgiven for concluding that sulphur mustard can interact with many biochemical processes within cells, either directly or via the generation of free radicals. One might ask: so what? It was known long ago that sulphur mustard could kill cells; it is interesting that in addition to killing cells outright, cells can be tipped towards apoptosis (assisted suicide) by sulphur mustard and it is both interesting, and confusing, to know that sulphur mustard can affect calcium entry into cells, release of calcium from stores in cells, signalling molecules and cytokine pathways *but* how does all this link with vesication? This we will now consider.

Basal cells are anchored to the basal lamina by hemidesmosomes. This complex structure can be imagined to span the basal membrane of the basal cells and to provide an intracellular anchorage point for microfilaments (discussed below) and an extracellular anchorage point for components of the basal lamina. Seven structural components have been identified in desmosomes and it is likely that these occur in hemidesmosomes. These components comprise three glycoproteins: desmoglein I (MW = 165 kD), desmocollin I (130 kD) and desmocollin II (130 kD) and four proteins: desmoplakin I (250 kD), desmoplakin II (215 kD), pakoglobin (83 kD) and a 75 kD basic polypeptide (Fawcett, 1994). The exact localization of these proteins is uncertain but the desmocollins are probably adhesive substances and are found in the narrow space between cell membranes (note that this refers to a desmosome where two cell membranes are opposed: less is known of the role of adhesive materials in the epidermal hemidesmosomes). Desmoplakins, plakoglobin and the 75 kD basic protein are all found in the dense plaque located against the inner surface of the cell membrane. Intermediate filaments (tonofilaments, keratin filaments) are linked to the adhesion plaque. Desmosomes are common in the deeper layers of the epidermis: the shrinkage caused by fixatives caused desmosomes to appear as the 'prickles' of the so-called prickle cell layer of classical light microscopy. In the more superficial layers, cell-to-cell linkages are fewer and intercellular lipids rather than junctional contacts provide the waterproofing of the epidermis. If the hemidesmosomes is complex, then so is the basal lamina. This is now divided into a lamina lucida adjacent to the cell membrane and a lamina densa between the lamina lucida and the underlying dermis. Each is about 40–50 nm in thickness. A sub-basal lamina is added by some authors – deep to the lamina densa. Laminin (and there are a number of sub-types) and fibronectin span the lamina lucida, running from the hemidesmosomes to the lamina densa. The latter contains the proteoglycan heparan sulphate and Type IV collagen: unique to the basal lamina. Type VII collagen runs in loops between the lamina densa and deeper anchoring plaques (again Type IV collagen) and ties fibres of Type III and Type I collagens to the basal lamina. Put simply – the basal cells are bound to the lamina densa by laminin and fibronectin and the lamina densa is bound to dermal collagen by Type VII collagen. This may sound sufficiently complicated but no mention has been made of other cell adhesion molecules: the integrins, the cadherins and the selectins, nor of the five classes of intermediate filaments: keratin being the major form in epidermal cells (vimentin, desmin, etc. are found in other tissues). Given that sulphur mustard alkylates proteins avidly, it is easy to imagine that the complex cell anchoring system might be affected by exposure to this compound. This is indeed the case. We will consider the problems in three parts.

(a) Effects on intracellular keratin filaments

Dillman *et al.* (2003) have recently shown that sulphur mustard causes keratin 14 and keratin 5 to clump into stable aggregates. These appear within 15 min of exposure to sulphur mustard. It was noted that a number of blistering disorders (epidermolysis bullosa simplex and epidermolytic hyperkeratosis) are associated with abnormalities of these keratin molecules (Fuchs *et al.*, 1994). The role of keratin 14 in linking to plectin (another component of the hemidesmosomes) and to the bullous pemphigoid antigen 'le' was also noted. Werrlein and Madren-Whalley (2000) demonstrated that sulphur mustard reduced expression of keratin 14 in cultured human keratinocytes. In addition, expression of an integrin ($\alpha_6 ß_4$) was reduced. This integrin has been shown to be abnormal in junctional epidermolysis bullosa. $\alpha_6 ß_4$ integrin has two components: α and ß – the ß component is attached to intracellular keratin 14 and the α component to extracellular laminin 5. Zhang and Monteiro-Riviere (1997) showed that the integrin could be found attached to the roof of sulphur mustard blisters: suggesting that pulling away of this molecule from extracellular laminin was an important feature of sulphur mustard blistering. It was suggested that intracellular calcium levels controlled the functioning of the integrin discussed above. The paper by Werrlein and Madren-Whalley (2000) should be consulted for further details.

(b) Effects on components of the basal lamina

Smith *et al.* (1997) have studied the effects of sulphur mustard on basal lamina proteins and have shown that laminin 5 was the major protein affected: immunofluorescent staining of this protein was consistently reduced in weanling pig skin exposed to sulphur mustard whereas other components (e.g. collagens IV and VII) were unaffected. Monteiro-Riviere *et al.* (1999) have studied the effects of sulphur mustard using a mouse ear model. Cleavage was precisely located to the upper lamina lucida but the cleavage plane was shown to run in and out of the basal cells – in some locations, the hemidesmosomes and basal cells were lifted – in others, the hemidesmosomes remained attached to the basal lamina and the cells were sheared away from them. The authors noted that the exact location of the cleavage plane differed from animal model to animal model. In most models, the cleavage plane is clearly placed in the lamina lucida. This was found to be the case in the isolated perfused pig skin flap (IPPSF) model; the one mammalian model that produces classical blistering as opposed to the formation of microvesicles only (Monteiro-Riviere and Inman, 1997). Further studies of the effects of sulphur mustard on basal lamina components have been contributed by Zhang *et al.* (1995) and by Monteiro-Riviere and Inman (1995): papers by these workers should be consulted for further details as should that by Lindsay and Rice (1995) who used a Yucatan mini pig model.

(c) Effects on subepidermal collagen

Millard *et al.* (1997) have focused on the effects of sulphur mustard on collagen in the subepidermal layer of the skin of the euthymic hairless guinea pig – another animal model. Two abnormal polypeptides were identified in extracts of the subepidermis after treatment with sulphur mustard: these were distinguishable from types I, III and IV collagens. These might be abnormal collagens produced as a result of exposure to sulphur mustard or breakdown products of normal collagens acted upon, for example, neutrophil collagenases. Very interestingly, the authors noted that the inherited blistering disease, bullous pemphigoid, was also characterized by the presence of a hybrid collagen called collagen XVII and that this is known to be associated with hemidesmosome adhesion complexes. This is a very suggestive finding. A number of matrix metalloproteinase enzymes are known to attack collagen: see the paper by Calvet *et al.* (1999) for evidence that these are up-regulated in the airways of guinea pigs exposed to sulphur mustard.

The account given above is only the briefest summary of the explosion of work on the mechanisms of effect of sulphur mustard that has been published during the past ten or so years. It is clear that Papirmeister's hypothesis linking DNA repair by poly(ADP-ribose) polymerase with a reduction in NAD^+ and activation of the hexose monophosphate shunt and subsequent activation of proteases is still plausible. However, so is the apoptosis theory linking as it does with increases in intracellular calcium levels. It has also been shown that sulphur mustard can affect, directly or indirectly, components of the epidermal–basal lamina linkage system and this may explain the blistering effect of this chemical. However, some questions remain:

- Why does only human skin, when compared with all other normal mammalian skins tested, blister as it does?
- Why do some chemically unrelated compounds cause blistering whereas other toxic substances do not?

Further work will be needed to answer these questions.

One recent paper by Grando (2003) is worth considering. The author has taken up a quite different line from those outlined above and has suggested that acetylcholine plays an important role in controlling epithelial cell adhesion. Nitrogen mustard is known to form a quaternary nitrogen grouping on cyclization and this is said to explain the cholinomimetic effects of this compound (Hunt and Philips, 1948). Grando (2003) showed that ligation of acetylcholine receptors with nitrogen mustard, in keratinocytes and bronchial epithelial cells, increased the activity of serine proteases. This effect was blocked by atropine (a muscarinic receptor blocker) and mecamylamine (a nicotinic receptor blocker). Further work on this interesting and novel hypothesis is needed.

HISTOPATHOLOGY OF SULPHUR MUSTARD EXPOSURE

The gross pathological effects of sulphur mustard exposure will be described in the section on clinical effects; here, the account will be limited to the light and electron microscopic appearances of damaged tissue. A very large number of experiments, using animal models and human volunteers, designed to investigate the effects of exposure to sulphur mustard, was undertaken during World War I. These have been described in detail by Ireland (1926). Ireland's account is particularly valuable for the photographic material presented.

Skin effects

In one series of experiments, sulphur mustard was applied in droplets of approximately 0.0004 cm^3 producing a 3–4 mm diameter contamination of the skin of the forearm in a series of volunteers. The progress of the lesions was observed and histopathological studies were made by means of biopsy.

Thirty minutes after contamination, the epidermis appeared generally shrunken with vacuolation of the deeper layers and nuclear changes in the granular layer. The cornified layer appeared rather thicker than usual and appeared to separate readily from the stratum lucidum. Occasional vacuoles were seen in the deepest layers of the epidermis. The dermis showed fewer changes than the epidermis.

Capillary damage in the papillary layer of the dermis was identified. The endothelial cells of capillaries in this region showed nuclear damage and vacuolation of their cytoplasm. Pericapillary oedema was noted. Leukocyte diapedesis was seen in some capillaries but near the centre of the lesion blood vessels appeared contracted and empty. Lymphatic vessels in the dermis were dilated.

The hair follicles and sebaceous glands showed changes similar to those described above, although sweat glands appeared normal.

By 18 h after contamination, a typical vesicle had appeared. On microscopy, liquefaction and hydropic changes in the epithelium of the centre of the lesion were noted. Small vesicles were seen at the epidermal–dermal border and in some places the epidermis was separated from the dermis by fluid. Epithelial cells of hair follicles and sebaceous glands also showed these changes.

The upper part of the dermis appeared oedematous and contained many degenerating nuclei. Capillary damage and local oedema was also seen in this region. Deeper in the dermis, blood vessels were congested and the lymphatic vessels appeared dilated and filled with proteinaceous fluid.

The subcutaneous tissues showed changes, particularly with relation to small blood vessels. Congestion of blood vessels and dilatation of lymphatics was commonly seen.

By 36 h after contamination, almost complete epidermal destruction was observed. Such epidermal cells as were left showed pyknotic nuclei. In some areas, only the deepest layer of epidermal nuclei remained. Leukocyte infiltration of the dermis was noted, particularly in relation to sweat glands, hair follicles and sebaceous glands. Small pockets of oedema and inflammatory cell infiltration were observed in the subcutaneous tissues.

The later progress of the lesion may be summarized as follows:

- 40–50 h: collapse of vesicles and progressive necrosis
- 72 h: eschar formation beginning
- 4–6 days: eschar beginning to slough; oedema and hyperaemia persisting locally
- 19 days: separation of eschar leaving a pigmented scar.

The examination of human postmortem material during World War I produced further observations:

- Infection was often found, the associated inflammatory response being marked.
- Striking increases in pigmentation were recorded. Cells were seen to be laden with melanin and this occurred not only in the basal layer of the epidermis.
- Deep burning was associated with increased dermal fibroblast activity.
- Thrombosis of blood vessels was seen only in instances of very severe burns.

Small lesions were observed to heal slowly by centripetal spread of peripheral epidermal cells and dermal fibrosis if dermal damage had occurred. Again, the increase in melanin production was characteristic.

In 1984, Vogt and colleagues reported a more detailed study using rabbits and guinea pigs (Vogt *et al.*, 1984). The light microscopic features described above were confirmed and extended by electron microscopy and the following observations added.

Histochemical observations. There was an increase in acid phosphatase and arylsulfatase activity in the region of basal epidermal cells, reaching a maximum at 3 h post-exposure. This was attributed to activation of lysosomal enzymes. At 8 and 19 h, post-exposure increased arylsulfatase staining of fibroblasts and histiocytes was noted. At 19 h, clusters of large ATPase positive mononuclear cells were identified just below the epidermis. These may have been derived from the Langerhans cells of the antigen presenting series.

Vascular permeability studies. Evans blue and horse radish peroxidase were used as markers of vascular permeability. At high doses of sulphur mustard (250 μg cm^{-2}), an early (30–60 min post-exposure) capillary fluid leak was noted. This appeared to be a reversible phenomenon. Later, and at all doses, a more marked leak occurred (8–48 h post-exposure). The proposal that the early effect may have been due to direct effects of sulphur mustard on capillary endothelial cells and that the later leak might be local vasoactive-metabolite-dependent was put forward.

Healing pattern. The inflammatory response was seen to reach a peak at 24–72 h post- exposure. Healing took place over 10 days.

The two-stage effect proposed could be summarized as follows:

(1) Immediate phase (within 1st hour) – direct damage to endothelium of capillaries and to superficial fibroblasts.
(2) Delayed phase – death of basal epidermal cells, generalized vascular leak, invasion by inflammatory cells.

It is interesting to note that the production of tissue oedema was inhibited by the application of topical steroids combined with systemic steroid therapy but that by 48 h post-exposure no difference could be detected between the lesions of control and steroid-treated animals. No evidence of enhanced healing of the lesions was obtained.

As has already been stated, animal skin does not blister as a result of contact with mustard. Papirmeister *et al.* (1984a,b) reported an extensive investigation of the effects of sulphur mustard on human skin using an interesting animal/human model. Athymic nude mice are immunologically incompetent and will accept skin grafts of human material. A similar rat model has also been developed. Human foreskin was used as the graft material. The most important findings reported from this study were probably those describing the pathological sequence at the electron microscopic level.

The following sequence of changes was defined:

(1) Condensation and margination of heterochromatin.
(2) Loss of euchromatin.
(3) Blebbing of the nuclear membrane.
(4) Appearance of perinuclear vacuoles.
(5) Swelling of endoplasmic reticulum.
(6) Progressive dissociation of rosettes of free ribosomes.
(7) Formation of cytoplasmic vacuoles.
(8) Loss of integrity of basal cell membrane.
(9) Leakage of cell contents and debris into lamina lucida of the basement membrane.
(10) Disruption of anchoring filaments of basal hemidesmosomes.
(11) Phagocyte infiltration into areas of damage.

It was also noted that not all basal epidermal cells were equally affected and suggested that the capacity for repair might be dependent on the stage of the cell cycle at the time of application of the mustard.

Sensitivity to sulphur mustard varies considerably from person to person. In addition, some individuals seem to become sensitized to sulphur mustard. This was investigated by Sulzberger *et al.* (1947) who reported a wide range of dermatological reactions to sulphur mustard. The authors' paper, which contains many interesting details relating to mustard gas and to other vesicants, should be consulted for details. Racial differences in sensitivity have been reported and disputed. Sulzberger *et al.* (1947) reported that negroes were less sensitive to sulphur mustard than whites and considered the suggestion that blondes were more sensitive than dark-haired whites. The latter suggestion they could not confirm. The greater sensitivity of whites than negroes has also been questioned by later workers.

Eye effects of sulphur mustard

Detailed studies of the effects of sulphur mustard and other mustards upon the eyes have been reported by Mann *et al.* (1948), Warthin and Weller (1918) and Friedenwald *et al.* (1948). The following account draws heavily upon Mann's seminal studies.

The effects of liquid sulphur mustard upon the eye mirror those upon the skin. Early corneal changes, including pyknosis in the epithelium and substantia propria, leading to corneal necrosis by 12 h were reported. Regeneration of the cornea occurred by approximately 65 h post-exposure in animals. Vascularization and scarring of the cornea followed in more severe cases. Vapour exposure was less likely to produce permanent eye changes than liquid exposure. Corneal necrosis was uncommon in cases of vapour-only exposure.

Conjunctival changes, including necrosis, desquamation and marked oedema, were noted. Petechial haemorrhages were commonly seen. In very severe cases of exposure during World War I, iritis and iridocyclitis were seen. The extension of changes into the posterior chamber of the eye was rare. Accessory structures, including the lachrymal gland, showed increased function and some of the smaller glands showed inflammation and necrosis. Periorbital tissues often showed congestion and oedema and a mild cellular infiltration of orbital muscles occurred. During World War I, secondary ophthalmic infections were comparatively common and panophthalmitis was seen.

One of the most distressing effects of exposure to sulphur mustard, which appears after a considerable delay, is late onset blindness associated with keratitis of the cornea. This was reported in cases of World War I mustard exposure during the 1920s and 1930s; the exact explanation for this effect seems unclear. Damage to the nerve endings of the cornea reflected in corneal anaesthesia seems to play a part.

The conjunctival (and presumably corneal) effects of sulphur mustard are generally believed only to be produced when the eye has been exposed to either vapour or liquid. Warthin and Weller (1918) reported Haldane as claiming that subcutaneous injection of mustard could 'cause conjunctivitis and death from pneumonia owing to the reabsorption of the gas into the circulation'. These observations appeared to confirm those made by Victor Meyer but could not be confirmed by Warthin and Weller (1918).

Eye effects of nitrogen mustards

As far as the effects of 'mustards' upon the eye are concerned, two groups may be defined; these were identified by Mann *et al.* (1948) as:

- The 'mustard gas group' comprising sulphur mustard and HN_3. HN_3 was observed to produce the same pattern of lesions as described for sulphur mustard above.
- The 'nitrogen mustard group' comprising HN_1 and HN_2.

The term 'nitrogen mustard group' was coined by Mann *et al.* (1948) during an extensive study of the eye effects of mustard compounds undertaken during World War II. It is confusing that HN_3 produced similar effects to sulphur mustard and therefore does not figure in the 'nitrogen mustard group'. The effects of the nitrogen mustard gas group were described as follows:

> *The typical reaction of the eye to injury with a compound of this class is characterized by the absence of a latent period and the depth of the injury due to a great power of penetration. The cornea is injured, the pupil contracts and a cellular exudate from the ciliary body appears within an hour.*

Later changes noted included:

- haemorrhagic iridocyclitis
- corneal oedema and vascularization
- slow recovery
- recurrent intraocular haemorrhages
- depigmentation of the iris and cataract formation.

Mann summed up as follows:

> *This reaction is typical and quite unlike that due to mustard gas, with its latent interval, absence of severe intraocular involvement and its typical relapsing vascularizing keratitis* (Mann *et al.*, 1948).

Respiratory changes

The extent of changes seen in the respiratory tract following exposure to sulphur mustard is dependent upon the duration of the exposure and the concentration of the agent in the inhaled air. Under warm environmental conditions, the respiratory effects of sulphur mustard vapour were observed to be increased. Data collected during World War I are often difficult to interpret because of the high incidence of 'superadded' infection in cases coming to post-mortem examination.

UPPER RESPIRATORY TRACT

Necrosis of the epithelium of the larynx, trachea and bronchi commonly occurred in cases of severe exposure, a 'diphtheritic' membrane being seen in severe cases. Hyperaemia of and petechial haemorrhages in the surface layers were common in cases of less severe exposure. Similar erosions in experimental animals persisted for some months post-exposure to the sulphur mustard. Occasionally during World War I, gangrenous changes in the trachea were reported. Under light microscopy, an exudate of epithelial cells, fibrin and mucus was seen. The basement membrane appeared swollen and poorly defined. Oedema of subepithelial tissues associated with inflammatory cell infiltration and dilation of blood vessels was common. In more severe cases, damage extended to the connective tissue and smooth muscle components of the walls of the airways. During the reparative phase, extensive squamous metaplasia was noted, the earliest changes being in the ducts of mucous glands. Cover of the entire damaged surface by a metaplastic stratified squamous epithelium was recorded. The later development of this epithelium is not well described and restoration of a pseudostratified ciliated columnar epithelium is not well documented.

CHANGES IN LUNG PARENCHYMA

Here, the high incidence of secondary infection makes interpretation of post-mortem findings particularly difficult. Even after low dose exposures, congestion and oedema appeared to be present in some cases and secondary infection often occurred. Classical changes were associated with those areas of lung parenchyma close to the airways. The peribronchial alveoli were seen to contain free red cells and often showed collapse. Further from the airways, a fibrinous exudate rather than free blood filled the alveoli. The red peribronchial zone, seen macroscopically, usually extended some 2–3 mm from the bronchial wall. The hypothesis of diffusion of sulphur mustard through the bronchial wall was put forward to explain these effects. Generalized areas of collapse, emphysema and oedema were all commonly seen at post-mortem. During the reparative phase, thickening of bronchial walls, organization of fibrinous oedema and proliferation of deeply staining cuboidal alveolar cells (Type II cells) was recorded. These cells were seen to grow over plugs of fibrinous exudate. The final picture was one of an organizing chemical pneumonitis.

Study of tissue samples from lungs of Iranian casualties who died as a result of exposure to mustard gas has revealed a pattern identical with that described above. In all, tissue from four patients was studied. Alveolar capillary congestion, haemorrhage, oedema, the formation of hyaline membranes and fibrosis were seen. The casualties died as a result of multi-system organ failure and the changes in the lung parenchyma were very similar to those seen in cases of Adult Respiratory Distress Syndrome (ARDS) (Maynard, unpublished observations).

Bone marrow changes

The effects of sulphur mustard upon the bone marrow were investigated during World War I by Pappenheimer and Vance (1920). Rabbits exposed to sulphur mustard at a level capable of inducing bone marrow depression revealed a general depletion of all elements of the bone marrow and a replacement by fat. The cells of the granulocyte series and the megakaryocytes appeared more susceptible to damage than those of the erythropoietic series.

During World War I, careful studies were made of the changes in white cell counts which occurred after exposure to mustard gas. The following phases were described:

(1) Day 1–Day 3: an increase in WBC count in peripheral blood was noted. This was accounted for mainly by a great increase in circulatory polymorphs. Lymphocytes were reduced in numbers during this period.

(2) From Day 4 onwards: in severe cases, a very rapid fall in the WBC count was recorded. Vedder (1925) reported a WBC count of 33 800 mm^{-3} on the second day, falling to 15 800 mm^{-3} on the third day and to 172 mm^{-3} on the seventh day, six hours before death. The early leukocytosis and subsequent leukopenia was observed in Iranian casualties seen during 1985 and 1986.

Table 5. Development of symptoms and signs following severe exposure to mustard gas

Time post-exposure	Symptoms and signs
20–60 min	Nausea, retching, vomiting and eye smarting have all been *occasionally* reported. Often no signs or symptoms are produced
2–6 h	Nausea, fatigue, headache. Inflammation of eyes. Development of intense pain in eyes, lacrimation, blepharospasm, photophobia, rhinorrhoea
6–24 h	General increase in severity of above effects. Inflammation of inner thighs, genitalia, perineum, buttocks and axillae followed by blister formation. Blisters may be large, pendulous and filled with a clear, yellow fluid. Death within 24 h of exposure is very rare
48 h	Condition generally worsened. Blistering more marked. Coughing appears: muco pus and necrotic slough may be expectorated. Intense itching of skin is common. Increase in skin pigmentation occurs

SYMPTOMS AND SIGNS OF SULPHUR MUSTARD POISONING

The use of mustard gas during the Iraq–Iran war furnished considerable clinical material and demonstrated that lessons forgotten since World War I regarding the management of CW casualties had to be relearnt. Colonel Jan L. Willems collected information on 65 casualties treated during the Iran–Iraq war in European hospitals. His report 'Clinical management of mustard gas casualties' should be studied by everyone who requires an up-to-date understanding of the effects of mustard gas (Willems, 1989). It is particularly valuable for the photographs of lesions excellently reproduced in colour which are more useful than black and white records of cases during World War I and are of perhaps even greater value than the coloured illustrations prepared by A. K. Maxwell during World War I. (A.K. Maxwell was a distinguished medical illustrator whose initials will be familiar to all who have studied the standard anatomical work, *Gray's Anatomy*.)

The hallmark of sulphur mustard exposure is the occurrence of a latent, symptom- and sign-free period of some hours post-exposure. The duration of this interval is dependent on the mode of exposure, environmental temperature and probably on the individual himself. Some people are markedly more sensitive to mustard than others.

Table 5 shows the evolution of symptoms and signs which might be expected following a severe exposure to sulphur mustard vapour. Liquid exposure of the eyes will produce more severe, possibly permanent, eye damage.

If mustard-contaminated food or water are ingested then the symptomatology will differ from that listed in Table 5 and the onset, after a few hours, of nausea, vomiting, abdominal pain, bloody vomiting and diarrhoea (in cases of severe poisoning), shock and prostration may be expected.

Symptoms and signs associated with skin lesions

The sequence of skin changes normally seen is as follows:

(1) Erythema (2–48 h post-exposure). This may be very striking and reminiscent of scarlet fever. Slight oedema of the skin may occur. Itching is common and may be intense. As the erythema fades, areas of increased pigmentation are left (this sequence is reminiscent of that seen in sunburn).

(2) Blistering. The appearance of the blisters has been described above. Blisters are not painful *per se*, although may be uncomfortable and may feel tense. Blisters at points of flexure, anterior aspects of elbows and posterior aspects of knees can seriously impede movement. Mustard blisters are delicate and may be easily rubbed off by contact with bed linen, bandages or during transport of casualties. Crops of new blisters may appear as

late as the second week post-exposure. The author has seen a blister about 1.5 × 0.5 cm in size on the thigh of an Iranian mustard gas casualty as late as approximately 3 weeks post-exposure. By this time, of the patient's earlier symptoms and signs, only mild photophobia and some darkening of the skin remained.

Blister fluid is not dangerous and does not produce secondary blistering if applied to skin (Sulzberger *et al.*, 1947). This observation has been a cause of considerable controversy. In 1943, Sulzberger and Katz performed a definitive experiment in which blister fluid from blisters produced upon volunteers by sulphur mustard and also lewisite, was aspirated and applied to normal skin both of the blistered individual and others. No blistering was produced by the application of the blister fluid.

(3) Deep burning leading to full thickness skin loss: this is particularly likely to occur on the penis and scrotum.

PROGRESS OF LESIONS

Though intact blisters are, as has already been pointed out, painless, once the blister cover is lost, lesions tend to be painful and some patients complain of very severe pain. Healing of skin lesions is slow. The areas which were markedly erythematous darken and may become very hyperpigmented. Brownish–purple to black discolouration of some areas may occur. This last change was particularly marked in some Iranian casualties and was commented upon by Willems (1989). These changes tend to disappear over a period of several weeks with desquamation leading to the appearance of areas of hypo-pigmentation. The appearance of such areas alongside those of hyper-pigmentation may be striking.

Experience in 1986 of Iranian mustard gas casualties suggested that hyper-pigmentation was most marked at the margins of affected areas. In some, the pigmentation seen was uneven and suggested, to some, droplet contamination. This was later doubted on the grounds that liquid contamination would be expected to produce more severe skin effects than were observed in these areas. The increases in pigmentation seen in several diseases and conditions are incompletely understood, e.g. chronic arsenic poisoning produces hyper-pigmentation and hypo-pigmentation which is classically described as having a 'rain-drop' pattern. Why the pigmentation should occur in such a pattern seems to be unknown.

Rubbing the damaged skin of patients exposed to mustard can lead to the production of secondary blisters (Nikolsky's sign: this sign can also be elicited in cases of pemphigus vulgaris). The late blistering described above may be a manifestation of this rather than a delayed effect of sulphur mustard itself.

Willems (1989) has commented upon attempts to correlate the extent of the skin lesions with the severity of the intoxication. This proved very difficult and the presence or absence of pulmonary damage and leukopaenia proved a better guide to the degree of poisoning.

Symptoms and signs associated with eye lesions

A marked conjunctivitis, local oedema, including oedema of the eyelids, blepharospasm and lachrymation are the classical signs. To these may be added early miosis, photophobia and severe eye pain (the headache sometimes described may be related to the eye pain and miosis). The onset of conjunctivitis may be delayed for up to 48 h in cases of mild exposure. Later adherence of the follicular margins with a muco-serous discharge occurs. As this dries, crusting is produced. If the damaged eye is opened for examination (this may be very painful), a hyperaemic band may be identified crossing the globe horizontally. This observation made during World War I was confined to cases seen in the early stages. Later, a white band reflecting severe damage replaces the hyperaemic band. Corneal ulceration is said to be unusual in cases of vapour exposure. However, histological examination of the eye of an Iranian casualty who had been exposed to mustard vapour in 1986 revealed stripping of the corneal epithelium. In cases of severe exposure, mustard may penetrate the anterior chamber and adhesion of the iris to the lens capsule may occur.

Infection of the eye is a very serious consequence of mustard gas exposure and may produce blindness.

Symptoms and signs associated with lesions of the respiratory tract

Rhinorrhoea, often profuse, is commonly seen after vapour exposure. Epistaxis may occur in severely affected patients. Inflammation and ulceration of the palate, nasopharynx, oropharynx and larynx follow, the voice becomes hoarse and temporary aphonia may occur. Reports from World War I suggest that laryngeal oedema and/or spasm sufficiently severe to necessitate tracheostomy occurred very seldom.

Willems (1989) reported findings on bronchoscopy in two patients: bilateral erythematous inflammation of the mucosa with bleeding and purulent secretions was reported. Casts of necrotic sloughed mucosa were also seen.

Coughing may be severe. As stated, a mucopurulent expectorate is produced. Necrotic slough may also be produced.

During World War I, factory workers involved in the production of mustard gas were particularly at risk. It was reported (Stockholm International Peace Research Institute, 1971) that:

> *At the main British factory, there were 1400 casualties among the plant workers, the accidentally burned and blistered exceeding 100 per cent of the staff every three months (presumably some members of staff were injured more than once). Conditions at the principal French plant, which supplied three quarters of the Allied-fired mustard gas, were equally unpleasant: the personnel ... is 90 per cent voiceless. About 50 per cent cough continuously .... By long exposure to the small amounts of vapour constantly in the air of the work rooms, the initial resistance of the skin is finally broken down. The chief result is that the itch makes sleep nearly impossible and the labourers are very much run down.*

Non-specific symptoms of mustard gas poisoning

Though the symptoms and signs listed above are the classical signs of sulphur mustard poisoning, clinicians should be aware of a group of symptoms and signs reported during World War I and not yet adequately explained. These included:

- diffuse skin pigmentation (this has been mentioned above)
- hypotension
- marked apathy and asthenia
- mental disturbance

The suggestion of adrenal damage was made during World War I to explain some of these effects. This has not been confirmed.

In 1986, Norris reviewed a number of aspects of the effects of mustard gas and reported work done in the late stages of World War I on the 'functional neurosis' which some authors claimed was recognised in 22% of all mustard cases. He commented:

> *They described an anxiety state in mild cases of gas poisoning and coughing or photophobia reinforced by hysteria and they produced the following breakdown of the 22% of casualties with neurosis:*
>
> *Functional photophobia 12.6% of all cases*
> *Functional aphonia 7.2%*
> *Functional vomiting 1.0%*
> *Effort syndrome 1.2%*
>
> *('functional' in this context implies absence of a clear pathological mechanism, at least of a physical kind).*

FIRSTHAND ACCOUNTS OF EFFECTS OF MUSTARD GAS

(1) Reported by Victor Lefebure (1921) from an account provided by a casualty.

> *I was gassed by dichlor-diethyl sulphide, commonly known as mustard stuff, on July 22 (1917). I was digging in (Livens Projectors) to fire on Lombartzyde. Going up we met a terrible strafe of HE (high explosive) and gas shells at Nieuport. When things quietened a little I went up with the three GS wagons, all that were left, and the carrying parties. I must say that the gas was clearly visible and had exactly the same smell as horseradish. It had no immediate effect on the eyes or throat. I suspected a delayed action and my party all put their masks on.*

On arriving at the emplacement we met a very thick cloud of the same stuff drifting from the front line system. As it seemed to have no effect on the eyes I gave orders for all to put on their mouthpieces and noseclips so as to breathe none of the stuff, and we carried on.

Coming back we met another terrific gas shell attack at Nieuport. Next morning, myself and all the eighty men we had up there were absolutely blind. The horrid stuff had a delayed action on the eyes, causing temporary blindness some seven hours afterwards. About 3000 were affected. One or two of our party never recovered their sight and died. The casualty clearing stations were crowded. On August 3, with my eyes still very bloodshot and weak and wearing blue glasses, I came home and went into Millbank Hospital on August 15.

(2) In October 1918, Adolph Hitler was exposed to mustard gas. He was at the time a runner in the 16th Bavarian Reserve Regiment. He included the following graphic account in *Mein Kampf* (Vol. 1, 1924).

During the night of October 13–14th (1918) the British opened an attack with gas on the front south of Ypres. They used the yellow gas whose effect was unknown to us, at least from personal experience. I was destined to experience it that very night. On a hill south of Werwick, in the evening of 13 October, we were subjected to several hours of heavy bombardment with gas bombs, which continued through the night with more or less intensity. About midnight a number of us were put out of action, some for ever. Towards morning I also began to feel pain. It increased with every quarter of an hour, and about seven o'clock my eyes were scorching as I staggered back and delivered the last dispatch I was destined to carry in this war. A few hours later my eyes were like glowing coals, and all was darkness around me.

This excellent description highlights the delayed effect of mustard gas upon the eyes and also the pain and incapacitation associated with such injuries. Hitler's reluctance to use chemical weapons during World War II has been attributed by some, perhaps fancifully, to his experience recorded above.

CLINICAL CHEMISTRY AND OTHER INVESTIGATIONS

Vesicles may be aspirated and the fluid obtained analyzed for thiodiglycol. Papers by Vycudilik (1985), Drasch *et al.* (1987), Wils *et al.* (1988), Black and Read (1988) and D'Agostino and Provost (1988) should be consulted for details of the techniques used. The same estimation may be performed on blood and urine. The paper by Wils *et al.* (1988) is important as it reports the results of investigations done on Iranian casualties. Of the casualties, more than 80% had urine thiodiglycol levels above the 95% confidence limit for levels calculated for the control group. These investigations may be of some use in differentiating blistering produced by mustard from that produced by other agents, e.g. lewisite.

Temperature increases occur during the early stages of mustard poisoning and these, plus the also typical early leukocytosis, should not be taken uncritically as evidence of infection. Later, the white cell count falls, the platelet count falls and finally the red cell count falls.

Culture of sputum and exudate from the eyes is important in order that appropriate antibiotic therapy may be given if infection supervenes. Infection of the chest is almost inevitable in a patient with extensive chemical lung damage and a very low white cell count. The dangers of infection are elaborated upon below.

MANAGEMENT OF CASES OF SULPHUR MUSTARD POISONING

Case management falls into two parts:

- First aid measures.
- Therapeutic measures.

First aid measures

These are of the greatest importance. Attendants should wear adequate protective clothing and respirators when dealing with contaminated casualties.

(1) Patients should be removed from the source of contamination.
(2) Areas of liquid contamination should be decontaminated using fullers' earth in liberal quantities. Washing with organic solvents, such as kerosene (paraffin), followed by soap and water is also valuable. Washing with organic solvents, if undertaken, should be continued for up to 30 min post-injury. The use of organic solvents was recommended by Vedder (1925). Chloramine solutions have also been extensively used. In recent years, there has been a move away from fullers' earth as a decontaminant and soap and water is now widely recommended.
(3) Liquid contamination of the eyes should be immediately rinsed out using normal saline, if available, or any source of water.

Therapy and medical management

There is no specific therapy for sulphur mustard or nitrogen mustard poisoning. Perhaps because of this a considerable number of palliative approaches have been suggested. Some lessons have been learnt in this area from the extensive use of alkylating agents as antineoplastic drugs. It will be obvious that a completely satisfactory evidence-based approach is unavailable: no clinical trials of alternative approaches have been undertaken.

SKIN EFFECTS

For areas of erythema and minor blistering, bland lotions (e.g. calamine) have been suggested. Silver sulphadiazine (Flamazine) 1% cream was used in the management (1986) of Iranian mustard gas casualties (Willems, 1989). This probably had value in reducing skin infection.

Dilute steroid preparations (hydrocortisone lotion) may be of symptomatic value in reducing irritation and itching. More powerful steroid preparations (beclomethasone dipropionate, i.e. 'Propaderm' cream) have also been used. Considerable symptomatic improvement was produced by the use of this preparation. Steroids have been said to delay healing and to enhance the likelihood of infection. No such ill-effects were observed in the Iranian casualties.

Severe pain and itching was reported by many of the Iranian mustard gas casualties. Therapy for pain ranged from paracetamol to morphine. It was found to be important to attempt to dissociate the pain from the panic exhibited by some patients, and diazepam in combination with a weak/mild analgesic proved, on several occasions, as effective as a potent analgesic. Very few patients were to require repeated doses of narcotic analgesics. Itching was troublesome in nearly all patients with extensive skin lesions and prevented sleep. Antihistamines, such as promethanzine and dimethindine maleate, proved effective. Carbamazepine was also used to great effect in one patient and allowed the use of narcotic analgesics to be stopped (Maynard, unpublished observations).

Willems (1989) (see above) reported a number of different patterns of management of skin lesions ranging from 'treating exposed' (at a burns unit) to treating by bathing and the use of wet dressings. Skin lesions of Iranian patients treated in London, were, on the whole, treated exposed. Silver sulphadiazine cream was used if evidence of skin infection was obtained. Regular swabbing was undertaken.

An older remedy to control itching – benzyl alcohol, 100 parts; ethyl alcohol, 96 parts; glycerine, 4 parts – prepared as a paint might also be tried.

Large full-thickness burns will not heal satisfactorily without grafting. One Iranian casualty with severe and extensive skin damage was grafted and the grafts were found to take well. This is the only case the author has seen where skin grafting has been used in the management of sulphur mustard burns.

EYE EFFECTS

Early decontamination is the key. Attempts to decontaminate the eye when more than 5 min have passed after liquid contamination are likely to be valueless. It has been stated (Ministry of Defence, 1987) that irrigation should:

> *Be immediate; if it is delayed for longer than five minutes the eye may have been damaged and washing may only serve to increase the injury.*

The concept of 'increasing the injury' presumably stems from the theory that any unabsorbed mustard may be spread by the washing to undamaged areas of the eye and subsequently cause damage there. Evidence for this is scanty and if copious quantities of fluids are used in the washing process it is unlikely. For damaged eyes the following are suggested:

(1) Saline irrigations.
(2) Use of vaseline on follicular margins to prevent sticking.
(3) If pain is severe, local anaesthetic drops, e.g. amethocaine hydrochloride (0.5%) should be used. Cocaine should be avoided as it may produce sloughing of the corneal epithelium. To some extent, all local anaesthetic preparations appear to damage the cornea. Some ophthalmologists advise the use of topical steroid preparations rather than of local anaesthetics. In the past, steroids have been stated to be contraindicated in cases of mustard damage to the eyes. They were, however, used on a number of Iranian patients and no-ill effects were reported. Certainly, before they are used expert ophthalmological opinion should be sought. If eye pain is very severe, systemic narcotic analgesics should be used.
(4) Chloramphenicol eye drops should be used to prevent infection. A number of different antibiotics were used in different centres by physicians managing Iranian chemical warfare casualties. These included chloramphenicol, tetracycline, oxytetracycline, bacitracin and polymyxin B. No conclusions regarding the most effective drug could be drawn. There seems little reason to abandon the use of chloramphenical eye drops.
(5) Mydriatics such as hyoscine eyedrops (0.5%) are useful in preventing sticking of the iris to the central area of the lens. In iritis, mydriatic drops reduce pain due to spasm of the iris.
(6) Dark glasses to alleviate photophobia.

Recently potassium ascorbate (10%) and sodium citrate (10%) drops have also been suggested. The ascorbate and citrate drops are to be given alternately every half hour for all the waking day. The patient will therefore be receiving each preparation at hourly intervals. These drops may be discontinued once a stable epithelial covering has formed (Wright, 1990).

Late corneal lesions are difficult to manage and blindness can occur. Contact lenses have proved very valuable for improving vision impaired by unevenness of the cornea (Mann *et al.*, 1948).

In all cases of mustard damage to the eyes, reassurance of severely frightened and often depressed patients is essential. During World War I, great emphasis was placed upon encouraging those with eye damage to return to normal activity as soon as possible and not to become dependent upon dark glasses and eye drops. Prolonged eye irritation with profuse lachrymation and photophobia was encountered in one Iranian patient seen by the author in London. Some 12 weeks post-exposure, she was still unable to face bright lights and tears poured almost continuously down her face. The cause of this very prolonged reaction is unknown.

As well as the above, a number of other methods has been proposed for managing mustard eye injuries (Foster, 1939). In cases of severe eye damage, an ophthalmological opinion must be sought.

RESPIRATORY EFFECTS

There is no specific therapy for mustard injuries of the respiratory tract. Severe coughing may be eased with codeine linctus. Antibiotic cover is recommended, the antibiotic chosen not being one liable to induce further bone marrow depression. Acetylcysteine was used in some patients from the Iran–Iraq War as a mucolytic. Evidence of its efficacy is scanty.

In severe cases, respiratory failure may ensue. The management of such patients is complex and cannot be dealt with here. Advice from physicians and, if ventilation is necessary, anaesthetists must be sought.

BONE MARROW DEPRESSION[1]

Bone marrow depression as a result of mustard poisoning is generally seen as therapeutically

[1] See Chapter 22 for a detailed account.

Table 6. Effect of thiosulphate and cysteine on LD_{50} of nitrogen mustards. Reproduced from the paper by Connors, T.A., Jeny, A and Jones, M. (1964), 'Reduction of the toxicity of "radiomimetic" alkylating agents in rats by thiol pretreatment-III. The mechanism of the protective action of thiosulphate.' *Biochem. Pharmacol.*, **13**, 1545–1550

	Merophan (IP)			HN_2 (IP)			HN_2 (SC)		
Pretreatment 95%	LD_{50} (mg kg^{-1})	Fiducial limits 95%	DRF	LD_{50} (mg kg^{-1})	Fiducial limits 95%	DRF	LD_{50} (mg k^{-1})	Fiducial limits 95%	DRF
None	3.67	3.24–4.16	1	1.28	1.08–1.51	1	2.06	1.74–2.43	1
Thiosulphate 2 g kg^{-1} IP, 30 min before	3.67	3.24–4.16	1	4.06	3.44–4.79	3.2	10.68	9.05–12.60	5.2
Cysteine 1 g kg^{-1} IP, 30 min before	15.24	13.65–16.71	4.2	6.33	5.37–7.46	4.9	10.68	9.39–12.14	5.2
Cysteine 1 g k^{-1} + thiosulphate 2 g kg^{-1}	5.30	4.50–6.23	1.4	12.27	11.31–13.31	9.6	17.34	15.16–19.84	8.6

irreversible. If the aplastic anaemia produced is severe, granulocyte and platelet transfusions and later red cell transfusions should be considered. Bone marrow transplantation has also been suggested.

A number of drugs are known to stimulate the bone marrow (much of the evidence of this comes from studies of the effects of drugs on normal marrow). The value of such drugs in mustard poisoning is quite unknown. However, in view of the paucity of other approaches, I list them as follows:

(1) Oxymethalone (a steroid) (Pharmaceutical Society of Great Britain, 1989)
(2) Lithium carbonate (Lyman *et al.*, 1980)
(3) Glucan (+ β-1, 3 polyglucose) (Di Luzio, 1983)

More important than these drugs are the colony stimulating factors which have recently become available. Granulocyte Colony Stimulating Factor and related factors should be considered in cases where bone marrow depression is marked.

It has been pointed out that cysteine can reduce the antitumour effects of alkylating agents (Bowman *et al.*, 1986). In 1964, Connors *et al.* reported a reduction in toxicity of nitrogen mustards by pretreating rats with a variety of thiols. Thiosulphate pretreatment (2 g kg^{-1} IP, i.e. maximum tolerated dose) produced an increase by a factor of 3.2 in the LD_{50} of HN_2 if given 30 min before the mustard. Cysteine hydrochloride (L-cysteine hydrochloride 1 g kg^{-1} IP, i.e. maximum tolerated dose) produced a five-fold increase in the LD_{50} of HN_2 if given 30 min before the mustard. Table 6 is reproduced from the paper by Connors *et al.* (1964).

The work reported by Connors *et al.* (1964) followed work done at CDE by Callaway and Pearce, reported in 1958. They investigated the effects of thiosulphate:trisodium citrate mixtures (10:1, 2.75 g kg^{-1}, IP) on the toxicity of sulphur mustard in rats. The combination was referred to as Thiocit and it was demonstrated that:

> *Thiocit afforded complete protection against greater than the median lethal dose of mustard gas whether given 10 min before or 10 min after mustard gas and raised the LD_{50} of mustard gas approximately three times.*

In 1986, Vojvodic and colleagues reported a study of the protective effects of a number of compounds and combinations of compounds on the toxicity of nitrogen and sulphur mustards (Vojvodic *et al.*, 1986). Protective indices similar to those obtained by Callaway and Pearce (1958) were reported. Work on the value of thiosulphate in the treatment of mustard gas poisoning has also been reported by Fasth and Sörbo (1973). Sodium thiosulphate has been recommended as a local treatment for accidentally extravasated doses of nitrogen mustards (Dorr *et al.*, 1988).

Despite the above work, thiosulphate and other thiols have not achieved an established place in the treatment of mustard gas poisoning. Interestingly (see later) Russian sources have recommended intravenous infusion of 30% sodium thiosulphate in the treatment of mustard gas poisoning.

Attempts have been made to learn from work done to reduce the unwanted effects of anticancer therapy involving exposure to either radiation of high doses of alkylating agents. A useful review by Glover *et al.* (1988) should be consulted for details. The compound WR-2721 has figured prominently in these studies.

Other measures which might be taken

If vomiting is a problem, then antiemetic drugs should be given. Phenothiazines would be a reasonable choice although recent work has shown that blockers of $5HT_3$ receptors may be of greater value.

Haemodialysis and haemoperfusion have also been suggested and the latter recently used. Haemoperfusion was used in a few patients from the Iran–Iraq war. There is no sound theoretical basis for such therapy as no active mustard has been identified in blood taken from known mustard gas casualties. As a measure of last resort, haemoperfusion might be justified although the grave attendant risk of infection should be considered.

A range of general supportive measures were undertaken in Western centres treating casualties from the Iran–Iraq war. These included:

(1) Use of H_2 antagonists to prevent stress ulceration.

(2) Use of heparin in order to prevent deep venous thrombosis.
(3) Use of Vitamins C and B_{12} and folic acid.
(4) Use of a single large dose of methyl prednisolone (2 g) as general protection against tissue damage.

The above advice on management has been culled from a study of the Western literature of mustard gas poisoning. Measures recommended by Eastern European sources differ substantially, in terms of drugs suggested, from the above. The following is taken from a Russian source.

For acute intoxication, intravenous infusion of:

- 40% glucose solution (20 ml; 1-2 daily)
- 30% sodium thiosulphate
- Calcium gluconate or chloride
- Cardiotonic drugs, e.g. camphor, caffeine
- Ascorbic acid (vitamin C)
- Thiamine (vitamin B_1)
- Pyridoxine (vitamin B_6)
- Blood transfusion

If leucopaenia and anaemia develop:

- Sodium nucleate, 0.5-1.0 g tds
- Methyl uracil, 1 g tds-qds
- Leucogen, 0.02 g tds
- Vitamin B_{12}, 0.01% solv, 1 ml, IV after 2–3 days

Note that sodium nucleate and leucogen are not recognized Western drugs.

PROGNOSIS FOR MUSTARD GAS CASUALTIES

As has already been pointed out, the great majority of mustard gas casualties survive. Resolution of specific problems can be difficult to predict but the following may provide a guide.

(1) Eye lesions: most are resolved within 28 days of exposure.
(2) Skin lesions: deep skin lesions may be expected to heal in up to 60 days. Superficial lesions heal in 14–21 days.
(3) Upper respiratory tract lesions: it is very difficult to define a time-course for complete recovery as patients from the Iran–Iraq conflict were often discharged while still coughing and complaining of expectoration. Lung function tests on patients with purely upper respiratory tract lesions were usually normal on discharge. Patients with parenchymal damage often showed an abnormal pattern on lung function testing.

Leukopaenia is a common finding in mustard gas casualties. It is often comparatively minor and the white cell count recovers within 14 days. A marked fall in the white cell count is a serious event and patients whose white cell counts fell to $< 200\ mm^{-3}$ did not survive despite artificial ventilation and intensive care. In all, the cause of death was overwhelming infection and multiple organ failure.

LONG-TERM EFFECTS OF MUSTARD GAS POISONING

The long-term effects of mustard may be divided into three groups.

Psychological effects

The syndrome described earlier may persist for some time.

Local effects

These include:

- Permanent blindness
- Visual impairment
- Scarring of the skin
- Chronic bronchitis
- Bronchial stenosis
- Sensitivity to mustard gas

Carcinogenic effects

Sulphur mustard is a known human carcinogen. The strongest evidence of induction of cancer by sulphur mustard comes from studies undertaken on mustard gas factory workers. The work of Wada *et al.* (1968), Nishimoto *et al.* (1988) and Yanagida *et al.* (1988) should be consulted for details. Exposures to mustard gas in these factories may have been considerable and prolonged. The study of British mustard gas workers revealed a clear increase in respiratory tract cancers (Easton *et al.*, 1988).

A more difficult question concerns the likelihood of developing cancer as a result of exposure to sulphur mustard on the battlefield. Here, the evidence is suggestive but not absolutely clear-cut. The study by Norman (1975) failed to demonstrate an increased risk of respiratory cancer in those exposed once to sulphur mustard. The study of Beebe (1960) suggested that the incidence of lung cancer was higher in those exposed to mustard gas during World War I.

REFERENCES

Aasted A, Darre E and Wulf HC (1987). Mustard gas: clinical, toxicological and mutagenic aspects based on modern experience. *Ann Plastic Surg*, **19**, 330–333.

Adair FE and Bagg HJ (1931). Experimental and clinical studies on the treatment of cancer by dichloroethylsulphide. *Ann Surg*, **93**, 190–199.

Aghanouri R, Ghanei M, Aslani J *et al.* (2004). Fibrogenic cytokine levels in bronchoalveolar lavage aspirates 15 years after exposure to sulfur mustard. *Am J Physiol Lung Cell Mol Physiol*, **287**, L1160–L1164.

Arroyo CM, Von Tersch RL and Broomfield CA (1995). Activation of alpha-human tumour necrosis factor (TNF-α) by human monocytes (THP-1) exposed to 2-chloroethyl ethyl sulphide (H-MG). *Human Exp Toxicol*, **14**, 547–553.

Arroyo CM, Broomfield CA and Hackley BE (2001). The role of interleukin-6 (IL-6) in human sulfur mustard (HD) toxicology. *Int J Toxicol*, **20**, 281–296.

Atkins KB, Lodhi IJ, Hurley LL *et al.* (2000). *N*-Acetylcysteine and endothelial cell injury by sulfur mustard. *J Appl Toxicol*, **20**, S125–S128.

Beebe GW (1960). Lung cancer in World War I veterans: possible relation to mustard gas injury and 1918 influenza epidemic. *J Natl Cancer Inst*, **5**, 1231–1251.

Bethlenfalvay NC and Bergin JJ (1972). Severe cerebral toxicity after intravenous nitrogen mustard therapy. *Cancer*, **29**, 366–369.

Bhat KR, Benton BJ, Rosenthal DS *et al.* (2000). Role of poly(ADP-ribose) polymerase (PARP) in DNA repair in sulfur mustard-exposed normal human epidermal keratinocytes (NHEK). *J Appl Toxicol*, **20**, S13–S17.

Black RM and Read RW (1988). Detection of trace levels of thiodiglycol in blood, plasma and urine using gas chromatography–electron capture negative-ion chemical ionisation mass spectrometry. *J Chromatogr*, **449**, 261–270.

Black RM, Brewster K, Clarke RJ *et al.* (1993). Metabolism of thiodiglycol (2,2'-thiobis-ethanol): isolation and identification of urinary metabolites following intraperitoneal administration to rat. *Xenobiotica*, **23**, 473–481.

Bowman WC, Bowman A and Bowman A (1986). *Dictionary of Pharmacology*. Oxford, UK: Blackwell Scientific.

Callaway S and Pearce KA (1958). Protection against systemic poisoning by mustard gas, (di(2-chloroethyl) sulphide), by sodium thiosulphate and thiocit in the albino rat. *Br J Pharmacol*, **13**, 395–398.

Calvet J-H, Planus E, Rouet P *et al.* (1999). Matrix metalloproteinase gelatinases in sulfur mustard-induced acute airway injury in guinea pigs. *Am J Physiol Lung Cell Mol Physiol*, **20**, L754–L762.

Cameron GR, Gaddum JH and Short RHD (1946). The absorption of war gases by the nose. *J Pathol*, **58**, 449–497.

Ciszewka-Popiolek B, Czerny K, Swieca M *et al.* (1989). Der Einfluß des Präparats Psoriazin auf das mophologische und das histochemische Leberbild (The effect of Psoriazin on the morphological and histochemical characteristics of the liver). *Gegenbaurs Morphol Jahrb Leipzig*, **135**, 875–880.

Connors TA, Jeny A and Jones M (1964). Reduction of the toxicity of 'radiomimetic' alkylating agents in rats by thiol pretreatment – III. The mechanism of the protective action of thiosulphate. *Biochem Pharmacol*, **13**, 1545–1550.

Cullumbine H (1947). The mode of penetration of the skin by mustard gas. *Br J Dermatol*, **58**, 291–294.

D'Agostino PA and Provost LR (1988). Gas chromatographic retention indices of sulfur vesicants and related compounds. *J Chromatogr*, **436**, 399–411.

Davison C, Rozman RS and Smith PK (1961). Metabolism of bis-ß-chlorethyl sulfide (sulphur mustard gas). *Biochem Pharmacol*, **7**, 65–74.

Di Luzio NR (1983). Immunopharmacology of glucan: a broad spectrum enhancer of host defense mechanisms. *Trends Pharm Sci*, **4**, 344–347.

Dillman JF, McGary KL and Schlager JJ (2003). Sulfur mustard induces the formation of keratin aggregates in human epidermal keratinocytes. *Toxicol Appl Pharmacol*, **193**, 228–236.

Dillman JF, McGary KL and Schlager JJ (2004). An inhibitor of p38 MAP kinase downregulates cytokine release induced by sulfur mustard exposure in human epidermal keratinocytes. *Toxicol in Vitro*, **18**, 593–599.

Dixon M and Needham DM (1946). Biochemical research on chemical warfare agents. *Nature*, **158**, 432–438.

Dorr RT, Soble M and Alberts DS (1988). Efficacy of sodium thiosulfate as a local antidote to mechlorethamine skin toxicity in the mouse. *Cancer Chemother Pharmacol*, **22**, 299–302.

Drasch G, Kretschmet E, Kauert E *et al.* (1987). Concentrations of mustard gas (bis(e-chloroethyl)sulfide) in the tissues of a victim of a vesicant exposure. *J Foren Sci* **32**, 1788–1793.

Easton DF, Peto J and Doll R (1988). Cancers of the respiratory tract in mustard gas workers. *Br J Ind Med*, **45**, 652–659.

Eldad A, Ben Meir P, Breiterman S *et al.* (1998). Superoxide dismutase (SOD) for mustard gas burns. *Burns*, **24**, 114–119.

Elsayed NM, Omaye ST, Klain GJ *et al.* (1992). Free radical-mediated lung response to the monofunctional sulfur mustard butyl 2-chloroethyl sulphide after subcutaneous injection. *Toxicology*, **72**, 153–165.

Fasth A and Sörbo B (1973). Protective effect of thiosulfate and metabolic thiosulfate precursors against toxicity of nitrogen mustard (HN_2). *Biochem Pharmacol*, **22**, 1337–1351.

Fawcett DW (1994). *The cell*. In: *A Textbook of Histology*, 12th Edition, pp. 1–53. London, UK: Chapman & Hall.

Foster J (1939). Ophthalmic injuries from mustard gas (D.E.S.). *Br Med J*, **2**, 1181–1183.

Fox M and Scott D (1980). The genetic toxicology of nitrogen and sulphur mustard. *Mutat Res*, **75**, 131–168.

Friedenwald JS, Scholz RO, Snell A *et al.* (1948). Studies on the physiology, biochemistry and cytopathology of the cornea in relation to injury by mustard gas and allied toxic agents. I. Introduction and outline. *Bull Johns Hopkins Hosp*, No. 2, 82–101.

Fuchs E, Coulombe P, Cheng J *et al.* (1994). Genetic bases of epidermolysis bullosa simplex and epidermolytic hyperkeratosis. *J Invest Dermatol*, **103**(Supplement 5), 25S–30S.

Glover D, Fox KR, Weiler C *et al.* (1988). Clinical trials of WR-2721 prior to alkylating agent chemotherapy and radiotherapy. *Pharmacol Ther*, **39**, 3–7.

Goodman L and Gilman A (eds) (1980). *The Pharmacological Basis of Therapeutics*. New York, London: MacMillan, Bailliere Tindall.

Grando SA (2003). Mucocutaneous cholinergic system is targeted in mustard-induced vesication. *Life Sci*, **72**, 2135–2144.

Haldane JBS (1925). *Callinicus: A Defence of Chemical Warfare*. London, UK: Kegan Paul, French, Tribner & Company Ltd.

Hayaishi L and Ueda K (1982). *ADP-Ribosylation Reaction, Biology and Medicine*. New York, NY, USA: Academic Press.

Hinshaw DB, Lodhi IJ, Hurley LL *et al.* (1999). Activation of poly (ADP-ribose) polymerase in endothelial cells and keratinocytes: role in an *in vitro* model of sulphur mustard-mediated vesication. *Toxicol Appl Pharmacol*, **156**, 17–29.

Hunt C and Phillips S (1949). The acute pharmacology of methyl-bis(2-chloroethyl)amine (NH_2). *J Pharmacol Exp Ther*, **95**, 131–144.

Ireland MM (1926). *Medical Aspects of Gas Warfare*, Volume XIV. The Medical Department of the United States in the World War. Washington, DC, USA: The Medical Department of the United States in the World War.

Kadar T, Fishbeine E, Meshulam Y *et al.* (2000). Treatment of skin injuries induced by sulfur mustard with calmodulin antagonists, using the pig model. *J Appl Toxicol*, **20**, S133–S136.

Kan RK, Pleva CM, Hamilton TA *et al.* (2003). Sulfur mustard-induced apoptosis in hairless guinea pig skin. *Toxicol Pathol*, **31**, 185–190.

Kaufmann SH, Desnoyers S, Ottaviano Y *et al.* (1993). Specific proteolytic cleavage of poly (ADP-ribose) polymerase an early marker of chemotherapy-induced apoptosis. *Cancer Res*, **53**, 3976–3985.

Kraker AJ and Moore CW (1988). Elevated DNA polymerase beta activity in a cis-diammine-dichloroplatinum(II) resistant P388 murine leukemia cell line. *Cancer Lett*, **38**, 307–314.

Lardot C, Dubois V and Lison D (1999). Sulfur mustard upregulates the expression of interleukin-8 in cultured human keratinocytes. *Toxicol Lett*, **110**, 29–33.

Lefebure V (1921). *The Riddle of the Rhine*. London, UK: W Collins Sons & Company, Ltd.

Levitt JM, Lodhi IJ, Nguyen PK *et al.* (2003). Low-dose sulfur mustard primes oxidative function and induces apoptosis in human polymorphonuclear leukocytes. *Int Immunopharmacol*, **3**, 747–756.

Lin P, Bernstein A and Vaughan FL (1994). Failure to observe a relationship between bis-(ß-chloroethyl)sulphide-induced NAD depletion and cytotoxicity in the rat kerinocyte culture. *J Toxicol Environ Health*, **42**, 393–405.

Lindahl T (1979). DNA glycolases, endonucleases for apurinic/apyrimidinic sites, and base excision-repair. *Prog Nucleic Acids Res Mol Biol*, **22**, 135–192.

Lindsay CD and Rice P (1995). Changes in connective tissue macromolecular components of Yucatan mini-pig skin following application of sulphur mustard vapour. *Human Exp Toxicol*, **14**, 341–348.

Lodhi IJ, Sweeney JF, Clift RE *et al.* (2001). Nuclear dependence of sulfur mustard-mediated cell death. *Toxicol Appl Pharmacol*, **170**, 69–77.

Lovatt Evans C (1970). *Biographical Memoirs of Fellows of the Royal Society*, Volume 16. London, UK: The Royal Society.

Lyman GH, Williams CC and Preston D (1980). The use of lithium carbonate to reduce infection and leukopenia during systemic chemotherapy. *New Engl J Med*, **302**, 257–260.

Mann I, Pirie A and Pullinger BD (1948). An experimental and clinical study of the reaction of the anterior segment of the eye in chemical injury, with special reference to chemical warfare agents. *Br J Ophthalmol*, **Monograph Supplement XII**, 1–171.

Meier HL and Millard CB (1998). Alterations in human lymphocyte DNA caused by sulfur mustard can be mitigated by selective inhibitors of poly(ADP-ribose) polymerase. *Biochim Biophys Acta*, **1404**, 367–376.

Meier HL, Millard C and Moser J (2000). Poly(ADP-ribose) polymerase inhibitors regulate the mechanism of sulfur mustard-initiated cell death in human lymphocytes. *J Appl Toxicol*, **20**, S93–S100.

Meyer V (1886). Ueber Thiodiglycol verbindungen. *Ber Dentach Chem Gesell (Berlin)*, **XIX**, 2359.

Millard CB, Bongiovanni R and Broomfield CA (1997). Cutaneous exposure to bis (2-chloroethyl)sulfide. Results in neutrophil infiltration and increased solubility of 180 000 M subepidermal collagens. *Biochem Pharmacol*, **53**, 1405–1412.

Ministry of Defence (1987). *Medical Manual of Defence Against Chemical Agents*, Ministry of Defence D/Med(F & S)(2)/10/1/1. London, UK: HMSO.

Mol MA and Smith WJ (1996). Calcium homeostasis and calcium signalling in sulphur mustard-exposed normal human epidermal keratinocytes. *Chem Biol Interact*, **100**, 85–93.

Monteiro-Riviere NA and Inman AO (1995). Indirect immunohistochemistry and immunoelectron microscopy distribution of eight epidermal–dermal junction epitopes in the pig and in isolated perfused skin treated with bis (2-chloroethyl) sulfide. *Toxicol Pathol*, **23**, 313–325.

Monteiro-Riviere NA and Inman, AO (1997). Ultrastructural characterization of sulfur mustard-induced vesication in isolated perfused porcine skin. *Microsc Res Techniq*, **37**, 229–241.

Monteiro-Riviere NA, Inman AO, Babin MC *et al.* (1999). Immunohistochemical characterization of the basement membrane epitopes in bis 2-chloroethyl) sulphide-induced toxicity in mouse ear skin. *J Appl Toxicol*, **19**, 319–328.

Nagy JM, Golumbic C, Stein WH *et al.* (1946). The penetration of vapours into human skin. *J Gen Physiol*, **29**, 441–469.

NATO (1985). *Handbook on the Medical Aspects of NBC Defensive Operations, AMedP-6, Part III: Chemical.* New York, NY, USA: North Atlantic Treaty Organization.

Neamati N, Fernandez A, Wright S *et al.* (1995). Degradation of lamin B1 precedes oligonucleosomal DNA fragmentation in apoptotic thymocytes and isolated thymocyte nuclei. *J Immunol*, **154**, 3788–3795.

Nicholson DW, Ali A, Thornberry NA *et al.* (1995). Identification and inhibition of the ICE/CED-3 protease necessary for mammalian apoptosis. *Nature*, **376**, 37–43.

Nishimoto Y, Yamakido M, Ishioka S, Watanabe S *et al.* (1988). Epidemiological status of lung cancer in Japanese mustard gas workers. In: *Unusual Occurrences as Clues to Cancer Aetiology* (RW Miller, Watanabe S, Fraumeui JF *et al.* eds), pp. 95–101. Tokyo, Japan: Japan Scientific Society Press, Taylor & Francis.

Norman JE (1975). Lung cancer mortality in World War I veterans with mustard-gas injury: 1919–1965. *J Natl Cancer Inst*, **54**, 311–317.

Norris K (1986). Salisbury, UK – Personal communication to author.

Orrenius S, McConkey DJ, Bellomo G *et al.* (1989). Role of Ca^{2+} in toxic killing. *Trends Pharmacol Sci*, **10**, 281–285.

Pappenheimer AM and Vance M (1920). The delayed action of mustard gas and lewisite. *J Exp Med*, **31**, 71–94.

Papirmeister B, Dorsey JK, Davison CL *et al.* (1970). Sensitization of DNA to endonuclease by adenine alkylation and its biological significance. *Fed Proc*, **29**, 726.

Papirmeister B, Gross CL, Petrali JP *et al.* (1984a). Pathology produced by sulfur mustard in human skin grafts on athymic nude mice. I. Gross and light microscopic changes. *J Toxicol Cutan Ocul Toxicol*, **3**, 371–391.

Papirmeister B, Gross CL, Petrali JP *et al.* (1984b). Pathology produced by sulfur mustard in human skin grafts on athymic nude mice. II. Ultrastructural changes. *J Toxicol Cutan Ocul Toxicol*, **3**, 393–408.

Papirmeister B, Gross CL, Meier HL *et al.* (1985). Molecular basis for mustard-induced vesication. *Fundam Appl Toxicol*, **5**, S134–S149.

Pharmaceutical Society of Great Britain (1977). *Martindale. The Extra Pharmacopoeia*, 27th Edition (JEF Reynolds, ed.). London, UK: The Pharmaceutical Press.

Pharmaceutical Society of Great Britain (1989). *Martindale. The Extra Pharmacopoeia*, 29th Edition (JEF Reynolds, ed.). London, UK: The Pharmaceutical Press.

Prentiss AM (1937). *Chemicals in War*. New York, NY, USA: McGraw-Hill Book Company Inc.

Pu Y, Lin P, Vaughan FL *et al.* (1995). Appearance of interleukin 1α relates DNA interstrand cross-links and cytotoxicity in cultured human keratinocytes exposed to bis-(2-chloroethyl)sulphide. *J Appl Toxicol*, **15**, 477–482.

Ray R, Legere RH, Majerus BJ *et al.* (1995). Sulfur mustard-induced increase in intracellular free calcium level and arachidonic acid release from cell membrane. *Toxicol Appl Pharmacol*, **131**, 44–52.

Renshaw B (1946). Mechanisms in production of cutaneous injuries by sulfur and nitrogen mustards. In: *Chemical Warfare Agents and Related Chemical Problems*, Volume 1, Chapter 23, pp. 479–518. Washington, DC, USA: US Office of Scientific Research and Development, National Defense Research Committee.

Robinson JP (1971). The rise of CB weapons. In: *The Problem of Chemical and Biological Warfare* (JP Robinson and M Leitenberg, eds), Chapter 2. New York, NY, USA: Stockholm International Peace Research Institute.

Rosenthal DS, Ding R, Simbulan-Rosenthal CMG *et al.* (1997). Intact cell evidence for the early synthesis and subsequent late apopain-mediated suppression, of poly (ADP-ribose) during apoptosis. *Exp Cell Res*, **232**, 313–321.

Rosenthal DS, Simbulan-Rosenthal CMG, Iyer S *et al.* (1998). Sulfur mustard induces markers of terminal differentiation and apoptosis in keratinocytes via a Ca^{2+}-calmodulin and caspase-dependent pathway. *J Invest Dermatol*, **111**, 64–71.

Rosenthal DS, Simbulan-Rosenthal CM, Iyer S *et al.* (2000). Expression of dominant-negative. Fas-associated death domain blocks human keratinocyle apoptosis and vesication induced by sulfur mustard. *J Biol Chem*, **20**, S43–S49.

Sabourin CLK, Petrali JP and Casillas RP (2000). Alterations in inflammatory cytokine gene expression in sulfur mustard-exposed mouse skin. *J Biochem Mol Toxicol*, **14**, 291–302.

Sabourin CLK, Danne MM, Buxton KL *et al.* (2002). Cytokine, chemokine and matrix metalloproteinase response after sulfur mustard injury to weanling pig skin. *J Biochem Mol Toxicol*, **16**, 263–272.

Schnyder J and Baggliolini M (1980). Induction of plasminogen activator secretion in macrophages by electrochemical stimulation of the hexose monophosphate shunt with methylene blue. *Proc Natl Acad Sci USA*, **77**, 414–417.

Shi Y, Schai BM and Green DR (1989). Cyclosporin A inhibits activation-induced cell death in T-cell hybridomas and thymocytes. *Nature*, **339**, 623–626.

Smith KJ, Graham JS, Hamilton TA *et al.* (1997). Immunohistochemical studies of basement membrane proteins and proliferation and apoptosis markers in sulfur mustard induced cutaneous lesions in weanling pigs. *J Dermatol Sci*, **15**, 173–182.

Stockholm International Peace Research Institute (1971). *The Problem of Chemical and Biological Warfare*. New York, NY, USA: Stockholm International Peace Research Institute.

Sulzberger MB and Katz JH (1943). The absence of skin irritants in the contents of vesicles. *US Navy Med Bull*, **43**, 1258–1262.

Sulzberger MC, Baer RI, Kanof A *et al.* (1947). Skin sensitization to vesicant agents of chemical warfare. *J Invest Dermatol*, **8**, 365–393.

Tewari M, Quan LT, O'Rourke K *et al.* (1995). Yama/CPP 32ß, a mammalian homolog of CED-3, is a cmA-inhibitable protease that cleaves the death substrate poly (ADP-ribose) polymerase. *Cell*, **81**, 801–809.

United Nations Security Council (1984). *Report of the Specialists Appointed by the Secretary-General to Investigate Allegations by the Islamic Republic of Iran concerning the use of Chemical Weapons*, United Nations Report S/16433. New York, NY, USA: United Nations Security Council.

United Nations Security Council (1986). *Report of the Mission Despatched by the Secretary-General to Investigate Allegations of the use of Chemical Weapons in the Conflict Between the Islamic Republics of Iran and Iraq*, United Nations Report

S/17911. New York, NY, USA: United Nations Security Council.

United Nations Security Council (1987). *Report of the Mission Despatched by the Secretary-General to Investigate Allegations of the use of Chemical Weapons in the Conflict Between the Islamic Republics of Iran and Iraq*, United Nations Report S/18852. New York, NY, USA: United Nations Security Council.

Vedder EB (1925). *The Medical Aspects of Chemical Warfare*. Baltimore, MD, USA: Williams and Wilkins Company.

Vogt RF, Dannenberg AM, Schofield BH *et al.* (1984). Pathogenesis of skin lesions caused by sulphur mustard. *Fundam Appl Toxicol*, **4**, S71–S83.

Vojvodć V, Milosavljević Z, Bošković B *et al.* (1980). The protective effect of different drugs in rats poisoned by sulfur and nitrogen mustards. *Fundam Appl Toxicol*, **5**, S160–S168.

Vycudilik W (1985). Detection of mustard gas bis(2-chloroethyl)-sulfide in urine. *Foren Sci Int*, **28**, 131–136.

Wada S, Miyanishi M, Nishimoto Y *et al.* (1968). Mustard gas as a cause of respiratory neoplasia in man. *Lancet*, **i**, 1161–1163.

Warthin AS and Weller CV (1918). The pathology of skin lesions produced by mustard gas (dichloroethylsulphide). *J Lab Clin Med*, **3**, 447–486.

Werrlein RJ and Madren-Whalley JS (2000). Effects of sulfur mustard on the basal cell adhesion complex. *J Appl Toxicol*, **20**, S115–S123.

West CJ (1920). The history of mustard gas. *Chem Met Eng*, **22**, 541–54.

Willems JL (1989). Clinical management of mustard gas casualties. *Ann Med Milit Belg*, **3** (**Supplement**), 1–61.

Wils ERJ, Hulst AG and van Laar J (1988). Analysis of thiodiglycol in urine of victims of an alleged attack with mustard gas, part II. *J Anal Toxicol*, 12, 15–19.

Wright P (1990). Moorfields Eye Hospital, UK – Personal Communication to author.

Yanagida, J., Hozawa, S., Ishioka, S *et al.* (1988). Somatic mutation in peripheral lymphocytes of former workers at the Okunojima poison gas factory. *Jpn J Cancer Res*, **79**, 1276–1283.

Zaniboni A, Simoncini E, Marpicati P *et al.* (1988). Severe delayed neurotoxicity after accidental high-dose nitrogen mustard. *Am J Hematol*, **27**, 304 (Letter).

Zhang Z and Monteiro-Riviere NA (1997). Comparison of integrins in human skin, pig skin and perfused skin: an *in vitro* skin toxicology model. *J Appl Toxicol*, **17**, 247–253.

Zhang Z, Peters BP and Monteiro-Riviere NA (1995). Assessment of sulfur mustard interaction with basement membrane components. *Cell Biol Toxicol*, **11**, 89–101.

20 DERMAL ASPECTS OF CHEMICAL WARFARE AGENTS

Robert P. Chilcott

Health Protection Agency, Chilton, UK

INTRODUCTION

The human integument is remarkably impermeable to a wide range of xenobiotics and this characteristic is mainly attributable to the outermost layer of skin, the stratum corneum. Normal skin is largely resistant to biological warfare agents (bacteria, viruses and toxins). However, a number of chemical warfare agents are able to diffuse freely through this barrier layer, resulting in local or systemic toxicity. The percutaneous toxicity of such agents is primarily related to the absorbed dose. This, in turn, is subject to a variety of factors such as environmental conditions and the physicochemical properties of the agent. Clearly, an understanding of such factors is essential for the successful development of topically administered medical countermeasures and therapies.

CHEMICAL WARFARE AGENTS OF RELEVANCE

The majority of substances listed on Schedule 1 of the *Chemical Weapons Convention* are considered to be percutaneous hazards (OPCW, 1993). In particular, sulphur mustard (HD) and VX are the most extensively studied agents of this genre and are representative of chemical warfare agents that exhibit mainly local or systemic percutaneous toxicity, respectively (Table 1).

With the exception of GB, the 'classic' percutaneous threat agents have relatively low volatilities (Figure 1). Consequently, these agents may persist in the environment for long periods and so their likely use in warfare would be to deny terrain and degrade supply operations at strategic installations. Volatility is also a principal determinant of the systemic toxicity of nerve agents.

From a military perspective, it is not necessarily the lethality of a percutaneous threat agent that is of primary concern but its ability to cause incapacitation. On this basis, HD can be considered to be equipotent to VX (the most toxic of nerve agents). In addition, the pathological consequences of HD exposure would likely impose a substantial burden on medical resources. Thus, while HD is not generally considered to be a lethal agent, its vesicant potency combined with its relative persistence and delayed effects contribute to its reputation as 'king of the war gasses'.

BASIC SKIN STRUCTURE AND FUNCTION

The skin is a multi-layered (veneered or stratified) structure comprising three principal layers, i.e. the epidermis, dermis (corium) and hypodermis (Figure 2). Human skin can be divided into two types – glabrous and non-glabrous. The former is mainly limited to the palms of the hand and soles of the feet (palmar–plantar skin) and comprizes a stratum corneum which is generally thicker and less permeable than non-glabrous skin.

The epidermis is predominantly (>90%) populated by keratinocytes (acanthocytes) which continuously undergo apical migration from the

Chemical Warfare Agents: Toxicology and Treatment (2nd Edition)
Edited by Timothy C. Marrs, Robert L. Maynard and Frederick R. Sidell

Table 1. Representative sample of chemical warfare agents capable of penetrating the skin (percutaneous hazards), categorized according to primary (local or systemic) effect

Primary effect	Standard designation	Alternate name(s)	Chemical name	Mw[a]
Local (Vesication)	CX	Phosgene oxime, nettle gas	Dichloroformoxime	114
	HD	Sulphur mustard, mustard gas, Yprite, Yellow Cross, Kampstoff 'Lost'	Bis-(2-chloroethyl) sulphide	160
	HL	Sulphur mustard/lewisite mixture	Bis-(2-chloroethyl) sulphide/ dichloro-(2-chlorovinyl)arsine	N/A[b]
	HN-1	Nitrogen mustard	2,2′-Dichlorotriethylamine	170
	L	Lewisite	Dichloro(2-chlorovinyl)arsine	207
	MD	Methyl-dick	Methyldichloroarsine	160
	Q	Sesqui-mustard	1,2-Bis (2-chloroethylthio) ethane	219
	T	—	Bis[2-(2-chloroethylthio)ethyl] ether	263
	T2	T2 Mycotoxin, 'yellow rain'	—	467
Systemic	GA	Tabun	Ethyl *N*,*N*-dimethylphosphoroamidocyanidate	162
	GB	Sarin	Isopropyl methylphosphonofluoridate	140
	GD	Soman	Pinacolyl methylphosphonofluoridate	182
	GF	Cyclosarin, CMPF	Cyclohexyl methylphosphonofluoridate	180
	VX	—	*O*-Ethyl *S*-(2-diisopropylamino-ethyl) methyl phosphonothiolate	267

Notes:
[a] Relative molecular mass.
[b] Not available.

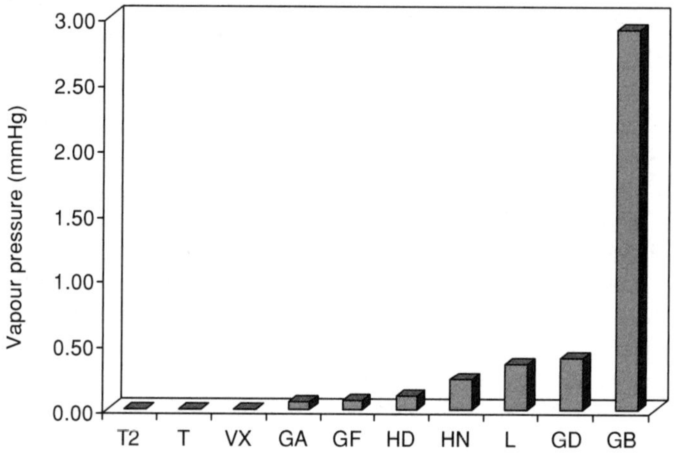

Figure 1. Vapour pressure of a selection of percutaneous threat agents

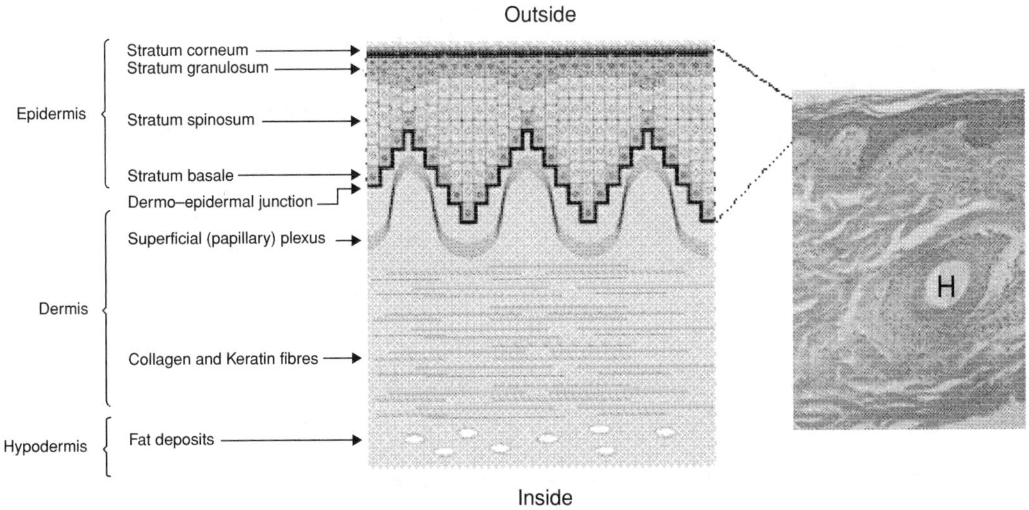

Figure 2. Schematic representation and corresponding histological section (stained with haematoxylin and eosin) of the main features of the skin, including a hair follicle (H)

stratum basale. During migration, keratinocytes undergo several stages of differentiation which can be identified histologically as the stratum spinosum, stratum granulosum and stratum corneum (Figure 2). Other cell-types present in the epidermis include Langerhans cells (which are involved with antigen presentation) and melanocytes (which synthesize the photo-protectant, melanin). The dendritic nature of these two cell-types enables them to migrate and populate the interstitial space between keratinocytes. Overall, the epidermis provides protection against xenobiotics, micro-organisms, some forms of radiation and limited mechanical trauma. However, the primary role of the epidermis is arguably to prevent water loss.

The epidermis is anchored to the dermis via a continuous, protein-rich region termed the dermo–epidermal junction (Figure 2). This structure is highly invaginated and forms characteristic ('rete') ridges on skin sections that are readily discernible under the optical microscope. The underlying blood supply (superficial plexus) interdigitates with the rete ridges, thus providing a large surface area for the transfer of nutrients, oxygen and waste products. Chemicals that are able to traverse the epidermis are generally subject to extensive systemic absorption by the superficial plexus at this anatomical region and so the dermis and hypodermis are not generally relevant to the percutaneous absorption kinetics of CW agents.

Terminally differentiated keratinocytes of the stratum corneum are known as corneocytes and are largely devoid of normal cellular functions, being predominantly composed of protein (keratin). The ultrastructure of the stratum corneum is described by the 'brick and mortar model' (Elias, 1983; Figure 3). The functional implication of this architecture is that some skin penetrants must diffuse via a long and tortuous route between adjacent corneocytes, thus reducing their rate of absorption. This is known as the intercellular route. In contrast, some chemicals may diffuse equally through both corneocytes and the lipid mortar, resulting in a transcellular route.

A third, potential route of entry across the skin involves diffusion down hair follicles and into sebaceous glands or via sweat glands. These are referred to as 'shunt pathways' and are the biological equivalent of intergalactic wormholes: both have been subject to intense debate but are of unknown practical relevance.

While corneocytes can be considered to be hydrophilic domains, they are surrounded by a lipid-rich matrix mainly comprising ceramides, free fatty acids and cholesterol (Downing *et al.*, 1987). Thus, the intercellular domain is predominantly a lipophilic environment. This combination imparts a degree of amphiphobicity upon

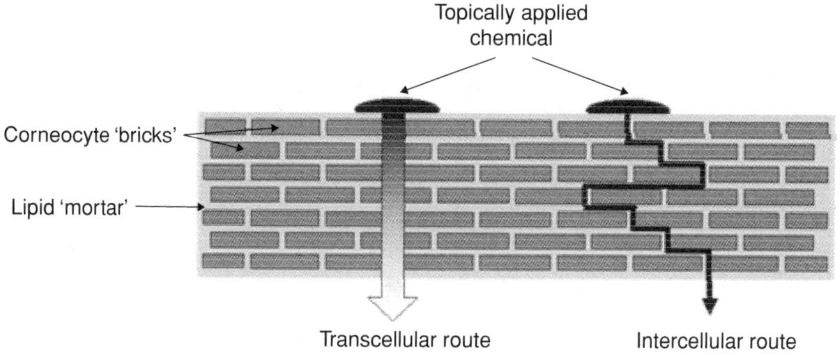

Figure 3. Schematic representation of the stratum corneum ('Brick and Mortar model') and corresponding routes of transcellular and intercellular penetration

the stratum corneum, providing limited protection against both lipophilic and hydrophilic penetrants.

The molecular packing of the lipid matrix effectively sets an upper limit on the physical size of molecules that may penetrate the stratum corneum (Figure 4). This is referred to as the 'rule of 500' (Bos and Meinardi, 2000) since few substances with a molecular weight above 500 Da are capable of passive diffusion through the skin (see Table 1).

PRINCIPLES OF SKIN ABSORPTION

Skin absorption occurs when a chemical undergoes passive diffusion through the stratum corneum into the viable layers of the skin. Diffusion is the transport phenomena by which matter moves from one part of a system to another as a result of random molecular motion and is characterized by Fick's laws (Crank, 1975).

Fick's first law of diffusion applies to 'steady-state' conditions. This situation may occur

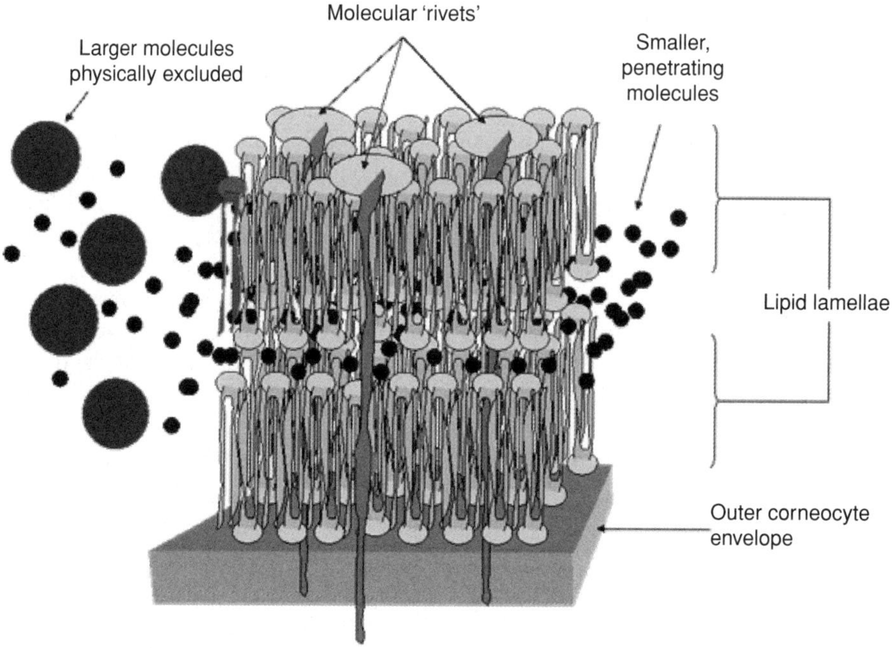

Figure 4. Representation of the arrangement of lipid lamellae between adjacent corneocytes and role as a 'molecular sieve'. Lipid layers are attached to adjacent corneocytes via long-chain ceramide (which act as a 'molecular rivet')

Table 2. Summary of factors influencing the skin absorption of chemical warfare agents based on Fick's laws of diffusion. The dimensions of each parameter are expressed in terms of those conventionally used for skin absorption kinetics

Factor	Symbol	Dimensions	Parameter
Partitioning	K_m	n/a[a]	Partition coefficient
Diffusivity	D	$cm^2\ h^{-1}$	Diffusivity coefficient
Distance	h	cm	Thickness
Concentration	C	$g\ cm^{-3}$	Concentration of penetrant on skin surface

Note:
[a] Not applicable.

following gross chemical skin contamination and is correspondingly referred to as an 'infinite dose'. Fick's second law applies when the amount of chemical on the skin surface is limited and so significantly decreases over time (due to percutaneous absorption or evaporation); so-called 'finite dose' conditions. Fick's laws relate to the four main factors that influence skin absorption (Table 2).

The relationship between these four parameters is given by a derivation of Fick's laws (Equation (1)), where J is the overall rate (flux) of diffusion across the skin (conventionally expressed in units of $g\ cm^{-2}\ h^{-1}$):

$$J = \frac{DCK_m}{h} \qquad (1)$$

A chemical's partition coefficient (K_m) quantifies the extent to which it moves into the stratum corneum from the skin surface. This is primarily determined by the relative solubility of the penetrant, commonly expressed as the octanol–water partition coefficient (K_{ow} or $\log P$) value.

Diffusivity (D) is a temperature-dependent parameter (Equation (2)) that essentially describes the mobility of a penetrating molecule within the stratum corneum. Diffusivity can be affected by a variety of factors, including the physical size of a penetrant and its potential interactions with the stratum corneum (through hydrogen bonding, electrostatic forces, etc.).

$$D = D_0 e^{-\frac{E_a}{RT}} \qquad (2)$$

where D_0 is the diffusivity coefficient (at infinite temperature), E_a is the activation energy, R is the molar gas constant and T is the absolute temperature.

'Distance' refers to how far a penetrant travels to traverse the stratum corneum. Theoretically, this should represent the average diffusional pathlength of the penetrating molecule. However, the thickness of the rate-limiting membrane (stratum corneum) is used as a more pragmatic alternative.

The concentration of an agent is assumed to be numerically equal to its liquid or vapour density (at a given temperature). Fick's laws assume that diffusion is driven by a concentration gradient, with molecules diffusing from areas of high to low concentration. However, it is important to understand that the actual driving force for diffusion is based on thermodynamic activity, not concentration (see factors affecting the percutaneous toxicity of CW agents – sweat).

The skin absorption kinetics of an agent may have implications for the medical management of exposed casualties. For example, an agent which undergoes relatively slow and prolonged skin absorption may have the potential to out-live an antidote administered by intramuscular bolus (Figure 5). Consequently, the antidote may required repeated administration to achieve clinical effectiveness.

SOME FACTORS AFFECTING THE PERCUTANEOUS ABSORPTION OF CW AGENTS

The central dogma of toxicology was formulated by Paracelsus (*aka* Theophrastus Phillippus Aureolus Bombastus von Hohenheim; ca. 1493–1541) who stated that it is the dose of a chemical which dictates its toxicity ('sola dosis facit venenum'; Oser, 1987). This applies equally to the

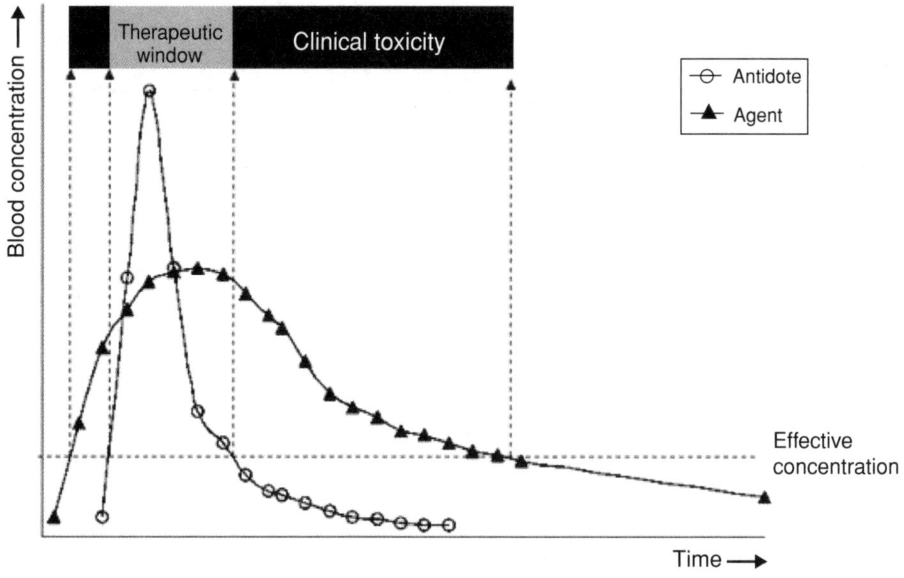

Figure 5. Theoretical representation of the toxicokinetics of a nerve agent (following dermal exposure) and pharmacokinetics of the corresponding antidote (i.m. administration). 'Effective concentration' refers to the threshold toxic or therapeutic blood concentrations of antidote, respectively

percutaneous toxicity of chemical warfare agents and so factors which affect skin absorption are necessarily factors that influence percutaneous toxicity.

Anatomical variation

The permeability of human skin can vary considerably according to anatomical location (Maibach *et al.*, 1971). The phenomenon of regional variation also applies to chemical warfare agents. One of the most extensive investigations of regional permeability in humans was conducted with VX in the late 1950s (Sim,1962). This study demonstrated that, for example, the skin of the scrotum is 350 times more permeable than that of the back of the hand (Figure 6).

Temperature

The diffusivity of a chemical through the skin is proportional to temperature (Equation (2)). Therefore, both skin and environmental temperatures may potentially affect skin absorption. Human volunteer studies involving the application of small ($\sim$ 0.2–0.6 μl) amounts of VX to cheek skin have clearly demonstrated a relation between environmental temperature and skin absorption (Figure 7).

Increased skin temperature is known to enhance the percutaneous absorption of HD (Figure 8) and results in more severe skin lesions (Renshaw, 1946), although this may also be attributable to sweating (see below).

Sweating

Sweating is a natural response to an increase in environmental temperature, physical exertion or stress and results in the deposition of an aqueous film over the skin surface. The influence of this water layer was investigated by Renshaw (1947) who observed that pre-wetting the skin surface significantly enhanced the severity of HD vapour-induced skin lesions. This effect decreased if the skin were allowed to dry before exposure (Figure 9).

One explanation for water having this effect involves a change in the thermodynamic activity of HD. The interaction of skin surface water and HD leads to the formation of an aqueous solution of agent. While this decreases the concentration of HD on the skin surface, the thermodynamic

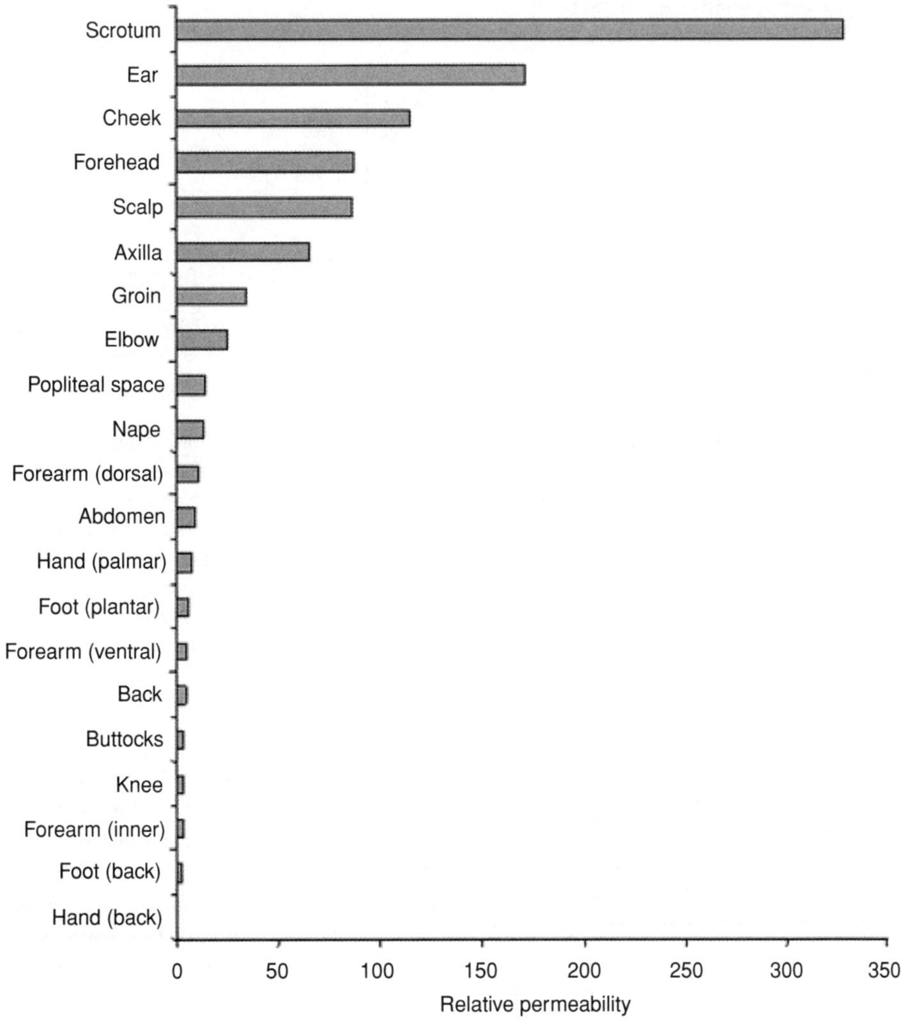

Figure 6. Anatomical (regional) variation in the permeability of human skin to cholinesterase, calculated from the rate of inhibition of cholinesterase (Sim, 1962). Permeability of each site expressed relative to the back of the hand (nominal permeability = 1)

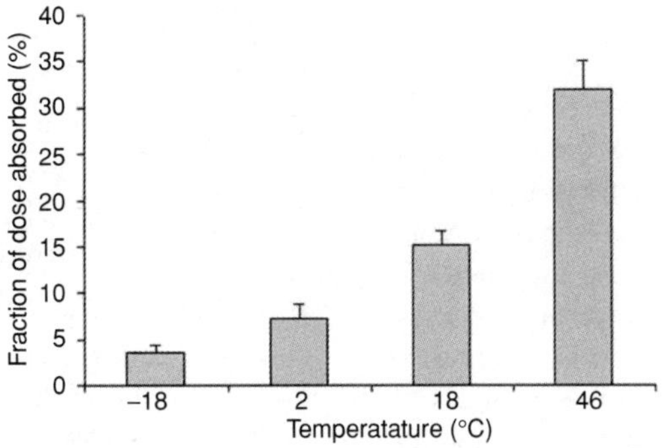

Figure 7. Effect of environmental temperature on the systemic absorption of VX in humans, following topical application to the face (cheek). Absorption is expressed as a percentage of the applied dose penetrated (average ± standard error of mean) 3 h post-exposure. Data from Craig *et al.* (1977)

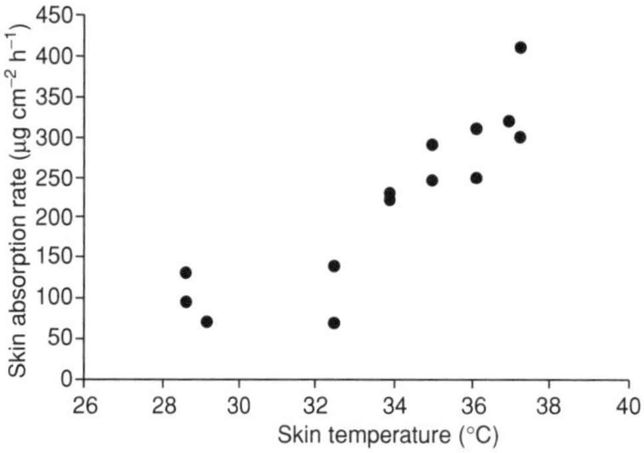

Figure 8. Effect of skin temperature on the human skin absorption of ^{14}C-radiolabelled HD. Absorption rates were based on the loss of radioactivity occurring from a dosing chamber containing liquid ^{14}C-HD following a 1 h exposure of abdominal skin. Data from Renshaw (1946)

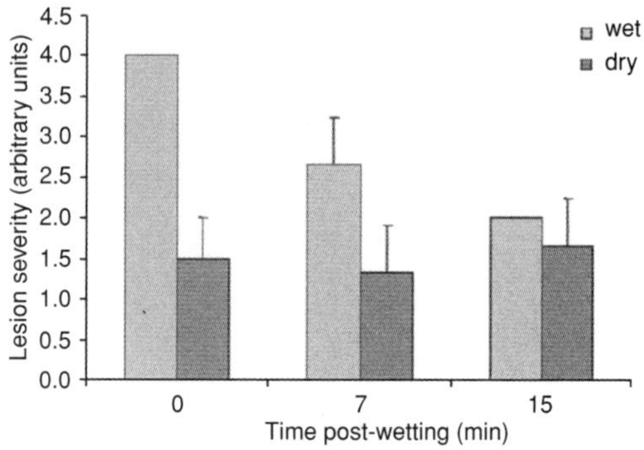

Figure 9. Severity of lesions resulting from exposure of dry or pre-wetted skin to HD vapour. 'Wet' skin was allowed to dry for increasing periods (up to 15 min) before exposure. Lesion severity was calculated from the original visual scoring index of faint erythema (score = 1), definite erythema (score = 2), raised (oedematous) erythema (score = 3) and blister (score = 4). Each bar represents the average lesion score of three subjects (± standard deviation). Data from Renshaw (1947)

activity (µ) increases. This is because HD is a lipophilic agent and so will have a higher affinity for the lipophilic domains of the stratum corneum rather than the aqueous solution. In other words, dilution in water increases the fugacity[1] of HD and this results in enhanced percutaneous toxicity. It is also conceivable that water may act by occluding HD on the skin surface, thus decreasing vapour loss of agent, which normally accounts for 80% of the applied dose (Renshaw, 1946; Chilcott *et al.*, 2000).

The lesion-enhancing effects of sweat may be a contributing factor to the relative sensitivity of the groin and axillae (armpits) to HD, although the thin stratum corneum, regional permeability and micro-environment (raised temperature) of these regions are also likely to be implicated.

Volatility

Vapour loss of a chemical warfare agent from the skin surface affects percutaneous toxicity by reducing the dose available for skin absorption. For agents of low volatility, such as VX, practically all of the applied dose may potentially be absorbed and so the percutaneous toxicity (LD_{50}) approaches that of an intravenous (IV) LD_{50}. In contrast, more volatile chemicals, such as G-agents, are subject to extensive vapour loss and so the percutaneous LD_{50} may be several hundred times the i.v. LD_{50} (Figure 10).

Clearly, factors which reduce or prevent the skin-surface vapour loss of a chemical warfare agent will tend to enhance skin absorption and percutaneous toxicity. Occlusion of contaminated skin (for example, with ointments or impermeable dressings) is an obvious means of impeding vapour loss and has implications for the

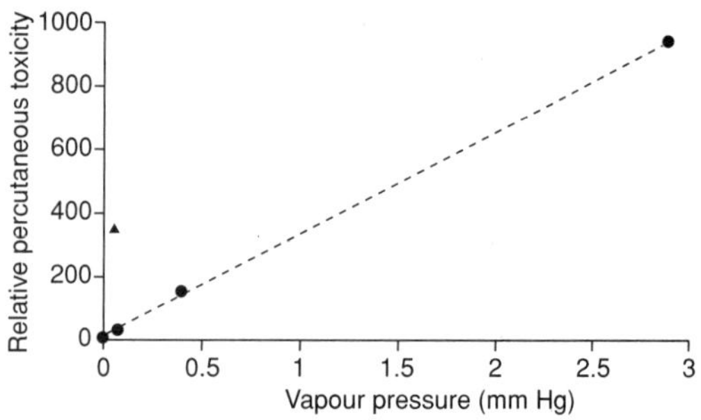

Figure 10. Relationship between vapour pressure and the relative percutaneous toxicity (expressed as the ratio of the percutaneous LD_{50} to the intravenous LD_{50}) of four nerve agents (VX, GF, GD and GB; indicated by solid circles). A notable exception to this relationship is GA (indicated with a solid triangle) which may undergo dermal metabolism prior to systemic absorption

[1] 'Fugacity' is derived from the Latin 'to escape' (literally 'flight' or 'flee') and is related to the thermodynamic activity of the penetrant in a given solution.

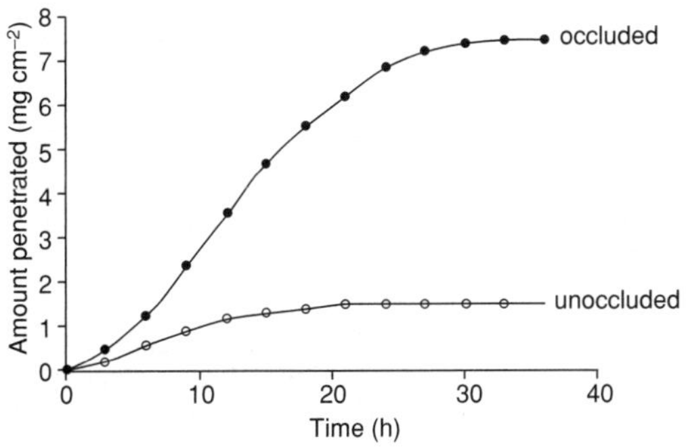

Figure 11. Skin absorption of HD (expressed as the cumulative amount penetrated per unit area) through human (breast) skin under occluded or unoccluded conditions. Penetration rates (J) of HD were 332 ± 96 and 38 ± 23 μg cm^{-2} h^{-1} under occluded and unoccluded conditions, respectively. Data from Chilcott *et al.* (2000)

medical management of dermal exposures to CW agents (see strategies to mitigate skin absorption). Indeed, animal studies conducted shortly after World War II demonstrated a significant increase in the percutaneous toxicity of GB if the contaminated skin were quickly covered with an occlusive 'antigas ointment' (Cullumbine *et al.*, 1954).

In vitro studies with human skin have quantified the influence of occlusion on the percutaneous absorption of HD. Under unoccluded conditions, up to 80% of an applied dose of HD evaporates from the skin surface. However, physical occlusion prevents this vapour loss and results in a five-fold increase in the total amount of absorbed HD (Figure 11) and a nine-fold increase in the skin absorption rate (Chilcott *et al.*, 2000).

STRATEGIES TO MITIGATE SKIN ABSORPTION

The most effective method for preventing skin exposure to chemical warfare agents is the correct and timely use of personal protective equipment (PPE), such as gloves, suits and boots. However, there are a variety of scenarios under which skin contamination may occur and so it is important to have effective medical countermeasures available to mitigate skin absorption.

Exposure of skin to chemical warfare agents can be reduced by the use of barrier creams, decontaminants or by removal of superficially absorbed chemicals ('catch-up therapies'). In short, prophylactic barrier creams seek to prevent skin contact whereas decontaminants and catch-up therapies are post-exposure measures that can only minimise skin absorption. Barrier creams may also be applied to skin under PPE to enhance protection at vulnerable areas, such as the groin and axillae.

Historical developments from a British perspective

Following World War I, British research efforts concentrated mainly on the development of anti-gas ointments (Figure 12) which could be used either as a decontaminant or prophylactic. These mainly consisted of a base containing soap, water and oil, with approximately 25% w/w active ingredient (Table 3).

The two most commonly deployed UK anti-gas ointments during World War II (A/G No. 5 and No. 6) contained a chloramine termed 'antiverm' (AV; 2,4-dichlorophenyl benzoyl chloroimide) to chemically neutralise vesicant agents. Both were efficacious decontaminants and prophylactics against vesicants, although A/G No. 6 was the preferred formulation for tropical conditions.

Substantial effort was placed upon improving the reactivity of the ointments towards HD and research focused on two types of compound; those which attacked the sulphur atom and those which underwent substitution reactions with the β-chlorine atoms of the sulphur mustard molecule. No significant improvement could be made and thus A/G No. 5 and No. 6 remained the ointments of choice until their withdrawal from service. During this time, the US army had also developed an anti-gas ointment called 'M5'. The active ingredient of this formulation was S-330 (tetrachloro-2,5-diimino-7,8-diphenylglycouril). In comparison with A/G No. 5 and No. 6, M5 ointment was slightly less efficacious. However, unlike A/G No. 6, it was not light sensitive and had the advantage of being effective in both hot and cold climates.

Table 3. Formulation of some anti-gas ointments, ca. 1918–1954

Ointment	Active ingredient	Emulsifier	Water (%)	Oil phase & excipient(s)
A/G No. 1	Bleaching powder (50%)	—	—	50% mineral jelly
A/G No. 2	Chloramine-T (26.7%)	6.7% ethylene diglycol stearate, 6.7% sodium stearate	59.9	—
A/G No. 3	Antiverm (25%)	—	—	27.5 % hardened palm kernel oil, 27.5% diethylphthalate, 20% magnesium carbonate
A/G No. 3a	Antiverm (25%)	—	—	17% hardened palm kernel oil, 26% diethylphthalate, 20% magensium carbonate, 12% beeswax
A/G No. 4	Chloramine-B (26.7%)	6.7% sodium distearate, 6.7% sodium stearate	59.9	—
A/G No. 5	Antiverm (25%)	5% sodium stearate, 1% potassium stearate	39	20% diethylphthalate, 10% hardened marine oil
A/G No. 6	antiverm (25%)	4% sodium stearate, 2% potassium stearate	39	20% diethylphthalate, 10% hardened ground-nut oil
M5	S330 (25%)	9% magnesium stearate	0	52% triacetin, 3% cellulose aceto-butyrate, 0.5% titanium oxide, 1.5% 'pigment'

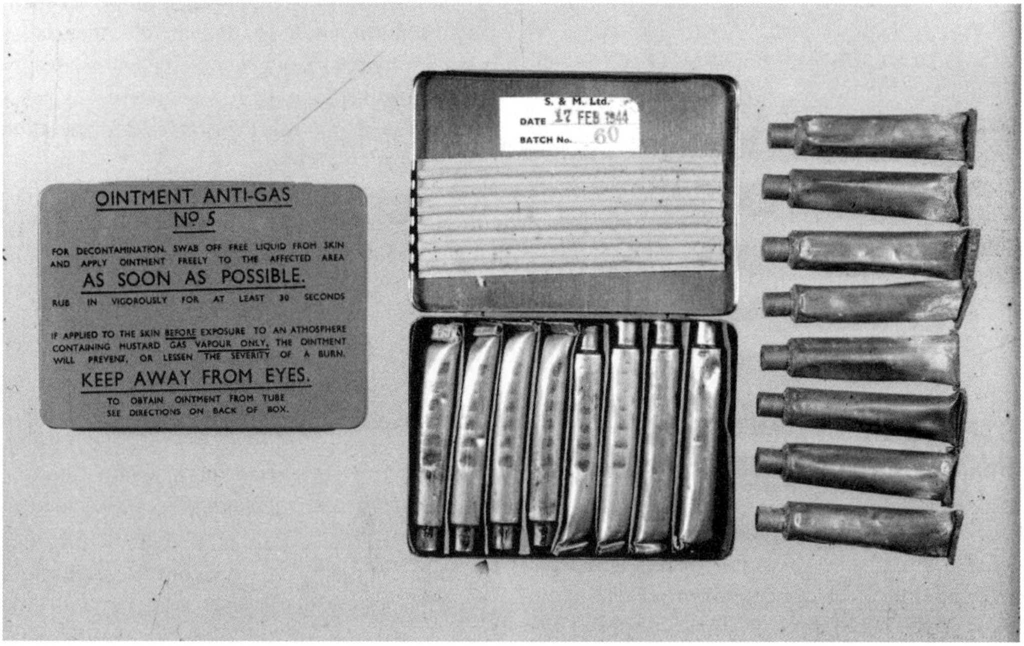

Figure 12. Tubes of Anti-gas No. 5 (ca. 1944) with corresponding instructions for use. Figure kindly provided by Dstl, Porton Down, Wiltshire, UK

Table 4. Nomenclature, presentation and composition of current (formally in-service) decontaminants

Country	Common designation	Class	Constituent(s)
UK	Fullers' earth	Passive powder	Montmorillonite clay
USA	Ambergard XE-555 resin (M291)	Reactive powder	Absorptive/ion-exchange resins
Canada	RSDL™[a]	Reactive liquid	Polyethylene glycol base, potassium butadiene monoximate

Note:
[a] RSDL, reactive skin decontamination lotion.

Further development of antigas ointments ceased in the 1950s because of their eight-fold enhancement in the percutaneous toxicity of G-agents (Cullumbine *et al.*, 1954). The UK military use of anti-gas ointments as decontaminants was formally discontinued in 1959, when it was recommended that the best means of mitigating skin absorption was by physical absorption onto powders. Instead, attention focused on the development of new decontaminants.

During the early 1960s, two methods of physical removal of agent were examined. The first used various types of fabric, such as asbestos, filter paper, woven viscose and woven nylon (Dennis and Errington, 1960). A second line of investigation examined absorbent powders (Bramwell and Green, 1962). Initially, 'Dutch powder' (a mixture of magnesium oxide and bleach powder) was shown to be efficient for the decontamination of H, G and V agents. However, subsequent experiments found that the effectiveness of Dutch powder was not associated with its bleaching properties but was instead related to its ability to passively absorb liquid agent. This work eventually led to the British military's adoption of fullers' earth as a universal skin decontaminant in 1963.

Current systems

Passive decontaminants, such as fullers' earth, are said to suffer from two disadvantages. First, the agent may remain active within the absorptive matrix and so may theoretically represent a secondary hazard. Secondly, liquid agent that has penetrated into the superficial skin layers or vapour exposures may not be amenable to decontamination with powders. A logical progression is to develop an active decontaminant that would degrade absorbed agent, both on the skin and, ideally, within the stratum corneum. Two countries have contributed significantly to such a development; Canada and the USA, both of which currently deploy 'reactive' skin decontamination kits (Table 4).

While Ambergard and RSDL™ represent significant technological advances over fullers' earth, all three are broadly equivalent in terms of efficacy and each have relative merits or disadvantages with regards to practicality, toxicity and economy.

The production and commercial availability of new materials possessing novel physicochemical properties has recently revived interest in the development of barrier creams. In particular, perfluorinated formulations such as AG-7™ and SERPACWA (skin exposure reduction paste against chemical warfare agents) are effective against a wide range of agents (Chilcott *et al.*, 2002,2005a; Liu *et al.*, 1999). Such contemporary formulations greatly enhance the protection afforded by PPE and may enhance the effectiveness of decontamination by retaining liquid agent on the surface.

Catch-up therapies

A topical catch-up therapy is one which seeks to extract and/or neutralise chemical warfare agents from within the skin. Ideally, a catch-up therapy should be efficacious several hours after dermal exposure. For rapidly absorbed agents, catch-up therapies are unlikely to be beneficial. However, slower skin penetrants or those which accumulate to form a superficial reservoir within the stratum corneum may be amenable to such countermeasures. The practical utility of catch-up therapies has been a matter of some controversy and two particular factors have been at the centre of the debate; skin reservoir formation and the role of the stratum corneum.

Research conducted during World War I indicated that HD was absorbed 'by some element on or immediately adjacent to the skin surface' where it remained for 'a considerable period' and that 'vapour loss from this element was still demonstrable after 45 minutes' (Smith *et al.*, 1919).[2] Furthermore, it was demonstrated that the human skin reservoir of HD could be extracted by the application of appropriate solvents, such as kerosene (Marshall *et al.*, 1918). This practical demonstration of the HD skin reservoir was refuted by subsequent (*ex vivo*) studies conducted during World War II which concluded that HD reacted immediately with proteins within human skin and thus 'all therapeutic procedures based on neutralisation of free penetrated HD must necessarily be valueless' (Renshaw, 1946). This interpretation was primarily based on histological evidence such as microautoradiographs of skin exposed to ^{14}C-HD or skin sections that had been specifically stained to indicate the presence of free (unreacted) HD (Axelrod and Hamilton, 1946; Cullumbine, 1946). However, this interpretation largely ignored the accumulation of material within the stratum corneum and instead focused on the lack of radiolabel or free HD in the dermis and viable epidermis: the stratum corneum was not generally recognized as being the main barrier layer until the 1950s.

More recent *in vitro* studies have supported the original conclusions of Smith *et al.* (1919) in that HD forms a considerable and persistent reservoir within the epidermal layers of human skin (Chilcott *et al.*, 2000) and it is also likely that other lipophilic agents, such as VX, may have similar skin absorption characteristics (Chilcott *et al.*, 2005b). Clearly, the potential formation of a cutaneous reservoir of agent has important implications for the medical management of casualties, since its efficient removal may prevent prolonged absorption of systemically active agents (see Figure 5) and the very presence of a mobile skin reservoir may represent a secondary hazard to medical staff.

Catch-up therapies may be based on three mechanistic approaches (or combinations thereof): physical removal of contaminated tissue, solvent extraction of agent and topical delivery of neutralising agents into the skin.

Debridement of the skin is arguably the most simple and effective option. A range of established techniques for mechanical or laser debridement exist (Graham *et al.*, 2005), although these have been largely untested for the purpose of removing superficial skin contaminated with chemical warfare agents and operator safety (via potential exposure to live agent) should be considered before use. Several chemical debriding agents (such as trichloroacetic or thioglycolic acids applied in kaolin pastes) have been specifically evaluated for removing HD-contaminated skin and were reportedly effective in preventing vesication up to 14 hours post exposure (Cullumbine, 1943). While chemical debridement clearly has potential for eliminating skin reservoirs of chemical warfare agents, it is difficult to control the depth of a chemical 'peel' and their use in cosmetic procedures has occasionally resulted in adverse effects (Cassano *et al.*, 1999).

Solvent extraction has been demonstrated to be effective against HD exposures (Smith *et al.*, 1919) but is less likely to completely remove contamination than debridement and the use of flammable solvents is perhaps not a practical option for battlefield conditions. Furthermore, no single solvent may be appropriate for all agents and the use of an inappropriate solvent may actually enhance skin absorption and toxicity.

The topical application of neutralising chemicals has one substantial drawback; a chemical designed to neutralise a particular agent may lack reactivity towards another and so a mixture of reactive chemicals may be required for a generic therapy. It is perhaps worth noting that chemicals which are reactive towards CW agents may also react with normal skin constituents, potentially resulting in sensitisation or irritancy. Indeed, many reactive topical countermeasures

[2] This description was remarkably prescient as the stratum corneum (the inferred 'element') was not considered to be implicated in skin barrier function until the 1950s due to its incoherent, flaky appearance under optical microscopy which was a direct result of histological processing techniques.

(such as anti-gas ointments) have provoked significant skin reactions.

SUMMARY

The percutaneous toxicity of chemical warfare agents is primarily related to the rate and extent of percutaneous absorption which, in turn, is subject to a variety of environmental, biological and physicochemical factors. Understanding the basic mechanisms which control skin absorption is essential for developing effective pre- and post-exposure medical countermeasures and progress in transdermal technology is likely to yield new challenges and opportunities in the future.

ACKNOWLEDGEMENTS

The author wishes to acknowledge the staff at Dstl, Porton Down, Wiltshire, UK for their help and assistance in providing reference material and John Dafydd Pritchard at the Health Protection Agency for his editorial assistance.

REFERENCES

Axelrod DJ and Hamilton JG (1946). Radio-autographic studies of the distribution of lewisite and mustard gas in the skin and eye tissues. *Am J Pathol*, **23**, 389–411.

Bos JD and Meinardi MM (2000). The 500 Dalton rule for the skin penetration of chemical compounds and drugs. *Exp Dermatol*, **9**, 165–169.

Bramwell ECB and Green DM (1962). *The use of powders in personal decontamination*. Porton Technical Paper 828. Porton Down, Wiltshire, UK: Dstl.

Cassano N, Alessandrini G, Mastrolonardo M *et al.* (1999). Peeling agents: toxicological and allergological aspects. *J Eur Acad Dermatol Venereol*, **13**, 14–23.

Chilcott RP, Jenner J, Carrick W *et al.* (2000). Human skin absorption of bis-2-(chloroethyl)sulphide (sulphur mustard) *in vitro*. *J Appl Toxicol*, **20**, 349–355.

Chilcott RP, Jenner J, Hotchkiss SAM *et al.* (2002). Evaluation of barrier creams against sulphur mustard: (I) *in vitro* studies using human skin, *Skin Pharmacol Appl Physiol*, **15**, 225–235.

Chilcott RP, Dalton CH, Hill I *et al.* (2005a). Evaluation of a barrier cream against the chemical warfare agent VX using the domestic white pig. *Basic Clin Pharmacol Toxicol*, **97**, 35–38.

Chilcott RP, Dalton CH, Hill I *et al.* (2005b). *In vivo* skin absorption and distribution of the nerve agent VX (*O*-ethyl-*S*-[2(diisopropylamino)ethyl] methylphosphonothioate) in the domestic white pig. *Human Exp Toxicol*, **24**, 347–352.

Craig FN, Cummings EF and Sim VM (1977). Environmental temperature and the percutaneous absorption of a cholinesterase inhibitor, VX. *J Invest Dermatol*, **68**, 357–361.

Crank J (1975). The diffusion equations. In: *The Mathematics of Diffusion*, Second Edition, pp. 1–10. Oxford, UK: Clarendon Press.

Cullumbine H (1943). *The prevention of vesication*. Porton Report 2518. Porton Down, Wiltshire, UK: Dstl.

Cullumbine H (1946). The mode of penetration of the skin by mustard gas. *Br J Dermatol*, **58**, 291–294.

Cullumbine H, Miles S, Callaway S *et al.* (1954). *Methods of personal cleansing against liquid H and G*. Porton Technical Paper 437. Porton Down, Wiltshire, UK: Dstl.

Dennis WL and Errington PC (1960). *Investigation of the removal of droplets from surfaces by absorbent materials*. Porton Technical Paper 729. Porton Down, Wiltshire, UK: Dstl.

Downing DT, Stewart ME, Wertz PW *et al.* (1987). Skin lipids: an update. *J Invest Dermatol*, **88**, S2–S6.

Elias PM (1983). Epidermal lipids, barrier function, and desquamation. *J Invest Dermatol*, **80**, S44–S49.

Graham JS, Chilcott RP, Rice P *et al.* (2005). Wound healing of sulphur mustard injuries: strategies for the development of improved therapies. *J Burns Wounds*, **4**, e1 (www.journalofburnsandwounds.com).

Maibach HI, Feldman RJ, Milby TH *et al.* (1971). Regional variation in percutaneous penetration in man. Pesticides. *Arch Environ Health*, **23**, 208–211.

Marshall EK, Lynch V and Smith HW (1918). On dichloroethylsulphide (mustard gas). II. Variations in susceptibility of the skin to dichloroethylsulphide. *J Pharmacol Exp Ther*, **12**, 291–301.

OPCW (1993). Convention on the prohibition of the development, production, stockpiling and use of chemical weapons and their destruction. The Hague, the Netherlands: Organization for the Prohibition of Chemical Weapons (www.opcw.org).

Oser BL (1987). Toxicology then and now. *Regul Toxicol Pharmacol*, **7**, 427–443.

Renshaw B (1946). *Mechanisms in production of cutaneous injuries by sulfur and nitrogen mustards*. In: Summary Technical Report of Division 9, NDRC,

Volume 1: *Chemical Warfare Agents and Related Chemical Problems*, Parts III–IV. Washington, DC, USA: Office of Scientific Research and Development.

Renshaw B (1947). Observations on the role of water in the susceptibility of human skin to injury by vesicants. *J Invest Dermatol*, **9**, 75–85.

Sim VM (1962). *Variability of different intact human skin sites to the penetration of VX.* Technical Report CRDLR 3122. Edgewood, MD, USA: US Army Chemical Corps Research and Development Command, Chemical Research and Development Laboratories, Army Medical Center.

Smith HW, Clowes GHA and Marshall EK (1919). On dichloroethylsulphide (mustard gas) i.v. The mechanism of absorption by the skin. *J Pharm Exp Ther*, **13**, 1–30.

21 SULPHUR MUSTARD INJURIES OF THE SKIN: PATHOPHYSIOLOGY AND CLINICAL MANAGEMENT OF CHEMICAL BURNS

Paul Rice

Dstl Porton Down, Salisbury, UK

INTRODUCTION

Sulphur mustard is a powerful vesicant (blistering agent) in man and the descriptions of the casualties from the Iran-Iraq War of 1984–1987, treated in various European medical centres, attest to the problems of the clinical management of both percutaneous and systemic sulphur mustard injuries (Mandl and Freilinger, 1984; Pauser *et al.*, 1984; Colardyn *et al.*, 1986; Willems, 1989). In addition to their association with large, pendulous fluid-filled blisters, the skin injuries differ in many other respects from thermal burns with which they are often compared (Figure 1). For example, the healing time is considerably longer than for comparable thermal injury and varies from species to species (Schrafl, 1938). Despite the obvious clinical importance of such lesions, the biochemical basis of vesication in human skin remains largely undefined (Gales *et al.*, 1989). Research to further elucidate the molecular mechanisms underlying vesication has been hampered by the lack of an appropriate animal model. In general, fur-bearing mammals do not produce blisters on challenge with either sulphur mustard liquid or vapour. This species variation in cutaneous response has long been known and several theories, none of which has yet been universally accepted, have been proposed to explain such differences (Flesch *et al.*, 1952; McAdams, 1956). The reasons for differences in the healing time remain equally elusive (Papirmeister *et al.*, 1984).

THE HISTORY OF THE USE OF SULPHUR MUSTARD IN WAR

Sulphur mustard or "mustard gas" is perhaps the most familiar member of the heterogeneous group of chemicals that are referred to as chemical warfare agents. The term "king of the battle gases", thought to have been coined by Foulkes in his commentaries on the use of chemical warfare agents in World War I (Foulkes, 1934), remains a relatively accurate and well-deserved description today.

The exact date of the first chemical synthesis of sulphur mustard is still disputed, but various dates between 1820 and 1854 have been quoted by a number of authors. In 1860, Guthrie reported the synthesis of sulphur mustard and noted some of its vesicant (blister-inducing) properties; this was followed by a report in 1886 detailing the synthesis of pure sulphur mustard (Meyer, 1886). There then followed a long period until the outbreak World War I, during which time little or no practical use for the compound was proposed. Although the British had successfully synthesized sulphur mustard by 1916, there were no plans to develop it as a chemical weapon. Germany, however, had recognized its potential military

Chemical Warfare Agents: Toxicology and Treatment (2nd Edition)
Edited by Timothy C. Marrs, Robert L. Maynard and Frederick R. Sidell © 2007 John Wiley & Sons, Ltd

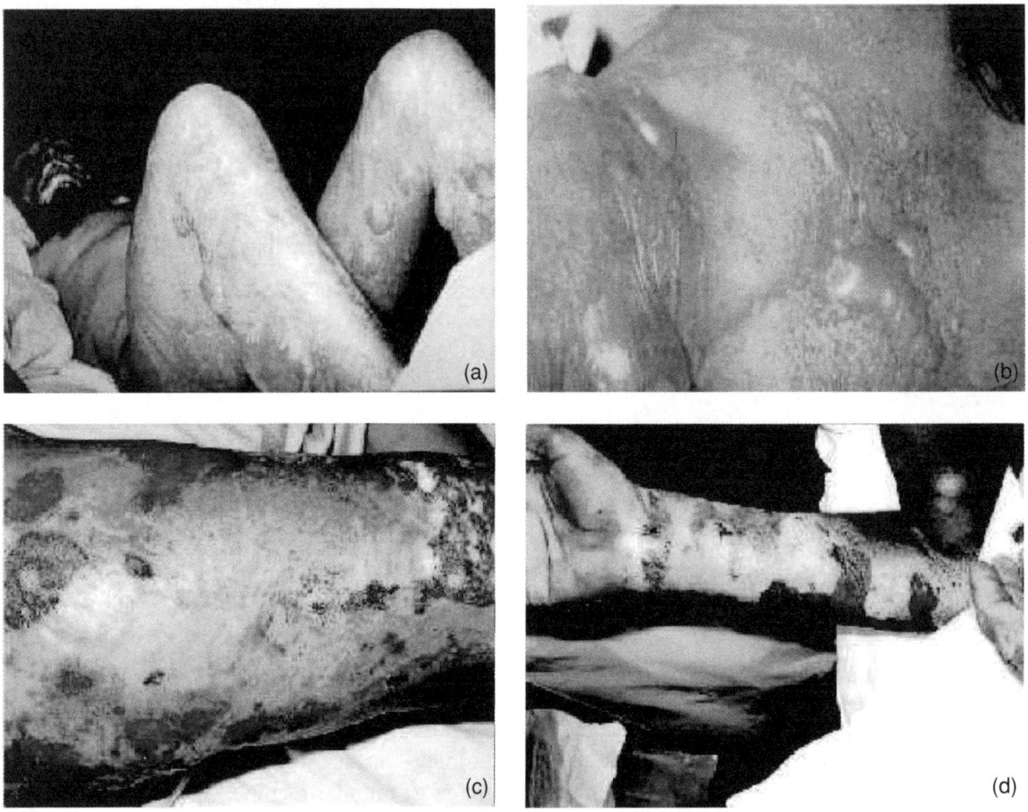

Figure 1. Illustrations of Iranian casualties from the Gulf War of 1984–1987. (a, b) Large fluid-filled blisters, characteristic of acute manifestation of sulphur mustard exposure of the skin. (c) A large sulphur mustard burn of the thigh following rupture of several large blisters and the early development of superficial infection of the resulting necrotic ulcers. (d) A partially healed mustard burn of the forearm showing the typical areas of peeling epidermis surrounded by zones of hypo- and hyper-pigmentation. © Crown Copyright, Dstl 2006

significance and, following initial development by Fritz Haber and industrial scale development by Lommel and Steinkopf (Papirmeister *et al.*, 1991), sulphur mustard was first deployed against the French troops at Ypres, Belgium on 12 July 1917 (Ireland, 1926).

During the first three weeks of its use, the British Army sustained more than 14 000 casualties, of who approximately 500 subsequently died (HMSO, 1923). By the end of the war, sulphur mustard was being extensively used by both sides and the overall casualty figure had risen to 400 000 according to some authorities; the British alone reported in excess of 140 000 casualties (Foulkes, 1934). Despite these appallingly high figures, the mortality rate did not exceed 2–3% (Haldane, 1925); the much higher figure of 13–14% reported following the Bari Harbour incident in 1943 is now thought to be atypical of sulphur mustard exposure and may have been compounded by fatalities as a result of direct blast injury (Alexander, 1947).

Since the use of sulphur mustard in the 1914–1918 war, there have been several allegations of its use around the world in numerous conflicts. Italian forces were reported to have used sulphur mustard against Abyssinian (Ethiopian) troops in 1936 and it was used, probably on several occasions, by the Japanese in China between 1937 and 1941 (Medema, 1986). Some evidence to support its use by Poland against Germany in 1939 and by Egypt against the Yemen in 1963 to 1967 has also been reported (Medema, 1986). Much more recently, well-founded allegations of

Table 1. Physicochemical properties of sulphur mustard (after Windholz and Budavari, 1983)

Parameter	Description/comments	
Chemical structure	$S(CH_2.CH_2.Cl)_2$	
	Molecular weight = 159	
Boiling point	215–217°C	
Melting point	14.4°C (for distilled mustard)	
Specific gravity	1.27	
Vapour pressure	Temperature (°C)	Pressure (mmHg)
	0	0.025
	14	0.070
	30	0.090
	40	0.450
Solubility	Sparingly soluble in water (0.68 $g\,l^{-1}$, $t_{0.5} = 5$ min)	
	In weak solutions, hydrolysed to thiodiglycol ($S(CH_2.CH_2.OH)_2$)	
Stability	Stable under normal conditions	
	Destroyed by strong oxidizing agents	

the use of mustard by Iraq against Iran (1984–1987) were made and largely substantiated by independent specialists acting on behalf of the UN Secretary General (United Nations Security Council, 1984,1986,1987). During the period of the Iran–Iraq conflict, numerous Iranian casualties whose injuries were compatible with exposure to sulphur mustard arrived for hospital treatment in several Western European countries, including the UK (Mandl and Freilinger, 1984; Pauser *et al.*, 1984; Colardyn *et al.*, 1986; Willems, 1989; Newman-Taylor, 1991).

CHEMICAL AND PHYSICAL PROPERTIES OF SULPHUR MUSTARD

The name 'mustard' was given to the compound by soldiers during World War I, apparently because of the smell perceived during gas attacks; since this time, the odour has been variously described as similar to that of garlic, mustard, horseradish and leeks. Sulphur mustard itself should not be confused with mustard oil (Parry, 1921; Reynolds, 1977); the latter is allyl isothiocyanate and, interestingly, is also a powerful vesicant. Mustard is often referred to by the letter 'H'; it can also be denoted as 'HS' (representing **H**un**S**toff or 'German stuff'), 'HD' to imply distilled mustard or 'LOST', believed to be a derivation of **LO**mmel and **ST**einkopf, the two German chemists responsible for its industrial-scale production. The compound remains referred to, particularly by the French, as 'Yperite' in view of its initial use at Ypres.

Sulphur mustard is an oily, colourless-to-brown liquid at room temperature and has the chemical name bis-(2-chloroethyl)sulphide; the chemical synonyms include 2,2′-dichlorodiethylsulphide, β,β'-dichloroethylsulphide, 1-chloro-2-(2-chloroethylthio)ethane and 1,1′-thiobis(2-chloroethane). Its physicochemical properties are summarized in Table 1. The vapour given off by a quantity of sulphur mustard has considerable penetrating ability; it rapidly passes through clothing to affect the underlying skin. Vapour will also penetrate substances such as wood and leather, albeit not as rapidly as cloth. Materials such as metal, glass and glazed ceramics are generally impervious; paint on metallic surfaces may, however, absorb mustard vapour and act as a potential source of a vapour hazard in the immediate vicinity.

THE INTERACTION OF SULPHUR MUSTARD WITH IMPORTANT BIOLOGICAL MACROMOLECULES

In order to fully appreciate the development of mustard skin injuries and the clinical problems associated with such lesions, it is important to have some understanding of the interactions between sulphur mustard and the important tissue macromolecules that result in cutaneous pathology. Sulphur mustard is a bifunctional alkylating agent capable of forming covalent linkages with nucleophilic groups in the cell (Mol *et al.*, 1989a). The ability of sulphur mustard to cross-link complementary strands of DNA has been extensively reviewed (Fox and Scott, 1980; Murname and Byfield, 1981), as well as its ability to bind to various important enzyme systems (Wheeler, 1962), collagen (Pirie, 1947) and keratin (Peters and Wakelin, 1947).

Reaction with DNA

The evidence that binding to DNA is a key mechanism underlying sulphur mustard's potential to cause cell injury has slowly accumulated since the elucidation of the structure and capacity for chemical reactions of DNA during the 1950s (Philips, 1950; Roberts *et al.*, 1968; Fox and Scott, 1980). The reaction with DNA is complex. As a bifunctional alkylating agent, sulphur mustard has two carbon chains that are capable of internal cyclization, a process which is necessary for alkylation to occur. Alkylation of complementary DNA bases by a single molecule of sulphur mustard is, therefore, possible and will lead to the formation of interstrand cross-links. The important chemical steps are summarized in Scheme 1. Alkylated guanine residues have a tendency to form base-pairs with thymine rather than cytosine (as normal), resulting in coding errors and inaccurate protein synthesis. This may ultimately lead to either non-production or excessive production of key metabolic enzymes and structural macromolecules. The effects on rapidly dividing tissues are particularly severe and have led to profound bone marrow suppression, gastrointestinal damage and spermatogenic arrest in human casualties (Willems, 1989).

Evidence is accumulating to suggest that damage to DNA is dependent on the mitotic state of the cell so that DNA has a differential sensitivity to sulphur mustard depending on what part of the cell cycle it has entered. Savage and Brekon (1981) have shown that 12–16 h after exposure to sulphur mustard, cultures of Syrian hamster fibroblasts produced a large number of chromatid aberrations due to substantial delay and disruption of the pre-S phase cells; most of these early S-phase cells failed to reach division within 36 h. The greatest depression in the rate of DNA synthesis occurs in the late G_1 and early S-phase, where the repair of DNA is severely reduced. Flow cytometric analysis of DNA has also shown blockage by sulphur mustard (and its monofunctional analogue, chloroethyl ethyl sulphide) in late G_1 and early S-phase of the cell cycle of cultured human keratinocytes (Gales *et al.*, 1989) and human peripheral blood lymphocytes (Sanders *et al.*, 1989).

It is evident that for DNA replication to occur, the supercoiled material constituting the chromosomes must unwind; this potentially increases the accessibility of the genetic bases to alkylating agents such as sulphur mustard, and involves numerous enzymes which are critical to this process. According to Pardee (1989), in late G_1-phase a number of enzymes involved in DNA synthesis appear; these include DNA polymerase (DP), cyclin (proliferative cell nuclear antigen), thymidine kinase (TK), ribonucleotide reductase (RR) and dihydrofolate reductase (DR). The enzymes are translated at cytoplasmic ribosomes and subsequently undergo translocation to the nucleus at the end of the G_1-phase. At the nucleus, the enzymes join together to form a multi-enzyme complex which contains RR and DR (required to catalyse precursor synthesis), TK (involved in thymidine salvage) and DP (required for the actual replication of the DNA strands). Finally, the DNA undergoes a conformational change at the replication origin. The G_1 to S-phase transition is, therefore, very sensitive to all processes which inhibit protein synthesis.

Once the cell has reached the S-phase, the DNA is replicated very precisely in a matter of a few hours by the initiation of bi-directional replication at numerous points in every chromosome; if replication is not completed in the S-phase, breakage occurs at subsequent mitosis (Lasky *et al.*, 1989). The replication of chromosomes also requires that the conformation (three-dimensional structure) as well as the activity/non-activity of genes within the chromosome are conserved. The alkylation of any of the highly sensitive machinery associated with the processes surrounding DNA replication may,

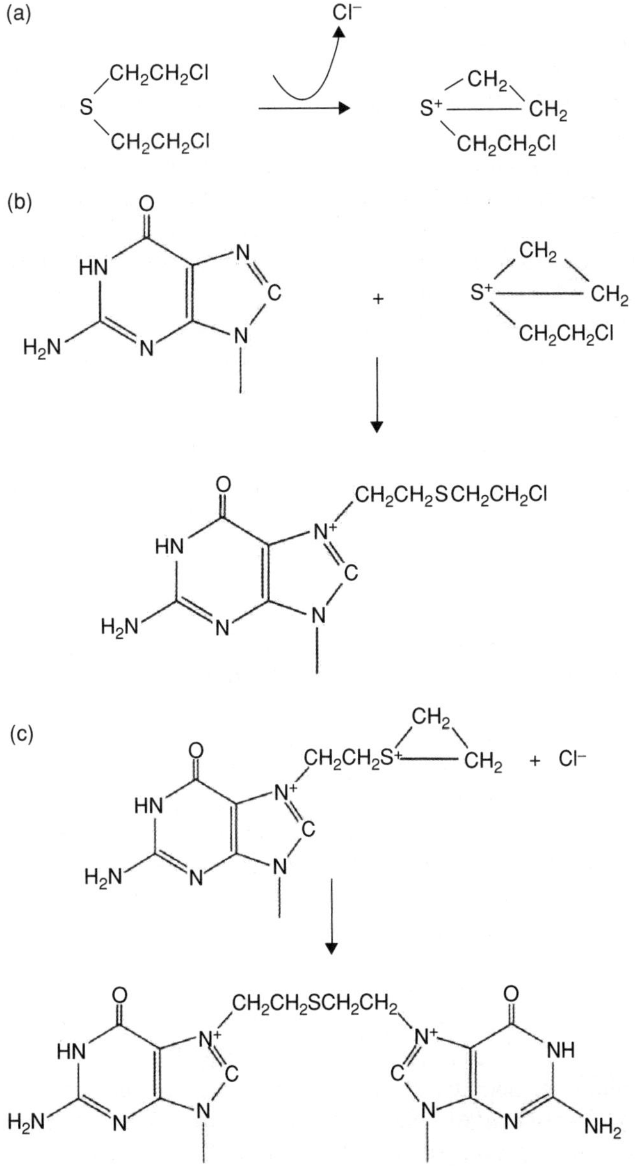

Scheme 1. Important chemical steps in the alkylation of complimentary DNA bases by a single molecule of sulphur mustard: (a) first cyclization reaction; (b) alkylation of DNA base; (c) second cyclization and alkylation of complimentary DNA base, resulting in a DNA cross-link

therefore, have a highly disruptive effect on the cell's metabolism.

Repair of DNA, whatever the initial cause, may also lead to further impairment of cellular function. An important aspect of the DNA repair mechanism involves the consumption of NAD^+ as a result of activation of poly (ADP-ribose)-polymerase (PRP), a DNA repair enzyme, in response to breaks occurring in the DNA backbone following alkylation of the DNA bases and the action of apurinic endonucleases. Other reports have also indicated that intracellular NAD^+ levels are decreased by DNA-damaging agents such as ionizing radiation, streptozotocin, neocarzinostatin and the mutagen, *N*-methyl-*N*′-nitro-*N*-nitrosoguanidine (Juarez-Salinas *et al.*, 1979).

PRP depletes cells of NAD^+ at vesicating doses of sulphur mustard, leading to the inhibition of glycolysis, stimulation of the NADP-dependent, hexose monophosphate shunt and cell death (Papirmeister *et al.*, 1984,1985). Stimulation of this latter enzyme pathway has also been associated with enhancement of protease synthesis and release, resulting in localized subepidermal blister formation in the skin (Smulson, 1989).

The seemingly central role of depletion of cellular NAD^+ in sulphur mustard-induced cutaneous injury has, however, recently been questioned. Gross *et al.* (1985) found that by pretreating an athymic nude mouse grafted with human skin with 3-aminobenzamide (an inhibitor of PRP), they could protect the graft from sulphur mustard challenge by maintaining cellular NAD^+ levels. In a similar experiment, Mol *et al.* (1989b) found that maintenance of NAD^+ levels in cultures of human epidermal cells did not protect the cells' energy metabolism, as measured by impairment of glucose uptake, when exposed to sulphur mustard. The situation has only become more confused by the results of an experiment conducted by Meier *et al.* (1987) who showed that treatment with nicotinamide (an NAD^+ precursor) could protect cultured human leucocytes from sulphur mustard challenge. This apparent dichotomy of results has been interpreted as being attributable to the relatively high rate of turnover of cultured epidermal cells when compared to leucocytes (Mol *et al.*, 1989b), and once again emphasizes the importance of the cell cycle to sulphur mustard-induced DNA damage.

Reaction with tissue proteases

The link between the biochemical effects of sulphur mustard – such as its reaction with DNA and the subsequent inhibition of glycolysis – and its vesicant action still remains poorly defined, although it is known that the local release of tissue proteases may damage the dermo–epidermal junction (Einbinder *et al.*, 1966; Kahl and Pearson, 1967; Briggamann *et al.*, 1984) and this idea has been incorporated into the mechanistic schema proposed by Papirmeister (Papirmeister *et al.*, 1985).

The involvement of proteases in the vesication process is becoming less speculative in the light of several papers that have established a crucial role for proteins such as plasminogen activator (PA) in pathological skin conditions such as psoriasis (Fraki *et al.*, 1983) and the blistering disorder of pemphigus (Hashimoto *et al.*, 1983; Singer *et al.*, 1985). Organ culture experiments have indicated that following exposure to sulphur mustard, tissue plasminogen levels increase and that the observed increases cannot be entirely explained in terms of either extravasation of serum or release from acute inflammatory cells (Woessner *et al.*, 1991). Furthermore, plasminogen has also been detected in the suprabasal, acantholytic cells of Darier's disease and in the degenerating basal cells of chronic cutaneous lupus erythematosis (Burge *et al.*, 1989); these observations suggest that the balance between protease activators and inhibitors, and between protease availability and location within the tissue, are important factors in the pathogenesis of these conditions.

Plasminogen activator (PA) is a serine protease which converts the proenzyme plasminogen into its active form plasmin by hydrolysing a single peptide bond in the precursor (Wun *et al.*, 1982). PA is found in a variety of cells and tissues, including keratinocytes. The hypothesis that keratinocyte PA may facilitate squame detachment (Myhre-Jensen and Astrup, 1971) is supported by the finding that cellular PA levels increase with increasing keratinocyte differentiation (Isserhof and Rifkin, 1983). According to Wilson and Reich (1978), PA secretion is a key mechanism underlying localized extracellular proteolysis.

PA is the main activator of latent collagenases in tissue (Werb and Gordon, 1975) and is able to induce the secretion of collagenase by fibroblasts as well as control the degradation of the extracellular matrix (ECM) and regulate the catabolism of tissue macromolecules (Liotta *et al.*, 1981). Human keratinocytes in culture synthesizes and secretes two inhibitors of PA (PA inhibitor I and II); the latter is able to prevent intraepidermal blister formation induced by pemphigus IgG autoantibody in human skin grown in organ culture (Hashimoto *et al.*, 1989). Given, therefore, that

PA is regulated by PA inhibitors (Leprince *et al.*, 1989), it follows that disruption of this negative control mechanism may lead to the loss of PA control *in vivo*. Any agent, such as sulphur mustard, which inhibits various components of cellular metabolism may also disrupt the proteolytic control mechanisms of the cell.

The role of inflammatory cells and the immune system in the release of tissue proteases

In addition to the direct effects which sulphur mustard may have in initiating proteolytic cascades, other sources of proteolytic enzymes have to be considered, the most important of these being the various inflammatory cell types that have been described as part of the sulphur mustard-induced cutaneous injury in various species (Papirmeister *et al.*, 1984; Vogt *et al.*, 1984; Mitcheltree *et al.*, 1989; Mershon *et al.*, 1990).

According to Sellers and Murphy (1981), in pathological situations involving significant infiltration of the tissues by neutrophils, elastase and cathepsin from these cells could contribute significantly to the lysis of collagen that is observed. Macrophages have the ability to degrade collagen intracellularly following phagocytosis of individual fibrils (Shoshan, 1981), and it is possible to speculate that these cells may play a role in the degradation of papillary dermal collagen that has been observed in rabbit skin following exposure to sulphur mustard (Knight and Rice, 1990, unpublished observations). Similar collagen degradation has also been observed in porcine skin following sulphur mustard vapour exposure using immunohistochemical techniques (Lindsay and Rice, 1995). Macrophages also secrete interleukin-1 (IL-l), which promotes the activation and multiplication of both B- and T-lymphocytes (Roitt *et al.*, 1988). On encountering the appropriate antigen, T-cells can secrete a preformed heparinase which degrades the heparin sulphate polysaccharide chains of the ECM proteoglycan scaffold; this allows T cells to traverse blood vessel walls and the endothelial ECM in order to reach their antigenic target sites.

Macrophages also possess a pericellular heparinase which facilitates their penetration of the ECM, allowing access to all extravascular compartments (Savion *et al.*, 1987). The exact roles for these cells and the immune system in the pathogenesis of sulphur mustard cutaneous injury remains largely unknown, although there is some evidence to suggest that the immunogenicity of various structural components of the skin, such as collagen, is altered following exposure to sulphur mustard (Berenblum and Wormall, 1939; Pirie, 1947; Jendryczko and Drozdz, 1985). Conversely, it is well known, that in high percutaneous and systemic doses, sulphur mustard is a powerful immunosuppressant (Vedder, 1925; Willems, 1989).

Reactions with the connective tissue matrices and the epidermal basement membrane

Having considered the potential role of proteases in sulphur mustard-induced injury, it is necessary to consider the likely targets for protease activity in the skin. The two most important targets in this respect are the connective tissue matrices of the upper papillary dermis and the basement membrane, and its associated fibrils that lie at the interface of the former and the overlying epidermis (the dermo–epidermal junction). The biochemical composition and ultrastructural organization of both of these structures have been extensively reviewed previously (Martinez-Fernandez and Amenta, 1983; Katz, 1984; Lunstrum *et al.*, 1986; Keene *et al.*, 1987; Bosman *et al.*, 1989) but have, to the author's knowledge, received little or no attention in relation to being possible targets following exposure to sulphur mustard.

THE CONNECTIVE TISSUE MATRIX

The connective tissue matrix of the papillary dermis is composed of collagens, elastin and other glycoproteins (Burgeson, 1988), as well as glycosaminoglycans (Lindahl and Hook, 1978). Normal adherent cells in culture, synthesize and deposit a connective tissue matrix which shows structural heterogeneity dependent on the cell phenotype and culture conditions (Leigh *et al.*, 1987). Type III collagen is essential for the normal tensile strength of the skin and accounts for

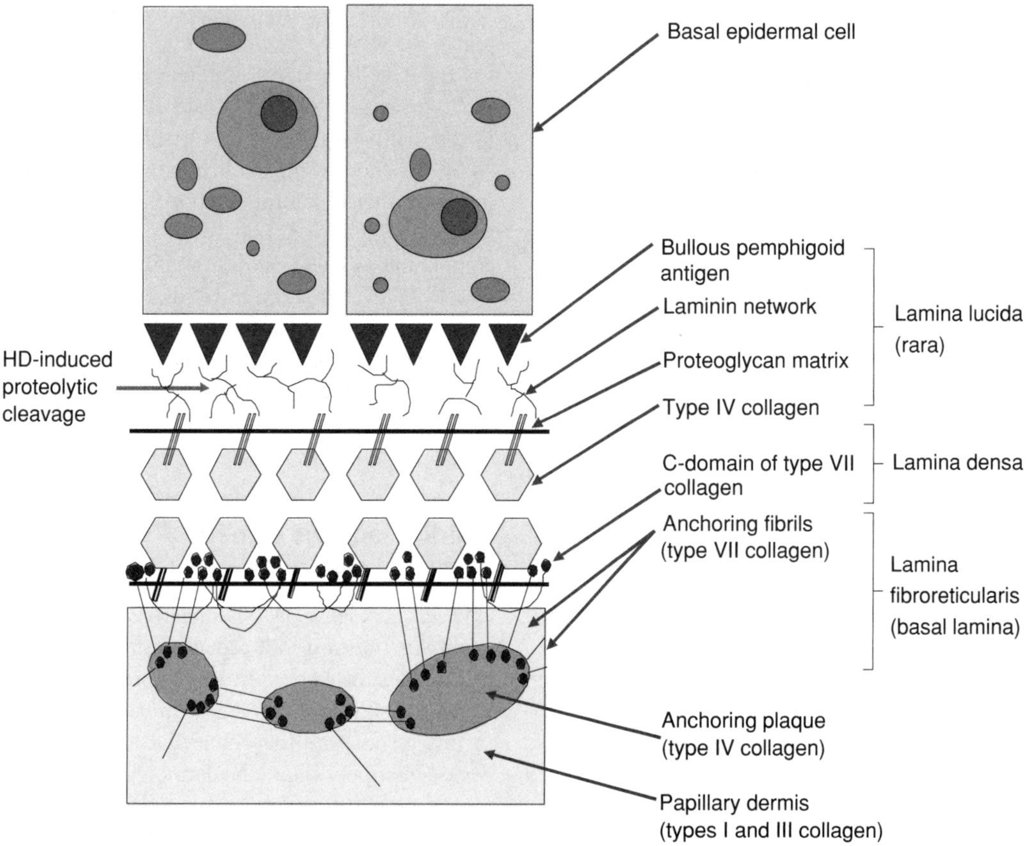

Figure 2. Diagrammatic representation of the molecular structure of the epidermal basement and the site of sulphur mustard-induced proteolytic cleavage that results in blister formation

10–50% of the total collagen content in adult tissues such as the skin, large arteries, skeletal muscle and the lung. Type I collagen is another major collagen type of the adult skin, and together with type III fibres, entrap the n-terminal domains of the anchoring fibrils and their associated plaques (Keene *et al.*, 1987).

THE BASEMENT MEMBRANE

Ultrastructurally, the basement membrane is composed of three distinct zones: the lamina lucida (or rara), the lamina densa and the lamina fibroreticularis (see Figure 2).

The lamina lucida is an electronlucent layer adjacent to the plasma membrane of the adherent basal epidermal cell and is composed largely of laminin. Bullous pemphigoid antigen (a disease-specific glycoprotein recognized by an antibody that circulates in the serum of patients with the disease) and a number of poorly defined antigens that are recognized by circulating antibodies in patients with herpes gestationis and scarring pemphigoid (Katz, 1984) have also been demonstrated in the lamina lucida. Delicate anchoring filaments traverse this layer and insert themselves into the attachment plaques of the hemidesmosomes of the overlying basal epithelial cells (Bosman *et al.*, 1989).

The lamina densa is an electron-dense layer situated at the stromal side of the lamina lucida and is composed of type IV collagen and the KF-1 antigen, a non-collagenous antigen which appears to be restricted to the basement membranes of stratified squamous epithelia and is completely absent in patients with the autosomal recessive condition of dystrophic epidermolysis bullosa (Katz, 1984; Burgeson, 1988). Recent evidence suggests that the antigen is the 450-kDa, globular C-domain of type VII collagen (Lunstrum *et al.*, 1986).

Finally, beneath the lamina densa is the lamina fibroreticularis or sub-basal lamina, a layer which appears to be restricted to basement membranes that support stratified, squamous epithelia (Bosman *et al.*, 1989). The layer is composed of elastic fibres that interdigitate with the type 1 and III collagen fibres of the underlying connective tissue matrix, and anchoring fibrils. The latter are composed predominantly of type VII collagen (Lunstrum *et al.*, 1986; Keene *et al.*, 1987; Leigh *et al.*, 1987) and stabilize the attachment of the basement membrane to the underlying stroma by forming a fibrous network; the C-terminal domains form globular structures which can be localized using immunoelectron microscopy of the inferior border of the lamina densa, cf. KF-1 antigen (Leigh *et al.*, 1987).

THE CONNECTIVE TISSUE MATRIX AND BASEMENT MEMBRANE AS TARGETS FOR SULPHUR MUSTARD ATTACK

The cells of the epidermis are anchored to the underlying dermis by a complex network of intra-cytoplasmic filaments, hemidesmosomes, anchoring fibrils and basement membrane components. Protease release into any region of the basement membrane could disrupt the components of this meshwork structure and, therefore, the structural integrity of the dermo–epidermal junction. This would, ultimately, lead to the separation of the epidermis from the dermis and vesicle (blister) formation. It is of interest to note that Mershon *et al.* (1990) have shown ultrastructural evidence of disruption of the anchoring fibrils, an observation that confirms the earlier reports of a similar lesion by Papirmeister *et al.* (1984). Careful inspection of the ultrastructural nature of these disruptions indicates that they are consistent with proteolytic cleavage of the type VII collagen molecule in the vicinity of the globular C-domain. Futhermore, studies performed in this author's laboratory that have examined the molecular changes in the epidermal basement membrane following mustard vapour exposure using immunohistochemical techniques, have confirmed that a key molecular event that leads to dermo–epidermal separation in early blister formation is the proteolytic cleavage of the laminin network of the lamina lucida (rara) region of the basement membrane (Lindsay and Rice, 1995) (Figure 2).

Sulphur mustard and its effects on important cutaneous enzyme systems

Most of the work on skin biochemistry in relation to sulphur mustard burns was stimulated by the enzyme theory of vesication (Peters, 1936). It remains to be determined whether (a) injury is primarily dependent upon a direct reaction of the agents with one or a few specific and highly important proteins or other molecular species, or (b) there is a more or less random reaction of the vesicant with reactive groups of numerous proteins producing the observed pathological effects.

Until fairly recently there was no convincing evidence that inactivation of any one enzyme or group of related enzymes has a causal relationship to the development of vesicant-induced burns. However, most of the earlier work concentrated on enzymes restricted to carbohydrate metabolism and many of the enzymes known today were undiscovered at the time of these early observations, e.g. adenosine triphosphatase (Jørgensen *et al.*, 1971). Even if it is assumed that vesicants act by alkylating or disrupting a single enzyme or group of similar enzymes, since the reaction time, i.e. exposure to full alkylation, is so short, therapy would need to be directed towards prophylaxis rather than treatment.

OXYGEN CONSUMPTION AND GLYCOLYSIS

In 1935, Berenblum showed that cutaneous applications of dichloroethyl sulphide, dichlorodiethyl sulphone and other vesicants inhibited the development of cutaneous tumours evoked by the application of coal tar. Subsequently, he demonstrated that sulphur mustard *in vitro* caused a limited inhibition of oxygen consumption by minced tumour tissue and a pronounced inhibition of its aerobic and anaerobic glycolysis (Berenblum *et al.*, 1936).

AEROBIC METABOLISM

Normal skin is an actively respiring tissue; in young rats, oxygen consumption of the skin can be as high as 4–5 $mm^3\ h^{-1}\ mg^{-1}$ of dry tissue (Needham and Dixon, 1941). Significant but limited inhibition of the basic respiratory rate does develop during the first 1–2 h after poisoning with 20% sulphur mustard in alcohol (Thompson, 1940). Oxygen consumption of untreated skin is not influenced by the addition of glucose (Needham and Dixon, 1941) but is markedly increased by succinate (Thompson, 1940). Aerobic glycolysis of normal rat skin is low and unaffected by pyruvate. About 15 h after treatment with sulphur mustard, there is no significant change in the rate of acid production in the presence of glucose; however, after 3 h a marked decrease has occurred which is out of proportion to the fall in oxygen consumption (Needham and Dixon, 1941).

ANAEROBIC GLYCOLYSIS

Under anaerobic conditions, normal excised rat skin and skin extracts in the absence of added substrate produce lactic acid at a low rate. After treatment with sulphur mustard *in vivo,* residual glycolysis, i.e. no added substrate, of excised but otherwise intact skin remains unaltered. However, if glucose is added to the system, its conversion to lactic acid is almost completely inhibited (Dixon, 1943).

Tests of a large number of vesicant and non-vesicant substances has demonstrated a good correlation between the skin-damaging action of mustard and its inhibitory effect on the glycolysis of glucose by rat skin (Wheeler, 1962).

OBSERVATIONS AND INTERPRETATIONS IN TERMS OF SKIN ENZYME SYSTEMS

The initial phosphorylation stages mediated by hexokinase are inhibited by sulphur mustard and other vesicant compounds (Dixon, 1943). Dixon demonstrated that treatment of skin with sulphur mustard *in vivo* results in the inhibition of the hexokinase system after a latency and according to a time-scale corresponding to that of glycolytic inhibition and the development of gross injury (12–24 h). This parallel between hexokinase inhibition and the fall of anaerobic glycolysis on the one hand, and the development of serious injury on the other remains striking, but it is possible that these alterations are co-phenomena rather than cause and effect (Renshaw, 1940).

Inhibition of glycolysis in the rat by sulphur mustard is reversed by the chelating agent 2,3-dimercaptopropanol (British Anti-Lewisite or BAL) (Barron *et al.*, 1948), but it is known that long continued application of BAL, although preventing blister formation on H-treated human skin, does not prevent cellular injury or death. The possible relation of the pyruvate oxidase system to vesication was investigated by Peters (1936). He showed that the oxidation of pyruvate in chopped brain preparations is strikingly inhibited by sulphur mustard and other vesicant agents, including nitrogen mustard (HN1), and that the effect did not depend upon inactivation of vitamin B_1 or glutathione. The significance of this enzyme system in the skin is as yet undetermined.

Cholinesterase is present in the skin and has been found to be inactivated by cutaneous applications of sulphur mustard and methyl *N*-(β-chloroethyl)-*N*-nitroso-carbamate (Thompson, 1942). The relevance of this to injury is not known.

THE HISTOPATHOLOGY OF VESICANT-INJURED SKIN

The gross and microscopic descriptions of human skin injured by vesicant agents, and sulphur mustard in particular, have not altered in detail since the original descriptions by Warthin and Weller in 1919 and Cullumbine in 1947. Application of the relatively new techniques of enzyme histochemistry, molecular biology and immunohistochemistry to the study of the pathogenesis and resolution of mustard injury have only just begun. Compared to the research into the aetiology of similar human blistering diseases, such as epidermolysis bullosa (Eady, 1987; Lin and Carter, 1989) and lichen planus pemphigoides

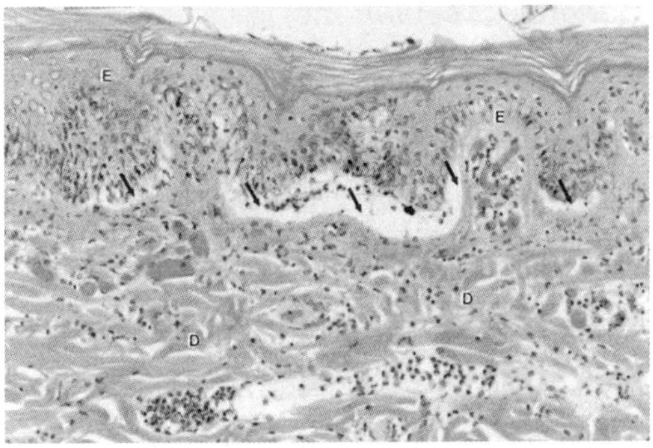

Figure 3. The early stages in the development of a sulphur mustard vapour injury in the Yucatan miniature pig at 24 h post-exposure. The cells comprising the epidermis (E) already show advanced degeneration, nuclear pyknosis and focal cytoplasmic vacuolation. The individual collagen bundles of the dermis (D) are forced apart by the accumulation of oedema fluid. Note the heavy acute inflammatory cell infiltrate throughout the dermis (D) and the epidermis (E) is focally separating from the underlying papillary dermis (indicated by arrows) (Harris' Haematoxylin & Eosin × 200). © Crown Copyright, Dstl 2006

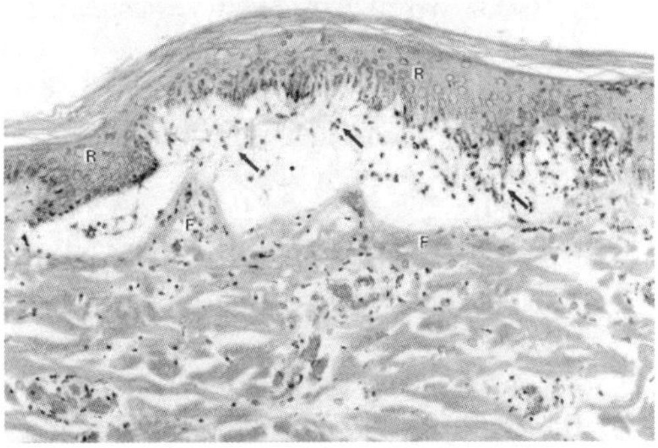

Figure 4. The fully developed microblister at 72 h post-exposure. The roof (R) of the blister is composed of intact but degenerate epidermal cells and the floor (F) of degenerate papillary dermal collagen and thrombosed, necrototic dermal papillary capillaries. The blister cavity contains strands of cytoplasmic debris, red cells and occasional acute inflammatory cells (indicated by arrows) (Harris' Haematoxylin & Eosin × 200). © Crown Copyright, Dstl 2006

(Gawkrodger *et al.*, 1989; Wilsteed *et al.*, 1991), the elucidation of the molecular pathology of vesicant-induced blistering is in its early stages (Lindsay and Rice, 1995).

Since the early descriptions of human casualties, there have been numerous studies of vesication in experimental animals, the hope being the discovery of a model system in which the development and healing of mustard-induced cutaneous injury can be studied. Recently, three animal models have been proposed that have renewed optimism in finding appropriate methods to study the interaction between mustard and the skin. Papirmeister *et al.* (1984) have described well-formed blisters in human skin grafted to athymic nude mice, whilst Mershon *et al.* (1990) have produced microblisters in the skin of euthymic, hairless guinea-pigs. The latter model responds in very similar ways to sulphur mustard challenge in domestic pigs, a model recently exploited by Mitcheltree *et al.* (1989), Lindsay and Rice (1995) and Brown and Rice (1998) (Figures 3 and 4). These three models are currently the only ones available.

Sulphur mustard injury in human skin

GROSS PATHOLOGY

Perhaps the best gross descriptions of human skin lesions are those of casualties from the World War I (Warthin and Weller, 1919; Schrafl, 1938; Cullumbine, 1947). More recently, this author has had the opportunity to observe the gross lesions and microscopy of several Iranian casualties injured during the recent conflict in the Iran-Iraq War (Rice and West, 1987 unpublished observations). The appearances of such injuries may be summarized as follows;

(a) The lesion is a chemical burn and with appearances unlike those seen in thermal, electrical or corrosive (acid/alkali) burns. There is little or no thrombosis of vessels, but a great degree of moistness of the affected area. The coagulated appearance of thermal injuries is not a feature of vesicant injury (Pearson, 1964; Takigawa and Ofuji, 1977).
(b) The skin at first is pale but then becomes erythematous within a few hours of exposure. Vesication is not usually seen until the second day and progresses, thereafter, for several more days.
(c) Scab formation begins within seven days once the early blisters begin to degenerate.

The skin may be made to vesicate in areas of erythema by slight trauma, e.g. on rubbing, and this phenomenon is known as 'Nikolsky's sign'. It does not imply the persistence of active vesicant (Sulzberger and Katz, 1943).

(d) Four to six days after exposure, necrosis is complete and separation of necrotic slough begins. The accompanying oedema and erythema may persist.
(e) By 16–20 days, separation of slough is complete and re-epithelialization has begun. Healing may take 3–8 weeks post-exposure to be complete and the casualty is often left with depigmented areas surrounded by zones of hyperpigmentation.

Based on these descriptions, it is possible to discern two striking differences between the lesions seen in animals and those observed in man. First, the production of grossly visible blisters occurs regularly in man but not in animals. These blisters may enlarge by coalescence and may appear several days after the removal of the casualty from the affected environment. Secondly, the healing time in man is considerably longer than that seen in animals (Schrafl, 1938; Renshaw, 1940). Several attempts have been made to explain the differences between animal and human skin with respect to vesication, and these have been alluded to previously (Flesch *et al.*, 1952; McAdams, 1956).

MICROSCOPY AND ULTRASTRUCTURAL OBSERVATIONS OF HUMAN VESICANT INJURIES

Although microscopic descriptions of World War I vesicant agent-induced lesions exist, more recently Papirmeister *et al.* (1985) have described a model (full-thickness human neonatal foreskin grafted to congenitally athymic nude mice) which appears to reproduce very accurately the findings previously described. This model has also allowed lesions to be studied at the ultrastructural level in an attempt to reveal additional information about the exact histogenesis of vesicant-induced cutaneous injuries. The observations made from this model can be summarised as follows.

General Observations

The severity of cutaneous injury appears to depend on the degree of alkylation occurring in the skin (Renshaw, 1940). Several dose regimens of liquid sulphur mustard were used, based on data derived from earlier studies by Renshaw (1947), viz:

$20\ \mu g\ cm^{-2}$ = mild injury
$60–120\ \mu g\ cm^{-2}$ = moderate injury
$635\ \mu g\ cm^{-2}$ = severe injury

At these doses, no systemic effects were noted as the maximum dose only represented approximately 25% of the percutaneous LD_{50} dose.

Microscopic Observations

The microscopic observations can be summarized as follows:

(a) There were minimal changes at 4 h, consistent with the known latency for sulphur mustard cutaneous injuries. By 7 h, there was evidence of vacuolation of basal keratinocytes which had progressed to focal necrosis accompanied by congestion and oedema of the dermis by 12 h.
(b) Widespread, multifocal necrosis of the epidermis accompanied by early acute inflammation was noted at 24 h. By 48 h, full-thickness necrosis had obliterated all normal cellular features, and there was a striking polymorph infiltrate at the base of the graft.
(c) Animals of the high-dose group showed dermo–epidermal separation at 24 h with the formation of a subepidermal blister at 48 h. The only deviation from true human injury was the impression of almost complete re-epithelialization of ulcerated lesions by 6 days post-exposure.

Ultrastructural Observations

The use of the electron microscope confirmed and extended the observations outlined above and affirmed that the development of the injury was both dose- and time-dependant. A definite sequence of morphological changes was described:

(a) Condensation of heterochromatin and loss of euchromatin.
(b) Blebbing of the nuclear membrane with the formation of perinuclear vacuoles.
(c) Swelling of the endoplasmic reticulum and disintegration of polysomes.
(d) Loss of the integrity of the cell membrane.
(e) Leakage of organelles into the extracellular space.
(f) Disruption of the anchoring filaments of the hemidesmosomes.

The authors concluded that sulphur mustard injury to human skin commences at the level of the basal keratinocyte and thus confirmed the original theories of McAdams (1956). They also drew some parallels with thermal injury, which is also thought to act by disrupting the basal epidermal layer (Cullumbine, 1947). These ultrastructural observations and the belief that blistering results from dermo–epidermal separation at the level of the epidermal basement membrane have now been confirmed by immunohistochemical studies (Lindsay and Rice, 1995).

CLINICAL MANAGEMENT

There is no specific therapy for sulphur mustard poisoning; the sole aim of clinical management in such cases is to maintain vital organ systems and alleviate symptoms. Skin burns can be severe and may involve extensive areas of the body surface. The naturally moist areas of the body, such as the genitalia, perineal regions, groins, lower back and axillae, often prove to be the most severely affected areas and crops of fresh blisters may appear at any time up to 2 weeks after exposure. The burns themselves tend to be superficial and will heal slowly without active treatment. However, experience in the clinical management of several Iranian casualties from the Iran–Iraq War (1984–1987) demonstrated that those with severe burns will require weeks of hospital care followed by lengthy convalescence and that, despite the superficial nature of the burn, it is all too easy to underestimate the period of care for such patients.

The current clinical management strategy

The current clinical management of sulphur mustard cutaneous injury is essentially that for a similar degree of thermal burn (Mellor *et al.*, 1991) but it is always important to bear in mind that the signs and symptoms of injury will not be evident for several hours after exposure. The overall management can be summarized as follows:

(a) For areas of erythema and minor blistering, bland lotions such as calamine are useful.
(b) Topical bacteriostatic agents such as 1% silver sulphadiazine (Flamazine) cream were used on Iranian casualties to reduce the incidence of secondary infection once the blisters had ruptured.
(c) Moderately severe pain and itching are common problems once blisters have developed and may be managed by the use of mild analgesics, antihistamines and small doses of diazepam. Occasionally, some cases experience severe pain and these may require narcotic analgesics such as morphine. Newman-Taylor (1991) reported that carbamazapine proved valuable in alleviating pain in one patient and that its use allowed the withdrawal of narcotic analgesics.
(d) Dilute topical steroids have proved beneficial in relieving irritation and reducing the attendant oedema at exposed sites; the use in human casualties appeared to have little or no effect on the subsequent rate of healing of the lesions, so confirming the earlier observations made by Vögt *et al.* (1984).
(e) Fluid replacement is calculated in the same way as for a thermal burn although unlike a thermal burn, large amounts of fluid loss will only occur once the blisters have formed, rather than in the first 24 h.
(f) Although the time to healing may be long, the evidence suggests that the eventual scar is softer and more pliable than that seen in thermal injuries. Wound contracture does not appear to be a major problem in this context, despite the predilection for the naturally moist areas, such as the axillae and groin.
(g) Numerous other drugs and regimes, including bathing in fresh human breast milk, have

been suggested (Hendrickx and Hendrickx, 1990), but there is no evidence that these have any therapeutic value in established cases.

Post-exposure surgical intervention in the management of mustard burns

Based on our previous experiences with Iranian casualties and the previous literature (Shrafl, 1938; Renshaw, 1940; Willems, 1989), it was recognized that large, full-thickness burns heal very slowly and may ultimately require skin grafting to achieve epithelial coverage of the ulcerated site. The success of grafting, assessed either by an improvement in the healing time or by the survival of the graft, will be determined by the efficiency of the tangential excision in cutting back to healthy tissue not showing sulphur mustard-induced damage. To ensure that excision is complete, both the depth and lateral extent of alkylation in the skin would need to be determined.

Histological examination of sulphur mustard-induced skin injuries in an established pig model (Brown and Rice, 1997,1998) has shown that the delayed rate of healing may be due, in part, to two distinct mechanisms:

(a) Alkylation of epidermal cells extends beyond the immediate region of exposure; although cells in this area may not ultimately die, the level of alkylation may be sufficient to delay or even prevent cell replication. Re-epithelialization of ulcerated lesions relies partly on the replication of cells from the undamaged epidermis at the edge of the lesion and partly from intact hair shafts in the base (Willems, 1989).

(b) In addition to achieving effective epidermal regeneration, re-epithelialization itself is dependent on the presence of an appropriate substrate on which regeneration can occur (Woodley *et al.*, 1985; Shakespeare and Shakespeare, 1987). The papillary dermis and basement membrane are vital in this respect and not only provide a structural scaffold for the epidermis but also act as signals for the subsequent differentiation of the overlying epidermis (Fleming, 1991). Immunohistochemical staining of the papillary dermis in pigs has shown that the collagen at this site is altered by exposure to sulphur mustard, and in this altered state, it may no longer function normally (Lindsay and Rice, 1995).

Over the last few years a number of wound-healing studies have been undertaken in the author's laboratory that have been directed towards establishing a more normal rate of wound healing by removing alkylated and sublethally injured epidermis and superficial papillary dermis. Rather than resorting to formal surgical excision, the technique of dermabrasion has been employed; this consists of simply scrubbing the wound surface under general anaesthesia with a sterile abrasive surface until punctate bleeding from the deeper, viable tissues is achieved. Currently, the technique is used in cases of severe facial acne and is well recognized in the management of deep dermal thermal burns (Holmes and Rayner, 1984).

Studies using a rotating abrasive surface on a hand-held electric drill [Figure 5(a)] to achieve superficial dermabrasion of established small, circular mustard vapour burns in our porcine model have shown that the healing time of such injuries can be reduced from 12 to 3–4 weeks (Rice *et al.*, 2000) [Figure 5(b)].

Following these initially encouraging results obtained through mechanical dermabrasion and the more recent advances in laser technology within the medical and surgical spheres, we have more recently performed (and reported) studies that have examined the role of surgical lasers to achieve enhancement of the healing rate of both sulphur mustard and Lewisite burns through vapourising the necrotic burn surface.

The studies employed both erbium:YAG and pulsed CO_2 surgical lasers [Figure 6(a)] to resurface established vesicant burns under general anaesthesia at between three and four days post-exposure. Sulphur mustard burns treated by pulsed CO_2 laser debridement showed a three-fold increase in healing rate compared to controls (untreated) at two weeks post exposure ($p = 0.061$) [Figure 6(b)]. This increase was achieved with only partial removal of necrotic tissue and may be due to disruption of the eschar

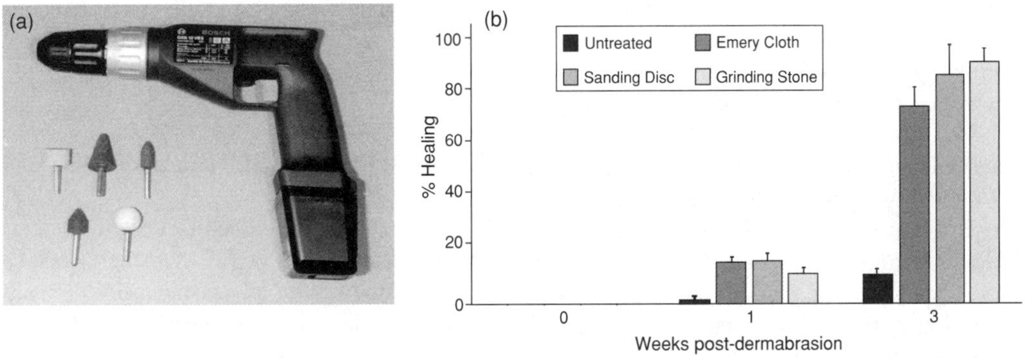

Figure 5. (a) An example of the hand-held drill used to perform mechanical dermabrasion. © Crown Copyright, Dstl 2006. (b) Comparative rates of epithermal healing for three different methods of dermabration of established mustard vapour burns in pigs

– wound bed interface either allowing earlier migration of regenerating epidermis or facilitation of phagocytosis of dead tissue as a result of a reduction in the eschar volume (Rice *et al.*, 2000). Similar results were achieved with lewisite vapour-induced burns, with a four-fold increase in healing rate at one week post-surgery (Lam *et al.*, 2002). Previous studies have shown that early surgical dermabrasion increases healing rates in mustard burns (Eldad *et al.*, 1998). Laser debridement offers the following advantages over other methods of physical debridement:

(a) Reduction of intra-operative bleeding
(b) Minimal debridement of normal skin surrounding the burn
(c) Reduced risk of aerosolized pathogens
(d) It is technically easier in relatively inaccessible parts of the body.

On the battlefield, the infection risk of an open burn-wound is very high. The zone of residual thermal damage induced by the CO_2 laser has the additional effect of sealing the wound surface, minimizing secondary infection. The zero infection rate in this study is encouraging when compared to previous dermabrasion studies of mustard burns in the same animal model in which a small number of infections occurred despite topical antimicrobials (Brown *et al.*, 1998).

Theoretically, the residual thermal damage produced by the laser may deepen a burn

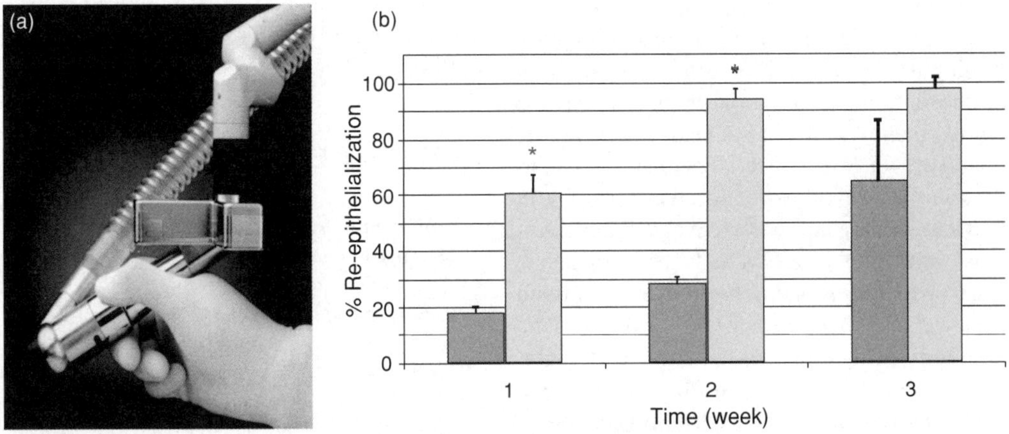

Figure 6. (a) Ultrapulse CO_2 laser. © Crown Copyright, Dstl 2006. (b) Heating rates of control and laser-ablated sulphur mustard burns lesions in the 'large white pig': *, $p < 0.061$: ▒, control; □, lasablation

sufficiently for it to require skin grafting. Should this be required, however, the zone of residual thermal damage produced by modern lasers has been shown to be sufficiently shallow to permit split skin grafting with excellent engraftment rates (Glatter *et al.*, 1998; Sheridan *et al.*, 1999).

Autoradiographic studies have shown that mustard localizes preferentially around hair follicles (Axelrod and Hamilton, 1946). This observation has led to the theory that mustard penetrates deeper along skin adnexae, damaging skin organelles more profoundly than elsewhere in the skin. However, in this study we found abundant evidence of epidermal regeneration from hair follicles. In this respect, this observation mirrors the slow epithelial regeneration from hair follicles in partial-thickness mustard burns in Iranian casualties from the Iran–Iraq War (Willems, 1989). A possible explanation is that as a result of its high lipid solubility, mustard may be retained in sebum rather than entering the epidermal cells, a solvent-partitioning effect.

The use of sulphur mustard in conflict is expected to produce large numbers of casualties with incapacitating partial-thickness burns, clogging up the medical support chain. These patients represent a huge potential wound-care problem. The increase in healing rates in this study was achieved through only a modest reduction in necrotic volume. In a mass chemical casualty situation, initial laser ablation of smaller lesions could potentially be safely delegated to appropriately trained paramedical staff, releasing medical and surgical staff to treat other casualties.

The control lesions in this study showed 65% healing at three weeks. This is in marked contrast to previous studies of partial-thickness mustard burns which were only 18% healed at this time-point (Brown *et al.*, 1998). This may be due to the semi-occlusive dressings used in this study. In previous studies, non-occlusive absorbant dressings were employed. Wounds re-epithelialize faster under moist conditions (Winter, 1962), and the use of occlusive dressings following cosmetic laser resurfacing promotes enhanced cell migration and diminished eschar formation (Collawn, 2000). Faster healing of the control lesions in this study suggests that early application of an occlusive dressing alone may accelerate healing without the requirement for additional surgery. We are currently planning studies to evaluate the effect of early application of occlusive dressings on the healing of mustard-induced burns.

SUMMARY AND CONCLUSIONS

Mustard gas was developed as a chemical warfare agent more than eighty years ago. Despite intensive study, little advancement in the management of skin burns has been made. The use of mustard in military conflicts and by terrorists remains a significant threat that if realized in practice would result in a large number of casualties with severely incapacitating, partial-thickness burns. Such injuries clearly present a huge potential wound-care problem.

Based on a number of relatively simple studies and careful, meticulous observation of the way blister agent burns develop and subsequently heal, a post-exposure strategy has been formulated which appears to overcome some of the clinical problems associated with this type of injury.

The techniques of mechanical dermabrasion and 'lasablation' represent notable advances in the management of chemical agent burns. In addition to their use in a military context, it seems likely that such procedures would similarly benefit the management of civilian chemical and thermal injuries to the skin.

The clinical value of such an approach will not be confirmed unless and until sulphur mustard is used again in anger. Despite significant progress under the terms of the Chemical Weapons Convention (CWC), such use cannot be ruled out.

REFERENCES

Alexander SF (1947). Medical report of the Bari harbour mustard casualties. *Milit Surg*, **101**, 1–17.

Axelrod DJ and Hamilton JG (1946). Radio-autographic studies of the distribution of Lewisite and mustard gas in skin and eye tissue. *Am J Pathol*, **23**, 389–411.

Barron ESG, Meyer I and Miller ZB (1948). The metabolism of the skin. Effects of vesicants. *J Invest Dermatol*, **11**, 97–118.

Berenblum I and Wormall A (1939). The immunological properties of proteins treated with b,b′-

dichlorodiethyl-sulphide and b,b′-dichlorodiethyl sulphone. *Biochem J*, **33**, 75–80.

Berenblum I, Kendal LP and Orr IW (1936). Tumour metabolism in the presence of anti-carcinogenic substances. *J Pathol Bacteriol*, **41**, 709–715.

Bosman FT, Cleutgens I, Beek C *et al.* (1989). Basement membrane heterogeneity. *Histochem J*, **21**, 629–633.

Briggamann RA, Schechter NM, Fraki I *et al.* (1984). Degradation of the epidermo–dermal junction by proteolytic enzymes from human skin and polymorphonuclear leucocytes. *J Exp Med*, **160**, 1027–1042.

Brown RFR and Rice P (1997). Histopathological changes in Yucatan minipig skin following challenge with sulphur mustard. A sequential study of the first 24 hours following challenge. *Int J Exp Pathol*, **78**, 9–20.

Brown RFR and Rice P (1998). Histopathological changes in large white pig skin following challenge with sulphur mustard. *J Cell Pathol*, **3**, 43–56.

Brown RFR, Rice P and Bennett NJ (1998). The use of laser Doppler imaging as an aid in clinical decision making in the treatment of vesicant burns. *Burns*, **24**, 692–698.

Burge SM, Williams-Cederholm S and Ryan TI (1989). Plasminogen and plasminogen activators in human skin in health and disease. *Br J Dermatol*, **120**, 307.

Burgeson RE (1988). Basement membranes. *Annu Rev Cell Biol*, **4**, 551–557.

Colardyn F, De Keyser H, Ringoir S *et al.* (1986). Clinical observation and therapy of injuries with vesicants. *J Exp Clin Toxicol*, **6**, 217–246.

Collawn SS (2000). Occlusion following laser resurfacing promotes re-epithelialisation and wound healing. *Plast Reconstruct Surg*, **105**, 2180–2189.

Cullumbine H (1947). Medical aspects of mustard gas poisoning. *Nature (London)*, **159**, 151.

Dixon M (1943). *The Phosphokinase Theory of Vesication; Its Present Position.* Report No 19 (Y7483). Cambridge, UK: Cambridge Biochemistry Laboratory.

Dixon M and Needham DM (1946). Biomedical research on chemical warfare agents. *Nature (London)*, **158**, 432–438.

Eady RA (1987). Babes, blisters and basement membranes : from sticky molecules to epidermolysis bullosa. *Clin Exp Dermatol*, **12**, 161–170.

Einbinder IM, Walzer RA, Mandl I (1966). Epidermodermal separation with proteolytic enzymes. *J Invest Dermatol*, **46**, 492–504.

Eldad A, Weinberg A, Breiterman S *et al.* (1998). Early non-surgical removal of chemically injured tissue enhances wound healing in partial thickness burns. *Burns*, **24**, 166–172.

Fleming S (1991). Cell adhesion and epithelial differentiation. *J Pathol*, **164**, 95–100.

Flesch P, Goldstone SB and Weidman FD (1952). Blister formation and separation of the epidermis from the corium in laboratory animals. *J Invest Dermatol*, **18**, 187–192.

Foulkes CH (1934). *Gas. The Story of the Special Brigade*. London, UK: William Blackwood and Son.

Fox M and Scott M (1980). The genetic toxicology of nitrogen and sulphur mustard. *Mutat Res*, **75**, 131–168.

Fraki IE, Lazarus GS, Gilgor RS *et al.* (1983). Correlation of epidermal plasminogen activator activity with disease activity in psoriasis. *Br J Dermatol*, **108**, 39–44.

Gales KA, Gross CL, Krebs RC *et al.* (1989). In: *Proceedings of the 1989 Medical Defence Bioscience Review (PMBR)*, USAMRDC, 15–17 August 1989, pp. 437–440. MD: USAMRICD.

Gawkrodger DI, Stavropoulos PG, McLaren KM *et al.* (1989). Bullous lichen planus and lichen planus pemphigoides – clinicopathological comparisons. *Clin Exp Dermatol*, **14**, 150–153.

Glatter RD, Goldberg JS, Shoemacker KT *et al.* (1998). Carbon dioxide laser ablation with immediate autografting in a full-thickness porcine burn model. *Ann Surg*, **228**, 257–265.

Gross CL, Meier HL, Papirmeister B *et al.* (1985). Sulfur mustard lowers nicotinamide adenine dinucleotide levels in human skin grafted to athymic nude mice. *Toxicol Appl Pharmacol*, **81**, 85–90.

Guthrie F (1860). On some derivatives from the olefins. *Chem Soc (London)*, **XII**, 109.

Haldane JBS (1925). *Callinicus, A Defence of Chemical Warfare*. London, UK: Kegan, Paul, Trench, Trubner.

Hashimoto K, Shafran KM, Webber PS *et al.* (1983). Anti-cell surface autoantibody stimulates plasminogen activator activity of human epidermal cells. *J Exp Med*, **157**, 259–272.

Hashimoto K, Wun T-C, Baird J *et al.* (1989). Characterisation of keratinocyte plasminogen activator inhibitors and demonstration of prevention of pemphigus IgG-induced acantholysis by a purified plasminogen activator inhibitor. *J Invest Dermatol*, **92**, 310–314.

Hendrickx A and Hendrickx B (1990). Management of war gas casualties. *Lancet*, **336**, 1248.

HMSO (1923). *History of the Great War – Medical Services. Diseases of the War*, Volume II, p. 291. London, UK: HMSO.

Holmes JD and Rayner CR (1984). The technique of late dermabrasion for deep dermal burns. Implications for planning treatment. *Burns*, **10**, 349–354.

Ireland MM (1926). Medical aspects of gas warfare. In: *The Medical Department of the United States Army in the World War*, Volume XIV, Washington, DC, USA: Government Printing Office.

Isserhof RR and Rifkin DB (1983). Plasminogen is present in the basal layer of the epidermis. *J Invest Dermatol*, **80**, 217.

Jendryczko A and Drozdz M (1985). Action of phenylalanine mustard on collagen *in vivo*. *Biomed Biochim Acta*, **44**, 497–501.

Jørgensen PL, Skou JC and Solomonson LP (1971). Purification and characterisation of ($Na^{+}+K^{+}$)-ATPase. *Biochim Biophys Acta*, **233**, 381–394.

Juarez-Salinas H, Sims JL and Jacobson MK (1979). Poly (ADP-ribose) levels in carcinogen-treated cells. *Nature (London)*, **282**, 740–741.

Kahl FR and Pearson RD (1967). Ultrastructural studies on experimental vesication. II. Collagenase. *J Invest Dermatol*, **49**, 616–631.

Katz SI (1984). The epidermal basement membrane zone: structure, ontogeny and role in disease. *J Am Acad Dermatol*, **11**, 1025–1037.

Keene DR, Sakai LY, Lunstrum GP *et al.* (1987). Type VII collagen forms an extended network of anchoring fibrils. *J Cell Biol*, **104**, 611–621.

Lam DG, Rice P and Brown RFR (2002). The treatment of Lewisite burns with laser debridement – 'lasablation'. *Burns*, **28**, 19–25.

Lasky RA, Fairman P and Blow JJ (1989). S-phase of the cell cycle. *Science*, **246**, 601–614.

Leigh IM, Purkis PE and Bruckner-Tuderman L (1987). LH-7.2 monoclonal antibody detects type-VII collagen in the sublamina densa zone of ectodermally derived epithelia, including skin. *Epithelia*, **1**, 17–29.

Leprince P, Rogister B and Moonen G (1989). A colorimetric assay for the simultaneous measurement of plasminogen activators and plasminogen activator inhibitors in serum-free conditioned media from cultured cells. *Ann Biochem*, **177**, 341–346.

Lin AN and Carter DM (1989). Epidermolysis bullosa: when the skin falls apart. *J Paediat*, **114**, 349–355.

Lindahl U and Hook M (1978). Glycosaminoglycans and their binding to biological macromolecules. *Ann Rev Biochem*, **47**, 385–417.

Lindsay CD and Rice P (1995). Changes in connective tissue macromolecular components of Yucatan minipig skin following application of sulphur mustard vapour. *Human Exp Toxicol*, **14**, 341–348.

Liotta LA, Goldfab RH and Brundage R (1981). Effect of plasminogen activator (urokinase), plasmin and thrombin on glycoprotein and collagenous components of basement membrane. *Cancer Res*, **41**, 4629–4636.

Lunstrum GP, Sakai LY, Keene DR *et al.* (1986). Large complex globular domains of type-VII protocollagen contribute to the structure of anchoring fibrils. *J Biochem Chem*, **261**, 9042–9048.

Mandl H and Freilinger G (1984). First report on victims of chemical warfare in the Gulf-war treated in Vienna. In: *Proceedings of the First World Congress on Biological and Chemical Warfare*, Ghent, Belgium: State University of Ghent and the National Science Foundation of Belgium, pp. 330–340.

Martinez-Fernandez A and Amenta P (1983). The basement membrane in pathology. *Lab Invest*, **48**, 656–677.

McAdams AJ (1956). A study of mustard vesication. *J Invest Dermatol*, **26**, 317–327.

Medema J (1986). Mustard gas : the science of H. *Nuc Biol Chem Def Technol Int*, **1**, 66–71.

Meier HL, Gross CL and Papirmeister B (1987). 2,2′-Dichlorodiethyl sulfide (sulfur mustard) decreases NAD^{+} levels in human keratinocytes. *Toxicol Lett*, **39**, 109–122.

Mellor SG, Rice P and Cooper GJ (1991). Vesicant bums. *Br J Plast Surg*, **44**, 434–437.

Mershon MM, Mitcheltree LW, Petrali JP *et al.* (1990). Hairless guinea-pig bioassay model for vesicant vapour exposure. *Fund Appl Toxicol*, **15**, 622–630.

Meyer V (1886). Versuche über die Haltbarkeit von Sublimatlosungen. *Ber Deutsch Chem Gesell*, **1**, 1725–1739.

Mitcheltree LW, Mershon MM, Wall HG *et al.* (1989). Microblister formation in vesicant-exposed pig skin. *J Toxicol (Cutan Oc Toxicol)*, **8**, 309–319.

Mol MAE, Van de Ruit A-M and Kuivers AW (1989a). In: *Proceedings of the 1989 Medical Defense Bioscience Review (PMBR)*, USAMRDC, 15–17 August 1989, MD: USAMRICD, pp. 57–61.

Mol MAE, Van de Ruit A-M and Kluivers AW (1989b). NAD^{+} levels and glucose uptake of cured human epidermal cells exposed to sulphur mustard. *Toxicol Appl Pharmacol*, **98**, 159–165.

Murnane JC and Byfield JE (1981). Irreparable DNA crosslinks and mammalian cell lethality with bifunctional alkylating agents. *Chem Biol Interact*, **38**, 75–86.

Myhre-Jensen O and Astrup T (1971). Fibrinolytic activity of squamous epithelium of the oral cavity and oesophagus of the rat, guinea-pig and rabbit. *Arch Oral Biol*, **16**, 1077.

Needham DM and Dixon M (1941). *The Metabolism of Normal and Vesicant Treated Skin*. Report No. 1

(U24815). Cambridge, UK: Cambridge Biochemistry Laboratory.

Newman-Taylor AJ (1991). Experience with mustard gas casualties. *Lancet*, **337**, 242.

Papirmeister B, Gross CL, Petrali JP *et al.* (1984). Pathology produced by sulphur mustard in human skin grafts on a thymic nude mice. I. Gross and light microscopic changes. *J Toxicol (Cutan Oc Toxicol)*, **3**, 371–391.

Papirmeister B, Gross CL, Meier HL *et al.* (1985). Molecular basis for mustard-induced vesication. *Fund Appl Toxicol*, **5** (**Supplement**), 134–149.

Papirmeister B, Feister AJ, Robinson SI *et al.* (eds) (1991). Historical and modern use of sulphur mustard in warfare. In: *Medical Defense Against Mustard Gas*, pp. 1–9. Boston, MA, USA: CRC Press.

Pardee AB (1989). G_1 events and regulation of cell proliferation. *Science*, **246**, 603–608.

Parry EJ (1921). In: *The Chemistry of Essential Oils and Artificial Perfumes,* Volume 1. London, UK: Scott, Greenwood and Son.

Pauser G, Alloy A, Carvena M *et al.* (1984). Lethal intoxication by war gases on Iranian soldiers. Therapeutic interventions on survivors of mustard gas and mycotoxin immersion. In: *Proccedings of the First World Congress on Biological and Chemical Warfare*, Ghent, Belgium: State University of Ghent and the National Science Foundation of Belgium, pp. 341–351.

Pearson RW (1964). Some observations on epidermolysis bullosa and experimental blisters. In: *The Epidermis* (W Montagna and WC Lobitz, eds), pp 611–626. New York, NY, USA: Academic Press.

Peters RA (1936). Effects of dichlorodiethyl-sulphone on brain respiration. *Nature (London)*, **138**, 127–128.

Peters RA and Wakelin RW (1947). Observations upon a compound of mustard gas and kerateine. *Biochem J*, **41**, 550–555.

Philips FS (1950). Recent contributions to the pharmacology of bis(2–haloethyl) amines and sulfides. *Pharmacol Rev*, **2**, 281–323.

Pirie A (1947). The action of mustard gas on ox cornea collagen. *Biochem J*, **41**, 185–190.

Renshaw B (1940). Mechanisms in the production of cutaneous injuries by sulphur and nitrogen mustards. In: *Chemical Warfare Agents and Related Chemical Problems*, Volume 1, pp. 479–518. Washington, DC, USA: US Office of Science Research and Development, National Defense Research Committee.

Renshaw B (1947). Observations on the role of water in the susceptibility of human skin to injury by vesicant vapours. *J Invest Dermatol*, **9**, 75–85.

Reynolds JE (ed.) (1977). *Martindale. The Extra Pharmacopoeia*, 27th Edition. London, UK: The Pharmaceutical Press.

Roberts JJ, Crathorn AR and Brent TP (1968). Repair of alkylated DNA in mammalian cells. *Nature (London)*, **218**, 970–972.

Rice P, Brown RFR, Lam DG *et al.* (2000). Dermabrasion – a novel concept in the surgical management of sulphur mustard injuries. *Burns*, **26**, 34–40.

Roitt I, Brostoff J and Male D (eds) (1988). Cell-mediated immunology. In: *Immunology*, pp. 11.1–11.12. Edinburgh, UK: Churchill Livingstone.

Sanders KM, Innace JK, Gross CL *et al.* (1989). In: *Proceedings of the 1989 Medical Defense Bioscience Review (PMBR)*, USAMRDC, 15–17 August 1989, MD: USAMRICD, pp. 419–422.

Savage JR and Brekon G (1981). Differential effects of sulphur mustard on S-phase cells of primary fibroblast cultures from Syrian hamsters. *Mutat Res*, **34**, 375–387.

Savion N, Disatnik M-H and Nero Z (1987). Murine macrophage heparanase: inhibition and comparison with metastatic tumour cells. *J Cell Physiol*, **130**, 77–84.

Schrafl A (1938). The symptoms, prophylaxis and the therapy of mustard injuries to the skin. *Protar*, **4**, 111–118.

Sellers A and Murphy G (1981). In: *International Review of Connective Tissue Research* (DA Hall and DS Jackson, eds). New York, NY, USA: Academic Press, pp. 151–190.

Shakespeare VA and Shakespeare PG (1987). Growth of cultured human keratinocytes on fibrous dermal collagen : a scanning electron microscopic study. *Burns*, **13**, 343.

Sheridan RL, Lydon MM, Petras LM *et al.* (1999). Laser ablation of burns : initial clinical trial. *Surgery*, **125**, 93–95.

Shoshan S (1981). In: *Internatianal Review of Connective Tissue Research* (DA Hall and DS Jackson, eds), p. 426. New York, NY, USA: Academic Press.

Singer KH, Hashimoto K, Lensen PK *et al.* (1985). Pathogenesis of autoimmunity in pemphigus. *Ann Rev Immunol*, **3**, 87.

Smulson ME (1989). In: *Proceedings at the 1989 Medical Defense Bioscience Review (PMBR)*, USAMRDC, 15–17 August 1989. MD: USAMRICD, pp. 361–371.

Sulzberger MB and Katz JH (1943). The absence of skin irritants in the contents of vesicles. *US Navy Med Bull*, **43**, 1258–1262.

Takigawa M and Ofuji S (1977). Early changes in human epidermis following thermal burn: an electron microscopic study. *Acta Dermatol (Stockholm)*, **57**, 187.

Thompson RHS (1940). *The Respiration of Rat Skin after Damage with Sulphur Mustard*, Report U9434. Oxford, UK: Department of Biochemistry, Oxford University.

Thompson RHS (1942). The action of chemical vesicants on cholinesterase. *J Physiol*, **105**, 370–381.

United Nations Security Council (1984). *Report of the Specialists Appointed by the Secretary General to Investigate Allegations by the Islamic Republic of Iran Concerning the Use of Chemical Weapons*, 26 March 1984. UN Report S\16433. New York, NY, USA: United Nations Security Council.

United Nations Security Council (1986). *Report of the Mission Dispatched by the Secretary General to Investigate Allegations of the Use of Chemical Weapons in the Conflict between Iran and Iraq*, 12 March 1986. UN Report S\17911. New York, NY, USA: United Nations Security Council.

United Nations Security Council (1987). *Report of the Mission Dispatched by the Secretary General to Investigate Allegations of the Use of Chemical Weapons in the Conflict between the Islamic Republics of Iran and Iraq*, 08 May 1987. UN Report S\18852. New York, NY, USA: United Nations Security Council.

Vedder EB (1925). Medical aspects of chemical warfare. In: *The Medical Aspects of Chemical Warfare*. Baltimore, MD, USA: Williams and Wilkins.

Vögt RF, Dannenberg AM, Schofield BH *et al.* (1984). The pathogenesis of skin lesions caused by sulphur mustard. *Fund Appl Toxicol*, **4** (Supplement), 71–83.

Warthin AS and Weller CV (1919). *The Medical Aspects of Mustard Gas Poisoning*. London, UK: Henry Kimpton.

Werb Z and Gordon S (1975). Secretion of a specific collagenase by stimulated macrophages. *J Exp Med*, **142**, 346–360.

Wheeler GP (1962). Studies related to the mechanisms of action of cytotoxic alkylating agents. *Cancer Res*, **22**, 651–688.

Willems JL (1989). Clinical management of mustard gas casualties. *Annal Med Milit (Belg)*, **3** (Supplement), 1–61.

Wilson EL and Reich E (1978). Plasminogen activator in chick fibroblasts: induction of synthesis by retinoic acid; synergism with viral transformation and phorbol ester. *Cell*, **15**, 385–392.

Wilsteed EM, Bhogal BS, Das AK *et al.* (1991). Lichen planus pemphigoides: a clinicopathological study of nine cases. *Histopathology*, **19**, 147–154.

Windholz M and Budavari S (1983). In: *The Merck Index: An Encyclopaedia of Chemicals, Drugs and Biologicals*, 9th Edition (M Windholz, S Budavari, LY Stroumtsos, *et al.*, eds). Rathway, NJ, USA: Merck and Company, Inc.

Winter GD (1962). Formation of the scab and rate of epithelialisation of superficial wounds in the skin of the young domestic pig. *Nature (London)*, **193**, 293–294.

Woessner JP, Dannenberg AM, Pula PJ *et al.* (1991). Extracellular collagenase, proteoglycanase and products of their activity, released in organ culture by intact dermal inflammatory lesions produced by sulphur mustard. *J Invest Dermatol*, **95**, 717–726.

Woodley DT, O'Keefe EJ and Prunieras M (1985). Cutaneous wound healing: a model for cell-matrix interactions. *J Am Acad Dermatol*, **12**, 420.

Wun T-C, Ossowski J and Reich E (1982). A proenzyme form of human urokinase. *J Biolog Chem*, **257**, 7262–7268.

22 THE NORMAL BONE MARROW AND MANAGEMENT OF TOXIN-INDUCED STEM CELL FAILURE

Jennifer G. Treleaven

Royal Marsden Hospital, Sutton, UK

BACKGROUND – THE NORMAL BONE MARROW

Sites of haematopoiesis

In human adults, the bone marrow involved with haematopoiesis is located in cavities within the pelvis, sternum, ribs, skull, vertebrae and proximal epiphyses of the femora and humeri (Figure 1). Bone marrow contains three components: trabecular bone, red marrow (haematopoietic), and yellow marrow (non-haematopoietic). The red marrow has an excellent blood supply and is concerned with the production of blood cells. These occupy approximately 60% of the space, the rest being composed of fat, water and protein, as can now be shown on magnetic resonance imaging of normal bone marrow (Vande Berg *et al.*, 1998). The yellow marrow has a sparse blood supply and is composed mainly of fat cells with approximately 15% water. Its role is to provide nutrition for the red marrow (Vogler and Murphy, 1988). The yellow marrow increases with age, as the trabecular bone becomes increasingly osteoporotic and is replaced by fat. Trabecular bone provides support for the red and yellow marrow.

Embryological development of the bone marrow

In the mammalian embryo, haemopoietic progenitors can first be detected in the vestigial yolk sac (Moore and Metcalf, 1970). Haemopoietic stem cells are then found in the liver and subsequently migrate to the bone marrow, which becomes the major site for generation of lineage-committed progenitor cells (Tavassoli, 1991). In infants and children, the long bones also contain haematopoietic tissue, but as the child matures these areas regress, and the previously cell-rich environment is replaced by fat cells (yellow marrow), which act as a source of nutrients for the haemopoietic bone marrow (red marrow). This conversion from red to yellow marrow takes place in a centripetal manner, staring with the distal bones of the feet and hands and progressing finally to the proximal bones, the humeri and femora. It is complete by 25–30 years of age.

Blood cell production

Over the last thirty years or so, our understanding of the complex mechanisms and interactions which regulate blood cell production has increased greatly. It is now recognized from experimental studies that cells are produced and mature within the bone marrow space from the relatively small population of pluripotent stem cells which comprise approximately 0.01–0.05% of the total nucleated marrow cell population and are established during embryogenesis (Metcalf and Moore, 1971). The two characteristic features of stem cells are that they have enormous ability to proliferate to produce more stem cells, a process known as self-renewal, and that they

Chemical Warfare Agents: Toxicology and Treatment (2nd Edition)
Edited by Timothy C. Marrs, Robert L. Maynard and Frederick R. Sidell

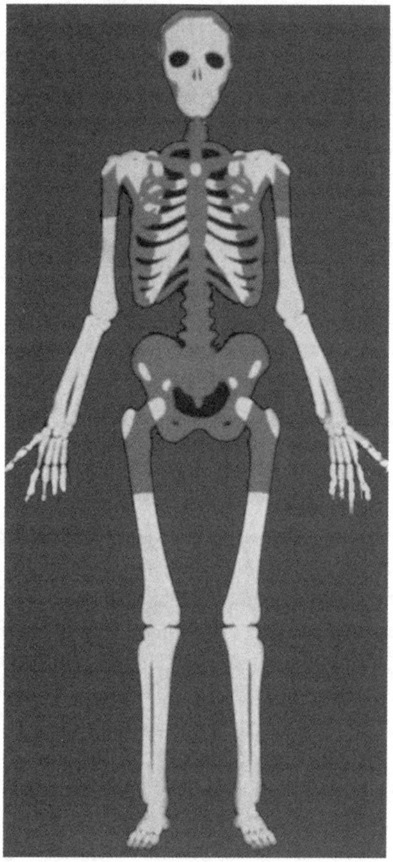

Figure 1. Sites in the adult skeleton where haematopoiesis takes place

are able to differentiate into specialized, mature 'end cell' types. As the stem cells differentiate and commit themselves to a particular cell line, they become less and less able to proliferate so that by the time the 'end cell' appears in the peripheral blood it has entirely lost its powers of self-replication (Figure 2). The mature cells into which such stem cells differentiate comprise osteoclasts, erythrocytes, platelets, and white blood cells which comprise T- and B-lymphocytes, monocytes/macrophages, neutrophils, eosinophils and basophils. Figure 3 depicts a simplified scheme of the stages of differentiation from stem cell to mature end cell, and mentions some of the growth factors involved in the differentiation processes of the various cells (see below).

Long-term effects of alkylating agents on the bone marrow; reduction of the stem cell pool

Most bone marrow progenitors are susceptible to dose-dependent reductions in viability and ability to proliferate after exposure to alkylating agents or radiation, but there is a proportion of cells which is relatively resistant to insult. Such cells play an important part in recovery of haematopoiesis, although they may subsequently

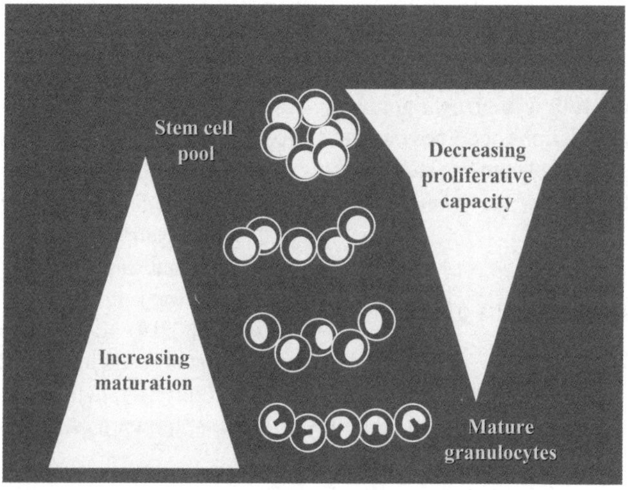

Figure 2. Changes in proliferative ability of stem cells through to mature cells

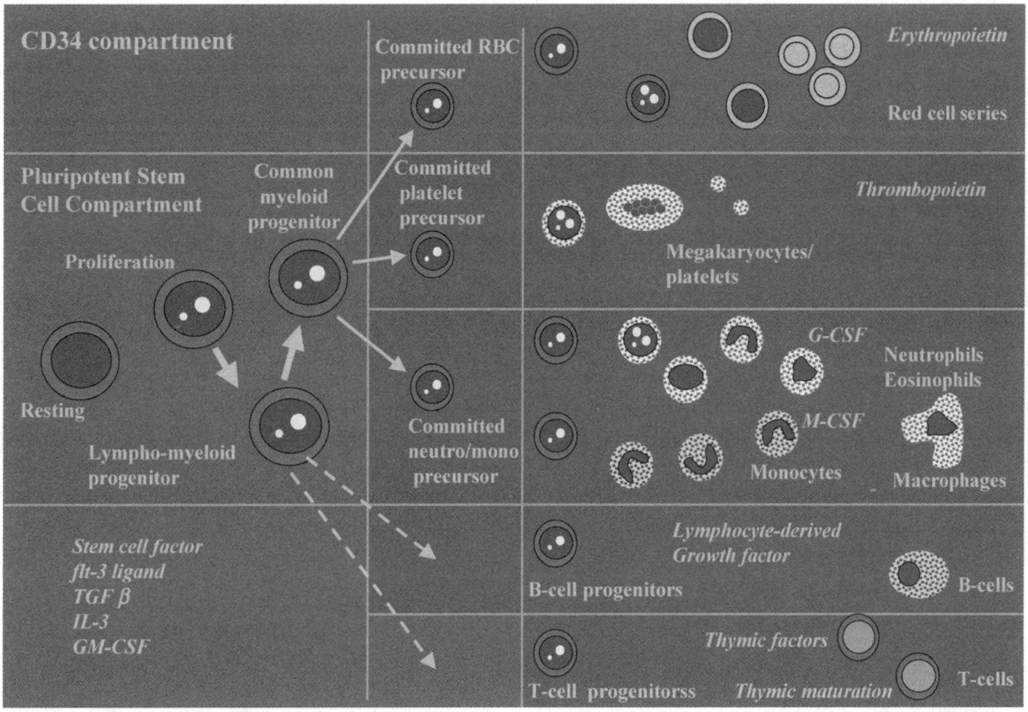

Figure 3. The various developmental stages from stem cell to mature end cells and the chief growth factors involved in differentiation

have a reduced capacity for self-renewal. This point can be clearly illustrated in the clinical oncology setting in patients who receive high-dose therapy with the alkylating agent, melphalan. If a marrow ablative dose of 200 mg m^{-2} is given to treat multiple myeloma and the patient is subsequently 'rescued' with previously harvested and cryopreserved autologous stem cells, for example, it is likely that after approximately 4 weeks the peripheral blood count will have reverted to complete normality. However, should the process be repeated subsequently with stem cells harvested after the first therapy is completed, it is likely that the peripheral blood cell counts will fail to revert to complete normality because the repeated treatment cycles have reduced the stem cell pool and its attendant potential for proliferation. In addition, repeated exposure to alkylating agents or to radiotherapy reduces the numbers of stem cells which can be harvested either from the bone marrow or peripherally, and so in a heavily pre-treated patient it may eventually prove impossible to obtain sufficient stem cells from a marrow harvest, or to 'mobilize' stem cells for peripheral harvest.

Early studies on the kinetics of haematopoiesis

In early animal studies, spleen colony forming units (CFU-S) were the cells most widely regarded as fitting the definition of a pluripotent stem cell, in that they also had the potential for proliferation, self-renewal and differentiation (Till and McCulloch, 1980). The kinetic heterogeneity of colony formation by CFU-S has been clearly shown and cell subsets have been isolated and identified which have proven capable of generating spleen colonies 8–12 days post transplantation (Visser *et al.*, 1984). The most primitive stem cells have been shown to have the capacity, when transplanted into irradiated mice, to effect regeneration of the lymphohaematopoietic cell population. Mice which have been thus repopulated can then donate cells to a second irradiated mouse which will also experience repopulation of its haematopoietic cells (Till and McCulloch, 1961). In such early experiments involving the retransplantation of regenerating CFU-S, it was observed that between the individual cells in the population, there was an extremely heterogeneous capacity for self-renewal (Worton *et al.*, 1975), which led to the idea of an age structure within the CFU-S population; some CFU-S were more mature than others and would take less time to develop as spleen colonies (perhaps 7–10 days) than would the less mature ones which would need 12–14 days (Schofield, 1978). However, even such multi-potential stem cells age and will eventually become exhausted by repeated transplantation or by damage by small amounts of irradiation, a fact which allowed them to be distinguished functionally from committed stem cells (Porteous and Lajtha, 1966). Progenitor cells were subsequently defined as cells which had the ability to give rise to colonies of morphologically recognizable progeny in semi-solid clonal cultures. The original cell culture assays that identified murine progenitor cells were characterized by colonies of mast cells, granulocytes and/or macrophages which were then demonstrated to be closely related to neutrophils and mononuclear phagocytes (Bradley and Metcalf, 1966). Since then, much work has been conducted on both animal and human stem cells and committed progenitor cells, designed to define the complex interactions which occur between stem cells and committed progenitor cells, the microenvironment in which they mature and proliferate and the various growth factors which aid maturation. Telomeres, which are essential genetic elements consisting of specific DNA repeats and associated proteins, have recently emerged as important regulatory elements which control the number of times normal somatic cells can divide. Granulocytes and lymphocytes, for example, show a significant decline in telomere length with age. The loss of individual stem cells because of telomere shortening should not influence overall haematopoiesis, however, although it may possibly contribute to the concept of stem cell 'exhaustion' as occurs in conditions such as aplastic anaemia. Exploitation of the mechanisms controlling regulation of telomere length by inactivation of down-stream signalling events may promote malignant expansion of cells or allow increased proliferation of an individual committed stem cell line (Antonchuk *et al.*, 2002).

Stem cell activity in physiological and adverse situations

The stem cells are the 'building blocks' responsible for renewing mature haematopoietic cells under physiological conditions and also for reconstituting haematopoiesis following damage to the haematopoietic system. Most commonly, this is after treatment with chemotherapeutic agents or radiation for malignant disease. Most mature blood cells only live for a period of hours or days, and so that there is a constant requirement for this loss to be replaced in the normal steady state. Should there be an increased requirement for a particular cell type as, for example, if haemorrhage or infection are present, this will prompt increased production of the required end cells, and more pluripotent stem cells will commit themselves to producing cells of a particular lineage. At any time there are haematopoietic cells at all stages of maturation, representing the various cell lines, present in the bone marrow space.

Under steady-state conditions, the average red blood cell takes approximately 5 days to develop from a committed stem cell, through the various stages of development from the early proerythroblast to the relatively mature reticulocyte and, finally, to the mature red cell seen in the peripheral blood. Its average life-span is 120 days if there are no abnormal extraneous circumstances, such as bleeding. Approximately 3×10^9 reticulocytes are delivered into the blood every day, representing approximately 1% of the circulating red cell pool.

Bone marrow granulocytes can be considered as existing either in the proliferative compartment, or the maturation storage compartment. Myeloblasts, promyelocytes and myelocytes are capable of replication and constitute the mitotic compartment, whereas metamyelocytes and mature granulocytes cannot divide, and constitute the storage compartment. Estimates as to the time taken for a granulocyte to proceed through the stages of development from myelocyte to mature blood granulocyte vary from 5 to 14 days, although this can be much more rapid in the presence of infection. Once in the blood, the cells survive only a matter of hours.

Megakaryocytes, which ultimately shed platelets into the peripheral blood, take two to three weeks to develop from committed stem cells, and platelets are shed from the megakaryocyte some 5 days after it has fully matured.

None of these early cells develops in isolation. They are dependent upon the presence of various growth factors to assist maturation.

The role of growth factors in haematopoiesis

The earliest growth factor to be identified was erythropoietin which was first described in 1906 by Carnot and Deflandre, who hypothesized that a circulating factor – then known as hemopoietene – regulated red cell production. Erythropoietin levels were subsequently noted to be higher in anaemic patients suffering from aplastic anaemia than in normal, non-anaemic subjects, and it was concluded that the erythropoietin was working to increase the levels of circulating red blood cells in the presence of the anaemia (Hasegawa *et al.*, 1968). Erythropoietin was then purified from the urine of patients suffering from aplastic anaemia and characterized as a 35 kDa glycoprotein (Miyake *et al.*, 1977).

In vitro clonal cultures documented that haematopoietic cells require positive stimulation for division to take place and they also provided a method for detecting and quantifying the regulatory molecules involved. It is now recognized that there are numerous haematopoietic growth factors which influence cells of several or specific haematopoietic lineages and which also affect cells outside the haematopoietic system. The first to be described were Macrophage-Colony stimulating factor (M-CSF), Granulocyte-Colony stimulating factor (G-CSF) and Granulocyte-Macrophage Colony stimulating factor (GM-CSF), for example, which influence predominantly the granulocyte and the granulocyte macrophage cell series, but which also have an effect on the pluripotent stem cells. Another example is Interleukin-3 (IL-3), initially known as multi-CSF, which has a very broad spectrum of activity and influences the development of all lineages derived from the pluripotent haematopoietic stem cell. Metcalf, in 1991, showed that G-CSF, GM-CSF and IL3 all had a clear effect on the proportion of different progenitor cells committed to the different lineages generated from the pluripotent stem cell in conjunction with colony stimulating factor. Several other Interleukins, numbered 4 through 18, have now been described and characterized, all of which activate one or more precursor cell lines, either alone or in conjunction with other growth factors. In addition, there are the Interferons, which are a heterogeneous group of proteins also involved in the modification of biological responses, the Transforming Growth Factor (TGFβ) family, Tumour Necrosis Factor (TNF), stem cell factor, thrombopoietin and FLK ligand.

Table 1. Some of the growth factors involved in haematopoiesis

GM-CSF
G-CSF
M-CSF
The Interleukins
Interferon
EP (erythropoietin)
Thrombopoietin
Stem cell factor
TNF (Tumour necrosis factor)
TGFb (Transforming growth factor)
FLK ligand

Recombinant human stem cell factor (rHuSCF) is a cytokine that stimulates both lineage-committed and non-lineage-committed haematopoietic progenitor cells (Broudy, 1997). There may also be a synergistic effect between rHuSCF and other cytokines with haematopoietic proliferative potential, including G-CSF, erythropoietin and Interleukin 3 and 6 (Bernstein *et al.*, 1991) and a number of studies have focused on the possibility of improving mobilization success by combining G-CSF with rHuSCF. However, the optimum mobilization schedule still remains to be identified. Table 1 lists the principle growth factors involved in the differentiation processes of the various cell lines, and Figure 3 depicts their interactions with the various end- cell precursors.

The role of the bone marrow microenvironment

In addition to a requirement for a pool of stem cells with the potential to multiply and mature and for the presence of the complex panel of various growth factors outlined above, the microenvironment within the bone marrow space with its various supporting cells must also be conducive to allowing optimum proliferation and differentiation of cells within the stem cell pool. The supporting cells, or stromal cells, include fibroblasts and stromal endothelial cells, all of which are mesenchyme-derived connective tissue-like cells which support the haematopoietic cells within the bone marrow space. If an innoculum of bone marrow is cultured, the supporting stoma cells are those which grow in culture, excluding the haematopoietic stem cells and their progeny. Abnormal interactions between the stoma cells and the haematopoietic precursor cells may result in abnormalities of progenitor cell differentiation and expression such as occurs in such conditions as cyclical neutropenia, where the progenitor cells may have a defective interaction with inhibitory aspects of the marrow microenvironment. This problem can be resolved permanently by stem cell transplantation, or temporarily by the use of G-CSF injections. If G-CSF or GM-CSF are used regularly, however, osteoporosis may be induced, implying that proliferation and release of haematopoietic stem cells requires cycles of bone marrow restructuring which are likely to be linked to osteoclast and osteoblast interactions (Bishop *et al.*, 1995; Takahashi *et al.*, 1996).

Cellular interaction also relies upon cell-to-cell adhesion. The molecules which mediate this are termed adhesion molecules or CAMs. A number of different adhesion molecules is now recognized, each of which is associated with a specific precursor cell line and many of which require the presence of calcium in order to associate with similar CAMs on neighbouring cell surfaces (Albeda and Buck, 1990).

TREATMENT APPROACHES IN SUBJECTS WITH BONE MARROW FAILURE

Background

The mustines are alkylating agents which act by interfering with the DNA strand unwinding necessary for transcription and replication. This, in turn, spatially disorientates the base sequence for the genetic code. Cross-linked DNA is susceptible to hydrolytic and ribonuclease-mediated attack, leading to deletion of the bis-purinylalkyl complex and scission of the DNA strands leading to miscoding. Thus, all mustines are mutagens and hence carcinogenic.

Mustard gas, or sulphur mustard, is closely related to the nitrogen mustards, for example, mustine hydrochloride which is used as a cytostatic agent in chemotherapy for treating malignant disease. It is a liquid at room temperature but evaporates to produce a vapour: as elaborated elsewhere in this book, both the liquid and the vapour cause blistering of the skin, severe irritation of the eyes and damage to the upper respiratory tract. These effects resolve over a period of weeks. However, the most serious effect of severe exposure can be bone marrow failure, which depending on the level of exposure may be either temporary or permanent. Sulphur mustard was used during World War I. In a small minority of soldiers exposed to mustard gas during World War I, keratitis appeared some twenty years post-exposure which in some cases led to blindness. However, the more immediate effect, and that which led to many deaths, was damage to the dividing cells of the reticuloendothelial system. Autopsies performed on soldiers killed in World War I indicated that sulphur mustard gas had an effect on rapidly dividing cells. Affected individuals had very low white blood cell counts (leukopenia), bone marrow aplasia (absence of maturing and mature cells), disruption of lymphoid tissues and ulceration of the gastrointestinal tract.

The first mustard anti-tumour agents were introduced in Europe as early as 1931. However, the sulphur mustard proved too toxic for systemic clinical use, and for clinical use, development turned towards nitrogen mustards as these were slightly less damaging. These agents were used for a number of years, largely in combination with other cytotoxic drugs, particularly in the treatment of Hodgkin's disease, a cancer of the reticuloendothelial system. The nitrogen mustards have largely fallen out of use now partly because other more effective cytotoxic drug combinations have become available, and partly because the mustards caused particularly severe and sometimes intractable vomiting as part of their side-effect profile.

General issues

In the event of a mustine gas attack, the victims will suffer a variety of serious medical problems in addition to the possible effects of the mustine gas on the bone marrow. In any victim, a fundamental problem is likely to be the difficulty of accurately assessing degree of exposure to which the bone marrow has been subjected; problems with bone marrow failure are going to occur after a period of days rather than immediately and hence the priority for any victims who are thought to have had severely toxic exposure will in all likelihood move from considerations of the bone marrow to centre on the more immediate problems of damage to skin, lungs and other vital organs. The confusion of having a terrorist attack and multiple victims will make it difficult to ensure that every victim is assessed fully and treatment administered in a logical and systematic manner. Although venous access may be difficult in persons suffering skin trauma, it is vital to establish good access so that antibiotics, fluids and analgesics can be administered. Of great importance, blood must be taken for human leucocyte antigen (HLA) typing in the event that it becomes necessary to identify a related or unrelated stem cell donor should the victim become aplastic. Since such typing relies upon the presence of white blood cells in the peripheral blood of the subject, such testing must be undertaken *before* the peripheral blood count falls significantly. It must be ensured, in all the confusion of a mass attack, that this vital step is not overlooked.

It may be that the victim will die of problems related to his other organs before bone marrow failure has a chance to manifest itself, or it may be that recovery from damage to all organs is complete and no signs of bone marrow failure develop. However, at the time of exposure it is impossible to know exactly what the outcome for any victim is likely to be and thus the safest approach is to obtain a sample for tissue typing from every victim and have the tests conducted so that the results are available in the event that marrow aplasia should develop.

Cytokine, blood product and antimicrobial support

CYTOKINE THERAPY

To date, only a few haematopoietic colony-stimulating factors (CSFs) are commercially available for the management of therapy-induced pancytopenias. These include GM-CSF, G-CSF and its pegylated form, PEG-G-CSF. Erthropoietin is also available in a short-acting and long-acting form. Thrombopoietin is currently not available commercially and neither are any factors which stimulate production of lymphocytes, although IL3 is under evaluation. The rationale for using these factors in the context of drug- or radiation-induced myelosuppression derives from the fact that they are known to reduce duration of neutropenia by a number of days in oncology patients treated with cytotoxic drugs or radiotherapy (Schiffer, 1996; Nemunaitis *et al.*, 1991). They may also speed peripheral blood cell recovery after exposure to otherwise lethal doses of irradiation in dogs (Schuening *et al.*, 1993; Nash *et al.*, 1994) and have certainly produced a survival benefit in prospective trials involving myelosuppressed primates (Neelis *et al.*, 1997). However, although they have been shown to shorten the duration of neutropenia in humans, there is no conclusive evidence that their use changes the rate of serious bacterial infection or produces a definitive reduction in morbidity and mortality, and hence expectations of their influence should be guarded (Vose and Armitage, 1995; Ozer *et al.*, 2000).

The value of these agents relies upon their ability to promote proliferation and differentiation of granulocyte progenitors to produce neutrophils. These are then primed, and their antimicrobial activity is enhanced (Mayer *et al.*, 1991) Thus, it can be appreciated that their use may be limited if an individual has received a large, marrow-ablative dose of toxin, since the CSFs will only be able to influence precursor cells which are already relatively far down the maturation pathway; they will not be able to exert any effect upon a seriously damaged or ablated stem cell pool. This can again be appreciated in the setting of clinical oncology, where CSFs have little or no effect until the effects of chemo or radiotherapy have abated to the extent that some progress

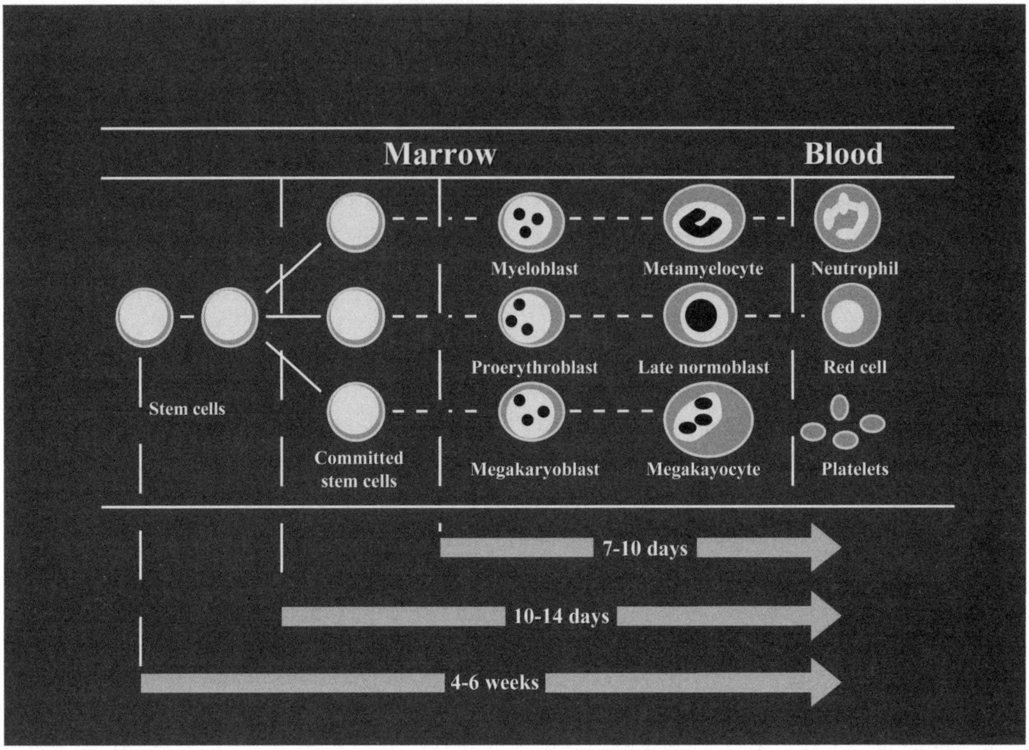

Figure 4. Aplasia times in relation to sites of action on cellular differentiation of the various groups of cytotoxic agents

with stem cell maturation has taken place, albeit the mature cells have not yet had sufficient time to appear in the peripheral blood. Studies have shown that recovery of neutrophils is no quicker if G-CSF is started on day 1 following an autologous PBSC, or if it is commenced after 7 days (Demirer *et al.*, 2002; Bence-Bruckler *et al.*, 1998). Depending on the cytotoxic agent used and its mode of action, this may take as long as 4–6 weeks if the toxic agent acts upon the stem cell pool as is the case with radiation and the alkylating agents, and no stem cell 'rescue' is provided. However, the aplasia time may be much shorter if an agent is used which acts on the more mature, committed stem cells. It is likely to be in this situation that CSFs are the most useful, or a few days after stem cell rescue has been given when the in-coming progenitor cells have had the opportunity to undergo some degree of maturation and proliferation. Figure 4 depicts a simplified overview of the effects of the various marrow toxins in terms of expected aplasia times in relation to the point in the stem cell maturation pathway at which they act.

G-CSF and GM-CSF

A few studies on irradiated non-human primates have examined the role and timing of G-CSF and GM-CSF administration. These suggest that when these agents are given the day after exposure and continued for up to 3 weeks, there is significant neutrophil enhancement which is not apparent if the agents are given later after exposure (Neelis *et al.*, 1997; MacVittie *et al.*, 1996). However, it is not clear from the studies whether the doses of radiation used would be expected to induce permanent aplasia in the experimental primate and there is no genuine parallel between these animal studies and the situation which may pertain in the human setting with regard to doses of radiation and response to SCF therapy. In the event of human exposure to marrow-toxic doses of mustine or other agents of chemical warfare, where the dose received will inevitably be unclear and in the absence of any conclusive animal or human data on the use of CSF, however, it would appear expedient to commence therapy with CSFs as soon as possible after exposure as there is likely to be some benefit in terms of an amelioration in both level and duration of neutropenia.

Whether the use of G-CSF or GM-CSF is preferable, or indeed use of the longer acting pegylated form of G-CSF, is also unclear and there are, again, only a few animal studies upon which to base such a judgement. These involved exposure to varying doses of radiation rather than a chemical bone-marrow toxin. For what it is worth, and results must be interpreted with caution, of dogs receiving G-CSF for 21 days after exposure to 4 Gy of radiation, 71% survived with complete haematopoietic recovery whereas only 3.6% of control animals survived with supportive care alone (Schuening *et al.*, 1989). However, in another study on dogs G-M-CSF was given, and this resulted in poor survivals of 10% (Nash *et al.*, 1994). If doses of radiation were increased beyond 4 Gy, survival rates fell even with administration of these agents, confirming that some degree of marrow reserve is necessary to enable them to have any positive effect on haematopoietic recovery. A parallel must exist with exposure to marrow-toxic agents of chemical warfare where the dose received will inevitably be unclear. In the light of the slim evidence available from animal studies relating to exposure to varying doses of radiation, if any parallel can be drawn, therapy with either G-CSF or GM-CSF should be commenced as soon as possible after exposure to the toxic agent.

Erythropoietin

The human gene for erythropoietin, a hormone produced in the kidney, was first cloned in the 1980s (Jacobs *et al.*, 1985), and the recombinant form was developed soon after that (Flaherty *et al.*, 1989). The recombinant alpha form of erythropoietin, epoetin, is now very widely used in the context of the anaemia associated with end-stage renal failure, and its potential has been investigated in many other clinical settings where anaemia is a significant problem. In the contexts of anaemia due to myeloablation and both allogeneic (Link *et al.*, 1994; Klaesson *et al.*, 1994; Biggs *et al.*, 1995) and autologous stem cell rescue (Chao *et al.*, 1994; Ayash *et al.*, 1994; Vannucchi *et al.*, 1996), a number of controlled, and mainly randomized studies have been conducted in an attempt to define the place of erythropoietin as opposed to red cell transfusion. In the allogeneic setting, erythropoietin consistently resulted in a faster time to independence from red cell transfusion, and the presence of increased reticulocyte numbers tended to confirm more rapid recovery of red cell function in the patients receiving erythropoietin. These results, however, were not seen in the studies where patients were receiving autologous stem cell support. In the situation of chemical warfare-induced bone marrow suppression, therefore, the place of erythropoietin remains unclear.

Stem cell factor

Recombinant human stem cell factor is not currently commercially available for use in humans or animals or for diagnostic purposes and can only be used for *in vitro* research. It will no doubt become available in the near future for use in animals and humans, as a bone marrow stimulant after chemotherapy, when it should expedite proliferation of erythroid, myeloid and lymphoid progenitors and also act synergistically with other colony-stimulating factors. In the context of its value after exposure of the bone marrow to large doses of mustard gas, however, the constraints concerning presence of residual viable stem cells amenable to stimulation will remain, as is also the situation with regard to potential value of the other growth factors which are already commercially available.

BLOOD PRODUCT SUPPORT

Blood products including red cells, plasma and platelets may be available for individuals needing such support after exposure to an agent of chemical warfare, although if a very large number of individuals require transfusion support, an unmeetable demand could be placed on blood transfusion centres and blood product donors. Transfusion of blood products may enable a person who has experienced a sub-ablative dose of marrow toxin to survive for days or weeks until marrow regeneration takes place, by circumventing lethal levels of anaemia or thrombocytopenia. Theoretically, it is possible for individuals unaffected by the chemical exposure to donate platelets and plasma by pheresis, thereby allowing much larger volumes of the commodity to be obtained than would be the case if, for example, they were retrieved from donation of a single unit of blood which is approximately 450 ml in volume. An individual who has been severely affected, and whose bone marrow has stopped producing cells either permanently or temporarily, may need transfusion with red cells once or twice a week. Platelet transfusions may be required daily, since the life of transfused platelets is short.

Transfusion of donated neutrophiles is a theoretical possibility, although in practice it is rarely possible to obtain these in sufficiently sustained, large numbers to enable a therapeutic effect. However, with pheresis of donors who have been primed with G-CSF with, or without dexamethasone, much larger numbers of neutrophils can be harvested than is possible when cells are taken without such prior stimulation or when the neutrophil layer is removed from a unit of red blood cells and plasma, the so-called 'buffy coat'. As yet, no definitive results have emerged to support or refute their use, however. Early studies suggested an advantage, but were uncontrolled and concerned a variety of underlying diseases, types of infection and differing doses of white cells (Strauss, 1995). More recent studies concerning cells collected after G-CSF and dexamethasone priming of the donor have suggested an advantage (Price *et al.*, 2000) in that transfusion recipients usually exhibit large post-transfusion neutrophil increments which are sustained for up to 24 h and where the neutrophils are capable of migrating to extravascular sites. Thus, transient benefit may result, particularly in individuals of low body weight, although evaluation of any such benefit remains very difficult. The current opinion is that neutrophil transfusions should be regarded as an experimental intervention and that if they are used, this should take place in the context of a randomized clinical trial.

It should be born in mind, however, that transfusion of any blood product is purely supportive and will not reverse the underlying problem of marrow aplasia or hypoplasia.

USE OF IRRADIATED BLOOD PRODUCTS

Prevention of transfusion-associated graft-versus-host disease

The standard UK definition of 'leucocyte–depletion' is less than 5×10^6 residual leucocytes per unit of red cells or platelets (Pamphilon *et al.*, 1999). Thus, if blood products are used, they should always be irradiated to 2 Gy prior to use, to ensure that the immunosuppressed recipient does not develop transfusion-mediated graft-versus-host disease because of the presence of residual viable lymphocytes in the donated blood product. Such graft-versus-host disease is usually fatal (Aoun *et al.*, 2003; Schroeder, 2002), and causes overwhelming skin desquamation, diarrhoea and liver failure secondary to fibrosis of the biliary system. Even if immunosuppression is given, this is usually insufficient to overcome the problem once allogeneic lymphocytes have engrafted in the recipient.

Prevention of cytomegalovirus infection

Blood products are now commonly leucodepleted at source, prior to being issued by the National Blood Service. In addition to reducing the chances of transfusion-related graft-versus-host disease developing in an immunosuppressed individual who receives unirradiated blood or platelets, this helps to avoid infection of the recipient by cytomegalovirus and other organisms which are mainly to be found in the leucocytes of infected individuals. The precise degree of leucocyte removal necessary to prevent CMV transmission is not known. If blood products are filtered to leave a residual leucocyte count of 10^5–10^6,this will leave less than 1000 latently infected leucocytes per unit, given that it is estimated that less than 0.2% of peripheral blood leucocytes in sero-positive individuals are latently infected (Hillyer *et al.*, 1999). Thus, although the likelihood of developing CMV infection or transfusion-related graft-versus-host disease will be reduced, these problems may still occur if blood products are not additionally irradiated. To date, there have been no randomized trials comparing the incidence of CMV infection in those receiving leucocyte-depleted blood which has not also been irradiated, with the incidence in individuals who have received leucocyte-depleted and irradiated blood products. The most relevant study is that by Ronghe *et al.* (2002), where immunosuppressed patients received CMV sero-negative red cell transfusions and leucocyte-depleted platelet transfusions. Patients who were sero-negative for CMV at the time of transplant did not experience a higher incidence of CMV infection than those who received irradiated platelet concentrates. Patients who were sero-positive for CMV at the time of transplantation experienced a 26% incidence of CMV infection, but this was no higher than in a historical group of patients who received CMV-negative platelets.

ANTI-MICROBIAL THERAPY

An individual who has been exposed to substantial doses of mustard gas will experience both neutropenia, which will predispose him/her to infection by bacterial agents, and lymphopenia which will result in defective cellular and humoral immunity, predisposing him/her to infection by fungi, viruses and other non-bacterial pathogens. The duration of leucopenia will correlate with the degree of exposure to the toxic agent. At worst, there will be no autologous recovery of neutrophil or lymphocyte function and an allogeneic stem cell transplant will be necessary. At best, the duration of such leucopenia will be for 1–6 weeks, before autologous marrow function returns.

The role of neutropenia as a major defect in host defence was defined by Bodey *et al.* in 1966 when they demonstrated that as the neutrophil count fell below 500–1000 mm^{-3}, the incidence of severe infection, the number of days spent taking antibiotics and the number of days of fever increased. If the neutrophil count was 500–1000 mm^{-3}, the incidence of infection was 14% and if it fell to below 100 mm^{-3} it was 24–60%. If the duration of granulocytopenia was more than 5 weeks, the incidence of infection was 100% (Dale *et al.*, 1979).

When sustained neutropenia develops after chemical or radiation exposure, problems with infection from bacterial, fungal and viral agents are very likely, particularly in view of the fact that the toxic exposure will, in all probability, have disrupted the integrity of the intestinal mucosal membrane, allowing both pathogens from the gut and mouth to enter the circulation and extraneous pathogens to breach the mucosal barrier. With little available in the way of a viable immune response against such invading organisms, prophylactic cover for accident victims should be provided with broad spectrum anti-bacterial, anti-fungal and anti-viral agents. The level of immunosuppression will parallel that seen in a patient suffering from a haematological malignancy who has received myeloablative doses of chemoradiotherapy. The difference is that in this situation the clinicians can predict the likely duration of such therapy-induced cytopenia, and they can therefore adapt their supportive therapy accordingly. In the situation of a mustard gas attack, degree of exposure is unlikely to be apparent, and therefore it will not be clear how long supportive care must last and whether there is any hope of bone marrow recovery.

ANTI-BACTERIAL PROPHYLAXIS

It was realized in the 1960s that sepsis from Gram negative organisms was associated with significant morbidity, if antibiotic therapy was delayed, and the prompt introduction of empirical antibiotics led to a great reduction in mortality rates in neutropenic cancer patients (Schimpff *et al.*, 1971; Love *et al.*, 1980). Since then, many empirical regimens have been evaluated in clinical trials. Probably the most popular even now is the combination of a beta-lactam such as ceftazidime, imipenem or meropenem, with an aminoglycoside such as gentamicin or amikacin (EORTC International Antimicrobial Therapy Co-operative Group, 1987; Hughes *et al.*, 1990). Such a combination, plus teicoplanin or vancomycin, should cover most bacterial pathogens, including *Pseudomonas aeriginosa*, and the common staphylococcal and streptococcal skin contaminants. However, problems may be encountered with emergence of resistant bacterial strains or with the nephrotoxicity associated with the aminoglycosides unless levels are carefully monitored.

Oral prophylaxis with a broad-spectrum agent such as ciprofloxacin may be appropriate in some cases although it should be borne in mind that no antibiotic will provide protection against all organisms and prolonged use may again encourage emergence of resistant strains, including *Pseudomonas aeruginosa* (Jones, 1999).

The role of gut decontamination

The normal gut harbours many bacilli which may cause overwhelming infection if the integrity of the gut mucosal barrier is disrupted. Over the years, a number of decontamination regimens have been tried, designed to rid the bowel of the Gram negative organisms which have the potential to cause severe sepsis, while retaining the anaerobic flora to prevent colonization (Levine *et al.*, 1973). Examples of such drug regimens include combinations of non-absorbable colistin, neomycin and amphotericin B, or an aminoglycoside with amphotericin B. Disadvantages include lack of compliance since the drugs can induce nausea and vomiting, or colonization by resistant strains.

FUNGAL INFECTIONS

Invasive fungal infections may also present a problem in those who remain immunosuppressed for more than a couple of weeks, and oral prophylaxis with one of the triazoles, either fluconazole – if candidiasis infection seems more likely, or itraconazole or the newer anti-fungal agent, voriconazole if infection with aspergillus is a significant possibility, is appropriate (Winston *et al.*, 2003; Pearson *et al.*, 2003). Fluconazole is ineffective against molds, although studies have shown that it reduces the incidence of superficial fungal infections such as occur with mucositis (Samonis *et al.*, 1990). Goodman *et al.* (1992) used 400 mg a day fluconazole as prophylaxis and they showed this to be clinically effective in reducing systemic fungal sepsis. Of major concern, however, is the theoretical selection of more resistant types of Candida such as *Candida krusei*, which are intrinsically resistant to fluconazole, and studies have shown a higher incidence of *C. krusei* infections in patients who had received fluconazole prophylaxis than in those who had not (Wingard *et al.*, 1991,1993). However, with the general reduction in other non-krusei Candida species causing infection, the benefits seem to outweigh the disadvantages. Since high-dose fluconazole is expensive, many units use lower doses for prophylaxis (Wakerly *et al.*, 1996), although to date there is no consensus concerning the ideal, cost-effective dose for use in this situation.

No prophylactic regimen has been proven to be clearly effective in the prevention of invasive mold infection, whether the species is *Aspergillus* or one of the newly emerging molds such as *Fusarium*, but itraconazole, voriconazole or one of the amphotericin preparations may provide some protection.

Both *Aspergillus* and *Candida* are very common nosocomial infections in the environment, although infections with *C. Tropicalis, albicans* and *C. parapsilosis* which used to be very prevalent have now decreased and there is a corresponding increase in infections with azole-resistant species such as *C. glabrata and krusei* (Dignani *et al.*, 2002). The incidence and severity of invasive fungal infections is closely related to the presence and severity of various other risk factors. These include:

(1) Previous fungal infection and colonization (unlikely in this population of previously fit individuals).
(2) State and degree of immunosuppression (probably difficult to ascertain with certainty in this patient population, although older age is a factor).
(3) Other organ dysfunction (renal failure, pulmonary dysfunction, severe mucositis).

In the event of certain fungal infection, intravenous anti-fungal therapy should be commenced with one of the polyenes, amphotericin B deoxycholate or one of its lipid formulations, all of which have a broad spectrum of activity against most of the *Candida* and *Aspergillus* species and *Trichosporon* (Ravankar and Graybill, 2002).

Other broad-spectrum antifungal agents include:

- ***Flucytosine*** – active against most *Candida* species and *C. neoformans* but should be used in combination to guard against emergence of resistant strains.
- ***Voriconazole*** – fungicidal against *Aspergillus* species and fungistatic against a wide variety of yeasts and molds, including some of the newly emerging fungi which are resistant to agents currently available.
- ***Posaconazole*** – a derivative of itraconazole which provides cover against *Aspergillus, Candida species* and *Cryptococcus* but is only available in an oral formulation.
- ***Ravuconazole*** – structurally similar to fluconazole and voriconazole and active against *Candida* species, *Aspergillus* species and *C. neoformans*.
- ***Caspofungin*** – a new Echinocandin, active against *Aspergillus* and *Candida* species but not effective against *Cryptococcus* and some other molds.

Prophylaxis against *Pneumocystis carinii*, classified as a fungus although in many ways it behaves more as a protozoal infection, should be with trimethoprim-sulphamethoxazole. However, it should be borne in mind that this can cause bone marrow suppression in its own right.

ANTI-VIRAL AGENTS

Herpes zoster and herpes simplex infections are common in individuals with impaired immunity and the use of prophylactic agents such as acyclovir has certainly proven beneficial in preventing these infections in stem cell transplant recipients. Infection with pneumocystis and cytomegalovirus (CMV) may also prove problematic in those suffering long-term immunosuppression, and prophylaxis with cotrimoxazole or pentamidine should be given to prevent pneumocystis infection, and valganciclovir can be used to prevent CMV reactivation. Valganciclovir, which also has activity against the herpes viruses, is the oral pro-drug of ganciclovir and it well absorbed from the gut, although ganciclovir must be administered intravenously. Unfortunately, however, both valganciclovir and ganciclovir can cause myelosuppression which could clearly present problems in the clinical setting of myelosuppression resulting from other causes. There are fewer agents available for viral infection prophylaxis and treatment, the main one being acyclovir which is active predominantly against the herpes viruses, or its newer derivatives, valaciclovir and famiciclovir. Other agents include cidofovir and foscarnet, which both have activity against the herpes viruses and CMV, but these agents are potentially nephrotoxic.

Respiratory virus infections (respiratory syncitial virus [RSV], parainfuenza viruses, influenza viruses and adenoviruses) are all highly contagious and symptomatic individuals should be isolated before virologic confirmation is available. To prevent progression of RSV from the upper to the lower respiratory tract, aerosolized ribavirin may be used alone or in combination with RSV-specific antibodies or palivizumab, an RSV-specific monoclonal antibody (Boeckh *et al.*, 2001).

Other supportive measures

HUMAN IMMUNOGLOBULIN

This is non-specific in its antimicrobial action but may offer protection against various bacterial and non-bacterial infections. It is commercially available from a number of sources. Intravenous immunoglobulin (IvIg) given every 3 weeks has been shown to be associated with a reduction in the number of bacterial infections suffered by patients with immunodeficiency resulting from chronic lymphocytic leukaemia, although the number of fungal and viral infections was not reduced and there was no survival advantage in the group of patients receiving IvIg (Cooperative Group for the Study of Immunoglobulin in Chronic Lymphocytic Leukaemia, 1988). As a short-term measure in the context of marrow depression secondary to mustine poisoning, its use may confer some advantages, although it is very costly.

GENERAL MEASURES TO PREVENT ACQUISITION OF INFECTION BY BACTERIAL, FUNGAL OR VIRAL PATHOGENS

These should include the following:

- Careful hand washing before coming into contact with a patient.
- Preventing dust accumulation.
- Maintaining negative pressure in high-risk areas of the environment.
- Avoiding patient exposure to tap water and food products likely to be contaminated.
- Using HEPA filtered accommodation which reduces exposure to air-borne particles.

Stem cell transplantation

HISTORICAL BACKGROUND

One of the earliest descriptions of an attempt to treat a patient suffering from aplastic anaemia was reported in 1939, when a patient was given a small quantity of bone marrow from his brother (Osgood *et al.*, 1939). Prior to this, Gloor (1930) described the cure of a patient with acute myeloid leukaemia by stem cell transplantation. However, stem cell transplantation as an approach for treating leukaemia and aplastic anaemia was not investigated seriously until the 1960s when it was found that dogs could survive 2–4 times the lethal exposure to irradiation (TBI) if they were given an infusion of bone marrow cells removed and stored prior to the TBI (Mannick *et al.*, 1960). However, almost all attempts to achieve allogeneic grafts in humans were unsuccessful. In 1958, six physicists were accidentally exposed to large doses of mixed gamma and neutron irradiation at Vinca in Yugoslavia (Mathé *et al.*, 1959). The most severely irradiated died, but of the remaining five, four were judged to have received a radiation dose of between 600–1000 rad. They were treated with allogeneic bone marrow infusions, and red cell antigen studies demonstrated that successful but temporary engraftment ensued. Of note was the fact that the pecipient red cell output paralleled the amount of marrow initially infused, perhaps indicating that the subjects who received the larger numbers of stem cells experienced enhanced haemopoietic recovery. Judging by the changes that the subjects who were not ABO and Rhesus compatible with their donors experienced, autologous reconstitution of haematopoietic activity occurred eventually, but the allogeneic bone marrow served to protect the patients until this had taken place – it was possible to document the transient changes in blood and Rhesus red cell grouping after the person exposed to radiation received the allogeneic marrow, but this eventually reverted to the original red cell group of the patient (Mathé *et al.*, 1959). Mathé *et al.*, (1965) subsequently went on to achieve the first durable allogeneic graft in a patient with leukaemia, but the patient died from problems probably related to graft-versus-host disease, or ‘secondary disease’, a condition which was at that time little understood.

Over the ensuing years, results from allogeneic transplantation improved. The need to use immunosuppressants to protect the recipient from graft-versus-host disease was recognized, and knowledge of human leukocyte groups and typing increased, allowing more accurate identification of a tissue-matched sibling donor. There followed the establishment of large panels of volunteer donors which made it possible in many cases to identify a donor outside of the immediate family despite the heterogeneity of the major histocompatibility complex.

BONE MARROW HARVESTING

Early on, bone marrow stem cells were harvested from the posterior and anterior iliac crests or the sternum, while the patient or donor was under a general anaesthetic. It was fairly rapidly established that a minimum number of nucleated cells was necessary to ensure engraftment, and that this related to the body weight of the recipient. At this point, it was not possible to identify the stem cells or early precursor cells immunologically, and a total nucleated cell count of the harvested material was obtained, into which the body weight of the recipient was divided in an attempt to ensure an adequate harvest. Normally, approximately 1 litre of harvested blood and marrow would be required to ensure the presence of a sufficient number of stem cells for an adult patient. The harvesting procedure would last for 1–2 h, depending on ease of aspiration, and

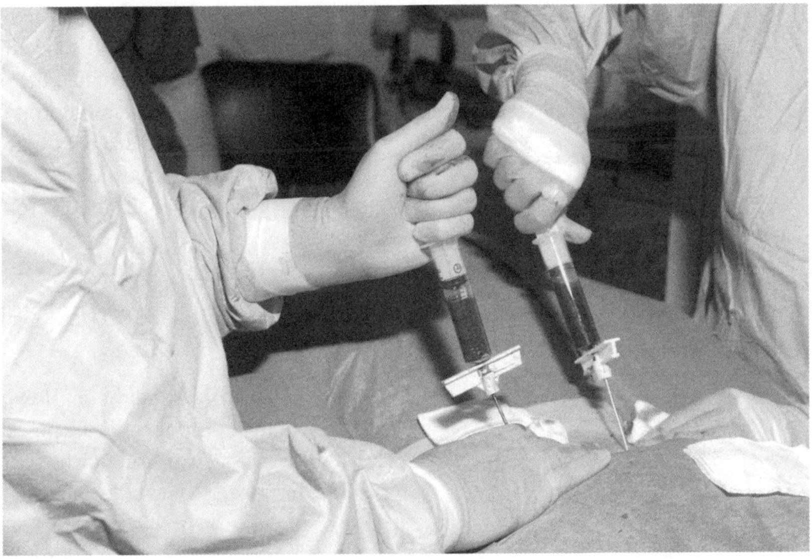

Figure 5. Bone marrow being harvested from the posterior iliac crests by two operators

two operators would remove marrow from the pelvis simultaneously (Figure 5), having inserted a needle containing a trocar into the bone marrow space, through the cortex of the bone (Figures 6 and 7). Once in position, the trochar was removed and marrow and blood were aspirated into heparinized 5 or 10 ml syringes which were then transferred into a sterile, heparinized blood collection bag, ready for either cryopreservation or for reinfusion into the patient. It was quickly realized that only small volumes of blood and marrow should be aspirated at each insertion of the needle, as this ensured removal of a relatively high number of nucleated cells in a relatively small volume of blood. As skin is fairly elastic, it was often possible to insert the needle into a single skin hole, while moving it around and inserting it into quite a large area of underlying bone through a number of individual holes made in the cortex.

The total volume of blood and marrow aspirated in order to obtain sufficient stem cells to support engraftment dictated whether or not the donor would require a blood transfusion after surgery, or whether it was sufficient to administer a course of iron and folic acid to help him make up the deficit. If patients were having their own

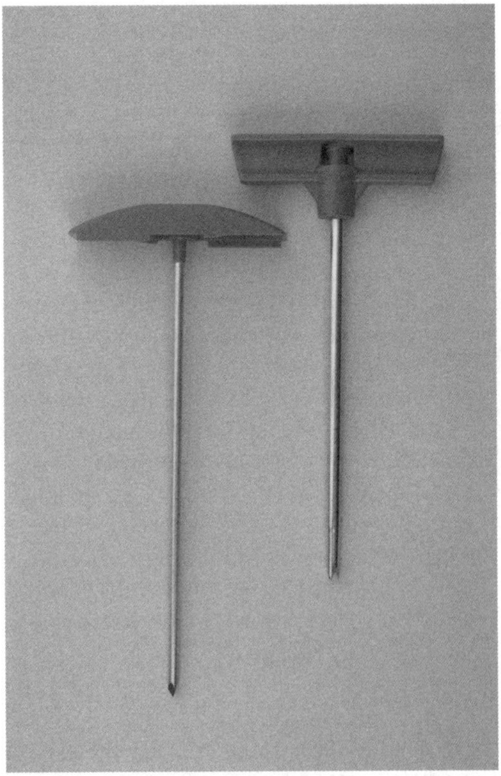

Figure 6. A bone marrow trephine biopsy needle such as may be used for bone marrow stem cell harvesting

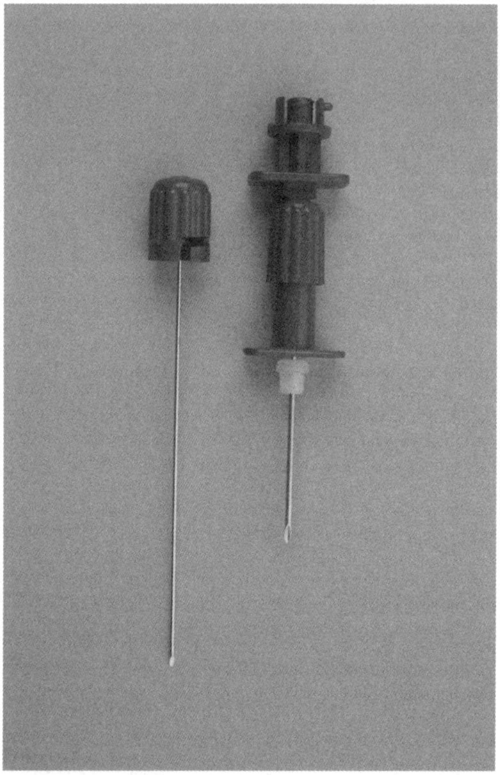

Figure 7. A bone marrow aspiration needle such as may be used for bone marrow stem cell harvesting

autologous marrow harvested for subsequent use as 'rescue' after high-dose chemoradiotherapy, it was usual to cross-match donated blood from the National Blood Service for them as in all probability they had already received numerous blood transfusions or would require transfusion of blood products during the course of their high-dose treatment. However, for sibling, or matched but unrelated stem cell donors it is more usual to attempt to avoid transfusion of donated blood because of the potential problem of causing infection in a normal donor, although the incidence of this in association with a blood transfusion is extremely small. For such healthy, normal donors it is usually adequate to provide them with a course of iron and folic acid post-donation, any deficit in haemoglobin level usually being corrected within a week or two of donation. If an allogeneic donor should require a normal haemoglobin level immediately after the harvesting procedure, it is possible for them to donate up to three units of their own blood which is stored in a blood refrigerator at weekly intervals before the stem cell harvesting procedure, and which can be returned to them following the procedure. By donating at intervals, they have had the opportunity to make up the loss in haemoglobin prior to the harvesting procedure so that they commence this with a normal haemoglobin level while having their autologous blood available for reinfusion afterwards.

PERIPHERAL BLOOD STEM CELL HARVESTING

As already described, very complex interactions between many components take place in both the normal, physiological situation and after the bone marrow has been subjected to a toxic insult. This requires regulation of the quantities of the various precursor and end cells circulating in the blood and present in the bone marrow. It was appreciated as early as the 1960s from animal studies that stem cells are present in the peripheral blood as well as in the bone marrow, albeit in very small numbers in the resting state. For the last fifteen years or so, it has become more the practice to harvest stem cells from peripheral blood rather than from the bone marrow itself. Higher yields of stem cells can be obtained which, in turn, lead to faster engraftment after reinfusion. In addition, the necessity for the donor to receive a general anaesthetic with its possible attendant risks is avoided, and in some cases, such as in the presence of bone marrow fibrosis or when the bone marrow space is extensively involved by metastatic tumour, it may be possible to obtain stem cells by peripheral harvesting when it would have been impossible to obtain them from the marrow itself.

STEM CELL MOBILIZING TECHNIQUES

Prior to peripheral blood stem cell harvesting, however, the stem cells must be mobilized or stimulated to enter the peripheral blood. The mobilization process is set into motion by stressing neutrophils and osteoclasts into activity by the use of cytokines with, or without chemotherapy. This results in the shedding and release of membrane-bound stem cell factor (SCF), proliferation of progenitor cells and activation or degradation of various adhesion molecules. Stem cells can be 'mobilized' to enter the peripheral blood in increased quantities either by the use of G-CSF alone, or by the use of G-CSF in combination with a cytotoxic drug, most commonly etoposide or cyclophosphamide. The mobilization process mimics an enhancement of the physiological release of progenitor cells from the bone marrow reservoir in response to stress such as may occur during injury or inflammation; a large, but non-myeloablative dose of a cytotoxic agent such as cyclophosphamide, iphosphamide or etoposide is administered, often with G-CSF stimulation. The latter may further enhance the number of stem cells entering the peripheral blood as bone marrow stem cell recovery takes place.

The cytotoxic drug has the effect of initially reducing the peripheral blood count substantially although the dose must, of course, be insufficient to permanently ablate the bone marrow stem cell pool. Approximately ten days later, when the effects of this are abating and bone marrow function is recovering, an enhanced number of stem cells enters the peripheral blood which can then be harvested by attaching the patient to a cell separator (Magagnoli *et al.*, 2001; Junghanss *et al.*, 2001).

In the situation of a mustard gas attack it may well be that, depending upon the dose that has been received, an enhanced number of stem cells will enter the blood if and when the peripheral blood count recovers. These will be the victims who do not need bone marrow support, however, since their count has recovered spontaneously. For those who have received a permanently myeloablative dose of the toxin, it will obviously not be possible to harvest autologous stem cells as these will have been eradicated, and only allogeneic stem cells will be able to reconstitute haematopoiesis. The difficulty will be early identification of individuals requiring stem cell support over those where the count will eventually recover spontaneously who will only need temporary blood product and antibiotic support.

THE CELL SEPARATOR

Various designs of cell separators have been used since the technique of stem cell harvesting from peripheral blood first became an option. In the main, these have relied upon the principle of centrifugation of the blood so that it separates into its respective components. The cell separators in use today rely upon a continuous flow of venous blood emerging under low pressure from the vein of one of the patient's arms and entering the centrifuge bowl. The heaviest part of the blood, namely the red blood cells, separates and sinks to the bottom of the bowl. The middle section, containing platelets, white cells and stem cells, is diverted into a separate collecting bag, and the lightest, acellular part of the blood, the plasma, remains uppermost. During collection, the red cells and plasma are diverted back into the patient via a vein on the opposite arm or into a centrally placed venous catheter, so that only the fraction of blood containing the stem cells is retained for immediate infusion or for cryopreservation. Figure 8 depicts the COBE Spectra Apheresis System (Gambro, Lakewood, USA) which is currently in very wide usage.

The volume of fractionated blood product containing the stem cells which is retained for cryopreservation is usually of the order of only 150–180 ml, so that patients do not experience the severe volume loss during collection as may occur when bone marrow stem cells are harvested. Thus, neither a replacement blood transfusion with its potential attendant problems, nor fluid replacement, are usually necessary after peripheral blood stem cell harvesting. In addition, the need for a general anaesthetic is avoided and the pain in the pelvic bones resulting from the trauma of aspiration during bone marrow harvesting, is avoided. However, peripheral blood stem cell harvesting usually requires two or three sessions on the separator to obtain sufficient stem cells, and the preparation of the donor with G-CSF causes him to experience bony aches and pains and fever, similar to the symptoms arising with a viral infection. These are however, limited to the time that the G-CSF is being administered.

PROBLEMS ASSOCIATED WITH BONE MARROW AND PERIPHERAL BLOOD STEM CELL HARVESTING

A number of publications (Jin *et al.*, 1985; Goldberg *et al.*, 1995) have characterized the problems associated with bone marrow and stem cell harvesting. With bone marrow harvesting, these

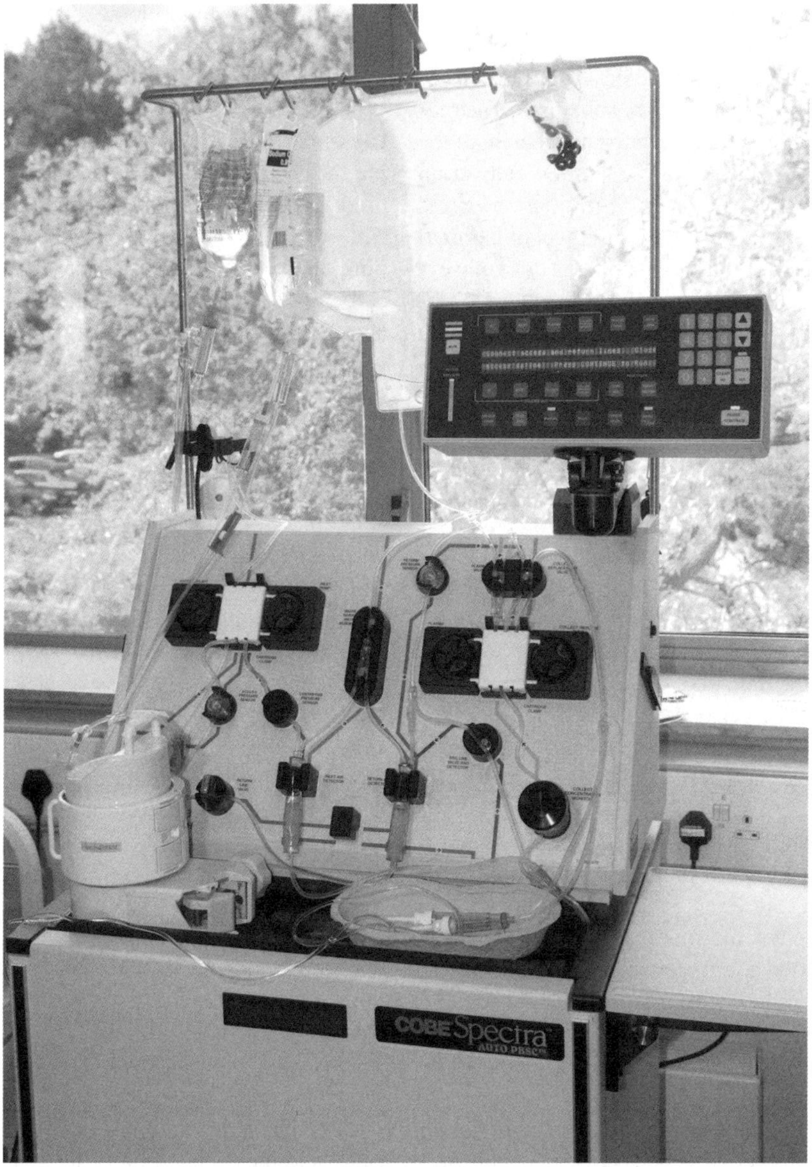

Figure 8. The COBE Spectra Apheresis System ready for use

largely relate to complications of general anaesthesia, pelvic bone pain in the harvested area or postoperative fever. Serious problems are fortunately rare, and in 2027 transplants reported to the International Bone Marrow Transplant Registry and 1263 transplants performed in Seattle, the incidence of major, i.e. life-threatening complications in the combined series was 0.27%. Neither death nor permanent problems occurred in any of the donors (Bortin and Buckner, 1983).

With peripheral blood stem cell harvesting, the most frequent problems relate to central venous catheter occlusion and bone pain in association with the G-CSF priming. Infectious complications during harvesting were experienced in 16% of patients during the PBSC harvesting.

The G-CSF commonly used to 'prime' patients and donors before peripheral blood stem cell harvesting can be given to normal, non-patient donors and has not been shown to have

Table 2. Steps to be taken after exposure to mustine gas in relation to possible bone marrow failure

(1) Check full blood count to use as a baseline
(2) Take blood sample for tissue-typing as early as possible and attempt to identify a matched sibling donor or a donor from the unrelated donor panels
(3) Start cytokine therapy if mustine dose is thought to be high enough to cause bone marrow suppression or if there is a fall in blood count:
- Stem cell factor (if available)
- G-CSF
- GM-CSF
- Erythropoietin

(4) Start supportive therapy:
- Anti-bacterials
- Anti-fungals
- Anti-virals
- Blood products (red cells and platelets)

(5) Monitor blood count on alternate days to detect any drop early
(6) Monitor closely for signs of infection

any long-term adverse side-effects. The use of cytotoxic drugs to stimulate stem cell production must obviously be limited to patients with malignant diseases who are being mobilized for their autologous stem cells to be used as 'rescue' after high-dose chemoradiotherapy. There is no role for their use to mobilize stem cells in the normal, non-patient donor because of the short-term and possible long-term side-effects.

SUMMARY OF STEPS TO BE TAKEN TO PROTECT AGAINST IRREVERSIBLE BONE MARROW DAMAGE AFTER EXPOSURE TO MUSTARD GAS

Table 2 summarizes the most important issues relating to bone marrow stem cell damage after mustine exposure.

Before the disaster

COLLECTION OF AUTOLOGOUS PERIPHERAL BLOOD STEM CELLS PROSPECTIVELY

Ideally, the entire population of potential gas victims should have their peripheral blood stem cells collected after simulation with G-CSF. These should be cryopreserved in liquid nitrogen where they will keep for many years if not required. If the entire population cannot undergo this procedure because of economic constraints and storage difficulties, then key individuals whose continuing existence is considered vital should undergo the procedure.

PROSPECTIVE TISSUE TYPING

All, or key individuals and those at particular risk in the event of an attack, should have their tissue type electively established along with that of any full siblings so that in the event of a disaster a donor has already been identified. If there is no sibling donor available, and any full sibling has only a 'one in four chance' of being a full match, a search through the unrelated donor panels, such as the Anthony Nolan Trust, can be instigated. This is based in the Royal Free Hospital in London and was established in 1974. It now holds data on more than 329 000 volunteer donors. In the UK, there is also the British Bone Marrow Registry, established in 1987, and holding information on approximately 150 000 donors and the Welsh Bone Marrow Donor Registry, established in 1989, and holding information on 33 000 donors. Worldwide, there are more than 8 000 000 volunteer donors registered in 52 donor registries in 37 countries.

After the disaster

RAPID BLOOD SAMPLING TO ESTABLISH THE VICTIM'S TISSUE TYPE IS VITAL

If the above steps have not been taken prior to the disaster, samples for tissue typing should be taken from victims as soon as is feasible and certainly within 24–48 h, in order to ensure that sufficient white blood cells remain in the peripheral blood to make tissue typing possible and subsequent identification of a donor.

BLOOD PRODUCT SUPPORT

Red blood cells

Blood counts should be monitored regularly so that blood and platelets can be transfused when appropriate. Broadly speaking, one unit of blood raises the haemoglobin level by approximately $1\ g\,dl^{-1}$. The normal range for haemoglobin level in an adult male is $13–18\ g\,dl^{-1}$ with significant symptoms of anaemia developing when the haemoglobin level falls below approximately $9\ g\,dl^{-1}$. The extent to which the bone marrow has been damaged will dictate the frequency of transfusions; if no red cell precursors at all are being produced, weekly transfusions will probably be necessary, whereas if some cells are being produced the haemoglobin is likely to remain at an acceptable level for a longer period of time.

Platelets

The normal platelet count is $150–400 \times 10^9\ l^{-1}$. Easy bruising and petechiae are likely to occur if the platelet count drops to below $20 \times 10^9\ l^{-1}$, and bleeding from the gums, nose and internal organs is likely at levels below this, particularly if skin trauma or trauma to other organs has occurred. Platelets are likely to need replacing on a daily basis as their survival after transfusion is only in the order of one or two days. Attempts should be made to maintain the platelet count at above $20–50 \times 10^9\ l^{-1}$ to minimize the chances of bleeding, particularly if there is concomitant trauma to other organs.

Leucocytes

As outlined earlier, neutrophils and monocytes cannot be replaced effectively by transfusion. Problems with bacterial infection are likely if the neutrophil count drops below $0.5 \times 10^9\ l^{-1}$, and fungal infections are more likely in the presence of neutopenia and monocytopenia. Lymphocytes cannot be replaced effectively by transfusion. Severe lymphopenia may persist for a year or more, long after the rest of the blood count has recovered. This results in a predisposition to viral and other non-bacterial infections.

GROWTH FACTOR SUPPORT

Recombinant erythropoietin and G-CSF injections can be administered either daily, or at less frequent intervals if the longer-acting, pegylated formulations are available. For short-acting G-CSF, a dose in the order of $10–15\ \mu g\,kg^{-1}$ daily may provide some protection. For short-acting erythropoietin, 50–100 International Units (IU) per kilogram body weight three times a week should be given either intravenously or subcutaneously.

ANTI-MICROBIAL THERAPY

Bacterial infections

As prophylaxis, a broad-spectrum anti-bacterial agent such as ciprofloxacin, 250 mg twice daily, should be administered orally. If fever develops, blood cultures should be taken in an attempt made to identify the infecting organism. Broad-spectrum intravenous antibiotics should then be commenced using a regimen combining agents effective against Gram negative, Gram positive and anaerobic organisms. Tazocin, gentamicin and teicoplanin is a commonly used regimen.

Fungal infections

Prophylaxis against fungal infections should also be commenced with daily oral itraconazole or fluconazole. If fever develops and the individual has had a poor white blood cell count for more than 2–3 weeks, intravenous amphotericin B or the liposomal formulation, Ambisome, should be given.

Prophylaxis against *P carinii* with either monthly pentamidine inhalations, or a small, daily dose of trimethoprim, should be commenced.

General prophylaxis against infection

Intravenous human immunoglobulin may be administered every 3–4 weeks at a dose of approximately 0.4 grams per kilogram body weight.

In patients who are neutropenic, careful handwashing should be undertaken before any form of physical contact with the patient.

REGULAR MONITORING FOR INFECTION AND BLOOD COUNT

This should include the following:

(a) In pancytopenic individuals a weekly chest X-ray should be taken to check for infective changes. Weekly nose swabs, throat swabs and urine samples should be taken for culture. In the event of a fever developing, these should be repeated before antibiotics are started or changed.

(b) Blood sampling should be carried out on alternate days to assess the need for blood product replacement and to monitor progress if growth factors are being given. Hepatic and renal function should be monitored at the same time.

REFERENCES

Albeda SM and Buck CA (1990). Integrins and other cell adhesion molecules. *FASEB J*, **4**, 2868–2880.

Aoun E, Shamseddine A, Chehal A *et al.* (2003). Transfusion-associated GVHD: 10 years' experience at the American University of Beirut-Medical Center. *Transfusion*, **43**, 1672–1676.

Antonchuk J, Sauvageau G and Humphries RK (2002). HoxB4-induced expansion of adult hematopoietic cells *ex vivo*. *Cell*, **109**, 39–45.

Ayash LJ, Elias A, Hunt M *et al.* (1994). Recombinant human erythropoietin for the treatment of anaemia associated with autologous bone marrow transplantation. *Br J Haematol*, **87**, 153–161.

Bence-Bruckler I, Bredeson C, Atkins H *et al.* (1998). A randomised trial of granulocyte-colony stimulating factor (Neupogen) starting day 1 vs day 7 post-autologous stem cell transplantation. *Bone Marrow Transplant*, **22**, 965–969.

Bernstein ID, Andrews RG and Zsebo KM (1991). Recombinant human stem cell factor enhances the formation of colonies by CD34+ and CD34+lin- cells, and the generation of colony-forming cell progeny from DC34+lin- cells cultured with Interleukin3, granulocyte-colony stimulating-factor, or granulocyte-macrophage colony stimulating factor. *Blood*, **77**, 2316–2321.

Biggs JC, Atkinson KA, Booker V *et al.* (1995). Prospective randomised double-blind trial of the *in-vivo* use of recombinant human erythropoietin in bone marrow transplantation from HLA-identical donors. The Australian bone marrow Transplant Study Group. *Bone Marrow Transplant*, **15**, 129–134.

Bishop NJ, Williams DM, Compston JC *et al.* (1995). Osteoporosis in severe congenital neutropenia treated with granulocyte colony stimulating factor. *Br J Haematol*, **89**, 927–928.

Bodey GP, Buckley M, Sathe YS *et al.* (1966). Quantitative relationships between circulating leukocytes and infection in patients with acute leukaemia. *Ann Int Med*, **64**, 328–340.

Boeckh M, Berrey MM, Bowden RA *et al.* (2001). Phase I evaluation of the RSV-specific humanised monoclonal antibody palivizumab in hematopoietic stem cell transplant recipients. *J Infect Dis*, **184**, 350–354.

Bortin MM and Buckner CD (1983). Major complications of marrow harvesting for transplantation. *Exp Hematol*, **11**, 916–921.

Bradley TR and Metcalfe D (1966). The growth of mouse bone marrow cells *in vitro*. *Aus J Exp Biol Med Sci*, **44**, 287–300.

Broudy VC (1997). Stem cell factor and haematopoiesis. *Blood*, **90**, 1345–1364.

Carnot P and Deflandre C (1906). Sur l'activite haemopoietique du serum au cours de la regeneration du sang. *Compt rend Acad Sci (Paris)*, **143**, 384–386.

Chao NJ, Schriber JR, Long GD *et al.* (1994). A randomised study of erythropoietin and granulocyte-colony stimulating factor (G-CSF) versus placebo and G-CSF for patients with Hodgkin's and non-Hodgkin's lymphoma undergoing autologous bone marrow transplantation. *Blood*, **83**, 2823–2828.

Co-operative Group for the Study of Immunoglobulin in Chronic Lymphocytic Leukaemia (1988). Intravenous immunoglobulin for the prevention of infection in chronic lymphocytic leukaemia. *N Eng J Med*, **6**, 902–907.

Dale DC, Guerry D IV, Wewerka JR *et al.* (1979). Chronic neurtopenia. *Medicine*, **58**, 128–144.

Demirer T, Ayli M, Dagli M *et al.* (2002). Influence of post-transplant recombinant human granulocyte colony stimulating factor administration on per-transplant morbidity in patients undergoing autologous stem cell transplantation. *Br J Haematol*, **118**, 1104–1111.

Dignani MC and Anaissie EJ. (2002). Candida. In: *Clinical Mycology*, Volume 1, 1st Edition (EJ Anaissie, M McGinnis and MA Pfaller, eds), pp. 195–239. Philadelphia, USA: Churchill Livingstone.

EORTC International Antimicrobial Therapy Cooperative Group. (1987). Ceftazidime combined with short and long course amikacin for empirical therapy of gram-negative bacteraemia in cancer patients with granulocytopenia. *New Eng J Med*, **317**, 1692–1698.

Flaharty KK, Grimm AM and Vlasses PH (1989). Epoetin: human recombinant erythropoietin. *Clin Pharm*, **8**, 769–782.

Gloor W (1930). Ein Fall von geheilter Myeloblastenleukaemie. *Münch Med Wochen*, **77**, 1096–1098.

Goldberg SL, Mangan KF, Klumpp TR *et al.* (1995). Complications of peripheral blood stem cell harvesting: review of 554 PBSC. *J Hematother*, **4**, 85–90.

Goodman JL, Winston DJ, Greenfield RA *et al.* (1992). A controlled trial of fluconazole to prevent fungal infections in patients undergoing bone marrow transplantation. *New Eng J Med*, **326**, 845–851.

Hasegawa M, Matsuki Y, Ozawa S *et al.* (1968). The role of erythropoietin in aplastic anemia – some aspects of the etiology and treatment of asplastic anemia. *Keio J Med*, **17**, 109–123.

Hillyer CD, Lankford KV, Roback JD *et al.* (1999). Transfusion of the HIV-seropositive patient: immunomodulation, viral reactivation and limiting exposure to EBV (HHV-4), CMV (HHV-5) and HHV-6,7 and 8. *Transfus Med Rev*, **13**, 1–17.

Hughes WT, Armstrong D, Bodey GP *et al.* (1990). From the Infectious Diseases Society of America. Guidelines for the use of antimicrobial agents in neutropenic patients with unexplained fever. *J Infect Dis*, **161**, 381–396.

Jacobs K, Shoemaker C, Rudersdorf R *et al.* (1985). Isolation and charactertisation of genomic and cDNA clones of human erythropoietin. *Nature*, **313**, 806–810.

Jin NR, Hill RS, Petersen FB *et al.* (1985). Marrow harvesting for autologous marrow transplantation. *Exp Hematol*, **13**, 879–884.

Jones RN (1999). Contemporary antimicrobial susceptibility patterns of bacterial pathogens commonly associated with febrile patients with neutropenia. *Clin Infect Dis*, **29**, 495–502.

Junghanss C, Leithauser M, Wilhelm S *et al.* (2001). High-dose etoposide phosphate and G-CSF mobilises peripheral blood stem cells in patients that previously failed to mobilise. *Ann Hematol*, 80, 96–102.

Klaesson S, Ringden O, Ljungman P *et al.* (1994). Reduced blood transfusion requirements after allogeneic bone marrow transplantation: results of a randomised, double-blind study with high-dose erythropoietin. *Bone Marrow Transplant*, **13**, 397–402.

Levine AS, Siegel SE, Schreiber AD *et al.* (1973). Protected environments and prophylactic antibiotics – a prospective controlled study of their utility in the therapy of acute leukaemia. *New Engl J Med*, **288**, 477–483.

Link H, Boogaerts MA, Fauser AA *et al.* (1994). A controlled trial of recombinant human erythropoietin after bone marrow transplantation. *Blood*, **84**, 3327–3335.

Love LJ, Schimpff SC, Schiffer CA *et al.* (1980). Improved prognosis of granulocytopenic patients with gram-negative bacteraemia. *Am J Med*, **68**, 643–648.

MacVittie TJ, Farese A, Herodin F *et al.* (1996). Combination therapy for radiation-induced bone marrow aplasia in nonhuman primates using synthetic SC-55494 and recombinant human granulocyte-colony stimulating factor. *Blood*, **97**, 4129–4135.

Magagnoli M, Sarina B, Balzarotti M *et al.* (2001). Mobilising potential of ifosphamide/vinorelabine-based chemotherapy in pretreated malignant lymphoma. *Bone Marrow Transplant*, **10**, 923–927.

Mannick JA, Lochte HL, Ashley CA *et al.* (1960). Autografts of bone marrow in dogs after lethal total body irradiation. *Blood*, **15**, 255–266.

Mathé G, Jammet H, Pendic B *et al.* (1959). Transfusions and grafts of homologous bone marrow in humans accidentally irradiated to high doses. *Rev Fr Etudes Clin Biol*, **4**, 226–238.

Mathé G, Schwarzenberg L, Amiel JL *et al.* (1965). Bone marrow Transplantation in the treatment of leukaemia. *Presse Med*, **10**, 1043–1046.

Mayer P, Schnntze E, Lam C *et al.* (1991). Recombinant murine granulocyte macrophage colony stimulating factor augments recovery and enhances resistance to infections in myelosuppressed mice. *J Infect Dis*, **163**, 584–590.

Metcalf D. (1991). Lineage commitment of hemopoietic progenitor cells in developing blast cell colonies: influence of colony-stimulating factors. *Proc Nat Acad Sci USA*, **88**, 1310–1314.

Metcalf D and Moore MA (1971). Haemopoietic cells. In: *Frontiers in Biology*, Volume 24 (Neuberger A & Tatum EL, eds), pp. 172–271. Amsterdam, The Netherlands: North-Holland.

Miyake T, Chung CK and Goldwasser E (1977). Purification of human erythropoietin. *J Biol Chem*, **252**, 5558–5564.

Moore MA and Metcalf D (1970). Ontogeny of the haemopoietic system; yolk origin of in-vivo and in-vitro colony forming cells of the developing mouse embryo. *Br J Haematol*, **18**, 279–296.

Nash RA, Schuening FG, Seidel K *et al.* (1994). Effect of recombinant canine granulocyte-macrophage colony-stimulating factor on hemopoietic recovery after otherwise lethal total body irradiation. *Blood*, **83**, 1963–1970.

Neelis KJ, Hartong SC, Egeland T *et al.* (1997). The efficacy of single-dose administration of thrombopoietin with coadministration of either granulocyte/macrophage colony-stimulating factor or granulocyte colony-stimulating factor in myelosuppressed Rhesus monkeys. *Blood*, **90**, 2565–2573.

Nemunaitis J, Rabinowe SN, Singer JW *et al.* (1991). Recombinant granulocyte-macrophage colony-stimulating factor after autologous bone marrow transplantation for lymphoid cancer. *New Eng J Med*, **324**, 1773–1778.

Osgood EE, Riddle MC and Matthews TJ (1939). Aplastic anaemia treated with daily transfusions and intravenous marrow; case report, *Ann Int Med*, **13**, 357–367.

Ozer H, Armitage JO, Bennett CL *et al.* (2000). 2000 update of recommendations for the use of hematopoietic colony-stimulating factors: evidence-based clinical practice guidelines. *J Clin Onc*, **18**, 3558–3585.

Pamphilon D, Rider J, Barbara J *et al.* (1999). Prevention of transfusion-transmitted cytomegalovirus infection. *Transfusion Med*, **9**, 115–123.

Pearson MM, Rogers PD, Cleary JD *et al.* (2003). Voriconazole: a new triazole antifungal agent. *Ann Pharm*, **37**, 420–432.

Porteous DD and Lajtha LG (1966). On stem cell recovery after irradiation. *Br J Haematol*, **12**, 177–188.

Price TH, Bowden RA, Boeckh M *et al.* (2000). Phase I/II trial of neutrophil transfusions from donors stimulated with G-CSF and dexamethasone for treatment of patients with infection in hemopoietic stem cell transplantation. *Blood*, **95**, 3302–3309.

Ravankar SG and Graybill JR (2002). Antifungal therapy. In: *Clinical Mycology*, Volume 1 1st Edition (EJ Anaissie, M McGinnis and MA Pfaller, eds), pp. 157–196. Philadelphia, USA: Churchill Livingstone.

Ronghe M, Foot A, Cornish J *et al.* (2002). The impact of transfusion of leucodepleted platelet concentrates on cytomegalovirus disease after allogeneic stem cell transplantation. *Br J Haematol*, **118**, 1124–1127.

Samonis G, Ralston K, Karl C *et al.* (1990). Prophylaxis of oropharyngeal candidiasis with fluconazole. *Rev Infect Dis*, **12**, S369–S373.

Schiffer CA (1996). Hematopoietic growth factors as adjuncts to the treatment of acute myeloid leukemia. *Blood*, **88**, 3675–3685.

Schimpff SC, Saterlee W, Young VM *et al.* (1971). Empirical therapy with carbenicillin and gentamicin for febrile patients with cancer and granulocytopenia. *New Eng J Med*, **284**, 1061–1065.

Schofield R (1978). The relationship between the spleen colony-forming cell and the haemopoietic stem cell. A hypothesis. *Blood Cells*, **4**, 7–25.

Schroeder ML (2002). Transfusion-associated graft-versus-host disease. *Br J Haematol*, **117**, 275–287.

Schuening FG, Storb R, Goehle S *et al.* (1989). Effect of recombinant human granulocyte colony-stimulating factor on hematopoiesis of normal dogs and on hematopoietic recovery after otherwise lethal total body irradiation. *Blood*, **74**, 1308–1313.

Schuening FG, Appelbaum FR, Deeg HJ *et al.* (1993). Effects of recombinant canine stem cell factor, a c-kit ligand, and recombinant granulocyte colony stimulating factor on hematopoietic recovery after otherwise lethal total body irradiation. *Blood*, **81**, 20–26.

Strauss RG (1995). Clinical perspectives of granulocyte transfusion: efficacy to date. *J Clin Apheresis*, **10**, 114–118.

Takahashi T, Wada T, Mori M *et al.* (1996). Over expression of the granulocyte colony stimulating factor gene leads to osteoporosis in mice. *Lab Invest* **74**, 827–834.

Tavassoli M (1991). Embryonic and fetal hemopoiesis: an overview. *Blood Cells*, **17**, 269–281.

Till JE and McCulloch EA (1961). A direct measurement of the radiation sensitivity of normal mouse bone marrow cells. *Rad Res*, **14**, 213–222.

Till JE and McCulloch EA (1980). Hemopoietic stem cell differentiation. *Biochim Biophys Acta*, **605**, 431–459.

Van de Berg BC, Malghem J, Lecouvet FE *et al.* (1998). Magnetic resonance imaging of normal bone marrow. *Eur Radiol*, **8**, 1327–1334.

Vannucchi AM, Bosi A, Ieri A *et al.* (1996). Combination therapy with G-CSF and erythropoietin after autologous bone marrow transplantation for haematological malignancies: a randomised trail. *Bone Marrow Transplant*, **17**, 527–531.

Visser JWM, Bauman JGJ, Mulder AH *et al.* (1984). Isolation of murine pluripotetic hemopoietic stem cells. *J Exp Med*, **50**, 1576–1590.

Vogler JB III and Murphy WA (1988). Bone marrow imaging. *Radiology*, **168**, 679–693.

Vose JM and Armitage JO (1995). Clinical applications of hematopoietic growth factors. *J Clin Onc*, **13**, 1023–1035.

Wakerly L, Craig A-M, Malek M *et al.* (1996). Fluconazole versus oral polyenes in the prophylaxis of immunocompromised patients: a cost-minimisation analysis. *J Hosp Infect*, **33**, 35–48.

Wingard JR, Merz WG, Rinaldi MG *et al.* (1991). Increase in *Candida krusei* among patients with bone marrow transplantation and neutropenia treated with fluconazole. *New Eng J Med*, **325**, 1274–1277.

Wingard JR, Merz WG, Rinaldi MG *et al.* (1993). Association of *Torulopsis glabrata* infections with fluconazole prophylaxis in neutropenic bone marrow transplant patients. *Antimicrob Agents Chemother*, **37**, 1847–1849.

Winston DJ, Maziarz RT, Chandrasekar PH *et al.* (2003). Intravenous and oral itraconazole versus intravenous and oral fluconazole for long-term antifungal prophylaxis in allogeneic hematopoietic stem cell transplant recipients. A multicenter randomized trial. *Ann Int Med*, **138**, 705–713.

Worton RG, McCulloch EA and Till JE (1975). Physical separation of haemopoietic stem cells differing in their capacity for self-renewal. *J Exp Med*, **130**, 91–102.

23 ORGANIC ARSENICALS

Timothy C. Marrs[1] and Robert L. Maynard[2]

[1] *Edentox Associates, Edenbridge, UK*
[2] *Health Protection Agency, Chilton, UK*

A number of organic arsenicals have been developed for use as chemical warfare agents, the majority during World War I. Of these, by far the most important is lewisite, isolated in pure form by Lee Lewis in 1918 (Table 1). Lewisite was developed in an attempt to provide a highly toxic, non-persistent, quick-acting vesicant compound. Mustard gas had been used on a large scale during World War I but its long persistence made it difficult for attacks to be launched on ground contaminated with that agent. Ethyldichlorarsine and diphenylchlorarsine were used by German troops although they were not as effective as had been expected.

LEWISITE

Lewisite (2-chlorovinyl dichloroarsine) was first synthesized in 1904 and rediscovered by Captain W. Lee Lewis in the USA, after whom the agent was named (Vilensky and Redman, 2003). Lewisite was first synthesized in bulk in the USA and, according to Prentiss (1937), the first shipment was on its way to Europe when the 1918 armistice was signed and apparently this shipment was destroyed at sea. Lewisite rapidly acquired a remarkable reputation as the 'dew of death' although there remains no proof of it having been used in war.

Lewisite is an odorless, colourless oily liquid of structural formula ClCH–CHAsCl2 (Table 2). Lewisite is said to darken on standing and the technical material is often blue–black to black in colour and is said to smell of geraniums. Lewisite, which is not soluble in water to any appreciable extent, nevertheless hydrolyzes rapidly when mixed with water and is rapidly hydrolysed by alkaline aqueous solutions such as sodium hypochlorite solution. The rapid hydrolysis of lewisite by water and especially by alkalis was seen to be advantageous to troops using the compound as a weapon. This property can nevertheless be a disadvantage in that it is unlikely that lewisite would be an effective weapon under wet conditions, because such conditions would render maintenance of effective field concentrations difficult. Prentiss (1937) gave the persistence of lewisite at 20° C as 9.6 times that of water whereas the persistence of sulphur mustard is 67 times that of water at the same temperature. The freezing point of lewisite is -18°C and thus no special preparations, of the type required with mustard, are needed for use under most winter conditions.

ABSORPTION

Lewisite is rapidly absorbed through the skin and mucous membranes. The distribution of lewisite has been assumed to follow that of other arsenicals, but this may well not be the case. The distribution of organomercurials and organotins is different from the inorganic salts of those metals and Inns *et al.* (1988) found that in rabbits, at doses of equal lethal toxicity, tissue levels of arsenic were much higher with sodium arsenite than with lewisite. There was a notable exception, namely the lungs, where tissue levels were slightly higher with lewisite. Bearing in mind that the intravenous lethal dose of lewisite is, on an

Chemical Warfare Agents: Toxicology and Treatment (2nd Edition)
Edited by Timothy C. Marrs, Robert L. Maynard and Frederick R. Sidell

Table 1. Organic arsenicals of chemical warfare importance

Chemical Name	Other names
2-Chlorovinyl dichlorarsine	Lewisite
Ethyldichlorarsine	Dick
Methylarsine	Methyl dick
Phenyldichlorarsine	Sneeze Gas
Diphenylchlorarsine	DA, Clark 1
Diphenylcyanarsine	DC, Clark 2
Diphenylamine chlorarsine	DM, Adamsite

Table 2. Physical properties of lewisite

Boiling point	190°C
Melting point	−13°C
Vapor pressure	at 0°C, 0.087 mmHg at 20°C, 0.395 mmHg

arsenic-content basis, much lower than that of sodium arsenite, the arsenic of lewisite clearly preferentially distributes to the lung. Inns *et al.* (1988) attributed the differences in toxic effects of lewisite and inorganic arsenic salts to differences in disposition. In fact comparatively little is known of the biotransformation of organic arsenicals (Klaassen, 2002), but it is known that trivalent arsenicals are converted to pentavalent derivatives and that these occur in the urine. Pentavalent arsenical compounds are less toxic than compounds of trivalent arsenic.

MODE OF EXPOSURE

As lewisite is a liquid it may be disseminated by shells, bombs or rockets or sprayed from aircraft. Lewisite can be mixed with sulphur mustard, which depresses the freezing point and renders sulphur mustard usable over a wider range of temperatures than sulphur mustard itself.

Individuals may be contaminated by contact with liquid lewisite or by inhalation of the vapor.

TOXICOLOGY

The toxicology of lewisite has been the subject of a comprehensive review (Goldman and Dacre, 1989).

Acute

Acute toxicity figures for man are not known but a lowest lethal concentration over 30 min of 6 ppm was quoted by Maynard (1989). The LD_{50} has been measured in a number of species (Table 3), while LCt_{50}s in a variety of species vary from 500 to 1500 mg min m^{-3} (Goldman and Dacre, 1989). The efficacy of lewisite, like that of mustard, depends partly upon its vesicant properties but lewisite is also a lethal systemic chemical weapon. About 30 drops (2.6 g), applied to the skin and not washed off or otherwise decontaminated would be expected to produce a fatal outcome in an average man.

Table 3. Acute toxicity of lewisite

Species	Route[a]	LD_{50} (mg kg^{-1}) (95% CL if available)	Reference
Rat	PO	50	US Army (1974)
Rat	PC	24	Maynard (1989)
Rabbit	IV	0.5	Maynard (1989)
Rabbit	IV	1.8 (1.6–2.1)	Inns *et al.* (1988)
Rabbit	PC	6	Maynard (1989)
Rabbit	PC	5.3 (3.5–8.5)	Inns and Rice (1993)
Guinea pig	PC	12	Maynard (1989)
Dog	PC	15	Maynard (1989)

Note:
[a] PC, percutaneous; IV, intravenous; PO, oral.

MODE OF ACTION

Arsenic, like most heavy metals, binds to a wide range of compounds, including macromolecules such as proteins, which contain sulphydryl groups (Ehrlich, 1909). Many of the effects of lewisite are thought to be due to binding to lipoic acid, a dithiol eight-carbon component of the pyruvate dehydrogenase complex (Peters *et al.*, 1945; Peters, 1948,1953). The pyruvate dehydrogenase complex in mammals catalyzes the oxidative decarboxylation of pyruvate to acetyl coenzyme A (Wieland, 1983). Binding of lewisite to lipoic acid prevents formation of acetyl coenzyme A from pyruvate and as pyruvate is a metabolite of glucose, inhibition of pyruvate decarboxylase prevents entry of glucose metabolites into the tricarboxylic acid cycle. Peters and colleagues (Peters *et al.*, 1945; Peters, 1948,1953) demonstrated that 15×10^{-6} M sodium arsenite produced 50% inhibition of the activity of the pyruvate dehydrogenase complex and observed that in animals treated with sodium arsenite, blood levels of pyruvate rose. In actual fact, the enzyme complex α-ketoglutarate dehydrogenase is similar to pyruvate dehydrogenase and would be expected to be similarly affected. More surprisingly, Platteborze (2005) found the subsequent citric acid cycle enzyme, succinate dehydrogenase to be up-regulated, in a system using microarrays containing sequence-verified human cDNAs screened with mRNA from human keratinocytes.

PATHOLOGY

SKIN

The pathological effects of lewisite have been less studied than those of other vesicants, particularly sulfur mustard. Ireland (1926) studied the progression of changes observed after the application of lewisite to the skin of horses. Five hours after exposure, marked edema of the skin was noted, extending into the dermis with separation of collagen fibres. More deeply placed blood vessels were surrounded by collections of polymorphs. By 24 h, there was thinning of the epidermis with nuclear pyknosis. The dermis was edematous and there were collections of fluid at the epidermal–dermal junction. At 48 h, there was a definite margin to the lesion and some repair was noted; damaged areas were heavily pigmented. In comparison with mustard, lewisite produced epidermal necrosis earlier, with more extensive edema and inflammation, and more vascular thrombosis. On the other hand repair with lewisite appeared to start earlier than with mustard. King *et al.* (1994) suggested that in lewisite-treated isolated perfused porcine skin flaps, epidermal–dermal seperation was localized in the lamina lucida. It was hypothesized by these workers that chemical modification of the glycoprotein adhesive, laminin, was responsible.

Lewisite is markedly vesicant when applied to human skin, but vesication has not been observed in animals that have been studied.

RESPIRATORY TRACT

The main effect of lewisite on the upper respiratory tract is to produce necrosis of the epithelium, which is accompanied by the formation of a false diphtheria-type membrane which consists of sloughed epithelial cells, together with inflammatory cells and mucus. The membrane may become detached and cause bronchial obstruction. Widespread edema and congestion of the lungs occurs and they may acquire a greyish-red to purple hue. Areas of atelectasis and secondary emphysema are common. Secondary bronchopneumonia is common and a frequent cause of death (Vedder, 1926).

Marked edema of mediastinal structures, including the pericardium, is seen.

EYE

Detailed pathological descriptions of the effects of lewisite on the eye are not available.

SYSTEMIC PATHOLOGICAL EFFECTS

In the studies of intravenously injected lewisite in the rabbit, changes in the lungs, including hemorrhage and edema, with lymphocytic infiltration, were seen (Inns *et al.*, 1988,1990). Damage was found in the biliary tree, including epithelial necrosis in the gall bladder. Inns and Rice

(1993), in a study in which lewisite was applied to the skin of rabbits, observed focal hepatocyte degeneration and transmural necrosis of the gall bladder. Small bile duct proliferation and early portal tract fibrosis was seen and there was focal mucosal necrosis in the duodenum; however, lung changes of the type seen with intravenous lewisite were not observed. Taken together, these data show clearly that the toxicity of lewisite is neither qualitatively nor quantitatively just that of the arsenic that it contains (see above).

Symptoms and clinical signs

Because lewisite has not been used in warfare, descriptions are mainly derived from accidental exposures or else clinical signs in man have been inferred from observations in animals. Findings with the related phenyldichlorarsine are also relevant (see below).

Immediately after exposure, there is eye irritation and coughing, sneezing, salivation and lachrymation rapidly follow. Contamination of the skin causes erythema at concentrations of 0.05–0.01 mg cm^{-2} (Inns *et al.*, 1990) with vesication after a few hours. Pain in the skin and eyes is immediate, a major point of difference from sulphur mustard. The effects in the eye and skin reach their greatest at 4–8 h post-exposure. The patient is seriously incapacitated, breathing with difficulty and unable to see. In severe cases, pulmonary edema follows and the patient may die of respiratory failure. In cases where skin contamination is extensive, there may be liver necrosis and the absorption of arsenic may be sufficient to cause death.

The eye lesions produced by lewisite are particularly serious: blindness will follow contamination of the eye with liquid lewisite unless decontamination is very prompt. It is possible to infer from animal experiments undertaken in the USA during World War II that exposure of the eye to liquid or vapour would produce blepharospasm and irritation. Severe ocular lesions can be produced by doses as low as 0.1 mg or exposure to the saturated vapor for 8 s at 23°C. The action on the eye is to produce rapid necrosis of the anterior parts of that organ (Friedenwald and Hughes, 1948).

It has been said that absorption of lewisite in food would be expected to give rise to the symptoms and signs of arsenical poisoning. These include severe stomach pain, vomiting, watery diarrhea, numbness and tingling, particularly in the feet, thirst and muscular cramps. Neuropathy, nephritis with proteinuria and/or encephalopathy may follow acute arsenical poisoning. Intravascular hemolysis and hemolytic anemia may occur, with, in extreme cases, renal failure (Stewart and Sullivan, 1992). Nevertheless the view that lewisite is simply inorganic arsenic poisoning is clearly false, and parenteral lewisite differs from inorganic arsenic in significant respects (see above): the same may be true of lewisite by the oral route.

Subchronic toxicity and reproductive toxicity

Studies to establish health and exposure criteria for lewisite have been undertaken. While these do not have much relevance to acute single-dose exposure, the type of exposure most likely in warfare or terrorist use, they are useful in hazard characterization for workers engaged in handling agents such as lewisite. In addition, such studies may be useful in characterising the hazard if food were contaminated during a chemical incident.

SUBCHRONIC TOXICITY

In a study of the subchronic toxicity of lewisite, rats (Sprague-Dawley) of both sexes (10 per dose group per sex) were dosed by gavage with lewisite in sesame seed oil at doses of up to 2.0 mg kg^{-1} bw day^{-1}, 5 days per week for 13 weeks. Some animals died during the study at doses of 0.5 mg kg^{-1} bw day^{-1} and above: these animals often had inflammatory changes in the respiratory tract. Body weight in females was decreased at the highest dose. At the highest dose, decreases in serum protein, creatinine and aminotransferase activity were seen in the male rats, and increases in lymphocytes and platelets in the blood of females. In both sexes, changes were seen in the forestomach at the highest dose, namely necrosis of the squamous epithelium with infiltration by neutrophils and macrophages accompanied by hemorrhage and fibroblast proliferation. Mild inflammation of the glandular stomach was observed at 1.0 and 2.0 mg kg^{-1} bw day^{-1}. The no-adverse effects level (NOAEL) was thought by the authors to be 0.5 mg kg^{-1} bw day^{-1}, although a more precautionary interpretation of decedency observed might suggest a NOAEL of 0.1 mg kg^{-1} bw day^{-1} (Sasser *et al.*, 1996). Were the data to be used for hazard characterization for dietary exposure, the decadency observed could probably be ignored, as due to instillation of lewisite into the respiratory tract, and 0.5 mg kg^{-1} bw day^{-1} be taken as the NOAEL.

REPRODUCTIVE AND DEVELOPMENTAL TOXICITY

Using dosing based upon the above study, a two-generation study of reproduction was undertaken in the rat (Sprague-Dawley), using lewisite in sesame oil. Administration was by gavage and doses of 0, 0.1, 0.25 and 0.6 mg kg^{-1} bw day^{-1} (5 days per week) were used and there was a control group of the same size. Each treatment group comprised 20 males and 25 females. Administration of test material was undertaken before, during and after mating until birth of the offspring, at which time the parental males were sacrificed, while dosing of the dams continued during lactation. At weaning, random male and female offspring (F_1) were selected to continue the study and lewisite was administered to them during adolescence. They were mated, during which lewisite was administered as it was after mating, until birth of the F_2 offspring, when the parental males were sacrificed. The dams received lewisite until weaning of the F_2 offspring at 3 weeks *post partum*. High mortality was seen at the highest dose in both generations of females and to a lesser extent in males. Excess early deaths were seen at both other test doses and this appeared dose-related. Lewisite had no adverse effect on reproduction, fertility and no organ-specific toxicity was noted in males or females: indeed in the F_0 generation, the fertility index of the females was significantly increased at the top dose. Biologically significant changes in pup survival and weight were not seen. Maternal toxicity (impaired body weight gain) was seen, however, particularly in the F_0 generation females; this was not clearly dose-related. No (gross) changes were seen in the forestomach or glandular stomach at necropsy. Decedents had changes in the respiratory system, consisting grossly of pleural effusion and pulmonary foci and microscopically of pulmonary edema, hemorrhage, inflammation of the airways and alveoli and inflammation of the pleura and mediastimum. The NOAEL for all effects, except maternal toxicity was thus the highest dose tested (0.6 mg kg^{-1} bw day^{-1}). The adverse effects observed in the parents may have been due to instillation of lewisite into the respiratory tract: if that were the case, and were the data to be used for hazard characterization for dietary exposure, the effects observed could be ignored and the NOAEL taken to be 0.6 mg kg^{-1} bw day^{-1}, the highest dose tested. Otherwise, a NOAEL was not observed (Sasser *et al.*, 1999).

Studies of developmental toxicity have been undertaken both in the rat and the rabbit (Hackett *et al.*, 1992). In the former exposure was from gestation day 6–15 and doses used were 0, 0.5, 1.0 and 1.5 mg kg^{-1} bw day^{-1} and in the latter exposure was from gestation day 6–19 and doses used were 0, 0.05 and 0.07 mg kg^{-1} bw day^{-1}. Maternal toxicity was seen but there no evidence of a teratogenic response.

MUTAGENICITY

Lewisite was not mutagenic in the Ames (Stewart *et al.*, 1989) or Chinese hamster ovary cell assay (HGPRT locus) (Jostes *et al.*, 1989a,b). Lewisite induced chromosomal aberrations in Chinese hamster ovary cells (Jostes *et al.*, 1989a). In a rat dominant lethal test *in vivo*, lewisite was administered to male CD rats at doses of 0.375, 0.75 or 1.5 mg kg^{-1} bw by gavage for 5 days, with appropriate vehicle and positive (ethyl methanesulphonate) controls. The male rats each mated with two virgin females. The females were killed on gestation day 14, while the males were killed at week 13. There was no indication of a dominant lethal mutagenic response or of any adverse effect on reproductive indices, except in the controls or on the male reproductive tract (Bucci *et al.*, 1993).

Management of poisoning

FIRST AID

The casualty must be removed from the source of contamination: rescue workers must be protected against liquid lewisite by protective clothing, gloves and boots and against the vapour by respirators. Because of the danger to both the casualty and his attendants, it is essential that standard drills be followed when decontaminating a casualty whose protective clothing is contaminated with lewisite. Fullers' earth is an effective decontaminant and dilute solutions of bleach would also be effective.

SPECIFIC THERAPY

Unlike vesicants of the mustard group, highly effective specific therapy is available for poisoning with organic arsenicals. Organic arsenicals are effectively antagonized by dithiol (divalent) chelating agents. In World War II, the United Kingdom Ministry of Supply funded studies at Oxford University into prospective antidotes for lewisite. The development of dimercaprol (British Anti-Lewisite, BAL, 2,3-dimercaptopropanol) was reported in 1940 (Ord and Stocken, 2000) but, for reasons of national security, not reported in the scientific literature until 1945, when the work on dimercaprol was published by Peters *et al.* (1945) and Stocken and Thompson (1946). Since World War II, dimercaprol has been the standard treatment for poisoning by arsenic compounds, including lewisite and other organic arsenicals and is effective for inorganic arsenic poisoning (Rittey, 1949). Dimercaprol binds to lewisite as follows:

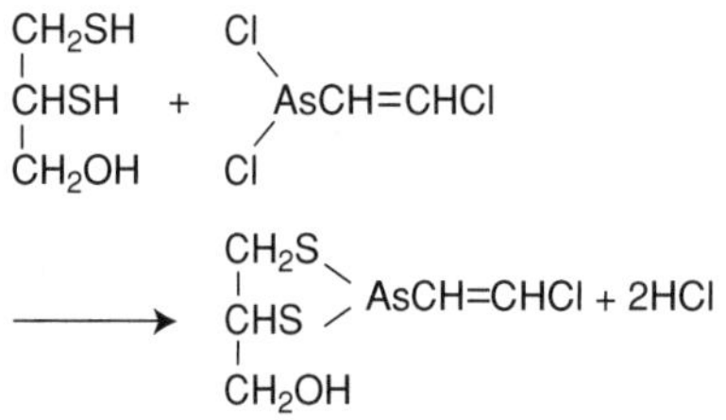

In animal studies, dimercaprol was able to protect against the effects of lewisite and reverse the enzyme inhibition produced by it. Dimercaprol was originally developed for parenteral use against systemic lewisite poisoning, and it was also available as an ointment for use against skin burns. However, dimercaprol has major disadvantages as an antidote. When used for systemic arsenic poisoning the dose of dimercaprol is limited by toxicity and the drug has to be given by intramuscular injection. Dimercaprol is dissolved in peanut oil and benzyl benzoate and injections are painful (Sulzberger *et al.*, 1946; Modell *et al.*, 1946). Furthermore, the material cannot be given intravenously, so that a loading dose cannot be administered.

A number of different dosing regimes have been recommended, and thus JSP 312 (1972) suggests:

- 2.5 mg kg^{-1} 4-hourly for four doses, then;
- 2.5 mg kg^{-1} twice daily.

Martindale (1993) gives the following regime:

- Day 1: 400–800 mg kg^{-1} IM in divided doses.
- Days 2 and 3: 200–400 mg kg^{-1} IM in divided doses.
- Days 4–12: 100–200 mg kg^{-1} IM in divided doses.

Within this range, the dose is determined by body weight and the severity of symptoms. Administration should be by deep intramuscular injection using multiple-injection sites. Intramuscular dimercaprol may produce alarming reactions. Pain at the injection site may last up to 24 h, while systemic reactions include increased blood pressure and tachycardia, nausea and vomiting, headache and feelings of constriction of the chest. Conjunctivitis, lachrymation, rhinorrhea, sweating, anxiety and agitation have also been reported. These effects pass in a few hours. It should also be noted that although there is a fair amount of knowledge about the theoretical basis of chelation, chelating agents are not usually totally specific (for review, see Bateman and Marrs, 1999). Thus, there is evidence that a dimercaprol will chelate zinc, an element essential for many biochemical processes (Emanuelli *et al.*, 1998). It is conceivable that such effects might contribute to ill-health following poisoning by lewisite, treated with chelating agents. Klaassen (2002) emphasized the need to maintain a high plasma level of the drug so that the rapidly excreted 2:1 complex of dimercaprol:arsenic should predominate over the 1:1 complex which is more slowly excreted.

Skin contamination should be treated with dimercaprol ointment. Eye contamination can be treated by application of dimercaprol (5–10% in vegetable oil) into the conjunctival sac. This should be performed as a matter of urgency. The eye drops produce pain on instillation in this way.

Two water-soluble analogs of dimercaprol have been studied as lewisite antidotes, namely *meso*-2,3-dimercaptosuccinic acid (DMSA) and 2,3-dimercapto-1-propane sulfonic acid (DMPS) (see review by Aposhian, 1993). Their structures are as follows:

DMSA	DMPS
COOH	CH_2SH
|	|
CHSH	CHSH
|	|
CHSH	CH_2SO^{3-} Na^+
|	
COOH	

These drugs circumvent two major disadvantages of dimercaprol discussed above, namely the need for intramuscular injection and the limitation of dose by toxicity. DMSA and DMPS are about 20 and 10 times less toxic than dimercaprol in the mouse (Aposhian *et al.*, 1984) and can be given orally. Inns *et al.* (1990) showed that, in rabbits poisoned with intravenous lewisite, dimercaprol, DMPS and DMSA were, on a molar basis, equieffective, but emphasized the limitations in dosage bestowed by the toxicity of dimercaprol. Studying the antidotal efficacy of dimercaprol, DMSA and DMPS against lewisite applied percutaneously to rabbits, Inns and Rice (1993) concluded that, at equimolar doses (40 μmol l^{-1}), there was little difference between the efficacy of dimercaprol, DMPS and DMSA; however, the low toxicity of DMPS and DMSA enabled the use of high doses of antidotes which produced improved survival, while dimercaprol at these molar doses would have been toxic. Protection ratios (LD_{50} with treatment/LD_{50} without treatment) were 13 and 16.9, respectively, for DMPS and DMSA at the higher dose of 160 μmol kg^{-1} IM. Hepatocellular damage (see above) appeared to be reduced by chelation therapy. Despite the evidence that at equimolar doses the three chelators are more or less equieffective, the two more novel drugs may have advantages other than just their low toxicity. Thus, Aposhian *et al.* (1984) showed that dimercaprol was the least effective of the three in reversing arsenite inhibition of the pyruvate dehydrogenase complex in the kidneys of mice or in a mouse kidney system *in vitro*. Another possible advantage of DMPS and DMSA against dimercaprol is that the latter drug increased the arsenic content of brains of rabbits poisoned with sodium arsenite, whereas the first two did not: however, the implications of this observation should be considered with some circumspection as it is possible that the arsenic was not biologically active, if chelated. DMPS is also effective in inorganic arsenic poisoning (Moore *et al.*, 1994) and DMS has been used in poisoning with heavy metals other than arsenic (Graziano, 1986; Fournier *et al.*, 1988).

DMPS and DMSA applied to the skin would probably be of value in lewisite-induced vesication. However, the disadvantages of dimercaprol largely relate to systemic treatment and the water-soluble analogues are unlikely to be better than dimercaprol ointment.

Other measures that may prove necessary include antibiotics to treat bronchopneumonia; the prophylactic use of antibiotics is controversial.

OTHER ORGANIC ARSENICALS

Phenyldichlorarsine

Phenyldichlorarsine is a less severe vesicant than lewisite, although Prentiss (1937) described its vesicant actions as 'not inconsiderable'. The effects of phenyldichlorarsine on an individual, exposed while working on it, were described in detail by Hunter (1978). There was severe blistering of one hand and two days after exposure, severe diarrhoea, vomiting and slight jaundice. This individual recovered, with intermittent vomiting and diarrhoea (which was not bloody) continuing for some days, while the skin had healed by the tenth day after exposure. Thus, the effects were not dissimilar to those of lewisite.

Phenyldichlorarsine was used on a large scale during the World War I. It was used as a compound capable of penetrating the respirators then available and also capable of producing severe irritation of the respiratory tract. It was apparently a very effective chemical weapon. Treatment should probably be as for lewisite and phenyldichlorarsine forms an analogous adduct to that produced by lewisite with dimercaprol (Dill *et al.*, 1987).

The 'Dicks'

Ethyl, methyl and phenyl dichlorarsine are known as the 'Dicks', more specifically ethyl dichlorarsine is 'dick' and methyl dichlorarsine is 'methyl dick'. They are vesicants similar to lewisite and dichlorarsine was used by the Germans in World War I to a limited extent. Phenyldichlorarsine was used in some percutaneous studies by Aposhian *et al.* (1984); these suggested that DMPS and DMSA would be useful in prophylaxis against skin burns.

Diphenylchlorarsine, diphenylcyanarsine and diphenylaminechlorarsine are irritants.

REFERENCES

Aposhian HV (1983). DMSA and DMPS – water soluble antidotes for heavy metal poisoning. *Annu Rev Pharmacol Toxicol*, **23**, 193–215.

Aposhian HV, Carter CD, Hoover TD *et al.* (1984). DMSA, DMPS and DMPA – as arsenic antidotes, *Fund Appl Toxicol*, **4**, S58–S70.

Bateman N and Marrs TC (1999). Antidotal studies. In: *General and Applied Toxicology.* (B Ballantyne, TC Marrs and T Syversen, eds), pp. 425–435. London, UK: MacMillan Reference Limited.

Bucci TJ, Parker RM, Dacre JC *et al.* (1993). *Dominant lethal study of lewisite in male rats*, AD-A290671. Fort Detrick, Frederick, MD, USA: US Army Medical Research and Materiel Command.

Dill K, Adams ER, O'Connor RJ *et al.* (1987). 2D NMR studies of the phenyldichloroarsine British anti-lewisite adduct. *Magn Reson Chem*, **25**, 1074–1077.

Ehrlich P (1909). Über den jetzigen Stand der Chemotherapie. *Ber Deutsch Chem Ges*, **42**, 17.

Emanuelli T, Rocha JBT, Pereira ME *et al.* (1998). δ-Aminolevulinate dehydratase inhibition by 2,3-dimercaptopropanol is mediated by chelation of zinc from a site involved in maintaining cysteinyl residues in a reduced state. *Pharmacol Toxicol*, **83**, 95–103.

Fournier L, Thomas G, Garnier R *et al.* (1988). 2,3-Dimercaptosuccinic acid treatment of heavy metal poisoning in humans. *Med Toxicol*, **3**, 499–504.

Friedenwald JS and Hughes WF (1948). *The Effects of Toxic Chemical Agents on the Eyes and their Treatment*, Advances in Military Medicine, Volume II. Boston, MA, USA: Little Brown and Company.

Goldman M and Dacre JC (1989). Lewisite: its chemistry, toxicology and biological effects. *Rev Environ Contam Toxicol*, **110**, 75–115.

Graziano JH (1986). Role of 2,3-dimercaptosuccinic acid in the treatment of heavy metal poisoning. *Med Toxicol*, **1**, 155–162.

Hackett PL, Sasser LB, Rommereim RL *et al.* (1987). *Teratology studies on lewisite and sulphur mustard agents: effects of lewisite in rats and rabbits*, AD-A198423, Fort Detrick, Frederick, MD, USA: US Army Research and Development Command.

Hackett PL, Sasser LB, Rommereim RL *et al.* (1992). Developmental toxicity of lewisite in rats and rabbits. *Toxicologist*, **12**, 198.

Hunter D (1978). *The Diseases of Ocuupations*, 6th Edition, pp. 368–369. London, UK: Hodder and Stoughton.

Inns RH and Rice P (1993). Efficacy of dimercapto chelating agents for the treatment of poisoning by percutaneously applied dichloro(2-chlorovinyl)arsine in rabbits. *Human Exp Toxicol*, **12**, 241–246.

Inns RH, Bright JE and Marrs TC (1988). Comparative acute systemic toxicity of sodium arsenite and dichloro(2-chlorovinyl)arsine in rabbits. *Toxicology*, **51**, 213–222.

Inns RH, Rice P, Bright JE *et al.* (1990). Evaluation of the efficacy of dimercapto chelating agents for the treatment of systemic organic arsenic poisoning in rabbits. *Human and Exp Toxicol*, **9**, 215–220.

Ireland MW (1926). *Medical Aspects of Chemical Warfare. The Medical Department of the US Army in World War 1*, Volume XIV. Washington, DC, USA: Government Printing Office.

Jostes RF, Rausch RJ and Sasser LB (1989a). *Toxicology studies on lewisite and sulphur mustard agents: mutagenicity study of sulphur mustard (HD) in Chinese hamster ovary cells.* Report AD-A216449. Fort Detrick Frederick, MD, USA: US Army Research and Development Command.

Jostes RF, Rausch RJ, Miller BM *et al.* (1989b). Genotoxicity of lewisite in Chinese hamster ovary cells. *Toxicologist*, **9**, 232.

JSP 312 (1972). *Medical Manual of Defence against Chemical Agents*. London, UK: HMSO.

King JR, Peters BP and Monteiro-Riviere NA (1994). Laminin in the cutaneous basement membrane as a potential target in lewisite vesication. *Toxicol Appl Pharmacol*, **126**, 164–173.

Klaassen CD (2002). Heavy metals and heavy metal antagonists. In: *Goodman and Gilman's The*

Pharmacological Basis of Therapeutics, 7th Edition (J Hardman and LE Liford, eds), pp. 1861–1875. New York, NY, USA: McGraw-Hill.

Martindale (1993). *The Extra Pharmacopoeia*, 30th Edition (JEF Reynolds, ed.), London, UK: Pharmaceutical Press.

Maynard RL (1989). *A review of chemical warfare agents*, CDE Technical Paper, 2nd Edition. Porton Down, Salisbury, UK: CDE.

Modell W, Gould H and Cattell M (1946). Clinical uses of 2,3-dimercapto-propanol (BAL). IV Pharmacologic observations on BAL by intramuscular injection in man. *J Clin Invest*, **25**, 480–487.

Moore DF, O'Callaghan CA, Berlyne G *et al.* (1994). Acute arsenic poisoning after treatment with 2,3-dimercaptopropanesulphonate (DMOS). *J Neurol Neurosurg Psychiat*, **57**, 1133–1135.

Ord MA and Stocken LA (2000). A contribition of chemical defence in World War II. *Trends Biochem Sci*, **25**, 253–256.

Prentiss AM (1937). *Chemicals in War*. New York, NY, USA: McGraw-Hill Book Company.

Peters R (1948). Development and theoretical significance of British Anti-Lewisite (BAL). *Br Med Bull*, **5**, 313–318.

Peters R (1953). Significance of biochemical lesions in the pyruvate oxidase system. *Br Med Bull*, **9**, 116–119.

Peters RA, Stocken LA and Thompson RHS (1945). British anti-lewisite (BAL). *Nature*, **156**, 616–691.

Platteborze PL (2005). The transcriptional effects of the vesicants lewisite and sulphur mustard on human epidermal keratinocytes. *Toxicol Mech Meth*, **15**, 185–192.

Rittey DAW (1949). Treatment of accidental arsenical poisoning with dimercaprol. *Lancet*, **2**, 719.

Sasser LB, Cushing JA, Mellick PW *et al.* (1996). Subchronic toxicity evaluation of lewisite in rats. *J Toxicol Environ Health*, **47**, 321–334.

Sasser LB, Cushing JA, Lindenmeier CW *et al.* (1999). Two-generation study of lewisite in rats. *J Appl Toxicol*, **19**, 229–235.

Stewart CE and Sullivan JB (1992). Military munitions and antipersonnel agents. In: *Hazardous Materials Toxicology* (JB Sullivan and GR Krieger, eds), pp. 986–1026. Baltimore, MD, USA: Williams and Wilkins.

Stewart DL, Sass EJ, Fritz LK *et al.* (1989). *Toxicology studies on Lewisite and sulphur mustard agents: mutagenicity of Lewisite in the salmonella histidine reversion assay*, Report AD-A213146. Fort Detrick, Frederick, MD, USA: US Army Medical Research and Development Command.

Stocken LA and Thompson RHS (1946). British antilewisite, arsenic and thiol excretion in animals after treatment of lewisite burns. *Biochem J*, **40**, 548–554.

Sulzberger MB, Boer RL and Kanof A (1946). Clinical uses of 2,3-dimercaptopropanol (BAL) III. Studies on the toxicity of BAL on percutaneous and parenteral administrations. *J Clin Invest*, **25**, 474–479.

US Army (1974). *Chemical Data Sheets*, Volume I, Report EO-SR-74001, pp. 65–72. Edgewood Arsenal, MD, USA: Development and Engineering Directorate.

Vedder EB (1926). *The Medical Aspects of Chemical Warfare*. Baltimore, MD, USA: Williams and Wilkins.

Vilensky JA and Redman K (2003). British antilewisite (dimercaprol): an amazing history. *Ann Emerg Med*, **41**, 378–383.

Wieland OH (1983). The mammalian pyruvate dehydrogenase complex: structure and regulation. *Rev Physiol Biochem Pharmacol*, **96**, 123–170.

24 PHOSGENE

Robert L. Maynard

Health Protection Agency, Chilton, UK

Phosgene is a colourless gas first prepared by John Davy in 1812. It was developed as a chemical weapon by German workers during World War I (WWI) and was first used, against British troops, near Ypres on 19 December 1915. Four thousand cylinders of phosgene (88 tons) were released causing 1069 casualties and 120 deaths. By 1915, the Allied Forces had also been experimenting with phosgene and the gas was later used on a substantial scale by both sides. Phosgene had been used as an intermediate in the dye industry for many years and was being produced, on a large scale, in Germany before WW1. The effect of the dominance of the German dye industry on the development of chemical warfare has been commented on by Lefebure (1921) and in Chapter 1.

Phosgene accounted for some 85% of all deaths attributed to chemical weapons during WWI (Ministry of Defence, 1987). Despite this, its efficacy as a chemical warfare agent was probably less than that of mustard gas. Phosgene acts by damaging the lungs and producing pulmonary oedema: it was thus classified, amongst other lung-damaging agents, with chloropicrin, trichloromethyl chloroformate and disulphur decafluoride. The lethality of phosgene and other lung-damaging agents led to an intensive search for more effective agents during WWI: an excellent account has been provided by Prentiss (1937).

CHEMICAL AND PHYSICAL PROPERTIES OF PHOSGENE

Phosgene is, at ordinary temperatures and pressures, a colourless gas with a suffocating smell said to be reminiscent of mouldy or sometimes freshly-mown hay. Its chemical and physical properties are shown in Table 1.

Early studies of the reaction between phosgene and water suggested that the hydrolysis of phosgene proceeded slowly (Ireland, 1926). This is untrue: Potts *et al* (1949) showed that hydrolysis occurred rapidly with the production of hydrochloric acid:

$$COCl_2 + H_2O \longrightarrow CO_2 + H_2O$$

The rapidity of hydrolysis was also noted by Prentiss (1937) who pointed out that phosgene could not be used successfully in wet weather. Production of hydrochloric acid in shells containing phosgene and some water was also identified as a problem by Prentiss.

It should be noted that, although the boiling point of phosgene (8.2°C) is below normal summer temperatures, the rate of evaporation of liquid phosgene is slow compared with that of chlorine (BP, −33.6°C) and to produce adequately lethal gas clouds chlorine was almost always mixed with phosgene: this was in fact done during the first phosgene attack referred to above.

Chemical Warfare Agents: Toxicology and Treatment (2nd Edition)
Edited by Timothy C. Marrs, Robert L. Maynard and Frederick R. Sidell

Table 1. Physical properties of phosgene

Formula	$COCl_2$
Melting point	−118°C
Boiling point	8.2°C
MW	99
Vapour pressure at 20°C	1215 mmHg
Vapour density	3.5

LIKELY MODE OF EXPOSURE TO PHOSGENE

Phosgene can be disseminated by aircraft, from pressurized containers and by bombs and shells. The gas is rapidly dispersed by the wind although, as it is heavier than air, it tends to accumulate in low-lying areas and dangerous concentrations may occur in cellars and trenches. Despite this, phosgene is classified as a non-persistent agent likely to be used to kill and incapacitate rather than to deny access to ground.

ABSORPTION OF PHOSGENE

Phosgene does not exert any significant effects on or via the skin. Local effects may be produced upon exposure of the eyes and lacrimation was reported during WWI (Ireland, 1926). It is often assumed that phosgene is absorbed across the lung. Nash and Pattle (1971) undertook a detailed analysis of the penetration of moist biological membranes by phosgene. They concluded that, given the rapidity of hydrolysis of phosgene, intact phosgene molecules would only penetrate a few tens of micrometers below such a surface. They argued that the upper respiratory tract would be significantly protected against the effects of phosgene by its surface layer of mucus but that the gas-exchange zone would be more susceptible to damage. Intact phosgene molecules might be expected to reach the pulmonary capillary blood but once there rapid hydrolysis would occur. This is discussed in more detail below.

TOXICITY OF PHOSGENE

A great deal of work on the toxicity of phosgene in a range of species was undertaken during WWI. The toxicity of phosgene in man is known and a figure of 3200 mg min m^{-3} is often quoted for the LCt_{50}. The origin of this figure is rather obscure and a figure of 3200 mg m^{-3} is sometimes given: this would be an LC_{50} rather than an LCt_{50}. The variation of LCt_{50} with duration of exposure has been discussed in Chapter 2; the general conformation to the Haber rule by phosgene is shown in Figure 1.

The apparent conformation to the predicted hyperbola shown in Figure 1 is, however, a little misleading: examine the data used to plot Figure 1, shown in Table 2.

These data are taken from American work on dogs and mice. The lethal index shown in Table 2 is the product of concentration (expressed in mg m^{-3}) and the duration of exposure (expressed in min). Prentiss (1937) commented that a constant lethal index should not be assumed for phosgene. Data from studies undertaken during WWI are not easy to interpret: today the LCt_{50} or the LC_{50} would be determined (see Chapter 2); during WWI the lethal dose may have been regarded as the dose needed to ensure killing of all the animals in the study.

The toxicity of phosgene may be perhaps more usefully illustrated by the following quotation from Prentiss (1937):

> *a concentration of 0.50 mg per litre being fatal after 10 minutes' exposure. In higher concentrations, which are often met in battle, one or two breaths may be fatal in a few hours.*

Vedder (1925) may also be quoted:

> *as little as 1 mg per litre may be lethal if exposure lasts more than a few minutes. . . .*

Haber's law (or Haber's rule) was discussed at some length in Chapter 2. A recent paper by Hatch *et al* (2001) has extended this discussion with special reference to phosgene. The authors focused on repeated and chronic exposures and showed that adaptation occurred if exposures were not overwhelming and were repeated daily. Adaptation was shown to wane over the period of a month or so. Repeated exposure at monthly intervals was associated with the development of chronic effects. This work shows that great care is needed in predicting responses from the *Ct* product if exposures are repeated and that the

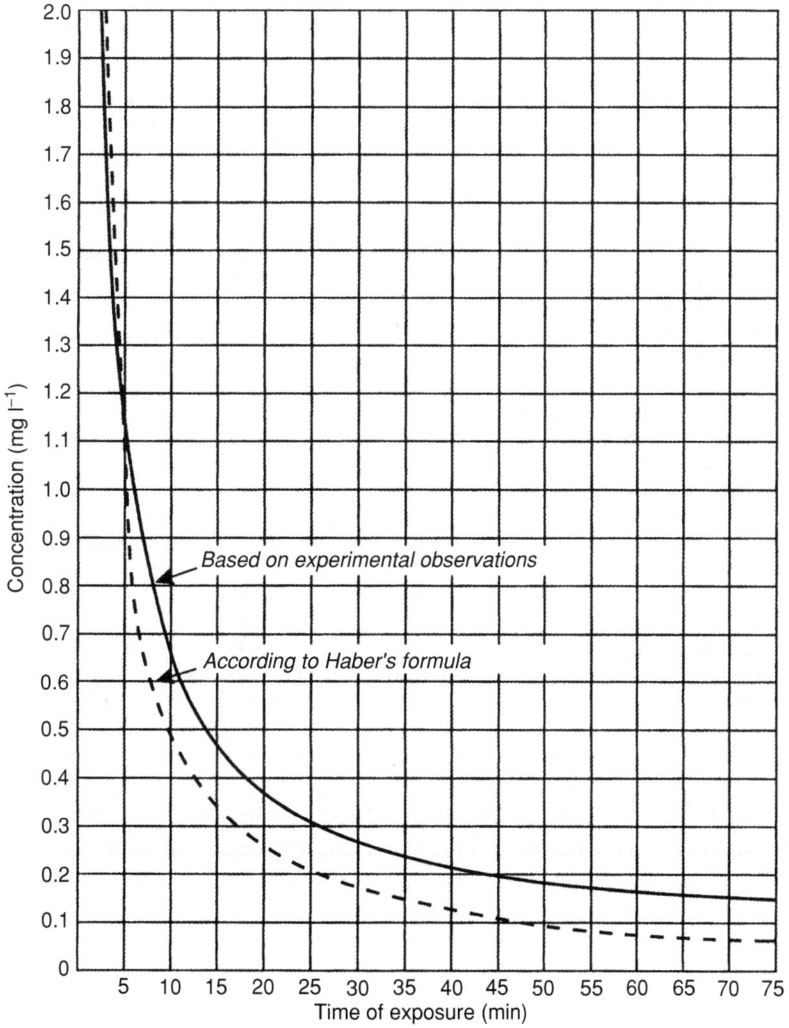

Figure 1. Toxicity curve for phosgene (in dogs)

interval between exposures can affect the pattern of response. In the chemical warfare or terrorist incident setting, repeated exposure is unlikely. The mechanism of adaptation is not clearly known although up-regulation of antioxidant defence systems is a possibility.

PHARMACOLOGICAL EFFECTS OF PHOSGENE

Inhalation of phosgene produces pulmonary oedema. In the years immediately following WWI, and still in some contemporary textbooks, this was assumed to be due to the liberation of hydrochloric acid in the lung and subsequent damage to the epithelial and endothelial surfaces. Potts *et al.* (1949) attempted to reproduce the effects of phosgene using hydrochloric acid: interestingly, a quite different toxicological picture was produced. Inhalation of hydrochloric acid led to immediate distress and rapid death; inhalation of phosgene led to few initial effects and a delayed death. These results obtained by Potts *et al.* (1949) confirmed those obtained by Winternitz in 1920. Nash and Pattle (1971) demonstrated that at a phosgene concentration of 25 ppm (110.5 mg m^{-3}) the HCl produced would easily be buffered.

Table 2. Lethal concentrations of gases (after 10 min exposure)

	American data		German data
Agent	mg l^{-1}	Lethal index	Lethal index
Phosgene	0.50	5 000	450
Diphosgene	0.50	5 000	500
Lewisite	0.12	1 200	1 500
Mustard gas	0.15	1 500	1 500
Chloropicrin	2.00	20 000	2 000
Ethylsulphuryl chloride	1.00	10 000	2 000
Ethyldichlorarsine	0.50	5 000	3 000
Ethylbromoacetate	2.30	23 000	3 000
Phenylcarbylamine chloride	0.50	5 000	3 000
Chloracetone	2.30	23 000	3 000
Benzyl iodide	3.00	30 000	3 000
Methyldichlorarsine	0.75	7 500	3 000
Acrolein	0.35	3 500	3 000
Diphenylchlorarsine	1.50	15 000	4 000
Diphenylcyanarsine	1.00	10 000	4 000
Bromacetone	3.20	32 000	4 000
Chloracetophenone	0.85	8 500	4 000
Benzyl bromide	4.50	45 000	6 000
Xylyl bromide	5.60	56 000	6 000
Brombenzyl cyanide	0.35	3 500	7 500
Chlorine	5.60	56 000	7 500
Hydrocyanic acid	0.20	2 000	1 000–4 000
Carbon monoxide	5.00	50 000	70 000

Using a mathematical model, they calculated that the 'maximum concentration of acid in a blood–air barrier of thickness 1 μm in contact with 25 ppm of phosgene is 7×10^{-10} M, which is negligible'. At very high concentrations of phosgene, the production of hydrochloric acid may play a part in its toxicity; at lower levels, other explanations should be sought.

Of modern theories, that of Potts *et al.* (1949), in which phosgene is described as forming diamides and that of Diller (1978), which proposed interactions between phosgene and a wide range of molecules have proved the most persuasive. Reactions proposed by Potts *et al.* (1949) and Diller (1978) are shown in Figures 2 and 3.

Frosolono and Pawlowski (1977) studied biochemical changes in various lung fractions prepared from rats exposed to phosgene at concentrations near to or above the LCt_{50}. A number of enzymes showed decreased activity in all fractions: these included *p*-nitrophenyl phosphatase, cytochrome *c* oxidase, ATPase and lactate dehydrogenase (LDH). The serum LDH rose. It was suggested that either inhibition of enzyme activity or loss of enzyme from cells would account for these changes. The data available did not allow

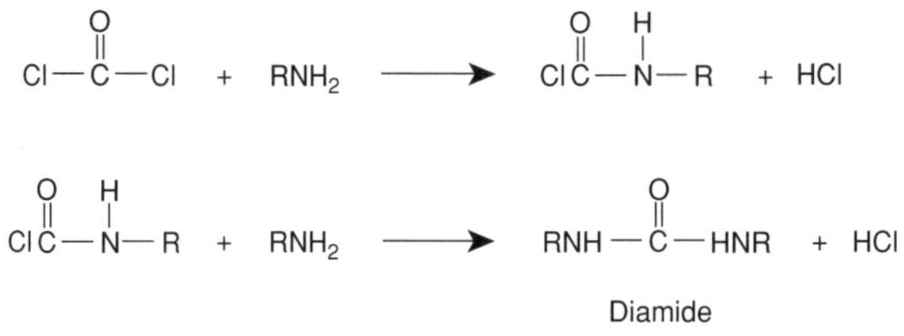

Figure 2. Reaction proposed for the formation of diamides from phosgene and diphosgene (Potts *et al.*, 1949)

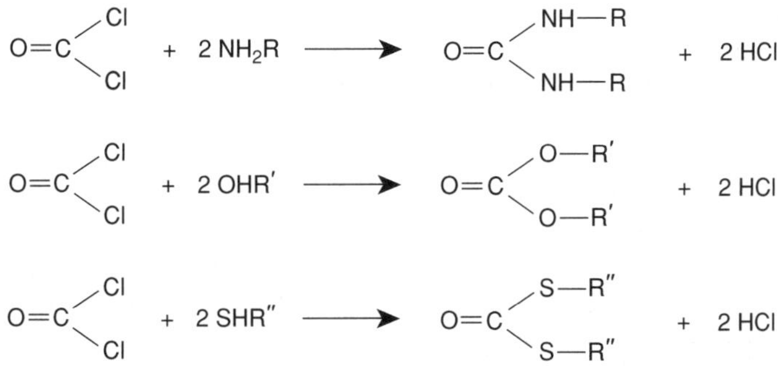

Figure 3. Interactions between phosgene and various molecules (Potts *et al.*, 1949; Diller, 1948)

a distinction to be drawn between these possible mechanisms. The view that phosgene binds to essential cell components and that this leads to cell damage is now generally accepted. In 1968, however, Everett and Overholt revived the hydrochloric acid hypothesis and drew into the debate the earlier work of Ivanhoe and Meyers (1964). It was suggested that the production of hydrochloric acid in the airways led to both severe reflex bronchoconstriction and that local irritation led to pulmonary oedema. This hypothesis does not appear to have attracted general support although the suggestion that precapillary vasoconstriction might lead to oedema has found a parallel in work on pulmonary oedema following head injury.

Exposure to very high levels of phosgene may lead to death before pulmonary oedema has developed. The cause of this is obscure, although the effect has been recorded in cases of exposure to high levels of other lung-damaging compounds, including chlorine. The hypothesis of reflex inhibition of respiration is often put forward in explanation of this effect.

Recent developments in understanding of the mechanism of action of phosgene and suggestions for therapeutic interventions

Recent work from the United States has shed new light on the possible mechanisms of action of phosgene. A series of papers by Sciuto and colleagues has focused on using therapeutic compounds of known mechanisms to probe the biochemical processes by which phosgene acts. The following account is based primarily on fourteen papers from this group; the original papers should be consulted for details. To this, references to other key studies have been added.

Work has developed along three major lines:

(a) Studies of changes in tissue levels of substances, such as glutathione (GSH), known to be important in defending against the effects of free radicals.
(b) Studies of active compounds, e.g. leukotrienes released as a result of exposure of the lung to phosgene.
(c) Studies of the effects of compounds such as *N*-acetylcysteine (NAC), and congeners, known to increase levels of GSH, compounds such as isoprenaline and aminophylline that are known to up-regulate production of 3,5-cyclic adenosine monophosphate (cAMP), and compounds such as ibuprofen that are known to interfere with the production of arachidonic acid (AA) derivatives, including the leukotrienes LTC_4, D_4 and E_4. These studies have been remarkably successful and it now seems possible to block phosgene-induced pulmonary oedema in animal models. Clinical demonstration of the efficacy of these interventions is, however, still lacking. Most encouraging is the fact that some of the interventions examined were effective when given after exposure of experimental animals to phosgene.

What, then, has been discovered?

(1) Kennedy *et al.* (1989) have shown that pre-treatment with a congener of cAMP, dibutyl cAMP (DBcAMP), with

aminophylline or with ß-adrenergic antagonists inhibit oedema development in isolated perfused rabbit lung exposed to phosgene. Post-exposure administration of aminophylline and the ß antagonist terbutaline also reduced oedema formation.

(2) Guo *et al.* (1990) working with an *in vivo* rabbit model showed that pre- or post-treatment with indomethacin, an inhibitor of thromboxane and prostacyclin production but not of leukotriene production, partially blocked phosgene-induced pulmonary oedema. The leukotriene receptor blockers FPL 55712 and LY 171883 dramatically reduced the oedema when given post-exposure to phosgene.

(3) Sciuto *et al.* (1995) showed that post-exposure treatment of rabbits with NAC prevented a reduction in GSH induced by phosgene. Production of leukotrienes LTC_4, D_4 and E_4 was also reduced.

(4) Sciuto *et al.* (1996a) showed that ibuprofen given both pre- and post-exposure to phosgene, in a rat model, inhibited oedema formation. Pentoxifylline (PTX) (an inhibitor of phosphodiesterase which can thus increase intracellular cAMP levels) given in the same way did not inhibit oedema production. The authors argued that the failure of PTX might be attributed to its haemodynamic properties leading to an increase in pulmonary blood flow. The authors expressed some surprise at this finding and noted that both aminophylline and methylxanthine, both known to increase cAMP levels had been shown to protect against the effects of phosgene.

(5) In 1996, Sciuto and coworkers showed that intratracheal administration of DBcAMP reduced oedema formation in an isolated, perfused rabbit lung model (Sciuto *et al.*, 1996b). Again, LTC_4, D_4 and E_4 production was inhibited. It was noted that intratracheal administration of DBcAMP was more effective than intravascular administration.

(6) Sciuto *et al.* (1997) showed that post-exposure treatment with aminophylline (known to increase intracellular cAMP levels) protected rabbits against phosgene-induced pulmonary oedema. Again, LTC_4, D_4 and E_4 production was reduced and measurement of the marker for free radical production, 'thiobarbituric acid-reactive substances' (TBARS) showed that aminophylline had reduced free radical production.

(7) In 1998, Sciuto and Stotts showed that post-treatment of guinea pigs with 5,8,11,14-eicosatetraynoic acid (ETYA) reduced pulmonary artery pressure after exposure to phosgene. ETYA is a competitive analogue of arachidonic acid and blocks the formation of arachidonic acid-derived mediators. Not all of the results of this study proved easy to interpret: ETYA decreased lipid peroxidation (as measured by TBARS formation) and decreased oedema formation but increased leukotiene release. The authors' account should be examined for details but the importance of possible species-specific effects is noted here.

(8) In 1998, Sciuto published a report of a study of early biochemical changes in lung lavage fluid and lung tissue following exposure of mice, rats and guinea pigs to phosgene. A decrement in tissue antioxidant levels (GSH) was reported although lavage levels of GSH were normal or slightly increased. The study focused on very early changes and it is difficult to know how closely these would represent likely later changes as oedema developed.

(9) In 2001, Sciuto and Morgan reported that dietary treatment with *n*-propyl gallate (nPG) but not with vitamin E, conferred increased resistance to phosgene in a mouse model. Lung GSH levels were increased by the pre-treatment. Again, paradoxical results were reported: the lower dose of nPG (0.75 wt% in rodent chow 5002) increased survival rate and lowered TBARS production but the higher dose (1.5 wt% in chow) did not.

(10) Sciuto *et al.* (2003) reported effects of phosgene on antioxidant enzymes and on GSH in lavage fluid from mice exposed to phosgene. All the compounds measured increased and then declined again over the

period 1 h to 7 days post-exposure with the exception of superoxide dismutase (SOD) which showed a sharp decline from 1 to 24 h post-exposure and then recovery by 7 days. The other enzymes measured included glutathione peroxidase, glutathione reductase and total glutathione. These results are not easy to interpret. The authors pointed out that it was known that exposure to phosgene reduced tissue levels of glutathione (Sciuto, 1998) and argued that the increased levels in lavage fluid might be part of a protective response. It may be noted that increased GSH levels say little about rate of GSH to GSSG conversion – both could be increased if the cycling rate between the oxidized and reduced forms was increased. The decrease in SOD levels was suggested to be due to increased levels of oxygen free radicals ($O_2^{\bullet -}$) and that these might be important in causing tissue damage. The possible conversion of the superoxide radical ($O_2^{\bullet -}$) to the hydroperoxy radical ($HO_2^{\bullet}$) and to the very toxic radical $OH^{\bullet}$ ($HO^{\bullet}$) was noted.

(11) In 2004, Sciuto and Hurt published a long paper touching on earlier work and extending this to look at a range of drugs (NAC, ibuprofen, aminophylline and isoprenaline), and many endpoints in both *in situ* perfused and *in vivo* rabbit models of phosgene-induced pulmonary oedema. Such an extensive study yielded many results worthy of discussion: the authors' main conclusions are given below.

(a) Intratracheal administration of NAC lowered pulmonary artery pressure, reduced the formation of LTC_4, D_4 and E_4 and increased the GSH/GSSG ratio.

Deduction: NAC protects against phosgene by acting as an antioxidant and by keeping up levels of the endogenous antioxidant GSH.

(b) Aminophylline reduced lipid peroxidation, reduced output of LTC_4, D_4 and E_4 and prevented a reduction in intracellular cAMP levels.

Deduction: aminophylline decreases lipid peroxidation and thus the output of leukotriene molecules that increase capillary permeability.

(c) Isoprenaline reduced pulmonary artery pressure, and also the increase in lung water seen in the model on exposure to phosgene. In addition, GSH levels were increased or maintained.

Deduction: isoprenaline prevents increased capillary leakage and had a useful effect on antioxidant levels in the lung.

(d) Ibuprofen reduced lipid peroxidation, increased or maintained GSH levels and reduced oedema formation.

All this leads to the following conclusions regarding the mode of action of phosgene.

(i) Phosgene produces damage to the lung by mechanisms similar to those of compounds that are or lead to the generation of free radicals: lipoperoxidation and leukotriene formation are hallmarks of such compounds.

(ii) A decrement in intracellular cAMP is produced and this leads to increased capillary permeability, probably as a result of effects on the endothelium cytoskeleton. It is known that a decrease in cAMP leads to a loosening of intercellular tight junctions and an increase in paracellular fluid leakage from capillaries.

(iii) The mechanism of effects on cAMP levels may involve increased activity of phosphodiesterase: this is inhibited by aminophylline. cAMP levels are increased by β_2 antagonists such as isoprenaline.

(iv) Decrements in tissue GSH levels are known to affect the regulation of intracellular calcium concentration: this is said to be mediated by effects on ion translocases in mitochondria, endoplasmic reticulum and the plasma membrane. An increase in free intracellular calcium concentration leads to blebbing of membranes and can cause cell death. Increased free calcium levels can also lead to activation of transcription factors such as $NF\kappa\beta$ (nuclear factor kappa beta) and thus to the activation of genes controlling production of interleukins. These locally acting transmitters or messenger molecules – some 23 individual species are known – are key factors in mediating the acute inflammation response.

Work by Ghio *et al.* (1991) showed that reduction of the neutrophil influx reduced lung injury and mortality following phosgene inhalation in a rat model.

Two further strands of evidence have also appeared. Werrlein *et al.* (1994) have shown that phosgene exposure affects F-actin levels in cultured lung cells (sheep and rat) and have argued that this is linked with changes in basal lamina morphology. Interestingly, these authors also reported phenotype switching in airway cells with an increased expression of the dendritic cell phenotype. The significance of this change is unknown. Disorganization of the basal lamina was associated with an increase in permeability of cell sheets.

In 1999, Jugg *et al.* reported that phosgene caused changes in the phospholipid components of surfactant. The authors suggested that a change in composition and in production rate (phosgene caused late increases in phospholipids) could affect the functioning of the surfactant layer and drew parallels with similar responses seen in lung injury leading to inflammatory reactions and the release of mediators of inflammation. The authors further suggested that drugs such as ambroxol, that are known to increase surfactant production, might have a place in the management of phosgene-induced lung injury.

The reader will now appreciate that our grasp of the mechanisms by which phosgene may injure the lung has increased dramatically during the last ten years or so. Some gaps in the theories of mechanism of action remain to be filled: does phosgene act as a free radical itself or is alkylation of key antioxidant species leading to an accumulation of endogenously produced radicals the key to the process? It is known that phosgene can be cleaved to produce the carbamoyl monochloride radical and Arroyo *et al.* (1993) have suggested that this may, in part, explain its reactivity with tissue components. Details of the chemical reactions between phosgene and a range of biologically important molecules have been provided by Babad and Zeiler (1972) – a valuable paper providing 358 references to the chemical literature.

It would be reasonable to infer from these mechanistic and experimental studies of the effects of well-known therapeutic agents that the treatment of pulmonary damage caused by phosgene could now be put on a firm footing. A combination of aminophylline, anti-inflammatory drugs (such as ibuprofen), isoprenaline and indeed *N*-acetylcysteine, would seem to be, at the least, a rational approach to therapy. We shall return to this point in discussing options for therapy.

THE PATHOPHYSIOLOGY OF PHOSGENE-INDUCED PULMONARY OEDEMA

The inhalation of phosgene in toxic quantities produces pulmonary oedema. The exact mechanisms involved remain remarkably obscure: in particular, the latent period, see below, is puzzling. No discussion of the effects of phosgene can be undertaken without a careful examination of current concepts of tissue fluid balance. These have changed substantially since WWI, although the early work of Starling (1896) has largely stood the test of time.

It is well-known that fluid tends to leak from the arterial and to be reabsorbed at the venous end of capillaries (Staub, 1974). Fluid not absorbed at the venous end of the capillaries is drained from the tissue spaces by the lymphatic system and returned to the vascular system via the lymphatic trunks which join the venous system at the root of the neck. Oedema is the accumulation of excess fluid in the tissues and may be produced by a number of causes. These include:

- an increase in hydrostatic pressure in the vascular system due perhaps to arterial hypertension or venous obstruction;
- a reduction in the colloid osmotic pressure of the blood, for example, in liver failure or malnutrition;
- an increase in the permeability of the capillary wall as, for example, occurs in the skin

as a result of the local production of, or introduction of, substances such as histamine and 5-hydroxytryptamine;

- a failure of the lymphatic drainage.

The balance of pressures, or forces, at the capillary wall was first described by Starling in 1896: the forces involved in controlling the movement of water across capillary walls are often referred to as 'The Starling Forces' (Starling, 1896). At most capillary walls, the major ions in the plasma move freely and thus do not exert an osmotic pressure. Capillary walls are, however, much less permeable to proteins – the concentration inside capillaries being greater than the concentration in the tissue fluid. An osmotic pressure differential is, therefore, established across the capillary wall. In addition, the unequal distribution of protein anions across the capillary wall causes a Gibbs–Donnan equilibrium of ions which pass readily across the wall to be established. The combination of the osmotic pressure exerted by the protein molecules (the colloid osmotic pressure) and that contributed by the concentration difference of freely permeable ions occurring as a result of the Gibbs–Donnan equilibrium, is described as the 'Oncotic Pressure' of the blood.

The balance of forces across the capillary wall is expressed succinctly by the Starling equation:

$$Q_f = K_f[(P_{mv} - P_{pmv}) - \theta(\pi_{mv} - \pi_{pmv})]$$

where

Q_f = water flow;
K_f = fluid conductance of the capillary wall;
P_{mv} = intracapillary hydrostatic pressure (microvascular pressure);
P_{pmv} = tissue fluid hydrostatic pressure (perimicrovascular pressure);
π_{mv} = microvascular osmotic pressure (colloid oncotic pressure);
π_{pmv} = perimicrovascular osmotic pressure (colloid osmotic pressure);
θ = Staverman reflection coefficient (if $\theta = 1$, then the membrane, capillary wall in this case, is completely impermeable to the molecule considered; if $\theta = 0$, then the membrane is freely permeable to the molecule considered).

In recent years, understanding of the flux of water and other substances across capillary walls has been clarified by the work of A.C. Guyton and his co-workers. An excellent account of fluid balance across the pulmonary capillaries may be found in recent editions of Guyton's *Textbook of Medical Physiology* (1986). Guyton's contributions include:

- demonstration that the interstitial hydrostatic pressure is sub-atmospheric (sometimes referred to as 'negative');
- the concept of a safety factor which protects against the formation of oedema.

The concept of the safety factor has caused some difficulty. In its simplest terms it may be thought of as the amount by which the capillary pressure has to be raised before oedema, i.e. swelling of the tissue spaces, occurs. This is a carefully constructed definition: note it says nothing about the flow of fluid across the capillary wall – this may be substantially increased but as long as no swelling, i.e. no accumulation of water, in tissue spaces occurs, then there is no oedema.

What are the components of the safety factor? There are in fact three key components: the negative interstitial pressure, the capacity of the lymphatic system to transport more fluid than it does under normal circumstances and the fact that increased lymph drainage tends to wash protein out of the interstitial spaces, thus reducing perimicrovascular osmotic pressure.

The safety factor in the pulmonary circulation is of the order of 21 mmHg, i.e. pulmonary capillary pressure has to be raised by 21 mmHg before fluid starts to accumulate in the interstitial spaces of the lung. One cause of confusion regarding the safety factor has been the confusion alluded to above between increase of fluid flow and increase of interstitial volume. Let us now examine the balance of forces across the pulmonary capillaries.

In Table 3, the pressures pushing or pulling fluid out of the capillaries are shown with a + sign, while those pushing or pulling fluid into the capillaries are shown with a – sign. Thus, as can be seen from this table, a small net outward force exists and a constant loss of fluid from lung capillaries is to be expected. The 'safety factor', 21 mmHg, relates to the changes in capillary pressure which can be sustained without an increase in interstitial volume. The component of the

Table 3. Balance of forces across the pulmonary capillaries

Force	Pressure (mmHg)
Forces tending to cause fluid to leave pulmonary capillaries and enter the interstitium	
• Capillary pressure	+7
• Interstitial fluid colloid osmotic pressure	+14
• Negative interstitial fluid hydrostatic pressure	+8
• **Total outward force**	**+29**
Forces tending to cause fluid to move from the interstitium into the pulmonary capillaries	
• Plasma colloid osmotic pressure	−28
• **Total inward force**	**−28**
• Total outward force	+29
• Total inward force	−28
• **Mean filtration pressure**	**+1**

safety factor contributed by the capacity for increased lymphatic drainage is fairly easily understood, as is the effect of washing protein out of the interstitial space. The contribution made by the negative interstitial pressure is less easy to understand. Guyton's analogy with a balloon may be helpful. Imagine a balloon connected to a vacuum pump. Let the pressure in the balloon be reduced to 10 mmHg below atmospheric. The balloon will be flat. Now turn down the vacuum pump a little until the pressure in the balloon is 5 mmHg less than atmospheric: the balloon will still be flat. Now switch off the vacuum pump and allow the balloon to equilibrate at atmospheric pressure: the balloon will still be flat. Now apply a small positive pressure (say +1 mmHg) via the tube connected to the balloon: the balloon will expand until all the creases present when it was collapsed have disappeared. To make the balloon expand any more, positive pressure will be needed. It will be appreciated that changing the pressure inside the balloon from −10 mmHg to 0 mmHg (i.e. atmospheric pressure) produced no change in the volume of the balloon. On the contrary, changing the pressure from 0 to +1 mmHg produced a sudden and substantial change in volume.

Consider now the *compliance* of the system, recalling that compliance is defined as change in volume for unit change in pressure (dV/dP). In the negative-pressure part of the curve, the system has an effectively zero compliance, i.e. the balloon can withstand a change of pressure of 10 mmHg without any change in volume. However, once the internal pressure becomes positive, the system then displays a huge compliance: a very substantial change in volume is produced by an increase in pressure of 1 mmHg.

Guyton has demonstrated, using an isolated limb preparation in which the blood vessels were perfused with fluid of known osmotic pressure and at a given hydrostatic pressure, that a curve as shown in Figure 4 could be recorded.

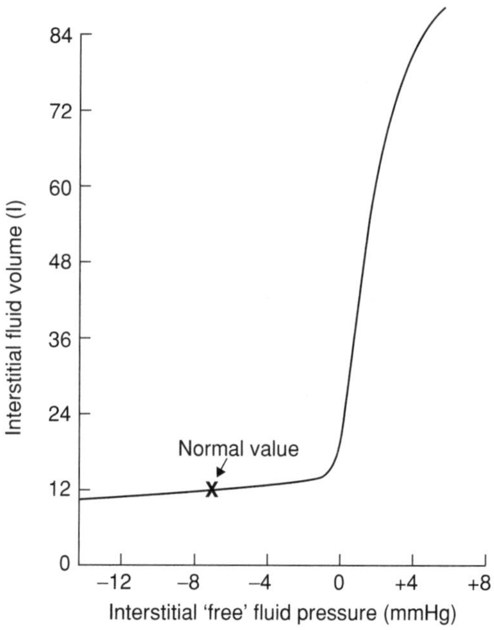

Figure 4. Pressure–volume curve of the interstitial spaces (extrapolated to the human being from data obtained from dogs)

It will also be appreciated that as fluid begins to leak from capillaries the pressure in the interstitial compartment rises rapidly and this will oppose the fluid leakage.

We have moved some way from the pathophysiology of phosgene poisoning – let us now consider the movement of fluid across the alveolar epithelium. Guyton (1986) has suggested that the capacity of the alveolar epithelium to resist fluid flow (i.e. from the interstitial space to the intra-alveolar space) is low; indeed, he has suggested that as soon as the interstitial pressure becomes positive and its volume suddenly expands (see above), fluid will flow into the alveoli. It will be understood that the negative interstitial pressure applies a draining force to the alveoli under normal conditions and the only movement of fluid into the alveoli from the interstitial space occurs as a result of 'capillary-creep' at intercellular junctions.

The presence of fluid in the alveoli, albeit only a thin layer, will exert a force on the interstitial space. This force may be calculated by use of the law of Laplace. This law is expressed for a single spherical curved surface by the following equation:

$$P = 2T/R$$

where P is the pressure exerted on the contents of the sphere by the curved surface, T is the surface tension and R the radius of curvature of the surface. In the lung we may think of P as representing the force, generated by the curved alveolar surface, pulling on the interstitium. In the alveolus the radius of curvature of the wall, and thus of the fluid lining layer, is not constant. In the highly curved corners of the alveoli the radius of curvature is small; along the flatter parts of the wall the radius of curvature is much larger. Thus the force exerted on the interstitium will vary: being greatest where the radius of curvature is small i.e. in the corners of the alveoli. Here fluid will be drawn into the alveolus from the interstitium and the corners of alveoli are often, at microscopy seen to contain small pools of fluid. It will be recalled that the alveolus is lined with surfactant: this lowers surface tension. Furthermore the surface tension exerted by a surfactant film is proportional to the area of the film: rising as the film is expanded. This allows the pressure within small and large alveoli to be equal and thus allows alveoli of different sizes to the connected together without the smaller spaces collapsing and the larger spaces expanding. Note that if surface tension were constant the Law of LaPlace would predict a higher pressure in the smaller alveoli than in their larger fellows. Now consider the effects of fluid leaking from the interstitium into an alveolus. As the alveolus fills the radius of curvature will fall and the force exerted on the interstitium will increase. Thus fluid filling will accelerate and sudden flooding occurs. Surfactant will reduce this effect by:

- reducing surface tension to well below that exerted by a water surface;
- further reducing the surface tension as the area (radius) falls.

These 'helpful' effects of surfactant will be reduced as the lipid film is diluted with water.

These mechanisms contribute to the sudden flooding of alveoli which occurs during the development of pulmonary oedema (Staub, 1974).

Although some readers may think the foregoing account of pulmonary oedema is long, it should be recalled that it is in fact a much simplified and abbreviated account. For an excellent and detailed account, the reader should consult Prichard's book *Edema of the Lung* (Prichard, 1982). Prichard and Lee (1987) have also contributed a valuable brief account of the topic in the *Oxford Textbook of Medicine*. This account stresses the importance of pulsatile flow in the pulmonary capillaries. Movement of fluid into the interstitial space is expected during systole but during diastole, movement back into the capillaries occurs. During exercise when the heart rate is increased, diastole is shortened, thus allowing less time for return of fluid from the interstitium. Interstitial fluid pressure rises and juxtacapillary receptors are stimulated, producing the sensation of breathlessness (see below). The concept of increased efflux of fluid during exercise is of great importance in phosgene poisoning (see below).

Phosgene damages the cells of the alveolar walls and thus the alveolar–interstitial fluid barrier. Damage to the pulmonary capillaries also occurs and the sequence of development of oedema described below supports this. Cameron and Courtice (1946) reported a series of studies on the effects of phosgene undertaken at the Chemical Defence Establishment at Porton Down (UK) during WWII. Rabbits, goats and dogs were exposed to phosgene; lymph flow from the lung was measured, as was the composition of the lymph. The concentration of oedema fluid was also monitored. It was found that dogs and goats showed a marked haemoconcentration during the development of pulmonary oedema: plasma volume fell. Rabbits, however, showed little haemoconcentration despite florid pulmonary oedema. Cameron commented:

> *The compensatory mechanism for rapidly restoring the plasma volume in rabbits appears to be detrimental to the life of the animal, for the more rapidly the plasma volume is restored, the more rapidly the oedema progresses. By partially dehydrating rabbits beforehand, the mortality can be decreased and haemoconcentration occurs. These observations would indicate that, in cases of haemoconcentration and fall in plasma volume due to a loss of fluid into the lungs, it would be better not to try to increase the plasma volume by transfusions, for anoxia due to pulmonary oedema is a more important factor in causing death than haemoconcentration and decreased plasma volume.*

These important conclusions have been quoted at some length, as they remain of fundamental importance in the management of phosgene poisoning (see below). Whitteridge (1948) studied the effects of phosgene using an anaesthetized cat preparation. A marked increase in respiratory rate and phrenic nerve activity was noted post-exposure. Later studies by Anand *et al.* (1986) and Paintal (1969) confirmed that stimulation of J receptors occurred after exposure to phosgene. More recent studies by Keeler *et al.* (1990) of the pathophysiology of phosgene in sheep have confirmed and extended the results of Cameron and Courtice (1946) referred to above. A range of other studies of the pathophysiological effects of phosgene have also been reported in recent years. Jaskot *et al.* (1989) demonstrated an increase in angiotensin-converting enzyme (ACE) activity in lung tissue obtained from rats exposed to 0.5 ppm phosgene for 4 h. Studies by Currie *et al.* (1985) demonstrated a reduction in levels of ATP and $Na^{+}K^{+}$-ATPase activity in the lungs of rats also exposed to low concentrations of phosgene.

Pathology of phosgene-induced lung damage

The optical microscopic changes following the inhalation of phosgene were extensively studied during and after WWI. An excellent description was provided by Winternitz (1920). The work of Meek and Eyester (1920) should also be consulted as it contains details of observations of the histopathological effects of phosgene on the dog lung made by the celebrated pulmonary anatomist William Snow Miller. Winternitz's observations have been confirmed and extended by the work of Pawlowski and Frosolono (1977). The chronological pattern of changes seen at the electron microscope level may be summarized as follows:

(1) The epithelium of the terminal bronchiole is the first area to be damaged. Intracellular vesiculation of both ciliated and Clara cells has been noted. This appears to be a relatively immediate effect.
(2) Oedema of the alveolar interstitium appears.
(3) Intracellular oedema of the cells of the interalveolar septae appears.
(4) Focal disruption of alveolar type I cells noted.
(5) Marked swelling of interalveolar septae observed.
(6) Appearance of frank intra-alveolar oedema.

The oedema fluid is eosinophilic at optical microscopy and contains much protein. It seems unlikely that phosgene attacks one particular cell type in the lung. The resistance of the thin part of the blood–air barrier to attack by phosgene has been noted but no deductions regarding protective mechanisms or mechanisms of damage seem to have been drawn.

The pattern described above is the common one seen in many types of permeability pulmonary oedema. This has been described and discussed in detail by Staub (1974), Teplitz (1979) and Robin (1979). The account provided by Prichard and Lee (1987) is of great value.

SYMPTOMS AND SIGNS OF PHOSGENE POISONING

Extensive descriptions of the symptoms and signs of phosgene poisoning may be found in the accounts of Ireland (1926), Vedder (1925), Diller (1978), Everett and Overholt (1968), Seidelin (1961), Fruhman (1974) and Cucinell (1974).

During or immediately after exposure to dangerous concentrations of phosgene the following symptoms and signs develop: eye irritation, respiratory tract irritation, lacrimation, coughing, a feeling of choking and tightness in the chest. In some cases, nausea, retching and vomiting have been reported. The severity of these early effects is, however, no guide to the severity of the poisoning: patients who may later die may show minimal or no symptoms or signs at the time of exposure. An effect frequently reported during WWI was the 'tobacco reaction': men who had inhaled only a small quantity of phosgene reported a flat metallic taste upon lighting a cigarette. This reaction has also been reported in cases of hydrogen cyanide and sulphur dioxide exposure but is ill-understood.

Following exposure to phosgene, a latent period ranging from 30 min to 24 h may occur before the onset of more serious symptoms and signs. The onset of these may be precipitated by exercise and collapse of patients exposed to phosgene on subsequently taking exercise was frequently reported during WWI (Ireland, 1926). Once the latent period is over, patients develop dyspnoea, a painful cough and cyanosis. Increasing quantities of frothy white or yellowish fluid may be expectorated. Later, the fluid tends to become pink-tinged; a mushroom-like efflux of pink foam ('Champignon d'ecume': mushroom of foam or froth) may appear at the mouth in the dying. On physical examination, diminished breath sounds, rales and rhonchi in all areas may be detected. The blood pressure may be low and the heart rate raised to more than 130 min^{-1}. Circulatory collapse and cardiac failure may occur.

Of fatal cases, 80% may be expected to die during the first 24–48 h post-exposure. Deaths during WWI in patients who had survived this initial period were often as a result of pneumonia. Selgrade *et al.* (1989) have shown that exposure of mice to low concentrations of phosgene (0.25 ppm for 4 h) significantly increased mortality rates on post-exposure challenge with aerosolized *Streptococcus zooepidimicus*. An account of a case of phosgene poisoning is reproduced below (Vedder, 1923).

> *February 3, 1917: A chemist was working at a new chemical product. A syphon of phosgene, required for the synthesis of this substance, burst on his table at 1.00 pm. A yellowish cloud was seen by a second person in the room to go up close to the chemist's face, who exclaimed, 'I am gassed', and both hurried out of the room. Outside, the patient sat down on a chair looking pale and coughing slightly.*
>
> *2.30 pm: In bed at hospital, to which he had been taken by car, having been kept at rest since the accident. Hardly coughing at all, pulse normal. No distress or anxiety and talking freely to friends for over an hour. During this time he was so well that the medical officer was not even asked to see the patient upon admission to hospital.*
>
> *5.30 pm: Coughing with frothy expectoration commenced and the patient was noticed to be bluish about the lips. His condition now rapidly deteriorated. Every fit of coughing brought up large quantities of clear, yellowish frothy fluid of which about 80 ounces were expectorated in one and a half hours. His face became of a grey ashen colour, never purple, though the pulse remained fairly strong. He died at 6.50 pm without any great struggle for breath. The symptoms of irritation were very slight at the onset; there was then a delay of at least four hours and the final development of serious oedema up to death took little more than an hour though the patient was continually rested in bed.*

MANAGEMENT OF PHOSGENE POISONING

A number of deductions regarding the proper management of phosgene poisoning may be drawn from the discussion of the pathophysiology of phosgene poisoning – these should be borne in mind in the following section.

As usual in cases of exposure to chemical warfare agents, management of phosgene poisoning may be divided into two parts, as follows:

First aid measures

The casualty should be removed from the source of the phosgene. Rescue workers should wear adequate respiratory protection. Decontamination of casualties is not necessary.

Therapy

A comprehensive account of the various approaches which have been used in the treatment of phosgene poisoning may be found in the papers presented at a recent symposium (Mehlman *et al.*, 1985). One of the standard rules laid down during WWI for the management of phosgene casualties was that the casualties should undertake no exercise and should be given complete rest as soon as possible. The effect of exercise on the efflux of fluid from pulmonary capillaries to the pulmonary interstitium has already been considered. Detailed experimental studies by Postel *et al.* (1946), however, indicated that at least in some species exercise did not have a marked effect on the outcome of phosgene poisoning. This was particularly the case in rats and mice although in dogs exercise after phosgene poisoning led to increased lung weight beyond that seen in the poisoned but rested controls. In clinical practice, bed rest is invariably recommended.

Patients presenting with cyanosis and PaO_2 levels of less than 50 mmHg require additional oxygen in their inspired air. Depending on the severity of lung damage, oxygen may be administered by face-mask or endotracheal tube. Positive-end expiratory pressure may be needed to achieve adequate oxygenation. An early recommendation that inspired gases should be passed through 90% ethyl alcohol to reduce frothing of airway secretions does not seem to be followed today. If mechanical ventilation is needed, low tidal volume should be used (Brown *et al.*, 2003).

During WWI, great emphasis was placed on the value of oxygen in the treatment of phosgene poisoning. The following extract is taken from *The Official History of the Great War: Medical Services; Diseases of War*, Volume II (Vedder, 1923):

> *Oxygen should always be given to casualties with serious pulmonary oedema, that is, to men with intense blue cyanosis or grey pallor. These need oxygen continuously and over a long period. If the supply permits of such use, it should be given also to milder cases of oedema in order to prevent their lapsing into a more serious state of asphyxia. It should be remembered that in every autopsy on early death from pulmonary irritant poisoning, extreme oedema of the lungs was found, that cyanosis is the main indication of such oedema, and that no case in whom it was possible to restore a pink colour by the proper use of oxygen died of this pulmonary oedema.*

Such therapy was recommended by those exceptionally experienced in treating phosgene poisoning on a large scale. Further details may be found in the works of Haldane (1919) and Barcroft (1920).

Rest and oxygen then are the main stays of management, while drugs are of secondary value.

DRUG TREATMENT

Steroids

The use of large doses of steroids has become standard in the management of permeability pulmonary oedema. There is considerable experimental evidence to suggest that steroids given to experimental animals before exposure to pulmonary oedema-inducing agents can ameliorate the oedema. This presumably is linked with the inhibition of the inflammatory response to the irritant. Evidence for the value of steroids after exposure to phosgene and other lung-damaging compounds is weak.

However, if steroids are to be given they should be given:

- as soon as possible after exposure
- in large doses
- discontinued fairly quickly.

Methyl prednisolone sodium succinate has been proposed in the following regime:

- 15 min or less post-exposure: 2000 mg IV or IM
- 6 h post-exposure: 2000 mg IV or IM
- 12 h post-exposure: 2000 mg IV or IM.

Continue at the same dose 12 hourly for 1–5 days, depending on the patient's condition (Ministry of Defence, 1987).

A number of reports of the value of steroids and some suggesting that they are of little value are noted below.

Everett and Overholt (1968) reported the successful use of hydrocortisone (200 mg per day) in a case of phosgene poisoning. Diller (1978) considered that the use of steroids was indicated in phosgene poisoning and suggested that inhibition of prostaglandin generation and release, inhibition of leukocyte release of kinins, inhibition of proteolysis and stabilization of mucopolysaccharides leading to maintenance of normal capillary permeability might underlie their effectiveness. Bradley and Unger (1982), on the other hand, felt that the evidence for the efficacy of steroids in phosgene poisoning was inconclusive although they used methylprednisolone in the case reported in their paper. The administration of steroids by inhalation has also been recommended (Ministry of Defence, 1987).

In 1987, Prichard and Lee summed up current opinion on the value of steroids in the management of permeability pulmonary oedema in their valuable chapter in *The Oxford Textbook of Medicine*. Their final paragraph on the point concluded:

> *[regarding the use of steroids in high-permeability pulmonary oedema] However, there is some logic in their use, for evidence is now very strong that granulocytes aggregate in the lungs of patients with ARDS as a result of complement activation. Steroids can not only reduce their aggregation but also diminish free radical production. Until a conclusive answer is produced, clinicians will continue to use steroids in large doses despite the absence of adequate supportive clinical data.*

The need for a clinical trial of the value of steroids in the management of phosgene poisoning is clear.

Antibiotics

The use of prophylactic antibiotics in cases of phosgene poisoning has often been recommended: penicillin G, amoxycillin and chloramphenicol have all been used (Diller, 1978; Everett and Overholt, 1968).

Other drugs which have been recommended

- Aminophylline (Everett and Overholt, 1968)
- Dopamine (Bradley and Unger, 1982)
- Codeine phosphate for alleviation of cough (Ministry of Defence, 1987)
- ε-aminocaproic acid (Diller, 1978)
- Phosgene antiserum (Ong, 1972).

The reader will recall the detailed discussion of the recent work on the mechanisms of action of phosgene and the suggestions made there regarding a rational approach to therapy. It is disappointing to record, therefore, that few of the compounds discussed have been shown to work in the clinical setting although it should be noted that they have not been exposed to clinical trials. Much of the evidence regarding clinical efficacy is anecdotal and based on a few cases. This may be inevitable as poisoning by phosgene is comparatively rare. Borak and Diller (2001) discussed the available evidence in detail and Table 4 has been compiled from their paper.

The series of entries in the right-hand column of Table 4 suggests that essentially none of the drugs discussed earlier and accepted as being effective in studies involving animal models has become established as a part of the clinical management of phosgene exposure. The need for organised therapeutic trials is thus acute.

Hexamethylenetetramine (HMT or methenamine or hexamine or urotropine)

The historical link between the use of HMT to alleviate or prevent the effects of phosgene is a long one. In designing respirators to defend against phosgene during WWI, a range of compounds was examined. For example, soda lime (a mixture of sodium hydroxide and calcium hydroxide) proved effective, while phenol and sodium phenolate were also effective and 'British Phenolate Helmet' (P Helmet) was first used on December 15, 1915. This was followed by the 'Phenolate Hexamine Helmet' (PH Helmet) from early 1916 onwards. The PH Helmets comprised a flannel sack which had been dipped in a mixture of

Table 4. Evidence for efficacy of various drugs in management of phosgene exposure

Therapeutic agent	Mechanistic evidence	Shown to be effective in animal models	Shown to be effective in human patients
Treatment given during the post-exposure latency period, i.e. before frank pulmonary oedema develops: 'post-exposure prophylactic therapy'			
Steroids	Yes	—	Doubtful
Ibuprofen	Yes	Yes	No
N-Acetylcysteine	Yes	Yes	No
Treatment of manifest pulmonary oedema			
Steroids	Yes	Yes	No
Aminophylline	Yes	Yes (but variable)	No
Diuretics	—	Mainly negative effects	No (except in cases of fluid overload and cardiac failure
Prostaglandin E_1	—	?	No
Surfactant	—	?	No
Antihistamines	—	?	No
ϵ-Amino caproic acid	—	?	No
Urease	—	?	No
Hexamethylenetetramine			
Pre-exposure		Yes	?
Post-exposure		No	No

sodium phenolate and hexamine. The discovery that hexamine neutralized phosgene was made in Russia (Prentiss, 1937). The addition of goggles to the PH Helmet provided considerable protection against lacrimatory agents as well as against phosgene; such helmets lasted for about 24 h of continuous use.

These chemically reactive masks were replaced by canister respirators fairly quickly: canisters containing lime and charcoal proved very effective in removing a wide range of chemical agents from the inspired air. Those interested in the development of respirators should consult Prentiss's excellent account (Prentiss, 1937).

Hexamine still figures in the British National Formulary (as methenamine hippurate) as a long-term treatment for urinary tract infections (Joint Formulary Committee, 1993). Hexamine was long considered as likely to be of little value in the management of phosgene poisoning: rather, as in the case of steroids, it was accepted that there was a case for its likely efficacy if given to experimental animals prior to exposure to phosgene but that it would not be likely to be effective when given after exposure had occurred. In 1971, Stravrakis challenged this view and reported several cases where hexamine had been given (20 ml of a 20% solution IV) as soon as possible after exposure to phosgene (Stravrakis, 1971). The author felt that the treatment had been of value and stated:

> *If given intravenously during the latent period soon after exposure, hexamethylenetetramine will adequately protect the victim of phosgene exposure.*

Diller held, on the contrary, that there was no evidence in the form of adequately controlled experiments to support the assertion that hexamine is of value when administered after phosgene exposure (Diller, 1978,1980). Diller (1980) is a valuable source of references to earlier papers on hexamine.

Few clinicians today would recommend the use of hexamine in the management of phosgene poisoning.

LONG-TERM EFFECTS OF EXPOSURE TO PHOSGENE

Chronic bronchitis and emphysema have been reported as consequences of phosgene exposure (Cucinell, 1974). Diller (1985) suggested that the great majority of patients who recovered after an exposure to phosgene would make a complete recovery. He noted that exertional dyspnoea had been reported and this sometimes took some time (months) to resolve. This has also been reported after exposure to mustard gas.

REFERENCES

Anand A, Paintal AS and Whitteridge D (1986). Phosgene stimulates J receptors and produces increased respiratory drive in cats. *J Physiol*, **371**, 1098.

Arroyo CM, Feliciano F, Kolb DL *et al.* (1993). Autoionization reaction of phosgene ($OCCl_2$) studied by resonance/spin trapping techniques. *J Biochem Toxicol*, **8**, 107–110.

Babad H and Zeiler AG (1972). The chemistry of phosgene. From: *The World's Knowledge*. Supplied by the British Library (www.bl.uk).

Barcroft J (1920). Discussion on the therapeutic uses of oxygen. *Proc R Soc Med London*, **XIII**, 59–99.

Borak J and Diller WF (2001). Phosgene exposure: mechanisms of injury and treatment strategies. *J Occup Environ Med*, **43**, 110–119.

Bradley BL and Unger KM (1982). Phosgene inhalation: a case report. *Texas Med*, **78**, 51–53.

Brown RFR, Jugg BJ, Harban FM, Platt J, Rice P (2003). Pathological improvement following protective ventilation strategy in an inhalationally-induced lung injury model. *American Journal of Respiratory and Critical Care Medicine*, **167**(7): A775, April 2003.

Cameron GR and Courtice FC (1946). The production and removal of oedema fluid in the lung after exposure to carbonyl chloride (phosgene). *J Physiol*, **105**, 175–185.

Cucinell SA (1974). Review of the toxicity of long-term phosgene exposure. *Arch Environ Health*, **28**, 272–275.

Currie WD, Pratt PC and Frosolono MF (1985). Response of pulmonary energy metabolism to phosgene. *Toxicol Ind Health*, **1**, 17–27.

Diller WF (1978). Medical phosgene problems and their possible solution. *J Occup Med*, **20**, 189–193.

Diller WF (1980). The methenamine misunderstanding in the therapy of phosgene poisoning. *Arch Toxicol*, **46**, 199–206.

Diller WF (1985). Late sequelae after phosgene poisoning: a literature review. *Toxicol Ind Health*, **1**, 129–133.

Everett ED and Overholt EL (1968). Phosgene poisoning. *JAMA*, **205**, 243–245.

Frosolono MF and Pawlowski R (1977). Effect of phosgene on rat lungs after single high-level exposure. I. Biochemical alterations. *Arch Environ Health*, **32**, 271–277.

Fruhmann G (1974). Die Behandlung des Lungenödems nach Reizgasinhalation. (Treatment of pulmonary edema following inhalation of irritant gas). *Ther Gegegnw*, **113**, 38–46.

Ghio AJ, Kennedy TP, Hatch GE *et al.* (1991). Reduction of neutrophil influx diminishes lung injury and mortality following phosgene inhalation. *J Appl Physiol*, **71**, 657–665.

Guo Y-L, Kennedy TP, Michael JR *et al.* (1990). Mechanism of phosgene-induced lung toxicity: role of arachidonate mediators. *J Appl Physiol*, **69**, 1615–1622.

Guyton AC (1986). *Textbook of Medical Physiology*, 7th Edition. Philadelphia, PA, USA: WB Saunders.

Haldane JS (1919). Lung-irritant gas poisoning and its sequelae. *J R Army Med Coll*, **33**, 494–507.

Hatch GE, Kodavanti U, Crissman K *et al.* (2001). An 'injury-time integral' model for extrapolating from acute to chronic effects of phosgene. *Toxicol Ind Health*, **17**, 285–293.

Ireland MM (1926). *Medical Aspects of Gas Warfare*, Volume XIV. Washington, DC, USA: Medical Department of the United States on the World War.

Ivanhoe F and Meyers FH (1964). Phosgene poisoning as an example of neuroparalytic acute pulmonary edema: the sympathetic vasomotor reflex involved. *Dis Chest*, **46**, 211–218.

Jaskot RH, Grose EC and Stead AG (1989). Increase in angiotensin-converting enzyme in rat lungs following inhalation of phosgene. *Inhal Toxicol*, **1**, 71–78.

Joint Formulary Committee (1993). *British National Formulary*, BNF Number 25. London, UK: British Medical Association and Pharmaceutical Society of Great Britain.

Jugg B, Jenner J and Rice P (1999). The effect of perfluoroisobutene and phosgene on rat lavage fluid surfactant phospholipids. *Human Exp Toxicol*, **18**, 659–668.

Keeler JR, Hurt HH, Nold JB *et al.* (1990). Phosgene-induced lung injury in sheep. *Inhal Toxicol*, **2**, 391–406.

Kennedy TP, Michael JR, Hoidal JR *et al.* (1989). Dibutyryl cAMP, aminophylline, and ß-adrenergic agonists protect against pulmonary edema caused by phosgene. *J Appl Physiol*, **67**, 2542–2552.

Lefebure V (1921). *The Riddle of the Rhine*. London, UK: W Collins Sons & Company, Ltd.

Meek WJ and Eyster JAE (1920). Experiments on the pathological physiology of acute phosgene poisoning. *Am J Physiol*, **51**, 303–320.

Mehlman MA, Fensterheim RJ and Frosolono MF (1985). Phosgene induced edema: diagnosis and therapeutic countermeasures. An international symposium. *Toxicol Ind Health*, **1**, 1–160.

Ministry of Defence (1987). *Medical Manual of Defence Against Chemical Agents*, Ministry of Defence D/Med(F & S)(2)./10/1/1. London, UK: HMSO.

Nash T and Pattle RE (1971). The absorption of phosgene by aqueous solutions and its relation to toxicity. *Ann Occup Hyg*, **51**, 303–320.

Ong SG (1972). Treatment of phosgene poisoning with antiserum anaphylactic shock by phosgene. *Arch Toxicol*, **29**, 267–278.

Paintal AS (1969). Mechanisms of stimulation of type J pulmonary receptors. *J Physiol*, **203**, 511–532.

Pawlowski R and Frosolono MF (1977). Effect of phosgene on rat lungs after single high-level exposure. II. Ultrastructural alterations. *Arch Environ Health*, **32**, 278–283.

Postel S, Tobias JM, Patt HM *et al.* (1946). The effect of exercise on mortality of animals poisoned with diphosgene. *Proc Soc Exp Biol Med*, **63**, 432–436.

Potts AM, Simon FP and Gerard RW (1949). The mechanism of action of phosgene and diphosgene. *Arch Biochem*, **24**, 329–337.

Prentiss AM (1937). *Chemicals in War*. New York, NY, USA: McGraw-Hill Book Company Inc.

Prichard JS (1982). *Edema of the Lung*. Springfield, IL, USA: Charles C. Thomas.

Prichard JS and Lee G de J (1987). Pulmonary oedema. In: *Oxford Textbook of Medicine* (DJ Weatherall, JGG Ledingham and DA Worrell, eds), Chapter 13, Oxford, UK: Oxford University Press.

Robin ED (1979). Permeability pulmonary oedema. In: *Pulmonary Oedema* (AP Fishman and EM Renkin, eds), pp. 217–228. Bethesda, MD, USA: American Physiological Society.

Sciuto AM (1998). Assessment of early acute lung injury in rodents exposed to phosgene. *Arch Toxicol*, **72**, 283–288.

Sciuto AM and Hurt HH (2004). Therapeutic treatments of phosgene-induced lung injury. *Inhal Toxicol*, **16**, 565–580.

Sciuto AM and Moran TS (2001). Effect of dietary treatment with *n*-propyl gallate or vitamin E on the survival of mice exposed to phosgene. *J Appl Toxicol*, **21**, 33–39.

Sciuto AM and Stotts RR (1998). Posttreatment with eicosatetraynoic acid decreases lung edema in guinea pigs exposed to phosgene: the role of leukotrienes. *Exp Lung Res*, **24**, 273–292.

Sciuto AM, Strickland PT, Kennedy TP *et al.* (1995). Protective effects of *N*-acetylcysteine treatment after phosgene exposure in rabbits. *Am J Respir Crit Care Med*, **151**, 768–772.

Sciuto AM, Stotts RR and Hurt HH (1996a). Efficacy of ibuprofen and pentoxifylline in the treatment of phosgene-induced acute lung injury. *J Appl Toxicol*, **16**, 381–384.

Sciuto AM, Strickland PT, Kennedy TP *et al.* (1996b). Intratracheal administration of DBcAMP attenuates edema formation in phosgene-induced acute lung injury. *J Appl Physiol*, **80**, 149–157.

Sciuto AM, Strickland PT, Kennedy TP *et al.* (1997). Postexposure treatment with aminophylline protects against phosgene-induced acute lung injury. *Exp Lung Res*, **23**, 317–332.

Sciuto AM, Cascio MB, Moran TS *et al.* (2003). The fate of antioxidant enzymes in bronchoalveolar lavage fluid over 7 days in mice with acute lung injury. *Inhal Toxicol*, **15**, 675–685.

Seidelin R (1961). The inhalation of phosgene in a fire extinguisher accident. *Thorax*, **16**, 91–93.

Selgrade MK, Starnes DM, Illing JW *et al.* (1989). Effects of phosgene exposure on bacterial, viral and neoplastic lung disease susceptibility in mice. *Inhal Toxicol*, **1**, 243–259.

Starling EH (1896). On the absorption of fluid from connective tissue spaces. *J Physiol*, **19**, 312–336.

Staub NC (1974). Pulmonary oedema. *Physiol Rev*, **54**, 678–811.

Stavrakis P (1971). The use of hexamethylenetetramine (HMT) in treatment of acute phosgene poisoning. *Ind Med Surg*, **40**, 30–31.

Teplitz C (1979). Pulmonary cellular and interstitial oedema. In: *Pulmonary Oedema* (AP Fishman and EM Renkin, eds), pp. 97–112. Bethesda, MD, USA: American Physiological Society.

Vedder EB (1923). Phosgene In: *History of The Great War. Medical Services. Diseases of the War*, Volume II (WG MacPherson, WP Herringham, TR Elliott *et al.*), pp. 78–108. London, UK: HMSO.

Vedder EB (1925). *The Medical Aspects of Chemical Warfare*. Baltimore, MA, USA: Williams & Wilkins Company.

Werrlein RJ, Madren-Whalley JS and Kirby SD (1994). Phosgene effects on F-actin organization and concentration in cells cultured from sheep and rat lung. *Cell Biol Toxicol*, **10**, 45–58.

Whitteridge D (1948). The action of phosgene on the stretch receptors of the lung. *J Physiol*, **107**, 107–114.

Winternitz MC (1920). *Pathology of War Gas Poisoning*. New Haven, CT, USA: Yale University Press.

25 CYANIDES: CHEMICAL WARFARE AGENTS AND POTENTIAL TERRORIST THREATS

Bryan Ballantyne[1], Chantal Bismuth[2] and Alan H. Hall[3]

[1]*Charleston, WV, USA*
[2]*Hôpital Fernard Widal, Paris, France*
[3]*Elk Mountain, WY, USA*

INTRODUCTION

The most frequently encountered free cyanides at normal temperature and pressure (NTP) are solids (notably sodium, potassium and calcium cyanides) or liquid hydrogen cyanide, which has a very high vapor pressure. They are used on a wide scale for the general benefit of mankind in industry, e.g. various manufacturing processes (synthesis of dyes, pigments, chelating agents, nitriles, monomers, resins and fibres), case hardening, electroplating, precious metal extraction, and for fumigation operations (Ballantyne, 1986,1988; Ballantyne and Salem, 2006; Homan, 1987). Additionally, and based on their rapid and high lethality to humans, they have been deliberately employed or suggested for intended use in various forensic, judicial, military, political and hate crime situations; these have been encountered as suicide, illegal euthanasia, homicide, judicial execution, politically motivated assassinations, antisocietal actions, and much has been written about the possible military use of cyanides in localized chemical warfare (CW) operations and by terrorists for psychological as well as deliberate lethal objectives (Ballantyne, 1987b,c; Bianco and Garcia, 2004; Gee, 1987). Hydrogen cyanide is generated during the combustion of materials containing nitrogen and carbon, and in this respect represents a hazard from exposure to the smokes from fires (Ballantyne, 1987a). In addition, cyanide-related toxicity and diseases may result from exposure to man-made and naturally occurring cyanogenic materials. The widespread known uses and potential uses/misuses of cyanides dictates the need for information on their physicochemical properties, acute and repeated exposure toxicity by various routes, organ and tissue specific toxicity, mechanism of toxic action, metabolism and detoxification, clinical toxicology, and the management of intoxication from cyanides. This review summarizes all these aspects of cyanides, and on the basis of the information presented discusses their possible uses in CW operations and as a weapon in the armamentarium of terrorists or terrorist organizations.

IDENTITIES AND PHYSICOCHEMICAL PROPERTIES OF CYANIDES

Hydrogen cyanide (HCN) is also chemically identified by the synonyms formonitrile, prussic acid and hydrocyanic acid. The physicochemical properties most related to the biological activity and uses of HCN are summarized in Table 1. Notable with HCN is that it is a low boiling point liquid (26.5°C). Related to its low MW (27.04) it has a high vapor pressure (600 mm Hg at 20°C) and low vapor density (0.947 at 31°C), Thus,

Chemical Warfare Agents: Toxicology and Treatment (2nd Edition)
Edited by Timothy C. Marrs, Robert L. Maynard and Frederick R. Sidell

Table 1. Physicochemical properties of hydrogen cyanide relevant to the toxicity and use[a]

Physical state at NTP	Liquid
Molecular weight	27.04
Density (liquid)	0.6884 g ml^{-1} (20°C)
Melting point	−13.3°C
Boiling point:	26.5°C
Vapor pressure	265.3 Torr (0°C)
	600 Torr (20°C)
	807 Torr (27.2°C)
Saturated vapor concentration	789 474 ppm (884 211 mg m^{-3}) (20°C)
Vapor density	0.947 (31°C)
Explosive limits:	61–41 vol% in air (100 kPa, 20°C)
Flash point (closed cup):	−17.8°C

Note:
[a] Data from Ballantyne, 1987b; Hathaway and Proctor, 2004; Homan, 1987.

although toxic vapor is readily generated from liquid HCN, there are difficulties in handling the material and the vapor is readily diffusible in free air conditions. In contrast, NaCN and KCN are solids (granules or powder) of relatively low vapor pressure. NaCN has a melting point of 564°C. Both are appreciably soluble in water. Although the salts are of very low vapor pressure, in moist conditions they hydrolyze with the liberation of HCN, which will volatilize and give them a characteristic CN odour. The release of HCN is markedly increased in the presence of acids.

BIOCHEMICAL MECHANISM OF THE ACUTE TOXICITY OF CYANIDE

CN can inhibit a large number of enzymes *in vivo* (see Table 2). Because of this, and since other biochemical and physiological functions may be adversely affected, the overall mechanism and presentation of CN intoxication may be very complex. However, the major mechanism for the acute toxic action of CN that has been extensively investigated is the inhibition of cytochrome c oxidase, the terminal oxidase of the respiratory chain, the functional result being a cytotoxic hypoxia. Thus, CN poisoning causes intracellular hypoxia by complexing with the ferric iron of mitochondrial cytochrome c oxidase, inhibiting the electron transport chain and oxidative phosphorylation, resulting in anaerobic metabolism with decreased ATP production and increased lactic acid production (Beasley and Glass, 1998). Those tissues having the greatest oxygen demand (myocardium and brain) are the most profoundly and rapidly affected. Often, stimulation followed by profound

Table 2. Enzymes additional to cytochrome c oxidase that are inhibited by cyanide

Enzyme	Reference
Acetoacetate decarboxylase	Autor and Fridovich (1970)
D-amino acid oxidase	Porter *et al.* (1972)
Carbonic anhydrase	Fenney and Brugen (1973)
Catalase	Kremer (1970)
Glutamate decarboxylase	Tursky and Sajter (1962)
2-Keto-4-hydroxyglutarate aldolase	Hansen and Dekker (1976)
Lipoxygenase	Aharony *et al.* (1982)
Nitrite reductase	Lafferty and Garrett (1974)
Ribulose diphosphate carboxylase	Marsho and Kung (1976)
Succinic dehydrogenase	Zanetti *et al.* (1973)
Superoxide dismutase	Borders and Fredivich (1985)
Tyrosine aminotransferase	Yasmamoto (1992)
Xanthine dehydrogenase	Coughlan *et al.* (1980)
Xanthine oxidase	Massey and Edmondron (1970)

depression is seen. Details of the mechanism are as follows.

Spectrophotometric evidence indicates that CN binds with both the reduced and oxidized forms of the cytochrome a_3 component of cytochrome c oxidase (Antonini *et al.*, 1971; Chance, 1952; Lemberg, 1969; Nicholls *et al.*, 1972; Van Buuren *et al.*, 1972). Since the rate of interaction of CN with the oxidized form of the enzyme is about two orders of magnitude lower than that for the reduced form, it has been suggested that the kinetically disruptive effect of CN on mitochondrial electron transport is at the reduced cytochrome a_3 level (Yonetani and Ray, 1965). The binding properties of CN on resting and pulsed cytochrome c oxidase was investigated in both their stable and transient turnover states. A model was developed that accounted for CN inhibition of the enzyme, the essential feature of which was the rapid and tight binding of CN to transient, partially reduced, forms of the enzyme populated during turnover. Computer experimental data fitted to kinetic predictions from the model indicated that the CN-sensitive form of the enzyme binds the ligand with combination constants in excess of 10^6 M^{-1} s^{-1} and K_D values of 50 nM or less. Kinetic difference spectra indicate that CN binds to oxidized cytochrome a_3, and that this occurs only when cytochrome a and Cu_A are reduced. It thus appears probable that CN reacts with the reduced form of cytochrome c oxidase, which may subsequently be converted to an oxidized enzyme–CN complex (Way, 1984). The oxidized form is relatively stable, but in the presence of reducing equivalents CN can dissociate from the enzyme-inhibitor complex to reactivate the enzyme (Ballantyne, 1987b). This reversible nature of the inhibition is the basis for the use of certain antidotal therapies which reactivate the enzyme by depleting intracellular CN, e.g. shifting the equilibrium to the extracellular (plasma) compartment by cyanmethaemoglobin (CNmetHb) formation, by chelation, or by conversion to thiocyanate (SCN). Although inhibition of cytochrome c oxidase appears to be the major feature in the overall toxicity of CN, other mechanisms may be operative. For example, Pettersen and Cohen (1985) found an equivalent degree of inhibition of cytochrome c oxidase in brain and heart following different subcutaneous (sc) doses of KCN (4 or 20 mg kg^{-1}); respective mortalities were 0 and 100%. At 4 mg kg^{-1}, cytochrome c oxidase inhibition was 40% in brain and 60% in myocardium at 10–20 min post-dosing. At 20 mg kg^{-1}, mice died within 5 min, and had 35% inhibition of cerebral and 60% inhibition of myocardial cytochrome c oxidase activity. In further studies, they compared the *in vitro* effects of CN on mouse brain mitochondrial respiratory and cytochrome oxidase activities (Pettersen and Cohen, 1986). Cytochrome oxidase activity was inhibited in a linear fashion with an IC_{50} of 4×10^{-4} M. Respiratory rates were slightly inhibited up to 10^{-4} M and 80% inhibited at 10^{-3} M. Thus, *in vitro*, large effects on respiration required >50% inhibition of cytochrome oxidase, and they proposed that 50% of cytochrome oxidase activity might be a functional reserve.

Since CN toxicity is mediated mainly by an intracellular mitochondrial mechanism, resulting in cytotoxic hypoxia, and because CN is sequestered by the erythrocyte (Vesey *et al.*, 1976; Vesey and Wilson, 1978) it is the plasma concentration of CN that it the prime determinant of cytotoxicity (Ballantyne, 1979,1987b; Vesey, 1976). A major cause for lethal toxicity by CN is a disturbance of central regulatory mechanisms for breathing, but experimental evidence indicates that direct myocardial toxicity may also be a significant actor in the lethal toxicity of CN (Ballantyne and Bright, 1979a; Susuki, 1968). The sensitivity of myocardial cytochrome c oxidase to CN is known from direct inhibitor studies (Camerino and King, 1966) and the low IC_{50} of 2.74 μM (Ballantyne, 1977).

METABOLISM AND DETOXIFICATION OF CYANIDE

Rapid detoxification of cyanide occurs in mammals; for example, rates of detoxification have been estimated at 0.076 mg CN kg^{-1} min^{-1} for guinea pigs (Lendle, 1964) and 0.017 mg CN kg^{-1} min^{-1} in humans (McNamara, 1976). Several pathways exist for the biodetoxification of cyanide of which the major one, responsible for conversion up to 80% of a dose of CN, is by the enzymatic conversion of cyanide to the significantly less acutely toxic thiocyanate (SCN)

(Okoh and Pitt, 1982; Silver *et al.*, 1982). Thus, rat LD_{50} values for NaCN by the peroral (po) and intraperitoneal (ip) routes are respectively 5.7 and 4.72 mg kg^{-1}, and the corresponding values for NaSCN are 764 and 540 mg kg^{-1}; thus, the conversion of NaCN to NaSCN results in a decrease in acute lethal toxicity by a factor of about 120-fold in the rat (Ballantyne, 1984). SCN is renally excreted with a half-life of 2.7 days in healthy subjects (Schulz *et al.*, 1983). Two enzyme systems are responsible for the transulphuration process (Ballantyne, 1987b; Lang, 1933; Sorbo, 1975): thiosulphate-cyanide transulphurase (EC 2.8.1.1; rhodanese) and β-mercaptopyruvate-cyanide transulphurase (EC 2.8.1.2). Thiosulphate-cyanide transulphurase is a mitochondrial enzyme that catalyzes the transfer of a sulphane sulphur atom from sulphur donors to sulphur acceptors:

$$CN^- + S_2O_3^- \longrightarrow SCN^- + SO_3^{2-}$$

The basic reaction involves transfer of sulphane sulphur from the donor (SCN) to the enzyme, forming a persulphide intermediate. The persulphide sulphur is transferred from the enzyme to a nucleophilic receptor (CN) to yield SCN. Although the enzyme activity varies between species and tissues, it is high for most species in liver, kidney and olfactory mucosa (Dahl, 1989; Himwich and Saunders, 1948). The nasal metabolism of CN may have a significant relevance to the toxicity of inhaled HCN.

ß-Mercaptopyruvate-cyanide transulphurases are present in blood, liver and kidney, and catalyze the reaction:

$$HS.CH_2.CO.CO.O^- + CN \longrightarrow SCN^- + CH_3.CO.CO.O^-$$

There is evidence that the thiosulphate sulphurtransferase system may not necessarily be the primary simplistic detoxification mechanism for CN as outlined briefly above, principally because little SCN penetrates the inner mitochondrial membrane to access the transferase system. A more general view of the role of sulphur in the detoxification process is that the supply of sulphane sulphur is from a rapidly equilibrating pool of potential sulphane sulphur donors, and these may include per- and polysulphides, thiosulphantes, polythionates, inorganic SCN and protein associated elemental sulphur. In this scheme the sulphurtransferases catalyze the formation, interconversions, and reactions of compounds containing sulphane sulphur atoms (Westley, 1981; Westley *et al.*, 1983). Overall, it is possible that sulphane sulphur is derived from mercaptopyruvate via β-mercaptopyruvate sulphurtransferase, and the various forms of sulphane sulphur are interconverted by thiosulphate sulphurtransferase. The sulphane carrier transporting the sulphur formed is albumin; the sulphane sulphur–albumin complex then reacts with CN. Pharmacokinetic studies indicate that the conversion of CN to SCN is predominantly in the central compartment with a volume of distribution approximating to that of the blood volume (Way, 1984). It is possible that the plasma albumin–sulphane complex is a primary detoxification buffer in normal metabolism (Westley, 1981; Vennesland *et al.*, 1982).

There is evidence for the possible participation of a third sulphurtransferase, cystathionase (cystathionine γ-lyase: EC 4.4.1.1), a cytosolic enzyme, in CN detoxification in the kidney and rhombencephalon (Wróbel *et al.*, 2004). A product of the γ-cystathionase reaction, bis(2-amino-2-carboxylethyl)trisulphide (thiocystine), may serve as a sulphur substrate donor for rhodanese. Another product of the reaction, 3-(thiosulpheno)-alanine (thiocysteine), may be an additional link between γ-cystathionase and CN biodetoxification. In addition to its function of generating sulphane sulphur compounds, γ-cystathionase also functions as a sulphane sulphur carrier.

Other pathways for the biodetoxification of CN which are quantitatively more minor, include:

(a) Exhalation of HCN, and as CO_2 resulting from its oxidative metabolism. For example, Boxer and Rickards (1952) found that rats dosed intravenously (iv) with $Na^{14}CN$ excreted 1.7% of the dose in expired air; 90% as CO_2 and 10% as CN. By a different route, Okoh (1983) found that 24 h after the sc injection of $Na^{14}CN$, 4.53% was excreted in expired air as 91% CO_2 and 9% CN. Traces of HCN can be detected in the expired air of normal humans, but this shows no correlation with blood CN concentrations. Most HCN in normal breath is derived from the oxidation of SCN by salivary peroxidase in the oropharynx (Lundquist *et al.*, 1988).

(b) Reaction with cystine produces β-thiocyanoalanine followed by ring closure to 2-aminothiazoline-4-carboxylic acid (ATC) or its tautomer 2-iminothiazoline-4-carboxylic acid (Wood and Cooley, 1956). After dosing rats ip and po with CN, Ruzo *et al.* (1978) found that SCN was the major metabolite (> 95%) and ATC a minor metabolite (< 5%). In the urine of healthy non-smokers, ATC was below the limit of detection (0.3 μM), but in cigarette smokers the urinary ATC concentrations ranged from < 0.3 to 1.1 μM (Lundquist *et al.*, 1995).

(c) Combination of CN with hydroxocobalamin yields cyanocobalamin, which is then excreted in urine or bile (Boxer and Rickards, 1951; Brink *et al.*, 1950; Herbert, 1975).

Additionally, erythrocytes have a high affinity for CN, and rapidly sequester CN from blood plasma (Barr, 1966; Schulz, 1984; Schulz *et al.*, 1983; Vesey *et al.*, 1976). This sequestration of CN by erythrocytes has been proposed as having a protective function in the detoxification of CN (Vesey and Wilson, 1978). The rapid sequestration of CN by erythrocytes was confirmed by McMillan and Svoboda (1982), who also showed that 4,4-di*iso*thiocyano-2,2-disulphonic acid, an inhibitor of ion transport, only slowed the uptake marginally, suggesting that CN passes through the erythrocyte membrane as HCN. They were also able to confirm that SCN is oxidized to CN and cyanate by erythrocytes, as proposed by other workers (Goldstein and Rieders, 1951; Pines and Crymble, 1952) and that haemoglobin (Hb) catalyzes SCN oxidation (Chung and Wood, 1971).

Okoh (1983) investigated the excretion of an acute po dose of ^{14}C-CN in rats; one group had unlabeled KCN in the diet (77 μmol rat^{-1} day^{-1}) for 6 weeks, and another group did not receive unlabeled KCN. In rats receiving a 6-week regime of unlabeled KCN, the main route of ^{14}C elimination was urine (83% in 12 h and 87% in 24 h), with SCN accounting for 71 and 79% of urinary activity in 12 and 24 h, respectively. The 24 h expired air excretion was 4% of the total dose, with 90% present as CO_2 and 10% as HCN. Rats not receiving dietary KCN had a similar urinary and expired air elimination of ^{14}C. The toxicokinetics of subchronic peroral dosing was also studied in rats receiving drinking water for 13 weeks containing KCN to achieve daily doses of 40, 80 and 160 mg kg^{-1} day^{-1} (Leuschner *et al.*, 1991). Blood CN concentrations remained relatively constant during the test weeks. SCN was detected in blood and urine in a dose-related manner. The CN:SCN ratio in urine was 1:1000. At 6 and 13 weeks of dosing, approximately 11% of the dosed CN was excreted as urinary SCN and only 0.003% as CN. The above findings indicated that subchronic peroral dosing with CN does not lead to substrate saturation of the detoxification pathway, and does not alter the mode of CN excretion.

Toxicokinetics in humans are poorly documented but the volume of distribution is probably between 0.4–0.5 l kg^{-1}; the half-life ($t_{1/2}$) is 30–60 min.

BIOCHEMICAL AND METABOLIC SEQUELAE OF CYANIDE INTOXICATION

Inhibition of cytochrome c oxidase and the associated disturbance of electron transport results in a decrease in mitochondrial O_2 utilization with decreased ATP levels (Olsen and Klein, 1947). Anaerobic metabolism results in lactic acid accumulation and lactate acidosis. The combination of cytotoxic hypoxia and lactate acidosis results in severe metabolic consequences, particularly in the central nervous system (CNS), causing disturbances of perception and consciousness. The endogenous buffering of lactate leads to a progressive fall in plasma HCO_3^-. In brains from CN-poisoned mice there is an increase in lactate, inorganic phosphate and ADP, with a decrease in ATP, phosphocreatine, glycogen and glucose (Estler, 1965; Isom *et al.*, 1975). Rats having an iv infusion of CN (4 mg kg^{-1} h^{-1}) had increases in the tricarboxylic acid intermediates succinate, fumarate and malate (Hoyer, 1984), indicating a disturbance of NAD^+- and FAD^+-dependent redox reactions, including pyruvate oxidation. Yamamoto and Yamamoto (1977) studied the acid–base changes produced by po NaCN intoxication in rats. There was a clear correlation between NaCN dose (7–20 mg kg^{-1}) and time to death, but no correlation between survival time and PO_2. As the NaCN dose decreased, and survival time increased, there was a significant fall in PCO_2 and plasma HCO_3^-, with increase in H^+ and lactate. Thus, as survival time increased the metabolic disturbance was more marked at the lower doses. Lactate acidosis with hyperglycaemia has been noted in dogs given iv KCN (Klimmek *et al.*, 1979) and rats given iv NaCN (Salkowsi and Penney, 1995). In addition, rats dosed with ip KCN had increased plasma lactate, glucose and oxypurines (Katsumata *et al.*, 1980). Since plasma allantoin was not significantly altered, the increase in plasma oxypurine was explained by degradation of tissue ATP during anoxia.

The effect of sublethal doses of CN (5 mg KCN g^{-1}) on glucose catabolism in the mouse was investigated by Isom *et al.* (1975), using ^{14}C-glucose, sodium glucuronate-1-^{14}C and sodium gluconate-1-^{14}C. They confirmed that in the mouse glucose is metabolized through three pathways: the Embden–Meyerhof–Parnas path and tricarboxylic acid cycle, the pentose phosphate shunt and the glucuronate path. CN was found to increase the catabolism of carbohydrate by the pentose phosphate shunt, and decrease the utilization of the Embden–Meyerhof–Parnas pathway, tricarboxylic acid cycle and the glucuronate pathway. They suggested that the increased catabolism of carbohydrate by the pentose phosphate shunt my produce a source NADPN that can reduce NAD by means of a transhydrogenase enzyme, and in this manner compensate for the aberrant redox state produced by CN intoxication. CN also blocks basal and glucagon-induced lipolysis (Camu, 1969). These findings indicate that CN can alter carbohydrate metabolism, resulting in an increased glycogenolysis and a shunting of glucose to the pentose phosphate pathway by decreasing the rate of glycolysis and inhibition of the tricarboxylic acid cycle.

GENERAL TOXICOLOGY OF CYANIDE BY ACUTE AND REPEATD EXPOSURE

Determinants of toxicity

For the exhibition of CN toxicity, the rate of accumulation and the absolute magnitude of free CN at the cellular target site(s) are prime determinants. To achieve this, many factors are important, which include bioavailability, biodistribution, detoxification and bioelimination of CN. Major factors that need to be considered are briefly discussed below.

FUNCTIONAL IMPLICATIONS OF DETOXIFICATION IN CYANIDE TOXICITY

As a result of the rapid absorption and biodistribution of CN, and because of the mechanism of toxic action by inhibition of intramitochondrial cytochrome c oxidase, cyanides are biologically rapidly reacting compounds. A major determinant to both the latency and severity of toxicity of CN is the balance between the quantitative rates of absorption versus that of its endogenous detoxification. This is such that after the end of an exposure, accumulation of CN will not occur. However, during exposure, as the dose and the quantitative rate of absorption increase the rate of availability of sulphur substrate is a determinant for detoxification, and a relative reduction in sulphurtransferase detoxification may occur with an increase in available free CN. When the rate of increase in toxicologically effective (free) CN is slow there will a delay to both the time to onset and progression of toxic effects. Within limits, this effect of the detoxification process produces a clear relationship between a given end point and the exposure dose of CN (Ballantyne, 1987b). With acute exposure to large doses of CN there may be a swamping of endogenous detoxification mechanisms, resulting in the relatively prompt onset of signs and their time to effect (including mortality). However, in general, the effective and efficient detoxification of CN prevents its long-term bioaccumulation. Thus, with an acute exposure to a sublethal dose of CN that produces signs during the absorption phase, after exposure and as the detoxification processes proceeds the signs ameliorate as CN is metabolized and excreted without any bioaccumulation.

RATE OF CYANIDE ABSORPTION

The rate and amount of CN absorbed through a primary exposure route (i.e. the absorbed dose) depends on:

(i) The physicochemical properties of the molecule. Thus, HCN is of low MW, non-ionized, and hence readily diffuses; in contrast, KCN has a higher MW and is ionized and hence absorbed to a significantly lesser extent. This is reflected in differences in LD_{50} values, e.g. the acute intramuscular (im) LD_{50} values (with 95% confidence limits) for HCN and KCN are, respectively, 0.018 (0.017–0.020) and 0.050 (0.042–0.063) mmol kg^{-1} (Ballantyne *et al.*, 1972).

(ii) The exposure dose. In particular the amount of CN available for absorption, which is a function of the exposure concentration, exposure time and number and timing of exposures.

(iii) Route of exposure. For example, HCN is readily absorbed through the pulmonary alveolar membrane, but skin presents a greater barrier. In addition, the integrity of the absorbing surface can be a significant determinant. Thus, cyanides are more readily absorbed through recently abraded skin than intact skin (Ballantyne, 1984,1994a).

Although the absorbed dose is a prime determinant of the amount of CN available for distribution to tissues, because of the sequestration of CN by erythrocytes and since CN in plasma is readily available for diffusion into intercellular and intracellular fluids, the concentration of free (unbound) CN is the major quantitative determinant of both the time to onset and severity of toxicity.

DIFFERENTIAL BIODISTRIBUTION OF CYANIDE

Clearly the post-absorption differential distribution of CN to the various systemic tissues will determine the relative proportions of CN present at detoxification and target tissue/cell sites. Thus, inhaled or percutaneously absorbed CN will initially enter the systemic circulation and only a small proportion will pass to the organs of detoxification, notably the liver. In contrast, a high proportion of CN dosed perorally will pass through the liver and undergo first-pass detoxification. However, there is evidence that the factors associated with hepatic transulphuration are more complex, since it has been demonstrated that dietary variations that cause a change in hepatic cyanide sulphurtransferase do not correlate with CN toxicity (Rutkowski *et al.*, 1985) and extensive injury to the liver, chemically or surgically induced, does not increase the susceptibility of the mouse to the lethal toxicity of CN (Rutowski *et al.*, 1986). The influence of route of exposure on the development of toxicity is likely to be due to the interactive relative effects of plasma transulphuration, erythrocyte sequestration, intracellular macromolecular binding and differential distribution to all tissues having detoxification biotransforming capacity.

MISCELLANEOUS FACTORS

Other determinants for the development of CN toxicity include:

(a) *Diurnal variation in toxicity*. The diurnal variation in the lethal toxicity of ip KCN (72.5 mg kg^{-1}) to mice was investigated by Baftis *et al.* (1981), using a 12-h light/dark cycle. Mortality peaked at 16.00 hours (83%) and was minimum at 08.00 hours (43%). A circadian pattern in time to death post-injection was also demonstrated, with latency being shortest at 20.00 hours and longest at 08.00 hours.

(b) *Age*. Fitzgerald (1954) found that for adult male mice the sc LD_{50} of NaCN was 5 mg kg^{-1}, in contrast with neonatal mice for which the LD_{50} ranged from 2.0 to 2.5 mg kg^{-1}.

(c) *Antidotes*. Clearly, the presence of antidotal substances will influence the development of toxicity. For example, binding to methaemogobin (metHb), chelation or the presence of sulphur donors will cause a shift in the equilibrium of CN from the intracellular to the extracellular compartment, and thus result in reactivation of cytochrome c oxidase.

Table 3. Acute lethal inhalation toxicity of HCN vapor as timed LC_{50} and $L(CT)_{50}$ values to various species

Species	Sex	Exposure time	Median lethal toxicity (with 95% confidence limits)	
			as LC_{50} (mg m^{-3})	as $L(CT)_{50}$ (mg min m^{-3})
Mouse[a]	M	30 min	176 (129–260)	5280 (3870–7880)
Rabbit[b]	F	45 s	2432 (2304–2532)	1824 (1728–1899)
Rabbit	F	5 min	409 (321–458)	2044 (1603–2288)
Rabbit	F	35 min	208 (154–276)	7283 (5408–9650)
Rat	F	10 s	3778 (3771–4313)	631 (562–719)
Rat	F	1 min	1129 (664–1471)	1129 (664–1471)
Rat	F	5 min	493 (372–661)	2463 (1861–3301)
Rat	F	30 min	173 (159–193)	5070 (4690–549)
Rat	F	60 min	158 (144–174)	9441 (8609–1399)
Rat[c]	NS[d]	5 min	553 (443–689)	2765 (2215–3445)

Notes:
[a] Matijak-Schaper and Alarie (1982).
[b] Ballantyne (1994b).
[c] Higgins *et al.* (1972).
[d] NS, not specified.

Acute inhalation toxicity of HCN vapor to laboratory mammals

In view of its low MW, poor ionization and thus ready diffusibility, HCN is rapidly absorbed in the lung; some typical acute lethality data are shown in Table 3. Examination of these data indicates that over the time-period studied, a few minutes to an hour, there is a disproportionate relationship between the exposure time required to produce mortality and the exposure concentration. Thus, as the lethal concentration is decreased the exposure time required to cause death increases, but not in proportion. The relationship between exposure time (T), exposure concentration (C), and the inhalation exposure dosage (CT) required to cause death was studied in detail by Ballantyne (1984,1987b,1994b). For shorter exposure periods to HCN vapor (a few seconds to few minutes), as the exposure concentration required to cause mortality decreases, only very short increases in exposure time are necessary to attain a 50% mortality. For example, the exposure time for a 1229 mg m^{-3} LC_{50} is 1 min, and for a 493 mg m^{-3} LC_{50} is 5 min. This accords with the fact that under both of these experimental conditions it is probable that there is near saturation of the endogenous detoxification mechanisms. However, with exposure times in excess of around 5 min, proportionately longer times are required to produced decreases in the LC_{50} values, e.g. a 173 mg m^{-3} LC_{50} requires that there be an exposure time of around 30 min. With higher inspired HCN vapor concentrations, there is a steep concentration gradient across the alveolar membrane, facilitating the transfer and absorption of HCN, which then enters the systemic circulation without a significant first-pass hepatic detoxification, and thus there is a rapid attainment of toxic tissue concentrations. With lower inhaled concentrations of HCN vapor, there is a slower rate of entry of HCN into the pulmonary circulation and a higher proportion of the inhaled dose is detoxified, resulting in a slower rise in the body burden of HCN, and thus a comparative delay to onset of signs and longer latency to death.

Typical signs of CN intoxication in experimental mammals exposed to HCN vapor are rapid breathing, weak and ataxic movements, loss of voluntary movement, convulsions, loss of consciousness, decrease in both breathing rate and depth and breathing irregularities. Necropsy findings are non-specific and consist mainly of congestion of various intra-abdominal viscera, pulmonary congestion and scattered pleural and alveolar haemorrhages (Ballantyne, 1994b). A concentration of 63 ppm for an exposure period of 30 min produced a 50% reduction in breathing rate (DC_{50}) recorded by plethysmography (Matijak-Schafer and Alarie, 1982). Below the DC_{50}, apnoea was seen intermittently between short periods of normal breathing, but as concentrations were increased apnoea became continuous. Purser *et al.* (1984) noted that in the early stages of an exposure to HCN vapor in the range 102–156 ppm (114–175 mg m^{-3}) there was a marked episode of hyperventilation with significant increase in respiratory minute volume, which may facilitate an increase in the absorbed dose.

An important practical feature of the inhalation toxicology of HCN is the development of incapacitating effects, since this may impede mobility and escape from a contaminated area. This could clearly be relevant to hazards from exposure to fire atmospheres, effectiveness of military operations and escape from enclosed areas where HCN has been deliberately released by terrorists. Over an HCN vapor concentration range of 102–156 ppm (114–175 mg^{-3}), Purser *et al.* (1984) found a linear relationship between exposure concentration and the time to the development of hyperventilation and subsequent incapacitation. The slope on the relationship was such that a doubling of the HCN vapor concentration from 100 to 200 ppm (112–224 mg m^{-3}) reduced the time to incapacitation from 25 to 2 min. Rat studies estimated that the HCN vapor concentrations producing incapacitating effects are about 65% of the lethal concentration (Levin *et al.*, 1987).

Human lethal inhalation toxicity estimates

A few cases of industrial exposure to HCN vapor have been documented that permit an indication of the lethal toxicity to humans of acute HCN vapor exposure. In one case, an individual who had intensive medial management survived a 3-min exposure to around 500 ppm (560 mg m^{-3}) HCN (Bonsall, 1984), but in other cases 270 ppm (302 mg m^{-3}) caused immediate death, 181 ppm (203 mg m^{-3}) was lethal within 1 min, and 135 ppm (151 mg m^{-3}) was lethal after 30 min (Dudley *et al.*, 1942). It has been stated that inhalation of high concentrations of HCN vapor (200–500 ppm or more) may cause sudden loss of consciousness after only one or two breaths (Hall and Rumack, 1998). For reference, a saturated vapor atmosphere of HCN is about 7.9×10^5 ppm (8.8×10^5 mg m^{-3}) at 20°C. Moore and Gates (1946) estimated the absorbed lethal inhalation dose of HCN for man as 1.1 mg kg^{-1}. This was based on available iv LD_{50} values (in mg kg^{-1}) for various species as 1.34 (dog), 0.81 (cat), 1.3 (monkey), 0.66 (rabbit), 1.43 (guinea pig), 0.81 (rat) and 0.99 (mouse). Using a detoxification rate of 0.017 mg kg^{-1} min^{-1}, LCt_{50} values were calculated according to the formula:

$$K = VaC - Dt$$

in which, K = iv lethal dose (mg kg^{-1}), C = total volume of air breathed (l kg^{-1}), a = fraction of inhaled dose absorbed and D = detoxification rate. Using this approach, timed LC_{50} values, shown in Table 4, were calculated for a 70 kg man having a breathing rate of 25 l min^{-1}. McNamara (1976) re-evaluated the lethal inhalation toxicity of HCN vapor and developed the values shown in Table 5. Based on available metabolic rate data, Hilado and Cummings (1977) suggested the following LC_{50}values for man, which are somewhat lower than the values calculated by

Table 4. Estimated acute inhalation lethal toxicity of HCN vapor to humans based on the Moore and Gates (1946) analysis. Calculated for a 70 kg man with a breathing rate of 25 l min^{-1} and a detoxification rate for cyanide of 0.017 mg kg^{-1} min^{-1}

	Calculated lethal inhalation toxicity	
Time (min)	as LC_{50} (mg m^{-3})	as LCt_{50} (mg min m^{-3})
1	4400	4400
3	1500	4500
10	504	5040
30	210	6300
60	140	8400

Table 5. Estimates of the acute human inhalation lethal toxicity of HCN vapor after McNamara (1976). Presented as LCt_{50} and timed LC_{50} values

Time (min)	LCt_{50} (mg min m^{-3})	LC_{50} (mg m^{-3})
0.5	2032	4064
1	3404	3404
3	4400	1466
10	6072	607
30	20632	688

Table 6. Comparative acute lethal toxicity of aqueous solutions of hydrogen, sodium and potassium cyanides to rabbits by different routes of exposure[a]

		LD_{50} (95% confidence limits) (mg kg^{-1})		
Route	Gender	HCN	NaCN	KCN
Intramuscular	M	0.52 (0.48–0.56)	1.61 (1.38–1.83)	3.06 (2.60–3.60)
	F	0.50 (0.45–0.55)	1.67 (1.51–1.84)	3.27 (270–4.10)
Peroral	F	2.49 (2.26–2.61)	5.11 (4.2–5.66)	5.82 (5.50–6.31)
Percutaneous (intact skin)	F	6.89 (6.43–7.52)	14.62 (13.7–15.35)	22.30 (20.40–24.00)
Percutaneous (abraded skin)	F	2.34 (2.02–2.61)	11.28 (9.17–12.67)	14.30 (13.27–25.09)
Transocular	F	1.04 (0.96–1.13)	5.06 (4.44–6.10)	7.87 (6.51–8.96)

Note:
[a] Data from Ballantyne, 1983a,b,1984;1994a; Ballantyne *et al.*, 1972.

McNamara (1976):

$$5\text{-min } LC_{50} = 680 \text{ ppm } (748 \text{ mg m}^{-3})$$
$$30\text{-min } LC_{50} = 200 \text{ ppm } (220 \text{ mg m}^{-3})$$

Taken together, the various estimates for human lethal toxicity to HCN vapor suggest that for an exposure of around 5 to 10 min a concentration of 500–600 mg m^{-3} would be fatal, but for a 1 min exposure the HCN vapor concentration would require to be around 4000 mg m^{-3}.

Contribution of percutaneous absorption to the lethal toxicity of HCN vapor

It is well known that cyanides applied to the skin as solids or in solutions can readily penetrate to the extent that percutaneous (pc) LD_{50} values can be calculated (Ballantyne, 1994a). Likewise, HCN vapor can be absorbed across the skin and additively contribute to the toxicity resulting from pulmonary absorption (Steffens, 2003). Fairley *et al.* (1934) demonstrated in guinea pigs and rabbits that exposure to HCN vapor resulted in the percutaneous absorption of the material sufficient to produce signs of toxicity, and with prolonged exposure to cause mortality. That there may be symptomatic percutaneous absorption of HCN vapor also comes from observations on occupationally exposed workers. Thus, employees working for 8 to 10 min in an atmosphere containing around 20 000 ppm HCN (22 400 mg m^{-3}), but wearing respiratory protective equipment, developed dizziness, weakness and headache (Drinker, 1932). Two reports of percutaneous HCN vapor intoxication, one in a fire fighter wearing self-contained breathing equipment, were described by Steffens (2003). According to Dugard (1987), the rate of absorption of HCN across human skin is proportional to the concentration of CN in the atmosphere. His studies lead to a conclusion that total body surface contact (18 500 cm^2 for a 70 kg individual) with 1 ppm (1.12 mg m^{-3}) HCN (by volume) can result in the absorption of 32 μg CN h^{-1}.

Acute lethal toxicity by non-inhalation routes of exposure

As discussed later, in addition to inhalation of HCN vapor there may be several other possible routes of exposure to solid cyanides or to solutions of cyanides in the context of military chemical warfare activities or their use by terrorist individuals or groups. Particularly relevant are exposure by the po, pc, im, and possibly transocular routes. A comparison of the acute lethal toxicity of solutions of HCN, NaCN and KCN by these routes for laboratory mammals is given in Table 6. It can be seen that by all routes the order of lethal toxicity is HCN > NaCN > KCN; this accords with the more diffusible nature of the un-ionized HCN. For the routes shown in Table 6, the cyanides are more lethally toxic by im injection than by po and transocular dosing, with larger LD_{50} values by the pc route. However, by the percutaneous route all cyanides had lower LD_{50} values (i.e. were more lethally toxic) when applied to abraded skin compared with intact skin. In addition, the LD_{50} values shown in Table 6 indicate a potential for systemic toxicity by contamination of the eye with cyanides. Biological determinants for transocular toxicity include induced conjunctival hyperaemia, blepharospasm, retaining material in contact with the vascular conjunctiva and drainage of material through the nasolacrimal duct onto the vascular nasal mucosa (Ballantyne, 1983a). By all routes, the cyanides exhibited a clear lethal potential.

Estimates of toxicity based on accidental and deliberate poisoning in humans indicates that NaCN has a lethal potential of around 2–3 mg kg^{-1} and KCN at 3–5 mg kg^{-1} (Ballantyne, 1987b). Ingestion of alkaline CN salts by humans (calcium, sodium and potassium cyanides) may cause symptomatology which is life-threatening in 30 to 60 min (Hall and Rumack, 1998). Rapid progression to coma, seizures, arrhythmias, intractable hypotension and apnoea may occur, and death is common.

Repeated-exposure general toxicity of cyanide

Several studies have been conducted in order to assess the toxic effects to various species by repeated exposure for different dosing periods. Some general aspects of the more significant investigations are briefly reviewed below: details are available in Ballantyne and Salem (2005, 2006). NaCN was incorporated in the diet (2.5 mg kg^{-1}) of Beagle dogs for 30–32 days; there were no significant findings with respect to clinical signs, food consumption, body weight, haematology or histology (American Cyanamid, 1959). Dogs receiving NaCN perorally by capsule at up to 6 mg kg^{-1} day^{-1} for 15 months had signs of toxicity immediately post-dosing, with increased erythrocyte count and decreased blood albumin. Degenerative changes were seen histologically in cerebrocortical neurones and cerebellar Purkinje cells (Herrting *et al.*, 1960). Baboons dosed with KCN (1 mg kg^{-1}) for 9 months had decreased Hb concentration and mean corpuscular Hb concentration (Crampton *et al.*, 1979). The potential for hepatorenal toxicity and thyroid injury was investigated in male rats given KCN in drinking water for 15 days (Sousa *et al.*, 2002). Although serum T_3 and T_4 were unaffected, thyroid gland histology showed a dose-related increase in the number of resorption vacuoles in the follicular colloid. There was cytoplasmic vacuolation of the hepatocytes, with degenerative changes at the higher doses. Renal histology revealed congestion and vacuolation of proximal tubular epithelial cells. Okolie and Osagie (1999) studied the effects of feeding KCN in the diet (702 ppm) of rabbits for 40 weeks. Increased lactate dehydrogenase activities were measured in serum, liver and kidney, consistent with a shift from aerobic to anaerobic metabolism. Biochemical evidence for hepatic toxicity was found as increases in serum and decreases in liver sorbitol dehydrogenase, alkaline phosphatase and glutamate-pyruvate transaminase. Nephrotoxicity was indicated by decreased renal alkaline phosphatase with increased serum activity, and accompanied by increased serum urea and creatinine. Histopathologically there were foci of congestion and necrosis in the liver and foci of renal tubular and glomerular necrosis. They also noted (Okolie and Osagie, 2000) that aspartate transaminase in activity in heart and serum was unaltered, but cardiac and pulmonary alkaline phosphatase activity was decreased. Serum amylase activity was unaltered. Histology of the pancreas and myocardium was normal, but lungs had foci of oedema and necrosis. Overall, the above findings indicate that long-term peroral dosing with CN produces hepatorenal and systemic pulmonary injury, but not cardiac or pancreatic toxicity. Soto-Blanco *et al.* (2001) in a multispecies study (rats, pigs and goats) found no biochemical or histological evidence for pancreatic exocrine or endocrine toxicity from subchronic peroral dosing with KCN. In a 13-week study, F334/N rats and $B6C3F_1$ mice received KCN in drinking water up to 300 ppm (Hébert, 1993). The only effects noted were a slight reduction in cauda epididymal weights in male rats and mice, and a reduced number of spermatid heads per testis in rats; female rats spent more time in proestrus and diestrus. In a 30-days study, Wistar rats were given KCN in drinking water up to 500 ppm (Rickwood *et al.*, 1987). Hepatic and cardiac mitochondrial respiration, and cardiac, hepatic and cerebral ATP concentrations were decreased in a dose-related manner. These findings accord with CN producing mitochondrial dysfunction.

SPECIFIC ORGAN AND TISSUE TOXICITY

CN intoxication may result in morphological and functional adverse effects in specific organ systems or tissues as a consequence of acute or repeated exposure to CN. These include both direct adverse reactions to and lesions of the respiratory, cardiovascular and central nervous systems. Secondary toxic effects, from SCN, may occur with the thyroid gland. These organ and tissue effects are summarized below.

Cardiac toxicity

In experimental acute CN intoxication, consistently high concentrations of CN were measured in the myocardium irrespective of whether CN was given by different routes or compared in different species (Ballantyne, 1983b,1984, 1987b). In addition, when cardiac mitochondria were exposed to ^{14}CN, there was a time- and concentration-dependent uptake of CN (Wisler *et al.*, 1991). Additionally, biochemical and enzyme cytochemical studies showed a significant inhibition of myocardial cytochrome c oxidase in vivo with acute CN poisoning (Ballantyne, 1977; Ballantyne and Bright, 1979a,b). In keeping is the fact that with rabbit tissue homogenate the IC_{50} for myocardial cytochrome oxidase was found to be 2.74 μM, and the calculated pIC_{50} value was 5.59 (Ballantyne, 1977).

Direct evidence for cardiotoxicity and pathophysiological effects of CN on the myocardium comes from morphological, biochemical and physiological studies in animals and from human clinical observations. Electrocardiographic (EEG) changes have been recorded by several observers. Leimdorfer (1950) compared ECG changes following iv NaCN in cats, monkeys and humans. Similar changes were noted, as an initial brief bradycardia followed by return of cardiac rate to pre-injection values. At 20 s post-injection there was increased T-wave amplitude, and S-T segment elevation; T-waves returned to normal within 15 min. Toxic doses in the cat (1–2 mg kg^{-1}) produced extreme and persistent bradycardia and arrhythmias; at 10 s post-injection there was an appearance of high biphasic T-waves followed by negative T-waves, Wenckebach periods, incomplete heart block, and then complete heart block. Voltage became low, extreme bradycardia developed, and ventricular fibrillation occurred. Rats dosed with ip KCN (10 mg kg^{-1}) showed sinus arrhythmia, disappearance of T-waves, elevation of S-T segment and terminal ventricular tachycardia with arrhythmias. With sublethal doses (2 mg kg^{-1}) there was slight bradycardia, T-wave suppression, and S-T segment elevation (Susuki, 1968).

Electrophysiological studies have demonstrated that CN causes a marked shortening of cardiac action potentials. This shortening can be counteracted by glucose, and is due to a marked increase in K^+ conductance (Van der Heyden *et al.*, 1985). Dogs dosed with iv CN (2.5 mg kg^{-1}) had an initial decrease in arterial blood pressure, hyperventilation, increased central venous pressure and bradycardia (Vick and Froehlich, 1985). This was followed by respiratory paralysis and increased blood pressure, and then by terminal apnoea, progressive hypotension, profound bradycardia and hypoxic ECG changes.

Rats given an acute ip injection of KCN showed minimal ultrastructural changes if animals died within 5 min. If mortality was delayed (10–15 min) there was sarcomere elongation, myofibril dissociation at the I-band, swelling and destruction of mitochondria, enlargement of sarcoplasmic reticulum vesicles, and capillary endothelial swelling (Susuki, 1968).

O'Flaherty and Thomas (1982) found an increase in cardiospecific creatine phosphokinase (CPK) activity at 2 h following a 5-min exposure to 200 ppm (224 mg m^{-3}) HCN vapor. Ganote *et al.* (1976) found that CPK release from rat heart perfused with O_2 medium containing KCN occurred 30 min later than was the case for hearts perfused with N_2 medium with KCN. Based on studies of the time-course of events using a Langendorff preparation with paced and unpaced guinea pig hearts, Baskin *et al.*, (1987) concluded that the time-dependent cardiac effects of CN appear to exhibit at least two components: (a) to exert an initial response on the β-adrenergic receptor, either directly or indirectly, and (b) to inhibit myocardial contractility through the inhibition of cytochrome oxidase. Kanthasamy *et al.* (1991) demonstrated that KCN produced a marked and sustained increase in catecholamines by sympathoadrenal stimulation. It was suggested that some of the cardiac and peripheral autonomic responses to CN are mediated partially by an elevation of plasma catecholamines.

Vascular toxicity and influences of cyanide on vascular reflexes

The ultimate effects of absorbed CN on the cardiovascular system is the result of complex interactions on various mechanisms which include, but are not limited to, direct effects on the myocardium, direct effects on the vascular system and adjustments in effector autonomic activity of the cardiovascular system. On the isolated aorta, CN can cause either contraction or relaxation depending on the CN concentration and the species investigated (Robinson *et al.*, 1984,1985a,b). The mechanism by which CN antagonizes noradrenaline-induced contractions of the rabbit aorta was investigated by Robinson *et al.* (1985b). Oubain did not alter the relaxant action, suggesting that stimulation of Na^+-K^+-ATPase is not involved. In addition, verapamil did not alter the CN-induced antagonism of noradrenaline-induced aortic strip contraction, and the CN-induced antagonism of high K^+ concentrations and noradrenaline contractions to the same extent, suggesting the effect was not related to the involvement of either intracellular or extracellular Ca^{++} stores. They found that ouabain and verapamil enhanced contractions, but atropine, pyrilamine, 2-bromolysergic acid diethylamide and pentolamine did not alter CN-induced aortic strip contractions. In contrast, 4,4′-di*iso*thiocyano-2,2′-stilbenedisulphonic acid (DIDS) or chlorpromazine partially reduced strip contractions. These findings may have the following relevance: (a) CN-induced vascular contractions are probably not a result of stimulation of muscarinic, serotoninergic or α-adrenergic receptors, (b) if effects on the coronary arteries are similar to those noted on the aorta, then hypoxia-induced depolarization could enhance CN-induced coronary artery vasoconstriction, and thus contribute to toxicity by increasing myocardial ischaemia, and (c) chlorpromazine or DIDS may be therapeutically effective in reducing lethality by inhibiting coronary vasoconstriction.

Paulet (1955) comparing normally innervated and bilaterally vagotomized animal preparations provided evidence for cardiac failure in acute CN poisoning, and implicated the following as major factors; direct myocardial toxicity, central vagal stimulation and inhibition of central sympathetic activity. Using cats, Tanberger (1970) confirmed a dose-dependent stimulation of parasympathetic activity, and also demonstrated a stimulation of sympathetic activity. Since changes in autonomic activity were observed after bilateral vagotomy and bilateral elimination of the carotid sinus, it was postulated that they represent effects primarily on the CNS. However, others described a reduction in vagal cardiovascular tone coupled with increased sympathoadrenal activity, resulting in increased cardiac output and arterial blood pressure (Krasney, 1970).

It is well appreciated that carotid and aortic chemoreceptors are important mechanisms for the reflex adjustment of autonomic activity leading to regulation and homeostasis in the cardiovascular system (Daly and Scott, 1964; Kahler *et al.*, 1962; Korner, 1959). Using intra-aortic injections of NaCN in dogs, Krasney (1971) found that CN caused abrupt increases in cardiac output, cardiac rate and arterial blood pressure, but systemic vascular resistance was unchanged. These effects were accompanied by a reflex hyperventilation. After carotid and aortic depressor nerve section, there were also increases in cardiac output and rate and arterial blood pressure; systemic vascular resistance was initially decreased and then returned to normal, but ventilation was unchanged. In contrast, with sinoaortic denervated anaesthetized dogs, there is an increase in blood pressure and vascular resistance. Thus, in the intact conscious dog, chemoreceptor reflexes are not essential for the increases in cardiac output and blood pressure that occur in response to CN-induced cytotoxic hypoxia. In addition, in the sinoaortic denervated animal the cardio-accelerator and vasoconstrictor responses to CN are abolished by surgical or pharmacological autonomic blockade , or by cervical cord transection (Krasney, 1967,1970; Krasney *et al.*, 1966). These findings suggest that the major sites for the initiation of circulatory responses to CN are outside the sinoaortic reflexogenic zone, and probably lie within the CNS (Krasney, 1971).

Effect of cyanide on the respiratory system and respiratory reflexes

Early and typical features of acute CN intoxication are tachypnoea and hyperpnoea, resulting in an increased tidal volume. This clearly may increase the inhaled dosage of HCN during the early stages of CN vapor exposure. The effect is generally believed to be due to stimulation of aortic and carotid body chemoreceptors following the accumulation of acid metabolites at these sites from the inhibition of cytochrome c oxidase (Comroe, 1974). Glomus cells, which are secretory cells in apposition to the afferent nerve endings in chemoreceptor zones, appear to have a role in the hypoxic transduction process. Thus, chemosensitivity is abolished following destruction of the glomus cells (Verna *et al.*, 1975), and chemosensitivity following sinus nerve section correlates with reinnervation of the glomus cells (Ponte and Sadler, 1989). Glomus cells of the rat carotid body are not homogeneous in their electrophysiological properties, and there are at least two populations (Donnelly, 1993) that differ in their voltage-dependent membrane current as well as resistance and capacitance, but neither generates repetitive action potentials. The subtypes rapidly respond to CN-induced cytotoxic hypoxia and may mediate separate roles in the organ response to chemostimulation. Acker and Eyzaguirre (1989) studied light absorbance changes in the mouse carotid body during CN intoxication, and obtained evidence for cytochromes other than a and aac which might have high affinity for PO_2 changes. Levine (1975) infused NaCN into the upper abdominal aorta (0.12 mg kg^{-1} min^{-1} over 10 min) and thus induced cytotoxic hypoxia in distal tissues without stimulation of aortic or carotid chemoreceptors. There were decreases in O_2 consumption (46 $\pm$ 6 SE %), increased arterial lactate (5.25 $\pm$ 0.2 mmol l^{-1}), increased arterial lactate/pyruvate ratio (50.4 $\pm$ 14.3), decreased arterial PCO_2 (21 $\pm$ 2 mmHg), and increased arterial pH (0.06 $\pm$ 0.01). In other experiments involving aorticocarotid denervation, metabolic and ventilatory changes also occurred after the aortic infusion of NaCN, suggesting that any recirculating CN capable of stimulating aorticocarotid chemoreceptors was limited. These and similar studies indicate that in addition to aorticocarotid chemoreceptor stimulation, intra-aortic CN may stimulate ventilation by other mechanisms.

Bhattacharya *et al.* (1994) compared KCN dosed by sc injection (0.5 and 1.0 LD_{50}) and inhalation of HCN vapor (55 ppm; 60.6 mg m^{-3}) for 30 min. By both routes, CN produced increased airflow, transthoracic pressure, and tidal volume accompanied by a significant decrease in pulmonary phospholipids. HCN also produced direct effects on the pulmonary cells as evidenced by decreased compliance.

Carotid body chemoreceptors may play a role in glucose homeostasis (Alvarez-Buylla and Alvarez-Buylla, 1988,1994), since stimulation of these chemoreceptors with CN results in a rapid hyperglycaemic response. Measurement of rat hepatic veno-arterial glucose difference demonstrated that stimulation of carotid body chemoreceptors with a bolus dose of NaCN (5 μg 100 g^{-1}) produced an immediate increase in hepatic glucose output. The same dose was not effective on glucose mobilization after bilateral adrenalectomy of neurohypophysectomy. Reflex glucose mobilization was maintained after adenohypophysectomy or in adrenalectomized rats after adrenal autotransplantation (Avarez-Buylla *et al.*, 1997). Thus, the neurohypophysis and adrenal glands are necessary for the hyperglycaemic response to carotid body chemoreceptor stimulation by CN.

Depression of respiration caused by CN may be mediated through the brain stem, notably the ventral medulla. Thus, microinjection of NaCN into the ventrolateral region of the cat medulla caused a depression of phrenic nerve activity and elevated cervical sympathetic tone and blood pressure, indicating that the respiratory depression and vasomotor excitation produced by CN is localized to discrete regions in the intermediate area ventrolateral medulla, with dissociation of respiratory and vasomotor responses (Mitra *et al.*, 1993). Haxhiu *et al.* (1993) also found that topical application of NaCN to the intermediate area of the ventral surface of the medulla oblongata decreased activity in the phrenic nerve and respiratory muscles, and increased blood pressure. In contrast, intrathecal (C5-T3) NaCN increased electrical activity of the respiratory muscles, but also caused increased arterial blood pressure. These data indicate that CN exerts site-specific qualitatively different responses along the neuraxis with respect to respiratory activity; at the ventral medullary surface it causes respiratory depression, but acting on spinal neurones it causes increased respiratory motor activity. However, at both levels (medulla oblogata and spinal cord) it causes increased sympathetic activity and increased blood pressure. Carroll *et al.* (1996) found that iv NaCN caused a dose-dependent rapid decline in intermediate ventral medullary surface neural activity, which was eliminated by bilateral carotid sinus denervation. This is consistent with a possibility that superficial neurone populations in the intermediate areas receive inhibitory influences from carotid chemoreceptors.

Neurotoxicity

Studies in laboratory animals and *in vitro* models, and clinical observations of exposed human subjects have demonstrated a neurotoxic potential for CN, notably CNS morphological and functional effects, including convulsions, loss of consciousness and perception, loss of central control functions and longer-term degenerative neuropathology.

ANIMAL STUDIES

In experimental acute CN intoxication, CN concentrations in the brain are high for all species and routes of dosing studied. CN concentration is high in the parenchyma as well as in the blood in intracerebral vessels (Ballantyne *et al.*, 1972), and there are generally no significant differences between CN in white and gray matter. The CN concentrations in brain are generally slightly lower than those in blood plasma and cerebrospinal fluid (Ballantyne, 1975,1987b). CN rapidly equilibrates across the neuronal plasma membrane and then slowly accumulates in mitochondria and membrane elements of the neurone (Borowitz *et al.*, 1994). Some *in vitro* studies have suggested that CN may produce direct effects on neurones, rather than by an inhibition of enzyme activity. For example, using slices of guinea pig hippocampus Aiken and Braitman (1989) found that CN (10–200 μM) rapidly depressed synaptic transmission between Schaffer collateral-commissural fibres and pyramidal cells. Analysis of input/output curves revealed both a decrease in excitatory postsynaptic potential generation and an increase in action potential threshold; these suggested that the rapidity of action was due to a direct effect on CNS neurones. However, most studies emphasize a role for biochemical mechanisms as being mainly responsible for neurotoxicity.

The contribution of the inhibition of cytochrome c oxidase to neurotoxicity has been stressed by several workers. Thus, CN produces a highly significant decrease in cerebral cytochrome c oxidase *in vivo* (Ballantyne, 1977; Ballantyne and Bright,1979b). Measurements of rabbit brain homogenate give an IC_{50} for CN-inhibited cytochrome c oxidase of 6.38 μM, and a calculated pIC_{50} of 5.20 (Ballantyne, 1977). *In vivo* measurements of cytochrome c oxidase activity by reflectance spectrometry showed that in the cerebral cortex of rats receiving sublethal doses of CN there was a noncumulative, transient and dose-dependent inhibition of the respiratory chain (Piantadosi *et al.*, 1983). The decreases in mitochondrial activity were accompanied by increases in regional cerebral HbO_2 saturation and blood volume. It was also shown that CN inhibition of cytochrome c oxide produced effects on the Hb spectrum by preventing the unloading of O_2 in capillaries. Further studies showed a dose-related suppression of electroencephalographic (EEG) activity with isoelectric conditions occurring usually after a 50% reduction in cytochrome c oxidase activity (Piantadosi and Silvia, 1986). Pretreatment with sodium thiosulphate resulted in a four-fold protection of brain cytochrome c oxidase activity from CN-induced redox changes.

It has been suggested that effects on cerebral enzyme systems other than cytochrome c oxidase may contribute to neurotoxic effects (Petterson and Cohen, 1993). Thus, inhibition of glutamate decarboxylase can result in a depletion of the inhibitory neurotransmitter γ-aminobutyric acid (GABA), possibly predisposing to the development of convulsions (Tursky and Sajter, 1962). Persson *et al.* (1985) found that CN (5–20 mg NaCN kg^{-1}, ip) increased glutamic acid concentrations in the cerebellum, striatum and hippocampus, but higher doses decreased both glutamic acid and GABA. Cassel *et al.* (1991) demonstrated that decreased brain GABA levels were associated with increased susceptibility to convulsions, and Yamamoto (1990) found that GABA concentrations were decreased by 31% in KCN-dosed mice exhibiting convulsions. The decrease in GABA concentrations and development of convulsions were abolished by administration of α-ketoglutarate.

Using hippocampal cultures, Patel *et al.* (1992) showed that NaCN-induced cytotoxicity is mediated mainly by activation of *N*-methyl-D-aspartate (NMDA) receptors. In addition, Yamamoto and Tang (1998) found that when cerebrocortical neurones were exposed to KCN or *N*-methyl-D-aspartate, lactate dehydrogenase efflux into the extracellular fluid increased in a dose-related manner. This was blocked by co-exposure to 2-amino-7-phosphonoheptanoic acid (a selective inhibitor of NMDA), melatonin (hydroxyl and peroxyl scavenger) or N^G-nitro-L-arginine (a nitric oxide synthase inhibitor). These observations suggest that activation of NMDA receptors and nitric oxide synthase and/or free radical formation may contribute to the neurotoxicity induced by cyanide or NMDA. Yamamoto and Tang (1996a) found the incidence of convulsions induced by CN was reduced by the intracerebroventricular pre-injection of carbatapentane (a glutamate release inhibitor) or sc melatonin, and cerebral lipid peroxidation was also abolished by predosing with melatonin (Yamamoto and Tang, 1996b). Mice dosed with KCN subcutaneously had elevated conjugated diene levels. Subcellular fraction studies showed that lipid peroxidation increased in the microsomal but not mitochondrial fractions (Ardelt *et al.*, 1994). *In vitro* studies showed that the increased peroxidized lipid in rat brain cortex slice following incubation with KCN was prevented by the omission of Ca^{++} from the medium or pretreatment with diltiazem (a Ca^{++} channel blocker). Such studies suggest that both free radical formation and increase glutamate release may contribute to CN-induced neurotoxicity.

The functional inter-relationships between CN and CNS dopamine (DA) and their role in neurotoxic effects is complex. Cassel and Persson (1985) found that NaCN produced dose-related decreases in rat striatum DA, but noradrenaline concentrations were not affected. In the olfactory tubercle the concentrations of both DA and noradrenaline increased. Mice dosed with KCN developed central dopaminergic toxicity as evidenced by decreased numbers of dopaminergic neurones in the basal ganglia and decreased DA in the striatum (Kanthasamy *et al.*, 1994a). In addition, Kiuchi *et al.* (1992) showed that perfusion of NaCN into the striatal region produced a transient but large increase in DA release associated with depletion of ATP. In severe acute CN intoxication there is decreased dopaminergic activity in the nigrostriatal area (Rosenberg *et al.*, 1989), and lethal doses of CN rapidly decrease striatal DA and increase L-dihydroxyphenylalanine (DOPA) (Cassel and Persson, 1992). It was subsequently shown that while high doses of CN decrease striatal DA and produce neurone damage of pathophysiological significance, low doses increase rat striatum DA but without significant neurone cytotoxicity (Cassel, 1995). Kanthasamay *et al.* (1994b) showed that CN reacts non-enymatically with the deaminated DA metabolite 3,4-dihydroxyphenylactaldehyde (DOPAL) to form a cyanohydrin adduct 2-hydroxy-3-(3,4-dihydroxyphenyl)propionitrile (HPN). *In vitro* and *in vivo* studies have shown the formation of HPN after exposure to CN. Incubation of pheochromocytoma PC12 cells with HPN resulted in 23% cytolethality and DA release increased by 39.8%; uptake was partially blocked by the catecholamine uptake inhibitor imipramine.

The cytotoxic hypoxia of acute CN intoxication affects the energy-dependent processes controlling cellular ionic homeostasis and the ionic disequilibrium normally maintained between the intracellular and extracellular fluid compartments (Maduh *et al.*, 1993). In isolated cell preparations, the cellular ionic disruption results in marked cellular acidosis and accumulation of cytosolic Ca^{++} (Bondy and Komulainen, 1988; Li and White, 1977; Nieminen *et al.*, 1988). This may result in disturbances of Ca^{++}-activated lipolytic enzyme activity, peroxidation of membrane phospholipids, changes in transmitter release and metabolism and effects on other Ca^{++}-modulating cell signaling systems. Johnson *et al.* (1986) found that CN significantly increased whole brain Ca^{++}, and demonstrated that centrally mediated tremors were correlated with the changes in brain Ca^{++}. Pretreatment with the Ca^{++} blocker diltiazem prevented the increase in brain Ca^{++} and attenuated the tremors. Yamamoto (1990) found that CN increased Ca^{++} concentrations in crude mitochondrial factions by 32%, and this increase was abolished by combined treatment with α-ketoglutarate and thiosulphate, but not with sodium thiosulphate alone. These findings indicate that Ca^{++} may have a significant role in mediating CN neurotoxicity. Further studies by Johnson *et al.* (1986) showed CN to cause a Ca^{++}-dependent increase in conjugated diene production in mice brain lipids, indicating peroxidation of membranes had occurred; this was blocked by diltiazem and by allopurinol. Other comparative studies with calmodulin inhibitors (trifloperazine, chlorpromazine and promethazine) given into the cerebral ventricles of mice before im CN, demonstrated they were inhibitory to CN-induced convulsions (Yamamoto, 1993a). These findings were interpreted as indicating that peroxidation of lipid membranes plays a role in CN neurotoxicity and that this action is related to altered regulation of neuronal Ca^{++} homeostasis and activation of xanthine oxidase.

Several other biochemical mechanisms have been implicated in causing, or contributing to, CN-induced encephalopathy. Thus, Yamamoto (1989,1993b) found that CN caused an increase in blood ammonia and brain neutral and aromatic amino acids (including leucine, isoleucine, tyrosine and phenylalanine). The CN-induced loss of consciousness, hyperammonaemia and increased brain amino acids were significantly abolished by α-ketoglutarate. In addition, the possible involvement of caspase-3-like activity in CN-induced neuronal apoptosis has been proposed (Gunaseker *et al.*, 1999). Yamamoto (1995) proposed a hypothesis by which CN-induced convulsions are dependent on Ca^{++}-calmodulin, nitric oxide, cyclic GMP and protein kinase C. It was suggested that CN inhibition of mitochondrial cytochrome oxidase leads to a depletion of ATP, which activates NMDA-sensitive glutamate receptors and causes increased synaptosomal Ca^{++}. This increased Ca^{++}, working in conjunction with the regulatory protein calmodulin, turns on the synthesis of nitric oxide, which diffuses to adjacent cells where it activates guanylate cyclase which synthesizes cyclic GMP. The latter activates protein kinase C. Possibly relevant was the finding (Maduh *et al.*, 1995) that an inhibitor of protein kinase C [1-(5-isoquinoline-sulphonyl)-2-methylpiperazine (H-7)] partially protected against experimental NaCN lethal toxicity.

Uncoupling protein 2 (UCP-2), a member of the mitochondrial anion carrier superfamily expressed in the inner mitochondrial membrane (Busquets *et al.*, 2001; Rousset *et al.*, 2004), modulates mitochondrial function by partially uncoupling oxidative phosphorylation, and has been reported to modulate cell death (Prabhakaran *et al.*, 2003; Shou *et al.*, 2003). In the mitochondrion, UCP-2 partially dissipates the proton electrochemical gradient by enhancing proton leakage from the inner membrane into the matrix. The uncoupling leads to a reduction of mitochondrial membrane potential, decrease of mitochondrial Ca^{++}-uptake, and a reduction of both cellular ATP production and reactive oxygen species (ROS) generation. Li *et al.* (2005) investigated the enhancement of cyanide-induced mitochondrial dysfunction and cortical cell necrosis by UCP-2. In primary rat cerebrocortical cells KCN produced an aptotic death at 200–400 μM; higher concentrations of KCN (500–600 μM) switched the mode of death from aptosis to necrosis. Their studies showed that increased expression of UCP-2 alters the response to a mitochondrial toxin by switching the mode of cell death from aptosis to necrosis. They concluded that UCP-2 levels influence cellular responses to cyanide-induced mitochondrial dysfunction.

Physiological studies have shown that marked cerebrovascular and blood perfusion changes occur with CN intoxication. Slow iv infusion of KCN increased cerebral blood flow in dogs by 130 to 200%, with respective blood CN concentrations of 1.0 and 1.5 $\mu g\ ml^{-1}$ (Pitt *et al.*, 1979). Cerebral O_2 consumption was unaffected at 1.0 μg CN ml^{-1}, but decreased to 75% of control vales at 1.5 μg CN ml^{-1}. The increased cerebral blood flow caused by CN was demonstrated in several species by Russek *et al.* (1963). By continuous infusion of NaCN, Funata *et al.* (1984) found an initial increase in blood flow in white and gray matter, after which acidosis developed, bradycardia occurred, and hypotension and decreased blood flow followed.

In addition to the above experimental biochemical and physiological studies which indicate that CN can access the CNS and produce pathophysiological effects, several studies have shown that CNS neuropathological changes, mainly of a degenerative nature, have developed following acute and subchronic dosing with CN. Haymaker *et al.* (1952) demonstrated almost exclusive gray matter necrosis in dogs if survival from acute CN poisoning was for more than 3 h. Lesions following repeated CN dosing have included degenerative changes in the CNS ganglion cells and Purkinje cells in dogs by 15-month peroral dosing (Hertting *et al.*, 1960); degeneration of cerebrocortical neurones and cerebellar Purkije cells with myelin pallor in the corpus callosum of rats after 22 weeks of subcutaneous KCN (Smith *et al.*, 1963); necrotic lesions in the optic nerve and corpus callosum after 3 months of sc NaCN (Lessel, 1971). Other studies have suggested a more selective involvement of the white matter in CN encephalopathy (Brierley *et al.*, 1976,1977; Funata *et al.*, 1984; Levine and Stypulkowski, 1959; Levine and Wenk, 1959). Lesions in the corpus callosum produced by acute exposure to HCN were studied ultrastructurally by Hirano *et al.* (1967), who described fenestration of the white matter. Acute CN leucoencephalopathy was defined histopathologically by Funata *et al.* (1984) as axonal swelling, destruction of myelin lamellae, intercellular oedema, astrocyte swelling and glial necrosis. In cats, continuous infusion of NaCN produced severe damage to the deep cerebral white matter, corpus callosum, pallidum and substantia nigra (Funata *et al.*, 1984). The authors postulated that the circulatory depression that develops during profound hypoxia decreases to a point where metabolic needs cannot be met. The topographic selectivity of CN-induced leucoencphalopathy appears to be related to the characteristics of the cerebral vascular pattern, and the severity of white matter lesions is a function of the intensity of both the degree of hypoxia and hypotension. The decreased venous pressure, combined with arterial hypotension, may also be a factor in the pathogenesis of CN leucoencephalopathy because of the decreased perfusion pressure resulting in cerebral hypoperfusion (Ballantyne, 1987b).

Neurobehavioral studies indicate that CN can alter behavior at doses that are not fatal. Mathangi and Namasivayam (2000) found that after a one-month treatment of Wistar rats with KCN (2 mg kg^{-1}, ip) there was memory deficit in a T-maze test, associated with decreased concentrations of DA and 5-hydroxytryptamine (5-HT) in the hippocampus. Available data indicates that motor and cognitive functions may be affected, but the exact nature of such changes and the conditions under which they appear are yet currently uncertain (D'Mello, 1987).

OBSERVATIONS ON CYANIDE-INTOXICATED HUMANS

The cerebral cytotoxic hypoxia, decreased brain ATP, increased brain ADP and increase in lactate and decrease in glycogen, are all factors contributing to some of the CNS signs and symptoms of acute CN intoxication (MacMillan, 1989). These include disturbances of consciousness and perception and loss of central control functions, including those for the respiratory and cardiovascular systems.

Several case reports, briefly summarized below, have described the clinical neurological and neuropathological sequelae of acute CN intoxication. Finelli (1981) described a 30-year-old male patient who attempted suicide with CN. Recovery was slow, and about 14 months after the poisoning incident he developed choreiform movements and dysdiadochokinesis of the left hand. Sixteen years after the original poisoning episode mental status was normal, but he was mildly dysarthric, muscle tone was decreased in all limbs, mild athetoid movements were present in the upper limb, and a left-hand dysdiadochokinesis was present. Computerized axial tomography (CAT) showed bilateral infarction of the globus pallidus and in the left cerebellar hemisphere. Utti *et al.* (1985) described an 18-year-old male who swallowed KCN (975–1300 mg) in an attempted suicide. Seven hours after sodium thiosulphate, sodium nitrite and oxygen he regained consciousness. Four months later, neurological examination revealed generalized rigidity and bradykinesia, with intermittent resting and postural tremor in the arms. Posture was unstable, with antero- and retro-pulsion. A diagnosis of Parkinsonism was made, and anticholinergics prescribed. Eighteen months after the original suicide attempt he successfully took his own life with an overdose of imipramine and alcohol. Autopsy showed destructive lesions in the globus pallidus, widespread lacunae in the striatum, with glial fibres and lipid-containing macrophages. The subthalamic nuclei were shrunken, and showed neurone loss with astrocyte proliferation. The zona reticularis of the substantia nigra showed complete loss of neurones with marked gliosis, but the area compacta was not affected. Varnell *et al.* (1987) described two cases of death due to acute CN poisoning from the ingestion of cyanide-adulterated Excedrin capsules. A CAT scan, carried out within 3 h of collapse, showed diffuse cerebral oedema with diffuse loss of gray–white discrimination. Borgohain *et al.* (1955) described a 27-year-old female who attempted suicide with KCN and subsequently developed persistent generalized dystonia. Cranial CAT showed bilateral putaminal lucencies, and magnetic resonance imaging (MRI) showed sharply delineated lesions corresponding to the two putamina. Carella *et al.* (1988) described a 46-year-old woman who ingested CN. On hospitalization, a CAT scan revealed no abnormalities. One year later she was hospitalized with speech and swallowing difficulties, and a CAT scan revealed moderate cortical atrophy. She had impairment of spatial and visual memory, poor visuo-perceptual performance and abstract reasoning. Five years following the poisoning incident, she was hospitalized with orolingual dystonia, with anarthria, dysphonia, reduced spontaneous blinking, moderate bradykinesia, slight spastic hypertonia, slight right hemiparesis and a positive Babinski sign. An EEG showed diffuse irritative activity, most prominent in the left temporo-occipital leads. CAT showed diffuse brain atrophy and two hypodense areas in the basal ganglia were present. MRI revealed atrophy of the cerebellum and cerebral hemispheres, with marked cerebral ventricular enlargement. Grandas *et al.* (1989) described a man who became comatose after ingesting NaCN. He regained consciousness but was apathetic with reduced speech and loss of balance. Dystonia and Parkinsonism developed during the subsequent years, and a CAT scan revealed lucencies in the putamen and external globus pallidus. Feldman and Feldman (1990) described a 28-year-old man who swallowed 800 mg of KCN in an attempted suicide. After intensive antidotal and psychiatric care, he developed severe Parkinsonian signs, including micrographia and hypersalivation. MRI revealed bilateral symmetrical basal ganglia abnormalities. A 29-year-old male who attempted suicide with 50 ml of 1% KCN was described by Messing (1991). He was apnoeic on hospitalization, but recovered after 7 h. Parkinsonism developed in the following weeks and then slowly regressed with residual dysarthria, bradykinesia of the upper limbs and brisk monosynaptic reflexes. Three weeks after the poisoning incident, a CAT scan was normal, but by 5 months there was hypodensity of the putamina. MRI at 8 weeks and 5 months sharply delineated signal elevation in T2 corresponding to the two putamina was detected. Parkinsonism was detected in two men who swallowed, respectively, 556 mg of CN (Utti *et al.*, 1985) and 8.6 mg CN kg^{-1} (Rosenberg *et al.*, 1989). A 48-year-old male electroplater who developed neurobehavioral disturbances after a significant splash of CN solution in the face was described by Kales *et al.* (1997). When seen 6 months later, he had paranoid psychosis with delusions of persecution and bilateral hand tremors. MRI demonstrated changes consistent with nigrostriatal degeneration.

Long-term and repeated exposure to CN have produced severe neurological effects, including hemiparesis and hemianopia (ATSDR, 1997; Sandberg, 1967). During long-term occupational exposure to 15 ppm HCN there was dizziness, fatigue, headache, disturbed sleep patterns, paraesthesiae and fainting (Blanc *et al.*, 1985). In addition, in workers exposed to HCN, Kumar *et al.* (1992) reported loss of immediate and delayed memory, decreased visual acuity and loss of psychomotor ability and visual learning.

Adverse effects of cyanide and thiocyanate on thyroid gland function

Animal studies and observations on occupationally exposed humans have suggested that CN and its detoxification product SCN may adversely affect the function of the thyroid gland. Thus, male rats dosed with KCN in drinking water for 15 days using a dose range of 0.0–9.0 mg kg^{-1} day^{-1} showed an increase in the number of resorption vacuoles in follicular colloid, but serum T_3 and T_4 concentrations were not significantly different from controls (Sousa *et al.*, 2002). Philbrick *et al.* (1979) found increases in plasma T_3 and T_4 secretion associated with increases in thyroid gland weights in weanling rats given diets containing 1500 ppm HCN or 2240 ppm SCN for 11 months. Kamalu and Agharanya (1991) fed HCN in balanced diets of rice to dogs for 14 weeks. They found a negative correlation between serum T_3 (decrease) and plasma SCN (increase). By 14 weeks, the serum T_4 concentration was 36% below the initial concentration and significantly different from the controls. Thyroid gland histology showed a large variation in the size of the lumina of the follicles, and colloid was sparse and pale staining. The follicular epithelium was thickened, cuboidal and tended to be multilayered.

Several occupational exposure studies have shown that repeated exposure to CN can have an adverse effect on the human thyroid gland. Thus, in workers exposed to 15 ppm HCN in a silver-reclaiming facility TSH levels were significantly higher than in controls (Blanc *et al.*, 1985). Serum SCN was evaluated in 35 non-smoking workers in a cable industry who worked in an electroplating process (Banerjee *et al.*, 1997). The mean serum SCN (± SD) was 316 ± 16 μmol l^{-1} compared with a value of 90.8 ± 9.02 μmol l^{-1} for 35 non-exposed controls. Serum thyroid hormone levels were: T_4 – controls 6.09 ± 0.601 μg dl^{-1}; exposed 3.81 ± 0.318 μg dl^{-1} ($P < 0.05$): T_3 – controls 111.0 ± 9.3 ng dl^{-1}; exposed 57.2 ng dl^{-1} ($P <$ 005). TSH – controls 1.20 ± 0.301 μU ml^{-1}; exposed 2.91 ± 0.29 μU ml^{-1} ($P < 0.05$).

Serum T_4 was negatively correlated with serum SCN ($r = -0.363$, $P < 0.05$) and serum TSH was positively correlated with serum SCN ($r = 0.354$, $P < 0.05$). The findings suggest that occupational exposure to CN results in impaired thyroid gland function. Several investigators believe that the thyroid effects resulting from CN exposure are mediated by its metabolite SCN, which inhibits both the uptake and utilization of iodine by the thyroid gland (Ermans *et al.*, 1972; Fukayama *et al.*, 1992; Solomonson, 1982).

SPECIFIC END-POINT TOXICOLOGIAL ASSESSMENTS

In addition to investigating the general and specific organ/tissue toxicological effects of CN, studies have also been conducted on specific end-points for assessing the potential toxicity of CN. These studies, part of general hazard evaluation processes, have included investigations for developmental, reproductive and genetic toxicology, and for oncogenic potential. They are briefly reviewed as follows.

Developmental toxicology of cyanide

Relatively few studies are available. Female albino rats dosed over gestational days (gd) 0–15 (KCN 3 mg kg^{-1} day^{-1}, ip) exhibited mortality (5%) and growth retardation, with a 3% incidence of meningocoele (Singh, 1982). Golden Syrian hamsters were given a slow infusion of NaCN from subcutaneously implanted osmotic minipumps (Doherty *et al.*, 1982). Preliminary observations indicated that 0.0125 mmol kg^{-1} h^{-1} did not produce anomalies, but at higher dose rates of 0.13 mg kg^{-1} min^{-1} and above, resorptions and maternal mortality occurred. Definitive studies, using a dose range of 0.126–0.1295 mmol kg^{-1} min^{-1} over gd 6–9, produced the following: (a) maternal toxicity, (b) increased resorptions, (c) reduced crown-rump length, (d) malformations mainly of the neural tube (nonclosure, encephalocoele, and exencephaly). While groups of antidotal controls (sodium thiosulphate) showed no maternal toxicity, no reductions in resorptions rates, crown-rump lengths were unaltered, and malformations significantly reduced compared with the groups not receiving thiosulphate. Thus, CN given by slow sc infusion at the period of maximum organogenesis is embryofetotoxic and teratogenic.

Reproductive toxicity of cyanide

KCN was incorporated into the diet (0.5 or 1.25 mg CN g^{-1}) of rats for 20 days before parturition, during pregnancy, through lactation, and in the postweaning period (Tewe and Maner, 1981a). There were no significant effects (compared with controls) with respect to gestational weight gain, litter size, pup birth weight, food consumption and body-weight changes during lactation, maternal liver and kidney weights, weanling weights and offspring mortality. Pigs were fed a diet containing 0.03, 0.28 or 0.52 mg CN g^{-1} throughout gestation and lactation. There were no differences with respect to litter size or birth weights (Tewe and Maner, 1981b). In a subchronic (13-week) drinking-water study, male and female F334/N rats and $B6C3F_1$ mice were dosed over a range of 0–300 ppm. Male rats and mice receiving 300 ppm had a slight reduction in cauda epididymal weight, and 300 ppm male rats had lower numbers of spermatid heads per testis, and over the range 3–300 ppm sperm motility was marginally reduced. Female rats receiving 100 and 300 ppm NaCN had a significantly longer time in proestrus and metestrus. It was concluded that these effects were not sufficient to decrease fertility (Hébert, 1993).

Genotoxicity

KCN did not cause reverse mutations in *Salmonella typhimurium* strains TA98, TA100, TA1535, TA1537 or TA1538 up to 3 × 10 mmol $plate^{-1}$, in the presence or absence of metabolic activation (DeFlora, 1981). In another *Salmonella typhimrium* study (TA97, TA98, TA100 and TA1535), with and without metabolic activation, CN did not increase the number of revertant colonies at 5 doses ranging 1.0–333 μg $plate^{-1}$ (Hébert, 1993). Negative bacterial mutagencity studies have also been reported for CN by Reitveld *et al.* (1983) and Owais *et al.* (1985). HCN was marginally mutagenic in *Salmonella typhimurium* TA100 (Kushi *et al.*, 1983). CN did not induce DNA strand breaks in cultures of mouse lymphoma cells (Garberg *et al.*, 1988).

Oncogenic potential

In a chronic study, male and female rats were give diets containing 0.07 or 0.09 mg HCN kg^{-1} for 104 weeks (Howard and Hanzel, 1955). There were no effects on growth rate, no signs of toxicity, and no histopathological findings. Philbrick *et al.* (1979) fed weanling rats with diets containing 1.5 g KCN kg^{-1} for 11.5 months (equivalent to 30 mg kg^{-1} day^{-1}). There were no signs or mortalities. Positive findings were reduced body weight gain, decreased plasma T_4 and decreased rate of T_4 secretion. There was no definitive neurohistopathology or oncogenic histopathology. The currently available information suggests that CN does not have an oncogenic potential (Ballantyne and Salem, 2005, 2006).

ON THE ODOUR OF HYDROGEN CYANIDE AND ITS RELEVANCE TO DETECTION OF AND BIOHAZARD FROM THE VAPOR

The odour of HCN vapor can be an initial clue as to the cause of poisoning or else a warning of the presence of the vapor in the atmosphere. However, this popular belief must be tempered with some important reservations. When detected, the odour is often likened to that of bitter almonds, since the latter releases HCN from the cyanogenic glucoside amygdalin in the seeds (Poulton, 1988). Although olfactory detection levels are often cited to be around 1.0 ppm (Guatelli, 1964), in those who can detect the odour of HCN there is a wide distribution of atmospheric-concentration detection levels between different individuals (Brown and Robinette, 1967). A range of 0.5–5.0 ppm (0.56–5.6 mg m^{-3}) has been cited by Kulig and Ballantyne (1993). It is important to be aware that some individuals are not able to detect the characteristic odour. The ability to detect the odour of HCN is a genetically determined trait that is absent in 2–45% of different ethnic populations (Ballantyne, 1987b). Kirk and Stenhouse (1953) suggested that CN anosmia is a sex-linked Mendelian recessive characteristic with males more affected than females. A higher proportion of CN anosmic males has also been found in other studies (Fukumoto *et al.*, 1957; Sayek, 1970). However, Huser *et al.* (1958) doubted that the trait was ether recessive or sex-linked, and Brown and Robinette (1967) did not consider that CN anosmia is a simple segregating genetically controlled trait. CN anosmia may cause several practical problems in various situations. Thus, in those incidents where CN odour cannot be detected there may not be a potentially helpful diagnostic indication of clinical poisoning or cause of death at post-mortem. In addition, CN anosmia will not allow an initial and early-warning indication of HCN in the atmosphere in occupational, military or potential terrorist situations. A potential biohazard exists for professionals with CN anosmia where HCN vapor may exist in analytical laboratories or autopsy rooms, as indicated by the following few briefly cited examples. The potential biohazard to autopsy staff involved in cases involving CN poisoning was demonstrated by the fact that with a case of suicide from swallowing CN, samples of blood taken from staff within 10 min of the completion of autopsy had increased blood CN concentrations (Andrews *et al.*, 1989). On the basis that some cases of acute CN poisoning could be missed because of CN anosmia, Fernando and Busuttil (1991) recommended that respirators should be worn during the autopsy of possible CN cases or that the stomach should be opened in a fume cupboard. For deliberate precautionary reasons (Nolte and Dasgupta, 1996) conducted an autopsy on a 32-year-old male who had committed suicide with HCN in a negatively pressurized isolation room and opened the stomach under a biosafety cabinet hood. CN measurements from the victim were: blood 5.7 $\mu g\ ml^{-1}$ and stomach contents 655 $\mu g\ ml^{-1}$. No CN was detected in blood specimens taken from three prosectors before and after the autopsy. It is essential that those who encounter, or may encounter, CN during their occupation should know if they are able to detect the odour of CN in the atmosphere. CN anosmia may be a source of worry or concern for those who need this sense from an occupational perspective. Indeed Nicholson and Vincenti (1994) described a case of phobic anxiety in a 20-year-old industrial process worker who was unable to detect the odour of HCN.

CLINICAL TOXICOLOGY

Symptomatology, signs and diagnostic criteria of acute cyanide exposure

In general medical practice, acute CN poisoning is a rare, potentially fatal but treatable condition (Hall and Rumack, 1998). However, CN and CN intoxication may be encountered in many settings: these include smoke inhalation, occupational exposures to cyanides, thermal degradation or acid contact with chemical compounds (e.g. cyanogen, cyanogen halides, calcium cyanide) or metabolic release after systemic absorption of laetrile, amygdalin or cyanogenic glycosides of plant origin (e.g. apricot, cherry or peach pits) and aliphatic nitriles (Ballantyne, 1987a; Beasley and Glass, 1998; Espinoza *et al.*, 1992; Geller *et al.*, 1991; Hall *et al.*, 1986; Meyer *et al.*, 1991). Sodium nitroprusside therapy may also result in cyanide poisoning (Hall and Rumack, 1987).

The initial symptoms and signs of acute CN poisoning are generally non-specific and may include central nervous system stimulation (giddiness, headache and anxiety), hyperpnoea, mild hypertension and palpitations (Hall and Rumack, 1998; Hall *et al.*, 1987). Later effects include nausea, vomiting, hypotension, generalized seizures, coma, apnoea, mydriasis (although the pupils may be sluggishly active to light), non-cardiogenic pulmonary oedema and various cardiac effects (tachycardia or bradycardia, supraventricular and ventricular arrhythmias, atrioventricular block, ischaemic electrocardiographic changes and, eventually, asystole) (Beasley and Glass, 1998; Hall and Rumack, 1998). Common symptoms that may be encountered in acute CN intoxication are listed in Table 7. Symptoms of early acute CN poisoning may be confused with anxiety or hyperventilation, which are common after non-lethal exposures. In contrast, if massive amounts of CN are absorbed rapidly, and particularly by the peroral or inhalation routes, collapse is often rapid in onset or almost instantaneous, and may be accompanied by convulsions, with death following rapidly. The number of the symptoms present, their order of appearance and their severity depends on a number of factors that include

Table 7. Common symptoms of acute cyanide intoxication in humans

Weakness
Fatigue
Headache
Anxiety
Restlessness
Palpitations
Confusion
Dizziness
Vertigo
Dyspnoea
Nausea
Nasal irritation (respiratory exposure)
Precordial pain

route of exposure, exposure concentration or dose, duration of exposure, rate of absorption and physical mode of presentation (Ballantyne and Salem, 2005, 2006). Typical physical signs of exposure to CN are listed in Table 8. Routine investigations of suspected cases of CN poisoning should include plasma lactate, serum electrolytes (with calculation of the anion gap), blood glucose concentrations, pulse oximetry, arterial blood gas analysis, chest radiography and 12-lead electrocardiography. If there is exposure to relatively low concentrations of HCN vapor, all of the symptoms and signs typical of CN intoxication may appear in progression (Ballantyne and Salem, 2005; 2006). However, as noted above, with massive doses absorbed over a brief period many of the characteristic symptoms and signs may not develop and there will be a prompt onset of intoxication with convulsions, coma, collapse and death. Thus, depending on the magnitude, degree and duration of exposure, the presentation of CN poisoning will be variable. Some patients, at one end of the clinical spectrum may be conscious and complain of only a few symptoms, with hyperpnoea being a presenting sign (Hall and Rumack, 1987), but at the other end of the spectrum patients may present in coma with areflexia, mydriasis and unresponsive pupils. Because of the mechanism of toxic action of CN by cytochrome c oxidase inhibition, cyanosis is usually not a presenting feature. However, if present it indicates that respiration has ceased or has been inadequate for several minutes. Thus, cyanosis is a late sign, noted at the stage of apnoea and circulatory collapse. If a diabetes insipidus-like condition occurs secondary to severe brain damage following marked acute CN poisoning, it is an ominous prognostic sign (Yen *et al.*, 1995). Rotenberg (2003) has drawn attention to the fact that children may show certain vulnerabilities in relation to acute CN intoxication. They could be exposed to higher inhaled doses of HCN because of their faster breathing rates and larger surface/volume ratios. In addition, due to a thinner stratum corneum, there may be proportionately greater percutaneous absorption of liquid or vapor.

Table 8. Physical signs of acute cyanide intoxication in humans

Initial increase in breathing rate and depth; later becomes slow and gasping
Vomiting
Diarrhea
Facial flushing
Transient increased blood pressure
Tachycardia followed by bradycardia
Cardiovascular collapse
Epistaxis
Convulsions
Loss of consciousness
Urinary and faecal incontinence
Cyanosis
Areflexia
Mydriasis and sluggish or non-responsive pupils
Decerebrate rigidity
Cardiac arrest

The early development of tachypnoea, usually accompanied by hyperpnoea, and resulting in an increased tidal volume, is frequently ascribed to stimulation of the carotid and aortic chemoreceptors as a result of local accumulation of acid metabolites following cytochrome c oxidase inhibition. However, the general blood acid–base balance changes, notably lactate acidosis, may also be an important factor in breathing changes.

Tachycardia is often initially present, followed by bradycardia. Blood pressure may be elevated at first, but often followed by developing hypotension. The ECG often shows increased T-wave amplitude, progressive shortening of the S-T segment, and eventual origin of the T-wave high on the R-wave (DeBush and Seidel, 1969; Wexler *et al.*, 1947). Third-degree heart block may occur (Lee-Jones *et al.*, 1970).

An important clinical biochemical feature that needs treatment is lactate acidosis, which when marked and sustained may account for several of the symptoms, signs and complications of acute CN intoxication (Graham *et al.*, 1977). Patients with significant CN poisoning usually have an elevated anion gap metabolic acidosis (Baud *et al.*, 1991; Hall and Rumack, 1998; LaPostolle, 2006). In relation to this, it should be noted that the following might require to be considered in the differential diagnosis of anion gap metabolic acidosis: methanol, paraldehyde, phenformin, iron, isoniazid, lactate, ethylene glycol, salicyclate and heroin intoxications, and uraemia and diabetic ketoacidosis (Centers for Disease Control, 2005; Chin and Calderon, 2000). Venous blood taken from victims of acute CN poisoning may be bright red in appearance, reflecting the cytotoxic hypoxia resulting in reduced O_2 extraction in tissues and a resultant decreased arteriovenous O_2 difference. For example, Johnson and Mellors (1988) recorded the case of a 30-year-old male who attempted suicide by swallowing about 3 g of NaCN, and who recovered with amyl nitrite, sodium thiosulphate and sodium nitrite. Blood gas analysis during the acute phase of intoxication (on oxygen) gave values of P_aO_2 256 mmHg (99.7% saturation), P_vO_2 84 mmHg (95.4% saturation), and an A–V O_2 difference of 1.4 ml dl^{-1}. The values during recovery (not on oxygen) were P_aO_2 81 mmHg (96.3% saturation), $P_vO_2$30 mm Hg (59.7% saturation) with an A–V O_2 difference of 8.1 ml dl^{-1}. The reduced A–V O_2 difference is regarded by some physicians as being an important clinical diagnostic feature, which may also be diagnosed clinically on ophthalmoscopy where retinal arteries and veins may appear nearly equally red. Thus, with routine laboratory analyses where CN intoxication is suspected there are elevated plasma lactate concentrations, an elevated anion gap metabolic acidosis, a relatively normal PO_2 in spontaneously ventilating patients, and an elevated peripheral venous blood PO_2 (>40 mmHg) or decreased arteriovenous O_2 saturation difference (central venous or mixed pulmonary artery O_2 saturation > 70% with a relatively normal co-oximeter measured arterial saturation) due to decreased oxygen extraction from the blood. Hydrogen sulphide and sodium azide poisoning may also cause this combination of laboratory findings (Hoidal *et al.*, 1986; Johnson and Mellors, 1988). Nakatani *et al.* (1993) found that the arterial ketone body ratio (acetoacetate/β-hydroxybutyrate), which reflects the redox state of hepatic mitochondria, is useful as a measure of the treatment of acute CN poisoning. The ultimate diagnosis of acute CN intoxication is by measurement of the whole blood or plasma CN concentration, as discussed in detail elsewhere (Ballantyne, 1973,1974,1976,1987b,c; Kulig and Ballantyne, 1993). Analyses should be carried out promptly after sampling or artifactually low values may be obtained (Ballantyne, 1983c). In addition, in severe acute CN ingestion poisoning, whole blood CN concentrations decrease rapidly after specific antidotal treatment, but may remain increased for prolonged periods when specific antidotes are not administered (Hall and Rumack, 1998). In the context of the emergency diagnosis of acute CN poisoning, there is a requirement for a rapid quantitative test for measurement of blood CN concentrations to confirm CN intoxication and to chemically monitor treatment (Lindsay *et al.*, 2004).

Late complications of acute CN poisoning have included pulmonary oedema (Graham *et al.*, 1977), acute renal failure (Mégarbane and Baud, 2003), rhabdomyolysis (Brivet *et al.*, 1983), CNS degenerative changes and early diffuse cerebral oedema (Fligner *et al.*, 1987; Varnell *et al.*, 1987) and neuropsychiatric manifestations including paranoid psychosis (Kales *et al.*, 1997).

The onset of symptoms of CN intoxication maybe delayed (up to 6–12 h) with exposure to cyanogenic compounds such as laetrile and plant cyanogenic glycosides or aliphatic nitrile compounds, such as acetonitrile in artificial 'glue-on' nail removers (Espinoza *et al.*, 1992; Geller *et al.*, 1991; Hall *et al.*, 1986). With long-term high-dose sodium nitroprusside infusion, elevated whole blood CN concentrations may occur and even signs of CN intoxication. However, some of the symptoms and signs that occur may be due to thiocyanate poisoning (a metabolite of sodium nitroprusside, not to be confused with CN poisoning, and not requiring a specific CN antidote therapy). With thiocyanate poisoning from nitroprusside therapy, clinical chemistry indications of CN poisoning (e.g. elevated lactate concentrations and metabolic acidosis) may not necessarily be present (Linakis *et al.*, 1991). CN poisoning in this setting can be prevented with concomitant administration of hydroxocobalamin and/or sodium thiosulphate (Cottrell *et al.*, 1978; Linakis *et al.*, 1991).

Asymptomatic CN-exposed patients should be kept under observation in a controlled setting such as an emergency/casualty department or intensive-care unit for a minimum of 4–6 h. When exposure has been to an aliphatic nitrile or cyanogenic glycoside, the observation period should be extended to at least 12 h (Geller *et al.*, 1991; Hall and Rumack, 1998).

Repeated exposure effects of cyanide

Symptoms and signs resulting from sequential repeated exposures to CN are often similar to those resulting from relatively low-dose acute exposure effects (Hathaway and Proctor, 2004). Notably, these include weakness, fatigue, nausea, headache, confusion, dizziness and vertigo. Blanc *et al.* (1985) reported on 36 workers formerly employed in a silver-reclaiming factory and who were chronically exposed to CN. Residual effects, up to months following cessation of exposure, included headache, eye irritation, fatigue, loss of appetite and epistaxis. Saia *et al.* (1970) compared 22 non-exposed control workers to 40 employees who worked near CN vats; the latter had twice the incidence of insomnia, tremors, dermatitis, epistaxis and vertigo. El Ghawabi *et al.* (1975) studied 36 male electroplaters from three factories in Europe; the respective mean HCN vapor concentrations in the workplace areas were 8.1, 6.4 and 10.4 ppm. The following were noted: headache, weakness, giddiness, throat irritation, vomiting, breathing difficulties, precordial pain, excess salivation and visual problems. 20 of the 36 had moderate enlargement of the thyroid gland. Radojicic (1973) investigated 43 CN-exposed electroplaters and annealers in Yugoslavia, most of whom complained of fatigue, headache and nausea.

Management of acute cyanide poisoning

Acute CN intoxication is an acute medical emergency. The management is best considered in the practical order of first-aid treatment, medical support measures and antidotal treatment. Acute CN poisoning is relatively uncommon and the majority of physicians will not see a case during their professional lifetime. It is therefore important that those who may encounter a suspect case should seek appropriate advice and guidance, ideally from their regional Poison Control Center, particularly with respect to diagnostic criteria and the decision as to whether to administer a specific antidote.

FIRST-AID TREATMENT

First-aid should be carried out by an appropriately trained person who has the necessary knowledge to understand the basis for the primary care. The first-aider (rescuer) should be wearing protective equipment, ideally skin protection (including gloves) and an absorbent filter or, preferably, air-supplied respirator self-contained positive pressure breathing apparatus (Ballantyne and Salem, 2005; Hall and Rumack, 1998). The first-aider should ensure the following:

(a) The affected individual is promptly transferred to a clean (uncontaminated) environment. Decontamination should be carried out (e.g. flushing the skin and eyes with water). Contaminated clothing should be removed, and isolated in double plastic bags.
(b) If breathing has stopped, or is labored, then artificial ventilation should be considered. This can be done by the Holger–Nielson method or, preferred, by using a mask with a manual ventilation bag (Ambu bag). Mouth-to-mouth ventilation should be avoided because of possible secondary intoxication in the first-aider (Lafin *et al.*, 1992; Sternbach, 1992; Thompson and Bayer, 1983). Rescuers should be careful to not inhale the victim's expired air.
(c) Activated charcoal has been demonstrated to decrease mortality in experimental acute CN poisoning (Lambert *et al.*, 1988) and 1 g may bind up to 35 mg CN (Anderson, 1946). An activated charcoal dose of 1 g kg^{-1} should be administered to patients known to have ingested CN.

(d) If breathing is difficult, 100% O_2 from a cylinder should be supplied by mask (see later).
(e) If ampoules of amyl nitrite are available, and the subject is breathing, one should be broken into a tissue and placed under the nose of the affected individual, or placed inside the ventilation bag for apnoeic patients for about 15–30 s out of each minute; this may be repeated every 3 to 5 min if necessary. Ampoules should be routinely checked on storage to ensure they are in-date. Outdated ampoules may *explode* (!) on crushing with spread of glass splinters, or there may have been loss of amyl nitrite due to decomposition. It has been recommended that ampoules should be stored below 15°C, and the shelf life taken as 6 months from the day of receipt (Beasley *et al.*, 1978). In addition, it should be remembered that if amyl nitrite is being used in the presence of 100% O_2, the mixture may be explosive. Otherwise, antidotes should not be administered until a physician is present.
(f) If cardiac arrest occurs, then external cardiac massage should be started, providing that the first-aider is appropriately and adequately qualified to do so.
(g) As early as possible, a sample of venous blood should be collected into an anticoagulated tube, which is then tightly sealed; this is required for subsequent analysis for the confirmation of a diagnosis of CN intoxication.

SUPPORTIVE MEDICAL MANAGEMENT

A physician should supervise and ensure the following:

(1) The airway is patent and aeration adequate. This may require endotracheal intubation. In addition, if ventilation is insufficient or breathing has stopped, than mechanical-assisted ventilation may be needed. Additionally, a large bore iv line should be inserted for therapeutic purposes.
(2) Although the basic pathophysiological mechanism in acute CN poisoning is cytotoxic hypoxia, there is clinical evidence that the use of O_2 is a valuable adjunct to treatment. By itself, the use of normobaric O_2 may have a minimal effect on acute CN poisoning (Litovitz, 1987) but it acts synergistically with other antidotes (Beasley and Glass, 1998; Holland and Koslowski, 1986; Kulig and Ballantyne, 1993). It has been stated to be particularly effective as an adjunct to sodium nitrite–sodium thiosulphate and to a lesser extent with sodium thiosulphate alone (Burrows and Way, 1977; Hart *et al.*, 1985; Litovitz, 1987; Sheehy and Way, 1968). Although it might be anticipated that hyperbaric O_2 (HBO) would be more effective than normobaric O_2, it is uncertain if HBO offers any clinical therapeutic advantage over normobaric O_2, either alone or in combination with other antidotes (Gorman, 1989; Kulig and Ballantyne, 1993; Litovitz, 1987; Salkowski and Penney, 1994; Tomaszewski and Thom, 1994). However, when antidotal treatment is not working, and where available, HBO should be considered as a treatment option (Goodhart, 1994). The arterial ketone body ratio (AKBR; acetoacetate/β-hydroxybutyrate) has been demonstrated to be a useful measure of the efficacy of treatment for CN poisoning (Nakatani *et al.*, 1993). It is closely correlated with electron transport and O_2 utilization. In CN poisoning and during recovery, as the P_vO_2 decreases, the AKBR increases.
(3) It is important to reverse the acid–base imbalance of lactate acidosis by the use of iv bicarbonate.
(4) With severe cases, it may be required to administer anticonvulsants and/or antiarrhythmic medication, when clinically indicated.
(5) Cardiovascular complications may require the use of atropine, iv fluids and vasopressors.
(6) In addition to periodic recording of physical signs and serial measurements of blood chemistry and arterial blood gases, there should be continual monitoring of blood pressure, electrocardiograms and pulse oximetry. However, pulse oximetry may

not be reliable following administration of methaemoglobin-inducing nitrite antidotes. Morphological neuroimaging is useful (Hantson and Deprez, 2006).

ANTIDOTAL TREATMENT

Acute CN poisoning is one of the relatively few chemical-induced intoxications for which specific antidotes are available. Indeed, a very large number of differing antidotes exist, of varying mechanisms of antidotal action, some of which are experimental but others have established clinical use and appropriate governmental approval for use in the treatment of acute CN poisoning. That a large number of antidotes exist is due, in part, to the fact that the mechanism of the lethal toxicity of CN is well understood, and thus work on the development of antidotal approaches is of considerable academic interest. In addition, CN is popularly known as a rapidly acting lethally toxic chemical and because of its potential use in chemical-warfare activities a readily produced antidote with limited side-effects is required for stockpiling purposes.

Patients with whole-blood CN concentrations higher than 3.0 μg ml^{-1} have rarely survived but specific antidotal therapy ensures survival if administered promptly enough (Hall and Rumack, 1998; Hall *et al.*, 1986). Some authors have noted that CN-intoxicated patients can recover spontaneously, rapidly and with minimal support measures, as judged by clinical findings and, on occasion, on blood-CN concentrations (Edwards and Thomas, 1978; Graham *et al.*, 1977). Based on such considerations, it has been suggested by some experienced physicians that a decision to administer antidotes should be based on the changing clinical condition of the patent, particularly the level of consciousness (Bryson, 1978; Peden *et al.*, 1986). However, this type of advice in general comes from physicians experienced in clinical toxicology or occupational health and it is given because some antidotes may themselves cause serious adverse health effects. In this respect, guidance to withhold antidotes is appropriate only for those who have the necessary clinical toxicology experience. A decision to give antidote(s) and which should be used should be taken by a general or emergency/casualty department physician only after consultation with, and receiving verbal advice from, a Poisons Control Center.

The main antidotes that have been investigated and developed experimentally, and also those currently accepted for use in the treatment of cases of human CN, poisoning are listed in Table 9. The following discussion briefly summarizes CN antidotes; more detailed reviews are to be found in Gracia and Shepherd (2004), Marrs (1987; 1988) and Meredith *et al.* (1993). The most convenient and meaningful way by which to classify antidotes is into the following three major groupings, based mainly on mechanistic grounds:

(a) Those enhancing endogenous biodetoxification mechanisms, predominantly agents facilitating SCN formation, either as sulphur donors or enzymes (mainly rhodanese).
(b) Agents which are direct capture and complexing agents or indirectly result in the formation of capture and complexing agents; both will result in the sequestration of CN and thus limit its availability to exert toxicity. Principal among these are the direct agents cobalt salts, cyanohydrin formers and methaemogobin (MetHb), and the indirect MetHb generators.
(c) Miscellaneous and adjunct antidotes, which have been developed or suggested for counteracting certain specific toxic mechanisms or signs/symptoms. These have included Ca^{++}-antagonists and anticonvulsants, and as such some are regarded as supportive.

The principal CN antidotes that have been clinically used and those developed experimentally are listed in Table 9 and reviewed briefly below.

Sulphur donors

These facilitate the endogenous sulphurtransferase mechanisms for the formation of SCN from CN. Sodium thiosulphate is frequently used in combination with other antidotes having different modes of antidotal action, e.g. with sodium nitrite or 4-dimethylaminophenol (4-DMAP). As

Table 9. Cyanide antidotes used clinically[a] and developed experimentally

Major class	Sub-class	Example
Biodetoxification enhancers	Sulphur donors	Sodium thiosulphate[a]
		Sodium ethanethiosulphonate
		Sodium propanethiosulponate
		Sodium tetrathionate
		Cystine
		Thiocystine
		Mercaptopyruvates
	Enzymes	Rhodanese
Direct complex and capture agents	Cobalt compounds	Dicobalt edetate[a]
		Cobalt histidine
		Cobalt chloride
		Cobalt acetate
		Sodium cobaltinitrite
		Cobamide
		Hydroxocobalamin[a]
	Cyanohydrin formers	Pyruvates
		α-Ketoglutarates
		D,L-Glyceraldehyde
		Glucose
		Mercaptopyruvates
	Methaemoglobin	Stroma-free methaemoglobin
Indirect complex and capture agents	Methemoglobin formers	Sodium nitrite[a]
		Amyl nitrite[a]
		4-Dimethylaminophenol[a]
		4-Aminopropriophenone
Adjuncts and miscellaneous	Ca^{++} antagonists	Flunarizine
		Diltiazem
		Verapamil
	Others	Chlorpromazine
		Phenoxybenzamine
		Centrophenoxine
		Etomidate
		Naloxone

a generalization, sodium thiosulphate is used as a subsidiary treatment on the basis that it is slow acting, possibly due to slow penetration into mitochondria. The alternative sulphur donors have only marginal advantage, and less is known of their human toxicity.

Rhodanese

The enzyme derived from liver is unstable, but that obtained from cultures of *Thiobacillus denitrificans* is more stable and has been studied in experimental animals (Meredith *et al.*, 1993). It has been used successfully as an antidote in experimental CN poisoning (Pronczuk de Garbino and Bismuth, 1981) but has not yet been employed in human antidotal therapy. The administration of rhodanese by encapsulation in carrier erythrocytes has been studied by Leung *et al.* (1991).

Ca^{++} antagonists

Based on the evidence indicating that CN intoxication (notably central neurotoxicity) may involve loss of mitochondrial energy metabolism and is associated with increase in cytosolic free Ca^{++}, the use of Ca^{++-}antagonists in the management of CN poisoning as been proposed. While there is experimental therapeutic and mechanistic evidence for this belief (Maduh *et al.*, 1993), clinical studies have not been undertaken.

Stroma-free MetHb

This is a directly acting CN capture agent, binding to give cyanmethaemoglobin (CNMetHb) and thus sequestering CN and reducing its bioavailability to exert toxicity. In this respect, the use of stroma-free MetHb avoids the problems associated with indirect MetHb capture agents that reduce the blood O_2-carrying capacity. Although MetHb is rapidly cleared from the vascular compartment, the overall evidence indicates that it may be an effective and relatively non-toxic antidote for acute CN poisoning (Boswell *et al.*, 1988).

Cyanohydrin formers

These also are direct capture agents for CN. The CN ion reacts with carbonyl groups to form cyanohydrins, resulting in a sequestering action. Among cyanohydrin formers, α-ketoglutaric acid has shown some promise experimentally as a CN antidote (Hume *et al.*, 1995; Moore *et al.*, 1986), and there is a possible additional antidotal activity by causing a decrease in convulsions (Yamamoto, 1990).

MetHb generators

These are indirectly acting CN antidotes that lead to the formation of MetHb that binds with, and sequesters, CN as CNMetHb. While MetHb does not have a greater affinity for CN than cytochrome c oxidase, there is a much larger potential source of MetHb than there is of cytochrome c oxidase and the efficacy of MetHb formation is thus primarily the result of mass action. A disadvantage of MetHb generation is the impairment of O_2 transport to tissues. Amyl nitrite generates only a small amount of MetHb by inhalation of the vapor (Bastian and Mercker, 1959) but is still recommended by several clinicians for the first-aid management of CN intoxication. Artificial ventilation with amyl nitrite ampoules broken into an Ambu bag was reported to be life-saving in dogs severely poisoned by CN, and before any significant MetHb formation occurred (Vick and Froelich, 1985). The vasogenic effect of amyl nitrite may be a factor in its antidotal effect in acute CN poisoning. Sodium nitrite given iv induces MetHb formation, which competes with cytochrome c oxidase for CN, as a result of which cytochrome c oxidase is reactivated.

A drawback of nitrites is the adverse effects they can cause on the cardiovascular system because of the vasodilation and hypotension they induce. It is generally recommended that the MetHb levels should be kept in a range of 5–40%: MetHb concentrations above 40% will significantly depress O_2 carriage (Hall and Rumack, 1986; Kirk *et al.*, 1993; Kulig and Ballantyne, 1993). Excess MetHb may be corrected by either methylene blue or toluidine blue, or exchange transfusion. Patients with cose-6-phosphate dehydrogenase deficiency are at risk from nitrite therapy because of the potential for haemolysis. Another MetHb generator is 4-dimethylaminophenol (4-DMAP) which sets up a catalytic cycle in the erythrocyte in which O_2 oxidizes DMAP to N,N-dimethylquinoneimine which oxidizes Hb to MetHb (Kiese, 1974). It is considered that 4-DMAP acts more rapidly in producing MetHb than does sodium nitrite and generates MetHb within a few minutes of injection (Weger, 1983). 4-DMAP antidote is also contra-indicated in those with glucose-6-phosphate dehydrogenase deficiency. *p*-Aminopropiophenone is also a MetHb generator but is slower in onset than 4-DMAP. However, both have been shown experimentally to be effective as CN antidotes (Bright and Marrs, 1987).

Cobalt compounds

These act as direct CN binding agents and thus sequester CN. The cobalt CN complexes are excreted in urine (Frankenberg and Sörbo, 1975), but inorganic cobalt compounds are possibly too toxic for clinical use (Paulet, 1961). They produce circulatory disturbance and may produce myocardial injury (Marrs, 1987). Hydroxocobalamin binds strongly to CN to form cyanocobalamin and does not interfere with O_2 transport. Several studies have confirmed the efficacy of hydroxocobalamin in experimental CN poisoning and its use has been advocated on the basis of low toxicity (Borron, 2006; Borron and Baud, 1996; DesLauriers *et al.*, 2006; Pontal *et al.*, 1982), although following its administration urticaria and vascular collapse have been reported (Dally and Gaultier, 1976). However, no significant allergic reactions have been reported following currently available hydroxocobalamin formulations. Since 1 mol of hydroxocobalamin binds only 1 mol of CN (Lovatt Evans, 1964) and has a high MW, large volumes of solution are necessary to treat substantial poisoning. Hydroxocobalamin (Cyanokit®) is available in France and is soon to be available in the USA and has a much better risk:benefit ratio than those of all other specific cyanide antidotes in current clinical use and is equally efficacious (Baud *et al.*, 1991; Beasley and Glass, 1998; Forsyth *et al.*, 1993; Hall and Rumack, 1987; Uhl *et al.*, 2006). Detoxification of 1 mmol CN (equivalent to 65 mg KCN) requires 1406 mg hydroxocobalamin (Meredith *et al.*, 1993) and therefore standard ampoules of hydroxocobalamin (containing 1–2 mg) are not effective. However, in some countries a formulation of 4 g hydroxocobalamin is available that can be reconstituted for iv use (Ballantyne and Salem, 2005). The French commercial preparation, Cyanokit®, contains 2.5 g hydroxocobalamin to be reconstituted in 100 ml of normal saline. Hydroxocobalamin has found use for treatment of cyanogen poisoning, notably sodium nitroprusside, and in CN intoxication resulting from inhalation of the products of combustion (Baud *et al.*, 1991; Forbin *et al.*, 2006; Houeto *et al.*, 1996). Although dicobalt edetate has been used in some countries there have been reports of severe adverse effects from its use, including vomiting, facial oedema, urticaria, collapse, chest pains, anaphylactic shock, hypotension, cardiac arrhythmias and convulsions (Hilmann *et al.*, 1974; Naughton, 1974; Tyrer, 1981). This has lead to a recommendation that dicobalt edetate should only be used antidotally with clearly established cases of acute CN intoxication and then only with caution (Meredith *et al.*, 1993; Pontal *et al.*, 1982).

Vasogenic compounds

Certain vasogenic compounds, by themselves not antidotal, may have a potentiating effect on mechanistically antidotal materials. For example, chlorpromazine potentiates the antidotal effect of thiosulphate (Way and Burrows, 1976), and phenoxybenzamine, an α-adrenergic blocking agent, also potentiates thiosulphate.

There is a significant geographical variation in the recommendations and use of specific CN antidotes. This is explained only partially on the basis of differing critical clinical analyses of the efficacy of antidotes versus potential adverse effects in a patient already compromised by acute CN poisoning. Other, often irrational, reasons for differences in attitudes and recommendations on the use of specific CN antidotes include the country of origin of the antidote, the differing experiences of clinicians in different countries, the differing attitudes and approaches (guidelines) of regional government departments having responsibilities for the registration and control of medicinal products and, indeed, in some locations by autocratic attitudes of some pompous dogmatic clinical toxicologists. Examples of geographical variations in (official) recommendations are as follows. Dicobalt edetate is used in the United Kingdom (Tyrer, 1981) and Australia (Worksafe Australia, 1989). In the United States of America, and partly related to the early historical development of cyanide antidotes in that country, sodium nitrite and sodium thiosulphate are used. In Germany, DMAP was developed and is used there (Steffens, 2003). Hydroxocobalamin is used in France on the logical basis of its effectiveness combined with low toxicity (Mégarbane and Baud, 2003) and is widely used (geographically) for the control of slow cyanogenesis during sodium nitroprusside infusions and for the treatment of CN intoxication in fire victims. Protocols for use of the various antidotes have been given elsewhere (Ballantyne and Salem, 2005). We stress the clear need for an international agreement about which CN antidotes are appropriate in given situations. A combined scientifically, pharmacologically and clinical toxicology rationalized critical analysis is needed based on, at least, the following considerations: (a) pharmacologically potent and potentially harmful antidotes may be mistakenly administered to a wrongly diagnosed patient, (b) certain adverse effects of the antidote may additively increase those due to CN intoxication, (c) toxicity of the antidote outweighs the benefit of a specific antidote, (d) the need for universal recommendations on treatment of CN poisoning that can be undertaken by general medical and support staff, (e) consideration of the risk – benefit of the antidote and its toxicity or lack thereof if the clinical impression of cyanide poisoning is not correct or uncertain, and (f) appropriate formulation and physicochemical characteristics that permit sustained stockpiling.

IMPLICATIONS OF CHEMICAL WARFARE AND TERRORIST ACTIVITIES WITH PARTICULAR REFERENCE TO THE USE OF CYANIDE AGAINST HUMAN TARGETS

Chemical warfare

By the standard military classification of chemical warfare (CW) agents, cyanides are referred to as lethal blood agents (HMSO, 1987; Maynard, 1999). In the context of CW operational situations, it is likely that CN in the form of atmospherically dispersed material would be applicable for two military situations, namely, the use of low concentrations to cause mental and physical incapacitation and the generation of high concentrations for lethal purposes. For the former purposes, it is well-known that HCN vapor produces disturbances of consciousness and perception; these coupled with muscle weakness and ataxia would lead to mental and physical incapacitation of troops and a marked reduction in their ability to conduct military tasks.

As a military lethal agent, HCN has been regarded as being of potential strategic usefulness because it exists as a colorless vapor, which has effects that are rapid in onset after the start of exposure. As noted above, the concentration range to produce lethality within 1 to 5 min is of the order of 500–4000 mg m^{-3}. For military purposes, HCN (coded AC) was used during World War I by France, with the first employment of HCN shells being on the Somme on 1 July 1916 (Prentiss, 1937). During World War II, CN was used as an agent of genocide in the German concentration camps and Japan allegedly used CN against China. In addition, Iraq allegedly used CN against the Kurds during the 1980s. HCN has an affinity for oxygen and is flammable (see Table 1) and is hence not efficient when dispersed by artillery shells (RAMC, 2002). In addition, because of its low vapor density and low MW it is readily diffusible and thus there are problems in achieving lethal concentrations against unprotected troops in open-battlefield conditions. In this respect, HCN acts as a non-persistent CW hazard. It is generally acknowledged that HCN vapor is only strategically useful as a lethal agent for localized and circumscribed circumstances. However, as noted above, less than lethal concentrations of HCN in the atmosphere cause mental and physical incapacitation, which may be of operational significance and are of value in reducing the efficiency and determination of enemy troops. Concentrations of the order of 100–200 ppm (112–224 mg m^{-3}) would be necessary to induce such incapacitating symptoms and signs.

The cyanogen halides, cyanogen chloride and cyanogen bromide, were also used in the World War I to release HCN on the battlefield but additionally they are strongly irritant to the eyes and other mucosal surfaces. Medical management is as for HCN inhalation but pulmonary irritation may also need to be treated additionally, as, for example, similarly to phosgene exposure.

Chemical terrorism

In the context of this discussion, a terrorist incident covers a wide-spectrum definition in which there is basically a desire to produce an assault on organized (law-abiding) societies in order to attempt to alter popular attitude, opinion, legislation or political dictate, by the use of techniques designed to make the terrorist motivation be fearfully known to the appropriate members or sectors of that society. The technique(s) frequently employed are usually designed to cause fear or panic in the targeted population but in some cases there may be a deliberate objective to cause random or targeted deaths. At one end of this definitional spectrum are individuals who have, for various and differing reasons, an aberrant behavior pattern against a civilized society or of individuals in that society and seek vengeance or a desire to frighten (terrorize) individual citizens or the general population at large in order to demonstrate, and make known, the reason(s) for their antisocial, and often destructive, physical and amoral activities. At the other end of the definitional spectrum are larger groups or organizations, often with evil motivation based on cult beliefs, extreme political/ethnic persuasions, religious dogma or ingrained abnormal behavioral attributes, and frequently a variable combination of all these factors.

The probable use of chemicals in some terrorist activities has been discussed and widely accepted as being likely to occur in some situations. The chemicals that might and indeed have been used, and their mode of delivery, vary widely. On a generic basis, chemical agents that could be used include irritant and/or disorienting materials (inducing panic), psychogenic materials (inducing panic and fear) and lethal agents (for panic and/or deliberate murderous purposes). It is assumed that most terrorists (or individuals using terrorist-like approaches) will be attracted to the use of chemicals that have the following attributes: (i) cheapness, (ii) can be readily obtained (in bulk if necessary) or readily manufactured (synthesized) in 'home' laboratories, (iii) capable of causing mass incapacitation/panic/fear/social disruption and/or mortality, (iv) will be comparatively easy to handle and use, (v) high biological activity and (for most circumstances) having effects that are rapid in onset, and (vi) chemicals and any associated dispersal systems procured without raising high levels of suspicion.

Specific chemicals or categories that have been used, or suggested as possibilities for use, across the definitional spectrum of terrorism include:

(a) Organophosphate anticholinesterases, including commercially available pesticidal and synthesized CW organophosphates. The most widely published account of organophosphate terrorism was the use of sarin by a Japanese cult organization (Aum Shinrykio (Supreme Truth)) against a mixed Japanese civilian population in a Tokyo subway during March 1995 and hence an otherwise random human target population (Okudera *et al.*, 1997; Simon, 1999). Twelve individuals were killed; had the sarin used been of greater purity and the delivery system more effective, casualties could have been in the thousands. Recognition as a terrorist incident would have been difficult since the cult members simply left punctured containers of sarin on baggage racks or the floor of five subway trains (see Chapter 13).

(b) Commercially available toxic industrial and agricultural chemicals; a large number of differing and readily available commercial chemicals which are potentially acutely lethal and/or harassing can be listed here and which could be administered by the inhalation and peroral routes. As examples are chloropicrin, phosgene (see Chapter 24), chlorine, paraquat, warfarin and even arsenicals (see Chapter 23).

(c) Cyanides, as HCN vapor or as cyanide salts either as solids or in solution. The ready availability of cyanides from laboratories, fumigation services and industrial organizations, coupled with their rapidly lethal and incapacitating toxic effects, makes the cyanides a likely group of substances in the terrorist armamentarium for lethal, panic and fear purposes. They are discussed in more detail below.

(d) The biotoxins ricin (see Chapter 27) and botulinum toxin (Tendler and O'Neill, 2005), the former having been used for political assassination purposes in public places (Simon, 1999) and discovered recently in security force raids on terrorist apartments (O'Neill, 2005).

The manner by which these various chemicals could be delivered to intended targets can vary markedly and depends primarily, but not exclusively, on the following factors: (a) the cause for the planned incident, (b) the target population (individuals or groups of persons), (c) whether the aim is to deliberately kill innocent people and/or to cause fear and panic, (d) whether the target(s) is intentionally selective or random, (e) whether the incident is planned and to be conducted by individual or group terrorist activity, and (f) the planned location. Individual dissemination methods are discussed below.

INDIVIDUAL/SMALL-GROUP TERRORIST INCIDENTS

An individual, or small group of individuals, can draw attention to their grievances(s) or vengeance mission by the use of chemicals. This loner/small-group activity is usually against a more specifically targeted population with circumscribed objectives, than is the larger national or international terrorist group activity. CN has been used in this respect on several occasions. This has included the repacking of medicinal capsules with CN and ensuring that a targeted individual receives that capsule, or else as a 'hate-crime' by substituting capsules on the shelves of a commercial establishment for non-targeted members of a community (Brahams, 1991; Centers for Disease Control, 1991; Curry, 1963; Dunea, 1983; Holland, 1983). A standard 250–500 mg capsule, repacked with Na or KCN, would be lethal to the majority of individuals in a population This activity has now been essentially stopped by the introduction of 'tamper-proof' wrappings. In Japan, the deaths of four individuals were reported following the consumption of curry laced with CN at a festival in Sonobe, Wakayama Prefecture; it was originally suggested that this incident could have been connected with a local vendetta over property rights, illegal parking and rubbish disposal (Watts, 1998). Despite the early reports that this mass food-borne poisoning incident was due to deliberate adulteration of curry with CN (Watts, 1998), subsequent clinical and laboratory evaluation revealed that arsenic was actually the causative substance. One of us (AHH) made the presumptive diagnosis of arsenic poisoning based on information relayed by the treating physicians through the Japan Center and the International Programme on Chemical Safety (IPCS, Geneva). Japanese authorities later confirmed the diagnosis of arsenic poisoning by laboratory analysis of food and clinical specimens. This emphasizes the care required in the interpretation of clinical toxicology findings and how inadequate or ill-informed qualitative testing meant to be confirmatory can be significantly misleading. Intramuscular injection of HCN from a concealed syringe has been considered a possibility for assassination purposes (Ballantyne *et al.*, 1972). Cult killings, as a means of conveying a sinister message, can be considered under this heading. A particularly notorious example was the mass killing in which over 900 people were encouraged to drink a CN solution in Jonestown (Thompson *et al.*, 1987). The use of CN for covert political reasons has included spraying HCN in the face from a canister, incorporation in cigarettes and administration by injection (Gee, 1987; Harris and Paxman, 1982; Seagrave, 1982).

MAJOR AND INTERNATIONAL TERRORIST THREATS

There have been multiple documented accounts about the discovery of information that has strongly suggested the existence of plans by some national and international terrorist groups for the possible use of cyanides in likely major terrorist activities; these are listed in Table 9. From the numbers of reports and the widespread variation in the geographical locations of possible sites for such terrorist activities, this potential use of cyanide is not to be ignored. Indeed, as this chapter is being written a suspected Al-Qaida terrorist, Kamel Bourgass, was imprisoned for life for the murder of a UK police Special Branch detective. Related to this, in a security force raid on a North London apartment during January 2003, anti-terrorist police discovered instructions and ingredients for the production of ricin and cyanide (Craven, 2005; Leading Article, 2005; O'Neill, 2005). CN would be extracted from apple pips and cherry stones. It had been planned to smear ricin on door handles in London and to use it to contaminate toiletries (including face creams, mouthwashes and toothbrushes) (Tendler and O'Neill, 2005). However, some reports need to be read and interpreted with care. For example, in an essay on cyanide as a chemical terrorism weapon it was stated that following the first World Trade Center bombing in 1993 investigators found traces of cyanide in vans where the explosion in the parking garage originated. This is not to necessarily to imply that CN may have been a possible use of chemicals in that incident (although the article stated that CN was used) but its presence was most probably the result of combustion of synthetic upholstery materials containing nitrogen and carbon, a process which is well-understood (Ballantyne, 1987a). In the USA, various governmental agencies, including the Centers for Disease Control and the Department of Homeland Security, consider cyanides among the most likely agents to be used for chemical terror (Khan *et al.*, 2000; NTARC, 2004).

The methods for delivery of CN are likely to be variable, depending on resources available to the terrorist organization, their knowledge and scientific/technical sophistication, the proposed location for the incident(s) and the intent and circumstances for the incident. Because of its low MW, low vapor density and ready diffusibility, HCN vapor is most likely to be effectively used in enclosed and/or confined spaces for maximum effect. These might include subways, small shopping centers, buildings with low ventilation rates or recirculating air and space-restricted indoor functions (e.g. cinemas, lectures theatres, etc.). It could be generated, depending on the volume of vapor needed, from cylinders of the liquid or from devices (maybe crude) for mixing Na or KCN with acidic fluids. Also likely is the use of CN, as solid salts or concentrated solutions, for contamination of various domestic, commercial and other publicly available sources of swallowed materials; this might include incorporation into pharmaceutical preparations, bottled drinks and injection into foodstuffs or food containers. Contamination of drinking water supplies would need to be at locations close to use. For example, massive amounts of solid cyanides, such as Na or KCN, would need to be dumped in public reservoirs in order for toxicologically effective concentrations to be still present at supply taps. Therefore, to be effective in drinking water supplies CN would need to be introduced into water storage tanks at the particular facility chosen as a location for terrorist activity.

PREPAREDNESS FOR MAJOR CHEMICAL TERROR INCIDENTS

Security forces and their support staff, who will have functions as first-responders, should be in a constant state of readiness in order to be able to immediately respond to an incident, which may come with little or no warning. There is, thus, a requirement for the following, in respect of any chemical terror incident including, in the context of this chapter, one specific to CN:

(1) Constant liaison with military, civil and political intelligence agencies in order to keep informed on the likelihood, nature, possible geographical location(s) and timing of potential incidents. However, one of the unfortunate lessons of following the New York World Trade Center tragedy of September 1991 was that informed discussion and exchange of information between government agencies was not optimum; this was partly related to political and departmental self-interests. It is hoped that this unfortunate communications hiatus is narrowing and interdepartmental co-operations are improving, although the current indications in the United States are not encouraging.

(2) Continual updating methods likely to be encountered in terrorist actions, expanding knowledge databases, appropriately revising techniques and maintaining equipment stores (including protective clothing and reliable respirators).

(3) Constant training sessions in dealing with potential terrorist incidents, and in all likely scenarios.

(4) Have technology available for, and develop skills to use, analytical methods for recognition and confirmation of threat chemicals. With CN as vapor, and as previously noted, odour is not a totally reliable recognition property because of the inability of some individuals to recognize the characteristic odour. Testing should be carried out to discover those who have CN anosmia. For the detection of atmospheric HCN, Draeger tubes are available but there is a need for portable instant read-out devices to quantitatively measure CN in the atmosphere at incident sites. Methods for the on-site and rapid detection of systemic CN intoxication by analysis of small blood samples need to be developed and refined.

(5) Security forces should maintain ongoing discussions with, and have frequent training exercises with, police, rescue organizations, emergency medical responders and healthcare providers.

THE ROLE OF HEALTH CARE INSTITUTIONS AND PROFESSIONAL HEALTH CARE PROVIDERS IN TERRORIST SITUATIONS INVOLVING CYANIDE

The US Centers for Disease Control (CDC) has prepared strategic recommendations relating to

Table 10. Recent examples of information discovered in relation to the possible use of cyanides in potential major terrorist incidents

Date: location: information	Reference
1995: Tokyo subway attack: precursors of CN were found in subway bathrooms	Sauter and Keim (2001)
2002: Arrests of four Moroccans, with ties to Al-Qaeda, who were plotting the use of cyanide to poison water supplies in areas around the American Embassy in Rome	BBC (2002)
December 2002: Recovery of a store of cyanide in Paris: linked to three suspected Al-Quaida operatives	Cloud (2004)
September 18, 2002: police raid on a North London (UK) apartment: discovery of instructions for manufacturing various poisons, including CN from fruit pips	O'Neill (2005)
January 5, 2003: police raid on a North London (UK) apartment: crude laboratory discovered for producing cyanide	O'Neill (2005)
May 2003: cyanide 'bomb' found in the possession of white supremacists in Texas	CNN (2004)
Early 2003: The US Central Intelligence Agency (CIA) uncovered a plot, apparently by an Al-Qaeda cell, in which hydrogen cyanide vapor would be released into the New York subway system, using a small device which would generate hydrogen cyanide vapor from sodium cyanide and hydrochloric acid	Suskind (2006)

preparedness for, and response to, biological and chemical terrorism (Khan *et al.*, 2000). A CDC task force stressed that preparedness for terrorist attacks constitutes a critical component in the US public health surveillance and response system. They identified five main (primary) activities that should be undertaken by public health organizations in order to enhance preparedness for terrorist chemical attacks. These are as follows:

(a) Epidemiological capacity should be enhanced for detecting and responding to chemical attacks.
(b) Awareness of chemical terrorism should be enhanced among emergency medical services (EMS) personnel, police officers, firefighters and nurses.
(c) Antidotes should be stockpiled.
(d) Bioassays should be developed and provided for the detection and diagnosis of chemical injuries.
(e) Educational materials should be prepared to inform the public during and after a chemical attack.

With respect specifically to CN terrorism, the following should be added and stressed to these particular recommendations. First, before antidotes are stockpiled there should be agreement, ideally at an international level, on what is the most appropriate antidote for the treatment of acute CN poisoning. As noted below, this decision should be based on a critical analysis of many criteria; currently we are of the opinion that hydroxocobalamin is most appropriate. Secondly, with respect to CN analyses, there is need for a portable method (kit) which is specific and, at least, semi-quantitative and can be used at the incident for a reliable bio-identification of acute CN intoxication. There is also a clear need for a reliable and sensitive method for the instantaneous measurement of atmospheric HCN concentrations and ideally continuous monitors with automatic warning devices for installation in locations where there is a potential that HCN vapor may be released. Thirdly, educational materials should be available for early distribution to the general population so they are prepared for what to expect in the event of a CN bioterroism event.

Hospitals have immense problems in maintaining their routine (day-to-day) medical, surgical and multiple other functions. It is, therefore, probably most appropriate for a small team to be appointed in regional hospitals, representing relevant specializations, which has the responsibility for developing and organizing arrangements needed in respect of the possibility of a terrorist incident in the hospital catchment area. This emergency response team should develop guidelines for dealing with the immediate and medical management of such an incident. These should include, across the board (chemical, biological and nuclear): (a) how to ensure that the necessary expertise can be summoned promptly, (b) triage procedures, (c) first-aid and medical management for all likely situations, this to include decisions on antidotes, (d) ensuring that there are readily available supplies of equipment and therapies for use at the incident site and in the hospital area, (e) arranging for decontamination areas in the hospital grounds, (f) ensuring the maintenance of essential free lines of communications with other advisory services (e.g. rescue groups, Poison Control Centres), (g) ensuring that appropriate protective equipment is readily available for use by staff who may be exposed, and (h) organizing educational and local training sessions and arranging for practical joint training with the local-responsible security and rescue services.

The short- and long-term emotional effects, and their consequent psychological evaluation and treatment, following a community exposure to a toxic material need to examined in detail and preparations for psychotherapy and mental care planned as part of the health advisory program (Greve *et al.*, 2005). It has been stressed that emergency medical care should be given in the context of a specific action plan unified with other emergency services. The provision of advanced life-support in the decontamination zone by protected and appropriately trained medical responders is now feasible technically (Baker, 2004).

In the context of the use of cyanides in a terrorist incident, specific considerations are as follows:

(1) As noted above, there is a need for a portable rapid blood test to screen for the presence of toxicologically significant concentrations of CN in blood, in order to aid in the immediate definitive diagnosis of CN intoxication on-site. Preliminary studies using Cyantesmo® strips, used by water treatment facilities, showed that the strips accurately and semi-quantitatively detected CN in solutions at concentrations $> 1\ \mu g\ ml^{-1}$ in 5 min (Rella *et al.*, 2003,2004).

(2) Protective equipment needed by those coming into contact with exposed victims should include impermeable clothing, gloves and respiratory protective equipment, either absorbent-canister type or, preferably, air-supplied. Because of its volatility and low MW, HCN is poorly absorbed by charcoal in the canister of a conventional cartridge respirator but the charcoal is made more reactive by impregnating it with metal salts. Modern NCB respirators are effective against an acute exposure to HCN vapor but should be changed immediately following an exposure (RAMC, 2002).

(3) Responders should be aware of the fact that, unlike many other toxic materials, symptoms, incapacitating signs and death can occur within seconds to minutes from the start of an exposure and thus rapid actions in the context of a well-prepared response is of high importance. The speed of recognition of the chemical nature of poisoning and fast intervention in the incident are critical. First-aid and medical managements stockpiles for use should include masks with manual ventilators for artificial ventilation, oropharyngeal airways, oxygen cylinders with masks, in-date ampoules of amyl nitrite (kept at, or below, 15°C), sodium bicarbonate for iv infusion and chosen antidote(s) (to be administered by a physician). Choice of antidote should be made in conjunction with relevant toxicology experts and Poison Control Centers. As noted above, there is a clear and pressing need for a rationalized international agreement on the most appropriate antidotes for acute CN poisoning, especially in the context of possible large numbers of CN intoxications during a terrorist incident and the fact that it will be needed to be used in a pre-hospital scenario (Thompson, 2004). It is our opinion that hydroxocobalamin is probably the most appropriate currently available antidote with an optimum risk–benefit ratio, and which can be used on-site at the incident scene but further work on the pharmacy of preparations is needed, particularly, to ensure sufficient potency and concentration of preparations, that adequate supplies can be made available for stockpiling purposes and costs. Much preparation work still needs to be

undertaken and according to Sauer and Keim (2001), the USA preparedness for a CN disaster is abysmal.

REFERENCES

Acker H and Eyzaguirre C (1989). Light absorbance changes in the mouse carotid body during hypoxia and cyanide poisoning. *Brain Res*, **409**, 380–385.

Aharony D, Smith JD and Smith MJ (1982). Inhibition of human platelet lipoxygenase by cyanide. *Experientia*, **38**, 1334–1335.

Aitken PG and Braitman J (1989). The effect of cyanide on neuronal and synaptic function in hippocampal slices. *Neurotoxicology*, **10**, 239–248.

Alvarez-Buylla R and Alvarez-Buylla ER de (1988). Carotid sinus receptors participate in glucose homeostasis. *Resp Physiol*, **72**, 47–60.

Alvarez-Buylla R and Alvarez-Buylla ER de (1994). Changes in blood glucose concentration in the carotid-body sinus modify brain glucose retention. *Brain Res*, **654**, 167–170.

Alvarez-Buylla R, Alvarez-Buylla E, Mendoza H *et al.* (1997). Pituitary and adrenals are required for hyperglycemic reflex initiated by stimulation of CBR with cyanide. *Am J Physiol*, **272**, R392–R399.

American Cyanamid (1959). *Report on sodium cyanide: 30-day repeated feeding study to dogs.* Cyanamid Report Number 59–14. Princeton, NJ, USA: Central Medical Department, American Cyanamid.

Anderson AH (1946). Experimental studies on the pharmacology of activated charcoal. *Acta Pharmacol*, **2**, 69–78.

Andrews JM, Sweeney ES, Grey TC *et al.* (1989). The biohazard potential of cyanide poisoning during postmortem examination. *J Forens Sci*, **34**, 1280–1284.

Antonini EA, Brunoori M, Greenwood C *et al.* (1971). The interaction of cyanide with cytochrome oxidase. *Eur J Biochem*, **23**, 396–400.

Ardelt KK, Borowitz JL, Maduh EU *et al.* (1994). Cyanide-induce lipid peroxidation in different organs: subcellular distribution and hydroperoxide generation in neuronal cells. *Toxicology*, **89**, 127–137.

ATSDR (1997). Toxicological profile for cyanide. Agency for Toxic Substances and Disease Registry, US Public Health Service. *ATSDR Toxicological Profiles on CD-ROM (Version 3:1; 2000).* Boca Raton, FL, USA: Chapman and Hall/CRC Press.

Autor AP and Fridovich I (1970). The interactions of acetoacetate decarboxylase with carbonyl compounds, hydrogen cyanide, and an organic mercurial. *J Biol Chem*, **245**, 5214–5222.

Baftis H, Smolens MH, His BP *et al.* (1981). Chronotoxicity of male BALB/cCr mice to potassium cyanide. In: *Proceedings of International Symposium on Chronopharmacology and Chronotherapy*, cited in Ballantyne (1987b).

Baker D (2004). Civilian exposure to toxic agents: emergency medical reponse. *Prehosp Disaster Med*, **19**, 174–178.

Ballantyne B (1973). An experimental assessment of decreases in measurable cyanide levels in biological fluids. *J Forens Sci Soc*, **13**, 111–117.

Ballantyne B (1974). The forensic diagnosis of acute cyanide poisoning. In: *Forensic Toxicology* (B Ballantyne, ed.), pp. 99–113. Bristol, UK: Wright.

Ballantyne B (1975). Blood, brain and cerebrospinal fluid cyanide concentrations in experimental acute cyanide poisoning. *J Forens Sci Soc*, **15**, 51–56.

Ballantyne B (1976). Changes in blood cyanide as a function of storage time and temperature. *J Forens Sci Soc*, **16**, 305–310.

Ballantyne B (1977). An experimental assessment of the diagnostic potential of histochemical and biochemical methods for cytochrome oxidase in acute cyanide poisoning. *Cell Molec Biol*, **22**, 109–123.

Ballantyne B (1979). Letter to the Editor. *Clin Toxicol*, **14**, 311–312.

Ballantyne B (1983a). Acute systemic toxicity of cyanides by topical application to the eye. *Cut Ocul Toxicol*, **2**, 119–129.

Ballantyne B (1983b). The influence of exposure route and species on the acute lethal toxicity and tissue concentrations of cyanide. In: *Developments in the Science and Practice of Toxicology* (AW Hayes, RC Schnell and TS Miya, eds), pp. 583–586. Amsterdam, The Netherlands: Elsevier Science Publishers.

Ballantyne B (1983c). Artifacts in the definition of toxicity by cyanide and cyanogens. *Fund Appl Toxicol*, **3**, 40–408.

Ballantyne B (1984). Comparative acute toxicity of hydrogen cyanide and its salts. In: *Proceedings of the Fourth Annual Chemical Defense Bioscience Review.* (RE Linstrom, ed.), pp. 477–501. Bethesda, MD, USA: US Army Medical Research Institute of Chemical Defense.

Ballantyne B (1986). Hazard evaluation of cyanide fumigant powder formulations. *Vet Human Toxicol*, **28**, 42.

Ballantyne B (1987a). Hydrogen cyanide as a product of combustion and a factor in morbidity and mortality from fires. In: *Clinical and Experimental Toxicology of Cyanides* (B Ballantyne and TC Marrs, eds), pp. 248–291. Bristol, UK: Wright.

Ballantyne B (1987b). Toxicology of cyanides. In: *Clinical and Experimental Toxicology of Cyanides* (B Ballantyne and TC Marrs, eds), pp. 41–126. Bristol, UK: Wright.

Ballantyne B (1987c). Post-mortem features and criteria for the diagnosis of acute lethal cyanide poisoning. In: *Clinical and Experimental Toxicology of Cyanides* (B Ballantyne and TC Marrs, eds), pp. 217–247. Bristol, UK: Wright.

Ballantyne B (1988). Toxicology and hazard evaluation of cyanide fumigation powders. *Clin Toxicol*, **26**, 325–335.

Ballantyne B (1994a). Acute percutaneous systemic toxicity of cyanides. *Cut Ocul Toxicol*, **13**, 249–262.

Ballantyne B (1994b). Acute inhalation toxicity of hydrogen cyanide vapor to the rat and rabbit. *Toxic Subst J*, **13**, 249–282.

Ballantyne B and Bright JE (1979a). Comparison of kinetic and end-point microdensitometry for the direct quantitative histochemical assessment of cytochrome oxidase activity. *Histochem J*, **11**, 173–186.

Ballantyne B and Bright JE (1979b). A kinetic histochemical method for the determination of cytochrome oxidase activity. *J Physiol London*, **300**, 1–2.

Ballantyne B and Salem H (2005). Experimental, clinical, occupational toxicology and forensic aspects of hydrogen cyanide with particular reference to vapor exposure. In: *Inhalation Toxicology*, Second Edition (H Salem and A Katz, eds), pp. 709–794. Boca Raton, FL, USA: CRC Press.

Ballantyne B, Bright J, Swanston DW *et al.* (1972). Toxicity and distribution of free cyanides given intramuscularly. *Med Sci Law*, **12**, 209–219.

Banerjee KK, Bishayee R and Marimutha P (1997). Evaluation of cyanide exposure and its effect on thyroid function of workers in a cable industry. *J Occup Environ Med*, **39**, 258–260.

Barr S (1966). The microdetermination of cyanide. Its application to the analysis of whole blood. *Analyst*, **91**, 268–272.

Baskin SI, Wilkerson G, Alexander K *et al.* (1987). Cardiac effects of cyanide. In: *Clinical and Experimental Toxicology of Cyanides* (B Ballantyne and TC Marrs, eds), pp. 138–155. Bristol, UK: Wright.

Bastian G and Mercker H (1959). Zur Frage der Zweckmäßigkeit der Inhalation von Amynitrit in der Behandlung der Cyanidvergiftung *Nauny-Schmiedebergs. Arch Exp Pathol Pharmakol*, **237**, 285–295.

Baud FJ, Barriot P and Toffis V (1991). Elevated blood cyanide concentrations in victims of smoke inhalation. *New England J Med*, **325**, 1761–1766.

BBC (2002). Cyanide plotters face terror charges. *British Broadcasting Corporation*, 21 February 2002 (http://news.bbc.uk/1/hi/world/europe/1833646.stm).

Beasley DMG and Glass WI (1998). Cyanide poisoning: pathophysiology and treatment recommendations. *Occup Med*, **48**, 427–431.

Beasley RWR, Blow RJ, Lunau FW *et al.* (1978). Amyl nitrite and all that. *J Soc Occup Med*, **28**, 142–143.

Bhattacharya R, Kumar P and Sachan AS (1994). Cyanide induced changes in dynamic pulmonary mechanics in rats. *Ind J Physiol Pharmacol*, **38**, 281–284.

Bianco PJMB and Garcia RA (2004). First case of illegal euthanasia in Spain: oral potassium cyanide poisoning. *Soud Lék*, **49**, 30–33.

Blanc P, Hogan M, Mallin K *et al.* (1985). Cyanide intoxication among silver reclaiming workers. *J Am Med Assoc*, **253**, 367–371.

Bondy SC and Komulainen H (1988). Intracellular calcium as an index of neurotoxic damage. *Toxicology*, **49**, 35–41.

Bonsall JJ (1984). Survival without sequelae following exposure to 500 mg m^{-3} hydrogen cyanide. *Human Toxicol*, **3**, 57–60.

Borders CL Jr and Fredivich I (1985). A comparison of the effects of cyanide, hydrogen peroxide, and phenylglyoxal on eukaryotic and prokaryotic Cu, Zn superoxide dismutases. *Arch Biochem Biophys*, **241**, 472–476.

Borgohain R, Singh AK, Radhakrishna H *et al.* (1955). Delayed onset generalized dystonia after cyanide poisoning. *Clin Neurol Neurosurg*, **97**, 213–215.

Borowitz JL, Kanthasamy A and Isom GE (1994). Toxicodynamics of cyanide. In: *Chemical Warfare Agents* (SM Somani, ed.), pp. 209–236. San Diego, CA, USA: Academic Press.

Borron S (2006). Blood cyanide and plasma cyanocobalamin concentrations after treatment of acute cyanide poisoning with hydroxocobalamin or vehicle in a canine model. *Clin Toxicol*, **44**, 630.

Borron SW and Baud FJ (1996). Acute cyanide poisoning: clinical spectrum, diagnosis and treatment. *Arh Hig Toksikol*, **47**, 307–322.

Boswell GW, Brooks DE, Murray AJ *et al.* (1988). Exogenous methemoglobin as a cyanide antidote in rats. *Pharmaceut Res*, **5**, 749–752.

Boxer GE and Rickards JC (1951). Chemical determination of vitamin B_{12}. IV. Assay of vitamin B_{12} in multivitamin preparations and biological materials. *Arch Biochem*, **30**, 392–401.

Boxer GE and Rickards JC (1952). Studies on the metablism of the carbon of cyanide and thiocyanate. *Arch Biochem*, **39**, 7–26.

Brahams D (1991). 'Sudafed capsules' poisoned with cyanide. *Lancet*, **337**, 968.

Brierley JB, Brown AW and Calverley J (1976). Cyanide intoxication in the rat physiological and neuropathological aspects. *J Neurol Neurosurg Psychiat*, **39**, 129–140.

Brierley JB, Prior PF, Calverley J *et al.* (1977). Cyanide intoxication in *Macaca mullata*. Physiological and neuropathological aspects. *J Neurol Sci*, **31**, 133–157.

Bright JE and Marrs TC (1987). Effect of *p*-aminopropiophenone (PAPP), a cyanide antidote, on cyanide given by intravenous infusion. *Human Toxicol*, **6**, 133–138.

Brink NG, Kuehl FA and Folkers K (1950). Vitamin B_{12}: The identification of vitamin B_{12} as a cyano-cobalt coordination complex. *Science*, **112**, 354.

Brivet F, Delfraissy JF, Duche M *et al.* (1983). Acute cyanide poisoning recovery with non-specific supportive therapy. *Intensive Care Med*, **9**, 33–35.

Brown KS and Robinette RR (1967). No simple pattern of inheritance in the ability to smell solutions of cyanide. *Nature*, **215**, 406–408.

Bryson DD (1978). Cyanide poisoning. *Lancet*, **i**, 92.

Burrows GE and Way JL (1977). Cyanide intoxication in sheep: therapeutic value of oxygen or cobalt. *Am J Vet Res*, **38**, 223–227.

Busquets S, Alvarez B, Van Royen M *et al.* (2001). Increased uncoupling protein-2 gene expression in brain of lipopolysaccharide-injected mice: Role of tumor necrosis factor-α? *Biochim Biophys Acta*, **1499**, 249–256.

Camerino PW and King TE (1966). Studies on cytochrome oxidase. 3. Reaction of cyanide with cytochrome oxidase insoluble and particulate forms. *J Biol Chem*, **241**, 970–979.

Camu F (1969). Effect of imidazole and cyanide upon rat tissue lipolysis. *Arch Int Physiol Biochem*, **77**, 663–666.

Carella F, Grassi MP and Savoiard M (1988). Dystonic-parkinsonian syndrome after cyanide poisoning: clinical and MRI findings. *J Neurol Neurosurg Psychiat*, **51**, 1345–1348.

Carroll JL, Gonzal D, Rector DM *et al.* (1996). Ventral medullary neuronal responses to peripheral chemoreceptor stimulation. *Neuroscience*, **73**, 989–998.

Cassel G (1995). Estimation of the convulsive effect of cyanide in rats. *Pharmacol Toxicol*, **77**, 259–263.

Cassel G and Persson SA (1985). Effect of acute cyanide intoxication on central catecholaminergic pathways. *Clin Toxicol*, **23**, 461–462.

Cassel G and Persson SA (1992). Effect of acute lethal cyanide intoxication on central dopaminergic pathways. *Pharmacol Toxicol*, **70**, 148–151.

Cassel G, Karlsson L and Sellström Å (1991). On the inhibition of glutamic acid decarboxylase and gamma-amino butyric acid transaminase by sodium cyanide. *Pharmacol Toxicol*, **69**, 238–241.

Centers for Disease Control (1991). Cyanide poisonings associated with over-the-counter medication – Washington State, 1991. *J Am Med Assoc*, **265**, 1806–1807.

Centers for Disease Control (2005). Atypical reactions associated with heroin use – five states, January–April, 2005. *Morbid Mortal Weekly Rep*, **54**, 793–796.

Chance B (1952). Spectra and related kinetics of respiratory pigments of homogenized and intact cells. *Nature*, **16**, 215–221.

Chin RG and Calderon Y (2000). Acute cyanide poisoning: a case report. *J Emerg Med*, **18**, 441–445.

Chung J and Wood JL (1971). Oxidation of thiocyanate to cyanide catalyzed by hemoglobin. *J Biol Chem*, **246**, 555–560.

Cloud DS (2004). Long in US signs, a young terrorist builds grim resume. *The Wall Street Journal (New York)*, February 10, p. 2.

CNN (2004). Cyanide arsenal stirs domestic terror fear. *CNN News*, January 30. (www.cnn.com/2004/US/southwest/01/30/cyanide.probe.ap/idex.html).

Comroe JM (1974). *Physiology of Respiration*, pp. 33–54. Chicago, IL, USA: Year Book Publishers.

Cottrell JE, Casthely P and Brodie JD (1978). Prevention of nitroprusside-induced cyanide toxicity with hydroxocobalamin. *New England J Med*, **298**, 809–811.

Coughlan MP, Johnson JL and Rajagopalan KV (1980). Mechanisms of inactivation of molybdoenzymes by cyanide. *J Biol Chem*, **255**, 2694–2699.

Crampton RF, Gaun IF, Harris R *et al.* (1979). Effect of low cobalamin diet and chronic cyanide toxicity in baboons. *Toxicology*, **12**, 221–234.

Craven N (2005). Assassin free to plot ricin error. *Daily Mail (London)*, Thursday, April 14, pp. 6–7.

Curry AS (1963). Cyanide poisoning. *Acta Pharmacol Toxicol*, **20**, 291–29.

Dahl AR (1989). The cyanide metabolizing enzyme rhodanese in rat nasal respiratory and olfactory mucosa. *Toxicol Lett*, **45**, 199–203.

Dally S and Gaultier M (1976). Choc anaphylactique dû à l'hydroxocobalamine. *Nouv Press Med*, **5**, 1917.

Daly M DeB and Scott MJ (1964). The cardiovascular effects of hypoxia in the dog with special reference to the contribution of the carotid body chemoreceptors. *J Physiol London*, **173**, 201–204.

DeBush RF and Seidel LG (1969). Attempted suicide by cyanide. *California Med*, **10**, 394–396.

DeFlora S (1981). Study of 106 organic and inorganic compounds in the Salmonella/microsome test. *Carcinogenesis*, **2**, 283–298.

DesLauriers CA, Burda AM and Whal M (2006). Hydroxocobalamin as a cyanide antidote. *Amer. J. Therap*, **13**, 161–165.

D'Mello GD (1987). Neuropathological and behavioral sequelae of acute cyanide toxicosis in animal species. In: *Clinical and Experimental Toxicology of Cyanides* (B Ballantyne and TC Marrs, eds), pp. 156–183. Bristol, UK: Wright.

Doherty PA, Ferm V and Smith RP (1982). Congenital malformations induced by infusion of sodium cyanide in Golden hamster, *Toxicol Appl Pharmaceut*, **64**, 456–464.

Donnelly DF (1993). Response to cyanide of two types of glomoid cells in mature rat carotid body. *Brain Res*, **30**, 157–168.

Drinker P (1932). Hydrocyanic gas poisoning by absorption through the skin. *J Ind Hyg*, **14**, 1–2.

Dudley HC, Sweeney TR and Miller JW (1942). Toxicology of acrylonitrile (vinyl cyanide). II. Studies of effects of daily inhalation. *J Ind Hyg Toxicol*, **24**, 255–258.

Dugard PH (1987). The absorption of cyanide through human skin *in vitro* from solutions of sodium cyanide and gaseous HCN. In: *Clinical and Experimental Toxicology of Cyanides* (B Ballantyne and TC Marrs, eds), pp. 127–137. Bristol, UK: Wright.

Dunea G (1983). Death over the counter. *Br Med J*, **286**, 211–212.

Edwards AC and Thomas ID (1978). Cyanide poisoning. *Lancet*, **i**, 92–93.

El Ghawabi SH, Gaffar MA, Esabarti AA *et al.* (1975). Chronic cyanide exposure: a clinical radioisotope and laboratory study. *Br J Ind Med*, **32**, 215–219.

Ermans AM, Delange F and Van der Velden M (1972). Possible role of cyanide and thiocyanate in the etiology of endemic cretinism. *Adv Exp Med Biol*, **30**, 455–486.

Espinoza OB, Perez M and Ramirez MS (1992). Bitter cassava poisoning in eight children: a case report. *Vet Human Toxicol*, **34**, 65.

Estler CJ (1965). Metabolic alterations in the brain during the course of non-lethal potassium cyanide poisoning and the influence of cyanide antagonists. *Naunyn-Schmeidebergs Arch Pharmacol*, **51**, 413–432.

Fairley A, Linton EC and Wild FE (1934). The absorption of hydrocyanic vapor through skin. *J Hyg*, **34**, 283–294.

Feldman JM and Feldman DD (1990). Sequelae of attempted suicide by cyanide ingestion: a case report. *Int J Psychiat Med*, **20**, 173–179.

Fenney J and Brugen ASV (1973). Cyanide binding to carbonic anhydrase. *Eur J Biochem*, **34**, 107–111.

Fernando GCA and Busuttil A (1991). Cyanide ingestion – case studies of four suicides. *Am J Forens Med Pathol*, **12**, 241–24.

Finelli PF (1981). Changes in the basal ganglia following cyanide poisoning. *J Comput Assist Tomog*, **5**, 755–756.

Fitzgerald LR (1954). Effect of injected sodium cyanide on newborn and adult mice *Am J Physiol*, **17**, 60–62.

Fligner CL, Luthi R, Linkaiyte-Weiss E *et al.* (1987). Cyanide poisoning from capsule tampering: clinical, pathologic and toxicological aspects. In: *Proceedings of the 39th Annual Meeting of the American Academy of Forensic Sciences*, pp. 106–107, San Diego, CA.

Forsyth JG, Mueller PD and Becker CE (1993). Hydroxocobalamin as a cyanide antidote: safety, efficacy and pharmacokinetics in heavily smoking normal volunteers. *Clin Toxicol*, **31**, 277–294.

Fortin J-L, Giocant J-P, Ruttimann *et al.* (2006). Prehospital administration of hydroxocobalamin for smoke inhalation associated cyanide poisoning: 8 years of experience in the Paris Fire Brigade. *Clin. Toxicol*, **44**(Suppl 1), 37–44.

Frankenberg L and Sörbo B (1975). Effect of cyanide antidotes on the metabolic conversion of cyanide to thiocyanate. *Arch Toxicol*, **3**, 81–89.

Fukayama H, Nasu M and Murakami S (1992). Examination of antithyroid effects of smoking products in cultured thyroid follicles: only thiocyanate is a potent antithyroid agent. *Acta Endocrinol*, **127**, 520–525.

Fukumoto Y, Nasu M and Murakami S (1957). A study on the sense of smell with respect to potassium cyanide solution and its hereditary transmission. *Jpn J Human Genet*, **2**, 7–16.

Funata N, Song SY, Funata M *et al.* (1984). A study of experimental cyanide encephalopathy in the acute phase – physiological and neuropathological correlation. *Acta Neuropathol*, **64**, 99–107.

Ganote GE, Worstell J and Kaltenlach JP (1976). Oxygen-induced enzyme release after irreversible myocardial injury. *Am J Pathol*, **84**, 2735.

Garberg P, Äkerblom, E-L and Bolcsfoldi G (1988). Evaluation of a genotoxicity test measuring DNA-strand breaks in mouse lymphoma cells by alkaline unwinding and hydroxyapatite elution. *Mut Res*, **203**, 155–176.

Gee DJ (1987). Cyanides in murder, suicide and accident. In: *Clinical and Experimental Toxicology of Cyanides* (B Ballantyne and TC Marrs, eds), pp. 209–216. Bristol, UK: Wright.

Geller RJ, Elkins BR and Iknoian RC (1991). Cyanide toxicity from acetonitrile-containing false nail remover. *Am J Emerg Med*, **9**, 268–270.

Goldstein F and Rieders F (1951). Formation of cyanide in dog and man following administration of thiocyanate. *Am J Physiol*, **167**, 47–50.

Goodhart GL (1994). Patient treated with antidote kit and hyperbaric oxygen survives cyanide poisoning. *Southern Med J*, **87**, 814–816.

Gorman DF (1989). Problems and pitfalls in the use of hyperbaric oxygen for the treatment of poisoned patients. *Med Toxicol Adverse Drug Exp*, **4**, 393–399.

Gracia R and Shepherd G (2004). Cyanide poisoning and its treatment. *Pharmacotherapy*, **24**, 1358–1365.

Graham DL, Laman D, Theodore J *et al.* (1977). Acute cyanide poisoning complicated by lactic acidosis and pulmonary edema. *Arch Int Med*, **137**, 1051–1055.

Grandas F, Artieda J and Obeso JA (1989). Clinical and CT scan findings in a case of cyanide intoxication. *Mov Disorders*, **4**, 188–193.

Greve KW, Bianchini KJ, Doane BM *et al.* (2005). Psychological evaluation of the emotional effects of a community toxic exposure. *J Occup Environ Med*, **47**, 5159.

Guatelli MA (1964). The toxicology of cyanides. In: *Methods of Forensic Sciences*, Volume 3 (AS Curry, ed.), pp. 233–265. New York, NY, USA: Academic Press.

Gunaseker PG, Ramesh GT, Borowitz JL *et al.* (1999). Involvement of caspase-3-like activity in cyanide-induced neuronal aptosis. *Toxicologist*, **48**, 87.

Hall AH and Rumack BH (1986). Clinical toxicology of cyanide. *Ann Emerg Med*, **15**, 1067–1074.

Hall AH and Rumack BH (1987). Hydroxycobalamin/sodium thiosulfate as a cyanide antidote. *J Emerg Med*, **5**, 115–121.

Hall AH and Rumack BH (1998). Cyanide and related compounds. In: *Clinical Management of Poisoning and Drug Overdose*, Third Edition (LM Haddad, MW Shannon and JF Winchester, eds), pp. 899–905. Philadelphia, PA, USA: WB Saunders.

Hall AH, Linden CH and Kulig KW (1986). Cyanide poisoning from laetrile: role of nitrite therapy. *Pediatrics*, **78**, 269–272.

Hall AH, Rumack BH and Schaffer MI (1987). Clinical toxicology of cyanide. In: *Clinical and Experimental Toxicology of Cyanides* (B Ballantyne and TC Marrs, eds), pp. 312–333. Bristol, UK: Wright.

Hansen BA and Dekker EE (1976). Inactivation of bovine liver 2-keto-4-hydroxyglutarate aldolase by cyanide in the presence of aldehydes. *Biochemistry*, **15**, 2912–2917.

Hantson J and Duprez T (2006). The value of morphological neuroimaging after exposure to toxic substances. *Toxicol. Rev*, **25**, 87–98.

Harris R and Paxman J (1982). *A Higher Form of Killing*, p. 199. London, UK: Chatto and Windus.

Hart GB, Strauss MB and Lennon PA (1985). Treatment of smoke inhalation by hyperbaric oxygen. *J Emerg Med*, **3**, 211–215.

Hathaway GJ and Proctor NH (2004). *Chemical Hazards of the Workplace*, Fourth Edition, pp. 346–347. Hoboken, NJ, USA: John Wiley & Sons, Inc.

Haxhiu MA, Erokwu B, Van Lunteren E *et al.* (1993). Central and spinal effects of sodium cyanide on respiratory activity. *J Appl Physiol*, **74**, 574–579.

Haymaker W, Ginzler JM and Ferguson RL (1952). Residual neuropathological effects of cyanide poisoning. *Milit Med*, **111**, 231–246.

Hébert CD (1993). *NTP Technical Report on toxicity studies of sodium cyanide (CAS No. 143-33-9). Administered in drinking water to F334/N rats and B6C3F1 mice*. NIH Publication 94-3386, November. Research Triangle Park, NC, USA: National Institutes of Health, National Toxicology Program.

Herbert V (1975). Drugs effective in megaloblastic anemias. In: *The Pharmacological Basis of Therapeutics*, Fifth Edition (LS Goodman and A Gilman, eds), pp. 1332–1346. New York, NY, USA: McMillan.

Herrting G, Kraupp O, Schnetz H *et al.* (1960). Untersuchungen über die Folgen einer chronischen Verabreichung akuttoxischer Dosen von Natriumcyanid an Hunden. *Acta Pharmacol Toxicol*, **17**, 27–42.

Higgins EA, Floea V, Thomas AA *et al.* (1972). Acute toxicity of brief exposures to HF, HCl, NO_2 and HCN with and without CO. *Fire Technol*, **2**, 120–130.

Hilado CJ and Cummings HJ (1977). A review of available LC_{50} data. *J Combust Toxicol*, **4**, 415–424.

Hilmann B, Bardham KD and Bain JTB. (1974). The use of dicobalt edetate (Kelocyanor) in cyanide poisoning. *Postgrad Med J*, **50**, 171–174.

Himwich WA and Saunders JP (1948). Enzymic conversion of cyanide to thiocyanate. *Am J Physiol*, **153**, 348–354.

Hirano A, Levine S and Zimmerman HM (1967). Experimental cyanide encephalopathy. Electron microscopic observations of early lesions of white matter. *J Neuropathol Exper Neurol*, **26**, 200–213.

HMSO (1987). *Medical Manual of Defence Against Chemical Agents*. London, UK: Her Majesty's Stationary Office.

Hoidal CR, Hall AH and Robinson MD (1986). Hydrogen sulfide poisoning from toxic inhalations of roofing asphalt fumes. *Ann Emerg Med*, **15**, 826–830.

Holland DJ (1983). Cyanide poisoning: an uncommon encounter.*J Emerg Nurs*, **9**, 138–140.

Holland MA and Koslowski IM (1986). Clinical features and management of cyanide poisoning. *Clin Pharm*, **5**, 737–741.

Homan ER (1987). Reactions, processes and materials with potential for cyanide exposure. In: *Clinical and Experimental Toxicology of Cyanides* (B Ballantyne and TC Marrs, eds), pp. 1–21. Bristol, UK: Wright.

Houeto P, Borron SW, Sandouk P *et al.* (1996). Pharmacokinetics of hydroxocobalamin in smoke inhalation victims. *Clin Toxicol*, **34**, 397–404.

Howard JW and Hanzal RF (1955). Chronic toxicity for rats of food treated with hydrogen cyanide. *J Agr Food Chem*, **3**, 325–329.

Hoyer S (1984). The effect of naftidroturyl on cyanide induced hypoxic damage to glucose and energy metabolism in brain cortex of rats. *Arzneim-Forsch*, **34**, 412–416.

Hume AS, Mozingo JR, McIntyre B *et al.* (1995). Antidotal efficacy of alpha-ketoglutaric acid and sodium thiosulfate in cyanide poisoning. *Clin Toxicol*, **33**, 721–724.

Huser HJ, Moor-Jankowski JK, Truog G *et al.* (1958). Clinical, genetic and coagulation-physiologic aspects of Hemophilia B in the bleeders of Tenna, with a contribution on the genetics of the coagulation factors. *Acta Genet*, **8**, 209–234.

Isom GE, Li DHW and Way JL (1975). Effects of sublethal doses of cyanide on glucose catabolism. *Biochem Pharmacol*, **24**, 81–85.

Johnson RP and Mellors JW (1988). Arteriolization of venous blood gases: a clue to the diagnosis of cyanide poisoning. *J Emerg Med*, **6**, 401–404.

Johnson JD, Meisenheimer TL and Isom GE (1986). Cyanide-induced neurotoxicity: role of neuronal calcium. *Toxicol Appl Pharmac*, **4**, 64–68.

Kahler RL, Goldblatt A and Braunmad E (1962). The effects of acte hypoxia on the systemic venous and arterial systems and on myocardial contractile force. *J Clin Invest*, **41**, 1553–1563.

Kales SN, Drinlage D, Dickey J *et al.* (1997). Paranoid psychosis after exposure to cyanide. *Arch Environ Health*, **52**, 245–246.

Kamalu, BP and Agharanya JC (1991). The effect of a nutritionally-balanced cassava (*Manihot esculenta Crantz*) diet on endocrine function using the dog as a model. 2. Thyroid. *Br J Nutrit*, **65**, 373–379.

Kanthasamy AG, Borowitz JL and Isom GE (1991). Cyanide-induced increases in plasma catecholamines: relationship to acute toxicity. *Neurotoxicology*, **12**, 777–784.

Kanthasamy AG, Borowitz JL, Pavlakovic G *et al.* (1994a). Dopaminergic neurotoxicity of cyanide: neurochemical, histological and behavioral characterization. *Toxicol Appl Pharmacol*, **126**, 156–163.

Kanthasamy AG, Borowitz JL, Pavlakovic G *et al.* (1994b). Interaction of cyanide with a dopamine metabolite: formation of a cyanohydrin adduct and its implications for cyanide-induced neurotoxicity. *Neurotoxicology*, **15**, 887–896.

Katsumata Y, Sato K, Oya M *et al.* (1980). Kinetic analysis of the shift of aerobic metabolism in rats during acute cyanide poisoning. *Life Sci*, **27**, 1509–1512.

Khan A, Levitt A, Sage M *et al.* (2000). Biological and chemical terrorism: strategic plan for preparedness and response. *Morbid Mortal Weekly Rep*, **49**, 1–14.

Kiese M (1974). *Methemoglobinemia: A Comprehensive Treatise*, p. 80, Cleveland, OH, USA: CRC Press.

Kirk RL and Stenhouse NS (1953). Ability to smell solutions of potassium cyanide. *Nature*, **171**, 698.

Kirk MA, Gerace R and Kulig KW (1993). Cyanide and methemoglobin kinetics in smoke inhalation victims treated with the cyanide antidote kit. *Ann Emerg Med*, **22**, 1413–1418.

Kiuchi Y, Inagaki M, Izumi J *et al.* (1992). Effects of local cyanide perfusion on rat striatal extracellular dopamine and its metabolites as studied by in vivo brain microdialysis. *Neurosci Lett*, **147**, 193–196.

Klimmek R, Fladerer M and Weger N (1979). Circulation, respiration and blood homeostasis of dogs during slow cyanide poisoning and after treatment with 4-dimethylaminophenol or cobalt compounds. *Arch Toxicol*, **3**, 121–133.

Korner PL (1959). Circulatory adaptations in hypoxia. *Physiol Rev*, **39**, 687–730.

Krasney JA (1967). Efficient components of the cardioaccelerator responses to oxygen lack and cyanide. *Am J Physiol*, **213**, 1475–1479.

Krasney JA (1970). Effects of sino-aortic denervation on regional circulatory responses to cyanide. *Am J Physiol*, **218**, 56–63.

Krasney JA (1971). Cardiovascular responses to cyanide in awake sino-aortic denervated dogs. *Am J Physiol*, **220**, 1361–1366.

Krasney JA, Hogan PM, Lowe RF *et al.* (1966). Comparison of the effects of hypoxia and cyanide on cardiac rate in spinal dogs. *Fed Proc*, **25**, 2.

Kremer ML (1970). Inhibition of catalase by cyanide. *Israel J Chem*, **8**, 790–807.

Kulig WK and Ballantyne B (1993). Cyanide toxicity. *Am Fam Phys*, **48**, 107–114.

Kumar P, Das M and Kumar A (1992). Health status of workers engaged in heat treatment (case hardening) plant and electroplating at cyanide bath. *Ind J Environ Protect*, **12**, 179–183.

Kushi A, Matsumoto T and Youmans WB (1983). Mutagen from the gaseous phase of protein hydrolysate. *Agr Biol Chem*, **47**, 1979–1982.

Lafferty MA and Garet RH (1974). Purification and properties of the neurospora crassa assimilatory nitrite reductase. *J Biol Chem*, **249**, 7555–7557.

Lafin SM, Lucchesi M, Fedak M *et al.* (1992). A need to revise the cyanide antidote package instructions. *J Emerg Med*, **10**, 623.

Lambert RJ, Kindler BL and Dschaeffer DJ (1988). The efficacy of superactivated charcoal in treating rats exposed to a lethal oral dose of potassium cyanide. *Ann Emerg Med*, **17**, 595–598.

Lang K (1933). Rhodanbildung im Tierkörper. *Biochem Z*, **259**, 243–256.

LaPostolle F, Borron S and Buad F (2006). Increased plasma lactate concentrations are associated with cyanide but not other types of acute poisoning. *Clin. Toxicol*, **44** (Suppl 1), 777.

Leading Article (2005). Poison plot. *The Times (London)*, Thursday April 14, p. 19.

Lee-Jones M, Bennett MA and Sherwell JM (1970). Cyanide self-poisoning. *Br Med J*, **2**, 780–781.

Leimdorfer A (1950). About anoxia of the heart roduced by intravenous sodium cyanide injections. *Arch Int Pharmacodyn*, **84**, 181–18.

Lemberg MR (1969). Cytochrome oxidase. *Physiol Rev*, **49**, 48–111.

Lendle L (1964). Wirkungsbedingngen und Blausäure und Schwefelwasserstoff und Möglichelten der Vergiftungsbehandlung. *Jpn J Pharmacol*, **14**, 215–224.

Lessel S (1971). Experimental cyanide optic neuropathy. *Arch Ophthalmol*, **86**, 194–204.

Leung P, Cannon EP, Petrkovics I *et al.* (1991). *In vivo* studies on rhodanese encapsulation in mouse carrier erythrocytes. *Toxicol Appl Pharmacol*, **110**, 268–274.

Leuschner J, Winkler A and Leuscchner F (1991). *In vivo* studies on rhodanese encapsulation in mouse carrier erythrocytes. *Toxicol Appl Pharmacol*, **10**, 268–274.

Levin BC, Gurman JL, Paabo M *et al.* (1987). Toxicological aspects of pure and mixed fire gases for various exposure times. *Toxicologist*, **7**, 201.

Levine S (1975). Nonperipheral chemoreceptor stimulation of ventilation by cyanide. *J. Appl. Physiol.*, **39**, 199–204.

Levine S and Stypulkowski W (1959). Experimental cyanide encephalopathy. *Arch Pathol*, **6**, 306–323.

Levine S and Wenk EJ (1959). Cyanide encephalopathy produced by intravenous route. *J Nerv Mental Dis*, **129**, 302–305.

Li PP and White TD (1977). Rapid effects of veratridine, tetrodotoxin, garamicidin D, valinomycin, and NaCN on the Na^+, K^+ and ATP contents of synaptosomes. *J Neurochem*, **28**, 967–975.

Li L, Prabhakaran K, Mills EM *et al.* (2005). Enhancement of cyanide-induced mitochondrial dysfunction and cortical cell necrosis by uncoupling protein-2. *Toxicol Sci*, **86**, 116–124.

Linakis JG, Lacouture PG and Woolf A (1991). Monitoring cyanide and thiocyanate concentrations during infusion of sodium nitroprusside in children. *Pediatr Cardiol*, **12**, 214–218.

Lindsay AE, Greenbaum AR and O'Hare D (2004). Analytical techniques for cyanide in blood and published blood cyanide concentrations from healthy subjects and fire victims. *Anal Chim Acta*, **51**, 185–195.

Litovitz T (1987). The use of oxygen in the treatment of acute cyanide poisoning. In: *Clinical and Experimental Toxicology of Cyanides* (B Ballantyne and TC Marrs, eds), pp. 467–472. Bristol, UK: Wright.

Lovatt Evans C (1964). Cobalt compounds as antidotes for hydrocyanic acid. *Br J Pharmacol*, **23**, 455–475.

Lundquist P, Roslig H and Sörbo B (1988). The origin of hydrogen cyanide in breath. *Arch Toxicol*, **61**, 270–274.

Lundquist P, Kagedal B, Nilsson L *et al.* (1995). Analysis of the cyanide metabolite 2-aminothiazoline-4-carboxylic acid in urine by high performance liquid chromatography. *Anal Biochem*, **22**, 1–8.

MacMillan VH (1989). Cerebral energy metabolism in cyanide encephalopathy. *J Cerebral Blood Flow Metab*, **9**, 156–162.

Maduh EU, Porter DW and Baskin SI (1993). Calcium antagonists. A role in the management of cyanide poisoning. *Drug Safety*, **9**, 237–248.

Maduh EU, Nealley EW, Song H *et al.* (1995). A protein kinase C inhibitor attenuates cyanide toxicity in vivo. *Toxicology*, **100**, 129–137.

Marrs TC (1987). The choice of cyanide antidotes. In: *Clinical and Experimental Toxicology of Cyanides* (B Ballantyne and TC Marrs, eds), pp. 383–401. Bristol, UK: Wright.

Marrs TC (1988). Antidotal treatment of acute cyanide poisoning. *Adv Drug Reacti Acute Poison Rev*, **4**, 178–206.

Marsho TV and Kung SD (1976). Oxygenase properties of crystallized fraction I protein from tobacco. *Arch Biochem Biophys*, **173**, 341–346.

Massey V and Edmondson D (1970). On the mechanism of inactivation of xanthine oxidase by cyanide. *J Biol Chem*, **45**, 6595–6598.

Mathangi C and Namasivayam A (2000). Effect of chronic cyanide intoxication on memory in albino rats. *Food Chem Toxicol*, **38**, 51–55.

Matijak-Schaper M and Alarie Y (1982). Toxicity of carbon monoxide hydrogen cyanide and low oxygen. *J Combust Toxicol*, **9**, 21–61.

Maynard RL (1999). Toxicology of chemical warfare agents. In: *General and Applied Toxicology*, Volume 3, Second Edition (B Ballantyne, TC Marrs and T Syversen, eds), pp. 2079–2109. London, UK: MacMillan.

McMillan DE and Svoboda AC (1982). The role of erythrocytes in cyanide detoxification. *J Pharmacol Exp Therap*, **21**, 37–41.

McNamara BP (1976). *Estimation of the toxicity of hydrocyanic acid vapors in man*. Edgewood Arsenal Technical Report, Number EB-TR76023. Bethesda, MD, USA: US Department of the Army.

Mégarbane B and Baud F (2003). Cyanide poisoning: diagnosis and atidote choice in an emergency situation. *Clin Toxicol*, **41**, 438–439.

Meredith TJ, Jacosen D, Haines JA *et al.* (1993). *Antidotes for Cyanide Poisoning*, Volume 2, *International Program of Chemical Safety/Commission of the European Communities, Evaluation of Antidotes Series*. Cambridge, UK: Cambridge University Press.

Messing B (1991). Extrapyramidal disturbances after cyanide poisoning (first MRI-investigation of the brain). *J Neural Transm*, **33**, 141–147.

Meyer GW, Hart GB and Strauss MB (1991). Hyperbaric oxygen therapy for acute smoke inhalation injuries. *Postgrad Med*, **89**, 221–223.

Mitra J, Dev NB, Trivedi R *et al.* (1993). Intramedullary sodium cyanide injection on respiratory and vasomotor responses in cats. *Resp Physiol*, **93**, 71–82.

Moore S and Gates N (1946). Cited by McNamara (1976).

Moore SJ, Norris JC, Ing KH *et al.* (1986). The efficacy of alpha-ketoglutaric acid in the antagonism of cyanide intoxication. *Toxicol Appl Pharmacol*, **81**, 265–273.

Nakatani T, Kosugi Y, Mori A *et al.* (1993). Changes in the metabolism of oxygen metabolism in a clinical course recovering from potassium cyanide. *Am J Emerg Med*, **11**, 213–217.

Naughton M (1974). Acute cyanide poisoning. *Anesth Intensive Care*, **4**, 351–356.

Neiminen A-L, Gores GJ, Wray BE *et al.* (1988). Calcium dependence of bleb formation and cell death in hepatocytes. *Cell Calcium*, **9**, 237–246.

Nicholls P, Van Burrn KJH and Van Gelder BF (1972). Biochemical and biophysical studies on the cytochrome aa_3. Binding of cyanide to cytochrome aa_3. *Biochem Biophys Acta*, **275**, 279–287.

Nicholson PJ and Vincenti GEP (1994). A case of phobic anxiety related to the inability to smell cyanide. *Occup Med*, **4**, 107–108.

Nolte KB and Dasgupta A (1996). Prevention of occupational cyanide exposure in autopsy prosectors. *J Forens Sci*, **41**, 146–148.

NTARC (2004). National terror alert: survival guide – facts about cyanide. National Terror Alert Resource Center (www.nationalterrorallert.com/readygude/cyanide.htm).

O'Flaherty EJ and Thomas WC (1982). The cardiotoxicity of hydrogen cyanide as a component of polymer hydrolysis smokes. *Toxicol Appl Pharmac*, **70**, 335–339.

Okoh PN (1983). Excretion of ^{14}C-labeled cyanide in rats exposed to chronic intake of potassium cyanide. *Toxicol Appl Pharmac*, **00**, 00–00.

Okoh PN and Pitt GAJ (1982). The metabolism in the rat of cyanide and gastrointestinal circulation of the resulting thiocyanate under conditions of chronic cyanide intake. *Can J Physiol Pharmacol*, **60**, 381–386.

Okolie NP and Osagie AU (1999). Liver and kidney lesions and associated enzyme changes induced in rabbits by chronic cyanide exposure. *Food Chem Toxicol*, **37**, 745–750.

Okolie NP and Osagie AU (2000). Differential effects of chronic cyanide intoxication on heart, lung and pancreatic tissues. *Food Chem Toxicol*, **38**, 543–548.

Okudera H, Morita H, Iwashita T *et al.* (1997). Unexpected nerve gas exposure in the city of Matsumoto: report of rescue activity in the first sarin gas terrorism. *Am J Emerg Med*, **15**, 527–528.

Olsen NS and Klein JR (1947). Effects of cyanide on the concentration of lactate and phosphates in brain. *J Biol Chem*, **167**, 739–746.

O'Neill S (2005). How high street poison plot ended in a bedsit bloodbath. *The Times (London)*, Thusday April 14, pp. 6–7.

Owais W, Janakat S and Hunaiti A (1985). Activation of sodium cyanide to a toxic but nonmutagenic metabolite by *Salmonella typhimurium*. *Mut Res*, **144**, 119–125.

Patel MN, Yim GKW. and Isom GE (1992). *N*-Methyl-*D*-aspartate receptors mediate cyanide-induced cytotoxicity in hippocampal cultures. *Neurotoxicology*, **14**, 35–40.

Paulet G (1955). On the relative importance of respiratory failure and cardiac failure in the lethal action of cyanides. *Arch Int Physiol Biochem*, **63**, 280–339.

Paulet G (1961). Nouvelles perspective dans le traitement de l'intoxication cyanhydrique. *Arch Mal Prof*, **22**, 120–127.

Peddy SB, Rigby MR and Shaffner DH (2006). Acute cyanide poisoning. *Pediatr Clin Care Med*, **7**, 79–82.

Peden NR, Taha R, McSorley PD *et al.* (1986). Industrial exposure to hydrogen cyanide: implications for treatment. *Br Med J*, **293**, 538.

Persson S-Å, Cassel G and Sellström, Å (1985). Acute cyanide intoxication and central transmitter systems. *Fund Appl Toxicol*, **5**, S150–S159.

Pettersen JC and Cohen SD (1985). Antagonism of cyanide poisoning by chlorpromazine and sodium thiosulfate. *Toxicol Appl Pharmac*, **81**, 265–273.

Pettersen JC and Cohen SD (1986). Comparison of the effects of cyanide (CN) on cytochrome oxidase (cyt ox) activity and respiration in mouse brain mitochondria (mito). *Toxicologist*, **6**, 199.

Pettersen JC and Cohen SD (1993). The effect of cyanide on brain mitochondrial cytochrome oxidase and respiratory activities. *J Appl Toxicol*, **13**, 9–14.

Philbrick DJ, Hopkins JB, Hil DC *et al.* (1979). Effects of prolonged cyanide and thiocyanate feeding in rats. *J Toxicol Environ Health*, **5**, 579–592.

Piantadosi CA and Silvia AL (1986). Cerebral cytochrome a, a_3 inhibition by cyanide in bloodless rats. *Toxicology*, **33**, 67–79.

Piantadosi CA, Sylvia DL and Jöbsis FF (1983). Cyanide-induced cytochrome a, a_3 oxidation-reduction responses in rat brain *in vivo*. *J Clin Invest*, **72**, 1224–1233.

Pines KL and Crymble MM (1952). *In vitro* conversion of thiocyanate to cyanide in the presence of erythrocytes. *Proc Soc Exp Biol Med*, **81**, 160–163.

Pitt BR, Radford EP, Gurtner GL *et al.* (1979). Interaction of carbon monoxide and cyanide on cerebral circulation and metabolism. *Arch Environ Health*, **34**, 354–359.

Pontal PG, Bismuth C, Garnier R *et al.* (1982). Therapeutic attitudes in cyanide poisoning. Retrospective study of 34 non-lethal cases. *Vet Human Toxicol*, **24** (**Supplement**), 90–93.

Ponte J and Sadler CI (1989). Studies on the regenerated carotid sinus nerve of the rabbit. *J Physiol London*, **410**, 411–424.

Porter DJT, Voet JG and Bright HT (1972). Reduction of D-amino oxidase by β-chloraniline: enhancement of the reduction rate by cyanide. *Biochim Biophys Res Commun*, **49**, 257–263.

Poulton JE (1988). Localization and catabolism of cyanogenic glycosides. In: *Cyanide Compounds in Biology*, Ciba Foundation Symposium Number 40, pp. 67–91. Chichester, UK: John Wiley & Sons, Ltd.

Prabhakaran K, Li L, Borowitz JL *et al.* (2003). Cyanide induces different modes of death in cortical and mesencephalic cells. *J Pharmacol Exper Therap*, **303**, 510–519.

Prentiss AM (1937). *Chemicals in War*. New York, NY, USA: McGraw-Hill.

Pronczuk de Garbino J and Bismuth C (1981). Propositions thérapeutiques actuelles en cas d'intoxification par les cyanures. *Toxicol Eur Res*, **III** (2), 690–756.

Purser DA, Grimshaw P and Berrill KP (1984). Intoxication by cyanide in fires: a study in monkeys using polyacrylonitrile. *Arch Environ Health*, **39**, 394–400.

Radojicic B (1973). Determining thiocyanate in urine of workers exposed to cyanide. *Arch Hig Rada Toksikol*, **24**, 227–232.

RAMC (2002). Cyanogen agents. *J R Army Med Corps*, **148**, 383–386.

Reitveld EC, Delbressine LPC, Waegenmaekers THJM *et al.* (1983). 2-Chlorobenzylmercapturic acid, a metabolite of the riot control agent 2-chlorobenzylidene malononitrile (CS) in the rat. *Arch Toxicol*, **5**, 139–144.

Rella J, Marcus S and Wagner BJ (2003). Rapid cyanide detection using the Cyantesmo® kit. *Clin Toxicol*, **41**, 736.

Rella J, Marcus S and Wagner BJ (2004). Rapid cyanide detection using the Cyantesmo® kit. *Clin Toxicol*, **42**, 897–900.

Rickwood D, Wilson MT and Darley-Usmar VM (1987). Isolation and characteristics of intact mitochondria. In: *Mitochondria: A Practical Approach* (VM Darley-Usmar, D Rickwood and MT Wilson, eds), pp. 1–16. Oxford, UK: IRL Press.

Robinson CP, Baskin SI and Franz DR (1984). The effects of cyanide on isolated aorta strips from the dog, rabbit and ferret. *Toxicologist*, **4**, 103.

Robinson CP, Baskin SI, Visnich N Jr *et al.* (1985a). The effect of cyanide and its interactions with norepinephrine on isolated aorta strips from the rabbit, dog and ferret. *Toxicology*, **35**, 59–72.

Robinson CP, Baskin SI and Franz DR (1985b). The mechanisms of action of cyanide on the rabbit aorta. *J Appl Toxicol*, **5**, 372–377.

Rosenberg NL, Myers JA and Martin WRW (1989). Cyanide-induced parkinsonism: clinical, MRI and 6-fluorodopa/PET studies. *Neurology*, **39**, 142–144.

Rotenberg JS (2003). Cyanide as a weapon of terror. *Pediat Ann*, **32**, 236–240.

Rousset S, Alves-Guerrra M-C, Mozo J *et al.* (2004). The biology of mitochondrial uncoupling proteins. *Diabetes*, **15**, S130–S135.

Russek M, Fernandez A and Vegas C (1963). Increase in cerebral blood flow produced by low dosages of cyanide. *Am J Physiol*, **204**, 309.

Rutkowski JV, Roebuck RJ and Smith RP (1985). Effect of protein free diet and food deprivation on hepatic rhodanese activity, serum proteins and acute cyanide lethality in mice. *J Nutrit*, **115**, 132–136.

Rutkowski JV, Roebuck RJ and Smith RP (1986). Liver damage does not increase the sensitivity of mice to cyanide given acutely. *Toxicology*, **38**, 305–314.

Ruzo LO, Unai T and Casda IE (1978). *J Agr Food Chem*, **26**, 916–925.

Saia B, DeRosa E and Penney DG (1970). Remarks on the chronic poisoning from cyanide. *Med Lav*, **61**, 580–586.

Salkowski AA and Penney DG (1994). Cyanide poisoning in animals and humans: a review. *Vet Human Toxicol*, **36**, 455–466.

Salkowski AA and Penney DG (1995). Metabolic, cardiovascular and neurologic aspects of acute cyanide poisoning in the rat. *Toxicol Lett*, **75**, 1–27.

Sandberg CG (1967). A case of chronic poisoning with potassium cyanide? *Acta Med Scand*, **181**, 233–236.

Sauter SW and Keim ME (2001). Hydroxocobalamin: improved public health readiness for cyanide disaster. *Ann Emerg Med*, **37**, 635–641.

Sayek I (1970). The incidence of the inability to smell solutions of potassium cyanide in the rural health center of Ortabereket. *Turk J Pediatr*, **12**, 72–75.

Schulz V (1984). Clinical pharmacokinetics of nitroprusside, cyanide, thiosulfate and thiocyanate. *Clin Pharmacokin*, **9**, 239–251.

Schulz V, Bonn R and Kindler J (1979). Kinetics of elimination of thiocyanate in 7 healthy subjects and 8 subjects with renal failure. *Klin Wochenschr*, **57**, 243–247.

Schulz V, Loffler A and Gheorghiu T (1983). The absorption of hydrocyanic acid from Linseed. *Leber Magen Dar*, **13**, 10–14.

Seagrave S (1982). *Yellow Rain*, p. 150. London, UK: Abacus.

Sheehy M and Way JL (1968). Effect of oxygen on cyanide intoxication: III. Mithridate. *J Pharmacol Exp Therap*, **161**, 163–168.

Shou Y, Li L, Prabhakaran K *et al.* (2003). p45 Mitogen-activated protein kinase contributes to Bax translocation and cytochrome c release in cyanide-induced apoptosis. *Toxicol Sci*, **75**, 99–107.

Silver EH, Kuttab SH, Hassan T *et al.* (1982). Structural considerations in the metabolism of nitriles to cyanide in vivo. *Drug Metab Dispos*, **10**, 495.

Simon JD (1999). The emerging threat of chemical and biological terrorism. In: *General and Applied Toxicology*, Volume 3, Second Edition (B Ballantyne TC Marrs and T Syversen, eds), pp. 1722–1730. London, UK: MacMillan.

Singh JD (1982). The lethality and teratogenicity of potassium cyanide in the rat. *Teratology*, **25**, 84A.

Smith ADM, Duckett S and Waters AH (1963). Neuropathological changes in chronic cyanide intoxication. *Nature*, **200**, 179–181.

Solomonson LP (1982). Cyanide as a metabolic inhibitor. In: *Cyanide in Biology* (B Vennesland, EE Conn, CJ Knowles *et al.*, eds), pp. 11–28. New York, NY, USA: Academic Press.

Sörbo B (1975). Thiosulphate sulfurtransferase and mercaptopyruvate sulfurtransferase. In: *Metabolic Pathways*, Volume 7, *Metabolism of Sulfur Compounds* (DM Greenberg, ed.), pp. 433–456. New York, NY, USA: Academic Press.

Sotto-Blanco B, Sousa B, Manzano JL *et al.* (2001). Does prolonged cyanide exposure have a diabetogenic effect? *Vet Human Toxicol*, **43**, 106–108.

Sousa AB, Soto-Blanco B, Guerra JL *et al.* (2002). Does prolonged oral exposure to cyanide promote hepatotoxicity and nephrotoxicity? *Toxicology*, **174**, 87–95.

Steffens W (2003). Percutaneous hydrocyanic acid poisoning. *Clin Toxicol*, **41**, 483.

Sternbach G (1992). Editor's comment. *J Emerg Med*, **10**, 623–624.

Suskind R (2006). *The One Percent Doctrine*, pp. 194–198. New York, NY, USA: Simon and Shuster.

Susuki T (1968). Ultrastructural changes of heart muscle in cyanide poisoning. *Tohoku J Exp Med*, **95**, 271–287.

Tanberger G (1970). The stimulating effect of cyanide on central nervous system, vagus and sympathetic activities and its importance on circulation. *Arch Int Pharmacodyn*, **187**, 61–66.

Tendler S and O'Neill S (2005). The al-Qaeda plot to poison Britain. *The Times (London)*, Thursday April 14, p 1.

Tewe OO and Maner JH (1981a). Long-term and carryover effect of dietary inorganic cyanide (KCN) in the life cycle performance of and metabolism of rats. *Toxicol Appl Pharmac*, **58**, 1–7.

Tewe OO and Maner JH (1981b). Performance and pathophysiological changes in pregnant pigs fed cassava diets containing different levels of cyanide. *Res Vet Sci*, **30**, 147–151.

Thompson JP (2004). The use of antidotes in the emergency treatment of cyanide poisoning. *Clin Toxicol*, **42**, 411.

Thompson SJ and Bayer MJ (1983). Access to cyanide antidote kits. *Ann Emerg Med*, **12**, 515–516.

Thompson RL, Manders WW and Cowan WR (1987). Postmortem findings of the victims of the Jonestown tragedy. *J Forens Sci*, **32**, 433–443.

Tomaszewski CA and Thom SR (1994). Use of hyperbaric oxygen in toxicology. *Emerg Med Clin N Am*, **12**, 437–459.

Tursky T and Sajter V (1962). The influence of potassium cyanide poisoningon the γ-aminobutyric acid level in rat brain. *J Neurochem*, **9**, 519–523.

Tyrer FH (1981). Treatment of cyanide poisoning. *J Soc Occup Med*, **31**, 65–66.

Uhl W, Nolting A, Golor G *et al.* (2006). Safety of hydroxocobalamin in healthy volunteers in a randomized placebo controlled study. *Clin Toxicol*, **44**, 17–28.

Utti RJ, Rajput AH, Ashenhurst EM *et al.* (1985). Cyanide-induced parkinsonism: a clinicopathologic report. *Neurology*, **35**, 921–925.

Van der Heyden G, Vereecka J and Carmliet E (1985). The effect of cyanide on the K-current in guinea pig ventricular myocytes. *Basic Res Cardiol*, **80** (Supplement 1), 3–96.

Van Buuren KJH, Nicholls P and Van Geler PF (1972). Biochemical and biophysical studies on cytochrome a,a_3. VI. Reaction of cyanide with oxidized and reduced enzyme. *Biochim Biophys Acta*, **256**, 257–276.

Varnell RM, Stimac GK and Fligner CL (1987). Ct diagnosis of toxic brain injury in cyanide poisoning: considerations for forensic medicine. *Am J Neuroradiol*, **8**, 1063–1066.

Vennesland B, Castric PA, Conn PE *et al.* (1982). Cyanide metabolism. *Fed Proc*, **41**, 263–264.

Verna A, Roumy M and Leitner LM (1975). Loss of chemoreceptive properties of the rabbit carotid body after destruction of the glomus cells. *Brain Res*, **100**, 13–23.

Vesey CJ (1976). Letter to the Editor. *Clin Toxicol*, **14**, 307–309.

Vesey CJ and Wilson J (1978). Red cell cyanide. *J Pharmac Pharmacol*, **30**, 20–26.

Vesey CJ, Cole PV and Simpson PJ (1976). Cyanide and thiocyanate concentrations following sodium nitroprusside infusion in man. *Br J Anaesthesiol*, **48**, 651–660.

Vick JA and Froelich HL (1985). Studies on cyanide poisoning. *Arch Int Pharmacody*, **273**, 314–322.

Watts J (1998). Japanese fear over festival cyanide poisoning. *Lancet*, **352**, 379.

Way JL (1984). Mechanism of cyanide intoxication and its antagonism. *Ann Rev Pharmacol Toxicol*, **24**, 451–481.

Way JL and Burrows GE (1976). Cyanide intoxication: protection with chlorpromazine. *Toxicol Appl Pharmacol*, **36**, 1–5.

Weger NP (1983). Treatment of cyanide poisoning with 4-dimethylaminophenol: experimental and clinical overview. *Fund Appl Toxicol*, **3**, 387–396.

Westley J (1981). Cyanide and sulfane sulfur. In: *Cyanide in Biology* (B Vennesland, EE Conn, CJ Knowles *et al.*, eds), pp. 61–76. London, UK: Academic Press.

Westley J, Adler H, Westley I *et al.* (1983). The sulfurtransferases. *Fund Appl Toxicol*, **3**, 377–382.

Wexler J, Whittenberger JL and Umke PR (1947). The effect of cyanide on the electrocardiogram of man. *Am Heart J*, **34**, 163–173.

Wisler JA, Dulaney MD, Pellicore LS *et al.* (1991). Transport of cyanide into guinea pig cardiac mitochondria. *Toxicol Lett*, **56**, 275–281.

Wood JL and Cooley SL (1956). Detoxification of cyanide by cystine. *J Biol Chem*, **21**, 449–457.

Worksafe Australia (1989). *Cyanide Poisoning*, National Occupational Health and Safety Commission, Publication Number WAP 89/032. Canberra, Australia: Australian Government Publishing Service.

Wróbel M, Jurkowska H, Śliwa L *et al.* (2004). Sulfurtransferases and cyanide detoxification in mouse liver, kidney and brain. *Toxicol Mech Methods*, **14**, 331–337.

Yamamoto H (1989). Hyperammonemia, increased brain neutral and aromatic amino acid levels, and encephalopathy induced by cyanide in mice. *Toxicol Appl Pharmac*, **9**, 415–420.

Yamamoto H (1990). Protection against cyanide-induced convulsions with α-ketoglutarate. *Toxicology*, **61**, 221–228.

Yamamoto H (1992). A possible mechanism for the increase in brain tyrosine levels induced by cyanide in mice. *Food Chem Toxicol*, **11**, 973–977.

Yamamoto H (1993a). Protective effect of calmodulin inhibitors against acute cyanide-induced lethality and convulsions in mice. *Toxicol Lett*, **66**, 73–79.

Yamamoto H (1993b). Relationship among cyanide-induced encephalopathy, blood ammonia levels and brain aromatic amino acid levels in rats. *Bull Environ Contamin Toxicol*, **50**, 249–281.

Yamamoto H (1995). A hypothesis for cyanide induced and toxic seizures with supporting evidence. *Toxicology*, **5**, 19–26.

Yamamoto H and Tang H-W (1996a). Effect of carbetapentane or melatonin on cyanide-induced neurotoxicity in mice. *Jpn J Toxicol Environ Health*, **6**, 488–491.

Yamamoto H and Tang H-W (1996b). Antagonistic effect of melatonin against cyanide-induced seizures and acute lethality in mice. *Toxicol Lett*, **7**, 19–24.

Yamamoto H and Tang HW (1998). Effects of 2-amino-7-phosphonoheptanoic acid, melatonin or N^G-nitro-L-arginine on cyanide or N-methyl-D-aspartate-induced neurotoxicity in rat cortical ells. *Toxicol Lett*, **94**, 13–18.

Yamamoto K and Yamamoto Y (1977). The blood acid–base changes in acute cyanide poisoning in the rat in comparison with those in acute anoxic anoxia. With special references to the relation with time to death. *Z Rechtsmed*, **79**, 125–135.

Yen D, Tsai ZJ and Wang M-L (1995). The clinical experience of acute cyanide poisoning. *Am J Emerg Med*, **13**, 524–528.

Yonetani T and Ray GS (1965). Studies on cytochrome oxidase. VI. Kinetics of the aerobic oxidation of ferro-cytochrome by cytochrome oxidase. *J Biol Chem*, **24**, 634–638.

Zanetti G, Galante YM, Arosio P *et al.* (1973). Interactions of succinate dehydrogenase with cyanide. *Biochim Biophys Acta*, **321**, 41–53.

26 RIOT CONTROL AGENTS IN MILITARY OPERATIONS, CIVIL DISTURBANCE CONTROL AND POTENTIAL TERRORIST ACTIVITIES, WITH PARTICULAR REFERENCE TO PERIPHERAL CHEMOSENSORY IRRITANTS

Bryan Ballantyne

Charleston, WV, USA

BACKGROUND

Historical aspects and recent uses

The human race has used chemicals, often dispersed as smokes or vapors, from recordable time for protection of themselves and for offensive purposes against other humans (SIPRI, 1971). The earliest recorded use of chemicals was probably of a burning mixture of pitch and sulphur (Greek fire), which was thrown at enemies to discourage them by virtue of its irritant effect on the eyes, respiratory tract and possibly skin. Janos Hunyadi used arsenical smokes to defend Belgrade against the Turks in 1456, and soldiers of the Bishop of Münster also used arsenical projectiles in 1672 as a siege weapon against Gronigen in the Netherlands. Plutach is said to have described an action by a Roman general in Spain during which a cloud of irritant smoke was used to drive the enemy out of concealment in caves (Ballantyne and Salem, 2004). More recently, in the late 20th Century, the USA was accused of this use of irritant smokes for 'tunnel denial' in the Vietnam War, i.e. to drive the enemy out of caves or tunnels, or to prevent the use of such spaces for concealment purposes. This led to strenuous political denials by the USA that this use of 'tear gas' is chemical warfare in the context of the Geneva Convention. The use of 2-chlorobenzylidene malononitrile (CS) for tunnel denial purposes by the USA in Vietnam has been documented (Ditter and Heal, 2004; Lewer and Schofield, 1997).

In contrast to the above, mainly military examples of the use of chemicals for subduing and harassing enemy soldiers, various chemicals have been used by security forces against civilian populations for peacekeeping operations in confrontational situations. This employment of chemicals has become an obvious feature during the 20th and 21st Centuries. Notable events in the recent history of the use of chemicals for peacekeeping operations are as follows. In 1912, the Parisian police used ethylbromoacetate (EBA) grenades to transiently disable lawless gangs based on the irritating effects of its smoke (Swearengen, 1966), and in the late 1920s 1-chloroacetophenone (CN) was used in French colonies to break up civil disorders. A rapid

Chemical Warfare Agents: Toxicology and Treatment (2nd Edition)
Edited by Timothy C. Marrs, Robert L. Maynard and Frederick R. Sidell

proliferation in the use of chemicals in peacekeeping operations occurred particularly in the USA during the mid-1920s as a consequence of the marked increase in both the crime rate and in urban gangster warfare. This resulted in the manufacture and marketing of anti-riot guns capable of firing tear gas projectiles into rioting crowds, or into buildings in order to evict armed besieged criminals. At about the same time, small tear gas pen guns became available on the market in the USA for general self-protection purposes. Later, around the mid-1950s, an increase in crimes of violence resulted in a proliferation in the sale of tear gas pen guns and the development for general sale of hand-held liquid irritant spray devices. However, legislation in many countries now prohibits their sale to and use by members of the general population, presumably to the advantage of the felon.

Since the 1950s the use of chemicals, mainly of an irritant nature, as an aid to restore law and order in demonstrations and riots has increased markedly on a worldwide basis, and such chemicals with appropriate delivery devices are now standard in the armamentarium of police and security forces. Their use, seen frequently on television news programs, sometimes appears to be what may be regarded as excessive and uncontrolled. For example, more than 5000 riot control canisters were used against demonstrators at the Summit of the Americas in Quebec City during April 2001 (Wier, 2001), and during June 1987 anti-political demonstrations in South Korea caused the government to use 351 200 tear gas canisters and grenades against civilian demonstrators (Hu *et al.*, 1989). This was against a lack of knowledge by the Korean medical community, mainly because the government withheld information on the nature of the agent(s) used, and did not issue any guidelines to the public or the health authorities (Hu *et al.*, 1989). The widespread use of 2-chlorobenylidene malononitrile (CS) in 1969 by the Royal Ulster Constabulary during sectarian riots in Northern Ireland led to widespread publicity and a public outcry. As a consequence the United Kingdom Secretary of State for the Home Office appointed an independent Committee of Enquiry to investigate this use of CS, particularly on the 13th and 14th August, 1969. A major conclusion of this Committee of Enquiry was that in spite of the extreme discomfort resulting from exposure to CS-containing smokes, it would be only under extreme and exceptional circumstances that doses could be received that might be injurious to health or cause death. However, they made a recommendation that when chemical agents are used for civil peacekeeping purposes, the effects of such agents should be studied more akin to that of a new drug rather than of a weapon (HMSO, 1969,1971).

Circumstances and politics

The circumstances of peacekeeping operations are variable with respect to the numbers of persons involved, the cause for the disturbance and the activities by demonstrators and by security forces at the site of the disturbance. As examples of the extremes of such disturbances are criminal assault by one member of the public on another, a prisoner or suspect individual resisting arrest, to large-scale riots by protestors/demonstrators and involving, by definition, the use of potentially physically damaging or injurious activities. Peacekeeping operations against members of a civilian population, usually as a protesting group, have been defined as planned and co-ordinated procedures to control civil disturbance involving, or between, members of a civilian population and where the level of physical violence could result in the destruction of private and/or public property and also where there is a risk of injury, or even death, to persons actively engaged in the disturbance or who, by a chance event, are present in the geographical area of the disturbance (Ballantyne and Salem, 2004). If many individuals are involved in such a disturbance, then this constitutes what is popularly defined as a riot. At the other end of the scale where only a few individuals, or a single malefactor, are involved then this usually falls under a description of self-protection for those involved in trying to control the disturbance or overpower a potentially dangerous individual. Although, when used for self-protection or riot control purposes, it is the planned desire to employ chemicals of a low health hazard potential and delivery methods that carry the minimum potential for injury, their use has been occasionally associated with instances of injury or even death. The nature of, and causes for these are discussed in detail below. Substances of a similar chemical nature are used for both the control of individuals (self-protection) and for riot control purposes, and for this reason they have often been loosely referred to as riot control agents (RCAs). It is usually the case in Westernized societies and it is the aim of the security forces, particularly in planned operations, to restore order with the minimum of physical violence using methods least likely to cause injury or to escalate the situation. Relatively recently it has been a major objective of several governments only to approve chemicals that are used in the context of civil disturbances and civilian disturbance which have the following characteristics: (a) rapid incapacitating effects against even the most motivated of malefactors, (b) are easy to disseminate and subsequently to decontaminate, (c) have a long shelf life, (d) are of low cost, and (e) do not produce significant short- or long-term adverse health effects when used against a heterogeneous population, which could include the young, elderly or otherwise infirmed (Maynard, 1999). Similar considerations apply to the equipments that are used to disseminate chemicals used in peacekeeping operations. Although there have been improvements in the safety characteristics of chemicals used and their delivery systems, there are still instances of physical injury from delivery systems and of direct or indirect adverse effects from the substances used. The majority of countries currently, but usually only to a degree, draw a distinction between agents that are used to control disturbances of the peace in the context of situations involving a civilian population, and agents that are intended for use in chemical warfare military operations. However, certain chemical agents that are used by the military for offensive purposes are generally classified, or at least referred to as RCAs; for example, the irritant chemicals employed by the military for tunnel denial are also those that have been used in the context of a civil disturbance. This overlap with specific chemicals for different functional uses has lead to development and expression of political opinions that amount to a cover-up in their ambiguity. Thus, the official view of the USA administration is that they do not consider RCAs to be chemical weapons as defined in the Geneva Convention of 1925, but there is still some lack of clarification and definition in this respect (NRC, 1999; Sidell, 1997; Takafuji and Kok, 1997; White House, 1994).

Civil disturbances differ in their magnitude, frequency, preplanning, organization, the cause for the disturbance, geographical localization, influence of extremists, confrontation with antagonistic groups (opposing the cause for the demonstration) and the threat to life and property. Experience over the past few decades with civil disturbances in which security forces have been used to control such situations by methods which include physical force and the use of chemicals has indicated that these approaches may result in some short-lived health effects, very occasionally to more severe injuries, and rarely, to deaths (Ballantyne, 1977a, 2006). The causes for these outcomes, which have occurred in active participants and bystanders, are many and variable, and some may be associated with factors such as participant age or state of health. Therefore, currently and particularly in Western (generally non-repressed) societies, these outcomes can lead to wide publicity, including immediate media coverage, accusations of excessive and unnecessary force by security authorities, claims for injury, litigation and subsequent public discussions and, rarely, official biased or non-biased enquiries. These post-event factors and the forensic implications of the use of riot control procedures have been discussed in detail elsewhere (Ballantyne and Salem, 2004).

Production of equipments and of chemicals for use in the control of civil disturbances is now a thriving business, and many of the involved companies have been contracted by law-enforcement agencies. This has the considerable disadvantage that such commercial firms, some of them small, have limited research resources and facilities for investigating the potential adverse health effects of chemicals and the injuring capability of physical equipments. Indeed, in spite of the recommendations of the UK Home Office Committee of Enquiry noted above, there are few countries that have any official regulations governing the health testing, manufacturing specifications and recommended safe use of riot control equipment and chemicals (Ballantyne, 2006). Personal attempts to obtain Material Safety Data Sheets (MSDSs) for riot control chemicals has yielded either no or pathetically inadequate responses. From time to time, individual law-enforcement agencies, or a few government departments, may review the employment of RCAs and issue recommendations and guidelines for their use, but these are not frequent, active, productive, nor necessarily impartial documents. Truly independent review and advisory bodies appointed by manufacturers or appropriate government departments are few in number. As a consequence, most laboratory and clinical toxicology studies on health issues pertaining to RCAs are conducted mainly by armed services research establishments under government control, and usually as a part of their CW programs.

PHYSICAL PROCEDURES AND CHEMICAL AGENTS USED IN PEACEKEEPING OPERATIONS

The physical equipments and chemical agents used by law-enforcement agencies (security forces) for the control of small disturbances and of larger scale riots can be categorized as follows.

Physical measures for the remote incapacitation of individuals

Under this group of procedures are included those physical measures employed to incapacitate at a distance from the intended target. These basically involve the ballistic and, now, electrical immobilization of individuals. This group includes the use of beanbags or other projectiles, such as plastic or rubber bullets, fired from gun dischargers at the targeted individual. Clearly these are devices meant to incapacitate active rioters by physically causing pain and immobilization. Equally clearly is the fact that depending on the body region struck by these projected devices, there is a potential for soft tissue and bone injury, and indeed deaths have been associated with the resultant trauma in a few cases. Tasers involve the discharge of electrode needles aimed at the trunk of an individual. The electrodes remain attached to the discharge device by fine wires, which carry high voltage pulses that cause transient muscular spasms, weakness and incapacitation. Tasers are used widely, and in the USA even against very young children in questionable circumstances, and are stated to carry the minimum potential for adverse health effects, although it is difficult to conceive that those with advanced cardiac disease and arrhythmias are free from risk.

Peripheral chemosensory irritant (PCSI) chemicals

These materials currently form the major group of chemicals used to harass and incapacitate individuals in a civil disturbance, and form the major discussion component of this chapter. PCSI materials interact with sensory nerve receptors in the skin and exposed mucosal surfaces, producing local discomfort or pain at the site of contact, together with related local and systemic reflexes. The local reflexes include excess lacrimation and blepharospasm in the eye, and with inhaled materials there occurs excess nasal and tracheobronchial secretions, sneezing, coughing and changes in breathing rate (Ballantyne, 1999a,2005a). The sensations and local reflexes cause harassment and a desire to vacate the contaminated zone, and form the basis for their use in peacekeeping operations. In view of the fact that the peripheral sensory effects develop shortly after exposure and they generally last for a short period (several minutes), when used in peacekeeping operations PCSIs are sometimes referred to as 'harassing agents' or 'short-term incapacitants'. They are the most effective and frequently used chemicals for peacekeeping and form the bulk of the presentation in this chapter. Those which have been employed, or are potential PCSIs include 1-chloroacetophenone (CN), 2-chorobenzylidene malononitrile (CS), dibenz(b.f)-1,4-oxazepine (CR), oleoresin capsicum (OC) and pelargonic acid vanillylamide (PAVA).

In addition to their use by security forces for civilian peacekeeping operations, PCSIs are used in military situations for enemy troop denial operations (e.g. to flush opponents out of caves or tunnels), training purposes to simulate CW operations, and to test for effectiveness of respirator fit. There also exists the possibility that PCSI materials could be employed by terrorists to create panic situations in limited or enclosed spaces, or for distraction purposes in or near locations where more serious and damaging effects have been planned to occur. This could involve the use of smoke or aerosol generated material, or spraying solutions of PCSIs. The panic and disturbance that may be caused by the discharge of riot control PCSIs is illustrated by the following example. During February 2003, a security guard attempted to stop a fight on a dance floor in a Chicago nightclub by using an OC spray device. As a consequence, a stampede started with several hundred persons charging for the front door, resulting in the deaths of 23 persons (USA Today, 2003). A less dramatic example occurred in a crowded nightclub in Leicester, UK, where a CS aerosol canister was discharged and resulted in 23 persons having to attend a local hospital accident and emergency department/receiving room (Brekell and Bodiwala, 1998).

Obscuring cloud smoke agents

Although used principally in military operations (Ballantyne and Salem, 2006; Burton *et al*, 1982; Mishra, 1984), obscuring (screening) clouds can also be used in peacekeeping operations to cause distraction, disorientation, and obscure potential target objectives from the rioter. They are usually dispersed as pyrotechnically generated smokes from grenades or canisters. Some smokes used for military purpose are not appropriate for use in civil situations based on toxicity studies demonstrating potential adverse health effects, notably respiratory tract injury (Ballantyne and Salem, 2006). These include phosphorus pentoxide (Ballantyne, 1998), titanium tetrachloride (Ballantyne, 1982), zinc oxide/hexachloroethane (Marrs *et al.*, 1983a,1988) and zinc chloride (Evans, 1945; Johnson and Stonehill, 1961). Smokes of significantly lower acute and repeated exposure toxicity are needed for use in peacekeeping operations; these could include cinnamic acid (Ballantyne and Clifford, 1978; Marrs *et al.*, 1989), and dyes such as Disperse Red 9, Solvent Green 3 and Solvent Yellow 33, although high concentrations and/or mixtures of these dyes may result in pulmonary dye retention and foreign body reactions (Marrs *et al.*, 1984; Sun *et al.*, 1987).

Visible and occult person markers

These substances are used to contaminate malefactors for immediate discouragement and for subsequent identification purposes. They can be dispersed from hand-held canisters for one-on-one use, or from water cannon for large-scale antiriot use. One series of markers are visible dyestuffs resulting in immediate staining of clothing and skin for recognition and/or identification. They are used in solution and delivered by liquid projection devices. These chemicals clearly need to be chosen on the basis of their ability to stain skin and clothing but also on lack of local toxicity to skin and eye, notably irritancy. For example, while Gentian violet (methylrosaniline chloride) produces immediate, intense, and prolonged staining of skin, it also causes severe and persistent eye injury (Ballantyne *et al.*, 1973a). In addition, they should not present environmental problems. Occult markers are colorless substances that will fluoresce under ultraviolet light, and thus used for identification purposes.

Malodorous substances

Contamination of persons with malodorous substances, such as amines or mercaptans, has been suggested as a means of deterring the less motivated rioters because of psychological and physiological effects such as olfactory repulsion and nausea (Witten *et al.*, 1970). They can be delivered by the use of frangible missiles, and addition of thickening agents will prolong adhesiveness. Clearly they should be devoid, particularly, of local adverse effects, not be detrimental to the environment and be capable of subsequent easy decontamination.

Low-friction polymers

It has been suggested, perhaps thoughtlessly, that dispersion of slippery viscous agents in the direction of rioters would impair their co-ordinated movements and make it difficult for them to undertake malicious tasks. However, not only would there be the likelihood for uncontrolled physical accidents, but also problems in the control of motorized vehicles in the area, including those of the security forces. Additionally, in public areas, there would be a need for prompt street decontaminations by environmentally acceptable methods.

Centrally acting neuropharmacological agents

There have been suggestions that drug-injecting dart guns could be used to systemically deliver pharmacological agents that could induce, for example, tranquilization or emesis. (Conner, 1967; Dean-Drummond, 1975; Security Planning Corporation, 1972; Swearengen, 1966). Clearly, this would necessitate security forces having additional skilled training in the use of such weapons to avoid serious penetration injures, and could result in litigation and possibly open political arguments. Agents having central neuropharmacological activities have been used more frequently in hostage situations, although not without cost to the hostages, For example, Russian security forces used a narcotic fentanyl analog (remifentanil) against Chechen rebels who held about 800 people hostage in a Moscow theatre during late October 2002. Pumping the vapor into the theatre for about 30 min resulted in the deaths of 119 hostages at the scene of the incident, and over 200 required hospital admission. To their disgrace, neither the security forces nor the responsible government officials informed the local hospitals and medical personnel of the nature of the agent used and it was some 5 days after the incident that the authorities openly declared the substance that was used (Ballantyne and Salem, 2004; Salem *et al*, 2005a). However, there is still some uncertainty about the conditions of use of the fentanyl – some have suggested that a mixture of fentanyl and halothane was used, and others believe that massive doses of carfentanil (a potent opioid) were used to saturate the theatre so that a maximum effect by inhalation could be achieved (Salem *et al.*, 2005a).

METHODS FOR THE GENERATION AND DISPERSAL OF RIOT CONTROL AGENTS

General considerations

Smokes used for military screening and training purposes and for civilian screening and deterrent purposes are usually generated pyrotechnically from grenades. Marking agents used by security forces against civilian targets are dispersed in solution by small- or large-scale dispensers, and sometimes mixed with PCSI materials. Depending on the operational circumstances, PCSI materials are used either as airborne dispersions or projected as solutions in the form of coherent liquid jets. Airborne PCSI materials may be widely disseminated and can result in many individuals being non-selectively affected, depending on how local meteorological conditions affect drift and persistence. Airborne dispersions are usually generated pyrotechnically from grenades as smokes, or else as aerosol droplets, mists or vapor generated from pressurized systems containing PCSIs in a volatile solvent. They may also be dispersed in the atmosphere as powder clouds from fogging devices. The use of the word 'gas' to describe airborne PCSIs (e.g. CS gas or tear gas) is technically incorrect, since a gas is a totally airborne dispersion of molecules and exists as such at NTP (STP). Solutions of PCSI materials projected as coherent fluid streams can be used to more selectively contaminate specific individuals or groups of persons. Depending on the volume of fluid to be dispersed, projection devices can range from small hand-held pressurized canisters for use against single or small numbers of individuals to water cannons to engage larger groups or crowds. Examples of specific methods for the generation and dispersion of RCAs are briefly discussed below. Detailed accounts have been written by Ditter and Heal (2004) and Swearengen (1966).

Aerosolization by thermal volatilization ('smoke generation')

This is the most frequently used method for generation of screening smokes for military operations and for the dispersion of PCSI materials for military situations and for the control of civilian crowds by security forces. It involves the pyrotechnic generation of smokes from chemicals that are generally thermostable. The active material is mixed with a base such as chlorate or lactate which, on ignition, causes volatilization of the intended reactive chemical which subsequently condenses into a cloud of solid or supercooled liquid droplets, usually in the respirable range (0.5–2.0 μm MMAD). In a short period the suspended particles or droplets settle or aggregate into larger agglomerations. On a calm day, a standard explosive pyrotechnic canister or grenade will generate a cloud, which is some 20–30 feet (*ca* 6–9 m) in diameter. The burning grenade has an igniting fuse that initiates after a delay of 1–5 s, after which the filler is ignited and pressure builds in the grenade body until sufficient to open gas ports and expel the pyrotechnic mix. Burning grenades are thrown or propelled by a launcher. The technology of pyrotechnic devices has been discussed by Swearengen (1966). A variant on the standard smoke generator, which produces a source of smoke which spreads from the unifocal generation system, is a grenade which on ignition disperses (projects) several burning sub-units over a large area, and thus produces a large cloud almost simultaneously from the multiple sub-units over a wide area, but with a reduced emission from each focal sub-unit. One example of such a device is the rubber bursting grenade, which consists of a rubber cylindrical casing packed with small pellets composed of PCSI chemicals mixed with a pyrotechnic composition. The internal pressure resulting from ignition of the mixture causes the rubber casing to burst and thus scatters slow burning pellets over a wide area, producing a multifocal source of smoke (Ballantyne and Johnston, 1974a). Cartridges can be fired from a shotgun, a rifle-mounted launcher or a gas/air gun. They allow for greater accuracy and distance of delivery than hand-thrown grenades, but many are high-velocity projectiles and can cause serious body injury or lethality if fired at too close a range. The contents may be pyrotechnic mix, non-pyrotechnic powder, or liquid. Aircraft or helicopter-mounted systems consist of a compressed air tank, agent tanks and hose and nozzle assemblies. Man-portable large dispensers use a compressed air tank, agent tanks and a hose and nozzle. Some systems dispense PCSI solutions, others PCSI powders.

Generation of powder clouds

PCSI materials can be blown into the atmosphere in fine form from fogging machines. This mode of dissemination avoids the possibility of toxic effects from inhaled combustion products which are a feature of grenade-generated smoke. For large-scale applications, and using a mobile fogger, PCSIs can be mixed with a material such as Neocil to improve flow properties. For example, hydrophobic anti-agglomerative powder CS formulations (CS1 and CS2) have been produced for use in fogging machines. CS1 is micronized powder containing 5% hydrophobic silica aerogel, which persists for about 2 weeks; CS2 is a siliconized microencapsulated form of CS1 with improved flow properties and greater weather resistance, and composed of 95% micropulverized CS with silica treated with hexamethyldisilazane. Since they persist in the environment, these formulations are not used in civilian peacekeeping operations, but used to produce a persistent hazard in military operations (Blumenfeld and Meselson, 1971; Ministry of Defence, 1972). On a smaller scale, and mainly for self-defence purposes, 'tear gas pen guns' or 'tear gas pistols' are available in some countries. These devices contain a small explosive charge that propels and discharges a cloud of PCSI material. They have a relatively high injury potential, and have caused severe local skin and eye injuries. In the USA, such devices became available to the general pubic in the late 1920s (Rengstorff, 1969a). One complication of the use of pyrotechnically generated smokes is that they contain combustion products formed during the burning of the mixture. Some of the products of combustion may be detrimental to the health of the exposed individuals following inhalation. As part of a hazard evaluation of pyrotechnically generated smokes it is necessary to perform chemical analyses to determine the nature and amounts of the various combustion products that may be formed during high-temperature dispersion processes (Kluchinsky *et al.*, 2001,2002). In addition, investigation of the role of smoke from the burning mixture used to produce a pyrotechnically generated, as well as the inhalation toxicity of the pyrotechnically generated RCA itself, may be valuable in determining relative contributions to inhalation toxicity. This was investigated, for example with CR generated smokes. A comparison of the acute inhalation toxicity of pure CR aerosols with that of pyrotechnically generated CR smoke suggested that components from the combustion of the burning mixture may have made a significant contribution to mortalities. The acute lethal

Table 1. Comparison of acute lethal inhalation toxicity to four species of pyrotechnically (grenade) generated CR smoke and that of smoke resulting from the combustion of the burning mix alone[a]

Exposure time (min)	Grenade	CR Dosage (mg min m^{-3})	Mortality (No. dying/No. exposed)			
			Rabbit	Rat	Guinea pig	Mouse
60	Blank[b]	0	9/10	5/5	5/5	9/10
60	CR	148 000	4/10	2/5	1/5	3/10
60	CR	182 000	10/10	5/5	2/5	2/10
120	Blank	0	10/10	4/5	5/5	10/10
120	CR	325 000	—	10/10	10/10	17/20

Notes:
[a] Data after Ballantyne (1977b).
[b] Burning mix without CR.

toxicity of smoke from grenades containing burning mix was compared with those from burning mix plus CR. The results (Table 1) show that a high proportion of mortalities occurred with exposure to smoke from burning mix alone, and that this may be a factor in mortalities from exposure to pyrotechnically generated CR smoke (Ballantyne, 1977b).

Dispersion as vapor

Pyrotechnically generated smokes from grenades or powder clouds may not be acceptable from a safety point of view in certain circumstances; for example, where fires may be produced, or in confined spaces with low rates of ventilation where asphyxial effects could develop. In order to reduce such potential hazards, a variety of approaches have been investigated. One of these is the use of highly volatile PCSI materials, which can be projected into enclosed spaces, but no material of sufficiently low toxicity has yet been found which is acceptable. An alternative approach is the use of a PCSI substance dissolved in a volatile solvent, and the mixture projected into the area by means of a pressurized aerosol canister or a frangible missile. This results in the development of an airborne vapor or aerosol of irritant (Ballantyne, 1979).

Dispersion as liquid solutions

Solutions of marking agents and PCSI substances can be projected as coherent liquid jets, allowing the selective engagement of specific targets. Using such coherent jets on a small scale avoids, or limits, the possibility for physical, kinetic or thermal injuries. However, with larger-scale use such as water cannon, the potential for kinetic injuries is still present, as is the possibility for physical injury resulting from individuals slipping on the relatively large volumes of fluid projected. With water cannon, a PCSI substance can be dissolved in the water at a low concentration in order to cause skin and eye irritation following drenching, and this avoids some of the potential complications from inhaling the products of grenade-generated smokes of irritant materials (Ballantyne *et al.*, 1976a).

Hand-held irritant liquid projection devices of various types have been manufactured commercially for use by the police and other security forces; additionally, some devices have also been sold for self-protection purposes to the general public, although this practice is illegal in some countries, including the UK, and all civil airlines prohibit them being carried as hand luggage. The devices are designed to permit the user to direct a spray of irritant solution at the face of an assailant, causing prompt but temporary incapacitation from severe eye and facial skin effects. These effects may be used to render an individual susceptible to be overpowered for arrest purposes. In some cases, where a highly volatile solvent is used in the pressurized canisters, it is recommended that the volatile solution be discharged at the upper torso, so that irritant vapor is produced and causes harassment to the individual from skin, eye and respiratory tract sensory irritant effects. Typical handheld canisters range from 0.3 to 10 oz (*ca* 8.5–285 g) in contents. Maximum spray range for an aerosol is 3 to 10 m, and

Table 2. Examples of hand-held personal protection irritant liquid dispersion devices[a]

Common name	PCSI[b]	Solvent[b]	Propellant
MACE	CN	Kerosene hydrocarbons,	Freon® 113
		1,1,1-trichloroethane	—
Paralyzer®	CS	Dichloromethane	Nitrogen
PIS[c]	CS	Methyl *iso*butyl ketone	Nitrogen
SPAD[d]	CR	50% Aqueous PEG 300[e]	Nitrogen
CapStun®	OC	*iso*Propanol	*iso*Butane/propane
Punch II	OC	*iso*Propanol	*iso*Butane
Sabre 5.5®	OC	TS	Dimel® 134a/P
Sabre®	OC/CS	TS	Dymel® 134a/P
Guardian	OC	2-(2-Butoxyethoxy)ethanol	Tetrafluoroethane
CapTor®	PAVA	50% aqueous ethanol	Nitrogen

Notes:
[a] Data from Ballantyne and Salem, 2004; Fisher, 1970; Olajos and Stopford, 2004; Shreenivason and Boese, 1970; Stefee *et al*., 1995.
[b] PCSI, peripheral chemosensory irritant; CN, 1-chloroacetophenone; CS, *o*-chlorobenzylidene malononitrile; OC, oleoresin capsicum; PAVA, pelargonic acid vanillylaminde; TS, trade secret (proprietary).
[c] PIS, Personal incapacitant spray used by UK police forces.
[d] SPAD, Self protection aid device (developed by UK Defence Ministry).
[e] PEG, polyethylene glycol.

for direct fluid spray jets is 2 to 5 m. Some devices still employ highly volatile propellants such as propane and isobutene, although many current devices use nitrogen pressurization; this has the disadvantage of reduction in overall pressure, resulting in a progressively weaker stream with use. Some devices now use Dymel® (1,1,1,2-tetraflurorethane) as propellant. This has the advantages of being non-ozone depleting, maintains canister pressurization, compatibility, and not flammable (Olajos and Stopford (2004)). Bigger aerosol spray devices have been produced for use against groups of individuals in larger-scale peacekeeping operations. Some examples of irritant liquid projection devices are listed in Table 2.

The choice of solvent or solvent formulation is of considerable importance, not only for performance characteristics but also to avoid the occurrence of local and systemic toxicity to the recipient. Thus, the solvent or solvent system used should meet the following criteria:

(i) It should have physicochemical characteristics that will not result in any physical hazards during the handling and dispersion of the irritant formulation; e.g. it should not be too highly volatile or flammable such that it could present a fire or explosive hazard, or result in concentration of the irritant formulation and hence increased potential for health risks in exposed individuals. In addition, the formulation should have an acceptable analytical lifetime (shelf-life) with respect to maintaining the composition of the formulation without deleterious reaction products being formed.
(ii) The solvent should be chemically compatible with the active ingredient (marker and/or PCSI), and other possible additives, such that potentially harmful reaction products are not formed.
(iii) Solvents should not decrease the physiological potency of PCSIs.
(iv) The solvent and/or formulation should not have undesirable effects on the general environment; e.g. persistence, odor or environmental toxicity.
(v) The solvent should not have significant local and/or systemic toxicity, and therefore does not present *per se* undesirable health hazards to exposed individuals, or have potentiating or synergistic effects on the known toxicity of PCSIs and markers.

The latter factor is particularly important. While the toxicity of riot control chemicals is usually well documented, the possible influence of and relevance of assessing health hazards from formulation constituents may receive less detailed attention (Ballantyne and Salem, 2004; Gray, 2004; Holopainen *et al.*, 2003). It is therefore of clear importance that formulating agents, and particularly solvents, that may constitute a major proportion of the in-use formulation, should be of acceptably low local and systemic toxicity, and not potentiate or synergize any potential adverse effects that are known to be produced by a PCSI material. They should be chemically compatible with the active constituent and not cause the formation of reaction products that could be deleterious to health. The importance of the need for detailed consideration to be given to the possibility of adverse health effects from exposure or overexposure to formulation of RCAs is illustrated by documented case-reports of chemically induced injury from formulation constituents. For example, moderate eye injury and elevated intraocular pressure (IOP) are produced by dichloromethane (Ballantyne *et al.*, 1976b), and methyl *iso*butyl ketone (MIBK) causes an irritant dermatitis (Gray, 2004). It has been demonstrated experimentally that with an aerosol generator of CS in dichloromethane, both materials contribute to the local ocular toxicity following contamination of the eye (Ballantyne, 1979). Local cutaneous toxicity from the use of CS in MIBK was reported in individuals exposed to this formulation in police personal incapacitant sprays (BBC, 2004a; Euripidou *et al.*, 2004; Varma and Holt, 2001). The skin lesions described were erythematous dermatitis and blisters that persisted for longer than anticipated from CS exposure alone, and thus MIBK probably enhanced the local cutaneous effects. One UK police liquid irritant spray device contains 5% CS (Varma and Holt, 2001), a concentration known to produce keratitis when dissolved in the non-irritant PEG 300 (Ballantyne *et al.*, 1974); the corneal injury is likely to be more severe with MIBK as solvent. In addition to causing local ocular injury, dichloromethane may also result in systemic toxicity following inhalation of the vapor and its hepatic metabolism to carbon monoxide (Horowitz, 1986; Rioux and Myers, 1988,1989). Thus, a case of systemic carbon monoxide intoxication was reported for a 39-year-old female after she was exposed to solution from a personal defence spray containing CS in dichloromethane; her carboxyhaemoglobin (COHb) concentration was 20.4%, and she recovered after treatment with 100% normobaric oxygen (Duenas *et al.*, 2000). The same authors also described cases of symptomatic carbon monoxide intoxication in three boys (aged 4, 5 and 9 years) who were exposed to CS in dichloromethane accidentally discharged from a personal protection device; the solution formulation was probably 0.8% CS in 49% dichloromethane. The respective COHb concentrations were 16, 12 and 9%, and the patients were successfully treated with 100% normobaric oxygen.

The above discussion reinforces recommendations that investigations designed to assess potential adverse effects from RCAs should include detailed studies and considerations not only of the active RCA but also any formulation constituents and the total formulation itself (Ballantyne, 1977a,1987; Ballantyne and Salem, 2004).

BIOMEDICAL CRITERIA FOR EVALUATING POTENTIAL IN-USE HUMAN HEALTH HAZARDS OF RIOT CONTROL AGENTS

As noted above, the Committee of Enquiry into the use of CS in Northern Ireland recommended that chemical agents intended for use in the control of a civil disturbance should be assessed from the viewpoint more of a medicinal drug than that of a military weapon (HMSO, 1969,1971). However, in reaching this conclusion the Committee did make the distinction that with new drugs there is a need for a balanced risk–benefit analysis and these are purely professional medical considerations. However, with a chemical intended for use against humans for peacekeeping purposes, while many of the hazard and risk assessments are of a medical nature, additional other questions may arise which are concerned with social policy and there may also be a need for political considerations. From a medical point of view an assessment of the in-use safety of RCA agents, and their formulations, requires information derived from several sources and specialties. A comprehensive study of the toxicology of the material is needed to determine the potential for chemically induced adverse effects; such studies are conducted in the laboratory using *in vivo* and in *vitro* approaches. This should be conducted for assessment of the effects of the active RCA, the formulation constituents and the formulation. It is also necessary to ensure that the delivery system, used to disseminate the RCA formulation to its intended target or target area, does not introduce additional or modifying factors. If toxicology studies indicate that the RCA material and formulation(s) have an appropriate degree of safety (absence from potential adverse health effects), these studies may be followed by carefully conducted and controlled human volunteer studies. Such studies should have the approval of an appropriately qualified and independent review group. These studies should be initially limited to particular aspects of possible health concerns raised by the preceding toxicology and ancillary studies. They may then progress to trials having the combined objectives of efficacy of agents/delivery systems and 'safety-in-use'. These aspects related to safe use are considered in more detail as follows.

Toxicology studies

Toxicology studies are designed to determine the potential of a substance, or mixture of substances, to produce adverse biological effects (toxicity); in this context, adverse effects are defined as those that are detrimental to the normal functioning or survival of the organism (Ballantyne, 1984; Ballantyne *et al.*, 1999a). If adverse effects are produced it is necessary to know their nature, tissue(s) affected, dosage–effect relationship and incidence, mechanism by which they are produced, factors influencing their induction, and reversibility (spontaneous and induced). The design of a toxicology testing program should allow for coverage of the active RCA, its formulations and the fact that a heterogeneous population will be exposed. The results of toxicology studies should permit for decisions on operational restrictions on the chemical agent, its formulation and delivery system. Clearly for those who participate in a civil disturbance, exposure to RCA chemicals will generally be for a single or a few irregularly but closely spaced intervals, and most exposures will be brief and exposure dosages low, although some individuals (particularly if in confined spaces) may experience higher exposure doses. Thus, initial toxicology studies should involve acute and short-term repeated exposures by the intended route(s), primary irritancy (skin and eye) and sensitization. For those exposed more frequently, including production workers and security forces, studies of longer duration are required in order to determine if the more frequent exposures are associated with adverse effects not seen in acute or short-term studies, and to assist in developing protective and precautionary measures, including the establishment of workplace exposure guidelines; e.g. threshold limit values (TLVs) have been suggested for CS (ceiling value of 0.05 ppm) and for CN ([time-weighted average over 8 h] TWA_8, 0.05 ppm) (ACGIH, 2006). In the heterogeneous population encountered with a civil disturbance there will be differences in age, sex, reproductive status and state of health. Thus, there are testing requirements with respect to developmental, reproductive and genetic toxicology. The oncogenic potential of RCAs needs to be investigated with respect to repeated occupational exposure, and also because this is an issue that may receive comment by those having concern over the sociopolitical aspects of peacekeeping operations and by the investigative media. The following sections briefly review and discuss what is generically regarded as needed in a toxicology testing program for chemicals used in peacekeeping operations. However, each testing program needs to be developed individually for specific RCAs, taking into account the nature and intended use of the material, foreseeable misuse and how toxicity may be modified by formulation and some modes of dispersion. In addition, special studies may be necessary on a case-by-case basis. Details of the methodologies used for *in vivo* and *in vitro* toxicology testing are available in detail in several texts; e.g. see Anderson and Conning (1993), Ballantyne *et al.* (1999b), Hayes (2001) and Niesink *et al.* (1996).

ACUTE TOXICITY

There is a clear requirement for single dosing studies by relevant route(s) of exposure to determine lethal toxicity (LD_{50} or timed LC_{50}) and sublethal injuring potential with dose–response relationships, including no-effects levels. In addition to investigating the pure material it is wise to study the technical material to include any impurities that may be in the material used in munitions and also to determine the influence of variants on acute toxicity; e.g. solvents, additives and the influence of pyrotechnic decomposition products. The latter should include chemical analysis of atmospheres generated from the combustion process.

PRIMARY IRRITATION

Whether dispersed as aerosol, smoke or in solution, RCAs will contaminate the skin and eye and may cause local acute inflammation at these sites. It is therefore a necessity to be aware if contact with skin and eye will have a local tissue injuring potential, and how formulation may influence the outcome. Supplemental to standard eye irritation tests, it is of practical value to undertake studies on the influence of RCAs on corneal thickness by pachymetry and on intraocular pressure (IOP) by tonometry (Ballantyne, 1999b; Ballantyne *et al.*, 1977a).

SENSITIZATION

Since the skin and the respiratory tract are usual routes of exposure to RCAs, there is a need to determine, especially in respect of those recurrently exposed, the potential for the induction of immune-mediated cutaneous and/or respiratory tract sensitization. *In vivo* and *in vitro* approaches are available for such investigations (Kimber and Dearman, 1999).

REPEATED EXPOSURE STUDIES

From both operational and occupational perspectives there is a requirement to study the effects of RCAs initially by short-term repeated exposures in order to determine the potential for cumulative toxicity. The need for subchronic and combined chronic/oncogenicity studies depends on a broad spectrum of considerations with decisions made on a case-by-case basis (Ballantyne, 1999c). The wide range of factors includes exposure patterns, nature and biological reactivity of the RCA, genetic toxicology, suspect biological activity and sociopolitical demands.

DEVELOPMENTAL AND REPRODUCTIVE TOXICOLOGY

Since one element of the heterogeneous population exposure pattern is that this may include females of childbearing age, it is necessary that information should be available on the potential for embryotoxic, teratogenic and reproductive hazards. These may require special design considerations. For example, with PCSI RCA materials to be dispersed as airborne material, the design and interpretation of investigations needs to take into consideration the possibility of stress caused by both sensory irritation and aerosol exposure during gestation (Upshall, 1973,1977a).

GENETIC TOXICOLOGY

As a general guide to potentially serious biological hazards, biological reactivity, and to assist in determining the need for chronic toxicity/oncogenicity studies, it is necessary to conduct studies to define the mutagenic and clastogenic potential of RCAs.

METABOLISM AND TOXICOKINETICS

Such studies, conducted by the intended route of exposure, can be of considerable value in assessing potential hazards, defining systemic, organ and tissue doses, and permitting quantitative risk assessments. Metabolism studies may give in formation on the possible contribution of metabolites to the toxicity of the parent material. Investigation on the absorption, biodistribution and excretion of parent compound and its metabolites will assist in defining the potential for cumulative toxicity, possible target organs and tissues, and allow quantitative assessments of hazards. Such studies also assist in the design and subsequent interpretation of repeated exposure investigations.

ADDITIONAL, SPECIAL AND COMPOUND SPECIFIC STUDIES

Because of the variety of substances and of chemical structures that are encountered as RCAs, and due to the differing modes of dispersion, special studies may be required on a case-by-case basis for a given RCA, formulation or delivery mode. In some instances, these can be predicted as being necessary and incorporated into the toxicology testing program during the planning phase, but in other cases the needs may arise as a consequence of experience and/or analysis of in-use situations and incidents. Some representative examples are as follows, but for any given RCA and/or delivery system the need for additional or non-standard studies requires to be discussed on a unique case-by-case basis. RCA materials are used in conditions conducive to the development of open wounds and therefore the need arises to determine if contamination of a superficial wound with an RCA results in adverse effects on the normal cutaneous healing processes (Ballantyne and Johnson, 1974b). The CS rubber bursting grenade, mentioned above, used to produce a multifocal source of pyrotechnically generated irritant smoke may result in thermal injury from the direct wounding capability of the individual scattered burning pellets. This additional specific source of trauma was investigated in separate studies (Ballantyne and Johnston, 1974a). An example of the need for investigation of a particular possible health effect from experience and investigation of incidents was provided by the enquiry into the use of CS during August 1969, in the Northern Ireland riots. Following the riots, reports appeared in newspapers describing cases of diarrhoea, particularly in infants and young children (The Observer, 1969; The Times, 1969). The investigating Committee of Enquiry into the use of CS in these riots commented that that there were certain features of the outbreak, including time to onset, that made them hesitant to ascribe the diarrhoea as being due to the effects of exposure to CS smokes (HMSO, 1969). In addition, in controlled human volunteer exposures to CS aerosols, diarrhoea was described as occurring in only a very small proportion (ca. 1%) of subjects (Punte *et al.*, 1963), and was not described at all in another study by different investigators (Beswick *et al.*, 1972). A few complaints of a need to defecate were made by volunteer subjects who ate food grossly contaminated with CS (Kemp and Wilder, 1972). In order to experimentally clarify the possibility of an association between exposure to CS and the occurrence of diarrhoea, several species of laboratory mammals were given CS daily by gavage over five consecutive days. This study gave no evidence for an increase in wet stool production, and no indication for an irritant (inflammatory) reaction in the alimentary tract (Ballantyne and Beswick, 1972).

Human volunteer studies and trials

Human volunteer studies, which should be conducted under strictly controlled medical conditions and having had the prior approval of an appropriately qualified independent panel (Institutional Review Board/Ethics Committee), may be needed for various reasons. These include: (a) confirming or otherwise the suspicion of the occurrence of a toxic or potentially adverse physiological/pharmacological effect suggested by the preliminary laboratory *in vivo* or *in vitro* toxicology studies, (b) quantitation of a known effect in human volunteers, (c) assessing the potency or usefulness of an RCA, (d) conducting small-scale clinical trials on effectiveness of RCA materials and delivery systems, and (e) conducting full-scale trials to simulate the likely situation to be encountered and confirm the effectiveness and usefulness under such situations. Items (d) and (e), particularly, can combine the major objectives of the trial with routine medical, physiological and biochemical monitoring to determine the effects of the trial procedures on physiological and biochemical homeostasis, and to detect unsuspected effects from the exposure conditions. This supportive monitoring often includes cardiovascular status (blood pressure, ECG), respiratory function tests, peripheral blood haematology, blood clinical chemistry, urinalysis and ophthalmic investigations (including pachymetry and tonometry). These supportive studies are of value not only in assessing safety-in-use for the healthy population and defining any operational restrictions on the RCA/formulation/delivery system, but also may indicate susceptible sub-populations, including those with established ill health (e.g. cardiovascular disease, respiratory disease, aneurysm, glaucoma, etc.). They clearly complement the toxicology studies and are of considerable value in hazard evaluations. Volunteer trials should ideally start at threshold levels (doses), be conducted with appropriate medical cover, only involve participation of subjects who have been fully informed of the nature of the human studies and any potential risks and permit withdrawal of participants at any stage in the trial. In addition, as noted previously, human studies (trials) should only be conducted after approval by the local (establishment) ethical medical safety committee, and by an independent advisory group (as discussed by Ballantyne, 2005b). These groups should be periodically informed of the progress and results of the studies, and should confirm that it is appropriate to continue (authorize) an extension of the trials. It is of the utmost importance that written and signed detailed records should be kept of all aspects of the trials, and on discussions of the local approval committee and of the independent approving body.

ASSESSMENT OF THE EFFECTIVENESS OF RIOT CONTROL PROCEDURES IN THE CONTEXT OF SAFETY CONSIDERATIONS

In assessing the operational usefulness and effectiveness of RCAs, especially PCSIs, some of the more important and relevant factors that need to be taken into consideration are as follows.

Effectiveness in intended use

Specific factors that need to be examined in determining the effectiveness and comparative effectiveness of PCSIs are threshold concentration, incapacitating concentration, effectiveness ratio and comparative potency, all of which are discussed in more detail below. In addition, considered in relation to effectiveness are latency to PCSI effect, duration of effects, performance against distracting effects, effects of avoidance and countermeasures, and the possible influence of formulation and formulating constituents. The majority of these practical considerations are based on information collected from carefully controlled and conducted laboratory and human volunteer studies; the latter can be conducted with integral biomedical monitors for safety-use-use assessments (see above). When in-use information is available, this may be of value, but clearly requires careful interpretation because of problems in quantitation of such data and the fact that it is obtained in situations generally beyond the control of the observer(s).

Toxicity and human adverse effects reporting

As discussed above, if toxicological studies indicate a potential to produce adverse effects it is a necessity to know their nature, incidence, dosage dependency, mechanism of production, factors which influence their development and reversibility; the latter to include both spontaneous reversibility (healing) and induced reversibility (antidotal and other treatments). As stressed, in addition to determining in detail the toxicity of a specific RCA, it is an essential for a complete assessment of potential for in-use adverse health effects that the possible modifying effects of impurities, formulation and delivery system be investigated; this to include additive, potentiating and synergistic effects, and the possible formation of new or unsuspected substances produced by reaction of the RCA with formulating constituents and also from the delivery mode (e.g. combustion products in pyrotechnically generated materials). The influence of such factors on toxicity may also have a modifying influence on the desired pharmacological effect of the active RCA.

Relationship of toxicity to effectiveness and hazard evaluation

Hazard evaluation involves a determination of whether the known toxicity of a material will develop and be exhibited under the specific conditions of use and exposure to the material. In the context of the use of most RCAs, the intended or incidental (accidental, bystanders) exposures will usually only be brief and for a single or a few irregularly spaced exposures to the material; additionally, and depending on circumstances, a few individuals may have single but sustained exposure to RCAs, possibly in confined spaces. Therefore, with respect to chemically induced injury, in these circumstances the main concerns are for acute toxicity, acute inflammation and single exposure long-term tissue injury. There are, of course, other and related health aspects resulting from participation in a civil disturbance that may need medical management, such as physical injuries and/or short-term psychological effects; these are considered later in this chapter. Those occupationally exposed to RCAs may have more frequent and sustained exposures; e.g. production workers, military training sessions, security forces and emergency services. With production workers, there should be no overexposure if recommended industrial hygiene measures are followed (e.g. protective clothing and adherence to workplace exposure guidelines such as TLVs). With those likely to be exposed at in-use sites, such as security forces and emergency service workers, there should be provision for appropriate protective clothing. The sporadic and unplanned use of hand-held protective irritant liquid projection devices, however, may result in repetitive exposures. Participants or innocent bystanders who are at the scene of a civil unrest incident and may be exposed to RCAs will constitute a heterogeneous collection of the population; thus, there will be old and young persons, males, females of fertile age and individuals with variable states of health. In relation to dosage considerations in hazard evaluation, it should be born in mind, as discussed in more detail below, that the shorter the latency to a PCSI effect and the more potent the material, the smaller will be the dose required to achieve a harassing effect and thus from this perspective will be associated with a lesser potential for toxicity. However, and as discussed later, for a given degree of incapacitation, airborne PCSI high concentration–short duration exposures may result in a larger inhalation exposure dosage than that resulting from low concentration–long duration exposure conditions. The RCA to be used should be capable of being delivered to its intended target by appropriate physical methods that combine safety (i.e. no enhancement of perceived hazard) with the required degree of selectivity, but without attenuation of the desired effect.

Environment and decontamination

When chemicals are dispersed over large areas, particularly in urban areas, a multiplicity of differing environmental effects need to be considered and, where necessary, appropriate actions planned and effected. These will depend, among other things, on the physicochemical characteristics of the active RCA and its formulation, stability and persistence, amount dispersed, and local hydrology, geography and geology. The diversity of considerations may need to cover, as examples: effects on sewage organisms; persistence and resultant contamination of public and private property and any necessary decontamination procedures; effects on domestic animals; phytotoxicity; fresh water aquatic toxicity; effects of accidental contamination of food and/or drinking water. In relation to the latter consideration, PCSI materials may be detectable by taste at very low concentrations; e.g. CR is detectable in drinking water at $< 3.16 \times 10^{-5}$ % (Kemp and Weatherell, 1975).

PHYSIOLOGICAL, HARRASSING AND INCAPACITING EFFECTS OF PERIPHERAL CHEMOSENSORY IRRITANTS

PCSI materials interact with sensory nerve-ending receptors in skin and mucosal surfaces, producing local sensation (itch, discomfort or pain) with associated local and systemic reflexes. Local effects on the eye are a burning sensation, marked discomfort or pain with blepharospasm and excess lacrimation. Contamination of the nasal mucosa causes stinging or discomfort with rhinorrhoea, and in the mouth there is stinging with excess salivation. Inhaled PCSIs cause stinging, discomfort or pain in the respiratory tract, with constricting sensations in the chest, sneezing, coughing, increased tracheobronchial secretions, difficulties with breathing and periods of voluntary breath holding. From a biological viewpoint, the above-noted effects are bioprotective in nature, warning of the presence of an irritant material and limiting exposure to that material. Clearly, from the riot control perspective, all of the effects are harassing, and make affected individuals desire to vacate the contaminated area, and are detrimental to the conduct of co-ordinated activities. PSCI effects are usually experienced at concentrations significantly less than those that cause adverse effects (toxicity) by acute exposure. This difference in concentrations between PCSI and toxic effects limits the likelihood for exposure to potentially harmful concentrations (doses) of a PCSI. In addition to the local reflexes induced, there are also systemic reflexes produced, of which the most relevant for medical complications are transient hypertension and bradycardia.

PCSI materials used in the context of atmospherically dispersed materials are sometimes classified as sternutators if their main action is on the upper respiratory tract, and lacrimators if the principal action is on the eye. For most currently used PCSIs, this is not a useful classification descriptor, since both effects are frequently present. Thus, exposure to an airborne PCSI RCA will cause effects in skin, eye and respiratory tract. When dispersed in solution, the effects are generally limited to the area of contact, and the reflexes elicited are a function of the afferent nerve involvement. In general, PCSI effects appear within seconds of contact, and subside within 10–60 min, depending on exposure concentration and site affected.

The incidence of a specific response in a population to a PCSI stimulus when plotted as a function of exposure concentration takes the form of a sigmoid curve typical for many biological phenomena (e.g. Figure 1). This implies the existence within the population of hyposensitive and hypersensitive individuals to the right- and left-hand sides, respectively, of the distribution curve, but with the majority of individuals responding over a relatively narrow concentration range in the center between the extremes. From such frequency distribution data, it is possible to calculate median effective concentrations (Ballantyne, 1999a,2005a; Ballantyne *et al.*, 1977b). For a just detectable sensation plotted as a function of

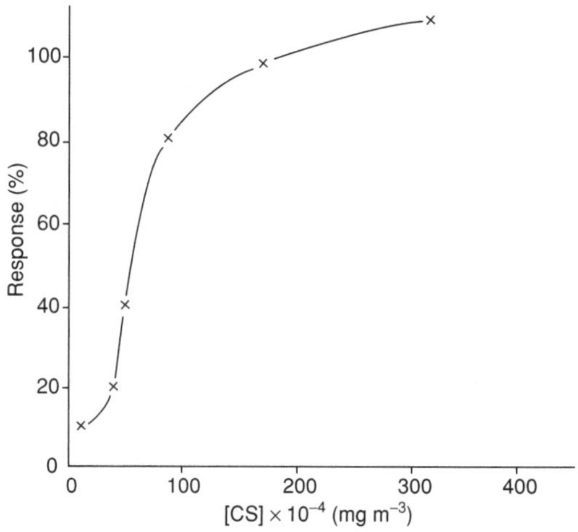

Figure 1. Concentration–population response curve for threshold peripheral chemosensory irritation to the human eye from an aerosol of 2-chlorobenzylidene malononitrile (CS). Probit analysis of this data gives a TC_{50} of 4.0×10^{-3} mg m^{-3}

exposure concentration, it is possible to calculate what is a just detectable (or threshold) sensation in 50% of the population studied (threshold concentration 50%; TC_{50}). This can similarly be calculated for incapacitating concentration, which is the concentration that cannot be voluntarily tolerated (usually for > 1 min). Clearly, incapacitating concentrations are relevant to effective concentrations in respect of peacekeeping operations. As with threshold effects, a median effective incapacitating concentration (IC_{50}) can be calculated from human studies data; however, in the practical context of a riot it is more meaningful to determine IC_{75} or IC_{90} values. The operational effectiveness of a specific a PCSI material with respect to its use in peacekeeping operations needs to be assessed in terms of both the absolute concentration of the material producing an incapacitating effect and the relationship between the incapacitating concentration and the TC_{50}. The closer these values, the smaller will be the IC_{50}/TC_{50} ratio and the more rapidly will incapacitating effects develop within a rising concentration of PCSI material. Thus, the ratio IC/TC can be referred to as the *effectiveness ratio*, and can be used for comparison with that of other PCSI materials. In addition, it is possible to compare the absolute potencies of different PCSI materials by examining the ratio of

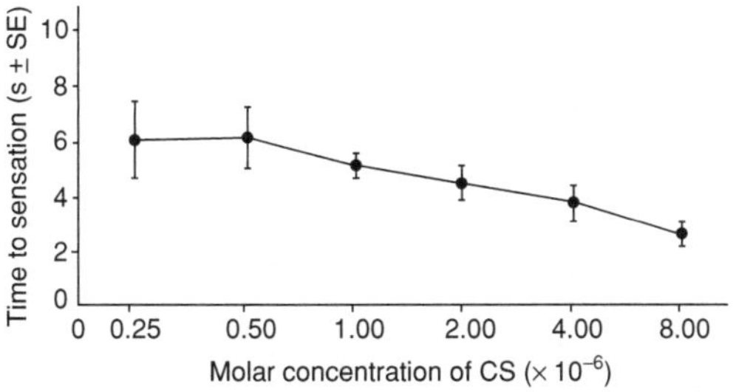

Figure 2. Effect of concentration of 2-chlorobenzylidene malononitrile (CS) dissolved in saline on the time to onset (latency) of sensation following the application of 0.01 ml droplets to the surface of the cornea of a human volunteer subject (data after Ballantyne and Swanston, 1973)

threshold or incapacitating effects at a given level (the comparative potency). An example of such a comparison is shown for CS and CR in Table 1. This shows that based on a consideration of the absolute PCSI concentrations, the relative (comparative) potency of CR is greater than that of CS at the TC_{50}, IC_{50} and IC_{75} levels. In addition, CR is more effective than CS based on the effectiveness ratios, which are smaller for CR than for CS at the IC_{50} and IC_{75}levels.

Additional features which need to be taken into account when assessing the effectiveness of PCSI materials are the time of onset (latency) of sensation and/or local reflex and the duration of the induced effects. Both of these are generally related to exposure concentration to a limiting value; as exposure concentration increases, the

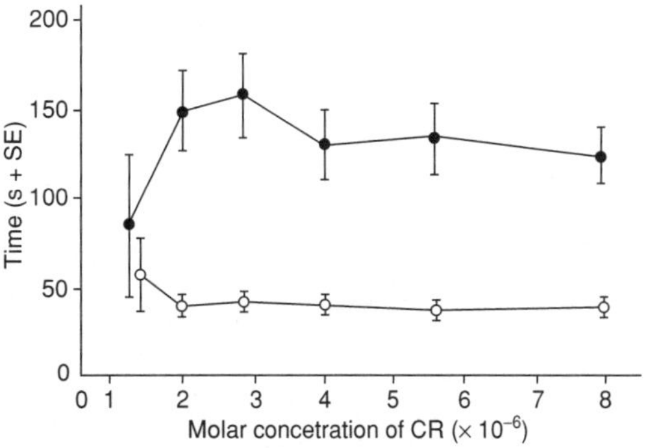

Figure 4. Effect of the concentration of dibenz(b.f)-1,4-oxazepine (CR) dissolved in saline on the time to onset (latency, o) and duration (•) of sensation produced after applying 0.01 ml droplets of solution to the corneal surface of a human volunteer subject (data after Ballantyne and Swanston, 1974)

latency decreases and the duration increases; examples are shown in Figures 2–6 for ocular and tongue irritation produced by solutions of the PCSIs CS and CN. Furthermore, as noted above, the proportion of a given population responding to a PCSI stimulus increases with exposure time, and this may have implications with respect to the inhalation exposure dose received with airborne PCSI materials. For example, Figure 7 shows the relationship between exposure concentration to CN aerosol and response to defined levels of incapacitating effects. As the exposure time increases, and at a specific exposure concentration, the proportion of the population responding to

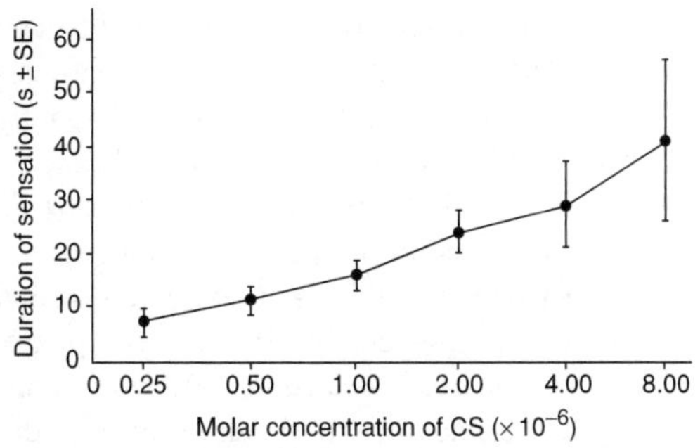

Figure 3. Effects of 2-chorobenzylidene malononitrile (CS) dissolved in saline on the duration of sensation produced after the application of 0.01 ml droplets to the surface of the cornea of a human volunteer subject (data after Ballantyne and Swanston, 1973)

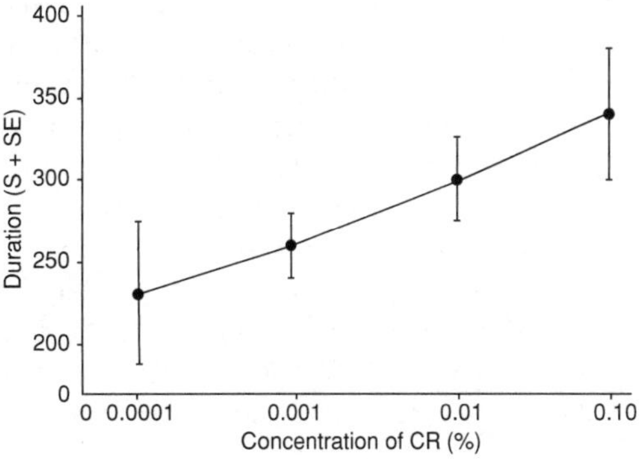

Figure 5. Effect of concentration of dibenz(b.f)-1,4-oxazepine (CR) in polyethylene glycol 300 on the duration of sensation in the human tongue following the application of 0.01 ml droplets of solution (data after Ballantyne and Swanston, 1974)

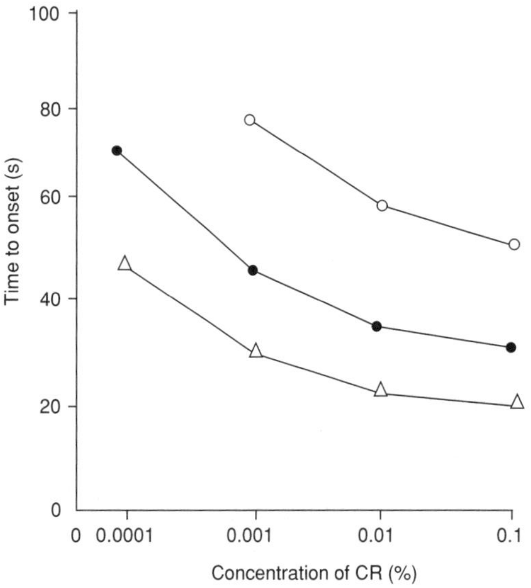

Figure 6. Effect of dibenz(b.f)-1,4-oxazepine (CR) dissolved in polyethylene glycol 300 on the time to onset (latency) of various types of perceived sensation on the human tongue following the application of 0.01 ml droplets: o, strong burning; •, detectable burning sensation; Δ, stinging sensation (data after Ballantyne and Swanston, 1974)

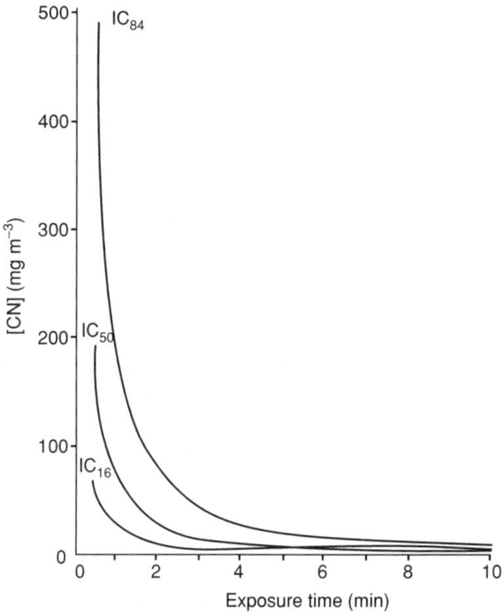

Figure 7. Relationship between exposure concentration and exposure time for aerosols of 1-choroacetophenone (CN) required to produce defined levels of harassment (incapacitating concentrations (ICs) of 16, 50 and 84%). Due to the non-linear relationships, the inhalation exposure dosages are greater for the high concentration–short exposure time than for the low concentration–long exposure time (data after McNamara *et al.*, 1968)

produce a given level of incapacitation (IC_{16}, IC_{50} and IC_{84} in this example) also increases, showing population-variability in the PCSI response. In addition, the relationship between increasing concentration and exposure time to produce given effect levels is not linear. Thus, the inhalation exposure doses (C*t* values) for a given degree of incapacitation (IC) are greatest for the high concentration–shorter exposure time (T) at a specified degree of incapacitation. Thus, while increasing the atmospheric concentration of a PCSI material decreases the latency to response, and increases the proportionate response in the population, it also increases the inhalation exposure dose and hence the potential for toxicity.

Several variable factors influence the latency to and potency of a PCSI response. Those of particular practical significance are as follows:

(1) *Concentration.* As discussed above, and within limits, increasing the exposure concentration to a PCSI will increase the proportionate response in a population, decrease latency to response, prolong the response and produce an overall more potent effect.
(2) *Particle size.* With aerosols of PCSI materials, the smaller respirable particles (ca. 1 μm MMAD) produce both ocular and respiratory effects, while larger particles (ca. 60 μm MMAD) cause predominantly ocular effects (Owens and Punte, 1963).
(3) *Vehicle.* With solutions, the use of surface active materials may enhance spread and penetration of a PCSI material into skin and mucosae, and hence facilitate the response.
(4) *Environmental conditions.* Elevated temperature and increased humidity may decrease tolerance to PCSI materials (Punte *et al.*, 1963).
(5) *Motivation.* In general, increased motivation and distracting influences will increase

the threshold for the induction of PCSI effects and increase tolerance to 'supra-threshold' concentrations of PCSI materials. This may clearly influence decisions on what constitute appropriate incapacitating concentrations in well-motivated rioters.

SPECIFIC PERIPHERAL CHEMOSENSORY IRRITANTS CONSIDERED FOR USE IN PEACEKEEPING OPERATIONS

General background comments

A variety of PCSI materials have been used, are currently in use and are being investigated for possible future use, by different modes of dissemination in the various aspects of peacekeeping operations. Organic arsenicals, such as the vomiting agents adamsite (DM; 10-chloro-5,10-dihydrophenarsazine) and diphenylchlorarsine (DA) in general have been banned by most countries because of their moderately high toxicity to humans. The arsenicals have been reviewed by Ballantyne (1977a,1987) and Salem *et al.* (2005b). 1-Chloroacetophenone (CN) has been used extensively, both as a grenade-generated smoke and in solution in devices for self-protection and security force-subduing devices. However, its use against humans has shown CN to produce several undesirable adverse health effects and, in a few cases, death (Ballantyne, 1977a). Therefore, several civilized countries, including the UK, have abandoned the use of CN against civilian populations, although it still finds use in the USA and some other countries. 2-Chlorobenzylidene malononitrile (CS) is used extensively both as a grenade-generated smoke and in solution for peacekeeping operations. As discussed below, the likelihood for serious short- or long-term adverse effects is low, and certainly less than with CN, and hence its widespread use by security forces, including the UK. Oleoresin capsicum (OC) is a complex mixture obtained from pepper plants (*Capsicum annuum* and *Capsicum frutescens*) and was introduced in the 1970s in the USA as an alternative to, particularly, CN. It is widely used by law-enforcement agencies in the USA, mainly in spray devices which are directed at the faces of assailants. OC has not been approved by UK authorities for use in aerosol spray devices because of the complexity of the mixture and the opinion that individual components would need to be examined with respect to human health hazard potential; coupled with the variability of the extract from batch to batch, this would make safety evaluations for different batches difficult and expensive. Nonivamide (pelargonic acid vanillylamide [PAVA]; 'synthetic capsaicin') is being investigated as a potential PCSI for use in personal protection and deterrent hand-held spray devices. It has the advantage over OC of being a single substance whose toxicology may be investigated in relation to the active PCSI, without complications from other constituents. Dibenz(b.f)-1,4-oxazepine (CR) is a potent PCSI material of low mammalian toxicity and not injurious to the eye and skin, and can be disseminated as both a grenade-generated smoke and in solution from liquid-projection devices. However, although human studies and trials have confirmed both the effectiveness of CR as an incapacitant and the low inflammatory potential, there are few authenticated reports of its use in operational situations. One disadvantage of CR is its chemical stability and persistence. Several PCSI substances that have initially been considered as having characteristics which would make them desirable as agents for use in peacekeeping operations, have been abandoned mainly because of toxicological considerations. For example, 1-methoxycycloheptatriene is a potent liquid PCSI with a high vapor pressure, which would make it suitable for easy delivery in frangible missiles. However, the vapor was shown to be centrally neurotoxic, producing cerebellar Purkinje cell injury (Marrs *et al.*, 1991).

The following sections briefly summarize the uses, chemistry, pharmacology, toxicology and effects on humans of various PCSIs, which are currently in use for peacekeeping operations or have a potential for such use. Individual PCSIs have also been reviewed in detail by Ballantyne (1977a,1987), Marrs *et al.* (1996), Olajos and Salem (2001) and Salem *et al.* (2001, 2005a,c).

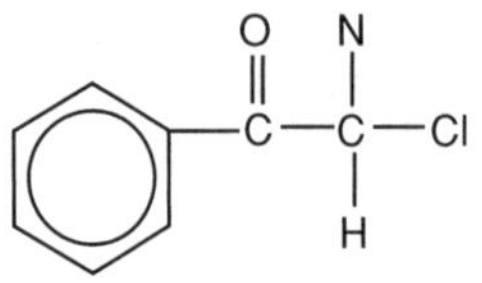

Figure 8. Structural formula of the RCA 1-chloroacetophenone (CN)

1-Chloroacetophenone (CN)

IDENTITIES AND PHYSICOCHEMICAL PROPERTIES

Synonyms: 2-chloro-1-phenylethanone; phenylacyl chloride; chloromethyl phenyl ketone; ω-chloroacetophenone; α- chloroacetophenone.
CAS number: 532-27-4
Molecular weight: 154.5
Molecular formula: C_8H_7OCl
Structural formula: see Figure 8
Physical state: white solid
Density: 1.31 g cm^{-3} (0°C)
Melting point: 59°C
Boiling point: 244–245°C
Vapor pressure (solid): 5.4×10^{-3} mmHg (20°C)
Vapor density: 5.3 (air = 1)
Solubility: water = 4.4×10^{-3} mol l^{-1} (∼ 68 mg dl^{-1})
Chemical identification: infrared spectrum (Ferslew *et al.*, 1986; Shreenivasan and Boese, 1970); UV spectrophotometry (Ferslew *et al.*, 1986); gas chromatography, flame ionization (Zerba and Ruveda, 1972); isothermal gas chromatography (Jane and Wheals, 1972); ion mobility spectrophotometry in the negative-ion acquisition mode (Allinson and McLeod, 1997a,b; Allinson *et al.*, 1998); NMR spectra (Ferslew *et al.*, 1986; Mesilaakso, 1996).

USES AND DISPERSION

As an incapacitant and harassing agent in riots and civilian peacekeeping operations as a grenade-generated smoke; as an incapacitant in solution for 'one-on-one' or small-group engagements; for enemy denial and dispersion purposes in military operations; used for respirator fit testing; smoke used for training purposes in military exercises; hostage and siege situations; civilian peacekeeping operations; personal protection devices. Hand-thrown grenades, projectile cartridges, aerosol and coherent liquid stream dispensers; cluster bombs; aircraft dispensers; portable powder disperser.

PCSI VALUES

Human Studies

Eye, aerosol TC_{50} = 0.3 mg m^{-3}
Respiration, aerosol TC_{50} = 0.4 mg m^{-3}
IC_{50} aerosol = 20–25 mg m^{-3}
ICt_{50} (aerosol in acetone) = 40 mg min m^{-3}
ICt_{50} (grenade smoke) = 20 mg min m^{-3}
(data after Ballantyne, 1977a; McNamara *et al.*, 1969)

ACUTE TOXICITY

Intravenous: LD_{50} values

Rabbit, male (in polyethylene glycol; 300 PEG 300) 31 mg kg^{-1} (slope = 8.3)
Rabbit, female (in PEG 300) 30 mg kg^{-1} (slope = 7.3)
Mouse, male (in PEG 300) 81 mg kg^{-1} (slope = 5.3)
Rat, female (in PEG 300) 41 mg kg^{-1} (slope = 8.4)

Animals exhibited convulsions lasting 15 s to 1 min within a few seconds of dosing. This was followed by limb tremors, intermittent muscle spasms and uncoordinated movements. Animals that died did so between 30 min and 23 h at the lower dosages and between 1 min and 1 h after the higher dosages. A characteristic sign was congestion of the vessels of the conjunctivae and iris, which persisted for 1 to 3 h. Animals that died had congestion of the liver, kidney, lung, thymus, spleen, conjunctivae and uveal tract, and haemorrhages in the lungs and thymus. Animals dying after 12 h had scattered irregular foci of renal cortical and medullary tubular necrosis. Survivors sacrificed 3 weeks after dosing had no abnormal findings.

Intraperitoneal: LD_{50} values

Rat, male (in PEG 300) 36 mg kg^{-1} (slope = 5.0)
Guinea pig, female (in PEG 300) 17 mg kg^{-1} (slope = 3.3)

Signs of toxicity developed within 10 min of dosing and consisted of pilo-erection, sluggish movements and evidence of peritoneal irritation on handling. In survivors, these effects gradually subsided and disappeared by 2 to 3 days. Animals that died developed progressive sluggishness of movements, showed shallow irregular breathing and became comatose, with death occurring 2 h to 5 days after dosing. Rectal temperature decreased in animals that died; in the hour before death, temperatures decreased from means values of 37.3°C to 27°C in the rat and 38.5°C to 28°C in guinea pigs. In animals that died, there was congestion of the stomach, small intestine, spleen, pelvic viscera and lungs. In those dying several days post-dosing there were also intra-abdominal adhesions with neutrophil infiltration of the mesentery. Small scattered haemorrhages were present in the pulmonary alveoli, thymus, and in a few animals in the small intestine and caecum. Survivors sacrificed 3 weeks after doing had multiple intra-abdominal adhesions.

Peroral: LD_{50} values

Rat, male (in PEG 300) 127 mg kg^{-1} (slope = 10.9)
Rabbit, female (in PEG 300) 118 mg kg^{-1} (slope = 5.1)
Guinea pig, female (in PEG 300) 158 mg kg^{-1} (slope = 5.7)

After a latent period of 15 to 30 min from dosing, the signs of toxicity were pilo-erection followed by reduced muscle tone and sluggish movements. In animals that survived, these effects gradually subsided over 2 or 3 days. Animals that died usually did so between 2 and 18 h post-dosing, although a few survived for 6–17 days. Animals dying within a few hours of dosing showed variable degrees of pulmonary congestion, and congestion and haemorrhages in the thymus. The stomach was markedly congested and had haemorrhagic erosions of the gastric mucosa. The small intestine was congested and in about half of the animals there was necrosis of the tips of the villi, and in some there was haemorrhagic necrosis of the mucosa. Kidneys of about half the animals that died had acute cortical and medullary tubular necrosis, and in a few animals there was hepatic centri- and mid-lobular necrosis. Animals that died days after dosing had gastric haemorrhagic necrosis with areas of mucosal regeneration and scar formation. About half of the surviving animals that were sacrificed 21 days post-dosing had no pathological abnormalities but the remainder demonstrated foci of gastric mucosal regeneration and intra-abdominal adhesions.

Inhalation: Pure CN Aerosols

Male rats, female rabbits, female mice and female guinea pigs were exposed to pure CN aerosols generated from molten CN in a Collison spray and introduced into the air stream to the exposure chamber. The ranges of average concentrations, exposure times and inhalation exposure dosages were, respectively, as follows: rats, 15–37 min, 417–749 mg m^{-3}, 6255–27 713 mg min m^{-3}; rabbits, 15- 60 min, 465–742 mg m^{-3}, 9300–33 060 mg min m^{-3}; mice, 15–37 min, 592–719 mg m^{-3}, 9000–26 603 mg min m^{-3}; guinea pigs, 15–60 min, 243–764 mg m^{-3}, 3645–37 620 mg min m^{-3}. The following LCt_{50} values were calculated from the dosage–mortality data:

Rat, male 8750 mg min m^{-3} (slope = 2.8)
Rabbit, female 11 480 mg min m^{-3} (slope = 11.5)
Guinea pig, female 13 140 mg min m^{-3} (slope = 9.8)
Mouse, male 18 200 mg min m^{-3} (slope = 4.6)

Most animals died during the first two days after exposure. The lungs of animals that died within the first 48 h were congested and oedematous, and many had multiple variable-sized haemorrhages. The trachea was congested and contained excess mucus. Histology showed moderate to marked congestion of alveolar capillaries and intrapulmonary veins, with intra-alveolar haemorrhages, and excess secretions in the bronchioles and intrapulmonary bronchi. In a few animals, there were areas of collapse distal to occluded bronchioles. In addition, there was patchy acute inflammatory cell infiltration of the trachea, bronchi and bronchioles, and a few instances of focal epithelial necrosis. Animals dying after 48 h also had bronchopneumonic changes. In those with the more extensive lung injury there were scattered circumscribed areas of acute renal cortical and medullary tubular necrosis, and some instances of cloudy swelling of centrilobular hepatocytes and a few instances of hepatic centrilobular necrosis. Animals sacrificed 14 days post-exposure generally showed no pathology, but a few had scattered pulmonary haemorrhages and patchy congestion of the tracheal mucosa (Ballantyne and Swanston, 1978).

For humans, the maximum safe inhaled dose of CN is estimated at 50 mg min m^{-3} (Punte *et al.*, 1962a). The human acute lethal inhalation dosage, extrapolated from animal data, has been estimated at 8500–25 000 mg min m^{-3} (United Nations, 1969; WHO, 1970). Based on a statistical analysis of numerous studies over a 47-year period, McNamara *et al.* (1968) consider the best estimate for the LCt_{50} for CN to man as 7000 mg min m^{-3} for pure aerosol and 14 000 mg min m^{-3} for a commercial grenade.

PRIMARY IRRITATION

Skin. The skin irritating effect of CN was studied in rabbits, guinea pigs and mice. A volume of 0.1 ml of 12.5% CN in either corn oil or acetone was spread over the shaven dorsal trunk skin. After 6 h of unoccluded contact, the sites were inspected at 5 h, and 1, 2, 4, 7, 14 and 21 days. There was little difference in the cutaneous response to CN whether dissolved in corn oil or acetone. At the end of the 6 h contact period there was slight to moderate erythema and slight to marked oedema. The effects slowly subsided, but resolved completely only after 7–14 days. Mild to moderate desquamation was seen at 4–7 days. From 24 h post-application, variable sized ecchymoses with scattered areas of necrosis developed, and resolved within 7–14 days. Histology of skin biopsies removed from a few animals 3 days after application of CN to the skin showed extensive necrosis of the epidermis and of the collagen in the outer dermis at the site of the original local contamination with CN. Oedema was present throughout most of the dermis with neutrophil infiltration of the dermis (Ballantyne and Swanston, 1978). Clinical evidence has shown that with humans CN can cause a primary contact dermatitis; erythema, oedema, vesiculation, petaechiae, purpura and necrosis have all been described (Holland and White, 1972; Jolly and Carpenter, 1968; Penneys, 1971; Schwartz *et al.*, 1957).

Eye. The ophthalmic toxicology of CN was investigated by a standard rabbit eye irritation test using solutions in PEG 300 in the range of 1–10%, as a solid, and with exposure to an aerosol (Ballantyne *et al.*, 1975). In solution, CN produced marked and persistent inflammatory effects whose severity and duration were dosage-related. Lacrimation was seen at all concentrations, varying between slight and moderate and resolving by 4 days with 1%, and mild to moderate and of 3 days duration with 5% CN, to a gross excess persisting up to 11 days with 10%. Blepharitis varied from mild and of 3–4 days duration with 1% to marked and lasting up to 16 days with 10%. Chemosis was usually detectable within 10 min post-instillation, and was mild and persisted for 4 days with 1%, and severe at 1 h with 10% and gradually subsided over the following 2 weeks. Hyperaemia of the conjunctivae and nictitating membrane was slight and resolving by 4 days with 1%, and severe which took up to 4 weeks to resolve with 5 or 10%; conjunctival haemorrhages were seen with animals receiving 2% and higher concentrations of CN. Iritis was just detectable with 1% (2 days duration), mild with 2% (4 to 7 days duration), and increasingly severe with 5 and 10% but became obscured by keratitis. Keratitis was as follows: 1%, not seen, 2%, just detectable by 24 to 48 h and up to 3 days in duration. Histology of eyes with active keratitis showed a thickening and loss of eosinophilia of the substantia propria with neutrophil infiltration. With 5% CN, keratitis was just detectable by 24 h, mild to moderate at 3 days, moderate to severe by 1 to 2 weeks, and thereafter slowly resolved but still detectable by 38 days; delayed corneal neovascularization was generally limited to the periphery of the cornea. After 10% CN, keratitis was detectable by 24 h and thereafter became progressively more severe, with only slow and partial resolution. Histological examination of eyes demonstrating a moderately severe keratitis demonstrated loss of the corneal epithelium, marked neutrophil infiltration of the thickened substantia propria and occasional hypopyon. Corneal neovascularization varied in extent from peripheral to affecting the whole cornea.

A comparison of the influence of solvent on the eye irritating potential of CN was undertaken by comparing the ocular effects of 5% CN in PEG 300, trichloroethane, corn oil and tri(2-ethylhexyl)phosphate. The most marked differences were seen with the cornea. With CN in trichloroethane and tri(2-ethylhexyl)phosphate, keratitis was just detectable and of a few days duration; CN in corn oil produced mild keratitis of about 2 weeks duration; CN in PEG 300 caused a moderate and persistent corneal injury.

The effect of solid CN applied to the cornea was investigated using 0.1, 0.25, 0.5, 1, 2 and 5 mg. Marked and sustained damage to the cornea, iris, conjunctivae, and eyelids was produced by 5 mg CN, being only slightly less with 2 mg. When 1 mg was applied, the effects were generally mild and usually resolved within 14 days. Amounts less than 1 mg produced just detectable effects of a few days duration. Keratitis produced by solid CN was more marked and persistent than that caused by equivalent amounts in solution, for example, while 1% CN in PEG 300 was without effect on the cornea, 1 mg solid CN caused a just detectable keratitis of several days duration. The no-effect level for solid CN on the cornea was between 0.1 and 0.25 mg.

The effect of a CN aerosol exposure was studied using a cloud generated by spraying 20% CN in dichloromethane through a nozzle into a 10 m^3 chamber. The mean atmospheric concentration of CN in the chamber was 719 mg m^{-3} with an exposure time of 15 min (Ct = 10 800 mg min m^{-3}). This caused a just detectable excess lacrimation and conjunctival hyperaemia, with mild blepharitis, of a few days duration.

The effect of solutions of CN in PEG 300 on corneal thickness was investigated *in vivo* in the rabbit using solutions in the concentration range 0.01–0.75% with measurements being taken before instillation and at 1 and 24 h, and 2, 4, 8 and 17 days post-treatment. Corneal thickness was measured by an optical pachymeter attached to a slit-lamp biomicroscope. Increases in corneal thickness were statistically significant for exposures to 0.03% CN and above, and both the magnitude and duration were related to the CN concentration. Times for corneal thickness to return to control values varied between 2 days for 0.02% and 14 days for 0.75%.

The effects of solutions of CN in PEG 300 on intraocular pressure (IOP) were studied *in vivo* in the rabbit using a concentration range of 0.0625 to 1.0% CN. Measurements were taken using an applanation tonometer before instillation of solution on to the cornea and at 10 min and 1 h post-instillation. At 10 min, there were statistically significant and concentration-related increases in IOP for concentrations at and above 0.125%, and which ranged from 16% for 0.125% to 98% with 1% CN. By 1 h, the IOP had decreased to control values, with the exception of 1% CN which was still elevated at 13%.

SENSITIZATION

That CN induces a type IV hypersensitivity reaction has been confirmed in animal studies (Rothberg, 1970) and in conventional skin patch testing in man (Marzulli and Maibach, 1974; Penneys, 1971; Penneys *et al.*, 1969). Induction concentrations of CN (0.5% and 1.0% in acetone) were applied to the skin of guinea pigs, and challenge doses applied 3 to 4 weeks later (0.1 ml of 0.2 or 1.0% CN); there was a 100% sensitizing response to both challenge concentrations (Chung and Giles, 1972). There is clinical evidence for allergic contact dermatitis in individuals exposed to CN (Fuchs and in der Wiesche, 1990; Penneys *et al.*, 1969; Queen and Stander, 1941).

REPEATED EXPOSURE TOXICITY

Mice were exposed to CN at 87.6 mg m^{-3} for 15 min a day for 10 days (daily Ct = 1314 mg min m^{-3}) (Kumar *et al.*, 1994). After 5 days, the lungs demonstrated diffuse haemorrhage, perivascular swelling, congestion of alveolar capillaries and bronchial inflammatory infiltrates. There was hepatic centrilobular necrosis and coagulative necrosis of renal tubules. At 10 days, more severe lung damage was apparent.

Rats and mice were exposed to CN aerosol concentrations ranging from 4.8 to 64 mg m^{-3} for 6 h a day for 14 days (daily Ct values 1728–23 040 mg min m^{-3}). All rats exposed to 19 mg m^{-3} and higher ($\geq$ 6840 mg min m^{-3}) died during the first week. During exposure, rats demonstrated partial blepharospasm, lacrimation and shortness of breath. All mice exposed to 10 mg m^{-3} CN and higher ($\geq$ 3600 mg min m^{-3}) died during the first week (NTP, 1990a). As an extension to these studies, a sub-chronic inhalation study was conducted in which rats and mice were exposed to concentrations of CN in the range

0.25 to 4 mg m^{-3} for 6 h a day, 5 days a week for 13 weeks (range of daily *Ct*s, 90 to 1440 mg min m^{-3}). All rats survived the 13 weeks of exposure. Eye irritation and corneal opacities were seen in rats exposed to 0.5 mg m^{-3} CN and higher. Mice deaths were as follows: 1/10 at 4 mg m^{-3} and 1/10 at 0.5 mg m^{-3}. No clinical chemistry changes or histopathology were found in rats and mice (NTP, 1990a).

BIOCHEMICAL MECHANISMS

CN is an SN2 alkylating agent with activated halogen groups that react readily at nucleophilic sites. Prime biochemical targets include SH-containing enzymes, leading to their inhibition. CN inhibits human plasma BChE by a non-SH interaction (Salem *et al.*, 2005d). The acute toxicity of CN may be related, at least in part, to the inhibition of SH-containing enzyme activity.

METABOLISM AND TOXICOKINETICS

The metabolism and metabolic fate of CN are poorly characterized.

DEVELOPMENTAL AND REPRODUCTIVE TOXICOLOGY

Studies conducted in the early 1960s indicated that CN can affect critical stages of embryonic development when instilled in alcoholic solution at mM concentrations directly into cultures of hen's eggs (Mankes and Mankes, 2004). The relevance of these findings to human developmental risks is uncertain, especially since conventional developmental toxicity studies *in vivo* have not been conducted.

GENETIC TOXICOLOGY

No information is available.

ONCOGENICITY

CN was investigated for oncogenic potential in a National Toxicology Program (NTP) 2-year bioassay in which rats were exposed to 0, 1 or 2 mg m^{-3} and mice to 0, 2 or 4 mg m^{-3}. Mortality rates were unaffected for rats, but survival of female mice was significantly decreased. Compound-related non-neoplastic lesions were seen in rats (squamous metaplasia of the olfactory epithelium, hyperplasia and metaplasia of nasal epithelium, and in females also inflammation, ulceration and squamous hyperplasia of the forestomach) and mice (hyperplasia and metaplasia of the respiratory epithelium). Neoplastic effects were not seen in male or female mice or male rats, but female rats had an increase in mammary gland fibroadenomas, indicating equivocal evidence of carcinogenicity (NTP, 1990a).

HUMAN VOLUNTEER STUDIES AND IN-USE OBSERVATIONS

Punte *et al.* (1962a) exposed volunteers to a maximum inhalation dosage of 350 mg min m^{-3}; median particle size 0.6 to 1.1 μm. The 1-min EC_{50} for irritation was 213 mg m^{-3}, the 2-min ECt_{50} was 119 mg min m^{-3} and 3-min ECt_{50} 93 mg min m^{-3}. Symptoms included lacrimation, eye discomfort, blurred vision, nasal discharge, burning sensation in the nose and throat, and difficulty with breathing.

Vaca *et al.* (1996) described the clinical findings in a case where CN spray was accidentally used as a room deodorant, 18 h after which the individual presented with shortness of breath and radiological evidence of pulmonary oedema. There was progression of the oedema, which was reversed after glucocorticoid treatment. Thornburn (1982) described the consequences of CN being released into 44 prisoner cells, resulting in 8 prisoners being hospitalized. Five had pseudomembranes in their throats, and three developed tracheobronchitis requiring the use of bronchodilators. Four had facial burns and 3 had ankle burns. One had an asthmatic-like reaction starting 1 to 2 days post-exposure.

Several deaths have been attributed to overexposure to CN, especially grenade-generated smokes with exposures in confined spaces (Chapman and White, 1978; Gonzales *et al.*, 1954; Stein and Kirwin, 1964; Thornburn, 1982). Death was associated with severe lesions of the airways (including necrosis of the laryngeal,

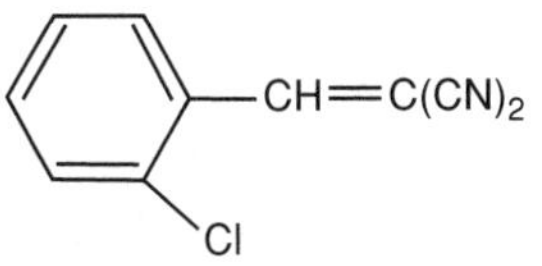

Figure 9. Structural formula of the RCA 2-chlorobenzylidene malononitrile (CS)

tracheal and bronchial epithelium with pseudomembrane formation, bronchiolar epithelium desquamation) and pulmonary oedema. Calculations of the exposure dosages in these lethal cases ranged 41 000 to 145 500 mg min m^{-3}.

2-Chlorobenzylidene malononitrile (CS)

IDENTITIES AND PHYSIOCHEMICAL PROPERTIES

Synonyms: *o*-chlorobenzylidene malononitrile; 2-chlorophenyl methylene propanedinitrile; β,β-dicyano-*o*-chlorostyrene.
CAS number: 2698-41–1
Molecular weight: 188.6
Molecular formula: $C_{10}H_5N_2Cl$
Structural formula: see Figure 9
Physical state: white solid
Melting point: 94°C
Boiling point: 310–315°C
Vapor pressure (solid): 3.4×10^{-3} mmHg
Solubility: water = 2.0×10^{-4} mol l^{-1} (~ 4 mg dl^{-1}); half-life in water = 14 min at pH 7.4/25°C; at pH 9, the half-life is ca. 1 min. CS is readily soluble in dichloromethane.
Chemical identification: infrared spectrum (Ferslew *et al.*, 1986; Shreenivasan and Boese, 1970); UV spectrophotometry (Ferslew *et al.*, 1986); fluorescence spectrophotometry (Ferslew *et al.*, 1986); gas chromatography (Zerba and Ruveda, 1972); isothermal gas chromatography (Jane and Wheals, 1972); ion mobility spectrometry (Allinson *et al.*, 1998); NMR spectra (Ferslew *et al.*, 1986; Mesilaakso, 1996); GC–MS (Smith *et al.*, 2002).

USES AND DISPERSION

For military training operations, respirator testing and enemy denial and dispersion use; hostage and siege situations; civilian peacekeeping operations; personal protection devices. Hand-thrown grenades, projectile cartridges, aerosol and coherent liquid stream dispensers; cluster bombs; aircraft dispensers; portable powder disperser. Hydrophobic antiagglomerative powders have been used in fogging machines.

PCSI VALUES

Animal Studies

Solution (in saline), blepharospasm EC_{50}, rabbit = 5.9 (3.8–10.0) $\times 10^{-5}$ M, guinea pig = 2.3 (1.9–2.4) $\times 10^{-5}$ M (Ballantyne and Swanston, 1973).

Human Studies

Aerosol, TC_{50} for ocular sensation = 4.0 (2.3–6.6) $\times 10^{-3}$ mg m^{-3}
Aerosol TC_{50} for respiratory tract = 23×10^{-3} mg m^{-3}
Solution (in saline), blepharospasm EC_{50} = 3.2 (2.1–5.1) $\times 10^{-6}$ M; corneal sensation TC_{50} 7.3 (4.2–11.2) $\times 10^{-7}$ molar; tongue sensation TC_{50} = 6.8 (5.0–10.6) $\times 10^{-6}$ M
IC_{50} for aerosol = 3.6 mg m^{-3}
(data after Ballantyne 1977a; Ballantyne and Swanston, 1973)

ACUTE TOXICITY

Intravenous: LD_{50} values

Rabbit, female (in PEG 300) 27 mg kg^{-1}
(slope = 13.8)
Mouse, male (in PEG 300) 48 mg kg^{-1}
(slope = 9.5)
Rat, female (in PEG 300) 28 mg kg^{-1}
(slope = 20.4)

Major signs of toxicity, seen within a few seconds of dosing, were convulsions, collapse and panting breathing. Blepharospasm, miosis and excess lacrimation also occurred in rabbits. Time to death varied between 1 and 24 h with the majority of animals dying within the first hour of dosing. In animals that died there was congestion of the liver, kidneys and lung, and occasional swelling of the centrilobular hepatocytes was seen. Animals dying a few hours post-dosing had small scattered irregular foci of acute renal cortical tubular necrosis. Survivors usually recovered from breathing difficulties within 30 min to 1 h, but muscle hypotonia and unsteadiness of gait persisted for 18 to 24 h, after which they were free from signs. There were no pathological findings in survivors sacrificed 21 day after dosing.

Intraperitoneal: LD_{50} values

Rat, male 48 mg kg^{-1} (slope = 10.8)
Guinea pig, female 73 mg kg^{-1} (slope = 21.7)

Animals became restless and developed piloerection within 5 min of dosing. There then developed salivation, rapid shallow breathing, extensor spasms and convulsions. In survivors, these effects subsided within 24 h and no abnormalities were seen on sacrifice at 21 days. Animals that died usually did so within 1 to 18 h, although a few survived for 2 to 9 days. Animals that died showed congestion of the liver, small intestine and lungs, with scattered patches of inter- and intra-alveolar haemorrhages. Small circumscribed foci of acute renal tubular necrosis were occasionally seen.

Peroral: LD_{50} values

Rat, male (in PEG 300) 1366 mg kg^{-1}
(slope = 7.2)
Rat, female (in PEG 300) 1284 mg kg^{-1}
(slope = 10.4)
Guinea pig, female (in PEG 300) 212 mg kg^{-1}
(slope = 8.3)
Rabbit, male (in PEG 300) 231 mg kg^{-1}
(slope = 4.5)
Rabbit, female 143 mg kg^{-1} (slope = 5.1)

Signs of toxicity appeared within 2 to 4 h of dosing, as piloerection, increased salivation and tremor, followed by rapid shallow breathing and reduced locomotion. Depending on the dosage, survivors recovered within 2 to 10 days. Animals that died showed a continued deterioration with death between 4 h and 7 days. Death was preceded by increased breathing difficulties, convulsions and collapse. Animals that died had multiple extensive haemorrhagic erosions of the gastric mucosa, with wall perforation in a few cases. The gastric mucosa and submucosa were congested and oedematous. Congestion of the spleen, small intestine, thymus and lungs was present. In a few animals, the small intestinal villi were congested with necrosis of their tips. Kidneys from a few animals had foci of acute renal cortical tubular necrosis. Survivors sacrificed 3 weeks after dosing showed areas of regeneration in the gastric mucosa (Ballantyne and Beswick, 1972; Ballantyne and Swanston, 1978).

Inhalation; Pyrotechnically Generated Smoke

(a) The comparative lethality of CS grenade-generated CS smoke to several species was determined by exposure to high concentrations of CS (around 4 g m^{-3}) for 5 to 20 min. LCt_{50} data (with 95% confidence limits) were as follows (Ballantyne and Callaway, 1972):

Guinea pig: 35 000 (25 000–45 000) mg min m^{-3}
Rabbit: 63 000 (50 000–80 000) mg min m^{-3}
Rat: 68 000 (61 000–77 000) mg min m^{-3}
Mouse: 76 000 (61 000–119 000) mg min m^{-3}

Lungs of animals that died were macroscopically oedematous and congested, with multiple variable sized haemorrhages. Histological examination of these lungs demonstrated moderate to severe congestion of alveolar capillaries and intrapulmonary veins, alveolar haemorrhages, haemorrhagic atelectasis and pulmonary oedema. Intrapulmonary bronchi and bronchioles had excess mucus. Survivors sacrificed 14 days post-exposure showed no residual histopathology.

(b) To define tissues lesions in detail, histopathology was conducted following single exposures of rats and hamsters to various concentrations of pyrotechnically generated CS smoke for different exposure times and sacrificed at various times after exposure. Animals that died or were sacrificed at 1, 10 and 28 or 29 days post-exposure had necropsy to examine for any gross pathology with the following tissues removed and fixed for subsequent histological examination: lung, heart, small intestine, liver, pancreas, spleen, kidney, adrenal glands, gonads, brain and bone marrow. Results for the various exposure concentrations were as follows.

(i) 750 mg m^{-3} CS for 30 min; Ct 22 500 mg min m^{-3}. There were no mortalities. At day 1 sacrificed hamsters and rats had minimal lung pathology (alveolar capillary congestion, few scattered alveolar haemorrhages) and one rat had a few small foci of renal tubular necrosis at the inner cortex. There was no histopathology in animals sacrificed at 10 and 28 days.

(ii) 480 mg m^{-3} CS for 60 minutes; Ct 28 800 mg min m^{-3}. Mortalities which occurred up to 14 days post-exposure, with the majority during the first 48 h, were as follows: male hamsters 34%, female hamsters 25%, male rats 10% and female rats 5%. The lungs of all rats that died had moderately severe congestion of the alveolar capillaries and intrapulmonary veins, scattered intra-alveolar haemorrhages and occasional patches of pulmonary oedema. The livers were depleted of glycogen and about half of them congested, and in a few there was centrilobular or inner midzonal necrosis. Kidneys showed moderately severe and extensive tubular necrosis, being particularly marked in the medulla. Adrenal glands were depleted of lipid. There was no histopathology in the heart, small intestine, pancreas, gonads, spleen, myeloid tissue and brain. Hamsters that died within 48 h of exposure had moderately severe pulmonary congestion, scattered intra-alveolar haemorrhages and minimal pulmonary oedema. There was depletion of liver glycogen, but otherwise this organ was normal. Kidneys showed tubular necrosis in both cortex and medulla. Adrenal glands showed lipid depletion, but were otherwise normal. Remaining tissues were normal. Hamsters that died 48 h post-exposure had lungs which showed either no pathology or minimal lesions. A few hamsters had hepatocellular necrosis, but kidneys from all hamsters had tubular necrosis of varying severity and extent. Survivors sacrificed 1 day post-exposure showed minimal pathology; a few rats had minor pulmonary congestion and a few scattered alveolar haemorrhages, and one had slight hepatic congestion with scattered necrotic hepatocytes; in hamsters, there was minimal congestion, few scattered haemorrhages and occasional oedema, and two hamsters had small foci of necrosis at the corticomedullary junction.

(iii) 150 mg m^{-3} for 120 min; $Ct = 18\,000$ mg min m^{-3}. Only two of the exposed animals died; these were hamsters which died at 10 and 16 days post-exposure, with bronchopneumonia as the only pathological finding. Of the animals sacrificed at 1 day, the only findings were a few scattered alveolar haemorrhages and (in hamsters) a few small scattered foci of acute renal tubular necrosis at the inner cortex.

Overall, the above findings indicate that high exposure concentration is the main determinant for acute toxicity (Ballantyne and Callaway, 1972).

Inhalation: Pure CS Aerosols

Animals were exposed to aerosols of CS by passing nitrogen through molten CS in a Collison spay and introducing into the air stream to the exposure chamber. The species used (with ranges of exposure times, exposure concentrations and dosages) were as follows: male rats, 10–60 min, 1802–2699 mg m^{-3}, 18 020–161 930 mg min m^{-3}; female rabbits, 5–60 min, 836–3066 mg m^{-3}, 4230–183 960 mg min m^{-3}; male mice, 15–30 min, 1432–2550 mg m^{-3}, 21 480–76 500 mg min m^{-3}; female guinea pigs, 10–45 min, 1302–2380 mg m^{-3}, 23 260–71 200 mg min m^{-3}. LCt_{50} values calculated from the inhalation exposure dosage-mortality data were as follows:

Rat, male 88 480 mg min m^{-3} (slope = 2.8)
Rabbit, female 54 090 mg min m^{-3} (slope = 3.1)
Guinea pig, female 67 200 mg min m^{-3} (slope = 3.8)
Mouse, male 50 010 mg min m^{-3} (slope = 3.4)

Exposed animals on removal from the exposure chamber had increased buccal and nasal secretions and increased breathing rates. These signs disappeared within 1 h of removal from the chamber. The lungs from animals that died during the first 48 h following exposure were congested and oedematous and had multiple variable sized pulmonary haemorrhages. Histology showed moderate to marked alveolar capillary congestion and intrapulmonary veins, inter- and intra-alveolar haemorrhages, and increased mucus in the lumen of bronchioles and intrapulmonary bronchi.

There was congestion of the liver, kidney, spleen and small intestine. A few animals with more extensive lung injury had foci of acute renal cortical and medullary tubular necrosis. Survivors sacrificed 14 days after exposure to CS aerosol did not have any gross or histological pathology (Ballantyne and Swanston, 1978).

Rats and hamsters were exposed acutely to CS at 480 mg m^{-3} for 1 h and 150 mg m^{-3} for 2 h ($Ct = 18\,000$ mg min m^{-3}) and sacrificed at 32 months. There were no effects on survival, and no CS-related toxicity (Marrs *et al.*, 1983b).

Estimates for the acute human lethal inhalation dosage for CS vary between 25 000 and 150 000 mg min m^{-3} (WHO, 1970), but there have been no authenticated reports of death from exposure to CS smokes. The Committee of Enquiry into the use of CS in Northern Ireland concluded that the likelihood of death after inhaling an exposure dose of pyrotechnically generated smoke that is one-tenth that which would lead to death of 50% of those exposed is less than one chance in 100 000, i.e. the accumulation of an inhalation exposure dose of pyrotechnically generated smoke of the order of 5000 mg min m^{-3} would constitute a virtually negligible risk to life.

PRIMARY IRRITATION

Skin. The potential to cause skin irritation was investigated in rats, guinea pigs and mice who had 0.1 ml 12.5% CS in either corn oil or acetone applied unoccluded to the shaven dorsal trunk skin for 6 h, and the area subsequently and periodically inspected up to 21 days for local reactions. Erythema was more marked than oedema, and both resolved by 7 days post-application. Histological examination of skin biopsies taken from a few animals at 3 days after CS application showed foci of epidermal necrosis in the contaminated area with spongiosis and acute inflammatory infiltration of the outer dermis (Ballantyne and Swanston, 1978).

In humans, contamination of the skin usually results in erythema, but the precise effects depend on concentration, formulation and environmental temperature. At higher ambient temperature and relative humidity the cutaneous irritant effects are more prominent, possibly because of excessive diaphoresis (Salem *et al.*, 2005e). Two phases of erythema have been described following contamination of the skin with solutions of CS; immediate erythema appearing within a few minutes and persisting for less than 1 h, followed about 2 h later by delayed erythema persisting for 24–72 h (Weigand, 1969). Occasionally, if exposures have been heavy and sustained there may be more marked cutaneous reactions, including oedema and vesication (Hellreich *el al.*, 1967; Salem *et al.*, 2005e). Primary contact dermatitis from CS exposure is less marked than that from CN (Holland and White, 1972).

Eye. The ophthalmic toxicology of CS has been investigated in rabbits using solutions, solid material and pyrotechnically generated smoke. Solutions of CS were tested in PEG 300, using 0.5, 1, 2, 5 and 10% CS. The following were noted, all showing concentration–effect relationships for both duration and severity. Excess lacrimation was seen in all treatment groups and ranged from mild and of 24 h duration at 0.5% to marked and taking 2 weeks to subside at 10%. Blepharitis was mild and 24 h in duration at 0.5% and moderate and persisted for 2 weeks at 10%. Chemosis and hyperaemia of the conjunctivae and nictitating membrane were mild and persisted 24 h at 0.5% and moderate and 2 weeks in duration at 10%. Iritis was not seen at 0.5%, but moderate and of 24 h duration with 1 and 2%, and just detectable and of 1 to 4 days duration at 5%. With 10% CS, iritis was mild to moderate, persisted for 2 to 7 days and histologically was seen as congestion of vessels with minimal neutrophil infiltration of the stroma. Keratitis was not observed at 0.5%, and was just detectable and of 1 to 3 days duration with 1% and 3 to 7 days duration with 2%. With 5% CS, keratitis was mild to moderate and in most animals persisted from 4 to 14 days. Peripheral neovascularization of the cornea was seen in a few animals, but regressed within a few days. With 10% CS, a just detectable keratitis was seen by 24 h that became more marked by 2 days, and thereafter slowly resolved. Histology of eyes from animals still demonstrating a gross keratitis showed patchy detachment of corneal epithelium, thickening of the substantia propria and neutrophil infiltration. These findings indicate that inflammatory effects will result from contamination of the eye with solutions of CS in PEG 300, with a threshold of around 0.5 % and just detectable transient keratitis occurring at 1% CS. A comparison of the eye irritation caused by 5% CS with that produced by 3.72% *o*-chlorobenzaldehyde or 1.75% malononitrile (the concentrations of the hydrolysis products of CS) suggested that any injury produced by CS is due to the parent material and not its products of hydrolysis. Solid CS applied to the cornea produced irritant effects less marked than with similar amounts of CS in solution. Even 5 mg of solid CS caused only a trivial and transient keratitis compared with the more marked injury resulting from a 5% solution. CS smoke was tested by exposing rabbits for 15 min to the pyrotechnic effluent from CS grenades detonated in a 10 m^3 chamber; average concentration was 6 g m^{-3}. The only ocular effects noted were transient slight excess of lacrimation, a just detectable blepharitis and hyperaemia of the conjunctival vessels of about 24 h duration. Histological examination of eyes 7 days after exposure revealed no abnormalities. The eye-irritating effects of a given CS concentration may be dependent on the solvent used. Thus, Gaskins *et al.* (1972) found that 10% CS in 1,1,1-trichloroethane did not induce cornel injury, but 10% CS in dichloromethane caused a moderate keratitis (Ballantyne, 1979).

The effect on IOP of solutions of CS in PEG 300 (0.125 to 5.0%) applied topically to the eye was investigated in rabbits receiving acute 0.1 ml volumes. Peak-concentration-dependent increases in IOP were measured at 10 min post-application, ranging from 7% with 0.125% CS to 52% with 5.0% CS; pressures returned to control values by 1 h (Ballantyne *et al.*, 1974).

Rengstorff and Mershon (1969) contaminated the eyes of human volunteers with 0.1 or 0.25% CS in water. This resulted in marked blepharospasm persisting for 10 to 135 s and transient conjunctivitis, but no evidence for corneal injury on slit-lamp biomicroscopy. They further examined CS by applying, as drops or spray, 0.1 or 0.25% CS solutions in water and 0.5% polysorbate 20 (Rengstorff and Mershon, 1971). After 1 to 5 s, there was intense eye pain and blepharospasm, the effects being worse with aerosol than with drops. Attenuation of effects began by 10 min and was resolved by 30 min. There was no evidence of edema or injury to the corneal epithelium or stroma. Gutentag *et al.* (1960) exposed human volunteers to CS in a wind tunnel at < 5 mg m^{-3}. There was instantaneous conjunctivitis, lacrimation and eye burning sensation and pain. After exposure, pain resolved in 2 to 5 min, lacrimation disappeared in 12 to 15 min and conjunctivitis cleared within 25 to 30 min.

SENSITIZATION

There is evidence from both animal studies (Rothberg, 1970) and human experience (Fuchs and in der Wiesche, 1990; Schmunes and Taylor, 1973) that CS is a hapten and causes allergic contact dermatitis, although less potent as a skin sensitizer then CN (Fisher, 1970). Chung and Giles (1972) studied the potential for skin sensitization by topical application of daily induction concentrations of 0.5 or 1.0% CS in acetone, and after a 3-week rest period applied challenges with 0.1 ml of 0.2 and 1.0% CS. Both concentrations demonstrated a sensitizing reaction.

REPEATED EXPOSURE TOXICITY

Short-Term Repeated Inhalation Exposure

Rats, rabbits, guinea pigs and mice were exposed to pyrotechnically generated CS smoke at a concentrations ranging 34.2–56.4 m^{-3} for 5 h a day, and from 1 to 7 successive daily exposures. Survivors were sacrificed 14 days after their final exposure. Analysis of the cumulative inhalation exposure dosage versus mortality data gave LCt_{50} values for these low concentration multiple-exposure conditions as follows:

Guinea pigs, 49 000 mg min m^{-3}
Rabbits, 54 000 mg min m^{-3}
Rats, 25 000 mg min m^{-3}
Mice, 36 000 mg min m^{-3}

Animals that died after the repeated exposures had moderate to marked congestion of the alveolar capillaries and intrapulmonary veins, and multiple variable sized areas of haemorrhage, both inter- and intra-alveolar; a few animals had moderate pulmonary oedema (Ballantyne and Callaway, 1972).

Pathology of Rats Having Short-Term Repeated Daily Exposures to Thermally Generated CS Aerosols

The following two experiments with rats were performed using high and low CS concentrations for short and long periods, respectively. CS aerosol was generated from molten CS in a Collison spray.

(a) Exposure for 5 min a day for 5 successive days to between 840 to 3050 mg m^{-3} CS. The 5-day cumulative *Ct*s ranged 41 080 to 44 100 mg min m^{-3}. There were no mortalities during or following exposure. Eighteen rats had lesions at sacrifice. During the first 2 days following the exposures, there was minimal alveolar capillary congestion with a few scattered alveolar haemorrhages. Rats sacrificed between 2 and 18 days after the final exposure showed scattered patches of bronchopneumonia.

(b) Exposures to average CS concentrations of 12.5–14.8 mg m^{-3} for 80 min a day for 9 successive days; total cumulative *Ct*s ranged 907–10 785 mg min m^{-3}. There were 5/50 (10%) mortalities; these occurred after the 7th or 8th exposure in 3 rats and 5 days after the final exposure with 2 rats. All had widespread acute bronchopneumonic changes. Several survivors sacrificed 1–6 h following the final exposure had alveolar capillary congestion and a few scattered alveolar haemorrhages. Those sacrificed 1–5 days after the final exposure had bronchopneumonic changes (Ballantyne and Callaway, 1972).

Punte *et al.* (1962b) exposed dogs to CS for 1 min per day, 5 days a week for 5 weeks, and rats for 5 min per day, 5 days a week for 5 weeks; the daily *Ct*s were 680 mg min m^{-3} (dogs) and 3640 mg min m^{-3} (rats), and the cumulative *Ct*s were 17 000 mg min m^{-3} and 91 000 mg min m^{-3}, respectively. No dogs died, but 3/60 rats died during exposure after cumulative exposure dosages of 25 000 and 68 000 mg min m^{-3}. There were no effects on clinical chemistry (electrolytes and creatinine) and no gross pathology. Marrs *et al.* (1983c) exposed mice, rats and guinea pigs to a 3–4 μm aerosol of CS at concentrations of 3, 30 and 192–236 mg m^{-3} for 1 h per day for 120 days, and sacrificed survivors at 1 year after the start of exposures. Excess mortality was noted in the high concentration group of each species and was related to concentration rather than cumulative inhalation exposure dosage. Acute alveolitis was seen in guinea pigs that died, but there were no pathological findings in rats or mice after 12 months in the medium- or low-exposure groups or in surviving animals in the high-dose group. In guinea pigs there was histopathological evidence of chronic laryngitis and tracheitis in the medium-exposure group and survivors of the high-concentration group.

In a study preliminary to an NTP carcinogenesis bioassay, mice and rats were exposed to CS2 up to a concentration of 100 mg m^{-3} for 6 h per day, 5 days a week for 2 weeks. CS2-related effects were blepharospasm, listlessness, nasal discharge and mouth breathing. This study was followed by a sub-chronic investigation in which rats and mice were exposed to graded concentrations of CS2 in the range 0.4 to 6.0 mg m^{-3} (maximum daily *Ct* was 2160 mg min m^{-3}); exposures were 6 h per day, 5 days a week for 13 weeks. In rats, CS2-related lesions were nasal mucosal erosions, hyperplasia, squamous metaplasia and inflammatory cell infiltration; inflammation and hyperplasia of the epithelium was seen in the larynx and trachea. In mice, there was squamous metaplasia and inflammation in the nasal mucosa, and inflammation and hyperplasia of the tracheal and laryngeal epithelium (NTP, 1990b).

BIOCHEMICAL MECHANISMS

Like CN, CS is an SN_2-alkylating agent that forms adducts with thiol groups and inhibits SH-containing enzymes which may form the basis for some aspects of the acute toxicity of CS. Some examples of probable enzyme biochemical interactions of CS are as follows: alkylation of GSH peroxidase causes GSH depletion with metabolic dysfunction and oxidative damage; alkylation of GSH S-transferase also results in GSH depletion and decreased detoxification; alkylation of lactate dehydrogenase leads to disturbed glycolysis; alkylation of dihydrolipoic acid causing lipoic acid depletion results in decreased acetyl coenzyme A (Olajos, 2004). In view of the alkylating properties of CS, and its potential to disrupt regulatory proteins, investigations have been conducted to assess the effect of CS on cell cycle. Weller *et al.* (1995) used bromodeoxyuridine/Hoechst flow cytometry with Chinese hamster embryo (CHE) and human amniotic fluid-derived fibroblast (AFFL) cells to analyze interference of CS with cell cycle dynamics. It was found that there was arrest in the G0/G1 phase in synchronized CHE cells. In synchronously growing CHE cells, CS treatment resulted in cell kinetic perturbations. Human asynchronous AFFL cells showed marked cell cycle perturbations, at concentrations less than those causing cell cycle perturbations in CHE cells. The investigators suggested that the CS-induced cell cycle perturbations were the result of interactions of CS with cell cycle regulatory proteins that are involved in traversing the transition point in the late G1 phase, in DNA replication and/or spindle formation.

Given by parenteral injection (intraperitoneal) cyanide is generated from the malononitrile component; however, this is not a significant factor by other routes of exposure (see p. 574). By the oral routes there are haemorrhages and fluid loss in addition to the intrinsic toxicity of CS. By inhalation, the lethal toxicity of CS is by lung damage, leading to asphyxia and circulatory failure, which may be compounded by bronchopneumonia. Although there is histopathological evidence for centrilobular hepatic necrosis and focal acute renal cortical tubular necrosis in animals dying after high inhalation doses of CS, these are secondary to asphyxia and circulatory failure and not a direct hepatonephrotoxic effect from CS.

METABOLISM AND TOXICOKINETICS

Major metabolites of CS are 2-chlorobenzyl malononitrile, *o*-chlorobenzaldehyde and *o*-chlorohippuric acid, the glycine conjugate of *o*-chlorobenzaldehyde (Feinsilver *et al.*, 1971; Leadbeater, 1973). Exposure of cats and rats to aerosols of CS results in absorption of the material with its presence being detectable in blood along with two metabolites, *o*-chlorobenzaldehyde and *o*-chlorobenzylmalononitrile (Leadbeater, 1973). Over a 1-h exposure of cats to 750 mg m^{-3} CS aerosol, the blood concentrations of CS and *o*-chlorobenzylmalononitrile rapidly reached plateau levels, but that of *o*-chlorobenzaldehyde continued to rise throughout exposure. Reducing the aerosol exposure concentration to cats by a factor of 10 did not produce a proportionate decrease in the blood concentrations of CS and its metabolites; CS was reduced by 4.5, *o*-chlorobenzylmalononitrile by 7.7 and *o*-chlorobenzaldehyde by 5.9. This suggests that the two metabolites found in blood are not derived solely from CS absorbed from the respiratory tract, but that the compounds are absorbed independently. In the respiratory tract some CS must be converted to *o*-chlorobenzaldehyde, probably by hydrolysis, and some CS is converted to *o*-chlorobenzylmalononitrile by enzymatic reduction with NADPH (Leadbeater *et al.*, 1973). To determine principal sites of absorption, studies were conducted in which tracheal cannulated cats received the exposure through an oronasal mask or by exposure through the tracheal cannula. The findings showed that CS was absorbed through both the upper and lower respiratory tract; the blood concentrations were, respectively, 30 and 80% of those found in the intact cat. Pre-exposure to CS aerosols for 5 min per day for 4 days reduced the absorption of CS, possibly because of its enhanced metabolism and/or excretion. When rats were exposed to CS aerosols ranging 14–245 mg m^{-3} for 5 min, CS and *o*-chlorobenzylmalononitrile were detected in blood immediately following exposure, but *o*-chlorobenzaldehyde was detected only in those rats exposed to concentrations > 100 mg m^{-3}. The blood concentration of CS increased linearly with increasing aerosol exposure concentration up to 100 mg m^{-3} and then remained constant. Pre-exposure of rats to 100 mg CS m^{-3} for 5 min per day for 4 consecutive days significantly reduced CS and *o*-chlorobenzylmalononitrile in blood. Regression analysis of the rat data showed that there was a threshold concentration below which CS or *o*-chlorobenzylmalononitrile did not appear in blood; namely, 7 mg m^{-3} for CS and 6 mg m^{-3} for *o*-chlorobenzylmalononitrile. Pre-exposure to CS raised the threshold concentration to 19 g m^{-3} for CS and 11 mg m^{-3} for *o*-chlorobenzylmalononitrile. Following intragastric dosing of cats with CS, the blood concentrations of CS, *o*-chlorobenzaldehyde and *o*-chlorobenzylmalononitrile reached maximum values after 30 min and then slowly decreased. Ninety minutes after dosing, blood CS had returned to zero, but those of *o*-chlorobenzaldehyde and *o*-chlorobenzylmalononitrile were still high. This is probably due to the conversion of CS into the two metabolites in the gastrointestinal tract, and all three materials are absorbed independently. In rats, CS was not detected in blood until high doses of CS were administered, but *o*-chlorobenzaldehyde and *o*-chlorobenzylmalononitrile were present at lower CS dosages. The blood concentrations of CS and metabolites were significantly lower than with the cat. The *in vitro* blood half-lives in the cat for CS, *o*-chlorobenzaldehyde and *o*-chlorobenzylmalononitrile were found to be short, being 5.5, 4.5 and 9.5 s, respectively. The *in vivo* half-lives of CS and its metabolites were measured in freshly drawn heparinized blood from cats, rats and humans. The half-life of CS was essentially the same from all species (cat and man, 5 s; rat, 7 s) but there were wide variations in the *in vivo* blood half-lives of *o*-chlorobenzaldehyde (cat, 70 s; man and rat, 15 s) and *o*-chlorobenzylmalononitrile (cat, 470 s; rat, 30 s; man, 660 s).

There is evidence that, due to its malononitrile component, the major cause for mortality following acute intraperitoneal dosing is cyanide intoxication. The evidence for acute lethal cyanogenesis by this route is as follows (Ballantyne, 1983; Cucinell *et al.*, 1971; Jones and Israel, 1970):

(a) After a lethal acute intraperitoneal injection of CS, the concentrations of cyanide measured in blood and various tissues are similar to those measured after an acute intraperitoneal injection of a cyanide salt.
(b) Intraperitoneal CS produces a significant inhibition of cytochrome c oxidase.
(c) Sodium thiosulphate protects from the early toxicity of an intraperitoneal CS injection.
(d) The signs of acute intraperitoneal poisoning and LD_{50} values are similar, compared on a molar basis, between CS and inorganic cyanide.
(e) Following equitoxic intraperitoneal dosages of CS, malononitrile and HCN there is increased urinary thiocyanate excretion of a similar magnitude for the three compounds; the thiocyanate yields and isotoxic molar levels for cyanide and CS indicate that only one of the two nitrile residues in the malononitrile radical is cyanogenic *in vivo*.

However, even when given by the intraperitoneal route, mechanisms other than cyanogenesis are responsible for the acute toxicity. Thus, although thiosulphate affords some protection against acute intraperitoneal toxicity, it is less effective than it is against inorganic cyanide or malononitrile. Alkylation reactions are probably also involved. Lethality from airborne CS is by lung damage leading to asphyxia or, with delayed deaths, bronchopneumonia secondary to respiratory tract infection. Although, as discussed above, CS can be absorbed across the respiratory tract, while cyanogenesis can occur by this route it is very limited Thus, while Frankenberg and Sörbo (1973) were able to demonstrate elevated urinary thiocyanate after exposure to CS aerosols, this required a high inhalation exposure dosage; 21 000 mg min m^{-3} (3.5 g m^{-3} for 6 min). In the context of human exposures to CS, the likelihood of toxicologically significant cyanide being generated is very remote for the following reasons. First, the PCSI effect of CS will cause exposures to be self-limiting. For example, the IC_{50} for CS is 3.6 mg m^{-3}. For an exposure of 10 mg m^{-3}, which the majority of a population will find intolerable, a minute volume of 20 l, and a 1-min exposure (Ct = 10 mg min m^{-3}), then assuming complete retention of the inhaled CS, the amount of CS absorbed would be about 1.05 μmol. Since only one nitrile radical of the CS molecule is effectively cyanogenic, the inhaled CS would yield 1.05 μmol of cyanide. This is about equivalent to the cyanide content of a 30 ml puff from a cigarette (Osborne *et al.*, 1956). This accords with the observation that urinary thiocyanate was not elevated in human volunteers exposed to CS aerosols for their tolerance times, with dosages in the range 0.03–9.0 mg min m^{-3} (Swentzel *et al.*, 1970), and the finding noted above that six men exposed to a Ct of CS of about 90 mg min m^{-3} only a trace of one metabolite, *o*-chlorobenzylmalononitrile was detected in one man (Leadbeater, 1973; Leadbeater *et al.*, 1973).

DEVELOPMENTAL TOXICOLOGY

Conventional animal developmental studies in rats and rabbits exposed to CS aerosols at a concentration of around 10 mg m^{-3} for 5 min have shown that CS is not embryotoxic and teratogenic in rats and rabbits (Upshall, 1973). However, the exposure conditions, meant to simulate riot control situations, do not allow for definitive statements regarding the absence of embryofetoxicity under other (more extreme) conditions (Olajos and Salem, 2001).

Human epidemiological evidence provides no indication that CS is teratogenic or embryotoxic. Abortion, stillbirth and congenital abnormality records from Londonderry after the 1969 riots showed no association between CS exposure and interference with human pregnancy (HMSO, 1971; Lancet, 1971). McElhatton *et al.* (2004) in a prospective study collected outcome data of a series of 30 women who were exposed to CS during pregnancy. Transient maternal symptoms of ear, nose and throat irritation were present in two-thirds of the females, but there was no significant increase in adverse pregnancy outcome.

REPRODUCTIVE TOXICOLOGY

No definitive studies are available.

GENETIC TOXICOLOGY

Overall, experimental studies indicate that while CS binds to proteins, its binding capacity to DNA is low. For example, ^{14}C-CS given by intraperitoneal injection results in radioactivity being associated with liver and kidney proteins and little (about one-hundredth) with DNA (Von Daniken *et al.*, 1981), and thus the potential for direct DNA damage is low. CS does not stimulate DNA repair in V79 cells, 3T3 human fibroblasts, or in A549 alveolar tumor cells (Ziegler-Skylakakis *et al.*, 1989). CS covalently interacts with thiols or amino groups on proteins. A dose-dependent increase in spindle disturbances was observed after a 3-h exposure of V79 cells to CS (Schmid *et al.*, 1989). In a follow-up study, when the exposure time was increased to 20 h (two cell cycles), a significant increase in aneuploid cells was observed after exposure to CS and *o*-chlorobenzaldehyde, but not malononitrile (Schmid and Bauchinger, 1991), which is consistent with damage of spindles only after cells have attempted to divide. Mitotic spindle disruption can lead to the formation of micronuclei, and in V79 cells a concentration-dependent effect on micronucleus formation was observed (Ziegler-Skylakakis *et al.*, 1989). Such results have also been reported by other workers, leading to the conclusion that CS is an *in vitro* aneugen (Miller and Nusse, 1993; Nusse *et al.*, 1992; Salassidis *et al.*, 1991). However, *in vivo* studies have failed to detect micronucleus formation in mice given CS intraperitoneally or perorally (Grawe *et al.*, 1997; Wild *et al.*, 1983).

Several *Salmonella typhimurium* bacterial reverse-mutation studies have been conducted with CS, most of which show an absence of mutagenic potential in the absence or presence of metabolic activation (Meshram *et al.*, 1992; Rietveld *et al.*, 1983; Wild *et al.*, 1983; Zeiger *et al.*, 1987). One study showed CS to be apparently mutagenic in TA 100, but only after 72 h incubation at a high dose of 2 mg plate^{-1} (Von Daniken *et al.*, 1981). Since Wild *et al.* (1983) reported that CS is cytotoxic at 1.5 mg plate^{-1}, Durnford (2004) suggested that the TA100 mutagenic response was probably a reflection of toxicity. CS and CS2 were not mutagenic in *Salmonella typhimurium* strains TA98, TA1535 and TA1537 with or without metabolic activation (NTP, 1990b).

Several mammalian *in vitro* genotoxicity studies have been conducted with CS. V79 Chinese hamster ovary (CHO) cells exposed to CS (3 h, 75 μM) had 4 to 5 times more revertants to 6-thioguanine (TG) resistance than controls (Ziegler-Skylakakis *et al.*, 1989). CS was weakly mutagenic in the mouse L5178 tk^{+}/tk^{-} lymphoma forward gene mutation assay, with and without metabolic activation (McGregor *et al.*, 1988). Chromosomal aberrations and increased sister chromatid exchanges were seen following *in vitro* exposure of V79 CHO cells to CS (Bauchinger and Schmid, 1992).

A review of the available genotoxic data by the US Committee on Toxicology of the National Research Council lead to a conclusion that taken in their totality the tests on CS for gene mutation and clastogenicity make it unlikely that CS poses a genotoxic hazard to humans (Salem *et al.*, 2005c,e).

ONCOGENICITY

In an NTP inhalation bioassay for carcinogenicity, involving 2-year exposures, rats were exposed up to 0.75 mg m^{-3} CS and mice up to 1.5 mg m^{-3}. Non-neoplastic lesions were seen in the nasal cavity; with rats there was hyperplasia, squamous metaplasia, focal inflammation and proliferation of the periosteum of the turbinate bones, and in mice there was inflammation, hyperplasia and squamous metaplasia. In neither species was there evidence for a carcinogenic response (NTP, 1990b).

ADDITIONAL STUDIES

In order to determine if exposure to CS can cause diarrhoea, rats, guinea pigs and rabbits were given CS in PEG 300 by gavage at the species' LD_1 (respectively, 648, 111 and 70 mg kg^{-1}) daily for 5 sequential days, and their stools examined for consistency over the following 7 days. Wet stool production was seen infrequently and there were no differences between the CS-dosed animals and controls given PEG 300 alone. In another study, the same species were given doses of CS by gavage at daily doses of 10% of the species LD_{50} for 5 successive days; these doses were 136.6 mg kg^{-1} for male rats, 21.2 mg kg^{-1} for female guinea pigs and 23.0 mg kg^{-1} for male rabbits. Survivors were sacrificed 72 h after the final dose, submitted to necropsy examination and tissues removed for histology. A few guinea pigs had minimal lesions of the gastric fundal mucosa (petaechial haemorrhages, slight congestion and shallow foci of ulceration). No lesions were seen in lungs, liver, kidney spleen, pancreas or adrenal glands. Rabbits were exposed to grenade-generated smoke 17–58 mg m^{-3} for 12 to 30 min). There was no evidence of an increase in faecal material or its water content (Ballantyne and Beswick, 1972).

The possibility that CS could modify the rate and/or mechanism of healing of cutaneous injuries was investigated in rats (Ballantyne and Johnson, 1974b). Compared to untreated control lesions, up to 10 mg CS applied to abrasions, full-thickness skin wounds, or burns, did not influence the rate of healing.

The production of conversion products from CS during its pyrotechnic generation and dispersion has been studied by Klutchinsky *et al.* (2002). They detected at least sixteen organic thermal degradation products through rearrangement and loss of cyano and chlorine substituents present in the parent CS. They subsequently detected the liberation of hydrogen cyanide and hydrogen chloride (Klutchinsky *et al.*, 2002). Values for HCN were around the ACGIH TLV value.

HUMAN VOLUNTEER STUDIES AND IN-USE OBSERVATIONS

Absorption of CS by the respiratory tract was studied in six male volunteer subjects who were exposed to grenade-generated CS smoke rising from 0.5 to 1.5 mg m^{-3} over 90 min; 2 men left the exposure chamber after 20 min. CS and *o*-chlorobenzaldehyde were not detected in blood samples, and a trace of *o*-chlorobenzylmalononitrile (0.02–03 μM) was detected in the blood of one man exposed for 90 min (Leadbeater, 1973).

Whole-body drenches of human volunteer subjects were conducted using 0.001 to 0.005% CS in aqueous 3.3% (v/v) glyceryl triacetate (Ballantyne *et al.*, 1976a). There was rapid onset stinging in the eye with blepharospasm and excess lacrimation, all of about 3 min duration. This was followed by a stinging sensation of the skin, progressing down from face to neck, back and genitalia. It diminished to mild stinging in a few minutes and resolved by about 10 min. Increases in blood pressure were measured following the irritant drench; with 0.005% CS, the peak rises were measured by 1–2 min with the mean systolic blood pressure (SBP) increase being 31 mmHg (range 5 to 50 mmHg) and a mean diastolic blood pressure (DBP) increase of 19 mmHg (range 0 to 40 mmHg). The blood pressure (BP) returned to control (pre-drench) values within 2 to 13 min of the drench. Prompt onset increases in BP have also been measured in volunteer subjects exposed to CS smokes (Beswick *et al.*, 1972). For 27 volunteers, the control (pre-exposure) mean SBP was 123 mmHg which increased to 142 mmHg on exposure and then decreased to 124 mmHg within 20 min of the start of exposure; the respective corresponding values for DBP were 73, 84 and 75 mmHg. Pulse pressures were 49 mmHg pre-exposure, 57 mmHg after the start of exposure and 49 mmHg by 20 min. Heart rate, measured by ECG, decreased from 80 to 67 beats min^{-1}. There were no significant changes in peripheral blood haematology or blood clinical chemistry.

Exposure of volunteer subjects under controlled conditions did not result in effects on lymphocyte chromosomal morphology (Holland and Seabright, 1971), and no significant change in the chest radiograph, peak airflow, tidal volume or vital capacity (Beswick *et al.*, 1972). Pulmonary gas transfer and alveolar volume were unchanged (Cotes *et al.*, 1972b), but a reduction in exercise ventilation volumes was recorded (Cotes *et al.*, 1972b), possibly due to stimulation of respiratory tract receptors (Cole *et al.*, 1975). Reactive airways dysfunction syndrome (RADS) has been described following exposure to CS (Bayeaux-Dunglas *et al.*, 1999; Hu and Cristiana, 1992; Worthington and Nee, 1999) and after spay exposure to Deep Freeze® containing a mixture of 1% CS and 1% OC. A case of persistent, multisystem hypersensitivity reaction was described in a male sprayed heavily with CS solution (Hill *et al.*, 2000).

A follow-up clinical study of 34 young adults (15 males, 19 females; age range 21–39 years) who were exposed to CS liquid irritant spray in an enclosed space (coach) was conducted (Karagama *et al.*, 2003). At 8 to 10 months post-exposure, clinical examination (including spirometry) did not reveal any specific abnormalities.

Human deaths directly attributable to CS have not been authentically recorded (Hill *et al.*, 2000; Salem *et al.*, 2005a,c,e).

Dibenz(b.f)-1,4-oxazepine (CR)

IDENTITIES AND PHYSICOCHEMICAL PROPERTIES

CAS number: 257-07-8
Molecular weight: 195.2
Molecular formula: $C_{13}H_9ON$
Structural formula: see Figure 10

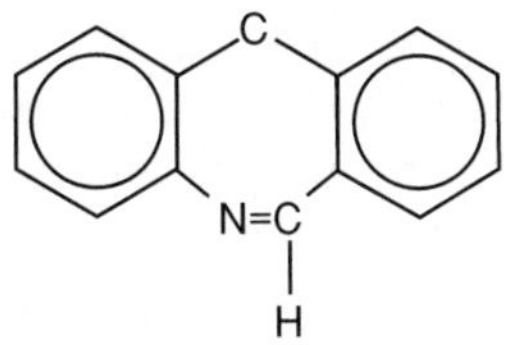

Figure 10. Structural formula of the RCA dibenz(b.f)-1,4-oxazepine (CR)

Physical state: pale yellow solid
Melting point: 72.5°C
Vapor pressure: 5.9×10^{-5} mmHg (20°C)
Vapor density: 6.7 (air =1)
pK_a : 2.97 (25°C)
Solubility: readily in most organic solvents; water – poor solubility = 3.5×10^{-4} mol l^{-1} (~ 7 mg dl^{-1}; 20°C)
Chemical identification: gas chromatography (D'Agostino and Provost, 1985); capillary column GC–MS (D'Agostino and Provost, 1995); thin layer chromatography (Makles *et al.*, 1999).

USES AND DISPERSION

Incapacitant as a solution and harassing agent as aerosol. Pressurized hand-held canister; pyrotechnic grenade. RCA as solution or grenade generated smoke.

PCSI VALUES

Animal Studies

Solution (in saline), blepharospasm EC_{50}, guinea pig = $3.5(2.8 - -4.3) \times 10^{-5}$ molar, rabbit = $7.9\ (5.1\text{–}12.5) \times 10^{-5}$ molar (Ballantyne and Swanston, 1974).

Human Studies

Pain threshold (blister base) = 5×10^{-6} M (Foster and Weston, 1986)
Aerosol, ocular sensation $TC_{50} = 4.0 \times 10^{-3}$ mg m^{-3}
Solution (in saline), blepharospasm $EC_{50} = 8.6\ (6.8\text{–}12.5) \times 10^{-7}$ molar, TC_{50} for ocular sensation = $4.9\ (3.8\text{–}6.5) \times 10^{-7}$ molar, tongue sensation $TC_{50} = 2.1(0.7 - -3.1) \times 10^{-6}$ molar
IC_{50} (aerosol) = 0.7 mg m^{-3}
(data after Ballantyne, 1977a; Ballantyne and Swanston, 1974)

ACUTE TOXICITY

Intravenous: LD_{50} values
Mouse, male (in propylene glycol) 130 mg kg^{-1} (slope = 5.5)
Mouse, female (in propylene glycol) 112 mg kg^{-1} (slope = 17.6)

Rat, male (in propylene glycol) 68 mg kg^{-1} (slope = 19.6)
Rat, female (in propylene glycol) 68 mg kg^{-1} (slope = 12.6)
Rabbit, female (in propylene glycol) 47 mg kg^{-1} (slope = 19.8)

In all species signs developed within a few seconds of injection, consisting of ataxia, tonic extensor spasms convulsions and (rabbits) miosis. Mortalities occurred within 10 min. Pathology seen in animals that died was congestion of liver sinusoids and alveolar capillaries. Survivors, sacrificed 14 days post-dosing, had no gross pathology or histopathological findings.

Intraperitoneal: LD_{50} values

Rat, male (in propylene glycol) 817 mg kg^{-1} (slope = 11.4)
Rat, female (in propylene glycol) 766 mg kg^{-1} (slope = 19.8)
Guinea pig, female: 463 mg kg^{-1}
(slope = 463 mg kg^{-1} (slope = 2.9)

Signs of toxicity generally appeared between 2 and 5 min after dosing and consisted of piloerection, indications of weakness, incoordination and sensitivity on handling. Deaths occurred between 20 h and 4 days post-dosing. Animals that died had mild to moderate congestion of alveolar capillaries, liver, small intestine and kidney, and occasional intra-alveolar haemorrhages. Survivors sacrificed 14 days after dosing had a few intra-abdominal adhesions, but otherwise no gross pathology or histopathology.

Peroral: LD_{50} values

Mouse, female (in propylene glycol) > 4000 mg kg^{-1}
Rat, male (in propylene glycol) 7500 mg kg^{-1} (slope = 7.2)
Rat, female (in propylene glycol) 5900 mg kg^{-1} (slope = 7.5)
Rabbit, female (in propylene glycol) 1760 mg kg^{-1} (slope = 4.40)
Guinea pig, female in propylene glycol) 629 mg kg^{-1} (slope = 9.4)

Animals that survived usually demonstrated an ataxic gait with reduced muscle tone and piloerection for 3 to 4 h, followed by mild sedation for 24 to 48 h. When sacrificed 14 days after dosing, there were no signs of gross pathology and no abnormal histopathology. Animals that died did so between 1 and 6 days post-dosing and demonstrated ataxia, muscle flaccidity, loss of consciousness and breathing difficulties. Necropsy, animals that died showed congestion of the gastric and small intestinal mucosa, a few petaechial haemorrhages in the gastric mucosa, congestion of liver and lungs and a few scattered foci of acute renal cortical tubular necrosis.

Percutaneous: LD_{50} values

Rabbit, female (in corn oil) > 450 mg kg^{-1}
Rabbit, female (in Vaseline paste) > 400 g kg^{-1}
Rabbit, female (in dimethyl sulphoxide) > 1500 mg kg^{-1}
There were no deaths or signs of systemic toxicity following application of CR to the skin. The only finding was mild local erythema. No gross pathology or histopathology was seen in animals sacrificed 14 days post-application.

Inhalation

The acute inhalation toxicity was studied for pure aerosols and grenade generated smoke; for the latter a comparison was made of the pyrotechnic smoke devoid of CR.

Aerosol Studies. CR was generated by blowing high-pressure air through molten CR in Collison or Dautreband aerolyzers. Rats were exposed to average CR concentrations ranging 870–2380 mg $^{-3}$ for between 15 and 180 min to give inhalation exposure dosages of 13 050 to 428 400 mg min m^{-3}. All rats showed blepharospasm and increased nasal and buccal secretions during exposure, but these signs disappeared shortly after exposure. There were no rat deaths during exposure or in a 14-day post-exposure observation period. For rabbits, guinea pigs and mice, the average CR concentrations (with exposure times and dosages) were 670 mg m^{-3} (14 min; 9380 mg min m^{-3}), 490 mg m^{-3} (120 min; 58 800 mg min m^{-3}), 570 mg m^{-3} (120 min; 68 400 mg min m^{-3}), 690 mg m^{-3} (193 min; 133 170 mg min m^{-3}) and 113 g m^{-3} (113 min; 169 500 mg min m^{-3}). Up to, and including an inhalation exposure dosage of 133 170 mg min m^{-3} there were no mortalities among rabbits and mice during exposure or in

Table 3. Peripheral chemosensory irritant potency and comparative potency of *o*-chlorobenzylidene malononitrile CS) and dibenz(b.f)-1,4-oxazepine (CR) aerosols[a]

Material	TC^{b}_{50} (mg m^{-3})	IC^{c}_{50} (mg m^{-3})	IC_{75} (mg m^{-3})	Effectiveness ratio	
				(IC_{50}/TC_{50})	(IC_{75}/TC_{50})
CS	4×10^{-3}	3.6	10.0	900	2500
CR	2×10^{-3}	0.7	1.1	350	550
CP[d]	2.0	5.1	9.1	—	—

Notes:
[a] Data after Ballantyne *et al.* (1977a).
[b] TC, threshold concentration.
[c] IC, incapacitating concentration.
[d] CP, comparative potency (CR/CS).

a 14-day post-exposure observation period, and only one of ten guinea pigs died. At the highest inhalation exposure dosage of 169 500 mg min m^{-3}, mortalities were as follows: rabbits 9/20, guinea pigs 5/20 and mice 2/40; thus, the LCt_{50} was above this value. Hamsters were exposed to CR aerosols as follows: 710 mg m^{-3} (60 min; 42 600 mg min m^{-3}), 1060 mg m^{-3} (220 min; 233 200 mg min^{-3}), 1850 mg m^{-3} (20 min; 370 000 mg min m^{-3}) and 2450 mg m^{-3} (18 min; 440 400 mg min m^{-3}). Mortalities were as follows: 370 000 mg min m^{-3} = 1/10; 440 400 mg min m^{-3} = 2/10.

Pyrotechnically Generated Smoke Studies. Grenades were functioned in a 10 m^3 chamber and rats, rabbits, guinea pigs and mice were exposed for between 20 and 120 min to average CR concentrations in the smoke ranging 2467–4025 mg m^{-3}, resulting in inhalation exposure dosages ranging 80 000–325 000 mg min m^{-3}. Animals did not die during exposure, but exhibited increased nasal and buccal secretions and rapid shallow breathing. These signs disappeared within an hour of the end of exposure. Animals that survived did not exhibit further signs of toxicity, but those that eventually died developed rapid breathing at 6 h before death leading to gasping breathing, decreased mobility and loss of consciousness. Probit analysis of the exposure–mortality data gave the following LCt_{50} values for the various species:

Rats, 139 000 mg min m^{-3} (slope = 6.5)
Rabbits, 160 000 mg min m^{-3} (slope = 6.2)
Guinea pigs, 169 000 mg min m^{-3} (slope = 7.6)
Mice, 203 600 mg min m^{-3} (slope = 6.1)

Histopathological effects seen in the lungs of animals that died up to 3 days post-exposure showed congestion of alveolar capillaries and intrapulmonary veins, multiple focal inter- and intra-alveolar haemorrhages and pulmonary oedema. Liver, spleen, adrenal gland and thymus were moderately congested. In a few animals, there were scattered small foci of acute renal cortical tubular necrosis. Animals that died after 4 days had pulmonary congestion, haemorrhages and oedema, but in addition some animals had bronchiolitis and neutrophil infiltration of the alveoli. Survivors sacrificed 14 days post-exposure had minimal congestion of alveolar capillaries with occasional small foci of intra-alveolar haemorrhage.

Toxicity of Pyrotechnic Smoke. Comparison of the mortality data for the pure CR aerosol exposures and those following exposure to grenade pyrotechnically generated smoke shows that the lethal toxicity of the aerosol was less then that for the CR generated as a smoke. This suggests that some of the products of combustion from the burning mixture may have contributed to the lethal toxicity of the pyrotechnic CR smoke. To conform this, several species were exposed to smoke generated from grenades containing only burning mixture and compared with the effects of mixture containing CR. The results shown in Table 3 indicate that a high proportion of mortalities result from exposure to smoke from the combustion of the burning mix alone which may be a significant factor in mortalities occurring from exposure to pyrotechnically generated CR smoke (Ballantyne, 1977b).

The human acute LCt_{50} for CR is probably in excess of 100 000 mg min m^{-3} (Ballantyne, 1977a).

PRIMARY IRRITATION

Skin. Weigand and Mershon (1970) patch-tested human subjects with CR solutions in the concentration range 0.01 to 1% for 5 to 30 min. Transient erythema was noted that persisted for 2 to 4 h.

Eye. The ophthalmic toxicology of CR was investigated as a solution in PEG 300 at concentrations of 1, 2, 5 and 10% instilled into the rabbit eye. Lacrimation was just detectable and about 1 h in duration with 1 and 2% CR, and mild persisting 2 to 3 days with 5 and 10% solutions. Blepharitis did not occur with 1%, and was only just detectable for 24 h with 2%. Solutions of 5 and 10% CR also produced mild blepharitis but it persisted for about a week. Chemosis did not occur with 1%, was just detectable for 1 h with 2%, and mild to moderate at 1 h with 5 and 10% and resolved over 48 h. Hyperaemia of the conjunctivae and nictitating membrane was just detectable over 24 to 48 h with 1 and 2%, and mild to moderate persisting for 3 days with 5 and 10%. Just detectable iritis was seen with 5 and 10% CR. Keratitis did not occur with 1 and 2% CR, and was just detectable with 5 and 10% persisting for 1 to 3 days. A few (10%) animals had a slightly more marked keratitis. Histology of an eye from a rabbit receiving 10% CR that developed a mild keratitis showed patchy stripping of the corneal epithelium and minimal neutrophil infiltration in a slightly thickened substantia propria.

With solid CR, 2 mg caused a just detectable lacrimation and conjunctival injection lasting less than an hour and 5 mg solid produced just detectable excess lacrimation, conjunctival injection, slight chemosis and minor blepharitis lasting about an hour (Ballantyne *et al.*, 1975).

Rabbits were exposed for 30 min to a CR aerosol of concentration of 360 mg m^{-3} (Ct 10 800 mg min m^{-3}) or 571 mg m^{-3} (Ct 17 130 mg min m^{-3}). The effects on the eye were similar for the two concentrations, consisting of just detectable excess lacrimation and conjunctival vessel injection of 1-h duration (Ballantyne *et al.*, 1975).

The effect of solutions of CR in PEG 300 on rabbit corneal thickness was studied *in vivo* (optical pachymeter) in the range 0.05 to 10%. Peak increases in thickness were measured between 1 and 6 h post-instillation and returned to normal between 2 and 10 days, depending on concentration. The effect on IOP of solutions of CR in PEG (0.5 to 5%) was investigated in the rabbit eye *in vivo* (applanation tonometry). Peak increases in IOP were measured at 10 min post-instillation for concentrations of 1% and above, ranging 20 to 40% and concentration-dependent (Ballantyne *et al.*, 1975).

SENSITIZATION

Studies on human volunteer subjects have shown that CR does not cause cutaneous contact sensitization (Holland, 1974).

REPEATED EXPOSURE TOXICITY

Short-Term Repeated Peroral Toxicity

Two studies were conducted (Ballantyne, 1977b). In one study, rats and guinea pigs were dosed daily with CR in PEG 300 at 10% of the species LD_{50}. There were no mortalities or clinical signs of toxicity, although weight gains did not occur over the dosing period. There was no statistically significant difference in the numbers of wet stools passed compared with PEG 300 controls. This and analysis of the total faecal material passed and its water content indicated that any changes in the fluid content of faecal material was due to dosing with PEG 300 and not due to CR. In a study with rabbits, similarly dosed, there was weight gain during the dosing period and again no effects on faecal material and its water content from dosed CR. Blood collected from rabbits 72 h after the final dose of CR did not show any evidence of CR-producing peripheral blood haematology changes (haemoglobin, haematocrit, erythrocyte count or leukocyte count) or any blood plasma clinical chemistry effects (bilirubin, creatinine, urea, lactate dehydrogenase and glutamate pyruvate transaminase). Necropsy, 72 h after the final dose did not show any gross pathology in the three species studied and no histopathology (lung, liver, kidney, adrenal gland, spleen, stomach, small intestine, pancreas, parotid salivary gland and gonads).

Short-Term Repeated Ocular Contact

A group of ten rabbits had daily applications of 0.1 ml 0.05% CR in PEG 300 to the surface of the cornea for ten successive days and were examined periodically for the following five months (Ballantyne *et al.*, 1975). The only effect seen was during the dosing period and was a just detectable conjunctival hyperaemia persisting for a few minutes after the instillation. Over the 5-month follow-up period, there were no adverse effects on the eye, and histology of the eye at 5 months showed no abnormalities.

Repeated exposure of the eye was studied by Rengstorff *et al.* (1975) who applied 0.025 ml of 5% CR suspension to the eye of rabbits for 5 days a week for 4 weeks, and kept under observation for 60 days. There was transient conjunctivitis, but no corneal effects on slit-lamp biomicroscopy and no fluorescein staining. Light and electron microscopy at 32 days after the final dose did not show any corneal injury.

Short-Term Repeated Inhalation Toxicology

Kumar *et al.* (1994) exposed mice to an average CR concentration of 1008 mg m^{-3} daily for 15 min a day for 10 days. At 5 days, lungs demonstrated minimal alveolar haemorrhage and after 10 days there was moderate alveolar haemorrhage and alveolar capillary congestion and a few foci of renal cortical necrosis.

Sub-chronic Inhalation Toxicology

Groups of hamsters and mice were exposed 5 days a week for 18 weeks to CR aerosols of MMAD 2.86 ± 1.17 μm generated from molten CR: exposure average concentrations and dosages were 5 min to 204 mg m^{-3} (daily *Ct* 1022 mg min m^{-3}, cumulative *Ct* 9200 mg min m^{-3}), 8.6 min to 236 mg m^{-3} (daily *Ct* 2033 mg min m^{-3}, cumulative *Ct* 183 000 mg min m^{-3}), and 15.8 min to 267 mg m^{-3} (daily *Ct* 4222 mg min m^{-3}, cumulative *Ct* 380 000 mg min m^{-3}). Survivors were sacrificed one year after the start of exposure. Mortality occurred at the highest dosage group, but no cause for death could be established although pneumonitis was present in several animals. The only exposure-related pathology found was an increased incidence of chronic laryngeal inflammation (Marrs *et al.*, 1983d).

Sub-chronic Repeated Cutaneous Contact

CR dissolved in acetone was applied to the skin of C3H and Porton-strain mice at 1 mg (equivalent to 40 mg kg^{-1}) daily for 12 weeks (Marrs *et al.*, 1982). 80 weeks after the end of dosing, animals were sacrificed. A high incidence of fatty infiltration of the liver might have been due to acetone exposure, but otherwise there was no abnormal histopathology.

METABOLISM AND TOXICOKINETICS

CR aerosols are rapidly absorbed from the respiratory tract. The plasma half-life of CR after inhalation exposure to CR aerosols is about 5 min, which is consistent with the plasma half-life following intravenous dosing (Upshall, 1977b). Absorption of CR from the gastrointestinal tract and its metabolic fate is similar to that following intravenous administration (French *et al.*, 1983a). Most blood radioactivity at 5 min was as the lactam derivative which decreased bi-exponentially and the parent compound which was barely detectable at 1 h. The sulphate conjugates of the hydroxylactam products of CR metabolism appeared quickly in the vascular compartment. The pattern of distribution was consistent with that of a highly lipophilic material undergoing hepatic bioconversion, biliary excretion, entrohepatic circulation and renal excretion. Comparative metabolic studies were conducted in rats, guinea pigs and monkeys, and it was found that similar metabolism and excretory patterns were noted between the species investigated, and that urinary excretion was the major route of elimination (French *et al.*, 1983a). The lactam derivative dibenz[b.f]-1:4-oxazepin-11-(10H)-one is a primary metabolite and the precursor of urinary hydroxylated metabolites. In the rat, metabolites of CR include dihydro-CR, the amino alcohol of CR (2-amino-2′-hydroxymethyldiphenylether), the lactam (dibenz[b.f]-1:4-oxazepin-11-(10H)-one), hydroxylactams

(4-, 7-, and 9-hydroxylactams) and possibly an arene oxide intermediate. The hydroxylactam intermediates, resulting from microsomal oxidation, undergo sulphate conjugation and are excreted in the urine. Biliary excretion of CR may also occur, as well as sulphate conjugation. Glucuronide conjugation may also occur with the hydroxylactams and the amino alcohol metabolite. *In vitro* metabolic studies with isolated liver preparations and isolated hepatocytes (Furnival *et al.*, 1983) demonstrated that Phase I metabolism of CR involves the following metabolic processes: (a) ring opening and reduction of CR to form 2-amino-2′-hydroxymethyldiphenyl ether, (b) the oxidation of CR at C-11 to form the cyclic lactam, and (c) hydroxylation of the lactam by microsomal mixed function oxidases to the 4-, 7-, or 9-hydroxylactams. The hydroxylactams and amino alcohol intermediates of CR metabolism undergo Phase II conjugation reactions. Additional *in vitro* and *in vivo* metabolic investigations (French *et al.*, 1983b) supported the conclusions that the major metabolic pathway of CR in the rat is oxidation to the lactam, subsequent ring hydroxylation, sulphate conjugation and renal excretion. With the exception of CR-lactam, Phase I metabolites of CR are less acutely toxic than CR.

DEVELOPMENTAL TOXICOLOGY

Studies with rats and rabbits have shown CR not to be embryotoxic or teratogenic under the conditions of dosing by gavage or aerosol exposure (Upshall, 1974). Aerosol exposures were to 2, 20 and 20 mg m^{-3} for 5 to 7 min.

REPRODUCTIVE TOXICOLOGY

No definitive studies are available.

GENETIC TOXICOLOGY

CR is stated not to be genotoxic in a *Salmonella* bacterial mutagenicity test, a CHO forward gene mutation test (HGPRT locus), mouse lymphoma cell assay (L5178Y/tk^+/tk^-), and a micronucleus test (Colgrave *et al.*, 1983).

ONCOGENICITY

No chronic studies are available to assess oncogenic potential.

ADDITIONAL STUDIES

The synthesis of CR from sodium phenoxide and 1-chloro-2-nitrobenzene involves a four-stage route, and three contaminants might occur in the technical CR so produced, namely, 2-dinitrophenyl ether, 2-aminodiphenyl ether and 2-formamidodiphenyl ether. To determine if these contaminants could result in acute hazards, they were individually investigated for acute intravenous toxicity, primary skin irritation studies and eye irritation (Ballantyne, 1977b). All three ether intermediate had intravenous toxicity LD_{50} values in male mice greater than that of CR; the respective values, in mg kg^{-1}, were CR 130, 2-dinitrophenyl ether 838, 2-aminodiphenyl ether 212 and 2-formamidodiphenyl ether 285. Given by gavage to male rats, there were neither mortalities nor signs of toxicity with doses up to 4000 mg kg^{-1} with the dinitrophenyl and aminodiphenyl ethers, and doses up to 2000 mg kg^{-1} with 2-formamidoddiphenyl ether. None of the ethers were irritant to rat skin over a 7-day observation period. A rabbit eye irritation test with 5% solutions showed CR and the three ethers produced mild to moderate excess lacrimation, blepharitis, conjunctival hyperaemia and chemosis, with the effects being slightly greater with CR than any of the ether intermediates. Thus, the three ether intermediate in technical CR should not present an acute toxic hazard.

HUMAN VOLUNTEER STUDIES AND IN-USE OBSERVATIONS

Ashton *et al.* (1977) exposed human volunteers to a 1–2 μm aerosol of concentration 0.25 mg m^{-3} for 1 h. Peak expiratory flow was decreased by 7% after 20 min of exposure, which the investigators considered was due to pulmonary irritant receptors causing bronchoconstriction.

Whole-body drenches of human volunteer subjects was carried out using 0.001 to 0.0025% CR in aqueous 3.3% dipropylene glycol monomethyl ether. There was almost immediate discomfort and pain in the eye with blepharospasm and increased lacrimation, which began to subside within 3 to 5 min. In addition, almost immediately there was stinging of the periorbital skin, which then spread to the remainder of the face and was often accompanied by excess salivation and rhinorrhea. Within two or so minutes, stinging would develop around the neck and irritation of the genitalia developed. Stinging of the shoulders and back would begin at 3 to 4 min, and by 5 to 6 min an intense burning sensation would be present. By 10 min, cutaneous sensations were reduced to mild tingling. In general, stronger symptoms were produced by 0.0025% CR compared with 0.001% CR, but individual variations were marked. Injection of the conjunctivae and cutaneous erythema developed within the first few minutes and both persisted for about 2 h. There was no subjective correlation between the degree of erythema and that of cutaneous sensation. Mean peak increases in SBP were 45 mmHg (range 30 to 70 mmHg) after 0.001 % CR, and 59 mmHg (range 30 to 80 mmHg) following 0.0025% CR. Peak increases in DBP were 23 mmHg (range 15 to 30 mmHg) with 0.001% CR and 20 mmHg (range 20 to 45 mmHg) with 0.0025% CR. The peak values for SBP and DBP for both concentrations of CR were statistically significantly greater than control (pre-drench) values ($p < 0005$). Peak values were attained between 1 and 3 min post-drenching and decreased to control values within 2 to 13 min of drenching. Pulse rates increased in all subjects, but with marked variation between individuals and there were no significant differences between the increases in heart rate for 0.001 and 0.0025% CR, with respective mean increases of 26 and 27 beats min^{-1}. A few subjects had an initial bradycardia, followed after a few minutes by tachycardia. Similar rapid onset increases in BP have been measured in volunteers having CR solutions (0.01 to 0.1%), topically applied to the surface of the cornea (Ballantyne, 1977a,1987).

The effects of solutions of CR on IOP in humans was studied in carefully controlled volunteer studies in which 0.04 ml of either 0.05% or 0.1% CR dissolved in PEG 300 was applied to the surface of the cornea. IOP was measured by applanation tonometry before application of CR solution and at 5 and 15 min and 3.5 and 24 h post-application; measurements were made on the treated and the contralateral untreated eye; blood pressure was measured at the times of tonometry. The results are shown graphically in Figure 11. Peak increases in IOP were measured at 5 min; with 0.05% CR a 40% increase in the treated eye and 16% in the contralateral eye, and with 0.1% CR 44% in the contaminated eye and 17% in the contralateral eye. In the CR-treated eyes IOP decreased to control values in 15 min with 0.05% CR and 3.5 h with 0.1% CR. Changes in IOP in the contralateral eye generally followed the time-course for the transient increase in DBP (Ballantyne *et al.*, 1977a).

Oleoresin capsicum (OC)

IDENTITIES AND PHYSIOCHEMICAL PROPERTIES

Source and Composition. OC is an oily reddish–brown extract of pepper plants of the genus *Capsicum*, principally from *Capsicum annuum* and *Capsicum frutescens*. Depending on the variety of the chilli pepper, OC contains from 0.01 to 1.0% capsaicinoids on a dry mass basis. The extract is a complex mixture of fat-soluble phenols (capsaicinoids), some of which are listed in Table 4. The composition of OC is highly variable and depends on factors such as the conditions of extraction, maturity of the fruit and environmental conditions in which the *Capsicum* plants were grown. Thus, the capsaicinoid content of extracts used in 'pepper spray' varies between manufacturers (1.2–12.6%; Smith and Stopford, 1999). Capsaicin and dihydrocapsaicin make up to 80–90% of the total capsaicinoids in OC extract.

Capsaicin is the most potent irritant component and quantitatively the major capsaicinoid of OC extract and has the following identities and characteristics:

CAS number: 404-86-4
Synonyms: *trans*-8-methyl-*N*-vanillyl-6-nonenamide; *N*-[4(4-hydroxy-3-methoxyphenyl)methyl]-8-methyl-6-nonenamide
Molecular weight: 305
Structural formula: see Figure 12

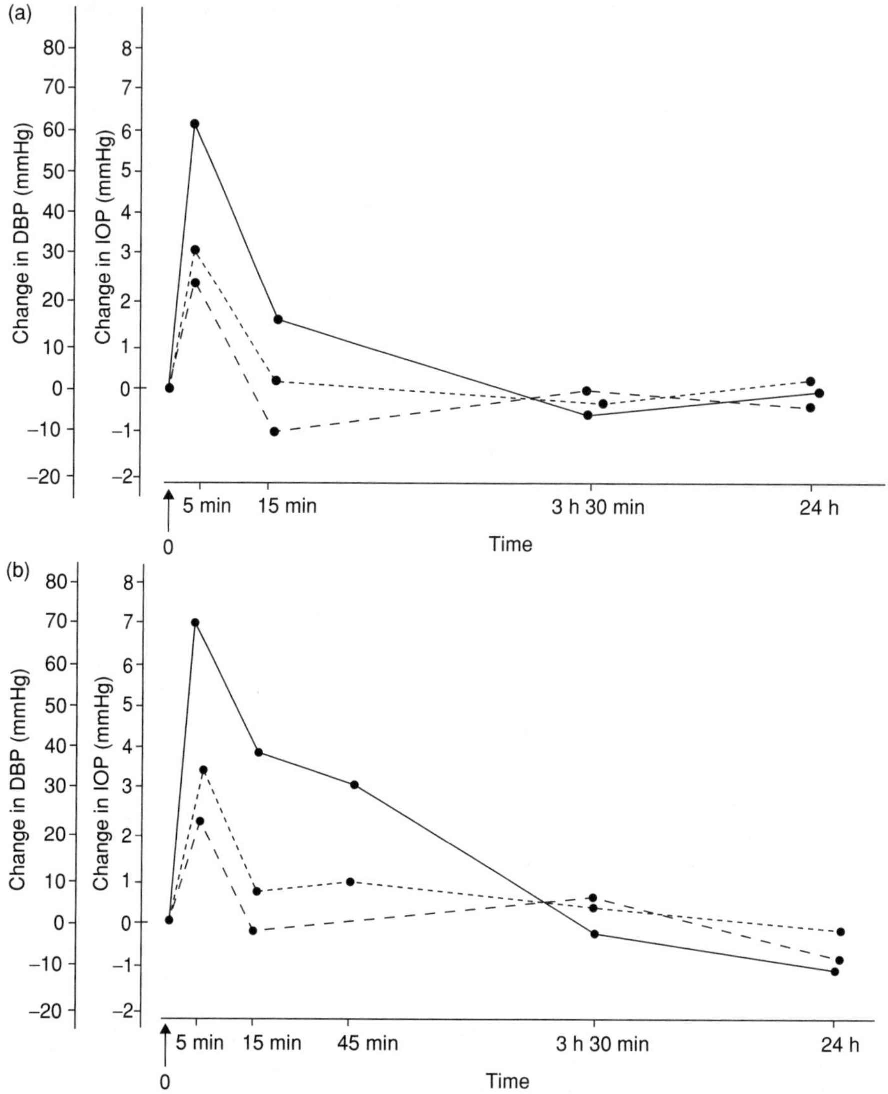

Figure 11. Effects of intraocular pressure (IOP) of the human eye and on systemic diastolic blood pressure (DBP) following the application of solutions of dibenz(b.f)-1,4-oxazepine (CR) in 50/50 vol% of polyethylene glycol 300 and water. A volume of 0.04 ml of either (a) 0.05% or (b) 0.1% CR was applied to one eye: results are given as average for six males in each treatment group: (——), treated eye; (– – –), untreated eye; (· · · · · · · · ·), diastolic blood pressure (data after Ballantyne *et al.*, 1997a)

Melting point: 64°C
Vapor pressure: 0.011 mmHg (20°C)
Identification: TLC (Fung *et al.*, 1982; Jane and Wheals, 1972); IR spectrophotometry (Fung *et al.*, 1982); HPLC (Fung *et al.*, 1982; Krebs *et al.*, 1982); GC–MS (Fung *et al.*, 1982; Hass *et al.*, 1997); liquid chromatography-tandem mass spectrometry (Reilly *et al.*, 2001, 2002a,b).

USES AND DISPERSION

Used as an additive spice in various foodstuffs. In analgesic creams it causes nerve fibre substance

Table 4. Some capsaicinoids in extract of pepper plants (*Capsicum annuum* and *Capsicum frutescenes*) (with approximate % in extract and CAS numbers)[a]

Capsaicin (*trans*-8-methyl-*N*-vanillyl-6-nonenamide; 70%; 404-86-4)
Dihydrocapsaicin (20%; 19408-84-5)
Norhydrocapsaicin (7%; 28789-35-7)
Homocapsaicin (1%; 58493-48-4)
Homodihydrocapsaicin (1%; 279-06-5)
Nonivamide (0.25%)

Note:
[a] After Constant and Cordell (1996), Cooper *et al.* (1991).

P reduction, especially in chronic painful conditions, e.g. postherpetic neuralgia. The active PCSI material in irritant spray devices used for antipersonnel and peacekeeping operations ('pepper spray' is 1–15% OC).

PCSI VALUES

Human Studies

Cough threshold, capsaicin = 6×10^{-6} M
Nasal irritation, capsaicin = 2×10^{-5} M
Nasal irritation (capsaicinoids) = 4 μg m^{-3}
Blister base model, capsaicin, pain at 0.5 μmol l^{-1}
(Chan *et al.*, 1990; Foster and Weston, 1986; Fujimura *et al.*, 1992; Sanico *et al.*, 1997).

For culinary purposes, the relative 'hotness' of capsaicinoids may be measured based on sensation to the tongue; this is usually measured in Scoville units which is the greatest dilution that can be detected by the tongue. For example, the 'hotness' of pure capsaicin is 15 000 000 Scoville units and 10% OC is 1 500 000 Scoville units.

ACUTE TOXICITY

Intravenous: capsaicin = 0.56 mg kg^{-1}
Intraperitoneal: capsaicin= 7.6 mg kg^{-1}
Intramuscular: capsaicin = 7.8 mg kg^{-1}
Subcutaneous: capsaicin = 9.0 mg kg^{-1}
Peroral: capsaicin = 190 mg kg^{-1}

Animals given OC perorally show gastric mucosal hyperaemia, oedema, focal bleeding and focal necrosis.

Percutaneous: 512 mg kg^{-1}

Inhalation:
Rat: LCt_{50} for OC = 835 000 mg min m^{-3}
Mouse: LCt_{50} for OC = 270 000 mg min m^{-3}

PRIMARY IRRITATION

Eye. In the rat, 1% capsaicin applied to the eye causes a neurogenic inflammation and loss of reaction to chemical and mechanical stimuli for up to a week.

REPEATED EXPOSURE TOXICITY

Repeated dosing with capsaicin produces systemic desensitization to chemogenic and thermal nociceptive stimulation (Hayes *et al.*, 1980,1981; Jansco *et al*, 1977; Miller *et al.*, 1980,1981) and has been stated to be the initial manifestation of the long-term neurotoxic action of capsaicin (Olajos and Lakoski, 2004). Experimental studies indicate that exposure to high doses of capsaicin and its analogs results in long-lasing

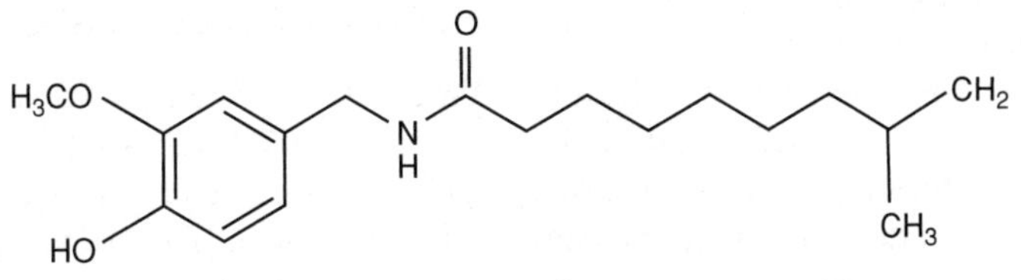

Figure 12. Structural formula of the RCA capsaicin

insensitivity to irritants, pain, and temperature (Govindarajan and Sathyanarayana, 1991). Long-term effects involving the pulmonary system are characterized by desensitization of the airways to chemical irritants and marked inhibition of vagal bronchoconstrictive effects (Lundberg and Saria, 1982).

Intragastric dosing with capsaicin (50 mg g^{-1} day^{-1}) for 60 days caused reduced body weight gains and reductions in plasma urea nitrogen, glucose, phospholipids, triglyceride, transaminase and alkaline phosphatase (Monsereenusorn, 1983).

MECHANISMS

The mechanistic effects of capsaicin on neurones is known to be due to an action on a subpopulation of neuropeptide-containing afferent neurones and involves the activation of a specific receptor (the 'vallinoid' receptor) that recognizes capsaicinoid molecules (Szallasi and Blumberg, 1990a,b,1999; Szallasi *et al.*, 1991). Structure–activity studies have demonstrated a requirement for both the vanilloid ring and acyl chain moieties for pharmacological activity (Caterina and Julius, 2001; Szallasi and Blumberg, 1999). In a more recent nomenclature for transient receptor potential (TRP) cation channels (Montell *et al.*, 2002), the capsaicin receptor (vallinoid receptor type-1; VR1) has been renamed TRPVR1. Activation of vallinoid receptors leads to opening of a particular type of receptor operated cation channel (Marsh *et al.*, 1987; Wood *et al.*, 1988). The influx of Ca^{++} and Na^{+} leads to depolarization which triggers local release of neuropeptides, central protective reflexes, as well as autonomic motor responses (Lundberg and Lundberg, 1984; Martling, 1987; Stjarne, 1991). A transient excitement of primary afferents is followed by a more prolonged condition of refractoriness, as a result of which the primary afferents become unresponsive to further application of capsaicin/capsaicinoids (desensitization). The influx of Ca^{++} and Na^{+} may lead to rapid cellular damage and eventual cell death by osmosis and Ca^{++}-dependent proteases (Jansco *et al.*, 1984). Thus, after subcutaneous capsaicin (50 mg kg^{-1}) to neonatal rats, upwards of 50% of dorsal root ganglion neurones are rapidly destroyed (Fitzgerald, 1983; Jansco *et al.* 1977).

In addition to stimulating afferent nerve fibre endings in mucosal surfaces and producing a typical PCSI effect, OC also induces the release of tachykinins or neuropeptides like substance P and neurokinin A. This induces neurogenic inflammation of the airway blood vessels, epithelial cells, glands and smooth muscle, leading to vasodilation, increased vascular permeability, neutrophil chemotaxis, mucus secretion and bronchospasm (Smith and Stopford, 1999).

METABOLISM AND TOXICOKINETICS

Capsaicin and capsaicinoids undergo Phase I metabolic conversion involving both oxidative and non-oxidative paths. The liver is the major site of this enzymatic activity. Lee and Kumar (1980) demonstrated the conversion of catechol metabolites via hydroxylation of vanillyl ring. In rats, dihydrocapsaicin is metabolized to products that are excreted in the urine as glucuronides (Kawada and Iwai, 1985). The generation of a quinone derivative occurs via *O*-demethylation at the aromatic ring with concomitant oxidation of the semiquinone and quinone derivatives or via demethylation of the phenoxy radical intermediate of capsaicin. Additionally, the alkyl side chain of capsaicin is also susceptible to oxidative deamination (Wehmeyer *et al.*, 1990). There is evidence that capsaicinoids can undergo aliphatic oxidation (ω-oxidation) (Surh *et al.*, 1995; Reilly *et al.*, 2003) which is a possible detoxification pathway. Non-oxidative pathways are also involved in the bioconversion of capsaicin, e.g. hydrolysis of the acid–amide bond to yield vanillylamine and fatty acyl moieties (Kawada *et al.*, 1984; Kawada and Iwai, 1985; Oi *et al.*, 1992).

DEVELOPMENTAL AND REPRODUCTIVE TOXICOLOGY

Pregnant rats were injected with single 10% alcoholic or 10% Tween-stabilized doses of 50 mg OC kg^{-1} on gestation days 14, 18 or 20, or to two doses of 50 mg OC kg^{-1} on days 15–16 and 16–17 of gestation (Kirby *et al.*, 1982). Reduced crown-rump length was noted in pups from dams given 50 mg OC kg^{-1} on gestation day 18, but all other reproductive and developmental monitors from OC groups were comparable to controls. Behavioral analysis of pup activity revealed a 40% depression in capsaicin-exposed animals. Acid phosphatase histochemistry of the spinal cord demonstrated a marked reduction in the substantia gelatinosa of the dorsal root.

Mice given intraperitoneal injections of capsaicin at 0.4, 0.8 and 1.6 mg kg^{-1} day^{-1} on 5 consecutive days did not have alterations in epididymal weights, caudal sperm counts, sperm morphology, testicular weights or testicular histology (Narasimhamurthy and Narasimhamurthy, 1988).

GENETIC TOXICOLOGY

Some studies on the genotoxicity of capsaicin are contradictory, due to the variability in purity of the material studied and also, as discussed under oncogenesis, to metabolism. Bacterial mutagencity studies have given variable results, but generally they were only weakly positive and often only with one strain. Toth *et al.* (1984) found capsaicin to be non-mutagenic in *Salmonella typhimurium* TA 97, TA100 and TA102, and positive with TA98 at near toxic doses with rat, but not mouse, S9 activation. Using an ethanol extract of *Capsicum annum* and *Capsicum frutescens*, positive results were obtained in the absence of S9 with *Salmonella typhimurium* TA98 and TA100 (Damhoeri *et al.*, 1985). An ether extract of *Capsicum annum* was weakly mutagenic with *Salmonella typhimurium* TA100 but not TA97 or TA98 (Azizan and Blevins, 1995). In a mouse bone marrow micronucleus test, fractionated capsaicinoids from *Capsicum frutescens* induced micronuclei (Villasensor and de Campo, 1994). Purified capsaicin induced micronuclei (Arceo *et al.*,1995; Nagabhushan and Bhide, 1985).

ONCOGENICITY

Capsaicin and capsaicinoids may be involved in complex interactions *in vivo* related to carcinogenesis, cocarcinogenesis and anticarcinogenesis, which are a reflection, in part, of metabolism and metabolite activity. Capsaicin is metabolized by the P450-dependent mono-oxygenases of the liver to a phenoxy radical intermediate that can bind to protein and nucleic acid. This activated metabolite can lead to toxic, mutagenic and carcinogenic activity, or interactions with other factors may inhibit such activity. Simultaneously capsaicin and dihydrocapsaicin inactivate cytochrome P450 IIE1 and other microsomal mono-oxygenases by binding irreversibly to the enzyme. This inactivation can have chemoprotective effects against other chemicals that require metabolic activation (Surh and Lee, 1995). In this way, capsaicin attenuates carcinogenicity by heterocyclic amines. It inhibits metabolism and covalent DNA binding of aflotoxin B, and modulates expression of c-*erb* and c-*myc*, two genes frequently activated in carcinogenesis in mice livers. It has been demonstrated that topical application of capsaicin prior to vinyl carbamate or *N*-nitrosodimethylamine inhibited skin tumorigenesis in mice; both materials require metabolic activation for carcinogenic activity, which is blocked by capsaicin (Suhr *et al.*, 1995). A rat cocarcinogenesis model demonstrated that capsaicin inhibited lung and liver carcinogenesis by dinitrosamine, *N*-methylnitrosourea and *N*-dibutylnitrosamine; alone, it was not carcinogenic (Jang *et al.*, 1991). The effect of capsaicin on the tumor promoter 12-*O*-tetradecanoylphorbol-13-acetate (TPA) was concluded by LaHann (1986) to be a facilitation of the onset of TPA-induced tumor formation, and capsaicin could increase the risk of skin cancer. Sasajimi *et al.* (1987) demonstrated that capsaicin induced ornithine decarboxylase activity, an enzyme used as an index of tumor promoting activity. A 79-week dietary feeding study with a mixture of capsaicinoids (64.5% capsaicin, 3.26% dihydrocapsaicin) was not carcinogenic in $B6C3F_1$ mice, and the hepatocellular tumor incidence was negatively correlated with capsaicin dose (Akagi *et al.*, 1998). Studies by Kim *et al.* (1985) suggest that capsaicinoids may act as cocarcinogens.

ADDITIONAL STUDIES

Studies were conducted in rodents to determine the influence of capsaicin on the pulmonary inflammation associated with respiratory infections (McDonald, 1992). It was found that exposure during *Parainfluenza* infection caused a 3- to 5-fold increase in neurogenic inflammation of the airways, and during *Mycoplasma pulmonis* infection a 30-fold increase in neurogenic plasma extravasation occurred that persisted for several weeks.

HUMAN VOLUNTEER STUDIES AND IN-USE OBSERVATIONS

As a spray, OC causes severe ocular pain/discomfort, lacrimation, blepharospasm, nasal irritation, bronchoconstriction, severe coughing and sneezing, shortness of breath, burning sensation of the skin; there may also be a hypotensive crisis and hypothermia (Olajos and Lakoski, 2004). A number of deaths have been reported when OC was used in conjunction with other forms of restraint or health factors (positional asphyxia, cocaine/PCP/or amphetamine intoxication, neuroleptic malignant syndrome) (Busker and van Helden, 1998; Granfield *et al.*, 1994; Lifschulz and Donaghue, 1991; McLaughlin and Siddle, 1998; Stefee *et al.*, 1995). Although, as yet, a causal relationship with OC exposure has not been established, since many reported deaths occurred within an hour of exposure the possibility of a direct relationship cannot be excluded. In particular, considerations and conclusions on the occurrence of deaths where OC is used in conditions of positional restraint should take into account the fact that OC exacerbates the physiological stress.

The potency (effectiveness) of OC ('pepper') sprays varies with the total concentration of extract used and the variability of the capsaicinoid composition of the extract used. Ten volunteer police officers were exposed to OC spray (5.5% OC) pointed at the face for 0.5–1.5 s; they were examined before exposure and at 30 min, 1 day, 1 week and 1 month post-exposure (Vesaluoma *et al.*, 2000). They reported mild to moderate facial and ocular stinging (mean duration 24 min; range 4–50 min). They developed mild to moderate facial erythema and half had nasal congestion. At 1 min, there was a tachycardia (increased from basal 80 bpm to 116 bpm) and decreased to control values by 10 min. At 20 min, four volunteers had focal corneal epithelial cell damage, which resolved by 24 h. Conjunctival hyperaemia was present at 20 min and persisted, on average, for 9.8 h. A transient decrease (24 h) of mechanical contact sensitivity occurred, but with non-contact gas aesthesiometry there was decreased mechanical sensitivity for one week. Chemical sensitivity to CO_2 was high for about 24 h and decreased below normal after a week, but sensitivity to cold was not affected. Confocal microscopy suggested temporary corneal epithelial swelling, but keratocytes, endothelial cells and sub-basal nerves were unchanged. The authors conclude that acute exposure produces minor corneal effects, but caution should be exercised with repeated exposures since long-lasting changes in corneal sensitivity could occur, possibly associated with injury of nerve terminals, mainly of unmyelinated polymodal nociceptive fibres.

Observations on persons exposed cutaneously to capsaicinoids show that they cause local tingling, burning sensations, erythema, edema and sometimes vesiculation. Nasorespiratory responses to capsaicin and OC include sneezing, rhinorrhea, burning sensation in the throat, wheezing, dry cough and shortness of breath (Fuller, 1990; Stefee *et al.*, 1995). In one study, 13/22 workers exposed to airborne capsaicinoids developed rhinorrhea and cough even at concentrations less than 1 $\mu g\ m^{-3}$ (Blanc *et al.*, 1991). In another study with controls and hot pepper workers, it was found that inhalation of dilute nebulized capsaicin caused dose-dependent cough without inducing tachyphylaxis or significant decrease in baseline pulmonary function in either group (Blanc *et al.*, 1991). Other studies have demonstrated that capsaicin causes transient (< 1 min) dose-dependant bronchoconstriction; there was a 20–50% increase in airways resistance at doses that did not induce cough (Fuller, 1990; Fuller *et al.*, 1985). There was no difference in the duration or magnitude of bronchoconstriction in normal subjects, smokers or asthmatics; it is considered that the mechanism is mediated either by substance P (acting directly or indirectly) or through reflex vagal bronchoconstriction resulting from stimulation of the C-fibres. Human volunteer studies with capsaicinoids given orally showed that they cause a burning sensation in the mouth, throat, chest and abdomen, nausea, vomiting and diarrhoea. Investigation has shown that in a few cases there may be gastric inflammatory changes with hyperaemia, oedema and focal petaechiae. In a review of 1531 exposures to OC sprays reported to the Texas Poison Network Center over the period 1998–2002 ingestion was a major route of exposure in 205 cases (19.9%). The most common symptoms were oropharyngeal irritation and nausea (Forrester and Stanley, 2003). OC contributed to the death of a prisoner with pre-existing bronchiolitis (Stefee *et al.*, 1995), and produced near fatal lung injury in an accidentally exposed infant (Billmere *et al.*, 1996).

Capsaicin nasal challenges were administered to 8 subjects (4 normal and 4 with perennial allergic rhinitis) every 10 min for 5 doses. Capsaicin 20 μM (0.5 μg per spray) when sprayed into the nose induced burning, nasal discharge and lacrimation, with an increase in total protein content of nasal lavage fluid in those with allergic rhinitis (Philip *et al.*, 1994). Morice *et al.* (1992) examined long-term tachyphylaxis to 1-min capsaicin inhalations at 10-min intervals for 40 min in 10 normal subjects. Cough was attenuated with capsaicin at 3, 10, 30 and 100 μM. At the highest concentration, cough was still attenuated at 180 min.

In an analysis of 100 cases collected over a 3-year period of prison inmates presenting to a jail ward emergency area following being sprayed with 'pepper spray' (10% OC), ocular findings included scleral injection and, in seven individuals, corneal abrasions (Brown *et al.*, 2000). A review of the medical records of 81 persons who presented at an emergency department after being exposed by spray device to OC in the conduct of law-enforcement action by the Kansas City Police Department was undertaken (Watson *et al.*, 1996). The major presenting effects were burning sensation in the eye and conjunctival hyperaemia. Skin effects (burning sensation and local erythema) were less frequent (about one-third of patients) and respiratory effects were documented in only 6 of the 81 subjects (including shortness of breath, wheezing and cough). Corneal abrasions were identified in 7/30 cases where fluorescein staining was performed. The authors concluded that only a small proportion of subjects exposed to OC spray will have clinically significant toxicity and that the potential for ocular and pulmonary toxicity should be considered in cases of prolonged exposure. Observations on humans exposed to capsaicin and OC have shown that the following clinically relevant effects may occur: by inhalation, an acute hypertension which may lead to headache; augmentation of allergic reactions; precipitation of bronchospasm and laryngospasm (Smith and Stopford, 1999; Stefee *et al.*, 1995). Considerations such as these lead to suspicions concerning remarks made to the effect that the use of OC spays in the context of peacekeeping operations are 'safe'; in fact, the evidence indicates that in a heterogeneous population there may be a few instances of serious adverse health sequelae.

CAUSES OF INJURES AND MORTALITIES IN PEACEKEEPING OPERATIONS

Circumstances of exposure to riot control agents

Riot control agents may be encountered in a variety of situations in addition to classical large-scale riots. The actual and possible situations are numerous and varied and may result in injuries of various types and severity to single individuals or to a multiplicity of persons who may require treatment at local hospital accident and emergency departments (receiving rooms) or the offices of medical practioners. The following are situations and conditions where exposure to RCAs and other forms of peacekeeping armory may be encountered:

(1) *Small Group Activity.* This covers the use of RCAs by security services (mainly police but sometimes security guards) during arrest or the quelling of disturbances by single or small groups of individuals. In these circumstances, it is most likely that the main equipment will be hand-held irritant liquid spray devices. In the USA these may contain CN, CS or OC, or combinations of these, and in the UK the majority of devices will use CS, but some police forces, including Leicestershire Police (BBC, 2004d),

Sussex Police (BBC, 2001a), North Wales Police (BBC, 2002a) and Hertfordshire Police (BBC, 2005c), have changed to PAVA (Captor), which speaks against a central authorization and approval system for self defence weaponry within the UK national Police organization. Statements, by police authorities and the Home Office, that PAVA is a safer alternative to CS should be interpreted cautiously, and comments that the formulation is totally safe or has no health risks (BBC, 2004d) reflect an ignorance of the available toxicology and human data. Rappert (2003), in respect of the use of CS sprays by police forces in England and Wales, has drawn attention to doubt about the robustness of precautions taken and a continuing failure for relevant government departments to respond and learn from problems identified. Although in the majority of instances where the police employ irritant sprays there is appropriate use and justification, in some instances there has clearly been disregard for situations and possible consequences. This has resulted in public complaints, media coverage and possible litigation because of apparent unnecessary and/or excessive use of irritant sprays (BBC, 2004a,2005b,d), and therefore the treating emergency medical personnel or physicians should record and retain detailed, signed and dated written notes of any incident. This is emphasized by the number of situations in which police have apparently used liquid irritant spray devices against young, old, infirmed and ill persons, sometimes with fatal outcomes. Typical examples include the following: (a) subduing a 73 year-old pensioner who required hospitalization and subsequently died after surgery for an aneurysm (BBC, 1998a,b) (a Home Office pathologist denied that there was an association between rupture of the aneurysm and the use of a CS spray – in the opinion of this author, a very questionable statement [see effects of CS exposure on blood pressure and risk factor for those with aneurysms]); (b) a 53-year-old diabetic male with coronary arteriosclerosis was sprayed with a CS device and subsequently had a sudden heart failure and died (BBC, 2003a,b) – although testifying pathologists stated that the heart attack may have been a consequence of the emotional stress, they made no mention of the well-established systemic arterial hypertension that follows the induction of a PCSI effect (see also (3) below); (c) use of CS spray against 14 year-old boys being held in custody (BBC, 2004f). Police or security guard uses against individuals or small group disturbances cover a multiplicity of situations and locations, as indicated by the following examples: in arrest situations, courtroom and inquest outbursts (BBC, 1998a, 2001b, 2005e), house parties (BBC, 2002b), sporting events (BBC, 2003d,g), public houses and bars (BBC, 2003g), night clubs (USA Today, 2003; BBC, 2005f) and university unions (BBC, 2005g).

(2) *Full-Scale Civil Disturbances.* In the control of full-scale riots (e.g. for politically motivated reasons) or clashes between rival groups (e.g. sporting matches), the following may be used by security forces to regain control of the situation: (a) PSCI irritants dispersed as pyrotechnically generated smokes from grenades or canisters (mainly currently CS, but some countries may still use CN); (b) there appears to be increasing use of irritant liquid/aerosol dispersers of various capacities (depending on the country, these may use CN, CS or OC).

(3) *Vandalization.* Release of PCSIs in public places by vandals or antisocial individuals to deliberately create stressful and frightening situations. The perpetrators often obtain liquid irritant spray devices by stealing them or purchasing in countries where they are readily available, such as Germany. Typical situations have included release of irritant in bars or public houses (BBC, 2004b), nightclubs (BBC, 2002c,2004c), London tube stations (subways) (BBC, 2005a), shops during robberies (BBC, 2004e) and schools (BBC, 2003e,f).

(4) *Use Against the Disabled.* The use of personal protection liquid irritant sprays by police and hospital security personnel to subdue clearly mentally ill patients is totally

unacceptable from medical, humane and sociopolitical points of view. There have, as examples, been reports of the use of such spray devices to (i) subdue an old-age pensioner with Alzheimer's disease at a care home (BBC, 1997), (ii) restrain patients in psychiatric wards as reported independently by the Mental Health Act Commission (BBC, 1999), and (iii) spraying CS at a paranoid schizophrenic while detained under the Mental Health Act – four police officers were subsequently sent for trial accused of assault with actual bodily harm (BBC, 2003c). Such examples raise questions about the welfare of mentally ill patents in custody (Trigwell, 1997).

(5) *Terrorist Activity.* It is conceivable that terrorists could obtain and use RCAs for primary or diversionary purposes. Terrorists, like vandals, could obtain liquid irritant spray devices or even RCA grenades by theft or by purchase in countries where they are freely available. Used in public areas, and particularly if confined, they would create fright and panic which could result in injuries secondary to individuals attempting to vacate the area rapidly. It is also conceivable that RCAs could be used to create a diversionary situation to the conduct of more sinister and serious activities being prepared or conducted in, or close to, the diversionary area.

(6) *Hostage Situations.* These can be regarded as an extension of civil unrest and terrorism and but may present special circumstances of limited information and the wide range of physical and chemical arms that may be used to release hostages. In addition to screening smokes, irritant smokes and irritant solutions, there may be the use of systemically active pharmacological agents (e.g. fentanyl), noise-distracting stun grenades and live ammunition. The extreme situation may result in extensive injuries and mortalities to both terrorists and hostages.

In several of the above situations where multiple persons are involved in a disturbance of the peace, the use of RCAs, notably liquid irritant spray devices and irritant smokes, may lead to fright, panic and physical injuries and in a few instances deaths have been reported where escape is attempted from confined spaces. Additionally, the security forces may also use physical methods for controlling activities of individuals or groups; these may include fists, batons, night sticks, projected rubber/plastic bullets and bean bags, horses, and tasers. Thus, emergency medical services and hospital emergency/casualty departments can expect disturbances of the peace to result in casualties ranging from single individuals to a multiplicity of persons, depending on the nature of the disturbance. The nature of the injuries encountered may include (a) direct (primary) chemical injury from the RCA used and the circumstances of exposure, (b) ballistic or thermal injury from the RCA delivery system or physical methods of restraint or dispersion, and (c) emotional/psychological reactions from the experiences of being involved in a civil disturbance. These are considered in more detail below.

Physical injuries

PROJECTILE INJURIES

These may result from either the deliberate use of physical force by security forces to control a given situation, or be secondary to the dispersion methods used to deliver RCAs. Injuries from projectiles such as rubber bullets or plastic bullets, used to disperse crowds where there is marked civil unrest, have included bruising, fractures of bones (including skull and face), visceral damage and eye trauma; deaths have occurred secondary to trauma (Marshall, 1976). They continue to be used with a high casualty rate, most recently in San Francisco, Los Angeles and Buenos Aires (BBC, 2003h). Smoke or tear gas grenades launched remotely from dischargers may also cause physical injury by direct contact.

PHYSICAL AND THERMAL INJURIES

RCA grenades or canisters, which are picked up before, or as, they detonate, may cause burns as well as explosive injuries. In addition, if they are projected into enclosed spaces (e.g. rooms) they may cause burns secondary to ignition of the area contents. Special devices such as the rubber bursting grenade may cause local burns by lodgment of smoldering pellets between clothing and skin. Tear gas pen guns or pistols discharged at close range to the body, particularly the head and neck, have caused severe and on occasion fatal injuries (Adams *et al.*, 1966; Ayers and Stahl, 1972; Hopping, 1969; Smialek *et al.*, 1975; Stahl and Davies,1969; Stahl *et al.*, 1968).

PANIC INJURIES

If physical or chemical methods of control are employed against civilians this may cause panic situations resulting in injuries and even mortalities. This was illustrated by an event in a Chicago nightclub during February 2003, when a security guard attempted to break up a fight on the dance floor using an OC spray device. This started a stampede involving several hundreds of persons towards the front door and resulted in the deaths of 21 persons (US Today, 2003). In addition to the panic effects in such situations, difficulty to escape from confined spaces may be compounded by blepharospasm and excess lacrimation.

INCIDENTAL PHYSICAL INJURIES AND ACCIDENTS

When water canon is used, injures may occur from the water force and also from difficulty in walking or running on the slippery wetted ground. If an airborne PCSI material drifts from its intended area due to a change in wind then if the cloud accesses an area where motorized vehicles are present this could lead to visual difficulties with drivers and resultant accidents. A similar situation could exist if low-viscosity polymers are dispersed on the ground. Arrest and in-custody deaths may result from restraint techniques, positional asphyxia, restraint in cases of cocaine intoxication, alcoholic intoxication and the neuroleptic malignant syndrome; these are all practical considerations and have been discussed in detail elsewhere (Bell *et al.*, 1992; Granfield *et al.*, 1994; Lifschultz and Donaghue 1991; Luke and Reay, 1992; McLaughlin and Siddle, 1998; Reay *et al.*, 1992; Stefee *et al.*, 1995). Additionally, and particularly in the USA, high-voltage tasers may be used to incapacitate and immobilize; these have been used somewhat indiscriminately, including against very young children. Contrary to what is stated by manufacturers, it difficult to believe that these devices are safe to use against those with advanced cardiac disease, including arrhythmias.

Structural and functional chemical injuries

Major sites of structural and functional injury and adverse effects caused by PCSIs and which are of clinical relevance are as follows.

EYE INJURIES

Contamination of the eye with a PCSI dispersed as aerosol, solid or in solution, in addition to causing a transient blepharospasm and increased lacrimation, will also result in a marked conjunctoblepharitis. Although with aerosols this usually resolves within minutes to a very few hours, prolonged exposure to CN aerosols may produce a more persistent and severe conjunctoblepharitis and possibly corneal injury (Thornburn, 1982) and heavy exposure to CN aerosol may produce necrosis of corneal epithelium (Grant, 1986). Injury to the eye has been reported with irritant liquid spray devices, particularly those containing CN. Thus, laboratory rabbit eye irritation studies have shown CN in polyethylene glycol (PEG) 300 produces threshold corneal injury at 1% with severe keratitis occurring at 2% (Ballantyne *et al.*, 1975). Eye injury from CN-containing MACE has been documented (Kling, 1969; MacLeod, 1969; Macrae *et al.*, 1970; Oksala and Salminen, 1975; Pearlman, 1969; Rose, 1969). Discharge of particulate irritant material from tear gas pen guns has also been associated with severe eye injury, notably again with devices containing CN (Hoffmann, 1967; Hopping, 1969; Oaks *et al.*, 1969). Effects of these tear gas pen injuries have included chemosis, corneal oedema, corneal epithelial stripping, necrotizing keratitis and iridocyclitis (Laibson and Oconor, 1979; Levine and Stahl, 1968). The initial severe eye injury is probably a consequence of physical damage from the blast and heat coupled with explosive deposition of solid irritant particles in the wound leading to chemical necrosis. Corneal abrasions have also been seen in humans exposed to OC sprays (Brown *et al.*, 2000).

In addition to causing structural injuries to the eye and its adnexa, local contamination of the eye with PCSI materials can result in functional ocular changes, some of which are pharmacologically induced and can have medical significance. One such effect is the induction of rapid onset increases in intraocular pressure (IOP). Many PCSI materials that cause an increase in IOP can, but often at much higher concentrations, also induce inflammatory changes in the eye (Ballantyne, 1999a; Ballantyne *et al.*, 1972). IOP changes have been demonstrated in both laboratory animals and in controlled human volunteer studies following ocular contamination with CN, CS and CR, and the findings are briefly summarized as follows. Using 0.1 ml droplets of CN in PEG 300 and measurement of IOP by applanation tonometry, concentration-related increases in IOP were detectable by 10 min post-application in the rabbit and generally returned to control (pre-treatment) values by 1 h; increases were statistically significant at 10 min with 0.125% CN and higher, with a peak increase of 98% with 1.0% CN (Ballantyne *et al.*, 1975). With CS in PEG 300, there were also dose-related increases in IOP at 10 min post-application but the increases were proportionately less than those for CN, with a maximum increase of 52% with 5.0% CS; pressures were down to control values by 60 min (Ballantyne *et al.*, 1974). With CR in PEG 300, increases in IOP also occurred at 10 min, with a maximum effect of 40% using 5.0% CR; increases were statistically significant at 1.0% CR and greater (Ballantyne *et al.*, 1975). Overall, the laboratory findings indicate that in the rabbit PCSI materials cause prompt onset concentration-related increases in IOP, reaching maximum values within a few minutes of contaminating the eye and returning to control values by about 1 h. The findings with CR were confirmed in human volunteer studies using solutions of CR in PEG 300. In one series, 0.4 ml of 0.05 and 0.1% CR was applied to one eye and IOP measured sequentially by applanation tonometry up to 24 h and blood pressure measured simultaneously (Ballantyne *et al.*, 1977a). Peak increases in IOP were measured at 5 min post-application; increases at this time for 0.05% CR were 40% in the treated eye and 16% in the contralateral (untreated) eye, and for 0.1% CR, 44% in the treated eye and 17% in the contralateral eye (see Figure 9). In CR-treated eyes, IOP decreased to control values by about 15 min with 0.05% CR and 3.5 h with 0.1%CR. The changes, shown in Figure 9, show a dose (concentration)-related effect for both the magnitude and duration of the IP changes in the contaminated eye; however, in the contralateral untreated eye there are smaller increases in IOP which are quantitatively similar for both concentrations of CR and which follow the time-course of a transient increase in diastolic blood pressure. These findings indicate that the rise in IOP is, in part, due to a generalized effect on both the treated and untreated eye. However, the more marked and sustained effect in the CR-contaminated eye is a consequence of a local ocular effect. Applied to the human eye, CR and other PCSI substances, cause local pain and an associated increase in systemic arterial and central venous pressures. The increased central venous pressure will elevate episceral venous plexus pressure and hence increase IOP in both the PCSI treated and untreated eye. Accompanying the local pain from the PCSI effect is hyperaemia of the conjunctival blood vessels and the vascular congestion will impede drainage of aqueous humor and further increase IOP. That the human eye is more sensitive to the ocular hypertensive effect of PCSIs than is a standard laboratory animal model is suggested by the fact that while 0.05% CR applied to the human eye causes a 40% increase in IOP at 10 min, when the same concentration is applied to the rabbit eye the IOP increased by only 4% at the same time period (Ballantyne *et al.*, 1977a). Comparable studies on IOP have not been conducted with OC and PAVA, but the fact that these materials cause both a PCS effect and, at higher concentrations, are capable of causing inflammatory effects on they eye strongly suggests OC and PAVA will produce prompt onset–short duration ocular hypertensive effects. From a clinical perspective, the increases in IOP are generally only briefly sustained and should not present a hazard to most individuals. However, there is a possibility that those with incipient narrow angle glaucoma may be precipitated into an attack, and those with established glaucoma may experience an exacerbation (Ballantyne *et al.*, 1973b). Since the incidence of glaucoma is around 2% and most cases occur in people over the age of 40 years, it is likely that the number of vulnerable individuals in a civil disturbance will be low. However, those having responsibility for triage and medical management should be aware of the potential for an ocular hypertensive effect and that such subjects should be referred for ophthalmic screening. When the concentration of RCA in contact with the eye is such that ocular injury (inflammation) occurs, then anterior segment damage occurs which will further increase and sustain elevated IOP (Ballantyne *et al.*, 1973a,1977a).

CUTANEOUS INJURIES

With harassing, but otherwise relatively low-exposure doses to PCSI materials, the only visible cutaneous effect is normally erythema. However, sustained exposure to high CN concentrations may result in severe erythema, edema and chemical burns of the skin (Thornburn, 1982). Both CN and CS may cause allergic contact dermatitis (Frazier, 1976; Fuchs and Weische, 1990; Holland and White, 1972; King *et al.*, 1995; Leenutaphong and Goerz, 1989; Penneys, 1971; Penneys *et al.*, 1969; Pfeiff, 1984; Rothberg, 1970).

RESPIRATORY TRACT INJURIES

The discomfort, breathing difficulties and cough induced by exposure to an airborne PSCI material usually resolve in 10 min to about 1 h. Those with pre-existing respiratory tract diseases, such as chronic obstructive pulmonary disease or asthma, may be at increased risk from the exposure to PCSI materials and chemicals having pulmonary neurogenic effects, such as OC (National Institute of Justice, 1994). Sustained exposure to high concentrations, as may occur in an enclosed space, may result in laryngotracheobronchitis (Thornburn, 1982). It is possible that exposure of hypersusceptible individuals may result in laryngospasm, possibly fatal. Lung injury is most likely to occur when there is exposure to high concentrations, particularly of pyrotechnically generated smoke, in confined poorly ventilated spaces from which escape is difficult; e.g. a barricaded room or cell, or an automobile (Greaves, 2000; Thornburn, 1982). Several cases of fatal respiratory tract injury have been recorded from such acute overexposure situations, particularly with DM and CN (Chapman and White, 1978; Gonzales *et al.*, 1954; Krapf, 1981; Ministry of Defence, 1972; Stein and Kirwin, 1964). Death usually occurs between 12 h and several days post-exposure. It is characterized pathologically by laryngeal and tracheobronchial necrosis with pseudomembrane formation, pulmonary oedema and alveolar haemorrhages (Chapman and White, 1978; Stein and Kirwin, 1964). If respiratory tract injury occurs, then secondary infection may develop.

One potential respiratory complication of exposure to high and potentially tissue-injuring concentrations of PCSI RCAs is 'Reactive Airways Dysfunction Syndrome' (RADS), which is a respiratory complication of pulmonary overexposure to an airborne irritant (inflammation-inducing) chemical, which was first described by Brooks *et al.* (1985). Symptoms, asthma-like in character, develop within a few hours of exposure nature. There is persistence of respiratory symptoms and airways hyperreactivity for years after exposure. All chemicals that have been shown to be aetiologic agents in the pathogenesis of RADS share the common characteristic of being biologically irritant in nature. It is therefore considered that PCSI RCAs which also have the potential to cause pulmonary inflammation and injury will cause RADS (Ballantyne and Salem, 2004). In fact, RADS has been described following exposure to CS (Bayeaux-Dunglas *et al.*, 1999; Hu and Cristani, 1992; Worthington and Nee, 1999). Although RADS has not been described as a complication of exposure to OC alone, it has been reported in an individual who was exposed for about 30 s in an enclosed space to Deep Freeze® (I% OC/1% CS). He developed the typical symptoms of cough, chest tightness, wheezing and shortness of breath, with pulmonary function tests demonstrating reversible and fixed obstructive pulmonary disease. Since there may be long-term sequelae associated with RADS, those who are responsible for triage and medical treatment of victims of exposure to PCSI RCAs need to be aware of this possibility, and where there are clinical indications refer the subject for a pulmonary function evaluation.

CARDIOVASCULAR CONSEQUENCES

Studies on human volunteer subjects have demonstrated that within a short period of exposure to PCSI RCAs there are increases in both SBP and DBP, often with an associated reflex bradycardia. These cardiovascular effects have been noted after following exposure to airborne irritants (aerosols and grenade-generated smokes), to dilute solutions topically applied to the eye (ca. 0.01–0.1%), and by whole-body drenches to even more dilute solutions of irritant (0.001–0.005% CS and 0.001–0.0025% CR (Ballantyne *et al.*, 1976a). With the whole-body drenches of dilute solutions of CS and CR, there was immediate discomfort to the eye, blepharospasm and profuse lacrimation of 3 to 5 min duration. Skin sensations followed shortly after the onset of eye irritation and were described as stinging or burning; on average, the cutaneous effects persisted for 5 to 10 min with CS and 15 to 20 min with CR. The shorter duration of cutaneous irritant effects with CS was probably related to hydrolysis in the dilute solutions and the lower intrinsic PCSI effect compared with CR (Ballantyne and Swanston, 1973,1974). There was erythema whose distribution was similar to that of the cutaneous sensory irritation but the degree of erythema was not strongly correlated with the intensity of the sensation. Increases in blood pressure were first measured within 1 min of the irritant drenches and were moderate to marked. With CR, peak increases in SBP ranged 30 to 70 mmHg (average 45 mmHg) following an 0.001% drench, and ranged 30 to 80 mmHg (average 59 mmHg) following 0.0025% CR. Peak increases in DBP ranged 15 to 30 mmHg (average 23 mmHg) with 0.001% CR and 20 to 45 mmHg (average 20 mmHg) after 0.0025% CR. These peak increases occurred between 1 and 3 min post-drenching, Times for DBP to return to within 10 mmHg above control (pre-exposure) values ranged 2 to 15 min; average times were 7.4 min for 0.001% CR and 12 min for 0.0025% CR. Blood pressure changes for the CS drenches were as follows. Peak SBP increases 5 to 50 mmHg (average 31 mmHg) and for DBP 0 to 40 mmHg (average 19 mmHg) and occurred between 1 and 2 min after drenching with return to control values by 2 to 13 min. Mechanistically, these increases in blood pressure could be the result of several, possibly interacting factors, including a direct systemic pharmacological effect from absorbed PCSI, a cold pressor effect from the drench *per se* and/or the pain and discomfort coupled with apprehension. A direct pharmacological effect seems unlikely for the following reasons. The onset of the BP rise is abrupt, and it is highly unlikely that sufficient PCSI material could be absorbed percutaneously to exert a systemic pharmacological effect over the 20-min period that the pressures were elevated. With CR, for example, *in vivo* and *in vitro* studies with rat skin using 0.005% CR dissolved in 3.3% dipropylene glycol monomethyl ether (DPM; the solvent system used in the drench studies) demonstrated a steady state absorption rate of 300 pg cm^{-2} min^{-1}; similar values have been demonstrated for human skin *in vitro* (Leadbeater and Creasey, 1973, personal communication). Assuming a constant reservoir of CR solution on the skin, a 70 kg man with body surface area of 1.8 m^2 would absorb CR percutaneously at a rate of about 5.4 μg min^{-1} from a 0.005% solution. Thus, over 20 min about 108 μg CR would be absorbed, equivalent to 1.5 μg kg^{-1} over this time period. Studies in cats have shown that the minimum dose of CR needed to produce a detectable pressor effect is 2.5 μg kg^{-1} when given as an acute bolus injection into the carotid artery (the most sensitive route of dosing). When infused into the right atrium or aortic arch, doses of 62.5 μg kg^{-1} are required to produce a comparable pressor response (Green and Muir, 1975, personal communication). Thus, the amount of CR that could be absorbed by a human male over a 20-min exposure to 0.005% CR solution is unlikely to induce a pressor effect. In addition, the peak increases in BP occurred not later than 4 min, when only 22 μg (0.3 μg kg^{-1}) could have been absorbed from the 0.005% CR solution. Similar considerations apply for the toxicokinetics of CS on BP (Brimblecombe *et al.*, 1972). The possible contribution of a cold pressor effect was studied in six volunteers who were subjected to a cold-water shower alone (Ballantyne *et al.*,

1976a). This produced a mean-peak SBP increase of 32 mmHg (range 0 to 85 mmHg) and a mean-peak DBP increase of 7 mmHg (range 0 to 25 mmHg); SBP returned to control values within 2 to 3 min. These findings suggest that the pressor effect of a cold drench may contribute to a small extent in the initial increases in BP but cannot account for the more sustained increases seen with the irritant drenches. Since the magnitude and duration of the BP increases was generally related to the degree of discomfort of the CR and CS drenches, it was considered appropriate to investigate the effects of a different pain-inducing procedure. This involved induction of ischaemic pain in the forearm muscles temporarily deprived of their blood supply by the application of a sphygmomanometer cuff at 200 mmHg and measuring BP in the contralateral arm (Ballantyne *et al.*, 1976a). This caused increases in both SBP and DBP; peak SBP increase was 37 mmHg (range 25 to 45 mmHg) and in DBP was 26 mmHg (range 15 to 40 mmHg). The increase in SBP caused by ischemic pain was statistically significantly less than that due to a 0.0025% CR drench ($p < 0.001$) but not significantly different from that resulting from 0.001% CR or 0.005% CS drenches ($p < 0.5$); however, there was no significant difference in the rise in DBP caused by ischaemic pain and the irritant drenches. These finding suggest that the increase in BP following irritant drenches is mainly a consequence of the intense and widespread cutaneous and ocular discomfort, possibly coupled with apprehension. Similar rapid onset increases in BP have also been recorded following the topical application of more concentrated solutions of CR (0.01 to 0.1%) to the surface of the cornea in human volunteers; the increased SBP and DBP in this situation are also considered to be a consequence of the severe ocular induced pain (Ballantyne, 1977a,1987). Additionally, exposure to PCSI smokes and aerosols results in a prompt increase in BP. For example, with CS smokes there were abrupt steep rises in SBP, DBP and pulse pressure following the start of exposure (Beswick *et al.*, 1972). Thus, for a group of volunteers the pre-exposure average SBP was 123 mmHg which increased to 142 mmHg on exposure to CS smoke and then decreased to 124 mmHg by 20 min; the corresponding values for DBP were 73 mmHg, increasing to 84 mmHg and resolving to 75 mmHg. Pulse pressures were 49 mmHg before exposure, 57 mmHg on CS exposure, decreasing to 49 mmHg by 20 min. Heart rate decreased from 80 to 67 beats per minute.

The above findings confirm that exposure to PSCI RCAs by airborne dispersion or by contamination in solution produce abrupt and marked increases in SBP and DBP, with resolution within about 0.5 h of the start of exposure. The magnitude and duration of the changes can be tolerated without significant medical hazards in healthy individuals. However, as with other stressful situations, some susceptible individuals may be at increased risk from the induced transient hypertensive episode; this will include those with essential hypertension, established myocardial infarction and coronary artery disease, cardiac arrhythmias and arterial aneurysms (Ballantyne, 1977a,1987; Ballantyne and Salem, 2004).

DECONTAMINATION, FIRST-AID AND MEDICAL MANAGEMENT OF CASUALTIES FROM PEACEKEEPING OPERATIONS

As stressed above, the circumstances and consequences of a civil disturbance (small or large) are very varied. The emergency medical services and physicians should expect casualties varying from simple irritant and emotional effects to severe chemical injury and/or physical injury and hospitalization may be required in the latter groups. Those seeking medical assistance will be a heterogeneous group; old, young, male, female, in prior good health, and some with established diseases. For anticipated large demonstrations, advance warning by security forces to the emergency services and local accident/emergency departments of the procedures likely to be adopted by the security forces would help greatly aid hospitals and health care services how to prepare. Unfortunately, this appears to be a rare occurrence. If civil unrest erupts into violent demonstrations and riots, this will enormously increase the requirements for health care and social services. For example, the use of CS grenades in a Vietnamese detention centre in

Hong Kong during 1995 resulted in 1500 people being transferred to a British Red Cross Medical Clinic for assessment and treatment (Anderson *et al.*, 1996). Likewise, in even so-called civilized countries the potential for objections and uprising of the electorate against viewpoints, actions and extreme legislation of and by their elected representatives who assume dogmatic and self-interest attitudes and opinions, is a clear possibility with resultant suppression by security forces; such scenes are common viewing on television and in newspapers. The need for advanced planning for such situations is clear and has been discussed in more detail elsewhere (Ballantyne and Salem, 2004; Ballantyne, 2006). Since, in part through political propaganda, many health-care facilities have developed advanced antiterrorist plans, education and coordinating groups (Baker, 1999,2004; Georgopoulos *et al.*, 2004; Kales and Christiani, 2004; Laurent *et al.*, 1999; Noeller, 2001; Okumura *et al.*, 1998), it seems reasonable to suggest that these facilities could also form a focus to discuss and develop plans and preparations for civil unrest, since several of the concepts, likely scenarios, chemicals and munitions, have commonalities. However, as commented elsewhere, there could be sensitive legal and sociopolitical restraints and constraints on a strictly combined functional approach, and on the development of policies, procedures and authorization. Although there have been critical comments on the organizational basis of some of these special health care groups (Boulton *et al.*, 2005; Morby *et al.*, 2000) they nonetheless could be used as a starting point for the development of agreed coordinated national plans (Ballantyne, 2006). The author regards it essential that there should be informed communications between security forces, intelligence agencies and the heath-care community on the tactics used in civil control situations and, where possible, advance warning of likely incidents in order for health-care institutions to develop necessary preparative plans. Information on the nature, actions, physiological–pharmacological–toxicological effects and complications should be made available to health-care providers to assist in advance planning and training sessions. Since there is 'common-sense reasoning' to believe that recent civil demonstrations and riots may represent an initial wave against dictatorial political efforts of capitalization, multicuturalization and globalization which may not be to the advantage of the majority of the electorate, and indeed sometimes civilization in general, it is likely that there may be an increase in the need for support services to deal with the consequences of civil unrest.

It is unfortunate, and indeed reprehensible, that several manufacturers of riot control agents, devices and related equipment do not issue readily available informative literature, and indeed when it is supplied it is often incomplete and technically questionable. Apparently, MSDSs are not considered as a necessary regulatory requirement for civil control munitions. Statements that formulation constituents of materials to be deliberately used against a human heterogeneous population are 'trade secrets' is not acceptable from medical and ethical points of view, and needs to be prohibited.

Decontamination

In order to relieve casualties of distress from the action of PCSIs and also to facilitate the unimpeded work of those treating casualties, it is essential that decontamination of the subject, in the absence of contamination of health-care workers, should be undertaken promptly. Ideally, a separate well-ventilated area remote from the hospital general building should be available to receive, examine and decontaminate patients; if it can be avoided, receiving rooms should not be contaminated at the inconvenience of other patents, and PCSIs may enter the general hospital atmosphere through closed circulation air-conditioning systems. In many cases where there has been only exposure to a PCSI RCA and this was briefly sustained, then decontamination in a fresh-air stream is the optimum first approach, and in many cases this is all that may be required (Blaho and Stark, 2000; Lee *et al.*, 1996). Contaminated clothing should be removed and stored in a sealed plastic bag. Disposable gloves and gowns should be worn, and if there is PCSI in the air the use of impermeable goggles will help prevent discomfort and inconvenience to emergency medical personnel; if available and needed because of an accumulation of PCSI in the air from the patient, respirators should be worn in order to permit free and continued treatment. If decontamination of PCSI requires showering or washing, patients should be advised that this may result in a temporary reprise of symptoms as water leaches a PCSI out of contaminated hair. If a patient is in need of specialist care or investigation then it is essential that the patient be decontaminated as completely as possible to ensure that there is no contamination of specialist units. In addition, if surgery is required decontamination and protective measures should be sufficient to ensure that there will not be secondary contamination of the anaesthetist, since this can cause problems for intubation (Barlow, 2000; Bhattacharya and Haywood, 1993).

Triage

Where many casualties are predicted or when it has become apparent to the health-care facilities 'not-given' advance warning, there should be arrangements for suitably qualified emergency medical staff to conduct triage on arriving patients. Clearly, triage staff themselves should be protected against secondary contamination. Triage will involve allotting priorities for clinical examination and treatment based on the presence or absence of physical injuries, severe chemical injuries and decisions on whether an individual falls into an 'at-risk' category who may develop secondary complications from the medical and psychological consequences of participating in a civil disturbance. These include, but are not limited to, individuals with coronary artery disease, myocardial ischaemia, cardiac arrhythmias, essential hypertension, arterial aneurysm, chronic pulmonary disease, asthma, reactive airways diseases, glaucoma and convulsions.

Psychological effects

For most, except the hardiest and most experienced demonstrators or malefactors, being a participant or an incidental bystander in a civil disturbance can be an emotionally disturbing experience, particularly if exposed to physical or irritant insults. Therefore, many people will present in a state of depression, fright, anxiety or even hysteria. Delayed psychological reactions may also occur. Psychological reactions may also be seen in those living in an area where the threat of violence exists (Frazer, 1971; Lyons, 1971). Severely emotionally affected individuals may require sedation and reassurance from psychiatric social workers, and with extreme and sustained reactions the advice of a psychiatrist may be needed. It is important to reassure individuals exposed to PCSI RCAs that recovery will occur promptly and without any residual sequelae.

Ocular effects

Acute chemically induced conjunctoblepharitis may be present from smoke, aerosol or liquid contamination of the eye. In mild cases of smoke or aerosol exposure, aeration in the open air or using an electrical fan may be all that is necessary for decontamination and relief (Blaho and Stark, 2000; Gray, 1995; Lee *et al.*, 1996; Yih, 1995). With more severe cases, gentle decontamination of the eye may be necessary by irrigation with cool water or saline for several minutes. Visual acuity might be slightly reduced immediately after contamination of the eye, and while the individual is symptomatic, but acuity rapidly returns to normal (Rengstorff, 1969b; Yih, 1995). If ocular discomfort is extreme, or if it persists, the use of local anaesthetic eye drops may give relief, e.g. propoxymetocaine hydrochloride. However, the use of local anaethesia should be restricted to essential cases and ideally be authorized by an ophthalmologist, since it may impair the regeneration of injured corneal epithelium and retard healing (Leopold and Leiberman, 1971). Those having marked and persistent eye discomfort with blepharospasm should have a detailed ophthalmic examination (including fluorescein staining and biomicroscopy) to confirm, or otherwise, the absence of corneal and/or anterior segment injury (Blaho and Stark, 2000; Brown *et al.*, 2000; Wier, 2000). As discussed in detail earlier, contamination of the eye with a PCSI RCA in solution causes a transient increase in IOP. While this is of no clinical significance in healthy subjects with normal eyes, persons with incipient or established glaucoma may be at risk and should receive appropriate expert ophthalmic investigation, advice, and if necessary, follow-up. With OC, superficial anaesthesia and loss of the blink reflex may lead to corneal abrasions from contact lenses or foreign bodies. Since capsaicin disrupts the corneal epithelium, those with impaired corneal integrity (exposure keratitis, keratomalacia or recurrent corneal erosion) may be more susceptible (Smith and Stopford, 1999).

There are differences of opinion concerning the relative advantages of wearing contact lenses during exposure to PCSI materials. Thus, and based on trials with CS aerosol generators, it has been stated that individuals wearing soft contact lenses keep their eyelids open more easily and longer, and have quicker and better orientation (Aalphen *et al.*, 1985). In addition, a limited report described two police officers wearing soft contact lenses who were able to conduct their duties effectively and without complications during training drills with CN and CS (Royall, 1977, cited by Aalphen *et al.*, 1985). However, others believe that the wearing of soft contact lenses could result in visual problems relating to effects from both PCSI and formulation constituents (Ballantyne and Salem, 2004). Soft contact lenses may lead to entrapment of material between the lens and surface of the cornea and increase contact time with the cornea; this may facilitate injury from PCSI or formulation constituents. It is generally believed that contact lenses of contaminated patients should be removed to ensure that their eyes can be adequately irrigated. It has been noted that hydrophilic soft contact lenses, secondarily to their network structure, may absorb chemicals and thus be a source of sustained exposure and enhance local toxicity (Loriot and Tourt, 1990). In this respect, it has been specifically noted that soft contact lenses contaminated with OC may absorb the material and it may be difficult to remove residual OC (Lee *et al.*, 1996). Additionally, it has been noted that some solvents of irritant solutions may led to solubilization or fragmentation of soft contact lenses (Ballantyne, 2006, personal unpublished data) or to hardening of soft contact lenses (Holopainen *et al.*, 2003). This may result in increased irritation and possible superficial corneal injury, which may be enhanced by the manual rubbing of eyes that characteristically accompanies topical sensory irritation of the eye.

Skin effects

Cutaneous erythema from exposure to a PCSI RCA is normally all that is seen on examination of the skin, and should be decontaminated by a copious soap and cool-water wash. With CS, and if the symptoms and signs are marked, advantage may be taken of its rapid hydrolysis in alkaline solutions. For example, Weigand (1969) recommended the following alkaline solution (pH 9.4) to produce rapid hydrolysis and relief of symptoms: aqueous 6% sodium bicarbonate, 3% sodium carbonate and 1% benzalkonium chloride. In most circumstances, however, either aeration or a simple water wash is sufficient. Sustained exposure to high concentrations of grenade-generated smokes, particularly CN, may cause a more severe primary irritant dermatitis, requiring therapeutic measures, such as topical corticosteroids. In addition, a hypersensitivity reaction may develop in a minority of individuals, although this is most likely to occur in frequently exposed security forces. Perforating injuries with embedded PCSI irritant material, notably CN, may provoke severe chronic suppurative and necrotizing reactions, resulting in degenerative changes and fibrosis affecting skin, nerve and muscle (Stahl *et al.*, 1968).

Respiratory tract

Respiratory effects from PCSI RCAs normally subside rapidly and completely if exposure was in the open air. However, if exposure to high concentrations has been sustained and particularly if it occurs in enclosed spaces, then the patient should be kept under observation and, as dictated by the clinical condition, radiological and pulmonary function studies conducted. Those with established pulmonary disease and asthma may have an exacerbation of their symptoms and signs. Those having sustained gross overexposure to grenade generated smokes may develop laryngotracheobronchitis, possibly requiring the use of bronchodilators, postural drainage, corticosteroids and prophylactic antibiotics (Thornburn, 1982). Individuals exposed to high concentrations of PCSIs, particularly if sustained, should be observed for several days (Park and Giammona, 1972; Sanford, 1976). The development of RADS in a few individuals having had acute high-dosage inhalation exposure to PCSI RCAs is a possibility, as discussed earlier.

Cardiovascular status

As discussed in detail above, those exposed to smokes, aerosols or solutions of PCSI materials may have transient increases in SBP and DBP, often accompanied by a reflex bradycardia. This could be compounded with that caused by the emotional experience of being involved in a civil disturbance. Those who give a history of cardiovascular disease (notably myocardial infarction, coronary artery disease, cardiac arrhythmias, essential hypertension, aneurysm) may require admission and further investigation.

Gastrointestinal effects

Direct oral contamination from irritant liquid sprays or the swallowing of contaminated saliva may experience stinging in the mouth and throat and increased salivation, and in a few cases there may be nausea and vomiting if sufficient amounts are swallowed; very rarely has haematemesis been reported (Anderson *et al.*, 1996). With CS, diarrhoea is not likely to be a complication of exposure. However, as noted previously, animal studies and controlled human feeding studies with natural pepper products have demonstrated that capsaicinoids can cause variable degrees of gastrointestinal irritation, and in keeping with this documented reports have been published on gastrointestinal signs and symptoms following exposure to OC in the context of civil disturbances. In an analysis of 1531 exposures to OC sprays reported to the Texas Poison Center Network over the period 1998–2002, ingestion was the primary route of exposure in 205 cases [19.9% – the most common symptoms were mouth and throat irritation and nausea (Forrester and Stanley, 2003)]. Ballantyne (unpublished data) also noted that exposure to OC sprays in the context of civil disturbances can result in throat irritation, nausea and vomiting. The nature and duration of the gastrointestinal consequences to OC spray exposure do not result in any clinically significant complications or long-term health effects.

Miscellaneous signs, symptoms and complications

Rarely have complications from CS exposure included haemoptysis and haematemesis (Anderson *et al.*, 1996). A case of multisystem hypersensitivity reaction characterized by delayed development of cutaneous rash, pneumonitis, hypoxaemia, hepatitis and hypereosinophilia, with rapid response to corticosteroids, was described in a man heavily exposed to CS solution spray. The authors ascribed this to a systemic allergic reaction; patch testing confirmed sensitization to CS (Hill *et al.*, 2000). The possible abuse and complications from illegal drugs should be considered where clinical suspicions exists (Hayman and Berkely, 1971). Physicians should be aware of the possible contribution from formulation constituents; some may exacerbate effects known to be produced by RCAs (e.g. eye injury and skin irritation) and others may introduce additional toxicological factors (such as the development of carbon monoxide intoxication from the absorption and metabolism of dichloromethane as a formulation solvent).

REFERENCES

Aalphen CCK-v, Visset R, van der Linden JE *et al.* (1985). Protection of the police against tear gas with soft lenses. *Mil Med*, **150**, 451–454.

ACGIH (2006). *2006 TLVs® and BEIs®*. Cincinnati, OH, USA: American Conference of Governmental Industrial Hygienists.

Adams JP, Fee N and Kenmore PI (1966). Tear gas injuries. *J Bone Joint Surg*, **48A**, 436–442.

Akagi A, Sano N, Uehara H *et al.* (1998). Non-carcinogenicity of capsaicinoids in B6C3F1 mice. *Food Chem Toxicol*, **31**, 1065–1071.

Allinson G and McLeod CW (1997a). Characterization of lachrymators by ambient temperature ion mobility spectrometry. *J Forens Sci*, **42**, 312–315.

Allinson G and McLeod CW (1997b). Characterization of tear gas residues by ion mobility spectrometry. *Appl Spectrom*, **51**, 1880–1889.

Allinson G, Saul C, McLeod CW *et al.* (1998). Identification of tear gases in suspect spray cans and cloth samples by ion mobility spectrometry. *J Forens Sci*, **43**, 845–849.

Anderson D and Conning DM (1993). *Experimental Toxicology*, Second Edition. Cambridge, UK: Royal Society of Chemistry.

Anderson PJ, Lau GSN, Taylor WRJ *et al.* (1996). Acute effects of the potent lacrimator *o*-chlorobenzylidene malononitrile (CS) tear gas. *Human Exp Toxicol*, **15**, 461–465.

Arceo SDB, Madrigal-Bujaidor E, Montellano EC *et al.* (1995). Genotoxic effects produced by capsaicin in mouse during subchronic treatment. *Mut Res*, **345**, 105–109.

Ashton I, Cotes JE and Holland P (1977). Acute effect of dibenz(b.f)-1,4-oxazepine aerosol upon lung function of healthy young men. *J Physiol London*, **275**, 85P.

Ayers KM and Stahl CJ (1972). Ballistic characteristics of wounding effects of tear gas gun loaded with ortho-chlorobenzalmalononitrile. *J Forens Sci*, **17**, 292–297.

Azizan A and Blevins RD (1995). Mutagenicity and antimutagenicity testing of six chemicals associated with the pungent properties of specific spices as revealed by the Ames Salmonella/microsomal assay. *Arch Environ Contamin Toxicol*, **28**, 248–258.

Baker DJ (1999). The pre-hospital management of injury following mass toxic release: a comparison of military and civil approaches. *Resuscitation*, **42**, 155–159.

Baker D (2004). Civilian exposure to toxic agents: emergency medical response. *Prehosp Disast Med*, **19**, 174–178.

Ballantyne B (1977a). Riot control agents. Biomedical and health aspects of the use of chemicals in civil disturbances. In: *Medical Annual*, (RB. Scott and J Frazer eds.) *1977*, pp. 7–41. Bristol, UK: Wright.

Ballantyne B (1977b). The acute mammalian toxicology of dibenz(b.f)-1,4-oxazepine. *Toxicology*, **8**, 347–379.

Ballantyne B. (1979). Evaluation of ophthalmic hazards from an aerosol generator of *o*-chlorobenzylidene malononitrile (CS). *Mil Med*, **144**, 691–694.

Ballantyne B (1982). Dosage levels associated with the induction of respiratory tract inflammation by acute exposure to titanium tetrachloride smokes. *Toxicologist*, **2**, 45.

Ballantyne B (1983). The cyanogenic potential of 2-chlorobenzylidene malononitrile. *Toxicologist*, **3**, 64.

Ballantyne B (1984). Toxicology. In: *Kirk-Othmer: Encyclopedia of Chemical Technology*, Supplemental Volume, Third Edition (M Grayson, ed), pp. 894–924. New York, NY, USA: John Wiley & Sons, Inc.

Ballantyne B (1987). Clinical toxicology and forensic aspects of riot control chemicals. In: *Abstracts of the 24th International Meeting, International Association of Forensic Toxicologists, Banff.* (GR Jones and PP Singer eds.) pp. 484–504. Alberta, Canada: University of Alberta Printing Services.

Ballantyne B (1998). Acute inhalation toxicity of phosphorus pentoxide smoke. *Toxic Sub Mech*, **17**, 251–266.

Ballantyne B (1999a). Peripheral sensory irritation: basics and applications. In: *General and Applied Toxicology,* Volume 2, Second Edition. (B Ballantyne, TC Marrs and T Syversen, eds), pp. 611–630. London, UK: MacMillan Reference.

Ballantyne B (1999b). Toxicology related to the eye. In: *General and Applied Toxicology,* Volume 2, Second Edition edited by (B Ballantyne, TC Marrs and T Syversen, eds), pp. 737–774. London, UK: MacMillan Reference.

Ballantyne B. (1999c). Repeated exposure toxicity. In: *General and Applied Toxicology*, Volume 1, Second Edition (B Ballantyne, TC Marrs and T Syversen, eds.), pp. 55–66. London, UK: Macmillan Reference.

Ballantyne B (2005a). Peripheral chemosensory irritation with particular reference to the respiratory tract. In: *Inhalation Toxicology,* Second Edition. (H Salem and A Katz, eds), pp. 269–306. Boca Raton, FL, USA: CRC Press.

Ballantyne B (2005b). The occupational toxicologist: professionalism, morality and ethical standards in the context of legal and non-litigation issues. *J Appl Toxicol*, **25**, 496–513.

Ballantyne B (2006). Medical management of the traumatic consequences of civil unrest incidents: causation, clinical approaches, needs, and advance planning criteria. *Toxicol Rev*, in press.

Ballantye B and Beswick F (1972). On the possible relationship between diarrhea and *o*-chlorobenzylidene malononitrile (CS). *Med Sci Law*, **12**, 121–128.

Ballantyne B and Callaway S (1972). Inhalation toxicology and pathology of animals exposed to *o*-chlorobenzylidene malononitrile (CS). *Med Sci Law*, **12**, 43–65.

Ballantyne B and Clifford EC (1978). Short term inhalation toxicology of cinnamic acid smoke. *J Combust Toxicol*, **5**, 253–260.

Ballantyne B and Johnson WG (1974a). Safety aspects of the rubber bursting grenade. *Med Sci Law*, **14**, 144–150.

Ballantyne B and Johnson WG (1974b). *o*-Chlorobenzylidene malononitrile (CS) and the healing of cutaneous injuries. *Med Sci Law*, **14**, 93–97.

Ballantyne B and Salem H (2004). Forensic aspects of riot control agents. In: *Riot Control Agents. Issues in Toxicology, Safety, and Health* (EJ Olajos and W Stopford, eds), pp. 231–258. Boca Raton, FL, USA: CRC Press.

Ballantyne B and Salem (2006). Screening smokes; applications, toxicology, clinical considerations, and medical management. In: *Chemical Warfare Agents: chemistry, pharmacology and therapeutics*. Second edition (H Salem and BJ Lukey, eds). Boca Raton, FL, USA: CRC Press, in press.

Ballantyne B and Swanston DW (1973). The irritant potential of dilute solutions of *ortho*-chlorobenzylidene malononitrile (CS) on the eye and tongue. *Acta Pharmacol Toxicol*, **32**, 266–277.

Ballantyne B and Swanston DW (1974). The irritant effects of dilute solutions of dibenzoxazepine (CR) on the eye and tongue. *Acta Pharmacol Toxicol*, **35**, 412–423.

Ballantyne B and Swanston DW (1978). The comparative acute mammalian toxicity of 1-chloroacetophenone (CN) and 2-chlorobenzylidene malononitrile (CS). *Arch Toxciol*, **40**, 75–95.

Ballantyne B, Gazzard MF and Swanston DW (1972). Effects of solvents and irritants on intraocular pressure in the rabbit. *J Physiol London*, **266**, 12P–14P.

Ballantyne B, Gazzard MF and Swanston DW (1973a). Eye damage caused by crystal violet. *Br J Pharmacol*, **49**, 181–182.

Ballantyne B, Beswick W and Price Thomas D (1973b). The presentation and management of individuals contaminated with solutions of dibenzoxazepine (CR). *Med Sci Law*, **13**, 265–268.

Ballantyne B, Gazzard MF, Swanston DW *et al.* (1974). The ophthalmic toxicology of *o*-chlorobenzylidene malononitrile (CS). *Arch Toxicol*, **32**, 149–168.

Ballantyne B, Gazzard MF, Swanston DW *et al.* (1975). The comparative ophthalmic toxicology of 1-chloroacetophenone (CN) and dibenz(b.f)-1,4-oxazepine (CR). *Arch Toxicol*, **34**, 183–201.

Ballantyne B, Gall D and Robson DC (1976a). Effects on man of drenching with dilute solutions of *o*-chlorobenzylidene malononitrile (CS) and dibenz(b.f)-1,4-oxazepine (CR). *Med Sci Law*, **6**, 159–170.

Ballantyne B, Gazzard MF and Swanston DW (1976b). The ophthalmic toxicology of dichloromethane. *Toxicology*, **6**, 173–187.

Ballantyne B, Gazzard MF and Swanston D (1977a). Applanation tonometry in ophthalmic toxicology. In: *Current Approaches in Toxicology* edited by (B Ballantyne, ed.), pp. 158–192. Bristol, UK: John Wright.

Ballantyne B, Gazzard MF and Swanston DW (1977b). Irritancy testing by respiratory exposure. In: *Current Approaches in Toxicology* (B Ballantyne, ed.), pp. 129–138. Bristol, UK: John Wright.

Ballantyne B, Marrs TC and Syversen T (1999a). Fundamentals of Toxicology. In: *General and Applied Toxicology*, Volume 1, Second Edition (B Ballantyne, TC Marrs and T Syversen, eds), pp. 1–32. London, UK: MacMillan Reference.

Ballantyne B, Marrs TC and Syversen T (eds) (1999b). *General and Applied Toxicology,* Volume 2, Part 2, Second Edition. London, UK: MacMillan Reference.

Barlow N (2000). Letter to the Editor. *Resuscitation*, **47**, 91–92.

Bauchinger M and Schmid E (1992). Clastogenicity of 2-chlorobenzylidene malononitrile (CS) in V79 Chinese hamster cells. *Mutat Res*, **282**, 231–234.

Bayeaux-Dunglas M-C, Depars P, Touati M-A *et al.* (1999). Occupational asthma in a teacher after repeated exposure to tear gas. *Rev Mal Resp*, **16**, 558–559.

BBC (1997). CS spray used to subdue pensioner. *British Broadcasting Corporation News*, UK Edition, dated 5 December 1997 (http://new.bbc.co.uk/1/hi/uk/37188.stm).

BBC (1998a). CS gas 'unprovoked', say campaigners. *British Broadcasting Corporaton News*, UK Edition, dated 29 June 1998 (http://news.bbc.co.uk/1/hi/uk/122548.stm).

BBC (1998b). Inquest into pensioner's death after CS spray. *British Broadcasting Corporation News*, UK Edition, dated 25 September 1998 (http://nwes.bbc.co.uk/1/hi/uk/180389.stm).

BBC (1999). CS gas used on psychiatric patients. *British Broadcasting Corporation News*, UK Edition, dated 3 August 1999 (http://news.bbc.co.uk/1/hi/health/41890.stm).

BBC (2001a). Police use pepper spray. *British Broadcasting Corporation News*, UK Edition, dated 16 March 2001 (http://new.bbc.co.k/1/hi/uk/1223786.stm).

BBC (2001b). Police defend use of CS spray. *British Broadcasting Corporation News*, UK Edition, dated 29 May 2001 (http://news.bbc.co.uk/1/i/scotland/1357325.htm).

BBC (2002a). First Welsh force adopts pepper spray. *British Broadcasting Corporation News*, UK Edition, dated 5 August 2002 (http://news.bbc.co.uk/1/hi/wales/2173434.stm).

BBC (2002b). Police called to noisy party. *British Broadcasting News*, UK Edition, dated 25 November 2002 (http://news.bbc.co.uk/1/hi/england/2512045.stm).

BBC (2002c). 45 hurt in nightclub gas incident. *British Broadcasting Corporation News*, UK Edition, dated 16 November 2002. (http://news.bc.co.uk/1/hi/england/2484407.stm).

BBC (2003a). Inquest told police used CS gas. *British Broadcasting Corporation News*, UK Edition, dated 20 January 2003 (http://new.bbc.co.uk/1/hi/england/2678423.stm).

BBC (2003b). Open verdict in 'CS gas death'. *British Broadcasting Corporation News*, UK Edition, dated 31 January 2003 (http://news.bbc.co.uk/1/hi/england/2714675.stm).

BBC (2003c). Schizophrenic sprayed with CS gas. *British Broadcasting Corporation News*, UK Edition, dated 18 February 2003 (http://news.bbc.co.uk/1/hi/england/2776637.stm).

BBC (2003d). Violence erupts at boxing match. *British Broadcasting Corporation News*, UK Edition, dated 24 November 2003 (http://news.bc.c.uk/1/hi/england/south_yorkshire/3233064.stm).

BBC (2003e). CS gas in school – 3 suspended. *British Broadcasting Corporation News*, UK Edition, dated 23 September 2003 (http://news.bbc.uk.co/1/hi/wales/north_east/3131668.stm).

BBC (2003f). Boy charged after gas incident. *British Broadcasting Corporation News*, UK Edition, dated 7 October 2003 (http://news.bbc.co.k/1/hi/scotland/3170912.stm).

BBC (2003g) Police defend use of gas. *British Broadcasting Corporation News*, UK Edition, dated 6 October 2003 (http://news.bbc.co.uk/1/hi/england/derbyshire/ 3169548.stm).

BBC (2003h). Baton rounds 'more dangerous'/Doctors urge rubber bullet ban. *British Broadcasting Corporation News*, World Edition (http://new.bbc.co.uk/2/hi/uk.new/northern.ireland/2926225stm>&health/20033999.stm).

BBC (2004a). Concern over effects of CS spray. *British Broadcasting Corporation News*, UK Edition, dated 23 August, 2004 (http://news.bbc.co.uk/hi/health/3584560.stm).

BBC (2004b). Man sought after CS gas incident. *British Broadcasting Corporation News*, UK Edition, dated 28 May 2004 (http://news.bbc.co.uk/1/hi/scotland/3756049.stm).

BBC (2004c). Jail term for nightclub CS attack. *British Broadcastng Corporation News*, UK Edition, dated 5 October 2004 (http://news.bbc.co.uk/1/hi/derbyshire/3717490.stm).

BBC (2004d). CS replaced with 'safer option'. *British Broadcasting Corporation News*, UK Edition, dated 21 December 2004 (http://bbc.news.uk.co/1/hi/england/leicestershire/4114761.stm).

BBC (2004e). Two shops in CS attacks. *British Broadcasting Corporation News*, UK Edition, Dated 13 February 2004 (http://news.bbc.co.uk/1/hi/england/derbyshire/3486649.stm).

BBC (2004f). CS spay 'used on teenage boys'. *British Broadcasting Corporation News*, UK Edition, dated 16 October 2004 (http://news.bbc.co.uk/1/hi/england/manchester/3749510.stm).

BBC (2005a). CS gas released in tube station. *British Broadcasting Corporation News*, UK Edition, dated 18 May 2005 (http://new.bbc.co.uk/1/ hi/uk/4560591.stm).

BBC (2005b). Police acused in CS spray claim. *British Broadcasting Corporation News*, UK Edition, dated 2 May 2005 (http://news.bbc.co.uk/1/hi/england/lincolnshire/4504069.stm).

BBC (2005c). 'Chilli' spray to replace CS gas. *British Broadcasting Corporation News*, UK Edition, dated 27 June 2005 (http://news.bbc.co.uk/1/hi/england/beds/bucks/herts/4626779.stm).

BBC (2005d). Police accused of CS spray claim. *British Broadcasting Corporation News*, UK Edition, dated 2 May 2005 (http://news.bbc.co.uk/1/hi/england/lincolnshire/4504069.stm).

BBC (2005e). Man, 23, held over CS gas attack. *British Broadcasting Corporation News*, UK Edition, dated 7 February 2005 (http://news.bbc.co.uk/1/hi/england/manchester/4243299.stm).

BBC (2005f). CS spray hits four in nightclub. *British Broadcasting Corporation News*, UK Edition, dated 10 April 2005 (http://news.bbc.co.uk/1/hi/england/southern_counties/4429769.stm).

BBC (2005g). CS spray used at student's union. *British Broadcasting Corporation News*, UK Edition, dated 4 May 2005 (http://news.bbc.co.uk/1/hi/wales/mid/4514243.stm).

Bell MD, Rao VJ, Wetli CV *et al.* (1992). Positional asphyxia in adults: a series of 30 cases from the Dade and Broward County Florida Medical Examiners Offices from 1982 to 1990. *Am J Forens Med Pathol*, **13**, 101–107.

Beswick FW, Holland P and Kemp KH (1972). Acute effects of exposure to *o*-chlorobenzylidene malononitrile (CS) and the development of tolerance. *Br J Indust Med*, **29**, 298–306.

Bhattacharya ST and Haywood AW (1993). CS gas – implications for the anaesthetist. *Anaesthesia*, **48**, 896–897.

Billmere DF, Vinicur C, Ginda M *et al.* (1996). Pepper-spray-induced respiratory treated with extracorporeal membrane oxygenation. *Pediatrics*, **96**, 961–963.

Blaho K and Stark MM (2000). CS is a particulate spray, not a gas. *Br Med J*, **321**, 46.

Blanc P, Lui D, Juarez C *et al.* (1991). Cough in hot pepper workers. *Chest*, **99**, 27–32.

Blumenfeld R and Meselson M (1971). The military value and political implications of the use of riot control agents in chemical warfare. In: *The Control of Chemical and Biological Weapons*, pp. 64–93. New York, NY, USA: Carnegie Endowment.

Boulton ML, Abellera J, Lemmings J *et al.* (2005). Brief report: terrorism and emergency preparedness in State and territorial public health departments – United States, 2004. *Morbid Mortal Week Rep*, **54**, 459–460.

Brekell A and Bodiwala GG (1998). CS-gas exposure in a crowded night club: the consequences for an accident and emergency department. *J Accid Emerg Med*, **15**, 56–64.

Brimblecombe RW, Green DM and Muir AW (1972). Pharmacology of *o*-chlorobenzylidene malononitrile. *Br J Pharmacol*, **29**, 298–306.

Brooks SM, Weiss MA and Bernstein L (1985). Reactive airways dysfunction syndrome. Case reports of persistent airways hyper-reactivity following high-level irritant exposures *J Occup Med*, 27, 473–476.

Brown I, Takeuchi D and Challoner K (2000). Corneal abrasions associated with pepper spray exposure. *Am J Emerg Med*, **18**, 271–272.

Burton EG, Clarke MI, Miller RA *et al.* (1982). Generalizations and characteristics of red phosphorus smoke aerosols for inhalation exposure of laboratory animals. *Am Indust Hyg Assoc J*, **43**, 767–772.

Busker RW and van Helden HPM (1998). Toxicological evaluation of pepper spray as a possible weapon for the Dutch police force. Risk assessment and efficacy. *Am J Forens Med Pathol*, **19**, 309–316.

Caterina MJ and Julius D (2001). The vanilloid receptor: a molecular gateway to the pain pathway. *Ann Rev Neurosci*, **24**, 487–517.

Chan OY, Lee CS, Tan TK *et al.* (1990). Health problems among spice grinders. *Jpn Soc Occup Med*, **40**, 111–115.

Chapman AJ and White C (1978). Death resulting from lachrymatory agents. *J Forens Sci*, **23**, 247–256.

Chung CW and Giles AL Jr (1972). Sensitization of guinea pigs to alpha-chloroacetophenone (CN) and *ortho*-chlorobenzylidene malononitrile (CS) tear gas chemicals. *J Immunol*, **109**, 284–293.

Cole TJ, Cotes JE, Johnson GR *et al.* (1975). Comparison of effects of ammonia and CS aerosol upon exercise ventilation and cardiac frequency in healthy men. *J Physiol London*, **252**, 28P–29P.

Colgrave HF, Lee C, Marrs TC *et al.* (1983). Repeated dose inhalation toxicity and mutagenicity-status of CR (dibenz(b.f)-1,4-oxazepine). *Br J Pharmacol*, **78**, 169.

Conner WC (1967). Human violence stopped by dart-shooting tranquilizer gun. *J Amer Med Assoc*, **201**, 34–35.

Constant HL and Cordell GA (1996). Nonivamide, a constituent of capsicum oleoresin. *J Nat Prod*, **59**, 425–429.

Cooper TH, Guzinski JA and Fisher C (1991). Improved high performance liquid chromatography method for the determination of major capsaicinoids in capsicum oleoresins. *J Agr Food Chem*, **39**, 2253–2256.

Cotes JE, Dabbs JM, Evans MR *et al.* (1972a). Effect of CS aerosol upon lung transfer and alveolar volume in healthy men. *Q J Expt Physiol*, **57**, 199–206.

Cotes JE, Evans RL, Johnson GR *et al.* (1972b). The effect of CS aerosol upon exercise ventilation and cardiac frequency in healthy men. *J Physiol London*, **222**, 77P–78P.

Cucinell SA, Swentzel K, Biskup R (1971). Biochemical interactions and metabolic fate of riot control agents. *Fed Proc*, **30**, 86–91.

D'Agostino PA and Provost LR (1985). Gas chromatographic retention indices of chemical warfare agents and simulants. *J Chromatogr*, **331**, 47–54.

D'Agostino PA and Provost LR (1995). Analysis of irritants by capillary column gas chromatography tandem mass spectrometry. *J Chromatogr Anal*, **695**, 65–73.

Damhoeri A, Hosono A, Itoh T *et al.* (1985). *In vitro* mutagenicity tests on capsicum pepper, shallot and nutmeg oleoresins. *Agr Biol Chem*, **49**, 1519–1520.

Dean-Drummond A (1975). *Riot Control*. London, UK: Royal United Services Institute for Defence Studies.

Ditter DM and Heal CS (2004). Application and use of riot control agents. In: *Riot Control Agents. Issues in Toxicology, Safety, and Health* (EJ Olajos and W Stopford, eds), pp. 17–24. Boca Raton, FL, USA: CRC Press.

Duenas A, Filipe S, Ruz-Mambrilla M *et al.* (2000). CO poisoning caused by inhalation of CH_3Cl contained in personal defense sprays. *Am J Emerg Med*, **18**, 120–121.

Durnford JM (2004). Genetic toxicity of riot control agents. In: *Riot Control Agents. Issues in Toxicology, Safety, and Health* (EJ Olajos and W Stopford, eds), pp. 183–200. Boca Raton, FL, USA: CRC Press.

Euripidou E, MacLehose R and Fletcher A (2004). An investigation into the short term and medium term health impacts of personal incapacitant sprays. A follow up of patients reported to the National

Poisons Information Service (London). *Emerg Med J*, **21**, 548–552.
Evans EH (1945). Casualties following exposure to zinc chloride smoke. *Lancet*, **ii**, 368–370.
Feinsilver L, Chambers HA, Vocci FJ *et al.* (1971). *Some metabolites of CS from rats*. Edgewood Arsenal Technical Report Series, Number 4521, dated May 1971. Edgewood Arsenal, MD, USA: Department of the Army.
Ferslew KE, Orcutt RH and Hagardorn AH (1986). Spectral differentiation and gas chromatographic/ mass spectrometric analysis of the lacrimators 2-chloroacetophenone and *o*-chlorobenzylidene malononitrile. *J Forens Sci*, **31**, 658–665.
Fisher AA (1970). Dermatitis due to tear gases (lachrymators). *Br J Dermatol*, **9**, 91–95.
Fitzgerald M (1983). Capsaicin and sensory neurons – a review. *Pain*, **15**, 109–130.
Forrester AB and Stanley SK (2003) The epidemiology of pepper spray exposures reported in Texas in 1998–2002. *Vet. Human Toxicol*, **45**, 27–33.
Foster RW and Weston KM (1986). Chemical irritant algesia assessed using the human blister base. *Pain*, **25**, 269–278.
Frankenberg L and Sörbo B (1973). Formation of cyanide from *o*-chlorobenzylidene malononitrile and its toxicological significance. *Arch Toxicol*, **31**, 99–108.
Frazer RM (1971). The cost of commotion: an analysis of the psychiatric sequelae of the 1969 Belfast riots. *Br J Psychiat*, **118**, 257–264.
Frazier CA (1976). Contact allergy to mace. *J Am Med Assoc*, **236**, 25–26.
French MC, Harrison JM, Inch TD *et al.* (1983a). The fate of dibenz(b,f)-1,4-oazepine (CR) in the rat, rhesus monkey, and guinea pig, part 1, metabolism *in vivo*. *Xenobiotica*, **13**, 345–359.
French MC, Harrison JM, Newman J *et al.* (1983b). The fate of dibenz(b.f)-1,4-oxazepine (CR) in the rat, part iii, the intermediary metabolites. *Xenobiotica*, **13**, 373–381.
Fuchs T and in der Wiesche M (1990). Contact dermatitis from CN and CS (tear gas) among demonstrators. *Z Haut*, **65**, 25–28.
Fujimura M, Kamio Y, Sakamoto S *et al.* (1992). Tachyphylaxis to capsaicin-induced cough and its reversal by indomethacin, in patients with the sinobronchial syndrome. *Clin Autonomic Res*, **2**, 397–401.
Fuller RW (1990). The human pharmacology of capsaicin. *Arch Int Pharmacodyn*, **303**, 147–155.
Fuller RW, Dixon CM and Barnes PJ (1985). Bronchoconstrictor response to inhaled capsaicin in humans. *J Appl Physiol*, **58**, 1080–1084.
Fung T, Jeffery W and Beveridge AD (1982). The identification of capsaicinoids in tear gas spray. *J Forens Sci*, **27**, 812–821.
Furnival B, Harrson JM, Newman J *et al.* (1983). The fate of dibenz(b.f)-1,4-oxazepine in the rat: part II, metabolism *in vitro*. *Xenobiotica*, **13**, 361–372.
Gaskins JR, Hehir RM, McCaulley DF *et al.* (1972). Lachrimating agents (CS and CN) in rats and rabbits. *Arch Environ Health*, **24**, 449–454.
Georgopoulas PG, Fedele P, Shade P *et al.* (2004). Hospital response to chemical terrorism: personal protective equipment, training, and operations planning. *Am J Indust Med*, **46**, 432–445.
Gonzales TA, Vance M, Helpern M *et al.* (1954). *Legal Medicine*. New York, NY, USA: Appleton-Century Crofts.
Govindarajan VS and Sathyanaravana MN (1991). Capsicum – production, technology, chemistry, and quality. Part V. Impact on physiology, pharmacology, nutrition, and metabolism: structure, pungency, pain and desensitizing sequences. *Crit Rev Food Sci Nutrit*, **23**, 207–288.
Granfield G, Onnen J and Petty CS (1994). Pepper spray and in-custody deaths. *Executive Brief*, March 1994. Alexandria, VA, USA: International Association of Chiefs of Police.
Grant WM (1986). *Toxicology of the Eye*, Third Edition, pp. 207–208. Springfield, IL, USA: Charles C. Thomas.
Grawe J, Nusse M and Adler I-D (1997). Quantitative and qualitative studies of micronucleus induction in mouse erythrocytes using flow cytometry. 1. Measurement of micronucleus induction in peripheral blood polychromatic erythrocytes by chemicals with known or suspected genotoxicity. *Mutagenesis*, **12**, 1–8.
Gray PJ (1995). Treating CS gas injuries of the eye. *Br Med J*, **311**, 871.
Gray PJ (2004). Formulation affects toxicity. *Br Med J*, **321**, 46.
Greaves JA (2000). The automobile as a confined space for toxic chemical hazards. *Am J Indust Med*, **38**, 481–482.
Gutentag PL, Hart J, Owens EJ *et al.* (1960). *The evaluation of CS aerosols as a riot control agent*. US Army Chemical Warfare Laboratories Technical Report, CWLR 2365. Bethesda, MD, USA: Army Chemical Center.
Hass JS, Whipple RE, Grant PM *et al.* (1997). Chemical and elemental comparison of two formulations of oleoresin capsicum. *Sci Justice*, **37**, 15–24.
Hayes AW (2001). *Principles and Methods of Toxicology*, Fourth Edition. Philadelphia, PA, USA: Taylor & Francis.

Hayes AG, Skingle M and Tyers MB (1980). The antinociceptive effects of single doses of capsaicin in the rodent. *Br J Pharmacol*, **70**, 96P–97P.

Hayes AG, Skingle M and Tyers MB (1981). Effects of single doses of capsaicin on nociceptive thresholds in the rodent. *Neuropharmacology*, **20**, 505–511.

Hayman CR and Berkely MJ (1971). Health care for war demonstrators in Washington, April-May, 1971. *Med Ann Distr Columbia*, **40**, 633–637.

Hellreich A, Goldman RH, Bottiglieri NG *et al.* (1967). *The effect of thermally-generated CS aerosols on Human skin*. Edgewood Arsenal Technical Report Series, Number 4075, dated January 1967. Edgewood Arsenal, MD, USA: Department of the Army.

Hill AR, Silverberg NB, Mayorga D *et al.* (2000). Medical hazards of the tear gas CS. A case of persistent, multisystem, hypersensitivity reaction and a review of the literature. *Medicine*, **79**, 234–240.

HMSO (1969). *Report of the enquiry into the Medical and Toxicological aspects of CS (Orthochlorobenzylidene malononitrile). Part 1 – Enquiry into the Medical Situation following the use of CS in Londonderry on 13th and 14th August, 1969.* Cmnd 4173. London, UK: Her Majesty's Stationary Office.

HMSO (1971). *Report of the enquiry into the Medical and Toxicological aspects of CS (Orthochlorobenzylidene malononitrile). Part 2. Enquiry into Toxicological Aspects of CS and its use for Civil Purposes.* Cmnd 4775. London, UK: Her Majesty's Stationary Office.

Hoffmann DH (1967). Eye burns caused by tear gas. *Br J Ophthalmol*, **51**, 265–268.

Holland P (1974). The cutaneous reactions produced by dibenzoxazepine (CR). *Br J Dermatol*, **90**, 657–665.

Holland P and Seabright M (1971). Effects of CS on lymphocyte chromosome structure. In: *Clinical and Preclinical Studies on Ortho Chlorobenzylidene Malononitrile (CS)*, Technical Note Number 82. Porton Down, Salisbury, UK: Chemical Defence Establishment.

Holland P and White RG (1972). The cutaneous reactions produced by *o*-chlorobenzylidene malononitrile and ω-chloroacetophenone when applied directly to the skin of human subjects. *Br J Dermatol*, **86**, 150–154.

Holopainen JM, Molnen JAO, Hack T *et al.* (2003). Toxic carriers in pepper sprays may cause corneal erosion. *Toxicol Appl Pharmacol*, **186**, 155–162.

Hopping W (1969). Lesions caused by close-range shots from from gas pistols. *Klin Monatsbl Augenheilkd*, **135**, 270–272.

Horowitz BZ (1986). Carboxyhemoglobinemia caused by inhalation of methylene chloride. *Am J Emerg Med*, **18**, 691–695.

Hu H and Cristani D (1992). Reactive airways dysfunction after exposure to tear gas. *Lancet*, **339**, 1535.

Hu H, Fine J, Epstein P *et al.* (1989). Tear gas – harassing agent or toxic chemical weapon? *J Am Med Assoc*, **262**, 660–663.

Jane I and Wheals BB (1972). Chromatographic characterization of lachrymatory agents in tear gas aerosols. *J Chromatogr Anal*, **70**, 151–153.

Jang JJ, Cho KJ, Lee YS *et al.* (1991). Different modifying responses of capsaicin in a wide-spectrum initiation model of F344 rat. *J Korean Med Sci*, **6**, 31–36.

Jansco G, Kiraly E and Jansco-Gabor A (1977). Pharmacologically induced selective degeneration of chemosensitive primary sensory neurones. *Nature*, **270**, 741–743.

Jansco G, Karcsu S, Kiraly E *et al.* (1984). Neurotoxin-induced nerve cell degeneration: possible involvement of calcium. *Brain Res*, **295**, 211–216.

Johnson FA and Stonehill RB (1961). Chemical pneumonitis from inhalation of zinc chloride. *Dis Chest*, **40**, 619–624.

Jolly HA and Carpenter CL (1968). Tear gas dermatitis. *J Am Med Assoc*, **203**, 808.

Jones JRN and Israel MS (1970). Mechanism of toxicity of injected CS gas. *Nature (London)*, **228**, 315–317.

Kales SN and Cristiani DC (2004). Acute chemical emergencies. *New England J Med*, **350**, 800–808.

Karagama YG, Newton JR and Newbegin CJR (2003). Short-term and long-term physical effects of exposure to CS spray. *J R Soc Med*, **96**, 172–174.

Kawada T and Iwai K (1985). *In vitro* and *in vivo* metabolism of dihydrocapsaicin, a pungent principle of hot pepper in rats. *Agr Biol Chem*, **49**, 441–448.

Kawada T, Suzuki T, Takahashi M *et al.* (1984). Gastrointestinal absorption and metabolism of capsaicin and dihydrocapsaicin. *Toxicol Appl Pharmacol*, **72**, 449–456.

Kemp KH and Weatherell A (1975). *The subjective detection of CR contamination in drinking water*. CDE Technical Note, Number 238, dated June 1975. Porton Down, Salisbury, UK: Ministry of Defence, Chemical Defence Establishment.

Kemp KH and Wilder WB (1972). Taste perception and palatability following CR exposure. *Med Sci Law*, **12**, 113.

Kim JP, Park JG, Lee MD *et al.* Co-carcinogenic effects of several Korean foods on gastric cancer induced by N-methyl-N-nitrosoguanidine in rats. *Jpn J Surg*, **15**, 427–437.

Kimber I and Dearman RJ (1999). Evaluation of respiratory sensitization potential of chemicals. In: *General and Applied Toxicology*, Volume 2, Second Edition. (B Ballantyne, TC Marrs and T Syversen, eds), pp. 701–720. London, UK: Macmillan Reference.

King K, Tunget CI, Turche B *et al.* (1995). Severe contact dermatitis from chemical mace containing 1-chloroaetophenone. *J Toxicol Cut Ocul Toxicol*, **14**, 57–62.

Kirby ML, Gale TF and Mattio TG (1982). Effects of prenatal capsaicin treatment on fetal spontaneous activity, opiate receptor binding, and acid phosphatase in in spinal cord. *Exp Neurol*, **76**, 298–308.

Kling RP (1969). Chemical mace – a controversy. *Eye Ear Nose Throat Monthly*, **48**, 13.

Kluchinsky TA Jr, Savage PB, Sheely MV *et al.* (2001). Identification of CS-derived compounds formed during heat-dispersion of CS riot control agent. *J Microcol Sep*, **13**, 186–190.

Kluchinsky TA Jr, Savage PB, Fitz R *et al.* (2002). Liberation of hydrogen cyanide and hydrogen chloride during high-temperature dispersion of CS riot control agent. *Am Ind Hyg Assoc J*, **63**, 493–496.

Krapf R (1981). Acute exposure to CS tear gas and clinical observations. *Schweiz Med Wschr*, **111**, 2056–2060.

Krebs J, Prime RJ and Leung K (1982). Rapid determination of capsaicin, CN and CS in tear gas by HPLC. *Can Soc Forens Sci J*, **15**, 29–33.

Kumar P, Flora SJ, Pant SC *et al.* (1994). Toxicological evaluation of 1-chloroacetophenone and dibenz[b.f]-1,4-oxazepine after repeated inhalation exposure in mice. *J Appl Toxicol*, **14**, 411–416.

LaHann TR (1986). Effect of capsaicin on croton oil and TPA induced carcinogenesis and inflammation. *J West Pharmacol Soc*, **29**, 145–149.

Laibson PR and Oconor J (1979). Explosive tear gas injuries of the eye. *Trans Am Acad Ophthalmol Otolaryng*, (July–August), 811–819.

Lancet (1971). Leading Article. Toxicity of CS. *Lancet*, **2**, 698.

Laurent JF, Richetr F and Michel A (1999). Management of victims of urban chemical attack: the French approach. *Resuscitation*, **42**, 141–149.

Leadbeater L (1973). The absorption of *o*-chlorobenzylidene malononitrile (CS) by the respiratory tract. *Toxicol Appl Pharmacol*, **25**, 101–110.

Leadbeater L, Sainsbury GL and Utley D (1973). *ortho*-Chlorobenzylmalononitrile: a metabolite formed from *ortho*-chorobenzylidene malononitrile (CS). *Toxicol Appl Pharmacol*, **25**, 111–116.

Lee SS and Kumar S (1980). Metabolism *in vitro* of capsaicin, a pungent principle of red pepper, with rat liver homogenates. In: *Microsomes, Drug Oxidation, and Chemical Carcinogenesis*, Volume 2. (MJ Coon, AH Conney, RW Estabrook, eds), pp. 1009–1012. New York, NY, USA: Academic Press.

Lee RJ, Yolton RL, Schnider C *et al.* (1996). Personal defense sprays: effects and management of exposure. *Am Optomet Assoc*, **67**, 548–560.

Leenutaphong V and Goerz G (1989). Allergic contact dermatitis from chloroacetophenone (tear gas). *Contact Derm*, **20**, 316.

Leopald IH and Lieberman TW (1971). Chemical injuries of the cornea. *Fed Proc*, **30**, 92–95.

Levine RA and Stahl CJ (1968). Eye injury caused by tear gas weapons. *Am J Ophthlmol*, **65**, 497–508.

Lewer N and Schofield S (1997). *Non-Lethal Weapons: A Fatal Attraction?* London, UK: Zed Books Ltd.

Lifschultz BD and Donoghue ER (1991). Deaths in custody. In: *Legal Medicine*, pp. 45–71. Philadelphia, PA, USA: Saunders.

Loriot J and Tourte J (1990). Hazards of contact lenses used by workers. *Int Arch Occup Environ Health*, **62**, 105–108.

Luke JL and Reay DT (1992). The perils of investigating and certifying deaths in police custody. *Am J Forens Med Pathol*, **13**, 98–100.

Lundberg JM and Saria A (1982). Bronchial smooth muscle contraction induced by stimulation of capsaicin-sensitive sensory neurons. *Acta Physiol Scand*, **116**, 473–476.

Lundberg L and Lundberg JM (1984). Capsaicin sensitive sensory neurons mediate the response to nasal irritation induced by the vapor phase of cigarette smoke. *Toxicology*, **33**, 1–7.

Lyons HA (1971). Psychiatric sequelae of the Belfast riots. *Br J Psychiat*, **118**, 265–273.

MacLeod IF (1969). Chemical mace: ocular affects in rabbits and monkeys. *J Forens Sci*, **14**, 34–37.

Macrae WG, Willinsky MD and Basu PK (1970). Corneal injury caused by aerosol irritant projectors. *Can J Ophthalmol*, **5**, 3–11.

Makles Z, Sliwakowski M and Nousiainen P (1999). Detection of dibenzo[b,f][1,4]oxazepine in the presence of other lacrimators with thin layer chromatography. *Chem Anal*, **44**, 257–262.

Mankes RF and Mankes KM (2004). Reproductive and developmental toxicology of riot control agents. In: *Riot Control Agents. Issues in Toxicology, Safety, and Health* (EJ Olajos and W Stopford, eds), pp. 161–181. Boca Raton, FL, USA: CRC Press.

Marrs TC, Gray MI, Colgrave HF *et al.* (1982). A repeated dose study of the toxicity of CR applied to the skin of mice. *Toxicol Lett*, **13**, 259–265.

Marrs TC, Clifford EC and Colgrave HF (1983a). Pathological changes produced by exposure of rabbits and rats to smokes from mixtures of hexachloroethane and zinc oxide. *Toxicol Lett*, **19**, 247–252.

Marrs TC, Clifford E and Colgrave HF (1983b). Late inhalation toxicology and pathology produced by exposure to a single dose of 2-chlorobenzylidene malononitrile (CS) aerosol in rats and hamsters. *Med Sci Law*, **23**, 257–265.

Marrs TC, Colgrave HF, Cross NL *et al.* (1983c). A repeated dose study of the toxicity of inhaled 2-chlorobenzylidene malononitrile (CS) aerosol in three species of laboratory animal. *Arch Toxicol*, **52**, 183–198.

Marrs TC, Colgrave HF and Cross NL (1983d). A repeated dose study of the toxicity of technical grade dibenz(b.f)-1,4-oxazepne in mice and hamsters. *Toxicol Lett*, **17**, 13–21.

Marrs TC, Colgrave HF, Gazzard MF *et al.* (1984). Inhalation toxicology of a smoke containing Solvent Yellow 33, Disperse Red 9 and Solvent Green 3 in laboratory animals. *Human Toxicol*, **3**, 298–308.

Marrs TC, Colgrave HF, Edginton JAG *et al.* (1988). The repeated dose toxicity of zinc oxide/hexachloroethane smoke. *Arch Toxicol*, **62**, 123–132.

Marrs TC, Colgrave HF, Edginton JAG *et al.* (1989). Repeated dose inhalation toxicity of cinnamic acid smoke. *J Hazard Mater*, **21**, 1–15.

Marrs TC, Allen IV, Colgrave HF *et al.* (1991). Neurotoxicity of 1-methoxycycloheptatriene – a Purkinje cell toxicant. *Human Exp Toxicol*, **10**, 93–101.

Marrs TC, Maynard RL and Sidell ER (1996). *Chemical Warfare Agents: Toxicology and Treatment*, pp. 221–230. Chichester, UK: John Wiley & Sons, Ltd.

Marsh SJ, Stansfeld CE, Brown DA *et al.* (1987). The mechanism of action of capsaicin on sensory c-type neurones and their actions *in vitro*. *Neuroscience*, **23**, 275–289.

Marshall TK (1976). Wounds and trauma. In: *Gradwohl's Legal Medicine*, Third Edition (FE Camps, AE Robinson and BGB Lucas, eds), pp. 294–296. Bristol, UK: John Wright.

Martling CR (1987). Sensory nerves containing tachykinins and CGRP in the lower airways. *Acta Physiol Scand*, **130**, **Supplement 563**, 1–57.

Marzulli FN and Maibach HI (1974). The use of graded concentrations in studying skin sensitizers: experimental contact sensitization in man. *Food Cosmet Toxicol*, **2**, 219–227.

Maynard RL (1999). Toxicology of chemical warfare agents. In: *General and Applied Toxicology*, Volume 3, Second Edition (B Ballantyne, TC Marrs and T Syversen, eds), pp. 2079–2109. London, UK: Macmillan Reference.

McDonald DM (1992). Infections intensify neurogenic plasma extravasation in the airway mucosa. *Am Rev Resp Dis*, **146**, S40–S44.

McElhatton PR, Sidhu S and Thomas SHL (2004). Exposure to CS gas in pregnancy. *J Toxicol Clin Toxicol*, **42**, 547.

McGregor DB, Brown A, Cattanach P *et al.* (1988). Responses of the L5178K tk+/tk-mouse lymphoma cell forward mutation assay II. 18 coded chemicals. *Environ Mol Mutagen*, **11**, 91–118.

McLaughlin V and Siddle B (1998). Law enforcement and custody deaths. *The Police Chief* (August), 38–41.

McNamara BP, Vocci FJ and Owens EJ (1968). *The toxicology of CN*. Edgewood Arsenal Technical Report Series, Number 4207, dated December 1968. Edgewood Arsenal, MD, USA: Department of the Army.

McNamara BP, Owens EJ, Weimer JT *et al.* (1969). *Toxicology of riot control chemicals – CS, CN, and DM*. Edgewood Arsenal Technical Report Series, Number 4309, dated March 1969. Edgewood Arsenal, Bethesda, MD, USA: Department of the Army.

Meshram GP, Malini RP and Rao KM (1992). Mutagenicity evaluation of riot control agent *o*-chlorobenzylidene malononitrile (CS) in the Ames/microsome test. *J Appl Toxicol*, **12**, 377–384.

Mesilaakso M (1996). Analysis of the ^{1}H and ^{13}C[^{1}H] NMR spectral parameters of tear gases, α-chloroacetophenone, bibenz[b,f][1,4]oxazepine and 2-chlorobenzylidene malononitrile. *Mag Reson Chem*, **34**, 989–994.

Miller BM and Nusse N (1993). Analysis of micronuclei induced by 2-chlorobenzylidene malononitrile (CS) using fluorescence *in situ* hybridization with telemetric and centrometric DNA probes and flow cytometry. *Mutagenesis*, **8**, 35–41.

Miller MS, Buck SH, Brendel K *et al.* (1980). Dihydrocapsaicin-induced hypothermia and substance P depletion in rats. *Pharmacologist*, **22**, 206.

Miller MS, Buck SH, Schnellmann R *et al.* (1981). Substance P depletion and analgesia induced by capsaicin analogs in guinea pigs. *Fed Proc*, **40**, 274.

Ministry of Defence (1972). *Medical Manual of Defence Against Chemical Agents*. London, UK: Her Majesty's Stationary Office.

Mishra PK (1984). Role of smokes in warfare. *Defence Sci J*, **44**, 173–179.

Monsereenusorn Y (1983). Subchronic toxicity studies of capsaicin and capsicum in rats. *Res Commun Chem Path Pharmacol*, **41**, 95–110.

Montell C, Birnbaumer L, Flockerzi V *et al.* (2002). A unified nomenclature for the superfamily of TRP cation-channels. *Mol Cell*, **9**, 229–231.

Morby P, Murray V, Cummins A *et al.* (2000). The capability of accident and emergency departments to safely decontaminate victims of chemical incidents. *J Accid Emerg Med*, **17**, 344–347.

Morice AH, Higgns KS and Yeo WW (1992). Adaptation of cough reflex with different types of stimulation. *Eur Resp J*, **5**, 841–847.

Nagabhushan M and Bhide SV (1985). Mutagenicity of chilli extract and capsaicin in short term tests. *Environ Mutagen*, **7**, 881–888.

Narasimhamurthy M and Narasimhamurthy K (1988). Non-mutagencity of capsaicin in albino mice. *Food Chem Toxicol*, **26**, 955–958.

National Institute of Justice (1994). Oleoresin capsicum: pepper spray as a force alternative. *NIJ, Technology Assessment Program*. Washington, DC, USA: US Department of Justice, Office of Justice Programs, National Institute of Justice.

Niesink RJM, De Vries J and Holliger MA (1996). *Toxicology. Principles and Applications*. Boca Raton, FL, USA: CRC Press.

Noeller TP (2001). Biological and chemical terrorism: recognition and management. *Cleveland Clinic J Med*, **68**, 1001–1006.

NRC (1999). *Chemical and Biological Terrorism*. Washington DC, USA: National Research Council, National Academy Press.

NTP (1990a). *Toxicology and carcinogenesis studies of 2-chloroacetophenone (CAS No. 532-27-4) in F344/N rats and B6C3F1 mice (inhalation studies)*. National Toxicology Program, Technical Report 379. Research Triangle Park, NC, USA: National Toxicology Program.

NTP (1990b). *Toxicology and carcinogenesis studies of CS2 (4% o-chlorobenzalmalononitrile) (CAS No. 2698-41-1) in F334/N rats and B6C3F1 mice (inhalation studies)*. National Toxicology Program, Technical Report 377. Research Triangle Park, NC, USA: National Toxicology Program.

Nusse M, Recknagel S and Beisker W (1992). Micronuclei induced by 2- chlorobenzylidene malononitrile contain single chromosomes as demonstrated by the combined use of flow cytometry and immunofluorescent staining with anti-kinetochore antibodies. *Mutagenesis*, **7**, 57–67.

Oaks IW, Dorman JE and Petty RW (1969). Tear gas burns of the eye. *Arch Opthalmol*, **63**, 698–706.

Oi Y, Kawada T, Watanabe T *et al.* (1992). Induction of capsaicin-hydrolyzing enzyme activity in rat liver by continuous oral administration of capsaicin. *J Agr Food Chem*, **40**, 467–470.

Oksala A and Salminen I (1975). Eye injuries caused by tear-gas hand weapons. *Arch Ophthalmol*, **63**, 908–913.

Okumura T, Suzuki K, Fukuda A *et al.* (1998). The Tokyo subway sarin attack: Disaster management, Part 2: Hospital response. *Acad Emerg Med*, **5**, 618–624.

Olajos EJ (2004). Biochemistry, biological interactions, and pharmacokinetics of riot control agents. In: *Riot Control Agents. Issues in Toxicology, Safety and Health* (EJ Olajos and W Stopford, eds), pp. 37–63. Boca Raton, FL, USA: CRC Press.

Olajos EJ and Lakoski JM (2004). Pharmacology/toxicology of oleoresin capsicum, capsaicin, and capsaicinoids. In: *Riot Control Agents. Issues in Toxicology, Safety and Health* (EJ Olajos and W Stopford, eds), pp. 123–143. Boca Raton, FL, USA: CRC Press.

Olajos EJ and Salem H (2001). Riot control agents: pharmacology, toxicology, biochemistry and chemistry. *J Appl Toxicol*, **21**, 355–391.

Olajos EJ and Stopford W (2004). Introduction and historical perspective. In: *Riot Control Agents. Issues in Toxicology, Safety and Health* (EJ Olajos and W Stopford, eds), pp. 1–24. Boca Raton, FL, USA: CRC Press.

Osborne JS, Adamek S and Hobbs ME (1956). Some components of gas phase of cigarette smoke. *Anal Chem*, **28**, 211–215.

Owens EJ and Punte CL (1963). Human respiratory and ocular irritation studies utilizing *o*-chlorobenzylidene malononitrile aerosols. *Am Indust Hyg Assoc J*, **24**, 262–264.

Park S and Giammona ST (1972). Toxic effects of tear gas on an infant following prolonged exposure. *Am J Dis Childhood*, **123**, 245–246.

Pearlman AI (1969). Nonlethal weapons for use by law enforcement agencies. *New Physic*, (August), 625–628.

Penneys NS (1971). Contact dermatitis due to chloroacetophenone. *Fed Proc*, **30**, 96–99.

Penneys NS, Isreal RM and Indgin SM (1969). Contact dermatitis due to 1-choroacetophenone and chemical mace. *New England J Med*, **281**, 413–415.

Pfeiff B (1984). Allergic contact dermatitis to chloroacetophenone (mace). *Z Hautkr*, **60**, 178–184.

Philip G, Baroody FM, Proud D *et al.* (1994). The human nasal response to capsaicin. *J Allergy Clin Immunol*, **94**, 1035–1045.

Punte CL, Ballard TA and Weimer JT (1962a). Inhalation studies with chloroacetophenone, diphenylaminochloroarsine and pelargonic morpholide. 1. Animal exposures. *Ind Hyg J*, **23**, 194–198.

Punte CL, Weimer JT, Ballard TA *et al.* (1962b). Toxicologic studies on *o*-chlorobenzylidene malononitrile. *Toxicol Appl Pharmacol*, **4**, 656–662.

Punte CL, Owens EJ and Gutentag PJ (1963). Exposure to *ortho*-chlorobenzylidene malononitrile. *Arch Environ Health*, **6**, 366–374.

Queen FB and Stander T (1941). Allergic contact dermatitis following exposure to tear gas (chloroacetophenone, CN). *J Am Med Assoc*, **117**, 1879.

Rappert B (2003). Health and safety in policing: lessons from the regulation of CS sprays in the UK. *Soc Sci Med*, **56**, 1269–1278.

Reay DT, Fligner CI, Stilwell AD *et al.* (1992). Positional asphyxia during law enforcement transport. *Am J Forens Med Pathol*, **13**, 90–97.

Reilly CA, Crouch DJ, Yost GS *et al.* (2001). Determination of capsaicin, dihydrocapsaicin, and nonivamide in self-defense weapons by liquid chromatography–mass spectrometry and liquid chromatography–tandem mass spectrometry. *J Chromatogr A*, **912**, 259–267.

Reilly CA, Crouch DJ, Yost GS *et al.* (2002a). Detection of pepper spray residues on fabrics using liquid chromatography–mass spectrometry. *J Forens Sci*, **47**, 37–43.

Reilly CA, Crouch DJ, Yost GS *et al.* (2002b). Determination of capsaicin, nonivamide and dihydrocapsaicin in blood and tissue by liquid chromatography–tandem mass spectrometry. *J Anal Toxicol*, **26**, 313–319.

Reilly CA, Ehlhardt WJ, Jackson DA *et al.* (2003). Metabolism of capsaicin by cytochrome P450 produces novel dehydrogenated metabolites and decreases cytotoxicity to lung and liver cells. *Chem Res Toxicol*, **16**, 336–349.

Reitveld EC, Delbressine LPC, Waegemaekers THJM *et al.* (1983). 2-Chlorobenzylmercapturic acid, a metabolite of the riot control agent 2-chlorobenzylidene malononitrile (CS) in the rat. *Arch Toxicol*, **54**, 139–144.

Rengstorff RH (1969a). Tear gas and riot control agents: a review of eye effects. *Optomet Weekly*, **60**, 25–28.

Rengstorff RH (1969b). The effect of the riot control agent CS on visual acuity. *Mil Med*, **134**, 219–221.

Rengstorff RH and Mershon MM (1969). CS in water: effects on human eyes. *Edgewood Arsenal Technical Report Series*, Number 4377, dated December 1969. Edgewood Arsenal, MD, USA: Department of the Army.

Rengstorff RH and Mershon MM (1971). CS in water. II. Effects on human eyes. *Mil Med*, **136**, 149–151.

Rengstorff RH, Petrali JP, Mershon MM *et al.* (1975). The effect of the riot control agent dibenz(b.f)-1,4-oxazepine (CR) in the rabbit eye. *Toxicol Appl Pharmacol*, **34**, 45–48.

Rioux JP and Myers RA (1988). Methylene chloride poisoning: a pragmatic review. *J Emerg Med*, **6**, 227–238.

Rioux JP and Myers RA (1989). Hyperbaric oxygen for methylene chloride poisoning: report of two cases. *Annu Emerg Med*, **18**, 691–695.

Rose I (1969). Mace – a dangerous police weapon. *Ophthalmologia*, **158**, 448–454.

Rothberg S (1970). Skin sensitizing potential of the riot control control agents BBC, DM, CN and CS in guinea pigs. *Mil Med*, **135**, 552–556.

Salem H, Olajos EJ and Katz SA (2001). Riot control agents. In: *Chemical Warfare Agents: Toxicity at Low Levels* (SM Somani and JA Romano Jr, eds), pp. 321–372. Boca Raton, FL, USA: CRC Press.

Salem H, Ballantyne B and Katz SA (2005a). Riot control agents. In: *Encyclopedia of Toxicology*, Second Edition (P Wexler, ed.), pp. 706–723. Oxford, UK: Elsevier.

Salem H, Ballantyne B and Katz SA (2005b). Arsenical vomiting agents. In: *Encyclopedia of Toxicology*, Second Edition (P Wexler, ed.) pp.171–173. Oxford, UK: Elsevier.

Salem H, Ballantyne B and Katz SA (2005c). Inhalation toxicology of riot control agents. In: *Inhalation Toxicology*, Second Edition. (H Salem and SA Katz, eds), pp. 485–528. Boca Raton, FL, USA: CRC Press.

Salem H, Ballantyne B and Katz SA (2005d). CN. In: *Encyclopedia of Toxicology*, Second Edition. (P Wexler, ed.), pp. 626–628. Oxford, UK: Elsevier.

Salem H, Ballantyne B and Katz SA (2005e). CS. In: *Encyclopedia of Toxicology*, Second Edition (P Wexler, ed.), pp. 686–690. Oxford, UK: Elsevier.

Sallasidis K, Schmid E and Bauchinger M (1991). Mitotic spindle damage induced by 2-chlorobenzylidene malononitrile (CS) in V79 Chinese hamster cells examined by different staining of the spindle apparatus and chromosomes. *Mutat Res*, **262**, 263–266.

Sanico AM, Atsuta S, Proud D *et al.* (1997). Dose-dependent effects of capsaicin nasal challenge: *in vivo* evidence of human airway neurogenic inflammation. *J Allery Clin Immunol*, **100**, 632–641.

Sanford, J.P. (1976). Medical aspects of riot control (harassing) agents. *Ann Rev Med*, **77**, 412–429.

Sasajimi K, Willey JC, Banks-Schlegel SP *et al.* (1987). Effects of tumor promoters and co-carcinogens on growth and differentiation of cultured cells of human esophageal epithelial cells. *J Nat Cancer Inst*, **78**, 419–458.

Schmid E and Bauchinger M (1991). Analysis of the aneuploidy inducing capacity of

2-chlorobenzylidene malononitrile (CS) and metabolites in V79 Chinese hamster cells. *Mutagenesis*, **6**, 303–305.

Schmid E, Bauchiner M, Ziegler-Skylakakis K *et al.* (1989). 2-Chlorobenzylidene malononitrile (CS) causes spindle disturbances in V79 Chinese hamster cells. *Mut Res*, **226**, 133–136.

Schmunes E and Taylor JS (1973). Industrial contact dermatitis. Effect of the riot control agent orthochlorobenzylidene malononitrile. *Arch Dermatol*, **107**, 212–216.

Schwartz L, Tulipan L and Birmingham DJ (1957). *Occupational Diseases of the Skin*. Philadelphia, PA, USA: Lea and Febiger.

Security Planning Corporation (1972). *Nonlethal weapons for law enforcement, research needs and priorities*. A Report to the National Science Foundation, Washington, DC, USA.

Shreenivason V and Boese BA (1970). Identification of lachrymators. *J Forens Sci*, **15**, 433–442.

Sidell FR (1997). Riot control agents. In: *Textbook of Military Medicine. Medical Aspects of Chemical and Biological warfare* (FR Sidell, ET Takefuji and DR Frantz, eds), Chapter 12, pp. 307–324. Washington, DC, USA: Office of the Surgeon General, US Army. TMM Publications, Borden Institute.

SIPRI (1971). Stockholm International Peace Research Institute. *Problems of Chemical and Biological Warfare. A Study of Historical, Technical, Military and Legal and Political Aspects of CBW*, Volume 1, *The Rise of CB Weapons*; Volume 10, *Humanities*. Stockholm, Sweden: SIPRI.

Smialek JE, Ratanoproeska O and Spitz WV (1975). Accidental death with a tear-gas pen gun: a case report. *J Forens Sci*, **20**, 709–713.

Smith CG and Stopford W (1999). Health hazards of pepper spray. *N Carolina Med J*, **60**, 268–274.

Smith PA, Kluchinsky TA, Savage PB *et al.* (2002). Traditional sampling with laboratory analysis and solid phase microextraction sampling with field gas chromatograph/mass spectrometry by military industrial hygienists. *Am Ind Hyg Assoc J*, **63**, 284–292.

Stahl CJ and Davis JH (1969). Missile wounds caused by tear gas Pen guns. *Am J Clin Pathol*, **52**, 270–276.

Stahl CJ, Young BC, Brown RJ *et al.* (1968). Forensic aspects of tear-gas pen guns. *J Forens Sci*, **13**, 442–469.

Stefee CH, Lantz PA, Flannagan LM *et al.* (1995). Oleoresin capsicum (pepper) spray and 'in custody deaths'. *Am J Forens Med Pathol*, **16**, 185–192.

Stein AA and Kirwin WE (1964). Chloroacetophenone (tear gas) poisoning: a clinico-pathological report. *J Forens Sci*, **9**, 374–382.

Stjarne P (1991). Sensory and reflex control of nasal mucosal blood flow and secretion: clinical implications in non-allergenic nasal hyperreactivity. *Acta Physiol Scand*, **142** (Supplement 600), 1–64.

Sun JD, Henderson RF, Marshall TC *et al.* (1987). The inhalation toxicity of two commercial dyes: solvent yellow 33 and solvent green 3. *Fund Appl Toxicol*, **8**, 358–371.

Surh Y-J and Lee SS (1995). Capsaicin, a double-edged sword: toxicity, metabolism and chemopreventive potential. *Life Sci*, **56**, 1845–1855.

Surh Y-H, Ahn SH, Kim K-C *et al.* (1995). Metabolism of capsaicinoids: evidence for aliphatic hydroxylation and its pharmacological implications. *Life Sci*, **56**, 305–311.

Swearengen TR (1966). *Tear-Gas Munitions*. Springfield, VA: USA: Thomas.

Swentzel KC, Merkey RP, Cucinell SA *et al.* (1970). *Unchanged thiocyanate levels in human subjects following exposure to CS aerosol*. Edgewood Arsenal Technical Memorandum, Number 100–8, dated June 1970. Edgewood Arsenal, MA, USA: Department of the Army.

Szallasi A and Blumberg PM (1990a). Specific binding resiniferatoxin, an ultra-potent capsaicin analog by dorsal root ganglion membranes. *Brain Res*, **524**, 106–111.

Szallasi A and Blumberg PM (1990b). Resiniferatoxin and its analogs provide novel insights into the pharmacology of the vanilloid (capsaicin) receptor. *Life Sci*, **47**, 1399–1408.

Szallasi A and Blumberg PM (1999). Vanilloid (capsaicin) receptors and mechanisms. *Pharmacol Rev*, **51**, 159–212.

Szallasi A, Szolcsanyi J, Szallasi Z *et al.* (1991). Inhibition of [3H] resiniferatoxin binding to dorsal root ganglion membranes as a novel approach in evaluating compounds with capsaicin-like activity. *Naunyn-Schmiedeberg's Arch Pharmacol*, **344**, 551–556.

Takafuji ET and Kok AB (1997). The chemical warfare threat and the military health care provider. In: *Textbook of Military Medicine. Medical Aspects of Chemical and Biological Warfare*. (FR Sidell, ET Takefuji and DR Frantz, eds), Chapter 4, pp. 86–94. Washington, DC, USA: Office of the Surgeon Generral, US Army. TMM Publications, Borden Institute.

The Observer (1969). 60 babies ill from riot gas. *The Observer (London)*, dated 24 August, p. 1.

The Times (1969). Defence ministry concern at Ulster riot gas. *The Times (London)*, dated 26 August, p. 1.

Thornburn KM (1982). Injuries after the use of the lachrymatory agent chloroacetophenone in a confined space. *Arch Environ Health*, **37**, 182–186.

Toth B, Rogan E and Walker B (1984). Tumorigenicity and mutagenicity studies with capsaicin of hot peppers. *Anticancer Res*, **4**, 117–119.

Trigwell P (1997). CS gas has been used as a chemical restraint in a mentally ill person. *Br Med J*, **314**, 444.

United Nations (1969). *Chemical and Bacteriological (Biological) Weapons and the Effects of their Possible Use*, Report of the Secretary General, Annexe 111. New York, NY, USA: United Nations.

Upshall DG (1973). Effects of *o*-chlorobenzylidene malononitrile (CS) and the stress of aerosol inhalation upon rat and rabbit embryonic development. *Toxicol Appl Pharmacol*, **24**, 45–59.

Upshall DG (1974). The effects of dibenz(b.f)-1:4-oxazepine (CR) upon rat and rabbit embryonic development. *Toxicol Appl Pharmacol*, **24**, 301–311.

Upshall DG (1977a). Embryonic development and inhalation stress. In: *Current Approaches in Toxicology* (B Ballantyne, ed.), pp. 79–85. Bristol, UK: John Wright.

Upshall DG (1977b). Riot control smokes: lung absorption and metabolism of peripheral sensory irritants. In: *Clinical Toxicology* (W Duncan and BJ Leonard, eds), pp. 121–129. Amsterdam, The Netherlands: Excerpta Medica.

USA Today (2003). Chicago club may loose its licenses. *USA Today*, dated 20 February, Section A, p. 4.

Vaca FE, Myers JH and Langdorf M (1996). Delayed pulmonary edema and bronchospasm after accidental lacrimator exposure. *Am J Emerg Med*, **14**, 402–405.

Varma S and Holt PJ (2001). Severe cutaneous reaction to CS gas. *Clin Exp Dermatol*, **26**, 248–250.

Vesaluoma M, Müller J, Lambiase A *et al.* (2000). Effects of oleoresin capsicum pepper spray on human corneal morphology and sensitivity. *Invest Ophthalmol Visual Sci*, **41**, 2138–2147.

Villesansor IM and de Campo EJ (1994). Clastogenicity of red pepper (*Capsicum frutescens* L.) extracts. *Mutat Res*, **312**, 151–155.

Von Daniken A, Frederich U, Lutz W *et al.* (1981). Tests for mutagenicity in Salmonella and covalent binding to DNA and protein in the rat of the riot control agent *o*-chlorobenzylidene malononitrile (CS). *Arch Toxicol*, **49**, 15–27.

Watson WA, Stremel KR and Westdorp EJ (1996). Oleoresin capsicum (Cap-Stun) toxicity from aerosol exposure. *Ann Pharmacotherap*, **30**, 733–735.

Wehmeyer KR, Kasting GB, Powell JH *et al.* (1990). Application of liquid chromatography with on-line radiochemical detection to metabolism studies on a novel class of analgesics. *J Pharmaceut Biomed Anal*, **8**, 177–183.

Weigand DA (1969). Cutaneous reaction to the riot control agent CS. *Mil Med*, **134**, 437–440.

Weigand DA and Mershon MM (1970). *Cutaneous reaction to EA 3547 in propylene glycol*. Edgewood Arsenal Technical Report, Number 4413, dated December 1970. Edgewood Arsenal, Bethesda, MA, USA: Department of the Army.

Weit E (2001). The health impact of crowd-control agents. *Canad Med Assoc J*, **164**, 1889.

Weller EM, Kubbies M and Nusse M (1995). Induction of cell cycle perturbations by tear gas 2-chlorobenzylidene malononitrile (CS) in synchronously and asynchronously proliferating mammalian cells. *Cytometry*, **19**, 334–342.

White House (1994). Riot control agents. Press Release, dated 23 June. Washington, DC, USA: Office of the Press Secretary, White House.

WHO (1970). *Health Aspects of Chemical and Biological Weapons*. Geneva, Switzerland: World Health Organization.

Wier E (2001). The health impact of crowd-control agents. *Can Med Assoc J*, **164**, 1889.

Wild D, Eckhardt K, Harnasch D *et al.* (1983). Genotoxicity study of CS (ortho-chlorobenzylidene malononitrile) in *Salmonella*, Drosophila and mice. *Arch Toxicol*, **54**, 167–170.

Witten B, Wagman W, Saffer R *et al.* (1970). *Malodorous substances as riot control and training agents*. Edgewood Arsenal Technical Report, Number 4370, dated 00000 1970. Edgewood Arsenal, MA, USA: Department of the Army.

Wood JN, Winter J, James IF *et al.* (1988). Capsaicin-induced ion fluxes in dorsal root ganglion cells in culture. *J Neurosci*, **8**, 320–322.

Worthington E and Nee PA (1999). CS exposure – clinical effects and management. *J Accid Emerg Med*, **16**, 168–179.

Yih P-H (1995). CS gas injury to the eye. *Br Med J*, **311**, 276.

Zeiger E, Anderson B, Howorth S *et al.* (1987). *Salmonella* mutagenicity tests. III. Results from the testing of 225 chemicals. *Environ Mutagen*, **9**, 1–109.

Zerba EN and Ruveda MA (1972). Gas chromatographic determination of riot control agents. *J Chromatogr Anal*, **68**, 245–247.

Ziegler-Skylakakis K, Summer KH and Andrae U (1989). Mutagenicity and cytotoxicity of 2-chlorobenzylidene malononitrile (CS) and its metabolites in V79 Chinese hamster cells. *Arch Toxicol*, **63**, 314–319.

27 RICIN AND ABRIN POISONING

Sally M. Bradberry[1], J. Michael Lord[2], Paul Rice[3] and J. Allister Vale[1]

[1] *National Poisons Information Service and West Midlands Poisons Unit, Birmingham, UK*
[2] *University of Warwick, Coventry, UK*
[3] *Dstl, Salisbury, UK*

RICIN

Ricin is a globular glycoprotein, which makes up 1–5% by weight of the beans of the castor oil plant *Ricinus communis*, an annual shrub common in warm climates. The beans are 0.5–2 cm in length, attractive and brightly coloured. Castor oil obtained from the beans by cold pressing is used as a purgative and laxative. Hot pressing of the beans, followed by solvent extraction, is used to produce specialist oils and lubricants. The residue is used as a cattle feed or as a fertiliser, after ricin has been destroyed by heating (Pevny, 1979).

Ricin can be extracted from the waste mash generated by castor oil production or from whole castor beans by a relatively simple and inexpensive process. The resulting product is a soluble white powder which is stable under ambient conditions, but can be detoxified by heating for 10 min at 80°C or for 1 h at 50°C (Burrows and Renner, 1999).

THE POTENTIAL USE OF RICIN AS A CHEMICAL WEAPON

The discovery of ricin in the Senate in Washington in 2004, at a Paris railway station in 2003 plus the finding in Afghanistan by a journalist, of a description of ricin purification, demonstrate the reality of its perceived potential as a chemical weapon. The suitability of ricin for this purpose derives from its extreme toxicity to mammalian cells, the fact that the source is naturally occurring and relatively easily prepared. The use of ricin to cause mass casualties would require either its aerosolization by means of a dispersal device, or its addition to food and beverages as a contaminant. It should be noted that by the oral route, ricin is approximately one thousand-fold less toxic than by either the inhalation or parenteral routes. Generating a large-scale aerosol would be best achieved with a dry powder consisting of very small particles. By inhalation (breathing in solid or liquid particles) and injection (into soft tissue or a vein), the lethal dose is approximately 5–10 μg/kg body weight, that is, for an adult weighing 70 kg, the lethal dose would be 350 to 700 μg.

RICIN STRUCTURE

Ricin is a prototype ribosome-inactivating protein (RIP). RIPs have been isolated from many sources including bacteria, fungi and the leaves, seeds and roots of plants (Stirpe *et al.*, 1992; Barbieri *et al.*, 1993; Hartley *et al.*, 1996). In general, most RIPs can be classified as either type 1 or type 2 (Peumans *et al.*, 2001). Type 1 RIPs are the most common ($\sim$ 100 described to date), and consist of a single enzymatic chain of $\sim$ 30 kDa. Type 2 RIPs, typified by ricin, are potent cytotoxins with an A-chain that is structurally and functionally equivalent to the type 1 RIPs, disulphide-linked to a lectin referred to as the B chain. The $\sim$ 30 kDa B-chain is able to bind terminal galactose residues on cell surface components. It follows that although type 1 and type 2 RIPs are

Chemical Warfare Agents: Toxicology and Treatment (2nd Edition)
Edited by Timothy C. Marrs, Robert L. Maynard and Frederick R. Sidell © 2007 John Wiley & Sons, Ltd

active against ribosomes *in vitro,* only the type 2 RIPs such as ricin, are cytotoxic owing to the presence of a B chain that mediates the binding and entry of holotoxin into susceptible cells.

Among the RIPs, the tertiary structure of ricin was the first to be determined (Montford *et al.*, 1987; Rutenber *et al.*, 1991; Rutenber and Robertus, 1991). It was shown to be a globular, glycosylated heterodimer in which the A-chain (**r**icin **t**oxin **A** chain, RTA) and the B-chain (**r**icin **t**oxin **B** chain, RTB) are joined by a single disulphide bond. The 262 amino acyl residues of RTB form a bilobal structure lacking α-helices or β-sheets that probably arose by gene duplication (Villafranca and Robertus, 1981). Each of the structurally homologous lobes contains three related subdomains and at least one of these in each lobe possesses a sugar-binding pocket. In contrast, RTA, comprising 267 amino acyl residues, has three structural domains with approximately 50% of the polypeptide arranged into α-helices or β-sheet (Katzin *et al.*, 1991; Mlsna *et al.*, 1993; Weston *et al.*, 1994).

RICIN BIOSYNTHESIS

Ricin is made in the fatty endosperm cells of maturing castor oil seeds where it accumulates in protein storage vacuoles to 5% of total particulate protein (Lord *et al.*, 1994). Although ricin is a heterodimer, its individual subunits are initially synthesised together in the form of a precursor (Butterworth and Lord, 1983). The first 26 residues constitute a signal peptide for translocation into the endoplasmic reticulum (ER) lumen of the plant cell (Ferrini *et al.*, 1995), followed by a 9 residue propeptide of no known function. Then follows mature RTA, which is joined to RTB by a 12 residue intramolecular propeptide (Lamb *et al.*, 1985). The signal peptide and the two propeptides are absent in mature ricin, with the former being removed during synthesis and segregation into the ER lumen, and the latter being removed in vacuoles. Within the ER, the nascent proricin protein undergoes glycosylation and folding (Roberts and Lord, 1981), before being transported to storage vacuoles via the Golgi complex (Lord, 1985a,b). Upon deposition in vacuoles, the mature disulphide-linked RTA-RTB heterodimer is finally generated by the endoproteolytic removal of the *N*-terminal and internal propeptides (Harley and Lord, 1985). The intramolecular propeptide has been shown to be both necessary (Frigerio *et al.*, 1998) and sufficient (Frigerio *et al.*, 2001; Jolliffe *et al.*, 2003) for vacuolar targeting.

Recombinant proricin has been produced and its properties examined (Richardson *et al.*, 1989). It is an active prolectin containing functional sugar binding sites. However, proricin lacks *N*-glycosidase activity, consistent with reports that in mature holotoxin, RTA must be reductively separated from RTB in order to have activity (Lewis and Youle, 1986; Wright and Robertus, 1987). It appears that in both the dimeric ricin holotoxin *and* in proricin, the RTB moiety can sterically obstruct the active site of RTA. Manufacture as a precursor perhaps explains why *Ricinus* cells are able to synthesize large amounts of ricin even though endogenous ribosomes are susceptible, albeit much less so than animal ribosomes (Cawley *et al.*, 1977; Harley and Beevers, 1982; Taylor *et al.*, 1994) to the action of RTA (Endo and Tsurugi, 1988). Since the endosperm cells of *Ricinus* show no signs of ribosome damage while storing large amounts of ricin, it can further be deduced that active RTA does not escape from low pH vacuoles to reach the cytosol. The generation of an active holotoxin only within the 'safe haven' of vacuoles therefore enables *Ricinus* seeds to synthesize and store large quantities of a potent toxin that can effectively function in plant defence against predators without compromising its own survival.

THE SUBSTRATE FOR RICIN

The way in which ricin inactivates ribosomes was first described by Endo and Tsurugi (1987) who showed that RTA cleaves a single specific glycosidic bond within the large ribosomal RNA (rRNA) of the 60S subunit of eukaryotic ribosomes. RTA must be separated from RTB to display this activity (Lewis and Youle, 1986; Wright and Robertus, 1987). The glycosidic bond is that of a specific adenine nucleotide that lies in a universally conserved sequence of 12 nucleotides, 5′-AGUACGAGAGGA-3′ (target adenine shown in bold) known as the sarcin-ricin loop (SRL) (Wool *et al.*, 2000). The SRL is associated with the binding of elongation factors (Ban *et al.*, 1999). Modification of this region by RTA thereby prevents the binding of such factors and inhibits the elongation cycle of protein synthesis (Sperti *et al.*, 1973; Nilsson and Nygård, 1986). The ribosome is a large, complex substrate that operates via structural transitions, and it is likely that the RTA polypeptide also interacts with sites outside the immediate vicinity of the SRL. The presence of particular ribosomal proteins may therefore allow or prevent access of RTA to the SRL and might thereby contribute to substrate specificity.

MODE OF ACTION OF RICIN A CHAIN

The catalytic activity of ricin lies exclusively within the A chain. RTA, along with other RIPs, contains several invariant residues. These include Tyr 80, Tyr 123, Glu 177 and Arg 180 that are located in the active site. Although the precise mechanism of *N*-glycosidase action for RTA remains unclear, Monzingo and Robertus (1992) have proposed a mechanistic model based upon the crystal structures of RTA in complex with small molecules and site-directed mutagenesis studies (Weston *et al.*, 1994; Monzigo and Robertus, 1992; Schlossman *et al.*, 1989; Frankel *et al.*, 1990; Ready *et al.*, 1991; Kim and Robertus, 1992; Chaddock and Roberts, 1993). The kinetic analysis of RTA variants with mutations at either Arg 180 or Glu 177 showed little change to the K_m (the binding affinity for the substrate) of the enzyme but the k_{cat} (a measure of the catalytic rate of the reaction) was altered, suggesting that these two residues are involved in the catalytic mechanism of the enzyme rather than in substrate binding. The rRNA substrate is bound in the active site cleft with the target adenine making stacking interactions with tyrosine residues Tyr 80 and Tyr 123. Arg180 is positioned so that it can protonate the leaving group at N-3 of the target adenine, facilitating an electron flow which results in the breaking of the bond between N-9 of the target adenine ring and the C-1′ of the ribose. Through this action, it has been reported that a single A chain molecule in the cytosol can depurinate ~1500 ribosomes per minute, leading to a rapid inhibition of protein synthesis.

BINDING OF RICIN TO MAMMALIAN CELLS

When ricin is applied to the exterior of most animal cells, it will bind to complex carbohydrates containing terminal *N*-acetyl galactosamine or β-1,4-linked galactose residues (Olsnes and Pihl, 1982). The abundance of cell surface galactolipids and glycoproteins provides an explanation for the diversity of cell lines that are sensitive to ricin. The relative contribution of glycoproteins and/or glycolipids in the uptake of ricin by target cells remains uncertain, although a role for glycosphingolipids as receptors has recently been ruled out (Spilsberg *et al.*, 2003). Native ricin also possesses the ability to bind and enter some cell types by virtue of its own carbohydrate side chains. A limited number of cell lines express mannose receptors on their surface, to which ricin, with its high mannose-type glycans typical of plant glycoproteins, can bind. Mannose receptor-mediated uptake of ricin has been described in both macrophages (Simmons *et al.*, 1986) and rat liver endothelial cells (Magnusson *et al.*, 1993).

ENDOCYTOSIS OF RICIN

Due to its promiscuous binding, ricin becomes localised to all types of membrane invaginations (Sandvig and van Deurs, 1996) and as such, is a valuable tool in the study of different endocytic mechanisms (Sandvig and van Deurs, 2000; Iversen *et al.*, 2001) (Figure 1). Ricin endocytosis via clathrin-coated pits has been demonstrated (van Deurs *et al.*, 1985). There is also evidence of clathrin-independent mechanisms of toxin uptake. For example, Hep-2 cells, in which coated pit formation has been blocked by hypotonic shock and potassium ion depletion, remain sensitive to ricin (Moya *et al.*, 1985).

Clathrin-independent endocytosis includes uptake by caveolae (specialized, invaginated

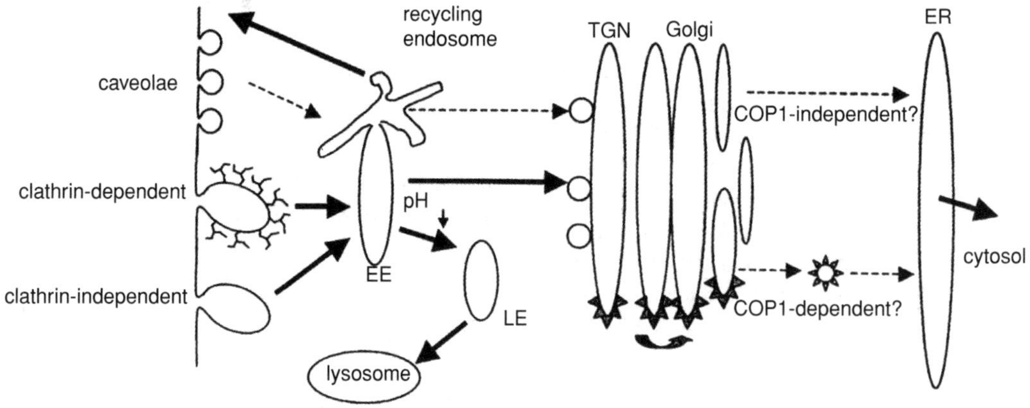

Figure 1. Schematic representation of the pathways that may be used by ricin to enter eukaryotic cells: TGN, trans-golgi network; ER, endoplasmic reticulum; EE, early endosomes; LE, late endosomes; COP1, coatomer protein 1

lipid rafts) and macropinocytosis (Nichols and Lippincott-Schwartz, 2001). When both caveolae and clathrin-dependent endocytosis are perturbed by the removal of membrane cholesterol, ricin uptake still occurs (Simpson *et al.*, 1998; Rodal *et al.*, 1999). The molecular machinery controlling this clathrin-independent endocytosis has yet to be elucidated, but uptake may occur by more than one mechanism (Sandvig and van Deurs, 1996; Llorente *et al.*, 1998; Herskovits *et al.*, 1993; van der Bliek *et al.*, 1993; Artalejo *et al.*, 1995; Damke *et al.*, 1995; Werbonat *et al.*, 2000; Lamaze *et al.*, 2001). In cells lacking functional clathrin-dependent uptake, the rate of ricin internalization is reduced to half of that in control cells, suggesting that ~50% of surface membrane uptake occurs by a clathrin-independent mechanism (Sandvig and van Deurs, 1996,2000; Ray and Wu, 1981).

TRANSPORT OF RICIN FROM ENDOSOMES TO THE ENDOPLASMIC RETICULUM

The majority of ricin that initially enters the cell is delivered to early endosomes from where most appears to be either recycled to the cell surface or delivered via late endosomes to lysosomes for degradation (Figure 1). The remainder, approximately 5% of internalized ricin, has been visualized within the perinuclear Golgi with 70–80% of this small fraction being located in the *trans*-Golgi network (TGN) (van Deurs *et al.*, 1988). The transfer of ricin from endosomes to the Golgi apparatus is complex and experimental studies suggest more than one route exist (Iversen *et al.*, 2001; Simpson *et al.*, 1998; Ghosh *et al.*, 1998; Mallard *et al.*, 1998; Mallet and Maxfield, 1999; Sandvig *et al.*, 2002; Lombardi *et al.*, 1993; Riederer *et al.*, 1994).

From the Golgi, ricin follows a retrograde path to the endoplasmic reticulum (Rapak *et al.*, 1977). The classic pathway of retrograde transport involves movement via so-called coatomer protein 1-coated vesicles (Cosson and Letourneur, 1994,1997; Girod *et al.*, 1999) but a second, less well characterized coatomer protein 1-*independent* pathway (Figure 1) may also be involved (White *et al.*, 1999; Chen *et al.*, 2002).

A pre-requisite for the activity of RTA is that it is reductively separated from RTB. At present it is unclear whether this occurs in the ER or whether it occurs in the reducing environment of the cytosol.

RELEASE OF RICIN INTO THE CYTOSOL

It is likely that ricin's complex intracellular journey through the Golgi and endoplasmic reticulum (ER) is essential for successful release of intact A chain to the cytosol. It has been proposed that ricin 'hijacks' a route from the ER to the cytosol that is normally reserved for faulty proteins destined for destruction there (Lord *et al.*, 2003). This ER-associated degradation pathway (ERAD) allows ricin access to the cytosol without the toxin succumbing to the normal fate of ERAD substrates, namely degradation in the cytosol by proteasomes (multi-enzyme complexes that digest and remove unwanted endogenous proteins). There is evidence (Hazes and Reed, 1997; Deeks *et al.*, 2002) that the reason ricin is not recognized by the proteasome destruction system is a paucity of lysine residues in its structure. Lysine residues serve as the attachment site for the small protein ubiquitin, which itself attaches to the proteasome to initiate destruction by unfolding. It appears that the low lysine content of RTA attenuates what might otherwise be a very efficient process for removing and thereby disarming this toxin in the cytosol. Interestingly, when proteasomes are inhibited, cells are typically sensitised 2–3-fold to ricin (Deeks *et al.*, 2002; Smith *et al.*, 2002; Day *et al.*, 2001). This suggests that some interaction between RTA and proteasomes occurs and offers partial protection against ricin toxicity. Partially unfolded RTA has been shown to refold in the presence of ribosomes to regain full catalytic activity (Argent *et al.*, 2000). Thus, ribosomes appear to act as suicidal 'chaperones' since, by the very act of refolding RTA, they are themselves inactivated.

Subsequent to protein synthesis termination a process of programmed cell death, apoptosis, ensues (Hughes *et al.*, 1996). The details of this process remain to be clarified but there is evidence that ricin A and B chains are involved and that the mechanism is *independent* of that which inhibits protein synthesis (Battelli, 2004).

TOXICOKINETICS OF RICIN EXPOSURE

As a relatively large protein, ricin is unlikely to be extensively absorbed from the gastrointestinal tract. In animal studies, most orally administered ricin was found in the large intestine after 24 h, with only limited systemic uptake (Ishigaro *et al.*, 1992). Dermal absorption of ricin through intact skin is most unlikely to occur, unless there are open cuts. In mice, intravenously administered ricin was distributed predominantly to the spleen, kidneys, heart and liver (Fodstad *et al.*, 1976). Ricin administered intramuscularly was found to localize in the draining lymph nodes (Griffiths *et al.*, 1986). Ricin was rapidly eliminated as degraded proteins in rats, predominantly in the urine, with only 11% of radiolabelled toxin remaining 24 h after intravenous injection (Ramsden *et al.*, 1989).

CLINICAL FEATURES OF RICIN POISONING

Ricin is toxic via all routes, although the features of poisoning and severity of toxicity vary markedly with the dose and route of exposure. The estimated LD_{50} of ricin in mice by intragastric administration was 20 mg/kg body weight (Franz and Jaax, 1997) although the frequently cited human fatal oral dose of 1 mg/kg body weight (Wedin *et al.*, 1986; Kopferschmitt *et al.*, 1983; Palatnick and Tenenbein, 2000) remains unconfirmed. In mice the LD_{50} by intravenous injection was 5 μg/kg body weight (Franz and Jaax, 1997). The fatal dose by injection in man has been suggested to be of the order of 1–10 μg/kg body weight (Crompton and Gall, 1980), but there are no reliable data to support this. Although a chemist is reported to have survived the alleged intramuscular injection of 150 mg ricin (Fine *et al.*, 1992), in reality as he injected the contents of only one castor bean, the maximum ricin content that he could have injected would have been about 10 mg, that is approximately 140 μg/kg body weight.

Many of the features observed in poisoning can be explained by ricin-induced endothelial cell damage, with fluid and protein leakage and tissue oedema, causing a so-called 'vascular leak syndrome'. Disseminated intravascular coagulation has been observed in experimental animals following intravenous ricin administration and this is also likely to reflect endothelial cell damage.

Hepatocellular and renal damage is at least partly secondary to vascular damage and impaired tissue perfusion, rather than a direct effect of the toxin itself. Animal studies have suggested that Kupffer cells (liver macrophages) are the primary targets in ricin-induced hepatotoxicity. Kupffer cell destruction impairs the liver's capacity to detoxify endogenous toxins, which may contribute to further damage (Bingen *et al.*, 1987). Mice administered intraperitoneal ricin show evidence of oxidative stress in the liver and kidneys with accumulation of lipid peroxidation products which may also arise from damaged Kupffer cells (Kumar *et al.*, 2003).

Ricin ingestion

Poisoning by ricin ingestion is an unlikely mode of chemical terrorism. To date, most cases of ricin poisoning by this route have involved eating castor beans. In these circumstances, the degree of seed mastication is important since beans swallowed whole may pass through the gastrointestinal tract intact, whereas the chewing of seeds facilitates ricin release.

In cases of substantial ingestion of castor beans, the onset of gastrointestinal features occurred typically within a few hours with oropharyngeal irritation (Challoner and McCarron, 1990; Möschl, 1938), vomiting (Wedin *et al.*, 1986; Kopferschmitt *et al.*, 1983; Palatnick and Tenenbein, 2000; Challoner and McCarron, 1990; Möschl, 1938; Bispham, 1903; Kaszás and Papp, 1960; Hutchinson, 1900; Despott and Cachia, 2004), abdominal pain (Wedin *et al.*, 1986; Kopferschmitt *et al.*, 1983; Challoner and McCarron, 1990; Möschl, 1938; Kaszás and Papp, 1960; Despott and Cachia, 2004; Aplin and Eliseo, 1997; Malizia *et al.*, 1977; Meldrum, 1900; Ingle *et al.*, 1966) and diarrhoea (Wedin *et al.*, 1986; Kopferschmitt *et al.*, 1983; Challoner and McCarron, 1990; Möschl, 1938; Hutchinson, 1900; Despott and Cachia, 2004; Malizia *et al.*, 1977; Ingle *et al.*, 1966; Spyker *et al.*, 1982). Haematemesis (Meldrum, 1900), bloody diarrhoea (Hutchinson, 1900; Malizia *et al.*, 1977) or melaena (Rauber and Heard, 1985) may occur. Subsequent features reflect fluid and electrolyte loss with hypotension (Despott and Cachia, 2004; Rauber and Heard, 1985), tachycardia (Wedin *et al.*, 1986; Kopferschmitt *et al.*, 1983; Challoner and McCarron, 1990; Despott and Cachia, 2004; Ingle *et al.*, 1966), tachypnoea (Wedin *et al.*, 1986; Bispham, 1903), sweating (Meldrum, 1900), dehydration (Kopferschmitt *et al.*, 1983; Challoner and McCarron, 1990; Malizia *et al.*, 1977; Ingle *et al.*, 1966) and peripheral cyanosis (Kopferschmitt *et al.*, 1983; Rauber and Heard, 1985; Koch and Caplan, 1942). Pre-renal impairment secondary to hypovolaemia is common in patients with substantial gastrointestinal fluid loss (Kopferschmitt *et al.*, 1983) and may progress to renal failure (Despott and Cachia, 2004; Rauber and Heard, 1985). In more severe cases, hypovolaemic shock, with oliguria or anuria, may ensue (Kopferschmitt *et al.*, 1983; Bispham, 1903; Meldrum, 1900). Proteinuria (Challoner and McCarron, 1990; Malizia *et al.*, 1977; Spyker *et al.*, 1982), haematuria (Challoner and McCarron, 1990; Malizia *et al.*, 1977; Spyker *et al.*, 1982) and hyaline casts on urine microscopy (Malizia, *et al.*, 1977) have also been reported.

Some degree of transient liver damage is likely in all but the mildest cases with increased hepatic transaminase and lactate dehydrogenase activities (Challoner and McCarron, 1990; Despott and Cachia, 2004; Levin *et al.*, 2000) and less commonly hyperbilirubinaemia (Palatnick and Tenenbein, 2000). Occasionally, liver function tests remain normal until several days after exposure (Palatnick and Tenenbein, 2000).

Other reported features include disorientation (Möschl, 1938; Kaszás and Papp, 1960; Malizia *et al.*, 1977; Levin *et al.*, 2000), drowsiness (Koch and Caplan, 1942), confusion (Wedin *et al.*, 1986; Koch and Caplan, 1942), lightheadedness (Challoner and McCarron, 1990), muscle cramps (Kopferschmitt *et al.*, 1983; Bispham, 1903; Rauber and Heard, 1985), convulsions (Möschl, 1938), intravascular haemolysis (Malizia *et al.*, 1977), bradycardia (Challoner and McCarron, 1990), hypertonia (Möschl, 1938; Rauber and Heard, 1985), miosis (Malizia *et al.*, 1977), mydriasis (Hutchinson, 1900; Balint, 1974) and blurred vision (Kopferschmitt *et al.*, 1983).

Investigations may show a metabolic acidosis (Challoner and McCarron, 1990; Levin *et al.*, 2000) leucocytosis (Kopferschmitt *et al.*, 1983; Challoner and McCarron, 1990; Despott and Cachia, 2004; Malizia *et al.*, 1977), hyperglycaemia (Challoner and McCarron, 1990; Despott and Cachia, 2004; Levin *et al.*, 2000) or hypoglycaemia (Challoner and McCarron, 1990), hypophosphataemia (Challoner and McCarron, 1990) and increased creatine kinase activity (Challoner and McCarron, 1990).

In a series of ten paediatric castor bean ingestions, seven had transient ECG abnormalities, including QT interval prolongation (in five cases), intraventricular conduction disturbance and repolarization changes. The authors suggested that these changes were probably secondary to metabolic disturbances (Kaszás and Papp, 1960). It is possible that these abnormalities may have resulted from ricin-induced apoptosis of key elements of the cardiac conduction system. Such an effect of ricin has been noted *in vivo* in rats (Leek, 1989).

Patients who receive prompt symptomatic and supportive care following castor bean ingestion are likely to survive, with a fatality rate for treated patients of approximately 2% (Rauber and Heard, 1985). In fatal cases, death usually occurs on the third day or later and is due to multi-organ failure (Franz and Jaax, 1997). The most common findings at autopsy are ulceration of the mucosa of the stomach and small intestine, necrosis of mesenteric lymph nodes, hepatic necrosis and nephritis (Franz and Jaax, 1977).

Intramuscular and subcutaneous ricin administration

Rats administered ricin 5 μg (33–50 μg/kg body weight) intramuscularly survived a maximum of 35 h. Post-mortem examination of the intestinal tract demonstrated severe haemorrhage with apoptosis of cells lining the small intestine, particularly the ileum, which showed marked lymphoid cell and macrophage infiltration of villi. The stomach and colon were largely unaffected (Leek *et al.*, 1989).

Ricin was allegedly used in the assassination of the Bulgarian defector Georgi Markov. Those investigating the case estimated the injected dose of ricin to be some 500 μg, although ricin was never isolated analytically (Crompton and Gall, 1980). There was immediate pain at the site of injection (the thigh), with fatigue, nausea, vomiting and fever developing over the next 24 h. On admission, some 36 h after the incident, local necrotic lymphadenopathy was present and gastrointestinal haemorrhage ensued with hypovolaemic shock and renal failure; death occurred on the third day. At autopsy, there was evidence of pulmonary oedema and haemorrhagic necrosis of the small bowel; haemorrhages were observed in the lymph nodes local to the injection site, in the myocardium, testicles and pancreas.

A 36-year-old chemist self-administered two intramuscular injections of ricin prepared from a single castor bean (Fine *et al.*, 1992). Although it was calculated that he had injected 150 mg ricin, this would be impossible from a single bean. The maximum amount of ricin that is extractable from a single bean is about 10 mg, which would equate to a maximum dose of no more than 140 μg/kg body weight. Ten hours after the injections, he complained of headache and rigors. He developed anorexia and nausea, a sinus tachycardia, erythematous areas around the puncture wounds and local lymphadenopathy at the sites of injection. Investigation showed mildly increased hepatic transaminase activities. He was discharged, well, after 10 days.

A third case of intramuscular ricin injection involved a 53 year-old male who injected part of 13 chewed castor beans into his thigh with suicidal intent (Passeron *et al.*, 2004). He developed necrotic cellulitis complicated by *Enterococcus faecalis* infection requiring emergency surgical debridement, but recovered after three months hospitalization.

A 20-year-old man was admitted to hospital 36 h after injecting castor bean extract subcutaneously (Targosz *et al.*, 2002). He complained of nausea, weakness, dizziness, chest and abdominal pain and myalgia with paraesthesiae of the extremities. Hypotension, anuria and a metabolic acidosis were noted on examination and fresh blood was present in the rectum, possibly related to the development of a bleeding diathesis. Hepatorenal and cardiorespiratory failure then developed and the patient died 18 h after admission following an asystolic arrest.

Intravenous ricin administration

When intravenous ricin 18–20 μg/m^2 (approximately 0.5 μg/kg body weight) was investigated as a potential chemotherapeutic agent it caused flu-like symptoms with marked fatigue and myalgia, and sometimes nausea and vomiting (Fodstad *et al.*, 1984).

In studies using ricin A chain-linked to a monoclonal antibody for anti-tumour immunotherapy, the principal dose-limiting side-effect was 'vascular leak syndrome'. This was characterized by hypoalbuminaemia (Engert *et al.*, 1997; Baluna *et al.*, 1996), oedema-associated weight gain (Engert *et al.*, 1997; Baluna *et al.*, 1996), pulmonary function abnormalities including reduced FVC, pleural effusion (Baluna *et al.*, 1996) and pulmonary oedema (Baluna *et al.*, 1996), renal insufficiency with oliguria and impaired creatinine clearance (Baluna *et al.*, 1996), cardiac failure and hypotension (Baluna *et al.*, 1996), sometimes in association with a pericardial effusion (Baluna *et al.*, 1996). Other common features included weakness (Engert *et al.*, 1997), nausea and vomiting (Engert *et al.*, 1997), myalgia, associated in some cases with increased total creatine kinase activity (Engert *et al.*, 1997), joint discomfort (Engert *et al.*, 1997), thrombocytopenia (Engert *et al.*, 1997) and occasionally 'allergic reactions' during the infusion (Engert *et al.*, 1997).

In a study of 56 patients treated with ricin A chain immunotoxin, 12 required interruption or termination of treatment due to the severity of side-effects, and two patients died as a result of vascular leak syndrome (Baluna *et al.*, 1996). Vascular leak syndrome was more common in patients who had received radiotherapy prior to immunotherapy (Schindler *et al.*, 2001).

Ricin inhalation

Non-human primates exposed to ricin by inhalation, developed necrotizing interstitial and alveolar inflammation with oedema and fibrinopurulent pneumonia. These manifestations typically occurred after a dose-dependent delay of 8–24 h (Wilhelmson and Pitt, 1993). Similar findings have been observed in rodents (Griffiths *et al.*, 1995; Kokes *et al.*, 1994; Brown and White, 1997).

An allergic syndrome has been observed in workers occupationally exposed to castor bean dust (Topping *et al.*, 1982; Figley and Elrod, 1928). Susceptible patients presented with acute onset conjunctivitis (Topping *et al.*, 1982), rhinitis (Topping *et al.*, 1982; Kanerva *et al.*, 1990), sneezing, urticaria and wheeze (Topping *et al.*, 1982), which responded to conventional measures and removal from exposure. Historically, castor bean dust has been the cause of endemic asthma in the locality of a castor oil mill (Figley and Elrod, 1928). The allergen identified as being responsible for this effect is now known not to be ricin itself but a separate protein.

Topical ricin exposure

Both type I and type IV allergic responses have been reported following dermal exposure to castor bean dust (Kanerva *et al.*, 1990; Metz *et al.*, 2001). A 21-year-old female had an anaphylactic-type response after a castor bean from her necklace disintegrated in her fingers. She immediately developed sneezing, rhinitis and periorbital oedema, with facial urticaria and erythema, requiring subcutaneous adrenaline (epinephrine) (Lockey and Dunkelberger, 1968). However, as suggested above, ricin is only one of several allergenic proteins in castor beans.

Ricin is severely irritating to the eye. In animal studies, pseudomembranous conjunctivitis occurred following application of ricin solutions in concentrations of 1:1000 – 1:10 000 (Grant and Schuman, 1993).

DIAGNOSIS OF RICIN POISONING

The detection of ricin to confirm exposure is difficult, not least because the toxin is metabolized and eliminated rapidly. The most commonly used method is the enzyme-linked immunosorbant assay (ELISA) which can be applied to environmental and biological samples and allows ricin detection for up to 48 h after exposure with a detection limit of approximately 200 pg/ml (Griffiths *et al.*, 1986; Leith *et al.*, 1988). More recently, an ELISA that uses monoclonal antibodies with distinct specificities for ricin A and B chains has been developed (Shyu *et al.*, 2002a) which offers high specificity but is too slow to be useful for rapid diagnosis. The same group has now developed a sensitive and rapid immunochromatographic assay in which a monoclonal antibody raised against ricin A chain is conjugated to colloidal gold which serves as a detection reagent (Shyu *et al.*, 2002b). This analysis is complete in less than 10 min and can detect ricin A chain at a concentration of 100 pg/ml. Alternatively, ricinine, an alkaloid present in crude preparations of ricin, can serve as a biomarker for ricin exposure. The method uses high performance liquid chromatography and mass spectrometry and can detect ricinine in urine for at least 48 h after exposure (Johnson *et al.*, 2005).

Detection of anti-ricin antibodies could potentially aid the diagnosis of ricin poisoning in those who survive for two to three weeks. However, humoral IgM responses would likely be of short duration only and no immunological memory would be anticipated without 'boosting'. Anti-ricin antibodies would not be detected in those dying soon after exposure.

MANAGEMENT OF RICIN POISONING

Ingestion

The benefit of gastric lavage or the administration of activated charcoal is uncertain but may be considered if patients present within the first hour following ingestion of ricin. Gastrointestinal fluid losses should be replaced. Cardiopulmonary, hepatic and renal function should be monitored. Organ dysfunction should be managed conventionally.

Parenteral ricin exposure

Symptomatic and supportive measures are the mainstay of management of this highly toxic route of exposure. There is some evidence that the intravenous administration of anti-ricin antibody shortly (within 1–2 h) after ricin exposure may improve survival (Houston, 1982) although this has to be confirmed and probably has no utility in a civilian population where there would be no near 'real-time' detection of ricin exposure.

Ricin aerosol inhalation

Removal from exposure and airway support with adequate ventilation are the priorities. Pulmonary oedema should be treated with continuous positive airway pressure (CPAP) or, in severe cases, with mechanical ventilation and positive end expiratory pressure (PEEP).

EXPERIMENTAL STRATEGIES FOR PROTECTING AGAINST RICIN TOXICITY

Since the mechanisms by which ricin binds to, is internalized by and subsequently paralyses protein synthesis within mammalian cells, have been clarified, experimental studies have focused on developing protective strategies to counteract ricin's effects. The two main approaches involve the design of therapeutic inhibitors (to prevent cell binding, intracellular routing or the enzymatic activity of ricin) and antitoxin vaccination to induce production of toxin-neutralizing antibodies (Marsden *et al.*, 2005). Of these options, vaccination development is the more promising since all therapeutic inhibitors investigated have either failed or have adverse effects that preclude their use (Marsden *et al.*, 2005).

Although no human anti-ricin vaccine is currently available, animal studies have demonstrated that both passive administration of anti-ricin antibodies and active immunization with inactivated ricin toxoid, have the potential to protect against a ricin challenge. Intravenous administration of antiricin IgG protects against an intravenous ricin exposure although it has a significantly lower protective effect against aerosolized ricin and has to be administered shortly after ricin exposure to be beneficial (Hewetson *et al.*, 1993; Foxwell *et al.*, 1985). *Pretreatment* with aerosolized antiricin IgG not only protects against a subsequent lethal dose of aerosolized ricin but also reduces lung damage (Poli *et al.*, 1996). These treatments are unlikely to offer any real therapeutic solution in a civilian context.

Active immunization with ricin toxoid is designed to provide long-lasting protection against a subsequent ricin challenge. Since inhalation is perceived to be the major route of potential exposure to biological agents, studies have concentrated on protection of the respiratory system to aerosolized ricin (Hewetson *et al.*, 1995; Griffiths *et al.*, 1995,1999; Kende *et al.*, 2002). Various routes of administration and toxoid delivery systems have been investigated. For example, intranasal or intratracheal immunization with ricin toxoid provided a protective antibody response, although did not prevent significant lung damage due to an inefficient local immune reponse (Poli *et al.*, 1996; Hewetson *et al.*, 1995; Griffiths *et al.*, 1995).

Localized respiratory immunity following vaccination with ricin toxoid has been improved by the delivery of toxoid encapsulated in a liposomal membrane, allowing delivery of antigen directly into the respiratory tract at a reduced dose and with fewer total inoculations required to achieve immune protection while minimizing local lung inflammation (Griffiths *et al.*, 1997,1998,1999; Morris *et al.*, 1994; Gupta and Siber, 1995). Encapsulated ricin toxoid also has the advantage of being a much more effective mucosal antigen than aqueous vaccine when administered orally, thus offering the potential for oral vaccination (Kende *et al.*, 2002).

A potential problem with ricin toxoid immunization is that at least some preparations have shown a tendency to revert to the toxic form when stored at room temperature or 4°C (Griffiths *et al.*, 1998). This has led to the examination of different vaccination candidates based upon ricin A chain (Griffiths *et al.*, 1998). The challenge is to develop a ricin A chain derivative that is sufficiently immunogenic to be used as a vaccine without having the enzymatic activity of the complete ricin A chain molecule.

The most promising solution seems to be the development of a genetically engineered ricin A chain which has been modified to eliminate enzyme activity (Roberts and Lord, 1992). Amino acid substitutions of sections of the ricin A chain sequence have resulted in a protein with reduced enzyme activity and which lacks the peptide sequence responsible for the vascular leak syndrome, so eliminating this adverse effect (Soler-Rodríguez *et al.*, 1993: Baluna *et al.*, 1999). The resulting vaccine protected mice from an intraperitoneal dose of ricin ten times the LD_{50} value (Smallshaw *et al.*, 2002,2003).

Other investigators (Marsden *et al.*, 2004) introduced an inhibitor peptide into the ricin A chain which completely eliminated the *in vivo* cytotoxicty of the protein, was non-toxic when injected into rats and elicited an immune response that protected the animals from an intratracheal ricin dose of five times the LD_{50}. Most recently, a fragment of the ricin A chain has been identified which has no enzymic activity, does not induce vascular leak syndrome, is very stable and completely protects mice from subsequent exposure to a lethal ricin challenge when the toxin is administered either by intraperitoneal injection or whole-body aerosol exposure (Olsen *et al.*, 2004; Lebeda and Olson, 1999; McHugh *et al.*, 2004).

ABRIN

Abrin is a plant toxin, which is closely related to ricin in terms of its structure and chemical properties. It is obtained from the seeds of *Abrus precatorius* (commonly known as 'jequirity bean' or 'rosary pea'), a tropical vine cultivated as an ornamental plant in many locations. Jequirity beans are usually scarlet in colour with a black spot at one end (though less common different coloured varieties exist) and are approximately 3 × 8 mm in size.

Abrin is soluble in water (Niyogi and Rieders, 1969) and can be extracted by soaking the shelled, ground seeds of *Abrus precatorius*. This crude extract contains several proteins besides abrin. Various methods for the isolation of pure abrin have been described (Olsnes, 1976,1978; Lin *et al.*, 1981). In summary, these involve the initial extraction of the toxin using an acidic aqueous solution, followed by several purification steps. Contaminating proteins are removed by dialysis against distilled water and by Sepharose 4B affinity chromatography. Abrin is separated from abrus agglutinin (another protein found in jequirity beans, which is capable of agglutinating red blood cells, but is non-toxic) by ion-exchange chromatography or sucrose gradient centrifugation. The yield of pure abrin powder is approximately 0.075 wt % of the jequirity bean (Olsnes, 1978).

At least theoretically, the similarity of abrin to ricin gives it a similar potential as a chemical weapon, although the relative low-scale cultivation of jequirity plants compared with castor oil plants make it less likely that abrin would be used in a large-scale attack.

STRUCTURE AND MECHANISMS OF TOXICITY OF ABRIN

Like ricin, abrin is a type 2 ribosome inactivating protein with A and B chains linked by a disulphide bond. The abrin A chain comprises 251 amino acid residues compared to 267 in the ricin A chain, both having three folding structural domains and a molecular weight of approximately 30 kDa. The abrin B chain has a molecular weight of 35 kDa with 60% of its amino acid residues identical to those of ricin's B chain. Both B chains have two saccharide binding sites which are highly conserved between the two toxins.

The mechanism of ribosome inactivation by abrin is essentially identical to ricin, save that *in vitro* studies suggest abrin is an even more potent toxin than ricin in this regard (Griffiths *et al.*, 1994). As with ricin, cell death is ultimately via apoptosis (Qu and Qing, 2004) and involves mitochondrial membrane damage and reactive oxygen species production (Shih *et al.*, 2001; Narayanan *et al.*, 2004). *In vitro* studies suggests abrin B chain triggers apoptosis while the A chain is responsible for its progression (Ohba *et al.*, 2004). Agglutination of red blood cells has been cited as a further mechanism of toxicity, but pure abrin has only a low haemagglutinating ability, the *in vitro* agglutination caused by jequirity bean extract being due rather to abrus agglutinin (Lin *et al.*, 1981).

TOXICOKINETICS OF ABRIN EXPOSURE

The structural similarity of abrin to ricin is reflected also in its toxicokinetic properties with limited absorption following ingestion. An *in vitro* study has confirmed that abrin is digested substantially by trypsin (Lin *et al.*, 1970a). In contrast, parenterally administered abrin distributes widely (Fodstad *et al.*, 1976; Lin *et al.*, 1970b). Elimination is almost exclusively renal as low molecular weight degradation products (Fodstad *et al.*, 1976).

CLINICAL FEATURES OF ABRIN EXPOSURE

Ingestion

As with ricin, poisoning by abrin ingestion is an unlikely mode of chemical terrorism and cases to date have involved eating jequirity beans. There is no case in which ingestion of a single jequirity bean has been proven to cause death and several where children remained asymptomatic after ingestion of at least one bean (Swanson-Biearman *et al.*, 1992; Hart, 1963). In such cases the hard, impermeable coating of intact seeds facilitates passage intact through the gastrointestinal tract without toxin release. In contrast, unripe abrin seeds may present more of an ingestion hazard since their shell is softer and can be chewed most easily (Hart, 1963; Davis, 1978; Fernando, 2001).

Features following substantial abrin ingestion have usually been very similar in nature and time-course to those described for ricin, with initial gastrointestinal manifestations (Davis, 1978; Fernando, 2001; Kobert, 1906; Guggisberg, 1968; Frohne *et al.*, 1984; Pillay *et al.*, 2005), followed in severe cases by neurological features and organ failure (Pillay *et al.*, 2005). There are, however, several reports of atypical presentation following jequirity bean ingestion. A 13-year-old boy first developed symptoms about five days after ingesting at least two jequirity beans (Fernando, 2001). He presented with drug-resistant pulmonary oedema and hypertension in addition to gastrointestinal features and tachycardia and required intermittent positive pressure ventilation for two days. In another case (reported only in abstract), a 15-month-old child who ingested over 20 jequirity beans experienced only 'mild' elevation of aspartate aminotransferase (AST), and minor hepatomegaly (Swanson-Biearman *et al.*, 1992).

Frohne *et al.* (1984) described the clinical course of a man who presented with severe gastroenteritis after ingesting a powder made from jequirity beans. He developed acute pancreatitis, leucocytosis, bloody ascites, increased serum amylase and lipase activities, hypocalcaemia and diabetes mellitus. Moreover, in addition to delirium, hallucinations, reduced consciousness and generalized seizures, he also developed a hemiparesis. Respiratory depression necessitated ventilation for 10 days but all symptoms improved gradually over the following three weeks. Neurological recovery was complete.

Death has occurred up to four days after the onset of symptoms (Davis, 1978). Autopsies performed on three patients revealed that ingestion of jequirity beans mainly affected the gastrointestinal tract, with haemorrhage, erythema and oedema (Davis, 1978). Moderate cerebral oedema was also noted.

Parenteral Abrin Administration

In the early 20th Century, abrin was used for homicide in several countries, including India and Sri Lanka (Hart, 1963; Byam and Archibald, 1921). The seeds of *Abrus precatorius* were ground, mixed to a paste with water and then formed into needles known as 'suis'. When the needles had hardened, they were used to pierce the skin of the victim. In at least one documented case, this resulted in death. The patient was 'pricked' in the cheek by a needle, which he extracted immediately; he died two days later. No clinical details were provided (Byam and Archibald, 1921).

The human minimum lethal dose of abrin by intravenous injection is more than 0.3 μg/kg (Gill, 1982) since in clinical trials, patients have tolerated this dose without serious adverse effects.

Dogs administered abrin intravenously showed anorexia and weight loss, weakness and fever (Fodstad *et al.*, 1979). Animals dying from intoxication expired within 15–40 h of dosing while those that survived recovered completely over 1–3 weeks. In this study, the minimum lethal dose was in the range 1.2–1.35 μg/kg. The authors determined corresponding values for rats (0.35–0.5 μg/kg), guinea-pigs (0.4–0.5 μg/kg) and rabbits (0.03–0.06 μg/kg), demonstrating significant 'inter-species' variation in susceptibility (Fodstad *et al.*, 1979).

Intraperitoneal injection in mice caused anorexia, drowsiness and diminished then absent reflexes (el-Shabrawy *et al.*, 1987). Guinea-pigs given intraperitoneal abrus seed extract showed similar features (Routh and Lahiri, 1971). The intraperitoneal dose which was lethal to 50% (LD_{50}) of treated mice has variably been reported as 8.34 mg/kg (el-Shabrawy *et al.*, 1987) and 0.02 mg/kg (Lin *et al.*, 1969). This discrepency is likely to reflect differences in purity of the abrin extract and/or experimental conditions.

Inhalation

Human toxicity after inhalation of abrin has not been reported. Rats exposed to varying abrin concentrations by inhalation remained clinically well for 18–24 h and then developed general malaise, anorexia, piloerection and respiratory distress. Examination showed pulmonary oedema, alveolitis and diffuse necrosis of the epithelium lining the lower respiratory tract but no physical injury outside the lungs. The LD50 in rats for abrin inhalation was 3.3 μg/kg (Griffiths *et al.*, 1995a). As inhalational toxins there is no significant difference in potency between abrin and ricin (Griffiths *et al.*, 1995b).

Ophthalmic and dermal exposure

Abrin applied to the eye causes severe inflammation of the conjunctiva with localized necrosis (Grant and Schuman, 1993). At the end of the 19th Century, the extract of jequirity bean was used therapeutically for its inflammatory properties, to treat various eye complaints including trachoma. However, the inflammation produced was very difficult to control and, in some cases, the use of this infusion resulted in permanent damage to the cornea, and occasionally blindness (Grant and Schuman, 1993).

In one case, a jequirity infusion applied to the eye ran through the tear duct and into the throat, causing pain and inflammation of the pharynx and larynx, with fever, unilateral swelling of the face and neck and cervical lymphadenopathy. Dysphagia and laboured breathing ensued but the patient eventually recovered (Kobart, 1906).

There are no reports of human or animal toxicity by skin contact with abrin.

DIAGNOSIS

As for ricin, abrin detection is possible by radioimmunoasay with little cross-reactivity between the two toxins (Godal *et al.*, 1981).

MANAGEMENT OF ABRIN POISONING

The principles of symptomatic and supportive care following abrin exposure by any route are identical to those described above for ricin. Similarly, experimental strategies have been pursued, aiming to protect against abrin toxicity. Animal data suggest that subcutaneous injection with increasing sublethal doses of aqueous *Abrus precatorious* seed extract can promote a degree of tolerance and enable mice to withstand an abrin dose 11 times greater than the LD_{50} (Niyogi and Rieders, 1969). This is most unlikely to have a clinical application.

Improved survival after a lethal dose of abrin has also been claimed with ascorbic acid pretreatment for 48 h prior to abrin administration (Clark *et al.*, 1981). However, the benefit was completely lost if ascorbate administration occured 15 min or more after abrin exposure, again limiting any clinical application.

As with ricin, antitoxin vaccination with either toxoid or antibodies offers the most promising means of effective prophylaxis against abrin poisoning. Rats administered three doses of subcutaneous abrin toxoid at three-week intervals were protected against a subsequent lethal aerolized abrin challenge three weeks after the last immunization, although they still developed significant lung damage (Griffiths *et al.*, 1995b). It is of particular interest that despite the close structural similarlity between ricin and abrin, the toxoids of each toxin do not provide 'cross-immunity' for each other.

CONCLUSIONS

The potential for ricin or abrin to be 'weapons of mass destruction' is limited by their need for delivery as an aerosol or widespread food or water contamination. Nevertheless, these toxins are exquisitely destructive to mammalian cells and pose sufficient threat to prioritize greater understanding of their mechansims of toxicity to facilitate the development of effective vaccines.

REFERENCES

Aplin PJ and Eliseo T (1997). Ingestion of castor oil plant seeds. *Med J Aust*, **167**, 260–261.

Argent RH, Parrott AM, Day PJ *et al.* (2000). Ribosome-mediated folding of partially unfolded ricin A-chain. *J Biol Chem*, **275**, 9263–9269.

Artalejo CR, Henley JR, McNiven MA *et al.* (1995). Rapid endocytosis coupled to exocytosis in adrenal chromaffin cells involves Ca^{2+}, GTP and dynamin but not clathrin. *Proc Natl Acad Sci USA*, **92**, 8328–8332.

Balint GA (1974). Ricin: the toxic protein of castor oil seeds. *Toxicology*, **2**, 77–102.

Baluna R, Sausville EA, Stone MJ *et al.* (1996). Decreases in levels of serum fibronectin predict the severity of vascular leak syndrome in patients treated with ricin A-chain containing immunotoxins. *Clin Cancer Res*, **2**, 1705–1712.

Baluna R, Rizo J, Gordon BE *et al.* (1999). Evidence for a structural motif in toxins and interleukin-2 that may be responsible for binding to endothelial cells and initiating vascular leak syndrome. *Proc Natl Acad Sci USA*, **96**, 3957–3962.

Ban N, Nissen P, Hansen J *et al.* (1999). Placement of protein and RNA structures into a 5 Å-resolution map of the 50S ribosomal subunit. *Nature*, **400**, 841–847.

Barbieri L, Battelli MG and Stirpe F (1993). Ribosome-inactivating proteins from plants. *Biochim Biophys Acta*, **1154**, 237–282.

Battelli MG (2004). Cytotoxicity and toxicity to animals and humans of ribosome-inactivating proteins. *Mini Rev Med Chem*, **4**, 513–521.

Bingen A, Creppy EE, Gut JP *et al.* (1987). The Kupffer cell is the first target in ricin-independent hepatitis. *J Submicrosc Cytol*, **19**, 247–256.

Bispham WN (1903). Report of cases of poisoning by fruit of *Ricinus communis*. *Am J Med Sci*, **12**, 319–321.

Brown RFR and White DE (1997). Ultrastructure of rat lung following inhalation of ricin aerosol. *Int J Exp Pathol*, **78**, 267–276.

Burrows WD and Renner SE (1999). Biological warfare agents as threats to potable water. *Environ Health Perspect*, **107**, 975–984.

Butterworth AG and Lord JM (1983). Ricin and *Ricinus communis* agglutinin subunits are all derived from a single-size polypeptide precursor. *Eur J Biochem*, **137**, 57–65.

Byam W and Archibald RG (eds) (1921). *The Practice of Medicine in the Tropics*. London, UK: Henry Frowde and Hodder & Stoughton.

Cawley DB, Hedblom ML, Hoffman EJ *et al.* (1977). Differential ricin sensitivity of rat liver and wheat germ ribosomes in polyuridylic acid translation. *Arch Biochem Biophys*, **182**, 690–695.

Chaddock JA and Roberts LM (1993). Mutagenesis and kinetic analysis of the active site Glu177 of ricin A-chain. *Protein Eng*, **6**, 425–431.

Challoner KR and McCarron MM (1990). Castor bean intoxication. *Ann Emerg Med*, **19**, 1177–1183.

Chen A, Hu T, Mikoryak C *et al.* (2002). Retrograde transport of protein toxins under conditions of COP1 dysfunction. *Biochim Biophys Acta*, **1589**, 124–139.

Clark CE, Keplinger JL, Riddle MR *et al.* (1981). Time dependency of ascorbate compound administration to alter the mean survival time of abrin intoxication. *Fed Proc Fed Am Soc Exp Biol*, **40**, 865.

Cosson P and Letourneur F (1994). Coatomer interaction with di-lysine endoplasmic reticulum retention motifs. *Science*, **263**, 1629–1631.

Cosson P and Letourneur F (1997). Coatomer (COP1)-coated vesicles: role in intracellular transport and protein sorting. *Curr Opin Cell Biol*, **9**, 484–487.

Crompton R and Gall D (1980). Georgi Markov – death in a pellet. *Med Legal J*, **48**, 51–62.

Damke H, Baba T, van der Bliek AM *et al.* (1995). Clathrin-independent pinocytosis is induced in cells overexpressing a temperature-sensitive mutant of dynamin. *J Cell Biol*, **131**, 69–80.

Davis JH (1978). Abrus precatorius (rosary pea). The most common lethal plant poison. *J Fla Med Assoc*, **65**, 188–191.

Day PJ, Owens SR, Wesche J *et al.* (2001). An interaction between ricin and calreticulin that may have implications for toxin trafficking. *J Biol Chem*, **276**, 7202–7208.

Deeks ED, Cook JP, Day PJ *et al.* (2002). The low lysine content of ricin A chain reduces the risk of proteolytic degradation after translocation from the endoplasmic reticulum to the cytosol. *Biochemistry*, **41**, 3405–3413.

Despott E and Cachia MJ (2004). A case of accidental ricin poisoning. *Malta Med J*, **16**, 39–41.

el-Shabrawy OA, el-Gengaihi S and Ali Ibrahim N (1987). Toxicity and tetratogenicity of abrin. *Egypt J Vet Sci*, **24**, 135–142.

Endo Y and Tsurugi K (1987). *N*-glycosidase activity of ricin A-chain. Mechanism of action of the toxic lectin ricin on eukaryotic ribosomers. *J Biol Chem*, **262**, 8128–8130.

Endo Y and Tsurugi K (1988). The RNA *N*-glycosidase activity of ricin A-chain. The characteristics of the enzymatic activity of ricin A-chain with ribosomes and with rRNA. *J Biol Chem*, **263**, 8735–8739.

Engert A, Diehl V, Schnell R *et al.* (1997). A phase-I study of an anti-CD25 ricin A-chain immunotoxin (RFT5-SMPT-dgA) in patients with refractory Hodgkin's lymphoma. *Blood*, **89**, 403–410.

Fernando C (2001). Poisoning due to *Abrus precatorius* (jequirity bean). *Anaesthesia*, **56**, 1178–1180.

Ferrini J-B, Martin M, Taupiac M-P *et al.* (1995). Expression of functional ricin B chain using the baculovirus system. *Eur J Biochem*, **233**, 772–777.

Figley KD and Elrod RH (1928). Endemic asthma due to castor bean dust. *JAMA*, **90**, 79–82.

Fine DR, Shepherd HA, Griffiths GD *et al.* (1992). Sub-lethal poisoning by self-injection with ricin. *Med Sci Law*, **32**, 70–72.

Fodstad Ø, Olsnes S and Pihl A (1976). Toxicity, distribution and elimination of the cancerostatic lectins abrin and ricin after parental injection into mice. *Br J Cancer*, **34**, 418–425.

Fodstad Ø, Johannessen JV, Schjerven L *et al.* (1979). Toxicity of abrin and ricin in mice and dogs. *J Toxicol Environ Health*, **5**, 1073–1084.

Fodstad Ø, Kvalheim G, Godal A *et al.* (1984). Phase I study of the plant protein ricin. *Cancer Res*, **44**, 862–865.

Foxwell BMJ, Detre SI, Donovan TA *et al.* (1985). The use of *anti*-ricin antibodies to protect mice intoxicated with ricin. *Toxicology*, **34**, 79–88.

Frankel A, Welsh P, Richardson J *et al.* (1990). Role of arginine 180 and glutamic acid 177 of ricin toxin A chain in enzymatic inactivation of ribosomes. *Mol Cell Biol*, **10**, 6257–6263.

Franz DR and Jaax NK (1997). Ricin toxin. In: *Medical Aspects of Chemical and Biological Warfare* (FR Siddell, ET Takafuji and DR Franz, eds), pp. 631–642. Washington, DC, USA: Office of the Surgeon General at TMM Publications.

Frigerio L, Vitale A, Lord JM *et al.* (1998). Free ricin A chain, proricin and native toxin have different cellular fates when expressed in tobacco protoplasts. *J Biol Chem*, **273**, 14194–14199.

Frigerio L, Jolliffe NA, Di Cola A *et al.* (2001). The internal propeptide of the ricin precursor carries a sequence-specific determinant for vacuolar sorting. *Plant Physiol*, **126**, 167–175.

Frohne D, Schmoldt A and Pfänder HJ (1984). Die Paternostererbse – keineswegs harmlos. *Dtsch Apothek Zeit*, **124**, 2109–2113.

Ghosh RN, Mallet WG, Soe TT *et al.* (1998). An endocytosed TGN38 chimeric protein is delivered to the TGN after trafficking through the endocytic recycling compartment in CHO cells. *J Cell Biol*, **142**, 923–936.

Gill DM (1982). Bacterial toxins: a table of lethal amounts. *Microbiol Rev*, **46**, 86–94.

Girod A, Storrie B, Simpson JC *et al.* (1999). Evidence for a COP1-independent transport route from Golgi complex to the endoplasmic reticulum. *Nat Cell Biol*, **1**, 423–430.

Godal A, Olsnes S and Pihl A (1981). Radioimmunoassays of abrin and ricin in blood. *J Toxicol Environ Health*, **8**, 409–417.

Grant WM and Schuman JS (1993). *Toxicology of the Eye. Effects on the Eyes and Visual System from Chemicals, Drugs, Metals and Minerals, Plants, Toxins and Venoms; also Systemic Side Effects from Eye Medications*, Fourth Edition. Springfield, IL, USA: Charles C. Thomas.

Griffiths GD, Newman H and Gee DJ (1986). Identification and quantification of ricin toxin in animal tissues using ELISA. *J Forens Sci Soc*, **26**, 349–358.

Griffiths GD, Lindsay CD and Upshall DG (1994). Examination of the toxicity of several protein toxins of plant origin using bovine pulmonary endothelial cells. *Toxicology*, **90**, 11–27.

Griffiths GD, Rice P, Allenby AC *et al.* (1995a). Inhalation toxicology and histopathology of ricin and abrin toxins. *Inhal Toxicol*, **7**, 269–288.

Griffiths GD, Lindsay CD, Allenby AC *et al.* (1995b). Protection against inhalation toxicity of ricin and abrin by immunisation. *Human Exp Toxicol*, **14**, 155–164.

Griffiths GD, Bailey SC, Hambrook JL *et al.* (1997). Liposomally-encapsulated ricin toxoid vaccine delivered intratracheally elicits a good immune response and protects a lethal pulmonary dose of ricin toxin. *Vaccine*, **15**, 1933–1939.

Griffiths GD, Bailey SC, Hambrook JL *et al.* (1998). Local and systemic responses against ricin toxin promoted by toxoid or peptide vaccines alone or in liposomal formulations. *Vaccine*, **16**, 530–535.

Griffiths GD, Phillips GJ, and Bailey SC (1999). Comparison of the quality of protection elicited by toxoid and peptide liposomal vaccine formulations against ricin as assessed by markers of inflammation. *Vaccine*, **17**, 2562–2568.

Guggisberg M (1968). A propos d'une curieuse intoxication par des grains de chapelet (abrus precatorius). *Rev Med Suisse Romande*, **88**, 206–208.

Gupta RK and Siber GR (1995). Adjuvants for human vaccines – current status, problems and future prospects. *Vaccine*, **13**, 1263–1276.

Harley SM and Beevers H (1982). Ricin inhibition of *in vitro* protein synthesis by plant ribosomes. *Proc Nat Acad Sci USA*, **79**, 5935–5938.

Harley SM and Lord JM (1985). *In vitro* endoproteolytic cleavage of castor bean lectin precursors. *Plant Sci*, **41**, 111–116.

Hart M (1963). Hazards to health. Jequirity-bean poisoning. *New Engl J Med*, **268**, 885–886.

Hartley MR, Chaddock JA and Bonness MS (1996). The structure and function of ribosome inactivating proteins. *Trends Plant Sci*, **1**, 254–259.

Hazes B and Read RJ (1997). Accumulating evidence suggests that several AB-toxins subvert the endoplasmic reticulum-associated protein degradation pathway to enter target cells. *Biochemistry*, **36**, 11051–11054.

Herskovits JS, Burgess CC, Obar RA *et al.* (1993). Effects of mutant rat dynamin on endocytosis. *J Cell Biol*, **122**, 565–578.

Hewetson JF, Rivera VR, Creasia DA *et al.* (1993). Protection of mice from inhaled ricin by vaccination with ricin or by passive treatment with heterologous antibody. *Vaccine*, **11**, 743–746.

Hewetson JF, Rivera VP, Lemley PV *et al.* (1995). A formalinized toxoid for protection of mice from inhaled ricin. *Vaccine Res*, **4**, 179–187.

Houston LL (1982). Protection of mice from ricin poisoning by treatment with antibodies directed against ricin. *J Toxicol Clin Toxicol*, **19**, 385–389.

Hughes JN, Lindsay CD and Griffiths GD (1996). Morphology of ricin and abrin exposed endothelial cells is consistent with apoptotic cell death. *Human Exp Toxicol*, **15**, 443–451.

Hutchinson LTR (1900). Poisoning by castor oil seeds. *Br Med J*, **1**, 1155–1156.

Ingle VN, Kale VG and Talwalker YB (1966). Accidental poisoning in children with particular reference to castor beans. *Ind J Pediatr*, **33**, 237–240.

Ishiguro M, Tanabe S, Matori Y *et al.* (1992). Biochemical studies on oral toxicity of ricin. IV. A fate of orally administered ricin in rats. *J Pharmacobiodyn*, **15**, 147–156.

Iversen T-G, Skretting G, Llorente A *et al.* (2001). Endosome to Golgi transport of ricin is independent of clathrin and of the Rab9- and Rab11-GTPases. *Mol Cell Biol*, **12**, 2099–2107.

Johnson RC, Lemire SW, Woolfitt AR *et al.* (2005). Quantification of ricinine in rat and human urine: a biomarker for ricin exposure. *J Anal Toxicol*, **29**, 149–155.

Jolliffe NA, Ceriotti A, Frigerio L *et al.* (2003). The position of the proricin vacuolar targeting signal is functionally important. *Plant Mol Biol*, **51**, 631–641.

Kanerva L, Estlander T and Jolanki R (1990). Long-lasting contact urticaria. Type I and type IV allergy from castor bean and a hypothesis of systemic IgE-mediated allergic dermatitis. *Dermatol Clin*, **8**, 181–188.

Kaszás T and Papp G (1960). Ricinussamen-Vergiftung von Schulkindern. *Arch Toxikol*, **18**, 145–150.

Katzin BJ, Collins EJ and Robertus JD (1991). Structure of ricin A-chain at 2.5 Å. *Proteins*, **10**, 251–259.

Kende M, Yan C, Hewetson J *et al.* (2002). Oral immunization of mice with ricin toxoid vaccine encapsulated in polymeric microspheres against aerosol challenge. *Vaccine*, **20**, 1681–1691.

Kim Y and Robertus JD (1992). Analysis of several key active site residues of ricin A chain by mutagenesis and X-ray crystallography. *Protein Eng*, **5**, 775–779.

Kobert R (1906). *Lehrbuch Der Intoxikationen*, Second Edition. Stuttgart, Germany: Ferdinard Enke.

Koch LA and Caplan J (1942). Castor bean poisoning. *Am J Dis Child*, **64**, 485–486.

Kokes J, Assaad A, Pitt L *et al.* (1994). Acute pulmonary response of rats exposed to a sublethal dose of ricin aerosol. *FASEB J*, **8**, A144.

Kopferschmitt J, Flesch F, Lugnier A *et al.* (1983). Acute voluntary intoxication by ricin. *Human Toxicol*, **2**, 239–242.

Kumar O, Sugendran K and Vijayaraghaven R (2003). Oxidative stress associated hepatic and renal toxicity induced by ricin in mice. *Toxicon*, **41**, 333–338.

Lamaze C, Dujeancourt A, Baba T *et al.* (2001). Interleukin-2 receptors and detergent-resistant membrane domains define a clathrin-independent endocytic pathway. *Mol Cell*, **7**, 661–671.

Lamb FI, Roberts LM and Lord JM (1985). Nucleotide sequence of cloned cDNA coding for preproricin. *Eur J Biochem*, **148**, 265–270.

Lebeda FJ and Olson MA (1999). Prediction of a conserved, neutralizing epitope in ribosome-inactivating proteins. *Int J Biol Macromol*, **24**, 19–26.

Leek MD (1989). *Pathological Changes Induced by Ricin Poisoning*, PhD Thesis. Leeds, UK: University of Leeds.

Leek MD, Griffiths GD and Green MA (1989). Intestinal pathology following intramuscular ricin poisoning. *J Pathol*, **159**, 329–324.

Leith AG, Griffiths GD and Green MA (1988). Quantification of ricin toxin using a highly sensitive avidin/biotin enzyme-linked immunosorbent assay. *J Forens Sci Soc*, **28**, 227–236.

Levin Y, Sherer Y, Bibi H *et al.* (2000). Rare Jatropha multifida intoxication in two children. *J Emerg Med*, **19**, 173–175.

Lewis MS and Youle RJ (1986). Ricin subunit association. Thermodynamics and the role of the disulfide bond in toxicity. *J Biol Chem*, **261**, 11571–11577.

Lin J-Y, Chen C-C, Lin L-T *et al.* (1969). Studies on the toxic action of abrin. *J Formos Med Assoc*, **68**, 322–324.

Lin J-Y, Kao C-L and Tung T-C (1970a). Study on the effect of tryptic digestion on the toxicity of abrin. *J Formos Med Assoc*, **69**, 61–63.

Lin J-Y, Ju S-T, Shaw Y-S *et al.* (1970b). Distribution of I^{131}–labeled abrin *in vivo*. *Toxicon*, **8**, 197–201.

Lin J-Y, Lee T-C, Hu S-T *et al.* (1981). Isolation of four isotoxic proteins and one agglutinin from jequiriti bean (Abrus precatorius). *Toxicon*, **19**, 41–51.

Llorente A, Rapek A, Schmid SL *et al.* (1998). Expression of mutant dynamin inhibits toxicity and transport of endocytosed ricin to the Golgi apparatus. *J Cell Biol*, **140**, 553–563.

Lockey SD Jr and Dunkelberger L (1968). Anaphylaxis from an Indian necklace. *JAMA*, **206**, 2900–2901.

Lombardi D, Soldati T, Riederer MA *et al.* (1993). Rab9 functions in transport between late endosomes and the trans Golgi network. *EMBO J*, **12**, 677–682.

Lord JM (1985a). Precursors of ricin and *Ricinus communis* agglutinin. Glycosylation and processing during synthesis and intracellular transport. *Eur J Biochem*, **146**, 411–416.

Lord JM (1985b). Synthesis and intracellular transport of lectin and storage protein precursors in endosperm from castor bean. *Eur J Biochem*, **146**, 403–409.

Lord JM, Roberts LM and Robertus JD (1994). Ricin: structure, mode of action and some current applications. *FASEB J*, **8**, 201–208.

Lord JM, Jolliffe NA, Marsden CJ *et al.* (2003). Ricin: mechanisms of cytotoxicity. *Toxicol Rev*, **22**, 53–64.

Magnusson S, Kjeken R and Berg T (1993). Characterization of two distinct pathways of endocytosis of ricin by rat liver endothelial cells. *Exp Cell Res*, **205**, 118–125.

Malizia E, Sarcinelli L and Andreucci G (1977). Ricinus poisoning: a familiar epidemy. *Acta Pharmacol Toxicol*, **41**, 351–361.

Mallard F, Antony C, Tenza D *et al.* (1998). Direct pathway from early/recycling endosomes to the Golgi apparatus revealed through the study of shiga toxin B-fragment transport. *J Cell Biol*, **143**, 973–990.

Mallet WG and Maxfield FR (1999). Chimeric forms of furin and TGN38 are transported with the plasma membrane in the trans-Golgi network via distinct endosomal pathways. *J Cell Biol*, **146**, 345–359.

Marsden CJ, Knight S, Smith DC *et al.* (2004). Insertional mutagenesis of ricin A chain: a novel route to an anti-ricin vaccine. *Vaccine*, **22**, 2800–2805.

Marsden CJ, Smith DC, Roberts LM *et al.* (2005). Ricin: current understanding and prospects for an antiricin vaccine. *Expert Rev Vaccines*, **4**, 229–237.

McHugh CA, Tammariello RF, Millard CB *et al.* (2004). Improved stability of a protein vaccine through elimination of a partially unfolded state. *Protein Sci*, **13**, 2736–2743.

Meldrum WP (1900). Poisoning by castor oil seeds. *Br Med J*, **1**, 317.

Metz G, Böcher D and Metz J (2001). IgE-mediated allergy to castor bean dust in a landscape gardener. *Contact Dermatitis*, **44**, 367.

Mlsna D, Monzingo AF, Katzin BJ *et al.* (1993). Structure of recombinant ricin A chain at 2.3 Å. *Protein Sci*, **2**, 429–435.

Montfort W, Villafranca JE, Monzingo AF *et al.* (1987). The three dimensional structure of ricin at 2.8 Å. *J Biol Chem*, **262**, 5398–5403.

Monzingo AF and Robertus JD (1992). X-ray analysis of substrate analogs in the ricin A-chain active site. *J Mol Biol*, **227**, 1136–1145.

Morris W, Steinhoff MC and Russell PK (1994). Potential of polymer microencapsulation technology for vaccine innovation. *Vaccine*, **12**, 5–11.

Möschl H (1938). Zur Klinik und Pathogenese der Rizinvergiftung. *Wien Klin Wochenschr*, **51**, 473–475.

Moya M, Dautry-Varsat A, Goud B *et al.* (1985). Inhibition of coated pit formation in Hep_2 cells blocks the cytotoxicity of diphtheria toxin but not that of ricin toxin. *J Cell Biol*, **101**, 548–559.

Narayanan S, Surolia A and Karande AA (2004). Ribosome-inactivating protein and apoptosis: abrin causes cell death via mitochondrial pathway in Jurkat cells. *Biochem J*, **377**, 233–240.

Nichols BJ and Lippincott-Schwartz J (2001). Endocytosis without clathrin coats. *Trends Cell Biol*, **11**, 406–412.

Nilsson L and Nygård O (1986). The mechanism of the protein-synthesis elongation cycle in eukaryotes. Effect of ricin on the ribosomal interaction with elongation factors. *Eur J Biochem*, **161**, 111–117.

Niyogi SK and Rieders F (1969). Toxicity studies with fractions from *Abrus precatorius* seed kernels. *Toxicon*, **7**, 211–216.

Ohba H, Moriwaki S, Bakalova R *et al.* (2004). Plant-derived abrin-a induces apoptosis in cultured leukemic cell lines by different mechanisms. *Toxicol Appl Pharmacol*, **195**, 182–193.

Olsnes S (1976). Abrin and ricin: structure and mechanism of action of two toxic lectins. *Bull Inst Pasteur*, **74**, 85–99.

Olsnes S (1978). Toxic and nontoxic lectins from *Abrus precatorius*. *Methods Enzymol*, **50**, 323–330.

Olsnes S and Pihl A (1982). Toxic lectins and related proteins. In: *Molecular Action of Toxins and Viruses* (P Cohen and S van Heyningen, eds), pp. 51–105. Amsterdam, The Netherlands: Elsevier.

Olson MA, Carra JH, Roxas-Duncan V *et al.* (2004). Finding a new vaccine in the ricin protein fold. *Protein Eng Des Sel*, **17**, 391–397.

Palatnick W and Tenenbein M (2000). Hepatotoxicity from castor bean ingestion in a child. *J Toxicol Clin Toxicol*, **38**, 67–69.

Passeron T, Mantoux F, Lacour J-P *et al.* (2004). Infectious and toxic cellulitis due to suicide attempt by subcutaneous injection of ricin. *Br J Dermatol*, **150**, 154.

Peumans WJ, Hao Q and van Damme EJM (2001). Ribosome-inactivating proteins from plants: more than RNA *N*-glycosidases? *FASEB J*, **15**, 1493–1506.

Pevny I (1979). Ricinusschrot-Allergie. *Dermatosen*, **27**, 159–162.

Pillay VV, Bhagyanathan PV, Krishnaprasad R *et al.* (2005). Poisoning due to white seed variety of *Abrus precatorius*. *J Assoc Physicians Ind*, **53**, 317–319.

Poli MA, Rivera VR, Pitt ML *et al.* (1996). Aerosolized specific antibody protects mice from lung injury associated with aerosolized ricin exposure. *Toxicon*, **34**, 1037–1044.

Qu X and Qing L (2004). Abrin induces HeLa cell apoptosis by cytochrome *c* release and caspase activation. *J Biochem Mol Biol*, **37**, 445–453.

Ramsden CS, Drayson MT and Bell EB (1989). The toxicity, distribution and excretion of ricin holotoxin in rats. *Toxicology*, **55**, 161–171.

Rapak A, Falnes PØ and Olsnes S (1997). Retrograde transport of mutant ricin to the endoplasmic reticulum with subsequent translocation to cytosol. *Proc Natl Acad Sci USA*, **94**, 3783–3788.

Rauber A and Heard J (1985). Castor bean toxicity reexamined: a new perspective. *Vet Human Toxicol*, **27**, 498–502.

Ray B and Wu HC (1981). Internalization of ricin in Chinese hamster ovary cells. *Mol Cell Biol*, **1**, 544–551.

Ready MP, Kim Y and Robertus JD (1991). Site-directed mutagenesis of ricin A-chain and implications for the mechanism of action. *Proteins*, **10**, 270–278.

Richardson PT, Westby M, Roberts LM *et al.* (1989). Recombinant proricin binds galactose but does not depurinate 28 S ribosomal RNA. *FEBS Lett*, **255**, 15–20.

Riederer MA, Soldati T, Shapiro AD *et al.* (1994). Lysosome biogenesis requires Rab9 function and receptor recycling from endosomes to the *trans*-Golgi network. *J Cell Biol*, **125**, 573–582.

Roberts LM and Lord JM (1981). The synthesis of *Ricinus communis* agglutinin, cotranslational and posttranslational modification of agglutinin polypeptides. *Eur J Biochem*, **119**, 31–41.

Roberts LM and Lord JM (1992). Cytotoxic proteins. *Curr Opin Biotechnol*, **3**, 422–429.

Rodal SK, Skretting G, Garred O *et al.* (1999). Extraction of cholesterol with methyl-beta-cyclodextrin perturbs formation of clathrin-coated endocytic vesicles. *Mol Biol Cell*, **10**, 961–974.

Routh BC and Lahiri SC (1971). Some actions of the seeds of Abrus precatorius. *Bull Calcutta Sch Trop Med*, **19**, 46–47.

Rutenber E and Robertus JD (1991). Structure of ricin B-chain at 2.5 Å resolution (1991). *Proteins*, **10**, 260–269.

Rutenber E, Katzin BJ, Ernst S *et al.* (1991). Crystallographic refinement of ricin to 2.5 Å. *Proteins*, **10**, 240–250.

Sandvig K and van Deurs B (1996). Endocytosis, intracellular transport and cytotoxic action of Shiga toxin and ricin. *Physiol Rev*, **76**, 949–966.

Sandvig K and van Deurs B (2000). Entry of ricin and Shiga toxin into cells: molecular mechanisms and medical perspectives. *EMBO J*, **19**, 5943–5950.

Sandvig K, Grimmer S, Lauvrak SU *et al.* (2002). Pathways followed by ricin and Shiga toxin into cells. *Histochem Cell Biol*, **117**, 131–141.

Schindler J, Sausville E, Messmann R *et al.* (2001). The toxicity of deglycosylated ricin A-chain-containing immunotoxins in patients with non-Hodgkin's lymphoma is exacerbated by prior radiotherapy: a retrospective analysis of patients in five clinical trials. *Clin Cancer Res*, **7**, 255–258.

Schlossman D, Withers D, Welsh P *et al.* (1989). Role of glutamic acid 177 of the ricin toxin A chain in enzymatic inactivation of ribosomes. *Mol Cell Biol*, **9**, 5012–5021.

Shih S-F, Wu Y-H, Hung C-H *et al.* (2001). Abrin triggers cell death by inactivating a thiol-specific antioxidant. *J Biol Chem*, **276**, 21870–21877.

Shyu H-F, Chiao D-J, Liu H-W *et al.* (2002a). Monoclonal antibody-based enzyme immunoassay for detection of ricin. *Hybrid Hybridomics*, **21**, 69–73.

Shyu R-H, Shyu H-F, Liu H-W *et al.* (2002b). Colloidal gold-based immunochromatographic assay for detection of ricin. *Toxicon*, **40**, 255–258.

Simmons BM, Stahl PD and Russell JH (1986). Mannose receptor-mediated uptake of ricin toxin and ricin A chain by macrophages. Multiple intracellular pathways for a chain translocation. *J Biol Chem*, **261**, 7912–7920.

Simpson JC, Smith DC, Roberts LM *et al.* (1998). Expression of mutant dynamin protects cells against diphtheria toxin but not against ricin. *Exp Cell Res*, **239**, 293–300.

Smallshaw JE, Firan A, Fulmer JR *et al.* (2002). A novel recombinant vaccine which protects mice against ricin intoxication. *Vaccine*, **20**, 3422–3427.

Smallshaw JE, Ghetie V, Rizo J *et al.* (2003). Genetic engineering of an immunotoxin to eliminate pulmonary vascular leak in mice. *Nat Biotechnol*, **21**, 387–391.

Smith DC, Gallimore A, Jones E *et al.* (2002). Exogenous peptides delivered by ricin require processing by signal peptidase for transporter associated with antigen processing-independent MHC class1-restricted presentation. *J Immunol*, **169**, 99–107.

Soler-Rodriguez A-M, Ghetie M-A, Oppenheimer-Marks N *et al.* (1993). Ricin A-chain and ricin A-chain immunotoxins rapidly damage human endothelial cells: implications for vascular leak syndrome. *Exp Cell Res*, **206**, 227–234.

Sperti S, Montanaro L, Mattioli A *et al.* (1973). Inhibition by ricin of protein synthesis *in vitro*: 60S ribosomal subunit as the target of the toxin. *Biochem J*, **136**, 813–815.

Spilsberg B, Van Meer G and Sandvig K (2003). Role of lipids in the retrograde pathway of ricin intoxication. *Traffic*, **4**, 544–552.

Spyker DA, Sauer K, Kell SO *et al.* (1982). A castor bean poisoning and a widely available bioassay for ricin. *Vet Human Toxicol*, **24**, 293.

Stirpe F, Barbieri L, Battelli MG *et al.* (1992). Ribosome-inactivating proteins from plants: present status and future prospects. *Biotechnology*, **10**, 405–412.

Swanson-Biearman B, Dean BS and Krenzelok EP (1992). Failure of whole bowel irrigation to decontaminate the GI tract following massive jequirity bean ingestion. *Vet Human Toxicol*, **34**, 352.

Targosz D, Winnik L and Szkolnicka B (2002). Suicidal poisoning with castor bean (*Ricinus communis*) extract injected subcutaneously – case report. *J Toxicol Clin Toxicol*, **40**, 398.

Taylor S, Massiah A, Lomonossoff G *et al.* (1994). Correlation between the activities of five ribosome-inactivating proteins in depurination of tobacco ribosomes and inhibition of tobacco mosaic virus infection. *Plant J*, **5**, 827–835.

Topping MD, Henderson RTS, Luczynska CM *et al.* (1982). Castor bean allergy among workers in the felt industry. *Allergy*, **37**, 603–608.

Van der Bliek AM, Redelmeier TE, Damke H *et al.* (1993). Mutations in human dynamin block – an intermediate stage in coated vesicle formation. *J Cell Biol*, **122**, 553–563.

van Deurs B, Pedersen LR, Sundan A *et al.* (1985). Receptor-mediated endocytosis of a ricin–colloidal gold conjugate in vero cells. Intracellular routing of vacuolar and tubulo-vesicular portions of the endosomal system. *Exp Cell Res*, **159**, 287–304.

van Deurs B, Sandvig K, Petersen OW *et al.* (1988). Estimation of the amount of internalized ricin that reaches the *trans*-Golgi network. *J Cell Biol*, **106**, 253–267.

Villafranca JE and Robertus JD (1981). Ricin B chain is a product of gene duplication. *J Biol Chem*, **256**, 554–556.

Wedin GP, Neal JS, Everson GW *et al.* (1986). Castor bean poisoning. *Am J Emerg Med*, **4**, 259–261.

Werbonat Y, Kleutges N, Jakobs KH *et al.* (2000). Essential role of dynamin in internalization of M_2 muscarinic acetylcholine and angiotensin AT_{1A} receptors. *J Biol Chem*, **275**, 21969–21974.

Weston SA, Tucker AD, Thatcher DR *et al.* (1994). X-ray structure of recombinant ricin A-chain at 1.8 Å resolution. *J Mol Biol*, **244**, 410–422.

White J, Johannes L, Mallard F *et al.* (1999). Rab6 coordinates a novel Golgi to ER retrograde transport pathway in live cells. *J Cell Biol*, **147**, 743–760.

Wilhelmsen C and Pitt L (1993). Lesions of acute inhaled lethal ricin intoxication in rhesus monkeys. *Vet Pathol*, **30**, 482.

Wool IG, Correll CC and Yuen-Ling C (2000). Structure and function of the sarcin-rich domain. In: *The Ribosome: Structure, Function, Antibiotics and Cellular Interactions* (J Garrett, L Douthwaite, PJ Lilias *et al.*, eds), pp. 461–473. Washington, DC, USA: American Society for Microbiology.

Wright HT and Robertus JD (1987). The intersubunit disulfide bridge of ricin is essential for cytotoxicity. *Arch Biochem Biophys*, **256**, 280–284.

28 THE TOTAL PROHIBITION OF CHEMICAL WEAPONS

Graham S. Pearson

Department of Peace Studies, University of Bradford, UK

INTRODUCTION

The prohibition of the use of poison in warfare is rooted in taboos found in moral and cultural traditions around the world. Ancient Greeks and Romans observed a prohibition on the use of poison and poisonous weapons and the 'Manu Law of War' in India in 500 BC banned the use of such weapons. A thousand years later, regulations on the conduct of war drawn from the Koran by the Saracens prohibited poisoning.

In more recent times, following an initiative by Czar Alexander II of Russia, the delegates of 15 European States met in Brussels on 27 July 1874. The conference adopted the Brussels Declaration of 1874 which included the following provision: 'Art. 13. According to this principle are especially "forbidden": (a) Employment of poison or poisoned weapons ... (e) The employment of arms, projectiles or material calculated to cause unnecessary suffering'. However, this declaration was not accepted by all of the governments as a binding convention and it was therefore not ratified.

In 1899, Russia proposed the holding of a Peace Conference at the Hague which would *inter alia* revise the Brussels Declaration of 1874. At this Peace Conference, a Declaration was agreed on 29 July 1899 in which 'The Contracting Powers agree to abstain from the use of projectiles the object of which is the diffusion of asphyxiating or deleterious gases'. This prohibition was subsequently included in the 1907 Hague Convention IV concerning the laws and customs of wars on land which in Article 23 states: 'In addition to the prohibitions provided by special Conventions, it is especially forbidden – To employ poison or poisoned weapons; ... To employ arms, projectiles, or material calculated to cause unnecessary suffering; ... '.

Following the extensive use of chemical weapons, such as chlorine, phosgene and mustard gas, during World War I, the Geneva Protocol of 1925, prohibiting the use of asphyxiating and other gases in war, was agreed. As many of the States that became Parties to the Geneva Protocol entered reservations that they would no longer be bound by the prohibitions should such weapons be used against them, the Geneva Protocol of 1925 was effectively a prohibition of 'first-use'.

During World War II, both chemical and biological weapons were acquired by the combatants for use in retaliation in kind should such weapons be used against them. Neither chemical nor biological weapons were used during World War II. Subsequent to World War II, nuclear weapons increased in strategic significance and led in the 1950s and 1960s to the abandonment of biological and toxin weapons. Although the initial aim in the 1960s had been to negotiate a treaty banning both chemical and biological weapons, reluctance by the Soviet Union and the United States, led to the negotiations focusing on biological and toxin weapons. These led to the 'Biological and Toxin Weapons Convention', which opened for signature in 1972 and entered into force in 1975.

The next two decades saw continued negotiation of a treaty to prohibit chemical weapons. The extensive use of chemical weapons in the

Chemical Warfare Agents: Toxicology and Treatment (2nd Edition)
Edited by Timothy C. Marrs, Robert L. Maynard and Frederick R. Sidell

Iran/Iraq war of the 1980s led to the negotiations being completed with the opening for signature of the Chemical Weapons Convention in January 1993 and its entry into force on 29 April 1997.

This chapter examines the prohibitions in the relevant treaties – the Geneva Protocol of 1925, the Biological and Toxin Weapons Convention and the Chemical Weapons Convention – and concludes that chemical weapons are totally prohibited. Consideration is given to the risk of use of chemical weapons posed in the 21st Century, both by states and by other organizations and individuals, such as terrorists, and to how these risks can be countered by the effective implementation of the treaties.

1925 GENEVA PROTOCOL

Following World War I, the Peace Treaty of Versailles between the Allied and Associated Powers and Germany was signed on 28 June 1919 and entered into force on 10 January 1920. This included, as Article 171, the requirement:

> *The use of asphyxiating, poisonous or other gases and all analogous liquids, materials or devices being prohibited, their manufacture and importation are strictly forbidden in Germany.*
>
> *The same applies to materials specially intended for the manufacture, storage and use of the said products or devices.*

This terminology, describing the prohibited agents as *asphyxiating, poisonous or other gases and all analogous liquids, materials or devices*, was used in the 1922 Washington Treaty and subsequently in the preamble to the Geneva Protocol of 1925.

The 1922 Washington Treaty in relation to the use of submarines and noxious gases in warfare was signed on 6 February 1922 between the USA, the British Empire, France, Italy and Japan. Article 5 of this treaty stated that:

> *The use in war of asphyxiating, poisonous or other gases, and all analogous liquids, materials or devices, having been justly condemned by the general opinion of the civilized world and a prohibition of such use having been declared in treaties to which a majority of the civilized Powers are parties.*
>
> *The Signatory Powers, to the end that this prohibition shall be universally accepted as a part of international law binding alike the conscience and practice of nations, declare their assent to such prohibition, agree to be bound thereby as between themselves and invite all other civilized nations to adhere thereto.*

This treaty did not enter into force because the treaty was not ratified by France.

These early treaties focused largely on chemical weapons and there is no mention of biological weapons in either the 1919 Peace Treaty of Versailles or the 1922 Washington Treaty. It has been pointed out (Mierzejewski and Moon, 1999) that this did not reflect indifference but an assumption voiced by an advisory group to the US Committee on Limitation of Armaments. This advisory group recommended that chemical weapons be prohibited by international agreement and that it be classed with methods of warfare such as poisoning wells, introducing disease and so on. This association of chemical warfare, with a form of warfare clearly regarded then as abhorrent, was intended to strengthen the argument for prohibition.

Following the signing of the 1922 Treaty of Washington in relation to the use of submarines and noxious gases, considerable pressure was mounted in the Assembly of the League of Nations to make the prohibition on noxious gases universal by securing adherence from member nations. This led to the 1925 Geneva Conference that culminated in the agreement on 17 June 1925 of the *Protocol for the Prohibition of the Use in War of Asphyxiating, Poisonous or other Gases, and of Bacteriological Methods of Warfare*.

The 1925 Geneva Protocol states:

> *The Undersigned Plenipotentiaries*, in the name of their respective Governments:
>
> *Whereas* the use in war of asphyxiating, poisonous or other gases, and of all analogous liquids, materials or devices, has been justly condemned by the general opinion of the civilised world.
>
> *Whereas* the prohibition of such use has been declared in Treaties to which the majority of Powers of the world are Parties.
>
> *To the end* that this prohibition shall be universally accepted as a part of International Law,

binding alike the conscience and the practice of nations.

Declare:

That the High Contracting Parties, so far as they are not already Parties to Treaties prohibiting such use, accept this prohibition, agree to extend this prohibition to the use of bacteriological methods of warfare and agree to be bound as between themselves according to the terms of this declaration.

The High Contracting Parties will exert every effort to induce other States to accede to the present Protocol. Such accession will be notified to the Government of the French Republic, and by the latter to all signatory and acceding Powers, and will take effect on the date of the notification by the Government of the French Republic.

The present Protocol, of which the French and English texts are both authentic, shall be ratified as soon as possible. It shall bear to-day's date.

The ratifications of the present Protocol shall be addressed to the Government of the French Republic, which will at once notify the deposit of such ratification to each of the signatory and acceding Powers.

The instruments of ratification and of accession to the present Protocol will remain deposited in the archives of the Government of the French Republic.

The present Protocol will come into force for each signatory Power as from the date of deposit of its ratification, and, from that moment, each Power will be bound as regards other Powers which have already deposited their ratifications.

It can be seen from the text of the 1925 Geneva Protocol that '*each Power will be bound as regards other Powers which have already deposited their ratifications*'. It was consequently not surprising that many States qualified their ratifications with reservations such as that made by the United Kingdom that:

(1) The said Protocol is only binding on His Britannic Majesty as regards those Powers and States which have both signed and ratified the Protocol or have finally acceded thereto.
(2) The said Protocol shall cease to be binding on His Britannic Majesty towards any Power at enmity with Him whose armed forces, or the armed forces of whose allies, fail to respect the prohibitions laid down in the Protocol.

These reservations meant that in World War II, States acquired chemical and biological weapons so that they would be able to retaliate in kind should chemical or biological weapons be used against them. Acquisition and stockpiling of such weapons was not prohibited under the Geneva Protocol of 1925 which applied only to the use of such weapons in war. It thus became regarded, because of the reservations, as a prohibition of 'first-use'.

Current position of the Geneva Protocol

As of February 2005, there are 134 States Parties to the 1925 Geneva Protocol. In addition, many of the States Parties, which entered reservations, have lifted those reservations as they are incompatible with the obligations under the later Biological and Toxin Weapons Convention and the Chemical Weapons Convention. However, there have been successive United Nations General Assembly resolutions on measures to uphold the authority of the 1925 Geneva Protocol, such as that adopted in October 2004 which include language that:

Calls upon those States that continue to maintain reservations to the 1925 Geneva Protocol to withdraw them;

Such reservations that the Geneva Protocol shall cease to binding on the State Party with respect to the use in war of asphyxiating, poisonous or other gases, and all analogous liquids, materials or devices, in regard to any enemy State if such State or any of its allies fails to respect the prohibitions laid down in the Protocol continue to be maintained by a number of States Parties. As of September 2005, these include Algeria, Angola, Bahrein, Bangladesh, Fiji, India, Iraq, Israel, Jordan, Kuwait, Libya, Nigeria, North Korea, Pakistan, Papua-New Guinea, the United States of America and Vietnam.

The maintenance of such reservations today are incompatible with the obligations that many of these States Parties have entered into as States Parties to the later treaties – notably the Biological and Toxin Weapons Convention and the Chemical Weapons Convention and it is for that reason that the successive UN General Assembly resolutions call upon those States that continue to maintain reservations to lift them.

BIOLOGICAL AND TOXIN WEAPONS CONVENTION

Following World War II and the increasing attention being given to nuclear weapons, a number of States abandoned their national chemical and biological weapons programmes, preferring to concentrate on nuclear weapons as the ultimate deterrent. This led to a situation in the late 1960s when consideration was given to whether an international treaty could be agreed to strengthen the 1925 Geneva Protocol by prohibiting all chemical and biological weapons. Agreement was eventually reached on negotiating a treaty totally prohibiting the development, production, stockpiling and acquisition of biological weapons.

A draft convention was submitted by the UK in 1969 to the Conference of the Committee on Disarmament and this gained support from first the USA and then from the Soviet Union. It is, however, evident that its content was considerably diluted in a bilateral negotiations between the USA and the Soviet Union who, consciously or unconsciously, 'gutted' the draft treaty of some of its more important components (Sims, 2001). Nevertheless, this led to the Biological and Toxin Weapons Convention (BTWC) which opened for signature on 10 April 1972 and entered into force three years later on 26 March 1975.

The central prohibition of the BTWC is in Article I:

> *Each State Party to this Convention undertakes never in any circumstances to develop, produce, stockpile or otherwise acquire or retain:*
> *(1) Microbial or other biological agents, or toxins whatever their origin or method of production, of types and in quantities that have no justification for prophylactic, protective or other peaceful purposes;*
> *(2) Weapons, equipment or means of delivery designed to use such agents or toxins for hostile purposes or in armed conflict.*

The prohibition is complete as the wording used, in what is referred to as the 'General Purpose Criterion' obliges States Parties to undertake never in any circumstances to develop, produce, acquire or retain microbial or other biological agents or toxins *whatever their origin or method of production, of types and in quantities that have no justification for prophylactic, protective or other peaceful purposes.*

The comprehensiveness of this prohibition has been reaffirmed in Final Declarations agreed by the States Parties to the BTWC at successive Review Conferences at approximately five-year intervals. At successive BTWC Review Conferences, the language in the Final Declaration in the Article I section regarding scientific and technological developments has been developed.

Thus, at the **First BTWC Review Conference** in 1980, the Final Declaration[1] in this respect in Article I simply stated that:

> *The Conference believes that Article I has proved sufficiently comprehensive to have covered recent scientific and technological developments relevant to the Convention.*

Unlike subsequent BTWC Review Conferences, no language was included in respect of the implications of scientific and technological developments in regard to either apprehensions about such developments or the scope of the Convention.

By the **Second BTWC Review Conference** in 1986, the Final Declaration[2] contained stronger language and had developed into two paragraphs – one addressing apprehensions:

> *The Conference, conscious of apprehensions arising from relevant scientific and technological developments, inter alia, in the fields of* ***microbiology, genetic engineering and biotechnology****, and the possibilities of their use for purposes inconsistent with the objectives and the provisions*

[1] United Nations, *Review Conference of the Parties to the Convention on the Prohibition of the Development, Production and Stockpiling of Bacteriological (Biological) and Toxin Weapons and on their Destruction*, 3–21 March 1980, Final Declaration, BWC/CONF.1/10, Geneva, 1980 (available at http://www.opbw.org).

[2] United Nations, *Second Review Conference of the Parties to the Convention on the Prohibition of the Development, Production and Stockpiling of Bacteriological (Biological) and Toxin Weapons and on their Destruction*, 8–26 September 1986, Final Declaration, BWC/CONF.II/13, Geneva, 1986 (available at http://www.opbw.org).

> *of the Convention, reaffirms that the undertaking given by the States Parties in Article I applies to all such developments* (emphasis added);

and the other making a clear reaffirmation as to the scope of the Convention:

> *The Conference reaffirms that the Convention* ***unequivocally*** *applies to* ***all*** *natural or* ***artificially created*** *microbial or other biological agents or toxins* ***whatever their origin or method of production.*** *Consequently, toxins (both proteinaceous and non-proteinaceous) of a microbial, animal or vegetable nature and* ***their synthetically produced analogues*** *are covered* (emphasis added).

The first paragraph addressed apprehensions arising from developments *inter alia* in the fields of microbiology, genetic engineering and biotechnology and reaffirmed that the undertakings in Article I, the basic prohibition, applied to all such developments. The second paragraph addressing the scope has two sentences. The first sentence reaffirmed that the Convention unequivocally applies to all natural or artificially created agents, whatever their origin or method of production, thereby emphasizing the all-embracing scope of the Convention. The second sentence was an explanatory one that makes it clear that toxins of a microbial, animal or vegetable nature and their synthetically produced analogues are covered.

At the **Third BTWC Review Conference** in 1991 the Final Declaration[3] contained similar language, on this occasion combined into a single paragraph:

> *The Conference, conscious of apprehensions arising from relevant scientific and technological developments, inter alia, in the fields of microbiology, genetic engineering and biotechnology, and the possibilities of their use for purposes inconsistent with the objectives and the provisions of the Convention, reaffirms that the undertaking given by the States Parties in Article I applies to all such developments. The Conference also reaffirms that the Convention unequivocally covers all microbial agents or toxins, naturally or artificially created* ***or altered****, whatever their origin or method of production* (emphasis added).

The first sentence repeated the apprehensions addressed at the Second Review Conference using identical language referring to microbiology, genetic engineering and biotechnology. The second sentence reaffirmed the scope which was further extended by addition of the words 'or altered'.

At the **Fourth BTWC Review Conference** in 1996 the Final Declaration[4] was broadened, and again was in two paragraphs, one addressing apprehensions:

> *The Conference, conscious of apprehensions arising from relevant scientific and technological developments, inter alia, in the fields of microbiology, biotechnology,* ***molecular biology****, genetic engineering and* ***any application resulting from genome studies****, and the possibilities of their use for purposes inconsistent with the objectives and the provisions of the Convention, reaffirms that the undertaking given by the States Parties in Article I applies to all such developments* (emphasis added).

and the other reaffirming the scope:

> *The Conference also reaffirms that the Convention unequivocally covers all microbial or other biological agents or toxins, naturally or artificially created or altered,* ***as well as their components,*** *whatever their origin or method of production, of types and in quantities that have no justification for prophylactic, protective or other peaceful purposes* (emphasis added).

In the apprehensions paragraph, the developments to which the undertaking in Article I applied *inter alia* in the fields of *microbiology, biotechnology, molecular biology, genetic engineering and any application resulting from genome studies,* were broadened by the addition of *molecular biology* and of *any application resulting from genome studies.* The scope

[3] United Nations, *Third Review Conference of the States Parties to the Convention on the Prohibition of the Development, Production and Stockpiling of Bacteriological (Biological) and Toxin Weapons and on their Destruction*, Geneva, 9–27 September 1991, BWC/CONF.III/23, Geneva 1991 (available at http://www. opbw.org).

[4] United Nations, *Fourth Review Conference of the States Parties to the Convention on the Prohibition of the Development, Production and Stockpiling of Bacteriological (Biological) and Toxin Weapons and on their Destruction*, Geneva, 25 November – 6 December 1996, BWC/CONF.IV/9, Geneva 1996 (available at http://www.opbw.org).

paragraph was also extended from that in 1986 by the addition of '*as well as their components*'.

Analysis

The central prohibition in Article I of the BTWC enshrined in the general purpose criterion prohibits all biological and toxin agents *whatever their origin or method of production, of types and in quantities that have no justification for prophylactic, protective or other peaceful purposes.* The successive Final Declarations of the Review Conferences have reaffirmed that *the Convention unequivocally covers all naturally or artificially created or altered microbial or other biological agents or toxins as well as their components.* The prohibition is thus comprehensive and covers all biological or toxin agents, past, present and future.

It is also to be noted from a chemical viewpoint that the BTWC prohibition covers all *naturally or artificially created or altered* toxins *as well as their components.* It consequently prohibits a range of chemicals and, because the BTWC prohibits use against humans, animals and plants, it prohibits the use of such chemicals against humans, animals and plants.

CHEMICAL WEAPONS CONVENTION

Article IX of the BTWC addresses the subject of the prohibition of chemical weapons:

> *Each State Party to this Convention affirms the recognized objective of effective prohibition of chemical weapons and, to this end, undertakes to continue negotiations in good faith with a view to reaching early agreement on effective measures for the prohibition of their development, production and stockpiling and for their destruction, and on appropriate measures concerning equipment and means of delivery specifically designed for the production or use of chemical agents for weapons purposes.*

The years following the agreement of the BTWC saw continued negotiation in Geneva of a treaty to prohibit chemical weapons. This received an impetus during the 1980s when chemical weapons were used extensively during the Iran/Iraq war and led to the completion of the Chemical Weapons Convention (CWC) in the early 1990s. Unlike the earlier BTWC which was a four- or five-page document, the CWC was considerably elaborated into a treaty of 45 pages and three annexes which bring the document to over 160 pages.

The CWC was opened for signature on 13 to 15 January 1993 and entered into force on 29 April 1997, 180 days after the 65th State Party had deposited its instrument of ratification.

The central prohibition of the CWC is in the first paragraph of Article I of the Convention:

> *1. Each State Party to this Convention undertakes never under any circumstances:*
> *(a) To develop, produce, otherwise acquire, stockpile or retain chemical weapons, or transfer, directly or indirectly, chemical weapons to anyone;*
> *(b) To use chemical weapons;*
> *(c) To engage in any military preparations to use chemical weapons;*
> *(d) To assist, encourage or induce, in any way, anyone to engage in any activity prohibited to a State Party under this Convention.*

In addition, Article I includes further prohibitions:

> *2. Each State Party undertakes to destroy chemical weapons it owns or possesses, or that are located in any place under its jurisdiction or control, in accordance with the provisions of this Convention.*
> *3. Each State Party undertakes to destroy all chemical weapons it abandoned on the territory of another State Party, in accordance with the provisions of this Convention.*
> *4. Each State Party undertakes to destroy any chemical weapons production facilities it owns or possesses, or that are located in any place under its jurisdiction or control, in accordance with the provisions of this Convention.*
> *5. Each State Party undertakes not to use riot control agents as a method of warfare.*

For the purposes of the CWC, the term 'chemical weapons' is defined in Article II as follows:

> *1. 'Chemical Weapons' means the following, together or separately:*
> *(a) Toxic chemicals and their precursors, except where intended for purposes not*

prohibited under this Convention, as long as the types and quantities are consistent with such purposes;
(b) Munitions and devices, specifically designed to cause death or other harm through the toxic properties of those toxic chemicals specified in subparagraph (a), which would be released as a result of the employment of such munitions and devices;
(c) Any equipment specifically designed for use directly in connection with the employment of munitions and devices specified in subparagraph (b).

and the term 'toxic chemical' is also defined in Article II:

2. 'Toxic Chemical' means:
Any chemical which through its chemical action on life processes can cause death, temporary incapacitation or permanent harm to humans or animals. This includes all such chemicals, regardless of their origin or of their method of production, and regardless of whether they are produced in facilities, in munitions or elsewhere.

It can thus be seen, that as in the BTWC, the chemical weapons are defined in the CWC under a general purpose criterion which makes it clear that the prohibition covers *any chemical which through its chemical action on life processes can cause death, temporary incapacitation or permanent harm* and that this *includes all such chemicals, regardless of their origin or of their method of production and regardless of whether they are produced in facilities, in munitions or elsewhere.* The prohibition thus covers **all** chemical weapons, past, present and future.

Riot Control Agents

However, unlike the BTWC which prohibits all uses of biological or toxin weapons *for hostile purposes or in armed conflict*, the CWC has a different treatment in regard to riot control agents where each State Party undertakes *not to use riot control agents as a method of warfare*. This is further amplified by the definition in Article II of the CWC of the words 'purposes not prohibited under this Convention' as follows:

9. 'Purposes Not Prohibited Under this Convention' means:
(a) Industrial, agricultural, research, medical, pharmaceutical or other peaceful purposes;
(b) Protective purposes, namely those purposes directly related to protection against toxic chemicals and to protection against chemical weapons;
(c) Military purposes not connected with the use of chemical weapons and not dependent on the use of the toxic properties of chemicals as a method of warfare;
(d) Law enforcement including domestic riot control purposes.

It is thus evident that the use of riot control agents for law enforcement, including domestic riot control purposes, is not prohibited under the Convention. The CWC requires under Article III Declarations that each State Party shall submit to the Organization, not later than 30 days after the Convention enters into force for the State Party, the following declaration:

(e) With respect to riot control agents: Specify the chemical name, structural formula and Chemical Abstracts Service (CAS) registry number, if assigned, of each chemical it holds for riot control purposes. This declaration shall be updated not later than 30 days after any change becomes effective.

First Review Conference

The First Review Conference of the CWC was held in The Hague on 28 April to 9 May 2003. Although, like the BTWC, the Convention requires that such Review Conferences *shall take into account any relevant scientific and technological developments* the Political Declaration[5] agreed by the States Parties simply stated:

2. The States Parties will continue to take account of developments in science and technology in the implementation of the Convention, in accordance with its provisions.

[5] Organization for the Prohibition of Chemical Weapons, *Political Declaration as Approved by the First Special Session of the Conference of the States Parties to Review the Operation of the Chemical Weapons Convention,* Conference of the States Parties, First Review Conference 28 April – 9 May 2003 (available at http://www.opcw.org).

The Report[6] of the Review Conference provided some additional information:

> *7.23 The First Review Conference* ***considered*** *the impact of developments in science and technology on the Convention's prohibitions. The definitions contained in Article II, in particular of the terms 'chemical weapons' and 'chemical weapons production facility', were found to adequately cover these developments and to provide for the application of the Convention's prohibitions to any toxic chemical, except where such a chemical is intended for purposes not prohibited by the Convention, and as long as the types and quantities involved are consistent with such purposes. The First Review Conference* ***noted****, however, that science is rapidly advancing. New chemicals may have to be assessed in relation to their relevance to the Schedules of Chemicals of the Convention. The First Review Conference* ***requested*** *the Council to consider the developments in relation to additional chemicals that may be relevant to the Convention, and assess, inter alia, whether these compounds should be considered in the context of the Schedules of Chemicals.*

Although not reflected in the Report or the Political Declaration, the report[7] of the Scientific Advisory Board of the Organization for the Prohibition of Chemical Weapons (OPCW) prepared for the First Review Conference makes it clear that consideration was given to a number of scientific and technological developments. The Scientific Advisory Board made a number of observations:

> *(a) over recent years, many new procedures have been developed to speed up the synthesis of new chemicals required, in particular, for biological evaluation by the pharmaceutical industry. Some relevant examples of this are combinatorial chemical techniques, together with other methods for rapid synthesis and screening;*
> *(b) advances in molecular biology (such as genomics and proteomics) are creating new opportunities both to design new biologically active chemicals and to develop processes to synthesise such chemicals using enzymes or cell-based systems. The rapid pace of developments in the biomolecular sciences, coupled with advances in chemical synthesis, increase the possibility that new toxic chemicals will be found that may have properties that would make them suitable candidates for CW. However, these advances do not significantly change the situation, in view of the large numbers of already known toxic compounds, many of which are not listed in the Schedules;*
> *(c) in particular, while the time required for the early stages of agent development may have shrunk considerably as a result of these developments, the subsequent stages in the development of such a new toxic chemical into an effective CW (in the traditional meaning of the term, ...) are much less affected by these developments in science and technology;*
> *(d) these developments underline, on the other hand, the importance of the assertion that the Schedules do not constitute a definition of CW. They also suggest the need to look beyond the Schedules in the future development of the OCAD (OPCW Central Analytical Database) by proposing the addition of analytical data of new toxic compounds that are directly relevant to the Convention.*

In addition, the Scientific Advisory Board (SAB) made additional observations:

> *3.13 Notwithstanding these scientific developments and the SAB's conclusion that they do not significantly change the situation with respect to CW, it must also be noted that experience has shown that there is a possibility that less sophisticated CW may be opted for, with little regard to agent stability and shelf life (i.e. weapons filled for immediate use). In that context, even toxic chemicals (as well as precursor chemicals) that would not normally be considered to pose a risk to the Convention may be relevant. The same would apply to terrorists using toxic materials as weapons.*
> *3.14 The SAB was also aware of concerns about the development of new riot control agents (RCAs), and other so-called 'non-lethal'*

[6] Organization for the Prohibition of Chemical Weapons, *Report of the First Special Session of the Conference of the States Parties to Review the Operation of the Chemical Weapons Convention (First Review Conference) 28 April – 9 May 2003*, Conference of the States Parties, First Review Conference 28 April – 9 May 2003, RC-1/5, 9 May 2003 (available at http://www.opcw.org).

[7] Organization for the Prohibition of Chemical Weapons, *Note by the Director-General: Report of the Scientific Advisory Board on Developments in Science and Technology*, Conference of the States Parties, First Review Conference 28 April – 9 May 2003, RC-1/DG.2, 23 April 2003 (available at http://www.opcw.org).

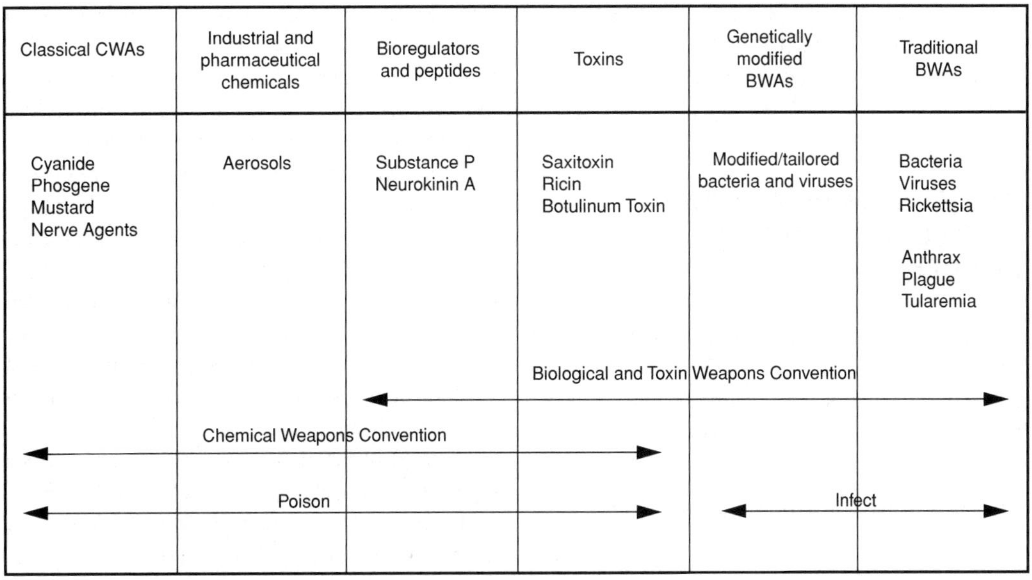

Figure 1. The 'spectrum' of chemical and biological agents

weapons utilising certain toxic chemicals, such as incapacitants, calmatives, vomiting agents, and the like. There are specific provisions in the Convention that deal with RCAs and the legitimate use of toxic chemicals for law enforcement purposes. The SAB noted that the science related to such agents is rapidly evolving, and that results of current programmes to develop such 'non-lethal' agents should be monitored and assessed in terms of their relevance to the Convention. However, based on past experience and the fact that many of these compounds act on the central nervous system, it appears unlikely from a scientific point of view that compounds with a sufficient safety ratio would be found.

The SAB rightly concluded by underlining the importance of the general purpose criterion which embraces **all** chemicals in their remarks that:

> *3.15 The SAB stressed the importance that* ***all new toxic chemicals****, no matter what their origin or method of synthesis,* ***are covered by the Convention's definition of CW****, unless they were intended for purposes not prohibited by it, and only as long as their types and quantities would be consistent with these purposes. The SAB underlined the importance of this aspect of the definition of CW as a safeguard for the validity of the Convention* (emphasis added).

Analysis

The general purpose criterion of the Chemical Weapons Convention makes it clear that all toxic chemicals are prohibited as this *includes all such chemicals, regardless of their origin or of their method of production, and regardless of whether they are produced in facilities, in munitions or elsewhere.* It is useful to consider the range of potential chemical and biological agents, which can be regarded as a 'spectrum' (Figure 1).

This figure shows that the materials prohibited under the two Conventions – the CWC and the BTWC – rightly overlap in regard to toxins as well as in the area of bioregulators and peptides with the CWC listing two toxins – ricin and saxitoxin – in Schedule 1. Furthermore, **both** Conventions address dual-use materials and technology, **both** totally prohibit a class of weapons which cause death or harm to humans and animals primarily through inhalation or ingestion and **both** have general purpose criteria in their basic prohibition.

There is consequently complete prohibition of **all** chemical and biological materials whatever their origin or method of production, of types and in quantities that have no justification for prophylactic, protective or other peaceful purposes. It is important to recognize that the prohibition is universal and embraces **all** toxic chemicals and is **not** limited to those that have previously been identified or used as potential chemical warfare agents.

IMPLEMENTATION OF THE CHEMICAL WEAPONS CONVENTION

As this book is focused on chemical warfare agents, it is necessary to consider the implementation of the prohibitions enshrined in the Chemical Weapons Convention (CWC).

National Implementing Legislation

Article VII of the CWC sets out the obligations for each State Party in regard to national implementation measures. The first three paragraphs of Article VII state the general undertakings:

> *1. Each State Party shall, in accordance with its constitutional processes, adopt the necessary measures to implement its obligations under this Convention. In particular, it shall:*
>
> *(a) Prohibit natural and legal persons anywhere on its territory or in any other place under its jurisdiction as recognized by international law from undertaking any activity prohibited to a State Party under this Convention, including enacting penal legislation with respect to such activity;*
>
> *(b) Not permit in any place under its control any activity prohibited to a State Party under this Convention;*
>
> *(c) Extend its penal legislation enacted under subparagraph (a) to any activity prohibited to a State Party under this Convention undertaken anywhere by natural persons, possessing its nationality, in conformity with international law.*
>
> *2. Each State Party shall cooperate with other States Parties and afford the appropriate form of legal assistance to facilitate the implementation of the obligations under paragraph 1.*
>
> *3. Each State Party, during the implementation of its obligations under this Convention, shall assign the highest priority to ensuring the safety of people and to protecting the environment, and shall cooperate as appropriate with other States Parties in this regard.*

Each State Party is thus obliged to implement penal legislation to prohibit any persons within the State or under its jurisdiction from undertaking any activity prohibited under the Convention. It is this requirement that prohibits any use of chemical weapons by sub-national groups or sub-State actors such as terrorist groups. Although not all of the States Parties to the CWC have yet implemented effective national penal legislation, the States Parties to the Convention recognized this shortcoming at the First Review Conference in 2003 and initiated an Action Plan aimed at all States Parties enacting the necessary national implementation legislation.

In addition, Article VII sets out obligations regarding the relationship between the State Party and the Organization for the Prohibition of Chemical Weapons. This includes the requirement to designate or establish a National Authority to serve as the national focal point for effective liaison with the Organization and other States Parties.

Scheduled and other Chemicals

Another key element in regard to national implementation is set out in Article VI which addresses Activities Not Prohibited under this Convention. This starts by setting out the right of States Parties to use toxic chemicals *for purposes not prohibited under the Convention.*

> *1. Each State Party has the right, subject to the provisions of this Convention, to develop, produce, otherwise acquire, retain, transfer and use toxic chemicals and their precursors for purposes not prohibited under this Convention.*

The Article then goes on to place the obligation on each State Party to adopt the necessary measures to ensure that toxic chemicals are only used *for purposes not prohibited under this Convention.*

> *2. Each State Party shall adopt the necessary measures to ensure that toxic chemicals and their precursors are only developed, produced, otherwise acquired, retained, transferred, or used within its territory or in any other place under its jurisdiction or control for purposes not prohibited under this Convention.*

In addition, it is required that:

> *To this end, and in order to verify that activities are in accordance with obligations under this Convention, each State Party shall subject toxic chemicals and their precursors listed in Schedules 1, 2 and 3 of the Annex on Chemicals, facilities related to such chemicals, and other facilities as specified in the Verification Annex, that are located on its territory or in any other place under its jurisdiction or control, to verification measures as provided in the Verification Annex.*

It is important to underline the fact that the obligation placed upon each State Party to adopt necessary measures applies to **all** toxic chemicals and their precursors. In **addition**, Article VI requires that verification measures as in the Verification Annex shall be applied to a particular subset of toxic chemicals and their precursors, namely those listed in the Schedules 1, 2 and 3 of the Annex on Chemicals to the CWC.

The 'Scheduled Chemicals' are primarily chemicals which have been used at some time or have been considered for use at some time as chemical weapons. Particular attention is given in the CWC to the verification of such chemicals. However, the designation of some chemicals in the Schedules is merely an aid to the verification of the Convention. It has to be underlined that the prohibitions in the Convention apply to **all** toxic chemicals as defined in Article II of the Convention and that each State Party is required to adopt the necessary measures to **ensure** that *toxic chemicals and their precursors are only developed, produced, otherwise acquired, retained, transferred, or used . . . for purposes not prohibited under this Convention.* Consequently the prohibitions are all embracing and the national measures to ensure that toxic chemicals are only used for purposes not prohibited under the Convention are equally all embracing. Chemical weapons are truly totally prohibited.

SCHEDULE 1 CHEMICALS

The Annex on Chemicals to the CWC sets out the guidelines for the different Schedules. Those for Schedule 1 are that the following criteria shall be taken into account in considering whether a toxic chemical or precursor shall be included:

> *(a) It has been developed, produced, stockpiled or used as a chemical weapon as defined in Article II;*
> *(b) It poses otherwise a high risk to the object and purpose of this Convention by virtue of its high potential for use in activities prohibited under this Convention because one or more of the following conditions are met:*
> *(i) It possesses a chemical structure closely related to that of other toxic chemicals listed in Schedule 1, and has, or can be expected to have, comparable properties;*
> *(ii) It possesses such lethal or incapacitating toxicity as well as other properties that would enable it to be used as a chemical weapon;*
> *(iii) It may be used as a precursor in the final single technological stage of production of a toxic chemical listed in Schedule 1, regardless of whether this stage takes place in facilities, in munitions or elsewhere;*
> *(c) It has little or no use for purposes not prohibited under this Convention.*

Schedule 1 includes the following toxic chemicals – the chemical classes for nerve agents which include GB (sarin), GD (soman), GA (tabun) and VX, a number of sulfur mustards, three lewisites, three nitrogen mustards, saxitoxin, ricin and a number of precursors.

SCHEDULE 2 CHEMICALS

The guidelines for Schedule 2 are that the following criteria shall be taken into account in considering whether a toxic chemical or precursor shall be included:

> *(a) It poses a significant risk to the object and purpose of this Convention because it possesses such lethal or incapacitating toxicity as well as other properties that could enable it to be used as a chemical weapon;*
> *(b) It may be used as a precursor in one of the chemical reactions at the final stage of formation of a chemical listed in Schedule 1 or Schedule 2, part A;*
> *(c) It poses a significant risk to the object and purpose of this Convention by virtue of its importance in the production of a chemical listed in Schedule 1 or Schedule 2, part A;*
> *(d) It is not produced in large commercial quantities for purposes not prohibited under this Convention.*

Schedule 2 includes the following toxic chemicals – amiton, PFIB (perfluoroisobutene) and BZ (3-quinuclidinyl benzilate) and a number of precursors.

SCHEDULE 3 CHEMICALS

The guidelines for Schedule 3 are that the following criteria shall be taken into account in considering whether a toxic chemical or precursor shall be included:

> *(a) It has been produced, stockpiled or used as a chemical weapon;*
> *(b) It poses otherwise a risk to the object and purpose of this Convention because it possesses such lethal or incapacitating toxicity as well as other properties that might enable it to be used as a chemical weapon;*
> *(c) It poses a risk to the object and purpose of this Convention by virtue of its importance in the production of one or more chemicals listed in Schedule 1 or Schedule 2, part B;*
> *(d) It may be produced in large commercial quantities for purposes not prohibited under this Convention.*

Schedule 3 includes the following toxic chemicals – phosgene, cyanogen chloride, hydrogen cyanide and chloropicrin and a number of precursors.

Implementation of the General Purpose Criterion

As already noted, a central provision of the Chemical Weapons Convention (CWC) is the general purpose criterion which prohibits *'Toxic chemicals and their precursors, except where intended for purposes not prohibited under this Convention, as long as types and quantities are consistent with such purposes'*. The implementation of this general purpose criterion is placed by Article VI on each State Party which *'shall adopt the necessary measures to ensure that toxic chemicals and their precursors are only developed, produced, otherwise acquired, retained, transferred, or used within its territory or in any other place under its jurisdiction or control for purposes not prohibited under this Convention'*.

During the first few years after the CWC entered into force, for quite understandable reasons, the OPCW and the States Parties have focussed correctly first on the destruction of chemical weapons and of chemical weapon production facilities and then on the verification of Scheduled chemical facilities. The Verification Annex to the CWC in Part IX states that the implementation of verification of the regime for other chemical production facilities – those producing more than 200 tonnes of unscheduled discrete organic chemicals or more than 30 tonnes of an unscheduled discrete organic chemical containing the elements phosphorus, sulphur or fluorine – shall start at the beginning of the fourth year after entry into force of the Convention. As the Convention entered into force on 29 April 1997, the start of the fourth year was in 2001.

Although the importance of implementing the general purpose criterion has been recognized by analysts of the CWC and the OPCW, little attention has yet been given to how this might be achieved. As Julian Perry Robinson pointed out[8] in July 2000, *'the OPCW Technical Secretariat is sighted only towards those 29 chemicals and 14 families of chemicals that are listed in the CWC Annex on Chemicals'* and *'It is the National Authorities therefore, not the OPCW Technical Secretariat, that are primarily responsible for implementing the general purpose criterion which ... is absolutely vital to the future of the treaty'*. It is encouraging to note that the 1999 Annual Report[9] by the UK National Authority includes mention of the application of the general purpose criterion and concludes that *'National authorities need to consider this situation further'*.

The UK National Authority 2001 Annual Report[10] states that a workshop had been held at which it was agreed that *'it was essential for States Parties to recognise the importance of the*

[8] Julian Perry Robinson, *Memorandum submitted by Professor J P Perry Robinson, University of Sussex*, Foreign Affairs Committee, Eighth Report, *Weapons of Mass Destruction*, 25 July 2000, Appendix 29, p. 203.

[9] Department of Trade and Industry, *1999 Annual Report on the operation of the Chemical Weapons Act 1996*, DTI/Pub 4913/2k/6/00/NP, June 2000.

[10] Department of Trade and Industry, *Annual Report for 2001 on the operation of the Chemical Weapons Act 1996* (available at http://www2.dti.gov.uk/nonproliferation/cwcna/2001-report.pdf).

GPC [general purpose criterion] *in requiring that* ***any*** *toxic chemical can only be used for permitted purposes'* and that *'this stipulation does not apply only to the chemicals listed in the Schedules'*. It goes on to say that States Parties should review their activities to implement the general purpose criterion in the wake of the terrorist attacks on 11 September, 2001.

The most recent UK National Authority Annual Report[11] for 2003 has a section entitled 'The General Purpose Criterion'. This recognizes that the negotiators of the Convention realized that *developments in science and technology could lead to the discovery of chemicals that, while not subject to verification under the CWC, could undermine its object and purpose*. The section goes on to note that *the CWC is not self-implementing and requires each State Party to ensure that its obligations have direct legal effect. The treaty requires that national legislation is enacted to cover all activities prohibited under the Convention involving the use of toxic chemicals and their precursors*. The report notes that in order to meet its obligations under the CWC, the National Authority has been encouraged to consider the overall regulation of chemicals within the UK and that a meeting had been held of all the relevant authorities and agencies to discuss the regulatory framework of chemicals.

In the following, an analysis is made first as to why the general purpose criterion should be seen as being of increasing importance and then of some current international initiatives that are addressing chemicals that are of potential risk to public health or to the environment in order to explore how these initiatives might be harnessed to implement the CWC general purpose criterion.

THE IMPORTANCE OF THE GENERAL PURPOSE CRITERION

It was noted above that the General Purpose Criterion is clearly recognized as being central to the health of the Convention and that it is incorrect to have perceptions that there are gaps in regard to chemicals such as novichoks (a binary nerve agent), etc. The point is that the prohibitions and definitions in Articles I and II of the Convention are **all** embracing and that the lists of chemicals making up the Schedules were never intended to be, and never could be, comprehensive. The Convention totally prohibits the development, production, acquisition, stockpiling retention or use of chemical weapons *'under any circumstances.'*

The Schedules to the Convention were essentially finalized nearly 20 years ago and understandably focused on those chemicals widely known then as having been used or developed for use as chemical weapons. In order to create a somewhat wider safety net, chemicals belonging to the same classes as the known chemical weapon agents were also included in the Schedules – sometimes referred to as 'families'. Since then there has been a greater appreciation that the risks to the Convention are posed by chemicals from a spectrum of potential agents.

The mid-spectrum region between chemical and biological agents includes substances such as bioregulators and toxins. These are all chemicals and almost all are not included in the Schedules – the two that are listed in Schedule 1 are ricin and saxitoxin. These mid-spectrum materials can now be readily produced in quantity – and for prohibited purposes, impurities are not a problem. The challenge to the Convention posed by such materials is further increased by the recent advances in drug-delivery techniques.

It is concluded that there are increasing risks from unscheduled chemicals as there are significant advances in technology and in biotechnology, and there were already a range of known unscheduled chemicals such as intermediate volatilitity agents (IVAs) or novichoks, mid-spectrum materials such as bioregulators, and calmatives, even though the General Purpose Criterion ensured that all such chemicals are embraced by the prohibitions of the CWC. The question can be debated as to how likely is it that Scheduled Chemicals would be chosen for 'break-out' from the Convention with the perception being that this was becoming less likely. Consequently, the First Review Conference of the CWC needed to be alert to the dangers which

[11] Department of Trade and Industry, *Annual Report 2003 Operations of the Chemical Weapons Act 1996* (available at http://www2.dti.gov.uk/non-proliferation/cwcna/2003-report.pdf).

might already be present but would certainly be there in future.

Although the General Purpose Criterion is the heart and soul of the Convention which provides the best protection against new agents in respect both of the Convention and in providing protection against chemical terrorism, the situation in regard to the enactment nationally by States Parties of the essential overarching penal legislation to implement the general purpose criterion and the Convention is grave. Attention needs to be given by States Parties to how best to implement nationally the general purpose criterion in a way that demonstrates both nationally and to the other States Parties that the State Party has an effective arrangement in place to build confidence that chemicals are not being developed, produced or used for prohibited purposes.

INTERNATIONAL INITIATIVES

As was noted in an article by Pearson (2000), the world growth in trade in the 1960s and 1970s has led to increasing attention being given to the potential risks to the environment and to public health from chemicals. Consequently, there are a number of international, regional and national initiatives that are addressing chemical safety and the potential risks to the environment and/or to the health of the general public or workers.

The international initiatives primarily arise from the United Nations promotion of global cooperation on issues relating to the environment and public health. There are several international treaties relating to the control of chemicals:

- *Stockholm Convention on Persistent Organic Pollutants (POPs).*[12] POPs are chemicals which remain in the environment for lengthy periods without degrading and thus cause damage when they arrive in the environment.
- *Montreal Protocol on Substances that Deplete the Ozone Layer.*[13] This protocol has introduced measures to restrict the production and use of chemicals that damage the ozone layer.
- *Rotterdam Convention on the Prior Informed Consent Procedure for Certain Hazardous Chemicals and Pesticides in International Trade.*[14] This requires the provision of information from risk assessments to importing nations prior to the import of such chemicals.
- *Basel Convention on the Control of Transboundary Movements of Hazardous Wastes and their Disposal.*[15] This requires that transboundary movements of hazardous wastes be reduced to a minimum at that such wastes should be disposed of as close as possible to their source of generation.

These Conventions have largely arisen from a number of initiatives in relation to chemicals undertaken over the years by the United Nations Environment Programme (UNEP). A discussion paper[16] prepared for the meeting of the Governing Council of UNEP on 3 to 7 February 2003 stated:

> I. *Chemicals Challenge*
> *2. Chemicals are essential for development and everyday life. Modern fertilizers and pesticides have been a boon to agriculture and helped us feed our growing populations. Chemicals have served medicine in many ways, ranging from pharmaceuticals to the equipment and materials used in hospitals. From transportation through information technology to entertainment – our quality of life would not be the same today without a healthy chemicals and manufacturing industry. Today, the pace of growth in the global chemicals industry is astonishing. There are some 70,000 different chemicals on the market with 1500 new ones introduced every year.*

[12] Stockholm Convention on Persistent Organic Pollutants (available at http://www.pops.int).

[13] Montreal Protocol on Substances that Deplete the Ozone Layer (available at http://www.unep.ch/ozone/montreal.shtml).

[14] Rotterdam Convention on the Prior Informed Consent Procedure for Certain Hazardous Chemicals and Pesticides in International Trade (available at http://www.pic.int/en/ViewPage.asp?id=101).

[15] Basel Convention on the Control of Transboundary Movements of Hazardous Wastes and their Disposal (available at http://www.basel.int).

[16] United Nations, Governing Council of the United Nations Environment Programme, *Background Paper for Consideration by the Plenary: State of the Environment, The chemicals work of the United Nations Environment Programme,* UNEP/GC.22/10/Add.1, 20 December 2002 (available at http://www.unep.org/gc/gc22/documents.asp).

3. As we have come to learn, however, chemicals are not all benign. Some chemicals have been implicated in various disorders and diseases, including cancer, reproductive disorders and failures, birth defects, neurobehavioural disorders and impaired immune functions. Many thousands of cases of accidental poisoning result from the inappropriate use of highly toxic pesticide formulations, or their use in locations where protective equipment is unavailable or unused. Chemicals deplete the ozone layer, cause climate change and affect biodiversity. They accumulate in poorly managed stockpiles and waste sites. Many persist in the environment and bioaccumulate, leading to ever increasing levels in humans and wildlife. These are just some of the effects we know – there are not enough data on most of the chemicals in use today to understand their risks. Furthermore, basic protection measures for consumers, workers and the environment are often lacking. Increasingly, the manufacture of chemicals is shifting from developed to developing countries where the capacity to provide such protection is limited.

4. Increasing globalization and the enormous market for chemicals and the products they are used means that chemical safety programmes must be strengthened and steps must be taken to integrate these programmes wisely into sustainable development. The Plan of Implementation of the World Summit on Sustainable Development,[17] *and the response by UNEP is a step in this direction.*

The UNEP chemicals programme has as its goal the making of the world to be a safer place from toxic chemicals. This is done by helping governments to take necessary global action for the sound management of chemicals, by promoting the exchange of information on chemicals and by helping to build the capacities of countries around the world to use chemicals safely. The UNEP programme takes forward several of the programmes identified at the 1992 United Nations Conference on Environment and Development (sometimes referred to as the Earth Summit, 1992) held in Rio de Janeiro which adopted Agenda 21 which in Chapter 19 identified six programme areas for the environmentally sound management of chemicals:

(a) expanding and accelerating the international assessment of chemical risk;
(b) harmonization of classification and labelling of chemicals;
(c) information exchange on chemicals and chemical risks;
(d) establishment of risk reduction programmes;
(e) strengthening of national capabilities and capacities for management of chemicals;
(f) prevention of illegal international traffic in toxic and dangerous products.

While most chemicals are benign in the concentration levels to which we are exposed to them, others present risks to human health or to the environment. Sustainable development requires the global capacity for the sound management of chemicals. National capacities exist within most developed countries, but to a more limited extent elsewhere. One aim in building global capacity is to extend the sound management of chemicals to all countries – that is, to take steps to ensure that all countries have the information necessary, expertise and resources to manage chemicals safely under the conditions of production or use in that country. A second aim of global capacity is ensuring that the necessary global actions are taken to address risks that are not dealt with by national actions alone.

Expanding access to information and information tools is one of the primary ways in which UNEP helps countries to develop their capabilities in assessing and managing chemical risks. A wide range of information products have been issued by UNEP Chemicals, such as the International Register of Potentially Toxic Chemicals (IRPTC), often with partner organizations such as the International Programme on Chemical Safety (IPCS), the Inter-Organizational Programme on the Sound Management of Chemicals (IOMC), the Organization for Economic Co-operation and Development (OECD) and the Intergovernmental Forum on Chemical Safety (IFCS). The IFCS

[17] Report of the World Summit on Sustainable Development, Johannesburg, South Africa, 26 August–4 September 2002, A/CONF.199/20, Chapter I, resolution 2, annex (available at http://www.johannesburgsummit.org/documents/summit_docs/131302_wssd_report_reissued.pdf).

was established[18] in 1994 as *'a non-institutional arrangement whereby representatives of governments meet to consider and to provide advice and where appropriate, make recommendations to governments, international organizations, intergovernmental bodies and non-governmental organizations involved in chemical safety on aspects of chemical risk assessment and environmentally sound management of chemicals'*. Its aim is the integration and consolidation of national and international efforts to promote chemical safety. The representatives of governments have the right to vote, whereas intergovernmental and non-governmental organizations participate without the right to vote. The IFCS secretariat is located in Geneva within the WHO.

The IFCS at its Forum III in Bahia, Brazil in October 2000 adopted the Bahia Declaration on Chemical Safety[19] which called for the promotion of *'global cooperation for chemicals management; for pollution prevention; for sustainable agriculture; and for cleaner processes, materials and products'*, and for ensuring that *'all countries have the capacity for sound management of chemicals, particularly through coordinated national policies, legislation and infrastructure'*. It also recognized that *'Many countries are still struggling to establish the essential infrastructure for chemical safety'* and that *'Standards of chemical safety across much of the world fall short of that needed to provide adequate protection of human health and the environment'*. The Declaration calls for the ratification and implementation of *'chemicals conventions and agreements and ensuring efficient and effective co-ordination between all chemical safety-related organizations and activities'*, as well as the promotion of *'the entry into force at the earliest possible time of international treaties and agreements concerning chemical safety that are under negotiation or not yet in operation'*.

[18] International Conference on Chemical Safety, *Resolution on the Establishment of an Intergovernmental Forum on Government Safety*, Stockholm, 25–29 April 1994, IPCS/IFGC/94.Res.1, 29 April 1994 (available at http://www.who.int/ifcs/fs_res1.htm).

[19] Intergovernmental Forum on Chemical Safety, *Bahia Declaration on Chemical Safety*, Third Session – Forum III Final Report, IPCS/FORUM III/23w (available at http://www.who.int/ifcs/forum3/final.htm).

The Forum III also agreed Priorities for Action beyond 2000[20] which set out a structured approach with dated 'milestones' to achieve the objectives of the Bahia Declaration.

A decision[21] calling for a strategic approach to international chemicals management was adopted by the UNEP Governing Council at its Seventh Special Session in February 2002. This noted the steps being taken to implement the Stockholm Convention on Persistent Organic Pollutants, the Rotterdam Convention on the Prior Informed Consent Procedure and the Basel Convention on the Control of Transboundary Movements of hazardous Wastes and their Disposal and decided that *'there is a need to further develop a strategic approach to international chemicals management and endorses the Intergovernmental Forum on Chemical Safety Bahia Declaration and Priorities for Action beyond 2000 as the foundation of this approach'*. The decision goes on to underline *'that the strategic approach to chemicals management should promote the incorporation of chemical safety issues into the development agenda and identify concrete proposals for strengthening capacity for the sound management of chemicals and the related technologies of all countries, taking into account the vast difference in capabilities between developed and developing countries in this field'*. This initiative was subsequently endorsed[22] by the World Summit on Sustainable Development in Johannesburg on 26 August to 4 September 2002. The Summit agreed to:

> *'23. Renew the commitment, as advanced in Agenda 21 to sound management of chemicals throughout their life cycle and of hazardous wastes, for sustainable development as well as*

[20] Intergovernmental Forum on Chemical Safety, *Annex 6 Priorities for Action beyond 2000*, Third Session – Forum III Final Report, IPCS/FORUM III/23w, Annex 6 (available at http://www.who.int/ifcs/forum3/final.htm).

[21] UNEP Governing Council, *Decision SS.VII/3 Strategic approach to international chemicals management*, Seventh Special Session, February 2002 (available at http://www.chem.unep.ch/irptc/strategy/default.htm).

[22] United Nations, *Report of the World Summit on Sustainable Development*, Johannesburg, South Africa, 26 August–4 September 2002. A/CONF.199/20, Resolution 2, Annex, pp. 19-20 (available at http://www.johannesburgsummit.org/html/documents/summit_docs/131302_wssd_report_reissued.pdf).

for the protection of human health and the environment, inter alia, aiming to achieve, by 2020, that chemicals are used and produced in ways that lead to the minimization of significant adverse effects on human health and the environment, using transparent science-based risk management procedures, taking into account the precautionary approach, as set out in principle 15 of the Rio Declaration on Environment and Development, and support developing countries in strengthening their capacity for the sound management of chemicals and hazardous wastes by providing technical and financial assistance. This would include actions at all levels to:

(a) Promote the ratification and implementation of relevant international instruments on chemicals and hazardous waste, including the Rotterdam Convention on the Prior Informed Consent Procedure for Certain Hazardous Chemicals and Pesticides in International Trade so that it can enter into force by 2003 and the Stockholm Convention on Persistent Organic Pollutants so that it can enter into force by 2004, and encourage and improve coordination as well as supporting developing countries in their implementation;

(b) Further develop a strategic approach to international chemicals management based on the Bahia Declaration and Priorities for Action beyond 2000 of the Intergovernmental Forum on Chemical Safety by 2005, and urge that the United Nations Environment Programme, the Intergovernmental Forum, other international organizations dealing with chemical management and other relevant international organizations and actors closely cooperate in this regard, as appropriate;

(c) Encourage countries to implement the new globally harmonized system of classification and labelling of chemicals[23] *as soon as possible with a view to having the system fully operational by 2008;*

(d) Encourage partnerships to promote activities aimed at enhancing environmentally sound management of chemicals and hazardous wastes, implementing multilateral environment agreements, raising awareness of issues relating to chemicals and hazardous waste and encouraging the collection and use of additional scientific data;

(e) Promote efforts to prevent international illegal trafficking of hazardous chemicals and hazardous wastes and to prevent damage resulting from the transboundary movement and disposal of hazardous wastes in a manner consistent with obligations under relevant international instruments, such as the Basel Convention on the Control of Transboundary Movements of Hazardous Wastes and their Disposal; ....'.

It is thus evident that there is a significant international effort addressing safety in chemicals which is seeking to promote the incorporation of the safe use of chemicals into the development agenda in countries around the world. This effort is essentially focused on capacity-building programmes in chemical safety for developing countries and countries with economies in transition.

OTHER INTERNATIONAL INITIATIVES

Organization of Economic Co-operation and Development (OECD)

The 30-nation[24] OECD in 1991 adopted a Council decision/recommendation[25] *considering that strengthened national and co-operative international efforts to investigate systematically and reduce the risks of hazardous existing chemicals will substantially alleviate threats of serious or irreversible damage to the environment and/or the health of the general public or workers ...DECIDES that Member countries shall co-operatively investigate high production volume (HPV) chemicals in order to identify those which*

[23] Work on the globally harmonized system of classification and labelling of chemicals is being led by the International Labour Organization together with the United Nations Institute for Training and Research.

[24] The 30 member countries of the OECD are Australia, Austria, Belgium, Canada, Czech Republic, Denmark, European Communities, Finland, France, Germany, Greece, Hungary, Iceland, Ireland, Italy, Japan, Korea, Luxembourg, Mexico, The Netherlands, New Zealand, Norway, Poland, Portugal, Slovakia, Spain, Sweden, Switzerland, Turkey, United Kingdom and the United States.

[25] OECD, *Decision-Recommendation of the Council on the Co-operative Investigation and Risk Reduction of Existing Chemicals*, C(90)163/Final, 31 January 1991 (available at http://www.oecd.org/ehs/CA90163.HTM).

are potentially hazardous to the environment and/or to the health of the general public or workers. In addition, the decision-recommendation *DECIDES that Member countries shall establish or strengthen national programmes aimed at the reduction of risk from existing chemicals to the environment and/or the health of the general public or workers* and *RECOMMENDS that, where appropriate, Member countries undertake concerted activities to reduce the risks of selected chemicals, taking into account the entire life cycle of the chemicals. These activities could encompass both regulatory and non-regulatory measures including the following: the promotion of the use of cleaner products and technologies; emission inventories; product labelling; use limitations; economic incentives; the phase-out or banning of chemicals*. The decision-recommendation also *INVITES the Secretary-General to take the necessary steps to ensure that this work is carried out in co-operation with other international organizations and, in particular, in collaboration with the UNEP/IRPTC and the IPCS*.

In order to make this task manageable, the OECD decided to concentrate on high production volume (HPV) chemicals – these are chemicals being produced or imported at levels greater than 1000 tonnes per year in at least one OECD country. The chemicals are listed in an OECD list of high production volume chemicals[26] which currently includes 5235 substances. In addition, the OECD has agreed a minimum set of data in order to determine its potential hazard – the Screening Information Data Set (SIDS).[27] This enables resources to be concentrated on carrying out further work on chemicals of concern.

Using the data from the SIDS, mainly provided by co-operation with the chemical industry, OECD Member countries prepare a SIDS Initial Assessment Report (SIAR) which highlights any potential risk and contains recommendations for further action, if any, on the chemical. The SIAR is discussed at a meeting (SIDS Initial Assessment Meeting (SIAM)) of experts from all Member countries, from other international organizations and from non-member countries, as nominated by the United Nations' IPCS, as well as representatives of the manufacturing companies. The SIAR, amended as appropriate, is made available worldwide by posting on the Internet and by provision to UNEP Chemicals for inclusion in their database and publication as a contribution to the Inter-Organizational Programme on the Sound Management of Chemicals (IOMC). The OECD HPV Chemicals programme was refocused in 1998 in order to increase transparency, efficiency and productivity. The current aim is to complete SIDS testing for the first tranche of some 1000 chemicals on the HPV list – which contains 5235 chemicals – by the end of 2004.

[26] The first OECD HPV chemicals list was compiled in 2000, containing 5235 substances and is based on the submissions of nine national inventories and that of the European Union. A further list was compiled in 2003 (available at http://www.oecd.org/EN/document/0,,EN-document-525-14-no-1-9998-0,00.html). The OECD Integrated HPV Database is available at the same website.

[27] Information on the SIDS and the evaluation and assessment process is provided in the Manual for the Investigation of HPV Chemicals (available at http://www.oecd.org/EN/document/0,,EN-document-525-nodirectorate-no-5-33255-12,00.html).

United Nations Institute for Training and Research (UNITAR)

The United Nations Institute for Training and Research (UNITAR) has mounted training and capacity building programmes in chemicals and waste management (CWM) to support developing countries and countries in economic transition in their efforts to ensure that dangerous chemicals and waste are handled safely without causing harm to human health and the environment. These programmes are closely linked to Chapter 19 of Agenda 21 and to related recommendations of the IFCS. The CWM programmes are implemented through partnerships with the participating organizations of IOMC and are funded through extra-budgetary funds provided by Member States and international organizations. Programmes have been supported by the governments of Australia, Austria, Canada, Denmark, Germany, The Netherlands, Switzerland and the United States, the European Commission, UNEP Chemicals and the Food and Agricultural Organization of the United Nations, with core funding being provided by Switzerland and the Netherlands. A particularly important element of the CWM programmes is in the elaboration of national profiles to indicate current capabilities and capacities for management of chemicals and the specific needs for improvements. As of August 2002, some 92 countries,[28] including several OECD Member States, have prepared or are preparing a National Profile following the guidelines laid down in the UNITAR/IOMC National Profile Guidance Document.[29] National Profiles are available on the internet for 45 countries[30] at a UNITAR/European Chemicals Bureau website.

In addition, UNITAR have organized a number of thematic workshops on priority topics of national chemicals management capacity building. Of particular interest, is one on 'Developing and Strengthening National Legislation and Policies for the Sound Management of Chemicals' which was held in Geneva in 1999 with funding provided by Switzerland and the Technical Secretariat of the OPCW. The Final Report[31] of this Workshop in its record of the perspective of international organization includes that the OPCW *'noted the close linkage of implementing obligations under the Chemical Weapons Convention and the general infrastructure for the management of chemicals, such as import and export control mechanisms, licensing data reporting, laboratory capacities'*. The workshop concluded that unacceptable risks to health, safety and environmental quality continually primarily because existing laws and regulations are fragmented across sectoral boundaries with no unifying policy mechanisms and governments are therefore urged to review their chemical legislation, including regulations and regulatory structures, to ensure that they efficiently and effectively promote the sound management of chemicals.

[28] Detailed information on the current status of the national profile for each country is available at: (http://www.unitar.org/cwm/homepage/a/np/globalstatus/frglobal-header.htm).

[29] UNITAR/IOMC, *Preparing a National Profile to Assess the National Infrastructure for Management of Chemicals: A Guidance Document,* 1996 (available at http://www.unitar.org/cwm/homepage/a/np/npdoc/index.htm).

[30] Available at: (http://www.unitar.org/cwm/nationalprofiles/English/national.htm).

[31] UNITAR/IOMC/IFCS, *Developing and Strengthening National Legislation and Policies for the Sound Management of Chemicals,* Observations and Conclusions of an International Expert Meeting, Geneva, Switzerland, 22–25 June 1999, Final Report (available at http://www.unitar.org/cwm/publications/pdf/tw3_(22_jan_02).PDF).

International Council of Chemical Associations (ICCA) Global Initiative on HPV Chemicals

The global chemical industry launched a global Initiative on High Production Volume (HPV) chemicals[32] on 3 October 1998 at the meeting of the Board of Directors of the ICCA. The goal of this initiative is to prepare harmonized, internationally agreed datasets and initial hazard assessments under the SIDS programme of the OECD. The key element of the ICCA initiative is the improvement of the current database of approximately 1000 OECD HPV chemicals based on information gathering and where necessary additional testing by the end of 2004. The ICCA HPV Working List, as of July 2002, lists 1257 chemicals[33] and shows which substances have already been assessed by the SIDS Initial Assessment Meeting (SIAM).

European Union

The European Union (EU) had identified the potential risks of chemicals as a policy priority in the 1970s and the 1980s which saw the drawing up of EINECS (European INventory of Existing Commercial Substances) which lists and defines some 100 000 chemicals which were deemed to be on the European Union market between 1 January 1971 and 18 September 1981; EINECS is an inventory containing 100 195 substances. Any new chemicals subsequently brought onto the market are included in ELINCS (European LIst of New Chemical Substances); this currently comprises some 4000 notifications in total, representing about 2000 substances, which have been notified since 1981, corresponding to about 400 notifications each year. The Fourth Community Action Programme on the Environment (1987–1992) underlined the need for a legislative instrument, which would provide a comprehensive structure for the evaluation of the risks posed by 'existing chemicals'. The development of the legal instruments in the European Union took

[32] ICCA Global Initiative on High Production Volume (HPV) Chemicals (available at http://www.cefic.org/activities/hse/mgt/hpv/hpvinit.htm).

[33] The ICCA HPV Working List, July 2002 Update (available at http://www.cefic.org/activities/hse/mgt/ hpv/ICCA Working List - July 2002 Update - 070103.xls).

Table 1. Data required for high production volume chemicals

Name and EINECS number of the substance
Quantity of the substance produced or imported
Information on the reasonably foreseeable uses of the substance
Data on the physicochemical properties of the substance
Data on the pathways and environmental fate
Data on the ecotoxicity of the substance
Data on the acute and subacute toxicity of the substance
Data on carcinogenicity, mutagenicity and/or toxicity for reproduction of the substance
Any other indication relevant to the risk evaluation of the substance

place in parallel with the development of new initiatives by the OECD which had led to the launching of an extensive programme in 1988 on existing chemicals, an area in which several EU Member States were already active.

European Union Directives require the evaluation and control of the risks to the environment and/or public health of both existing and new chemicals. The European Chemicals Bureau (ECB) located in Ispra, Italy provides technical support for the development of EU chemicals policy and its website[34] provides information on both existing and new chemicals. The Existing Substances Regulation[35] provides for the evaluation and control of risks posed by existing chemicals in four steps:

Step I	Data collection
Step II	Priority setting
Step III	Risk assessment
Step IV	Risk reduction

The data reporting is divided into two broad categories – first, data on high production volume (HPV) substances produced or imported in quantities exceeding 1000 tonnes per year, and secondly, data on low production volume (LPV) substances which have been produced or imported in quantities between 10 and 1000 tonnes per year.

[34] European Chemicals Bureau website (available at http://ecb.jrc.it/).

[35] European Community, *Council Regulation (EEC) No 793/93 of 23 March 1993 on the evaluation and control of the risks of existing substances* (available at http://ecb.jrc.it/existing-chemicals).

Table 2. Toxicity Data required for high production volume chemicals

5.1	Acute toxicity
5.1.1	Acute oral toxicity
5.1.2	Acute inhalation toxicity
5.1.3	Acute dermal toxicity
5.1.4	Acute toxicity (other routes of administration)
5.2	Corrosiveness and irritation
5.2.1	Skin irritation
5.2.2	Eye irritation
5.3	Sensitization
5.4	Repeated dose toxicity
5.5	Genetic toxicity in vitro
5.6	Genetic toxicity in vivo
5.7	Carcinogenicity
5.8	Toxicity to reproduction
5.9	Other relevant information
5.10	Experience with human exposure

The data required for HPV chemicals are specified in Table 1. The toxicity data requirements are comprehensive (Table 2.)

The EU Directive makes it clear that industrial and commercial secrecy shall not apply *inter alia* to the name of the substance, the name of the manufacturer and the summary results of the toxicological and ecotoxicological tests.

On the basis of the information submitted and on the basis of national lists of priority substances, the Commission shall regularly draw up lists of priority substances or groups of substances *requiring immediate attention because of their potential effects on man or the environment.* These lists are published by the Commission; four such lists have so far been published, totalling 141 chemicals.[36] The main motivations for establishing the EU working list are two-fold: first as the basis for the priority lists,

[36] European Community, *Commission Regulation (EC) No 1179/94 of 25 May 1994 concerning the first list of priority substances as foreseen under Council Regulation (EEC) No 793/93*. European Community, *Commission Regulation (EC) No 2268/95 of 27 September 1995 concerning the second list of priority substances as foreseen under Council Regulation (EEC) No 793/93*. European Community, *Commission Regulation (EC) No 143/97 of 27 January 1997 concerning the third list of priority substances as foreseen under Council Regulation (EEC) No 793/93*. European Community, *Commission Regulation (EC) No 2364/2000 of 25 October 2000 concerning the fourth list of priority substances as foreseen under Council Regulation (EEC) No 793/93* (available at http://ecb.jrc.it/existing-chemicals).

and secondly because industry is encouraged to include substances on the working list as by doing so, HEROs (High Expected Regulatory Outcome substances) can be better identified and possible NEROs (No Expected Regulatory Outcome substances) can be removed from the working list if convincing evidence is brought forward by industry.[37]

The notification schemes for new substances,[38] manufactured or imported within the EU, were first introduced during the 1970s by individual Member States. The current version is the 7th Amendment[39] to Directive 67/548/EEC which requires the provision of data, with increasing detail, according to the quantity of the substance placed on the market, viz 10 kg, 100 kg and 1000 kg per year per manufacturer with further toxicological and ecotoxicological testing required at quantities exceeding 100 and 1000 tonnes per year. Since the 7th Amendment, the European Chemicals Bureau has received about 400 to 450 notifications per annum referring to about 300 to 350 new substances.[40] Notifications from the UK contribute about 28% of the cumulative total, followed by Germany (25%), France (12%), The Netherlands (9%) and Italy (7%). Foreign imports, particularly from the USA, Japan and Switzerland, represent about half of the new notified substances (Table 3).

61% of new chemicals are notified for production volumes between 1 and 10 tonnes per year (Annex VIIA), 28% for production in smaller quantities (Annex VIIB and VIIC) and about 10.5% in larger volumes (Level 1 and 2).

As an example of the additional data required as the quantity placed on the market increases, the toxicological data requirements are summarised in Table 4.

Table 3. Types of notification and annual quantities

Type of notification	Annual quantity
Level 2 (1000 tonnes)	> 1000 tonnes
Level 1 (100 tonnes)	> 100 tonnes
VIIA	> 1 tonne
VIIB	> 100 kg and < 1 tonne
VIIC	> 10 kg and < 100 kg

As the quantity of a new substance increases through Level 1 to Level 2, so the additional toxicological data required converges with the data required for High Production Volume existing substances. The Directive also requires that the substances shall be classified as very toxic, toxic or harmful, according to the criteria shown in Table 5.

The data provided in the new substances notification procedure are used to assign one of the following risk assessments[41] to the new substance:

(a) The substance is of no immediate concern.
(b) The substance is of concern – assessment revision deferred to tonnage threshold attainment.
(c) The substance is of concern – assessment to be reviewed immediately.
(d) The substance is of concern – recommendations for risk reduction to be instigated immediately.

FUTURE EU CHEMICALS POLICY

The European Union chemicals policy is currently being redeveloped as the current system of assessing chemicals on the market has made only very slow progress. The European Commission, because of concerns about the lack of information about chemicals, in February 2001 published a white paper[42] outlining ideas on future

[37] European Community, *Priority Setting* (available at http://ecb.jrc.it/Priority-Setting/).

[38] European Community, *New Chemicals* (available at http://ecb.jrc.it/new-chemicals/).

[39] European Community, *Council Directive 92/32/EEC of 30 April 1992 amending for the seventh time Directive 67/548/EEC on the approximation of the laws, regulations and administrative provisions relating to the classification, packaging and labelling of dangerous substances* (available at http://europa.eu.int/eur-lex/en/lif/dat/1992/en_392L0032.html).

[40] European Community, *New Chemicals* (available at http://ecb.jrc.it/new-chemicals/).

[41] European Community, *Commission Directive 93/67/EEC of 20 July 1993 laying down the principles for assessment of the risks to man and the environment of substances notified in accordance with Council Directive 67/548/EEC* (available at http://europa.eu.int/eur-lex/en/lif/dat/1993/en_393L0067.html).

[42] European Commission, *The White Paper on the strategy for a future Chemicals Policy,* COM (2001) 88 final, 27 February 2001 (available at http://europa.eu.int/comm/enterprise/chemicals/chempol/whitepaper/whitepaper.htm).

Table 4. Toxicological data requirements

Level	Toxicological testing	Type of notification
4.1	Acute Toxicity[a]	—
4.1.1	Administered orally	VIIC, VIIB, VIIA
4.1.2	Administered by inhalation	VIIC, VIIB, VIIA
4.1.3	Administered cutaneously	VIIA
4.1.5	Skin irritation	VIIB, VIIA
4.1.6	Eye irritation	VIIB, VIIA
4.1.7	Skin sensitization	VIIB, VIIA
4.2	Repeated dose[b]	—
4.2.1	Repeated dose toxicity	VIIA
4.3	Other effects	—
4.3.1	Mutagenicity	VIIB, VIIA
4.3.2	Screening for toxicity related to reproduction	VIIA
4.3.3	Assessment for toxicokinetic behaviour	VIIA

Note:
[a] For acute toxicity testing at VIIC or VIIB, one route of administration is sufficient. Gases should be tested by inhalation. Substances other than gases should be tested by oral administration. At VIIA, substances other than gases shall be administered by at least two routes, one of which should be the oral route. The choice of the second route will depend on the nature of the substance and the likely route of human exposure. Gases and volatile liquids should be administered by the inhalation route.
[b] For repeated dose testing, the route of administration should be the most appropriate having regard to the likely route of human exposure, the acute toxicity and the nature of the substance. In the absence of contra-indications, the oral route is usually the preferred one.

chemicals strategy known as the New European Chemicals Strategy (NECS). This introduces a new system of chemicals control for both new and existing substances which has been called the REACH system:

Registration of basic information of substances to be submitted by companies, in a central database.
Evaluation of the registered information to determine hazards and risks.
Authorization requirements imposed on the use of high-concern substances. This process will be used for both new and old.
CHemicals.

The broader aims of NECS include compliance with the various United Nations and other international agreements on the use and control of chemicals, as well as the provision of assistance to developing countries, so that their capability and capacity for managing chemicals can be strengthened. The European Commission is expected to release legislative proposals for the introduction of NECS by April 2003.

National Initiatives

Individual countries, such as the United Kingdom and the United States of America, have adopted particular national strategies to augment the regional and international initiatives into the evaluation of the risk assessment of chemicals. As an example of a national approach, the United Kingdom in December 1999 published a

Table 5. Classification of substances according to various criteria

Parametar	Very toxic	Toxic	Harmful
LD_{50} oral in rat (mg/kg body weight)	< 25	25 to 200	200 to 2000
LD_{50} dermal in rat (mg/kg body weight)	< 50	50 to 400	400 to 2000
LC_{50} (inhalation) rat (mg/l/4 h)	< 0.25	0.25 to 1	1 to 5

chemical strategy[43] setting out policies to avoid harm to the environment or to human health through environmental exposure to chemicals. This strategy includes the need for precautionary action for chemicals which are likely to cause serious or irreversible damage to the environment and identifies environmental persistence, tendency to bioaccumulate and toxicity as the properties that are especially important. A Stakeholder Forum established in September 2000 has advised the UK government on criteria for concern rapidly identifying those chemicals which need a risk management strategy as a matter of urgency. These criteria for concern[44] were developed in order to trigger a structured review process and provide a fast-track procedure for high risk chemicals. The strategy states that all documents considered by the Stakeholder Forum and all records of its meetings will be made available to the public; these are available on the web.[45]

The United States of America in 1998 announced the Chemical Right-to-Know (RTK) Initiative[46] which was the US government response to an Environmental Protection Agency (EPA) study that found that very little basic toxicity information is publicly available on most of the HPV chemicals made and used in the USA. It should be noted that the US definition of HPV chemicals is different from that used in the rest of the world, as the US definition is a chemical produced in or imported into the USA in amounts of over a million pounds a year – approximately 444 tonnes. The RTK initiative aims to rapidly test chemicals – using the same tests as in the OECD SIDS – and make the data available to scientists, policy makers, industry and the public. An EPA Chemical Hazard Data Availability Study[47] showed that the US produces or imports close to 3000 chemicals at over 1 million pound a year, yet there was no basic toxicity information publicly available for 43% of the HPV chemicals produced in the USA and that a full set of basic toxicity information is only available for 7% of these chemicals. The EPA has invited industry chemical manufacturers and importers to participate in a voluntary challenge programme to provide the basic toxicity data on the HPV chemicals that they produce. EPA intends that chemicals not adopted in the voluntary programme be tested under the HPV Test Rule. Some 2080 of the 2800 HPV chemicals were adopted by the deadline of 1 December 1999. Detailed information on much of this programme is available on the EPA website.

Notification of new chemicals is required in the USA under the TSCA (Toxic Substances Control Act) Inventory Update Rule[48] which requires the reporting of basic data every four years on chemicals produced or imported in an amount exceeding 10 000 lb (4540 kg ~ 4.5 tonnes). Typically, data is provided on approximately 9000 organic substances each four years. However, unlike the EU notification of new substance requirements, the US requirement does not require provision of toxicity data although proposals are currently being considered[49] to modify the US requirement so as to require the collection of a broad-based database of use and exposure information on chemicals produced or imported in quantities exceeding 25 000 lb.

43 Department of the Environment, Transport and the Regions, *Sustainable production and use of chemicals – a strategic approach, The Government's Chemicals Strategy,* London, December 1999 (available at http://www.detr.gov/environment/chemistrat/index.htm).

44 Department of the Environment, Food and Rural Affairs, Chemicals Stakeholder Forum, Criteria for Identifying Chemicals of Concern (available at http://www.defra.gov.uk/environment/chemicals/csf/criteria/index.htm). See also The Chemicals Stakeholder Forum, Criteria for Concern of the Chemicals Stakeholder Forum (available at http://www.defra.gov.uk/environment/chemicals/csf/criteria.htm).

45 Chemicals Stakeholder Forum, Criteria for Identifying Chemicals of Concern (available at http://www.defra.gov.uk/environment/chemicals/csf/papers.htm).

46 Environmental Protection Agency, *Chemical Right-to-Know Initiative* (available at http://www.epa.gov/chemrtk).

47 Environmental Protection Agency, *Chemical Hazard Data Availability Study,* prepared by EPA's Office of Pollution Prevention and Toxics, April 1998 (available at http://www.epa.gov/opptintr/chemtest/hazchem.htm).

48 Environmental Protection Agency, *The TSCA Inventory Update Rule (IUR)* (available at http://www.epa.gov/opptintr/iur98/).

49 Environmental Protection Agency, *Fact Sheet: Proposed IUR Amendments,* 26 July 1999 (available at http://www.epa.gov/opptintr/iuramend/iurafact.htm).

Other Initiatives

Although particular attention has been given above to the UNEP, OECD, ICCA and European Union initiatives demonstrating how there is a concerted effort to obtain data both on existing chemicals and on new chemicals placed on the market, it is evident that there are several global activities which are aimed at taking forward the six priority programme areas of Agenda 21, Chapter 19 so that there is sound management of chemicals worldwide. These include the following:

(a) The International Programme on Chemical Safety (IPCS)[50] was established in 1980 with the WHO as its executing agency. The two main roles of IPCS are to:
- establish the scientific health and environmental risk assessment basis for safe use of chemicals;
- strengthen national capabilities and capacities for chemical safety.

IPCS products include Health and Safety Guides, Environmental Health Criteria documents, International Chemical Safety cards. Activities include:

- the global harmonization of approaches to risk assessment through increased understanding;
- responses to chemical incidents and emergencies which are usually accidental and unexpected but may be caused deliberately, for example, as a result of terrorist action.

(b) The Intergovernmental Forum on Chemical Safety (IFCS)[51] established in 1994, which has as one of its functions the identification of priorities for co-operative action on chemical safety, particularly taking into account the special needs of developing countries. The IFCS in 1994 established Priorities for Action[52] for the implementation of the six priority programme areas of Agenda 21 Chapter 19. As already noted above, Forum III agreed[53] the Bahia Declaration on Chemical Safety and Priorities for Action Beyond 2000.

(c) The Inter-Organization Programme for the Sound Management of Chemicals (IOMC),[54] established in 1995, provides a mechanism to co-ordinate the efforts of intergovernmental organizations (UNEP, ILO, FAO, WHO, UNIDO, UNITAR and OECD) in the assessment and management of chemicals. IOMC compiles summary reports of ongoing activities categorized by the six priority programme areas of Agenda 21 Chapter 19. Capacity-building has been given a high priority with a comprehensive review being issued in 1998.[55]

(d) The Global Information Network on Chemicals (GINC)[56] was initiated in 1994 to foster generation and circulation of chemical-related information among all countries and international organizations for the promotion of chemical safety. Its website includes a useful guide with links to the principal sites providing chemical safety information around the world.

EVALUATION

There are already mechanisms in place within nations and regions, such as the European Union, which are also reflected in other areas of the world, notably through the OECD and UNEP Chemicals programmes, to respond to the Agenda 21 Chapter 19 priority programme area to expand and accelerate the international assessment of chemical risks. These programmes ensure that **data regarding the risks** to public health and to the environment are available for **both** existing and new chemicals.

The data required increase with the quantity of chemical – using the EU situation as a model, the data requirements are shown in Table 6.

[50] Information on IPCS is available at: (http://www.who.int/pcs/).

[51] Information on IFCS is available at: (http://www.who.int/ifcs/).

[52] Available at: (http://ww.who.int/ifcs/fs_res2.htm).

[53] Available at: (http://ww.who.int/ifcs/forum3.final.html).

[54] Information on IOMC is available at: (http://www.who.int/iomc).

[55] Available at: (http://www.who.int/iomc/capacity/caprep.html#toc).

[56] Information on GINC is available at: (http://www.nihs.go.jp/GINC/other/aboutginc.htm).

Table 6. Data requirements for evaluation

Annual quantity	Existing chemicals	New chemicals
>10 kg and <100kg	—	VIIC
>100 kg and < 1 tonne	—	VIIB
> 1 tonne	—	VIIA
10 to 1000 tonnes	Low Production Volume	—
> 100 tonnes	—	Level 1 (100 tonnes)
> 1000 tonnes	High Production Volume	Level 2 (1000 tonnes)

It is noted that the EU scheme is intended to identify HEROs (High Expected Regulatory Outcome substances) as well as possible NEROs (No Expected Regulatory Outcome substances) and that national schemes, such as that in the UK, includes the establishment of a 'fast-track' procedure for chemicals that present a high risk to public health or to the environment.

Given that the EU has expanded to include many of the Central and Eastern European states and that international trade in chemicals will continue to increase, it is reasonable to expect that the EU requirements for toxicity information on both existing and new chemicals will come to be applied to an increasing extent around the world.

In addition, it should be noted that there is considerable emphasis throughout the programmes described above in making information on the risks posed by chemicals available to the public.

THE CWC REQUIREMENTS

The general purpose criterion within the CWC in Article II.1(a) states that 'chemical weapons' include *'Toxic chemicals and their precursors, except where intended for purposes not prohibited under this Convention, as long as the types and quantities are consistent with such purposes'*. As chemical weapons, by their nature, involve toxic chemicals which cause death, temporary incapacitation or permanent harm to humans or animals, there is clearly a parallel between chemicals which might be used as chemical weapons and existing or new chemicals which are highly toxic – and are the subject of the ongoing national, regional and international initiatives aimed at ensuring the sound management of chemicals and the reduction of risks to human health or the environment. The implementation of the general purpose criterion in the CWC is clearly placed upon the States Parties by the requirement in Article VI that:

> *2. Each State Party shall adopt the necessary measures to ensure that toxic chemicals and their precursors are only developed, produced, otherwise acquired, retained, transferred, or used within its territory or in any other place under its jurisdiction or control for purposes not prohibited under this Convention.*

In considering how National Authorities in the States Parties to the OPCW might implement the general purpose criterion, it is considered that use should be made of the ongoing national and international programmes addressing the safe management of chemicals, as these programmes are focused on those chemicals that present the **greatest** dangers to health and to the environment. Particular attention should be addressed to those chemicals that present the greatest risks to public health and that are available in quantity for purposes not prohibited under the Convention. As traditionally, it has been recognized that for a single attack using chemical weapons, a quantity of about 1 tonne of agent is required, it follows that for a militarily significant capability, a quantity of 300 tonnes or more of agent would be needed. Consequently, it would be appropriate for National Authorities to utilize in respect of **existing** chemicals, the data emerging from the ongoing international HPV chemicals programme (for chemicals in the USA in excess of 444 tonnes per annum and elsewhere in excess of 1000 tonnes per annum) and, in respect of **new** chemicals, the data relating to new substances being placed on the market in quantities in excess of 1 tonne, in order to identify those chemicals that presented the greatest risk to public health. National Authorities could then determine what further action was appropriate and necessary to ensure that the national obligations under Article VI.2 of the CWC are being met.

The General Purpose Criterion also applies to newly encountered hazardous chemicals which might be judged to lack market potential and so fail to enter the reporting systems. Such chemicals could be more toxic than the traditional chemical weapon agents – and thus smaller quantities than 300 tonnes may present a risk to the Convention. It is, however, noted that in the UK, the Health and Safety Executive guidance[57] on the notification of new substances states that the regulations apply to anyone who supplies a new substance which '*includes selling it, lending it to someone else, passing it on, giving it away or importing it*' into the EU. Furthermore, the EU requirements for the notification of new substances do require provision of toxicity information for any new chemical produced in quantities in excess of 10 kg. While it is possible that a significant military quantity (300 tonnes or more for a traditional CW agent – or a smaller quantity for a more toxic novel chemical) of a new chemical that has not been placed on the market could be produced – and thus present a risk to the CWC – it is recognized that the overall trend is increasingly to require the provision for health and safety reasons of toxicity information on chemicals being produced in a facility and for the provision of such information on new chemicals being placed on the market in quantities in excess of 10 kg. National Authorities implementing the General Purpose Criterion will also need to consider other chemicals, both known and novel, which have not entered the reporting chains in the chemical safety regimes.

From the point of view of the effective implementation of the CWC, there is much to be said for the States Parties individually encouraging both the implementation and extension of the international HPV chemicals programme and the EU notification of new substances.

As the General Purpose Criterion is a central provision in the CWC, it is important that both the fact and the method of its implementation is made generally known. It would be important for National Authorities to report to the OPCW, as well as nationally both what action they have taken and the nature of this action to implement the General Purpose Convention, thereby strengthening the implementation of the CWC and ensuring its continued health and effectiveness in totally preventing chemical weapons.

INCAPACITATING CHEMICALS

A particular risk to the prohibitions in the Chemical Weapons Convention arises from the continuing interest being shown in incapacitating chemicals by some of the States Parties. International attention on this was heightened by the incident on 23 October 2002 when a group of some 50 Chechens took over 700 people hostage in a Moscow theatre. Three days later, Russian forces stormed and retook the building after two hostages were killed. The action by Russian forces was preceded by large quantities of 'gas' being pumped into the building with the intention of incapacitating the hostage-takers. The hostage-takers and over 120 hostages died, and many others were hospitalized because of the effects of the gas.[58]

The 'gas' used was stated by the Russian Health Minister to be a derivative of fentanyl, an opiate chemical related to morphine,[59] perhaps mixed with other agents (Van Damme, 2002). Reports suggesting that atropine, an antidote to anticholinergic nerve agents, did not reverse the effects of the gas whereas naloxone, an antidote

[57] Health & Safety Executive, *The NONS Regulations* (available at http://www.hse.gov.uk/hthdir/noframes/nons/nons2.htm).

[58] The risk to the Chemical Weapons Convention from incapacitating chemicals is addressed in Malcolm R. Dando, *The Danger to the Chemical Weapons Convention from Incapacitating Chemicals*, University of Bradford, Department of Peace Studies, First CWC Review Conference Paper No. 4, March 2003 (available at http://www.brad.ac.uk/acad/scwc).

[59] CBS News.com (2002) *Russia IDs Theater Gas: Fentanyl* (available at http://www.cbsnews.com/stories/2002/10/31/world/main527614.shtml).

to opiates, did supported the conclusion that a fentanyl derivative was used in the attempt to incapacitate the hostage-takers.[60,61]

In regard to the Chemical Weapons Convention (CWC), it might, at first sight, appear that there is little to be discussed. The Russian Health Minister stated that the hostage deaths from the gas had been the result of their weakened state and insisted that the Chemical Weapons Convention had not been violated. It might also be argued that the use of an agent that was believed to be an anaesthetic which *'could not cause death'*[62] falls under the 'Purposes not Prohibited' exemption of paragraph 9(d) of Article II of the CWC which allows use of chemicals for *'Law enforcement including domestic riot control purposes'*.[63]

However, interest on the part of security and military agencies in regard to such materials may have been given significant enhancement by the scientific and technological developments during the 1990s that might be seen to allow improved differentiation between the incapacitating effects of some chemical agents and their other, more dangerous, effects. Clearly, if such opportunities are seen to exist, and are widely exploited, erosion of the General Purpose Criterion at the heart of the Convention is highly probable. For that reason, the issue of novel non-lethal chemicals cannot be ignored by the Review Conference which, according to paragraph 22 of Article VIII:[64]

> *'... shall take into account any relevant scientific and technological developments...'*

It is evident[65] that the US military is still attempting to discover new forms of chemical incapacitant. Moreover, some might argue that the peaceful purpose exemption of Article II.(9)(d) of the Chemical Weapons Convention, which allows for 'Law enforcement including domestic riot control purposes', would allow quite new law enforcement chemicals with complex physiological effects on humans to be developed – particularly as no definition is offered for what chemicals are permitted for law enforcement other than that Schedule 1 chemicals may not be used.

At the time of the negotiation of the Convention, an Editorial in the *Chemical Weapons Convention Bulletin* (1994) pointed out the dangers:

> *'The Chemical Weapons Convention in no way limits use of tear gas or other temporarily disabling chemicals by police forces for purposes of domestic riot control.* ***But the language used to exempt other law enforcement purposes has created ambiguity in the heart of the Convention****...'* (emphasis added).

In particular, the editorial noted that:

> ***'What is at stake is the ability of the treaty regime to withstand technical change****. For new chemical agents and technologies have begun to emerge whose attractions for weapons purposes may eventually drive them through the loopholes which the ambiguity has created'* (Emphasis added).

The evidence presented[66] by Dando shows that we have now reached that point as, in regard to a number of systems that could well be of interest to those with a malign intent, it is clearly now possible to create specific chemicals that target specific receptor sub-types and thereby cause specific behavioural effects.

It is certainly possible to find strong advocates of non-lethal chemical weapons who believe that

[60] Fox News (2002) US *Questions Use of Chemical Gas* (available at http://www.foxnews.com/story/0,2933,66911,00.html).

[61] Fox News (2002) *Gas Russia Used in Hostage Siege was Fentanyl, US Officials Say* (available at http://www.foxnews.com/story/0,2933,669,88,00.html).

[62] CBS News.com (2002) *Russia IDs Theater Gas: Fentanyl* (available at http://www.cbsnews.com/stories/2002/10/31/world/main527614.shtml).

[63] The Chemical Weapons Convention (text) (available at http://www.opcw.org/html/db/cwc/eng /cwc_article_II.html).

[64] The Chemical Weapons Convention (text) (available at http://www.opcw.org/html/db/cwc/eng /cwc_article_VIII.html).

[65] Joint Non-Lethal Weapons Directorate (2003) *Front End Analysis for Non-lethal Chemicals* (available at http://www.sunshine-project.org/publications/nlwdpdt/feachemical.jpg).

[66] Malcolm R. Dando, *The Danger to the Chemical Weapons Convention from Incapacitating Chemicals,* University of Bradford, Department of Peace Studies, First CWC Review Conference Paper No. 4, March 2003 (available at http://www.brad.ac.uk/acad/scwc).

it is necessary to consider selective changes to current international legal agreements. As Fidler (2001) has noted:

'The selective change perspective uses changes in military operations and technologies as a basis for advocating selective, case-by-case reforms in international law to allow NLW (Non-Lethal Weapons) development and use...'.

However, he argued, inherent in that selective change position is a much more radical position that could upset the current international legal system that we have developed over centuries to constrain warfare. He noted, for example, that:

'Arguments in favour of developing and deploying NLWs often rely on the new capabilities such weapons give military forces and suggest that such capabilities affect how we evaluate the ethics of weapons' use...'.

As an example, at present soldiers are clearly not allowed directly to target civilians with their lethal weapons. If it was to be agreed that civilians could be targeted directly with non-lethal weapons (although with such weapons there will always be a risk of deaths[67]) where does that leave the principle of discrimination between combatants and non-combatants?

It was argued strongly prior to the First Review Conference of the CWC in 2003 that with the extremely rapid current rate of development in the life sciences it would be **dangerous** to leave this matter unattended to at the 2003 Review Conference. It is likely that the scientific possibilities will be even more tempting to those seeking new weapon systems after five more years. The bioregulators and synthetic analogues, such as fentanyl, considered here are mid-spectrum agents covered, and correctly so, by both the Biological and Toxin Weapons Convention and the Chemical Weapons Convention. A NATO Advanced Research Workshop held in the run-up to the 2001 Review Conference of the Biological and Toxin Weapons Convention considered the scientific changes carefully and recognized that the scope of the Convention should be reaffirmed as at previous Review Conferences by a consensus statement in regard to Article I, along the following lines:[68]

'The Conference... reaffirms that the Convention unequivocally covers all microbial or other biological agents or toxins, naturally or artificially created or altered, as well as their components, whatever their origin or method of production, of types and in quantities that have no justification for prophylactic, protective or other peaceful purposes. ***Consequently, prions, proteins and bioregulators, and their synthetically produced analogues and components are covered****'* (emphasis added).

Unfortunately, no Final Declaration was agreed by the 2001–2002 Review Conference of the Biological and Toxin Weapons Convention and consequently the opportunity for such a consensus statement was missed.

A system of reaffirmation and extended understandings of the scope of Biological and Toxin Weapons Convention has come about through the Final Declarations agreed at successive Review Conferences since 1980. There is much to be said for the Review Conferences of the Chemical Weapons Convention to gain similar benefits from extended understandings agreed in their Final Declarations.[69] It was suggested that appropriate language would be to state that:

'The Conference also reaffirms that the Convention unequivocally covers all chemicals, regardless of whether they are produced in facilities, in munitions or elsewhere, of types and in quantities that are consistent with purposes not prohibited under this Convention'.

[67] Klotz L, Furmanski M and Wheelis Mark (2003) *Beware the Siren's Song: Why 'Non-Lethal' Incapacitating Chemical Agents are Lethal* (available at http://microbiology/ucdavis.edu.faculty/mwheelis/sirens_song.pdf).

[68] Graham S. Pearson, *New Scientific and Technological Developments of Relevance to the Fifth BTWC Review Conference,* Review Conference Paper No. 3, University of Bradford, October 2002 (available at http://www.brad.ac.uk/acad/sbtwc).

[69] Graham S. Pearson, *Relevant Scientific and Technological Developments for the First CWC Review Conference: The BTWC Review Conference Experience.* CWC Review Conference Paper, No. 1, University of Bradford, August 2002 (available at http://www.brad.ac.uk/acad/scwc).

Furthermore, in order to avoid any possible misunderstanding, it was suggested that an explanatory sentence should be added to state that:

> ***'Consequently, toxins, prions, proteins, peptides and bioregulators and their biologically or synthetically produced analogues and components are covered'*** (emphasis added).

It was argued that the CWC Review Conference must address this issue to prevent a dangerous erosion of the purpose and objective of the Convention. The risks are real, as a recent US military legal review noted (Department of the Navy, 1997):

> *'Convulsives and calmatives may rely on their toxic properties to have a physiological effect on humans. If that is the case, and these two NLWs (Non-Lethal Weapons) are not considered RCAs (Riot Control Agents), in order to avoid being classified as a prohibited chemical weapon, they would have to be used for the article I(9)(d) "purpose not prohibited" the law enforcement purpose. As discussed...* ***the limits of this 'purpose not prohibited' are not clear and will be determined by the practice of states****'* (emphasis added).

If the Review Conference failed clearly to address this issue, novel non-lethal weapons based on the new understanding of the nervous system and its chemical neurotransmitters and receptors could well have been deployed and used before there is an opportunity for the next CWC Review Conference to address the issue. The events in Moscow in late 2002 would then be seen as a 'harbinger' of a much more dangerous future with a seriously eroded chemical weapons prohibition regime rather than as an isolated 'hangover' from the military developments of the Cold War period which has served as a useful signal to strengthen the understanding that all such chemicals are prohibited under the Chemical Weapons Convention.

Although it is true that the CWC is, in a sense, under continuous review through the annual Conferences of the States Parties and the regular Executive Council meetings, it was argued that it would be irresponsible if the States Parties at the forthcoming Review Conference in April 2003 – given the mandate of the Review Conference specified in the Convention to *convene in special sessions to undertake reviews of the operation of this Convention: such reviews shall take into account any relevant scientific and technological developments* – were to fail to address the real and present danger to the Chemical Weapons Convention from incapacitating chemicals. The outcome should be a clear reaffirmation of the comprehensive prohibition of toxic chemicals in the Convention, making it clear in the reaffirmation that all incapacitating chemicals are covered. Consideration could also be given to an action placed on the annual Conference of States Parties and the Executive Council to be vigilant to ensure that there is no erosion of the chemical weapons prohibition regime through incapacitating chemicals.

Unfortunately, although there was some mention of the risk from incapacitating chemicals in a few of the statements made by States Parties at the first CWC Review Conference, the Review Conference did not address this issue. As an Editorial in *The CBW Conventions Bulletin* noted (Editorial, 2003a), following the Review Conference '*Several issues seem not to have engaged the Review Conference. No mention of so-called "non-lethal" agents or "law enforcement" ended up in the final version of the Review Document. States parties' reluctance to address this issue was evident prior to the Review Conference, and the timing was perhaps not right to tackle this controversial issue, but it simply should not be simply ignored now for another five years. Indeed,* ***it is hard to think of any other issue having as much potential for jeopardizing the long-term future of the CWC regime****'* (Emphasis added). The subsequent issue of *The CBW Conventions Bulletin* in its editorial (Editorial, 2000b) addressed the danger posed by such 'non-lethal' weapons to both the CWC and the BTWC. There has been no indication thus far that the States Parties to the CWC are prepared to address the risks posed by such materials. It is, however, true that the General Purpose Criterion in both the CWC and the BTWC does totally prohibit all toxic chemicals – past, present and future – whether incapacitating or lethal, unless they are for purposes not prohibited under the Conventions.

CONCLUSIONS

Chemical weapons are totally prohibited under the provisions of the Chemical Weapons Convention. The General Purpose Criterion ensures that all toxic chemicals, past, present and future, are prohibited unless they are for purposes not prohibited under the Convention. The regime against chemical weapons will become more effective as the Chemical Weapons Convention approaches universality and international initiatives in regard to toxic chemicals become more widely applied throughout the world. National measures to ensure that toxic chemicals do not present a risk to health and safety can and should be harnessed to ensure effective implementation of the obligations under the Chemical Weapons Convention to ensure that such chemicals are only used for purposes permitted under the Convention.

REFERENCES

Department of the Navy (1997). *Preliminary Legal Review of Proposed Chemical-Based Nonlethal Weapons*, 30 November. Alexandria, VA, USA: Office of the Judge Advocate General (International and Operational Division).

Editorial (1994). New technologies and the loophole in the Convention. *CBW Conventions Bull*, 23(March), 1–2 (available at http://www.sussex.ac.uk/spru/hsp).

Editorial (2003a). Where do we go from here? The first CWC conference and the next five years. *CBW Conventions Bull*, 60(June), 1–5 (available at http://www.sussex.ac.uk/spru/hsp).

Editorial (2003b). 'Non-lethal' weapons, the CWC and the BMW. *CBW Conventions Bull*, 61(September), 1–2 (available at http://www.sussex.ac.uk/spru/hsp).

Fidler DP (2001). 'Non-lethal' weapons and international law: three perspectives on the future. *Med Conflict Survival*, **17**, 194–206.

Mierzejewski JW and Moon JEvan C (1999). Poland and biological weapons in *Biological and Toxin Weapons: Research, Development and Use from the Middle Ages to 1945* (ed. Geissler E. and Moon, J. E. van C.), 63–69, Stockholm International Peace Research Institute, Oxford University Press, Oxford.

Pearson GS (2000). The CWC General Purpose Criterion: how to implement? *CBW Conventions Bull*, 49(September), 1–7 (available at http://www.sussex.ac.uk/spru/hsp).

Sims NA (2001). *The Evolution of Biological Disarmament*, 4, Stockholm International Peace Research Institute, Oxford University Press, Oxford.

Van Damme B (2002). Moscow theatre siege: a deadly gamble that nearly paid off. *Pharmaceut J*, **269**, 723–724.

29 AN A–Z OF COMPOUNDS OF INTEREST IN RELATION TO CHEMICAL WARFARE AND OTHER MALEVOLENT USES OF POISONS

Philippa Edwards and Robert L. Maynard

Health Protection Agency, Chilton, UK

INTRODUCTION

In assembling this A–Z, we have included many toxins that have never been used in association with chemical warfare. Some have been used for the purpose of murder or assassination. Many of these poisons are natural substances, of which most are in a form not readily suited to use in conventional warfare or terrorist attacks. However, given the rapid advances in biotechnology and formulation technology, the potential for such uses in the future cannot be excluded.

We have also included some substances that are useful in the protection from, or treatment of the adverse effects of chemical warfare agents.

Although the use of agents in aggression is often uncertain, we have indicated where possible if particular agents have or may have been used, or prepared and stocked with a view to such use.

A number of agents are known by a variety of names, which has caused confusion. We have therefore included the most common synonyms for a particular agent.

We hope that this compendium will be a useful reference text for those working in the field and for those responsible for taking steps to safeguard the safety of the public. The entries include some of the history and natural history of many of the agents to provide information beyond the minimum required for such purposes in order to interest, as well as inform, the reader.

MAJOR SOURCES

Beadnell CM (1943). *An Encyclopaedic Dictionary of Science and War*. London: CA Watts and Company.

Blyth AW and Blyth MW (1920). *Poisons: their Effects and Detection*, Fifth Edition. London: Charles Griffin and Company.

Bowman WC and Rand MJ (1980). *Textbook of Pharmacology*, Second Edition. London, UK: Blackwell Science.

Bowman WC, Bowman A and Bowman A (1986). *Dictionary of Pharmacology*, London, UK: Blackwell Science.

CDC website (http://www.bt.cdc.gov/chemical).

Edstrom A (1992). *Venomous and Poisonous Animals*. Malabar, FL, USA: Krieger Publishing Company.

Farrow C, Wheeler H, Bates N *et al.* (2000). *The Chemical Incident Management Book*. London, UK: HMSO.

The information provided in this chapter is the responsibility of the authors and has not been verified by the Health Protection Agency.

Chemical Warfare Agents: Toxicology and Treatment (2nd Edition)
Edited by Timothy C. Marrs, Robert L. Maynard and Frederick R. Sidell

Gossell IA and Bricker JD (1994). *Principles of Clinical Toxicology*. NY: Raven Press.
Henderson Y and Haggard HW (1943). *Noxious Gases*, Second Edition. New York, NY, USA: Reinhold Publishing Corporation.
Klaassen CD, Amdur MO and Doull J (1986–1996). *Casarett and Doull's Toxicology*, Third–Fifth Editions. NY: Macmillan.
Laessøe T and Del Conte A (1996). *The Mushroom Book*. London, UK: Dorling Kindersler Ltd.
Perry Robinson J (1971). *The Problem of Chemical and Biological Warfare*, Volume 1. Stockholm, Sweden: Stockholm International Peace Research Institute, Amsqvist and Wiskell.
Polson CJ, Green MA and Lee MR (1983). *Clinical Toxicology*, Third Edition. London: Pitman.
Prentiss AM (1937). *Chemicals in War*, Fourth Edition. London & NY, USA: McGraw Hill Book Company.
Sartori M (1939). *The War Gases*. London: Churchill.
Budavari S (1989). *The Merck Index*, Eleventh Edition. Rahway, NJ, USA: Merck and Company, Inc.

COMPOUNDS

A-stoff

Chloroacetone *qv*.

Abrin

Toxin derived from the seeds of the Rosary pea or Jequirity bean (*Abrus pecatorius*). Powdered abrin is yellowish-white; it is soluble in water and stable. Two glycoprotein chains: acidic chain (30 000D) – inhibits protein synthesis; neutral chain (35 000D) – binds to cell wall and facilitates entry. Effects similar to ricin *qv*, but it is more toxic.

Absinthe

Spirit containing 68 vol% ethanol, flavoured with herbs from the Swiss Jura: main flavouring wormwood (*Artemisia absinthium*). The most toxic component of the essential oil of wormwood is the terpenoid, thujone: a convulsant in animals in large doses. Over indulgence: delirium, hallucinations, brain damage. Banned in Switzerland in 1908 and in France in 1915. Imitation absinthe: Pernod, flavoured with liquorice.

AC

Hydrogen cyanide *qv*.

Acetic acid

Flammable colourless liquid: gives vinegar its characteristic smell. Irritant vapour: eyes and respiratory tract. Can be fatal on ingestion. 25 mg/m^3: conjunctivitis. 60 mg/m^3: extreme eye and nose irritation: 125 mg/m^3 intolerable.

Acetonitrile

Colourless volatile liquid: sweet ether-like odour. Can be absorbed through skin as well as by inhalation. Slowly metabolized to cyanide. Clinical effects delayed. >850 mg/m^3: weakness, nausea, convulsions, death.

Ackee fruit

Fruit of *Blighia sapida*: known in Jamaica as ackee; known as 'isin' in Nigeria. Causes 'vomiting sickness' but probably only in the undernourished. Vomiting is followed by convulsions. Ackee contains two hypoglycaemic amino acids: hypoglycin A and B. A metabolite blocks fatty acid oxidation, resulting in increased glucose consumption as an energy source; this combined with inhibition of gluconeogenesis by reduction of NADH and acetyl CoA, results in hypoglycaemia.

Aconite (Aconitine)

Used by Moors in Europe in the 15th Century and earlier in India and China. Used by elephant hunters in India and said to have been tested in small arms ammunition at Buchenwald. Aconitine and delphinine: alkaloids from the root of *Aconitine napellus* (aconite, monkshood, wolfsbane) and from the seeds of *Delphinium staphisagria* (starvesacre seeds). Delays repolarization due to impaired closure to Na channels. Weak diaphoretic: was used in fevers. Local anaesthetic: used as a liniment for toothache, rheumatism. Absorption via the skin can produce systemic toxic effects. Toxic effects: depression of respiration, myocardial fibrillation. Marked abdominal pain, projectile vomiting, fear of drinking water (as per hydrophobia in rabies), cold feeling, sweating, giddiness, muscular weakness. Blyth and Blyth (1920) give details on use in murders.

Acrilet

Acrylonitrile *qv*.

Acrolein

Papite, acrylic aldehyde. Used by the French in 1916. Clear liquid, bp 52°C: impure form is greenish-yellow. Irritant and lachrymator at $> 7\ mg/m^3$.

Acrylamide

Vinyl monomer, many industrial uses. White powder, very soluble in water. Dermal, oral and inhalation absorption. Repeated exposure: polyneuritis, sensory losses in limbs, weakness, ataxia.

Acrylonitrile

Vinyl cyanide, Acritet, Fumigrain, Ventox. Colourless, explosive, flammable liquid. Yellow on standing. Reacts violently with oxidizing agents. Absorbed: all routes. Eye and respiratory tract irritant, corrosive to skin, sensitizer. Partly metabolized to cyanide. Effects: CNS, liver, kidney, cardiovascular system, gut. Headache, nausea, vomiting, weakness, jaundice, coma, convulsions and cardiac arrest can occur without warning.

Adamsite

Diphenylamine chloroarsine *qv*.

Aflatoxins

Toxins produced by *Aspergillus flavis* and *A. parasiticus*, fungi that occur naturally in several foodstuffs, including peanuts. Liver damage and cancer. Turkey X disease: major outbreaks of poisoning in turkeys in the USA and UK in 1960.

Agent 15

See BZ.

Agent Orange

Agent Orange was a 50–50 mix of 4-(2,4-dichlorophenoxy) butanoic acid (2,4-D) and 2,4,5-trichlorophenoxy acetic acid (2,4,5-T) mixed with kerosene or diesel fuel and used as a defoliant herbicide in Vietnam by the USA forces. An estimated 19 million gallons of Agent Orange were used in South Vietnam during the war. 'Orange' because of orange ring painted on the containers. 2,4,5-T contains small quantities of dioxins (see TCDD), which can cause chloracne, birth abnormalities and possibly cancer.

Aggregoserpentin

Trimucytin, a polypeptide/polysaccharide complex isolated from the venom of the viper *Trimeresurus mucrosquamatus* activates collagen receptors and causes aggregation mainly through phospholipase C-phosphoinositide pathway. Similar toxins found in the venom of related viper species.

Algal blooms

This subject is complicated by the fact that algal blooms can be produced by two types of algae. Those produced by blue – green algae and those (red tides) produced by dinoflagellates. For red tides, see saxitoxin; other blooms can be green, yellow, blue and milky. They produce rashes and blisters on the skin, lips and genitalia of swimmers. The blue – green algal toxins are water-soluble peptides: a number of different types have been described as being produced by different species: neurotoxins (see anatoxins), hepatotoxins and cytotoxins.

Amanita muscaria

Fly agaric mushroom. Widespread and common in north temperate zones. Dried preparation used by certain North Asian communities as a social euphoriant and intoxicant. In addition, causes CNS depression, sweating, lachrymation, salivation, bradycardia. Contains cholinergic agonists and GABA agonists, including muscarine, muscaridine, bufotenine and related indole alkaloids, muscimol and ibotenic acid. Mushrooms used by Norse warriors to produce a 'berserk rage' may have been the related species, *A. pantherina*, which is associated with induction of mania.

Amanita phalloides

Death cap mushroom. Responsible for > 90% of 'mushroom deaths' in the UK. Contains the peptides, amanitine and phalloidine. Hepatotoxic: α-amanitine is the major toxic component. Direct attack on hepatocyte nuclei. Vomiting 8–12 h post-ingestion; cramping abdominal pain; diarrhoea. 2–3 days latent period then: jaundice, circulatory collapse, haemorrhage, death. Death rate 30% in 'best hands'.

Amines (alkyl)

Methylamine, ethylamine and their respective di- and tri-forms, propylamine and isoamylamine. Irritant gases or low-boiling-point liquids with pungent ammoniacal or fishy odours. Trimethylamine (putrefaction of hawthorn flowers and other plant and animal matter), triethylamine, isopropylamine and isoamylamine (produced from putrefaction of yeast) are the most toxic with rat oral LD_{50} values between 0.5 and 1 g/kg bw.

Amiton

Early V type OP. *S*-2-diethylaminoethyl *O*,*O*-diethylphosphorothioate. Colourless liquid, absorbed by inhalation or dermal route. Available as the quaternary oxalate salt as Tetram. Introduced in 1955 as an insecticide: effective against mites.

Amygdalin

Laetrile. Cyanogenic glycoside found in apple, peach, plum, apricot, cherry, almond stones. Hydrolyzed to HCN.

Anatoxins

Anatoxins are potent toxins derived from a species of cyanobacteria (blue–green algae) called *Anabaena flos-aquae*. The first published report of the potentially lethal effects appeared in the journal *Nature* in 1878. Three common anatoxins have been described: anatoxin-a and homoanatoxin-a are secondary amines and anatoxin-a(s) is a phosphate ester of a cyclic *N*-hydroxyguanine structure. Anatoxin-a (also known as very fast death factor, VFDF) and homoanatoxin-a cause irreversible activation of the nicotinic acetylcholine receptor and rapid death by respiratory arrest (mouse LD_{50} is approximately 250 μg/kg ip). Anatoxin-a(s) is a potent cholinesterase inhibitor (mouse LD_{50} 20–40 μg/kg). They are inactivated by heat. These toxins can enter the body by ingestion, inhalation of aerosols and through damaged skin.

Ancrod

Anticoagulant thrombin-like enzyme found in venom of Malayan pit viper, *Agkistrodon rhodostoma*.

Angel dust

Angel's mist, peace pills, goon, hog, T and many other street names. See phencyclidine.

Aniline

Benzenamine, phenylamine, aminobenzene, aminophen, kyanol. Oily liquid, colourless when fresh, darkening on exposure to air and light, characteristic smell and burning taste; bp 184–186°C; volatile with steam. Absorbed by oral, dermal or inhalation routes. 1 g can be fatal; aniline and nitrobenzene produce methaemoglobin. Slate blue–brown skin. CNS anoxia. Renal tubular necrosis, hepatic necrosis. Haemolytic in large doses.

Antimony and its compounds

Cause dermatitis, conjunctivitis and ulceration. Several salts (e.g. tartar emetic: potassium antimony tartrate) have been used therapeutically as parasiticides. The dark cosmetic Kohl is finely powdered antimony sulphide. Stibine: poisonous gas SbH_3, produced by action of acid on antimony residues, such as can occur in storage batteries. Nausea, vomiting, colic, can be fatal. Stibine produces liver damage and jaundice, as does arsine.

Apamin

Octadecapeptide (MW 2039) from bee venom. Blocks Ca^+ gated K^+ channels. Unco-ordinated movements; hyper-excitability. Crosses the blood–brain barrier.

Aquinite

French name for chloropicrin *qv*.

Arecoline

The major active alkaloid obtained from the betel nut *qv*. Agonist at muscarinic and nicotinic cholinergic receptors. Euphoria, CNS stimulation, vertigo, salivation, retching, miosis, tremor, bradycardia.

Arsenic

The account by Blyth and Blyth (1920) contains much historical information on the use of As as an aphrodisiac, as a way on improving the appearance of horses, as a parasiticide in sheep dip and in fireworks to produce different colours. Many old preparations listed. Poisoning: Acute form: death in 24 h, vomiting, diarrhoea, cold extremities, weak pulse, drop in temperature. Sub acute: coated tongue, great thirst, abdominal pain, pain in loin, swallowing painful. Breath smells of arsine (garlic), convulsions. Slow poisoning (wallpaper used to be coloured green with arsenical dyes: bacteria growing on the damp paper released arsenic): malaise, jaundice, runny nose, convulsions, 'hectic-fever' death.

Arsenic trichloride

Marsite, butter of arsenic. Colourless liquid, decomposes in moist air to HCl and arsenious oxide. Mainly irritant effects. Very toxic; more than would be predicted from arsenic content. Toxicity as a war gas reported as 1.5 times that of chlorine, possibly due to deeper penetration into the lung.

Arsine

Arsenic trihydride. Colourless inflammable gas, smells like garlic, 2.7 × density of air, very poisonous even at 0.5 ppm. Produces haemolysis. Symptoms can be delayed by 2–24 h: nausea, malaise, shivering, intense headache, vomiting, pain in loin, urine: port-wine coloured, skin becomes copper coloured, or shows 'rain-drop' pigmentation. Odd changes in eye colouration are reported: brown eyes can go grey, said to be due to effects on melanin metabolism.

Arum maculatum

Cuckoo pint. The red berries are poisonous. Contains an acrid irritant.

Asterosaponin L

See astichoposide C.

Astichoposide C

Related to thelenotoside B and asterosaponin L. Starfish neurotoxins, steroidal glycosides, affect membrane ion permeability and inhibit Na^+/K^+ ATPase, depolarize membranes, block cholinergic neuromuscular transmission. Act like ouabain.

Atraxotoxin

Also known as atraxin. Toxic component of the funnel web spider *Atrax robustus*. Disrupts ACh vesicles and depletes the nerve terminals of ACh. MW: 15 000–20 000. Activates Na channels.

Atropine

Atropine is racemic hyoscyamine. Hyoscine and hyoscyamine are the main active components of *Atropa belladona* (deadly nightshade), *Hyoscyamus niger* (black henbane) and *Datura stramonium* (thorn apple). Anticholinergic: dilated pupils, thirst, flushing, mania, collapse, tachycardia. Flushing can look like scarlet fever: due to wide blood vessel dilatation in response to inhibition of heat loss due to inhibition of sweating. Rabbits, some only, make atropinase and can eat leaves of deadly nightshade without suffering ill-effects. Eating such rabbits that have recently been feeding on deadly nightshade can produce atropine poisoning. Stramonium cigarettes used to be prescribed for asthma.

ATX II

Produced by the Wax Rose sea anemone (*Anemonia sulcata*). 47 amino acid basic polypeptide, MW 4770. Prolongs opening of voltage-gated Na^+ channels. Reports of human deaths due to exposure in the Mediterranean.

Azides

Compounds of hydrogen or a metal and the monovalent N_3 radical. Some azides are unstable and explode spontaneously. Act like cyanide: block cytochrome oxidase. Hyperkinesia seen in animals. Sodium azide: potent vasodilator. Highly toxic: hypotension, tachycardia, tachypnoea, severe headache and convulsions. Rat oral LD_{50}: 45 mg/kg bw. Used in automatic blood counters, as a propellant for inflating air-bags in cars, nematocide, rot control in fruit.

B-Stoff

German abb. for bromoacetone *qv*.

BA

British and USA abb. for bromoacetone *qv*.

Barium

Barium chloride and carbonate are rat poisons. Death can occur rapidly: < 1 h, but is usually delayed. Gut upset, vomiting, colic, diarrhoea, blurred vision, bradycardia, areflexia and paralysis. Lethal dose 0.8 g, but recovery has been seen with much larger doses.

Batrachotoxin

Most toxic of the steroidal alkaloids present in the skin of dart-poison frogs, *Phyllobates*. Used as an arrow poison in western Colombia. Opens Na^+ channels and depolarizes nerve fibres: irreversible depolarization leading to paralysis; ventricular fibrillation, cardiac and respiratory failure and death. Effect abolished by tetrodotoxin. Only toxic via damaged skin or digestive tract. Mice: lethal at 2–3 μg/kg SC and 0.1 μg/kg IV.

BB

British abb. for dichlordiethylene sulphide *qv*.

BBC gas

Bromobenzyl cyanide *qv*.

Be-stoff

Austro–Hungarian abb. for bromoacetone *qv*.

Benactyzine

Centrally acting anticholinergic. Has an antidepressant effect. Has been considered as an alternative or addition to atropine in the treatment of nerve-agent poisoning. Do not confuse with BZ *qv*, which also has anti-cholinergic effects.

Benzyl bromide

Cyclite, T-stoff, G *qv*. Some sources also give benzyl bromide the French name Fraissite (see benzyl iodide). Used briefly in WWI: given up when more effective compounds appeared. Clear liquid, bp 199°C, aromatic odour. Decomposes in contact with iron: put in lead canisters in projectiles. Used as a mixture with castor oil, alcohol, thiosulphate and glycerol: irritant above 4 mg/m^3, salivation, nausea, CNS-depressant. Absorbed well by charcoal.

Benzyl chloride

Used briefly in WWI. Colourless liquid, bp 179°C. Attacks iron, tin and copper vigorously. Irritant, CNS-depressant.

Benzyl iodide

Fr: Fraissite, named after Dufraisse, a pharmacist seriously poisoned while carrying out research on this agent. Fraisinite also given by some sources. Used by French, March 1915: *before* the first chlorine attack by Germany. Colourless crystals, mp 24°C, bp 226°C. Gives off a vapour. Very strong irritant, lachrymator at 2 mg/m^3.

Berger mixture

Zinc, carbon tetrachloride, zinc oxide and kieselguhr. Used in smoke candles by the French in WW1. The finely divided zinc reacts with the organic chloride to produce zinc chloride and carbon to produce a high temperature (1200°C) and a dense smoke. The kieselguhr is used to produce a paste of the components to prevent the zinc settling at the bottom of the mixture.

Bertholite

French term for chlorine.

Betel nuts

Seed of the tropical palm *Areca catechu*. Betel nuts may be chewed solely, but also in combination with lime and betel leaves (*Piper betel*). The betel leaves, which contain phenols, probably produce synergistic effects in combination with betel nuts. Lime ($Ca(OH)_2$) quickens the absorption of the main psycho-active ingredients, arecoline *qv* and structurally related alkaloids. The practice of betel-chewing is widespread throughout Asia and dates from over 7000 years ago. 8–10 g of pulverized nut may be lethal.

BIBI

Dibromomethyl ether *qv*.

Bicuculline

Convulsant alkaloid found naturally in several species, including *Corydalis* sp. Blocks the hyperpolarizing effects of GABA: molecule contains a GABA-like moiety.

Black Bryony

Tamus communis. Poisonous berries: can kill. Gastroenteritis in large doses. Paralysis of legs reported.

Black widow spider venom

A–Latrotoxin is the major protein neurotoxin in the venom of the black widow spider (*Latrodectus mactans*). The 'red back' spider, *L. nasseltii*, produces the same toxic effects. Pain, sweating, circulatory collapse, muscular rigidity, respiratory failure, death. Depletes nerve endings of ACh, adrenaline and glutamate: acts on vesicle membrane.

Blue cross gases

Solids of low volatility with great irritant properties, e.g. diphenyl chloroarsine, diphenyl cyanoarsine *qv*.

BM mixture

Zinc dust 35.4%, CCl_4 41.6%, sodium chlorate 9.3%, NH_4Cl 5.4%, $MgCO_3$ 8.3%. American **B**ureau of **M**ines improvement on Berger Mixture. Very effective smoke, various starter mixtures, some including sulphur, were also used with this mixture.

BN-stoff (Bn stoff)

Bromomethyl ethyl ketone *qv*.

Boric acid

Water-soluble, volatile with steam. Gut effects: vomiting and diarrhoea, meningitis-like symptoms. Bright red skin. Boiled lobster appearance: bright red nails, fingers and toes. Skin exfoliates later. Lethal dose 15–20 g adults; 3–5 g infants.

Botulinum toxins

Seven antigenicaly distinct (A–G) protein toxins produced by anaerobic bacteria of the genus *Clostridium,* which are the cause of the paralytic food poisoning known as botulism (Latin for sausage-related condition because this was the food associated with the initial identification of the condition). Block ACh release at neuromuscular junction: irreversible binding: nerve fibres degenerate; also cause agglutination of red blood cells. Nerve recovery by sprouting at terminals and establishment of new endings. Botulinum toxin A is the most toxic compound known. Death due to respiratory muscle paralysis.: LD_{50} (human) given as 0.01–0.001 μg/kg. Used therapeutically in neurology and cosmetically to reduce facial wrinkles.

Bretonite

French abb. for iodoacetate *qv*.

British Type S smoke

Used as a filling for smoke candles: smoke torch Mark 1 Type S. KNO_3 45%, sulphur 12%, pitch 30%, borax 9%, glue 4%. Later, the Mark 2 appeared: KNO_3 40%, sulphur 14%, pitch (hard) 29%, borax 8%, coal dust 9%. Yellow–brown smoke. Not a good obscurant but cheap and much used in WWI.

Brodifacoum

White powder. Rat poison, anticoagulant; rat oral LD_{50} 270 μg/kg bw.

Bromine

Heavy dark red liquid, vapour pressure at 20°C: 172 mmHg, bp 59°C. Limited use as a war gas: used with chlorine to increase persistence in WWI. Irritating vapour, eye irritant. Fatal at 220 mg/m^3 for 30–60 min.

Bromlost

Dibromoethyl sulphide *qv.*

Bromomethyl ethyl ketone

German terms: Bn-stoff. French Abb, homomartonite. Pale yellow liquid, bp 145°C, volatility at 20°C: 34 000 mg/m^3. Powerful lachrymator.

Bromoacetone

German term: B-stoff; USA and UK: BA. Used by Germany in 1915: shells and bombs. Colourless liquid, pungent odour, commercial product is yellow–brown, bp 24°C, vapour pressure at 20°C: 9 mmHg. Strong irritant: very effective lachrymator at 1.5 mg/m^3; lethal at 3.2 mg/m^3 for 10 min. Liquid causes blisters on skin: heal quickly but painful on sensitive parts of the body.

Bromoacetophenone

White crystals: greenish on exposure to light. Mp: 50°C; bp ca. 250°C; easily volatile with steam; soluble in organic solvents. Lachrymator.

Bromobenzyl cyanide

Camite, CA, BBC. Used by French in solution in chloropicrin, also used by USA. Very effective war gas: great persistence and lachrymatory power. Yellowish–white crystals, turn pink as they decompose. Technical product: oily brown liquid, bp 242°C; vapour pressure at 20°C: 0.012 mmHg. Difficult to break down; attacks metals but not glass. Very irritant: minimal effective concentration 0.2 mg/m^3 insupportable from 5 mg/m^3. Said to be the most powerful lachrymator used in WWI: about as potent as chloroacetophenone. Last lachrymator introduced in WWI: 1918. Very persistent in soil.

Bromomethyl ethyl ketone

Colourless pale yellow liquid. Bp 146°C. Irritant, lachrymator. Limit for eye irritation 1.6 mg/m^3.

Bromophosgene

See carbonyl bromide.

Bromopicrin

Crystals: melt at 10°C. Irritant but less so than chloropicrin *qv*.

Bromovinyl dibromoarsine

Bp 108°C. Irritant.

Brucine

Alkaloid related to strychnine; occurs naturally with strychnine. Best source: false angustura bark: this contains very little strychnine. White crystals, very bitter taste. Action similar to strychnine but less toxic and convulsions may be lacking: oral LD_{50} rat 1 mg/kg bw.

BTX

Botulinum toxin *qv*.

Bufotoxin, Bufotalin and Bufotenine

Usually isolated from toad skin: psychostimulants: hallucinogenic effects. Also found in certain plants and fungi.

Bungarotoxins

Protein components of the venom of the banded krait (*Bungarus multicinctus*). Two major components: α- and β-bungarotoxins. α-Bungarotoxin binds irreversibly to ACh receptor causing neuromuscular blockade and muscle paralysis similar to effects of curare. β-Bungarotoxin contains several components: prevents ACh release at skeletal neuromuscular junction. Crude venom LD_{50} SC mouse, 0.16 mg/kg bw.

Burdock root

Sold as burdock tea. Atropine-like effects: hallucinogenic, dilated pupils.

BZ

3–Quinoclinidinyl benzilate. Atropine-like CNS depressant. White powder. Effective by all routes: percutaneous uptake poor although good for related compounds. Onset: 1 h. Increased heart rate, dizzy, ataxia, vomiting, dry mouth, blurring of vision. 4–12 h: unable to move around. 12–96 h: random behaviour, delusions, hallucinating. Heat stroke is a risk. Agent 15 may well be identical with BZ. BZ should not be confused with benactyzine.

C-stoff

Different sources give differing accounts of the composition of the gas with this German name. Some indicate that the irritant component was methyl chlorosulphonate *qv*, used mixed 1:3 with dimethyl sulphate. This agent is also given the French name Villantite. Most sources indicate the toxic component as monochloromethyl chloroformate *qv*, also known by the French name Palite. Both methyl chlorosulphonate and monochloromethyl chloroformate were first used by the Germans in June 1915, which has perhaps given rise to the confusion.

CA

USA term for bromobenzyl cyanide *qv*.

Cadmium

Metal: inhalation is the greatest hazard. Within several hours: irritation of the airways, nausea, vomiting, chest pain, dizziness, pulmonary oedema, death. Survivors: emphysema, nephrotoxicity, osteomalacia.

Calcium cyanamide

Carbamide, nitrolime. White to grey-black powder. Skin-caustic. Releases cyanide. Used as fertilizer, defoliant, herbicide, pesticide and veterinary anthelmintic. Citrated form used as an anti-alcohol treatment as per disulfiram.

Calcium peroxide

White–yellowish powder. Strong oxidizer/ respiratory and dermal irritant. Decomposes to H_2O_2. 10% solution can cause skin burns. Ingestion: irritation and gas embolism.

Camite

French : bromobenzyl cyanide *qv*.

Camphor

Occurs in all parts of the camphor tree *Cinnamomum camphora*. White solid, penetrating odour; sublimes appreciably at ambient temperature; readily volatile in steam. Teaspoonful of camphorated oil said to produce serious toxic effects in an adult. Children: 1 g camphor: profuse dermal, gastric and renal haemorrhaging. CNS deterioration, death. Liver and kidney: fatty degeneration. Rapid absorption from gut: lipid-soluble. Convulsions: status epilepticus.

Campiellite

French term for 25% cyanogen bromide *qv*, 25% bromoacetone *qv*, 50% benzene.

Campillit

Italian term for cyanogen bromide *qv*.

Cantharides

Spanish fly *Cantharis vesicatoria*. Allegedly aphrodisiac. Active component – cantharadin. Encountered also as the powdered insect. Effects: blistering of skin and corrosive effect on mouth, abdominal pain, vomiting, diarrhoea, blood passed in vomit and urine. Early increase in WBC. Severe illness: recovery likely but deaths recorded. Lethal dose: 60 mg cantharadin, 162 mg Spanish fly.

CAP gas

USA abb.: chloroacetophenone *qv*, later known as CN.

Capsaicin

Pepper spray, oleoresin capsicum. Extracted from cayenne pepper and paprika: proposed as a harassing agent in WWI. Used in pepper sprays. Irritant. Damages sensory nerve endings.

Carbamates

Reversible anticholinesterases widely used as insecticides. Absorbed: all routes. Usual signs of anti-AChE poisoning: salivation, sweating, bradycardia, gut spasm, involuntary micturition, convulsions, coma.

Carbon disulphide

Liquid, bp 46.5°C. Pure gas reputedly has a pleasant odour but industrial preparations smell foul. Solvent used in rubber and rayon industries. Explosive; poisonous by all routes. Acute toxicity – euphoria, restlessness, mucous membrane toxicity, nausea, vomiting, convulsions. Chronic exposure: psychosis, tremor, polyneuropathy lower limb weakness and parasthesiae. Dermal contact may cause burning sensation, erythema and exfoliation.

Carbonyl bromide

Bromophosgene. Colourless heavy liquid. Bp 65°C. Toxic effects similar to phosgene.

Carotatoxin

A neurotoxic polyacetylenic alcohol found in carrots.

Cassava

Cassava shrub: *Manihot utilissima*. Important food: root eaten in tropics. High starch content. Tapioca is prepared from cassava. Leaves and outer parts of roots contain linamarin: enzymatic hydrolysis to acetone and HCN. Peeling and drying important. Boiling drives off the HCN. Acute poisoning: abdominal pain, vomiting, confusion, weakness, respiratory depression. Chronic poisoning: demyelination. Optic nerve damage: amblyopia.

CB

British abb. cyanogen bromide *qv*.

CBR

British abb. 50% phosgene *qv*, 50% arsenic trichloride *qv*.

CC

British abb. cyanogen chloride *qv*.

CC-2

Used to impregnate clothing to protect against chemical weapons: Sym-bis-(chloro-2,4,6-trichlorophenyl)urea.

CDA

Clarke II, diphenyl cyanoarsine *qv*. The abb. CDA is also used for an unrelated cationic germicidal detergent.

Ce

Austrian abb. cyanogen bromide *qv*.

Cedentite

French term for mixture of *o*-nitrobenzyl chloride and *p*-nitrobenzyl chloride *qv*.

CG

UK/USA term for phosgene *qv*.

Charybdotoxin

Venom of Israeli scorpion (*Leiurus quinquestriatus*). Protein toxin MW 7000. Blocks calcium-activated potassium channels in mammalian skeletal muscle.

Chlorine

French abb. Bertholite. Ypres 22 April 1915: 168 tons released from cylinders by German forces: 15 000 casualties: 5000 dead. Yellow–green gas, pungent odour. Liquefied by 6 atm at 70°F. Sensory nerves affected, pain in chest, pulmonary oedema. Limit of supportability 1 min 100 mg/m^3; LC_{50} 1 h inhalation, rat 0.42 mg/m^3. High chemical reactivity makes it easy to defend against.

Chloroacetone

Tonite. A-stoff, CAP gas. Used by French in WWI. Clear liquid bp 119°C. Irritant, lachrymator. Odour of HCl, intolerable at 0.018 mg/l. Well absorbed by charcoal. Minor role in WWI.

Chloroacetophenone

USA tear gas codename CN. Colourless to yellow crystals. Mp 59°C. Soluble in organic solvents. Solution evaporates to leave a fine solid aerosol. Lachrymator at 0.3 mg/m^3; skin irritant at 100 mg/m^3. Rapidly reversible. Dogs and horses relatively insensitive.

Chloroacetic acid

Colourless-white deliquescent crystals. Smells of acetic and hydrochloric acids. Irritant to skin and mucous membranes.

o-Chlorobenzylidinemalononitrile

CS, tear gas. Skin and eye irritant; rapidly reversible. White solid, low vapour pressure; used as aerosol. Rapidly hydrolyzed in water.

1,2-bis(2-Chloroethylthio)ethane

Sesqui-mustard, agent Q, QN2. Solid, Mp 56°C. volatility <1 mg/m^3 at 20°C. British laboratory reported vesicant activity to be five times that of mustard gas and this may have been an underestimate.

Bis(2-Chloroethylthioethyl) ether

Agent T. Vesicant: Mp 10°C. Like sesqui-mustard a low-volatility vesicant. Persistent.

Chloroform

Sweet-smelling, volatile liquid, bp 61°C. Acute effects: CNS depression, myocardial depression, respiratory arrest; chronic effects liver damage and carcinogenesis.

Chloroformoxine

Crystalline needles. Small quantities volatilize at ordinary temperatures. Lachrymator and vesicant.

Chlorophenothane

DDT *qv*.

Chloropicrin

Klop; acquinite; PS; NC, Larvacide 100; Picfume. Russians used it in hand grenades in 1916: dissolved the chloropicrin in sulphuryl chloride (50%). Insecticide, fungicide, used to kill rats in ships. Oily liquid, yellowish in impure form, intense odour; bp 112°C, attacks lead energetically, attacks iron and copper more slowly. Absorbed well by activated charcoal. Strong irritant and toxic asphyxiant.

Chlorostyryl dichloroarsine

Arsenical CW agent developed by the Allies at the end of WWI.

Chlorosulphonic acid

Used by France and Germany in WWI. Colourless corrosive liquid; bp 153°C. Fumes in contact with air, producing HCl and H_2SO_4. Strong irritant. Used as a smoke-producing agent.

Chlorovinyl dichloroarsine

Lewisite, L. Liquid bp 190°C, colourless when pure: brown when impure. Smells of geraniums. Rapidly decomposed by damp air and water. Irritant vesicant, systemic poison due to As. As little as 0.5 ml on skin can cause severe toxicity and 2 ml, death.

Chlorovinyl methyl chloroarsine

Liquid, bp 115°C. Less irritant than methyl dichloroarsine but also vesicant. Blisters said to be difficult to heal.

Cholera toxin

Protein toxin secreted by the bacterium *Vibrio cholerae*. Activates adenyl cyclase. Leads to prolonged loss of Na and water from the gut.

ClCl gas

Dichloromethyl ether *qv*.

Cicutoxin

Solid unsaturated diol from water hemlock *Cicuta virosa*: umbelliferus plant. Extremely poisonous. Pain and burning in stomach, nausea, vomiting, convulsions. As in strychnine poisoning. Swallowing impossible. Respiratory failure. Deadly: 31 cases: 14 died. Fatal dose unknown.

Ciguatoxin

From ciguateric Moray eels. Acts like batrachotoxin *qv*. Ciguatera is also the term applied to the syndrome (headache, paraesthesiae, lassitude, haematuria, pruritis, cardiac dysrhythmias, vomiting, diarrhoea) caused sporadically by the consumption of a range of reef-dwelling fish. In both types, the source of the poison may be the blue–green algae or other components of the diet but detailed descriptions of the chemical identity of the poison(s) are not available.

CK

Formerly CC: cyanogen chloride *qv*.

Clairsite

Perchloromethyl mercaptan *qv*.

Clark I

Diphenyl chloroarsine *qv*.

Clark II

Diphenyl cyanoarsine *qv*.

CN

Formerly known as CAP: chloroacetophenone *qv.*

CNS

Chloroacetophenone solution or mixture of chloropicrin, chloroacetophenone and chloroform.

Cocaine

Major component of eight alkaloids in dried coca leaves (*Erythroxylon spp*). Small doses excite the CNS, large doses: paralysis/convulsions. Local anaesthetic effect. Failure of accommodation, dilated pupils. Early effects: dry nose and throat, difficulty in swallowing, fainting, vomiting. Hyperaesthesia followed by diminution of sensation. Sweating and pallor. Death can occur as a result of respiratory failure.

Coelenterates (Cnidarians)

Many very toxic nematocyst toxins in this group of marine invertebrates, which include the Portuguese man-of-war (*Physalia*). Intense pain, shock, paralysis, nausea and vomiting, delirium, often fatal.

Colchicine

Major alkaloid extracted from seeds of *Colchicum autumnale*: common meadow saffron. Nausea, vomiting, convulsions, paralysis, death. Used as a treatment for gout and in experimental biology as an arrestor of mitosis to facilitate examination of chromosomes.

Collongite

French name for phosgene *qv.*

Coniine

Alkaloid extracted from hemlock *Conium maculatum*. Used to execute Soccrates. Colourless liquid, bp 166°C, volatile with steam. Acts like nicotine: large doses produce paralysis and respiratory depression.

Conotoxins

A large number of toxic peptides are found in cone snails (Conidae). They include four classes of small (13–29 amino acid) basic peptides, known as conotoxins, that have paralytic effects: ω-conotoxins block presynaptic calcium channels, α-conotoxins bind to nicotinic ACh receptors in muscle, μ-contoxins bind to muscle sodium channels, κ-conotoxins may affect potassium channels. Other cone snail toxins include 'sleeper' that induces sleep in mice, conopressin that produces scratching behaviour in mice and increases blood pressure in mammals and a larger (ca. 100 amino acid) toxin that produces convulsions in mice.

Convulxin

Platelet-aggregating factor found in the venom of the snake *Crotalus durissus.*

CR

Dibenzoxazepine *qv.*

Crotalocytin

Platelet-aggregating factor found in the venom of the snake *Crotalus horridus.*

Crotamine

Basic polypeptide toxin from rattlesnake *Crotalus terrificus*. Activates sodium channels in muscle: depolarization and contracture, blocked by tetrodotoxin. See also crotoxin.

Croton oil

From seeds of *Croton tiglium*: a West-Indian plant. Symptoms and signs: pain in abdomen, purging, vomiting. Death uncommon.

Crotoxin

Mixed polypeptide complex neurotoxin from rattlesnake venom. Prevents ACh release and blocks muscle ACh receptor. Systemic effects: dizziness, sensory and motor depression, collapse; may be fatal.

Crude Oil Smoke

The smoke is a dense aerosol of carbon particles. Most effective forms produced by incomplete combustion of oil: each carbon particle has an unburnt oil film around it. Described as 'slightly suffocating when dense'.

CS

See *o*-chlorobenzylidenemalononitrile.

Cuckoo Pint

See *Arum maculatum.*

Cucumarioside G

See Holothurin.

Curare (Wookari)

Plant extract used by South American Indians to prepare poison arrows. Sold with different names (e.g. Tube-curare) depending on the container used for packaging. Toxicologically active principle is the alkaloid tubocurarine chloride. Muscle relaxant, toxic by respiratory paralysis and hypotension; competitive blocker ACh receptor in muscle.

CX

See phosgene oxime.

Cyanides

HCN, KCN, etc. HCN formed from polyurethane by combustion. Cyanogenic plants (*qv*) and cyanogen halides (*qv*) are all poisonous by virtue of cyanide produced. NaCN and KCN hydrolyze in air to produce HCN. Adding HCl to KCN was the method used in judicial execution in the USA. Absorption by inhalation, ingestion and via skin. Effects: watering of eyes, headache, salivation, palpitations, difficulty in breathing, numbness, 'floating feeling'. CN ions bind to cytochrome oxidase system: blocks internal respiration. Recovery is rapid following appropriate treatment.

Cyanogen $(CN)_2$

Colourless poisonous gas; bp –21°C; almond-like odour, pungent at lethal concentrations. Acts as per cyanide *qv*.

Cyanogen bromide

Used by Austrians, September 1917: as a mixture with benzene, bromacetone and benzene: Campiellite. Crystalline, mp 52°C, vapour pressure at 25°C: 119.5 mmHg, volatility at 16°C: 155 000 mg/m^3. Soluble in organic solvents, sparingly soluble in water. Corrosive to metals. Irritant to eyes and airways at 6 mg/m^3. Effects as per cyanide.

Cyanogen chloride

Used by French, October 1916. Colourless liquid: very volatile, bp 12.5°C, vapour pressure at 20°C: 1000 mmHg, volatility at 20°C 3.3 kg/m^3. Usually contains 2–5% HCN. Added to Zyklon B (*qv*) to prevent polymerization of HCN and to provide a warning by irritant properties. Irritant at 2.5 mg/m^3. Lethal 40 mg/m^3 for 10 min.

Cyanogen fluoride

Colourless gas at room temperature. Sublimes at –72°C. Not much seems to be known.

Cyanogen iodide

Not used in WWI. White crystals mp 146°C. Dissolves well in hot water. Pungent odour; lachrymator, causes convulsions, paralysis and death by respiratory failure following ingestion but not inhalation.

Cyanogenic foods

Several foods contain cyanogenic glycosides, such as amygdalin. Kernels of stone fruits, bitter almonds (250 mg HCN per 100 g). Kernels appear to become toxic when mixed with water, e.g. in chewing. Boiling destroys the poison: Plants: cassava tubers: Nigerian poisoning. Choke cherries, arrow grass: cattle poisoning. Laurel leaves: chopped leaves used in killing bottles by entomologists, tips of immature bamboos (800 mg HCN/100 g). Coloured Java and black Puerto Rico varieties of Lima beans. CN is metabolized to thiocyanate, which is goitrogenic and can cause foetal cretinism.

Cyanuric triazide

Used in war-gas mixtures: effects similar to those of cyanide.

Cycasin

Toxic component of badly prepared starch from nuts of the cycad tree. Paralysis: arms and legs similar to amyotrophic lateral sclerosis. Death in 5 years. Also carcinogenic. Possible cause of dementia on the Pacific Island of Guam.

Cyclite

French abb. for benzyl bromide *qv*.

Cyclon

German abb. for cyanoformate esters.

Cytisine

See laburnum.

D-stoff

Dimethyl sulphate *qv*; also used as abb. for phosgene.

DA

Diphenyl chloroarsine *qv* Clark I.

Dart-poison frogs

130 species, often brightly coloured. 200 alkaloids isolated from skin secretions. Batrachotoxins *qv*, are the most potent poisons, other toxins include pumilotoxin B, which affects intracellular Ca levels causing muscle spasms and convulsions.

Datura

Datura stramonium: Jimson weed, devil's weed, Thorn apple. Contains atropine, hyoscine, hyoscyamine. Stramonium tea and cigarettes used to be used for asthma. Dilated pupils, mania, red skin colour (dilated vessels and no sweating) thirst, tachycardia, vomiting, convulsions. 100–125 seeds lethal. Death uncommon. Datura was apparently used in India by poisoners to remove 'idiots' from positions of high rank.

DC

Belgian abb., diphenylcyanoarsine *qv*.

DDT

Dicophane, chlorophenothane. Solid organochlorine insecticide. Delays closure of sodium channels in nerve membrane and partially blocks K^+ channels. Convulsant: respiratory paralysis. Absorbed following oral, inhalation or dermal exposure. Very stable.

DEET

Diethyltoluamide. Liquid insect repellant often used to protect troops in conflict scenarios. Generally considered safe, irritant to eyes; CNS effects: light-headedness. Nausea, vomiting, diarrhoea, hypotension.

Delphinine

See aconite.

Dendrotoxin

59 Amino acid polypeptide toxin from Eastern green mamba (*Dendroaspis augusticeps*). Inhibits potassium channels and facilitates release of ACh at nerve terminals: increases evoked release but does not produce spontaneous release of ACh.

Dermorphin

Heptapeptide isolated from skin of South American frogs of the *Phylomedusae* family. Opioid activity: many times more potent than met-enkephalin.

o-Dianisidine chlorosulphonate

The first irritant compound used in shells by Germany in WWI. Shells contained lead balls embedded in this compound. Called dianisidine niespulver: (Niespulver is German for sneeze powder).

Diazobenzol

Tyrotoxicon. A toxin reported in the early 1900s in putrid cheese and other dairy products, and producing violent diarrhoea and dehydration, symptoms similar to cholera. Chemical identification – possibly diazobenzol.

Diazomethane

Highly toxic yellow gas, mp –35°C; soluble in ether; concentrated solutions are explosive; strong irritant with a delayed inflammatory effect; carcinogenic.

Dibenzoxazepine

Tear gas codename, CR. White solid; low vapour pressure; used as aerosol. Irritant. More potent than CS but less toxic. Low solubility and not easily hydrolyzed by water and thus less easy to remove from the skin and hair and from surroundings than CS.

Dibromoacetylene

Colourless heavy liquid, bp 76°C. Unstable: inflames spontaneously in air, burns with a red flame, heating produces explosion. Inhalation of vapour: violent and prolonged headache.

1,2-Dibromomethane

Colourless viscous liquid with a chloroform-like odour. Rapidly absorbed: all routes. Strong respiratory irritant. Inhalation: bronchospasm, laryngeal oedema, chemical pneumonitis, pulmonary oedema. Skin: erythema, blistering. Oral: nausea, vomiting, weakness, headache, brady/tachycardia, oliguria, jaundice, confusion, coma. Metabolic acidosis. Shock. CNS: depression, skeletal muscle necrosis.

Dibromoethyl sulphide

White crystals, mp 31–34°C, volatility at 20°C: 400 mg/m^3. Vesicant.

Dibromomethyl ether

Colourless liquid, bp 154°C. Lachrymator. Mortality rate given as 400-fold greater than that of phosgene. Known as BIBI and as X gas.

Dichlorodimethyl ether

Bis(chloromethyl) ether, BCME, dichloromethyl ether, chloromethyl ether, Cici. Colourless liquid, bp 106°C, suffocating odour; decomposes in moist air to HCl and formaldehyde; eye and respiratory irritant, acts on labyrinth, producing staggering. Used as a war gas and as a solvent for ethyl dichloroarsine and mustard gas.

1,2-Dichloroethane

Colourless oily liquid: chloroform-like odour. Sweet taste. Vapour can be ignited. Irritant to eyes and gut. CNS depressant. Sensitizes myocardium to endogenous catecholamines. Headache, ataxia, dizziness, dilated pupils, convulsions, coma, renal failure (tubular necrosis), liver damage (jaundice), extrapyramidal effects can occur. Death: arrhythmia.

Dichloroethyl sulphide

Sulphur mustard, mustard gas, H, HS, Yperite, Hunstoff, LOST. The abbreviation HD, distilled mustard is also used, especially by American authors. German use of mustard gas for the first time in WWI, July 1917: took the Allies a year to produce enough to respond. Also used by Italy in the Abyssinian campaign in the 1930s and by Iraq in the Iran–Iraq war in the late 1980s. Oily liquid, colourless if pure, crude state: brown liquid, bp 217.5°C, mp 14.4°C, vapour pressure at 20°C: 0.115 mmHg, volatility at 20°C: 625 mg/m^3. Odour: garlic, leeks, onions, mustard. Reacts with oxidizing agents: dichloroethyl sulphoxide and sulphone. The sulphone is vesicant and irritant. Eyes, skin, lung effects. Blister fluid not dangerous. Conjunctivitis and chemosis. Blisters appear for up to 24–48 h: damp warm areas of skin badly affected. Coughing: sloughing of airway epithelium. Bone marrow depression can be lethal. Carcinogenic.

Dichloroformoxime

See phosgene oxime.

Dichloromethane

Methyl chloride, methylene dichloride. Colourless liquid, bp 40–41°C. Industrial solvent and paint remover. Fumigant. Rapidly metabolized to carbon monoxide: produces carboxy-Hb. Irritant to eyes and skin: blistering on prolonged contact. Anaesthetic effects on CNS: coma, death.

Dichloromethyl chloroformate

Colourless liquid, bp 110°C, irritant: less so than the mono form but more toxic. Used as a war gas mixed with monochloromethyl chloroformate *qv*.

Dichlorovinyl arsenious sulphide

Formed by reaction between hydrogen sulphide and dichlorovinyl chloroarsine. Viscous yellow–brown liquid, nauseating smell. Irritant to mucous membranes.

Dichlorovinyl chloroarsine

Clear yellow/yellow–brown liquid, bp 230°C. Very reactive. Hydrolyzed by water. Similar effects to lewisite *qv* but less potent.

Dichlorovinyl methyl arsine

Liquid, bp 140°C, irritant at > 2 mg/m^3, vesicant. Similar effects to chlorovinyl methyl chloroarsine *qv*.

Dick

Used by Germany as an abb. for ethyl dichloroarsine *qv*. Also methyl dick and phenyl dick: the 'Dicks'.

Dicophene

See DDT.

Diethyl arsine

Liquid, bp 105°C. Said to be produced by the mould *Penicillium brevicaule* growing on wallpaper coloured green with arsenical green pigments. Causes chronic exposure to arsenic.

S-2-Diethylaminoethylphosphorothioate

See Amiton.

Diethyltoluamide

See DEET.

Digitalis and similar compounds

Cardiotoxic glycosides in *Digitalis purpurea*: foxglove. Oral: malaise, vomiting, early tachycardia, auditory and visual disturbances, gastric irritation, swollen tongue, foul breath, hiccough, convulsions, death. Cardiotoxic glycosides and alkaloids are also found in a number of other plants, such as the yew (*Taxus cuspidata)* and false hellebore (*Veratrum viride*), dog bane (*Apocynum spp*). Some of these cardiotoxins have been used as arrow poisons.

Diiodoacetylene

White crystals, mp 78.5°C. Strong, unpleasant odour, explodes on rubbing in a mortar. Vigorous irritant: eyes affected.

Diiodoethyl sulphide

Developed in 1920 as a war gas but not used in warfare. Bright yellow crystals, mp 62°C, insoluble in water and common organic solvents. Vesicant action similar to dichloroethyl sulphide:

noted as having greater toxic power than sulphur mustard.

Dimethyloxybenzidine

Dianisidine: mp 170°C. Developed by Germany as a war gas.

Dimethyl sulphate

Used as a war gas mixed with methylchlorosulphonate (German D-stoff) or chlorosulphonic acid (French Rationite). Colourless odourless liquid, bp 188°C, volatility at 20°C: 3300 mg/m^3, decomposed by water and damp air. Delayed appearance of symptoms may result in lethal exposures; liquid produces severe skin damage and can result in systemic toxicity; vapours – delayed severe inflammation and necrosis of eyes and respiratory tract; a range of CNS symptoms and fatal pulmonary damage. 0.5 mg/l for 30 min said to be fatal. Leaves a peculiar analgesia of the skin that is said to last for 6 months.

Dioscorine

Found with dihydrodioscorine in yams (e.g. *Dioscorea hirsute*): painful burning of mouth, GI upset, convulsions.

Diphenyl bromoarsine

White crystals, mp 55°C. Irritant, less active than the chloro compound below.

Diphenyl chloroarsine

Clark I. Used by Germany in 1917. Canister penetrator: surprised Allied forces. Dark brown liquid: semi-solid viscous mass. Smells of shoe polish. Crystalline in pure state, mp 41°C, low vapour pressure: 0.0005 mmHg at 20°C. Sprayed in solvent to produce aerosol of particles 0.1–1 μm diameter. Dispersed also by explosion and from burning candles: these were described as 'irritant candles'. Vesicant, irritant at 0.1 mg/m^3, lethal at high concentration. Headache, vertigo, trembling.

Diphenyl chlorstibine

White crystals, mp 68°C. Irritant, causes sneezing.

Diphenyl cyanoarsine

Clark II, CDA. Used in 1918 alone and mixed with diphenyl chloroarsine. Colourless crystals: odour of garlic and bitter almonds, mp 35°C. Dispersed as an aerosol. Odour detectable at 0.01 mg/m^3, limit of supportability 0.25 mg/m^3 for 1–2 min. Toxic effects similar to those of diphenyl chloroarsine but slightly more potent.

Diphenyl cyanostibine

Crystals, mp 116°C. Irritant, causes sneezing.

Diphenylamine chloroarsine

Adamsite, phenarsazine chloride, DM. Used by Allied forces in WWI. Crystalline solid, mp 190°C, canary yellow when pure, dark green–brown when impure. Irritant at 0.1 mg/m^3. Not endurable at 0.4 mg/m^3. Low lethality. Envelopes containing Adamsite reportedly sent in June 2003 to embassies in Belgium. Skin and eye irritation: no serious effects.

Disulfiram

Produced by the inky cap mushroom *(Coprinus atramentarius)*; see mushrooms.

Disulphur decafluoride

See Z.

Dithiophosgene

Formed from thiophosgene by exposure to light.

Ditran

Trade name. One of the glycolate esters that have 'atropine-like' effects but more marked actions on the CNS. Disorientation, depersonalization, hallucinations, paranoid feelings. Related to BZ and phencyclidine.

DJ

British abb. for phenyl dichloroarsine *qv*.

DM

Adamsite, diphenylamine chloroarsine *qv*.

DNOC

Dinitro-*o*-cresol. Moderately volatile with steam; weed killer: acts by uncoupling oxidative phosphorylation. Nausea, sensation of heat, flushing, tachycardia, cyanosis, coma, death. Yellow conjunctivae. May be confused with OP poisoning: do NOT give atropine. Severity of symptoms increased by higher ambient temperature.

ED

Ethyl dichloroarsine *qv*.

Eledoisin

Peptide toxin produced by the posterior salivary glands of some cephalopoda. (e.g. *Eledone moschata*). Vasodilator: hypotension, increased capillary permeability.

Equinatoxin

Produced by the purple rose sea anemone (*Actinia equina*). 147 amino acid, MW ca. 20 000. LD_{50} rat: 33 μg/kg IV. Cardiotoxin: bradycardia, respiratory depressant. 100 μg/kg: pulmonary oedema. Damages cell membranes.

Erabutoxin

Peptide neurotoxins from the sea snake *Laticauda semifasciata*: resembles bungarotoxin: blocks nicotinic receptors.

Ergot alkaloids

Products of the fungus *Claviceps purpurea*: ergot of rye. Many ergot alkaloids known, with a range of biological activities from vasoconstrictor and oxytocic to hallucinogenic effects: used therapeutically in migraine and obstetrics. Outbreaks of poisoning ('Saint Anthony's Fire') relatively common in the Middle Ages but now rare. The convulsive form of ergotism is well described by Blyth and Blyth (1920) and Bowman and Rand (1980). Bowman *et al.* (1986) note a family who all lost feet as a result of the vasoconstrictor activity.

Erucic acid

A long-chain unsaturated fatty acid found in plants, especially rapeseed and mustard seed oil. Uncouples oxidative phosphorylation in rat heart mitochondria: cardiac damage in rats. Effects in man not reported.

Eserine

See physostigmine.

Ethyl bromide

Colourless liquid, bp 38°C, lachrymator producing eye pain.

Ethyl bromoacetate

The first harassing agent used during WWI. Used by the French in August 1914. Clear colourless liquid, mp –13.8°C, bp 168°C, volatility at 20°C: 21 000 mg/m^3. Irritant and lachrymator at > 10 mg/m^3.

Ethyl carbazol

Carbazole. White flaky solid, mp 68°C, bp 190°C. Insoluble in and little affected by water. Harassing agent used by Germany in WWI. Vapour is seven times heavier than air, respiratory irritant.

Ethyl chloroacetate

Colourless liquid, bp 144°C. Odour of fruit. Lachrymator but used mainly in the production of the bromo and iodo forms.

Ethyl-bis(2-chloroethyl)amine

See HN1.

Ethyl chlorosulphonate

Used by the French in late 1915. Colourless oily liquid, pungent odour, bp 153°C, rapidly decomposed in cold water. Irritant at > 2 mg/m^3. Used mixed with bromoacetone.

Ethyl cyanoformate

Powerful lachrymator, decomposed by water.

Ethyl dichloroarsine

Ethyl Dick. Introduced in March 1918 by Germany. Thought to be more effective than mustard gas in offensive operations because of immediate vesicant effects and low persistence. Mobile liquid, bp 156°C, volatility at 20°C: 20 000 mg/m^3; vapour pressure at 21.5°C: 2.29 mmHg, odour when diluted is of fruit, hydrolyzes in water to ethyl arsenious oxide with nauseating smell of garlic. Irritant at 1.5 mg/m^3, vesicant at 1.0 mg/cm^2 on skin.

Ethyl fluoroformate

Liquid, bp 57°C. Powerful lachrymator.

Ethyl iodoacetate

SK: developed at **S**outh **K**ensington, London (Imperial College). Used by the Allies in WWI, often as a mixture with chloropicrin: 90% SK and 10% chloropicrin. Dense colourless liquid: turns brown in air. Bp 179°C, low volatility. Eye irritant. Lachrymator at 1.4 mg/m^3; toxic at 1.5 g/m^3 for 10 min. About a third as toxic as phosgene. Serious toxicity unlikely in field because of low volatility. One of the most powerful lachrymators used during WWI. Beadnell (1943) says it produced great eye pain.

Ethylidenediamine

One of a number of toxic diamines produced by putrefaction of meat or fish. When administered SC to guinea pigs: abundant secretions from nose and mouth, dilated pupils, projection of eyeballs, dyspnoea, death takes 24 h.

Ethyl nitrite

Sweet spirits of nitre: 4% ethyl nitrite in 70% ethanol used as a diuretic. Overdose produces methaemoglobinemia, hypotension and narcosis.

Ethylsulphuryl chloride

Ethyl chlorosulphonate, Sulvanite. Colourless liquid, bp 135°C, fumes in air. Pungent smell, more lachrymatory and toxic than the methyl compound but less volatile. Intolerable at 50 mg/m^3 and toxic at 1g/m^3.

Ethylene glycol

Clear colourless sweet tasting liquid. Highly toxic by ingestion. Poorly absorbed across skin. Metabolized via glycoaldehyde to glycolic acid. The metabolites cause acidosis, block oxidative phosphorylation and respiration and cause hypocalcaemia and renal damage. CNS depression, nystagmus, ophthalmoplegia, hyporeflexia, metabolic acidosis, pulmonary oedema, renal failure.

Eucalyptus

Active principle: eucalyptol (cineole), gastrointestinal irritant and CNS depressant. Lethal dose reported as between 3.5 and 31 ml. Vomiting, faintness, ataxia, coma, pinpoint pupils.

F gas

A nerve agent of the V series developed in Sweden.

F-stoff

German abb. for titanium tetrachloride *qv*.

Fire corals

Fire corals are coelenterates belonging to the genus *Millepora*. On touch, toxins in nematocysts are released and cause serious local and sometimes systemic effects. Toxins have been purified from various species and include quaternary ammonium compounds, proteins, 5-hydroxytryptamine, catecholamines, diphenylamine, histamine and histamine liberators.

Fluorides and fluoroacetates

Insecticides, rat poisons: sodium fluoride, sodium fluoroacetate: crystalline. Block citric acid cycle by forming fluorocitrate, which inhibits the next enzyme in the cycle. Effects: vomiting, convulsions, coma, respiratory and cardiac arrest. Deadly at ca 5 mg/kg oral.

FM

USA abb. for titanium tetrachloride *qv*.

Forestite

French term for hydrogen cyanide *qv*.

Fraissite

French term: 50/50 mixture of benzyl iodide and benzyl chloride. Some sources indicate this term for benzyl bromide *qv* or benzyl iodide *qv* alone.

Fraisinite

Given by some sources as the French term for benzyl iodide *qv*.

French 4

Hydrogen cyanide *qv*.

French 4B

Cyanogen chloride *qv*.

FS

USA abb. for sulphur trioxide/chlorsulphonic acid mixture.

Fullers' earth

Name given to a range of highly adsorbent clay-like substances consisting of hydrated aluminum silicates, used predominantly as absorbants. Adsorption reversible. Used as a decontaminant. Traditionally used to remove grease and oil from wool: this process is called fulling, hence the name 'fullers' earth'. In the USA, two varieties of fullers' earth are mined, montmorillonite (mainly aluminium silicate) and palygorskite (attapulgite, mainly magnesium aluminium silicate).

Fumergerite

Beadnell (1943) uses this term for 'smoke cloud agent': titanium tetrachloride *qv*.

Furocoumarins

Photosensitizing and phototoxic chemicals found in more than two dozen plant sources, including celery, parsnips and parsley. Major compounds: psoralen, methoxsalen, trioxsalen, bergapten and imperatorin. Stimulate melanocyte response to UV.

Fusariotoxin

See mycotoxin T2.

G agents

Term used originally to describe the anti-cholinesterase nerve agents discovered to have been produced by Germany just before and during World War II. The 'G' stands for Germany. See GA, GB, GD, GE, GF.

GA

Ethyl N,N-dimethylphosphoramidocyanidate. Tabun. First of the nerve agents synthesized by Schrader in 1937. Goebels called for tabun and sarin to be used against invading troops. Given the large stocks held in Germany, it is not easy to see why this was not done. Mp –50°C; vapour pressure at 20°C: 0.057 mmHg, Volatility: 490 mg/m^3 at 25°C. Cholinesterase inhibitor;

toxic by inhalation and by absorption through skin and eyes. Lethal dose for humans of the order of 0.01 mg/kg bw.

Galantamine

Reversible carbamate anticholinesterase and allosteric modulator at nicotinic cholinergic receptor sites potentiating nicotinic neurotransmission. Originally derived from the bulbs of the Caucasian snowdrop (*Galanthus woronowii*), now prepared synthetically for clinical use in the treatment of Alzheimer's disease.

GB

Isopropyl methylphosphonofluoridate. Sarin. Most volatile of the better-known nerve agents. Mp –57°C; vapour pressure at 25°C: 2.9 mmHg; volatility at 25°C: 22 000 mg/m^3. Anticholinesterase. Used in terrorist attack on the Tokyo subway in 1995. Lethal dose for humans similar to GA *qv*.

GD

Pinacolyl methylphosphonofluoridate. 1,2,2-trimethylpropyl methyl phosphonofluoridate. Soman. Colourless liquid; bp 167°C; vapour pressure at 25°C: 0.4 mmHg; volatility at 25°C: 3900 mg/m^3, gives off an odour of rotting fruit when vaporizing. The vapour is colourless. The lethal dose for soman through inhalation is about half that of sarin. It is also a far more persistent agent than sarin so that it can easily remain in a particular area for a day or longer, depending on the atmospheric conditions. Also toxic by skin absorption. Anticholinesterase.

GE

Isopropyl ethylphosphonofluoridate. Ethyl sarin: anticholinesterase nerve agent similar to GB *qv*.

Gelsemine

Derived from Carolina jasmine (*Gelsemium sempervirens*). Large doses have strychnine-like effects. Respiratory failure, dilated pupils, paralysis of eye movements, convulsions, death. 11 mg said to have killed a healthy woman.

GF

Cyclohexyl methylphosphonofluoridate. Cyclosarin. Anticholinesterase nerve agent similar to sarin, see GB. Liquid, vapour pressure at 25°C: 0.07 mmHg; volatility at 25°C: 680 mg/m^3.

α-Glycerotoxin

Produced by a Mediterranean polychaete (*Glycera convoluta*). Globular protein, MW 300 000. Triggers ACh release from nerve endings.

Glycol

Ethylene glycol, antifreeze. Intoxication produces nausea and vomiting, convulsions, papilloedema, focal and general seizures. 100 ml minimum lethal dose; 200–400 ml certain death.

Glycyrrizin

Triterpene glycoside responsible for the sweetness in liquorice. Promotes sodium retention and an increase in blood pressure.

Gonyautoxin II and III

Gonydiatoxin. Toxins isolated from red tide algal blooms. Act like saxitoxin *qv*.

Goon

See Angel dust.

Grayanotoxins

Derived from rhododendron leaves. Open sodium channels: act like batrachotoxin. Alleged to be one of the earliest known chemical warfare agents in the form of rhododendron honey given to Roman invaders. The intoxication by honey still occurs occasionally. Rarely fatal; salivation, dizziness, weakness, excessive perspiration, nausea, vomiting low blood pressure or shock, heart abnormalities, paresthesia, loss of co-ordination, muscular weakness, convulsions.

Green cross shells

German classification for shells containing compounds with a high vapour pressure and 'great toxic power on the respiratory tract': phosgene, trichloromethyl chloroformate (diphosgene), chloropicrin, etc.

H

USA term for Hunstoff, sulphur mustard, mustard gas, dichloroethyl sulphide *qv*.

Harmine

Alkaloid isolated from the seeds of *Pergonum harrmala*: a plant from the Asian steppes now spread to the Mediterranean and South-East Asia. Similar material can be derived from *Banisteriopsis caapi*: an American vine-like plant. Caapi: a snuff, is prepared from the stem. Other obscure names: *Ayahuasca* or *Yage*: from the root of *Haemadictyon amazonicum*. All said to contain dimethyltryptamine. Snuff or drink prepared from these sources produces violent inebriation, vomiting, tremor, convulsions.

HC mixture

Finely divided zinc plus hexachloroethane plus zinc oxide: very effective smoke producer.

HD

USA term for distilled sulphur mustard; dichloroethyl sulphide *qv*.

Heloderma

Polypeptide poisons from venomous lizards of the *Heloderma* species. Lethal dose is 10 mg SC for a rabbit.

Hemlock

See coniine.

Henbane

Poisonous plants *Hyscyamus niger* and *albus*: black and white henbane. Main toxic alkaloids are hyoscyamine *qv* and scopolamine *qv*. Similar toxins in several species of *Solanaceae*. Oils and tincture poisonous. Unlike atropine, hyoscine does not seem to produce delirium.

Hexachloroethane

Colourless crystals, sublimates at 184–187°C, suffocating odour. Used in smoke barrages.

Hexachloromethyl carbonate

Triphosgene. White crystals, mp 79°C. Odour of phosgene. Effects similar to phosgene *qv*.

Hexanitroethane

Poison gas and explosive.

Hexanitro phenylamine

Aniline explosive used in floating mines: exceedingly poisonous.

Histrionicotoxin

Major toxin from the skin of the Colombian tree frog: *Dendrobates histrionicus*. Blocks cation channels associated with ACh receptors.

HL

Mustard/lewisite mixture. Used to lower the mp of mustard to facilitate use in cold conditions.

HN1

Bis (2-chloroethyl) ethylamine, *N*-ethyl 2,2′ di(chloroethyl)amine, ethylbis(2-chloroethyl) amine, one of the nitrogen mustards. Oily colourless–yellow liquid, mp 34°C; bp 85°C, vapour pressure at 25°C: 0.722 mmHg, volatility at 25°C, 2290 mg/m^3. Similar effects to those of sulphur mustard but more damaging to the eye: rapidly penetrates to the interior of the eyeball.

HN2

Bis (2-chloroethyl) methylamine, 2-chloro-*N*-(2-chloroethyl)-*N*-methylethanamine. One of the nitrogen mustards. Mustine hydrochloride or mechlorethamine: used as a cytotoxic drug. Oily colourless–yellow liquid, mp 60°C; bp 75°C, vapour pressure at 25°C: 1.25 mmHg. Similar effects to those of sulphur mustard but more severe effects on the eye: rapidly penetrates to the interior of the eyeball.

HN3

Tris (2-chloroethyl)amine. A nitrogen mustard. Oily colourless–yellow liquid, mp –4°C; bp 138°C; vapour pressure at 25°C: 0.0382 mmHg. Effects similar to sulphur mustard. Differs from HN1 and HN2 in that the effects on the eye are similar to sulphur mustard: limited penetration.

Hog

See Angel dust.

Holothurins and Cucumariosides

Triterpene glycosides produced by sea cucumbers. Several actions, including block of Na/K-activated ATPase. Depolarizing.

Homomartonite

French term for bromomethyl ethyl ketone.

Hordenine

Sympathomimetic amine found with mescaline *qv* in peyote nuts.

HS

USA term for sulphur mustard, **H**un**S**toff, dichloroethyl sulphide *qv*.

HT

Mixture of 60% H and 40% di(β-chloroethylthio) diethylether.

Hunstoff

See dichloroethyl sulphide.

Hydrazine

Colourless, oily, fuming liquid: ammoniacal smell. Mp 2°C, bp 120°C. Absorbed by all routes. CNS depressant and stimulant, liver and renal damage, delayed eye irritation, vesicant.

Hydrocarbons

Methane and ethane: simple asphyxiants. Propane, butane: anaesthetic; Pentane–decane: anaesthetic and irritant.

Hydrogen bromide

Colourless gas, bp −67°C, used as a poison gas, liquid is yellow, fumes in moist air. Irritant and corrosive.

Hydrogen cyanide

Blausäure, prussic acid. Used in WWI only by France. Clear colourless liquid, odour of bitter almonds. Bp 25°C, mp –13.4°C vapour pressure at 15°C: 500 mmHg. Lethal concentration: 120–150 mg/m^3 (1 h); 200 mg/m^3 (10 min); figures vary with source; < 30 mg/m^3 said to be safe: eliminated rapidly. Blocks cytochromes in mitochondria: rapid collapse after a short period of hyperpnoea, dizziness and headache. CNS damage possible in some who recover from acute effects.

Hydrogen fluoride

Colourless fuming gas/liquid. Bp 19.4°C. Extremely corrosive. Inhalation: pulmonary oedema which may be delayed. Electrolyte imbalances: calcium and magnesium fall: arrhythmias.

Hydrogen selenide

Gas, bp – 41°C; disagreeable odour; irritating to eyes and mucosa; causes dizziness, nausea, 'garlic breath'.

Hydrogen sulphide

Colourless gas; 'rotten-egg smell'. Lethal within minutes at 0.1–0.2% in air. Stink bombs beloved of school boys: H_2S dissolved in water, banned in the UK in 1981. Irritant and lachrymator. Early hyperpnoea due to direct effect on carotid body as per cyanide. Sulphmethaemoglobin formed.

Hyoscine

Scopolamine. Anticholinergic, chemically and pharmacologically similar to hyoscyamine. Present in several poisonous plants, including *Solanaceae* and *Datura qv*.

Hyoscyamine

Pharmacologically active enantiomer of atropine. Anticholinergic. Present in several poisonous plants, including *Solenaceae* and *Datura qv*.

Hypoglycin

See Ackee fruit.

Iboga

A central African term for a substance prepared from the cortex of the roots of *Tabernanthe iboga*. Large doses are said to produce catalepsy and death. Main active component is ibogaine and twelve other alkaloids. Psychological effects like those of tricyclic antidepressants, plus monoamine oxidase inhibitors and anticholinesterases.

Ibotenic acid

Excitotoxic amino acid derived from mushrooms *Amanita pantheria* and *A. muscaria*. Insecticidal effects. CNS actions, and taste-enhancing qualities similar to glutamate.

Ichthyocrinotoxin

Found with tetrodotoxin in ovaries, liver and skin of puffer fish and toadfish.

Ictrogen

Derived from lupins. In large doses, a 'nerve poison'. Acts like phosphorus: yellow atrophy of the liver, intense jaundice, somnolence, fever, paralysis, degeneration of heart, muscles, kidney. Mainly an animal poison: sheep, etc.

Iminodipropionitrile

IDPN. Colourless liquid, bp 134°C. Absorbed by oral, inhalation and dermal routes but relatively high doses (of the order of g/kg bw) required for toxic effects. Produces 'waltzing syndrome' in rodents. Damages axons in spinal cord and brain stem by interaction with neuronal cytoskeleton.

Iodoacetone

French abb. Bretonite. Pale yellow liquid, bp 102°C, pungent odour, lachrymator.

Iprite

Mustard gas; dichloroethyl sulphide *qv*.

Iron pentacarbonyl

Liquid, bp 103°C: vapour burns with a brilliant flame. Decomposes to produce carbon monoxide: this is the toxic hazard.

Isin

See Ackee fruit.

Japanese star anise

Illicium religiosum, I. anisatum. Highly toxic Japanese variant of the safe plant, Chinese star anise (*I. verum*). All parts of the plant, particularly the seeds, are poisonous, causing vomiting and epileptiform convulsions of the type caused by picrotoxin, with dilated pupil and cyanosis. Kills by effects on respiratory and cardiovascular centres in medulla. Regarded as a sacred plant and found growing around Buddhist temples and graveyards.

JBR

British abb. for a mixture: 50 parts hydrogen cyanide, 25 parts arsenic trichloride, 25 parts chloroform.

JL

British abb. for a mixture: 50 parts HCN, 50 parts chloroform.

Joro spider toxin

JSTX. A group of arylamide toxins from the venom glands of the Japanese joro spider, *Nephila clavate*, and related species. Blocks glutamate receptors in both crustacea and mammalia.

K-stoff

Palite. Monochloromethyl chloroformate and dichloromethyl chloroformate. Also used, according to Beadnell (1943) and Sartori (1939), as an abbreviation for phenylcarbylamine chloride.

Kainic acid

Excitotoxic amino acid, stimulates the CNS by acting on glutamate receptors.

KJ

British abb. for tin tetrachloride *qv.*

Klop

German term for chloropicrin *qv.*

Kratom

See Mitragyna.

KSK

British term for a mixture of 70% ethyl iodoacetate and 30% ethanol and ethyl acetate.

L

See lewisite.

Laburnum

Laburnum anagyroides: seeds, flowers, bark contain the highly toxic alkaloid, cytisine. Produce: dilated pupils, gastro-intestinal upset and vomiting, weakness, inco-ordination, asphyxia, death. Rabbits and hares are apparently unaffected.

Lacrimite

Thiophosgene *qv.*

Laetrile

Amygdalin *qv.*

Lathyrism

Two forms, neuro- and angio/osteo-lathyrism, known in humans, cattle and horses, to result from consumption of the vetch, *Lathyrus sativus*, and related species. Neurotoxicity with stiffness of legs, increased reflexes, spastic paralysis and death from asphyxia. The toxins change the elasticity of the aorta. Aneurysms (weak ballooning in the arteries) develop and burst. Osteolathyrism affects skeletal development, cartilages and bones grow abnormally leaving the body deformed. Toxicity related to presence of toxic amino acids, 3-*N*-oxalyl-L-2,3-diaminopropionic acid or ß-ODAP, is associated with neurolathyrism, while β-aminopropionitrile, which interferes with cross-linking in collagen, is associated with angio and osteolathyrism.

Latrotoxin

See Black widow spider.

Lead

Acute poisoning rare: gut upset, black faeces due to lead sulphide formed in the gut. Decreased urinary output due to renal damage. Cardiovascular collapse possible. Chronic poisoning: demyelination, lead palsy. 'Wrist drop'. Encephalopathy:

restlessness, lethargy, insomnia, convulsions, delirium, projectile vomiting. 'Lead line' ('Burton's lead line') on gums due to deposition of lead sulphide. Stippling of red cells. Constipation. Anaemia: haem synthesis impaired. Seen in children eating flakes of lead paint and putty: pica. The word 'pica' is a reference to the common magpie (*Pica pica*): a bird of omnivorous eating habits.

Leptodactyline

(*m*-Hydroxyphenethyl) trimethylammonium. Particularly abundant in the skin of South American tropical frogs of the genus *Leptodactylus*. Neuromuscular blockade; oral LD_{50} mouse: 300 mg/kg; acts like coniine *qv*.

Lewisite

Chlorovinyl arsines; chlorovinyl dichloroarsine, dichlorovinyl chloroarsine and trichlorovinyl arsine and mixtures of these *qv*. It is interesting that several authors refer to the many varieties of lewisite. The chlorovinyl dichloroarsine variant *qv* is said to be typical.

Lewisite B

Described by Beadnell (1943), without the chemical name, as a variety of lewisite with 'especial lung-irritant effects'.

Lindane

99% γ-Hexachlorocyclohexane. White crystalline powder; vapour pressure at 20°C, 9×10^{-6} mmHg. Insecticide. Inhalation of powder produces CNS depression and sensitizes the myocardium to catecholamines. Ingestion can produce convulsions in children; 45 g said to produce convulsions in an adult, 392 g: status epilepticus. Adult: 840 mg/kg lethal; 180 mg/kg lethal in a child.

Lobeline

Piperidine-like alkaloid from Indian tobacco (*Lobelia inflata*). Partial agonist at nACh receptors and alters dopamine storage and release. Therapeutic uses: respiratory and peripheral vascular stimulant, emetic and smoking cessation. Poor oral absorption but can cause convulsions at very high doses.

Lophotoxin

LTX. White crystalline neurotoxin purified from the Sea Whip, a coelenterate. LD_{50} in mice: 8.9 mg/kg SC. Produces irreversible post-junctional block at the neuromuscular junction. Synthetic analogues with similar properties have been synthesized.

LOST

German name for mustard gas, dichloroethyl sulphide *qv*): **Lo**mmel and **St**einkopf developed the compound.

LSD

An ergot alkaloid *qv*: hallucinogen lysergic acid diethylamide, a salt (**L**yserg**s**äure **D**iethylamide tartrate) of which was first synthesised in Germany in 1937. Produces psychosis. 40 μg effective: inhalation or ingestion: effects peak at 2 h and last up to about 8 h. Half-life: 3 h. Interacts with endogenous transmitters including 5HT. See also mescaline.

M-1

Lewisite; chlorovinyl dichloroarsine *qv*.

Maculotoxin

Produced by the blue ringed octopus and now known to be identical with tetrodotoxin *qv*.

Madagascar ordeal poison

Fruit from the tree *Tanghinia venenifera* used in Madagascar as a 'trial by ordeal' poison. The Asian upas tree is similar and legend had it that sleeping under this tree was lethal. Both fruits contain cardiac glycosides.

Magnesium cyanide

White powder: releases HCN on contact with water. The commercial poison is CYMAG, a rat poison.

Male fern

Aspidium filixmas. Native plant of Europe. Extract of the root used to treat tapeworm: 45 g extract said to be fatal for an adult, 7–10 g for a child. Active component: filicic acid. Pain, heaviness of limbs, faintness, somnolence, dilated pupils, albuminuria, lockjaw, collapse. In animals: dragging of legs, salivation, dyspnoea. Widespread oedema and congestion of lung, brain, liver, spleen.

Mambog

See Mitragyna.

Manganite

See hydrogen cyanide.

Marsite

See arsenic trichloride.

Martonite

Bromoacetone *qv* and chloroacetone (80:20) used by the French in WWI. Sometimes also mixed with stannic chloride.

Mastoparans

A series of fourteen amino acid peptide components of the venoms of different species of wasp. Degranulate mast cells. Not found in bee venom.

Mauguinite

Sometimes given as Mauginite. Cyanogen chloride *qv*.

MD

Methyl dichloroarsine *qv*, 'Methyl Dick'.

Mellitin

Pain-producing 26 amino acid peptide component of bee venom. Four molecules join to produce a tetrameric pore in cell membranes. Depolarizing.

Mercaptans

Many individual compounds. Colourless liquids: rotten cabbage/garlic/skunk-like smell. Used as a stench gas: odorant and as the starting point for synthesis of plastics, insecticides, etc. Irritants. Large exposure: drowsiness, chemical pneumonitis.

Mercury

Elemental mercury, inorganic and organic mercury compounds are toxic for the nervous system and kidneys and have caused significant poisoning in humans but the symptoms predominating vary with the type of mercury and the characteristics of exposure. Elemental mercury: damage by inhalation; volatile at room temperature; readily passes biological membranes. Mercurous salts less toxic than mercuric. The phrase 'as mad as a hatter' arose from the effects of inorganic mercury salts seen among hatters and furriers who used mercuric nitrate in the felting process as a means of rendering resilient hair soft. Aryl, alkyl and alkoxyalkyl mercury compounds are of toxicological significance. The most serious report of organic mercurial poisoning resulted from the ingestion of methyl mercury, which passed through the food chain to man through fish as a result of contamination of the sea at Minamata Bay in Japan. About 1200 persons were poisoned, 45 died. Typical acute symptoms of oral inorganic mercury salt poisoning are: metallic taste, marked salivation, diarrhoea, renal failure. Chronic poisoning: salivation, damage to lips, gums and teeth. Tremor. Sudden attacks of temper, irritability.

Mescaline

3,4,5-Trimethoxyphenethylamine: similar structure to neurotransmitter amines. One of a number of alkaloids found in the peyote cactus *Lophophora williamsii* of Mexico, where it was used in religious ceremonies from the time of the Inca civilization: visual hallucinogen: colour effects made famous by Aldous Huxley in his book *Doors of Perception*. Effects similar to those of LSD *qv*.

Metaldehyde

Metacetaldehyde. Slug poison and solid fuel. CNS effects: rigidity, convulsions, opisthotonus, pupils may be constricted or dilated. Death 24 h after ingestion. Metabolic acidosis: acetic acid produced. Recovery likely.

Methyl-bis(2-chloroethyl)amine

Nitrogen mustard. Liquid, bp 87°C. Vesicant, also causes nausea, vomiting and haemorrhage. Volatility: 2000 mg/m^3 at 20°C. Casualty producing exposure (100 mg min/m^3) is given as less than that for mustard gas (200 mg min/m^3).

Methyl bromide

Colourless liquid: faint agreeable smell. Fumigant, used in fire extinguishers. Skin lesions look like 2nd-degree burns. Inhalation causes headache, vertigo, diplopia, convulsions, bronchospasm, pulmonary oedema, death.

Methyl chloroformate

Clear liquid, bp 71.4°C, powerful lachrymator. Zyklon A: 90% methyl cyanoformate and 10% methyl chloroformate. Developed as a CW agent. Also used as an insecticide.

Methyl chlorosulphonate

Likely component of C-stoff, French name – Villantite. Used by Germany in cloud attacks in April 1915. Colourless liquid, bp 134°C, pungent odour, volatility at 20°C: 60 000 mg/m^3. Irritant.

Methyl cyanoformate

Colourless liquid, bp 100°C. Powerful lachrymator, decomposed by water.

Methyl dichloroarsine

German name: methyl dick. Colourless liquid, bp 37°C, volatility at 20°C: 74 440 mg/m^3. Fruity smell. Irritant at 2 mg/m^3. Vesicant.

Methyl ethyl ketone

Colourless volatile liquid: acetone-like odour, used as a solvent. Absorbed all routes: rapidly by inhalation. Causes headache, dizziness, confusion, CNS depression at high concentrations.

Methyl ethyl ketone peroxide

Colourless liquid: acetone-like odour. Explosive when pure: usually used as a 60% solution in dimethyl phthalate. Strong oxidizing agent: reacts violently with salts of heavy metals, acids and alkalis. Corrosive, produces liver and kidney damage.

Methyl fluoroacetate

MFA. Examined as a possible CW agent but humans are less susceptible than laboratory animals. The fluoroacetates inhibit the mitochondrial enzyme aconitase and disrupt the citric acid cycle. The first signs of poisoning are nausea and apprehension, followed by convulsions. Death is due to ventricular fibrillation.

Methyl fluoroformate

Liquid, bp 40°C. Powerful lachrymator.

Methyl fluorosulphonate

Liquid, bp 92°C, smells of ether. Attacks glass but not rubber. Vesicant and toxic.

Methyl formate

Colourles liquid, bp 31.5°C. Mainly used as an intermediate for producing chlorinated derivatives but is itself toxic, causing irritation of mucous membranes, retching, narcosis and death from pulmonary damage.

Methyl guanidine

Found in putrefying meat. Exposure leads to rapid respiration, dilated pupils, paralysis and death.

Methyl methacrylate

Colourless volatile liquid, vapour pressure at 20°C: 29.3 mmHg, acrid fruity odour. Irritant: produces pulmonary oedema in animals in large doses. Moderately toxic by ingestion and inhalation: headache, fatigue, irritability.

Methyl sulphurylchloride

Transparent viscous liquid, bp 133°C. Heavy vapour. Lachrymator, produces pulmonary oedema.

Metridiolysin

Cytolytic toxin produced by the Sea Pink sea anemone (*Metridium senile*). Produce 33 nm diameter ring like structures in cell membranes.

Mezereon

From the shrub *Daphne mezereum*. Berries and bark poisonous, vesicant. A girl died of eating 12 berries: burning sensation in mouth, vomiting, giddy, narcosis, convulsions, death.

MFA

See methyl fluoroacetate.

Mitragyna

From a South-East Asian plant *Mitragyna speciosa*. Related to the coffee plant. Alkaloids include mitragynine similar to ibogaine. Euphoria, vomiting, vertigo, arrhythmias, twitching. The dried leaf is called 'Kratom' and is smoked, the distillate is called 'Mambog' and is imbibed.

Modeccin

A lectin from the African plant, *Adenia digitata*, a succulent. Effects similar to those of ricin and abrin: ribosome attack. Seeds commercially available.

Monkshood

See aconite.

Monochloromethyl chloroformate

War gas used alone and mixed with dichloromethyl chloroformate, known as Palite, K-stoff, C-stoff. Colourless liquid, bp 107°C. Irritant odour. Powerful lachrymator: 2 mg/m^3 effective, produces great eye pain according to Beadnell (1943). Systemic toxicity low.

MPPP

1-Methyl 4-phenyl 4-propionoxy-piperidine. A street drug. Acts like pethidine, 'synthetic heroin'. Easily converted to and may contain MPTP *qv*: produces Parkinsonian effects.

MPTP

1-Methyl-4-phenyl-1,2,3,6 tetrahydropyridine. Destroys neurones in the nigrostriatal pathways: irreversible Parkinsonism. A protoxin: oxidized in two steps to the MPP$^+$ ion. Toxic by inhalation or skin contact, as well as ingestion.

Murexine

Neuromuscular blocker; choline derivative produced by *Murex trunculus* and related marine molluscs. Derived from molluscs: acts like coniine *qv*.

Muscarine

Cholinergic alkaloid found in *Amanita Muscaria* *qv*.

Muscimol

Agarin. Found in *Amanita muscaria qv*. GABA agonist; CNS depressant.

Mushrooms

See also *Amanita phalloides* and *A. muscaria*. *A. virosa*: the 'destroying angel'/'angel of death' mushroom, which used to be more common than *A.phalloides* in the USA; *A. pantherina*: panther cap: *Clitocybe rivulosa*: lawn funnel cap and many *Galerina* species are all highly toxic. Other dangerous mushrooms: *Flammulina velutipes*, velvet shank or winter fungus; widespread and common in Northern zones; cultivated in the Far East for use in cooking but contains a potent cardiotoxin. *Pluvatus oleatus*: mediterranean: necrosis of the liver, kidney and gut. *Coprinus atramentarius*: inky cap contains disulfiram, *qv*., and causes palpitations and nausea if alcohol is consumed even several days later. *Omphalotus olearius*, jack o'lantern, *Paxillus involutus*, brown roll rim. *Cortinarius orellanus and C. rubellus*, web caps lfc: severe and irreversible kidney damage typically appearing long after ingestion.

Mustard gas

See dichloroethyl sulphide.

Mustard oil

Allyl isothiocyanate. Distilled from mustard seeds. Colourless or pale yellow volatile liquid. Powerful irritant and vesicant.

Mustard sulphoxide

Non-toxic reaction product of mustard gas and $Ca(OCl)_2$.

Mustard sulphone

A very toxic reaction product of mustard gas and strong oxidizing agents.

Mycotoxin T2

Better known as *Fusarium* mycotoxin. LD_{50} (rat) 4 mg/kg. Irritant, produces aleukia in birds. Bread contaminated with this toxin killed 1000s of Russians in the period after WWI. Poisonous by all routes. Produces vomiting and haematemesis.

Mydaleine

A toxic alkaloid (ptomaine) obtained from putrid flesh and from herring brines. Guinea pig: 50 mg lethal. Dilated pupils, increased secretions, increased temperature. Cats: profuse diarrhoea and vomiting.

Mydatoxine

Another substance allegedly found in putrefying meat: lachrymation, diarrhoea, convulsions.

Myotoxin a

A myonecrotic toxin found in the venom of the prairie rattlesnake: *Crotalus viridis*. Basic polypeptide, MW 4600. Binds to sarcoplasmic reticulum, affects Ca^{++} levels: produces cellular swelling and necrosis.

Myristicin

Methoxysafrole; aromatic hallucinogen found in nutmeg *(Myristica fragrans)*. Myristicin is also found in several members of the carrot family (Umbelliferae). Nutmeg has been used as a snuff mixed with betel, *qv*. Myristin, was also used as an insecticide. Elemicin is another active component. Produces a fall in blood pressure, probable neurotoxic effects on dopaminergic neurons, brain damage. Myristicin is a weak inhibitor of monoamine oxidase. Toxic by ingestion of ca. 5 g of nutmeg (1–2 mg myristicin/kg bw).

Naphthalene

Fire lighters, moth balls. Children susceptible to toxic effects. Oral: nausea, vomiting, diarrhoea, excitement, coma, convulsions. Hepatic necrosis. If there is a deficiency of glucose-6-phosphate dehydrogenase, then rapid red cell haemolysis with renal tubular blockage and anuria can occur. Vapour causes headache, diplopia and confusion.

NC

UK and USA abb. for mixture: 80 parts chloropicrin and 20 parts stannic chloride.

Nerve agents

See GA, GB, GD, GE, GF, VX , VE, VM, etc.

Neurine

Trimethylvinylammonium hydroxide. Decomposition product from choline. 46 mg/kg SC lethal in mice: effects as per muscarine.

Nga or Ngwa

'Bushman' term for the beetle *Diamphidia nigro-ornata*. Used as a source of arrow poison. Juice from 10 pupae enough for one arrow. Protein toxin, MW 54 000. A few μg said to be lethal IP in mouse. Depolarizing agent, possibly haemolytic.

Nickel carbonyl

Colourless liquid with high vapour pressure at room temperature. Exposure can occur during electroplating and nickel refining. Vapour is insoluble in water and penetrates deep into the lung, causing pulmonary oedema and cancer.

Nicotine

3-(1-Methyl-2-pyrrolidinyl)pyridine, the main alkaloid present in tobacco leaves. Colourless-to-pale yellow liquid, bp 125°C, volatile with steam. Used as insecticide fumigant. Massive doses produce immediate collapse: vomiting, convulsions, death. Cyanosis and dilated pupils may be seen. In less severe cases: nausea, vomiting, headache, and later diarrhoea. Bradycardia followed by tachycardia. Temperature depressed: cold sweat. Lethal dose for an adult: 40–60 mg. 4 mg can produce alarming symptoms. Does not need much to produce effects in children: blowing soap bubbles with an old pipe can allegedly do so. Some insecticides contain > 95% nicotine.

Niespulver

See *o*-dianisidine chlorosulphonate.

Nitrobenzene

Pale yellow oily liquid. Smells of bitter almonds. Sometimes known as 'Oil of Mirbane'. Used in cheap perfumes. Toxic by inhalation or ingestion. Haemoglobin converted to methaemoglobin: cyanosis, anoxia, death.

Nitrobenzyl chloride

o-Nitrobenzyl chloride. Harassing agent used by the French in 1915. Crystalline, mp 48.5°C. Vesicant and lung irritant properties at > 1.8 mg/m^3.

Nitrogen mustard

See methyl-bis(2-chloroethyl)amine.

Nivalenol

Tricothecene toxin from *Fusarium* species; haemorrhagic agent causing blistering and necrosis, dizziness, vomiting, may be lethal.

Notexin

Polypeptide neurotoxin from the Australian tiger snake (*Notechis scutatus scutatus*). Prevents release of ACh from nerve endings: paralyses muscles.

Noxiustoxin

A scorpion (*Centruroides noxius*) toxin; blocks K^+ channels.

Nutmeg

Contains myristicin *qv*.

Ochratoxins

Ochratoxin A is the most potent toxin isolated from the mould *Aspergillus ochraceus*. LD_{50}, oral, rat: 20 mg/kg. Acute effects mainly on kidneys, also toxic to neural, reproductive and immune systems and carcinogenic.

Octopamine

Sympathomimetic toxin found in citrus fruits.

Oleoresin capsicum

OC, pepper spray, capsaicin *qv*.

Oleum

Fuming sulphuric acid. Used as a smoke producer in WWI. Dripped onto a bed of quick lime: much heat generated: evaporation of the acid: condensation into fine droplets, the acid is hygroscopic. Dense fog produced.

Ololiuqui

Also known as Pieule. Extracted from the seeds of species of vine-like creepers: Morning glory, convolvulus or bindweed. Especially *Rivea corymbosa* and *Ipomoea violacea*. Contains analogues and precursors of lysergic acid. Small doses: increased brightness of visual perception without hallucinations. Large doses: hyperactivity and anxiety.

Opacite

French term for tin tetrachloride *qv*.

Organomercury compounds

Various organomercury compounds (e.g. ethylmercuric chloride, methyl mercury, dimethyl mercury), found as environmental contaminants or used as fungicides, are irreversible neurotoxins. Damage cell bodies of neurones in the dorsal root ganglia and brain. See also mercury.

Ostracitoxin, pahutoxin, homopahutoxin

Haemolytic poisons secreted in mucous by box fishes *(Ostraciontidae)*. Pahutoxin is the choline chloride ester of 3-acetoxypalmitic acid. Ostracitoxin IP causes ataxia, laboured breathing, coma and death in mice (lethal dose 0.2 mg/g bw). Pahutoxin has a haemolytic effect on vertebrate red cells *in vitro* and this is correlated with its lethal properties.

Ouabaine

White powder or crystals; poisonous cardiac glycoside (g-strophanthin) from the seeds of the African trees *Strophanthus gratus* and *Acokanthera ouabaio*. Used as a heart stimulant and by some African peoples as a dart poison. Blocks the Na^+/K^+-ATPase.

Oxalic acid

Crystalline colourless solid. Found in many plants, particularly the leaves of the rhubarb plant and in the juice of wood sorrel. Salts of Sorrel, Salts of Lemon. Used as a bleach and ink stain remover. The concentrated solution is corrosive. Harmful if inhaled or absorbed through skin; 15 g oral: agonizing pain, nausea and vomiting with haematemesis. Lowers plasma calcium: convulsions, tetany, tingling of mucous membranes. Renal damage: tubular necrosis. Smallest lethal dose reported: 3.8 g in a 16-year-old boy.

Oxalyl chloride

Colourless liquid, bp 64°C. Vapour attacks the respiratory tract strongly.

Pahutoxin

See ostracitoxin.

Palite

French term for monochloromethyl chloroformate mixed 74:25 with stannic chloride or with dichloromethyl chloroformate (see K-stoff).

Palytoxin (PTX)

Neurotoxin found in soft coral and believed to be produced by a flagellate and concentrated by the coral (*Palythoa toxica*). Most deadly non-protein toxin known: LD_{50} (mice): 0.15 μg/kg. 20–50 times as toxic as tetrodotoxin. Intense vasoconstrictor: 0.06 μg/kg IV in dogs produced a transient rise in blood pressure followed by rapid hypotension and death within 5 min.

Papite

Acrolein. Liquid, bp 52°C. Lachrymatory at > 7 mg/m^3. Used in mixtures as a war gas, for instance with titanic chloride.

Paraquat and diquat

Bipyridilium herbicides. Paraquat: death caused by ingestion and SC dosing in humans. Widespread cellular proliferation in the lung: pulmonary fibrosis. Early neurotoxic signs: increased excitability, convulsions. Also: liver and kidney damage. Possible myocardial damage. LD_{50}, human: about 40 mg/kg. Diquat: kidney the main target.

Pardaxins

Polypeptide toxins of about 33 amino acid derived from the Red Sea flat fish *Pardachirus marmoratus* and other *P. species*. LD_{50}, mice: 25 mg/kg IP. Pardaxins are ionophores that have neurotoxic and haemolytic effects. The toxins also act as shark repellents.

Parsley: oil of

Apiol. Ingestion: gastroenteritis, liver and kidney damage, can be fatal. Dark olive-green liquid. Adulteration with TOCP can occur. Used as an arbortifacient.

Peace pills

See Angel dust.

Pelargonic acid morpholide

Pelargonyl vanillamide, PAVA. Synthetic analogue of capsaicin *qv*.

Perchloromethyl mercaptan

Clairsite. Used by French in September 1915: first gas shell used by France. Not found to be very efficient. Clear, yellow, oily liquid, bp 148°C. Offensive smell. Irritant at 10 mg/m^3; intolerable at 70 mg/m^3.

Pereirine

One of three alkaloids extracted from the bark of the Brazilian tree pereira (*Geissospermum laeve*). Said to act like strychnine: details lacking. Geissospermine causes death by paralysis, pereirine causes paralysis, fever, and death, Vellosine is also present but its toxicity is not reported.

Perstoff

Trichloromethyl chloroformate, diphosgene *qv*.

Peyote

See mescaline.

Pfiffikus

German name for phenyl dichloroarsine *qv*, phenyl dick.

PG

British abb. for 50% phosgene and 50% chloropicrin.

Phalaris tuberosa

A grass found in East Africa. At certain seasons this plant contains bufotenine and related tryptamine alkaloids with hallucinogenic effects. Sheep eat the grass: motor disturbances: bufotenine acting on spinal pathways. Bufotenine blocks uptake of norepinephrine. Ventricular fibrillation can occur if animals become excited.

Phalloidine

See *Amanita phalloides*.

Phencyclidine

1-(1-Phenylcyclohexylpiperidine). PCP. Angel dust. Hallucinogen. Bizarre, aggressive behaviour simulating schizophrenic/manic depressive psychosis, numbness of fingers and toes, anaesthesia, respiratory depression, open-eye coma, adrenergic crisis, malignant hyperthermia.

Phenol

Carbolic acid. Toxic by ingestion or through skin. Corrosive. Death generally due to respiratory depression. Renal damage can occur. Oesophageal stricture may be produced by drinking. Lethal dose adult: 10–30 g.

Phenyl carbylamine chloride

Used to disguise the smell of mustard gas. Oily, pale yellow liquid, bp 210°C, low volatility: 2100 mg/m^3 at 20°C. Onion-like odour. Strong irritant: 3 mg/m^3. Very persistent. Little used in WWI.

Phenyl dibromoarsine

Harassing agent used by Germany in 1918. Lethal at 2000 mg min m^3.

Phenyl dichloroarsine

Phenyl dick, Pfiffikus. Colourless liquid, turns yellow slowly, bp 255°C, volatility at 20°C: 404 mg/m^3. Lung irritant, vesicant. Used as a solvent for diphenyl cyanoarsine in war gases and mixed with 40% diphenyl chloroarsine as Sternite.

Phosgene

Carbonyl chloride, CG, Collongite, D-stoff. Used by Germany as a mixture with chlorine at Nieltje in Flanders on December 19th 1915. 88 tons released: 1069 casualties, 129 fatal. 80% of WWI gas fatalities due to phosgene. Colourless gas, bp 8°C with odour variably described as: new-mown hay or mouldy hay. Tobacco reaction produced: cigarette after exposure produces a metallic and unpleasant taste: Graves notes this in 'Goodbye to All That'. This reaction was also produced by HCN and SO_2. Ten times as toxic as chlorine but producing comparatively little irritation, being characterized by a latent period (h) and then producing permeability pulmonary oedema. Physical effort can trigger the oedema. Lethal at 0.5 g/m^3 for 10 min. Long-term lung damage can occur in survivors.

Phosgene oxime

Dichloroformoxime. CX. White crystalline powder, mp 40°C, bp 129°C, high vapour pressure, unpleasant odour. Violent irritant, lachrymator, vesicant. Few mg on skin: severe irritation, intense pain, necrotizing wound, 'very few compounds are as painful and destructive of tissues'. Eye exposure: irritation, blindness. Lungs: pulmonary oedema. Skin: white patch with erythematous ring around it. Necrosis later: turns yellowish.

Phosphine

Colourless gas that smells of decaying fish or garlic. Irritant. Grain fumigant. Rat paste: zinc phosphide. Damp can produce phosphine from aluminium and zinc phosphides. Effects: lung and liver damage, circulatory collapse. Green fluorescent sputum produced.

Phosphorus

White phosphorus. Most effective of all military smokes. Spontaneously inflammable at room temperature: kept under water. Waxy solid, mp 44°C. Upon exposure to air: luminous glow and then bursts into flames. Vapour toxic, although there is little evidence of toxicity due to exposure to phosphorus smokes. Smoke: phosphorus pentoxide formed and this is converted into phosphoric acid. Solid produces severe burns: slow to heal.

Red phosphorus: formed by heating white phosphorus to 250–300°C out of contact with air. Reddish-brown amorphous powder. More stable than white phosphorus: not luminous, not spontaneously inflammable. Need to heat to 260°C before it ignites. Vapour not said to be toxic. Used mixed with white phosphorus to produce smoke.

Phosphorus pentachloride

White fuming crystals. Otherwise similar properties to phosphorus trichloride *qv*.

Phosphorus trichloride

Colourless fuming liquid, bp 76°C. Vapour decomposes in damp air to HCl and phosphoric acids. This may be slow enough to allow penetration deep into the lung. Highly irritant and corrosive: 50 ppm dangerous even for short periods, 600 ppm rapidly fatal.

Physostigmine

Eserine. Reversible anticholinesterase. Derived from Calabar (*Physostigma venenosum*) beans, the 'ordeal bean'.

Picrotoxin

Drug obtained form the seeds of 'fish beans' (used to poison fish) of the *Anamirta* species. Convulsant. Blocks GABA transmission: blocks presynaptic inhibition. 120 mg apparently lethal for cat, 16 mg lethal in guinea pig, 100–200 mg lethal in man; GI upset, diarrhoea, fever, convulsions.

Pilocarpine

Major cholinergic alkaloid derived from the Jaborandi plant: from its leaves. Acts on muscarinic receptors: sweating, salivation, flushing, lachrymation, tachycardia.

Pituri

Dried leaves of *Duboisia hopwoodi*: chewed or smoked by Australian Aborigines. Contains hyoscine *qv* and nornicotine. Delays hunger.

Platypus

Ornithorhynchus anatinus: the duckbill platypus. Males have a venom gland connected to a 13 mm spur close to the heel. Venom containes a mixture of chemicals; produces extreme pain, shock, weakness and perhaps cytolytic and haemolytic effects.

PS

See chloropicrin.

Pseudotritontoxin

Found in the skin secretions of the red salamander (*Pseudotriton ruber*). A high-MW toxin: ca. 200 000. Effects like tetrodotoxin *qv*.

Psilocybin and psilocin

Hallucinogens produced from the Mexican mushroom: *Psilocybe mexicana*.

Psoralen

See furocoumarins.

Ptomaine

An old word used to describe the mixture of toxins produced in decaying meat. Not one compound; can include bacterial toxins such as cadaverine, a fuming liquid biogenic polyamine.

Pukateine

Alkaloid from the New Zealand tree: the Pukatea (*Laurelia novae-zelandiae*). Local action on lips and tongue like that of aconitine. Large doses cause convulsions.

Pyrethrins

Viscous liquid insecticides extracted from flowers of *Chrysanthemum*: more effective in insects than mammals partly because of lower insect body temperatures. Skin effects in man. CNS effects in mammals at high doses.

Pyridine

Colourless to yellow liquid, bp 115° with a distinctive odour. Volatile with steam. Irritant to skin. CNS depressant leading to respiratory depression. Ingestion can produce liver and kidney damage. On ingestion: nausea, headache, insomnia, 'nervousness'. Chronic exposure: chronic bronchitis and emphysema.

Pyriminil

Trade name: Vacor. Yellow or yellow–green powder. Rat poison, an alternative to warfarin. 5.6 mg/kg dangerous in man. Interferes with nicotinamide metabolism: rats die from paralysis and respiratory arrest. Destroys beta cells in pancreatic islets: diabetes produced. Acute exposure to pyriminil may produce the following signs and symptoms: abdominal pain, nausea, vomiting, diarrhoea, anorexia, dehydration, urinary retention, and urinary tract irritation, muscular weakness, aching, chest pain, irregular heartbeat, tremors, mental confusion, and central nervous system depression. Pyriminil dust may produce irritation of the eyes and mucous membranes, dilated pupils and visual disturbances.

Q

1,2-bis-(2-Chloroethylthio) ethane *qv*, sesquimustard, QN2. Analogues with similar structures and properties.

Quebracho alkaloids

From the bark of *Aspidosperma quebracho blanco*. Six alkaloids identified: quebrachine, aspidospermine, aspidospermatine, aspidosamine, hypoquebrachine. Quebrachine is said to be poisonous with effects like those of strychnine.

Rationite

French CW agent: chlorosulphonic acid *qv* and dimethylsulphate.

Red squill

Bulbs contain a mixture of glycosides: Scillaren A+B. Cardiotoxic effects like those of digitalis. Blurred vision, arrhythmias, convulsions. Used as a rodenticide: respiratory failure and convulsions. Low toxicity in man because of powerful emetic effects.

Ricin

Protein toxin produced from castor oil seeds. *Ricinis communis*. Castor oil itself is safe: the ricin is destroyed by heating during preparation. Four dimeric lectins identified. Two of these cause haemagglutination and two interfere with RNA functioning. Oral effects: gastric haemorrhage, bloody diarrhoea, fever. Later: renal and liver failure, convulsions, death. Georgi Markow stabbed with an umbrella that delivered a metal pellet containing ricin (about 500 μg) on Waterloo Bridge, London in 1978. Died of what appeared to be septicaemia. A case in Winchester (UK) of 'self-poisoning'. Lethal dose in an adult: three seeds. Lethal dose by injection: μg–mg quantity. Related to abrin.

Rotenone

Derris. An insecticide of plant origin. Can be used as a fish poison. Oral: nausea and vomiting. Chronic exposure: liver and kidney damage. Massive exposure: respiratory paralysis and death. By inhalation: intense respiratory stimulation and then depression, convulsions, death. Inhibits mitochondrial electron transport.

S

British abb. for methylbis(2-chlorothyl)amine *qv* (nitrogen mustard).

Salamander

Amphibian. Skin glands contain poison that cannot be discharged by the animal but is released on pressing the skin. Effects said to be like rabies: excitable stage with increased reflexes, rapid respiration and dilated pupils, increased nasal and buccal secretions. Convulsive phase: dyspnoea. Paralytic phase: respiratory failure and death. Once symptoms begin in animals, death is said to be certain. Lethal dose in a dog: 0.7 mg/kg SC. The secretions from the water salamander (*Triton cristatus*) cause haemolysis of red cells, increase in BP, cardiac arrest.

Salamandarin

The toxin produced by salamanders.

Sanguinarine

Compound derived from poppy seed oil. Inhibits oxidation of pyruvic acid, which accumulates in the blood. Used as antimicrobial to control dental disease but also toxic – cause of epidemic dropsy.

Santonin

Derived from the seeds of *Artemesia* species. Blyth and Blyth (1920) reported 18 cases of poisoning when the drug had been used as a worming agent. Disturbance of colour perception, taste and smell. Convulsions, dilated pupils, heart slow and weak, collapse, death.

Saponin

Term applied to a range of sapogenin glycosides widely distributed in plants, all poisonous, and all producing froth on being shaken with water. Used for killing fish. Effects on man: oral administration increased mucus secretion, nausea, low toxicity; intravenous – haemolysis at extremely low doses.

Sapotoxins

Toxic component of the Soap Bark tree (*Quillaja saponaria*). Powerful irritant, local anaesthetic and muscular poison, vomiting, diarrhoea and gastroenteritis if taken in large doses internally, headache, vertigo, vomiting, hot skin, rapid feeble pulse, progressive muscular weakness and finally coma and death.

Sarafotoxin(s)

Components of the venom of the Israeli burrowing asp (*Atractaspis engaddenisis*). Remarkable structural similarities to endothelins: potent vasoconstrictors. 21 amino acid residue peptides.

Sarin

Name suggested by Schrader for GB *qv*. Said to be derived from the names of those involved in its production: **S**chrader, **A**mbros, **R**udriger, van der **Lin**de.

Saurine

See scombroid.

Sauvagine

Linear polypeptide: 40 amino acid from skin of South American frog *Phyllomedusa sauvagei*. Vasodilator.

Savin oil

From the sarin, formerly classified as a juniper: *Sabina communis* (*Juniperus sabina*). Irritant. Blyth and Blyth (1920) reported violent vomiting, great pain, diarrhoea and death.

Saxitoxin

Non-protein toxin produced by the dinoflagellate *Gonyaulax catanella*, found in shellfish. The toxin was first extracted from the Alaskan butter clam : *Saxidomus giganteus*, hence the name. Cause of paralytic shellfish poisoning. Lethal in mice at 10 μg/kg. Blocks influx of sodium ions: stabilizes the cell membrane. Large-scale production is feasible as the organisms can be cultured. Neosaxitoxin is produced by *Gonyaulax tamarensis*. Red tide colour produced by the red dye peridinin. Toxins present: saxitoxin, neosaxitoxin, gonydiatoxin (several sub-types). Skin exposure produces irritation and occasionally vesication.

Scombroid fish poisoning

Scombroid poisoning is caused by ingestion of food that contains high levels of histamine and other amines, such as saurine. Symptoms: nausea, vomiting, abdominal cramps, diarrhoea, flushing, headache, urticaria, burning in mouth. The high levels of histamine are caused by the action of certain bacteria on normally safe foods, including cheese and fish. Most outbreaks have resulted from the ingestion of fish of the sub-order *Scombrodei*, particularly tuna, also mackarel.

Scopolamine

Hyoscine *qv*.

Scorpion poisons

Over 80 different scorpion toxins known. About 40 low-MW protein toxins isolated from scorpion venoms: generally prevent closure of sodium channels and block potassium channels: prolong action potential and increase transmitter release. Last joint of the tail: two oval glands, a bladder and a sting that injects the venom. Large scorpions can kill. Effects: local inflammation, gangrene, vomiting, diarrhoea, weakness, fever, fainting, delirium, coma, convulsions, death. Little effect by mouth: dogs eat scorpions.

Selenium

Industrial hazard. Used to 'blue' the barrels of guns. Effects: breath smells of garlic/rotten onions, gastrointestinal disturbance, spasms of limbs, peripheral cyanosis, death due to depression of CNS cardiovascular and respiratory centres. Lethal at 4 mg/kg in man. Dimercaprol *enhances* toxicity.

Selenium dioxide

Toxic, intense irritant and easily generated as an aerosol.

Selenium oxychloride

Colourless or yellowish fuming liquid. Vesicant, causes pulmonary oedema. 0.01 ml applied dermally to a rabbit causes death.

Senf gas

Mustard gas, dichloroethyl sulphide *qv*.

Sesqui-mustard

See Q and 1,2-bis(2-chloroethylthioethyl) ethane.

Silicon tetrachloride

Smoke-producing agent. Colourless liquid, bp 60°C. Fumes in moist air producing silicon hydroxide and hydrochloric acid. Neutralizing the HCl with ammonia keeps the smoke generation going. If used with ammonia the generating power is greater than that of phosphorus. Looks like ordinary fog.

SK

See ethyl iodoacetate.

Snake neurotoxins

α-Toxins: post-synaptic neurotoxins. Many identified: polypeptides: 61–74 amino acid, 4–5 disulphide bridges, act like curarae *qv*, e.g. α-bungarotoxin (see bungarotoxins).

β-Toxins: pre-synaptic neurotoxins, polypeptides, all chemically related to phospholipase A but no link between the phospholipase activity and toxicity. Effects similar to α-glycerotoxin and latrotoxin *qv*, e.g.: crotoxin, notexin, taipoxin and β-bungarotoxin.

Sodium chlorate

Colourless–white crystals. Saline taste. Ingestion: ataxia, dizziness, anoxia, cyanosis, convulsions, coma, brown–black urine, liver and kidney damage. Produces MetHb: not responsive to methylene blue: death from ingestion of more than 150 g due to hypoxic damage to tissues. Effects may be delayed 12 h post-ingestion.

Sodium fluoride

White, crystalline powder, ingestion produces collapse of cardiovascular system. More than 8 mg fluoride/kg bw is potentially toxic. Chelates calcium.

Solanaceous alkaloids

Alkaloids from various solanaceous plants. The main alkaloids are (−)-hyoscine (scopolamine) and (−)-hyoscyamine. Atropine is racemic hyoscyamine.

Solanine

Alkaloid related to atropine, hyoscine. Found in *Solanum* and similar species, including potatoes, tomato plant leaves, Jerusalem cherry plant. Effects: gastroenteritis, vomiting, collapse, twitching, coma. Confusion and hallucination can last for several days.

Solenodon

A large insectivore: looks like a squirrel-sized shrew. Two genuses: *Solenodon paradoxus* (found only on Haiti) and *Atopogale cubanus* (found only on Cuba). Salivary glands produce venom which is delivered via a deeply grooved second incisor. Venom produces paralysis and convulsions: nature of venom unknown.

Soma

Legendary psychoactive drug of the ancient Aryans. Possibly of plant or fungal origin.

Soman

See GD.

Sparteine

A quinolizidine alkaloid isolated from several brooms and lupins (*Fabaceae*), including *Lupinus, Spartium* and *Cytisus*. It has been used as an oxytocic and an anti-arrhythmia agent. Cardiac stimulant, higher doses – muscular trembling, inco-ordination, emesis, catharsis and finally paralysis of the respiratory and motor centres, cardiac arrest.

Staphylococcus enterotoxin B (SEB)

Heat-stable peptide toxin produced by the bacterium *Staphylococcus aureus*. Developed as a CW agent. Non-volatile, dispersed as an aerosol. Inhalation exposure to < 1mg min/m^3 produces fever, chills, myalgia, septic shock, death.

Starvesacre seeds

See aconite.

Sternite

60% phenyl dichloroarsine; 40% diphenyl chloroarsine. Irritant used by French in WWI.

Stibine

SbH_3: poisonous gas: see antimony.

Strychnine

Alkaloid extracted from plants of the *Strychnos* family. *S. nux-vomica*: East Indies and India. Was used in tonics: no longer. Bitter: detectable at 1 in 10^5 dilution. Absorbed across gut, metabolized by liver, excreted by kidneys. Glycinergic antagonist, CNS stimulant. Lethal dose: adult 32 mg, child 15 mg. Effects: rapid onset of tremor, twitching, convulsions, great pain, opisthotonus, sweating, respiratory paralysis. Convulsions can be triggered by any disturbance.

Suberitine

The haemolytic and neurotoxic protein isolated from *Suberites domuncula*, a silicaceous sponge. Lethal in dogs at 10 mg/kg. Non-toxic by mouth. Effects: vomiting, diarrhoea, dyspnoea, haemorrhages into gut, peritoneum and pericardium.

Sulphur monochloride

Reddish-yellow liquid used as a solvent and fumigant. Fumes in moist air producing sulphur dioxide, sulphur and HCl. Irritant to upper respiratory tract. Lethal to mice at 150 ppm for 1 min.

Sulphur trioxide

Sulphuric trioxide. After phosphorus, the most effective smoke developed in WWI despite needing a moist atmosphere to produce full effects. Liquid, bp 45°C, mp 18°C, transparent solid. Fumes in moist air producing droplets of sulphuric and sulphurous acid, which are hygroscopic. 10 mg/m^3: coughing, 30 mg/m^3: produces complete obscurity at 7 m.

Sulphuryl chloride

Colourless liquid, bp 69.2°C. Smoke-producing agent, pungent odour, corrosive to skin and mucous membranes. Used by the British and French in WWI as a mixture with phosgene and chloropicrin.

Sulvanite

Sulvinite, ethyl chlorosulphonate, colourless oily liquid, bp 152°C, pungent odour, irritant at > 2 mg/m^3. Used by the French with bromoacetone in WWI.

Surpalite

French: trichloromethyl chloroformate *qv*.

T

Bis (2-chloroethylthioethyl) ether. A higher mustard. Liquid. Said to have three times the vesicant power of mustard gas. Also used as an abb. for Angel dust (phencyclidine) *qv*.

T-stoff

Xylyl bromide *qv*. According to Prentiss (1937), also used as a term for benzyl bromide.

Tabun

See GA.

Taipoxin

Polypeptide neurotoxin produced by the Australian Taipan snake (*Oxyuranus scutellatus*). Prevents release of ACh from nerve endings, similar to bungarotoxin *qv*.

Taxine

Derived from the leaves and berries of the yew tree (*Taxus baccata*). Effects: fainting, pallor, contracted pupils, weak pulse, convulsions, cardiac and respiratory depression: death.

TCDD

Tetrachlorodibenzo-*p*-dioxin. Contaminant of PCPs and PCBs and of Agent Orange *qv*. Effects: chloracne, nephritis, haemorrhagic cystitis, polyneuropathy, birth defects. Seveso, Italy, July 1976: ca. 2 kg TCDD released by a factory explosion; birds, cats and dogs died.

Tedania ignis

Sponge: produces two neurotoxins: both affect ACh release, one increasing, the other blocking.

Tellurium, tellurates and tellurites

Absorbed by dermal inhalation and oral routes, results in 'garlic breath'. Damages nervous system like lead. Mild effects reported in workers. Peripheral neuropathy and paralysis in rats.

Teonanacatl

In the Aztec tongue: 'flesh of God'. Mushrooms of the Psilocybe group, especially *Ps. mexicana*. See psilocybin. Acts like lysergide but with only 1% of the potency.

TEPP

Tetraethyl pyrophosphate. Very toxic anticholinesterase originally synthesized as a potential insecticide. Lethal dose oral 1.1 mg/kg, dermal 2.4 mg/kg.

Tetanine and tetanotoxine

Tetanus toxin. Protein neurotoxin produced by the bacterium *Lostridium tetani*. Blocks inhibitory synapses in the spinal cord: spastic paralysis, followed by block of ACh release in periphery, resulting in flaccid paralysis.

Tetrachloro dinitroethane

Crystals, decompose at 130°C, producing NO_2. Melts at 142°C. Powerful irritant and toxicant. Eight times as powerful as a lachrymator and six times as toxic as chloropicrin.

Tetrodotoxin

The puffer fish toxin. Captain Cook severely affected in 1774. The fish is prepared by trained cooks and eaten as 'fugu' in Japan. Also found in ovaries and liver of related fish species and some cephalopods. Non-protein MW 319. Effects: numbness and tingling of lips, vomiting, fall in BP, weakness, paralysis, respiratory failure, death. Lethal dose in mice 5 μg/kg: among the most potent of toxins. Blocks sodium channels and prevents depolarization. Believed to be produced by bacteria in the fish. Looked at by Japan in WWII as a potential CW agent.

Thallium

Widely used in optical glass: raises the refractive index. Rodenticide, insecticide, depillatory creams, 'Zelio paste' used as a rat poison: seems to have been a favourite with murderers. Effects: gastrointestinal upset, colic, sensory disturbances: extreme pain. Spreads from feet to legs to trunk. Pain in toes like gout. Motor paralysis: ascending paralysis looks like Landry's ascending paralysis. Phrenic and vagus affected: diaphragmatic and cardiac effects. Delirium, rhinitis, conjunctivitis, hair loss. Diabetic glucose curve due to damage to beta cells in islets. Irritant effects delayed by 1 or 2 days, neurological and other effects seen even later. Lethal at 12 mg/kg adult.

Thelenotoside B

See Astichoposide C.

Thiophosgene

Lacrimite. Used by the Austrians and French in WWI. Oily yellow–orange liquid. Pungent odour, fumes in air, bp 74°C. Exposure to light produces dithiophosgene. Lethal at 4000 mg/m^3 for 30 min.

Thrombocytin

Serine protease platelet-aggregating factor found in venom of the snake *Bothrops atrox*.

Tin tetrachloride

Stannic chloride, KJ, Opacite, Smoke-producing agent: less dangerous than phosphorus. Liquid, bp 114°C, fumes in moist air: stannic hydroxide and HCl. Less-dense smoke than some but penetrated WWI gas mask canisters. Irritant. Used as a mixture with phosgene and chloropicrin. Very expensive for large-scale use in war.

Titanium tetrachloride

F-stoff, FM. Colourless liquid, bp 136°C. Reacts vigorously with water in moist air to produce the hydroxide and HCl. Acrid due to the HCl.

Tityustoxin

Polypeptides TsTX-Kα and TsTX-Kβ from Brazilian scorpion toxin (*Tityus serrulatus*). Selectively block voltage-gated non-inactivating K^+ channels in nerve membranes.

Toad

Toxins bufotenin, bufotenidin, bufoviridin, bufotalin, bufotoxin, bufinin found in skin glands of *Bufo vulgari*: common toad and related species. All hallucinogens: affect Na/K ATPase.

TOCP

Tri-*o*-cresyl phosphate. Most toxic of the tricresyl phosphate isomers. Poor AChE inhibitor but classic agent for producing delayed polyneuropathy in hens and humans. 'Ginger Jake paralysis' in the USA in 1930 due to contamination of an illegal alcoholic ginger beverage with cresyl phosphates used in the extraction process. Mass outbreaks have also occurred as a result of using contaminated cooking oil. A single dose can cause a largely irreversible axonal degeneration in the peripheral nerves and spinal cord as a result of interaction with neuropathy target esterase. See also tricresyl phosphate.

Toluene diisocyanates

Several isomers. Colourless crystals or liquids turning pale yellow and darkening on exposure to air and light. Irritant vapour from 0.05 ppm. Respiratory sensitization. Neurological damage produced: euphoria, ataxia, drowsiness, coma. Delayed and lingering effects: poor memory, personality changes, instability, depression.

Tonite

French name for chloroacetone *qv*.

C-Toxiferines I and II

Principal paralytic alkaloids present in an alternative form of curare, calabash curare, prepared from the bark of *Strychnos toxifera*. Used to be packed in calabashes or gourds, hence the C (for calabash) in the name. Used as an arrow poison. Actions similar to tubocurarine.

Toxin Var 3

Typical structure for polypeptide scorpion toxin. 65 amino acid peptide with four disulphide bridges. The bridges confer resistance to deactivation by pH and temperature. This toxin delays closure of Na channels.

Trialkyltins

Butyl, ethyl and methyl forms. Neurotoxins. Damage myelin. Developing brain particularly susceptible. Use as anti-fouling agent on ships now prohibited. Affect mitochondrial metabolism and inhibit oxidative phosphorylation. Myelin damaging: peripheral neuropathy. Damage throughout CNS: oligodendrocytes affected.

Trichloromethyl chloroformate

Diphosgene, Surpalite, Perstoff. Colourless liquid, odour of phosgene, bp 128°C. Less irritant than the mono- and di-forms but has the lung-damaging properties of phosgene. Used in shells in WWI.

Trichloronitroso methane

Nitro-chloroform, chloropicrin. Dark-blue liquid, bp 127°C. Disagreeable smell. Lachrymator: sternutator.

Trichlorovinyl arsine

Crystalline, mp 21.5°C, bp 260°C. No irritant action but said by Beadnell (1943) to be poisonous.

Trichothecene mycotoxin

Nivalenol. Powerful toxin derived from *Fusarium nivale*. Haemorrhagic, vesicant, necrosis producing, nausea, vomiting. Suspected constituent of 'yellow rain' deposits during conflict in South-East Asia, although deposits also said to contain bee excreta.

Tricresyl phosphate

Tritolyl phosphate, TCP. Mixture of isomers that have varying potencies as AChE inhibitors and inducers of delayed polyneuropathy. 'Lindol' is the trade name for a mixture of isomers with a reduced content of the most toxic form, TOCP *qv*. Use now largely restricted to use as an additive to lubricants, specialist fuels and hydraulic fluids.

Tutin and coriamyrtin

Glycosides present in extracts of *Cararia* species in New Zealand and Europe, respectively: both are convulsants with chemistry and pharmacology similar to picrotoxin. Both berries and the leaves are poisonous in large doses, and several instances of death are on record from eating the fruit. Toxic honey is produced as a result of bees feeding on honeydew produced by the sap-sucking vine hopper insect (*Scolypopa sp.*) feeding on native New Zealand tutu bushes. A number of people have been killed, incapacitated and hospitalized over the years from eating toxic honey. Symptoms include vomiting, delirium, giddiness, increased excitability, stupor, coma and violent convulsions.

Tyrotoxicon

See diazobenzol.

Urotropine

Hexamethylenetetramine, HMT, HMTA, hexamine. Urinary antiseptic: releases formaldehyde in urine and this kills bacteria. Used with sodium phenate and NaOH as an absorbant of poisonous gases. Tried unsuccessfully as an antidote to phosgene.

V agents

American name given to a series of organophosphorus compounds containing a basic nitrogen side chain. Low-volatility nerve agents, active via the skin. Effects as per sarin: classic anticholinesterase effects: miosis, increased secretions, bradycardia, respiratory paralysis, death. 1 drop on skin can be fatal. See VX.

VE

O-Ethyl-*S*-[2-(diethylamino)ethyl] ethylphosphonothioate, V agent *qv*.

Veratrine

Mixture of alkaloids extracted from *Veratrum* species, including green hellebore and lentils. Veratridine: powerful sensory irritant: violent sneezing. Delays sodium channel closure: effects resemble those of hypocalcaemia. Germine, has been used in myasthenic gravis. Effects: gastrointestinal upset, vomiting, diarrhoea, giddy, dilated pupils. Deaths recorded.

Verotoxin

Shiga toxin. Toxic protein produced by *E. coli* and *Shigella dysenteriae*. Produces haemolytic-uremia syndrome: renal failure, dysentery, haemorrhagic colitits.

Vetch

See lathyrism.

VG

O,*O*-Diethyl-*S*-[2-(diethylamino)ethyl] phosphorothioate, V agent *qv*.

Vicine and convicine

Glycosides found with DOPA in Broad (Fava) beans (*Vicia faba*). Produces haemolytic anaemia in patients with G-6-P dehydrogenase deficiency.

Villantite

French: methyl chlorosulphonate *qv*.

Vincennite

French: 50% hydrocyanic acid, 30% arsenic trichloride, 15% stannic chloride, 5% chloroform.

Viscumin lectin 1

ML-1. Lectin found in mistletoe. Comparable in toxicity with ricin and abrin: inhibits protein synthesis as does ricin.

Vitrite/Vivrite

French: mixture of cyanogen chloride and arsenic trichloride.

VM

O-Ethyl-*S*-[2-(diethylamino)ethyl] methylphosphonothioate. V agent *qv*.

Volkensin

Lectin from *Adenia volkensii*: the Kilyambiti plant, an African succulent. Comparable to ricin and abrin: same mechanism and effects on protein synthesis.

Vomiting gas

Chloropicrin *qv*.

Vomitoxin

A *fusarium* mycotoxin.

VX

O-Ethyl-*S*-[2-(diisopropylamino)ethyl] methylphosphonothioate. Exceedingly toxic nerve agent. Low volatility: percutaneous hazard. MW: 267.4, mp −51°C; vapour pressure at 25°C: 0.0007 mmHg.

W

Agent W: USA code name for ricin.

Warfarin

Rat poison: inhibits production of clotting factors II (prothrombin), VII, IX and X. Bleeding: joints, bruising, shock. Death can be delayed by 2 weeks.

White cross gases

Lachrymators: bromoacetone, chloroacetophenone *qv*.

Wolfsbane

See aconite.

X-gas

All the following have been called X-gas: dibromo(di)methyl ether, methylsulphuryl chloride, dibromomethyl ether, dichloromethyl ether, ethyl dibromoarsine, phenyl dibromoarsine, ethyl bromoacetate.

Xylene bromide

Harassing agent. Used by Germany in 1915.

Xylyl bromide

T-stoff. Germans used on January 31st, 1915: first major gas used in war. Colourless liquid, bp 210–220°C, odour of elder blossom. Irritant.

Yellow cross gases

Low vapour pressure, toxic vesicants. Sartori (1939) lists mustard and lewisite as examples although the latter was not available in Germany in WWI.

Yellow star gas

British for: chloropicrin and chlorine mixture.

Yew

See taxine.

Yohimbine

Alkaloid: α-adrenergic receptor antagonist. Vasodilator: said to be an aphrodisiac. Local anaesthetic about as potent as cocaine. Releases antidiuretic hormone from posterior pituitary. Related to reserpine.

Yperite

French name for sulphur mustard.

Z

Disulphur decafluoride. Made in the UK in 1934. High toxicity, no odour. In rhesus monkey, was one tenth as toxic as phosgene.

Zinc chloride

Butter of Zinc, smoke known as hexite: respiratory and eye irritant. Common mixture is the zinc chloride smoke mixture (HC), consisting of hexachloroethane, fine-grained aluminium and zinc oxide. The smoke consists mainly of zinc chloride, zinc oxychlorides and hydrochloric acid. Its toxicity is caused mainly by the content of strongly acidic hydrochloric acid, but also to thermal effects of the reaction of zinc chloride with water. Lesions of the mucous membranes of the airways, dyspnea, retrosternal pain, hoarseness, stridor, lachrymation, cough, expectoration, and in some cases haemoptysis. Delayed pulmonary edema, cyanosis or bronchopneumonia may develop.

Zusatz

German name for phosgene *qv*.

Zyklon A

90% methyl cyanoformate and 10% methyl chloroformate.

Zyklon B

Liquid HCN plus cyanogen chloride or bromide. The additions prevented polymerization of the HCN and thus removed a hazard of explosion. The irritant was also said to provide a warning in case the HCN could not be detected. Developed by the German conglomerate IG Farben: produced chemical weapons in WWI. Was used for mass murder in concentration camps.

INDEX

Note: Figures and Tables are indicated by *italic page numbers*, and footnotes by suffix 'n'. 'CW' = 'chemical warfare', 'CWA' = 'chemical warfare agent', and 'RCA' = 'riot control agent'